# HANDBOOK OF ROBOTIC SURGERY

# HANDBOOK OF ROBOTIC SURGERY

*Edited by*

### STÊNIO DE CÁSSIO ZEQUI
*Urology Division, Referral Center for Urological Tumors and Graduate School, A.C. Camargo Cancer Center, São Paulo, São Paulo, Brazil*
*Surgery-Anatomy-Urology, Faculty of Medicine-USP-FMRP, São Paulo University-Ribeirão Preto, Ribeirão Preto, São Paulo, Brazil*
*Department of Urology, Graduate School, Federal University of São Paulo-UNIFESP, São Paulo, São Paulo, Brazil*

### HONGLIANG REN
*Department of Electronic Engineering, The Chinese University of Hong Kong, Sha Tin, Hong Kong*
*Shun Hing Institute of Advanced Engineering (SHIAE), The Chinese University of Hong Kong, Sha Tin, Hong Kong*

ELSEVIER

**ACADEMIC PRESS**
An imprint of Elsevier

Academic Press is an imprint of Elsevier
125 London Wall, London EC2Y 5AS, United Kingdom
525 B Street, Suite 1650, San Diego, CA 92101, United States
50 Hampshire Street, 5th Floor, Cambridge, MA 02139, United States

**Notices**

Knowledge and best practice in this field are constantly changing. As new research and experience broaden our understanding, changes in research methods, professional practices, or medical treatment may become necessary.

Practitioners and researchers must always rely on their own experience and knowledge in evaluating and using any information, methods, compounds, or experiments described herein. In using such information or methods they should be mindful of their own safety and the safety of others, including parties for whom they have a professional responsibility.

To the fullest extent of the law, neither the Publisher nor the authors, contributors, or editors, assume any liability for any injury and/or damage to persons or property as a matter of products liability, negligence or otherwise, or from any use or operation of any methods, products, instructions, or ideas contained in the material herein.

ISBN 978-0-443-13271-1

For information on all Academic Press publications
visit our website at https://www.elsevier.com/books-and-journals

Publisher: Mara Conner
Acquisitions Editor: Sonnini Yura
Editorial Project Manager: Isabella Silva
Production Project Manager: Surya Narayanan Jayachandran
Cover Designer: Matthew Limbert

Typeset by STRAIVE, India

# Dedication

To my wife, Patricia, and my sons Victor and Vinícius, with love—Stenio.
To my wife, Lily, my daughter Eva and my son Evan, with love—Hongliang.
To all the mentors, surgeons, colleagues, students, and patients, who have inspired us—Stenio and Hongliang.

# Contents

## I

## Introduction, technology basics, evolution and next frontiers of surgical robots

### 1. Introduction
STÊNIO DE CÁSSIO ZEQUI AND HONGLIANG REN

### 2. History of robotic surgery: Trends in autonomy levels and enabling technologies
HONGLIANG REN AND STÊNIO DE CÁSSIO ZEQUI

### 3. Needle inserts into soft tissue: Basic technical aspects
MINGYUE LU, CHWEE MING LIM, AND HONGLIANG REN

### 4. Kinematic concepts in minimally invasive surgical flexible robotic manipulators: State of the art
JIEWEN LAI, BO LU, AND HONGLIANG REN

### 5. Bioinspired flexible and compliant robotic manipulators for surgery
TAO ZHANG AND HONGLIANG REN

### 6. Haptic feedback technology in robot-assisted surgery
LAILU LI, WENCHAO YUE, AND HONGLIANG REN

### 7. Robust magnetic tracking and navigation in robotic surgery
SHIJIAN SU AND HONGLIANG REN

## 8. 3D reconstruction of deformable tissues in robotic surgery

MENGYA XU, TIEBING TANG, ZIQI GUO, AN WANG, BEILEI CUI, LONG BAI, AND HONGLIANG REN

## 9. Deep reinforcement learning in surgical robotics: Enhancing the automation level

CHENG QIAN AND HONGLIANG REN

## 10. Ultrasound guidance and robotic procedures: Actual and future intelligence

LONG BAI, LEI ZHAO, AND HONGLIANG REN

## 11. Robotic staplers

JEFFERSON LUIZ GROSS AND JOÃO PAULO DE OLIVEIRA MEDICI

## 12. Actual competitive and new models of surgical robots

DANIEL COSER GOMES AND STÊNIO DE CÁSSIO ZEQUI

## 13. Enabling intelligent procedures: Endoscopy dataset collection trends and pipeline

LYUXING HE, HUXIN GAO, AND HONGLIANG REN

## 14. Pathway of robotic learning to perform autonomous sewing

YUNKAI LV, HAO ZHANG, HUAICHENG YAN, AND HONGLIANG REN

## 15. Telesurgery applications, current status, and future perspectives in technologies and ethics

THIAGO CAMELO MOURÃO, SHADY SAIKALI, EVAN PATEL, MISCHA DOHLER, VIPUL PATEL, AND MÁRCIO COVAS MOSCHOVAS

# II

# Education, safety, ethical and administrative aspects for a successful robotic program

# III

# The current and future clinical applications of robotic surgery among medical specialties

## 48. Head and neck and transoral robotic surgery
JOSÉ GUILHERME VARTANIAN, RENAN BEZERRA LIRA,
AND LUIZ PAULO KOWALSKI

## 49. Robotic neck dissection
RENAN BEZERRA LIRA, JOSÉ GUILHERME VARTANIAN,
AND LUIZ PAULO KOWALSKI

## 50. Otolaryngology: Sleep apnea and benign diseases robotic surgery
WEI LI NEO, CHWEE MING LIM, AND SONG TAR TOH

## 51. Robotic cardiac surgery: Advancements, applications, and future perspectives
BURAK ERSOY AND BURAK ONAN

## 52. Robotic thoracic surgery
JENNIFER PAN, AMMARA WATKINS, AND ELLIOT SERVAIS

## 53. Esophagus/foregut and pancreatic robotic surgery
FELIPE J.F. COIMBRA, REBECA HARA NAHIME, SILVIO MELO TORRES,
AND IGOR CORREIA FARIAS

## 54. Bariatric robotic surgery
FRANCISCO GUERRA, JR., SHINIL K. SHAH, ERIK B. WILSON,
AND MELISSA M. FELINSKI

# Contributors

Diego Abreu   Department of Urology, Hospital Pasteur, Montevideo, Uruguay

Phillipe Abreu   Division of Transplant Surgery, Department of Surgery, University of Colorado Anschutz Medical Center, Aurora, CO, United States

Samuel Aguiar, Jr.   Colorectal Cancer Reference Center, A.C. Camargo Cancer Center; Department of Colorectal Surgery, Hospital AC Camargo, São Paulo, SP, Brazil

Tarek Ajami   Desai Sethi Urology Institute, Miller School of Medicine, University of Miami, Miami, FL, United States

Vanessa Alvarenga-Bezerra   Department of Gynecological Oncology, Hospital Israelita Albert Einstein, São Paulo, SP, Brazil

Maurice Anidjar   Department of Surgery, Division of Urology, McGill University, Montréal, Quebec, Canada

Fabio Antonellis   Department of General Surgery, Colleferro, Italy

Armen Aprikian   Department of Surgery, Division of Urology, McGill University, Montréal, Quebec, Canada

Raphael L.C. Araujo   Department of Digestive Surgery, Universidade Federal de São Paulo; Department of Oncology, Hospital Israelita Albert Einstein, São Paulo, SP, Brazil

Pier Paolo Avolio   Department of Surgery, Division of Urology, McGill University, Montréal, Quebec, Canada; Department of Biomedical Sciences, Humanitas University, Milan, Italy

Long Bai   Department of Electronic Engineering, The Chinese University of Hong Kong, Sha Tin, Hong Kong

Glauco Baiocchi   Department of Gynecologic Oncology, AC Camargo Cancer Center, São Paulo, Brazil

Alex Barbosa   Department of Surgical Oncology, Santa Casa de Maceió, Alagoas, Brazil

Andrés Barrios   Urology Department, Hospital de San José, Bogota, Colombia

A. Betancourt   General Surgery Department, Good Samaritan Medical Center, West Palm Beach, FL, United States

Romina Bianchi   Queens University, Kingston, Ontario, Canada

Bianca Bianco   Department of Urology, Hospital Israelita Albert Einstein; Discipline of Sexual and Reproductive Health and Population Genetics, Department of Collective Health, Faculdade de Medicina Do ABC, São Paulo, SP, Brazil

Carlos Leonardo Malta Braga   Santa Casa de Belo Horizonte; Hospital Unimed, Belo Horizonte, Brazil

Jefferson Braga Silva   Hospital Moinhos de Vento, Porto Alegre, Brazil

Éder Silveira Brazão, Jr.   Urology Division, A.C. Camargo Cancer Center, São Paulo, São Paulo, Brazil

Nicolo Maria Buffi   Department of Biomedical Sciences, Humanitas University; Department of Urology, IRCCS Humanitas Research Hospital, Milan, Italy

Catarina Vellinho Busnello   Hospital de Clínicas Porto Alegre, Porto Alegre, Brazil

Rodrigo Sousa Madeira Campos   Urology Division, A.C. Camargo Cancer Center, São Paulo, São Paulo, Brazil

Ziyan Cao   The Milton and Carroll Petrie Department of Urology at The Icahn School of Medicine at Mount Sinai, New York, NY, United States

Arie Carneiro   Department of Urology, Hospital Israelita Albert Einstein, São Paulo, SP, Brazil

Paolo Casale   Department of Biomedical Sciences, Humanitas University; Department of Urology, IRCCS Humanitas Research Hospital, Milan, Italy

Leandro Totti Cavazzola   Universidade Federal do Rio Grande do Sul, Porto Alegre, Brazil

Antonio Caycedo-Marulanda   Orlando Health Colon and Rectal Institute; University of Central Florida, Orlando, FL, United States; Queens University, Kingston, Ontario, Canada

Valeria Celis   Catherine and Joseph Aresty Department of Urology, USC Institute of Urology, Keck School of Medicine, University of Southern California, Los Angeles, CA, United States

Steven Lee Chang   Department of Urology, Brigham and Women's Hospital, Harvard Medical School; Lank Center for Genitourinary Oncology, Dana-Farber Cancer Institute, Boston, MA, United States

Jaya S. Chavali   Glickman Urological and Kidney Institute, Cleveland Clinic, Cleveland, OH, United States

Aline Yuri Chibana   Department of Anesthesiology, A.C. Camargo Cancer Center, São Paulo, Brazil

Manish Kumar Choudhary   The Milton and Carroll Petrie Department of Urology at The Icahn School of Medicine at Mount Sinai, New York, NY, United States

Rafael Clavijo   Urology Department, Hospital de San José, Bogota, Colombia

Felipe J.F. Coimbra   Department of Abdominal Surgery, Reference Center on Upper GI & HPB Oncology, São Paulo, Brazil

Paulo Roberto Corsi   National Directory, Brazilian College of Surgeons (CBC); Brazilian Chapter of the American College of Surgeons, Rio de Janeiro; Department of Surgery, Faculty of Medical Sciences of Santa Casa de São Paulo (FCMSCSP), São Paulo, Brazil

**Beilei Cui** Department of Electronic Engineering, The Chinese University of Hong Kong, Sha Tin, Hong Kong

**Rafael Cunha de Almeida** Department of Ophthalmology, ABC Foundation School of Medicine, Santo André; Department of Ophthalmology, University of Santo Amaro, School of Medicine, São Paulo, Brazil

**Walter Henriques da Costa** Urology Division, A.C. Camargo Cancer Center, São Paulo, São Paulo, Brazil

**Leonardo Emilio da Silva** Department of Surgery, Faculty of Medicine, Federal University of Goias, Goiânia; Robotic Surgery, Hospital Israelita Albert Einstein, Goiania, Goias, Brazil

**Cecilia da Silva Angelo** Surgical Center Nursing, A.C. Camargo Cancer Center, São Paulo, Brazil

**Daniel DaJusta** Nationwide Children's Hospital Division of Pediatric Urology, The Ohio State University Department of Urology, Columbus, OH, United States

**Everson Luiz de Almeida Artifon** Surgery Department, Hospital das Clínicas of University of São Paulo Medical School; Department of Surgery Post-Graduate Program, São Paulo Medical School, São Paulo, SP, Brazil

**Giovana Abrahão de Araújo Moriya** Brazilian Association of Perioperative Nurses (SOBECC), São Paulo, SP, Brazil

**Vanessa de Brito Poveda** School of Nursing of University of São Paulo; Brazilian Association of Perioperative Nurses (SOBECC), São Paulo, SP, Brazil

**Stênio de Cássio Zequi** Urology Division, Referral Center for Urological Tumors and Graduate School, A.C. Camargo Cancer Center; Department of Urology, Graduate School, Federal University of São Paulo-UNIFESP, São Paulo; Surgery-Anatomy-Urology, Faculty of Medicine-USP-FMRP, São Paulo University-Ribeirão Preto, Ribeirão Preto, São Paulo, Brazil

**Héber Salvador de Castro Ribeiro** Department of Abdominal Surgery, A.C. Camargo Cancer; Reference Center on Upper GI & HPB Oncology, A.C. Camargo Center, São Paulo, Brazil

**Lucio Flavio de Magalhães Brito** Pontifical Catholic University of São Paulo, Exact and Technological Sciences Faculty—Biomedical Engineering, São Paulo, Brazil

**Camila Mendonça de Moraes** Federal University of Rio de Janeiro (UFRJ), Rio de Janeiro, Brazil

**Renato Almeida Rosa de Oliveira** Urology Department, Beneficência Portuguesa de São Paulo; Instituto de Urologia, Oncologia e Cirurgia Robótica, São Paulo, SP, Brazil

**Rita de Cássia Burgos de Oliveira** University of São Paulo, School of Nursing, São Paulo, Brazil

**João Paulo de Oliveira Medici** Department of Thoracic Surgery, A.C. Camargo Cancer Center, São Paulo, Brazil

**Yasmin Russo de Toledo** Surgical Center Nursing, A.C. Camargo Cancer Center, São Paulo, Brazil

**Pedro Debieux** Hospital Israelita Albert Einstein, São Paulo, SP; Hospital Beneficência Portuguesa de São Paulo, São Paulo, Brazil

**Celeste Del Basso** SCDU General Surgery, Surgical Oncology, Minimally Invasive, Robotic and HBP Surgery, AOUAL SS. Antonio e Biagio e Cesare Arrigo, Alessandria, Italy

**Mischa Dohler** VP Emerging Tech, Ericson Inc., Silicon Valley, CA, United States; Advisory Board, FCC (TAC) & Ofcom (Spectrum); King's College London, London, United Kingdom

**Taurino dos Santos Rodrigues Neto** Associate Physician in RetinaPro Ophthalmological Clinic, São Paulo, Brazil

**Burak Ersoy** Department of Cardiovascular Surgery, University of Health Sciences, Istanbul Mehmet Akif Ersoy Thoracic and Cardiovascular Surgery Training and Research Hospital, Istanbul, Turkey

**Marcio Roberto Facanali Junior** Department of Gastroenterology, Hospital das Clínicas of University of São Paulo Medical School, São Paulo, SP, Brazil

**Marco Faria-Correa** Plastic Surgery, Mount Elizabeth Novena Specialist Centre, Singapore, Singapore

**Igor Correia Farias** Department of Abdominal Surgery, Reference Center on Upper GI & HPB Oncology, São Paulo, Brazil

**Melissa M. Felinski** Division of Minimally Invasive and Elective General Surgery, Department of Surgery, McGovern Medical School at UT Health, Houston, TX, United States

**Ricardo B.V. Fontes** Rush University Medical Centre, Chicago, IL, United States

**Molly Fuchs** Nationwide Children's Hospital Division of Pediatric Urology, The Ohio State University Department of Urology, Columbus, OH, United States

**Huxin Gao** Department of Biomedical Engineering, National University of Singapore, Faculty of Engineering, Singapore, Singapore; National University of Singapore (Suzhou) Research Institute (NUSRI), Suzhou, China

**Guglielmo Gazzetta** Division of Breast Surgery, European Institute of Oncology, Istituto di Ricovero e Cura a Carattere Scientifico (IRCCS), Milan, Italy

**Desrene Gibson** Orlando Regional Medical Center, Orlando Health Robotic Surgery Program, Orlando, FL, United States

**Camilo Giedelmann** Urology Department, Hospital de San José, Bogota, Colombia

**Juliana Rizzo Gnatta** School of Nursing of University of São Paulo; Brazilian Association of Perioperative Nurses (SOBECC), São Paulo, SP, Brazil

**Howard Goldman** Glickman Urological and Kidney Institute, Cleveland Clinic, Cleveland, OH, United States

**Daniel Coser Gomes** Urology Division, A.C. Camargo Cancer Center, São Paulo, São Paulo, Brazil

**Marcos Gómez Ruiz** National Directory, Brazilian College of Surgeons (CBC), Rio de Janeiro, Brazil; Marques de Valdecilla University Hospital; Valdecilla Biomedical Research Institute (IDIVAL), Santander, Spain

**Omaira Rodríguez González** Chief of Surgical Department, Clínicas Caracas Hospital, Faculty of Medicine, Central University of Venezuela, Caracas, Venezuela

Vitagliano Gonzalo Department of Urology, Hospital Alemán, Buenos Aires, Argentina

Jefferson Luiz Gross Department of Thoracic Surgery, A.C. Camargo Cancer Center, São Paulo, Brazil

Francisco Guerra, Jr. Division of Minimally Invasive and Elective General Surgery, Department of Surgery, McGovern Medical School at UT Health, Houston, TX, United States

Gustavo Cardoso Guimarães Hospital Israelita Albert Einstein; Urology Department, Beneficência Portuguesa de São Paulo; Instituto de Urologia, Oncologia e Cirurgia Robótica, São Paulo, SP, Brazil

Ziqi Guo Department of Electronic Engineering, The Chinese University of Hong Kong, Sha Tin, Hong Kong

Ashraf S. Habib Division of Women's Anesthesia, Department of Anesthesiology, Duke University Medical Center, Durham, NC, United States

Camila De Souza Hagui Department of Anesthesiology, A.C. Camargo Cancer Center, São Paulo, Brazil

C. Hartmann Department of Hernia and Abdominal Wall Reconstruction, Good Samaritan Medical Center-TENET Health, West Palm Beach, FL, United States

Junya Hata Department of Urology, Fukushima Medical University School of Medicine, Fukushima, Japan

Yutaro Hayashi Department of Pediatric Urology, Nagoya City University Graduate School of Medical Sciences, Nagoya, Japan

Lyuxing He Department of Biomedical Engineering, National University of Singapore, Faculty of Engineering, Singapore, Singapore; National University of Singapore (Suzhou) Research Institute (NUSRI), Suzhou, China

Alexandre Kyoshi Hidaka Urology Division, Centro Universitario FMABC, Santo André, SP, Brazil

Mohammud Shakeel Inder Imperial College Healthcare NHS Trust, London, United Kingdom

Leanne Iorio Orlando Health Colon and Rectal Institute, Orlando, FL, United States

Cian L. Jacob Department of Urology, University of Kansas, Kansas City, KS, United States

Eduardo Giroud Joaquim Department of Anesthesiology, A.C. Camargo Cancer Center, São Paulo, Brazil

Gisele Maia Jünger Institute of Robotic Surgery at Faculty of Medical Sciences of Minas Gerais, Belo Horizonte, Minas Gerais, Brazil

Camila Cohen Kaleka Hospital Israelita Albert Einstein, São Paulo, SP; Hospital Beneficência Portuguesa de São Paulo, São Paulo, Brazil

Gil Kamergorodsky Department of Gynecology, Federal University of São Paulo (Dr. Kamergorodsky), São Paulo, Brazil

Jihad Kaouk Glickman Urological and Kidney Institute, Cleveland Clinic, Cleveland, OH, United States

Joshua Karas Orlando Health Colon and Rectal Institute, Orlando, FL, United States

Wassim Kassouf Department of Surgery, Division of Urology, McGill University, Montréal, Quebec, Canada

Ryan Kelly Department of Neurosurgery, Rush University Medical Center, Chicago, IL, United States

Andrew Kerner Orlando Health Colon and Rectal Institute, Orlando, FL, United States

Archan Khandekar Desai Sethi Urology Institute, Miller School of Medicine, University of Miami, Miami, FL, United States

Yoshiyuki Kojima Department of Urology, Fukushima Medical University School of Medicine, Fukushima, Japan

Fernando Korkes Urology Division, Centro Universitario FMABC, Santo André; Hospital Municipal da Vila Santa Catarina; Hospital Israelita Albert Einstein, São Paulo, SP, Brazil

Luiz Paulo Kowalski Department of Head and Neck Surgery and Otorhinolaryngology, A.C. Camargo Cancer Center; Head and Neck Surgery Department and LIM 28, University of São Paulo Medical School, São Paulo, Brazil

Jiewen Lai Department of Electronic Engineering, The Chinese University of Hong Kong, Sha Tin, Hong Kong

Mario M. Leitao, Jr. Gynecology Service, Department of Surgery, Memorial Sloan-Kettering Cancer Center and Department of Obstetrics and Gynecology, Weill Cornell Medical College New York, New York, NY, United States

Giovanni Battista Levi Sandri Digestive Surgery Unit, Fondazione Policlinico Universitario Agostino Gemelli IRCCS, Rome, Italy

Lailu Li Department of Electronic Engineering, The Chinese University of Hong Kong, Sha Tin, Hong Kong

Chwee Ming Lim Department of Otorhinolaryngology—Head and Neck Surgery, Singapore General Hospital; Surgery Academic Clinical Programme, Duke-NUS Graduate Medical School, Singapore, Singapore

Vagner Loduca Lima Department of Ophthalmology, ABC Foundation School of Medicine, Santo André, Brazil

Trenton Lippert University of South Florida Morsani College of Medicine Tampa, FL, United States

Renan Bezerra Lira Department of Head and Neck Surgery and Otorhinolaryngology, A.C. Camargo Cancer Center; Robotic Surgery Program, Hospital Israelita Albert Einstein, São Paulo, Brazil

Andressa Cristina Sposato Louzada Albert Einstein Israeli College of Health Sciences (FICSAE), São Paulo, Brazil

Bo Lu Robotics and Microsystem Center, Soochow University, Suzhou, China

Mingyue Lu The Department of Biomedical Engineering, National University of Singapore, Singapore, Singapore

Giovanni Lughezzani Department of Biomedical Sciences, Humanitas University; Department of Urology, IRCCS Humanitas Research Hospital, Milan, Italy

Yunkai Lv East China University of Science and Technology; Tongji University, Shanghai, China; The Chinese University of Hong Kong, Sha Tin, Hong Kong

Fabiana Baroni Alves Makdissi Department of Mastology, A.C. Camargo Cancer Center, São Paulo, Brazil

Ankur Malpani Desai Sethi Urology Institute, Miller School of Medicine, University of Miami, Miami, FL, United States

Asher Mandel The Milton and Carroll Petrie Department of Urology at The Icahn School of Medicine at Mount Sinai, New York, NY, United States

Joao Manzi Miami Transplant Institute, Jackson Memorial Hospital, University of Miami, Coral Gables, FL, United States

Tomas Mansur Duarte de Miranda Marques Department of Colorectal Surgery, A.C. Camargo Cancer Center, São Paulo, Brazil

Ringa Maximiliano Department of Urology, Hospital Alemán, Buenos Aires, Argentina

Luis G. Medina Catherine and Joseph Aresty Department of Urology, USC Institute of Urology, Keck School of Medicine, University of Southern California, Los Angeles, CA, United States

Rafael Ribeiro Meduna Urology Division, A.C. Camargo Cancer Center, São Paulo, São Paulo, Brazil

Kentaro Mizuno Department of Pediatric Urology, Nagoya City University Graduate School of Medical Sciences, Nagoya, Japan

Wilson R. Molina Department of Urology, University of Kansas, Kansas City, KS, United States

Luiz Antônio Mondadori Department of Anesthesiology, A.C. Camargo Cancer Center, São Paulo, Brazil

Renato Moretti-Marques Department of Gynecological Oncology, Hospital Israelita Albert Einstein, São Paulo, SP, Brazil

Márcio Covas Moschovas AdventHealth Global Robotics Institute, Celebration; University of Central Florida (UCF), Orlando, FL, United States

Thiago Camelo Mourão Department of Urology, A.C. Camargo Cancer Center, São Paulo, Brazil

Miguel Nacul Service of Surgery, Hospital Moinhos de Vento, Porto Alegre, RS, Brazil

Bruno Nahar Desai Sethi Urology Institute, Miller School of Medicine, University of Miami, Miami, FL, United States

Rebeca Hara Nahime Department of Abdominal Surgery, Reference Center on Upper GI & HPB Oncology, São Paulo, Brazil

Wei Li Neo Department of Otorhinolaryngology—Head and Neck Surgery, Singapore General Hospital, Singapore, Singapore

Hidenori Nishio Department of Pediatric Urology, Nagoya City University Graduate School of Medical Sciences, Nagoya, Japan

José Ignacio Nolazco Department of Urology, Brigham and Women's Hospital, Harvard Medical School, Boston, MA, United States; Servicio de Urología, Hospital Universitario Austral, Universidad Austral, Pilar, Argentina

Paul J. Oh Glickman Urological and Kidney Institute, Cleveland Clinic, Cleveland, OH, United States

Burak Onan Department of Cardiovascular Surgery, University of Health Sciences, Istanbul Mehmet Akif Ersoy Thoracic and Cardiovascular Surgery Training and Research Hospital, Istanbul, Turkey

Jennifer Pan Department of General Surgery, Beth Israel Deaconess Medical Center, Boston, MA, United States

E. Parra Davila Department of Hernia and Abdominal Wall Reconstruction, Good Samaritan Medical Center-TENET Health, West Palm Beach, FL, United States

Dhruti M. Patel The Milton and Carroll Petrie Department of Urology at The Icahn School of Medicine at Mount Sinai, New York, NY, United States

Evan Patel AdventHealth Global Robotics Institute, Celebration, FL, United States

Vipul Patel AdventHealth Global Robotics Institute, Celebration; University of Central Florida (UCF), Orlando, FL, United States

Adriana M. Pedraza The Milton and Carroll Petrie Department of Urology at The Icahn School of Medicine at Mount Sinai, New York, NY, United States

Sérgio Podgaec Department of Obstetrics and Gynecology, Medical School Hospital, University of São Paulo, São Paulo, Brazil

Jaime Poncel Catherine and Joseph Aresty Department of Urology, USC Institute of Urology, Keck School of Medicine, University of Southern California, Los Angeles, CA, United States

Giada Pozzi Division of Breast Surgery, European Institute of Oncology, Istituto di Ricovero e Cura a Carattere Scientifico (IRCCS), Milan, Italy

Cheng Qian Department of Electrical Engineering and Information Technology, Technical University of Munich, Munich, Germany

Savitha Ramachandran Department of Plastic and Reconstructive Surgery, Singapore General Hospital, Singapore, Singapore

Leonardo O. Reis UroScience, State University of Campinas, Unicamp, and Urologic Oncology Department, School of Life Sciences, Pontifical Catholic University of Campinas, Campinas, São Paulo, Brazil

Hongliang Ren Department of Electronic Engineering; Shun Hing Institute of Advanced Engineering (SHIAE), The Chinese University of Hong Kong, Sha Tin, Hong Kong

Maurício Murce Rocha Urology Division, A.C. Camargo Cancer Center, São Paulo, São Paulo, Brazil

Omaira Rodriguez Robotic Surgery Program, University Hospital of Caracas, Medicine Faculty, Central University of Venezuela, Caracas, Venezuela

Sergio Roll Department of Surgery, Faculty of Medical Sciences of Santa Casa de São Paulo (FCMSCSP); Abdominal Wall Group of the Department of Surgery at the Faculty of Medical Sciences of Santa Casa de São Paulo—(FCMSCSP); Hernia Surgery Center at German Hospital Oswaldo Cruz, São Paulo, Brazil

Alexandre Antonio Marques Rosa Department of Ophthalmology, ABC Foundation School of Medicine, Santo André; Department of Ophthalmology, Federal University School of Pará, Pará, Brazil

Alexander Rosemurgy  Digestive Health Institute, AdventHealth Tampa, Tampa, FL, United States

Sharona Ross  Digestive Health Institute, AdventHealth Tampa, Tampa, FL, United States

Mariana Costa Rossette  Gynecologic Division, Universidade Federal de São Paulo, Escola Paulista de Medicina, São Paulo, SP, Brazil

Shady Saikali  AdventHealth Global Robotics Institute, Celebration, FL, United States

Idvaldo Salazar-M-Messias  Department of Urology, Dr Arnaldo Vieira de Carvalho Cancer Institute; Catherine and Joseph Aresty Department of Urology, USC Institute of Urology, Keck School of Medicine, University of Southern California, Los Angeles, CA, United States

José Gadú Campos Salcedo  Urology Department, Hospital Central Militar; Urology Center, Hospital Angeles, Mexico City, Mexico

Rubens A. Sallum  Department of Digestive Surgery, Universidade Federal de São Paulo, São Paulo, SP, Brazil

Alexis Sánchez  Orlando Regional Medical Center, Orlando Health Robotic Surgery Program, Orlando, FL, United States

Rafael Sanchez-Salas  Department of Surgery, Division of Urology, McGill University, Montréal, Quebec, Canada

Sepehr Sani  Department of Neurosurgery, Rush University Medical Center, Chicago, IL, United States

Yuichi Sato  Department of Urology, Fukushima Medical University School of Medicine, Fukushima, Japan

Aref S. Sayegh  Catherine and Joseph Aresty Department of Urology, USC Institute of Urology, Keck School of Medicine, University of Southern California, Los Angeles, CA, United States

Lalithkumar Seenivasan  Department of Biomedical Engineering, National University of Singapore, Singapore, Singapore; Department of Electronic Engineering; Shun Hing Institute of Advanced Engineering (SHIAE), The Chinese University of Hong Kong, Sha Tin, Hong Kong

Elliot Servais  Division of Thoracic and Cardiovascular Surgery, Lahey Hospital & Medical Center, Burlington, MA, United States

Shinil K. Shah  Division of Minimally Invasive and Elective General Surgery, Department of Surgery, McGovern Medical School at UT Health, Houston; Michael E. DeBakey Institute for Comparative Cardiovascular Science and Biomedical Devices, Texas A&M University, College Station, TX, United States

Taimur Shah  Imperial College Healthcare NHS Trust, London, United Kingdom

Marina Sonagli  Department of Mastology, A.C. Camargo Cancer Center, São Paulo, Brazil

Nicolas A. Soputro  Glickman Urological and Kidney Institute, Cleveland Clinic, Cleveland, OH, United States

René Sotelo  Keck Medical Center at the University of Southern California; Catherine and Joseph Aresty Department of Urology, USC Institute of Urology, Keck School of Medicine, University of Southern California, Los Angeles, CA, United States

Paulo Roberto Stevanato Filho  Colorectal Cancer Reference Center, A.C. Camargo Cancer Center, São Paulo, SP; Brazilian Society of Surgical Oncology–SBCO, Rio de Janeiro, Brazil

Shijian Su  Department of Electronic Engineering, The Chinese University of Hong Kong, Sha Tin, Hong Kong

Iswanto Sucandy  Digestive Health Institute, AdventHealth Tampa, Tampa, FL, United States

Chandru P. Sundaram  Indiana University School of Medicine, Indianapolis, IN, United States

Hon Sen Tan  Department of Women's Anaesthesia, KK Women's and Children's Hospital, Singapore, Singapore

Tiebing Tang  Department of Electronic Engineering, The Chinese University of Hong Kong, Sha Tin, Hong Kong

Simon Tanguay  Department of Surgery, Division of Urology, McGill University, Montréal, Quebec, Canada

Saulo Borborema Teles  Department of Urology, Hospital Israelita Albert Einstein, São Paulo, SP, Brazil

Ashutosh Tewari  The Milton and Carroll Petrie Department of Urology at The Icahn School of Medicine at Mount Sinai, New York, NY, United States

Yu Tian  Department of Electronic Engineering; Shun Hing Institute of Advanced Engineering (SHIAE), The Chinese University of Hong Kong, Sha Tin, Hong Kong

Marcos Tobias-Machado  Department of Urology, Dr Arnaldo Vieira de Carvalho Cancer Institute, São Paulo; ABC Medical School, Santo André, SP, Brazil

Antonio Toesca  Division of Breast Surgery, European Institute of Oncology, Istituto di Ricovero e Cura a Carattere Scientifico (IRCCS), Milan, Italy

Song Tar Toh  Department of Otorhinolaryngology—Head and Neck Surgery, Singapore General Hospital; SingHealth Duke-NUS Sleep Centre, Singapore, Singapore

Laís Gonçalves Tolentino  Vila da Serra Hospital, Nova Lima, Minas Gerais, Brazil

Flavio Daniel Saavedra Tomasich  Department of Surgery, Federal University of Paraná, Curitiba, Paraná; National Directory, Brazilian College of Surgeons (CBC), Rio de Janeiro; Committee on Minimally Invasive Surgery and Robotic Surgery, Brazilian College of Surgeons (CBC), Rio de Janeiro, Brazil; Brazilian Society of Oncologic Surgery (SBCO), Rosemont, IL, United States; Department of Surgery, Erasto Gaertner Cancer Center, Curitiba, Paraná, Brazil

Silvio Melo Torres  Department of Abdominal Surgery, Reference Center on Upper GI & HPB Oncology, São Paulo, Brazil

Adam Vacek  University of Edinburgh Medical School, Edinburgh, United Kingdom

José Guilherme Vartanian  Department of Head and Neck Surgery and Otorhinolaryngology, A.C. Camargo Cancer Center, São Paulo, São Paulo, Brazil

Lucas B. Vergamini  Department of Urology, University of Kansas, Kansas City, KS, United States

**Rodrigo Vianna** Miami Transplant Institute, Jackson Memorial Hospital, University of Miami, Coral Gables, FL, United States

**Luiz Carlos Von Bahten** Department of Surgery, Federal University of Paraná, Curitiba, Paraná; National Directory, Brazilian College of Surgeons (CBC), Rio de Janeiro; Department of Surgery, Pontificia Universidade Católica do Paraná (PUC-PR), Curitiba; Clinica Cirurgica, UFPR, Curitiba, Parana, Brazil

**Vinayak Wagaskar** The Milton and Carroll Petrie Department of Urology at The Icahn School of Medicine at Mount Sinai, New York, NY, United States

**An Wang** Department of Electronic Engineering, The Chinese University of Hong Kong, Sha Tin, Hong Kong

**Ammara Watkins** Division of Thoracic and Cardiovascular Surgery, Lahey Hospital & Medical Center, Burlington, MA, United States

**Bristol B. Whiles** Department of Urology, University of Kansas, Kansas City, KS, United States

**Erik B. Wilson** Division of Minimally Invasive and Elective General Surgery, Department of Surgery, McGovern Medical School at UT Health, Houston, TX, United States

**Nelson Wolosker** Albert Einstein Israeli College of Health Sciences (FICSAE), São Paulo, Brazil

**Mengya Xu** Department of Electronic Engineering, The Chinese University of Hong Kong, Sha Tin, Hong Kong; Department of Biomedical Engineering, National University of Singapore (NUS), Singapore, Singapore; NUS (Suzhou) Research Institute, Suzhou, China

**Huaicheng Yan** Tongji University, Shanghai, China

**Guilherme Yazbek** Vascular and Endovascular Surgery Department at AC Camargo Cancer Center, São Paulo, Brazil

**Courtney Yong** Indiana University School of Medicine, Indianapolis, IN, United States

**Wenchao Yue** Department of Electronic Engineering, The Chinese University of Hong Kong, Sha Tin, Hong Kong

**Rafael Ribeiro Zanotti** Urology Division, A.C. Camargo Cancer Center, São Paulo, São Paulo, Brazil

**Hao Zhang** East China University of Science and Technology, Shanghai, China

**Tao Zhang** Department of Electronic Engineering, The Chinese University of Hong Kong, Sha Tin, Hong Kong

**Lei Zhao** Department of Electronic Engineering, The Chinese University of Hong Kong, Sha Tin, Hong Kong; College of Computer Science and Electronic Engineering, Hunan University, Changsha, China

**Bruno Zilberstein** Service of Digestive Surgery, Sao Leopoldo Mandic School of Medicine, Campinas, SP, Brazil

# Foreword 1

I am honored to be invited to write this foreword. As a urological oncologist, I have watched the clinical evolution of robotics for almost 25 years. This followed 20 years of transformation with the introduction of minimally invasive surgery (MIS) for stones (percutaneous nephrolithotomy) and renal surgery (laparoscopic nephrectomy). I well remember attending the first workshop on laparoscopic radical prostatectomy in Paris, where the French introduced us to minimally invasive radical prostatectomy using the laparoscope. At the time, it was exciting and innovative but in retrospect relatively primitive primarily because of two-dimensional imaging to guide the surgeons. Robotic-assisted radical prostatectomy with stereoscopic visualization which provided a third dimension followed quickly. This was the first widespread application of surgical robotics in high-income countries; the majority of radical prostatectomies are now done with robotic assistance. Throughout the evolution of robotic surgery, surgeons have attempted to replicate the steps of open surgery, but frequently, time-saving shortcuts and other opportunities to improve on the classical approaches have emerged. The changes have not only been technical. MIS and robotics have brought massive change to the established order including for the generation of surgeons who exclusively trained in and practiced open surgery. They have had to adapt. Some continued to do open procedures despite increasing evidence that MIS with or without robotic assistance was safe, precise, and achieved equal or better outcomes, most notably less morbidity.

It is important to note that for surgeons, robotic surgery is currently robotic-assisted surgery and not autonomous robotics, at least for now.

Dr. Zequi and colleagues have assembled a monumental treatise of 77 chapters that will interest all surgeons, robotic engineers, and technicians as well as allied health disciplines and health administrators. It will be the go-to HTA (health technology assessment) manual for robotic surgery.

From an academic perspective, this text tells the story of robotics in most surgical disciplines, but there are common elements for all the subspecialties. There is so much to cover in this new and evolving field, which is hard to know in what order to present the issues.

The introductory chapters cover the basics that surgeons encounter with all minimally invasive surgery: haptics, navigation including the use of imaging, telesurgery, simulation, training, and the recently recognized application of artificial intelligence, among others. A bit later in the text, there is an overview of the advantages and disadvantages of the robotic approach.

Simulation and training have been major credential and administrative issues, although the widespread practice of MIS, more recently with robotic assistance, has made this a core competency of most training programs. Appropriately, the editors have included extensive contributions about building multidisciplinary clinical teams, managing complications, and health economics.

The final half of the chapters are devoted to the present state of robotics in all surgical subspecialties. There is something for all surgeons, especially urologists.

Congratulations to Dr. Zequi for conceiving this unique contribution to the field of surgery.

*Michael A.S. Jewett*
**Princess Margaret Cancer Centre, University Health Network,**
**University of Toronto, Toronto, ON, Canada**

# Foreword 2

Guided by the expertise of Stênio de Cássio Zequi Sr. and Hongliang Ren, this handbook offers a detailed overview of the field of robotic surgery, mapping its progression from early concepts to modern practice and potential future developments. The initial chapters provide a deep dive into bioengineering, thoroughly examining how the fusion of robotics, augmented reality, simulation, and artificial intelligence has shaped the landscape of surgery. Included in this section is a detailed analysis of the technical breakthroughs that have propelled the field to its current state.

The subsequent segment shifts focus to the humanistic elements of robotic surgery, integrating the pillars of education, training, and leadership. It addresses the quintessential elements of mentoring, teaching, accreditation, and the multifaceted process of certification in robotics. Ethical, legal, and economic considerations are also explored, shedding light on patient safety, risk management, and the effective incorporation of these technologies in healthcare settings. This section is tailored for a broad audience, comprising doctors, students, residents, nursing staff, and administrative personnel, all of whom play a pivotal role in the ecosystem of healthcare.

The concluding part of this book presents an in-depth assessment of the clinical applications of robotic surgery across a diverse range of medical specialties. Contributions from renowned leaders and experts around the globe bring unique perspectives to this work. More than just an academic resource, this book acts as a vibrant guide through the progressive field of robotic surgery. It inspires readers to pursue innovative solutions and further advancements, resonating with Thomas Edison's timeless words, "There's a way to do it better. Find it." With each page, the reader will be equipped with the knowledge to navigate the new era of surgical care, one that promises greater precision, efficiency, and, above all, an unwavering commitment to patient well-being. Welcome to the vanguard of surgical innovation. Welcome to the "Handbook of Robotic Surgery."

*Ashutosh Tewari*
*Adriana M. Pedraza*
*Vinayak Wagaskar*
**The Milton and Carrol Petrie Department, Icahn School of Medicine at Mount Sinai,**
**New York, NY, United States**

# Preface

The era of robotics began after the Second World War, with the first industrial models being used in the mid-1950s and 1960s. Since then, robotics has developed in the most diverse areas of human knowledge, both in basic industry as well as in the manufacturing industry, means of transport, agriculture, geological, maritime, aerospace sciences, etc. This not only impacted and irreversibly transformed production chains but also provided effectiveness, precision, and safety for the most complex and risky tasks for human beings. With the development of computing and the Internet, the influence of robotics is inseparable from the daily lives of citizens, especially in developed and developing countries. Imagining a current world without robots seems almost impossible.

In medicine and other health sciences, it could not be different.

With endoscopic and laparoscopic procedures and percutaneous interventions, minimally invasive surgeries became popular in the 1980s and 1990s. However, these advancements came at the expense of limitations in visualization (two-dimensional), poor dexterity, limited range of movement, and exposure to ionizing radiation and organic fluids, associated with great ergonomic wear and tear on surgeons and their assistants. Even so, access to delicate anatomical structures in difficult-to-reach regions remained a challenge to overcome.

Faced with these limitations and the need to offer surgical treatment in remote geographic regions or conflict zones, among other reasons, around three decades ago, the era of robotic surgery started. Robotic surgery has been rapidly adopted, since it provides greater dexterity with articulated surgical tools, different degrees of freedom, expanded three-dimensional vision, tremor filtration, associations with surgical navigation devices, advances in ultrasound, virtual reality, electrosurgery, staplers, sealers, etc.

With the arrival of the Internet of Things, 5G, virtual reality, artificial intelligence, machine learning, and telecommunications, it seems to present unimaginable possibilities, such as telesurgery over long distances, both on Earth and in remote regions of outer space.

However, when we look at the current situation, we see that the implementation and development of robotic surgery systems require several professionals: surgeons, nurses, healthcare professionals, and hospital managers. Before these, bioengineers and clinical engineers must develop the technology and facilitate its installation.

Although the acquisition of robotic surgical platforms can increase a hospital's reputation and strengthen its market share, it also increases costs for acquisition, maintenance, depreciation, etc., influencing the healthcare market and ecosystem (patients, hospitals, industry, dealers, payers, etc.).

After centuries, robotic surgery has also profoundly changed the "status quo" of a surgery team: for the first time, there is now an interface between the surgeons and their patients and the absence of intraoperative tactile sensation, making it necessary to compensate for visual feedback. For the first time, communications between surgeons, their assistants, scrub nurses, and anesthetists take place indirectly. There is a need to remodel roles and leadership levels among the various surgical professionals, aiming for patient safety.

The secular teaching of surgery is also changing, with the industry as an important player and the possibility of training in simulated models and virtual reality, with local proctor action or even though by the teleproctoring. With this, a bioethical benefit is established: the need for training in animal or cadaveric models and "in vivo" patients is reduced. However, new bioethical challenges raise questions such as the following: What are the levels of responsibility of local and remote proctors? How are the professional credentialing and accreditation processes defined? Will new accreditations be necessary for health professionals to work on the different robotic models for each new device and each manufacturing company? These issues highlight the evolving landscape of robotic technology in healthcare and the importance of addressing ethical considerations to ensure safe and responsible implementation.

From a socioeconomic point of view, the benefits of robotic surgery provide patients with rapid recovery, fewer complications, and a rapid return to their professional activities, with long-term savings, but at the expense of a significant outlay at the time of surgery. The following questions arise: How to equalize cost-effectiveness and training in developed and developing countries? How to expand robotic surgery for public health patients? What would be the procedures in which robotic surgery is more advantageous or could be waived? How to train an entire institution in safety protocols, in avoiding near-errors? How to be prepared for factors associated with complications, depending on patient characteristics, dependent on the surgical team, or dependent on machinery statements?

How quickly will we be able to incorporate all upcoming technologies? How to prepare leadership teams, mentoring and training among surgeons, nurses, and bioengineers? What are the medical specialties in which robotic surgery is already established, and which are those in which it is still taking its first steps?

A robotic surgery program depends on numerous professionals involved in its design, installation, pre-, intra-, and postoperative. Therefore, to acquire the essential foundations of an activity that involves health, bioengineering professionals, administrators, managers, etc., we understand that it is necessary to have a work rarely found in the currently available literature.

A publication that contains all of these basic principles allows readers to be prepared to delve deeper into the specific areas of robotic surgery. Therefore, we planned a handbook with almost 80 chapters, which is divided into three sections: Section I is dedicated to the introduction and history of robotic surgery, robotic bioengineering, and technology. Section II is focused on Education, training, administrative, nurse, medical and administrative teams, and bioethical aspects. Section III describes the clinical use of robotic surgery, in all medical specialties that practice robotic surgeries. To this task, we invited several international key opinion leaders and asked them to prepare basic and intermediate-level chapters, being intelligible to both students and healthcare professionals (surgeons, students, residents, and nurses), as for professional and students of bioengineering, and health managers.

The authors were given full intellectual freedom to express their opinions, offering the broadest view of the current scenario without editorial biases. To better consolidate knowledge, at the end of each chapter, there are two succinct sections: *Future scenario*, in which, in a few paragraphs, the authors discuss future trends and achievements in their area, and the *Key points*: up to six bullet sentences, which summarize the essential information of each chapter.

Videos will be accessible in a dedicated repository: https://data.mendeley.com/datasets/yzhtwf469w/1. We would like to express our sincere thanks and appreciation to all the authors, who devoted their time and efforts to disseminate holistic knowledge of all areas and actors involved in the theater of contemporary robotic surgery.

*Stênio de Cássio Zequi*
*Hongliang Ren*

# Acknowledgments

Stênio de Cássio Zequi and Hongliang Ren:

We would like to extend our sincerest thanks to the people who contributed to this publication:

- The authors who dedicated their time to knowledge in the production of their chapters.
- Miguel Romanello Joaquim, Research Assistant and Masters in Bioengineering/Pre-Med student at the University of Pennsylvania, for his help in suggesting topics related to technology.
- Dr. Rafael Ribeiro Zanotti, MD, MSc, robotic surgeon at our institution, for his full support in the production of our video repository.
- To Dr. Rene Sotelo for his valuable collaboration in suggesting and contacting some authors of chapters related to patient safety.
- The Elsevier's team: Sonnini Ruiz Yura, Senior Acquisitions Editor, for the invitation to this production; Isabella Conti Silva, Editorial Project Manager, for the uninterrupted monitoring; and Jeromel Tenorio, Editorial Assistant II, for contact with the authors and Surya Narayanan Jayachandran, Senior Project Manager.
- To the Urology Department and all surgeons, anesthesiologists, residents, fellow nurses, and administrative and back-office staffs at the A.C. Camargo Cancer Center, for their commitment and for helping to build our successful institutional robotic surgery program.
- Everyone who directly or indirectly contributed to this work.
- And finally, to all our patients, who are the greatest motivation of our mission.

# Introduction, technology basics, evolution and next frontiers of surgical robots

# 1

# Introduction

*Stênio de Cássio Zequi*[a,b,c] *and Hongliang Ren*[d,e]

[a]Urology Division, Referral Center for Urological Tumors and Graduate School, A.C. Camargo Cancer Center, São Paulo, São Paulo, Brazil [b]Surgery-Anatomy-Urology, Faculty of Medicine-USP-FMRP, São Paulo University-Ribeirão Preto, Ribeirão Preto, São Paulo, Brazil [c]Department of Urology, Graduate School, Federal University of São Paulo-UNIFESP, São Paulo, São Paulo, Brazil [d]Department of Electronic Engineering, The Chinese University of Hong Kong, Sha Tin, Hong Kong [e]Shun Hing Institute of Advanced Engineering (SHIAE), The Chinese University of Hong Kong, Sha Tin, Hong Kong

## Introduction

Over the past three decades, robotic surgery (RS) has undergone rapid and exponential advancement, transitioning from its conceptualization and early prototypes to the practical implementation of "master-slave" robotic systems in operating rooms worldwide.

RS is widely used in daily clinical practice in several medical specialties and has been investigated for many other medical applications. In this way, RS is achieving more and more acceptance in the surgical community, often before being approved as superior by the scientific literature [1].

This development has occurred in the face of its recognized advantages, such as more-precise and less-invasive procedures, minimal rates of surgical bleeding and blood transfusions, reduced risks of surgical site infections and deep vein thrombosis [2], reduced surgical complications, and the ability to reach challenging surgical zones in anatomical areas with difficult access or those close to delicate and vital neurovascular or endocrine structures, which historically required aggressive open surgical access. Additionally, RS leads to proportionately lower postoperative pain and reduced consumption of analgesics as well as shorter hospital stays and quicker recovery for normal activities [2–4].

However, many of these advantages have also been long obtained with minimally invasive surgical procedures (such as video laparoscopic endoscopic surgeries). Thus, more advantages were necessary to justify the exponential growth of RS. We can say that for surgeons, surgical robots allow working under better ergonomic conditions, resulting in fewer occupational joint articular diseases and a longer professional career in comparison with traditional open or video laparoscopic surgeries [5–7].

Other RS advantages are the safety systems that can interrupt undesired movements such as tremors, hierarchical levels of safety alarms, enhanced vision with an advanced three-dimensional optical lens, virtual and augmented reality, articulated surgical movements under surgeon command through the "endowrist" with degrees of freedom similar or superior to humans. Software allowing patient positioning motions or changing surgical ports during surgery are some time-saving characteristics of RS. Other advances that can save time for health professionals include picture-in-picture views, joining images of previous scans or real-time ultrasound or radioscopy, microscopic images, surgical navigation, and live teleinteractions with proctors or other colleagues. For patients, advances can offer better functional and aesthetic results, such as fluorescence surgery can lead to a better evaluation of tissular vascularization for the anastomosis of suspicious lymphatic areas as well as more precise movements in limited surgical spaces and reduced port or natural orifice access. Robots advance concomitantly, and dedicated surgical devices have progressed, resulting in new technologies of electric surgery with more potent vessel sealers, surgical staplers, clip applicators, etc. [8].

Regarding bioengineering, the transformations required to develop these modern equipments were due to the incorporation of advances in several areas of human knowledge related to the robotics industry, as: agriculture, reality, naval, aerospatial, maritime, military, geological, etc. and also, in the computational, optical, and game industries, among others. The progressive incorporation of artificial intelligence (AI), augmented virtual reality, simulators, and teaching revolutionized surgical training and teaching protocols. This reduced the need for need for or training in human patients or in cadaveric models, with its inherent bioethical dilemmas and legal risks and the high costs involved [9–16].

One of the original motivations for RS was for the military to allow surgeries to be directed at long distances from patients; the era of telementoring and proctoring has arrived. This innovation has permitted investigational surgeries and, at the same time, allowed tutoring and splitting of duties. These challenges will surely be solved through the modern Internet and high-speed and high-quality data traffic, etc. [13,14,16–18]. Autonomous robots are being tested in several areas, including urban cars, aerospatial medicine, and some surgical prototypes [19,20]. AI linked with robots may help surgeons perform their procedures or even replace them in some surgical steps [16,19,20].

Much of this discussion above has occurred in an era dominated by one company through the Da Vinci Robotic models. In this "Da Vinci era", an unedited scenario in medical education was presented in which surgeons might be formed and certified by the industry instead of traditional surgical teaching certified by health institutions. Additionally, these procedures were limited to centers with surgical platforms installed (this problem was more prominent in underdeveloped countries, with fewer installed robots and limited access per professional). Another debate was born that was similar to the aviation industry in which a pilot certified to fly in a specific model is not allowed to fly in another model. This might also occur in the surgical robotic industry: surgeons certified in a specific robotic model, might be authorized to perform surgeries in robotic machines created by distinct companies? Or would it be allowed to them practice surgeries with different robotic models developed by the same company that they were previously certified in?

Several new surgical platforms have been developed and launched in Asia, Europe, and North America in recent years [12,21]. Gradually, national and international health authorities have authorized these new models, followed by global commercialization. This leads to a new scenario requiring bioengineering and health professionals. For engineers, these new job opportunities are opened, and increasingly, multiple novice competencies and subspecialties are required. It has moving them ahead for unpreceded avenues of knowledge and illimited discovering. For surgeons, nurses, anesthesiologists and medical residents or students, the multiplicity of these machines promote overture for several medical specialties. However, it offers them challenges in using these new machines at a speed that is difficult to achieve. In the near future, we will have doctors or nurses skilled in Robot "A" for radical prostatectomy but not specialized at using robots "B" or "C" for the same surgery. How do we get autonomy or overcome the learning curve for each machine for the same procedures? How do we compare the surgical outcomes and cost effectiveness between Machines "A," "B," "C," etc.? What about the costs for teaching, training, and simulating hours for each specific robot for surgeons or nurses? How will we pay for it? How will special medical societies determine the "bar levels" or the test "cut-offs" to accredit someone as "a specialist" in each robotic machine? Must hospitals have several robotic models? In the future, will we have polyvalent robots acting in several specialties?

Beyond the engineers, the doctors and nurses, and the patients, we must not forget other actors. The market of RS moves billions of dollars: Patients seek robots and robotic surgeons. Surgeons are more and more indicating robotic procedures. Health payers must be prepared to reimburse hospitals, health professionals. Nosocomial institutions must pay for technical assistance, suffer for machine depreciation, and must continuously upgrade their robotic installations. The surgical industry always develops new robotic instruments and surgical goods to be commercialized. Although the minimally invasive surgeries result for the patients in a quick recovery of professional activities, and lessen their incoming losses, these processes are accompanied by the progressive increase in medical inflation [22,23].

For this challenging scenario, hospital administrative teams, managers, and health insurance companies must adapt at the same velocity as these transformations. The cost effectiveness of RS has not been totally understood in the face of the scarce literature, suggesting that the higher the caseload procedures, the lower the surgical complications and the more cost-effective robotic surgery is [24–26]. However, due to extreme variability according to distinct surgical procedures, each surgeon's expertise, each country's environmental and economic realities, and regional competition, the center must find its own solutions regarding the viability of these surgical modalities [24–26].

In this way, robotic centers usually enhance their reputation and can attract patients and customers from their market competitors [27,28].

To achieve excellence under the pillars of value-based healthcare, in their robotic programs, healthcare institutions must offer maximum safety to patients during their surgeries. Its hospitality teams must provide the best possible experience for patients who seek this technology. The promotion of mentoring and leadership programs among doctors and nurses must be constantly encouraged. It is also necessary to offer adequate infrastructure to clinical and biomedical engineering teams, to maximize performance and reduce equipment depreciation [29–31].

The impact of actual robotic surgery scenario must be viewed optimistically, similarly that was verified with the advent and popularization of videolaparoscopy and other minimally invasive proccedures as: surgeries, digestive endoscopy, percutaneous image-guided biopsies and ablations, microsurgeries, and laser surgery etc. Although provocative, the impact of actual robotic surgery must be viewed optimistically, similarly that was verified with the advent and popularization of videolaparoscopy and several minimally invasive procedures as: endoscopies, percutaneous image-guided biopsies and ablations, microsurgeries, and laser ablations, etc.

In the face of these issues and with the goal of providing a comprehensive resource for individuals seeking to delve into almost all aspects of robotic surgery, this book was created as a "handbook." It targets a wide range of professionals, including surgeons, bioengineers, nurses, medical students, residents, and healthcare administrators. The volume aims to cover the fundamental principles of diverse areas within the realm of robotic surgery.

Esteemed global thought leaders in the field were invited to contribute as authors, tasked with preparing chapters that offer basic to intermediate levels of understanding in medicine. These chapters are designed to be accessible to engineers while also ensuring that the bioengineering sections can be comprehended by doctors (including general surgeons, residents, and academics), nurses, other healthcare professionals, hospital managers, administrators, stakeholders, and similar individuals.

After finishing this book, readers will probably be interested in more specialized literature in their respective areas.

This book is divided into three sections: **Section I**, introduces the history and evolution of robotic surgery. There are 15 chapters focused on robotic bioengineering of actual and future robotic machines, basic technical aspects (kinematics, robotic manipulators, haptics, navigation and calibration, fluorescence, ultrasound guidance surgeries, robotic staplers, artificial muscles, new robotic models, new technological frontiers, image-guided robotic therapies, AI systems, virtual reality, simulators, cooperation and remote mode operations systems, etc.). The **Section II—Education**, promotes an overview on the administrative processes of robotic surgery, focusing on the training, leadership, mentoring, proctoring, of robotic surgical teams. Also, we visit issues such as: optimization of robotic surgical rooms rotativity, anesthesiology and perioperative care. In this section, also, we cover the main aspects of patient safety, evolving health professionals, and institutional infra-structures. The second part of this Section II, is dedicated to preventing complications and surgical accidents; identifying factors related to the machine, or related to the patients, to the positioning, or due the patient' previous surgeries, etc.; and reinforces the role of proactive prevention of complications. In the end of Section II, the chapters discuss the cost effectiveness of robotic surgery, including this scenario in developing countries. Ethical and medicolegal aspects of robotics, and the certification of professionals and credentialing of robotic centers in developed and developing nations, and general advantages and disadvantages of RS are also debated in these manuscripts.

The **Section III**, reviews and updates the clinical use of robotic surgery, in all medical specialties. There are almost 50 chapters prepared by worldwide recognized key opinion leaders surgeons, debating the most realized robotic surgeries, as radical prostatectomy, per example, and the less realized robotic procedures. These chapters discuss the main technical aspects: positioning, trocars insertions, tips and tricks, results, and complications. The specialties are Urology (kidney and upper tract, prostate, bladder, urinary diversion, pelvic and retroperitoneal lymphadenectomy, lithotripsy and urodynamics); Head and Neck (head neck and transoral surgery, neck lymphadenectomies), and Otolaryngology (sleep apnea and benign diseases); Thoracic and Cardiovascular Surgery, General Surgery and Digestive system, (esophagus, gastric, hepatobiliary, foregut, pancreatic surgery, hernias and abdominal wall surgery, Colorectal, Bariatric, robotic surgery); and Gynecology (general and endometriosis surgeries). The actual frontiers of RS, as the single port use and controverted areas of use or robotics, such as Breast surgery and Pediatric RS, are included, too. Abdominal Organs and Kidney Transplantations are in separated texts. RS in Orthopedy (joint and hip), and in Neurosurgery (cranial and spine) chapters are included. More recent specialties in which RS is in the initial steps or under investigation, such as ophthalmology, endoscopy, microsurgery, and plastic surgery, are also discussed.

Each chapter will conclude with an examination of future perspectives related to the subject matter, accompanied by three to six bullet points encapsulating the Key points of the text. These succinct sentences will highlight the most crucial and pertinent information discussed in the chapter.

Additionally, to reinforce learning for our readers, many of our authors produced surgical videos or podcasts regarding their respective chapters. These materials can be freely accessed at our links to our audiovisual repository: https://data.mendeley.com/datasets/yzhtwf469w/1.

Our objective is to provide readers with the essential foundations for their advancements in robotic surgery while also offering researchers comprehensive and up-to-date yet concise answers to their inquiries across all aspects and future trends within the realm of robotic surgery.

# References

[1] Horn D, Sacarny A, Zhou A. Technology adoption and market allocation: the case of robotic surgery. J Health Econ 2022;86, 102672. https://doi. org/10.1016/j.jhealeco.2022.102672. Epub 2022 Sep 14 PMID: 36115136.

[2] JWF C, Khetrapal P, Ricciardi F, Ambler G, Williams NR, Al-Hammouri T, Khan MS, Thurairaja R, Nair R, Feber A, Dixon S, Nathan S, Briggs T, Sridhar A, Ahmad I, Bhatt J, Charlesworth P, Blick C, Cumberbatch MG, Hussain SA, Kotwal S, Koupparis A, McGrath J, Noon AP, Rowe E, Vasdev N, Hanchanale V, Hagan D, Brew-Graves C, Kelly JD, iROC Study Team. Effect of robot-assisted radical cystectomy with intracorporeal urinary diversion vs open radical cystectomy on 90-day morbidity and mortality among patients with bladder cancer: a randomized clinical trial. JAMA 2022;327(21):2092–103. https://doi.org/10.1001/jama.2022.7393. PMID: 35569079. PMCID:PMC9109000.

[3] Liu L, Lewis N, Mhaskar R, Sujka J, DuCoin C. Robotic-assisted foregut surgery is associated with lower rates of complication and shorter post-operative length of stay. Surg Endosc 2023;37(4):2800–5. https://doi.org/10.1007/s00464-022-09814-6. Epub 2022 Dec 7 PMID: 36477641.

[4] Steffens D, McBride KE, Hirst N, Solomon MJ, Anderson T, Thanigasalam R, Leslie S, Karunaratne S, Bannon PG. Surgical outcomes and cost analysis of a multi-specialty robotic-assisted surgery caseload in the Australian public health system. J Robot Surg 2023;17(5):2237–45. https://doi.org/10.1007/s11701-023-01643-6. Epub 2023 Jun 8 PMID: 37289337. PMCID:PMC10492768.

[5] Diana M, Marescaux J. Robotic surgery. Br J Surg 2015;102(2):e15–28. https://doi.org/10.1002/bjs.9711. PMID: 25627128.

[6] Wee IJY, Kuo LJ, Ngu JC. A systematic review of the true benefit of robotic surgery: ergonomics. Int J Med Robot 2020;16(4), e2113. https://doi. org/10.1002/rcs.2113. Epub 2020 May 6 PMID: 32304167.

[7] Hayashi MC, Sarri AJ, Pereira PASV, Rocha MM, Zequi SC, Machado MT, de Souza AH, Magno LAV, Faria EF. Ergonomic risk assessment of surgeon's position during radical prostatectomy: laparoscopic versus robotic approach. J Surg Oncol 2023;128(8):1453–8. https://doi.org/10.1002/jso.27419. Epub 2023 Aug 21 PMID: 37602508.

[8] Williamson T, Song SE. Robotic surgery techniques to improve traditional laparoscopy. JSLS 2022;26(2), e2022.00002. https://doi.org/10.4293/JSLS.2022.00002. PMID: 35655469. PMCID: PMC9135605.

[9] Vasey B, Lippert KAN, Khan DZ, Ibrahim M, Koh CH, Layard Horsfall H, Lee KS, Williams S, Marcus HJ, McCulloch P. Intraoperative applications of artificial intelligence in robotic surgery: a scoping review of current development stages and levels of autonomy. Ann Surg 2023;278 (6):896–903.

[10] Gholizadeh M, Bakhshali MA, Mazlooman SR, Aliakbarian M, Gholizadeh F, Eslami S, Modrzejewski A. Minimally invasive and invasive liver surgery based on augmented reality training: a review of the literature. J Robot Surg 2023;17(3):753–63. https://doi.org/10.1007/s11701-022-01499-2. Epub 2022 Nov 28 PMID: 36441418.

[11] Park C, Shabani S, Agarwal N, Tan L, Mummaneni PV. Robotic-assisted surgery and navigation in deformity surgery. Neurosurg Clin N Am 2023;34(4):659–64. https://doi.org/10.1016/j.nec.2023.05.002. Epub 2023 Jun 26 PMID: 37718112.

[12] Alip SL, Kim J, Rha KH, Han WK. Future platforms of robotic surgery. Urol Clin North Am 2022;49(1):23–38. https://doi.org/10.1016/j.ucl.2021.07.008. Epub 2021 Oct 25 PMID: 34776052.

[13] Pandav K, Te AG, Tomer N, Nair SS, Tewari AK. Leveraging 5G technology for robotic surgery and cancer care. Cancer Rep (Hoboken) 2022;5 (8), e1595. https://doi.org/10.1002/cnr2.1595. Epub 2022 Mar 9 PMID: 35266317. PMCID:PMC9351674.

[14] Howard KK, Makki H, Novotny NM, Mi M, Nguyen N. Value of robotic surgery simulation for training surgical residents and attendings: a systematic review protocol. BMJ Open 2022;12(6), e059439. https://doi.org/10.1136/bmjopen-2021-059439. PMID: 35701063. PMCID: PMC9198707.

[15] Siqueira-Batista R, Souza CR, Maia PM, Siqueira SL. Robotic surgery: bioethical aspects. Arq Bras Cir Dig 2016;29(4):287–90. https://doi.org/10.1590/0102-6720201600040018. PMID: 28076489. PMCID:PMC5225874.

[16] O'Sullivan S, Nevejans N, Allen C, Blyth A, Leonard S, Pagallo U, Holzinger K, Holzinger A, Sajid MI, Ashrafian H. Legal, regulatory, and ethical frameworks for development of standards in artificial intelligence (AI) and autonomous robotic surgery. Int J Med Robot 2019;15(1), e1968. https://doi.org/10.1002/rcs.1968. PMID: 30397993.

[17] Ballantyne GH. Robotic surgery, telerobotic surgery, telepresence, and telementoring. Review of early clinical results. Surg Endosc 2002;16 (10):1389–402. https://doi.org/10.1007/s00464-001-8283-7. Epub 2002 Jul 29 PMID: 12140630.

[18] Nakanoko T, Oki E, Ota M, Ikenaga N, Hisamatsu Y, Toshima T, Kanno T, Tadano K, Kawashima K, Ohuchida K, Morohashi H, Ebihara Y, Mimori K, Nakamura M, Yoshizumi T, Hakamada K, Hirano S, Ikeda N, Mori M. Real-time telementoring with 3D drawing annotation in robotic surgery. Surg Endosc 2023. https://doi.org/10.1007/s00464-023-10521-z. Epub ahead of print PMID: 37935920.

[19] Saeidi H, Opfermann JD, Kam M, Wei S, Leonard S, Hsieh MH, Kang JU, Krieger A. Autonomous robotic laparoscopic surgery for intestinal anastomosis. Sci Robot 2022;7(62), eabj2908. https://doi.org/10.1126/scirobotics.abj2908. Epub 2022 Jan 26 PMID: 35080901. PMCID: PMC8992572.

[20] Pantalone D. Surgery in the next space missions. Life (Basel) 2023;13(7):1477. https://doi.org/10.3390/life13071477. PMID: 37511852. PMCID: PMC10381631.

[21] Wilson M, Badani K. Competing robotic systems: a preview. Urol Clin North Am 2021;48(1):147–50. https://doi.org/10.1016/j.ucl.2020.09.007. Epub 2020 Nov 5 PMID: 33218589.

[22] Rodrigues Martins YM, Romanelli de Castro P, Drummond Lage AP, Alves Wainstein AJ, de Vasconcellos Santos FA. Robotic surgery costs: revealing the real villains. Int J Med Robot 2021;17(6), e2311. https://doi.org/10.1002/rcs.2311. Epub 2021 Aug 21 PMID: 34268880.

[23] Okhawere KE, Shih IF, Lee SH, Li Y, Wong JA, Badani KK. Comparison of 1-year health care costs and use associated with open vs robotic-assisted radical prostatectomy. JAMA Netw Open 2021;4(3), e212265. https://doi.org/10.1001/jamanetworkopen.2021.2265. PMID: 33749767. PMCID: PMC7985723.

[24] Trinh QD, Bjartell A, Freedland SJ, Hollenbeck BK, Hu JC, Shariat SF, Sun M, Vickers AJ. A systematic review of the volume-outcome relationship for radical prostatectomy. Eur Urol 2013;64(5):786–98. https://doi.org/10.1016/j.eururo.2013.04.012. Epub 2013 Apr 19 PMID: 23664423. PMCID:PMC4109273.

[25] Van den Broeck T, Oprea-Lager D, Moris L, Kailavasan M, Briers E, Cornford P, De Santis M, Gandaglia G, Gillessen Sommer S, Grummet JP, Grivas N, Lam TBL, Lardas M, Liew M, Mason M, O'Hanlon S, Pecanka J, Ploussard G, Rouviere O, Schoots IG, Tilki D, van den Bergh RCN, van der Poel H, Wiegel T, Willemse PP, Yuan CY, Mottet N. A systematic review of the impact of surgeon and hospital caseload volume on

oncological and nononcological outcomes after radical prostatectomy for nonmetastatic prostate cancer. Eur Urol 2021;80(5):531–45. https://doi.org/10.1016/j.eururo.2021.04.028. Epub 2021 May 5 PMID: 33962808.

[26] Forsmark A, Gehrman J, Angenete E, Bjartell A, Björholt I, Carlsson S, Hugosson J, Marlow T, Stinesen-Kollberg K, Stranne J, Wallerstedt A, Wiklund P, Wilderäng U, Haglind E. Health economic analysis of open and robot-assisted laparoscopic surgery for prostate cancer within the prospective multicentre LAPPRO trial. Eur Urol 2018;74(6):816–24. https://doi.org/10.1016/j.eururo.2018.07.038. Epub 2018 Aug 22 PMID: 30143383.

[27] Kaye DR, Mullins JK, Carter HB, Bivalacqua TJ. Robotic surgery in urological oncology: patient care or market share? Nat Rev Urol 2015;12 (1):55–60. https://doi.org/10.1038/nrurol.2014.339. Epub 2014 Dec 23 PMID: 25535000.

[28] Kuklinski D, Vogel J, Henschke C, Pross C, Geissler A. Robotic-assisted surgery for prostatectomy—does the diffusion of robotic systems contribute to treatment centralization and influence patients' hospital choice? Heal Econ Rev 2023;13(1):29. https://doi.org/10.1186/s13561-023-00444-9. PMID: 37162648. PMCID: PMC10170785.

[29] Lawrie L, Gillies K, Duncan E, Davies L, Beard D, Campbell MK. Barriers and enablers to the effective implementation of robotic assisted surgery. PLoS One 2022;17(8), e0273696. https://doi.org/10.1371/journal.pone.0273696. PMID: 36037179. PMCID: PMC9423619.

[30] Randell R, Honey S, Hindmarsh J, Alvarado N, Greenhalgh J, Pearman A, Long A, Cope A, Gill A, Gardner P, Kotze A, Wilkinson D, Jayne D, Croft J, Dowding D. A realist process evaluation of robot-assisted surgery: integration into routine practice and impacts on communication, collaboration and decision-making. Southampton (UK): NIHR Journals Library; 2017. PMID: 28813131.

[31] Lucas SR, Schabowsky CN. Clinical engineering in robotic surgery programs. Biomed Instrum Technol 2016;50(6):415–20. https://doi.org/10.2345/0899-8205-50.6.415. PMID: 27854493.

# History of robotic surgery: Trends in autonomy levels and enabling technologies

*Hongliang Ren*[a,b] *and Stênio de Cássio Zequi*[c,d,e]

[a]Department of Electronic Engineering, The Chinese University of Hong Kong, Sha Tin, Hong Kong [b]Shun Hing Institute of Advanced Engineering (SHIAE), The Chinese University of Hong Kong, Sha Tin, Hong Kong [c]Urology Division, Referral Center for Urological Tumors and Graduate School, A.C. Camargo Cancer Center, São Paulo, São Paulo, Brazil [d]Surgery-Anatomy-Urology, Faculty of Medicine-USP-FMRP, São Paulo University- Ribeirão Preto, Ribeirão Preto, São Paulo, Brazil [e]Department of Urology, Graduate School, Federal University of São Paulo-UNIFESP, São Paulo, São Paulo, Brazil

## Introduction

Robotic surgery has revolutionized the practice of surgery, by performing complex procedures with improved outcomes. The history of robotic surgery is characterized by significant advancements in autonomy levels [1], with the integration of enabling technologies playing a crucial role in this progression.

The early stages of robotic surgery were characterized by the development of telemanipulation systems, which allowed surgeons to remotely control robotic arms with enhanced precision and dexterity. These systems laid the foundation for the birth of robotic surgery, providing surgeons with improved control, stability, and range of motion. The success of telemanipulation systems sparked further advancements in the field, leading to the integration of more advanced autonomous features that have transformed the practice of surgery.

## Early stages: Telemanipulation and the birth of robotic surgery

The early stages of robotic surgery were marked by the development of telemanipulation systems, which laid the foundation for the birth of robotic surgery as we know it today. Telemanipulation systems allowed surgeons to remotely control robotic arms with increased precision and dexterity, overcoming the limitations of hand tremors and providing an improved range of motion.

The concept of telemanipulation in early robotic surgical system [2] consisted of a robotic arm equipped with surgical instruments that could be controlled by a surgeon from a remote console. It allowed surgeons to perform precise movements with improved accuracy. The robotic arm faithfully replicated the surgeon's hand movements, isolated unintended hand tremors or imprecise hand motions. This telemanipulation capability opened up new possibilities for surgical interventions by providing surgeons with enhanced control and stability.

Early telemanipulation systems laid the foundation for the evolution of robotic surgery. Surgeons and engineers identified the potential for incorporating advanced autonomous features. Telemanipulation systems enabled precise remote control of robotic arms, eliminating hand tremors and improving range of motion.

## Advancements in autonomy levels

## Level 1 autonomy: Enhanced telemanipulation

The initial advancements in robotic surgery focused on enhancing telemanipulation capabilities. Systems such as the da Vinci Surgical System (Intuitive Surgical Inc.) in the late 1990s provided surgeons with improved dexterity, 3D visualization, and ergonomic benefits.

The level 1 autonomy surgical system enhanced surgeons' capabilities through improved dexterity and 3D visualization. The system enabled precise maneuvers, reducing errors. The da Vinci Surgical System's high-definition 3D camera provided detailed views, enhancing depth perception and spatial awareness. Surgeons could navigate complex anatomical structures more effectively. Furthermore, level 1 autonomy systems addressed the issue of surgeon fatigue by providing ergonomic benefits. The robotic arms were designed to be highly maneuverable, allowing surgeons to operate with increased comfort and reduced physical strain. The system also offered adjustable controls and ergonomic consoles, enabling surgeons to maintain a comfortable posture during lengthy procedures. These advanced telemanipulation systems had a profound impact on surgical outcomes. The elimination of hand tremors and improved range of motion provided by level 1 autonomy systems significantly reduced the risk of errors and complications during surgery.

Level 1 autonomy systems enhanced surgical outcomes by reducing the limitations of traditional laparoscopic surgery. Surgeons could remotely control robotic arms with increased precision, eliminating hand tremors and providing improved range of motion. These systems also facilitated improved ergonomics, allowing surgeons to operate comfortably and safely for extended periods with active mechanical guiding constraints (e.g., virtual fixtures [3]).

## Level 2 autonomy: Surgical task-level assistance

The next-stage level 2 autonomy in the evolution of robotic surgery marked a significant advancement with the incorporation of surgical assistance features at certain task levels. These features aimed to provide surgeons with real-time feedback, guidance, and enhanced situational awareness during specific tasks, e.g., biopsy or needle insertion [4], discretely commanded by human operators. The integration of preoperative planning, intraoperative imaging, and augmented reality technologies played a crucial role to provide real-time feedback and guidance in this stage of development [5,6].

One of the key advancements in level 2 autonomy was the utilization of advanced imaging techniques. Computed tomography (CT), magnetic resonance imaging (MRI), and ultrasound were employed to generate detailed anatomical models of the patient's anatomy [7]. These models could be precisely registered with the surgical field, providing surgeons with a comprehensive understanding of the patient's anatomy before and during the procedure.

During surgery, these anatomical models were overlaid onto the surgeon's view using augmented reality technologies. This real-time guidance allowed surgeons to visualize critical structures, such as blood vessels, nerves, and tumors, that may not be readily visible to the naked eye. Surgeons could navigate complex anatomical regions with greater confidence and accuracy.

The integration of augmented reality also facilitated the visualization of important surgical data, such as vital signs, anesthesia information, and instrument tracking. This real-time feedback further enhanced the surgeon's situational awareness, enabling them to make informed decisions and adjustments during the procedure. The incorporation of surgical assistance features in level 2 autonomy systems significantly reduced human error and improved surgical accuracy. Surgeons could rely on the real-time guidance provided by the augmented reality overlays, ensuring precise instrument placement and minimizing the risk of unintentional damage to surrounding tissues.

Moreover, these automated features allowed for better preoperative planning. Surgeons could simulate the surgical procedure on the anatomical models derived from imaging data, identifying potential challenges and optimizing the surgical approach. This preoperative simulation enhanced surgical efficiency and reduced operating time.

Level 2 autonomy in robotic surgery introduced surgical assistance features that incorporated preoperative planning, intraoperative imaging, and augmented reality technologies. These advancements provided surgeons with real-time feedback, guidance, and enhanced situational awareness during procedures. The integration of advanced imaging techniques and augmented reality overlays improved surgical accuracy, reduced human error, and facilitated better preoperative planning. Level 2 autonomy marked a significant leap forward in the field of robotic surgery, setting the stage for further advancements in autonomous surgical systems.

## Level 3 autonomy: Collaborative surgery

Collaborative surgical robots are capable of working alongside surgeons for both interactive task planning and collaborative robot-assisted tasks [8,9]. These systems analyzed surgical data, made predictions, and provided collaborative assistance in decision-making processes. Collaborative surgical robots employed algorithms to analyze real-time surgical data, such as vital signs, anatomical structures, and instrument positions. These systems could provide feedback, suggestions, and warnings to assist surgeons during critical phases of the procedure.

Equipped with perceptual capabilities, they can understand the surgical scenario, plan and execute tasks, and adapt their plan as required. Similar to level 2, the machine assumes control during the execution of tasks [10]. The collaboration between surgeons and autonomous robots resulted in improved surgical outcomes and reduced surgical complications.

## Level 4 autonomy: Supervised autonomy

Level 4 represents high autonomy systems capable of interpreting preoperative and intraoperative information, task planning, and robot-assisted task executions under human supervision [11]. They can devise an interventional plan consisting of a sequence of tasks, execute the plan autonomously, and adjust it if necessary. A surgeon oversees the system using discrete control. Recent developments in robotic surgery have led to the emergence of supervised autonomy. These systems leverage artificial intelligence (AI) and machine learning (ML), and computer vision (CV) to perform complex surgical procedures with minimal human intervention. Surgeons act as supervisors, ensuring patient safety while the robotic system carries out the procedure.

Supervised autonomy systems combine the capabilities of AI, ML, and computer vision to analyze surgical data, make predictions, and perform specific tasks autonomously. These systems continuously learn from past experiences and adapt their behavior to improve performance. Surgeons oversee the actions of the robotic system, maintaining control and accountability throughout the procedure.

## Level 5 autonomy: Full autonomy

Full autonomy in robotic surgery [11], where robots can perform entire procedures independently, is an ongoing area of research and development. Achieving level 5 autonomy requires addressing technical, ethical, and legal challenges to ensure patient safety and accountability.

Full autonomy systems would be capable of independently performing surgical procedures from start to finish, without the need for human intervention. These systems would rely on advanced AI algorithms, ML models, and sophisticated sensor technologies to analyze surgical data, make decisions, and execute precise maneuvers. However, achieving full autonomy in surgical procedures poses significant challenges and demands extensive research and validation.

## Enabling technologies

### Robotics and mechatronics

The combination of robotics and mechatronics has been instrumental in the development of surgical robots. Actuators, sensors, and robotic arms enable precise and controlled movements, while haptic feedback systems provide surgeons with a sense of touch. The continuous refinement of these technologies has contributed to the increased autonomy levels in robotic surgery.

Robotic arms are a key component of surgical robots, designed to mimic the movements of a surgeon's hands with improved accuracy and stability. These arms are equipped with various joints and mechanisms that allow them to rotate, bend, and articulate with multiple degrees of freedom. This flexibility enables surgeons to reach challenging anatomical locations and perform delicate maneuvers with greater ease and precision [12].

Actuators, such as electric motors or hydraulic systems, provide the necessary power to drive robotic arms. These actuators convert electrical or hydraulic energy into mechanical motion, enabling the arms to move in a controlled manner. By carefully controlling the actuators, surgeons can manipulate the robotic arms with fine movements, enabling precise surgical interventions.

Sensors are another crucial component of robotic surgery systems. These sensors provide feedback to the system, allowing it to perceive and respond to its environment. For example, force sensors can measure the amount of force applied during a surgical procedure, providing valuable information to ensure safe and accurate interactions between the robot and the patient's tissues.

Haptic feedback systems have also been integrated into surgical robots to provide surgeons with a sense of touch. These systems use sensors and actuators to simulate the tactile sensations that a surgeon would experience during manual surgery. By receiving haptic feedback, surgeons can better perceive the forces exerted on tissues, assess tissue characteristics, and perform delicate tasks with enhanced precision.

The continuous development and refinement of robotics and mechatronics technologies have contributed to the advancement of surgical robots and the increasing levels of autonomy. These advancements have allowed for more complex and autonomous tasks to be performed by robotic systems, reducing the reliance on direct human control and enabling the execution of tasks with greater efficiency, accuracy, and safety.

## Imaging technologies

The advancement of imaging technologies [13] has significantly contributed to the capabilities of robotic surgical systems. These technologies, such as high-resolution imaging modalities and fluorescence imaging, have revolutionized the visualization of anatomical structures and enhanced surgical precision.

One key imaging technology that has greatly impacted robotic surgery is three-dimensional (3D) visualization. Traditional two-dimensional imaging can sometimes limit surgeons' spatial orientation and depth perception. However, with 3D visualization, surgeons can have a more immersive and realistic view of the surgical field. This technology provides a three-dimensional representation of the anatomy, allowing surgeons to better understand the spatial relationships between structures. The improved depth perception facilitates more accurate tissue dissection, suturing, and precise manipulation of surgical instruments.

Intraoperative ultrasound is another imaging technology commonly used in robotic surgery. It involves the use of ultrasound probes to provide real-time imaging of internal structures during the procedure. Surgeons can visualize anatomical details, identify critical structures, and monitor changes in real-time. Intraoperative ultrasound is particularly valuable in procedures that require precise localization, such as tumor resections or organ preservation surgeries.

Fluorescence imaging is another imaging technology that has found valuable applications in robotic surgery. By utilizing fluorescent contrast agents, surgeons can visualize specific tissues or structures during surgery. This is particularly useful in procedures such as tumor resections, where the differentiation between healthy and diseased tissue is critical, or between well-vascularized virus tissues with reduced blood perfusion. Fluorescent contrast agents are selectively taken up by targeted tissues, making them stand out from the surrounding tissue when illuminated with specific wavelengths of light. Real-time feedback from fluorescence imaging systems enables surgeons to identify tumor margins, localize vital structures, and ensure complete tumor removal. This technology enhances surgical accuracy and reduces the risk of leaving behind residual tumor tissue.

Furthermore, advancements in imaging technologies have facilitated the integration of intraoperative imaging with robotic surgical systems. This integration allows surgeons to acquire real-time imaging data during the procedure, providing valuable information for decision-making and surgical guidance. For example, image-guided robotic systems can register preoperative data, with the live surgical field. This enables surgeons to navigate complex anatomical regions with greater precision, avoid critical structures, and improve the overall safety and outcomes of the procedure.

The evolution of imaging and sensing technologies has significantly enhanced the capabilities of robotic surgical systems. Three-dimensional visualization provides surgeons with a more immersive and realistic view of the surgical field, improving spatial orientation and depth perception. Fluorescence imaging enables the visualization of specific tissues or structures, aiding in procedures such as tumor resections. The integration of intraoperative imaging with robotic systems further enhances surgical guidance and decision-making.

## Sensing technologies

Sensing technologies play a crucial role in robotic surgery by providing surgeons with real-time feedback and enhancing their ability to interact with tissues and navigate the surgical field. These technologies include force sensors, tactile sensors, electromagnetic tracking, optical tracking, and shape sensing.

Force sensors are used to measure the forces and pressures applied during surgical procedures [14]. By integrating force sensors into robotic surgical instruments, surgeons can receive feedback on the amount of force exerted on tissues. This feedback allows for better control over delicate tissues, reducing the risk of damage or injury. Force sensing technology provides surgeons with a sense of touch, enabling them to perform tasks that require a delicate touch, such as suturing or tissue manipulation. Incorporating force feedback into robotic surgery has demonstrated positive impacts on surgical outcomes, reducing unintended damage and improving the success rate of procedures. Additionally, force sensing technology enables the detection of tumors during palpation.

Tactile sensors are another type of sensing technology used in robotic surgery. These sensors are designed to mimic the sense of touch, allowing surgeons to perceive the texture, stiffness, and compliance of tissues. By providing tactile feedback, these sensors enhance the surgeon's ability to assess tissue characteristics and make informed decisions during the procedure.

However, accurately sensing instrument-tissue interaction forces poses challenges due to limited space and the need for biocompatibility during surgical procedures. Researchers have developed sensor-based methods utilizing elastic structures and sensing elements to measure interaction and gripping forces. These methods can be broadly classified into two categories: electrical signal-based and optical signal-based.

Electrical signal-based sensors utilize principles like resistance change, capacitance change, and piezoelectric effect. Resistance change-based sensors were initially integrated into surgical instruments for force sensing in robotic surgery. Recent advancements in micro-electro-mechanical systems (MEMS) and materials science have expanded the exploration of capacitance change and piezoelectric-based sensors. However, these electrical signal-based methods are susceptible to measurement errors and interference in environments with electrical and magnetic disturbances.

On the other hand, optical signal-based sensors offer inherent advantages in mitigating interference issues. They have gained popularity in robotic surgery for force sensing purposes. Two commonly employed optical sensor types in surgical sensing are intensity modulation-based and wavelength modulation-based sensors. Intensity modulation-based sensors have a simple design and are easily implemented to calculate physical parameters such as force and displacement. However, wavelength modulation-based sensors face challenges related to miniaturization, limiting their scalability and affecting resolution and measurement range. One promising optical sensor element utilized for wavelength modulation is fiber Bragg gratings (FBG) [15,16]. FBG sensors have garnered considerable attention in the field of robotic surgery force sensing due to their miniaturization advantages and compatibility with electromagnetic environments. Nonetheless, optical sensors encounter limitations related to the bending optical fibers, which can lead to signal attenuation or loss.

Electromagnetic tracking and optical tracking are technologies used to monitor the position and movement of surgical instruments in real-time. Electromagnetic tracking utilizes electromagnetic fields to track the position and orientation of instruments equipped with electromagnetic sensors. Optical tracking, on the other hand, relies on cameras and markers to track the movement of instruments. These tracking technologies enable accurate spatial tracking, alignment with preoperative planning, and precise navigation within the surgical field. Surgeons can visualize the position and movement of instruments on a display, allowing them to make precise movements and adjustments during the procedure.

Intraoperative instrument shape sensing technologies [17] provide real-time feedback on the shape, position, and orientation of surgical instruments within the body. These technologies typically involve the integration of sensors, such as strain gauges or fiber optic systems, directly into the instruments. By monitoring the deformation and bending of the sensors, surgeons can accurately determine the shape and curvature of the instruments during the procedure. This information is then processed and displayed on a visualization system, allowing surgeons to visualize the exact shape and position of the instrument in relation to the surrounding anatomy. Intraoperative instrument shape sensing technologies enable surgeons to navigate complex anatomical structures with enhanced precision, manipulate instruments with greater accuracy, and avoid potential collisions or damage to adjacent tissues. This real-time feedback enhances the overall safety and efficacy of robotic surgical procedures, ultimately leading to improved patient outcomes. Continued advancements in instrument shape sensing technologies hold tremendous potential for further enhancing the capabilities and performance of robotic surgical systems.

Overall, sensing technologies in robotic surgery provide surgeons with real-time feedback and navigation capabilities. Force sensors and tactile sensors enable surgeons to have a sense of touch and better control over delicate tissues. Electromagnetic tracking, optical tracking, and intraoperative shape sensing technologies provide real-time feedback and visualization, enhancing surgical guidance, precision, and decision-making. These sensing technologies are instrumental in advancing the field of robotic surgery, improving surgical outcomes, and expanding the capabilities of surgeons.

## Artificial intelligence and machine learning

The integration of AI and ML algorithms has revolutionized robotic surgery by enabling autonomous decision-making and adaptive learning. These technologies analyze large datasets, extract meaningful insights, and assist in surgical planning, assistance, and prediction [18–22].

AI and ML techniques are used to process and interpret a vast amount of medical data, including patient demographics, preoperative imaging, surgical plans, and intraoperative sensor data. These algorithms can identify patterns, make predictions, and generate surgical recommendations based on the analysis of these data. For example, ML models can predict the risk of complications, estimate blood loss, or identify potential anatomical anomalies.

Furthermore, AI and ML algorithms enable adaptive learning in robotic surgical systems [21,22]. By continuously analyzing and learning from surgical data, these systems can improve their performance over time. This adaptive learning capability allows robotic systems to refine their surgical techniques, optimize outcomes, and adapt to variations in patient anatomy.

## Current applications and future prospects

### Clinical applications of robotic surgery

Robotic surgery has found applications in various surgical specialties, including otorhinolaryngology [23–25], urology [26], gynecology [27], cardiothoracic surgery [28], and general surgery. Procedures, such as prostatectomy, hysterectomy, and colorectal surgery, have benefited from the enhanced capabilities of robotic surgical systems.

In urology, robotic-assisted radical prostatectomy with visualization, precise tissue dissection, and improved suturing capabilities led to reduced blood loss, shorter hospital stays, and improved functional outcomes for patients. In gynecology, robotic surgery has been widely adopted for procedures such as hysterectomy and myomectomy. Robotic systems enable surgeons to perform complex cardiac surgeries through small incisions, minimizing trauma to the chest and reducing the risk of complications.

General surgery has also seen applications of robotic surgery in procedures such as colorectal surgery, bariatric surgery, and hernia repair. The robotic platform provides surgeons with improved access and precision, facilitating complex maneuvers and enabling minimally invasive approaches.

### Challenges and future directions

Despite the significant progress in robotic surgery, several challenges remain. Regulatory considerations, ethical concerns, data privacy, and liability issues need to be addressed for the widespread adoption of autonomous robotic systems. Furthermore, further research is required to refine autonomy algorithms, improve training data, and ensure the safety and reliability of autonomous surgical interventions.

One of the key challenges is establishing the safety and efficacy of autonomous surgical systems. Extensive testing, validation, and regulatory approvals are necessary to ensure that these systems meet the required standards and can be trusted to perform surgical procedures autonomously.

Ethical considerations surrounding the use of autonomous systems in surgery need to be carefully addressed. The responsibility for patient outcomes, accountability for errors or complications, and the role of human supervision in autonomous procedures are important ethical considerations that require careful deliberation.

## Future scenario

The future trajectory of robotic surgery hinges upon the ongoing development and advancement of autonomy levels. This trajectory necessitates the refinement of algorithms to optimize performance, the integration of adaptive learning capabilities to enhance system intelligence, and the incorporation of advanced imaging and sensing technologies for improved perception and decision-making. Furthermore, the successful implementation of autonomous robotic surgery calls for the diligent addressing of concerns pertaining to patient safety, data privacy, and liability.

Autonomous robotic surgery holds great potential in improving patient outcomes, reducing surgical complications, and enhancing the efficiency of healthcare delivery. However, challenges, such as the need for robust training data, regulatory hurdles, and ethical considerations, pose significant barriers to the widespread adoption of fully autonomous systems. Balancing the benefits of autonomy with the importance of human expertise and judgment remains a crucial aspect to be addressed.

While the path forward is not without its challenges, the potential benefits offered by fully autonomous systems are vast, with the capacity to usher in a transformative paradigm shift in surgical care. As technology continues to progress and collaborations between researchers and surgeons flourish, the vision of fully autonomous robotic surgery draws ever nearer, offering the promise of safer, more precise, and highly efficient surgical interventions.

# Key points

- Robotic surgery has evolved from basic telemanipulation to advanced autonomous capabilities. The history of robotic surgery is a testament to the remarkable progress achieved in the field. From the early telemanipulation systems to the current level of supervised autonomy, robotic surgery has revolutionized surgical practice.
- The evolution of autonomy levels in robotic surgery includes enhanced telemanipulation, surgical assistance, collaborative surgery, supervised autonomy, and the future goal of full autonomy.
- The trends in autonomy levels highlight the enabling technologies to enhance surgical robotic capabilities.
- Robotic surgery has found applications in various surgical specialties and offers advantages in terms of improved patient outcomes and healthcare efficiency.

# References

[1] Yang G-Z, et al. Medical robotics—Regulatory, ethical, and legal considerations for increasing levels of autonomy. vol. 2. American Association for the Advancement of Science; 2017. eaam8638.
[2] Kwoh YS, Hou J, Jonckheere EA, Hayati S. A robot with improved absolute positioning accuracy for CT guided stereotactic brain surgery. IEEE Trans Biomed Eng 1988;35(2):153–60.
[3] Park S, Howe RD, Torchiana DF. Virtual fixtures for robotic cardiac surgery. In: International conference on medical image computing and computer-assisted intervention. Springer; 2001. p. 1419–20.
[4] Lu M, Zhang Y, Lim CM, Ren H. Flexible needle steering with tethered and untethered actuation: current states, targeting errors, challenges and opportunities. Ann Biomed Eng 2023;51(5):905–24. https://doi.org/10.1007/s10439-023-03163-8.
[5] Nadeau C, Ren H, Krupa A, Dupont P. Intensity-based visual servoing for instrument and tissue tracking in 3D ultrasound volumes. IEEE Trans Autom Sci Eng 2014;12(1):367–71.
[6] Wu K, et al. Safety-enhanced model-free visual servoing for continuum tubular robots through singularity avoidance in confined environments. IEEE Access 2019;7:21539–58.
[7] Ren H, Kazanzides P. AISLE: an automatic volumetric segmentation method for the study of lung Allometry. Stud Health Technol Inform 2011;163:476–8. Submitted for publication.
[8] Schleer P, Drobinsky S, de la Fuente M, Radermacher K. Toward versatile cooperative surgical robotics: a review and future challenges. Int J Comput Assist Radiol Surg 2019;14:1673–86.
[9] Padoy N, Hager GD. Human-machine collaborative surgery using learned models. In: 2011 IEEE international conference on robotics and automation. IEEE; 2011. p. 5285–92.
[10] Shademan A, Decker RS, Opfermann JD, Leonard S, Krieger A, Kim PC. Supervised autonomous robotic soft tissue surgery. Sci Transl Med 2016;8(337). 337ra64–337ra64.
[11] Han J, Davids J, Ashrafian H, Darzi A, Elson DS, Sodergren M. A systematic review of robotic surgery: from supervised paradigms to fully autonomous robotic approaches. Int J Med Robot Comput Assist Surg 2022;18(2), e2358.
[12] Li Z, Wu L, Ren H, Yu H. Kinematic comparison of surgical tendon-driven manipulators and concentric tube manipulators. Mech Mach Theory 2017;107:148–65. Available: https://www.sciencedirect.com/science/article/pii/S0094114X16302580.
[13] Lee Y-J, van den Berg NS, Orosco RK, Rosenthal EL, Sorger JM. A narrative review of fluorescence imaging in robotic-assisted surgery. Laparoscopic Surg 2021;5.
[14] Kumar KS, et al. Stretchable and sensitive silver nanowire-hydrogel strain sensors for proprioceptive actuation. ACS Appl Mater Interfaces 2021;13(31):37816–29. https://doi.org/10.1021/acsami.1c08305.
[15] Li T, King NKK, Ren H. Disposable FBG-based Tridirectional force/torque sensor for aspiration instruments in neurosurgery. IEEE Trans Ind Electron 2020;67(4):3236–47. https://doi.org/10.1109/tie.2019.2905829.
[16] Li T, Shi C, Ren H. A high-sensitivity tactile sensor Array based on Fiber Bragg grating sensing for tissue palpation in minimally invasive surgery. IEEE/ASME Trans Mechatron 2018;23(5):2306–15. https://doi.org/10.1109/tmech.2018.2856897.
[17] Song S, Li Z, Yu H, Ren H. Shape reconstruction for wire-driven flexible robots based on Bézier curve and electromagnetic positioning. Mechatronics 2015;29:28–35.
[18] Qiu L, Ren H. RSegNet: a joint learning framework for deformable registration and segmentation. IEEE Trans Autom Sci Eng 2021.
[19] Qiu L, Ren H. U-RSNet: an unsupervised probabilistic model for joint registration and segmentation. Neurocomputing 2021;450:264–74.
[20] Van M, Wu D, Ge SS, Ren H. Fault diagnosis in image-based visual servoing with eye-in-hand configurations using Kalman filter. IEEE Trans Industr Inform 2016;12(6):1998–2007.
[21] Xu M, Islam M, Lim CM, Ren H. Class-incremental domain adaptation with smoothing and calibration for surgical report generation. Springer International Publishing; 2021. p. 269–78.
[22] Xu M, Islam M, Ming Lim C, Ren H. Learning domain adaptation with model calibration for surgical report generation in robotic surgery. IEEE; 2021. https://doi.org/10.1109/icra48506.2021.9561569. Available: https://doi.org/10.1109/ICRA48506.2021.9561569.
[23] Gu X, Li C, Xiao X, Lim CM, Ren H. A compliant transoral surgical robotic system based on a parallel flexible mechanism. Ann Biomed Eng 2019;47(6):1329–44. https://doi.org/10.1007/s10439-019-02241-0.
[24] Li C, Gu X, Xiao X, Lim CM, Ren H. A robotic system with multichannel flexible parallel manipulators for single port access surgery. IEEE Trans Industr Inform 2019;15(3):1678–87. https://doi.org/10.1109/tii.2018.2856108.
[25] Li C, Gu X, Xiao X, Lim CM, Ren H. Flexible robot with variable stiffness in transoral surgery. IEEE/ASME Trans Mechatron 2020;25(1):1–10. https://doi.org/10.1109/tmech.2019.2945525.
[26] Checcucci E, et al. 3D imaging applications for robotic urologic surgery: an ESUT YAUWP review. World J Urol 2020;38:869–81.
[27] Moon AS, Garofalo J, Koirala P, Vu M-LT, Chuang L. Robotic surgery in gynecology. Surg Clin 2020;100(2):445–60.
[28] Chitwood Jr WR. Historical evolution of robot-assisted cardiac surgery: a 25-year journey. Ann Cardiothorac Surg 2022;11(6):564.

# 3

# Needle inserts into soft tissue: Basic technical aspects

*Mingyue Lu[a], Chwee Ming Lim[b], and Hongliang Ren[c,d]*

[a]The Department of Biomedical Engineering, National University of Singapore, Singapore, Singapore [b]Department of Otorhinolaryngology—Head and Neck Surgery, Singapore General Hospital, Singapore, Singapore [c]Department of Electronic Engineering, The Chinese University of Hong Kong, Sha Tin, Hong Kong [d]Shun Hing Institute of Advanced Engineering (SHIAE), The Chinese University of Hong Kong, Sha Tin, Hong Kong

## Basic procedure for needle insertion

From the mechanics arising from the needle-tissue interface, three fundamental phases (surface deformation, tip insertion, and tip and shaft insertion) of this needle-tissue interaction can be described [1]. The respective characteristics of each phase are as follows.

## Phase 1: Surface deformation

The surface deformation phase begins when the needle inserts just onto the surface of the tissue (Fig. 1A) and ends with a rupture of the tissue surface, which corresponds to the maximum insertion force (Fig. 1B, E). During the surface deformation phase, as the loading pressure at the needle tip increases, so does the stress in the soft tissue around the puncture point. However, the tissue is not penetrated (the tissue surface moves with the needle).

## Phase 2: Needle tip insertion

Phase 2 begins once a puncture occurs in the soft tissue and the needle tip insertion phase begins. It ends once the tissue surface slides from the needle tip onto the needle shaft. At this stage, the needle pushes against the soft tissue, which is momentarily pierced, and the needle advances inside the tissue (Fig. 1C). The actual event (where the boundary breaks) is called the puncture. Tissue breakage continues until the strain energy level is reduced sufficiently enough for crack growth to proceed steadily, and phase 3 begins.

## Phase 3: Tip and shaft insertion

The tip and shaft insertion phase begins when the needle shaft penetrates deeper into the soft tissue and ends once the needle stops moving or encounters a new layer of the tissue surface (Fig. 1D). The needle-soft tissue interaction results in load distribution along the needle-tissue contact area (Fig. 1F). During phase 3, the needle tip will be subjected to cutting forces. Simultaneously, the needle shaft sustains various increasing frictional forces because of the increased needle shaft-soft tissue contact area as the needle advances. Therefore, the needle tip and shaft must be able to sustain these forces in order to avoid breakage of the needle within the tissue actuation.

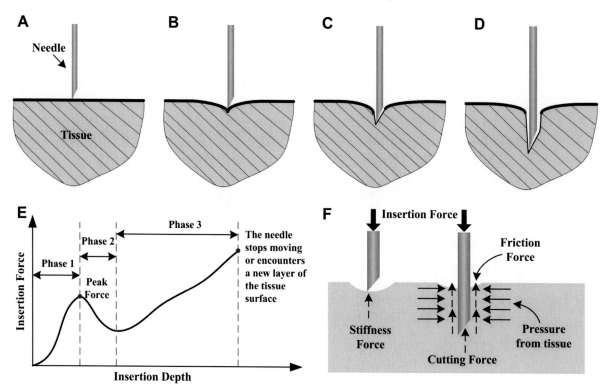

FIG. 1  The basic procedure and force distribution for needle insertion: (A) The moment of needle-tissue contact; (B) Surface deformation; (C) Needle tip insertion; (D) Tip and shaft insertion; (E) A typical force curve of needle insertion; (F) The force distribution along the needle-tissue contact area.

## Needle insertion mechanical analysis

### Stiffness-friction-cutting force modeling

The force generation mechanism should be elucidated and modeled to simulate the insertion force. Many surveys have stated that it is essential to research the insertion force (axial force). Okamura et al. proposed that the needle insertion force ($F_{insertion}$) is the combined result of stiffness force ($F_{stiffness}$) caused by the internal stiffness of the tissue, the cutting force ($F_{cutting}$) of the needle tip is used to cut tissue, and the changing friction force ($F_{friction}$) because of the increased needle shaft-soft tissue contact area [2]. Based on the needle-soft tissue interaction mechanism previously, the total applied insertion force can be expressed accurately by:

$$F_{insertion} = F_{stiffness} + F_{friction} + F_{cutting} \qquad (1)$$

The stiffness force is the first to occur. In the surface deformation phase (phase 1), the insertion force equals the stiffness force. The stiffness force only exists before the soft tissue is penetrated, and this contact is entirely reversible from the undeformed tissue surface to the maximally deformed tissue surface. Considering the insertion force from the point of view of fracture mechanics, the peak force represents the puncture applied to the needle, a steady increase followed by a sharp decrease. This is because when the tissue surface is pierced, the energy accumulated in the surface deformation phase is usually too great, which leads to a sudden rupture. Therefore, the force drop is relatively significant, and the stored strain energy is devoted to propagating the crack (Fig. 1E). This is an irreversible process. In the tip and shaft insertion phase (phase 3), the insertion force can be obtained based on the friction force and cutting force, and friction force is the most significant contributor to the insertion force [3].

Inspired by Okamura et al., modeling stiffness, cutting, and friction force attracts much attention. The nonlinear spring and contact model are often used to describe the stiffness force [4]. The cutting force is essentially constant under the same conditions. However, there may be some fluctuations in rupture, depending on how heterogeneous the tissue is [5]. In various materials, there are indications that as the insertion depth increases, the friction force increases almost linearly.

## Multilayer needle insertion force modeling

In clinical percutaneous puncture, the needle usually needs to penetrate several layers of skin, fat, muscle, and other tissues or organs to reach the target location (deep in the body). After conducting many force analyses on single-layer homogeneous materials and tissues, some scholars focus on modeling and simulating multilayer or different materials and tissues. Each layer of biological tissue has variable mechanical properties, which lead to varying levels of friction force on the needle shaft, cutting forces on the needle tip, and stiffness forces induced by the tissue stiffness. An example of two-layer tissue is shown in Fig. 2. The phases of force analyses in a single layer will repeat as the needle encounters changes in internal structure or tissue properties (Fig. 2A). For example, when inserted through two-layer tissue, the change process of insertion force is shown in Fig. 2B.

As in the previous analysis, there are still three main insertion phases (surface deformation phase, needle tip insertion, and tip and shaft insertion). They repeat the same processes in the multilayer soft tissue model. This is because once the needle tip reaches the interface between two different tissue layers, its further insertion will push the tissue instead of piercing it. Therefore, the force model during the complete insertion of two layers of tissue can be concluded as follows:

$$F_{insertion} = \begin{cases} F_{stiffness1} & 0 \leq x \leq d_1 \\ F_{friction1} + F_{cutting1} & d_1 x \leq d_2 \\ F_{friction1} + F_{stiffness2} & d_2 x \leq d_3 \\ F_{friction1} + F_{friction2} + F_{cutting2} & d_3 x \leq d_4 \end{cases} \tag{2}$$

The previous analysis is also applicable to soft tissue insertion with more than three layers. Typically, a general force modeling was proposed by Carra et al. to research the needle insertion into multilayered tissue to reach the liver. They have modeled stiffness force by a nonlinear model, friction force by a modified Dahl model, and cutting force by a

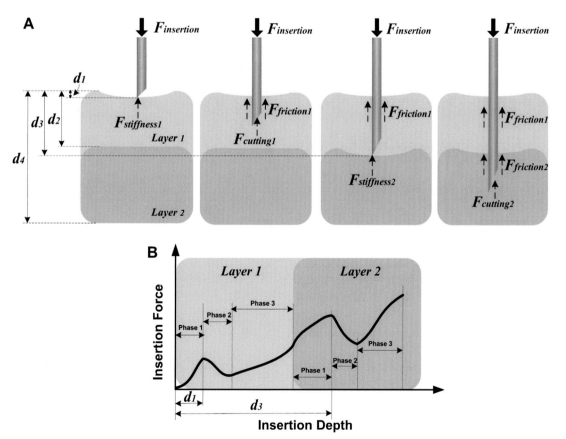

FIG. 2    Needle insertion into a two-layer tissue. (A) The load distribution along the needle-tissue contact area in two-layer tissue. (B) Force on the needle when inserted through two-layer tissue.

constant [6]. Similarly, Jiang et al. proposed appropriate substitutions for several tissue layers of humans consisting of polyvinyl alcohol (PVA) hydrogel and silicone rubber. A contact model was built to calculate the stiffness force, and the nonlinear Winkler foundation model was selected to obtain the friction force and cutting force [7]. Considering the inconsistent needle-tissue interaction forces, Lee et al. proposed a needle deflection model for two-layered tissue [8]. The multilayer needle insertion force model could show the force characteristic change when the needle punctures different layers of soft tissues. Compared with the previously proposed single-layer model, the multilayer model is more practical in modeling needle insertion throughout robot-assisted MIS because it is closer to the natural structure of actual soft tissue.

## Effect factors of insertion force

During percutaneous puncture surgery, the degree of tissue deformation increases with increasing insertion force, thus reducing the accuracy of needle puncture surgery, leading to a higher degree of tissue damage and increasing pain to the patient [9]. Therefore, this fact has motivated many researchers to investigate ways to reduce or minimize puncture force to facilitate more accurate procedures for better medical treatment. The insertion force of needles during soft tissue puncture has been extensively studied over the past decade by experiments with needles inserted into artificial and biological tissues. Needle properties, tissue properties, and insertion methods influence the insertion force. However, how these factors affect insertion force remains to be determined. This section studies the correlation between effect factors and insertion force by retrieving the experimental results in the literature to provide a reference for the research on the precise control of needle insertion force in robot-assisted needle intervention surgery. The effect factors of needle insertion force are analyzed and discussed in Table 1.

TABLE 1   Correlation analysis between effect factors and insertion force.

| Effect factors | | | The correlation to insertion force | | | |
|---|---|---|---|---|---|---|
| Items | Vital component | Details | Stiffness force | Friction force | Cutting force | Supplemental instruction |
| Needle characteristics | The geometry of the needle tip | Such as the blunt, beveled, tapered, Sprotte, diamond, and Tuohy | ☑ | ☐ | ☑ | • The needle insertion force is closely related to the tip sharpness. • The power of puncture needles made of different materials will also be different. |
| | The geometry of the needle body | Some bio-inspired needle | ☐ | ☑ | ☑ | |
| | Bevel angle | 8°–85° | ☑ | ☐ | ☑ | |
| | Diameter | 14G–25G | ☑ | ☑ | ☑ | |
| Tissue properties | Biological tissues | Alive or dead; In vivo or ex vivo; single-layer or multilayer tissue | ☑ | ☑ | ☑ | • Biological tissues exhibit inhomogeneity, nonlinearity, and anisotropy. • The axial load distribution of needle inserts into artificial material is substantially uniform. |
| | Artificial tissues | PVA hydrogels, silicone rubber, and gelatin | ☑ | ☑ | ☑ | |
| | Individual difference | Age, water content, experimental pretreatment | ☑ | ☑ | ☑ | |
| Needle insertion method | Rotation-assisted needle insertion | The unidirectional needle rotation; the bidirectional needle rotation | ☐ | ☐ | ☑ | • The cutting performance can be improved because of the drilling effects. • The unidirectional needle rotation causes winding of tissue, which can be improved by the bidirectional rotation. |

TABLE 1 Correlation analysis between effect factors and insertion force—cont'd

| Items | Vital component | Details | Stiffness force | Friction force | Cutting force | Supplemental instruction |
|---|---|---|---|---|---|---|
| | Vibration-assisted needle insertion | The effect of different vibration frequencies | ☐ | ☑ | ☑ | • Applying high-frequency vibration is also an effective method of increasing the local insertion speed of the needle. |
| | Hybrid control of rotation and vibration | Combination of rotation and vibration | ☐ | ☑ | ☑ | • Using high-frequency vibration to apply rotation can reduce torsional friction and minimize tissue damage. |
| | Others | The "push-pull" motion; needle surface coating; The needle shooting procedure; water-jet technology; intermittent needle insertion; needle insertion location; needle puncture angle | ☑ | ☑ | ☑ | • Different motion and drive modes, such as interrupt or continuous and manual or robot-assisted, also affect insertion force. |

"☑" means there is a significant correlation between effect factors and insertion force, and "☐" stands for the related research is blank or the correlation needs further determination.

## Needle characteristics

When considering needle properties, the focus of most research is the geometry of the needle. The needle insertion force relies on the geometry of the needle tip and the needle body. Several needle tip configurations have been developed, such as the blunt, beveled, tapered, Sprotte, diamond, and Tuohy (Fig. 3A). The needle insertion force is closely related to the tip sharpness. The 8°–85° beveled needles are the most commonly used clinical tool. When using beveled needles, the direction the needle tip faces due to asymmetrical cutting forces during insertion into the tissue can cause needle deflection. Currently, many variations of beveled tip needles are on the market. There are three typical examples, as shown in Fig. 3B. Some indications indicate that smaller bevel angles (e.g., standard bevel) produce lower insertion forces, and larger bevel angles (e.g., true short bevel) result in higher cutting forces. Tapered needles produce higher peak force than beveled needles, and cannulae and trocars affect needle-tissue contact area and cutting force. Blunt needle geometry is used in clinical procedures to reduce accidental stick injuries. However, the blunt geometry of the needle tip will require greater insertion force to penetrate the tissue.

Moreover, drawing inspiration from nature, some researchers proposed several bio-inspired surgery needle designs to control the insertion force [10], such as mosquito-inspired needles [9] and honeybee stinger-inspired needles with barbs [3]. Such organisms in nature often have complex structural compositions. Therefore, these bio-inspired needle designs have considered the geometric of both the needle tip and the needle body. For example, the needle tip and body of the mosquito-inspired needle imitate the geometry of the labrum and maxilla of mosquitoes, respectively. Due to the bionic needle tip and needle body geometry design, the needle-tissue contact area is reduced, and the friction force can be minimized, thereby reducing the insertion force. The study showed mosquito-inspired needles could reduce the maximum insertion force by up to 39% [9]. The honeybee stinger-inspired needle is also a similar principle,

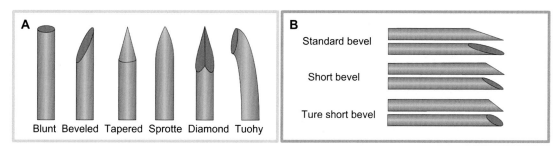

FIG. 3 Different geometries of the needle. (A) Basic geometry of the needle tip (from the left): blunt, beveled, tapered, Sprotte, diamond, Tuohy. (B) Needles with different bevel angles.

and it has been tested to decrease the total insertion force by up to 50% [11]. These bio-inspired surgery needles usually have complex geometry of needle bodies. Meanwhile, they have tremendous potential to advance the development of surgical needles for more efficient MIS procedures [12].

In addition, typical clinical needle diameters range from 14G (Ø 2.1mm) for tissue biopsy to 30G (Ø 0.3mm) for vaccination procedures. Larger needle diameters result in greater penetration, cutting, and friction forces [5]. As the needle diameter increases, the peak insertion force increases dramatically. The power of puncture needles made of different materials will also be different.

## Tissue properties

The insertion force during needle puncture is also affected by tissue properties, such as tissue material, individual differences, and experimental pretreatment.

The most commonly used materials in needle puncture soft tissue experiments can be divided into biological tissues and artificial materials. At the same time, using biological tissue is only sometimes a viable option regarding reproducibility, visibility, and ethics. Therefore, many researchers have proposed artificial phantoms to replace biological tissue. Biological soft tissues are often viscoelastic or hyperelastic. For a more accurate needle insertion simulation process, the viscoelasticity of soft tissues should be considered. People tend to choose tissue-mimicking materials that are stable and transparent so that deformation information can be easily obtained. PVA hydrogels [4], silicone rubber [13], and gelatin can exhibit viscoelastic behavior mimicking biological tissues and have been widely used as phantom materials to mimic the mechanical behavior of soft biological tissues. Artificial materials are often easier to model, can be customized, and are more durable than biological tissues. Although artificial tissues have many advantages, due to the differences between artificial tissues and biological tissues, we must make a reasonable choice according to the actual situation. Biological tissues also exhibit inhomogeneity, nonlinearity, and anisotropy. For example, when inserting a needle into a homogeneous isotropic artificial material, it is found that the axial load distribution is substantially uniform.

Different states (alive or dead) of the same biological tissue will produce various stiffness forces. Meanwhile, in vivo and in vitro experiments represent two completely different conditions. However, measuring insertion force in vivo is much more complex than measuring force in vitro, both in a practical and ethical sense. The presence of skin was found to significantly affect the peak force, with tissues with higher water content more easily punctured.

## Needle insertion method

The needle insertion method affects the needle-tissue interaction. Therefore, the surgeon must possess considerable skill and experience to control the needle to reach a determined target within the tissue. The robotic-assisted needle insertion is proposed to improve the effectiveness of puncture interventions that are gaining acceptance. From a force analysis perspective, robotic-assisted needle insertion has the potential to control the insertion force and improve precision by implementing novel insertion strategies, such as regular needle rotation and vibration, which could not be performed reliably manually.

### Rotation-assisted needle insertion

Rotation-assisted needle insertion is expected to affect the insertion force. The cutting performance can be improved because of the drilling effects. The lower cutting force also means that the tissue deforms less during needle insertion. At the same time, because the asymmetric force can be eliminated, the deflection of the needle can be minimized through axial rotation. Therefore, the puncture accuracy will be improved. In recent years, many researchers have paid attention to the studies of needle insertion with unidirectional rotation.

Needle rotation typically reduces friction by 10% in chicken breasts and up to 50% in porcine gelatin and beef. However, despite the benefits of reduced cutting force, a histological study suggested that axial rotation may pose a potential risk to patients [14]. Needle rotation with high speed could cause serious winding due to torsional friction (Fig. 4A), which enlarges the size of cracks along the needle path, exacerbating tissue damage and leading to complications such as bleeding and hematoma. Moreover, it was observed that when the rotating needle was inserted into the rat, the connective tissue was wound, and the force increased during the needle extraction. Therefore, the researchers aim to find a method to reduce tissue damage and insertion force, which is impossible to achieve with unidirectional needle rotation.

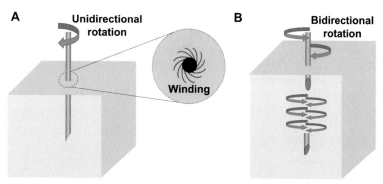

FIG. 4    Rotation-assisted needle insertion. (A) The unidirectional needle rotation can cause tissue winding. (B) The bidirectional needle rotation.

A potentially effective solution is the bidirectional needle rotation, which can reduce the tissue winding and damage caused by unidirectional needle rotation, while maintaining good tissue cutting performance (Fig. 4B). Lin et al. demonstrated that when the needle rotation speed is high, bidirectional needle rotation is more suitable for reducing tissue damage [15]. The bidirectional rotation method of rotating every 360° clockwise/counterclockwise shows that the deflection and tissue damage are reduced [14].

### Vibration-assisted needle insertion

Adding vibration to needle puncture procedures effectively reduces patient pain and improves puncture accuracy. Therefore, vibration-assisted needle insertion is also an important research direction. Shin-ei et al. first proposed that the needle insertion force could be reduced by introducing the vibration into the needle insertion [16]. In their work, the insertion force was reduced by up to 69%. In this process, the amplitude is controlled by the driving voltage of the piezoelectric actuator. The vibration-assisted needle insertion schematic demonstration is shown in Fig. 5A.

Applying high-frequency vibration is a method to improve the local needle insertion speed, while fast insertion can reduce friction force. Moreover, needle vibratory tissue cutting has been shown to be a method that can improve cutting performance and thereby improve the precision of needle placement in vivo. Therefore, the vibration-assisted needle insertion method reduced the insertion force of phase 2 and phase 3 (Fig. 5B), which has been proven by Barnett

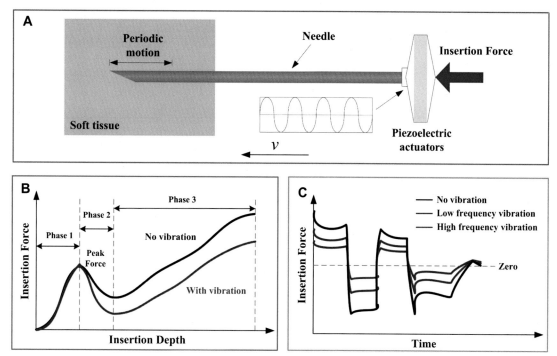

FIG. 5    Vibration-assisted needle insertion. (A) The schematic illustration of the vibration-assisted needle insertion. (B) The vibration method decreases the insertion force. (C) The vibration frequency affects the insertion force.

et al. [17]. To further study the mechanism of vibration affecting the insertion force, Huang et al. [18] used a subcutaneous injection needle (27 G needle) for vibration-assisted puncture to reduce the peak insertion force by 28%. Clement et al. [19] reduced the peak force by 73% after using the subcutaneous injection needle (25 G needle) for vibration-assisted puncture. In addition, Khalaji et al. demonstrated that friction could be reduced by introducing vibration low amplitude motion on the conventional insertion profile using a close proximity needle (18 G needle). Fig. 5C vividly shows the effect of vibration frequency on the insertion force between needle and tissue. The results of several studies showed that when the vibration frequency is increased, the total insertion force is significantly reduced [20].

### Hybrid control of rotation and vibration

Robot-assisted needle insertion makes complex control strategies possible. Based on considering the impact of rotation and vibration needle insertion methods on the insertion force, some researchers proposed a hybrid puncture method that combines rotation and vibration needle insertion. Using high-frequency vibration to apply rotation can reduce torsional friction and minimize tissue damage [21]. Tsumura et al. applied the hybrid control of vibration and bidirectional rotation to lower abdominal needle insertion. Comparing and analyzing the insertion force, torque, and hole area of the tissue due to needle insertion, they demonstrated that the hybrid control of vibration and bidirectional rotation could minimize tissue damage and reduce the insertion force [22].

### Other needle insertion force control methods

Some researchers have proposed many other novel methods to control insertion force in the past few years. We have mentioned that the bionic needle has excellent potential in minimally invasive interventional surgery. Some researchers imitate the shape of the mouthparts of insects, and some mimic the needle insertion method of mosquito vibration piercing. Here, a self-propelled needle insertion method that imitates the wasp ovipositor is introduced. Bloemberg et al. have designed a self-propelled needle for prostate interventions based on the "push-pull" motion inspired by the ovipositor [23]. Different motion and drive modes, such as interrupt or continuous and manual or robot-assisted, also affect insertion force. The intermittent needle insertion method is proposed for lung surgery to overcome the internal tissue motion caused by disturbances such as breathing motions. Moving and stopping the needle motion at a specific time frame is repeated during the breathing cycle [24]. The insertion force is also affected by the insertion location and needle puncture angle, and the smaller the puncture angle, the greater the insertion force. Based on the force effect of the insertion velocity, a fast needle insertion technique is proposed to reduce the insertion force [25]. However, high velocity could be dangerous (the needle insertion velocity should usually be kept at less than 10.0 mm/s), and the security of such a "needle shooting" method for essential or sensitive tissue cannot be guaranteed. In this case, maybe high-frequency vibration is a premium solution, as mentioned previously. The polydopamine (PDA) coating has been used to reduce the friction force of the needle surface, and the total insertion force can be reduced by 20%–25% [26]. The feasibility of using water-jet technology for reducing insertion force has been proved [27].

## Future scenario

The future tendencies in the field of needle insertion in surgical procedures are likely to revolve around advancements in technology and methodologies to further enhance precision, reduce tissue damage, and improve patient outcomes.

Needle design will continue to evolve, incorporating bio-inspired features and novel materials and geometries to reduce insertion forces and improve tissue interactions. Robotic-assisted needle insertion systems will offer greater precision, real-time feedback, and the ability to perform complex maneuvers with minimal tissue disruption. Advances in imaging technologies, such as augmented reality and intraoperative imaging, will provide surgeons with enhanced visualization during needle insertion. Another aspect is that we must develop new needle insert approaches that can help lower insertion forces, making the procedures more comfortable and less traumatic for patients.

## Key points

- Understanding and modeling needle insertion forces and the factors that influence them are crucial for improving surgical outcomes, minimizing tissue damage, and enhancing patient comfort during needle-based procedures, especially in MIS.

- The needle insertion process involves three phases: surface deformation, needle tip insertion, and tip and shaft insertion, with varying force components at each stage.
- Needle insertion force is influenced by stiffness, cutting, and friction forces, and multilayer needle insertion modeling is important for understanding force characteristics in different tissue layers.
- Factors affecting insertion force include needle characteristics (geometry, diameter), tissue properties (biological vs artificial, viscoelasticity), and insertion methods (rotation, vibration, hybrid control).
- Rotation-assisted and vibration-assisted needle insertion methods, as well as a hybrid approach, offer promising strategies to reduce tissue damage and improve precision in minimally invasive surgery.

## Acknowledgments

This work was supported by Hong Kong Research Grants Council (RGC) Collaborative Research Fund (C4026-21G), Research Grants Council (RGC) Research Impact Fund (RIF R4020-22), and the key project 2021B1515120035 (B.02.21.00101) of the Regional Joint Fund Project of the Basic and Applied Research Fund of Guangdong Province.

## References

[1] Jushiddi MG, Mulvihill JJ, Chovan D, Mani A, Shanahan C, Silien C, Tiernan P. Simulation of biopsy bevel-tipped needle insertion into soft-gel. Comput Biol Med 2019;111, 103337.
[2] Okamura AM, Simone C, O'leary MD. Force modeling for needle insertion into soft tissue. IEEE Trans Biomed Eng 2004;51(10):1707–16.
[3] Sahlabadi M, Khodaei S, Jezler K, Hutapea P. Insertion mechanics of bioinspired needles into soft tissues. Minim Invasive Ther Allied Technol 2018;27(5):284–91.
[4] Jiang S, Li P, Yu Y, Liu J, Yang Z. Experimental study of needle-tissue interaction forces: effect of needle geometries, insertion methods and tissue characteristics. J Biomech 2014;47(13):3344–53.
[5] Van Gerwen DJ, Dankelman J, van den Dobbelsteen JJ. Needle-tissue interaction forces-a survey of experimental data. Med Eng Phys 2012;34(6):665–80.
[6] Carra A, Avila-Vilchis JC. Multi-layer needle insertion modeling for robotic percutaneous therapy. In: 2010 4th international conference on bioinformatics and biomedical engineering. IEEE; 2010. p. 1–4.
[7] Li P, Jiang S, Yu Y, Yang J, Yang Z. Biomaterial characteristics and application of silicone rubber and PVA hydrogels mimicked in organ groups for prostate brachytherapy. J Mech Behav Biomed Mater 2015;49:220–34.
[8] Lee H, Kim J. Estimation of flexible needle deflection in layered soft tissues with different elastic moduli. Med Biol Eng Comput 2014;52(9):729–40.
[9] Gidde STR, Acharya SR, Kandel S, Pleshko N, Hutapea P. Assessment of tissue damage from mosquito-inspired surgical needle. Minim Invasive Ther Allied Technol 2022;1–10.
[10] Ma Y, Xiao X, Ren H, Meng MQH. A review of bio-inspired needle for percutaneous interventions. Biomimetic Intell Robot 2022;2(4), 100064.
[11] Gidde STR, Ciuciu A, Devaravar N, Doracio R, Kianzad K, Hutapea P. Effect of vibration on insertion force and deflection of bioinspired needle in tissues. Bioinspir Biomim 2020;15(5), 054001.
[12] Lu M, Zhang Y, Lim CM, Ren H. Flexible needle steering with tethered and untethered actuation: current states, targeting errors, challenges and opportunities. Ann Biomed Eng 2023;51(5):905–24.
[13] Li T, Pan A, Ren H. Reaction force mapping by 3-axis tactile sensing with arbitrary angles for tissue hard-inclusion localization. IEEE Trans Biomed Eng 2020;68(1):26–35.
[14] Tsumura R, Takishita Y, Fukushima Y, Iwata H. Histological evaluation of tissue damage caused by rotational needle insertion. In: 2016 38th annual international conference of the IEEE engineering in medicine and biology society (EMBC). IEEE; 2016. p. 5120–3.
[15] Lin CL, Huang YA. Simultaneously reducing cutting force and tissue damage in needle insertion with rotation. IEEE Trans Biomed Eng 2020;67(11):3195–202.
[16] Ei T, Yuyama K, Ujihira M, Mabuchi K. Reduction of insertion force of medical devices into biological tissues by vibration. Jpn J Med Electron Biol Eng 2001;39:292–6.
[17] Barnett AC, Lee YS, Moore JZ. Needle geometry effect on vibration tissue cutting. Proc Inst Mech Eng Pt B J Eng Manuf 2018;232(5):827–37.
[18] Huang YC, Tsai MC, Lin CH. A piezoelectric vibration-based syringe for reducing insertion force. In: IOP conference series: Materials science and engineering. IOP Publishing; 2012. p. 012020. Vol. 42, No. 1.
[19] Clement RS, Unger EL, Ocón-Grove OM, Cronin TL, Mulvihill ML. Effects of axial vibration on needle insertion into the tail veins of rats and subsequent serial blood corticosterone levels. J Am Assoc Lab Anim Sci 2016;55(2):204–12.
[20] Khalaji I, Hadavand M, Asadian A, Patel RV, Naish MD. Analysis of needle-tissue friction during vibration-assisted needle insertion. In: 2013 IEEE/RSJ international conference on intelligent robots and systems. IEEE; 2013. p. 4099–104.
[21] Tsumura R, Takishita Y, Iwata H. Needle insertion control method for minimizing both deflection and tissue damage. J Med Robot Res 2019;4(01):1842005.
[22] Tsumura R, Iordachita I, Iwata H. Fine needle insertion method for minimising deflection in lower abdomen: in vivo evaluation. Int J Med Robot Comput Assist Surg 2020;16(6):1–12.
[23] Bloemberg J, Trauzettel F, Coolen B, Dodou D, Breedveld P. Design and evaluation of an MRI-ready, self-propelled needle for prostate interventions. PLoS One 2022;17(9), e0274063.
[24] Tsumura R, Kakima K, Iwata H. Intermittent insertion control method with fine needle for adapting lung deformation due to breathing motion. In: 2020 IEEE/RSJ international conference on intelligent robots and systems (IROS). IEEE; 2020. p. 3192–9.

I. Introduction, technology basics, evolution and next frontiers of surgical robots

[25] Mahvash M, Dupont PE. Fast needle insertion to minimize tissue deformation and damage. In: 2009 IEEE international conference on robotics and automation. IEEE; 2009. p. 3097–102.

[26] Patel KI, Gidde ST, Li H, Podder T, Ren F, Hutapea P. Insertion force of Polydopamine-coated needle on phantom tissues. In: Frontiers in biomedical devices, Vol. 41037. American Society of Mechanical Engineers; 2019. V001T06A005.

[27] Babaiasl M, Yang F, Swensen JP. Towards water-jet steerable needles. In: 2018 7Th IEEE international conference on biomedical robotics and biomechatronics (biorob). IEEE; 2018. p. 601–8.

# 4

# Kinematic concepts in minimally invasive surgical flexible robotic manipulators: State of the art

*Jiewen Lai[a], Bo Lu[b], and Hongliang Ren[a]*

[a]Department of Electronic Engineering, The Chinese University of Hong Kong, Sha Tin, Hong Kong [b]Robotics and Microsystem Center, Soochow University, Suzhou, China

## Introduction

Minimally invasive surgery (MIS) allows the surgeon to perform surgical operations through small incisions by administering the instruments inserted into the patient's body, thereby reducing the patient's wound healing time, associated pain and suffering, and risk of infection. The use of surgical robots in MIS has been enhancing the precision of the intraoperative MIS task and the dexterity of surgical instrument maneuvering. With the prevailing 5G technology, robot-assisted MIS surgery leads to the possibility of remote surgery [1]. Different types of MIS have been semi-robotized or robotized [2], such as surgery in ophthalmology [3], heart [4], thoracic [5], gastrointestinal [6], gynecology [7], orthopedics [8], etc. Even though many surgical robots are commercially available in the market, the development of novel surgical robots and medical mechatronic instruments that better fit specific medical procedures has continued in academia. Among all novel surgical robots, flexible robots, or soft/continuum robots [9], have stood out from the crowds of many other rigid-limbs robots in robot-assisted minimally invasive surgery and interventions (RAMIS) [10]—for their inherent compliance, dexterity, and safe human-robot interaction. State-of-the-art medical robots that employ flexible manipulators can be found in some big brands of commercial surgical robots, such as the Da Vinci SP platform [11], the SHURUI system [12], and the Medrobotics Flex Robotic System [13].

According to a recent metaanalysis [14], many interdisciplinary challenges of MIS flexible robotic manipulators have been tackled, such as robotic design, materials selection, fabrication, actuation methods, sensing, force exertion, modeling, and stiffness variability, among others. One of the most well-studied fields is the kinematics of these flexible continuum robots. Flexible robots for MIS kinematically distinguish themselves from conventional industrial robots in terms of reduced dimensions (therefore, the type of joints and transmission mechanisms vary), motion constraints, joint perceptions, geometry, use of materials, etc. The continuum structure of the manipulator body physically guarantees essential flexibility. In Ref. [15], the authors summarize continuum robot kinematics into two mappings, namely, robot-independent mapping and robot-dependent mapping. The former mapping exists independently from the actuation mechanism as it concerns only the task space and the configuration space of the robot, while the latter focuses on the relationship between the actuation mechanism and the geometrical resultant of the robot. Different actuation methods also constitute the variety in kinematics modeling [16]. These kinematics concepts are mostly applicable to the MIS flexible manipulators designed for surgical manipulation, endoscopic applications, and force sensing [9]. However, for the novel soft material-based flexible MIS robots, modification in kinematics shall be given with the consideration of including but not limited to the soft material's properties, hysteresis upon the actuation, mechanics, nonlinear deformation, and robot-environment interaction. All these factors diversify the kinematics concepts of MIS flexible robots. Alternatively, with the help of machine learning algorithms and robotic perception, the mappings can also be coupled and become data-driven [17], neglecting conventional kinematics.

TABLE 1   Databases and the corresponding search entries.

| Database | Search query syntaxes[a] |
|---|---|
| Web of Science (WOS) | TI = (soft OR continuum OR flexible) AND TI = (robot* OR manipulator*) AND TI = (surg* OR medic* OR MIS*) AND TI = (survey OR review OR advances OR overview) |
| Scopus | TITLE("Soft" OR "Continuum" OR "Flexible") AND TITLE("Robot*" OR "Manipulator*") AND TITLE("Surg*" OR "Medic*" OR "MIS*") AND TITLE("survey" OR "review" OR "overview") |
| PubMed | (("robot*"[All Fields]) AND ("soft"[title] OR "continuum"[title] OR "flexible"[title])) AND ("surg*"[Title] OR "MIS"[Title] OR "medic*" [Title]) AND ("survey"[All Fields] OR "review"[Title] OR "overview"[Title] OR "advances"[Title]) |

[a] *Filtered with the range of 2012–22.*

FIG. 1   Number of review articles on soft continuum robots for MIS in the past 10 years.

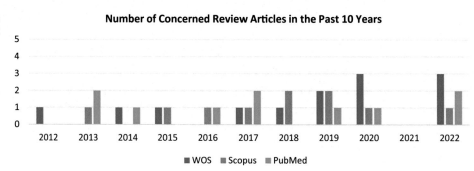

**Number of Concerned Review Articles in the Past 10 Years**

■ WOS   ■ Scopus   ■ PubMed

To exhaust the published review articles of the concerning fields, we searched the keywords on different databases. The databases and the corresponding search entries are given in Table 1, and the result is given in Fig. 1. Even though not many, recent decades of review articles on MIS-oriented (or other medical) flexible robots can be found in various scopes, such as medical applications [9,14,18,19], shape sensing techniques [20], specific surgeries and medical interventions [21,22], retrospectives and prospects [23,24], control strategies [25], etc. In-depth kinematic modeling surveys on soft continuum robots are available in the literature [15,26–29], focusing on the mathematical modeling of different kinds. However, there lacks a review of how the field utilizes the kinematics concepts in their MIS flexible robotic applications. The chapter herein introduces the kinematic concepts in MIS flexible robotic manipulators through case studies of the recent decade (2012–22). Typical state-of-the-art works are listed and compared regarding kinematics modeling and kinematic-based applications with minimally invasive surgical initiatives. By raising case studies, the kinematic concepts would have a stronger connection to medical considerations. The discussion is restricted to rigid-link continuum manipulators and soft manipulators designed for MIS or other similar medical applications—therefore, large robots are excluded from the case studies. The robots we discuss are referred to the manipulators with cylindrical or conical structures with an aspect ratio (length to cross-sectional diameter) greater than 1, and generally with 3D steerability that can be kinematically modeled. Note that concentric tube robots, dielectric elastomer actuators, and other uncommon soft robots are out of our scope.

The rest of this chapter is organized as follows: "Kinematic frameworks for MIS flexible manipulators" section briefly introduces the popular kinematics models for flexible robots—with the classification of rigid-link and soft manipulators. "Kinematics-based applications for MIS flexible robots" section demonstrates some kinematics-based applications for MIS robots through case studies. "Conclusions" section discusses the kinematics concepts in MIS flexible robots, and "Future scenario" section concludes this chapter.

## Kinematic frameworks for MIS flexible manipulators

Since the MIS flexible robotic manipulators would be used in the confined intracavity and surgical site, where the sensing techniques may not always be available, developing a referable kinematic model would be a baseline for any tangible MIS robots. The investigation of the kinematics of MIS manipulators is the prerequisite of quasistatic robotic motion control and robotic automation control, such as trajectory following and motion control.

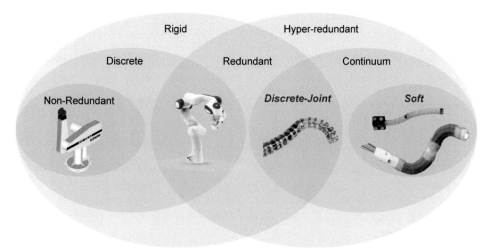

FIG. 2   Venn diagram of the rigid and hyperredundant robots based on the classification of materials of DOFs. The continuum robots can be further categorized as discrete-jointed and soft robots. (Images from left to right: 6-DOF PUMA 560, 7-DOF Franka Emika robot arm, an extensible segment tendon-driven continuum robot (Continuum Robotics Lab, University of Toronto), a cable-driven soft robot (Biomimetic Robotics Lab of PolyU), and the STIFF-FLOP soft medical robot.)

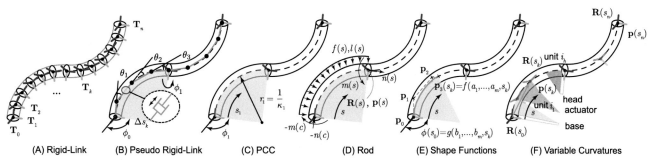

FIG. 3   Schematic of various kinematic frameworks used to describe the manipulator's geometry of rigid-link and soft continuum robots. The continuum robot with physical rigid-link can be represented by (A) Hyperredundant rigid-link structure with each physical joint represented by homogeneous transformation matrices; (B) Pseudo-rigid-link model to approximate a continuum body with a number of virtual links and rotational joints. Sometimes, passive prismatic joints are used. (C) Piecewise constant curvature (PCC) model is often used for rigid-link and soft continuum robots. It simplifies robot geometry by using constant curvature arcs. (D) Rod models, such as Cosserat rod and beam theory, are often used to represent robots. (E) Shape functions, such as the Bézier Curve, can be used to approximate the robot configuration in space. (F) Variable curvatures that consider the prototypical fact can sometimes better fit the robot's shape.

This section categorizes flexible manipulators by continuum robots with rigid links (i.e., backbone or notched-tube structure) and soft continuum manipulators (see Fig. 2). Such categorization is based on the fact that the continuum robots with discrete joints share some common kinematic knowledge with serial-link robots, while the soft continuum robots sometimes exhibit more complicated kinematics considering the nature of materials and soft mechanics.

Thus far, several kinematic frameworks have been proposed and utilized in MIS flexible robots. One can use different geometrical or algebraic models to describe flexible robot kinematics. Fig. 3 sketches a couple of model-based kinematic frameworks for discrete-jointed and soft continuum robots. The following subsections will briefly introduce these kinematic frameworks in robot-independent mapping, where the robot-dependent actuator space is not primarily considered. Since there are extensive reviews in the literature, this part provides only a concise description without rewinding the detailed mathematics derivation. As a handbook, useful literature for corresponding models will be provided.

## Discrete-jointed continuum manipulators

A discrete-jointed continuum robot is a type of flexible robot, usually characterized by multiple degrees of freedom (DOF) and a finite number of rigid joint-link serial structures. Some may define them as "Semisoft" continuum robots [30], or "Rigid segment" continuum robots [31]. Table 2 lists some typical discrete-jointed continuum manipulators based on different kinematic models for MIS for the below case studies and discussions.

TABLE 2    Sample discrete-jointed and multibackbone continuum manipulators for MIS.

| Ref. | Features | Main materials | Actuation methods | Kinematic models | Medical applications | Dimensions |
|------|----------|----------------|-------------------|------------------|---------------------|------------|
| [32] | 13 rigid links with spherical joints | Polyether ether ketone (PEEK) | Cable driven | Discrete rigid-link model, CC | Manipulator | 20 mm OD, 4 mm ID, 140 mm length |
| [33] | Proximally constrained bending motion, rigid vertebras | – | Cable driven | Discrete rigid-link model, CC | Manipulator | 7.5 mm OD, length about 65 mm |
| [34] | 17 rigid links with spherical joints and a gripper | NiTi backbone | Cable driven | Discrete rigid-link model, D-H | Maxillary sinus surgery/manipulator | 4 mm OD, 30 mm length |
| [35] | Multiple 1D rolling joints, variable stiffness | ABS rolling joints | Cable driven | Discrete rigid-link model | Manipulator | 20 mm OD, 12 mm ID |
| [36] | Notched structure | Metal | Cable driven | Pseudo-rigid-link (curve), D-H | Osteonecrosis, bone drilling | 6 mm OD, 35 mm length |
| [37] | Hybrid motion/force control multibackbone | NiTi | Cable driven | PCC | Manipulator, haptic, tissue stiffness mapping | 5 mm OD, 17.5 mm length for each segment |
| [38] | Tension propagation model with tendon friction considered | Stainless steel | Cable driven | PCC | Manipulator, endoscope | 3.4 mm OD, 120 mm length |
| [39] | Three-segment continuum robot with multiple working channels | PTFE | Tendon-driven (pushing/pulling) | PCC | Transurethral bladder tumor resection | 5 mm OD, three 1.8 mm ID working channels, 18, 20, 15 mm length, respectively |
| [40] | Two-segment continuum robot with shape-sensing capability | – | Cable driven | PCC/shape function | Manipulator | – |
| [41] | A rigid segment between two continuum segments with notched structures | Metal | Tendon driven (pushing/pulling) | Cosserat rod | Manipulator | – |
| [42] | With extensible sections and path-following motion | NiTi, magnets | Cable driven | Cosserat rod | Manipulator | 5 mm OD, 15–55 mm segment length |
| [43] | RCM motion with a commercial robot arm | NiTi tube with notches | McKibben muscles (pneumatic) | Cosserat rod | Manipulator | 7 mm OD, 6 mm ID, 66 mm length |
| [44] | FBG sensing | NiTi tube with notches | Cable-driven | Jacobian-approximated curvatures | Manipulator | 6 mm OD, 34 mm length |
| [45] | 3D-printed spring-like flexible joints with presettable stiffness | Metal | Tendon driven (pushing/pulling) | PCC with FEM-based correction | Manipulator | 5 mm OD, 1.2 mm ID, 43.15 mm length |
| [46] | 3-DOF, master-slave control | Titanium alloy | Cable driven | Learning based | Transoral laryngeal surgery/manipulator | 6 mm OD, dual 2.1 mm ID, 52.5 mm length |
| [47] | Disc vertebras | PTFE tube backbone | Tendon driven | Learning based | Manipulator | 8 mm OD, 100 mm length |

### Rigid-link model

Although the medical continuum robots listed in Table 2 are primarily assembled using rigid components, they exhibit compliant and flexible behavior upon actuation and interaction. Similar to the conventional kinematics models for rigid-link robots, one can use a homogeneous transformation matrix to represent the position and orientation of each link in space as

$$\mathbf{T}_i = \begin{bmatrix} \mathbf{R}_i & \mathbf{p}_i \\ \mathbf{0}_{1\times3} & 1 \end{bmatrix} \in SE(3)$$

where $\mathbf{T}_i$ represents the "pose" of the $i$-th link with respect to some global coordinates, $\mathbf{R}_i \in SO(3)$ denotes the rotation matrix, and $\mathbf{p}_i \in \mathbb{R}^3$ is the position vector of a point on the link. When considering a Cartesian frame attached to the $i$-th link moves along with it, one can use $\mathbf{T}_i^{i-1}$ to describe the transformation from the $(i-1)$-th link to the $i$-th link. If one describes the pose of the $i$-th link with respect to a reference frame of $\{0\}$, it can be denoted by taking the continued multiplication of the successive frame as

$$\mathbf{T}_i^0 = \prod_{k=1}^{i} \mathbf{T}_k^{k-1}(\mathbf{q}_k)$$

where $\mathbf{q}_k \in \mathbb{R}^m$ represents the actuator variables, with $m$ being the number of actuators. Fig. 3A provides a schematic of how the traditional serial-link kinematics can be applied to the modeling of continuum robots. It is often used by flexible robots with discrete rigid-link structures where they can be geometrically modeled. For MIS-oriented continuum robots, various joint-link structures have been proposed, such as those with spherical joints [32,34], serial vertebras [33], and rolling joints [35]. In particular, Piltan et al. [34] model a 4-DOF continuum robot composed of 17 discrete rigid links by using the Denavit-Hartenberg (D-H) convention [48]. However, hyperredundancy complicates the kinematics. With some proper assumptions on the robot geometry, the model can be simplified.

### PCC

In fact, many experiments suggest that the continuum manipulators deform their shape continuously along the axial direction—which is mathematically equivalent to a rigid-link robot with infinitesimally tiny links and an infinite number of joints [24]. Therefore, early continuum robotics researchers established a simplified kinematic representation to describe the robots by using multiple serially connected circular arcs with constant curvatures, while each arc can be expressed by a finite set of parameters. This is usually referred to as a piecewise-constant curvature (PCC) model. It is termed based on the assumption that the curvature in each segment is constant over its longitudinal length upon bending. In PCC model, the forward kinematics can be written as

$$\mathbf{T}_i^0 = \prod_{k=1}^{i} \mathbf{T}_{cc,k}^{k-1}(\mathbf{\Psi}_k)$$

where $\mathbf{T}_{cc,k}^{k-1}(\mathbf{\Psi}_k)$ represents the local transformation (i.e., the $i$-th segment's) matrix with respect to its previous segment, and $\mathbf{\Psi}_k$ denotes the configuration variables in a vector form defining the geometry ("arc parameters") of the local curvature as

$$\mathbf{\Psi}_k(\mathbf{q}) = [\kappa_k \ \psi_k \ l_k]^\top$$

where $\kappa_k, \psi_k, l_k$ represent the local curvature, angle of bending plane, and arc length, respectively (see Fig. 3C). The configuration of the local link is also a function of the actuator variables [15].

The arc geometry can also be represented by Frenet-Serret formulas (for example, in Ref. [49]) that parameterize a curve in terms of arc length $s$ by defining a local coordinate frame, which moves along the curve in terms of a unit vector tangent to the curve $t(s)$, a unit vector $n(s)$, and a unit binormal vector $b(s) = t(s) \times n(s)$, such that a coordinate along the arc can be obtained by taking the integration of the tangent curve as $\mathbf{p}(s) = \int_0^s t(s)\mathrm{d}s$. Similar results can be obtained by using the exponential coordinates based on the Lie group theory [39,50,51] that decompose the homogeneous transformation of a constant curvature arc into rotation transformation and in-plane transformation. Note that those different representations are only diverse in mathematical forms but lead to the same CC assumptions. With many other variants and modified forms, the board-PCC model plays an essential role in soft and continuum robot kinematics modeling.

### Cosserat rod

Modeling continuum robots using the Cosserat rod theory was first introduced by Trivedi et al. [52]. The method involves equilibrium relationships relating the internal and external forces and moments along the elastic robot backbone or robot body. As shown in Fig. 3D, the local rotation matrix and position vector along with the arc length with respect to a reference frame can be derived based on the differential kinematic relationship as

$$\frac{\mathrm{d}\mathbf{R}(s)}{\mathrm{d}s} = \mathbf{R}(s) \times u(s)$$

$$\frac{d\mathbf{p}(s)}{ds} = \mathbf{R}(s) \cdot v(s)$$

where $u(s) \in \mathbb{R}^3$ and $v(s) \in \mathbb{R}^3$ are the curvature vector containing angular rates of change about the rotation matrix and the linear velocity of the position vector, respectively. Considering the static equilibrium, the rod model consists of nonlinear ordinary differential equations (ODEs) for the internal force and moment vectors as [9]

$$\frac{dn(s)}{ds} + f(s) = 0$$

$$\frac{dm(s)}{ds} + \frac{d\mathbf{p}(s)}{ds} \times n(s) + l(s) = 0$$

where $n(s)$, $m(s)$, $f(s)$, and $l(s)$ are the vectors of internal force and moment, and vectors of external distributed force and moment, respectively. These equilibrium equations are generally coupled to the differential kinematic relationships through some material constitutive laws based on the material properties and mechanics. As it can satisfactorily describe the nonlinear strain-stress relationship, the method applies to many walks of the continuum and soft robots that obey elastic deformations upon actuation. In discrete-jointed continuum robots, the rod method is often used in modeling the elastic primary backbone that represents the continuum robot's curve. Recent examples that model medical discrete-jointed continuum robots using rod theory can be seen in Ref. [41, 42, 53].

### Shape functions

Shape functions, usually in the form of curve parameters, can be used to approximate the continuum robot geometry and model the kinematics. One can describe the robot curve using linear combinations of mode shape functions as [29]

$$c(s) = \sum_{r=1}^{n} b_r \mathbf{M}_r(s)$$

where $b_r$ denotes the $r$-th coefficient and $\mathbf{M}_r$ is the $r$-th shape functions. By principle, two curve parameters, namely, the curvature $\kappa(s)$ and angle of bending plane $\psi(s)$, are necessary to fully represent an in-plane bending. It is useful in the venue of continuum robotic sensing, where the general sensing techniques provide discrete instead of continuous data. One can employ the shape functions to estimate the geometry of continuum robots in space. For example, in Ref. [40], quadratic Bézier curves are used to reconstruct the wire-driven flexible robots with multiple bending segments by reading the discrete joint position and direction. In Ref. [54], piecewise cubic Bézier curves are employed to model a discrete-jointed continuum robot.

### Jacobian approximation

Model-free approaches are also available to approximate the prototype-dependent kinematics of discrete-jointed continuum robots. In general, the velocity of a robot manipulator can be described by an input-output system of $\dot{x} = \dot{f}(q)$ where the linear velocity of the end-effector $\dot{x}$ is a first derivative function of the actuator inputs $\mathbf{q}$. The continuous function can be linearly approximated in an instantaneous status such that $\dot{x} \approx \mathbf{J}\dot{q}$, where $\mathbf{J}$ denotes the Jacobian matrix calculated by taking the partial derivatives of the $f(\mathbf{q})$. The Jacobian matrix that maps the actuator inputs and robot end-effector motion output can be empirically estimated—since the output can be physically measured using various methods, such as vision [55], FBGs [44], etc. However, when the system is redundant (i.e., there are multiple possible solutions given the same output), the method could be ineffective unless with more measurable outputs or higher dimensional output for the control. Even though the approximation is robot dependent, it provides a rapid numerical solution of the kinematics of a tangible continuum robot system.

### Finite element

The variety of notched and hinge structures of continuum robots diversifies their kinematics. Finite element methods (FEM) are typically used to model the continuum robots in the venues where the robot morphology, internal/external force, and actuation are required to be included. FEM-enhanced kinematics could provide the correction terms subject to the new continuum robot design [45], design comparison and optimization [56], etc. Since there are quite a lot of rigid components in discrete-jointed continuum robots, the computational cost of FEM could be less expensive than that for soft-bodied robots.

## Learning

It can be challenging to precisely model the continuum robots' kinematics due to the nonlinear factors such as friction and backlash in actuation, assembly error, robot gravity, payload, etc. Machine learning-based methods were long proposed to cope with the limitations of model-based kinematics, especially with the recent advancement of computational hardware and intelligent algorithms. Similar to the Jacobian approximation, machine learning-based methods are widely used in continuum robot kinematics by mapping the relationship between the input and output of the system. Nonlinear factors can be included in the learning, such as external payload [47], disturbances [57], master-side motion [46], etc. However, tedious data collection and offline training would become inevitable. Especially, reinforcement learning requires extensive interactions with the environment, which may cause damage to some types of continuum robots [58].

Here, we summarize the abovementioned kinematic models for discrete-jointed continuum robots in terms of robot dependency and model continuity using Fig. 4. Robot dependency shows the extent of the model's generality and rapid implementation among prototypes, while model continuity indicates the compliance of the robot description in different scenes. Given the same type of robots with known parameters, explicit mathematical methods often depend less on the prototypes from one to another. Robot kinematics can be rapidly implemented for control by adapting the parameters. On the contrary, the Jacobian approximation, FEM, and learning-based kinematics are more robot dependent, meaning that possible prototype differences cannot be easily identified by parameterization but by a re-run on either the simulation or real-world system—especially the learning-based kinematics modeling, which usually involves tedious data collection and offline training. This is a trade-off as the "model-less" methods often out weighted the "model-based" methods in terms of kinematics and statics accuracy. For medical applications, the major efforts on modeling have addressed the forward kinematics and static deflection, while the control has generally been solved from a quasistatic inverse kinematics perspective using traditional robotics knowledge like analytical and numerical (optimization, heuristics, and data-driven) methods.

## Soft continuum manipulators

Soft continuum manipulators are primarily made from soft materials with low Young's modulus [59] comparable to human tissues. The soft nature guarantees interaction safety even in the worst-case scenarios where the robot-environment collision occurs; therefore, soft manipulators are becoming promising choices for developing RMIS tools [14]. Soft continuum manipulators demonstrate weak physical joint-link structures, higher deformability, and lower Poisson's ratio, making the modeling approaches not necessarily consistent with that of the discrete-linked continuum robots. The nonlinear nature of most soft materials—usually excluded from modeling rigid-bodied robots—will contribute to one of the significant factors that differentiate the kinematics of soft robots. With that being said, most of the kinematics models for discrete-linked continuum robots, except for the rigid-link model, are generally applicable to the soft manipulators by posing some assumptions. Table 3 lists some typical discrete-jointed continuum manipulators based on different kinematic models for MIS for the below case studies and discussions.

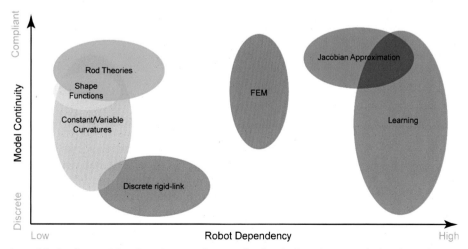

FIG. 4   The kinematic models for discrete-jointed continuum robots can be sketched out in terms of robot dependency and model continuity.

TABLE 3   Sample soft continuum manipulators for MIS.

| Ref. | Features | Main materials | Actuation methods | Kinematic models | Medical applications | Dimensions |
|------|----------|----------------|-------------------|------------------|----------------------|------------|
| [60] | Single segment with four magnetic rings | Elastomer (PEBAX) | Magnetically actuated | PRB | Surgical catheter | 2 mm OD, 1.2 mm ID, 52 mm length |
| [61] | Two segments, capable of elongation motion and variable stiffness. Cadaver study | Ecoflex 00-50 | Fluidic driven | PCC | Manipulator | 14.5 mm OD, 50 mm length |
| [62] | Two segments, visual-enhanced, configuration constraints | Agilus30 polymer | Cable driven | PCC | Blood suction tube | 9 mm OD, 3.6 mm ID, 50 mm length per segment |
| [63] | Two segments, null space motion | Agilus30 polymer | Cable driven | PCC (variant) | Manipulator | 9 mm OD, 3.6 mm ID, 50 mm length per segment |
| [64] | Single segment, capable of elongation and variable stiffness | Silicone rubber | Pneumatic driven | CC | Manipulator | 25 mm OD, 13 mm ID, 50 mm length |
| [65] | Single segment with three channels for medical instruments | Silicone rubber | Cable driven | CC | Manipulator | 10 mm OD, 45 mm length |
| [66] | Single segment, single actuation, variable stiffness | Ecoflex 00-50 | Tendon and pneumatic | Cosserat rod | Manipulator | 15 mm OD, 3 mm ID, 75 mm length |
| [67] | Two-segment, variable stiffness | Elastosil M4601 A/B | Cable driven | Cosserat rod | Manipulator | 20 mm OD, 1.5 mm ID, 160 mm length |
| [68] | Single-segment, interactive navigation | Ecoflex 00-50 | Fluidic driven | FEM-based Jacobian estimation | Manipulator | 20 mm OD |
| [69] | Two segments, visual servoing with planar motion | PDMS | Cable driven | Visual servoing-based Jacobian estimation | Manipulator | 9 mm OD, 50 mm length per segment |
| [70] | Single segment | Agilus30 polymer | Fluidic driven | Learning/FEM | MRI-guided transoral laser microsurgery | 9.2 mm OD, 1.6 mm ID |
| [71] | Single segment with vision-based online learning | Ecoflex 00-50 | Fluidic driven | Learning/FEM | Manipulator, endoscope | 13 mm OD, 93 mm length |
| [72] | Two segments, sim-to-real visual servoing | Agilus30 polymer | Cable-driven | Learning/FEM (SOFA) | Manipulator, endoscope | 6 mm OD, 50 mm length per segment |
| [73] | Single segment with locomotion | Ecoflex 00-30 | Pneumatic driven | Empirical | Manipulator | 25 mm OD, 15 mm ID |
| [74] | Single segment | PDMS | Pneumatic driven | Euler-Bernoulli Beam | Micromanipulator | 1 mm OD |
| [75,76] | Single-segment, extrinsic tip controlled, extremely small | TPU, NdFeB (30% by volume) | Magnetically actuated | Euler-Bernoulli Beam | Surgical catheter | 0.5–0.6 mm OD |

## *Pseudo-rigid-body*

As soft manipulators are, in most cases, composed of multiple pieces of soft elastic segments that deform in continuous configurations, the rigid-link model would be inapplicable. However, one can assume the continuum segment as a pseudo-rigid-body (PRB, or pseudo-rigid-link) structure, a widely used modeling assumption in compliant mechanics. Although the PRB models generally consist of a finite number of serially connected pseudo-links, their modeling conventions can vary depending on the accuracy requirement. There are several PRB models reported in the literature. For example, in Ref. [60], the authors model a single soft segment using a 6-DOF PRB model with four pseudo-rigid links connected by three rotational joints. The highly customizable PRB links fit the soft magnetic catheters with heterogeneously distributed magnets. In Ref. [77], a soft segment is simplified as two pseudo-links with passive variable length connected by a rotational joint. Several PRB conventions for soft robots, such as RPR, RPPR, and RPRPR, are raised and discussed in Ref. [78]. The method is also applicable to describe the elastic backbones of the discrete-jointed continuum robots.

### PCC

The PCC modeling method is also applicable to the soft-bodied continuum robots following the same assumptions [61,62,65], usually with virtual backbones to represent the soft body's geometry. However, the original PCC convention does not primarily consider the soft materials property, external load, or physical contact with the environment. Common phenomena like elongation, shrinkage, and nonlinear inflation that occur on soft manipulators would be neglected by the PCC assumptions unless with model modification. There are several modified PCC models that enhance the model adaptiveness for soft manipulators. The use of different methods for PCC modification has been reported, such as incorporating FEM [79,80] or augmented PRB model [81], considering the soft body mechanics [63], mode shape function-based variable curvature [82], etc. Therefore, there is still plenty of room to modify the CC convention for different soft manipulators based on the robot designs and prototypes.

### Cosserat rod

Cosserat rod theory frameworks are one of the ideal choices for soft-bodied robot modeling, as they describe a material-dependent homogeneous transformation frame to express the "backbone" as a function of the arc length along the robot. The rod theory frameworks are friendly to the description and investigation of variable stiffness soft robots for medical applications [66,67] related to the material's nonlinearity, external payload, and actuation mechanics. Variant rod methods have also been developed. For instance, a compromise method called piecewise constant strain (PCS) was proposed by Renda et al. [83], which discretizes the Cosserat rod method and assumes piecewise constant deformation along the soft body (inspired by the PCC) to achieve simplification, benefiting the model adaptability.

### Others

Other kinematics modeling methods for soft manipulators are similar to that of the discrete-jointed continuum robots, including FEM, mechanics beam, Jacobian estimation, machine learning, empiricism, etc. Among them, FEM and mechanics methods can be established independently from prototypes, while the others are robot-dependent unless with some fusions. For example, the finite element analysis (FEA) can partially become a source for the Jacobian estimation [68] or machine learning [70]. In Ref. [71], the FEA-based model of a single-bending soft robot is used to initialize the online learning supported by the ground truth measurement from the prototype—which greatly reduces the computational complexity and makes the Jacobian initialization easier. For soft robots with simple nonlinear bending motion, one can empirically map the kinematic relationship [73]. Millimeter-grade soft robotic catheters are usually magnetically tip-driven to navigate through complex environments like blood vessels and bronchia. These types of soft robots are partially bent at their tip to "switch" direction by the amplitude of the electromagnetic field and directly dragged by the locatable magnetic source. The intrinsic actuation allows symmetrical force distribution. Hence, the catheters can be ideally modeled as short cantilever beams [75,76].

In this section, we would like to summarize the soft robots' kinematics by their conveniences and competency in describing the nonlinearity of the robot's motion. Fig. 5 demonstrates an overview of the summary. Geometry-based methods like PCC and PRBs are relatively limited in expressing the robot's nonlinear behavior but benefit rapid kinematics modeling with available parameters. Their nonlinearity can be further enhanced by fusing other methods. Rod and beam theories are capable of illustrating the soft robot's physical behaviors related to internal and external forces. Model-free methods that employ fine meshes or experimental data generally neglect geometry complexity and nonlinearity. Therefore, they can derive the kinematics of some highly customized soft manipulators.

## Kinematics-based applications for MIS flexible robots

There is a growing trend that flexible manipulators are being adopted as the end-effectors of RMIS because of the inherent interaction safety. For flexible robots, kinematics is fundamental for many MIS applications that require precise control. When there is a new MIS flexible robotic tool being designed, a thorough study of kinematics would provide a clear prediction of how the robot behaves, how we can have better control, and even suggest how we can take advantage of the compliant structures. In this part, we introduce several kinematics-based RMIS applications for flexible robots by case study.

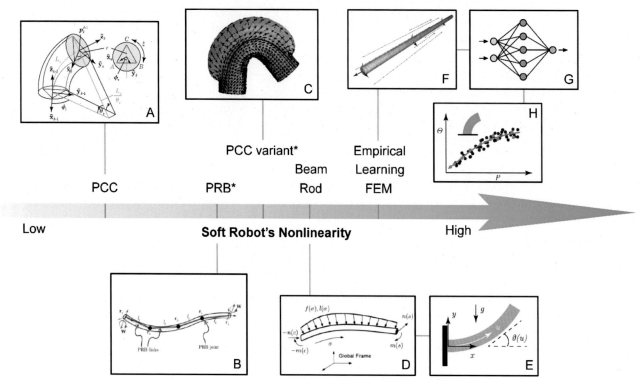

FIG. 5    Different kinematics modeling methods for soft robots in terms of the conveniences in expressing nonlinearity: (A) PCC is often used in soft robotic kinematics; (B) PRBs can describe complicated soft robots' motion with sufficient control point; (C) PCC can be further improved by incorporating other modeling factors; (D) Cosserat rod theory includes forces in the modeling; (E) Euler-Bernoulli beam theory are often used in describing soft robots with simple bending with some ideal assumptions; (F) FEM, (G) machine learning, and (H) empirical methods are competent of describe highly nonlinear soft robot behavior despite of the model complexity.

## Workspace optimization

The workspace is typically defined as a set of endpoints that can be approached by the end-effector of the manipulators. Generally, the robot's end-effector is repositioned by bending the continuum structure in a confined environment without excessively pivoting around the incision/trocar. As shown in Fig. 6A, this could significantly reduce the motion of the robot's footprint (operational space) and improve the end-effector's workspace, which is helpful to the maturation of specific RMIS and improvement of interaction safety. While using flexible robots with simple bending structures and RCM (remote-center-of-motion), mechanism can improve the end-effector's workspace, employing multisegment flexible tools can even reduce the operation space that prevents the tools from pivoting around the incision.

In this regard, model-based kinematics can be employed to investigate and optimize the end-effector workspace prior to robot fabrication. It allows the researchers to tailor design flexible robotic tools for different surgical venues. For example, Li et al. [84] explicitly analyzed the workspace of a discrete-jointed cable-driven continuum manipulator with different longitudinal constraints based on the PCC model. By partially constraining the proximal segment from bending, single-segment robots could attain a wider workspace without adding extra DOF to the system. In Ref. [85], the authors optimized the multisegment continuum robot design according to the reachable workspace and desired functional volume with the help of PCC modeling. In addition to the kinematics, static analysis can be incorporated into the PCC in estimating the end-effector's workspace [86], such that different forces would be included in the modeling. For soft material-based robots, FEM modeling [71,87] could provide an intuitive sensation of workspace that allows one to optimize the parameters in not only robot geometry but also physical properties like morphology and the use of materials. In Ref. [88], Cosserat rod models were employed to analyze a hybrid rigid-flexible soft robot with the consideration of material constitutive law. Other kinematic models are also often used in estimating the soft robot's workspace that fits biomedical applications. In Ref. [89], a beam model was used to compute the soft robotic end-effector's spatial position and further bioprinting control in MIS. Supported by the field data, Lin et al. [90] reported a flexible magnetic-driven continuum robot with enlarged effectiveness in constrained conditions for retrograde intrarenal surgery. Fig. 6B summarizes the general workflow of how workspace optimization works with kinematic analysis during the design stage of a standalone flexible robotic tool for MIS.

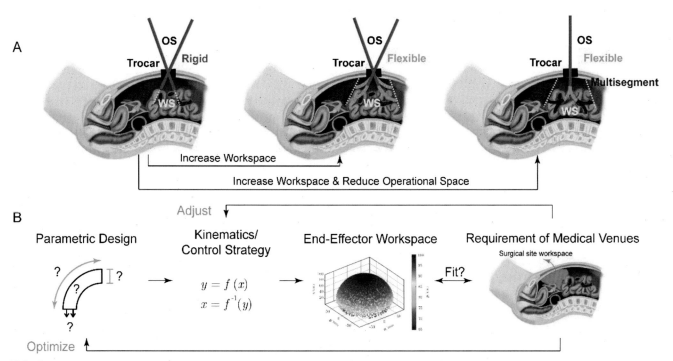

**FIG. 6** (A) Compared to rigid RMIS tools, using flexible RMIS tools can increase the end-effector's workspace and reduce the footprint's operational space. (B) General workflow of workspace optimization for RMIS flexible robotic tools.

## Motion constraints

Instead of operating in free space, MIS flexible robots are often expected to work in confined, obstructed environments that could possibly constrain the robot's motion. In this regard, the kinematics of flexible robots can be modified and further investigated to fit the MIS scenes. Several kinematics-based flexible robot motions considered constraints have been published in the past decade. For example, Bajo and Simaan [91] derived the constrained differential kinematics based on the PCC model to detect and localize the contacts along a surgical multisegment continuum robot. The methods were further applied to the transurethral bladder resection task [92]. In another work from them [37], they presented a framework that compromises the PCC-based robot motion and force controller to estimate the stiffness and shape of the external objects and environment on a multibackbone continuum robot for MIS. In Ref. [62], a configuration constraint was assigned to a two-segment PCC-based soft robot to regulate the end-effector's orientation with respect to the task surface for effective blood suction. Follow-the-leader motion, which is desirable for many MIS applications, was developed for tendon-driven continuum robots based on the Cosserat rod model [93]. Sensor-based online Jacobian estimation methods were introduced to control continuum manipulators to work in constrained environments [55] and even perform soft object manipulation [94,95].

Except for making use of the kinematics of standalone miniature multisegment robots, one can associate the macro-robot arm and attain extra degrees of freedom that compensate the microrobots with simple 2-DOF bending motion (i.e., single-segment continuum robots). This is often seen when the macro-micro MIS robots are coupled as one piece. The task space operations can be performed through the RCM constraints enabled by redundant kinematics—which raises worthy optimization problems in multiple solutions for the joint space control. For example, a spherical linkage RCM mechanism with a single-segment discrete-jointed continuum robot was proposed by Qi et al. [96]. The end-effector's motion of the PCC-modeled robot was mechanically constrained, and the tip motion could be kinematically modeled by extending the base frame. In Ref. [97], the authors attached a cable-driven continuum robot with a single bending segment onto a commercial robot arm (UR5) to obtain RCM motion. The RCM constraints formulated a QP problem with multiple cost functions to realize multi-DOF kinematic control. Similar works can be found in Ref. [98], which proposes a convex optimization method to control a micro-macro robotic motion with alternating direction method of multipliers (ADMM) and damped least square (DLM) algorithms. Recently, Alambeigi et al. [99] report using such a robotic system for orthopedic MIS, showing the competency and performance of employing motion constraint strategy on flexible robots for RMIS applications.

## Conclusions

This chapter provides a concise overview of how kinematics concepts are utilized in RMIS with flexible robotic manipulators. We limit our focus to the recent decade and subjects of discrete-jointed and soft continuum robots for potential RMIS applications. Through state-of-the-art case studies, we cover kinematic modeling methods with multidimensional comparisons and their practical applications. Our discussion is confined to the literature on MIS flexible robots, which provides realistic and reproducible references on how flexible robots are designed, considered, and adopted in medical applications. In terms of applications, we address workspace optimization and motion constraints, which can be resolved by investigating robot kinematics. Overall, this chapter serves as a valuable resource for researchers, medical robot designers, and developers by providing a comprehensive overview of the recent advances in the kinematic modeling of flexible robotic manipulators in RMIS and their practical applications.

## Future scenario

The authors contend that there is still considerable potential for the development and refinement of flexible robotic kinematics. As flexible robots become increasingly softer, lighter, and smaller, they are becoming more versatile and multifunctional in terms of robotic motion and control. This presents a significant challenge to the efficacy of kinematics modeling for RMIS applications. Moreover, comprehensive RMIS scenarios still pose significant challenges to the kinematics and statics of continuum robots. Meanwhile, new types of flexible robots are evolving, which may require model revisions or even new models. However, the authors remain optimistic that ongoing research efforts will lead to the continued advancement and evolution of flexible robotic manipulators in RMIS, ultimately improving patient outcomes and medical procedures.

## Key points

- Kinematic concepts in MIS flexible robotic manipulators through case studies of the recent decade (2012–22) are described.
- Two types of MIS flexible robotic manipulators have been extensively investigated in the field, namely, the flexible manipulators and the soft continuum manipulators. Both of them share similar kinematic modeling concepts, but the latter demonstrates more room for exploration.
- The kinematics-based applications for MIS flexible robots are challenging to fit the intended clinical scenarios. More medical critiques can be considered in the robot design and kinematics.

## Acknowledgment

This work was supported by Hong Kong Research Grants Council (RGC) Collaborative Research Fund (C4026-21G), Research Grants Council (RGC)—Research Impact Fund (RIF R4020-22), General Research Fund (GRF 14211420), Research Grants Council (RGC)—NSFC/RGC Joint Research Scheme N_CUHK420/22, Shenzhen-Hong Kong-Macau Technology Research Programme (Type C) STIC Grant SGDX20210823103535014 (202108233000303) the key project 2021B1515120035 (B.02.21.00101) of the Regional Joint Fund Project of the Basic and Applied Research Fund of Guangdong Province, Hong Kong Research Grants Council (RGC) Collaborative Research Fund (CRF C4063-18G and C4026-21G), General Research Funds (GRF 14216022, GRF 14203323), CUHK IdeaBooster Fund Award (IDBF23ENG06), and CUHK Faculty Direct Grant (4055213).

## References

[1] Sachs J, Andersson LAA, Araújo J, Curescu C, Lundsjö J, Rune G, Steinbach E, Wikström G. Adaptive 5G low-latency communication for tactile internet services. Proc IEEE 2019;107:325–49.
[2] Taylor RH, Menciassi A, Fichtinger G, Fiorini P, Dario P. Medical robotics and computer-integrated surgery. In: Springer handbook of robotics. Springer; 2016. p. 1657–84.
[3] de Smet MD, Naus GJL, Faridpooya K, Mura M. Robotic-assisted surgery in ophthalmology. Curr Opin Ophthalmol 2018;29:248–53.
[4] Sharifi M, Salarieh H, Behzadipour S, Tavakoli M. Beating-heart robotic surgery using bilateral impedance control: theory and experiments. Biomed Signal Process Control 2018;45:256–66.
[5] Li S, Ai Q, Liang H, Liu H, Yang C, Deng H, Zhong Y, Zhang J, He J. Nonintubated robotic-assisted thoracic surgery for tracheal/airway resection and reconstruction: technique description and preliminary results. Ann Surg 2022;275, e534.
[6] Kinross JM, Mason SE, Mylonas G, Darzi A. Next-generation robotics in gastrointestinal surgery. Nat Rev Gastroenterol Hepatol 2020;17:430–40.

[7] Yip HM, Wang Z, Navarro-Alarcon D, Li P, Cheung TH, Greiffenhagen C, Liu Y-h. A collaborative robotic uterine positioning system for laparoscopic hysterectomy: design and experiments. Int J Med Robot Comput Assist Surg 2020;16, e2103.

[8] Min Z, Wang J, Pan J, Meng MQ-H. Generalized 3-D point set registration with hybrid mixture models for computer-assisted orthopedic surgery: from isotropic to anisotropic positional error. IEEE Trans Autom Sci Eng 2020;18:1679–91.

[9] Burgner-Kahrs J, Rucker DC, Choset H. Continuum robots for medical applications: a survey. IEEE Trans Robot 2015;31:1261–80.

[10] Kwok K-W, Wurdemann H, Arezzo A, Menciassi A, Althoefer K. Soft robot-assisted minimally invasive surgery and interventions: advances and outlook. Proc IEEE 2022;110:871–92.

[11] Dobbs RW, Halgrimson WR, Madueke I, Vigneswaran HT, Wilson JO, Crivellaro S. Single-port robot-assisted laparoscopic radical prostatectomy: initial experience and technique with the da Vinci® SP platform. BJU Int 2019;124:1022–7.

[12] Chen Y, Zhang C, Wu Z, Zhao J, Yang B, Huang J, et al. The SHURUI system: a modular continuum surgical robotic platform for multiport, hybrid-port, and single-port procedures. IEEE/ASME Trans Mechatron 2022;27(5):3186–97. https://doi.org/10.1109/TMECH.2021.3110883.

[13] Lang S, Mattheis S, Hasskamp P, Lawson G, Güldner C, Mandapathil M, Schuler P, Hoffmann T, Scheithauer M, Remacle M. A European multicenter study evaluating the flex robotic system in transoral robotic surgery. Laryngoscope 2017;127:391–5.

[14] Runciman M, Darzi A, Mylonas GP. Soft robotics in minimally invasive surgery. Soft Robot 2019;6:423–43.

[15] Webster III RJ, Jones BA. Design and kinematic modeling of constant curvature continuum robots: a review. Int J Robot Res 2010;29:1661–83.

[16] Zhong Y, Hu L, Xu Y. Recent advances in design and actuation of continuum robots for medical applications. Actuators 2020;9(4):142. https://doi.org/10.3390/act9040142.

[17] Thuruthel TG, Shih B, Laschi C, Tolley MT. Soft robot perception using embedded soft sensors and recurrent neural networks. Sci Robot 2019;4, eaav1488.

[18] Zhang Y, Lu M. A review of recent advancements in soft and flexible robots for medical applications. Int J Med Robot Comput Assist Surg 2020;16, e2096.

[19] da Veiga T, Chandler JH, Lloyd P, Pittiglio G, Wilkinson NJ, Hoshiar AK, Harris RA, Valdastri P. Challenges of continuum robots in clinical context: a review. Prog Biomed Eng 2020;2:032003.

[20] Shi C, Luo X, Qi P, Li T, Song S, Najdovski Z, Fukuda T, Ren H. Shape sensing techniques for continuum robots in minimally invasive surgery: a survey. IEEE Trans Biomed Eng 2016;64:1665–78.

[21] Cao Y, Shi Y, Hong W, Dai P, Sun X, Yu H, Xie L. Continuum robots for endoscopic sinus surgery: recent advances, challenges, and prospects. Int J Med Robot Comput Assist Surg 2022; e2471.

[22] Gifari MW, Naghibi H, Stramigioli S, Abayazid M. A review on recent advances in soft surgical robots for endoscopic applications. Int J Med Robot Comput Assist Surg 2019;15, e2010.

[23] Dupont PE, Nelson BJ, Goldfarb M, Hannaford B, Menciassi A, O'Malley MK, Simaan N, Valdastri P, Yang G-Z. A decade retrospective of medical robotics research from 2010 to 2020. Sci Robot 2021;6, eabi8017.

[24] Dupont PE, Simaan N, Choset H, Rucker C. Continuum robots for medical interventions. Proc IEEE 2022;110:847–70.

[25] George Thuruthel T, Ansari Y, Falotico E, Laschi C. Control strategies for soft robotic manipulators: a survey. Soft Robot 2018;5:149–63.

[26] Li Z, Wu L, Ren H, Yu H. Kinematic comparison of surgical tendon-driven manipulators and concentric tube manipulators. Mech Mach Theory 2017;107:148–65.

[27] Armanini C, Messer C, Mathew AT, Boyer F, Duriez C, Renda F. Soft robots modeling: A structured overview. arXiv preprint arXiv:2112.03645; 2021.

[28] Barrientos-Diez J, Dong X, Axinte D, Kell J. Real-time kinematics of continuum robots: modelling and validation. Robot Comput Integr Manuf 2021;67:102019.

[29] Rao P, Peyron Q, Lilge S, Burgner-Kahrs J. How to model tendon-driven continuum robots and benchmark modelling performance. Front Rob AI 2021;7:630245.

[30] Li S, Hao G. Current trends and prospects in compliant continuum robots: a survey. Actuators 2021;10(7):145. https://doi.org/10.3390/act10070145.

[31] Tan N, Gu X, Ren H. Design, characterization and applications of a novel soft actuator driven by flexible shafts. Mech Mach Theory 2018;122:197–218.

[32] Ji D, Kang TH, Shim S, Lee S, Hong J. Wire-driven flexible manipulator with constrained spherical joints for minimally invasive surgery. Int J Comput Assist Radiol Surg 2019;14:1365–77.

[33] Li Z, Feiling J, Ren H, Yu H. A novel tele-operated flexible robot targeted for minimally invasive robotic surgery. Engineering 2015;1:073–8.

[34] Piltan F, Kim C-H, Kim J-M. Adaptive fuzzy-based fault-tolerant control of a continuum robotic system for maxillary sinus surgery. Appl Sci 2019;9:2490.

[35] Kim Y-J, Cheng S, Kim S, Iagnemma K. A stiffness-adjustable hyperredundant manipulator using a variable neutral-line mechanism for minimally invasive surgery. IEEE Trans Robot 2013;30:382–95.

[36] Alambeigi F, Wang Y, Sefati S, Gao C, Murphy RJ, Iordachita I, Taylor RH, Khanuja H, Armand M. A curved-drilling approach in core decompression of the femoral head osteonecrosis using a continuum manipulator. IEEE Robot Autom Lett 2017;2:1480–7.

[37] Bajo A, Simaan N. Hybrid motion/force control of multi-backbone continuum robots. Int J Robot Res 2016;35:422–34.

[38] Kato T, Okumura I, Song S-E, Golby AJ, Hata N. Tendon-driven continuum robot for endoscopic surgery: preclinical development and validation of a tension propagation model. IEEE/ASME Trans Mechatron 2014;20:2252–63.

[39] Sarli N, Del Giudice G, De S, Dietrich MS, Herrell SD, Simaan N. TURBot: a system for robot-assisted transurethral bladder tumor resection. IEEE/ASME Trans Mechatron 2019;24:1452–63.

[40] Song S, Li Z, Meng MQ-H, Yu H, Ren H. Real-time shape estimation for wire-driven flexible robots with multiple bending sections based on quadratic Bézier curves. IEEE Sensors J 2015;15:6326–34.

[41] Zhao B, Zeng L, Wu Z, Xu K. A continuum manipulator for continuously variable stiffness and its stiffness control formulation. Mech Mach Theory 2020;149:103746.

[42] Amanov E, Nguyen T-D, Burgner-Kahrs J. Tendon-driven continuum robots with extensible sections—a model-based evaluation of path-following motions. Int J Robot Res 2021;40:7–23.

[43] Smoljkic G, Borghesan G, Devreker A, Poorten EV, Rosa B, De Praetere H, De Schutter J, Reynaerts D, Sloten JV. Control of a hybrid robotic system for computer-assisted interventions in dynamic environments. Int J Comput Assist Radiol Surg 2016;11:1371–83.

[44] Sefati S, Murphy RJ, Alambeigi F, Pozin M, Iordachita I, Taylor RH, Armand M. FBG-based control of a continuum manipulator interacting with obstacles. In: 2018 IEEE/RSJ international conference on intelligent robots and systems (IROS); 2018.

[45] Feng F, Hong W, Xie L. Design of 3D-printed flexible joints with presettable stiffness for surgical robots. IEEE Access 2020;8:79573–85.

[46] Feng F, Zhou Y, Hong W, Li K, Xie L. Development and experiments of a continuum robotic system for transoral laryngeal surgery. Int J Comput Assist Radiol Surg 2022;17:497–505.

[47] Wang Z, Wang T, Zhao B, He Y, Hu Y, Li B, Zhang P, Meng MQ-H. Hybrid adaptive control strategy for continuum surgical robot under external load. IEEE Robot Autom Lett 2021;6:1407–14.

[48] Corke PI. A simple and systematic approach to assigning Denavit–Hartenberg parameters. IEEE Trans Robot 2007;23:590–4.

[49] Ros-Freixedes L, Gao A, Liu N, Shen M, Yang G-Z. Design optimization of a contact-aided continuum robot for endobronchial interventions based on anatomical constraints. Int J Comput Assist Radiol Surg 2019;14:1137–46.

[50] Murray RM, Li Z, Sastry SS. A mathematical introduction to robotic manipulation. CRC Press; 2017.

[51] Grazioso S, Gironimo GD, Siciliano B. From differential geometry of curves to helical kinematics of continuum robots using exponential mapping. In: International symposium on advances in robot kinematics; 2018.

[52] Trivedi D, Lotfi A, Rahn CD. Geometrically exact models for soft robotic manipulators. IEEE Trans Robot 2008;24:773–80.

[53] Mitros Z, Thamo B, Bergeles C, Da Cruz L, Dhaliwal K, Khadem M. Design and modelling of a continuum robot for distal lung sampling in mechanically ventilated patients in critical care. Front Rob AI 2021;8:611866.

[54] Guo H, Ju F, Chen B. Preliminary study on shape sensing for continuum robot affected by external load using piecewise fitting curves. In: 2019 IEEE international conference on robotics and biomimetics (ROBIO); 2019.

[55] Yip MC, Camarillo DB. Model-less feedback control of continuum manipulators in constrained environments. IEEE Trans Robot 2014;30:880–9.

[56] Tian J, Wang T, Fang X, Shi Z. Design, fabrication and modeling analysis of a spiral support structure with superelastic Ni-Ti shape memory alloy for continuum robot. Smart Mater Struct 2020;29:045007.

[57] Cui Z, Li J, Li W, Zhang X, Chiu PWY, Li Z. Fast convergent antinoise dual neural network controller with adaptive gain for flexible endoscope robots. IEEE Trans Neural Netw Learn Syst 2022.

[58] Tan N, Yu P, Zhong Z, Zhang Y. Data-driven control for continuum robots based on discrete zeroing neural networks. IEEE Trans Ind Inform 2022.

[59] Rus D, Tolley MT. Design, fabrication and control of soft robots. Nature 2015;521:467–75.

[60] Venkiteswaran VK, Sikorski J, Misra S. Shape and contact force estimation of continuum manipulators using pseudo rigid body models. Mech Mach Theory 2019;139:34–45.

[61] Abidi H, Gerboni G, Brancadoro M, Fras J, Diodato A, Cianchetti M, Wurdemann H, Althoefer K, Menciassi A. Highly dexterous 2-module soft robot for intra-organ navigation in minimally invasive surgery. Int J Med Robot Comput Assist Surg 2018;14, e1875.

[62] Lai J, Huang K, Lu B, Zhao Q, Chu HK. Verticalized-tip trajectory tracking of a 3D-printable soft continuum robot: enabling surgical blood suction automation. IEEE/ASME Trans Mechatron 2021;27:1545–56.

[63] Lai J, Lu B, Zhao Q, Chu HK. Constrained motion planning of a cable-driven soft robot with compressible curvature modeling. IEEE Robot Autom Lett 2022;7:4813–20.

[64] Ranzani T, Cianchetti M, Gerboni G, De Falco I, Menciassi A. A soft modular manipulator for minimally invasive surgery: design and characterization of a single module. IEEE Trans Robot 2016;32:187–200.

[65] Li R, Chen F, Yu W, Igarash T, Shu X, Xie L. A novel cable-driven soft robot for surgery. J Shanghai Jiaotong Univ (Sci) 2022;1–13.

[66] Roshanfar M, Sayadi A, Dargahi J, Hooshiar A. Stiffness adaptation of a hybrid soft surgical robot for improved safety in interventional surgery. In: 2022 44th annual international conference of the IEEE Engineering in Medicine & Biology Society (EMBC); 2022.

[67] Xiao Q, Musa M, Godage I, Su H, Chen Y. Kinematics and stiffness modeling of soft robot with a concentric backbone. J Mech Robot 2022;1–13.

[68] Lee K-H, Leong MCW, Chow MCK, Fu H-C, Luk W, Sze K-Y, Yeung C-K, Kwok K-W. FEM-based soft robotic control framework for intra-cavitary navigation. In: 2017 IEEE international conference on real-time computing and robotics (RCAR); 2017.

[69] Lai J, Huang K, Lu B, Chu HK. Toward vision-based adaptive configuring of a bidirectional two-segment soft continuum manipulator. In: 2020 IEEE/ASME international conference on advanced intelligent mechatronics (AIM); 2020.

[70] Fang G, Chow MCK, Ho JDL, He Z, Wang K, Ng TC, Tsoi JKH, Chan P-L, Chang H-C, Chan DT-M, et al. Soft robotic manipulator for intraoperative MRI-guided transoral laser microsurgery. Sci Robot 2021;6, eabg5575.

[71] Lee K-H, Fu DKC, Leong MCW, Chow M, Fu H-C, Althoefer K, Sze KY, Yeung C-K, Kwok K-W. Nonparametric online learning control for soft continuum robot: an enabling technique for effective endoscopic navigation. Soft Robot 2017;4:324–37.

[72] Lai J, Ren T-A, Yue W, Su S, Chan JYK, Ren H. Sim-to-real transfer of soft robotic navigation strategies that learns from the virtual eye-in-hand vision. IEEE Trans Ind Inform 2024;20:2365–77.

[73] Zhang B, Fan Y, Yang P, Cao T, Liao H. Worm-like soft robot for complicated tubular environments. Soft Robot 2019;6:399–413.

[74] Gorissen B, Vincentie W, Al-Bender F, Reynaerts D, De Volder M. Modeling and bonding-free fabrication of flexible fluidic microactuators with a bending motion. J Micromech Microeng 2013;23:045012.

[75] Wang L, Kim Y, Guo CF, Zhao X. Hard-magnetic elastica. J Mech Phys Solids 2020;142:104045.

[76] Kim Y, Parada GA, Liu S, Zhao X. Ferromagnetic soft continuum robots. Sci Robot 2019;4, eaax7329.

[77] Lai J, Huang K, Chu HK. A learning-based inverse kinematics solver for a multi-segment continuum robot in robot-independent mapping. In: 2019 IEEE international conference on robotics and biomimetics (ROBIO); 2019.

[78] Santina CD, Katzschmann RK, Bicchi A, Rus D. Model-based dynamic feedback control of a planar soft robot: trajectory tracking and interaction with the environment. Int J Robot Res 2020;39:490–513.

[79] Runge G, Wiese M, Günther L, Raatz A. A framework for the kinematic modeling of soft material robots combining finite element analysis and piecewise constant curvature kinematics. In: 2017 3rd international conference on control, automation and robotics (ICCAR); 2017.

[80] Caasenbrood B, Pogromsky A, Nijmeijer H. Control-oriented models for hyperelastic soft robots through differential geometry of curves. Soft Robot 2023;10(1):129–48.

[81] Katzschmann RK, Della Santina C, Toshimitsu Y, Bicchi A, Rus D. Dynamic motion control of multi-segment soft robots using piecewise constant curvature matched with an augmented rigid body model. In: 2019 2nd IEEE international conference on soft robotics (RoboSoft); 2019.

[82] Godage IS, Medrano-Cerda GA, Branson DT, Guglielmino E, Caldwell DG. Modal kinematics for multisection continuum arms. Bioinspir Biomim 2015;10:035002.

[83] Renda F, Cacucciolo V, Dias J, Seneviratne L. Discrete Cosserat approach for soft robot dynamics: a new piece-wise constant strain model with torsion and shears. In: 2016 IEEE/RSJ international conference on intelligent robots and systems (IROS); 2016.

[84] Li Z, Ren H, Chiu PWY, Du R, Yu H. A novel constrained wire-driven flexible mechanism and its kinematic analysis. Mech Mach Theory 2016;95:59–75.

[85] Xu K, Zhao J, Zheng X. Configuration comparison among kinematically optimized continuum manipulators for robotic surgeries through a single access port. Robotica 2015;33:2025–44.

[86] Yuan H, Li Z. Workspace analysis of cable-driven continuum manipulators based on static model. Robot Comput Integr Manuf 2018;49:240–52.

[87] Amehri W, Zheng G, Kruszewski A. Fem based workspace estimation for soft robots: a forward-backward interval analysis approach. In: 2020 3rd IEEE international conference on soft robotics (RoboSoft); 2020.

[88] Altuzarra O, Solanillas DM, Amezua E, Petuya V. Path analysis for hybrid rigid–flexible mechanisms. Mathematics 2021;9:1869.

[89] Zhou C, Yang Y, Wang J, Wu Q, Gu Z, Zhou Y, Liu X, Yang Y, Tang H, Ling Q, et al. Ferromagnetic soft catheter robots for minimally invasive bioprinting. Nat Commun 2021;12:1–12.

[90] Lin D, Wang J, Jiao N, Wang Z, Liu L. A flexible magnetically controlled continuum robot steering in the enlarged effective workspace with constraints for retrograde intrarenal surgery. Adv Intell Syst 2021;3:2000211.

[91] Bajo A, Simaan N. Kinematics-based detection and localization of contacts along multisegment continuum robots. IEEE Trans Robot 2011;28:291–302.

[92] Bajo A, Pickens RB, Herrell SD, Simaan N. Constrained motion control of multisegment continuum robots for transurethral bladder resection and surveillance. In: 2013 IEEE international conference on robotics and automation (ICRA); 2013.

[93] Neumann M, Burgner-Kahrs J. Considerations for follow-the-leader motion of extensible tendon-driven continuum robots. In: 2016 IEEE international conference on robotics and automation (ICRA); 2016.

[94] Mo H, Ouyang B, Xing L, Dong D, Liu Y, Sun D. Automated 3-D deformation of a soft object using a continuum robot. IEEE Trans Autom Sci Eng 2020;18:2076–86.

[95] Lai J, Lu B, Huang K, Chu HK. Gesture-based steering framework for redundant soft robots. IEEE/ASME Trans Mechatron 2024;1–13.

[96] Qi P, Zhang C, Li J, Li Z, Dai JS, Althoefer K. A compact continuum manipulator system with enhanced steering abilities for robot-assisted surgery. In: 2016 6th IEEE international conference on biomedical robotics and biomechatronics (BioRob); 2016.

[97] Zhang X, Li W, Chiu PWY, Li Z. A novel flexible robotic endoscope with constrained tendon-driven continuum mechanism. IEEE Robot Autom Lett 2020;5:1366–72.

[98] Alambeigi F, Sefati S, Armand M. A convex optimization framework for constrained concurrent motion control of a hybrid redundant surgical system. In: 2018 annual American control conference (ACC); 2018.

[99] Alambeigi F, Bakhtiarinejad M, Sefati S, Hegeman R, Iordachita I, Khanuja H, Armand M. On the use of a continuum manipulator and a bendable medical screw for minimally invasive interventions in orthopedic surgery. IEEE Trans Med Robot Bionics 2019;1:14–21.

# 5

# Bioinspired flexible and compliant robotic manipulators for surgery

*Tao Zhang and Hongliang Ren*

**Department of Electronic Engineering, The Chinese University of Hong Kong, Sha Tin, Hong Kong**

## Abbreviations

| | |
|---|---|
| **CM** | compliant manipulator |
| **DOFs** | degrees of freedom |
| **FBG** | fiber Bragg grating |
| **MIS** | minimally invasive surgery |
| **NOTES** | natural orifice transluminal endoscopic surgery |

## Introduction

In modern medical treatment, robot-assisted MIS plays an increasingly important role in confined surgical areas [1,2]. To ensure the safety of the surgical process and complete complex operations in the narrow surgical environment, bioinspired robots have become an essential part of the current research [3–5].

As shown in Fig. 1, there are different bioinspired CMs: snake-inspired [6,7] discrete CMs, octopus- [8,13] or elephant trunk-inspired [9,10] continuum CMs, and bioinspired (such as wasps) medical tools [11,12]. Inspired by snake skeletons, the discrete manipulators have a series of the same joints; each of the joints is equivalent to a snake skeleton joint, and the driving tendons are equivalent to snake muscle. Continuum CMs have a continuum backbone, equivalent to the continuum structure of octopuses or elephant trunks, and they are also equipped with driving tendons equivalent to the muscle. The properties of discrete and continuum manipulators are shown in Table 1. In addition, there are several bioinspired medical tools. For example, inspired by the wasp ovipositor, a flexible needle is proposed, and the needle body is divided into four parts, equivalent to the multipart wasp ovipositor.

In surgeries, bioinspired CMs can make a difference. Discrete and continuum manipulators can be used with other instruments, such as electric knives and clamps, to provide a multidegree of freedom base or a free steering channel for the instruments. Bioinspired surgical tools, such as the steering needle, can be used alone to complete several surgical tasks, such as tissue puncture. However, improvements still need to be made when applied in a specific surgery, such as NOTES. In NOTES, the surgical robot needs to complete the injection, clamping, lifting, cutting, and other surgical operations after passing through the winding and narrow path. A strict working environment puts higher requirements for surgical manipulators, such as size, flexibility, compliance, and bearing capacity. The size of the surgical instruments used for laparoscopy should be controlled at 5–10 mm [14], while for NOTES, due to the limitation of the existing endoscope, the endoscopic instruments should be 3.7 mm or less [15]. During the percutaneous puncture, the size of the puncture needle will directly affect the degree of injury to the human body [11]. In addition, the maximum tension provided by the operating arm should be at least 300 g during endoscopic tissue removal [15]. Therefore, minimizing the size, maintaining sufficient DOFs, and ensuring the tool-tip stiffness is the leading research goal of the compliant manipulators. In addition, there are apparent modeling and driving errors in existing compliant devices due to the nonlinear properties of the compliant manipulators. Reducing errors through optimizing the structure or improving the modeling method is also one of the issues that should be considered.

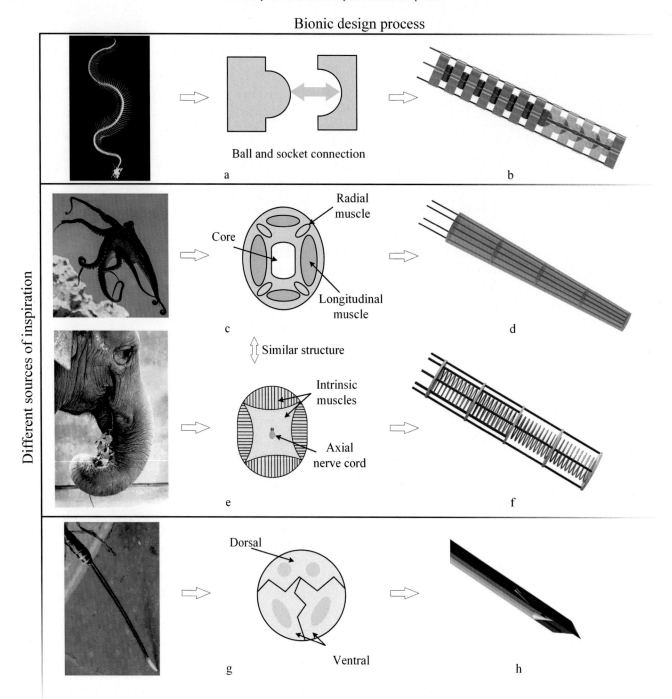

FIG. 1   Bioinspired manipulators: (A) Snake skeleton and structure analysis [6]; (B) snake-inspired discrete manipulator [7]; (C) Octopus and structure analysis [8]; (D) Octopus-inspired continuum manipulators [8]; (E) elephant trunks and structure analysis [9]; (F) elephant trunk-inspired continuum manipulators [10]; (G) Wasp ovipositor and structure analysis [11]; (H) Wasp ovipositor-inspired flexible needle [12].  *Panel A: Image by Ludovic Charlet from Pixabay, https://pixabay.com/photos/skeleton-snake-bones-reptile-white-3813465/; Panel C: Image by Thomas from Pixabay, https://pixabay.com/photos/octopus-tentacles-underwater-sea-6121101/; Panel E: Image by Liselotte Brunner from Pixabay, https://pixabay.com/photos/elephant-pachyderm-old-frown-folds-181062/; Panel G: Image by Francisco Corado Rivera from Pixabay, https://pixabay.com/photos/insect-yard-nature-wasp-4558264/;*

TABLE 1 Discrete CMs and continuum CMs.

| | Categories | Advantages | Disadvantages | Research focus |
|---|---|---|---|---|
| Discrete manipulators | 1. Socket joints<br>2. Rolling joints<br>3. Hinge joints<br>4. Composite joint | 1. High axial stiffness | 1. Friction between joints<br>2. Complex structure | 1. Smaller size<br>2. High stiffness<br>3. More DOFs |
| Notch continuum manipulators | 1. Single-edge notch<br>2. Two-edge notch<br>3. Cross-notch | 1. No inner friction<br>2. Different stiffness can be achieved by flexibly adjusting the notches<br>3. Large interior space, and can be used as a working channel for surgical instruments | 1. Hard to model the elastic deformation | |
| Backbone continuum manipulators | 1. Central backbone<br>2. Multiple backbones | 1. No inner friction<br>2. Simple structure<br>3. Different stiffness can be achieved by flexibly adjusting the backbones | 1. Low stiffness<br>2. Hard to model the thin backbone | |

# Discrete compliant manipulators

Arising from snake-like robots, discrete CM [16,17] is a hyperredundant serial manipulator with a series of same joints. The adjacent joints are connected by the driving tendons passing through the holes in joints, and the contact surface is spherical [7] or cylindrical [18–20].

Flexible robots with discrete joints have been used in endoscopes [15] and flexible manipulators [21]. A well-known one is the flexible wrist of the Da Vinci Robotic Surgical system [22]. Snake manipulators can finish complex operations, such as cutting and suturing. Then, another similar flexible wrist has also been proposed [23]. In addition, a 6-DoF endoscope is proposed for single-port laparoscopic surgery [24]. The adjacent joints are connected by cylindrical surfaces and driving tendons. A 3-DoF SMA-tendon-driven flexible endoscope is also proposed in Ref. [25], which can adjust the total length of the endoscope.

When applied in the endoscope, as the size limitation is not strict, and there tend to be one or more manipulators inside the endoscope, the stiffness of the flexible endoscope requires no special consideration. However, the size limitation is strict when applied to endoscopic instruments. The diameter of working channels is typically just 3.5 mm [26] or even less. The stiffness of the manipulators will reduce with the reduction of the size. To enhance the stiffness of such instruments, optimization of mechanical structure and antagonistic control strategy has been studied [30].

First, simple spherical or rolling joints will cause low stiffness and load capacity, while stiffness and load capacity are essential evaluation standards of a manipulator. Then, two approaches are proposed: the first is to enhance the stiffness of the mechanical structure itself, and the second is to adjust the arrangement of the driving wires.

To prevent unwanted relative sliding of adjacent joints and enhance the strength of the manipulator when maintaining a small structural size, elastic fixtures are added to link two adjacent joints [31,32]. A pair of elastic fixtures are inserted into the holes of two joints. When the joints bend to an angle, elastic fixtures can provide extra stiffness. An approach from another aspect was proposed by [27] (Fig. 2F). A pair of half-gear and a rigid link (dumbbell) links two adjacent joints. The half-gears could avoid relative sliding, while the rigid link can ensure that the joints will not displace relatively in the axial direction. In addition, the use of rigid components also prevents torsion deformation. A snake-like manipulator with a riveted structure is proposed by Ref. [29] (Fig. 2H). The riveted structure limits the motion between adjacent joints to one DOF, relative rotation, and improves the overall strength of the manipulator. However, the extra structures need more space, which is in contradiction with strict size requirements.

The literature [28] combines the advantage of gear joints and elastic fixtures and adjusts the position of the driving tendons to enhance manipulator stiffness (Fig. 2H). The driving tendons only connect with the manipulator in the base and distal tip. When the manipulator bends, the arm of the actuation force becomes longer, and the load capacity has also been enhanced. However, when bending, the driving wire beyond the original size range of the operating arm will become an obstacle to movement during the surgical operation.

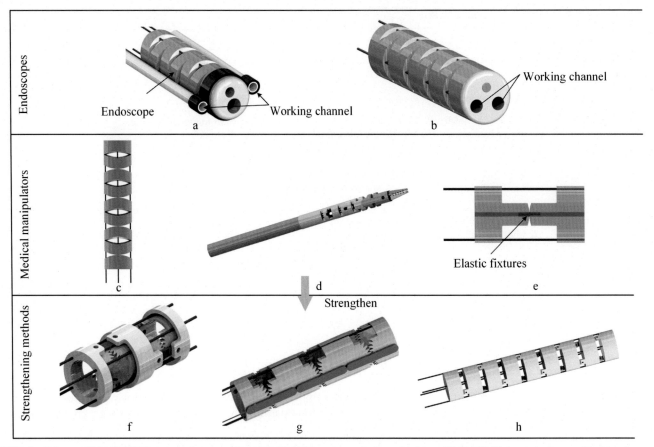

**FIG. 2** Discrete CMs: (A–B) Discrete endoscopes [15,24,26]. (C–E) Discrete surgical manipulators [18,19,21]. (F) A manipulator with riveted structure [27]. (G) A manipulator with gear joints, elastic fixtures, and adjusted driving tendons [28]. (H) A manipulator with riveted [29].

In addition, although the discrete joint manipulators have been applied in many robots and shown good performance, the friction between the adjacent joints still impacts the motion accuracy, and the structure of joints also limits the size reduction.

## Continuum compliant manipulators

As shown in Fig. 1, arising from octopuses or elephant trunks, manipulators with continuum elements are the main type of flexible manipulators, such as notch CMs [33] and backbone CMs [34]. Compared with the discrete CMs, there is no inner friction of the continuum manipulators except the friction between the driving wires and other elements. However, the computation time of the model of continuum manipulators will be much longer, as the deformation of the elastic elements is complex [35].

### Notch manipulators

The notch manipulators consist of a notch elastic tube and driving wires (as shown in Fig. 3). The notch can be divided into single-edge notch [36,37], two-edge notch [38–40], and cross-notch [41,42]. For notch continuum manipulators, the main research content is optimizing the shape, size, location, and number of notches to meet different requirements, such as stiffness and workspace.

In some applications, bending in a single direction can meet the DOF requirements. A single-edge notch surgical robot is designed to navigate around sharp corners when carrying out otolaryngology surgery [36]. As navigating around sharp corners requires enough driving torque to force the flexible manipulator to bend to a large-angle, a single-edge notch is designed to provide a long driving force arm when the total size of the manipulator is strictly limited. In addition, to meet the stiffness requirements of the neurosurgical applications, Castigliano's second theorem is used to establish the relationship between the notches and the stiffness and the workspace of a 1.3-mm notch manipulator [37], and the cut depth and duty cycle are optimized to obtain a higher stiffness.

However, in other fields, such as single-port abdominal surgery and NOTES, to get more DOFs, there need to be more notches in different directions, although they will cause low stiffness. A two-segment asymmetric triangular-notch continuum manipulator is designed to realize two-DOF bending, and the manipulator shape is also accurately described using a mechanics-based kinematic model [43]. In addition, an asymmetric rectangular notch manipulator is also proposed for lesions of the lateral skull [44]. The tube can be divided into flexible links and rigid links, and the two driving tendons are fixed in different rigid links to realize both "C" and "S" shape motion rely on only two tendons and the elastic force of the tube. Notches can also be arranged crossed to give the ability to bend in 3D space to the manipulators.

There are also studies in the combination of the single-edge notch and two-edge notch. For example, to finish brain surgery, a two-segment notch working channel is designed [45]. The proximal segment has single-edge notches, and the distal segment has two-edge symmetric notches. As the stiffness of the proximal segment is higher, the coupling when driving the distal segment will be reduced.

Besides the structure optimization of the notch, the establishment of an accurate kinematic model is also the main research direction, as the manipulator shape is not a constant curvature due to the different force conditions of each flexible part. The most popular is choosing a proper beam theorem and using the Euler-Lagrangian equations to get the relationship between the driving force and the manipulator shape. For example, a series of 2-node Timoshenko beam elements are used to describe the flexible parts [38]. However, the super elastic property of the tube is not considered. To solve this problem, the Euler-Bernoulli bending moment-curvature mechanics model [42] and Cosserat rod theory [46] are also studied.

Then, a notched tube has also been used as a part of other medical manipulators. For example, a 2-DOF notch manipulator is designed for invasive intracerebral hemorrhage evacuation [47]. The adjacent notches are

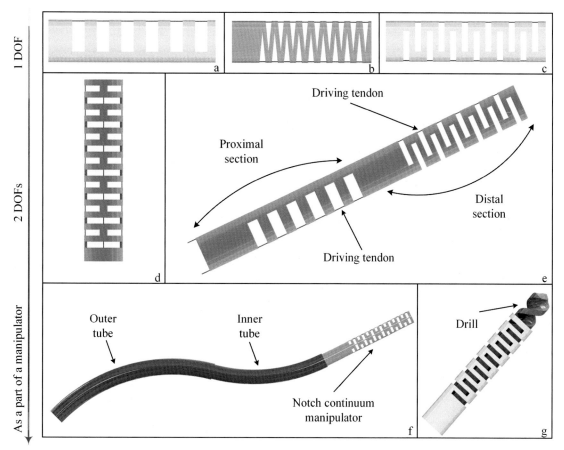

FIG. 3   Different types of notch manipulators: (A) A single-edge notch manipulator [36]. (B–C) Two-edge notch manipulators [38,39]. (D) A cross-notch manipulator [41,42]. (E) A two-segment notch manipulator [45]. (F) A manipulator that works as a part of a concentric tube robot [47]. (G) A manipulator that works as the channel of a bone drill [48].

perpendicular to each other. To meet the DOF requirements, the notched elastic tube has a prebending curvature and forms a concentric tube robot with a layer of outer tubes. Similarly, for osteolysis treatment, a notch manipulator is used as a working channel of the bone drill to raise the workspace to reach behind the acetabular cup [48]. The CM has also been equipped with an FBG sensor to obtain the nonconstant curvature shape, which is also the primary method to solve the nonconstant bending of the notch manipulator, except for the mechanical model method.

## Backbone manipulators

Backbone manipulators have also been applied in different medical robots, which could be divided into single-backbone and multibackbone manipulators [49] (as shown in Fig. 4). The backbone of a single-backbone manipulator is located in the center of the manipulator, which is linked to each disk of the manipulator. The manipulators are driven by four of three wires, which are only connected to one specific disk (usually the last one). However, the multibackbone manipulator has three or more driving elastic rods acting as backbones and driving rods. The elastic rods are linked in the same disk, and more backbones give the manipulator higher stiffness.

For a backbone manipulator, the backbone could be an elastic rod [50–52], tube [53], or spring [54]. Recently, 3D printing technology is also been applied to build a flexible backbone. A 3D-printed helical elastic backbone with a series of metal cylinders is designed in Ref. [55], which can avoid axial compression. In Ref. [56], a 3D-printed rigid and flexible coupled backbone is built, too. 3D print technology provides a new direction for designing and building flexible manipulators, such as embedding sensors into the backbones to realize sensing functions.

Compared with the notch manipulator, the backbone manipulator has a simpler structure. However, the long and thin backbone also leads to low stiffness, and the inconstant curvature shape of the backbone also makes modeling and control difficult. The most helpful method of improving the stiffness of the manipulator is increasing the stiffness of the backbone or directly increasing the size or number of the backbones.

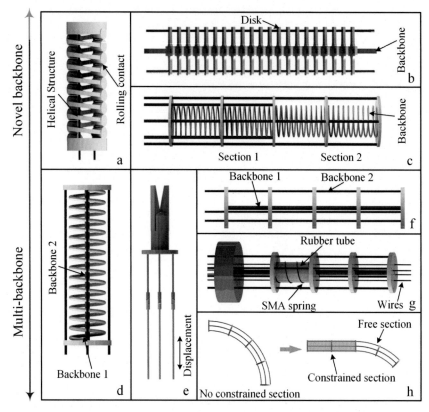

FIG. 4    Different types of backbone manipulators: (A) 3D-printed helical elastic backbone [55]. (B) 3D-printed rigid-flexible coupled backbone [56]. (C) A manipulator with three types of springs [10]. (D) A spring-backbone manipulator [57]. (E) A manipulator that can adjust the stiffness by changing the effective length of the backbone [58]. (F) A multibackbone manipulator [59]. (G) A manipulator that can adjust the stiffness by an SMA spring [60]. (H) A manipulator with a constraint joint [61].

When an external force is applied, the inner torque of the manipulator gradually increases toward the manipulator base. A multisegment backbone manipulator [10] enhanced the stiffness of the base segments. Three types of springs with variable hardness constants are adopted. The hardness constants of the springs gradually decrease toward the tips of the manipulators. Therefore, when the tip of the manipulator has applied an external force, the base will not appear to be a larger deformation. In addition, the twisting stiffness is also improved because of the use of the springs. Another spring-backbone manipulator is proposed in Ref. [57]. The manipulator has an inner backbone and an outer backbone; the inner one is a stretchable spring, and the outer one is a compressible spring. The two backbones improve the stiffness of the manipulator. However, although the springs can improve the stiffness, their shapes are more challenging to model. And the use of spring will also affect the miniaturization of a manipulator.

From another aspect, a multibackbone manipulator consisting of four backbone tubes is designed [59]. One tube is located at the center of each disk and fixed with all disks. The other three backbones are like the driving tendons and are only fixed with the final disk. These four backbones improved the stiffness, especially the twisting stiffness. Then, similar to the method of increasing the internal friction to improve the stiffness of the discrete manipulator, an approach that can enhance the tension of the backbones and improve the stiffness is proposed in Ref. [60]. An SMA spring covers a rubber tube, which encloses the backbone of the manipulator. When the stiffness needs to be enhanced, the temperature of the spring can be adjusted by adding a current. The diameter of the spring will decrease, and the rubber tube will contact the backbone, which will constrain the motion of the backbone and the deflection of the manipulator. As a result, the stiffness is improved. However, the improved inner friction will also impact the motion accuracy of the manipulator. Then, a constraint joint is used to enhance the stiffness of the manipulator further [61]. The structure introduced two concentric rigid curvature-constraining (CC) rods. By inserting the CC rods, the joints could be divided into two segments: a constrained segment, and an effective segment. The constrained segment can be seen as a rigid segment, and due to the shorting of the flexible segment, the deflection of the manipulator, when applied an external force, will decrease, too. In addition, adjusting the stiffness of a manipulator by changing the effective length of the backbone is studied in Ref. [58]. Three NiTi rods with a universal joint are adopted as the backbone and the driving rods. The universal joints could reduce the bending radius, which improves dexterity. The effective length of the three rods can be adjusted. When the size becomes shorter, the stiffness increases. Similar to the multibackbone manipulator, a flexible parallel manipulator is designed [62]. There are only two disks and six elastic rods in the manipulator. Due to the multiple elastic rods, the manipulator has compactness, compliance, and high stiffness and strength.

To improve the motion accuracy of the backbone manipulator, there are two directions. The first one is to adjust the structure of the manipulators and make the shape of a manipulator closer to an arc. The second one is to improve the accuracy of the model.

A bioinspired fishbone manipulator with a rigid-flexible-soft coupling structure can bend in a constant curvature curve [56]. When finding a more accurate model to describe a manipulator, the Bernoulli-Euler beam [63] is used in the initial research. However, this theory cannot describe the large deflection when a manipulator bends to a specific angle, especially when an external torque is applied. Then, the Kirchhoff-Love beam [64] and the Cosserat beam [44,65] are used to model the long and thin backbone. They tend to divide the whole backbone into minor elements and use constitutive equations to model the large deformation of the backbone. However, these methods require high computational power and can only supply a numerical solution instead of an analytical solution.

There are also methods that use special curves to fit the shape of a manipulator. A polynomial curvature models a soft underwater tentacle [66]. Using an affine curvature, an analytical shape function and kinematical model have been obtained, making it possible to get a dynamic analytical model. Euler curve is also chosen to describe the shape of a soft pneumatic robot [67], and the experimental results show that this curve can fit different types of soft manipulators. In [68], the Euler curve is used for a tendon-driven backbone manipulator. It focuses on the real shape of the backbone manipulator when an external force is applied to the tip of the manipulator. In addition, to reduce the computational complexity, the curvatures of different sections of the backbone manipulator are assumed to be in arithmetic progression. However, the study did not take torsion into consideration. Then, a more profound study is carried out in Ref. [69]. It proved that the torsion of the continuum robot is also in arithmetic progression and can approximate the 3D shape of the manipulator accurately.

In addition, there are also many researchers that try to explore more possibilities of the flexible continuum manipulators. A manipulator with extensible sections is studied in Ref. [70]. The manipulator consists of a three-segment concentric tube backbone, disks, and driving tendons. By adopting the concentric tube backbone, the manipulators can extend the length of each segment of the whole continuum manipulators. In addition, the same magnets are fixed in each disk to realize the uniform distribution of the disks. A novel manipulator which can realize micromotion and macroscopic motion is designed in Ref. [71]. The manipulator is a type of multibackbone manipulator. However, the

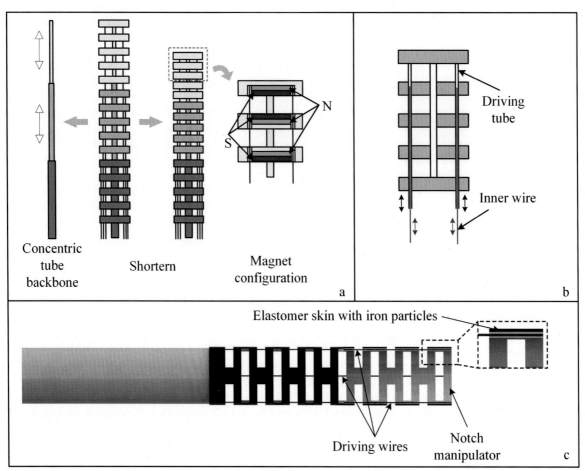

FIG. 5    Novel CMs: (A) A manipulator with extensible sections [70]. (B) A manipulator that can realize micromotion and macroscopic motion [71]. (C) A manipulator combining the tendon- and magnetic-driven modes [72].

backbones are double-deck tubes, which consist of an elastic tube and an inner rod. After bending to a specific angle by driving the tubes, the stiffness of the tube can be changed by inserting or pulling the rods. As a result, the mechanical equilibrium of the whole manipulator is changed too, which contributes to the micromotion of the tip of the manipulator. Combining the tendon- and magnetic-driven mode, a notch flexible manipulator coated with magnetic material is proposed in Ref. [72]. The manipulator can realize large-angle steering, driven by tendons, and high-precision manipulation, driven by the magnetic field. Combining the traditional flexible manipulators with different mechanical structures or driving methods to get novel work performance has caught the notation of many researchers and has become a new research direction for bioinspired manipulators (Fig. 5).

## Compliant tools

There are also compliant medical tools that are inspired by creatures [11]. For example, inspired by wasps, honeybees, or mosquitos, researchers proposed different types of flexible needles. Inspired by Hymenoptera insects, researchers proposed new clamps [73]. Crawling robots inspired by inchworms and earthworms are proposed by researchers separately [74,75]. In addition, manipulators inspired by plants are proposed [76,77]. However, considering the current level of processing and manufacturing, the motivation of many natural creatures cannot be fully reproduced by existing mechanical mechanisms. As a result, many biomimetic instruments have yet to be applied to the clinic. Only some simplified structure, such as flexible needles, clamps, and crawling robots, has been successfully used [78] (Fig. 6).

For bioinspired needles, reducing the insertion force is the key to reducing patients' pain. The ovipositor of a wasp is also a popular bionic object of low-insertion force flexible needle. Inspired by wasps, a multivalve puncture needle is proposed [80]. The body of the needle is divided into four lobes, which is like the bifurcated structure of the ovipositor

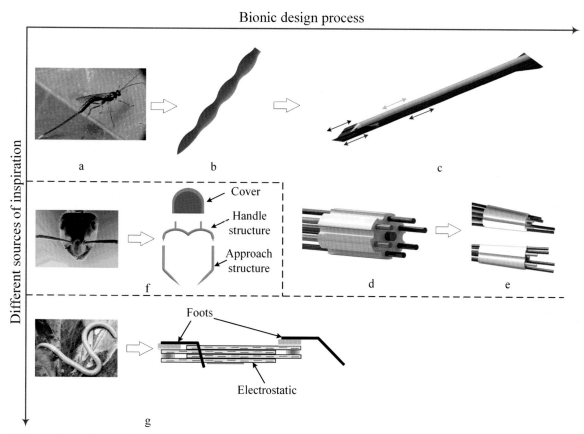

FIG. 6    Bioinspired complaint tools: (A) A wasp. (B) A mosquito-inspired needle [79]. (C) A wasp-inspired needle [80]. (D) A mosquito-inspired needle [81]. (E) Improvement of the mosquito-inspired needle [82]. (F) An ant and ant-inspired clip [83]. (G) An earthworm and earthworm-inspired crawling robot and endoscopic applications [75]. *Panel A: Image by Francisco Corado Rivera from Pixabay, https://pixabay.com/photos/insect-yard-nature-wasp-4558264/; Panel F: Image by Virvoreanu Laurentiu from Pixabay, https://pixabay.com/photos/stack-insect-ant-macro-animal-2129070/; Panel G: Image by Eukalyptus from Pixabay, https://pixabay.com/photos/letter-s-earthworm-lumbricidae-202404/.*

of a wasp. When inserting a tissue phantom, there is a cyclic drive mode, one of the four lobes is pushed in turn, while others are pulled to a lower value simultaneously. This insertion strategy can reduce the damage to the tissue phantom. However, the diameter is 4 mm, which will aggravate the injury. Therefore, it is a critical problem to reduce the insertion force while reducing the geometric size. A soft mosquito-inspired needle inspired by mosquitos is proposed [79], whose diameter is only 1.65 mm. The shape of the needle comes from the mosquito's mouthparts, which mainly include two parts: the shape of the maxilla and the labrum of the mosquito. In addition, the asymmetric structure of the mosquito proboscis is adjusted to a symmetrical structure to avoid being applied unbalanced external force during insertion. A longitudinal vibration at an appropriate frequency [84] is also adopted to reduce the insertion force.

However, a needle also needs to bypass obstacles during insertion, such as vessels. To achieve a controlled bending of the needle, a 2.5-mm 3-D steering needle is proposed in [12], which also consists of four parts linked with each other. It finds that the offset of the four parts is related to the steering angle of the needle and uses a multibeam approach based on Euler-Bernoulli beam theory to model the bending property when the four parts of the needle are in different configurations. A needle consisting of seven NiTi wires is also proposed [81]. One of the wires is in the center, while the other six are arranged evenly around it. By changing the relative position of the wires during insertion, different beveled planes can be formed to control the turn of the needle tip. Due to the simplification of the structure, the needle size is reduced to 1.2 mm. However, there are errors in the steering function, and the steering curvature of the needle is also too poor. To solve this problem, the NiTi wires with parallel configuration at the tip were changed to a centripetal or centrifugal configuration to enhance the steering performance of the needle [82]. The result shows that the centrifugal configuration can contribute to more significant curvature. The out diameter also changes from 1.2 mm to 1.55 mm, which is still not conducive to clinical applications. Therefore, the researchers still need to focus on the simpler structure, smaller size, and larger curvature steering needle.

The other bioinspired medical manipulators are clamps and minor crawling robots. An ant-inspired absorbable suture clip is proposed in Ref. [83]. The clip consists of a cover, a handle structure, and two approach structures.

The handle is elastic, and the cover is used to open the clip. When the approach structures are inserted into the tissue on either side of the wound, the cover can be removed, and the clip will hold the wound together. An earthworm-inspired robot is proposed in [75]. The robot consists of 4 ft and an electrostatic actuator. The feet are equivalent to the bristles on the belly of an earthworm, and the electrostatic actuator is equivalent to the muscle of the earthworm. The electrostatic includes several films which are formed by multiple electrodes. The electrodes can generate a driving force when applied an electrostatic field, which is the source of the actuator motion. In addition, the actuator can also perform different functions when combined with other devices. For example, when a blade is attached to the actuator, it can perform endoscopic tissue cutting. The use of the electrode-driven method also gives us a new direction to simulate the actions of creatures, which is different from using the traditional driving model, such as tendon-driven or hydraulic drive, and will get positive results.

# Future scenario

Presently, the bioinspired CMs applied in the medical field can be mainly divided into the slender manipulators inspired by the spine of a vertebrate or soft-bodied organisms and the compliant tools inspired by biological organs, such as ovipositor of wasps. During their development, these manipulators have played a role in laparoscopic surgery and NOTES. However, due to the extremely strict requirements of the medical robot system, the current medical manipulators need to be further studied to reduce the geometric size and improve dexterity, mechanism strength, and motion accuracy. Therefore, optimizing the structure to improve the dexterity and load capacity and developing the high-precision control model are still the main contents of the research.

In addition, combining the bioinspired structure with other types of manipulators, such as the concentric tube robot, and different driving modes, such as magnetic control, can often achieve expanding motion ability of manipulators, which can realize multiscale precision motion and enhance the performance of bioinspired structures.

# Key points

- Bioinspired flexible and compliant manipulators have played a key role in robotic surgery. The manipulators can be divided into discrete manipulators, continuum manipulators, and compliant tools.
- The discrete manipulators and the continuum manipulators are mainly inspired by snakes, octopus, or elephants. They have large length-to-diameter ratio and can be used to perform surgery that requires passing through natural orifices.
- The compliant tools are mainly inspired by varied insects. Their shape and function differ greatly. Typical examples are puncture needles, inside-crawling robots, etc.
- The main research content of these bioinspired manipulators are mechanical design and optimization, static and dynamic modeling, and controlled motion.

# Acknowledgments

This work was supported in part by Hong Kong Research Grants Council (RGC) Collaborative Research Fund (CRF C4026-21GF and CRF C4063-18G), and General Research Fund (GRF 14216022), NSFC/RGC Joint Research Scheme N\_CUHK420/22; Shenzhen-Hong Kong-Macau Technology Research Programme (Type C) Grant 202108233000303; Guangdong Basic and Applied Basic Research Foundation (GBABF) \#2021B1515120035.

# References

[1] Ren H, Lim CM, Wang J, et al. Computer-assisted transoral surgery with flexible robotics and navigation technologies: a review of recent progress and research challenges. Crit Rev Biomed Eng 2013;41(4–5).
[2] Zhou YUE, Ren H, Meng MQH, Tse ZTH, Yu H. Robotics in natural orifice transluminal endoscopic surgery. J Mech Med Biol 2013;13(02).
[3] Ma G, Wu C. Microneedle, bio-microneedle and bio-inspired microneedle: a review. J Control Release 2017;251:11–23.
[4] Omisore OM, Han S, Xiong J, Li H, Li Z, Wang L. A review on flexible robotic systems for minimally invasive surgery. IEEE Trans Syst Man Cybern Syst 2022;52(1):631–44.
[5] Yang Y, He Z, Jiao P, Ren H. Bioinspired soft robotics: how do we learn from creatures? IEEE Rev Biomed Eng 2022;PP.
[6] Liljebäck P, Pettersen KY, Stavdahl Ø, et al. Snake robots: Modelling, mechatronics, and control. London: Springer; 2013.
[7] Li Z, Du R. Design and analysis of a bio-inspired wire-driven multi-section flexible robot. Int J Adv Robot Syst 2013;10(4).
[8] Zheng T, et al. Model validation of an Octopus inspired continuum robotic arm for use in underwater environments. J Mech Robot 2013;5(2).

[9] Yeshmukhametov A, Koganezawa K, Yamamoto Y. Design and kinematics of cable-driven continuum robot arm with universal joint backbone. In: Presented at the 2018 IEEE international conference on robotics and biomimetics (ROBIO); 2018.

[10] Yeshmukhametov A, Buribayev Z, Amirgaliyev Y, Ramakrishnan RR. Modeling and validation of new continuum robot backbone design with variable stiffness inspired from elephant trunk. IOP Conf Ser: Mater Sci Eng 2018;417.

[11] Ma Y, Xiao X, Ren H, Meng MQH. A review of bio-inspired needle for percutaneous interventions. Biomimetic Intell Rob 2022;2(4).

[12] Watts T, Secoli R, Baena FRY. A mechanics-based model for 3-D steering of programmable bevel-tip needles. IEEE Trans Robot 2019;35(2):371–86.

[13] Zheng T, Branson DT, Kang R, et al. Dynamic continuum arm model for use with underwater robotic manipulators inspired by octopus vulgaris. In: 2012 IEEE international conference on robotics and automation. IEEE; 2012. p. 5289–94.

[14] Zidane IF, Khattab Y, Rezeka S, El-Habrouk M. Robotics in laparoscopic surgery - a review. Robotica 2022;41(1):126–73.

[15] Hwang M, Kwon DS. K-FLEX: a flexible robotic platform for scar-free endoscopic surgery. Int J Med Robot 2020;16(2), e2078.

[16] Hirose MMS. Biologically inspired snake-like robots. In: International conference on robotics and biomimetics; 2004.

[17] Sun Y, et al. A triboelectric tactile perception ring for continuum robot collision-aware. In: Presented at the 2021 21st international conference on solid-state sensors, actuators and microsystems (transducers); 2021.

[18] Kim Y-J, Cheng S, Kim S, Iagnemma K. A stiffness-adjustable hyperredundant manipulator using a variable neutral-line mechanism for minimally invasive surgery. IEEE Trans Robot 2014;30(2):382–95.

[19] Zhang D, Sun Y, Lueth TC. Design of a novel tendon-driven manipulator structure based on monolithic compliant rolling-contact joint for minimally invasive surgery. Int J Comput Assist Radiol Surg 2021;16(9):1615–25.

[20] Gao H, et al. GESRsim: gastrointestinal endoscopic surgical robot simulator. In: Presented at the 2022 IEEE/RSJ international conference on intelligent robots and systems (IROS); 2022.

[21] Li C, et al. A miniature manipulator with variable stiffness towards minimally invasive transluminal endoscopic surgery. IEEE Rob Autom Lett 2021;6(3):5541–8.

[22] Suh J-W, Kim K-Y. Design of a discrete bending joint using multiple unit PREF joints for isotropic 2-DOF motion. Int J Control Autom Syst 2017;15(1):64–72.

[23] Pierre Berthet-Rayne KL, Kim K, Carlo JS, Seneci A, Yang G-Z. Rolling joint design optimization for tendon driven snake-like surgical robots. In: Presented at the international conference on intelligent robots and systems, Madrid, Spain; 2018.

[24] Jusuk Lee JK, Lee K-K, Hyung S, Kim Y-J, Kwon W, Roh K, Choi J-Y. Modeling and control of robotic surgical platform for single-port access surgery. In: International conference on intelligent robots and systems; 2014.

[25] Mandolino M, Britz R, Goergen Y, Rizzello G, Motzki P. Development of an SMA driven articulation and autofocus mechanism for endoscope applications. In: Presented at the international conference and exhibition on new actuator systems and applications; 2022.

[26] Mandapathil M, Greene B, Wilhelm T. Transoral surgery using a novel single-port flexible endoscope system. Eur Arch Otorrinolaringol 2015;272(9):2451–6.

[27] Chu X, Yip HW, Cai Y, Chung TY, Moran S, Au KWS. A compliant robotic instrument with coupled tendon driven articulated wrist control for organ retraction. IEEE Rob Autom Lett 2018;3(4):4225–32.

[28] Hwang M, Kwon D-S. Strong continuum manipulator for flexible endoscopic surgery. IEEE/ASME Trans Mechatron 2019;24(5):2193–203.

[29] Kong Y, Song S, Zhang N, Wang J, Li B. Design and kinematic modeling of in-situ torsionally-steerable flexible surgical robots. IEEE Rob Autom Lett 2022;7(2):1864–71.

[30] Li S, Hao G. Current trends and prospects in compliant continuum robots: a survey. Actuators 2021;10(7).

[31] Suh J-W, Kim K-Y, Jeong J-W, Lee J-J. Design considerations for a hyper-redundant pulleyless rolling joint with elastic fixtures. IEEE/ASME Trans Mechatron 2015;20(6):2841–52.

[32] Suh J-W, Lee J-J, Kwon D-S. Underactuated miniature bending joint composed of serial pulleyless rolling joints. Adv Robot 2013;28(1):1–14.

[33] Thomas TL, Kalpathy Venkiteswaran V, Ananthasuresh GK, Misra S. Surgical applications of compliant mechanisms: a review. J Mech Robot 2021;13(2).

[34] Bajo A, Simaan N. Hybrid motion/force control of multi-backbone continuum robots. Int J Rob Res 2015;35(4):422–34.

[35] Li Z, Wu L, Ren H, Yu H. Kinematic comparison of surgical tendon-driven manipulators and concentric tube manipulators. Mech Mach Theory 2017;107:148–65.

[36] York PA, Swaney PJ, Gilbert HB, Webster III RJ. A wrist for needle-sized surgical robots. In: Presented at the 2015 IEEE international conference on robotics and automation. Seattle, WA: Washington State Convention Center; 2015.

[37] Eastwood KW, Azimian H, Carrillo B, Looi T, Naguib HE, Drake JM. Kinetostatic design of asymmetric notch joints for surgical robots. In: Presented at the 2016 IEEE/RSJ international conference on intelligent robots and systems (IROS), Daejeon, Korea; 2016.

[38] Du Z, Yang W, Dong W. Kinematics modeling and performance optimization of a kinematic-mechanics coupled continuum manipulator. Mechatronics 2015;31:196–204.

[39] Gao A, Murphy RJ, Liu H, Iordachita II, Armand M. Mechanical model of dexterous continuum manipulators with compliant joints and tendon/external force interactions. IEEE ASME Trans Mechatron 2017;22(1):465–75.

[40] Wang H, Liang W, Liang B, Ren H, Du Z, Wu Y. Robust position control of a continuum manipulator based on selective approach and Koopman operator. IEEE Trans Ind Electron 2023;1–10.

[41] Zeng W, Yan J, Huang X, Shin Cheng S. Motion coupling analysis for the decoupled design of a two-segment notched continuum robot. In: Presented at the 2021 IEEE international conference on robotics and automation (ICRA); 2021.

[42] Zeng W, Yan J, Yan K, Huang X, Wang X, Cheng SS. Modeling a symmetrically-notched continuum neurosurgical robot with non-constant curvature and superelastic property. IEEE Rob Autom Lett 2021;6(4):6489–96.

[43] Wenlong Y, Wei D, Zhijiang D. Mechanics-based kinematic modeling of a continuum manipulator. In: 2013 IEEE/RSJ international conference on intelligent robots and systems. IEEE; 2013. p. 5052–8.

[44] Gao A, Carey JP, Murphy RJ, et al. Progress toward robotic surgery of the lateral skull base: Integration of a dexterous continuum manipulator and flexible ring curette. In: 2016 IEEE international conference on robotics and automation (ICRA). IEEE; 2016. p. 4429–35.

[45] Chitalia Y, Jeong S, Yamamoto KK, Chern JJ, Desai JP. Modeling and control of a 2-DoF meso-scale continuum robotic tool for pediatric neurosurgery. IEEE Trans Robot 2021;37(2):520–31.

I. Introduction, technology basics, evolution and next frontiers of surgical robots

[46] Wang H, Wang X, Yang W, Du Z. Design and kinematic modeling of a notch continuum manipulator for laryngeal surgery. Int J Control Autom Syst 2020;18(11):2966–73.

[47] Yan J, et al. A continuum robotic cannula with tip following capability and distal dexterity for intracerebral hemorrhage evacuation. IEEE Trans Biomed Eng 2022;69(9):2958–69.

[48] Sefati S, et al. A surgical robotic system for treatment of pelvic osteolysis using an FBG-equipped continuum manipulator and flexible instruments. IEEE ASME Trans Mechatron 2021;26(1):369–80.

[49] Gu X, Li C, Xiao X, Lim CM, Ren H. A compliant transoral surgical robotic system based on a parallel flexible mechanism. Ann Biomed Eng 2019;47(6):1329–44.

[50] Qi F, Chen B, Gao S, She S. Dynamic model and control for a cable-driven continuum manipulator used for minimally invasive surgery. Int J Med Robot 2021;17(3), e2234.

[51] Gao H, Xiao X, Yang X, et al. A miniature 3-dof flexible parallel robotic wrist using NiTi wires for gastrointestinal endoscopic surgery. arXiv preprint arXiv:2207.04735; 2022.

[52] Zhu X, Xiao X, Ren H, Meng MQH. Design, modeling and simulation of an omnidirectional steerable surgical forceps for laparoscopic surgery. In: Presented at the 2021 27th international conference on mechatronics and machine vision in practice (M2VIP); 2021.

[53] Li Z, et al. Design of a novel flexible endoscope—cardioscope. J Mech Robot 2016;8(5).

[54] Li M, Kang R, Geng S, Guglielmino E. Design and control of a tendon-driven continuum robot. Trans Inst Meas Control 2017;40(11):3263–72.

[55] Hu Y, Zhang L, Li W, Yang G-Z. Design and fabrication of a 3-D printed metallic flexible joint for snake-like surgical robot. IEEE Rob Autom Lett 2019;4(2):1557–63.

[56] Zhou P, Yao J, Zhang S, Wei C, Zhang H, Qi S. A bioinspired fishbone continuum robot with rigid-flexible-soft coupling structure. Bioinspir Biomim 2022;17(6).

[57] Liang X, et al. Finite-time observer-based variable impedance control of cable-driven continuum manipulators. IEEE Trans Hum-Mach Syst 2022;52(1):26–40.

[58] Li C, Gu X, Xiao X, et al. Flexible robot with variable stiffness in transoral surgery. IEEE/ASME Trans Mechatron 2019;25(1):1–10.

[59] Wei W, Xu K, Simaan N. A compact two-armed slave manipulator for minimally invasive surgery of the throat. In: The first IEEE/RAS-EMBS international conference on biomedical robotics and biomechatronics, 2006. BioRob 2006. IEEE; 2006. p. 769–74.

[60] Jeon H, et al. Towards a Snake-like flexible robot with variable stiffness using an SMA spring-based friction change mechanism. IEEE Rob Autom Lett 2022;7(3):6582–9.

[61] Zhao B, Zeng L, Wu Z, Xu K. A continuum manipulator for continuously variable stiffness and its stiffness control formulation. Mech Mach Theory 2020;149.

[62] Bryson CE, Rucker DC. Toward parallel continuum manipulators. In: 2014 IEEE international conference on robotics and automation (ICRA). IEEE; 2014. p. 778–85.

[63] Grazioso S, Di Gironimo G, Siciliano B. A geometrically exact model for soft continuum robots: the finite element deformation space formulation. Soft Robot 2019;6(6):790–811.

[64] Yuan H, Zhou L, Xu W. A comprehensive static model of cable-driven multi-section continuum robots considering friction effect. Mech Mach Theory 2019;135:130–49.

[65] Chen Y, Wu B, Jin J, Xu K. A variable curvature model for multi-backbone continuum robots to account for inter-segment coupling and external disturbance. IEEE Rob Autom Lett 2021;6(2):1590–7.

[66] Stella F, Obayashi N, Della Santina C, et al. An experimental validation of the polynomial curvature model: identification and optimal control of a soft underwater tentacle. IEEE Rob Autom Lett 2022;7(4):11410–7.

[67] Gonthina PS, Kapadia AD, Godage IS, et al. Modeling variable curvature parallel continuum robots using euler curves. In: 2019 International conference on robotics and automation (ICRA). IEEE; 2019. p. 1679–85.

[68] Rao P, Peyron Q, Burgner-Kahrs J. Using Euler curves to model continuum robots. In: Presented at the 2021 IEEE international conference on robotics and automation (ICRA); 2021.

[69] Rao P, Peyron Q, Burgner-Kahrs J. Shape representation and modeling of tendon-driven continuum robots using Euler arc splines. IEEE Rob Autom Lett 2022;7(3):8114–21.

[70] Amanov E, Nguyen T-D, Burgner-Kahrs J. Tendon-driven continuum robots with extensible sections—a model-based evaluation of path-following motions. Int J Rob Res 2019;40(1):7–23.

[71] Del Giudice G, Orekhov AL, Shen JH, Joos K, Simaan N. Investigation of micro-motion kinematics of continuum robots for volumetric OCT and OCT-guided visual servoing. IEEE ASME Trans Mechatron 2021;26(5):2604–15.

[72] Zhang T, Yang L, Yang X, Tan R, Lu H, Shen Y. Millimeter-scale soft continuum robots for large-angle and high-precision manipulation by hybrid actuation. Adv Intell Syst 2021;3(2):2000189.

[73] Brito TO, Elzubair A, Araújo LS, Camargo SADS, Souza JLP, Almeida LH. Characterization of the mandible Atta Laevigata and the bioinspiration for the development of a biomimetic surgical clamp. Mater Res 2017;20(6):1525–33.

[74] Shi Z, Pan J, Tian J, Huang H, Jiang Y, Zeng S. An inchworm-inspired crawling robot. J Bionic Eng 2019;16(4):582–92.

[75] Wang H, et al. Biologically inspired electrostatic artificial muscles for insect-sized robots. Int J Rob Res 2021;40(6–7):895–922.

[76] Putzu F, Abrar T, Althoefer K. Plant-inspired soft pneumatic eversion robot. In: 2018 7th IEEE international conference on biomedical robotics and biomechatronics (Biorob). IEEE; 2018. p. 1327–32.

[77] Hu F, Lyu L, He Y. A 3D printed paper-based thermally driven soft robotic gripper inspired by cabbage. Int J Precis Eng Manuf 2019;20(11):1915–28.

[78] Bloemberg J, Stefanini C, Romano D. The role of insects in medical engineering and bionics: towards entomomedical engineering. IEEE Trans Med Rob Bionics 2021;3(4):909–18.

[79] Gidde STR, Acharya SR, Kandel S, Pleshko N, Hutapea P. Assessment of tissue damage from mosquito-inspired surgical needle. Minim Invasive Ther Allied Technol 2022;31(7):1112–21.

[80] Leibinger A, Oldfield MJ, Rodriguez YBF. Minimally disruptive needle insertion: a biologically inspired solution. Interface Focus 2016;6(3):20150107.

[81] Scali M, Pusch TP, Breedveld P, Dodou D. Ovipositor-inspired steerable needle: design and preliminary experimental evaluation. Bioinspir Biomim 2017;13(1), 016006.

[82] Knez M, et al. Design and evaluation of a wasp-inspired steerable needle. In: Presented at the bioinspiration, biomimetics, and bioreplication 2017; 2017.

[83] Brito TO, Dos Santos BB, Araújo LS, De Almeida LH, Da Costa MF. Analysis of biomimetic surgical clip using finite element modeling for geometry improvement and biomaterials selection. In: Proceedings of the 3rd pan american materials congress (The minerals, metals & materials series); 2017. p. 3–9.

[84] Gidde STR. Bioinspired surgical needle insertion mechanics in soft tissues for percutaneous procedures. Temple University; 2021.

# 6

# Haptic feedback technology in robot-assisted surgery

*Lailu Li\*, Wenchao Yue\*, and Hongliang Ren*

Department of Electronic Engineering, The Chinese University of Hong Kong, Sha Tin, Hong Kong

## Introduction

When compared to the visual and auditory senses, haptic perception, or the sense of touch, can improve task performance efficiency and immersion [1–5]. Since haptic perception corresponds to a complex human perception system that includes various nerve layers, tactile receptors (Meissner corpuscle, Pacinian corpuscle, Merkel cell, Ruffini ending, etc.), and organ systems, it is a comprehensive process with low fidelity [6]. Consequently, the potential of haptic technology for medical specialties that depend on sensory information, such as minimally invasive surgery (MIS), should always be considered [7–9].

Haptic feedback in open surgery refers to both conscious and unconscious touch sense that a surgeon feels while conducting surgical interventions. It is well recognized that haptic perception is limited in MIS compared to open surgery because the surgeon cannot directly access and touch the patient [10]. Haptics provides awareness of a wide range of surgical techniques, which vary depending on the structure and type of applied force and are associated with tissue damage, operation precision, and task completion time [11]. As a result, in teleoperated medical applications, providing haptic feedback to the surgeon is essential to safeguarding delicate internal body tissues from excessive force application.

Early teleoperation solutions offer the operator primarily visual feedback [12]. However, because the operator was forced to rely exclusively on visual input from the interface area and received no haptic feedback from the robot-environment interaction, this frequently resulted in significant cognitive processes. To compensate for this limitation, haptic interfaces were developed in the mid-1990s, first as remote controllers and subsequently as more comprehensive devices that can only provide a single force or haptic feedback [13]. In the subsequent decades, haptics perception, as well as interface stability, in virtual surgical simulators and MIS, has attracted the attention of researchers [14]. Utilizing haptic feedback to achieve closed-loop control has lately acquired popularity, particularly in medical applications. Haptic feedback has been reported to reduce remote operation errors while decreasing environmental harm and operating time [15]. Additionally, haptic information holds the potential to improve surgical navigation efficiency in vivo by guiding and remaining the instrument on the appropriate route and avoiding collisions between the surgical tool and tissue.

Fig. 1 depicts the teleoperation surgical robotic platform mainly consisting of a remote operation platform and a robotics interaction site [16]. The former is where the operator receives the surgical robotics system's feedback signal and generates control commands, and the latter is which contains the surgical robotics platform that interacts with the environment and responds to commands. Surgeons can send commands to the robot using a variety of techniques, including joysticks, keyboards, and mouse, head movements, voice commands, and so forth. Meanwhile, surgeons obtain visual, audio, or haptic feedback information from the remote operation platform.

The teleoperated surgical robotics system capable of providing haptic feedback to the surgeon contains two components: one is a tactile sensing or force estimation component on the manipulator side for quantifying the interaction state between the end effector and the target tissue (physic environment), and the other is a haptic device on the master side for transferring the perceived signals (audio, video, and haptics) to the surgeon [17]. Each of these sections has

---

*These authors contributed equally.

FIG. 1   A teleoperated surgical robotics feedback system has three main components: The master side, the follower side, and the physical world. The surgeon operator is the master side, which sends control signals and receives feedback information; the robotic surgical system is the follower side, which can realize interaction action and information extraction with the environment; and the physical world refers to the human body in our case.

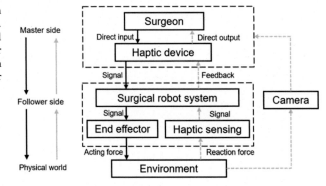

garnered much attention in the literature. There have been numerous attempts to precisely characterize the end effector-environment interaction and feedback to the surgeon in the most immersive way possible. Despite the development of highly precious force sensing technologies, easy-to-use haptic devices, and other significant progress in this field, most existing commercial systems consider few high-fidelity haptic rendering elements or still need to include haptic feedback [18]. Users are reluctant to use haptic feedback because the current systems can not realize haptic rendering with accuracy, realism, and comfort for the user for further performance improvement without loss of system stability. The estimation of interaction force at the end-effector tooltip and the identification of the most effective feedback types for various applications remain important research areas.

This chapter covers the literature on haptic feedback in teleoperated medical interventions from two decades until the present, with the objective of presenting the most noteworthy accomplishments, prevailing research trends, and prospective future directions. It investigates haptic devices related to robot-assist surgery and provides a new interface in depth, in addition to the influence of haptic feedback on user performance.

## Surgical haptic feedback interfaces

Surgical haptic feedback systems involve two key aspects, as shown in Fig. 2: sensing and perception. Sensing refers to the system's ability to sensitively detect and accurately measure the forces exerted during surgical tasks. Perception refers to the ability of the system to transmit this force information to the surgeon intuitively and informatively. Together, these two aspects form the basis of force haptic feedback, which has the potential to improve surgical operations with high precision and safety dramatically.

### Sensing types

Surgical force sensing force haptic feedback devices can be categorized into several types based on the underlying technology used. These types can be broadly classified into contact, noncontact, force-estimated, and hybrid force sensing.

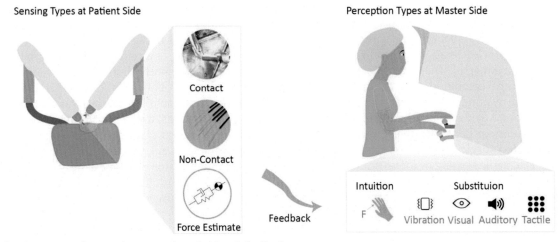

FIG. 2   Sensing types and perception types of surgical haptic feedback.

Contact-based sensing relies on direct contact between the surgical instrument and the tissue to provide force feedback. Contact-based sensors, including capacitive [19], piezoelectric [20], piezoresistive [21], and optical-based [22], can be integrated into the surgical instrument or attached to the surgical tool externally. When the external force is applied, the sensors measure the deformation increment of the core sensing elements and convert it into an electrical signal that can be processed and displayed to the surgeon. Contact-based force sensing has some benefits: high accuracy and precision, low noise, robustness, and reliability. However, it also can cause tissue damage or interfere with surgical procedures, requiring direct contact with the surgical tool or tissue.

Noncontact-based sensing uses sensors that do not require direct contact with the tissue to provide force feedback. Noncontact-based sensors can use various technologies, such as visual [23] or optical sensors [24]. These sensors can quantify the distance between the instrument and the tissue and provide feedback on the applied force based on the distance. The sensors can measure forces without interfering with the surgical procedure, with no risk of tissue damage, but with lower accuracy and precision than contact force sensing, susceptible to noise and environmental factors.

Force estimation is another method to compute the contact force between the medical instrument and tissue. This sensing type relies on the dynamic model of the system [25], neural network approach [26], and disturbance observer [27] are usually adopted. Those models generally input joint angles, velocities, and acceleration to estimate the force. It can provide real-time information, but the model is sensitive to the model parameters.

Hybrid sensing combines both contact-based and noncontact-based sensing methods to provide a more comprehensive haptic feedback experience [28]. Hybrid sensing can be used to detect both the force applied by the surgical instrument and the tissue properties, such as texture and hardness. It can improve the accuracy and precision of force measurements and overcome the limitations of individual sensing technologies, but the system is more complex and expensive than single sensing technologies.

## Perception types

Force haptic feedback devices for surgical applications can also be classified based on their feedback methods. These methods include intuition perception and substitution perception.

Intuition perception, also known as direct force perception, refers to the surgeon's capabilities to directly quantify the force being applied to tissues or organs during the operation procedure without relying solely on feedback from the haptic device [29]. However, due to the passive degree of freedom on the master side, this type of perception may lose the direction of the feedback force [30]. Generally, the force is scaled to the proper range within the perceptual ability of human beings and the actuator's ability.

The substitution perceptions use vibration, tactile, visual, and auditory as haptic feedback mediums. Vibrational motion can be generated and used to provide haptic feedback and can express various force levels or tissue properties, such as texture or stiffness. Vibration feedback can be delivered through the surgical instrument or a separate device [31]. Tactile feedback can also give the surgeon a sense of touch using pressure or texture and can be delivered through the surgical instrument or a separate device [17]. Visual feedback employs visual signals to bring the surgeon tactile information [32]. The visual cues can be in the form of a graphical interface or augmented reality display. This method provides a direct presentation of the forces being applied by the surgical instrument. Acoustic signals are also used as a medium to provide the surgeon with haptic information. The sound can be varied to convey different force levels or tissue properties. Auditory feedback can be delivered through headphones or speakers.

## Current force haptic interfaces in research stage

Actuation manners in force feedback haptic surgical devices are varied and can be categorized into different types based on the mechanism used to generate the force feedback. This section digs into the different force haptic devices from the actuation aspects. The common actuation forms of surgical force haptic are motor, cable-driven, pneumatic, magneto-rheological fluid (MRF), granular jamming, and piezoelectric (PZT). The comparison of these actuators under indexes, including output capacity, compact, fast response, lightweight, economical, and safety, is presented in Fig. 3. In addition, there are other forms of force haptic feedback, like auditory and visual cues. The summary of the current research stage of force haptic devices is demonstrated in Table 1. Fig. 4 presents the classification of the force haptic interfaces according to their actuation forms.

### *Motor*

Motors are commonly used in force feedback haptic surgical devices to generate force feedback. The motors can be either linear or rotary, providing precise control over the amount and direction of force feedback. The motor can

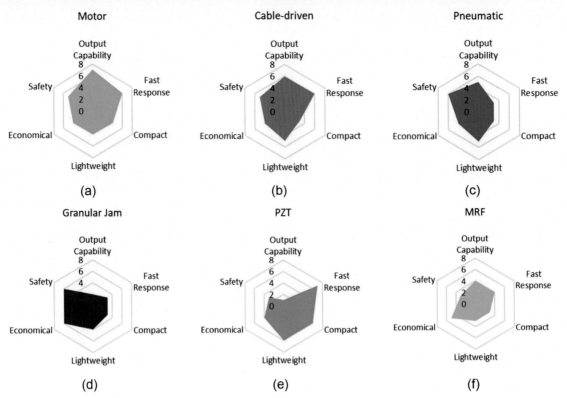

FIG. 3 Comparison results of the six indexes among different actuators.

TABLE 1 Haptic interfaces in research stage.

| # | Application scenario | Year | Actuator | Capacity | Sensing type | Perception type | Mounted type | Ref. |
|---|---|---|---|---|---|---|---|---|
| 1 | Needle insertion | 2016 | Vibrating motor array | Necessary maneuvers | NA | Substitution | Wrist wearable | [33] |
| 2 | Forceps grasp | 2006 | Voice coil motor | 800 Hz<br>6 N<br>Stiffness > 20 N/mm | NA | Intuition | Desktop | [34] |
| 3 | Slip sensation | 2016 | Linear movement of a base | LDP perception | NA | Intuition | Desktop | [35] |
| 4 | Force feedback | 2021 | Magnetic particle brake<br>Linear voice motor | NA | Force sensor | Intuition | Desktop | [36] |
| 5 | Needle insertion | 2011 | Cable-driven | 10 N | Force sensor | Intuition | Desktop | [37] |
| 6 | Grasp force | 2019 | Cable-driven | NA | NA | Intuition | Wearable | [38] |
| 7 | Laparoscopic forceps | 2019 | Motor<br>Cable-driven | NA<br>3 DOFs | Motor current | Intuition | Desktop | [39] |
| 8 | Force | 2011 | 4 micromotor | NA | NA | Intuition | Wearable | [40] |
| 9 | Contact location,<br>sliding, rolling | 2005 | Motor, push-and-pull wire | NA | NA | Intuition | Wearable | [41] |
| 10 | Grasp force | 2008 | Pneumatic | 20 Hz<br>3 N (unit) | Force sensor | Intuition | Desktop | [42] |
| 11 | Force | 2002 | Pneumatic | 0.014–16 N at fingertip | NA | Intuition | Wearable | [43] |
| 12 | Cutaneous | 2021 | Pneumatic balloon<br>Vibration motor | NA | Force sensor | Intuition | Desktop | [44] |
| 13 | Needle insertion<br>Palpation | 2015 | MRF | 14 N<br>18 kN/m | NA | Intuition | Desktop | [45] |
| 14 | Catheter insertion force | 2016 | MRF | Minimum 0.01 N | Force sensor | Intuition | Desktop | [46] |
| 15 | Stiffness | 2014 | Granular jamming | NA | NA | Intuition | Desktop | [47] |
| 16 | Stiffness | 2017 | Granular jamming | Maximum 7 N<br>5 mm displacement | NA | Intuition | Wearable | [48] |
| 17 | Palpation | 2020 | PZT | Maximum 1.16 N | Force sensor | Intuition | Desktop | [49] |
| 18 | Force feedback | 2019 | Bone-conduction headphone | NA | PZT | Substitution | Wearable | [50] |

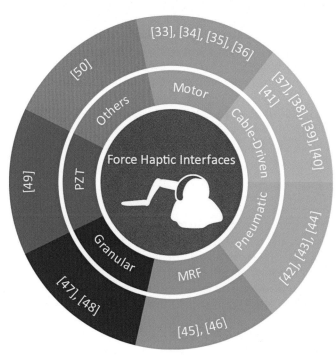

FIG. 4    A summary of the force haptic interface according to the actuation forms.

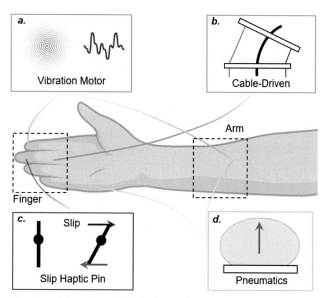

FIG. 5    (A) Vibrating motor actuates force haptic interface; (B) cable-driven force haptic feedback hand exoskeleton and force haptic feedback for fingertip; (C) pins triggered force haptic interface; (D) pneumatic actuated balloon array interface.

generate high levels of force and achieve fast response times. However, motors can be relatively bulky, heavy, and expensive, which may limit their use in some surgical instruments.

Multiactuator haptic interface in Fig. 5A utilizes the vibrating motor array to interpret the necessary needle steering operation to users, including acceleration, rotation, push, arrived, and withdrawal information [33]. Experimental findings imply that certain needle steering algorithms can be supported by the interface to assist surgeons in achieving high-quality implants and practicing needle steering techniques.

A forceps haptic device consisting of a voice coil actuator and a position sensor mounted on the master robot hand controller is proposed to render force and stiffness feedback [34]. The force feedback output is up to 6 N, and the stiffness feedback starts at 20 N/mm. Operators sense the grasp force and discriminate the grip of a vessel, hard tissue, to the pathological tissue.

A 3-degree-of-freedoms (DOFs) force feedback haptic interface is used in a 4 DOFs master-slave intraocular surgical robot system [35]. Force information at the patient side is described by regulating the currents of the two magnetic particle brakes and one linear voice motor. Experiments with force feedback illustrate more than 30% contact force decrease between the instrument and tissue.

In surgery, soft contact may be a forceps grasp tool or a body tissue. In these situations, operators commonly struggle to adjust the grasping force of the remote forceps to avoid dropping or damaging the held objects. The operator cannot accurately perceive slippage through cutaneous tactile feedback, especially incipient slippage (or preslide) at the contact surface. Honda and Hirai proposed a tactile feedback apparatus based on a bundle of pins under the fingertip of the operator [36] shown in Fig. 5C. The initial experiment results support that the pins' displacements can describe localized displacement phenomena (LDP) and improve the slip perception of the operator.

### Cable-driven

Cable-driven actuation systems use cables or wires to transmit force feedback to the hand of the surgeon. These systems can be lightweight, compact, and provide high force feedback levels [51,52]. However, they can be less precise than other actuation systems, and the cables can create friction and wear over time, requiring frequent maintenance.

Epidural anesthesia is the typical form of anesthesia before an operation. The anesthetists must hold the needle to penetrate ligamentum flavum and infiltrate the drug before dura matter [37]. However, the needle and tissue friction hinder the ability of the anesthetists to sense the insertion procedure. A cable-driven feedback device is proposed to provide anesthetists perception of the force at the tip of the needle [53,54].

Fig. 5B shows a grasp perception haptic feedback hand exoskeleton actuated by the motors developed to be used with the da Vinci surgery system [38,55]. The tendon-driven system renders the grasp feedback of the endo wrist to the operators by the grip of their thumb and index fingers. The tasks to grasp two different objects with the haptic equipment illustrate the improvement of the transparency of the robot aid surgery.

Zhao and Nelson utilized a cable-driven system to reflect the force information in laparoscopic surgery to the operators [39]. The interface conveys the position command of the operator to the surgical grasp and applies force feedback to the operator in accordance with the detected motor current.

Graham et al. developed a multimodal haptic feedback system for dexterous robotic telemanipulation [40]. The device provides force feedback to the operator's hand through a glove-based interface that includes fingertip sensors and vibrotactile actuators. The system also includes a display screen for visual feedback and an audio system for auditory feedback. The force feedback is achieved through impedance control, which measures the applied force by the operator and adjusts the interaction system's impedance accordingly. The device improves the dexterity and situational awareness of the operator during telemanipulation tasks.

A moving centroid beneath the finger of the user renders tactile feedback [41]. It gives the user simultaneous feedback of contact location and interaction forces in a thimble-sized container installed atop a haptic force feedback device. The two push-and-pull wires move the centroid along the thimble for a maximum length of 2 mm to portray contact location (sliding or rolling) and contactless, respectively [56]. Thus, a configuration generates a genuine experience of making and breaking contact [57].

### Pneumatic

Pneumatic actuation systems use compressed air or gas to generate force feedback as shown in Fig. 5D. These systems can provide high levels of force, are relatively compact and lightweight, and can be quiet during operation. However, they require a compressed air or gas source, which may not be readily available in some surgical settings.

Culjat et al. developed a tactile feedback system for robot-assisted surgery, consisting of a custom-made haptic device that provided surgeons using the da Vinci surgical system with real-time force and vibration feedback. A pneumatically actuated balloon array interface mounted on the da Vinci surgery system provides grasp force feedback perception for surgeons [42]. The device could achieve a response frequency of up to 20 Hz and an output force (balloon unit) of up to 3 N. The system was tested in experiments and shown to improve the surgeon's ability to detect tissue differences and perform precise movements during surgery. Rutgers Master II-ND hand exoskeleton is a pneumatically actuated haptic feedback device that provides force feedback to the fingers of the operator, including the thumb, index, middle, and ring fingers [43]. The system can render a force of up to 16 N at each fingertip under a weight of less than 100 g, which can be applied in surgical simulation. A bio-inspired controlling model is proposed to provide surgeons with force and vibration feedback using a pneumatic balloon and vibration motor [44]. The single pair of actuators mimics the population of cutaneous afferents by the weighted stimulating function.

## Magneto-rheological fluid

The unique characteristic of magneto-rheological fluid (MRF) is that the external magnetic field influences the fluid viscosity and shear stress. Thus, the transmission feature of the MRF is controllable, which is a vital benefit of the MRF [58]. MRF-based actuation systems use magnets to manipulate the fluid and generate force feedback. These systems can provide a wide range of force feedback options and be relatively compact and lightweight. However, MRF costs are expensive, and the magnetic field can interfere with other electronic devices in the surgical setting.

The MRF actuator is an alternative to haptic devices [59]. A 3-DOF (one of them is passive) haptic device that adopts four MRF clutches was proposed by Patel et al. [45]. The phantom needle insertion and tissue palpation experiments perform better than commercial haptic devices.

Guo et al. developed an MRF-based catheter operation system to provide the insertion friction sensation to surgeons [46]. Based on the detected force from the follower's end-effector, the insertion force is mimicked by regulating the viscosity of the MRF by altering the electromagnetic field. Thus, the surgeon can discriminate the collision and immediate surgery to be successful [60]. In the following study, comparison experiments demonstrate a shorter completion time and lower collision times when the haptic device is involved [61].

## Granular jamming

Granular jamming is a mechanical actuation system that uses granular materials, such as sand or beads, to generate force feedback. The materials can be compressed or released to simulate different levels of force. These systems can be relatively low-cost and provide a wide range of force feedback options. However, they can be less precise than other actuation systems, and the granular materials can wear out over time, requiring frequent maintenance.

A multifingered haptic palpation system that uses granular jamming stiffness feedback actuators is described in [47]. The device comprises a series of cylindrical chambers filled with granular material and linked by a vacuum source. By applying or releasing vacuum pressure, the granular material transitions between a jammed state (stiff) and an unjammed state (compliant), providing variable levels of force rendering effect onto the user's fingertips. The palpation device is designed to improve the accuracy and sensitivity of robotic palpation systems used in medical applications.

Zubrycki and Granosik proposed a glove-based control interface that uses the jamming principle to provide kinesthetic feedback [48]. The device comprises a glove filled with granular material and connected to a vacuum pump. When a vacuum is applied, the granular material jams together, creating a rigid structure that provides resistance to movement and a sense of force feedback. The device is designed to enhance the realism and effectiveness of virtual reality and teleoperation systems.

## Piezoelectric (PZT) actuator

Piezo-based actuation systems use piezoelectric materials to generate force feedback. These materials can change shape when an electrical voltage is applied, generating force feedback. These systems can be relatively compact and lightweight and provide fast response times. However, they can be relatively expensive and require complex control systems.

Yeh et al. depicted a haptic feedback device that uses a piezoelectric actuator to generate precise and rapid movements, providing users with realistic feedback in response to virtual interactions [49]. Various studied applications, such as video games and virtual reality, prove the ability of the haptic device in the aspects of interaction and user experience,

## Other forms

A bone-conduction headphone was paired with a force sensor to give the user minimal auditory distraction [50]. The sensor measures the distal force information in robotic surgery and transmits the data to the operator via the bone-conduction headphones. Suture trials with this sensation substitution feedback demonstrate the high performance of the operators.

# Current commercial force haptic interfaces

The haptic feedback device market is expected to expand significantly over the next few years because of the growing demand for more immersive gaming and virtual reality experiences and the increasing adoption of haptic technology in medical applications. Table 2 shows the summary of the current commercial surgical force haptic device.

TABLE 2    Commercial force haptic interfaces.

| # | Haptic interface | Workspace translation (mm$^2$), Rotation (yaw, pitch, roll in degrees) | Capacity Maximum feedback force, torque, and stiffness | Force feedback DOFs | Positional sensing DOFs | Ref. |
|---|---|---|---|---|---|---|
| 1 | CyberGrasp | NA | 12 N each finger | NA | NA | [62] |
| 2 | PHANTOM 1.5 High Force/6 DOF | Translation: $381 \times 267 \times 191$ Rotation: $297 \times 260 \times 335$ | 37.5 N 3.5 N/mm | 6 | 6 | [63–65] |
| 3 | Novint Falcon | Translation: $101 \times 101 \times 101$ | 8.9 N | 3 | 6 | [66] |
| 4 | Sigma 7 | Translation: $\varnothing 190 \times 130$ Rotation: $235 \times 140 \times 200$ | 20 N 0.4 Nm | 6+1 | 6+1 | [67] |
| 5 | Virtuose 6D | Translation: $1300 \times 658 \times 1080$ Rotation: $300 \times 100 \times 250$ | 35 N 3.1 Nm | 6 | 6 | [68] |
| 6 | HD2 | Translation: $800 \times 250 \times 350$ Rotation: $180 \times 180 \times$ Continuous | 19.71 N in $x$ 19.71 N in $y$ 13.94 N in $z$ 1.72 Nm | 6 | 6+1 | [69] |
| 7 | Senhance surgical system | NA | NA | NA | NA | [70] |
| 8 | MAKO | NA | NA | NA | NA | [71] |

The CyberGrasp (CyberGlove Systems, San Jose, CA, United States) is a wearable haptic interface that provides force feedback to the user's fingers and hand, allowing them to feel virtual objects and surfaces as if they were real. CyberGrasp routes low-friction tendons through an exoskeleton and electrical actuators far from the hand to send forces to the fingertips and help users discriminate the size and stiffness of the object [62]. Combined with the Cybertouch, straightforward sensations like pulses or persistent vibrations can be rendered to create detailed force feedback.

3D Systems (Rock Hill, SC, United States) provides a series of haptic feedback solutions, including Touch, Touch X, and PHANTOM premium. Among them, PHANTOM 1.5 High Force/6 DOF performs best. The device offers 6 DOFs, with force feedback up to 37.5 N and a stiffness range of 0–3.5 N/mm. It can achieve position resolution within 0.007 mm. The device's haptic feedback aids in improving the accuracy and precision of surgical movements, reducing the risk of complications and improving patient outcomes [63–65]. It has also been proven to be the best commercial haptic device [72].

The Novint Falcon (Novint Technologies, Albuquerque, United States) haptic device is a force feedback device used in surgical simulations to give users a realistic sense of touch and force. It uses advanced haptic technology to create real sensations of touch and resistance, allowing users to feel like they are interacting with a physical object. In [66], Novint Falcon was used as a haptic device to provide feedback signals to the surgeon while teleoperating with a flexible robotic surgical tool. The device provides force feedback to the surgeon, allowing for more robust control of the flexible surgical instrument during the procedure.

Force Dimension (Nyon, Switzerland) develops Omega, Delta, and Sigma series of haptic interfaces. The Force Dimension Sigma 7 is a haptic device that provides force feedback for surgical and other applications. The device uses a 6 DOFs force sensor to measure forces and torques in all directions, and it can exert forces of up to 7 N in any direction. The device also features high-resolution encoders that provide precise position and velocity information, allowing for accurate force control and feedback. The device is used as a haptic interface in the MiroSurge console to give the surgeon's hands force feedback during the surgical procedure [67]. The device is integrated with the console to provide a seamless user experience and to allow the surgeon to control the surgical instruments with high accuracy and precision. A study also shows tissue damage decreased when integrating a Sigma 7 with da Vinci Research Kit [73].

Another family of haptic feedback interfaces is the products provided by Haption (Aachen, Germany), including Scale 1, HapticMaser, Desktop 3D and 6D, Virtuose 3D, and 6D. The Virtuose 6D is a haptic device that uses six degrees of freedom to simulate the movements of objects in a virtual environment [68]. The device uses a combination of motorized and friction-based actuators to provide force feedback. The motorized actuators allow precise force control and generate forces up to 35 N. Howard and Szewczyk evaluated their proposed method in a simulated laparoscopic surgery task. They found that it significantly improved the precision of the instrument navigation compared to using only visual feedback [74].

Two high-fidelity haptic devices are offered by Quanser (Markham, Canada). The initial device is a Haptic Wand with five degrees of freedom, which employs a dual-pantograph kinematic design to offer three translational and two rotational DOFs [69]. The maximum force output of the HD2 force haptic feedback device is 19.71 N, and the maximum torque output is 1.72 N m. It has a 6 DOFs end effector, which allows for a wide range of motion and manipulation.

The Senhance surgical system (Asensus Surgical, Durham, NC, United States) is a robotic surgery platform that utilizes haptic force feedback technology to provide surgeons with a sense of touch and pressure during MIS [18,70]. The system's haptic feedback technology enables surgeons to feel the amount of force being applied to tissue, which helps to improve accuracy and control during surgical procedures.

The haptic feedback in the MAKO (Stryker, Portage, MI, United States) surgical system is provided by the RIO (Robotic Arm Interactive Orthopedic System) robotic arm [71]. The haptic feedback is designed to give the surgeon a sense of resistance as they move the surgical tools and make cuts, enabling them to quantify the exerted force precisely. The system is also integrated with a visual display, which allows the surgeon to see real-time feedback on the amount of force being applied by the robotic arm. This visual feedback helps the surgeon adjust as needed, ensuring that the procedure is performed with the highest accuracy and precision possible.

## Current issues with surgical force haptic feedback technology

In a master-slave manipulator system, haptic feedback with an ideal response renders operators the best sensation of the patient site's status, thus enhancing performance [75,76]. The haptic feedback interface intensifies the immersion of the operators as if they are operating directly on the other end [77]. However, delays induced by communication latency and disturbances commonly exist in an automatic system, severely creating barriers to the operator and degrading their performance [18]. That can cause a lag between the surgeon's movement and the corresponding feedback. Time delay can make it difficult for the surgeon to accurately gauge the force being applied, leading to potential complications and reduced accuracy.

Stability and transparency are vital concerns when using surgical haptic feedback interface applications. In a haptic system that is 100% transparent, force detection would be completely accurate with no errors, time delays, or differences in impedance [78]. Due to high instability, a "perfect" kinesthetic feedback system can similarly result in harmful oscillations in a "closed loop" [79]. Therefore, there must be a balance between transparency and stability. Instability also can be caused by factors, such as system noise, vibration, or drift, which can cause the feedback to be inconsistent or inaccurate. Instability can reduce the effectiveness of the haptic feedback and impact the surgeon's ability to perform the procedure accurately.

Haptic feedback systems may have technical limitations that impact their effectiveness in surgical applications. For example, systems may have a limited range of motion or force, making it challenging to simulate the surgical environment accurately. Additionally, some systems may not be able to accurately simulate certain types of tissue, such as soft tissue, which can impact their usefulness in specific surgical procedures. Sterilization or material compatibility is also an inevitable challenge [80]. Haptic feedback devices must be compatible and integrate seamlessly with existing surgical equipment and systems. This can be challenging, particularly in multivendor environments where different devices and systems may use different communication protocols and interfaces.

Using haptic feedback interfaces in surgery requires specialized training and skill acquisition. Surgeons must learn to interpret the device's force feedback and integrate it into their surgical technique.

## Conclusions

This chapter discusses the state-of-art force haptic feedback in surgical systems, including its sensing types, perception types, current interfaces, and challenges. Force haptic feedback can potentially improve surgical procedures accurately and safely, allowing surgeons to perceive and control the forces exerted during surgical tasks more accurately. Significant progress has been summarized in developing force haptic feedback systems specialized for surgical applications. Several commercial systems have been designed with force feedback capabilities. Additionally, research efforts have focused on developing more sophisticated control algorithms and sensor technologies and integrating haptic feedback with other sensory modalities. However, there are several challenges, including issues related to time delay, instability, technical limitation, and compatibility, which must be overcome to realize the technology's benefits fully.

## Future scenario

Looking to the future, several promising directions exist for developing force haptic feedback in surgical systems. Material science and sensor technology advances can help create more realistic and accurate simulations of surgical environments. Additionally, collaborations between engineers, clinicians, and researchers can help to identify and address the key challenges facing the development and implementation of force haptic feedback in surgical systems. By addressing these challenges and continuing to innovate in this field, haptic force feedback has the potential to revolutionize the field of surgical procedures and improve patient outcomes.

## Key points

- Covering perception types, interfaces, and challenges, this discussion explores the state-of-the-art force haptic feedback in surgical systems.
- Force haptic feedback enhances surgical procedures by improving accuracy and surgeons' dexterity and situational awareness in perceiving and controlling exerted forces during surgical teleoperation.
- Significant progress has been made in developing specialized force haptic feedback systems for surgical applications, with commercial systems available.
- Challenges such as the need for high-fidelity and accurate force haptic feedback in surgical environments need more attention.

## Acknowledgments

This work was supported by the Hong Kong Research Grants Council (RGC) Collaborative Research Fund (C4026-21G), Research Grants Council (RGC)—Research Impact Fund (RIF R4020-22) General Research Fund (GRF 14211420), Research Grants Council (RGC)—NSFC/RGC Joint Research Scheme N_CUHK420/22, Shenzhen-Hong Kong-Macau Technology Research Programme (Type C) STIC Grant SGDX20210823103535014 (202108233000303), the key project 2021B1515120035 (B.02.21.00101) of the Regional Joint Fund Project of the Basic and Applied Research Fund of Guangdong Province, Hong Kong Research Grants Council (RGC) Collaborative Research Fund (CRF C4063-18G and C4026-21G), and General Research Funds (GRF 14216022, GRF 14203323).

## References

[1] Lécuyer A. Simulating haptic feedback using vision: a survey of research and applications of pseudo-haptic feedback. Presence Teleop Virt 2009;18(1):39–53.
[2] Biswas S, Visell Y. Haptic perception, mechanics, and material technologies for virtual reality. Adv Funct Mater 2021;31(39):2008186.
[3] Cooper N, Milella F, Pinto C, Cant I, White M, Meyer G. The effects of substitute multisensory feedback on task performance and the sense of presence in a virtual reality environment. PLoS One 2018;13(2), e0191846.
[4] Westebring-van der Putten EP, Goossens RH, Jakimowicz JJ, Dankelman J. Haptics in minimally invasive surgery—a review. Minim Invasive Ther Allied Technol 2008;17(1):3–16.
[5] Ren H, Rank D, Merdes M, Stallkamp J, Kazanzides P. Multisensor data fusion in an integrated tracking system for endoscopic surgery. IEEE Trans Inf Technol Biomed 2011;16(1):106–11.
[6] Lucarotti C, Oddo CM, Vitiello N, Carrozza MC. Synthetic and bio-artificial tactile sensing: a review. Sensors 2013;13(2):1435–66.
[7] Xu L, et al. Information loss challenges in surgical navigation systems: from information fusion to AI-based approaches. Inf Fusion 2022;92:13–36.
[8] Yue W, et al. Dynamic piezoelectric tactile sensor for tissue hardness measurement using symmetrical flexure hinges and anisotropic vibration modes. IEEE Sensors J 2021;21(16):17712–22.
[9] Ju F, Ling S-F. A micro whisker transducer with sensorless mechanical impedance detection capability for fluid and tactile sensing in space-limited applications. Sensors and Actuators A Phys 2015;234:104–12.
[10] Puangmali P, Althoefer K, Seneviratne LD, Murphy D, Dasgupta P. State-of-the-art in force and tactile sensing for minimally invasive surgery. IEEE Sensors J 2008;8(4):371–81.
[11] Okamura AM. Methods for haptic feedback in teleoperated robot-assisted surgery. Ind Robot Int J 2004;31(6):499–508.
[12] Lichiardopol S. A survey on teleoperation. vol. 20. Technische Universitat Eindhoven; 2007. p. 40–60. DCT report.
[13] Hoeckelmann M, Rudas IJ, Fiorini P, Kirchner F, Haidegger T. Current capabilities and development potential in surgical robotics. Int J Adv Robot Syst 2015;12(5):61.
[14] Zhao Z, et al. Engineering functional and anthropomorphic models for surgical training in interventional radiology: a state-of-the-art review. Proc Inst Mech Eng H J Eng Med 2023;237(1):3–17.
[15] Wagner CR, Howe RD, Stylopoulos N. The role of force feedback in surgery: analysis of blunt dissection. In: International symposium on haptic interfaces for virtual environment and teleoperator systems. IEEE Computer Society; 2002. p. 73.
[16] Wu Y, Balatti P, Lorenzini M, Zhao F, Kim W, Ajoudani A. A teleoperation interface for loco-manipulation control of mobile collaborative robotic assistant. IEEE Robot Autom Lett 2019;4(4):3593–600.

[17] Schostek S, Schurr MO, Buess GF. Review on aspects of artificial tactile feedback in laparoscopic surgery. Med Eng Phys 2009;31(8):887–98.

[18] Patel RV, Atashzar SF, Tavakoli M. Haptic feedback and force-based teleoperation in surgical robotics. Proc IEEE 2022;110(7):1012–27.

[19] Kim U, Lee D-H, Yoon WJ, Hannaford B, Choi HR. Force sensor integrated surgical forceps for minimally invasive robotic surgery. IEEE Trans Robot 2015;31(5):1214–24.

[20] Sokhanvar S, Packirisamy M, Dargahi J. A multifunctional PVDF-based tactile sensor for minimally invasive surgery. Smart Mater Struct 2007;16(4):989.

[21] Kalantari M, Ramezanifard M, Ahmadi R, Dargahi J, Kövecses J. A piezoresistive tactile sensor for tissue characterization during catheter-based cardiac surgery. Int J Med Robot Comput Assist Surg 2011;7(4):431–40.

[22] Noh Y, Han S, Gawenda P, Li W, Sareh S, Rhode K. A contact force sensor based on S-shaped beams and optoelectronic sensors for flexible manipulators for minimally invasive surgery (MIS). IEEE Sensors J 2019;20(7):3487–95.

[23] Su Y-H, Sosnovskaya Y, Hannaford B, Huang K. Securing robot-assisted minimally invasive surgery through perception complementarities. In: 2020 fourth IEEE international conference on robotic computing (IRC). IEEE; 2020. p. 41–7.

[24] Puangmali P, Liu H, Seneviratne LD, Dasgupta P, Althoefer K. Miniature 3-axis distal force sensor for minimally invasive surgical palpation. IEEE/ASME Trans Mechatron 2011;17(4):646–56.

[25] Li Y, Miyasaka M, Haghighipanah M, Cheng L, Hannaford B. Dynamic modeling of cable driven elongated surgical instruments for sensorless grip force estimation. In: 2016 IEEE international conference on robotics and automation (ICRA). IEEE; 2016. p. 4128–34.

[26] Marban A, Srinivasan V, Samek W, Fernández J, Casals A. A recurrent convolutional neural network approach for sensorless force estimation in robotic surgery. Biomed Signal Process Control 2019;50:134–50.

[27] Liang W, Huang S, Chen S, Tan KK. Force estimation and failure detection based on disturbance observer for an ear surgical device. ISA Trans 2017;66:476–84.

[28] Yilmaz N, Bazman M, Alassi A, Gur B, Tumerdem U. 6-axis hybrid sensing and estimation of tip forces/torques on a hyper-redundant robotic surgical instrument. In: 2019 IEEE/RSJ international conference on intelligent robots and systems (IROS). IEEE; 2019. p. 2990–7.

[29] Okamura AM. Haptic feedback in robot-assisted minimally invasive surgery. Curr Opin Urol 2009;19(1):102.

[30] Verner LN, Okamura AM. Effects of translational and gripping force feedback are decoupled in a 4-degree-of-freedom telemanipulator. In: Second joint EuroHaptics conference and symposium on haptic interfaces for virtual environment and teleoperator systems (WHC'07). IEEE; 2007. p. 286–91.

[31] Pacchierotti C, Prattichizzo D, Kuchenbecker KJ. Cutaneous feedback of fingertip deformation and vibration for palpation in robotic surgery. IEEE Trans Biomed Eng 2015;63(2):278–87.

[32] Horeman T, Rodrigues SP, van den Dobbelsteen JJ, Jansen F-W, Dankelman J. Visual force feedback in laparoscopic training. Surg Endosc 2012;26:242–8.

[33] Rossa C, Fong J, Usmani N, Sloboda R, Tavakoli M. Multiactuator haptic feedback on the wrist for needle steering guidance in brachytherapy. IEEE Robot Autom Lett 2016;1(2):852–9.

[34] Rizun P, Gunn D, Cox B, Sutherland G. Mechatronic design of haptic forceps for robotic surgery. Int J Med Robot Comput Assist Surg 2006;2 (4):341–9.

[35] Zuo S, Wang Z, Zhang T, Chen B. A novel master–slave intraocular surgical robot with force feedback. Int J Med Robot Comput Assist Surg 2021;17(4), e2267.

[36] Honda H, Hirai S. Development of a novel slip haptic display device based on the localized displacement phenomenon. IEEE Robot Autom Lett 2016;1(1):585–92.

[37] Koseki Y, De Lorenzo D, Chinzei K, Okamura AM. Coaxial needle insertion assistant for epidural puncture. In: 2011 IEEE/RSJ international conference on intelligent robots and systems. IEEE; 2011. p. 2584–9.

[38] Secco EL, Tadesse AM. A wearable exoskeleton for hand kinesthetic feedback in virtual reality. In: International conference on wireless mobile communication and healthcare. Springer; 2019. p. 186–200.

[39] Zhao B, Nelson CA. A sensorless force-feedback system for robot-assisted laparoscopic surgery. Comput Assist Surg 2019;24(sup1):36–43.

[40] Graham JL, Manuel SG, Johannes MS, Armiger RS. Development of a multi-modal haptic feedback system for dexterous robotic telemanipulation. In: 2011 IEEE international conference on systems, man, and cybernetics. IEEE; 2011. p. 3548–53.

[41] Provancher WR, Cutkosky MR, Kuchenbecker KJ, Niemeyer G. Contact location display for haptic perception of curvature and object motion. Int J Robot Res 2005;24(9):691–702.

[42] Culjat MO, et al. A tactile feedback system for robotic surgery. In: 2008 30th annual international conference of the IEEE Engineering in Medicine and Biology Society. IEEE; 2008. p. 1930–4.

[43] Bouzit M, Burdea G, Popescu G, Boian R. The Rutgers Master II-new design force-feedback glove. IEEE/ASME Trans Mechatron 2002;7 (2):256–63.

[44] Ouyang Q, et al. Bio-inspired haptic feedback for artificial palpation in robotic surgery. IEEE Trans Biomed Eng 2021;68(10):3184–93.

[45] Najmaei N, Asadian A, Kermani MR, Patel RV. Design and performance evaluation of a prototype MRF-based haptic interface for medical applications. IEEE/ASME Trans Mechatron 2015;21(1):110–21.

[46] Yin X, Guo S, Hirata H, Ishihara H. Design and experimental evaluation of a teleoperated haptic robot–assisted catheter operating system. J Intell Mater Syst Struct 2016;27(1):3–16.

[47] Li M, et al. Multi-fingered haptic palpation utilizing granular jamming stiffness feedback actuators. Smart Mater Struct 2014;23(9):095007.

[48] Zubrycki I, Granosik G. Novel haptic device using jamming principle for providing kinaesthetic feedback in glove-based control interface. J Intell Robot Syst 2017;85(3):413–29.

[49] Yeh C-H, et al. Application of piezoelectric actuator to simplified haptic feedback system. Sensors and Actuators A Phys 2020;303:111820.

[50] Mikic M, Francis P, Looi T, Gerstle JT, Drake J. Bone conduction headphones for force feedback in robotic surgery. In: 2019 41st annual international conference of the IEEE Engineering in Medicine and Biology Society (EMBC). IEEE; 2019. p. 7128–33.

[51] Li L, Zhang L, Wang B, Xue F, Zou Y, Song D. Running experimental research of a cable-driven astronaut on-orbit physical exercise equipment. Machines 2022;10(5):377.

[52] Zhang L, Li L, Zou Y, Wang K, Jiang X, Ju H. Force control strategy and bench press experimental research of a cable driven astronaut rehabilitative training robot. IEEE Access 2017;5:9981–9.

[53] De Lorenzo D, Koseki Y, De Momi E, Chinzei K, Okamura AM. Coaxial needle insertion assistant with enhanced force feedback. IEEE Trans Biomed Eng 2012;60(2):379–89.

[54] Koseki Y, Kawai M, De Lorenzo D, Yamauchi Y, Chinzei K. Coaxial needle insertion assistant for epidural puncture effect of lateral force on needle. In: 2013 35th annual international conference of the IEEE Engineering in Medicine and Biology Society (EMBC). IEEE; 2013. p. 6683–6.

[55] Secco EL, Tadesse AM. Kinesthetic feedback for robot-assisted minimally invasive surgery (Da Vinci) with two fingers exoskeleton. In: International conference on wireless mobile communication and healthcare. Springer; 2019. p. 212–25.

[56] Provancher WR, Kuchenbecker KJ, Niemeyer G, Cutkosky MR. Perception of curvature and object motion via contact location feedback. In: Robotics research. The eleventh international symposium. Springer; 2005. p. 456–65.

[57] Springer SL, Ferrier NJ. Design and control of a force-reflecting haptic interface for teleoperational grasping. J Mech Des 2002;124(2):277–83.

[58] Jolly MR, Bender JW, Carlson JD. Properties and applications of commercial magnetorheological fluids. J Intell Mater Syst Struct 1999;10(1):5–13.

[59] Bicchi A, Raugi M, Rizzo R, Sgambelluri N. Analysis and design of an electromagnetic system for the characterization of magneto-rheological fluids for haptic interfaces. IEEE Trans Magn 2005;41(5):1876–9.

[60] Yin X, Guo S, Xiao N, Tamiya T, Hirata H, Ishihara H. Safety operation consciousness realization of a MR fluids-based novel haptic interface for teleoperated catheter minimally invasive neurosurgery. IEEE/ASME Trans Mechatron 2015;21(2):1043–54.

[61] Song Y, et al. Design and performance evaluation of a haptic interface based on MR fluids for endovascular tele-surgery. Microsyst Technol 2018;24(2):909–18.

[62] CyberGlove Systems. http://www.cyberglovesystems.com/cybergrasp. [Accessed 2023].

[63] Soleymani A, Li X, Tavakoli M. A domain-adapted machine learning approach for visual evaluation and interpretation of robot-assisted surgery skills. IEEE Robot Autom Lett 2022;7(3):8202–8.

[64] Soleymani A, Li X, Tavakoli M. Deep neural skill assessment and transfer: application to robotic surgery training. In: 2021 IEEE/RSJ international conference on intelligent robots and systems (IROS). IEEE; 2021. p. 8822–9.

[65] Cui Z, et al. A study of force feedback master-slave teleoperation system based on biological tissue interaction. In: Intelligent life system modelling, image processing and analysis. Springer; 2021. p. 134–44.

[66] Li Z, Feiling J, Ren H, Yu H. A novel tele-operated flexible robot targeted for minimally invasive robotic surgery. Engineering 2015;1(1):073–8.

[67] Tobergte A, et al. The sigma. 7 haptic interface for MiroSurge: a new bi-manual surgical console. In: 2011 IEEE/RSJ international conference on intelligent robots and systems. IEEE; 2011. p. 3023–30.

[68] Garrec P, Friconneau J-P, Louveaux F. Virtuose 6D: a new force-control master arm using innovative ball-screw actuators. In: ISR 2004-35th international symposium on robotics; 2004.

[69] Stocco LJ, Salcudean SE, Sassani F. Optimal kinematic design of a haptic pen. IEEE/ASME Trans Mechatron 2001;6(3):210–20.

[70] Asensus Surgical. https://www.senhance.com/us/digital-laparoscopy. [Accessed 2023].

[71] Babu BC, Thilak J. Total knee arthroplasty using robotics (MAKO). In: Knee arthroplasty: New and future directions. Springer; 2022. p. 491–505.

[72] Zareinia K, Maddahi Y, Ng C, Sepehri N, Sutherland GR. Performance evaluation of haptic hand-controllers in a robot-assisted surgical system. Int J Med Robot Comput Assist Surg 2015;11(4):486–501.

[73] Saracino A, et al. Haptic feedback in the da Vinci Research Kit (dVRK): a user study based on grasping, palpation, and incision tasks. Int J Med Robot Comput Assist Surg 2019;15(4), e1999.

[74] Howard T, Szewczyk J. Improving precision in navigating laparoscopic surgery instruments toward a planar target using haptic and visual feedback. Front Robot AI 2016;3:37.

[75] De Gersem G, Van Brussel H, Tendick F. Reliable and enhanced stiffness perception in soft-tissue telemanipulation. Int J Robot Res 2005;24(10):805–22.

[76] Niemeyer G, Slotine J-JE. Telemanipulation with time delays. Int J Robot Res 2004;23(9):873–90.

[77] Hirche S, Bauer A, Buss M. Transparency of haptic telepresence systems with constant time delay. In: Proceedings of 2005 IEEE conference on control applications, 2005. CCA 2005. IEEE; 2005. p. 328–33.

[78] Yamamoto T, Abolhassani N, Jung S, Okamura AM, Judkins TN. Augmented reality and haptic interfaces for robot-assisted surgery. Int J Med Robot Comput Assist Surg 2012;8(1):45–56.

[79] Pacchierotti C, Meli L, Chinello F, Malvezzi M, Prattichizzo D. Cutaneous haptic feedback to ensure the stability of robotic teleoperation systems. Int J Robot Res 2015;34(14):1773–87.

[80] Marbán A, Casals A, Fernández J, Amat J. Haptic feedback in surgical robotics: still a challenge. In: ROBOT2013: First Iberian robotics conference: Advances in robotics, vol. 1. Springer; 2014. p. 245–53.

# 7

# Robust magnetic tracking and navigation in robotic surgery

*Shijian Su and Hongliang Ren*

Department of Electronic Engineering, The Chinese University of Hong Kong, Sha Tin, Hong Kong

## Introduction

Robotic surgery refers to a surgical procedure that uses robotic equipment to enable surgeons to perform delicate and intricate procedures that may be challenging or impossible with other traditional surgical techniques [1]. This advanced technology facilitates not only minimally invasive surgery (MIS) through smaller incisions but also offers other significant advantages such as higher precision of movement, a greater range of motion (360° rotation for motor), better visualization (3D high-definition imaging of the surgical area), and the ability to operate inside the human body. Robotic surgery has developed rapidly in recent years and has been successfully utilized in various clinical procedures, such as spine surgery [2], fracture repair [3], neurosurgery [4], and cardiovascular surgery [5], among others.

### The role of tracking and navigation in robotic surgery

Surgical navigation systems play crucial roles in robotic surgery, specifically in trajectory planning, localization, and determining spatial relationships between medical images, surgical tools, and target lesions [6]. Surgical navigation systems indicate where the surgical instruments are and how to reach the target site safely [7,8] with core tracking components, which enable intraoperative and continuous tool tracking with respect to the surrounding anatomy [9]. More specifically, the tracking system will estimate the six-degree-of-freedom (6-DoF) pose of each surgical instrument and then output the pose information to track the motion of these surgical instruments in real time.

### Surgical instrument tracking technologies

Currently, the most used technologies for tracking surgical instruments are mainly based on mechanics, optical, magnetic, and medical imaging principles.

Mechanical digitizer tracking systems (MDTSs), also known as articulated arm coordinate measuring machines, consist of parallel or serial linkages with rotary joint encoders [10]. The position and orientation of the robotic end-effector can be determined using the forward kinematics model. MDTSs could achieve submillimeter accuracy and are ever widely employed in neuro- [11] and orthopedic [12] surgery. The MDTSs, such as the commercial coordinate measuring machine by Ferranti (British Ltd.) [13], have the mechanical structure of being bulky, difficult to sterilize, and only tracking one instrument at one time [14]. These inherent deficiencies make it not feasible in increasingly complex surgical procedures.

Optical tracking systems (OTSs) are a widely used tracking method in intra-operative procedures because of their high precision and stability [15]. OTSs based on camera and computer technology can be traced back to the 1980s [16]. In 1983, Maxwell at MIT presented a technology called "Graphic Marionette." It used two cameras and a series of LEDs attached to a human body to capture body motion [17]. The Polaris optical tracking system, still widely used in robotic surgery today, was introduced by Northern Digital Inc. (NDI, Canada) in 1996 [18]. Depending on the camera's sensitivity to light, OTSs can be classified into two categories: visible light-based and infrared tracking systems.

Visible light-based tracking systems use computer vision and image processing technology to recognize and track passive markers [19]. Infrared tracking systems generally utilize passive retro-reflective spheres or active infrared light-emitting diodes (LEDs) as tracking markers, while the former has the advantage of wireless tracking [20,21]. Since both categories of OTSs highly rely on the line-of-sight for tracking, tracking failure might trigger if surgical instruments are moved out of the camera's field of view or occluded by the surgeon's hand during surgical procedures. Thus, the applications of OTSs are subject to limitations in crowded operating rooms [22].

Magnetic tracking systems (MTSs) generally consist of magnetic field generators and sensors. According to the magnetic field generation type, MTSs can be divided into electromagnetic tracking systems (EMTSs) and permanent magnet tracking systems (PMTSs). For EMTSs, quasi-static electromagnetic fields are generated by transmitting coils excited by alternating current signals. Then the pose of the induction coils (EM sensors) with respect to the transmitting coils is estimated based on the magnetic induction intensity measured by the induction coils [23,24]. The tracking principle of PMTSs is similar to that of EMTSs, except that the static magnetic field is generated by permanent magnets (PMs), and the magnetic induction intensity is measured by a magnetometer array [25,26]. In 1969, Polhemus (Burlington, USA) first introduced electromagnetic technology for the head tracking of pilots in the military [27]. In addition, other EMTSs commonly used in surgical navigation are Aurora (Northern Digital Inc., Canada) and 3D Guidance (Ascension Technology Corp, an affiliate company of NDI since 2012) [28]. The magnetic field of the MTSs can penetrate through human tissues without any harm, making it an ideal technology for tracking surgical instruments in medical [29–32].

Medical image-guided tracking systems (MITSs) are utilized to perceive the shape and tip pose of surgical instruments during clinical interventions (interventional surgery) using medical imaging techniques, such as magnetic resonance imaging (MRI) [36], computed tomography (CT) [37], ultrasound (US) [34]. Computers were involved in medical imaging in the early 1970s with the advent of CT, US, and MRI [38]. Medical imaging techniques can image surgical instruments and tissue structures simultaneously, but the disadvantages include the limited real-time imaging capabilities for MRI and CT, the ionizing radiation for CT, and the low spatial resolution for US.

Fiber Bragg grating (FBG)-based tracking systems (FBGTSs) integrate multiple FBG sensors into a flexible instrument for tip pose prediction and shape sensing, where the FBG sensors function as measuring the curvature at the location of the sensor [33,39,40]. In 2000, Gander et al. first demonstrated curvature measurements with Bragg gratings embedded in multi-core fibers, and later Miller et al. reconstructed the two-dimensional shape of the fiber based on distributed curvature sensing [41]. FBGTSs are physically robust to electromagnetic interference but might suffer cumulative errors due to poor torsion and curvature sensing at discrete points [42].

The abovementioned tracking techniques are summarized in Fig. 1. Penetrability and surgeon-friendliness are key considerations for intrabody instruments in robotic surgery. MTSs (i.e., EMTSs and PMTSs) and MITSs can penetrate human tissue and track flexible (with multiple free-form curvatures) instruments inside the human body, reducing the invasiveness of medical interventions. Compared with MITSs, MTSs are less harmful and more friendly for the patient and surgeon.

## Significance of robust tracking and navigation

Highly accurate and robust tracking and navigation systems are indispensable to robotic surgery. Deterioration in the tracking system's performance might lead to failure or termination of the procedure, which would be catastrophic for the patient. Each tracking technique discussed above has specific advantages and limitations, and integrating multiple tracking techniques is a common way to improve tracking robustness [15,43]. For example, aiming at the drawbacks of OTSs and MTSs, Vaccarella et al. [44] presented a fusion algorithm based on an unscented Kalman filter (UKF) to enhance the robustness of the tracking system in scenarios where optical markers are occluded or when electromagnetic interference is present. Nevertheless, improving the robustness of mono-modal tracking systems remains a top priority, which, in turn, will enhance the performance of multimodal tracking systems.

Among MTSs (i.e., EMTSs and PMTSs), EMTSs require a wired connection to each surgical instrument to be tracked, whereas PMTSs utilize power-free permanent magnets as the tracking targets. PMTSs show considerable potential for applications within the human body, given its wireless and cost-effective advantages.

Considering the importance of robust tracking and navigation in robotic surgery, as well as the potential benefits of PMTSs, this chapter will focus on the survey of robust permanent magnet tracking.

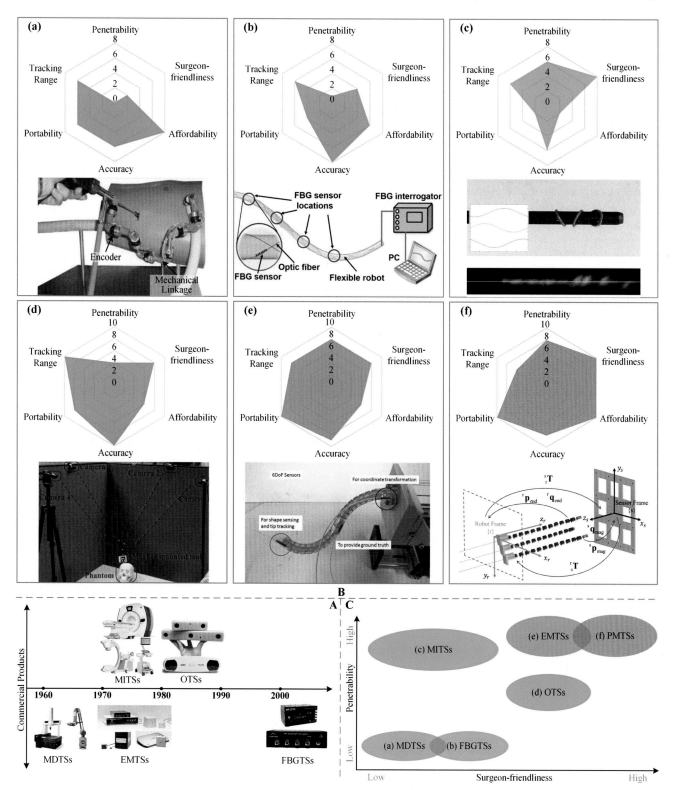

FIG. 1   Overview of various tracking techniques for robotic surgical navigation. (A) History of various tracking technologies. (B) Performance profiling. (a) Mechanical digitizer tracking systems (MDTSs) [14]; (b) Fiber Bragg grating (FBG)-based tracking systems (FBGTSs) [33]; (c) Medical image-guided tracking systems (MITSs) [34]; (d) Optical tracking systems (OTSs) [22]; (e) Electromagnetic tracking systems (EMTSs) [29]; (f) Permanent magnet tracking systems (PMTSs) [35]. (C) Comparison of penetrability and user-friendliness levels. Penetrability refers to the ability of the tracking system to track a target placed inside the human body wirelessly.

## Fundamentals of permanent magnet tracking system

### Mathematical model and tracking principle

Monitoring wireless capsule endoscopes (WCEs) inside the human gastrointestinal (GI) tract and the tips of flexible instruments in MIS are the most common applications of permanent magnet tracking in robotic surgical navigation.

The permanent magnet tracking system generally considers PMs as the tracking targets and employs a magnetometer array to observe the magnetic flux density distribution around the PMs. As illustrated in Fig. 2, a PM is embedded in the WCE or fixed at the distal end of the flexible robot.

Assuming that the sensor array consists of $N$ tri-axis magnetometers, the $l^{\text{th}}$ magnetometer is located at the position $\mathbf{p}_l = (x_l, y_l, z_l)$. The PM is placed at position $(a, b, c)$ with orientation $\mathbf{H}_0 = (m, n, p)^{\text{T}}$. Based on the magnetic dipole model, the magnetic flux density $\mathbf{B}_l$ generated by the PM at the $l^{\text{th}}$ magnetometer can be represented as

$$\mathbf{B}_l = B_{lx}\mathbf{i} + B_{ly}\mathbf{j} + B_{lz}\mathbf{k}$$
$$= \frac{\mu_r \mu_0 V M_0}{4\pi} \left( \frac{3\left(\mathbf{H}_0^{\text{T}} \cdot \mathbf{F}_l\right)\mathbf{F}_l}{R_l^5} - \frac{\mathbf{H}_0}{R_l^3} \right), \quad (l = 1, 2, \ldots, N), \tag{1}$$

where $\mu_0$ is the magnetic permeability of the vacuum and $\mu_r$ refers to the relative magnetic permeability of the medium (e.g., air); $V$ and $M_0$ represent the volume and the magnetization of the PM; $\mathbf{F}_l$ indicates the vector starting from the PM position $(a, b, c)$ to the magnetometer position $(x_l, y_l, z_l)$, and $R_l$ is the Euclidean norm of $\mathbf{F}_l$, i.e., $R_l = \|\mathbf{F}_l\|_2$.

Suppose the expected observation at the $l^{\text{th}}$ magnetometer is $\mathbf{B}_l$, while the real observation from the magnetometer is assumed $\mathbf{B}_l' = (B_{lx}', B_{ly}', B_{lz}')^{\text{T}}$. A well-established method to estimate the magnet pose is to minimize the following objective function by optimization algorithms:

$$E(\mathbf{v}) = \sum_{l=1}^{N} \left\| \mathbf{B}_l' - \mathbf{B}_l \right\|_2. \tag{2}$$

Among various optimization algorithms, the Levenberg-Marquart (LM) algorithm based on an initial guess and gradient descent is widely employed due to its outstanding execution speed and calculation accuracy.

### Analysis of robust magnetic tracking system

Apart from the magnetic fields generated by the PMs, the environment is also surrounded by geomagnetic fields. In addition, the geomagnetic and PM fields will inevitably suffer from hard and soft iron effects in the environment [17], as shown in Fig. 3A. The hard-iron effects refer to noise sources that generate magnetic fields actively. Different metallic objects generate hard-iron effects, such as motors and electrified wires. Ferromagnetic materials cause soft iron effects in different directions. The magnetic field distributions will be distorted after passing through the ferromagnetic

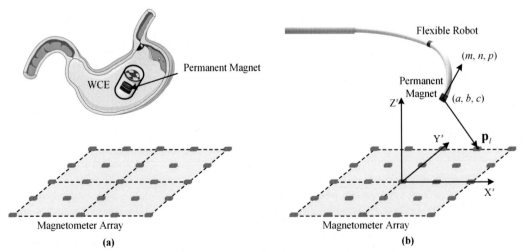

FIG. 2    Permanent magnet tracking system in medical applications. (A) Tracking the WCE pose in the GI tract; (B) Estimating the tip pose of the flexible robot.

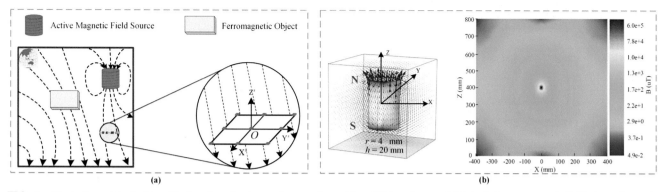

FIG. 3  Illustration of the factors affecting the robustness of magnetic tracking. (A) Interference fields include geomagnetic fields from the earth, hard-iron effect fields from active magnetic field sources, and soft-iron effect fields from ferromagnetic objects. (B) Magnetic induction intensity distribution around the PM calculated by the finite element analysis software COMSOL shows that the magnetic induction intensity decays rapidly with distance.

objects, such as reinforced concrete buildings and surgical instruments. These interference sources (geomagnetic field and iron effects) are considered noise during the magnetic tracking process.

Su et al. [45] demonstrated in their study that the performance of the PMTS deteriorated when the distance between the magnet and the planar magnetometer array was below 36 mm or above 126 mm. The degraded accuracy at the distal location is mainly due to the rapid attenuation of the magnetic field generated by PMs, as shown in Fig. 3B. In contrast, the approximate modeling error of the magnetic dipole model causes decreased accuracy at the proximal location.

Furthermore, the LM algorithm for minimizing the objective function (2) is a local optimization algorithm that demands an initial guess to start searching for the optimal solution. The LM algorithm might diverge or fall into a local optimum if the initial guess is far from the globally optimal solution. Divergence or falling into a local optimum often means tracking failure, which is fatal in robotic surgery.

## Existing methods for improving the robustness of magnetic tracking

As analyzed in Section "Analysis of robust magnetic tracking system", there are several underlying reasons to affect the robustness of magnetic tracking systems: magnetic field interference, limited tracking range, and dependence on initial guesses. The following section will discuss the associated solutions to overcome these problems.

### Antiinterference

Table 1 lists the state-of-the-art solutions for antiinterference, with most studies concentrating on the interference introduced by geomagnetic and actuating magnetic fields. These two interfering magnetic fields are naturally introduced when the magnetometer array is moved, or the EPM actuates the tracking target. However, each of these solutions has specific prerequisites for application. For example, Song et al. [48] assume that the geomagnetic vectors between adjacent magnetometers are equal for the geomagnetic field. However, this solution will be ineffective if some magnetometers are disturbed by unknown interference sources like tiny magnets or energized wires. To our knowledge, the existing literature on overcoming unknown interference sources is limited. Although Su et al. [55] have made a noteworthy contribution in this area, further research is warranted to improve the robustness of magnetic tracking systems further.

### Extending tracking range

Extending the tracking range of PMTSs is done primarily through hardware and algorithms. The relevant studies are listed in Table 2. The performance deterioration at the proximal location is primarily attributed to the inherent limitation of the magnetic dipole model, which cannot compensate for physical and geometry imperfections. Therefore, an intuitive solution is to correct the near-field model of the permanent magnet [59,60]. The magnetic field generated by the PM decays rapidly with distance. The low signal-to-noise ratio (SNR) at the distal location is the main

TABLE 1    Overview of antiinterference technology for robust magnetic tracking.

| Authors (year) | Interference sources | Approaches | Application scenarios |
|---|---|---|---|
| Dai et al. [46] | Geomagnetic field | An inertial measurement unit (IMU) was utilized to capture the tri-axial rotation of the magnetometer array. As a result, the variation of the tri-axial geomagnetic vectors was estimated and separated. | The magnetometer array was limited to small linear movements to maintain the geomagnetic vector as a constant |
| Shao et al. [47] | Geomagnetic field | Introduced a geo-noise cancellation sensor to measure the geomagnetic field. | It requires the installation of an additional sensor that is far away from the magnet and not subject to interference. |
| Song et al. [48] | Geomagnetic field | Proposed a tracking algorithm based on differential signals between adjacent magnetometers, which can reduce the disturbance from the geomagnetic field without additional sensors. | The geomagnetic vectors between adjacent magnetometers are assumed to be equal. |
| Wang et al. [49] | Geomagnetic field | Took the difference of the magnetic field values at two locations to remove the background geomagnetic noise. | It is based on the assumption that the magnetometer array is kept static throughout the sampling at the two locations. |
| Su et al. [50] | Geomagnetic field | Proposed a geomagnetic-vector-separation method based on the environmental magnetic model and optimization algorithm, where the magnet pose and geomagnetic vector are estimated simultaneously. | The proposed method assumes that the geomagnetic vectors measured by all magnetometers on the sensor array are consistent. |
| Xu et al. [51] | Actuating magnet | Eliminated the actuating magnetic field by integrating the observed magnetic fields in one period. | The tracking target and actuating magnet need to be kept rotating. |
| Shi et al. [52] | Actuating magnet | The external permanent magnet's (EPM) pose was estimated using an optical tracking system. Subsequently, the magnetic field components originating from the EPM were decoupled from the mixed field. | Rely on an OTS to separate the external actuating magnetic field from the composite magnetic field. |
| Song et al. [53] | Actuating magnet | The magnet to be tracked and the actuating magnet was estimated simultaneously using a multitarget tracking algorithm. | It increases the computational complexity and the risk of divergence. |
| Son et al. [54] | Electromagnet | Described the magnetic field of an electromagnet using the dipole model and subtracted it from the measured data. | It needs to establish the magnetic field model generated by the electromagnet precisely. |
| Su et al. [55] | Unknown interference sources like small magnets | Introduced graph optimization to solve the magnetic tracking problem and used a Huber cost function to reduce the weight of anomaly edges. | It is suitable for one or a few magnetometers subjected to interference sources. |
| Hu et al. [56] | Human body movements | Proposed a three-magnet tracking method, in which two magnets established a reference coordinate system and tracked the target relative to the reference system. | This approach results in computational complexity and deployment sophistication. |

reason for limiting the tracking range. Although Lv et al. [61] adopted tunnel magnetoresistance (TMR) magnetometers with more sensitivity and artificial neural network (ANN) to estimate the magnet pose, the positioning error at the distal location ($19.86 \pm 12.83$ mm in the tracking range between 206 and 306 mm) is still much higher than that at the proximal location ($2.59 \pm 0.86$ mm in the tracking range between 16 and 126 mm). Thus, it remains a challenge to extend the tracking range of the PMTS.

## Overcoming the dependence on initial guesses

Methodologies for overcoming the initial value dependence problem can be divided into three categories: analytical method, swarm intelligence, and deep learning, as shown in Table 3. The analytical method can obtain a closed-form analytical solution with low computational complexity. However, the analytical method might encounter a singularity problem due to measurement noises and the higher-order nonlinear relationship between the magnetic induction intensity and the PM poses [63]. This problem can be alleviated in some specific applications involving low-DoF motion [64,65].

TABLE 2 Comparison of extending tracking range for robust magnetic tracking.

| Authors (year) | Range | Approaches | Results |
|---|---|---|---|
| Hu et al. [57] | Full range | Utilized four planar arrays to form a cubic internal space with the dimensions of $0.5\,\mathrm{m} \times 0.5\,\mathrm{m} \times 0.5\,\mathrm{m}$. | The tracking system achieves an accuracy of 1.8 mm and 1.6° within the cubic space. |
| Sun et al. [58] | Proximal location | Employed an inertial measurement unit (IMU) array to construct a relationship with a single dipole landmark in the near magnetic field. | It could estimate the 3-DoF position from the single dipole landmark with an error lower than 0.5 mm. |
| Wu et al. [59] | Proximal location | Proposed a hybrid ANN (artificial neural network)-DM (dipole model) field model, in which ANN is used to characterize the near-end field, and DM is to model the far-end field. | The hybrid model's modeling error is about an order of magnitude lower than that of the magnetic dipole model. |
| Ren et al. [60] | Proximal location | Proposed a near-field correction method, i.e., dividing the field into several regions and finding a correction coefficient for each region. | The proposed method was validated on three different types of magnets, and its near-field modeling error was reduced from 3.32% to 2.8%. |
| Lv et al. [61] | Distal location | Developed a magnetometer array based on tunnel magnetoresistance (TMR) and proposed an LM-ANN-based tracking method, where ANN is used for the far-field zone. | The tracking distance of the proposed method was enhanced to 396 mm with a position error of under 25 mm. |
| Dai et al. [62] | Distal location | Presented an adaptive weighted fusion algorithm, which combines IMU and magnetic tracking data. | Due to the significant position error of IMU sensing, only the orientation accuracy at the distal location was improved. |

TABLE 3 Overview of resolving the dependency on initial guesses.

| Authors (year) | Algorithm types | Approaches | Results |
|---|---|---|---|
| Hu et al. [63] | Analytical method | Presented a linear algorithm based on algebra and matrix calculation to determine the 5-DoF magnet pose. | The linear approach has a singularity problem and lower accuracy than the nonlinear algorithm but less computational complexity. |
| Fan et al. [64] | Analytical method | Introduced an efficient linear algorithm based on an optically pumped magnetometer array. | Only the 1-DoF position of the target can be obtained from the derived cubic equation. |
| Dai et al. [65] | Analytical method | Proposed an analytical method-based magnetic tracking method. The simplified analytical expression is derived from the magnetic dipole model. | The 2-DoF position of the magnet can be calculated according to the measurement of a single tri-axis magnetometer. |
| Yang et al. [66] | Swarm intelligence | Utilized the particle swarm optimization (PSO) algorithm to estimate the 6-DoF pose of a rectangular magnet. | It took about 1 s to estimate the rectangular magnet pose. |
| Song et al. [67] | Swarm intelligence | Derived a closed-form analytical model for an annular magnet and estimated its 6-DoF pose by the PSO algorithm. | It took about 830 s to estimate the annular magnet pose. |
| Lv et al. [68] | Swarm intelligence | Proposed an improved whale optimization algorithm (WOA) based on the individual memory strategy to increase the convergence rate of multitarget tracking. | With the improvement, the localization success rate of WOA increased from 83.82% to 98.28%. |
| Sebkhi et al. [69] | Deep learning | Developed a data collection setup and trained a 5-layer fully connected network model to regress the magnet pose. | They spent a week gathering a dataset composed of ~1.7 million samples. |
| Fu et al. [70] | Deep learning | Divided the working space into 27 subregions and trained a classification model with 66,000 samples. | This classification model outputs discrete results and only can be used to provide initial guesses to the LM algorithm. |
| Qin et al. [71] | Deep learning | Trained a regression model with 20,691 samples to provide initial guesses for the LM algorithm. | The localization results from the regression model need to be further fine-tuned with the LM algorithm. |
| Su et al. [72] | Deep learning | Proposed a pose estimation network (called AMagPoseNet) based on dual-domain few-shot learning from a prior mathematical model. | The AMagPoseNet outperforms the optimization-based method regarding accuracy, robustness, and computational latency. |

Swarm intelligence algorithms use randomly distributed populations to explore the optimal solution according to a specific search strategy. Due to the stochastic nature of these algorithms, it is challenging to ensure the attainment of the globally optimal solution with a limited number of populations (for real-time considerations). In addition, new swarm intelligent algorithms, such as white shark optimizer (WSO) [73] and snake optimizer (SO) [74], are continually emerging and can also be leveraged to estimate initial guesses for magnetic tracking problems.

Deep neural networks possess a powerful nonlinear fitting capability. By training an end-to-end neural network with high-precision localization, it is possible to avoid the problem that the LM localization algorithm relies on initial guesses. However, the pose regression problem requires large labeled training sets [69]. If the localization model is trained using limited training data, it is typically only capable of providing initial guesses to the LM algorithm due to its low localization accuracy [70,71]. Thus, the primary issue in learning-based magnetic tracking is how to train a high-precision localization model with a few real-world samples.

## Discussion and future research trends

### Further improving the robustness of mono-modal PMTS

Although many studies have been conducted, much work remains to be done to achieve satisfactory robustness and accuracy before PMTS is applied in clinical applications. Besides the potential interferences from magnetic fields generated by small magnets or energized wires, unknown interference sources, such as sensor damage or saturation, can lead to tracking failures. Another significant shortcoming of PMTS is the limited tracking range caused by the rapid decay of the magnetic field. Adopting more sensitive magnetometers, such as fluxgate magnetometers [75] and optically pumped magnetometers [76], is a straightforward and practical approach, but their higher cost should also be taken into account.

Recently, deep learning has demonstrated its benefits in various fields [77,78], but its application to magnetic tracking is still limited. Future work can leverage deep learning techniques to eliminate the impacts of unknown interference sources and extend the tracking range. Researchers should also consider reducing the dependence on a large amount of training data, which is crucial for learning-based magnetic tracking. By addressing these challenges and limitations, PMTS can significantly improve the accuracy and robustness of surgical instrument tracking, leading to better clinical outcomes.

### Tightly coupled fusion of multimodal tracking systems

Each tracking technique has its drawbacks, such as the line-of-sight constraint and occlusion problem for OTS, the sensitivity of MTS to ferromagnetic objects, and the low SNR for ultrasound imaging. Combining multiple tracking techniques is a viable solution to compensate for the shortcomings of mono-modal tracking systems. Combinations of various tracking techniques have been employed to track surgical instruments during the pre-, intra-, and postsurgery. For example, Bharat et al. [79] utilized the EMTS to map the pose and shape of the catheter to the ultrasound images during ultrasound-guided prostate brachytherapy.

However, these combinations have been performed in a loosely coupled fusion manner, i.e., the output data of one modal tracking system is simply fused with the output data of another modal tracking system for complementary advantages. To achieve an integrated tracking system with high accuracy and reliability, a tightly coupled fusion of different tracking modalities from the basic system architecture and raw sensor data level will be a promising future research trend.

### Morphological perception

Typically, multiple EM sensors or FBG sensors are mounted at discrete joints to enable shape sensing of a continuum surgical robot. Through curve fitting techniques, such as the quadratic Bessel curve [29], the position and orientation information at these specific joints can be used to reconstruct the shape of the continuum surgical robot. Similarly, for PMTSs, shape sensing can be achieved by replacing the EM or FBG sensors mounted at the joints with permanent magnets.

Besides the widely used continuum surgical robots, medical instruments with innovative forms and technologies are emerging. For instance, deployable origami-based medical devices that start in a compact form, pass through the body to previously unreachable regions, and then unfold into their functional form when they arrive at the destinations. Such devices have the potential to improve the performance of drug delivery, encapsulation and micro-surgery, and cardiac catheterization [80]. Thus, exploring how to sense their position and morphology using magnetic tracking technology is a worthy research topic in the future. By leveraging magnetic tracking technology, researchers can further develop and optimize these novel medical devices, contributing to medical science and practice advancement.

## Novel application research

Many researchers have applied magnetic tracking technology to various scenarios beyond surgical instruments' pose tracking and shape sensing. Taylor et al. [81] demonstrate the real-time monitoring of muscle motion for intuitive control over prostheses or wearable robots. Dai et al. [82] employed three implanted permanent magnets to enable real-time dynamic target tracking of lung tumors, thereby improving the efficacy of robotic radiotherapy. The ability of magnetic fields generated by PMs to penetrate human tissue noninvasively opens up more possibilities for medical applications in the future.

## Conclusion

As surgical robots become increasingly involved in surgeries, the safe cooperation of multiple surgical robots in the surgical environment poses a significant challenge to their tracking and navigation systems. In this chapter, we first review the common tracking techniques for robotic surgery. Since robustness is a crucial consideration for the tracking system, we further summarize and analyze the robust tracking of PMTS from antiinterference, tracking range extension, and overcoming the dependence on initial guesses. Although significant breakthroughs have been made, some problems or practical limitations remain unresolved. Therefore, we discuss future research trends and potential schemes to improve tracking robustness. Overall, this paper provides a comprehensive review and forward-looking analysis of PMTS tracking for surgical robots.

## Future scenario

Magnetic tracking technology has shown great potential in surgical navigation, but there is still room for improvement in terms of robustness and accuracy. In the future, advancements in magnetic sensor technology could lead to more advanced and sensitive magnetometers that can detect smaller magnetic fields and provide more accurate tracking information. Additionally, new tracking principles could be developed that utilize specific magnetic fields generated in space to improve tracking accuracy and range.

Another potential future scenario for magnetic tracking technology is the development of novel tracking methods that combine magnetic tracking with other sensing modalities, such as optical or acoustic sensing. This could lead to more robust and accurate tracking systems that are less susceptible to interference and can provide more comprehensive tracking information. Additionally, the development of new materials and fabrication techniques could enable the creation of more compact and lightweight magnetic tracking systems that are easier to integrate into surgical instruments and devices.

## Key points

- Provide a comprehensive overview of the current state of tracking technologies in robotic surgery and highlight the importance of robust tracking and navigation for successful surgical outcomes.
- Magnetic tracking systems are widely used in surgical navigation, and the chapter explains the mathematical model and tracking principle behind their operation.
- Survey existing methods for improving the robustness of magnetic tracking, including antiinterference, extending tracking range, and overcoming the dependence on initial guesses.
- Discuss future research trends to improve tracking robustness, such as further improving the robustness of mono-modal PMTS, tightly coupled fusion of multimodal tracking systems, morphological perception for emerging medical devices, and novel application research.

## Acknowledgments

This work was supported in part by the Hong Kong Research Grants Council through the Collaborative Research Fund (CRF C4063-18G, CRF C4026-21G), the General Research Fund (GRF 14216022, GRF 14211420, GRF 14203323), and the Research Impact Fund (RIF R4020-22), in part by the Chinese University of Hong Kong Mainland Postdoctoral Sponsorship Program, in part by the Shenzhen-Hong Kong-Macau Technology Research Program (Type C) under Grant 202108233000303, in part by the NSFC/RGC Joint Research Scheme under Grant N_CUHK420/22, and in part by the key project of the Regional Joint Fund Project of the Basic and Applied Research Fund of Guangdong Province under Grant 2021B1515120035 (B.02.21.00101).

## References

[1] Schreuder H, Verheijen R. Robotic surgery. BJOG 2009;116(2):198–213.

[2] Huang M, Tetreault TA, Vaishnav A, York PJ, Staub BN. The current state of navigation in robotic spine surgery. Ann Transl Med 2021;9(1).

[3] Liebergall M, Ben-David D, Weil Y, Peyser A, Mosheiff R. Computerized navigation for the internal fixation of femoral neck fractures. JBJS 2006;88(8):1748–54.

[4] Archip N, et al. Non-rigid alignment of pre-operative MRI, fMRI, and DT-MRI with intra-operative MRI for enhanced visualization and navigation in image-guided neurosurgery. Neuroimage 2007;35(2):609–24. https://doi.org/10.1016/j.neuroimage.2006.11.060.

[5] Razavi R, et al. Cardiac catheterisation guided by MRI in children and adults with congenital heart disease. Lancet 2003;362(9399):1877–82.

[6] Xu L, et al. Information loss challenges in surgical navigation systems: from information fusion to AI-based approaches. Inf Fusion 2022.

[7] Wang TY, et al. Robotic navigation in spine surgery: where are we now and where are we going? J Clin Neurosci 2021;94:298–304.

[8] Mezger U, Jendrewski C, Bartels M. Navigation in surgery. Langenbecks Arch Surg 2013;398(4):501–14. https://doi.org/10.1007/s00423-013-1059-4.

[9] Ren H, et al. Computer-assisted transoral surgery with flexible robotics and navigation technologies: a review of recent progress and research challenges. Crit Rev Biomed Eng 2013;41(4–5).

[10] Geist E, Shimada K. Position error reduction in a mechanical tracking linkage for arthroscopic hip surgery. Int J Comput Assist Radiol Surg 2011;6(5):693–8. https://doi.org/10.1007/s11548-011-0555-7.

[11] Li C, King NKK, Ren H. A skull-mounted robot with a compact and lightweight parallel mechanism for positioning in minimally invasive neurosurgery. Ann Biomed Eng 2018;46(10):1465–78. https://doi.org/10.1007/s10439-018-2037-3.

[12] Li Q, et al. Effect of optical digitizer selection on the application accuracy of a surgical localization system—a quantitative comparison between the OPTOTRAK and flashpoint tracking systems. Comput Aided Surg 1999;4(6):314–21.

[13] The History of Coordinate Measuring Machines. https://www.cmmxyz.com/blog/the-history-of-coordinate-measuring-machines-cmmxyz/; 2020 (accessed).

[14] Monahan E, Shimada K. Computer-aided navigation for arthroscopic hip surgery using encoder linkages for position tracking. Int J Med Robot 2006;2(3):271–8.

[15] Wang J, Song S, Ren H, Lim CM, Meng MQ-H. Surgical instrument tracking by multiple monocular modules and a sensor fusion approach. IEEE Trans Autom Sci Eng 2018;16(2):629–39.

[16] Sturman DJ. A brief history of motion capture for computer character animation. *SIGGRAPH94, Course9*; 1994.

[17] Maxwell DR. Graphical marionette: a modern-day Pinocchio. Massachusetts Institute of Technology; 1983.

[18] NDI History. https://www.ndigital.com/company/ndi-history/ (accessed).

[19] Stefanelli LV, DeGroot BS, Lipton DI, Mandelaris GA. Accuracy of a dynamic dental implant navigation system in a private practice. Int J Oral Maxillofac Implants 2019;34(1).

[20] Cai K, Yang R, Lin Q, Wang Z. Tracking multiple surgical instruments in a near-infrared optical system. Comput Assist Surg 2016;21(1):46–55. https://doi.org/10.1080/24699322.2016.1184312.

[21] Decker RS, Shademan A, Opfermann JD, Leonard S, Kim PC, Krieger A. Biocompatible near-infrared three-dimensional tracking system. IEEE Trans Biomed Eng 2017;64(3):549–56.

[22] Wang J, Meng MQ-H, Ren H. Towards occlusion-free surgical instrument tracking: a modular monocular approach and an agile calibration method. IEEE Trans Autom Sci Eng 2015;12(2):588–95.

[23] Paperno E, Sasada I, Leonovich E. A new method for magnetic position and orientation tracking. IEEE Trans Magn 2001;37(4):1938–40. https://doi.org/10.1109/20.951014.

[24] Schlageter V, Besse PA, Popovic RS, Kucera P. Tracking system with five degrees of freedom using a 2D-array of hall sensors and a permanent magnet. Sens Actuators A: Phys 2001;92:37–42.

[25] Guo X, Yan G, He W. A novel method of three-dimensional localization based on a neural network algorithm. J Med Eng Technol 2009;33 (3):192–8.

[26] Nara T, Suzuki S, Ando S. A closed-form formula for magnetic dipole localization by measurement of its magnetic field and spatial gradients. IEEE Trans Magn 2006;42(10):3291–3.

[27] Polhemus, https://polhemus.com/company/history/. [accessed].

[28] Franz AM, Haidegger T, Birkfellner W, Cleary K, Peters TM, Maier-Hein L. Electromagnetic tracking in medicine—a review of technology, validation, and applications. IEEE Trans Med Imaging 2014;33(8):1702–25.

[29] Song S, Li Z, Meng QH, Yu H, Ren H. Real-time shape estimation for wire-driven flexible robots with multiple bending sections based on quadratic Bézier curves. IEEE Sens J 2015;15(11):6326–34.

[30] Song S, Li Z, Yu H, Ren H. Electromagnetic positioning for tip tracking and shape sensing of flexible robots. IEEE Sens J 2015;15(8):4565–75. https://doi.org/10.1109/jsen.2015.2424228.

[31] Bianchi F, et al. Localization strategies for robotic endoscopic capsules: a review. Expert Rev Med Devices 2019;16(5):381–403.

[32] Mateen H, Basar R, Ahmed AU, Ahmad MY. Localization of wireless capsule endoscope: a systematic review. IEEE Sens J 2017;17(5):1197–206. https://doi.org/10.1109/jsen.2016.2645945.

[33] Wei J, Wang S, Li J, Zuo S. Novel integrated helical design of single optic fiber for shape sensing of flexible robot. IEEE Sens J 2017;17 (20):6627–36.

[34] Stoll J, Ren H, Dupont PE. Passive markers for tracking surgical instruments in real-time 3-D ultrasound imaging. IEEE Trans Med Imaging 2011;31(3):563–75.

[35] Song S, Ge H, Wang J, Meng MQ-H. Real-time multi-object magnetic tracking for multi-arm continuum robots. IEEE Trans Instrum Meas 2021;70:1–9.

[36] Menten MJ, Wetscherek A, Fast MF. MRI-guided lung SBRT: present and future developments. Phys Med 2017;44:139–49.

[37] Schullian P, Widmann G, Lang TB, Knoflach M, Bale R. Accuracy and diagnostic yield of CT-guided stereotactic liver biopsy of primary and secondary liver tumors. Comput Aided Surg 2011;16(4):181–7. https://doi.org/10.3109/10929088.2011.578367.

[38] Bradley WG. History of medical imaging. Proc Am Philos Soc 2008;152(3):349–61.

[39] Li T, Qiu L, Ren H. Distributed curvature sensing and shape reconstruction for soft manipulators with irregular cross sections based on parallel dual-FBG arrays. IEEE/ASME Trans Mechatron 2019;25(1):406–17.

[40] Shi C, et al. Shape sensing techniques for continuum robots in minimally invasive surgery: a survey. IEEE Trans Biomed Eng 2016;64(8):1665–78.

[41] Floris I, Adam JM, Calderón PA, Sales S. Fiber optic shape sensors: a comprehensive review. Opt Lasers Eng 2021;139, 106508.

[42] Lu Y, Lu B, Li B, Guo H, Liu Y-H. Robust three-dimensional shape sensing for flexible endoscopic surgery using multi-Core FBG sensors. IEEE Robot Autom Lett 2021;6(3):4835–42. https://doi.org/10.1109/lra.2021.3067279.

[43] Lee Y, Do W, Yoon H, Heo J, Lee W, Lee D. Visual-inertial hand motion tracking with robustness against occlusion, interference, and contact. Sci Robotics 2021;6(58). eabe1315.

[44] Vaccarella A, De Momi E, Enquobahrie A, Ferrigno G. Unscented Kalman filter based sensor fusion for robust optical and electromagnetic tracking in surgical navigation. IEEE Trans Instrum Meas 2013;62(7):2067–81. https://doi.org/10.1109/tim.2013.2248304.

[45] Su S, et al. Investigation of the relationship between tracking accuracy and tracking distance of a novel magnetic tracking system. IEEE Sens J 2017;17(15):4928–37.

[46] Dai H, Hu C, Su S, Lin M, Song S. Geomagnetic compensation for the rotating of magnetometer Array during magnetic tracking. IEEE Trans Instrum Meas 2018;68(9):3379–86. https://doi.org/10.1109/TIM.2018.2875965.

[47] Shao G, Tang Y, Tang L, Dai Q, Guo Y-X. A novel passive magnetic localization wearable system for wireless capsule endoscopy. IEEE Sens J 2019;19(9):3462–72. https://doi.org/10.1109/jsen.2019.2894386.

[48] Song S, Wang S, Yuan S, Wang J, Liu W, Meng MQ-H. Magnetic tracking of wireless capsule endoscope in mobile setup based on differential signals. IEEE Trans Instrum Meas 2021;70:1–8.

[49] Wang M, Song S, Liu J, Meng MQ-H. Multipoint simultaneous tracking of wireless capsule endoscope using magnetic sensor array. IEEE Trans Instrum Meas 2021;70:1–10.

[50] Su S, Dai H, Zhang Y, Yuan S, Song S, Ren H. Magnetic tracking with real-time geomagnetic vector separation for robotic Dockable charging. IEEE Trans Intell Transp Syst 2023.

[51] Xu Y, Li K, Zhao Z, Meng MQ-H. A novel system for closed-loop simultaneous magnetic actuation and localization of wce based on external sensors and rotating actuation. IEEE Trans Autom Sci Eng 2021;18(4):1640–52.

[52] Shi Q, Liu T, Song S, Wang J, Meng MQ-H. An optically aided magnetic tracking approach for magnetically actuated capsule robot. IEEE Trans Instrum Meas 2021;70:1–9.

[53] Song S, Qiu X, Wang J, Meng MQH. Real-time tracking and navigation for magnetically manipulated untethered robot. IEEE Access 2016;4:7104–10. https://doi.org/10.1109/access.2016.2618949.

[54] Son D, Yim S, Sitti M. A 5-D localization method for a magnetically manipulated untethered robot using a 2-D Array of hall-effect sensors. IEEE/ASME Trans Mechatron 2015;21(2):708–16.

[55] Su S, Dai H, Cheng S, Lin P, Hu C, Lv B. A robust magnetic tracking approach based on graph optimization. IEEE Trans Instrum Meas 2020;69 (10):7933–40. https://doi.org/10.1109/TIM.2020.2986843.

[56] Hu C, Ren Y, You X, Yang W. Locating intra-body capsule object by three magnet sensing system. IEEE Sens J 2016;16(13):5167–76.

[57] Hu C, Li M, Song S, Yang WA, Zhang R, Meng QH. A cubic 3-Axis magnetic sensor Array for wirelessly tracking magnet position and orientation. IEEE Sens J 2010;10(5):903–13.

[58] Sun KC, Lo SW, Hu JS. A single dipole-based localization method in near magnetic field using IMU array. In: IEEE workshop on advanced robotics and ITS social impacts. IEEE; 2016. p. 152–7.

[59] F.Y. Wu, S. Foong, and Z. Sun, "A hybrid field model for enhanced magnetic localization and position control," IEEE/ASME Trans Mechatron, vol. 20, no. 3, pp. 1278–1287, 2015/06 2015, https://doi.org/10.1109/tmech.2014.2341644.

[60] Ren Y, Chao H, Sheng X, Feng Z. Magnetic dipole model in the near-field. In: IEEE International Conference on Information & Automation; 2015.

[61] Lv B, Chen Y, Dai H, Su S, Lin M. PKBPNN-based tracking range extending approach for TMR magnetic tracking system. IEEE Access 2019;7:63123–32.

[62] Dai H, Song S, Hu C, Sun B, Lin Z. A novel 6D tracking method by fusion of 5D magnetic tracking and 3D inertial sensing. IEEE Sens J 2018;18 (23):9640–8.

[63] Hu C, Meng QH, Mandal M. A linear algorithm for tracing magnet position and orientation by using three-Axis magnetic sensors. IEEE Trans Magn 2007;43(12):4096–101.

[64] Fan L, et al. A fast linear algorithm for magnetic dipole localization using total magnetic field gradient. IEEE Sens J 2017;18(3):1032–8.

[65] Dai H, Guo P, Su S, Song S, Zhao S, Cheng S. A simplified magnetic positioning approach based on analytical method and data fusion for automated guided vehicles. IEEE/ASME Trans Mechatron 2021.

[66] Yang WA, Hu C, Meng QH, Song S, Dai H. A six-dimensional magnetic localization algorithm for a rectangular magnet objective based on a particle swarm optimizer. IEEE Trans Magn 2009;45(8):3092–9.

[67] Song S, et al. 6D magnetic localization and orientation method for an annular magnet based on a closed-form analytical model. IEEE Trans Magn 2014;50(9):1–11.

[68] Lv B, Qin Y, Dai H, Su S. Improving localization success rate of three magnetic targets using individual memory-based WO-LM algorithm. IEEE Sens J 2021.

I. Introduction, technology basics, evolution and next frontiers of surgical robots

[69] Sebkhi N, Sahadat N, Hersek S, Bhavsar A. A deep neural network-based permanent magnet localization for tongue tracking. IEEE Sens J 2019;19(20):9324–31.

[70] Fu Y, Guo Y-X. Guideline on initial point finding and search bounds setting for biomedical magnetic localization. IEEE Sens J 2022.

[71] Qin Y, Lv B, Dai H, Han J. An hFFNN-LM based real-time and high precision magnet localization method. IEEE Trans Instrum Meas 2022;71:1–9. https://doi.org/10.1109/tim.2022.3165806.

[72] Su S, Yuan S, Xu M, Gao H, Yang X, Ren H. AMagPoseNet: real-time 6-DoF magnet pose estimation by dual-domain few-shot learning from prior model. IEEE Trans Industr Inform 2023.

[73] Braik M, Hammouri A, Atwan J, Al-Betar MA, Awadallah MA. White shark optimizer: a novel bio-inspired meta-heuristic algorithm for global optimization problems. Knowl-Based Sys 2022;243, 108457. https://doi.org/10.1016/j.knosys.2022.108457.

[74] Hashim FA, Hussien AG. Snake optimizer: a novel meta-heuristic optimization algorithm. Knowl-Based Sys 2022;242, 108320.

[75] Wang Z, et al. Highly-sensitive MEMS Micro-fluxgate magnetometer. IEEE Electron Device Lett 2022;43(8):1327–30. https://doi.org/10.1109/led.2022.3187447.

[76] Borna A, Carter TR, DeRego P, James CD, Schwindt PD. Magnetic source imaging using a pulsed optically pumped magnetometer array. IEEE Trans Instrum Meas 2018;68(2):493–501.

[77] Frey M, Nau M, Doeller CF. Magnetic resonance-based eye tracking using deep neural networks. Nat Neurosci 2021;24(12):1772–9. https://doi.org/10.1038/s41593-021-00947-w.

[78] Jumper J, et al. "highly accurate protein structure prediction with AlphaFold," (in eng). Nature 2021;596(7873):583–9. https://doi.org/10.1038/s41586-021-03819-2.

[79] Bharat S, et al. Electromagnetic tracking for catheter reconstruction in ultrasound-guided high-dose-rate brachytherapy of the prostate. Brachytherapy 2014;13(6):640–50. https://doi.org/10.1016/j.brachy.2014.05.012.

[80] Johnson M, et al. "fabricating biomedical origami: a state-of-the-art review," (in eng). Int J Comput Assist Radiol Surg 2017;12(11):2023–32. https://doi.org/10.1007/s11548-017-1545-1.

[81] Taylor CR, Srinivasan SS, Yeon SH, O'Donnell M, Roberts T, Herr HM. Magnetomicrometry. Sci Robotics 2021;6(57). eabg0656.

[82] Dai H, Dong L, Lv B, Chen Y, Su S. Feasibility study of permanent magnet-based tumor tracking technique for precise lung Cancer radiotherapy. IEEE Trans Instrum Meas 2020;70:1–10.

# 3D reconstruction of deformable tissues in robotic surgery

*Mengya Xu*[a,b,c], *Tiebing Tang*[a], *Ziqi Guo*[a], *An Wang*[a], *Beilei Cui*[a], *Long Bai*[a], *and Hongliang Ren*[a,d]

[a]Department of Electronic Engineering, The Chinese University of Hong Kong, Sha Tin, Hong Kong [b]Department of Biomedical Engineering, National University of Singapore (NUS), Singapore, Singapore [c]NUS (Suzhou) Research Institute, Suzhou, China [d]Shun Hing Institute of Advanced Engineering (SHIAE), The Chinese University of Hong Kong, Sha Tin, Hong Kong

## Introduction

Three-dimensional (3D) deformable and soft tissue reconstruction in robotic surgery is critical for several applications, including intraoperative navigation and image-guided robotic surgery automation. This study aims to systematically review the application of deformable tissue simulation, neural radiance fields (NeRF), and depth estimation in the 3D reconstruction of deformable tissue in robotic surgery.

### Deformable tissue simulation

Deformable tissues are models that are used in computer graphics to create things with deformable behavior. These have interaction and motion realism, two contradictory qualities. The many deformable models created to date have mostly emphasized one at the expense of the other. For the surgical simulation, to create a realistic environment, an accurate description model is needed for the interaction between surgical instruments and human organs. The complexity of deformation is in using real tissue data from patients, creating patient-specific models, and simulating real-time responses from the organ. The deformation of objects is mainly divided into two approaches such as the heuristic approach and continuum mechanical approach. In terms of the modeling deformable tissue, the physical properties of the organs, such as elasticity, must be considered. The physics model could be classified into two bases such as linear elasticity and nonlinear elasticity. In robotics and haptics, linear elasticity is the most used method for simulating soft tissues. Most soft tissues are naturally viscoelastic, meaning they respond to changes in both position and velocity. The characteristics of viscous fluids and elastic solids are present in viscoelastic materials.

### NeRF-based reconstruction

In facial structure reconstruction, a new technology called NeRF [1] could significantly impact related works in the coming years. The primary function of the algorithm is to take multiple photographs of a static scene from various perspectives and enable the rendering of additional views of that scene, specifically for view synthesis.

Unlike most of the previous works, NeRF chooses to reconstruct static 3D scenes based on a technology called neural implicit representation instead of explicit representation. In this way, they have made impressive progress on the high-resolution new view generation with higher efficiency. However, though implicit representation rendering is not a new concept, previous works related to view synthesis adopt it and suffer from the quality of these views: unacceptable distortion often occurs when rendering new views. However, NeRF somehow took a big step forward in solving this problem. By applying classic volume rendering to implicit representation scenes, they eliminate the drawback of poor figure quality of new view generation. All the previously mentioned advantages seem to tell us NeRF has broad application prospects in surgical soft tissue reconstruction.

## Depth estimation-based reconstruction

Depth and pose estimation is crucial in 3D reconstruction for applications like virtual reality in robotic surgery. Estimating the depth of a scene and the pose of the camera that captured it from a set of images is currently a well-established technique. However, many successful models depend on specific prerequisites, such as scenes with sufficient rigidity and texture, stable lighting conditions, and significant camera movements [2]. When dealing with medical images and precisely endoscopic procedures, these tasks become more challenging due to the lack of textures and unstable illumination. Nevertheless, creating 3D models within the human body and accurately localizing the camera is extremely valuable for applications like virtual augmentations or virtual reality in surgical procedures.

## Deformable tissue simulation

### Heuristic approach

Deformable spline: The first deformable models to be created and used in the surgical simulation were deformable splines, often referred to as active contours. Deformable splines then define potential energy proportional to the magnitude of elastic deformation using the fundamental theorems of differential geometry on the equivalence of forms. The Lagrange technique minimizes this energy for the displacements imposed at particular control points. Deformable splines are generally more complicated and computationally expensive than models of the spring-mass kind without really providing more realism.

Spring-mass model: Meshes of discrete mass locations and springs are the foundation for spring-mass models. They specify a set of mass-free springs that are welded in places or nodes to which a discrete mass is assigned and that are dispersed around the surface of a modeled object (Fig. 1).

### Continuum approach

Two basic models are currently used for the continuum approach of deformable tissue: finite-element method (FEM) and tensor precomputation.

*The finite-element method (FEM):* FEM was created to roughly resolve differential equations that were specified for a particular domain and with specific related boundary conditions. A limited number of subdomains or elements make up the discretized form of the item. The precomputation can take several hours, but it is quick enough to enable real-time simulation of a reasonable scale. Regarding physical realism, the FFE closely adheres to the initial theories of the mechanical behavior upon which they are founded. To get a linear connection, these presumptions must be entirely restricted. As a result, neither significant or quick deformations nor the partly viscous behavior of tissues can be accurately modeled.

*Tensor precomputation:* It is based on the idea of superposition. Additionally, it is presumed that the reaction to applying several loads at once will be identical to the total number of reactions to the corresponding single loads. The limitation to surface nodes is simple and does not call for additional compression when using the tensor approach. Furthermore, when there is a changeable set of contact nodes, it is ultimately feasible to dynamically modify the type of each node. This can happen when contact is lost, or new instruments come into contact with a virtual organ and slide down the surface. As long as there aren't too many contacts, numerous contacts can be readily managed. However, its applicability to large or sudden deformations is severely constrained because it is predicated on the linear elasticity theory. Topological changes, such as those that result from cutting, can still not be represented in real time since doing so would require recompiling all of the tensors, which is an expensive process.

*Linear and nonlinear elasticity:* To precisely specify the connection between stress and strain, constitutive equations based on continuum mechanics are used to describe the physical laws. The stiffness matrix may be precomputed, inverted, and stored, which makes linear elasticity a computationally efficient method for studying soft tissue deformation. However, a technique must employ completely nonlinear formulations to accommodate substantial deformation of biological tissues to be effective in clinical simulation, where better physical accuracy is required, assuming the additional computing cost is acceptable.

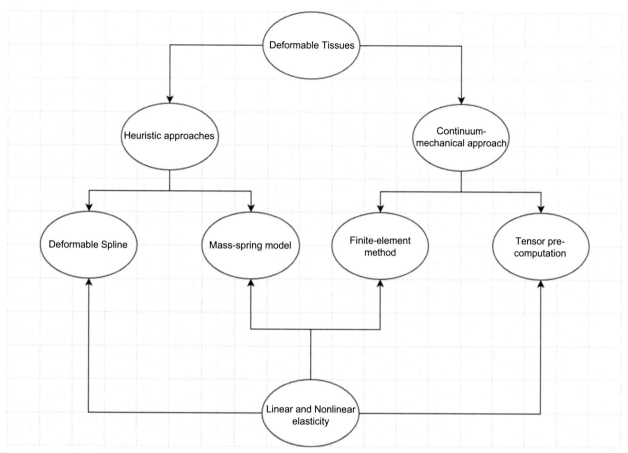

FIG. 1   Introduction to the simulation approaches of deformed tissue.

*Linear elasticity:* Most commonly utilized methods employ the principle of linear elasticity to elucidate the distortion of supple tissues in response to the need for real-time processing in surgical simulation. Adopting linear elasticity as the fundamental model entails certain assumptions that limit the accuracy of describing the actual material, albeit reducing computational time. Additionally, this approach is restricted to modest deformations and cannot cope with substantial soft tissue distortion. The embodiment of significant distortions, including global rotations, is often simulated using the framework of linear elasticity, resulting in an unrealistic augmentation of the model's volume.

*Nonlinear elasticity:* The deformable models must be consistent with the finite deformation theory to manage the massive deformation of soft tissues, which involves both geometric and material nonlinearities.

## Neural rendering for reconstructing natural scenes

Two critical aspects of the NeRF [1] approach are implicit representation and volume rendering. Fig. 2 showcases the quantitative improvement of NeRF in generating multiview 3D scenes compared to older technologies, using metrics such as PSNR, SSIM, and LPIPS.

*Implicit representation:* A continuous 5D function is used to represent the model in NeRF construction. The function takes five variables as input: the 3D position information $(x, y, z)$ of dots in the virtual space of the target model and two variables indicating the viewpoint angle. The function outputs the color and density of these dots. To achieve this representation, a deep fully connected neural network was constructed and optimized without employing convolutional layers.

*Volume rendering:* In essence, volume rendering is a set of techniques used to display 2D projections based on a 3D discretely sampled dataset. In the case of NeRF, volume rendering is employed through a method called "ray marching."

A camera ray is a hypothetical ray that traverses our model from a specific starting point and angle, and it is used to obtain the color for each pixel when synthesizing a new view of the target. We obtain the aforementioned dots with their properties by sampling along these camera rays. During the rendering process, for each pixel, we can trace a

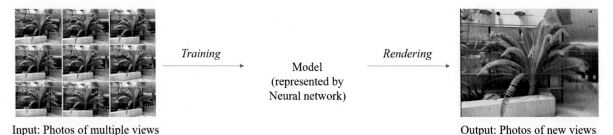

FIG. 2    Working process of NeRF [1]. Inputting photos of different views to train a neural network containing model, and then generating new views by rendering.

camera ray based on the desired view. By calculating the color and density using the position of each dot and the angle of incidence for the camera ray, we can overlay this information to obtain RGB values for the pixel. Combining all these pixels forms a complete image of the new view. Another noteworthy improvement introduced by NeRF is the implementation of a new positional encoding technique. This encoding maps each 5D input to a higher dimensional space, allowing for a more effective transformation of the 5D function and producing highly promising results. This enhancement enables NeRF to handle scene contents with higher frequency.

The original version of NeRF [1] has several drawbacks despite its impressive performance. Some follow-up work has significantly addressed these drawbacks and improved NeRF. One major drawback is the slow rendering speed of NeRF. Training and rendering a video using the original NeRF can take hours, even for low-resolution and low-frame-rate videos. This is primarily due to the complex rendering process that involves calculating each dot of the model. Many works have focused on addressing this issue by incorporating NeRF with frameworks like PyTorch or TensorFlow.

In this field, Nvidia's Instant NeRF [3] is currently considered the best choice. As the most competitive AI hardware company until now, Nvidia claims that it redesigned the structure of MLP, and it is possible to reconstruct a scene within a few seconds without much loss of model fineness. Second, the universality of NeRF is also a problem. For each dataset, NeRF needs to do training separately. Some current works get hands-on solving this problem by combining the convolution algorithm and source code of NeRF, like GRF [6] and pixelNeRF [15], but all the outcomes are decent but with limitations. The third problem is that NeRF focuses on reconstructing a static scene, which is unrealistic in most scenes. Due to this reason, some researchers began to consider the probability of adding time as the sixth dimension to achieve the goal of executing NeRF-related operations in videos. Some of these works have made solid progress in this field, like deformable neural radiance fields [7] and NeRF for dynamic scenes (D-NeRF) [8]. One of these works, EndoNeRF [5], focuses on minimally invasive surgery like laparoscopy in surgical modeling. They optimized the scenery for a single-viewed camera and made the deformation of the body tissue possible based on the work of D-NeRF. Although it may be challenging to implement in real surgical procedures, the concept of NeRF retains its significance in envisioning the future of surgery. Additionally, various advancements have been made in other domains, such as GRAFFE [9] and Mip-NeRF [10], serving as constant reminders of the remarkable ingenuity displayed by researchers in the field of NeRF.

## Neural rendering for reconstructing deformable tissues in robotic surgery

In robotic surgery, EndoNeRF [5] introduced the use of NeRF for deformable tissue reconstruction, setting a precedent for integrating NeRF into robotic surgery. Table 1 provides an overview of the representative works. The results in Table 1 display the training time of optimization and the peak signal-to-noise ratio (PSNR) value, which indicates the image quality.

Since the initial implementation of deformable tissue reconstruction, there has been a significant increase in the number of studies surpassing the performance of EndoNeRF [5]. Two notable works, namely, EndoSurf [11] and Neural LerPlane [12], specifically address two concerns associated with EndoNeRF. EndoSurf prioritizes tissue restoration with a smoother and more precise surface, while neural LerPlane greatly enhances the reconstruction speed.

### Stereo 3D reconstruction of deformable tissues

Based on D-NeRF [8] that reconstructs moving objects from multiple views, EndoNeRF [5] was presented to solve several challenges, including reconstruction of nonrigid deformable tissues, limited 3D clues resulting from the endoscopic camera, and partial occlusion by surgical instruments. It first increases the sampling probability of a ray with a

TABLE 1 The application of NeRF in robotic surgery.

| References | Aim | Method | Novelty | Results |
|---|---|---|---|---|
| EndoNeRF (2022) [5] | Deformable tissue | NeRF + time | Single-view deformation reconstruction | Time: 9h, PSNR: 26.311 |
| EndoSurf (2023) [11] | Smooth surface | EndoNeRF + SDF | SDF field generates smoother surface | Time: 9h, PSNR: 26.882 |
| LerPlane (2023) [12] | Fast reconstruction | Linear interpolation | Decomposes 4D space to reduce optimization time | Time: 10min, PSNR: 35.504 |
| LightNeuS (2023) [13] | Colon reconstruction | NeuS + illumination modeling | Model illumination decline on NeuS | Reconstruction error RMSE: 3.49, whereas NeuS: 6.40 |
| REIM (2023) [14] | Colon reconstruction | Illumination modeling | Locate light source in modeling | PSNR: 31.662 |

high occluded frequency during the whole video, and then, samples point on the ray by the Gaussian distribution. This strategy uses the complementary among video frames to remove surgical tool occlusion. Finally, import D-NeRF to reconstruct the dynamic scene and render a novel view. EndoNeRF is a significant starting point that expands the application of NeRF in robotic surgery.

## Surface reconstruction of deformable tissues

The density field used in EndoNeRF ambiguously estimates depth, resulting in its inability to accurately locate the surface. While a signed distance function (SDF) performs well in reconstructing continuous surfaces, it is limited to static scenes with multiple viewing directions, which restricts its application in endoscopic reconstruction. To address these issues, EndoSurf [11] was created to reconstruct smoother surfaces in deformable endoscopic scenes with a limited view angle. This is achieved by representing the dynamic scene with a canonical field and utilizing three multilayer perceptrons. First, a deformation network converts observation space points into canonical space. Second, an SDF network describes surface and geometry, enabling the reconstruction of tissue geometry instead of relying on the density field. At last, a radiance field, also known as the appearance network, learns the color of points on the surface, separate from the geometry. With these improved network structures, EndoSurf offers a better reconstruction of deformable tissues, enhancing the usability of NeRF in robotic surgery.

## Fast 4D reconstruction of deformable tissues

The novel method EndoSurf significantly improves reconstruction performance. However, to achieve real-time and accurate reconstruction during surgery, the reconstruction speed must also be significantly increased. By addressing this issue, instead of dealing with the extensive 4D data of 3D space and time, LerPlane [12] separates the 4D scene into two fields: a static field independent of time and a dynamic field related to time. Then, it further breaks down each field into three 2D planes: for the static field, it uses $XY$, $YZ$, and $XZ$ planes to represent spatial points of the scene, and for the dynamic field, it utilizes $XT$, $YT$, and $ZT$ planes to capture the changes of spatial points with time. This decomposition simplifies the projection of each spatial point onto six planes and facilitates the fusion of their features. As a result, the difficulty of deformation reconstruction is reduced, while the use of memory and computing resources is also minimized, leading to a significant reduction in training time. This method offers hope for real-time reconstruction in robotic surgery.

## Illumination modeling

Colon reconstruction during endoscopic procedures presents a formidable challenge. Despite the static nature of the scene, a major issue arises from the significant variation in color on the colon's internal surface depending on the positioning of the light source or camera. To address this concern, two innovative methods, LightNeuS [13] and REIM [14], propose similar approaches for modeling illumination decay. LightNeuS achieves this by estimating the distance between the light source and the target point. At the same time, REIM incorporates the location of the light source as an additional factor in predicting color.

Based on two properties of the colon image dataset, namely, the colon's watertight internal surface and the decay of illumination that can be represented by a function of $1/t^2$, LightNeuS [13] utilizes the SDF, which is suitable for reconstructing this surface. It also integrates a photometric model to optimize the photometric loss. Inspired by NeuS [15], LightNeuS incorporates the factor $1/t^2$ into the rendering equation of the primary NeuS network, enabling the neural network to learn how color changes depending on the distance. Although a simple improvement is made, LightNeuS represents the first work on colon-dense reconstruction and demonstrates the effectiveness of this enhancement.

Due to an issue where the movement of an endoscope can affect the surface illumination, resulting in blurred color, REIM proposes a variant of depth-supervised NeRF to model the illumination variation based on the light source location. This variant builds upon NeRF's multilayer perceptron, which takes spatial point coordinates and viewing direction as inputs, with the added parameter of the light source location to estimate point color. Experimental results demonstrate its effectiveness in synthesizing a novel, nonperfect, deformable model for the training dataset, thereby addressing the problem of insufficient data in surgical procedures.

## Depth estimation

Depth estimation using deep convolutional networks was first proposed in [16], where depth sensors were used to train the model in a supervised manner. Later, self-supervised methods were proposed, mainly based on multiview photometric consistency [17]. In-body monocular reconstructions depend on self-supervised depth estimation because it does not require redundant sensors and complicated rendering. Fig. 2 shows a main network structure for monocular depth estimation, which requires a DepthNet to estimate the Depth map and a PoseNet to estimate the pose translation from the Key Frame to the Target Frame. Then, the Reconstructed Frame is built from the pose and the Key Frame, where the DepthNet is optimized by minimizing the reconstruction error between the Target Frame and the Reconstruction Frame.

## Challenge and future work

### Deformable tissue simulation

Currently, no perfect deformable model meets all the diverse requirements for surgical simulation. The existing approaches tend to prioritize certain aspects while neglecting others. Anticipated advancements in CPUs are expected to result in improved memory and computational capacity. This enhanced capability will enable more complex surgical simulation scenarios, potentially leading to more realistic deformable models.

However, the currently considered best models may not have the most significant potential for further development. The spring-mass model has nearly exhausted all of its potential. Considering its significantly superior volumetric behavior, the linked volume technique can potentially become the most widely adopted approach.

To justify the need for additional work, more advanced methods must outperform the spring-mass or connected volume models. This includes the tensor-mass model or continuum-mechanical approaches. The BEM-based model likely holds the most promise because it requires a mesh almost as simple as the spring-mass model. However, its ability to overcome the linearity constraint, which mainly affects the realism of large deformations, will significantly influence its future. Nevertheless, we expect that over time, continuum-mechanical techniques will be capable of accurately simulating the deformation of human tissues in surgical simulations.

### NeRF-based reconstruction

As a nascent technology, NeRF has already impressed us with its impressive results and boundless potential. NeRF will play a significant role in disease diagnosis and surgical planning. There are two primary directions for NeRF's advancement. The first direction is enhancing the reliability of NeRF's rendering, which would facilitate the reconstruction of human structures. The second direction involves developing real-time NeRF reconstruction to aid in auxiliary diagnosis or remote surgery.

## Depth estimation

Conventional depth estimation techniques based on neural networks often overlook the spatiotemporal continuity of endoscopic videos, resulting in unreliable accuracy of deep learning algorithms [6]. Moreover, applying deep learning-based methods to endoscopic image analysis is challenging due to the unique in vivo environment [4]. A potential future direction could involve the development of a fully self-supervised depth estimation network.

## Conclusion

In conclusion, this chapter examines the application of deformable tissue simulation, NeRF, and depth estimation in the 3D reconstruction of deformable tissue during robotic surgery. The chapter explores different approaches to modeling deformable tissue, including heuristic and continuum mechanical approaches. It also discusses using NeRF to reconstruct static 3D scenes based on neural implicit representation, showcasing impressive advancements in high-resolution new view generation with improved efficiency. Furthermore, the chapter emphasizes the significance of depth and pose estimation in the 3D reconstruction for virtual reality applications in robotic surgery. Overall, this chapter offers a comprehensive overview of the current methods and techniques utilized for the 3D reconstruction of deformable tissues in robotic surgery.

## Future scenario

Examining three methods for reconstructing surgical scenes using robotics (deformable tissue simulation, NeRF, and depth estimation), we anticipate future advancements and challenges. As technology progresses, we can anticipate improved accuracy and real-time reconstruction, empowering surgical robots to navigate more effectively and aiding doctors in making informed decisions during procedures. Future techniques will prioritize addressing individual variations, necessitating personalized reconstruction plans based on patients' lesion type, anatomical structure, and physiological characteristics. Furthermore, future research may explore the fusion of different data types, such as CT scans or biological signals.

There are also several issues to be solved in the future. Robotic surgery scene reconstruction requires a significant amount of medical image data. Thus, protecting sensitive information and preventing data leakage becomes an urgent concern. Efficiently labeling medical images and verifying reconstruction accuracy should be the focus of future research, as deep learning models rely on labeled data. Additionally, implementing widely applicable real-time surgical navigation presents a challenge in achieving high-quality reconstruction with limited computing resources.

The future of 3D reconstruction for deformable tissues in robotic surgery shows promise, but there is still a need for sustained research efforts and interdisciplinary collaboration to address these challenges.

## Key points

- Deformable tissue simulation will become an important tool for surgical planning and training, helping doctors simulate surgical procedures and optimize operations.
- NeRF-based reconstruction can reconstruct 3D deformable tissues in robotic surgery, facilitating preoperative planning and surgical navigation.
- Depth estimation can work in real-time surgical navigation, which assists doctors in accurately measuring the location and distance of tissue during surgery.

## Acknowledgments

The work was supported by the Hong Kong Research Grants Council (RGC) Collaborative Research Fund (CRF C4026-21GF), the General Research Fund (GRF #14211420 and GRF #14216022), and the Shenzhen-Hong Kong-Macau Technology Research Programme (Type C) Grant 202108233000303.

# References

[1] Mildenhall B, Srinivasan PP, Tancik M, Barron JT, Ramamoorthi R, Ng R. Nerf: representing scenes as neural radiance fields for view synthesis. Commun ACM 2021;65(1):99–106.

[2] Recasens D, Lamarca J, Fácil JM, Montiel J, Civera J. Endo-depth-and-motion: reconstruction and tracking in endoscopic videos using depth networks and photometric constraints. IEEE Robot Autom Lett 2021;6(4):7225–32.

[3] Müller T, Evans A, Schied C, Keller A. Instant neural graphics primitives with a multiresolution hash encoding. ACM Trans Graph (ToG) 2022;41(4):1–15.

[4] Yu A, Ye V, Tancik M, Kanazawa A. pixelnerf: neural radiance fields from one or few images. In: Proceedings of the IEEE/CVF conference on computer vision and pattern recognition; 2021. p. 4578–87.

[5] Wang Y, Long Y, Fan SH, Dou Q. Neural rendering for stereo 3d reconstruction of deformable tissues in robotic surgery. In: Medical image computing and computer assisted intervention—MICCAI 2022: 25th international conference, Singapore, September 18–22, 2022, proceedings, Part VII. Springer; 2022. p. 431–41.

[6] Trevithick A, Yang B. Grf: learning a general radiance field for 3d representation and rendering. In: Proceedings of the IEEE/CVF international conference on computer vision; 2021. p. 15182–92.

[7] Park K, Sinha U, Barron JT, Bouaziz S, Goldman DB, Seitz SM, Martin-Brualla R. Nerfies: deformable neural radiance fields. In: Proceedings of the IEEE/CVF international conference on computer vision; 2021. p. 5865–74.

[8] Pumarola A, Corona E, Pons-Moll G, Moreno-Noguer F. D-nerf: Neural radiance fields for dynamic scenes. In: Proceedings of the IEEE/CVF conference on computer vision and pattern recognition; 2021. p. 10318–27.

[9] Niemeyer M, Geiger A. Giraffe: representing scenes as compositional generative neural feature fields. In: Proceedings of the IEEE/CVF conference on computer vision and pattern recognition; 2021. p. 11453–64.

[10] Barron JT, Mildenhall B, Tancik M, Hedman P, Martin-Brualla R, Srinivasan PP. Mip-nerf: A multiscale representation for anti-aliasing neural radiance fields. In: Proceedings of the IEEE/CVF international conference on computer vision; 2021. p. 5855–64.

[11] Zha R, Cheng X, Li H, Harandi M, Ge Z. Endosurf: neural surface reconstruction of deformable tissues with stereo endoscope videos. In: International conference on medical image computing and computer-assisted intervention. Springer; 2023. p. 13–23.

[12] Yang C, Wang K, Wang Y, Yang X, Shen W. Neural lerplane representations for fast 4d reconstruction of deformable tissues. arXiv preprint arXiv:2305.19906; 2023.

[13] Batlle VM, Montiel JM, Fua P, Tardós JD. Lightneus: neural surface reconstruction in endoscopy using illumination decline. In: International conference on medical image computing and computer-assisted intervention. Springer; 2023. p. 502–12.

[14] Psychogyios D, Vasconcelos F, Stoyanov D. Realistic endoscopic illumination modeling for NERF-based data generation. In: International conference on medical image computing and computer-assisted intervention. Springer; 2023. p. 535–44.

[15] Wang P, Liu L, Liu Y, Theobalt C, Komura T, Wang W. Neus: learning neural implicit surfaces by volume rendering for multi-view reconstruction. In: NeurIPS; 2021.

[16] Eigen D, Puhrsch C, Fergus R. Depth map prediction from a single image using a multi-scale deep network. In: Advances in neural information processing systems, vol. 27; 2014.

[17] Zhan H, Garg R, Weerasekera CS, Li K, Agarwal H, Reid I. Unsupervised learning of monocular depth estimation and visual odometry with deep feature reconstruction. In: Proceedings of the IEEE conference on computer vision and pattern recognition; 2018. p. 340–9.

# Deep reinforcement learning in surgical robotics: Enhancing the automation level

*Cheng Qian[a] and Hongliang Ren[b]*

[a]Department of Electrical Engineering and Information Technology, Technical University of Munich, Munich, Germany [b]Department of Electronic Engineering, The Chinese University of Hong Kong, Sha Tin, Hong Kong

## Introduction

The use of surgical robots has significantly increased in the past decade, driven by the need for precision, safety, and efficiency in surgeries [1]. Since the appearance of da Vinci robotic-assisted surgical system in 2000 [2], surgical robots have proven to help perform minimally invasive surgeries (MIS), providing better visualization, higher precision, and reduced invasiveness, and helping reduce surgeons' fatigue [3]. However, the full potential of surgical robots has yet to be realized, and there is still a need to improve their autonomy. In the past decade, more and more studies have been conducted on autonomous surgical robots [4]. To achieve autonomy in surgery, it is crucial for robots to understand the surgical task objectives, perceive complex physical environments, and autonomously make decisions. One of the biggest challenges in the autonomy of surgical robots is the high variance of surgical tasks [5], which is hard to address by explicitly modeling and planning. Therefore, artificial intelligence (AI) solutions emerged due to the model-free property and learning capability.

Deep reinforcement learning (DRL), a deep learning-based planning method that allows robots to learn from interaction with environments in a semi-supervised fashion without a predefined model, is one of the most promising approaches [6]. DRL has been increasingly highlighted in recent years, since its success in Atari [7]. It has demonstrated the possibility of enabling intelligent agents to outperform human experts in multiple fields. Compared to conventional planning methods, DRL provides advantages, e.g., end-to-end learning, complex decision-making, generalization, and transferability, handling uncertainties, and continuous learning. These properties enable DRL to handle high-dimensional inputs from cameras and sensors in surgeries, apply its acquired knowledge and skills to different patients, handle unforeseen variations, and continuously learn and refine their performance during surgery procedures. For these reasons, in the context of surgical robots, DRL provides a powerful model-free framework and a set of tools for learning various complicated surgical tasks with complex physical environments, which are hard to model [8]. Many studies have utilized DRL on robots under abundant surgical scenarios, e.g., ultrasound scanning, cutting and sewing, tissue retraction, needle steering, and catheterization. Currently, there have been some reviews on DRL in the scope of medical imaging, e.g., radiation therapy, image registration [9], healthcare application, e.g., clinical decision support [10–12], medicine, e.g., medicine treatment, or development [13]. However, there still lacks a review specifically on the applications of DRL in medical robots. To fill this gap, this chapter will present a literature review highlighting the typical state-of-art works in the past 5 years (2018–2023) that utilize DRL in autonomous surgical robots, and comparing their methodologies, limitations, and results. We divide these works into three categories according to their access modes, namely,

(1) Extra-body skin-interfaced procedure
(2) Intrabody procedure
(3) Percutaneous procedure

which are three main procedures that we find the combination of DRL and surgical robot is mainly applied to. The various surgical tasks learned by the robot in this review are illustrated in Fig. 1, including steerable needle planning in keyhole neurosurgery, needle insertion in ophthalmic microsurgery, neck vessel and spine US scanning, tissue cutting and retraction, and wound suturing.

To exhaust the published review articles of the concerning fields and extract the most relevant ones, we searched keywords on the database and excluded the irrelevant articles. The literature screening pipeline is shown in Fig. 2. We also counted the number of studies that applied DRL in medical scenarios in the past 5 years. In Fig. 3, the number of articles on the application of DRL in medical imaging, medical robotics, and dynamic treatment regimes in the past 5 years are listed.

We can see that the number of studies combining DRL with different medical fields has quickly emerged in recent years, which indicates a growing trend.

The rest of the chapters are organized as follows: Section "Basics of reinforcement learning" briefly introduces the fundamental theories in RL. Sections "Deep reinforcement learning in surgical robotics," "Preoperative procedure," and "Intrabody procedure" discuss the latest work of DRL in the fields of preoperative scanning, intrabody surgery, and image-guided autonomous robotic surgery, respectively.

FIG. 1   The seven different robotic surgical tasks contained in this review.

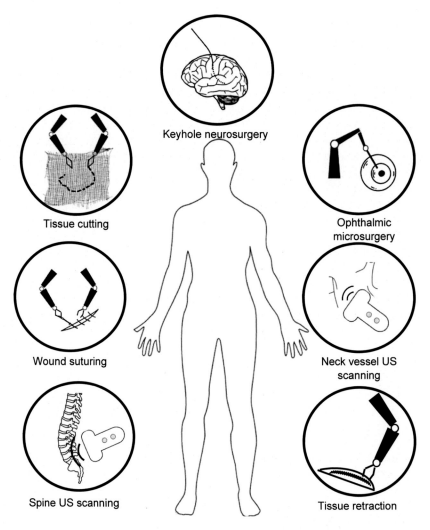

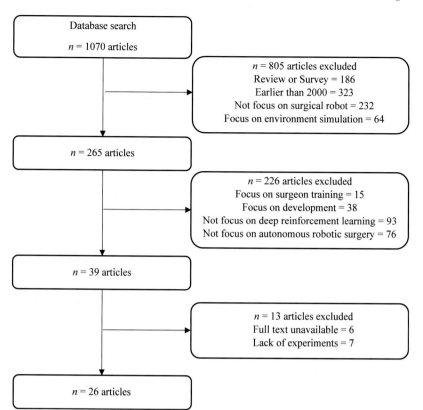

FIG. 2 Articles selection pipeline with keywords "surgical robot," "autonomous," and "deep reinforcement learning" according to PRISMA [14].

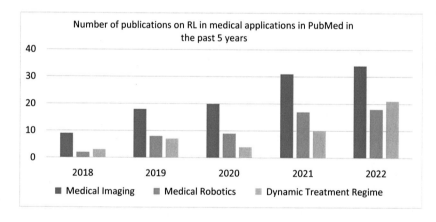

FIG. 3 Statistics of the number of publications on RL in three main medical applications in PubMed in the past 5 years. The combination of DRL with autonomous surgical robots and other medical fields has been a rising trend in the past 5 years.

## Basics of reinforcement learning

Before discussing the state-of-art works of DRL in surgical robotics fields, we will first give a general introduction to the fundamental knowledge in reinforcement learning (RL). Learning through interaction with the environment is the essence of RL [15], which means an agent learns to take action through rewards and penalties and refines its policy accordingly.

The fundamental of RL includes five essential elements: agent, environment, action, state, and reward. In the context of surgical robots, an example of it can be illustrated in Fig. 4, where the robot (agent) works at the surgical site of a human body (environment), moving the probe to find a feasible scan plane for the sacrum, and obtaining the current position information of probe via real-time ultrasound (US) images. At each time step, the robot possibly gets a positive or negative reward based on the current US image, which guides the robot toward the standard scan plane.

FIG. 4   An illustration of agent-environment interaction in RL under the context of surgical robots.

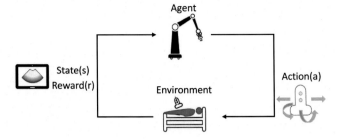

## Markov decision process

Markov decision process (MDP) is always used to formally describe the above-mentioned agent-environment interaction, which consists of [16,17]:

- State space ($S$): The set of possible states the agent can be. Each state represents a particular configuration or situation in the environment. In surgical robot scenarios, the state is often chosen as the robot's pose and target.
- Action space ($A$): The set of possible actions that the agent can take. Actions are the choices available to the agent in each state. Actions can be discrete or continuous depending on the accuracy requirements. It is preferred to be chosen as continuous in some safety-critical scenarios, which need high accuracy of control, e.g., in intrabody surgery.
- Transition ($T$): The transition probabilities that describe the dynamics of the environment. They specify the probability of transitioning from one state $s$ to another state $s'$, if the agent takes action $a$, which is represented as $T(s'|s,a)$. It includes important information about the robot-environment interaction, e.g., the interaction between the US probe and tissue.
- Reward ($R$): The immediate feedback that the agent receives from the environment for its action. It quantifies the desirability or value associated with transitions between states. The reward function is typically denoted as $r = R(s', a, s)$. It is commonly designed to guide the robot to achieve its goal. For example, for standard scan plane navigation in robotic ultrasound scanning, the reward is commonly designed to be the pose improvement of the probe to the target pose.
- Discount factor ($\gamma$): A value between 0 and 1 that determines the importance of future rewards compared to immediate rewards. It determines the preference of the agent for immediate rewards or long-term cumulative rewards.

Given MDP($S, A, T, R, \gamma$), the agent chooses the action at state $s$ with the observation $o$ it receives according to the policy $\pi(a|s)$. When the policy is deterministic, $\pi$ is a mapping from state $s$ to action $a$; when the policy is stochastic, $\pi$ represents the possibility of selecting action $a$ at state $s$.

The goal of RL is to find an optimal policy $\pi^*$ that maximizes the expectation of cumulative return, which is denoted as

$$\pi^* = \max_{\pi} \mathbb{E}\left[\sum_{t=0}^{T} \gamma\, r_t\right],$$

where $r_t$ is the reward at time $t$, and $T$ is the time horizon.

To be noticed that sometimes the full state information is not available for the agent, but only a part of it, instead. The agent has to predict the state information given an observation. For example, the ultrasound scanning robot has to detect its current position according to the real-time US image. In this case, the process is a partially observable MDP (POMDP) [18]. The set of the state information that is observable for the agent is called Observation. In this case, the policy $\pi$ is dependent on observation $o$ instead of state $s$.

Besides, in this review, only model-free RL algorithms are focused on. Therefore, the transition is assumed to be unknown.

## Deep reinforcement learning in surgical robotics

This section highlights the state-of-the-art studies of DRL for surgical robotics applications and discusses them in three parts: preoperative scanning, intrabody surgery, and percutaneous surgery. We will focus on how they formulate the problem in a DRL-based framework and different methodologies applied to augment RL to meet some surgery-specific requirements, such as risk analysis.

## Preoperative procedure

Surgical images are obtained using various imaging modalities, and ultrasound (US) scanning is the one that has been widely studied in combination with robotics. Over the past two decades, researchers have begun exploring the potential of robotics in applying US scanning. By equipping the robot arm with a probe, the robot can move the probe to perform US scanning on the patient. The accuracy, consistency, skill, and maneuverability of robotic manipulators can be used to improve the acquisition and utility of real-time ultrasounds [19]. However, to obtain high-quality ultrasound images, it is crucial to navigate the US probe to the correct scan plane [20] and maintain reasonable and consistent probe-skin contact force [21], as illustrated in Fig. 5. Therefore, standard scan plane localization and contact force control are two main challenges in robotic US, and so far, there have been several studies utilizing DRL to address them. In Table 1, the methodologies, metrics, and results used in the nine reviewed papers in this section are listed.

### Standard scan plane navigation

A standard scan plane in ultrasound imaging refers to a recommended imaging plane or view commonly used for a specific anatomical structure or diagnostic purpose. Finding the appropriate scan planes is crucial to obtaining good-quality US images. To enable autonomous robotic ultrasound scanning, the robot should be capable of detecting its own position and finding the way toward the standard scan plane of the specific anatomy with the real-time US image it obtains.

Hase et al. [22] proposed a framework in 2020 to train the robot to autonomously navigate to the standard scan plane of the sacrum with the information of a sequence of history US images with Deep Q-Network (DQN) [7]. The agent is trained on the 2D US images acquired by grid covering and moves with 2-DOF (degree of freedom) actions, namely, moving forward or backward. When moving closer to or further from the desired scan plane, the agent receives positive or negative rewards. A binary classifier that determines whether the robotic probe is at the standard scan plane, depending on the current US image, is used to make the agent stop at the correct position.

One of the limitations in Ref. [22] is that the probe is assumed to find the scan plane by moving in a 2D space, which is unavailable in real US scanning scenarios, where the relative pose between the probe and sacrum is not static. Therefore, in Li et al. [23], the state and action space is designed to be the 6D-pose and -twist of the probe so that the learned policy is no longer restricted to the collected data. The agent receives a reward proportional to the pose improvement of the probe at each time step. Besides, the quality of the US image is also considered in this work by evaluating the pixelwise confidence and giving corresponding rewards to the agent.

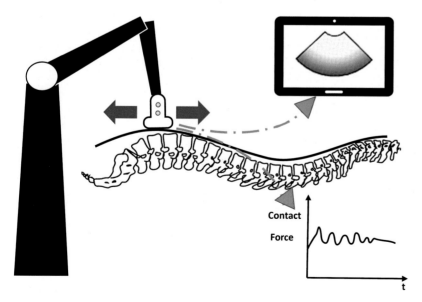

FIG. 5 The robot moves the probe to find the standard scan plane while keeping the contact force in a suitable range.

TABLE 1   The formulation, methodologies, and results of the reviewed papers in the section on preoperative planning.

| Ref. | Description | Algorithm | Observation | DOF of action | Reward | Result |
|------|-------------|-----------|-------------|---------------|--------|--------|
| [22] | Navigation toward the standard scan plane of sacrum | DQN | Sequential US images | 1 | + Moving closer<br>− Moving further | Policy correctness of 79.53% and reachability of 82.91% |
| [23] | Navigation toward the standard scan plane of the spine with consideration of US image quality | DQN | Sequential US images | 6 | + Pose improvement or image quality improvement<br>− Unallowable pose | 92% and 46% success rate in intra- and interpatient settings, respectively |
| [24] | Navigation toward different standard scan planes of spinal anatomies | DQN | Sequential US images | 6 | + Pose improvement or image quality improvement<br>− Unallowable pose | Pose error $\sim$5.18 mm/5.25° for intrapatient settings; Pose error $\sim$2.87 mm/17.49° for interpatient settings. |
| [25] | Navigation toward the standard scan plane of carotid vessels | A2C | Sequential segmented US images + Sequential vessel area changes | 3 | + Vessel area improvement<br>− Too small vessel area | 91.5% and 80% success rate in simulated and real environment, respectively |
| [26] | Guidance for novice operators in moving TEE probe toward the standard scan plane of heart with pressure awareness | DQN | Sequential US images | 3 | + Pose improvement<br>− Unallowable pose | Pose error of 2.72 mm/2.69° and 8.15 mm/5.58° without and with pressure awareness, respectively |
| [27] | Guidance for novice operators in obtaining correct US images of anatomy of interest | DQN | Sequential US images | 4 | + Moving closer<br>− Moving further | 86.1% success rate in giving correct guidance |
| [28] | Force control between probe and phantom | PPO | 6-D contact force | 2 | + Small horizontal force<br>− Big horizontal force or too big or too small vertical force | Difference of skin area in US images within 3%±0.4% from the hand-free scanning approach |
| [28] | Force control between probe and phantom | PPO | Single encoded RBG scene image | 3 | + Moving closer to the target surface, good US image quality, correct relative position<br>− Otherwise | 93% success rate in getting feasible US images |
| [29] | Force control between probe and phantom | PPO +Inverse RL | Contact force and torque, and corresponding linear and angular speed | 6 | Reward shaping via inverse RL | Posture error of 2.3±1.3° and 1.9±1.2° in X and Y axis compared to manual operation |

An agent that can recognize different anatomies and find the nearest one according to its current position can be advantageous for its flexibility in a real US scanning scenario. In Li et al. [24], an agent is trained to find three different spinal anatomies with a given standard view recognizer for different spinal anatomies.

Real US images as observations given to the agent can be noisy, which makes the agent hard to predict the real state correctly. One of the methods addressing this issue is presented in Bi et al. [25], which navigates the agent to the scan plane of the neck vessel. The study segments the US images with a pretrained U-Net [30] in advance and provides the segmented mask to the agent, where the area of interest has pixel values of one, while other areas have values of zero. Compared to real US images, it is much easier for the agent to extract information from segmented masks, which only contain binary values.

Besides, Li et al. [26] and Milletari et al. [27] proposed DRL-based frameworks for training an agent that guides a novice operator to find the standard scan planes in transesophageal echocardiography and chest sonography, respectively.

## Pose and force control

The way in which the ultrasound probe is positioned and controlled can have a significant impact on both the quality of the resulting ultrasound images and the overall safety of the robotic ultrasound system. It is essential to carefully control the pose and force used when operating the probe, as any errors or inconsistencies can compromise the quality of the imaging and potentially cause harm to the patient or the system itself. A system can ensure imaging quality while minimizing potential risks or complications by taking a deliberate approach to probe control. Unlike rigid objects, force control of the US probe should consider the compliance of the patient body. Besides, errors caused by target movement have to be compensated. Both of them are hard to model accurately. However, by taking advantage of DRL's model-free and end-to-end properties, the control of the US probe can be solved without explicitly modeling.

In Milletari et al. [27], an agent is trained with Proximal Policy Optimization (PPO) [31] to autonomously control the pose and force of the US probe with a force-to-displacement admittance controller. The agent has to provide proper 2D input command for the controller, namely, the desired torque of the US probe in long- and short-horizontal-direction, as illustrated in Fig. 5, to keep the vertical between the probe and the scanned surface with suitable contact force. A 6-D force sensor is attached to the robot end-effector to give the force feedback to the agent. A positive reward is given when the vertical contact force is suitable and the horizontal contact force is small enough, which means the probe is approximately vertical to the scanned surface. A similar work is done in Ref. [32], however, with an inverse RL method to study the reward function.

Differing from Refs. [27,32], in Ning et al. [28], the scene image captured by a RGB camera is also provided to the agent as observation. From the scene image, the agent can extract the information of both its own pose and the target pose. In the study [33], a convolutional autoencoder (CAE) and a reward prediction network are employed to achieve two objectives simultaneously. First, the CAE is used to decrease the dimensionality of the observation space, allowing for more efficient data processing. Second, the reward prediction network encodes force and ultrasound image information into the scene image, enhancing the resulting image's quality. By utilizing these techniques, the researchers improved the overall efficiency of the system.

## Intrabody procedure

Recently, flexible surgical robotic systems have been developed to improve intrabody surgery in the narrow areas of the human body. However, the teleoperation of surgical robots can be exhausting and needs long-term training time. More and more researchers have been working on the possibility of automating difficult surgical handling tasks, e.g., tissue cutting, suturing, knot tying, and tissue retraction, to reduce the surgeons' workload [34]. However, the large quantity of soft tissue in surgeries, including organs, blood vessels, and muscles, possess inherent compliance and deformability, making their manipulation challenging and requiring modeling and planning with high accuracy and complexity. Therefore, some studies have tried to unleash the model-free property of DRL in automating the tissue manipulation tasks in MIS, including tensioning, suturing, and retraction. In Table 2, the methodologies, metrics, and results used in the eight reviewed papers in this section are listed.

### Tensioning

Robotic surgery has revolutionized the medical field, and an electric knife is an effective tool for cutting and removing thin tissues. However, the electric knife alone may not be enough to cut effectively when it comes to deformable soft tissue. This is because soft tissue needs to be held in tension to be cut most effectively. Therefore, a second tool is required to pinch and tension the material while cutting. This technique is illustrated in Fig. 6, which demonstrates the use of two tools cooperatively to cut soft tissue. The first tool, the knife, cuts the tissue, while the second tool, which pinches and tensions the material, helps the knife cut more effectively. This technique is particularly important in robotic surgery, where precision is critical, and using multiple tools can help ensure the surgery is successful. To let the robot autonomously assist the surgeon in cutting, the robot has to learn the tensioning policies for different cutting contours.

In Thananjeyan et al. [35], a finite-element model is first developed for simulating the deformation and cutting of tissue. Then, considering the kinematics constraints of the surgical robot arm, the cutting outline is divided into several subdivision segments in advance. The agent is lastly trained in trust region policy optimization (TRPO) [43] to learn the optimal tensioning policy to minimize the cutting error with a single fixed pinch point. The agent receives sparse rewards at the end of each episode according to the final cutting error.

There needs to be more than a single tensioning point to assist cutting, when the cutting pattern is complex, for example, the cutting contours are zigzag and have to be divided into many segments. Therefore, in Refs. [36,37],

TABLE 2    The formulation, methodologies, and results of the reviewed papers in the section of intrabody surgery.

| Ref. | Description | Algorithm | Observation | DOF of action | Reward | Result |
|---|---|---|---|---|---|---|
| [35] | Tensioning policy for the selected pinch point | TRPO | Cutting trajectory and fiducial points locations | 2 | Sparse reward according to the final cutting error, when episode ends | Improvement of 43.3% compared to nontension baseline in term of cutting error |
| [36,37] | An improved pipeline enabling selecting multiple pinch points for different cutting segments | TRPO | Cutting trajectory and fiducial points locations | 2 | Sparse reward according to the final cutting error, when episode ends | Improvement of 50.6% compared to nontension baseline in term of cutting error |
| [38] | Autonomous collaborative needle hand-off task of PSM | Q-Learning | 3D position of robotic tip | 3 | Sparse reward according to the data points on user-defined trajectory | Dissimilarity between learned trajectory and reference trajectory with mean and standard deviation of (2.857, 1.488, 0.774mm) and [3.388, 2.286, 1.808mm) |
| [39] | Autonomous collaborative needle hand-off task of PSM in ego-centric spaces | DDPG +BC | Relative position and quaternion | 6 | + Reaching target pose − Collision | 97% and 73.3% success rate in simulation and real-world environment, respectively |
| [40] | Robotic tissue retraction learning from expert demonstration | PPO +GAIL | Gripper state, End-effector location, Target location | 3 | + Moving closer to tumor or target position + Moving further from tumor or target position | Average tumor exposure percentage of 84% and 90% in simulation and real-world environment, respectively |
| [41] | Robotic tissue retraction considering safety properties | PPO +Formal verification | Gripper state, end-effector location, target location | 3 | + Moving closer to tumor or target position + Moving further from tumor or target position or violating safety constraints | Safety violation rate of 3.07% and violation rate reduction of 24%, compared to nonsafety method |
| [42] | Robotic tissue retraction with sim-to-real | PPO+DCL | Sequential translated scene image | 2 | + Moving closer to tumor or target position + Moving further from tumor or target position | 50% success rate in real-world environment with raw camera images as input |

FIG. 6    The blue manipulator pinches the gray point and tensions the tissue, while the green manipulator is responsible for cutting (e.g., in endoscopic submucosal dissection [ESD]).

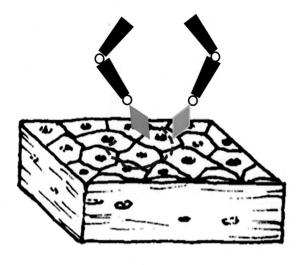

an improved pipeline is proposed to address this limitation. Specifically, a pinch point is chosen for each cutting segment instead of the whole contour and the agent learns different tensioning policies for each pinch point in a similar way as in Ref. [35]. Compared to Ref. [35], the improved method shows more accurate and robust performance, when handling complex cutting contours.

FIG. 7   The main manipulator (*left*) is operated by the surgeon, which inserts the needle through the tissue, while the assistive manipulator (*right*) pulls the needle out and hands it to the main manipulator.

TABLE 3   The formulation, methodologies, and results of the reviewed papers in the section of percutaneous surgery.

| Ref. | Description | Algorithm | Observation | DOF of Action | Reward | Result |
|---|---|---|---|---|---|---|
| [44] | Needle path planning for ophthalmic microsurgery | DDPGfD | Height maps of two corneal surfaces | 3 | Sparse reward when reaching the target position | Perforation-free percent depth of 84.75% ± 4.91% |
| [45] | RCM recommendation for keyhole neuro surgery | PPO | Target position, robot joints position | 2 | Sparse reward according to positioning accuracy, solvable inverse kinematics and mechanical joint motion | 93% Success rate of finding optimal RCM |
| [46] | Navigation of steerable needle toward the target position in brain in 2D space | DQN | 2D map based on segmented MRI images | 2 | + Position improvement | 93.6% Success rate of achieving target position |
| [47] | Navigation of steerable needle toward the target position in brain in 3D space | DDPG | 3D map based on segmented MRI images | 2 | + Achieve goal or in safe area − In unsafe area | Outperforms RRT* under different quantiles of constraints |
| [48] | Navigation of steerable needle toward the target position in brain in 3D space | GAIL | 3D map based on segmented MRI images | 6 | + Achieve goal − Obstacle collision | Average targeting error of 1.34 ± 0.52 mm in position and 3.16 ± 1.06 degrees in orientation |
| [49] | Navigation of steerable needle toward the target position in brain in 3D space | GA3C | Sequential frames of 3D map based on segmented MRI images | 6 | + Achieve goal − Obstacle collision | Outperforms RRT* under different quantiles of constraints |
| [50] | Optimization of catheterization trajectory obtained from demonstration | $PI^2$ | Catheter tip pose, Target position | 2 | + Position improvement, Vessel centerline alignment | Average targeting error of 2.92 mm and path length of 258.67 mm |
| [51] | Navigation of catheter tip to the target position | GAIL + PPO | Catheter tip pose, Target position | 3 | + Position improvement − Obstacle collision | 82.4% success rate of achieving target position |
| [52] | Adaptation of PID controller parameters for catheterization | DQN | Catheter tip pose | 3 | + Position improvement | Average error of 0.003 ± 0.0058 mm with respect to setting point |

## Suturing

Suturing is a critical step in wound closure during surgeries and in robot-assisted surgeries. However, robotic suturing can be laborious for novice operators. A collaborative robot that autonomously assists surgeons in performing some sub-tasks in robotic surgeries can effectively operators' fatigue. So far, utilizing DRL on surgical collaborative robots can learn how to autonomously collaborate with surgeons in the teleoperated suturing process, as illustrated in Fig. 7. Table 3 lists the methodologies, metrics, and results used in the nine reviewed papers in this section.

In Varier et al. [38], an agent is trained to use an assistive patient side manipulator (PSM) to pull the needle, translate it to the next suture point, and hand it to the surgeon after the needle is inserted through the tissue with the main PSM by the operator, as illustrated in Fig. 7. To address varying suture styles of users, the users are instructed to perform a running suture without a collaborative robot. The trajectory of a single hand-off task is collected, and an algorithm is designed to generate sparse rewards on the trajectory. Then, the agent imitates the operator's hand-off trajectory by maximizing the cumulative collected reward.

However, in Ref. [38], the state of the agent is respect to a fixed frame, which makes the learned policy strongly depend on the selection of the frame. In Chiu et al. [39], an improved method is proposed to address this issue by designing action spaces with respect to the ego-centric frame, which means the policy depends on the relative position and orientation of the assistive PSM relative to the main PSM and therefore can be directly applied to different robot configurations.

## Retraction

Another kind of tissue manipulation task in robotic surgery is tissue retraction. That is, to uncover the underlying anatomical region, the tissue is repeatedly held and pulled back in MIS [53]. To autonomously perform the tissue retraction task, the robot has to find the position of the tissue, move closer to it, and grasp it to the target position.

In Pore et al. [40], a robot learns to approach the tumor from its initial position and retract it to the target position from human demonstrations. The position of the tumor is assumed to be known from the preoperative data. The agent gets the reward based on whether it moves closer to the tumor or target position before or after grasping. Human demonstrations are collected to enable imitation learning. The agent is trained with generative adversarial imitation learning (GAIL) and PPO, where PPO acts as the action generator.

Safety is always the priority in surgeries, especially in tissue retraction, where the robot directly interacts with the tissue. However, due to the model-free property of DRL, the safety of the learned policy is always hard to verify, which leads to significant potential risks in surgery. In another work by Pore et al. [41], a framework for robotic tissue retraction incorporates the safety constraints during the DRL training with formal verification, which adds a penalty term in the reward function for unsafe actions and evaluates the safety of the learned policy [54]. The proposed method shows a large reduction in the safety violation rate, compared to Ref. [40].

In Refs. [40,41], the agents are assumed to have access to the full-state information, e.g., the robot joint angles and tissue position, which makes the policy largely depend on the accuracy of state extraction and lack robustness against the patient movements. In Scheikl et al. [42], a vision-based framework for robotic tissue retraction is proposed. The agent is trained with the simulated RGB scene image, and a translation model is trained to translate the observation function in simulation to the one in reality using domain adaptation. The trained agent achieves a success rate of 50% in real surgical scenarios.

## Percutaneous procedure

Percutaneous techniques are increasingly used in many surgical scenarios, including neurosurgery and ophthalmic surgery. The practical advantages include lower complexity rates and faster recovery time. It involves the precise insertion of a thin, hollow needle into specific anatomical structures for diagnostic, therapeutic, or monitoring purposes. This technique allows surgeons to access internal body tissues, organs, or vessels without the need for open surgery, minimizing patient discomfort, reducing the risk of complications, and promoting faster recovery.

## Needle insertion

Needle insertion surgery is common in various medical fields, including keyhole neurosurgery [55] and ophthalmic surgery [56]. Conventionally, a rigid needle is commonly used in these procedures. There has been some research on

DRL-based needle path planning. In Keller et al. [44], an agent is trained to control the yaw, pitch, and depth of the needle to achieve the goal position in ophthalmic surgery with optical coherence tomography (OCT) image as observation. In Gao et al. [45], an agent is trained to provide a remote center of motion (RCM) [57] recommendation in brain surgery. The author considers three aspects to evaluate the quality of RCM, namely, clinical obstacle avoidance (COA), mechanically inverse kinematics (MIK), and mechanically less motion (MLM), and the reward function is also designed based on these three aspects.

However, it can be challenging for rigid needles to find safe trajectories to insert toward the target without touching some critical anatomies, e.g., blood vessels, especially when the structure is complicated. Therefore, steerable needles have attracted much attention in the past decade due to their flexibility. Accurate path planning is one of the most crucial factors for successful steerable needle insertion, where the tissue-needle interaction has to be considered. In Lee et al. [46], an agent is trained to perform preoperative path planning for steerable needles in keyhole neurosurgery with DQN. The environment is simulated by segmenting 2D MRI images into obstacles and obstacle-free areas. The agent controls the bevel direction rotation and insertion depth to insert the needle toward the target area. The agent is rewarded when achieving the goal and punished when entering an unsafe area, e.g., a blood vessel. In Kumar et al. [47] and Segato et al. [48,49], similar frameworks are proposed, however, with 3D MRI images to enable 3D path planning and continuous actions.

Preoperative path planning provides initial planning, including the insertion point and a rough global plan. However, due to unexpected anatomical movements or needle-tissue interactions, the preplanned path can be violated, and therefore, intraoperative replanning is needed. Furthermore, it is also essential for the surgeon to easily detect the risk potential (possibility of the needle entering unsafe areas) of the re-planned path. In Tan et al. [58], a universal distributional Q-learning (UDQL) [59] based training framework is proposed to enable fast replanning and risk management. In UDQL, the expected Q-value is parameterized with a value distribution so that only a distribution with a high mean Q-value and low variance can be considered a safe plan.

## Catheterization

Catheterization is one of the most commonly used procedures in endovascular intervention. The catheter is guided to the target of the disease in the vasculature along with treatments such as stenting, embolization, and ablation, as illustrated in Fig. 8. However, the guidance of the catheter is not trivial. The surgeon needs to manipulate the catheter with limited 2D fluoroscopic information and minimize unwanted excessive tissue contacts. Due to the difficulty of

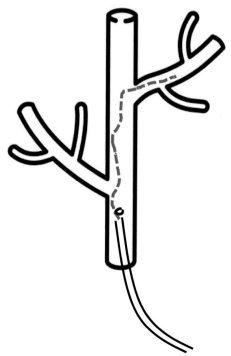

FIG. 8  The catheter is guided to the target position in vessel. The blue dot line is the planned path.

manually operating the catheter, robot-assisted catheterization has been researched in the past decade, and DRL is one of the promising methods for its path or trajectory planning.

In Chi et al. [50], an agent is trained to optimize the catheterization trajectory demonstrated by the experts. The expert demonstrations are first parameterized with dynamic motion primitives (DMP). The agent adjusts the parameters of DMP to optimize the trajectory, guiding the catheter tip toward the desired vessel plane while matching the trajectory with the vessel centerline as much as possible. The agent is trained with Path Integral ($PI^2$) [60] algorithm, which is a robust RL implementation based on trajectory rollouts. The environment is based on vascular models with flow simulation. Furthermore, Chi et al. [51] proposed a closed-loop catheter control framework based on GAIL to imitate the expert catheterization demonstration. An electromagnetic (EM) tracking sensor is attached to the catheter tip to take into account its real-time pose to enable intraoperative control. The agent is trained to imitate the five-motion primitive of the expert's hand demonstration, namely, pull, push, clockwise rotation, anticlockwise rotation, and stand-by. Besides, in Omisore et al. [52], DRL is utilized to tune the parameters of a PID controller in real time, to let it adapt to different blood flow settings.

## Future scenario

As surgical robotics continues to advance, the integration of deep reinforcement learning holds immense potential for enhancing the automation level in surgical procedures. With further research and development in this area, we can anticipate several future scenarios:

1. Autonomous Surgical Robots: Deep reinforcement learning algorithms can enable surgical robots to autonomously perform complex surgical tasks with high precision and accuracy. This could potentially reduce the need for human intervention during surgeries.
2. Personalized Surgical Procedures: By utilizing reinforcement learning, surgical robots can adapt and learn from patient-specific data and surgical outcomes. This could lead to personalized surgical approaches tailored to individual patients, resulting in improved surgical outcomes.
3. Real-time Decision-making: Reinforcement learning algorithms can enable surgical robots to make real-time decisions during surgeries based on dynamic feedback. This capability can enhance the robot's ability to adapt to changing surgical conditions and optimize surgical outcomes.
4. Collaborative Surgical Robotics: Reinforcement learning can facilitate the collaboration between surgical robots and human surgeons. By learning from human expertise and incorporating it into their decision-making processes, robots can become valuable assistants to surgeons, enhancing their capabilities and reducing surgical errors.
5. Continuous Learning and Improvement: Deep reinforcement learning enables surgical robots to continuously learn and improve their performance over time. By gathering and analyzing data from previous surgeries, robots can refine their skills and adapt to new challenges, ultimately enhancing their overall performance in surgical procedures.

## Conclusion

This literature review has provided an overview of the application of DRL in surgical robots. We divided the state-of-art works that applied DRL on surgical robots into three main fields: skin-interfaced, intrabody, and percutaneous, discuss how they formulate the problem based on RL-framework, and compare their methodologies and limitations. Based on the outstanding performance of DRL in these works, the integration of DRL algorithms into surgical robotic systems has the potential to revolutionize the field of robotic-assisted surgery by enhancing the autonomy and decision-making capabilities of these systems.

The technology of DRL is in its youth and still suffers from some limitations, e.g., low data efficiency, expensive to train in the real world, and lack of safety guarantee. Looking forward, further research is needed to refine and optimize DRL algorithms for surgical applications. This includes the following points: First, more efficient training methodologies. Currently, most DRL algorithms are very sample inefficient compared to other deep learning methods. Second, a more accurate simulation environment. In surgical scenarios, there exist a lot of deformable bodies interaction, which are much harder to simulate, compared to rigid bodies interaction. Third, addressing safety concerns. Safety is always the priority in surgeries. This could include risk analysis or interpretability of the model. Last, conducting clinical trials to evaluate the effectiveness and reliability of DRL-based surgical robots because, so far, few DRL-based robots have been tested in real clinical scenarios.

# Key points

- Surgical robotics is a rapidly evolving field that offers numerous benefits, including enhanced precision, reduced invasiveness, and reduced surgeon fatigue.
- Reinforcement learning, a type of machine learning, shows promise in enhancing the automation level in surgical robotics.
- Applications of reinforcement learning in surgical robotics include preoperative, intrabody, and percutaneous procedures.
- Reinforcement learning can teach surgical robots complex tasks such as suturing and tissue manipulation, improving their autonomy and accuracy.
- The integration of deep reinforcement learning can lead to future scenarios such as autonomous surgical robots, personalized surgical procedures, real-time decision-making, collaborative surgical robotics, and continuous learning and improvement.

# References

[1] Peters BS, et al. Review of emerging surgical robotic technology. Surg Endosc 2018;32:1636–55.
[2] Freschi C, et al. Technical review of the da Vinci surgical telemanipulatorhe. Int J Med Robotics Comput Assist Surg 2013;9(4):396–406.
[3] Gomes P, editor. Medical robotics: minimally invasive surgery. Elsevier; 2012.
[4] Moustris GP, et al. Evolution of autonomous and semi-autonomous robotic surgical systems: a review of the literature. Int J Med Robotics Comput Assist Surg 2011;7(4):375–92.
[5] Haidegger T. Autonomy for surgical robots: concepts and paradigms. IEEE Trans Med Robotics Bionics 2019;1(2):65–76.
[6] Ibarz J, Dogangil D, Moustris H, Freschi F, Peters A, et al. How to train your robot with deep reinforcement learning: lessons we have learned. Int J Robotics Res 2021;40(4–5):698–721.
[7] Mnih V, et al. Playing atari with deep reinforcement learning. arXiv Preprint 2013. arXiv:1312.5602.
[8] Nguyen H, La H. Review of deep reinforcement learning for robot manipulation. In: 2019 third IEEE international conference on robotic computing (IRC). IEEE; 2019.
[9] Zhou SK, et al. Deep reinforcement learning in medical imaging: a literature review. Med Image Anal 2021;73, 102193.
[10] Yu C, et al. Reinforcement learning in healthcare: a survey. ACM Comput Surv (CSUR) 2021;55(1):1–36.
[11] Coronato A, et al. Reinforcement learning for intelligent healthcare applications: a survey. Artif Intell Med 2020;109, 101964.
[12] Liu S, et al. Reinforcement learning for clinical decision support in critical care: comprehensive review. J Med Internet Res 2020;22(7), e18477.
[13] Jonsson A. Deep reinforcement learning in medicine. Kidney Dis 2019;5(1):18–22.
[14] Page MJ, et al. The PRISMA 2020 statement: an updated guideline for reporting systematic reviews. Int J Surg 2021;88, 105906.
[15] Arulkumaran K, et al. Deep reinforcement learning: a brief survey. IEEE Signal Process Mag 2017;34(6):26–38.
[16] Sigaud O, Buffet O, editors. Markov decision processes in artificial intelligence. John Wiley & Sons; 2013.
[17] Sutton RS, Barto AG. Reinforcement learning: an introduction. MIT Press; 2018.
[18] Spaan MTJ. Partially observable Markov decision processes. In: Reinforcement learning: state-of-the-art. Berlin, Heidelberg: Springer Berlin Heidelberg; 2012. p. 387–414.
[19] Priester AM, Natarajan S, Culjat MO. Robotic ultrasound systems in medicine. IEEE Trans Ultrason Ferroelectr Freq Control 2013;60(3):507–23.
[20] Baumgartner CF, et al. SonoNet: real-time detection and localisation of fetal standard scan planes in freehand ultrasound. IEEE Trans Med Imaging 2017;36(11):2204–15.
[21] Virga S, et al. Automatic force-compliant robotic ultrasound screening of abdominal aortic aneurysms. In: 2016 IEEE/RSJ international conference on intelligent robots and systems (IROS). IEEE; 2016. p. 508–13.
[22] Hase H, et al. Ultrasound-guided robotic navigation with deep reinforcement learning. In: 2020 IEEE/RSJ international conference on intelligent robots and systems (IROS). IEEE; 2020.
[23] Li K, et al. Autonomous navigation of an ultrasound probe towards standard scan planes with deep reinforcement learning. In: 2021 IEEE international conference on robotics and automation (ICRA). IEEE; 2021.
[24] Li K, et al. Image-guided navigation of a robotic ultrasound probe for autonomous spinal sonography using a shadow-aware dual-agent framework. IEEE Trans Med Robotics Bionics 2021;4(1):130–44.
[25] Bi Y, et al. VesNet-RL: simulation-based reinforcement learning for real-world us probe navigation. IEEE Robotics Autom Lett 2022;7(3):6638–45.
[26] Li K, et al. RL-TEE: autonomous probe guidance for transesophageal echocardiography based on attention-augmented deep reinforcement learning. IEEE Trans Autom Sci Eng 2023;21:1526–38.
[27] Milletari F, Birodkar V, Sofka M. Straight to the point: reinforcement learning for user guidance in ultrasound. In: Smart ultrasound imaging and perinatal, preterm and paediatric image analysis: first international workshop, SUSI 2019, and 4th international workshop, PIPPI 2019, held in conjunction with MICCAI 2019, Shenzhen, China, October 13 and 17, 2019, Proceedings 4. Springer International Publishing; 2019.
[28] Ning G, Zhang X, Liao H. Autonomic robotic ultrasound imaging system based on reinforcement learning. IEEE Trans Biomed Eng 2021;68(9):2787–97.
[29] Lillicrap TP, et al. Continuous control with deep reinforcement learning. In: The international conference on learning representations (ICLR); 2015.

[30] Ronneberger O, Fischer P, Brox T. U-net: Convolutional networks for biomedical image segmentation. In: Medical image computing and computer-assisted intervention—MICCAI 2015: 18th international conference, Munich, Germany, October 5–9, 2015, Proceedings, Part III 18. Springer International Publishing; 2015.

[31] Schulman J, et al. Proximal policy optimization algorithms. arXiv Preprint 2017. arXiv:1707.06347.

[32] Ning G, et al. Inverse-reinforcement-learning-based robotic ultrasound active compliance control in uncertain environments. IEEE Trans Ind Electron 2023.

[33] Badrinarayanan V, Kendall A, Cipolla R. Segnet: a deep convolutional encoder-decoder architecture for image segmentation. IEEE Trans Pattern Anal Mach Intell 2017;39(12):2481–95.

[34] Mayer H, et al. Automation of manual tasks for minimally invasive surgery. France: Gosier; 2008. p. 260–5.

[35] Thananjeyan B, et al. Multilateral surgical pattern cutting in 2d orthotropic gauze with deep reinforcement learning policies for tensioning. In: 2017 IEEE international conference on robotics and automation (ICRA). IEEE; 2017.

[36] Nguyen T, et al. A new tensioning method using deep reinforcement learning for surgical pattern cutting. In: 2019 IEEE international conference on industrial technology (ICIT). IEEE; 2019.

[37] Nguyen ND, et al. Manipulating soft tissues by deep reinforcement learning for autonomous robotic surgery. In: 2019 IEEE international systems conference (SysCon). IEEE; 2019. p. 1–7.

[38] Varier VM, et al. Collaborative suturing: a reinforcement learning approach to automate hand-off task in suturing for surgical robots. In: 2020 29th IEEE international conference on robot and human interactive communication (RO-MAN). IEEE; 2020.

[39] Chiu Z-Y, et al. Bimanual regrasping for suture needles using reinforcement learning for rapid motion planning. In: 2021 IEEE international conference on robotics and automation (ICRA). IEEE; 2021.

[40] Pore A, et al. Learning from demonstrations for autonomous soft-tissue retraction. In: 2021 international symposium on medical robotics (ISMR). IEEE; 2021.

[41] Pore A, et al. Safe reinforcement learning using formal verification for tissue retraction in autonomous robotic-assisted surgery. In: 2021 IEEE/RSJ international conference on intelligent robots and systems (IROS). IEEE; 2021.

[42] Scheikl PM, et al. Sim-to-real transfer for visual reinforcement learning of deformable object manipulation for robot-assisted surgery. IEEE Robot Autom Lett 2022;8(2):560–7.

[43] Schulman J, et al. Trust region policy optimization. In: International conference on machine learning. PMLR; 2015.

[44] Keller B, et al. Optical coherence tomography-guided robotic ophthalmic microsurgery via reinforcement learning from demonstration. IEEE Trans Robotics 2020;36(4):1207–18.

[45] Gao H, et al. Remote-center-of-motion recommendation toward brain needle intervention using deep reinforcement learning. In: 2021 IEEE international conference on robotics and automation (ICRA). IEEE; 2021.

[46] Lee Y, et al. Simulation of robot-assisted flexible needle insertion using deep Q-network. In: 2019 IEEE international conference on systems, man and cybernetics (SMC). IEEE; 2019. p. 342–6.

[47] Kumar J, Raut CS, Patel N. Automated flexible needle trajectory planning for keyhole neurosurgery using reinforcement learning. In: 2022 IEEE/RSJ international conference on intelligent robots and systems (IROS). IEEE; 2022.

[48] Segato A, et al. Inverse reinforcement learning intra-operative path planning for steerable needle. IEEE Trans Biomed Eng 2021;69(6):1995–2005.

[49] Segato A, et al. Ga3c reinforcement learning for surgical steerable catheter path planning. In: 2020 IEEE international conference on robotics and automation (ICRA). IEEE; 2020.

[50] Chi W, et al. Trajectory optimization of robot-assisted endovascular catheterization with reinforcement learning. In: 2018 IEEE/RSJ international conference on intelligent robots and systems (IROS). IEEE; 2018.

[51] Chi W, et al. Collaborative robot-assisted endovascular catheterization with generative adversarial imitation learning. In: 2020 IEEE international conference on robotics and automation (ICRA). IEEE; 2020.

[52] Omisore OM, et al. A novel sample-efficient deep reinforcement learning with episodic policy transfer for PID-based control in cardiac catheterization robots. arXiv Preprint 2021. arXiv:2110.14941.

[53] Patil S, Alterovitz R. Toward automated tissue retraction in robot-assisted surgery. In: 2010 IEEE international conference on robotics and automation. IEEE; 2010.

[54] Corsi D, et al. Formal verification for safe deep reinforcement learning in trajectory generation. In: 2020 fourth IEEE international conference on robotic computing (IRC). IEEE; 2020. p. 352–9.

[55] Reisch R, et al. The keyhole concept in neurosurgery. World Neurosurg 2013;79(2):S17. e9.

[56] Berry S, Ligda KO. Ophthalmic surgery. New York: Springer; 2015.

[57] Aksungur S. Remote center of motion (RCM) mechanisms for surgical operations. Int J Appl Math Electron Comput 2015;3(2):119–26.

[58] Tan X, et al. Robot-assisted flexible needle insertion using universal distributional deep reinforcement learning. Int J Comput Assist Radiol Surg 2020;15:341–9.

[59] Tan X, et al. Robust path planning for flexible needle insertion using Markov decision processes. Int J Comput Assist Radiol Surg 2018;13:1439–51.

[60] Theodorou E, Buchli J, Schaal S. A generalized path integral control approach to reinforcement learning. J Mach Learn Res 2010;11:3137–81.

# 10

# Ultrasound guidance and robotic procedures: Actual and future intelligence

*Long Bai[a],\*, Lei Zhao[a,b],\*, and Hongliang Ren[a]*

[a]Department of Electronic Engineering, The Chinese University of Hong Kong, Sha Tin, Hong Kong
[b]College of Computer Science and Electronic Engineering, Hunan University, Changsha, China

## Introduction

Robotic technology has been used in minimally invasive surgery (MIS) for nearly 40 years [1–3]. Patients can benefit from MIS via smaller incisions and faster recovery times, while robotics can further improve the accuracy and precision of MIS [4]. Meanwhile, in the context of the COVID-19 pandemic, robot-assisted surgery can further help medical staff reduce exposure risks [5,6]. Various imaging technologies are used to provide real-time dynamic perception to surgical robots [7–9]. Ultrasound (US) imaging technology is a standard medical imaging modality that can provide 2D or 3D imaging. Its potential applications can be anatomical detection, positioning, and relative spatial posture of targets or lesions, as well as tracking the shape of medical instruments and their real-time relative spatial position within surrounding tissues [10]. Figure 1 presents a typical US-guided robotic surgery system. US imaging can also be combined with other medical imaging techniques (e.g., RGB cameras [12] and magnetic resonance imaging (MRI) [13]). Some functional ultrasonography (e.g., Doppler [14] and Elastography [15]) can provide additional information and guidance for surgery. US image is cost-effective, noninvasive, and can conduct real-time volumetric imaging. Therefore, US guidance has been widely used in robotic surgery [1].

This chapter will begin with an overview of the different US modalities, as shown in Fig. 2, followed by an overall review of current US-guided robotic surgery. Finally, we will further discuss the current challenges and potential prospects of US-guided robotic surgery.

## US technologies in evolution

### Diagnostic US

With the characteristics of convenient inspection and intuitive images, diagnostic US can perform noninvasive imaging of internal organs in the human body [16]. Specifically, US technology is mainly used to diagnose liquid and substantive lesions. Its detectable areas cover the brain, heart, blood vessels, liver, gallbladder, pancreas, spleen, chest cavity, kidney, ureter, bladder, urethra, uterus, pelvic appendages, prostate, seminal vesicles, eyes, thyroid, breast, salivary glands, testes, peripheral nerves, and tendons of limbs. However, the diagnostic value of lesions in the stomach, lungs, and gastrointestinal tract is limited, because these areas are filled with air [19]. Typical forms of ultrasound imaging include 2D, 3D, and Dynamic 3D.

---

\* Equal contribution.

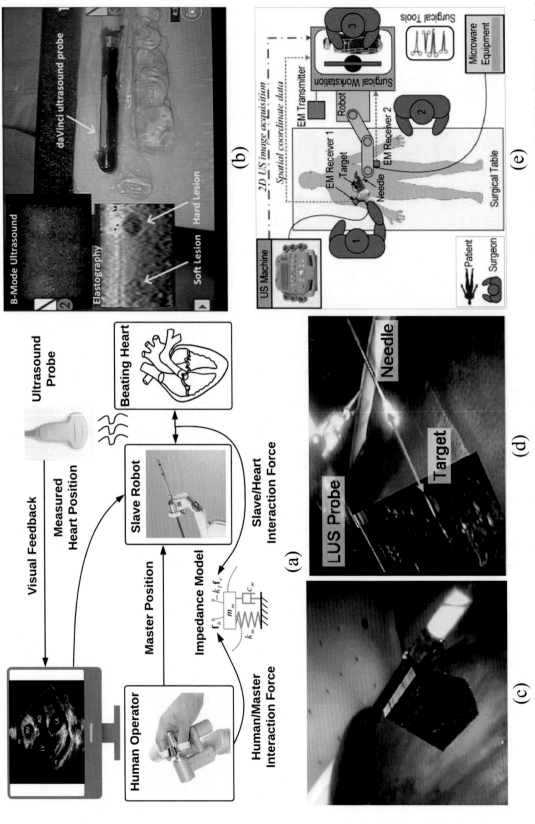

FIG. 1  Integration of US and surgical robot systems. (A) a typical US-guided robotic surgery system [11]; (B) two different stiffness values detected by US [1]; (C) 3D image obtained by laparoscopic US [2]; (D) 2D US image of the target lesion [2]; (E) concept of ultrasound-based ablation robotic system [1].

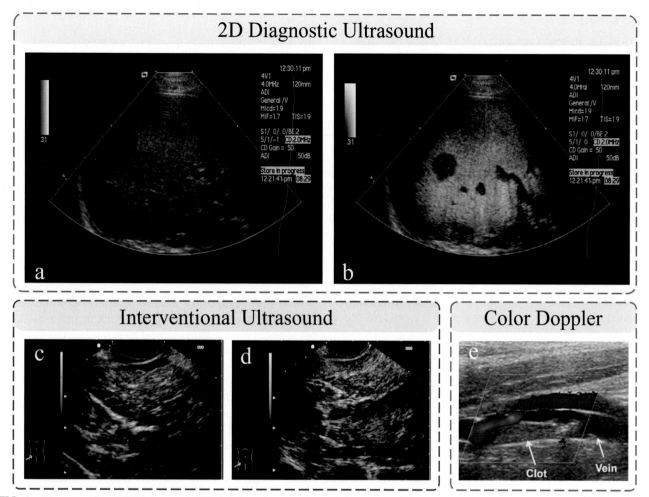

FIG. 2 Demonstration of different types of US. (A) Conventional 2D US scan [16]. (B) 2D US scan of the same patient following the administration of a US contrast agent. This lesion is clearly visible [16]. (C) An endoscopic US image of the left kidney and the adrenal gland. A T-tag anchor has been ejected with the endoscopic US needle to guide the instrument movement [17]. (D) The white reflection close to the adrenal gland is the anchor for guidance [17]. (E) The color Doppler US is detecting a clot blocking blood flow [18].

## Functional US

Functional US contains Doppler US, color Doppler US, and elastography. Doppler US uses the Doppler effect to calculate the relative velocity of a structure (e.g., blood flow) and detect whether the structure is moving toward or away from the probe [20]. Doppler US can be processed by autocorrelation technology, color-coded, and superimposed on the 2D US image [21]. Then, it can provide both morphological and hemodynamic information. Elastography is a method that can show the relative firmness of tissue and can distinguish tumors from healthy tissue. Elastography can display the relative stiffness map for the human tissue, e.g., a black-and-white map showing a high-contrast tumor image compared to an anatomical image, or a color-coded map overlaid on the anatomical image [22,23].

## Interventional US

Routine ultrasonography is conducted on the surface of the body. Some organs or lesions are located in the deep part of the pelvic cavity, such as vaginal examination of the uterus, and appendages. Endoscopic US can be inserted or injected into the target lesion, cyst cavity, body cavity, pipeline, and other specific parts to achieve the purpose of diagnosis and treatment, called interventional US [24,25].

## Therapeutic US

Therapeutic US shall focus high-level acoustic on targeted tissues to heat or ablate them [26]. High-intensity focused US (HIFU) refers to a type of therapeutic US using highly targeted, high-intensity sound beams, which can be used to change or eliminate diseased tissue (e.g., a tumor) [27] and break up clots in blood vessels [28].

## US guidance in robotic surgery

To overcome the limitations of US imaging, the innovative US robotic has been developed to integrate a robotic system with a US station. This unique setup involves attaching a US probe to the robot's end-effector, enabling enhanced imaging capabilities and expanding the possibilities of US-based procedures. This section primarily highlights the latest research findings on cutting-edge US robotic systems within the past 5 years and explores potential directions of US robotics in surgical guidance. We organize the section according to the level of robot autonomy (LORA) into tele-operated, human-robot cooperative, and autonomous systems [29].

### Tele-operated

A master–slave architecture is frequently utilized for the remote operation of the US probe in tele-operated robot-assisted US imaging. The sonographer controls the slave robot on the master side using a joystick or force feedback device, adjusting the US probe's direction and force to acquire the real-time US flowing and transfer them to the monitor. At the slave side, the robot fixed force sensor at the end performs an examination for the patient under the supervision of the sonographer.

An overview of the recent representative works is shown in Table 1. Arent et al. [30] presented a robotic system called ReMeDi for remote medical examination, which supported remote interview, observation, auscultation, and US examination. Guan et al. [31] designed a novel six-DOF robot, with a force sensor controlled by a dummy probe to reserve doctors' habits in ultrasound diagnosis. Adams et al. [32] assessed the acceptance of a telerobotic US system by users and patients in the routine prenatal sonographic examination. Kurnicki et al. [33] designed a kinematically redundant robotic arm and an additional force-torque sensor to reproduce all necessary movements of the examiner. Huang et al. [34] exploited a 6-DOF robot arm without force sensors to remotely manipulate the probe via a joystick for the subsequent 3D volume reconstruction and visualization. To reduce infection risk for the physician, a feasible tele-examination scheme for COVID-19 patients was investigated in Wang's study [35]. Giuliani et al. [36] developed a

TABLE 1 Overview of tele-operated US robotic systems.

| First author | Procedure/target | Key features | Robot type | LORA | Year |
|---|---|---|---|---|---|
| Arent et al. | US examination | User requirement-centered design methodology | Serial; 7 active DOF; custom design | Tele-operation | 2017 |
| Guan et al. | US examination | Transmit the US image information in real-time | Serial; 6 DOF; custom design | Tele-operation | 2017 |
| Adams et al. | Prenatal sonographic examinations | Clinically test for thirty participants in obstetric | Serial; 6 DOF; custom design | Tele-operation | 2018 |
| Kurnicki et al. | US examination | Reproduce all necessary movements of the examiner | Serial; 7 active DOF; custom design | Tele-operation | 2019 |
| Huang et al. | Vascular | 3D construction of human body parts | Serial; 6 DOF; C4 [Epson] | Tele-operation | 2019 |
| Wang et al. | COVID-19 pneumonia | Eliminated infection risk for the physician | Serial; 6 DOF; custom design | Tele-operation | 2020 |
| Giuliani et al. | echocardiography examination | User-centered design methodology | Serial; 7 active DOF; custom design | Tele-operation | 2020 |
| Geng et al. | US examination | Filtering haptic commands and improve safety | Serial; 6 DOF; UR5 [Universal Robots] | Tele-operation | 2020 |
| Zhou et al. | US examination | A good sense of operation and real-time | Serial; 7 DOF; Panda [Franka Emika Robts] | Tele-operation | 2021 |

TABLE 2  Overview of human-robot cooperated robotic US systems.

| First author | Procedure/ target | Key features | Robot type | LORA | Year |
|---|---|---|---|---|---|
| Virga et al. | Soft-tissue applications | Merely need for the tracked US images and force information | Serial; 7 DOF; LBR iiwa [KUKA] | Human-robot cooperated | 2018 |
| Esteban et al. | Spinal needle insertion | Visualize soft-tissue landmarks to guide the target anatomy | Serial; 7 DOF; LBR iiwa [KUKA] | Human-robot cooperated | 2018 |
| Li et al. | Liver tumors | A respiratory motion calibration and real-time multimodality imaging | Serial; 6 DOF; UR5 [Universal Robot] | Human-robot cooperated | 2018 |
| Jiang et al. | Orthopedic applications | Estimating the optimal probe orientation | Serial; 7 DOF; LBR iiwa [KUKA] | Human-robot cooperated | 2020 |
| Ipsen et al. | US examination | Integrating dynamic configuration changes and the robot's kinematics | Serial; 7 DOF; LBR iiwa [KUKA] | Human-robot cooperated | 2021 |

robot to assist doctors with remote echocardiography examinations, successfully executing auscultation on 14 patients. Geng et al. introduced a groundbreaking master–slave velocity mapping and control mechanism. This pioneering approach establishes a solid foundation for enabling precise and reliable US imaging remotely, opening up new possibilities for medical diagnostics, interventions, and treatment planning. By introducing real-time position tracking and dynamic guidance techniques, Zhou et al. [37] pushed the boundaries of the field even further. This notable contribution expands the possibilities for precise and interactive control of robotic systems, particularly in areas such as surgical navigation and medical interventions. Nowadays, tele-operated US systems have evolved to accommodate remote examination at varying distances supported by commercial systems.

Human-robot cooperated. The operator and the robot share the control degrees of freedom under a human-robot cooperated system. Several noteworthy human-robot cooperated systems are listed in Table 2. Virga et al. [38] proposed a correction technique for soft-tissue distortion induced by pressure in 3D ultrasound images. In a pivotal clinical investigation by Esteban et al. [39], robotic-guided needle insertion was performed by placing the needle holder along a planned trajectory within a 3D ultrasound (US) spine volume. Li et al. [40] proposed an innovative system that combines real-time multimodality imaging compatibility with a respiratory motion calibration module, which was specifically designed to facilitate the accurate insertion and guidance of needles during liver malignancy ablation procedures. Jiang et al. [41] enhanced the quality of the US images by full orientation optimization of the US probe. Ipsen et al. [42] suggested a system with the potential for diagnostic and therapy-guiding purposes, which consisted of a collaborative robot and a 4D ultrasound system to facilitate long-term data acquisition over extended imaging sessions. Current human-robot cooperated systems aim to assist physicians throughout the US examination through probe placement, navigation, and more intuitive visualizations. While increasing comfort and lessening the physician's mental strain, these collaborative aiding systems enhance the quality of US image acquisitions at the same time.

## Autonomous

Autonomous US-guided robotic systems have been created during the past 5 years, most of them at an experimental stage in Table 3. Autonomous robotic US imaging systems commonly comprise three key components: a US device, a robot arm, and a tracking system. The tracking system might be a passive mechanically encoded system or an optical/electromagnetic tracking system. The precise execution of the predetermined route for the ultrasound (US) scan is accomplished through the direct actuation of the robotic arm, ensuring accurate positioning and controlled movement during the imaging process. Hennersperger et al. [43] designed a US robot system that enabled direct patient-specific trajectory planning by choosing the start and end points on the condition of patients holding their breath. An et al. [44] adapted the HIFU transducer to identify the tumor size and location in the US images. Unlike previous systems, the robot controls the treatment transducer instead of the US probe. The study conducted by Kojcev et al. [45] showcased the superior measurement performance achieved with the US robotic system, surpassing that of expert-operated sonography. Huang et al. [45] designed a 6-DOF robot for full automatic determination of the scan range and path planning, which could adapt to the 3D contour of the skin surface. Langsch et al. [46] developed a real-time autonomous catheter-tracking trajectory and visual feedback system focusing on endovascular aneurysm repair. A method created by Von Haxthausen et al. [47] allows the robot to track peripheral arteries once the probe is manually

TABLE 3    Overview of autonomous robotic US systems.

| First author | Procedure/target | Key features | Robot type | LORA | year |
|---|---|---|---|---|---|
| Hennersperger et al. | Vascular | Robotic 3D US acquisitions | Serial; 7 DOF; LBR iiwa [KUKA] | Autonomous | 2017 |
| An et al. | Tumor | Treatment transducer controlled by robot | Serial; 4 DOF; YK400XG; [YAMAHA] | Autonomous | 2017 |
| Kojcev et al. | Thyroid | The reproducibility of measurements | Serial; 7 DOF; LBR iiwa [KUKA] | Autonomous | 2017 |
| Huang et al. | US examination | Fully automatic scanning and 3D imaging. | Serial; 6 DOF; C4 [EPSON] | Autonomous | 2018 |
| Langsch et al. | Vascular | Real-time catheter tracking and navigation | Serial; 7 DOF; LBR iiwa [KUKA] | Autonomous | 2019 |
| Von Haxthausen et al. | Peripheral arterial disease | The treatment transducer controlled by robot | Serial; 7 DOF; LBR iiwa [KUKA] | Autonomous | 2020 |
| Li et al. | Spine | Navigate a virtual US probe | Serial; 6 DOF; LBR iiwa [KUKA] | Autonomous | 2021 |
| Jiang et al. | Vascular | An end-to-end workflow | Serial; 6 DOF; LBR iiwa [KUKA] | Autonomous | 2021 |

placed initially, and convolutional neural networks are used to detect the vessels. Li et al. [48] successfully devised a deep reinforcement learning framework that autonomously navigated the virtual 6-DOF US probe to detect standard planes in real-time image feedback. Jiang et al. [49] investigated an end-to-end process of autonomous robotic US screening for tubular structures, and simultaneously estimated vessel radius in real-time US sweeps. The significance of autonomous systems in robot US systems cannot be understated, as they possess the potential to remove operator dependency and ultimately enable the development of fully automatic US-guided intervention systems.

## Trends of future directions

Image analysis, robot navigation, and virtual reality are the three key application areas to boost the autonomy of US robotic systems in the future. The intelligent image analysis systems can aim for specific disease diagnosis of the 2D/3D US images [50], and the identification of the individually optimal therapy [51]. Deep reinforcement learning has recently achieved a breakthrough in autonomous robot navigation for landmark detection in US images [52]. Virtual reality extends the robot for treatment guidance [53], simulation/verification of the robot setup [54,55], and fully immersed head-mounted displays. These approaches are of great significance for solving the task of US probe placement, which is still one of the open challenges in autonomous US robotic system development.

## Discussion and challenges

This section discusses further advancements in US-guided medical robotics and explores possible development directions and challenges. Figure 3 presents various US-assisted robotic surgery systems.

## Robotic-assisted surgery

For applications in robotic-assisted surgery, devices such as robotic arms shall replace surgeons operating on patients, which reduces instability and provides high-precision motion and manipulation [2,58]. Automation has been slow to advance in this application, with the major development being to provide additional feedback to the robot operator or the surgeon. This is because surgeons cannot achieve the same comprehensive perception (such as tactile perception and 360-degree viewing angle) as in-person surgery when operating remotely [56]. Existing studies include tracking surgical instruments and tissues based on US imaging [59–62], using elastography to provide tactile feedback to sense tumor localization [63], and introducing enhanced visual interfaces to aid surgeon perception [64]. In robotic-assisted surgery, the decision-making and execution of surgical robots require high accuracy, precision, and dexterity, and current fully automated products cannot thoroughly cover these requirements [58]. Meanwhile, the quantitative

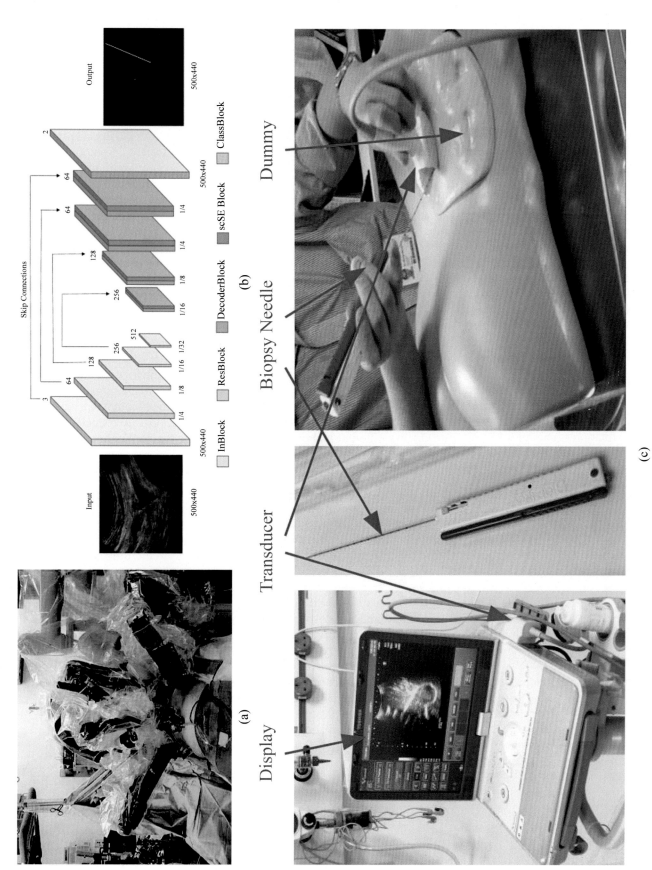

FIG. 3   Various US-assisted robotic surgery systems. (A) floor-standing da Vinci robotic platform for clinical use [56]; (B) excitation network with encoder–decoder architecture for US needle tracking [57]; (C) a simple experimental platform based on the handheld US and biopsy needle [57].

human anatomy that US imaging can provide is relatively limited, and the system needs to work more than partially autonomously based on this information alone [1]. In summary, surgeon interaction is still necessary. The fully automatic robotic surgery system still needs further research and clinical verification.

## Percutaneous surgery

Numerous automated/semi-automated procedures have been developed for percutaneous puncture surgery [65–71]. However, only a few semi-autonomous systems have been clinically tested, and most have still experimented in simple simulated scenarios [65,66,69]. Therefore, in these puncture procedures, the overall system does not require complex decision-making procedures, the number of steps is limited, the surgical site is also apparent and easy to reach, and the degree of automation of these systems is verified in these relatively easy environments. For example, the system presets the current and entry point of the biopsy needle. The biopsy needle only needs to reach the target along a straight line and exit along the same line [1]. Nevertheless, these systems underestimate the consideration of the intraoperative motion of the organ. Furthermore, these systems tested on phantoms/static images do not consider the tissue inhomogeneities and potentially complicated deformations, which oversimplifies the procedure of identifying and tracking targeted structures [69,70]. In this case, some semi-autonomous systems usually do not consider the task of object tracking under the primary assumption. Whenever the organ moves or deforms, real-time imaging can directly provide this feedback to the clinician, who can then compensate for the puncture position based on this feedback [67,68]. However, only a few studies considered correcting the puncture needle route under organ motion or deformation among the fully autonomous puncture systems [66,69,72].

Currently, the US images still have the following limitations: (i) the field of view (FOV) is limited, and (ii) the probe manipulation is sensitive to the imaging output. Therefore, it is challenging to effectively track soft tissue using US images [73,74]. In addition to hardware-based US adjustment strategies [75], various artificial intelligence-based segmentation and object detection algorithms have been proposed to automatically identify and track surgical instruments or human tissues, thereby improving the automation of robotic systems [57,76]. Artificial intelligence-based methods can effectively identify significant variations in images, providing automated intraoperative tracking and image interpretation capabilities [77–80]. Introducing different imaging modalities may be a potential improvement strategy, which we will discuss later.

## Endoscopic US

The endoscopic US system integrates ultrasonic and optical imaging. The system can monitor the surface morphology of mucosa and achieve histological information. Therefore, it provides the depth resolution that conventional electronic endoscopes lack. Meanwhile, endoscopic US significantly shortens the acoustic circuit utilizing internal detection, avoiding the interference of other organs on the US signal, and then using a higher-frequency range for detection compared with surface US imaging, significantly improving the accuracy and vertical resolution of ultrasonic images [81]. The increasing significance of endoscopic US lies in its ability to assist in assessing lesion extent, invasion depth, and lesion characteristics [82]. Combining endoscopic ultrasonography with robotic surgery, the endoscopic US can provide guidance information that cannot be provided by conventional RGB endoscopy and body surface US. For example, the endoscopic US can provide closer and more direct image-guided information when a surgical target is covered by fatty tissue or extended in muscle. However, the manipulation of the endoscopic US itself presents challenges. The endoscopic US itself is also a medical robot. Under the limited wound of MIS, it is difficult for the endoscopic US to reach the target manually through a specific area. The combination of the endoscopic US and natural orifice transluminal endoscopic surgery (NOTES) has proven safe and feasible [83]. In laparoscopy, mediastinoscopy/thoracoscopy, gastrojejunostomy, and adrenalectomy, endoscopic US has all been shown to be helpful in simulations and animal experiments [17,84]. More clinical trials shall be further conducted and researched to demonstrate the clinical value of the endoscopic US.

## Multimodel perception

Intraoperative navigation for US volume imaging remains a challenging goal. Various research works focus on combining the US with MRI or RGB endoscopy to provide a more comprehensive perception and FOV [13,70,85–88]. The combination of US and MRI can be divided into intraoperative combined navigation and two-step application. Several works have achieved joint tracking via rigid/deformable registration of heterogeneous images [13,70,85]. Other work has taken a two-step format, using information from US data for preoperative scene simulation and intraoperative MRI-based navigation for real-time navigation [86]. In the combination of US and RGB camera navigation, the most successful attempt so far is to combine RGB endoscopic cameras with US for laparoscopic surgery. These two modes can form a good complement. US images provide an intuitive visualization of the structure of human tissue and surgical instruments, while endoscopic videos can provide decision-makers with easy-to-interpret images of the body. The problem with the cooperation of these two images is that the fusion of the two images in different coordinate systems needs to be considered. The 2D plane presented by the endoscope is required to be registered to the US image [87]. Some techniques based on image processing or artificial intelligence have been developed to capture and register similar structures of images from the two modalities [87,88].

Another concern of US imaging for clinical application is that the currently widely used 2D US only shows one plane, and the operator must manually align it with the target structure. Therefore, if there is significant anatomical deformation outside this plane, it is challenging to visualize this on 2D US imaging. The method of volume imaging can perform 3D reconstruction on the 2D plane of the target area. However, the slow reconstruction still cannot avoid the geometric aberration caused by the possible deformation in the process. The emerging fast 3D/4D US imaging has the potential to help with this significantly [89]. Rapid volumetric imaging can provide real-time deformed structures in vivo for intraoperative navigation. Besides, internal US probes are used in some applications (e.g., laparoscopic surgery), but their FOV is usually limited [90]. Augmented reality technology may provide avenues to overcome these restrictions and provide decision-makers with a better operating environment [91].

## Conclusion

Accurate, safe, reliable, less harmful, and real-time US guidance is widely used in robotic surgery. Various tasks in robotic surgery, such as recognition, segmentation, object detection, and object tracking, can be performed on US images. Furthermore, the different forms of US (3D, 4D, color Doppler, and elastography) with image processing and artificial intelligence technologies can develop multimodal visualization and tracking techniques to provide more detailed feedback for the robot itself or clinicians. The properties of the US make it an essential component of semi-autonomous/autonomous surgical robotic systems.

## Future Scenario

Currently, many semi-autonomous US-guided surgeries have been clinically validated. New technologies are enhancing surgical visualization and guidance, with advances like contrast-enhanced US, ultra-high frequency US, and photoacoustic imaging, often combined with AR, becoming essential for better intraoperative anatomy visualization. In addition, US-assisted surgery can extend care to remote and unique environments, such as space or disaster-stricken areas. Furthermore, a comprehensive dynamic surgical map may be generated using the US to clinically verify the feasibility and safety of fully automated robotic surgery.

Looking ahead, the integration of ultrasound guidance with robotic surgery is poised to undergo significant advancements, driven by rapid technological innovations and an increasing demand for precision and safety in surgical procedures. One prominent direction is the enhancement of AI capabilities in robotic systems. Future developments are likely to focus on the implementation AI algorithms that can analyze ultrasound data in real time to provide dynamic surgical guidance. This could lead to improvements in the accuracy of diagnostics and the adaptability of surgical robots to changing conditions during operations.

Another anticipated trend is the evolution of multimodal imaging systems that seamlessly integrate ultrasound with other imaging modalities such as MRI and CT scans. This integration will provide a more comprehensive view of the surgical field, enabling more informed decision-making and precise manipulations by robotic systems. Additionally, the development of miniaturized and more flexible ultrasound probes could expand the usability of ultrasound-guided robotic systems in minimally invasive surgeries, further reducing recovery times and improving patient outcomes.

## Key Points

- We emphasize significant advances in ultrasound imaging technologies that have improved the quality and scope of visual data available during robotic surgeries. Innovations such as 3D and dynamic imaging have made it possible for surgeons to achieve better precision and outcomes.

- The integration of artificial intelligence with ultrasound-guided robotic systems has led to the development of more autonomous and efficient surgical robots. AI's role in interpreting ultrasound data and making real-time adjustments enhances the capabilities of surgical robots, leading to more consistent and reliable outcomes.
- The importance of human-robot interaction in surgical settings is underscored, highlighting advancements in teleoperation and cooperative systems that allow for finer control and flexibility during operations, ultimately improving patient safety and surgical efficacy.
- We outline current limitations, such as the need for higher resolution in ultrasound imaging and better sensory feedback in robotic systems. Addressing these challenges is critical for the next generation of robotic surgical systems, which will require more sophisticated sensor integration and data-processing capabilities.

# References

[1] Antico M, Sasazawa F, Wu L, et al. Ultrasound guidance in minimally invasive robotic procedures. Med Image Anal 2019;54:149–67.
[2] Leven J, Burschka D, Kumar R, et al. DaVinci canvas: a telerobotic surgical system with integrated, robot-assisted, laparoscopic ultrasound capability. In: International conference on medical image computing and computer-assisted intervention; 2005. p. 811–8.
[3] Shi C, Luo X, Guo J, et al. Three-dimensional intravascular reconstruction techniques based on intravascular ultrasound: a technical review. IEEE J Biomed Health Inform 2017;22(3):806–17. 2168–2194.
[4] Elgezua I, Kobayashi Y, Fujie MG. Survey on current state-of-the-art in needle insertion robots: open challenges for application in real surgery. Procedia CIRP 2013;5:94–9.
[5] Gates B. Responding to Covid-19—a once-in-a-century pandemic? N Engl J Med 2020;382(18):1677–9.
[6] Zemmar A, Lozano AM, Nelson BJ. The rise of robots in surgical environments during COVID-19. Nat Mach Intell 2020;2(10):566–72.
[7] Lee S-L, Lerotic M, Vitiello V, et al. From medical images to minimally invasive intervention: computer assistance for robotic surgery. Comput Med Imaging Graph 2010;34(1):33–45.
[8] Ho H, Yuen J, Mohan P, et al. Robotic transperineal prostate biopsy: pilot clinical study. Urology 2011;78(5):1203–8.
[9] Ren H, Sun Y, Tan KL. Preliminary development of a soft robotic ultrasound steering system. In: 2016 IEEE international conference on consumer electronics (ICCE); 2016. p. 567–8. 1467383643.
[10] Fenster A, Downey DB, Cardinal HN. Three-dimensional ultrasound imaging. Phys Med Biol 2001;46(5):R67.
[11] Cheng L, Tavakoli M. Ultrasound image guidance and robot impedance control for beating-heart surgery. Control Eng Pract 2018;81:9–17.
[12] Azizian M, Khoshnam M, Najmaei N, et al. Visual servoing in medical robotics: a survey. Part I: endoscopic and direct vision imaging–techniques and applications. Int J Med Robot Comput Assist Surg 2014;10(3):263–74.
[13] Ukimura O, Desai MM, Palmer S, et al. 3-Dimensional elastic registration system of prostate biopsy location by real-time 3-dimensional transrectal ultrasound guidance with magnetic resonance/transrectal ultrasound image fusion. J Urol 2012;187(3):1080–6.
[14] Adebar TK, Okamura AM. 3D segmentation of curved needles using Doppler ultrasound and vibration. In: International conference on information processing in computer-assisted interventions; 2013. p. 61–70.
[15] Deshmukh NP, Kang HJ, Billings SD, et al. Elastography using multi-stream GPU: an application to online tracked ultrasound elastography, in-vivo and the da Vinci surgical system. PLoS One 2014;9(12), e115881.
[16] Wells PN. Ultrasound imaging. Phys Med Biol 2006;51(13):R83.
[17] Fritscher-Ravens A, Ghanbari A, Cuming T, et al. Comparative study of NOTES alone vs. EUS-guided NOTES procedures. Endoscopy 2008;40(11):925–30.
[18] National Institute of Biomedical Imaging and Bioengineering. Ultrasound. National Institutes of Health; 2016.
[19] Sanches JM, Laine AF, Suri JS. Ultrasound imaging[M]. Springer; 2012.
[20] Routh HF. Doppler ultrasound. IEEE Eng Med Biol Mag 1996;15(6):31–40.
[21] Taylor KJ, Burns PN, Well PN. Clinical applications of Doppler ultrasound. Raven Pr; 1987.
[22] Gennisson J-L, Deffieux T, Fink M, et al. Ultrasound elastography: principles and techniques. Diagn Interv Imaging 2013;94(5):487–95.
[23] Sigrist RM, Liau J, El Kaffas A, et al. Ultrasound elastography: review of techniques and clinical applications. Theranostics 2017;7(5):1303.
[24] Holm HH, Skjoldbye B. Interventional ultrasound. Ultrasound Med Biol 1996;22(7):773–89.
[25] Rösch T, Lorenz R, Braig C, et al. Endoscopic ultrasound in pancreatic tumor diagnosis. Gastrointest Endosc 1991;37(3):347–52.
[26] Miller DL, Smith NB, Bailey MR, et al. Overview of therapeutic ultrasound applications and safety considerations. J Ultrasound Med 2012;31(4):623–34.
[27] Kennedy JE. High-intensity focused ultrasound in the treatment of solid tumours. Nat Rev Cancer 2005;5(4):321–7.
[28] Jo J, Forrest ML, Yang X. Ultrasound-assisted laser thrombolysis with endovascular laser and high-intensity focused ultrasound. Med Phys 2021;48(2):579–86.
[29] Beer JM, Fisk AD, Rogers WA. Toward a framework for levels of robot autonomy in human-robot interaction. J Hum Robot Interact 2014;3(2):74.
[30] Arent K, Cholewiński M, Domski W, et al. Selected topics in design and application of a robot for remote medical examination with the use of ultrasonography and ascultation from the perspective of the REMEDI project. J Autom Mob Robot Intell Syst 2017;11(2):82–94.
[31] Guan X, Wu H, Hou X, et al. Study of a 6DOF robot assisted ultrasound scanning system and its simulated control handle. In: 2017 IEEE international conference on cybernetics and intelligent systems (CIS) and IEEE conference on robotics, automation and mechatronics (RAM); 2017. p. 469–74.
[32] Adams SJ, Burbridge BE, Badea A, et al. A crossover comparison of standard and telerobotic approaches to prenatal sonography. J Ultrasound Med 2018;37(11):2603–12.
[33] Kurnicki A, Stańczyk B. Manipulator control system for remote USG examinantion. J Autom Mob Robot Intell Syst 2019;13.
[34] Huang Q, Lan J. Remote control of a robotic prosthesis arm with six-degree-of-freedom for ultrasonic scanning and three-dimensional imaging. Biomed Signal Process Control 2019;54, 101606.

[35] Wang J, Peng C, Zhao Y, et al. Application of a Robotic Tele-Echography System for COVID-19 Pneumonia. J Ultrasound Med 2021;40 (2):385–90.

[36] Giuliani M, Szczęśniak-Stańczyk D, Mirnig N, et al. User-centred design and evaluation of a tele-operated echocardiography robot. Health Technol 2020;10(3):649–65.

[37] Zhou J, Gao B, Xue B, et al. Real-time Interaction of a 7-DOF Robot for Teleoperated Ultrasonic Scanning. In: 2021 6th IEEE international conference on advanced robotics and mechatronics (ICARM); 2021. p. 483–6.

[38] Virga S, Göbl R, Baust M, et al. Use the force: deformation correction in robotic 3D ultrasound. Int J Comput Assist Radiol Surg 2018;13 (5):619–27.

[39] Esteban J, Simson W, Requena Witzig S, et al. Robotic ultrasound-guided facet joint insertion. Int J Comput Assist Radiol Surg 2018;13 (6):895–904.

[40] Li D, Cheng Z, Chen G, et al. A multimodality imaging-compatible insertion robot with a respiratory motion calibration module designed for ablation of liver tumors: a preclinical study. Int J Hyperthermia 2018;34(8):1194–201.

[41] Jiang Z, Grimm M, Zhou M, et al. Automatic normal positioning of robotic ultrasound probe based only on confidence map optimization and force measurement. IEEE Robot Autom Lett 2020;5(2):1342–9.

[42] Ipsen S, Wulff D, Kuhlemann I, et al. Towards automated ultrasound imaging—robotic image acquisition in liver and prostate for long-term motion monitoring. Phys Med Biol 2021;66(9), 094002.

[43] Hennersperger C, Fuerst B, Virga S, et al. Towards MRI-based autonomous robotic US acquisitions: a first feasibility study. IEEE Trans Med Imaging 2016;36(2):538–48.

[44] An CY, Syu JH, Tseng CS, et al. An ultrasound imaging-guided robotic HIFU ablation experimental system and accuracy evaluations. Appl Bionics Biomech 2017;2017.

[45] Huang Q, Lan J, Li X. Robotic arm based automatic ultrasound scanning for three-dimensional imaging. IEEE Trans Industr Inform 2018;15 (2):1173–82.

[46] Langsch F, Virga S, Esteban J, et al. Robotic ultrasound for catheter navigation in endovascular procedures. In: 2019 IEEE/RSJ international conference on intelligent robots and systems (IROS); 2019. p. 5404–10.

[47] von Haxthausen F, Hagenah J, Kaschwich M, et al. Robotized ultrasound imaging of the peripheral arteries–a phantom study. Curr Dir Biomed Eng 2020;6(1).

[48] Li K, Wang J, Xu Y, et al. Autonomous navigation of an ultrasound probe towards standard scan planes with deep reinforcement learning. In: 2021 IEEE international conference on robotics and automation (ICRA); 2021. p. 8302–8.

[49] Jiang Z, Li Z, Grimm M, et al. Autonomous robotic screening of tubular structures based only on real-time ultrasound imaging feedback. IEEE Trans Ind Electron 2021;69(7):7064–75.

[50] Al-Dhabyani W, Gomaa M, Khaled H, et al. Deep learning approaches for data augmentation and classification of breast masses using ultrasound images. Int J Adv Comput Sci Appl 2019;10(5):1–11.

[51] Gerlach S, Hofmann T, Fürweger C, et al. AI-based optimization for US-guided radiation therapy of the prostate. Int J Comput Assist Radiol Surg 2022;1–10.

[52] Kang SH, Jeon K, Kang S-H, et al. 3D cephalometric landmark detection by multiple stage deep reinforcement learning. Sci Rep 2021;11(1):1–13.

[53] Wake N, Nussbaum JE, Elias MI, et al. 3D printing, augmented reality, and virtual reality for the assessment and management of kidney and prostate cancer: a systematic review. Urology 2020;143:20–32.

[54] McKendrick M, Yang S, McLeod G. The use of artificial intelligence and robotics in regional anaesthesia. Anaesthesia 2021;76:171–81.

[55] Piana A, Gallioli A, Amparore D, et al. Three-dimensional augmented reality–guided robotic-assisted kidney transplantation: breaking the limit of Atheromatic plaques. Eur Urol 2022;82(4):419–26.

[56] Marohn CMR, Hanly CEJ. Twenty-first century surgery using twenty-first century technology: surgical robotics. Curr Surg 2004;61(5):466–73.

[57] Lee JY, Islam M, Woh JR, et al. Ultrasound needle segmentation and trajectory prediction using excitation network. Int J Comput Assist Radiol Surg 2020;15(3):437–43.

[58] Lanfranco AR, Castellanos AE, Desai JP, et al. Robotic surgery: a current perspective. Ann Surg 2004;239(1):14.

[59] Stoll J, Ren H, Dupont PE. Passive markers for tracking surgical instruments in real-time 3-D ultrasound imaging. IEEE Trans Med Imaging 2011;31(3):563–75. 0278-0062.

[60] Nadeau C, Ren H, Krupa A, et al. Intensity-based visual servoing for instrument and tissue tracking in 3D ultrasound volumes. IEEE Trans Autom Sci Eng 2014;12(1):367–71. 1545-5955.

[61] Ren H, Dupont PE. Tubular structure enhancement for surgical instrument detection in 3D ultrasound. Annu Int Conf IEEE Eng Med Biol Soc 2011;7203–6. 1457715899.

[62] Ren H, Dupont PE. Tubular enhanced geodesic active contours for continuum robot detection using 3d ultrasound. In: IEEE int conf robot autom; 2012. p. 2907–12. 1467314056.

[63] Schneider C, Baghani A, Rohling R, et al. Remote ultrasound palpation for robotic interventions using absolute elastography. In: International conference on medical image computing and computer-assisted intervention; 2012. p. 42–9.

[64] Rao AR, Gray R, Mayer E, et al. Occlusion angiography using intraoperative contrast-enhanced ultrasound scan (CEUS): a novel technique demonstrating segmental renal blood supply to assist zero-ischaemia robot-assisted partial nephrectomy. Eur Urol 2013;63(5):913–9.

[65] Mallapragada VG, Sarkar N, Podder TK. Robot-assisted real-time tumor manipulation for breast biopsy. IEEE Trans Robot 2009;25(2):316–24.

[66] Mallapragada VG, Sarkar N, Podder TK. Robotic system for tumor manipulation and ultrasound image guidance during breast biopsy. In: 2008 30th Annual international conference of the IEEE engineering in medicine and biology society; 2008. p. 5589–92.

[67] Neshat HRS, Patel RV. Real-time parametric curved needle segmentation in 3D ultrasound images. In: 2008 2nd IEEE RAS & EMBS International conference on biomedical robotics and biomechatronics; 2008. p. 670–5.

[68] Barva M, Uhercik M, Mari J-M, et al. Parallel integral projection transform for straight electrode localization in 3-D ultrasound images. IEEE Trans Ultrason Ferroelectr Freq Control 2008;55(7):1559–69.

[69] Hungr N, Troccaz J, Zemiti N, et al. Design of an ultrasound-guided robotic brachytherapy needle-insertion system. In: 2009 Annual international conference of the IEEE engineering in medicine and biology society; 2009. p. 250–3.

I. Introduction, technology basics, evolution and next frontiers of surgical robots

[70] Zhang S, Jiang S, Yang Z, et al. An ultrasound image navigation robotic prostate brachytherapy system based on US to MRI deformable image registration method. Hell J Nucl Med 2016;19(3):223–30.

[71] Guo J, Shi C, Ren H. Ultrasound-assisted guidance with force cues for intravascular interventions. IEEE Trans Autom Sci Eng 2018;16(1):253–60. 1545-5955.

[72] Long J-A, Hungr N, Baumann M, et al. Development of a novel robot for transperineal needle based interventions: focal therapy, brachytherapy and prostate biopsies. J Urol 2012;188(4):1369–74.

[73] Noble JA. Ultrasound image segmentation and tissue characterization. Proc Inst Mech Eng, Part H J Eng Med 2010;224(2):307–16.

[74] Ren H, Gu X, Tan KL. Human-compliant body-attached soft robots towards automatic cooperative ultrasound imaging. In: 2016 CSCWD. IEEE; 2016. p. 653–8. 1509019154.

[75] Ren H, Anuraj B, Dupont PE. Varying ultrasound power level to distinguish surgical instruments and tissue. Med Biol Eng Comput 2018;56 (3):453–67. 1741-0444.

[76] Venkatesh SS, Levenback BJ, Sultan LR, et al. Going beyond a first reader: a machine learning methodology for optimizing cost and performance in breast ultrasound diagnosis. Ultrasound Med Biol 2015;41(12):3148–62.

[77] Lempitsky V, Verhoek M, Noble JA, et al. Random forest classification for automatic delineation of myocardium in real-time 3D echocardiography. In: International conference on functional imaging and modeling of the heart; 2009. p. 447–56.

[78] Carneiro G, Georgescu B, Good S, et al. Detection and measurement of fetal anatomies from ultrasound images using a constrained probabilistic boosting tree. IEEE Trans Med Imaging 2008;27(9):1342–55.

[79] Chen AI, Balter ML, Maguire TJ, et al. Deep learning robotic guidance for autonomous vascular access. Nat Mach Intell 2020;2(2):104–15.

[80] Antico M, Sasazawa F, Dunnhofer M, et al. Deep learning-based femoral cartilage automatic segmentation in ultrasound imaging for guidance in robotic knee arthroscopy. Ultrasound Med Biol 2020;46(2):422–35.

[81] Guo J, Li H, Chen Y, et al. Robotic ultrasound and ultrasonic robot. Endosc Ultrasound 2019;8(1):1.

[82] Săftoiu A. State-of-the-art imaging techniques in endoscopic ultrasound. World J Gastroenterol: WJG 2011;17(6):691.

[83] Voermans RP, van Berge Henegouwen MI, Bemelman WA, et al. Feasibility of transgastric and transcolonic natural orifice transluminal endoscopic surgery peritoneoscopy combined with intraperitoneal EUS. Gastrointest Endosc 2009;69(7):e61–7.

[84] Saftoiu A, Bhutani MS, Vilmann P, et al. Feasibility study of EUS-NOTES as a novel approach for pancreatic cancer staging and therapy: An international collaborative study. Hepatogastroenterology 2013;60(122):180–6.

[85] Sonn GA, Margolis DJ, Marks LS. Target detection: magnetic resonance imaging-ultrasound fusion–guided prostate biopsy. Urol Oncol Semin Ori Investig 2014;903–11.

[86] Preiswerk F, Toews M, Cheng CC, et al. Hybrid MRI-Ultrasound acquisitions, and scannerless real-time imaging. Magn Reson Med 2017;78 (3):897–908.

[87] Ewertsen C, Săftoiu A, Gruionu LG, et al. Real-time image fusion involving diagnostic ultrasound. Am J Roentgenol 2013;200(3):W249–55.

[88] San José Estépar R, Westin C-F, Vosburgh KG. Towards real time 2D to 3D registration for ultrasound-guided endoscopic and laparoscopic procedures. Int J Comput Assist Radiol Surg 2009;4(6):549–60.

[89] Leung K-Y. Applications of advanced ultrasound technology in obstetrics. Diagnostics 2021;11(7):1217.

[90] Edgcumbe P, Singla R, Pratt P, et al. Augmented reality imaging for robot-assisted partial nephrectomy surgery. In: International conference on medical imaging and augmented reality; 2016. p. 139–50.

[91] Singla R, Edgcumbe P, Pratt P, et al. Intra-operative ultrasound-based augmented reality guidance for laparoscopic surgery. Healthc Technol Lett 2017;4(5):204–9.

# 11

# Robotic staplers

*Jefferson Luiz Gross and João Paulo de Oliveira Medici*

Department of Thoracic Surgery, A.C. Camargo Cancer Center, São Paulo, Brazil

## Introduction

Technological advances across many scientific disciplines have produced many surgical devices and instruments that are used during surgery.

The surgical stapler is an example of a device that is commonly used during surgical procedures and, at the same time, is in an almost constant state of developmental evolution. Although these devices are highly versatile and efficient, there have been well-documented incidences of staple line leaks leading to postoperative complications that often result from issues not attributable to ischemia. Of these, technical errors can play a significant role, potentially increasing the risk of bleeding, transfusions, and unplanned proximal diversions, in surgical procedures.

The concept of using staples for surgical procedures dates to the early 1900s. Surgeons and inventors experimented with assorted designs and materials for staples to close wounds.

In the 1900s, various inventors and medical professionals developed and patented early surgical stapler designs, but these early attempts were often troublesome and not widely adopted.

The modern surgical stapler we are more familiar with today can be attributed to the work of several key individuals and companies [1].

Hümér Hültl, a Hungarian surgeon, is often credited with inventing the first practical surgical stapler in 1921. His stapler was a circular design used for gastrointestinal surgeries [2].

Companies like US Surgical Corporation (now part of Medtronic) and Ethicon (a subsidiary of Johnson & Johnson) made significant contributions to the development and commercialization of surgical staplers in the mid-20th century.

Over the years, surgical stapler technology has advanced considerably. There are several types of surgical staplers, including linear, circular, and skin staplers, each designed for specific surgical applications [3].

The integration of robotic technology into surgery began in the early 2000s. Companies like Intuitive Surgical, with their da Vinci Surgical System, pioneered the use of robotic systems in minimally invasive procedures. These robotic systems allow for greater precision and control during surgery.

As robotic surgery became more prevalent, the need for robotic surgical staplers appeared. Manufacturers started developing stapling instruments compatible with robotic surgical systems, such as the da Vinci. These robotic staplers offer even more precise control and are used in various surgical specialties, including general surgery, urology, gynecology, thoracic, bariatric, and more.

## Advantages of robotic stapler

Robotic staplers supply high precision and accuracy, reducing the risk during surgery. Surgeons can control the stapler with great precision, which is especially important when working in delicate or hard-to-reach areas. It can perform repetitive tasks with minimal hand fatigue, which can be a significant advantage during lengthy surgical procedures.

Robotic systems often come equipped with high-definition cameras and three-dimensional imaging, offering surgeons a clear and detailed view of the surgical field. This can help in better decision-making and more precise stapling.

Surgeons can use robotic staplers from a console with ergonomic controls, reducing physical strain and allowing for more comfortable and extended use [4].

Robotic systems have a wide range of motion, allowing for precise positioning, suturing, and stapler placement. Some studies show that robotic staplers reduce the risk of complications such as air leaks, bleeding, or fistulas [5].

Surgeons can customize the stapling process by adjusting parameters like staple size, spacing, and depth, depending on the specific requirements of the surgery.

## Use of staplers among different specialties

The adoption of surgical staplers grew in the 20th century as they offered advantages over traditional sutures in terms of speed, precision, and reduced risk of infection. Surgical staplers became particularly popular in procedures involving the gastrointestinal tract, thoracic surgery, and bariatric surgery.

### Characteristics of robotic staplers

1. Size
2. Charges
3. Cannula diameter
4. Angulation

The main robotic staplers currently available are Endowrist and Sureform.

The difference between these devices is their size. Both have a 45-mm staple line, but Endowrist has a 30-mm stapler charge while Sureform has a 60-mm stapler charge [6].

### Endowrist

The EndoWrist Stapler for the da Vinci Xi System is an endoscopic linear stapler. The EndoWrist Stapler supplies fully wristed articulation in both horizontal and vertical directions and is controlled by the surgeon through the da Vinci Xi Surgeon Console. SmartClamp is a technology that provides intraoperative feedback to the surgeon and reduces the guesswork normally associated with stapling. For each reload color, SmartClamp technology detects whether the jaws can adequately close on the target tissue for the given staple height. If not, the EndoWrist Stapler will notify the user. The EndoWrist Stapler System also prevents firing when it detects that no reload is installed, or that a spent reload is installed, and notifies the surgeon and OR staff.

The EndoWrist Stapler 30 and 45 instruments and reloads are intended to be used with the da Vinci Xi surgical system for resection, transection, and/or creation of anastomosis in general, thoracic, gynecologic, and urologic surgery.

The EndoWrist stapler 30 and 45 are indicated for adult use, and the EndoWrist stapler 30 is indicated for pediatric use. The device can be used with staple-line or tissue-buttressing materials. The use of tissue like the liver or spleen is contra-indicated, because the tissue compressibility would be destructive. The use of the aorta is also contra-indicated.

The EndoWrist stapler 30 and 45 for the da Vinci Xi System are not compatible for use with the da Vinci, da Vinci S, or da Vinci Si surgical systems.

Reloads from Endowrist 45 mm: White, Blue, Green [6].

### Sureform

Also, from Intuitive, this device has 60-mm and 45-mm staple line, an anvil-centric design, straight tip, and curve tip anvil and works through a 12-mm cannula.

Reloads from Sureform 60 mm: White, Blue, Green, and Black reload 6 rows of staples.

Reloads from Sureform 45 mm: Gray, White, Blue, Green, and Black reload 6 rows of staples.

New knife with every reload [6].

### Angulation

Sureform supplies a 120-degree articulation, compared to 45-degree Signia and Echelon articulation.

Typically, white loads are used for vascular structures, such as pulmonary veins and artery, mesentery. Blue loads are used for small bowel, colon and pulmonary parenchyma, and green loads, for thick tissue such as the stomach and bronchus.

## Technical aspects of using robotic staplers

The robotic staplers provide the surgeon with complete control of performing sutures in different anatomic structures, and this is important because stapling is one of the most crucial and potentially hazardous steps of the operation. Modern staplers come with an intelligent integrated technology that makes automatic adjustments to the firing process; however, some technical aspects must be observed by robotic surgeons.

The surgeon must keep in mind that better outcomes from staplers come from observing some characteristics such as minimizing the trauma, avoiding tension in the structures, and careful and appropriate stapler location. These procedures depend on the surgeon respecting some technical aspects of stapler use to obtain better outcomes. The adequate use of robotic staplers presents certain challenges, and the transition from manual to robotic stapler requires appropriate training, and robotic surgeons must dedicate some period of training and understanding of robotic staplers to obtain better results. The technology of robotic staplers benefits procedures across different surgical specialties, mainly video-assisted thoracic surgery (VATS) lobectomies, low rectal, gastric, and esophageal surgeries.

The adequate use of robotic staplers maximizes the efficiency and safety of this modern device. Next, we will make some comments on how to properly use robotic staplers to obtain the best results with these high-tech devices [7].

### Adequate port placements

The port for the robotic stapler must be 12 mm, as other robotic instruments fit through 8 mm ports, a plastic reducer should be used to insert both robotic staplers and robotic instruments.

Considering that the end effector of a robotic stapler is longer than robotic instruments, it should be placed as far as possible from the surgical target. This considerable distance is necessary to allow the greatest degree of maneuverability in the cavity. Port placement for other robotic instruments and cameras must be planned to avoid collision and to allow a wide range of movements and easy access to surgical targets.

### Creating a sufficient space for robotic stapler placement

To avoid tension in the structure it is very important to create enough space on the opposite side of the structure that will be divided. Adequate dissection, clearing lymph nodes, and connective tissue around the anatomic structure should be performed before stapler placement.

### Using a stapler guide

Attaching a device to the anvil of the robotic stapler might facilitate the passage of the stapler behind the anatomic structure that will be fired with less tension. A PVC or latex catheter might be attached to the anvil of the robotic stapler to softly pass the anatomic structure and must be removed before firing.

### Simulate stapler placement before engaging the anatomic structure

The degree of movement of the robotic stapler is smaller than other robotic instruments. Because of this limited movement, it is important not to attempt to reposit the anvil of the stapler when it is under the anatomical structure. The recommendation is to simulate the position of the stapler end effector in the plane just in front of the anatomical structure that will be fired, and then the stapler can be moved back and advanced in the appropriate plane behind the anatomical structure, without additional movements of the end effector.

The robotic stapler represents a critical technical advancement; however, the surgeon who is operating the robot should observe some technical recommendations described above to increase the effectiveness and the safety.

## Safety of robotic stapler

Da Vinci released robotic staplers in 2014, since then its use has been increasing among different surgical specialties replacing traditional laparoscopic staplers. Safety is one of the concerns of recent use of any medical device.

In general, we can say that robotic staplers are safe. Surgeon control of staplers is one of the greatest advantages and contributes to its safety. High technology is incorporated in robotic staplers such as SmartFire technology, which monitors tissue compression before and during firing, making automatic adjustments to optimize the staple line; more maneuverability that leads to a better stapler placement, reducing the need to pull structures toward the stapler; Smart Clamp feedback that helps to detect if staplers jaws are adequately closed on tissue selected reload color prior to firing; microchip that will not let the system fire an incorrectly loaded or spent reload; among others. This advanced technology helps to improve the safety of robotic staplers [8].

We have no data on the current percentage of reported complications attributed to robotic staplers. In the United States of America, complications of robotic staplers can be reported spontaneously to FDA (Food and Drug Administration), which catalogs complications related to medical devices.

In 2021, Giffen et al. published a review of reported robotic stapler complications to FDA. From January 2014 to February 2020, they observed 225 complete reports of complications attributed to robotic staplers. It is not possible to estimate the percentage of complications since we have no data of the total number of robotic staplers that were used in this period. The most common reported complications were difficult to open and close (20.4%), failure to form staple line (18.2%), material fragmentation (13.8%), misfire (13.3%), failure to cut (3.6%), and another malfunction (30.7%). Of the reported adverse events, 31 (13.7%) resulted in patient harm, including bleeding, anastomotic leak, and bronchopulmonary fistula requiring blood transfusion, hospital readmission, and reoperation. One complication led to a patient's death. Robotic staplers were replaced by laparoscopic ones to finish the surgical procedure in 31 cases, and conversion to open surgery due to robotic stapler complication was reported in 14 (6.2%) cases. The specialties that more frequently reported adverse events with robotic staplers were colorectal, thoracic, and bariatric [9,10].

As the report of adverse events is voluntary, probably many complications of robotic staplers are not reported, and the true incidence is unknown.

Laparoscopic staplers have been used longer than robotic ones, and analyzing the safety aspects of laparoscopic staplers could bring some information of their reported complications. Kwaznezski and cols published a survey with questions about stapler's malfunctions among some minimally invasive surgeons. Personal or peer experience with laparoscopic stapler malfunction was reported in 60%–70% of the surgeons, and the surgical strategy must be changed in 25% of the cases. This study suggested that problems with laparoscopic staplers are very common and probably are underreported [11].

Due to the recent release of robotic staplers, just a few studies compare laparoscopic staplers with robotic ones in surgeries performed with the da Vinci platform. Gutierrez and cols reported a systematic review of operative outcomes of robotic surgical procedures performed with endoscopic linear staplers or robotic staplers. The author identified only three studies of direct comparison (two in colorectal surgery and one in gastric bypass surgery) and concluded that both staplers are safe and effective [12].

More recently, Zervos and cols performed a study to compare clinical outcomes and costs between hand-held and robotic staplers in a real-world clinical practice in a group of patients who were submitted to robotic lobectomy. They found that robotic staplers were associated with a reduced risk of bleeding, conversion to open surgery, air leak, and overall complications. On the other hand, the total index hospitalization costs were comparable between the two types of staplers [5].

The robotic staplers are safe and effective, however, there is little data about the complications of using robotic staplers. We should encourage the surgeons to report any adverse events to better understand the safety of robotic staplers. Furthermore, cost-effectiveness studies should compare robotic staplers with manual ones.

## Future scenario

Since they were invented more than 100 years ago, surgical staplers have evolved a lot in their mechanical aspects and technology. Surgical staplers made minimally invasive surgery possible, just as minimally invasive surgery forced the evolution of staplers to cut and seal tissues and vessels through small incisions. The advent of robotic surgery also brought an increase in the technological evolution of surgical staplers.

The mechanical aspects of robotic staplers need to evolve, such as improving the angle of articulation, more ergonomic designs, shaft and reload with lesser diameter to be used through the same portals as other robotic instruments. Technological and intelligence developments will lead to the development of different staplers, and the surgeon will choose which stapler will be most appropriate for each procedure.

Another aspect that we must increasingly care about is the possible environmental damage from medical devices. Therefore, more durable staplers made with materials that cause less environmental damage should be developed.

## Key points

- Staplers have made minimally invasive surgery possible and are constantly evolving. Robotic surgery has brought a great technological advance to staplers.
- Robotic staplers replaced the laparoscopic ones handled by the assistant in robotic surgery and returned control of the staplers to the surgeon at the console.
- Robotic staplers incorporate new and advanced technologies that provide real-time feedback to surgeons, they detect if the staple height is compatible with the tissue thickness, and they have safety systems to detect if the staplers are not able to fire.
- Proper use of robotic staplers requires observing some technical details so that their use is safer and more efficient.
- The robotic staplers are safe, with few reported complications.

## References

[1] Gaidry AD, Tremblay L, Nakayama D, Ignacio RCJ. The history of surgical staplers: a combination of Hungarian, Russian, and American innovation. Am Surg 2019;85(6):563–6.

[2] Akopov A, Artioukh DY, Molnar TF. Surgical staplers: the history of conception and adoption. Ann Thorac Surg 2021;112(5):1716–21.

[3] Ghosh S, More N, Kapusetti G. Surgical staples: current state-of-the-art and future prospective. Med Nov Technol Devices 2022;16, 100166. Available from: https://www.sciencedirect.com/science/article/pii/S2590093522000534.

[4] Galetta D, Casiraghi M, Pardolesi A, Borri A, Spaggiari L. New stapling devices in robotic surgery. J Visc Surg 2017;3:45.

[5] Zervos M, Song A, Li Y, Lee SH, Oh DS. Clinical and economic outcomes of using robotic versus hand-held staplers during robotic lobectomy. Innovations (Phila) 2021;16(5):470–6.

[6] Intuitive. Intuitive, 2023. Available from: https://www.intuitive.com/en-us/products-and-services/da-vinci/stapling.

[7] Chekan E, Whelan RL. Surgical stapling device-tissue interactions: what surgeons need to know to improve patient outcomes. Med Devices (Auckl) 2014;7:305–18.

[8] Pearlstein DP. Robotic lobectomy utilizing the robotic stapler. Ann Thorac Surg 2016;102(6):e591–3.

[9] Giffen Z, Ezzone A, Ekwenna O. Robotic stapler use: is it safe?-FDA database analysis across multiple surgical specialties. PLoS One 2021;16(6), e0253548.

[10] Johnson CS, Kassir A, Marx DS, Soliman MK. Performance of da Vinci stapler during robotic-assisted right colectomy with intracorporeal anastomosis. J Robot Surg 2019;13(1):115–9. Feb.

[11] Kwazneski 2nd D, Six C, Stahlfeld K. The unacknowledged incidence of laparoscopic stapler malfunction. Surg Endosc 2013;27(1):86–9.

[12] Gutierrez M, Ditto R, Roy S. Systematic review of operative outcomes of robotic surgical procedures performed with endoscopic linear staplers or robotic staplers. J Robot Surg 2019;13(1):9–21.

# Actual competitive and new models of surgical robots

*Daniel Coser Gomes*[a] *and Stênio de Cássio Zequi*[b]

[a]Urology Division, A.C. Camargo Cancer Center, São Paulo, São Paulo, Brazil [b]Urology Division, Referral Center for Urological Tumors and Graduate School, A.C. Camargo Cancer Center, São Paulo, São Paulo, Brazil

## Abbreviations

| | |
|---|---|
| **AI** | Artificial intelligence |
| **CT** | Computed tomography |
| **MIS** | Minimally invasive surgery |
| **ML** | Machine learning |
| **MRI** | Magnetic resonance imaging |
| **NOTES** | Natural orifice transluminal endoscopic surgery |
| **RAS** | Robotic assisted surgery |

## Introduction

Since the first prototypes were used almost three decades ago, research and development of surgical robots have been increasing exponentially. Initially dominated by a single major platform, the field has recently witnessed an explosion of new models across various medical specialties. These range from major surgeries to microsurgery, endoscopy, vascular surgery, and ophthalmology. Over time, progressively new platforms are launched, and, in their initial phases, they receive approvals from local regulatory agencies before gaining wider acceptance. Devices are increasingly becoming less invasive, more mobile, and with greater addition of technological resources.

In the following, we provide a brief description of the main platforms currently in use around the world. We believe that more and more, new robotic models will be produced and used in daily surgical practice, especially when they become widespread and their acquisition and maintenance costs decrease.

## Background

The PUMA 200 was the first surgical robot used in a clinical setting [1]. It was initially used for stereotaxic brain biopsies, where the surgeon would position the robot's arms to carry out the desired job. These robots could not carry out many consecutive tasks on human beings unattended at the time, and the robotic technology required human surgical assessment at the conclusion of every phase of robotic-assisted surgery (RAS) (Fig. 1).

The concept of RAS has expanded to its greatest extent since the introduction of the second generation of surgical robots in 1998–2000. This was prompted by the market need for highly precise robotic surgical systems that could complement the existing minimally invasive surgery (MIS) technology by extending its reach to established stereo-endoscopic platforms like thoracoscopy and laparoscopy.

In 1989, the company Computer Motion introduced a groundbreaking innovation: AESOP (Automated Endoscopic System for Optimal Positioning), a voice-controlled robotic arm equipped with an endoscope. The first iteration of this

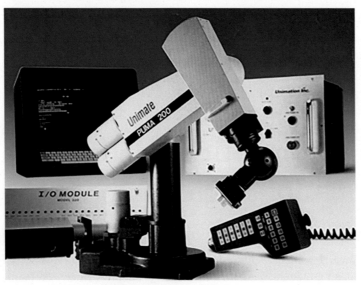

FIG. 1    Puma 200, the first robot used for assisting human neurosurgery in 1985 [32].

FIG. 2    AESOP (Automated Endoscopic System for Optimal Positioning) [2].

robotic arm, the AESOP 1000, received approval in 1994 and operated via foot pedals. Just 2 years later, its successor, the AESOP 2000, brought a significant advancement by replacing the pedals with a voice control system [2] (Fig. 2).

Despite initial successes, the evolving needs of surgical procedures called for more than just telemanipulation of the video camera; they required the surgeon's movements to be mirrored as well. Addressing this, Computer Motion unveiled the ZEUS system in 1998. This system featured arms and surgical instruments directly controlled by the surgeon. In this setup, the surgeon (master) directly commanded the robot (slave) [2] (Fig. 3).

FIG. 3    The ZEUS robot consisted of three arms, each independently attached to a surgical table [2].

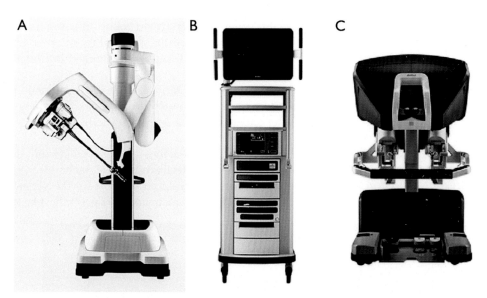

FIG. 4    da Vinci SP surgical system. (A) Patient cart; (B) Vision cart; (C) Surgeon console [33].

Many new technical advancements are now being developed on top of these second-generation platforms, which are the most popular and well-established systems in clinical usage today (e.g., Da Vinci by Intuitive Surgical) [3].

Bioinspired technologies were adopted by MIS and endoscopic techniques throughout the creation and evolution of second-generation robots. A preference for scarless surgery led to the development of the NOTES (Natural Orifice Transluminal Endoscopic Surgery) [4] platform and numerous MIS ports were converted into a SP (single port) platform [5], in an effort to reduce surgical exposure even more (Fig. 4).

Right now, there are more than 35 robots on the market for different types of surgical applications, and more than 150 are currently being developed by companies and start-ups [6], the majority at an early stage of funding. The pipeline analysis shows a strong interest in the development of procedure-specific robots, with the idea of making more affordable and compact solutions without compromising performance. Below, we discuss some of the mainly robotic platforms there is currently on the market and the ones that might come in the near future.

During the reading of this book chapters, it will be possible to find descriptions of many other surgical robots, dedicated to some specific specialties.

# Robotic platforms

## Abdominal and thoracic surgery

- Da Vinci Surgical System (Intuitive Surgical—USA): Extensively used for a variety of surgeries, including urological, gynecological, thoracic, cardiological, general, and even head and neck surgery. Ranging from its multiport platform "DaVinci X" and "Xi" to the single port "DaVinci SP" [3,5], they pioneered new capabilities in the operating room and dominated the market for nearly three decades, for a total of more than 12 million procedures performed all over the globe. In March 2024, the company obtained FDA clearance for da Vinci 5, its next-generation multi-port robotic system, with built in force feedback technology.
- Versius (CMR Surgical—GBR): Suitable for a range of abdominal and thoracic surgeries. The system entered the robotic market in 2021 and claims to overcome obstacles to the widespread adoption of robotic minimal access surgery, namely robot and instrument size, versatility, port placement, cost, and ease of use, allowing the system to be highly utilized and ultimately cost comparable to manual laparoscopic surgery. With an ergonomic console, which allows the surgeon to operate seated or standing, full camera control without foot switches, and individual arm docking, the Versius platform [7] has now been used in more than 5000 surgical procedures worldwide [8] (Fig. 5).
- Hugo RAS System (Medtronic—USA): With the very first procedure being performed on June 2021 in Chile, this robot is also a modular, multi-quadrant, open-console platform designed for a broad range of surgical procedures, the Hugo RAS System combines wristed instruments, 3D visualization, and has received regulatory approvals and is commercially available in various countries globally [9]. The compatibility with the globally trusted Medtronic surgical instrumentation may make Hugo a more accessible and successful competitor (Fig. 6).
- Toumai Laparoscopic Surgical Robot (MicroPort—CHN): Launched in November 2019, this robot-assisted laparoscopic surgery system includes a video system trolley, a four-armed robot unit, and a control console. The robot's proportions may all be changed to fit the patient's body habitus, surgical techniques, and posture of the surgeon, with either a multi or single-port platform [10]. In 2023, Toumai was utilized in China's first 5G ultra telerobotic hepatobiliary surgery [11] (Fig. 7).
- Hinotori Surgical Robot System (Medicaroid Corporation—JPN): Derived from the combination of the industrial robot technology of Kawasaki Heavy Industries and the medical knowledge of Sysmex, the Japanese robot launched in 2020 [12] also made its debut in telesurgery after succeeding in a demonstration of remote surgery between Singapore and Japan, 5000 km apart [13] (Fig. 8).
- Senhance Surgical System (Asensus Surgical—USA): Being a semi-active laparoscopy system, its sales model encompasses renting and reusable instruments to improve cost-effectiveness. The Senhance System [14] digitizes the interface between the surgeon and the patient to increase control and comfort for the surgeon and to reduce variability in laparoscopic surgery. Is the first platform to offer 3 mm instruments, the smallest instruments available

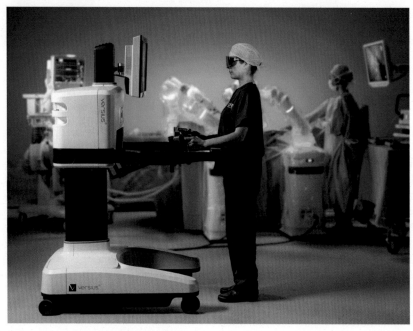

FIG. 5    Versius with 3D glasses and an open console design [34].

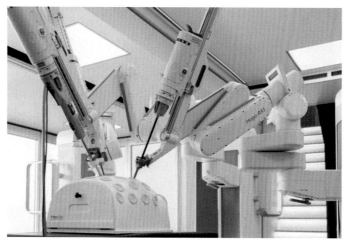

FIG. 6    Hugo: not an acronym, just a friendly name. [35].

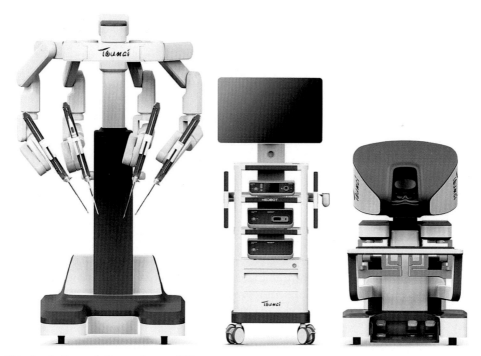

FIG. 7    Toumai: China's robotic revolution on its way [10].

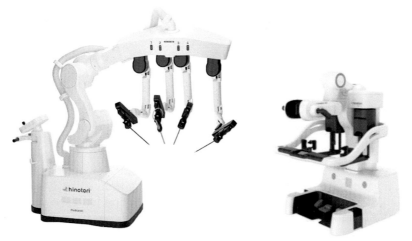

FIG. 8    Hinotori: the Japanese robot [13].

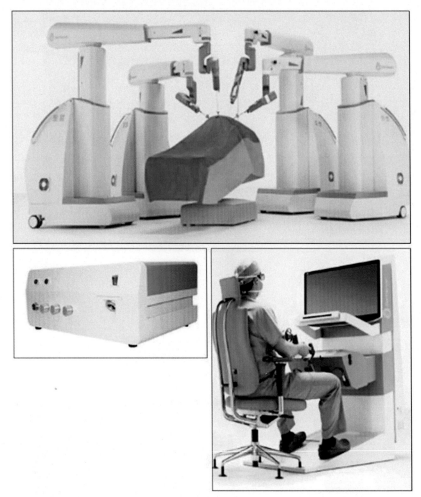

FIG. 9    Senhance: when laparoscopy fuses with robotics [36].

in the world on a robotic surgical platform. The manufacturer claims to remove the economic limitations of current robotic systems by time and cost-per-procedure comparable to manual laparoscopy. The platform has been receiving Conformité Européenne (CE) and Food and Drug Administration (FDA) approvals since 2017, being their last one in January 2023 (Fig. 9).

## Orthopedics

- Mako SmartRobotics (Stryker—USA): Specialized in orthopedic surgeries, particularly knee and hip replacements. The Mako Surgical system is a semi-active robot that provides surgeons with intraoperative haptic guidance for bone preparation and implant placement and personalizes a patient's surgical experience based on a specific diagnosis and anatomy [15] (Fig. 10).

## Neurosurgical

- NeuroBlate System (Monteris Surgical—USA) A minimally invasive, robotic, laser thermotherapy that uses magnetic resonance imaging (MRI)-guided surgical ablation technology designed specifically for use in the brain. NeuroBlate performs brain lesion ablation for patients without the invasiveness of an open neurosurgical procedure [16].
- Rosa One Brain (Zimmer Biomet Robotics—FRA): Specifically developed for stereotactic neurosurgery, this robotic platform can assist in a variety of neurosurgical procedures, ranging from epilepsy surgery and functional neurosurgery, including stereo electroencephalography, deep brain stimulation, stereotactic biopsy, ventricular endoscopy, and transnasal endoscopy [17] (Fig. 11).

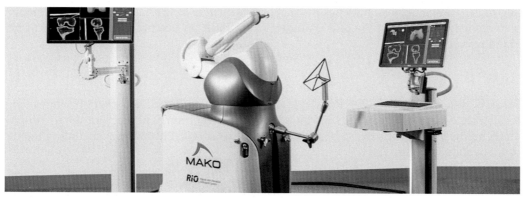

FIG. 10   Mako, orthopedic robot [37].

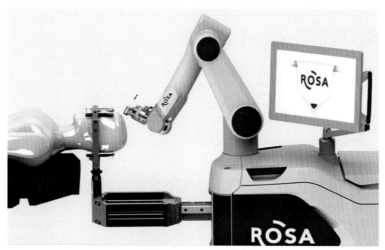

FIG. 11   Rosa One Brain, the French neurosurgeon robot [17].

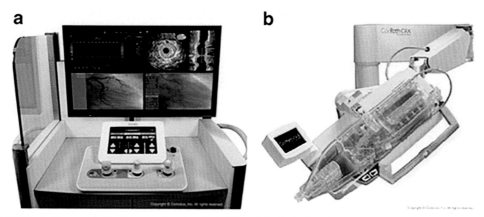

FIG. 12   CorPath GRX robotic system. The interventional cockpit (A) and the bedside robotic arm (B) [38].

## Vascular

- CorPath System (Corindus Vascular Robotics—USA): Specialized in vascular interventions, including coronary and peripheral vascular surgeries. It enables the placement of coronary guidewires and balloon/stent devices from the safety of a radiation-protected, interventional cockpit [18] (Fig. 12).
- Niobe ES Magnetic Navigation System (Stereotaxis—USA): Used in electrophysiology, which can include certain vascular procedures. It utilizes permanent magnets mounted on pivoting arms and positioned on opposing sides of

I. Introduction, technology basics, evolution and next frontiers of surgical robots

the operating table. These magnets influence the magnetic catheters in the heart to make micromovements of the catheter tip to navigate throughout the four chambers of the heart and complete the diagnosis and treatment of cardiac arrhythmias [19].

## Ophthalmology

- Preceyes Surgical System (Zeiss—NLD): Robotic assistant for eye surgery. The system provides surgeons with a precision better than 20 µm to position and hold instruments. This high surgical precision aims to improve treatment outcomes and empowers surgeons to establish new innovative surgical techniques, including the delivery of advanced therapeutics [20] (Fig. 13).

## Endoscopy and NOTES (Natural Orifice Transluminal Endoscopic Surgery)

- Monarch (Ethicon/J&J—USA): The Monarch Platform is a robotic-assisted bronchoscopy (RAB) that reaches deeper into the lung than traditional devices. Physicians can use the scope with both direct vision and software guidance to

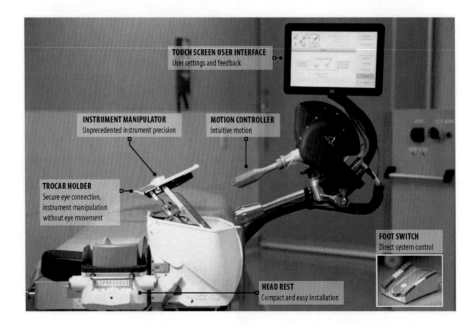

**PRECEYES Surgical System**

The PRECEYES Surgical System is a robotic assistant for vitreoretinal surgery. It supports surgeons in inserting and manipulating instruments inside the eye. The system is guided by positioning commands that the surgeon provides through an intuitive motion controller. The dedicated head rest facilitates easy installation onto existing OR tables.

The system provides surgeons with a precision better than 20 µm to position and hold instruments steady for an extended period of time. The high surgical precision aims to improve treatment outcomes. The system empowers surgeons to establish new innovative surgical techniques and to deliver advanced therapeutics.

**Precision**
Scaling and filtering of hand tremors yield unprecedented steadiness and precision of the instrument position. The standby function freezes any motion and allows to relax and reposition for optimal hand position.

**Safety**
A hybrid manual/assisted setup allows the surgeon to maintain patient contact. Residual eye movements are minimized by holding the trocar and safety boundaries are employed to limit instrument movements.

**Workflow optimization**
During highly demanding surgical steps, the robot is easily engaged to assist in specific tasks. Instruments are easily exchanged and their movements recorded for post-surgical evaluation and training purposes.

FIG. 13   Vein canulation with the ophthalmologic robot Preceyes [39].

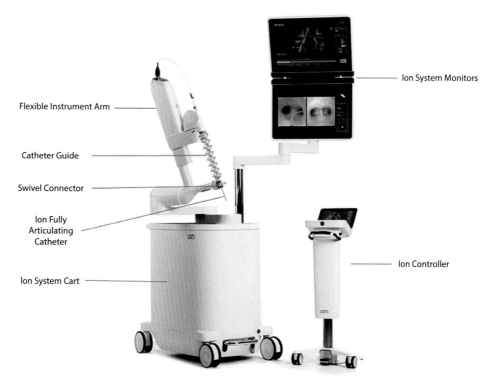

Flexible Instrument Arm

Catheter Guide

Swivel Connector

Ion Fully
Articulating
Catheter

Ion System Cart

Ion System Monitors

Ion Controller

FIG. 14    Ion, the robotic bronchoscope for lung biopsy [40].

hard-to-reach areas of the lung, leveraging airway optical recognition, electromagnetics, and robotic kinematic information [21].

- Ion Endoluminal System (Intuitive Surgical—USA): A robotic-assisted platform for minimally invasive biopsy in the lung. From the patient computed tomography (CT) scan, its software generates 3D airway trees and automatically creates a path and anatomy borders once a target is identified. It was FDA-approved in 2019, and in the first human use study, targets with a mean size of approximately 14 mm were reached in 96.6% of cases; The overall diagnostic yield was 79.3% with no reported incidence of pneumothorax or bleeding [21,22] (Fig. 14).

## Interventional radiology

- Ace Robotic System (XACT Robotics—ISR): The Israeli and world's first hands-free robotic system that combines advanced image-based procedure planning and navigation with robotic instrument insertion and nonlinear steering capabilities. Cleared to market in the U.S. in July 2020, for (CT)—guided percutaneous procedures [23] (Fig. 15).

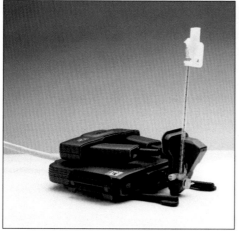

FIG. 15    Ace, the robotic platform for interventional radiology [41].

I. Introduction, technology basics, evolution and next frontiers of surgical robots

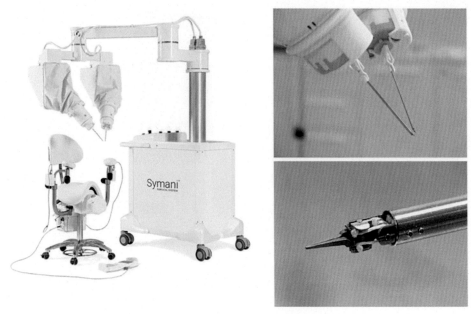

FIG. 16    Robotic microsurgery system Symani from MMI [42].

## Microsurgery

- Symani Surgical System (MMI Micro—ITA): Developed as a master–slave robotic system for microsurgery applications, designed to improve a surgeon's ability to access and suture small, delicate anatomy. The system has obtained the CE mark of approval and plans for FDA approval for US commercialization in the future [24] (Fig. 16).

## Upcoming platforms

- Enos (Titan Medical—CAN): Now on its 2.0 version, this SP platform, with a smaller footprint than its competitors, promises quicker draping and patient positioning with an open console. It has a 25-mm cannula with an integrated 2-D camera and a steerable 3-D endoscope. The endoscope works alongside the other instruments and both cameras have high-definition resolution and illumination. Still not FDA-approved and is currently in development [25] (Fig. 17).
- Ottava (Johnson & Johnson—USA): This general surgical robotics platform will employ Ethicon instrumentation (up to six robotic arms) with advanced visualization, machine learning, and a connected ecosystem. In Italian, "Ottava" means to play music an octave higher, as claimed by its manufactures. J&J MedTech aims to put Ottava up for US clinical trials in late 2024 [25,26].
- Mira Miniaturized RAS Platform (Virtual Incision—USA): First "miniaturized robot," Miras's in-a-tray form factor promises to allow booking RAS procedures in any available operating room and set up in minutes. Its design eliminates the need to drape or dock, minimizing RAS downtime. Virtual Incision is collaborating with NASA

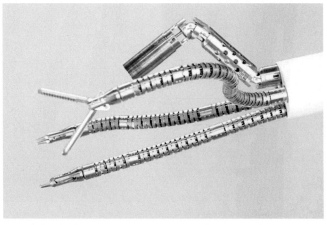

FIG. 17    Enos 2.0 prototype makes room for a third instrument arm by eliminating the integrated camera and using the 3-D scope [43].

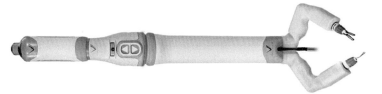

FIG. 18 Mira Miniaturized RAS Platform: going to space in 2024 [27,44].

FIG. 19 Beta2 Prototype with 9 degrees of freedom per arm and 360-degree visualization [45].

and the University of Lincoln-Nebraska to lay the foundations for telesurgery in remote locations, even space [27] (Fig. 18).

- Beta2 (Vicarious Surgical—USA): With a focus on abdominal access and visualization through a single 1.5 cm port. A camera and two robotic instruments can be passed through this port to maximize visualization, precision, and control of instruments for the surgeon in control of the device. The Company's technology was granted Breakthrough Device Designation by the FDA, although not yet been cleared for commercial sale in the United States [28] (Fig. 19).
- Avatera (Avatera Medical GmbH—DEU): The German company developed a single-patient cart with a four-arm robotic surgical system. It is currently conducting clinical trials. Avatera received the CE mark in 2019 and the company planned to perform a couple of hundred surgeries before commercialization. The next target geographies are Russia, India, the Middle East, and China. In a further stage, market entry in the US is planned [6,29] (Fig. 20).

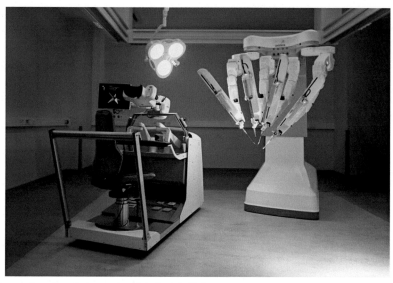

FIG. 20 Avatera: The German robot already in use in clinical trials [46].

I. Introduction, technology basics, evolution and next frontiers of surgical robots

## Future scenario

The implementation of artificial intelligence, augmented reality, miniaturization of systems, and tele-surgery are technology trends that are expected to impact the market and further facilitate the adoption and capabilities of robotic surgical systems soon. The development of augmented reality devices in surgery has allowed physicians to incorporate data visualization into diagnostic and treatment procedures to improve work efficiency, intraoperative visualization, and to enhance surgical training [30].

The types of machine learning (ML) relevant to train autonomous surgical devices can be divided into "unsupervised" and "supervised learnings" and can be applied to continuous or categorical data sources. Some ML techniques are rooted in traditional statistical principles (e.g., regression), whereas others involve decision mathematics and computer science principles. Reinforcement learning is rooted in psychological principles, where the agent (robot) performs its actions within its environment to achieve an end goal. Each action that brings it closer to its goal yields a positive reward (and vice versa). We are still not there yet, but certainly in a not-so-far future, we will be facing a real-life challenge side by side with the machine [6,31].

## Conclusions

The landscape of robotic surgery is witnessing a transformative shift with the introduction of new robotic platforms, challenging the long-held dominance of well-established systems. This recent surge of novel robotic platforms in the surgical field, not only emphasizes their role in thoracoabdominal surgeries but also in other specialized areas, such as orthopedics, neurosurgery, vascular, micro, endoscopic, and interventional radiology. These developments promise to mitigate the traditional barriers of high costs and limited access, democratizing robotic surgery and aiming for a future where robotic surgery is more advanced, efficient, and accessible for all.

## Key points

- RAS has progressed from early stereotaxic devices to advanced systems aiming for more affordable, compact, and easy-to-use platforms.
- There are more than 35 surgical robots on the market and more than 150 are currently being developed.
- Many other countries besides US, like China, Japan, UK, and Germany have their own RAS system now, and we can expect many more to come.
- Telesurgery, artificial intelligence, and machine learning are here to stay.

## References

[1] Kwoh YS, Hou J, Jonckheere EA, Hayati S. A robot with improved absolute positioning accuracy for CT guided stereotactic brain surgery. IEEE Trans Biomed Eng 1988;35(2).
[2] Morrell ALG, Morrell-Junior AC, Morrell AG, Mendes JMF, Tustumi F, De-Oliveira-e-silva LG, et al. The history of robotic surgery and its evolution: when illusion becomes reality. Rev Col Bras Cir 2021;48.
[3] Rosen J, Hannaford B, Satava RM. Surgical robotics: systems applications and visions. Los Angeles: University of California; 2011.
[4] Visconti TAdC, Otoch JP, Artifon ELdA. Robotic endoscopy. A review of the literature. Acta Cir Bras 2020;35(2).
[5] Morelli L, Guadagni S, Di Franco G, Palmeri M, Di Candio G, Mosca F. Da Vinci single site© surgical platform in clinical practice: a systematic review. Int J Med Robot Comput Assist Surg 2016;Vol. 12.
[6] Uffer D. Robotic-assisted surgery (RAS) review: clinical landscape, commercial arena, and future outlook. Alira Health HVM; 2021.
[7] Thomas BC, Slack M, Hussain M, Barber N, Pradhan A, Dinneen E, et al. Preclinical evaluation of the Versius Surgical system, a new robot-assisted Surgical device for use in minimal access renal and prostate surgery. Eur Urol Focus 2021;7(2).
[8] CMR Surgical. https://cmrsurgical.com/wp-content/uploads/2022/07/5000-Surgeries-Completed-Using-Versius.pdf. 2022. 5000 Surgeries completed using Versius.
[9] Prata F, Ragusa A, Tempesta C, Iannuzzi A, Tedesco F, Cacciatore L, et al. State of the art in robotic surgery with hugo RAS system: feasibility, safety and clinical applications. J Pers Med 2023;13.
[10] Wang Y, Qu M, Mei N, Jiang X, Lu X, Nian X, et al. A phase III randomized controlled study of a domestic endoscopic robot used in radical prostatectomy. Chinese. J Urol 2021;42(7).
[11] Microport Scientific Corporation. https://microport.com/assets/blog/Microport-e-2023-Interim-Report_2023-09-22-054538_djaz.pdf. 2023. Microport interim report 2023.
[12] Hinata N, Yamaguchi R, Kusuhara Y, Kanayama H, Kohjimoto Y, Hara I, et al. Hinotori Surgical robot system, a novel robot-assisted surgical platform: preclinical and clinical evaluation. Int J Urol 2022;29(10).

[13] Medicaroid Corporation. https://www.medicaroid.com/en/release/pdf/231011_en.pdf. 2023. Demonstration of remote surgery using the hinotoriTM surgical robot system successfully performed between Singapore and Japan.

[14] Stephan D, Sälzer H, Willeke F. First experiences with the new Senhance® Telerobotic system in visceral surgery. Visc Med 2018;34.

[15] Lin J, Yan S, Ye Z, Zhao X. A systematic review of MAKO-assisted unicompartmental knee arthroplasty. Int J Med Robot Comput Assist Surg 2020;16.

[16] Sloan AE, Ahluwalia MS, Valerio-Pascua J, Manjila S, Torchia MG, Jones SE, et al. Results of the NeuroBlate system first-in-humans phase I clinical trial for recurrent glioblastoma. J Neurosurg 2013;118(6).

[17] Rubino F, Eichberg DG, Cordeiro JG, Di L, Eliahu K, Shah AH, et al. Robotic guidance platform for laser interstitial thermal ablation and stereotactic needle biopsies: a single center experience. J Robot Surg 2022;16(3).

[18] Smitson CC, Ang L, Pourdjabbar A, Reeves R, Patel M, Mahmud E. Safety and feasibility of a novel, second-generation robotic-assisted system for percutaneous coronary intervention: first-in-human report. J Invasive Cardiol 2018;30(4).

[19] Feng Y, Guo Z, Dong Z, Zhou XY, Kwok KW, Ernst S, et al. An efficient cardiac mapping strategy for radiofrequency catheter ablation with active learning. Int J Comput Assist Radiol Surg 2017;12(7).

[20] De Smet MD, Naus GJL, Faridpooya K, Mura M. Robotic-assisted surgery in ophthalmology. Curr Opin Ophthalmol 2018;29.

[21] Kalchiem-Dekel O, Connolly JG, Lin IH, Husta BC, Adusumilli PS, Beattie JA, et al. Shape-sensing robotic-assisted bronchoscopy in the diagnosis of pulmonary parenchymal lesions. Chest 2022.

[22] Yarmus L, Akulian J, Wahidi M, Chen A, Steltz JP, Solomon SL, et al. A prospective randomized comparative study of three guided Bronchoscopic approaches for investigating pulmonary nodules: the PRECISION-1 study. Chest 2020.

[23] Fiorini P, Goldberg KY, Liu Y, Taylor RH. Concepts and trends in autonomy for robot-assisted surgery. Proc IEEE 2022;110(7).

[24] Barbon C, Grünherz L, Uyulmaz S, Giovanoli P, Lindenblatt N. Exploring the learning curve of a new robotic microsurgical system for microsurgery. JPRAS Open 2022;34.

[25] Millan B, Nagpal S, Ding M, Lee JY, Kapoor A. A scoping review of emerging and established Surgical Robotic platforms with applications in urologic surgery. Société Internationale d'Urologie Journal 2021;5.

[26] C. Hale. https://www.fiercebiotech.com/medtech/kidney-monitor-maker-potrero-medical-files-chapter-11-bankruptcy. 2023. J&J MedTech aims to put Ottava surgical robot up for US clinical trials in late 2024.

[27] Eugene Demaitre. https://www.robotics247.com/article/virtual_incisions_mira_surgical_robot_to_show_skills_on_international_space_station. 2022. Virtual incision's MIRA surgical robot to show skills on international space station.

[28] S. Whooley. https://www.massdevice.com/vicarious-surgical-wins-fda-breakthrough-designation-for-surgical-robot/. 2019. Vicarious surgical wins FDA breakthrough designation for surgical robot.

[29] Liatsikos E, Tsaturyan A, Kyriazis I, Kallidonis P, Manolopoulos D, Magoutas A. Market potentials of robotic systems in medical science: analysis of the Avatera robotic system. World J Urol 2022;40(1).

[30] Panesar S, Cagle Y, Chander D, Morey J, Fernandez-Miranda J, Kliot M. Artificial intelligence and the future of Surgical robotics. Ann Surg 2019;270.

[31] Andras I, Mazzone E, van Leeuwen FWB, De Naeyer G, van Oosterom MN, Beato S, et al. Artificial intelligence and robotics: a combination that is changing the operating room. World J Urol 2020;38.

[32] Gasparetto A, Scalera L. A brief history of industrial robotics in the 20th century. Adv Hist Stud 2019;08(01).

[33] Shaear M, Russell JO, Steck S, Liu RH, Chen LW, Razavi CR, et al. The intuitive da Vinci single port surgical system and feasibility of transoral thyroidectomy vestibular approach. Vol. 5. Ann Thyroid; 2020.

[34] Mayor N, Coppola AS, Challacombe B. Past, present and future of surgical robotics. Trends Urol Men's Health 2022;13(1).

[35] D.M. Marchini. https://formiche.net/2022/11/robot-hugo-chirurgia-sommella/. 2022. Vi presento Hugo, il robot per la chirurgia assistita. Parla Sommella.

[36] Sasaki M, Hirano Y, Yonezawa H, Shimamura S, Kataoka A, Fujii T, et al. Short-term results of robot-assisted colorectal cancer surgery using Senhance digital laparoscopy system. Asian J Endosc Surg 2022;15(3).

[37] Carolinas Center for Surgery. https://www.cc4surgery.com/hip-and-knee-replacement-what-is-mako-robotic-orthopedic-surgery/. 2019. What is MAKO robotic orthopedic surgery?.

[38] Lo N, Gutierrez JA, Swaminathan RV. Robotic-assisted percutaneous coronary intervention. Curr Treat Options Cardiovasc Med 2018;20(2):14.

[39] De Smet MD, Stassen JM, Meenink TCM, Janssens T, Vanheukelom V, Naus GJL, et al. Release of experimental retinal vein occlusions by direct intraluminal injection of ocriplasmin. Br J Ophthalmol 2016;100(12).

[40] Biswas P, Sikander S, Kulkarni P. Recent advances in robot-assisted surgical systems. Biomed Engin Adv 2023;100109.

[41] Kutsenko O, Narayanan G, Gentile N, Gandhi R. Robotics in interventional oncology: the next frontier in image-guided interventions. Endovascular Today 2023;22.

[42] Aitzetmüller MM, Klietz ML, Dermietzel AF, Hirsch T, Kückelhaus M. Robotic-assisted microsurgery and its future in plastic surgery. J Clin Med 2022;11.

[43] J. Hammerand. https://www.massdevice.com/titan-medical-strategic-buyers-enos-2-surgical-robot/. 2023. Titan Medical meets with strategic buyers and investors as surgical robot developer considers its future.

[44] OR Today Magazine. https://ortoday.com/virtual-incision-mira-surgical-robotic-system/. 2020. Virtual incision MIRA surgical robotic system.

[45] C. Newmarker. https://www.massdevice.com/vicarious-surgical-stock-continues-to-rise. 2021. Vicarious Surgical stock continues to rise.

[46] B. Oppermann. https://medizin-und-technik.industrie.de/markt/aus-der-branche/avatera-op-roboter-aus-jena-als-alternative-zum-davinci-system. 2022. Avatera: OP-Roboter aus Jena als alternative zum davinci-system.

# 13

# Enabling intelligent procedures: Endoscopy dataset collection trends and pipeline

*Lyuxing He[a,b], Huxin Gao[a,b], and Hongliang Ren[c,d]*

[a]Department of Biomedical Engineering, National University of Singapore, Faculty of Engineering, Singapore, Singapore [b]National University of Singapore (Suzhou) Research Institute (NUSRI), Suzhou, China [c]Department of Electronic Engineering, The Chinese University of Hong Kong, Sha Tin, Hong Kong [d]Shun Hing Institute of Advanced Engineering (SHIAE), The Chinese University of Hong Kong, Sha Tin, Hong Kong

## Introduction

A gastrointestinal (GI) examination involves evaluating the structure and function of the GI tract, which includes the esophagus, stomach, small intestine, and large intestine (colon), to diagnose or monitor diseases, evaluate symptoms, and screen for conditions such as colorectal cancer. However, traditional GI examinations have intrinsic limitations [1], including lesion detection rate, spatial complexity, etc. Therefore, there is an urgent need to incorporate automation into those methods, and the ability to perform scene reconstruction is highly beneficial for both autonomous medical robotics and experienced surgeons. This is yet challenging compared to the traditional 3D reconstruction setting due to several factors [2], such as limited size and number of carried hardware, insufficient training dataset, the lack of surface texture features, complex reflective surfaces, etc. To address the aforementioned problems, we propose GI-Replay and summarize two major contributions we have made to the research community:

1. We present a complete pipeline to construct a GI tract dataset for dense 3D reconstruction application. Our experimental setup can provide monocular and stereo images with training ground truth information, including depth and camera poses within an in vivo in-cavity porcine GI tract.
2. We introduce a neural radiance field (NeRF)-based 3D-reconstruction method that works with the above data collection pipeline. It takes a monocular RGB image training dataset. It produces accurate in-body scene reconstruction free from pose priors and supervised by geometric cues of implicit representations predicted by a coupled pretrained depth estimation network.

## Related work

### Dataset survey

The first part of the survey delves into the available datasets that have been pivotal in 3D reconstruction for the GI environment. With the rise in deep learning techniques, numerous public datasets have emerged to support research and development of various advanced diagnostic features across various tasks, including segmentation, disease classification, localization, tissue deformation and motion detection, and depth estimation. Table 1 gives an overview of the widely used such datasets, including dataset names, segments of the GI tract covered, ground truth data types, dataset sizes, and other related information in the following order: [13,11,7,5,4,9,6,8,12,10].

The following discussion will center around datasets for dense 3D reconstruction applications since those datasets typically provide enough size of data and various types of ground truth measurements due to the natural requirements

TABLE 1 A brief survey of the widely used datasets for applications related to GI tract.

| Name | Target | Object condition | | Organs (stomach.) | GT data – method/device | | | | | | | |
| | | In-vivo or ex-vivo | In cavity or dissected | | Image | Depth | Pose | 3D model | # of views | Varied lighting | Datasets size | Application |
|---|---|---|---|---|---|---|---|---|---|---|---|---|
| Hamlyn dataset | Human | In-vivo | In cavity | Colons (with Polyp), kidney, liver, ureter, kidney, abdomen | Olympus narrow-band imaging and Pentax i-Scan | Calculated using stereo matching software Libelas | N/A | N/A | 2 (stereo) | No | 10 (videos) + 7834 (images) | Segmentation, classification, disparity estimation, tissue deformation tracking |
| EndoSLAM | Animal (Porcine) | Ex-vivo | Dissected | Colons (with Polyp), small intestine, stomach | MiroCam Regular MC1000-W endoscope video capsule, Pillcam COLON2, High Resolution Endoscope Camera (YPC-HD720P), low resolution endoscope 3 in1 camera | 3D scanners (EinScan Pro 2X, Artec 3D Eva) | Calculated from Franka Emika Panda robotic arm joint trajectory | 3D scanner | 4 (multi-view) | Yes | 42700 (images) | Pose estimation, dense 3D reconstruction |
| C3VD | Phantom (physical simulator) | Ex-vivo | Dissected | Colons | Mono Olympus CF-HQ190L video colonoscope | Calculated using phantom 3D model | UR3 pose log | Sculpted | 1 (mono) | No | 22 (videos) | Pose estimation, dense 3D reconstruction |
| EndoMapper | Human + virtual | In-vivo | In cavity | Simulated colon using VR-CAPs and real complete routine endoscopies/colonscopies | EVIS EXERA III CF-H190 colonoscope/EVIS EXERA III GIF-H190 gastroscope | N/A | N/A | N/A | 1 (mono) | No | 59 (videos) | Segmentation, VSLAM evaluation |
| SCARED | Animal (Porcine) | Ex-vivo | Dissected | Complete routine of laparoscopies | da Vinci stereo | Structured light pattern projection to compute dense stereo reconstruction | da Vinci Xi surgical robot pose log | N/A | 5–10 (multiview) | No | 22843 (images) | Stereo correspondence, dense 3D reconstruction |

| | | | | | | | | | | | | |
|---|---|---|---|---|---|---|---|---|---|---|---|---|
| ERS | Human | In-vivo | In cavity | Gastrointestinal tract (colonoscopy and capsule endoscopy) | N/A | N/A | N/A | N/A | 1 (mono) | No | 6000 (images) | Segmentation, classification, multilabel image classification |
| MESAD | Human + Phantom | In-vivo | In cavity | Prostate | da Vinci Xi robotic system with a diameter of 8 mm (Intuitive Surgical Inc.) | N/A | N/A | N/A | 1 (mono) | No | 23,366 (MESAD-Real) + 22,609 (MESAD-Phantom) | Action detection during prostatectomy |
| HyperKvasir | Human | In-vivo | In cavity | Gastrointestinal tract | N/A | N/A | N/A | N/A | 1 (mono) | No | 10662 (labeled images), 99417 (unlabeled images), 1000 (segmentad) | Segmentation, classification, bounding box annotation, video annotation |
| EndoAbS | Phantom (physical simulator) | Ex-vivo | Dissected | Liver, kidneys, spleen | Stereo endoscope | Generated using a laser scanner | Generated using a laser scanner | N/A | 2 (stereo) | Yes (different light levels; presence of smoke) | 120 pair of images | Dense 3D reconstruction |
| StereoMIS | Animal (Porcine) + Human | In-vivo | In cavity | N/A | N/A | Stereo video feeds from the da Vinci Xi surgical robot | Camera forward kinematics | N/A | 2 (stereo) | Yes | 16 (videos) | SLAM in endoscopic surgery |
| EndoVisSub2018-F | Animal (Porcine) | Ex-vivo | Dissected | Kidney, small bowel | da Vinci Xi stereo structured light pattern projection | N/A | N/A | N/A | 2 (stereo) | No | 16 (procedures) + 5960 (frames) | Segmentation, detection of man-made and anatomical objects |

for traditional dense 3D reconstruction training or validation, making the datasets sufficient to be used for most of all other tasks.

The type of datasets for 3D reconstruction for the GI environment can be categorized into three classes: humans, animals, and phantoms (physical or virtual). Datasets like [13,5,9,6,8] are examples of using humans as data sources. The datasets are usually made from actual GI procedures of human patients and, therefore, are in vivo and in-cavity, which well-capture the real-time variability of living organs and spatial geometry of confined spaces of the GI tract. Refs. [9,8] both provide endoscopy image and video data. In addition to that, Ref. [13] provides corresponding depth maps for each stereo image pair, but the depth maps are estimated using stereo matching software LIBELAS using the stereo image pairs; Refs. [5,6] claim to provide ground truths for reconstruction purposes in addition to video and image data from humans, but the ground truths are collected from phantoms. More specifically, Ref. [5] uses computed tomography (CT) results and VR-Caps7 as simulator to generate ground truths, and virtual model built from CT with finite resolutions will not capture as many details; Ref. [8] incorporated data collected from artificial anatomies used for the training of surgeons, but this phantom is mainly designed for surgeon action detection during prostatectomy. It is not obvious that, due to ethnicity and safety concerns, the primary source of data presented in those datasets is videos and images of actual GI procedures of human patients only, and the only sensor used to build the dataset is a camera(s). This implies that the only trustworthy data are those images and videos, and all other information included in the datasets is, in fact, secondary or estimated.

Datasets like [11,4,3,10] are examples of using animals (mainly porcine) as data sources and are the most common and widely used ones among those for dense 3D reconstruction applications. Since ethnicity and safety are not the main concerns anymore, organs are typically brought outside of the body and dissected for the convenience of using advanced machinery. As a result of this, datasets collected from animals tend to cover more types of ground truth data. Ref. [11] uses four endoscope video cameras to record video data, Franka Emika Panda robotic arm to track the trajectory and quantify the pose values, and high-precision 3D scanners for ground truth organ shape (including depth) measurement; Ref. [4] uses da Vinci Xi surgical robot stereo camera to collect video data, robot pose log to compute ground truth pose, and structured light pattern projection to compute depth ground truths; Ref. [3] also uses da Vinci Xi surgical robot, but the dataset contains only video data, as the dataset is mainly used for segmentation and robotic surgical instruments detection tasks. As an outlier, Ref. [10] uses da Vinci Xi surgical robot in an in vivo and in-cavity environment of both animals and humans to collect datasets, and it can provide stereo video feeds as well as ground truth pose calculated using camera forward kinematics. The dataset does not include ground truth depths, although theoretically, it could have obtained them using structured light pattern projection. Despite the variety of ground truth measurements in datasets collected in ex vivo dissected animal organs, the real-time variability and spatial geometry are no longer preserved in the data, given that the targets are already dead and dissected during the experiment. Moreover, using robot joint history to calculate camera ground truth is intrinsically inaccurate. This is because errors in the forward kinematics will eventually lead to errors in the absolute camera pose [10].

Datasets like [7,12] are examples of using phantoms as data sources. The advantage of phantoms is that, ground truth measurements are directly available during manufacturing or modeling, and artificial details can be added to phantoms to simulate some specific scenarios that are not common in actual surgical scenes. Features of real-time variability and spatial geometry can also be reproduced somehow by animating virtual organs or choosing the appropriate materials for the physical model. Ref. [7] is acquired using a high-definition clinical colonoscopy in high-fidelity colon models. Each colon model is digitally sculpted in Zbrush (Pixologic) by a board-certified anaplastologist (JRG) using reference anatomical imagery from colonoscopic procedures. Ref. [12] builds physical phantoms through a moulding process similar to Ref. [14] using 3D model generated from anonymized DICOM CT medical images. The polyurethane organ phantom of the abdominal organ presented in the work also aims to mimic deformations of actual tissues. However, phantoms typically come with the simplification of complex interactions in the actual GI environment. Also, virtual phantoms usually lack haptic feedback, which is crucial for many medical procedures, while the static nature of physical phantoms prevents them from replicating biological processes. These drawbacks make the datasets collected from phantoms less realistic and applicable to surgical scenarios.

As the table suggests and the above discussions point out, there are currently no datasets capable of providing all accurate ground truth depth, camera pose, and camera views simultaneously within the in vivo and in-cavity environment of a human or animal GI tract. The success of 3D dense reconstruction often hinges on the accuracy of depth estimation, camera pose estimation, and the quality and quantity of camera views, and thus, providing enough size and variety of the data mentioned above can better help 3D dense reconstruction algorithms to train and validate their models. In vivo and in-cavity are also crucial, as the in vivo environment ensures the dataset captures real-time and dynamic observations. In contrast, the in-cavity environment preserves spatial relationships and effects of single lightening conditions, including specular reflections and ambiguity in surface normals. As described in Section 3, GI-Replay can collect all ground truth depth, camera pose, and camera views in an in vivo and in-cavity environment. This is because, first of all, the sensors used by GI-Replay are small in size and can all be conveniently installed in a typical endoscopic duodenoscope. Second, GI-Replay's methodologies of collecting ground truth data are relatively more accurate, as pose ground truths are measured by tracking specialized sensors in generated magnetic fields instead of using forward kinematics, and depth ground truths are aligned to the actual depths instead of pure vision approaches. All of these distinguish the collected dataset by GI-Replay from the currently used datasets listed in the table.

## 3D reconstruction approach survey

The second part of the survey discusses the most recent 3D reconstruction algorithms in the GI tract and the differences with our approach. Ref. [13] leverages depth convolutional networks to create pseudo-RGBD keyframes, estimates the camera motion using photometric methods, and fuse the registered pseudo-RGBD keyframes as truncated signed distance function (TSDF) volumetric representation. The issue with this method is that the rendering of TSDF does not provide proper shading, lighting, and texture mapping. In addition, defects, including gaps, missing parts, or inaccuracies, will be directly reflected in the synthesized views. Ref. [2] focused on deformable surgical scene reconstruction from single-viewpoint stereo input using NeRF additionally guided with depth-supervision loss. This work demonstrates superior performance in most of the robotic surgery datasets.

However, the model mentioned in the work is mostly derived from D-NeRF [15] that requires stereo image inputs and camera poses as prior during training, which are sometimes unavailable or expensive to obtain during the actual surgery.

Our approach attempts to densely 3D reconstruct the GI tract using monocular images only while producing view synthesis that is smooth and accurate without the need for camera pose priors.

## GI-Replay dataset

This section introduces the GI-Replay dataset collection pipeline. It starts with the equipment used for building the GI-Replay. It then demonstrates how the sensor nodes are coordinated and ground truth information is recorded. It concludes with an overview of the dataset architecture built by GI-Replay.

## Equipment

The main devices used are described below:

1. Olympus duodenoscope (1): TJF-Q190V Olympus Duodenoscope is the main carrier that equips and enables manipulations with all the sensors, as it has additional insertion tubes that carry external sensors and surgical tools to move around together with the probe head. In addition to that, the duodenoscope has the following properties:
   (a) Optical system: The duodenoscope comes with an advanced medical imaging system with a field of view of 100°, the direction of view of backward side viewing 15°, and a depth of field of 5–60 mm. It is coupled with the Olympus video processor discussed below.
   (b) Four-way angulation capability: The duodenoscope has angulation ranges of upward to 120° downward to 90° right to 110° and left to 90°.
   (c) Manual cleaning: Dirt or body liquid on the camera lens during the procedure on the optics of the duodenoscope can be manually flushed by water injected through the inner tube using the water/air nozzle.
2. EVIS LUCERA endoscopic video processor (1): As mentioned earlier, the video signal from TJF-Q190V Olympus Duodenoscope optical system is processed and output via EVIS LUCERA endoscopic video processor. The endoscopic video processor has the following components:
   (a) CV-260 video system center: convert signal input to video out in HDTV or SDTV, structure enhancement, electronic zoom, prefreeze image capture, automatic iris control, automatic white balance.
   (b) CLV-260 xenon light source: automatic 16-level brightness control, adjustable aeration (high/medium/low settings), emergency lamp cut-off.
3. USB cameras (2): Two OV6946 USB cameras with the same camera intrinsics essentially provide a stereo live of the probe head. Each camera has the following specifications:
   (a) resolution: $400 \times 400$
   (b) FPS: 30
   (c) depth of field: 3–100 mm
   (d) field of view: $120° \pm 15\%$
   (e) LEDs: 4, with a minimum illumination of 0.03 lux
   (f) passive sensor head size: module diameter as small as 2.0 mm

**4.** TrakSTAR EMF sensor (1): The TrakSTAR system by NDI uses an electromagnetic field generator to dynamically track specialized sensors' position and orientation in real time by interpreting variations in detected electromagnetic fields, therefore providing the position and orientation of the probe head. It does not require a line of sight between the sensor and the tracker and, therefore, is especially useful in medical applications where the sensor might be inside or obscured by the human body. The specific model we use is Model 180, and it has the following specifications:

  **(a)** 6 (position and orientation) degrees of freedom measurements

  **(b)** translation range: 58 cm

  **(c)** angular range: ±180° azimuth and roll and ±90° elevation

  **(d)** static accuracy: position of 1.4 mm RMS, orientation of 0.5° RMS

  **(e)** static resolution: position of 0.5 mm at 30.5 cm, orientation of 0.1° at 30.5 cm

  **(f )** passive sensor head size: module diameter as small as 2.0 mm

One important note is that ferromagnetic objects and stray magnetic fields in the operation volume may degrade performance.

Below are images of the complete GI-Replay dataset collection pipeline setup and a simple usage demonstration of probing inside a silicone pig stomach model. In Fig. 1, the left and right cables correspond to the USB cameras, and the middle cable corresponds to the TrakSTAR EMF sensor. Similarly, in Fig. 2, the side lighting sources represent the USB cameras, and the middle blond head represents the TrakSTAR EMF sensor. Fig. 3 illustrates how the Olympus Duodenoscope enters the cavity. First, a TJF-Q190V Olympus Duodenoscope prototype without the optical system is used;

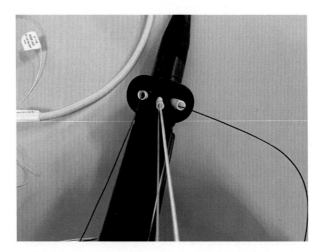

FIG. 1   Handle view.

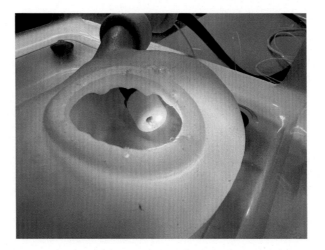

FIG. 2   Probe view.

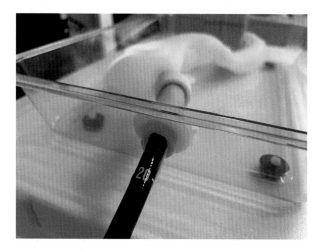

FIG. 3 Entry view.

therefore, the optical lenses presented in the figures are only USB cameras. Second, although the demonstration is presented in an ex vivo and dissected environment for simplicity, the pipeline can work the same in an in vivo and in-cavity environment, as the only requirement for GI-Replay to successfully collect data and construct a dataset is being able to enter the GI tract and explore inside of it manually. The rest of the process is done autonomously.

## Data flow

Robot operating system (ROS) is the middleware allowing onboard hardware to communicate, interact, and store data locally. As shown in Fig. 4, each hardware is abstracted as an ROS node and publishes the collected data at its rate through its topics. Since these sensors' data are published at different rates and times, it is crucial to synchronize them before storing them as a dataset, thus ensuring temporal consistency. To do so, a message filter filters and buffers data

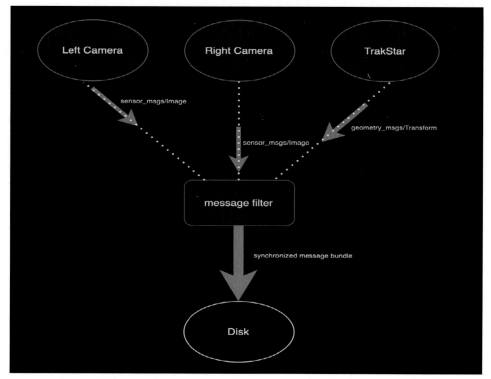

FIG. 4 Flow chart of how data are synchronized and stored by GI-Replay.

with different timestamps. Only data from sensors received with the same timestamp will be bundled and passed through the filter, and data buffered for a specified timeout will be discarded. A master data collection rate can also be set via buffering the bundled synchronized data for a lower limit of

$$\frac{1}{R_{\text{master}}} \geq LCM\left(\frac{1}{R_1}, \frac{1}{R_2}, \dots, \frac{1}{R_N}\right)$$

where $R_{\text{master}}$ is the lower limit of the master data collection rate, and $R_i$ is the publish rate of each sensor node.

## Depth ground truth

### *Depth map computation*

Different from the other ground truth measurements like pose and images that can be directly measured by sensors, the ground truth depth map is indirectly derived from the stereo camera pairs. To do so, images will be first rectified by obtaining the intrinsic parameters and performing stereo calibration. This will give intrinsic and extrinsic matrices and, therefore, the fundamental and essential matrices, allowing us to compute the rectification transforms that wrap the images into rectified pairs. The rectified stereo pairs are then passed to StereoSGBM [16] to produce the corresponding disparity maps, which give the following depth maps:

$$Z(x, y) = \frac{f \times B}{D(x, y)}$$

where $Z$ is the depth (distance of the point from the camera), $f$ is the focal length of the camera (usually in pixels), $B$ is the baseline or the distance between the two cameras, and $D$ is the disparity at that point.

### *Anchor point refinement*

This depth map's accuracy is further improved by iteratively aligning it with point depth ground truth measurements. To obtain a single-point depth ground truth, GI-Replay will need the pose information when the sensor head touches the wall. To do that, when the probe head is exploring within the cavity at any point during data collection, the probe head can reach the electromagnetic fields (EMF) sensor out toward the cavity wall while monitoring it through the camera feed. When the EMF sensor collides with the wall without much deformation, one can manually force a data collection action without waiting for the next data synchronization cycle by pressing an attached button or inputting a space key into the terminal.

Suppose $n$ such forced synchronization operations are taken along with data frames $D$ captured during the data collection process, then each frame $d_i \in D$ with camera pose $\mathcal{P}_i \in \mathcal{P}_{\text{data}}$ will potentially have $m(m \leq n)$ ground truth point depth measurements as anchor points for depth refinement. In other words, for each data frame $d_i$ with pose $\mathcal{P}_i = (R_i, T_i)$, there is a fixed array of camera origins captured at the cavity wall with poses $\mathcal{P}_{\text{wall}} = \{\mathcal{P}_j | j = 0, 1, \dots, n-1\} = \{(R_j, T_j) | j = 0, 1, \dots, n-1\}$ that can be related to data frame $d_i$, depending on whether those camera origins can be projected onto the current data frame.

To make use of those camera origins captured at the cavity wall, we will first convert the world coordinate of the $j$th traversed camera origin with $\mathcal{P}_j$ to camera coordinate. We first write its homogeneous coordinate as $C_j = [T_j, 1]^T = [t_{j,x}, t_{j,y}, t_{j,z}, 1]^T$. Then, the homogeneous coordinate of the camera origin is projected into the current data frame $C_i$ and transformed into the camera coordinate as follows:

$$\begin{bmatrix} X_w \\ Y_w \\ Z_w \\ \sigma_w \end{bmatrix} = C_i = \begin{bmatrix} R_i & T_i \\ 0 & 1 \end{bmatrix} \cdot C_j$$

$$\begin{bmatrix} X_c \\ Y_c \\ Z_c \end{bmatrix} = \frac{1}{\sigma_w} \begin{bmatrix} X_w \\ Y_w \\ Z_w \end{bmatrix}$$

Note that $Z_c$ is also the ground truth point depth measurement for the $i$th data frame imposed by the $j$th camera origins captured at the cavity wall.

The next step is to map the camera coordinate to the pixel coordinate. This is done by multiplying the camera intrinsics matrix with the camera coordinate, and the resulting $[u_j, v_j]^T$ is the $j$th pixel coordinate of the camera origin's projection onto the data frame:

$$\begin{bmatrix} X_p \\ Y_p \\ \sigma_p \end{bmatrix} = K \cdot \begin{bmatrix} X_c \\ Y_c \\ Z_c \end{bmatrix}$$

$$\begin{bmatrix} u_j \\ v_j \end{bmatrix} = \frac{1}{\sigma_p} \begin{bmatrix} X_p \\ Y_p \end{bmatrix}$$

If the resulting $[u_j, v_j]^T$ contains undefined values (e.g., infinity) that camera origin should be unrelated to the current data frame and should be discarded together with the corresponding point depth measurement. This is because the undefined projected pixel coordinate indicates the specific camera origins at the cavity are not captured by the current data frame and, therefore, will not affect the corresponding depth map. This will yield, for each data frame $di$ in the dataset $D$, there will exist $m$ related point depth ground truth measurement $\{Z_{g,k}|k=0,1,...,m-1\}$ at the corresponding pixel locations $\{[u_j, v_j]^T|k=0,1,...,m-1\}$ on the depth map, where $m \leq n$.

Using these single-point ground truth measurements and pixel locations as known anchor points, we can perform anchor point iterative minimization [17] in the energy function formulation similar to StereoSGBM to refine each computed depth map $d_i$ (abusing the variable $d_i$ to represent both data frames collected during data acquisition and computed depth maps). More specifically, this is done by minimizing the following sum:

$$\min_{d_i} E(d_i) = E_{\text{data}}(d_i) + E_{\text{smooth}}(d_i)$$

with $E_{\text{data}}$ measuring how far the depth values at the anchor points are away from the point ground truth depths, where $\lambda$ is the step coefficient:

$$E_{\text{data}}(d_i) = \sum_{k=1}^{m} \lambda \times (d_i(u_k, v_k) - Zg, k)^2$$

and $E_{\text{smooth}}$ measures the similarity of the neighboring pixels, where $\mathcal{N}(x,y)$ represents the neighboring pixels at $(x,y)$, $w(x,y,x',y')$ serves as a weighting factor that grows with the similarity between the neighboring pixels' intensities with the value proposed in Ref. [18], and truncated $L1$ distance is used as a discontinuity preserving cost function:

$$E_{\text{smooth}}(d_i) = \sum_{(x,y)} \sum_{(x',y')\in\mathcal{N}(x,y)} w(x,y,x',y') \cdot \|d_i(x,y) - d_i(x',y')\|_1$$

## Dataset architecture

The directory tree of dataset collected at each data acquisition process will contain the following four folders: the "img1" folder that stores the synchronized image data from the left camera of the stereo pair as .jpg, the "img2" folder that similarly stores the right camera image data, the "emf" folder that stores the data synchronization timestamp, camera pose, and single-point depth measurements, and the "depth" folder that stores computed depth maps refined using anchor point iterative minimization. Fig. 5 illustrates how the data are recorded in a .txt file, where *Translation* and *Orientation* together describe the camera pose, *Time stamp* records the exact timestamp each data synchronization happens, and *Depth* shows the start pose and end pose used to calculate the single-point depth measurement and the calculated value. Note that the .jpg and .txt files are named by numerical numbers that indicate the synchronization count, e.g., "5.txt" and "5.jpg" represent TrakSTAR sensor data and camera image data synchronized to the same timestamp at the fifth synchronization cycle. If the synchronization is forced manually for point depth ground truth measurement, the string "manual" will append to the front of the file names, e.g., "manual 5.txt" and "manual 5.jpg" represent TrakSTAR sensor data and camera image data of a forced synchronization at the fifth synchronization cycle.

## GI-Replay 3D reconstruction

This section describes how pose joint optimization and geometric cues supervision are added to traditional NeRF framework [19]. The intuition is that joint optimization of camera pose is needed since pose information is most likely unavailable for actual GI procedures. The geometric cues are also beneficial, as photoconsistency cues proposed in

```
Translation:
x: 0.914288
y: 0.578532
z: 0.144214

Rotation:
x: 0.790221
y: -0.0423354
z: 0.0472156
w: 0.609532

Time stamp:
2023-08-14 04:25:56.6131

Depth:
start pose:
translation:
  x: 0.914288
  y: 0.578532
  z: 0.144214
rotation:
  x: 0.790221
  y: -0.0423354
  z: 0.0472156
  w: 0.609532
end pose:
translation:
  x: 0.914288
  y: 0.6032
  z: 0.119881
rotation:
  x: 0.794671
  y: 0.00696116
  z: 0.03505
  w: 0.605987
calculated depth: 0.034649
```

FIG. 5  "emf" folder file content.

traditional NeRF can establish globally accurate 3D geometry in textured regions. Geometric cues like depth maps and normal maps provide local geometric information [20], enabling the reconstruction method to deal with large texture-less regions like GI track human tissues where traditional NeRF prone to fail [21].

## From the Traditional NeRF

NeRF represents the volumetric scene function $f_\theta(\mathbf{x}, \mathbf{d}) = (\sigma, \mathbf{c})$ with a neural network parameterized by $\theta$ that maps a 5D vector, composed of a 3D spatial location $\mathbf{x}$ and a 2D viewing direction $\mathbf{d}$, to a 4D output specifying the volume density $\sigma$ and RGB color $\mathbf{c}$ at that spatial location. Note that the 2D viewing direction encapsulates information about the camera poses $O$.

The expected color $C(r)$ along a camera ray $r(t)$ with near and far bounds $t_n$ and $t_f$ is given by

$$\hat{C}(r) = \int_{t_{\text{near}}}^{t_{\text{far}}} T(t)\, \sigma(r(t)) \mathbf{c}(r(t), d)\, dt$$

where the transmittance along the ray up to $t$ is given by

$$T(t) = \exp\left(-\int_{t_{\text{near}}}^{t} \sigma(s) ds\right)$$

Rendering a view from a continuous NeRF requires estimating using a stratified sampling approach that partitions $[t_n, t_f]$ into $N$ evenly-spaced bins, with each sample drawn uniformly at random within each bin. Let $\delta_i = t_{i+1} - t_i$, and the discretized estimation of predicted color is written as

$$\hat{C}(r) = \sum_{i=1}^{N} T_i(1 - \exp(-\tau_i \delta_i)) c_i$$

where the transmittance now becomes

$$T_i = \exp\left( - \sum_{j=1}^{i-1} \tau_j \delta_j \right)$$

And last, recall that NeRF is optimized via

$$L_{\text{NeRF}} = \sum_{x} \left| \hat{C}(x) - C(x) \right|$$

$$\theta^* = \arg \min_{\theta} L_{\text{NeRF}}\left( \hat{C} | C, O \right)$$

To throw camera pose prior $O$ away and boost reconstruction accuracy under complex textures and less-observed environment of the GI environment, we perform joint optimization [22] and include additional supervision terms using monocular geometric cues, where $\lambda_1$, $\lambda_2$, and $\lambda_3$ represent the weighting factors for each of the loss term, respectively:

$$L_{\text{total}} = L_{\text{NeRF}} + \lambda_1 L_{\text{depth}} + \lambda_2 L_{\text{norm}} + \lambda_3 L_{\text{pc}}$$

$$\theta^*, \phi^* = \arg \min_{\theta, \phi} L_{\text{total}}\left( \hat{C}, \hat{O} | C \right)$$

## Depth constraint $L_{\text{depth}}$

Geometric cues can be supervised via depth priors with the approach inspired by [23]. We take advantage of the pre-trained monocular depth estimation network Monodepth2 [24] and fine-tune it on monocular images and corresponding depth maps from Hamlyn dataset [13], a vivo endoscopy stereo video dataset of the Hamlyn Center Laparoscopic at Imperial College London. The depth estimation is incorporated into the NeRF pipeline such that each input image's corresponding depth map is predicted by Monodepth2, and the bundled monocular image and depth map are fed into training loss calculation and network optimization.

More specifically, the depth loss is the absolute difference between depth maps predicted by the neural radiance representation $\hat{D}_i$ and Monodepth2 $\overline{D}_i$, normalized by the depth variance of the neural radiance representation predictions to discourage weights with high uncertainty:

$$L_{\text{depth}} = \sum_{i=1}^{n} \frac{\left| \hat{D}_i - \overline{D}_i \right|}{\sqrt{var\left( \hat{D}_i \right)}}$$

with $\hat{D}_i$ rendered by trained NeRF is given by

$$\hat{D}_i = \sum_{i=1}^{N} T_i(1 - exp\left( -\tau_i \delta_i \right)) t_i$$

## Surface normal constraint $L_{\text{norm}}$

As a bi-product of the depth prior, we can also derive a surface normal map and impose surface normal consistency similar to Ref. [20]. More specifically, surface normal priors $\overline{n}$ are derived from the depth prior $\overline{D}$ by first approximating its partial derivatives in the x and y directions, respectively (or approximate it using Sobel operator):

$$\frac{\partial \overline{\mathbf{D}}}{\partial x} \approx \overline{\mathbf{D}}(x+1,y) - \overline{\mathbf{D}}(x-1,y)$$

$$\frac{\partial \overline{\mathbf{D}}}{\partial y} \approx \overline{\mathbf{D}}(x,y+1) - \overline{\mathbf{D}}(x,y-1)$$

Given the changes in depth in the $x$ and $y$ directions, the normal vector can be constructed. The negative of these derivatives gives the $x$ and $y$ components of the surface normal, and the $z$ component is set to 1 to indicate the upward direction. Thus,

$$\overline{\mathbf{n}}(x,y) = \left( -\frac{\partial \overline{\mathbf{D}}}{\partial x}, -\frac{\partial \overline{\mathbf{D}}}{\partial y}, 1 \right)$$

and normalizing it to a unit vector gives the surface normal prior:

$$\hat{\mathbf{n}} = -\frac{\mathbf{n}}{\|\mathbf{n}\|_2}$$

We can then use this surface normal map to form the surface normal constraint enforced by the following surface normal loss:

$$L_{\text{norm}} = \sum_{i=1}^{n} \|\hat{\mathbf{n}} - \overline{\mathbf{n}}\|_1 + \|1 - \hat{\mathbf{n}}^T \overline{\mathbf{n}}\|_1$$

where the surface normal map $\hat{\mathbf{n}}$ rendered by trained NeRF is given by the normalized negative gradient of the density field [25]:

$$\hat{\mathbf{n}} = -\frac{\nabla_x \sigma}{\|\nabla_x \sigma\|_2}$$

## Point cloud constraint $L_{\text{pc}}$

The intuition of the pose joint optimization scheme using depth priors comes from Ref. [22]. The pose dependency of the total loss function is realized by constraining pairwise point clouds back-projected from the depth maps predicted by Monodepth2. More specifically, depth maps $D$ are back-projected using the known camera intrinsics to point clouds $P = \{P_i \mid i = 0, 1, \dots, N-1\}$, which is then used to optimize the relative pose between consecutive point clouds by minimizing the following loss:

$$L_{\text{pc}} = \sum_{i=1}^{n} \sum_{j=1}^{n} d_{dc}\left(P_i, T_j T_i^{-1} P_i\right)$$

where $i$ and $j$ are consecutive pairs of image instances, $T_j T_i^{-1}$ represents the related pose that transforms point cloud $P_i$ to $P_j$, and $d_{cd}$ represents Chamfer Distance [26] that measures the distance between two point clouds as

$$d_{\text{cham}}(A,B) = \frac{1}{|A|} \sum_{a \in A} \min_{b \in B} \|a - b\|^2 + \frac{1}{|B|} \sum_{b \in B} \min_{a \in A} \|a - b\|^2$$

## 3D reconstruction results

### Depth prediction

The pretrained model used from Monodepth2 is mono640x192. Images from Hamlyn dataset sequences 5, 6, and 8 with halved resolution and reflection padding are used to fine-tune the pretrained model. The best performance is achieved with 10 training epochs. When evaluating the Hamlyn dataset sequences 1 and 4, the fine-tuned Monodepth2 achieved an average MSE of 10.721 on sequence 1 and 4.188 on sequence 4. Visual results are shown in Fig. 6.

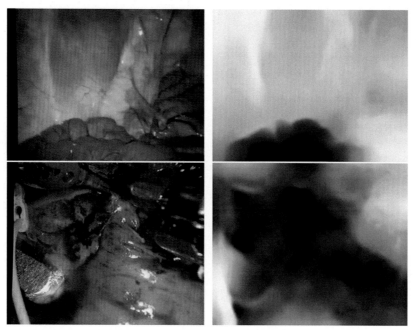

FIG. 6  Selected depth prediction input RBG image (left column) and corresponding depth map (right column) estimated by Monodepth2 fine-tuned and evaluated on Hamlyn dataset.

### Network training and evaluation

Sequence 1 from the Hamlyn dataset is used for training the GI-Replay 3D reconstruction network. The first 300 RGB images from sequence 1 and the corresponding depth maps predicted by the coupled depth network are fed as a training set, with pose parameters initialized with pose estimation generated by COLMAP. The training batch size is set to 1, and the sample rate is set to 8. The training curves are shown in Fig. 7. View synthesis results and the corresponding depth map estimated by the trained neural radiance representation are shown in Fig. 8.

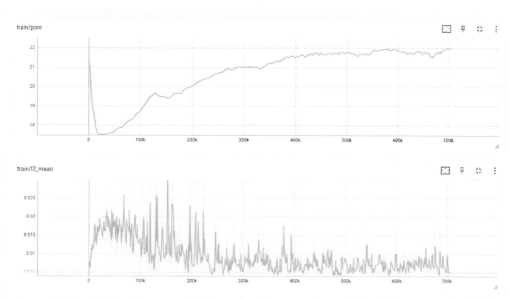

FIG. 7  Training loss and training PSNR of the proposed 3D reconstruction network over 700 k training steps on first 300 images from Hamlyn dataset sequence 1.

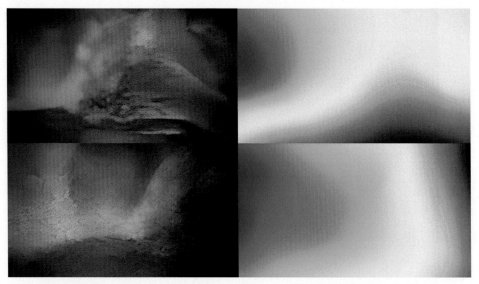

FIG. 8    Selected novel view synthesis (left column) and corresponding depth map (right column) generated by the proposed 3D dense reconstruction network after 700 k training steps.

## Conclusions

This work presents a semi-automated data collection pipeline with easy-access hardware that can collect synchronous monocular and stereo images with training ground truth information, including depth and camera pose within an ex-vivo in-cavity porcine GI tract. We also attempted to tackle the challenge of 3D reconstruction of a GI environment using endoscopy stereo video data similar to the GI-Replay dataset by introducing additional constraints inspired by novel pose-free and depth-supervised NeRF frameworks.

If granted additional time, we propose replacing the current ground truth depth measurement technique with the structured light method. This method involves projecting a known pattern, such as a grid or series of lines, onto a scene and utilizing distortions in the pattern to compute depth maps. The current technique is burdensome, as it necessitates the manual setting of anchor points for fine-tuning, and it is computationally demanding due to the local cost computation required by StereoSGBM for each pixel across multiple potential disparities. Regarding the 3D reconstruction aspect of our experiment, we aim to train the network until it converges and evaluate its performance on expanded datasets. This evaluation will allow us to examine the effects of the added loss function on rendered novel view synthesis, predicted depth maps, and camera poses. Additionally, we intend to conduct experiments with various constraints that can be incorporated into the NeRF framework to address the challenges associated with dense 3D reconstruction in the GI tract. For instance, these constraints would account for complex reflections caused by liquids and mucous membranes in the GI environment, considerations that are typically overlooked in traditional dense 3D reconstruction problem settings.

## Future scenario

The adoption of artificial intelligence (AI) in intelligent procedures is set to grow, with machine learning models being trained on increasingly large and complex datasets to improve the accuracy of sophisticated tasks, just as human surgeons. More specifically, these AI systems will learn to identify and segment various anatomical structures within the GI tract, differentiate between normal and pathological tissues, predict areas of concern that require closer examination, or even potentially foresee complications or suggest the most effective intervention strategies [27]. The continuously evolving nature of AI demands the creation of more nuanced and intricately detailed GI tract datasets, which are essential for the training of algorithms to match the discernment and precision of skilled human surgeons. These datasets must capture the vast variability in anatomy, pathology, and physiological responses encountered within the GI tract [8]. By encompassing a broader range of conditions, including rare anomalies and stages of disease progression, the datasets will enable AI systems to provide more accurate and personalized diagnostic insights. Furthermore, as those machine learning models advance, the datasets themselves should not become the bottleneck [27] and must also evolve to include multimodal data points such as dynamic response to intervention, real-time tissue characterization, and patient-specific anatomical features [8], thereby enhancing the AI's predictive capabilities and decision-making processes in real-time clinical scenarios.

In parallel, the future of GI tract imaging for endoscopic procedures is on the cusp of a transformative leap, primarily driven by the amalgamation of advanced imaging technologies and machine learning techniques. Innovations in endoscopic equipment and imaging techniques are anticipated to yield higher quality imaging, which, when coupled with sophisticated algorithms, will enable high-accuracy fast-computing 3D reconstruction technologies of the GI tract [28]. This will significantly enhance the surgeon's ability to navigate and understand the complex anatomy during procedures, leading to more accurate diagnoses and targeted treatments. The integration of augmented reality systems in endoscopy is also expected, providing clinicians with an overlay of 3D reconstructed images directly onto their field of view during procedures, further enriching the available visual information. All these benefits allude to safer, more efficacious, and more intelligent endoscopic procedures, paving the way for personalized therapeutic approaches and minimally invasive surgeries [29], ultimately leading to improved patient health and recovery experiences.

## Key points

- Comprehensive survey of current widely used datasets of GI tract, centering on datasets collection environment conditions, datasets collection techniques and equipment used, and the pros and cons.
- Short listings and analysis of state-of-art 3D reconstruction approaches in the GI tract.
- Introduction to a novel pipeline for constructing a GI tract dataset for dense 3D reconstruction applications, capable of providing monocular and stereo images with ground truth data, including depth and camera pose.
- Introduction to a NeRF-based 3D-reconstruction method paired with this dataset shows promise for accurate in-body scene reconstruction without reliance on pose priors, instead supervised by geometric cues.
- Brief description of future tendencies of GI tract datasets and 3D reconstruction methods development.

## Acknowledgments

The work was supported by the Hong Kong Research Grants Council (RGC) Collaborative Research Fund (CRF C4026-21GF), the General Research Fund (GRF #14211420 and GRF #14216022), and the Shenzhen-Hong Kong-Macau Technology Research Programme (Type C) Grant 202108233000303.

## References

[1] Wang S, et al. Toward automatic detection of gastric lesion for upper gastrointestinal endoscopy with neural network. J Med Robot Res 2022;7(01):2141003.

[2] Wang Y, et al. Neural rendering for stereo 3D reconstruction of deformable tissues in robotic surgery. In: International conference on medical image computing and computer-assisted intervention. Springer; 2022. p. 431–41.

[3] Allan M, et al. 2018 robotic scene segmentation challenge. ArXiv Preprint arXiv:2001.11190; 2020.

[4] Allan M, et al. Stereo correspondence and reconstruction of endoscopic data challenge. ArXiv Preprint arXiv:2101.01133; 2021.

[5] Azagra P, et al. Endomapper dataset of complete calibrated endoscopy procedures. Scientific Data 2023;10(1):671.

[6] Bawa VS, et al. Esad: endoscopic surgeon action detection dataset. ArXiv Preprint arXiv:2006.07164; 2020.

[7] Bobrow TL, et al. Colonoscopy 3D video dataset with paired depth from 2D-3D registration. In: Medical Image Analysis. Elsevier BV; 2023. p. 102956.

[8] Borgli H, et al. HyperKvasir, a comprehensive multi-class image and video dataset for gastrointestinal endoscopy. Scientific Data 2020;7(1):283.

[9] Cychnerski J, Dziubich T, Brzeski A. ERS: a novel comprehensive endoscopy image dataset for machine learning, compliant with the MST 3.0 specification. ArXiv Preprint arXiv:2201.08746; 2022.

[10] Hayoz M, et al. Learning how to robustly estimate camera pose in endoscopic videos. Int J Comput Assist Radiol Surg 2023;1–8.

[11] Ozyoruk KB, et al. EndoSLAM dataset and an unsupervised monocular visual odometry and depth estimation approach for endoscopic videos. Med Image Anal 2021;71:102058.

[12] Penza V, et al. Endoabs dataset: endoscopic abdominal stereo image dataset for benchmarking 3D stereo reconstruction algorithms. Int J Med Robot Comput Assist Surg 2018;14(5), e1926.

[13] Recasens D, et al. Endo-depth-and-motion: reconstruction and tracking in endoscopic videos using depth networks and photometric constraints. IEEE Robot Automat Lett 2021;6(4):7225–32.

[14] Condino S, et al. How to build patient-specific synthetic abdominal anatomies. An innovative approach from physical toward hybrid surgical simulators. Int J Med Robot Comput Assist Surg 2011;7(2):202–13.

[15] Pumarola A, et al. D-NeRF: neural radiance fields for dynamic scenes. CoRR abs/2011.13961; 2020. arXiv: 2011.13961.

[16] Hirschmuller H. Stereo processing by semiglobal matching and mutual information. IEEE Trans Pattern Anal Mach Intell 2007;30(2):328–41.

[17] Zarpalas D, et al. Anchoring graph cuts towards accurate depth estimation in integral images. J Display Technol 2012;8(7):405–17.

[18] Boykov Y, Veksler O, Zabih R. Fast approximate energy minimization via graph cuts. IEEE Trans Pattern Anal Mach Intell 2001;23(11):1222–39.

[19] Mildenhall B, et al. Nerf: representing scenes as neural radiance fields for view synthesis. Commun ACM 2021;65(1):99–106.

[20] Zehao Y, et al. MonoSDF: exploring monocular geometric cues for neural implicit surface reconstruction. In: Koyejo S, et al., editors. Advances in neural information processing systems, vol. 35. Curran Associates, Inc.; 2022. p. 25018–32.

[21] Remondino F, et al. A critical analysis of NeRF-based 3D reconstruction. Remote Sensing 2023;15(14):3585.

[22] Bian W, et al. Nope-nerf: optimising neural radiance field with no pose prior. In: Proceedings of the IEEE/CVF conference on computer vision and pattern recognition; 2023. p. 4160–9.

[23] Dey A, Ahmine Y, Comport AI. Mip-NeRF RGB-D: depth assisted fast *neural* radiance fields. ArXiv Preprint arXiv:2205.09351; 2022.

[24] Godard C, et al. Digging into self-supervised monocular depth estimation. In: Proceedings of the IEEE/CVF international conference on computer vision; 2019. p. 3828–38.

[25] Boss M, et al. Nerd: neural reflectance decomposition from image collections. In: Proceedings of the IEEE/CVF international conference on computer vision; 2021. p. 12684–94.

[26] Hajdu A, Hajdu L, Tijdeman R. Approximations of the Euclidean distance by chamfer distances. ArXiv Preprint arXiv:1201.0876; 2012.

[27] Chadebecq F, Lovat LB, Stoyanov D. Artificial intelligence and automation in endoscopy and surgery. Nat Rev Gastroenterol Hepatol 2023;20 (3):171–82.

[28] García-Vega A, et al. Multi-scale structural-aware exposure correction for endoscopic imaging. In: 2023 IEEE 20th international symposium on biomedical imaging (ISBI). IEEE; 2023. p. 1–5.

[29] Lin B, et al. Video-based 3D reconstruction, laparoscope localization and deformation recovery for abdominal minimally invasive surgery: a survey. Int J Med Robot Comput Assist Surg 2016;12(2):158–78.

# 14

# Pathway of robotic learning to perform autonomous sewing

*Yunkai Lv[a,b,c], Hao Zhang[a], Huaicheng Yan[c], and Hongliang Ren[d,e]*

[a]East China University of Science and Technology, Shanghai, China [b]The Chinese University of Hong Kong, Sha Tin, Hong Kong [c]Tongji University, Shanghai, China [d]Department of Electronic Engineering, The Chinese University of Hong Kong, Sha Tin, Hong Kong [e]Shun Hing Institute of Advanced Engineering (SHIAE), The Chinese University of Hong Kong, Sha Tin, Hong Kong

## Introduction

Since the 1970s, minimally invasive surgery (MIS) has become a better option than open surgery, which can avoid excessive blood loss and reduce postoperative recovery time [1]. The doctor inserts surgical instruments into the body through multiple small incisions on the patient's body surface and completes the surgical operation under the guidance of endoscopic images in MIS [2]. Compared with traditional open surgery, MIS has the advantages of less trauma, less pain, and quicker postoperative recovery. Therefore, it is widely used in neurosurgery, thoracic and abdominal surgery, orthopedics, and many other fields [3].

Sewing represents a fundamental surgical task within the realm of minimally invasive surgery, and it is known for its time-consuming nature. It should be pointed out that there are also many problems in minimally invasive surgery: (1) In MIS, the endoscope is manipulated by the assistant doctor, and the instruments used to perform the surgical operation are manipulated by the chief surgeon, so the orientation of the lens is hard to satisfy the requirements of the chief surgeon, which increases the difficulty of the operation [4]. (2) Holding the instrument for a long time will aggravate the fatigue of the doctor and affect the surgical effect [5]. These problems increase the complexity and difficulty of sewing and knotting to some extent.

Sewing and knotting is one of the basic operations that every surgeon should learn and master. However, sewing and knitting techniques are relatively difficult, and novice doctors must undergo a long study and training to carry out actual surgical operations. Even if a doctor has rich experience in knotting and sewing, fatigue caused by long operations may still lead to nonstandard sewing and unreliable knotting.

With the development of robotics and image processing technology, its application in medicine has become more and more extensive [6,7]. Surgical robots revolutionize medicine by seamlessly integrating the expertise of doctors with the remarkable capabilities of robotic technology. With its heightened sensitivity, precise positioning, and delicate movements, this innovative surgical tool offers numerous benefits. It accelerates treatment duration, alleviates patient discomfort, and lowers medical expenses. The fusion of MIS and robotic assistance propels a transformative shift in traditional medicine, fostering the advancement of novel theories and cutting-edge technologies.

In recent years, more scientific research institutions, commercial institutions, and scientific researchers have joined in the research of surgical robots. Some new technologies and theories have been developed, such as surgery instrument tracking based on image guidance [8], surgery instrument detection and segmentation [9], and autonomous endoscope control based on visual servoing [10]. The da Vinci surgical robot [11] and the research of minimally invasive flexible surgical robots have also attracted considerable research interest over recent years [12]. The main development and application of surgical robots are shown in Fig. 1.

Although robot-assisted MIS effectively addresses the limitations of traditional MIS, the current surgical robot system remains intricate, relying heavily on doctors to coordinate and control multiple robotic arms during surgical

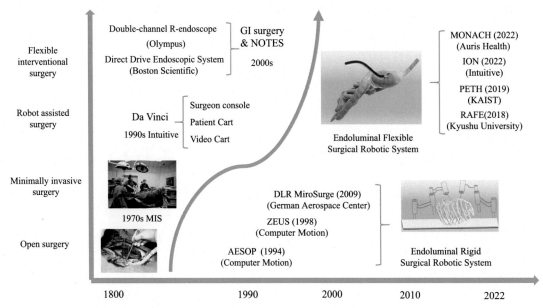

FIG. 1    Development and application of surgical robotic systems.

procedures. Furthermore, the lack of intelligence in surgical robots results in only marginal improvements in completion time for certain tasks involving intricate multi-arm cooperation [13,14]. This is particularly evident in complex sewing and knotting operations, where many conventional surgical techniques and technologies are no longer suitable for robot-assisted procedures. Consequently, achieving autonomous sewing and knotting represents a substantial challenge for surgical robots. In recent years, autonomous sewing technology for robots has significantly enhanced the surgical outcomes of robot-assisted minimally invasive procedures.

## Autonomous sewing robots

### Application of autonomous surgical robot in sewing

Suturing, a widely prevalent and time-consuming task in surgical procedures, involves three standard actions: needle insertion, needle withdrawal, and wire pulling. Presently, research efforts in autonomous suturing for robot-assisted MIS primarily concentrate on predefined surgical scenarios. From a kinematic perspective, robots can autonomously perform single or multiple suturing actions. The existing control methods primarily include visual servoing [15], teaching and learning [16], and surgical interaction motion modeling [17]. Visual servoing involves the use of auxiliary reference points for feature recognition, enabling closed-loop point-to-point positioning with the assistance of surgical images. The teaching and learning method controls the robot by collecting the surgeon's operational data to replicate the actions. The interactive motion modeling method achieves robot motion planning by establishing a mathematical relationship between surgical operation displacements and organ deformations.

### Development of autonomous sewing robot

In the past few decades, autonomous sewing robots have garnered significant research attention for their applications in MIS and have gained increasing prominence. Table 1 provides an overview of some autonomous sewing robots designed for MIS.

In Ref. [21], the development of the Endo-PAR surgical robot system has led to the realization of autonomous needle insertion using the visual servo method. This involves using a laser transmitter mounted on a robotic arm to indicate the insertion point for sutures. In Ref. [18], Nageotte et al. devised a path-tracking algorithm that relies on marked point information from surgical instruments and suture positions visible in endoscopic images. They employed visual servo control to guide surgical instruments, enabling them to complete suture needle insertion and extraction at specified marking points. In Ref. [20], a teaching method is introduced to govern a double-arm robot's knotting operation. This method utilized a stereo camera to record the robot's joint movements. Color representation and image coding are

TABLE 1  Development of autonomous sewing robot for MIS.

| Description | Organization | Type/Refs. |
| --- | --- | --- |
| Stitching path tracking based on visual servoing | Louis Pasteur University | Single arm [18] |
| Autonomous suture knot winding using recurrent neural network | Technical University Munich | Three arms [19] |
| Autonomous knotting based on visual servoing | The Queen's University of Belfast | Dual arm [20] |
| Autonomous needle insertion based on visual servoing | Technical University Munich University of Tokyo German Heart Center | Single arm & multi-arm [21] |
| Local autonomous suture using one-master two-slave structure that encloses the human intervention | Tokyo Medical and Dental University Shibaura Institute of Technology | Dual arm [22] |
| Autonomous rolling arc cycle knotting based on the visual 3D reconstruction method | Case Western Reserve University | Dual arm [23] |

applied to extract the end position and rotation information of surgical instruments. Task decomposition is then carried out based on the video-captured robot movements, effectively controlling the robot to complete the knotting operation. A local autonomous suture method is proposed for the process of "inserting needle-grasping needle-pulling needle" in Ref. [22].

In Ref. [19], the teaching and learning approach is employed to gather motion data from the joints of the Endo-PAR robot during knotting procedures. A cyclic neural network is introduced to enable the autonomous generation of knotting trajectories by the robot. In Ref. [23], two autonomous knotting methods for robotic-assisted MIS are developed to enhance knotting efficiency. These methods are designed by analyzing the most common knotting techniques in surgery. In 2014, a visual 3D reconstruction technology was used to obtain the positions of the small claw and the suture. A trajectory planning method is then applied to enable the robot to autonomously perform the rolling arc cycle knotting [24].

## Integration of iterative learning control and autonomous sewing robot

In this section, we will discuss the feasibility of integrating the iterative learning control (ILC) method into the control of autonomous sewing robot.

## Characteristics of autonomous sewing surgical robot

The development of personalized stent-graft suturing and the creation of a comprehensive one-stop solution framework has garnered substantial attention from researchers in recent years. In [25], a multirobot system is introduced to fabricate personalized medical stent grafts. Additionally, Ref. [26] describes the design of a multifunctional robotic system tailored for stitching 3D structured objects. A one-stop solution for suturing personalized stent grafts is created through a customized robotic suturing device and a closed-loop visual servo control system. In Ref. [27], an innovative modular design-based framework is developed to manufacture personalized medical stent grafts. This framework leverages demonstration learning, integrating 3D vision and multirobot collaboration algorithms. It effectively guides robots to learn and execute tasks seamlessly without interruptions and can adapt to various graft geometries.

It is worth noting that both the robot arm's motion and stitching actions exhibit repetitive characteristics. The robotic arm carries out identical actions in a repetitive manner during the sewing process, resulting in a periodic motion trajectory. This suggests that sewing robot systems can be described in terms of two-dimensional dynamics: (1) time-dimensional dynamics and (2) iteration-dimensional dynamics. These observations have inspired the integration of ILC into the development of the robot arm control algorithm.

## An overview of iterative learning control

ILC [28] involves learning from historical operational data, and it has demonstrated exceptional capabilities in achieving real-time responsiveness and precise tracking control. An illustrative example is presented in Fig. 2. ILC can be likened to an ideal control algorithm that mimics the human learning process. The fundamental process of ILC involves studying the inherent repetition factor of the system based on information from previous operations, with the aim of progressively enhancing control performance. The trajectory control problems can be divided into two types: (a) Tracking an entire reference trajectory; (b) Tracking multiple specified reference points (Fig. 3).

Drawing inspiration from the idea of ILC, both time-dimensional and iteration-dimensional dynamics can be effectively harnessed in the design of sewing action control for the robot system. In the preset scheme, the control input for each sewing action is adjusted based on the control input and the system error from preceding sewing actions, facilitating precise sewing. Next, we will briefly introduce two classical structures of ILC: feed-forward structure and feedback structure. For more in-depth information, please refer to Figs. 4 and 5. In a feed-forward structure, the control input $u_{i+1}(t)$ of $(i+1)$th iteration is calculated based on the control input $u_i(t)$ and output error $e_i(t)$ of the $i$th iteration. In the feedback structure, the control input is updated by $u_{i+1}(t) = u_i(t) + k(e_{i+1}(t))$.

ILC stands as a crucial subfield within learning control. Compared with the traditional control methods, ILC can handle high-uncertainty dynamic systems with an easy-to-understand control structure. It only requires less prior system information and has low computational complexity. ILC exhibits remarkable adaptability, finding applications in a wide array of fields. Crucially, ILC does not depend on the presence of a precise mechanism model for the dynamic systems it addresses. The exploration of ILC holds extraordinary significance for controlling nonlinear, complex, challenging-to-model systems and high-accuracy trajectory tracking controls.

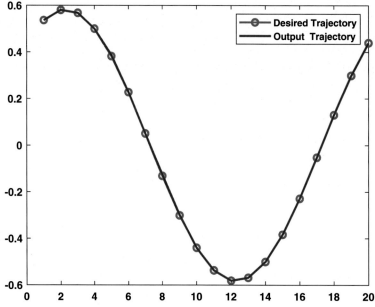

FIG. 2   Perfect tracking performance of ILC.

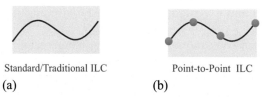

FIG. 3   The trajectory tracking of ILC. (A) Standard/traditional ILC (B) Point-to-point ILC.

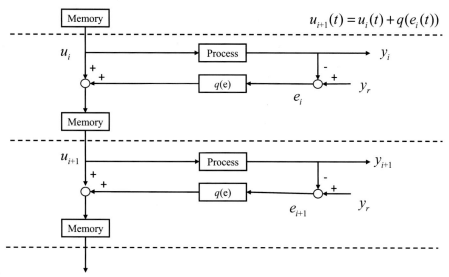

$$u_{i+1}(t) = u_i(t) + q(e_i(t))$$

FIG. 4    Feed-forward structure of ILC.

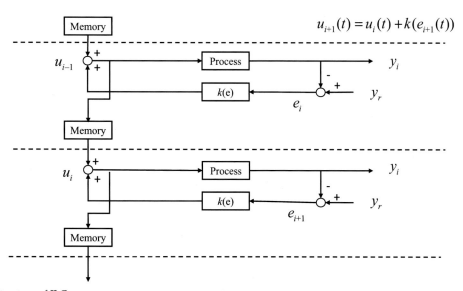

$$u_{i+1}(t) = u_i(t) + k(e_{i+1}(t))$$

FIG. 5    Feedback structure of ILC.

## Integration of ILC and autonomous sewing

Sewing is a repetitive process characterized by the continuous back-and-forth movement of the needle through the target object, driven by the terminal. To ensure the quality of the sewing, it is crucial that the needle's motion accurately adheres to the intended reference trajectory. However, if the sewing process is unexpectedly halted due to a mechanical arm failure, the system loses the data for the remaining duration of the ongoing operation. The current learning process is prematurely terminated, resulting in an undesirable control issue referred to as "iteration-varying length," as illustrated in Fig. 6.

Over the past few decades, several research efforts have put forth viable design and analysis approaches for addressing the iteration-varying length problem [29,30]. In Ref. [31], a framework is developed to tackle the control issue in repetitive systems with varying passing lengths. The concept of maximum-pass-length error is introduced to mitigate system output errors with variable dimensions. The mathematical model for iteration-varying length in discrete-time systems is introduced in Ref. [32], with a subsequent extension to continuous-time systems discussed in Ref. [33]. In Ref. [32], an ILC control algorithm with an iterative averaging operator is proposed, leveraging historical data to compensate for missing system information. Furthermore, in Ref. [33], a dynamic, iterative averaging operator is designed

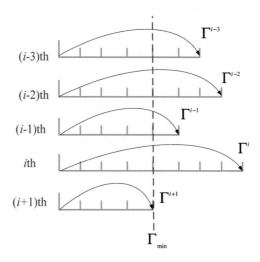

FIG. 6    Iteration-varying length.

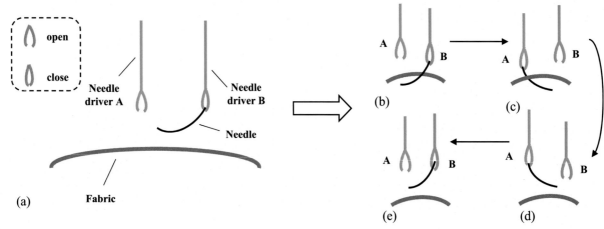

FIG. 7    The main steps included in one stitch cycle (A–E).

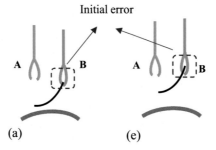

FIG. 8    The initial iteration error.

to minimize the impact of older historical data on the control system. In Ref. [34], two enhanced ILC methods are devised, utilizing iterative moving average operators. These methods ensure faster convergence by implementing a search mechanism that collects valuable data while eliminating redundant historical tracking data.

In the context of the bimanual sewing task, precisely tracking the reference trajectory by the sewing needle during the piercing motion is paramount. For instance, in Ref. [26], the completion of one stitch is considered a cycle, referred to as an iteration in this context, as illustrated in Fig. 7. It is worth noting that in steps b and d, the final positions of bent needles clamped by the needle driver "A" may exhibit some variations, as depicted in Fig. 8. In practical bimanual

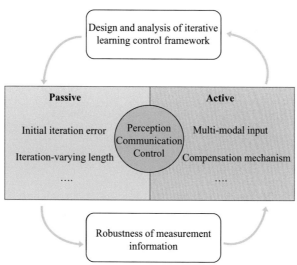

FIG. 9    Main structure of the integration of ILC and autonomous sewing robot.

sewing tasks, even a slight deviation in the initial localization can result in unexpected discrepancies between the actual sewing trajectory and the desired path.

In recent years, several relevant studies focusing on iteration-varying initial errors have been published [35–38]. These works delve into the impact of initial errors on system performance and offer various solutions. In Ref. [35], the study addresses the influence of initial errors on system performance. It introduces an ILC algorithm that utilizes multimodal input to mitigate the impact of initial errors. In Ref. [36], an ILC algorithm based on the two-dimensional system theory is proposed to solve the control problem of linear discrete-time multivariable systems with initial error. In Ref. [37], a novel ILC method is introduced. This method employs an average operator to effectively eliminate the effects of variable initial state errors on trajectory tracking, contributing to improved system performance. Reference [38] focuses on studying the intrinsic relationship between varying initial errors and the corresponding convergence characteristics in a specific class of repetitive nonlinear systems.

The main structure of this part is shown in Fig. 9.

Based on the above analysis and discussion, the repetitive operation characteristics of the autonomous sewing robot should receive more attention in the design of the control algorithm. Integrating ILC and autonomous sewing robots will be promising work.

## Conclusion

In this chapter, the related technologies of teleoperation surgical robots are surveyed briefly. The promising future of teleoperation surgical robots in telemedicine is discussed. Furthermore, the development of endoscopic minimally invasive and autonomous surgical robots is also reviewed.

## Future scenario

Generally, robotic sewing technology of vision-based is relatively mature. In contrast, autonomous robot sewing is still in its infancy, focusing on the control algorithm design. For the development of sewing control algorithms for autonomous robots, much research is needed with endless attempts, and much work is fundamental and challenging since little systematic research has been done so far.

Considering the repetitive operation characteristics of autonomous sewing, it is novel and challenging to incorporate ILC into the algorithm design of robotic autonomous sewing. The control algorithms of autonomous sewing robots need to improve, such as combining offline training with online self-tune to adapt to more sewing scenarios, and the surgeon can choose the control algorithm which most appropriate for current sewing task.

## Key points

- Autonomous sewing robots incorporate advanced algorithms that can perform the real-time regulation of controller parameters, they can safely perform sewing task.
- The use of autonomous sewing robots requires strict adherence to technical regulations in order to be safer and more efficient.
- The autonomous sewing robots are safe, which can be widely promoted.

## Acknowledgments

This work was supported by Hong Kong Research Grants Council (RGC) Collaborative Research Fund (C4026-21G), Research Grants Council (RGC)—Research Impact Fund (RIF R4020-22), General Research Fund (GRF 14211420), Research Grants Council (RGC)—NSFC/RGC Joint Research Scheme N_CUHK420/22, Shenzhen-Hong Kong-Macau Technology Research Programme (Type C) STIC Grant SGDX20210823103535014 (202108233000303), the key project 2021B1515120035 (B.02.21.00101) of the Regional Joint Fund Project of the Basic and Applied Research Fund of Guangdong Province, Hong Kong Research Grants Council (RGC) Collaborative Research Fund (CRF C4063-18G and C4026-21G), and General Research Funds (GRF 14216022, GRF 14203323).

## References

[1] Wamala I, Roche ET, Pigula FA. The use of soft robotics in cardiovascular therapy. Expert Rev Cardiovasc Ther 2017;15(10):767–74.

[2] Mei F. Research on key technologies about a medical robot for celiac minimally invasive surgery. Harbin Institute of Technology; 2012.

[3] Bo P. Research on laparoscopic robot and its key technology. Harbin Institute of Technology; 2009.

[4] Breedveld P, Stassen H, Meijer D. Theoretical background and conceptual solution for depth perception and eye-hand coordination problems in laparoscopic surgery. Minim Invasive Ther 1999;8(4):227–34.

[5] Lee Y, Kim J, Ko S. Design of a compact laparoscopic assistant robot. In: Proceedings of the International Conference on Control Automation and Systems; 2003. p. 2648–53.

[6] Islam M, Seenivasan L, Ming L, Ren H. Learning and reasoning with the graph structure representation in robotic surgery, medical image computing and computer assisted intervention. Springer International Publishing; 2020. p. 627–36.

[7] Pang W, Islam M, Mitheran S, Seenivasan L, Xu M, Ren H. Rethinking feature extraction: gradient-based localized feature extraction for end-to-end surgical downstream tasks. IEEE Robot Autom Lett 2022;7(4):12623–30.

[8] Islam M, Vibashan VS, Lim C, Ren H. ST-MTL: Spatio-temporal multitask learning model to predict scanpath while tracking instruments in robotic surgery. Med Image Anal 2021;67:101837.

[9] Islam M, Vibashan VS, Ren H. AP-MTL: Attention pruned multi-task learning model for real-time instrument detection and segmentation in robot-assisted surgery. In: IEEE international conference on robotics and automation; 2020. p. 8433–9.

[10] Gao H, Fan W, Qiu L, Yang X, Li Z, Zuo X, Li Y, Meng M, Ren H. SAVAnet: surgical action-driven visual attention network for autonomous endoscope control. IEEE Trans Autom Sci Eng 2022;20(4):2655–67.

[11] Ashrafian H, Clancy O, Grover V, Darzi A. The evolution of robotic surgery: surgical and anaesthetic aspects. Br J Anaesth 2017;119(1):i72–84.

[12] Patel N, Cundy T, Darzi AW, Yang GZ, Teare J. A novel flexible snake robot for endoluminal upper gastrointestinal surgery. Gastrointest Endosc 2014;79(5). AB147.

[13] Ding J, Goldman RE, Xu K, Allen PK, Fowler DL, Simaan N. Design and coordination kinematics of an insertable robotic effectors platform for single-port access surgery. IEEE/ASME Trans Mechatron 2013;18(5):1612–24.

[14] Stoyanov D, Scarzanella M, Pratt P, Yang G. Real-time stereo reconstruction in robotically assisted minimally invasive surgery. Lect Notes Comput Sci 2010;63(61):275–82.

[15] Kim S, Tan Y, Deguet A, Kazanzides P. Real-time image-guided telerobotic system integrating 3D slicer and the da vinci research kit. In: IEEE international conference on robotic computing; 2017. p. 113–6.

[16] Okamura A. Haptic feedback in robot-assisted minimally invasive surgery. Curr Opin Urol 2009;19:102–7.

[17] Nageotte F, Zanne P, Doignon C, DeMathelin M. Stitching planning in laparoscopic surgery: towards robot-assisted suturing. Int J Robot Res 2009;28:1303–21.

[18] Nageotte F, Zanne C, Mathelin M. Visual servoing-based endoscopic path following for robot-assisted laparoscopic surgery. In: IEEE/RSJ international conference on intelligent robots and systems; 2006. p. 2364–9.

[19] Mayer H, Gomez F, Wierstra D, Nagy I, Knoll A, Schmidhuber J. A system for robotic heart surgery that learns to tie knots using recurrent neural networks. Adv Robot 2008;22:13–4.

[20] Hynes P, Dodds G, Wilkinson A. Uncalibrated visual-servoing of a dual-arm robot for mis suturing. In: IEEE/RAS-EMBS international conference on biomedical robotics and biomechatronics; 2009. p. 420–5.

[21] Staub C, Knoll A, Osa T, Bauernschmitt R. Autonomous high precision positioning of surgical instruments in robot-assisted minimally invasive surgery under visual guidance. In: International conference on autonomic and autonomous systems; 2010. p. 64–9.

[22] Watanabe K, Kanno T, Kawashima K. Human-integrated automation of suturing task with one-master two-slave system for laparoscopic surgery. In: IEEE international conference on advanced intelligent mechatronics; 2016. p. 1180–5.

[23] Chow D, Newman W. Improved knot-tying methods for autonomous robot surgery. In: IEEE international conference on automation science and engineering; 2013. p. 461–5.

[24] Chow D, Jackson R, Newman W. A novel vision guided knot-tying method for autonomous robotic surgery. In: IEEE international conference on automation science and engineering; 2014. p. 504–8.

[25] Huang B, Ye M, Hu Y, Vandini A, Lee S, Yang G. A multirobot cooperation framework for sewing personalized stent grafts. IEEE Trans Industr Inform 2018;14(4):1776–85.

[26] Hu Y, Zhang L, Li W, Yang G. Robotic sewing and knot tying for personalized stent graft manufacturing. In: IEEE/RSJ international conference on intelligent robots and systems; 2018. p. 754–60.

[27] Huang B, Yang Y, Tsai Y, Yang G. A reconfigurable multirobot cooperation workcell for personalized manufacturing. IEEE Trans Autom Sci Eng 2022;19(3):2581–90.

[28] Arimoto S, Kawamura S, Miyazaki F. Bettering operation of robots by learning. J Robot Sys 1984;1(2):123–40.

[29] Seel T, Werner C, Schauer T. The adaptive drop foot stimulator-multivariable learning control of foot pitch and roll motion in paretic gait. Med Eng Phys 2016;38(11):1205–13.

[30] Lv Y, Zhang H, Wang Z, Yan H. Distributed localization for dynamic multiagent systems with randomly varying trajectory lengths. IEEE Trans Ind Electron 2022;69(9):9298–308.

[31] Seel T, Schauer T, Raisch J. Monotonic convergence of iterative learning control systems with variable pass length. Int J Control 2017; 90(3):393–406.

[32] Li X, Xu J, Huang D. An iterative learning control approach for linear systems with randomly varying trial lengths. IEEE Trans Autom Control 2014;59(7):1954–60.

[33] Li X, Xu J, Huang D. Iterative learning control for nonlinear dynamic systems with randomly varying trial lengths. Int J Adapt Control Signal Process 2015;29(11):1341–53.

[34] Li X, Shen D. Two novel iterative learning control schemes for systems with randomly varying trial lengths. Syst Control Lett 2017;107:9–16.

[35] Lee H, Bien Z. Study on robustness of iterative learning control with non-zero initial error. Int J Control 1996;64:345–59.

[36] Fang Y, Chow T. 2-D analysis for iterative learning controller for discrete-time systems with variable initial conditions. IEEE Trans Circuits Syst Part I: Fundam Theory Appl 2003;50:722–7.

[37] Park KH. An average operator-based pd-type iterative learning control for variable initial state error. IEEE Trans Automat Contr 2005;50:865–9.

[38] Wang Y, Chien C, Teng C. Direct adaptive iterative learning control of nonlinear systems using an output-recurrent fuzzy neural network. IEEE Trans Syst Man Cybern B Cybern 2004;34:1348–59.

# 15

# Telesurgery applications, current status, and future perspectives in technologies and ethics

*Thiago Camelo Mourão*[a], *Shady Saikali*[b], *Evan Patel*[b], *Mischa Dohler*[c,d,e], *Vipul Patel*[b,f], *and Márcio Covas Moschovas*[b,f]

[a]Department of Urology, A.C. Camargo Cancer Center, São Paulo, Brazil [b]AdventHealth Global Robotics Institute, Celebration, FL, United States [c]VP Emerging Tech, Ericson Inc., Silicon Valley, CA, United States [d]Advisory Board, FCC (TAC) & Ofcom (Spectrum), London, United Kingdom [e]King's College London, London, United Kingdom [f]University of Central Florida (UCF), Orlando, FL, United States

## Introduction

One of the initial reports of Telesurgery was described in 2001 when Jacques Marescaux made medical history by conducting the inaugural remote full operation using a robotic platform. This pioneering event involved a cholecystectomy performed between New York City, USA, and Strasbourg, France, with a mere 155 ms lag time [1]. This milestone achievement was properly named "The Lindberg Operation," drawing an analogy from the historic aviator who successfully crossed the Atlantic Ocean [2].

Since this pioneering event, several new technological improvements have been created in robotic surgery and telecommunication. However, until now, the main challenges emerged from the need to convert video images and surgical movements into electronic signals, compounded by the limitations in bandwidth and time delay within current telecommunication lines [3]. In this context, several variables impact transmission speed, including factors, such as distance, the inherent capabilities of the transmission system, the effectiveness of computer interfaces in compressing and decompressing data, and even cyberattacks [2,4].

Recently, several groups have described successful procedures performed with 5G technology with optimal signal transmission and operative outcomes [5,6]. Therefore, the ability to perform complex surgical manipulations from remote locations would eliminate geographical limitations while making surgical expertise available worldwide, improving patient treatment and surgical training. In this context, our chapter covers the humanitarian, ethical, and technical aspects involving the current and future scenarios of Telesurgery.

## Humanitarian implications of telesurgery

The healthcare landscape is shifting with rapid technological advancement, and Telesurgery, also known as remote surgery, is critical for humanitarian implications. This innovative approach holds the potential to deliver surgical expertise beyond geographical boundaries, transcending limitations, and offering support to underserved populations. Some critical aspects need to be considered:

## Enhanced access to specialized care in regions with limited or nonexistent specialized surgical experience

Remote surgical interventions enable patients to receive expert medical attention from expert surgeons located miles away. This is especially significant for communities in remote or underserved areas, where individuals might otherwise face barriers to accessing life-saving procedures.

## Equity in healthcare delivery

Telesurgery has the potential to level the playing field in healthcare. By enabling patients in underserved regions to receive the same quality of surgical care as those in well-equipped medical centers, Telesurgery promotes equity in healthcare delivery. This shift can potentially reduce health disparities and ensure that individuals, regardless of their geographic location, have an equal chance of receiving timely and proficient surgical interventions.

## Knowledge transfer and skill enhancement

Telesurgery could be used as an instrument for knowledge dissemination and skill enhancement among medical professionals worldwide. Surgeons in remote or developing areas can collaborate with experts from around the globe, engaging in training, mentorship, and even real-time guidance during procedures. This exchange of expertise strengthens the capabilities of local medical teams and contributes to sustainable healthcare capacity-building.

## Disaster response and humanitarian aid

In eventual disasters or emergencies, Telesurgery can be helpful when medical infrastructure is compromised. Medical teams from distant locations can provide critical surgical support remotely, aiding on-site relief efforts and bolstering the capacity of local medical personnel. This capacity to offer immediate assistance, even in the most challenging circumstances, highlights the humanitarian potential of Telesurgery.

## Maintaining constant surgical routine

Telesurgery enhances the resilience of healthcare systems against unforeseen challenges. In situations where travel restrictions, pandemics, or other disruptions delay traditional surgical practices, Telesurgery remains a viable solution to guarantee a constant routine of surgeries and healthcare, especially in oncologic patients. This adaptability contributes to maintaining essential surgical services during crises, ensuring that patient's needs are met regardless of external circumstances.

In summary, the humanitarian implications of Telesurgery are transformative. This innovative surgical care approach offers medical solutions to individuals and communities with restricted healthcare access, ensuring that all individuals, regardless of their location, have the opportunity to receive safe, timely, and proficient surgical interventions.

## Historical landmarks

Emerging as Cybersurgery, telerobotics began with solid funding and investments by the US Defense Advanced Research Project Agency (DARPA), driven by the imperative to offer remote surgical interventions for battlefield casualties. The primary objective was to respond rapidly to critical injuries such as profuse bleeding from major vessels, which account for a significant 90% of combat fatalities before reaching a medical facility. In this scenario, the battle concept of the golden hour (saving the soldier in the first hour of injury) was replaced by the golden minute. The injured soldier becomes the focal point of a telepresence surgical unit, where their image is projected through 3D imaging to a surgeon situated on an aircraft or a distant location [7]. However, one of the main limitations at that time was the transmission delay. According to the investigation performed by Marescaux's group, the threshold for an acceptable time delay, as perceived by a surgeon's sense of safety, was approximately 330ms [3].

During that timeframe, investigators performed laparoscopic robotic cholecystectomies on six porcine subjects using the ZEUS robotic system (Computer Motion, California, USA). The connection between the two locations was facilitated by a high-speed terrestrial optical-fiber network (FranceTelecom/Equant), enabling data transmission through dedicated connections utilizing asynchronous transfer mode (ATM) technology. A designated bandwidth of 10 megabits per second was assigned within an interconnected network that incorporated a network termination unit (NTU), providing a flexible conduit for diverse applications at both endpoints [3].

More recently, new transmission technologies and the implementation of a fifth-generation wireless system (5G) have marked a new era in Telesurgery history, driving remote medical care to unparalleled advancements. The attributes of the 5G network, distinguished by swift communication, negligible latency, and ample bandwidth, assume a critical role in steering this transformative advancement [8].

Within this context, recent investigations into modern Telesurgery have demonstrated the viability of standard general procedures even when transmission delays remain under 200 ms; notably, delays of less than 100 ms exhibit minimal repercussions. This supposition considers the synchronization of contemporary information processing and communication technologies with these established benchmarks [9,10]. In this scenario, Japan took a significant step in 2021 by officially endorsing remote surgical assistance, permitting surgeries to be executed remotely under the stipulation that an on-site surgeon is physically present [9].

## Telesurgery in practice: Technical aspects and case studies

The key to implementing a successful Telesurgery program, other than providing the necessary surgical expertise to conduct the surgery, is providing the necessary network infrastructure. The components involved in establishing this framework include communication systems, capable robotic platforms, redundant and reliable connections, latency mitigation, data security systems, and protocols. In a 2008 study, Nguan et al. conducted 18 porcine robotic pyeloplasties. Six surgeries happened in real-time, six via IP-VPN over landline, and six using satellite telesurgical connections. Network transport latencies were $66.3 \pm 1.5$ ms for landline and $560.7 \pm 16.5$ ms for satellite, calculated from point-to-point fiber network packet delays. Despite latency-related asynchrony in visual and motor functions, the operating surgeon reported smooth robot motion and accurate visual rendering. No network failures occurred. Procedural times for real-time (latency 0), IP-VPN landline, and satellite were comparable, and they did not significantly differ between groups: real-time ($41.3 \pm 15.0$ min), landline ($47.0 \pm 24.1$ min), and satellite ($51.8 \pm 4.7$ min) [11].

Potential applications of teleanesthesia were also performed in experiments, such as remote vital sign monitoring, automated anesthetic drug titration and administration, and procedures related to regional anesthesia [12].

With the introduction of 5G networks, more intricate surgical procedures can be performed remotely. This new and improved network infrastructure offers a high data transfer rate, reaching 10 GB/s. This allows the seamless broadcast of high-quality data with minimal latency. The design of 5G networks also allows for ultra-low latency communication (ULLC), which is crucial for real-time performance communication such as Telesurgery. The quality of service (QoS) attributes within the 5G network are crucial for instilling confidence in the system's usability. These attributes encompass a collection of techniques and mechanisms meticulously crafted to ensure that diverse network traffic types receive the appropriate level of performance and resources, tailored to their specific needs. This includes improved mobile broadband (eMBB), extensive machine-type communications (mMTC), and ULLC. Numerous parameters are fine-tuned to sustain the QoS offered by 5G networks, such as data throughput, packet loss rates, latency, and jitter. In the context of Telesurgery QoS, redundant communication plays a pivotal role. By furnishing at least one backup channel, it has been demonstrated that interruptions in the connection during robotic Telesurgery do not impact the transmission of images during surgery [13].

An important consideration is cybersecurity and data privacy. Broadcasting haptics, audiovisual communication, and patient data subjects it to potential cyberattacks or unauthorized access to the information. The network must be robust and be able to prevent illegal or unauthorized access to the connection. Threats, such as distributed denial of service (DDOS), blocking surgeons from transmitting commands to the robot, and man-in-the-middle (MITM) attacks, which basically allow attackers to become intermediaries between the surgeon and the robot, are serious concerns that need to be addressed [14]. Iqbal et al. have described a secure framework (SecureSurgiNET) that provides a safe working environment for Telesurgery. They propose several protocols to minimize the threat of cyberattacks or unauthorized access during the intraoperative phase of the procedure [15].

Another key technical barrier to the widespread adoption of Telesurgery is the interoperability of different systems. Extrapolating from other device manufacturers, devices and their cloud/connection services, as well as their multiple components, are provided by the same vendor, which creates a seamless integration of the surgeon experience. This interoperability guarantees commercial benefits to the vendor and creates exclusivity in terms of device usage. For example, buying a mobile telephone device and a SIM card from a different company along with third company for cloud storage will not interrupt the phone service. Allowing device interoperability in robotic surgery will allow for easier clinical adoption by medical centers. This will also create a healthy sense of competition, affecting prices and QoS [16].

In June 2019, a preliminary trial was conducted on swine using the "MicroHand" robotic system (WEGO Group, Weihai, China) over a 100 Mb/s Internet connection, spanning 300 km. The total intraoperative latency recorded was 170 ms. Building on this, in September 2019, the team achieved 5G-enabled remote surgery utilizing this same robotic system over a distance of 3000 km. Robot-assisted telesurgeries, including left nephrectomy, partial hepatectomy, cholecystectomy, and cystectomy, were sequentially performed by two surgeons in China. The average latency time measured was 264 ms, comprising a mean round-trip transit delay of 114 ms, alongside a data packet loss ratio of 1.20% [17]. This experiment demonstrated the feasibility of performing Telesurgery using local 5G networks with locally produced robotic platforms.

Other notable trials were conducted in China to further confirm the capability of locally produced robotic platforms in procedures using 5G networks. In 2021, three additional experimental procedures were performed on animals at 300 and 700 km distances. Innovative heterogeneous multilink network converged transmission technology was employed in these operations to enhance network communication and update the network architecture. Data packet transmission exhibited a flawless 0% packet loss rate, with an average round-trip signal transmission delay of under 60 ms. Moreover, the mean total delay across all three telesurgeries was under 250 ms [18]. The same author's team, using the MicroHand robot system, ventured into ultra-remote radical cystectomy across 3000 km. Throughout the procedure, real-time network attack monitoring through a security awareness system was in place. IPSEC VPN encryption and advanced firewalls effectively thwarted simulated network attacks, upholding network security. Importantly, robot master–slave operation remained consistently unaffected [19].

Another study presents perioperative outcomes from 29 patients who underwent robot-assisted radical nephrectomy conducted through remote means. While using a 5G network infrastructure, the authors observed a median total delay of 176 ms across a median distance of 187 km between the two integral components of the robotic system in China [6].

Robotic Telesurgery is not limited to urology. For instance, Tian et al. discussed telerobotics in stereotactic neurosurgery. They used the CAS-BH5 frameless robotic system in 10 cases between Beijing and Yan'an in late 2005 [20]. The primary surgeon and patient were over 1500 km apart, with no signal latency mentioned via the "Digital Data Network" for telecommunication. In 2019, Patel et al. reported successful long-distance telerobotic cardiology procedures. Five telerobotic-assisted percutaneous coronary interventions were conducted over 32 km. Employing a CorPath GRX robotic system (Corindus Robotics, Waltham, MA, USA), the procedures went smoothly with a 53 ms observed time delay [21].

Orthopedic surgery is another area with advancements in Telesurgery, which has more recently integrated robotic technology into its procedural landscape. In spinal fusion, computer-assisted robotic systems markedly enhance pedicle screw fixation accuracy [22]. A study involving 12 patients utilized the 5G telerobotic surgery system in conjunction with the TiRobot System. Notably, this study marks the first instance of telerobotic spinal surgery via the 5G network and employing a "one-to-many" surgical approach. The "one-to-many" approach enables a single surgical expert to remotely care for multiple isolated patients, optimizing surgical expertise resource utilization for enhanced medical services [23]. The outcomes demonstrated precise pedicle screw placement, with deviations from planned positions measuring less than 0.8 mm. No intraoperative adverse events were observed, affirming the safety and reliability of telerobotic spinal surgery rooted in the 5G network.

In another Telesurgery experience, the Japanese Hinotori robotic system (Medicaroid Corporation, Kobe, Japan) was employed. Despite spanning roughly 4000 km round-trip, the communication network flawlessly conveyed 3D 2 K images with a 30 ms round-trip time, leaving surgical performance mostly unaffected. Even at a 145 Mbps communication bandwidth, tasks proceeded smoothly without image compromise or delay, enabled by image compression [24]. Recent studies evidenced Hinotori's commercial feasibility in Telesurgery [24–27].

Furthermly, the viability of employing the Robotic Slave Micromanipulator Unit (RSMU) in Ophthalmology for remote photocoagulation of the ciliary body using a diode laser was evaluated in freshly unoperated, enucleated human eyes [28].

A recent study validated a novel robotic platform known as The Double Delta (DDelta), a research-oriented master–slave telesurgical robot based on parallel kinematics. Developed by the National Technical University of Athens in collaboration with Accrea Engineering, a robotics company based in Lublin, Poland, the platform was assessed through surgical exercises performed by a surgeon on a robotic surgery training phantom. The master controllers were linked to a local site via a 5G network, enabling remote teleoperation of the robot in a hospital setting. All tasks were successfully accomplished. The 5G network's low latency and high bandwidth contributed to a motion command latency of 18 ms, with a video delay of approximately 350 ms [29].

## Telementoring in robotic surgery

Telementoring offers promising solutions to address challenges such as underutilizing expert surgeons' time and the potential for establishing a novel payment model for this type of service. Several centers across different countries have already shared successful experiences in remotely guiding sought-after laparoscopic procedures. This emerging practice could also be crucial in educating surgical residents about laparoscopic or robotic surgery [7].

An example of technology facilitating Telementoring was Socrates, a robotic telecollaboration device developed by Computer Motion, CA. Socrates was specifically designed to streamline the Telementoring process. Utilizing Socrates, a telementor located at a remote site can connect with an operating room, enabling the seamless exchange of audio and video signals. Through Socrates' telestrator feature, the telementor can provide real-time annotations on anatomy or surgical instructions, enriching the guidance process [7].

Another resource for Telementoring is the metaverse. The convergence of extended reality and artificial intelligence (AI) within a 5G context transports medical practitioners into this virtual realm. Within the metaverse, three-dimensional models, holographic displays, augmented reality visuals, radiomics, and genomics collaborate to enhance healthcare through heightened information exchange facilitated by these novel technologies [30].

In this scenario, technical and clinical aspects of two 5G-assisted Telementoring procedures were evaluated. Synchronized high-precision chronometers tracked signal latency between emission and reception. Surgeons and mentors communicated in real-time via an audiovisual system, sharing identical internal and operative field visuals. Transmission speeds fluctuated between 95 and 106 MB/s across two procedures. The average latency for the first procedure stood at 202 ms, while the second procedure exhibited a lower average latency of 146 ms. Remarkably, in this study, the correlation between question and instruction transmission earned a flawless 10 out of 10 rating in both instances [31].

In conclusion, Telementoring applications in Telesurgery represent a transformative advancement with the potential to revolutionize surgical practice. This approach has effectively reduced learning curves, improved surgical outcomes, and expanded access to specialized surgical care, particularly in remote or underserved areas. As technology evolves, Telementoring has become an indispensable tool for surgical education and collaboration. However, its successful implementation demands careful consideration of technological infrastructure, regulatory frameworks, and ethical considerations.

## Ethical challenges

Despite its potential benefits, Telesurgery is not devoid of concerns and limitations. As this innovative field expands to multiple directions, Telesurgery's ethical dimensions are a topic of concern and discussions [32].

The initial ethical dilemma lies in the surgeon-patient relationship in Telesurgery, a domain wherein remote interventions hold immense promise while posing the challenge of upholding this connection across physical divides. The inherent absence of face-to-face interactions may compromise the interpersonal bond, potentially impeding the cultivation of a profound relationship with patients.

Further accentuating ethical considerations, the threat of dehumanization and objectification emerges. Telesurgery, driven by technical precision, risks inadvertently obscuring the patient's needs, experiences, and expectations. The sophistication of advanced robotic platforms paradoxically threatens to detach the surgeon from the patient's unique identity, necessitating a solid ethical commitment to ensuring that each patient remains a distinct individual rather than a mere surgical subject or clinical case.

The realm of patient vulnerability also transforms the context of Telesurgery. The remote nature of these procedures introduces novel challenges in addressing complications and safeguarding patient well-being, thus heightening the ethical imperative for well-informed decision-making and comprehensive informed consent. Furthermore, the indispensability of Telesurgery underscores the limitations of local surgeons in handling specific procedures.

Another concern regards conflict of interest involving financial and professional considerations in robotics and Telesurgery. These conflicts can potentially compromise the surgeon's primary commitment to the patient, highlighting the ethical demand for transparency, disclosure, and an unwavering dedication to patient interest.

Informed consent, one of the pillars of patient autonomy, is redefined in the face of remote surgical interventions. The unique challenges posed by such procedures necessitate a reimagining of comprehensive and sufficient informed consent. Patients must be empowered with a detailed comprehension of the procedure, its limitations, potential ramifications, contingency plans, and the surgeon's affiliations with entities. The roles of local and remote surgeons should be articulated in the consent process.

With Telesurgery transcending geographical boundaries, formulating legal and jurisdictional guidelines is paramount to ensuring patient safety and ethical integrity [33]. Determining responsibility in adverse events, addressing licensure across jurisdictions, and fostering international collaborations underscore the ethical need for a unified legal framework that supports patient welfare and professional accountability. The domains of surgeons' and hospitals' malpractice liabilities and reimbursement regulations emerge as critical areas in establishing a robust global Telesurgery network.

Lastly, the specter of cyberattacks and safeguarding patient confidentiality demand ethical contemplation in Telesurgery. The convergence of technology and patient data exposes vulnerabilities necessitating vigilance against cyber threats, thereby protecting surgical precision and patient privacy. Previous studies have elucidated diverse forms of cyberattacks with manipulation of surgeon actions, delays, and even movement obstructions [4].

In summary, the assurance of patient autonomy, transparent communication, and the preservation of trust in the surgeon-patient alliance stand as ethical imperatives across all medical fields. In this context, it remains crucial to ensure that Telesurgery's potential to enlarge access and enhance care is accompanied by an unwavering commitment to patient well-being, trust, and ethical responsibility.

## Future scenario

### Directives and advancements

A crucial point to consider is that Telesurgery is an extension of conventional surgery, allowing pioneering technology to reach underserved areas with restricted access to healthcare. Therefore, any advancements in conventional surgery should easily crossover into Telesurgery. Integrating AI in Telesurgery represents a transformative leap in healthcare. AI technologies are enhancing the precision and safety of remote surgical procedures. One of the key applications of AI in Telesurgery regards imaging analysis. Advanced AI algorithms can swiftly process and interpret medical images, such as MRI and CT scans, providing surgeons with real-time insights during procedures [34]. This assists in identifying and navigating through intricate anatomical structures, minimizing the risk of errors, and improving the overall surgical outcome.

Trustworthy AI (TAI) is paramount in the context of Telesurgery. Patients and healthcare providers rely heavily on AI-driven systems to ensure patient safety and optimize surgical outcomes. Key concepts of TAI include transparency, accountability, fairness, and robustness [35]. TAI encompasses the principles of transparency, accountability, fairness, and robustness. In Telesurgery, transparency is achieved by providing clear information about how AI algorithms operate and make decisions, fostering trust among surgeons and patients. Accountability ensures that in the event of an AI-related issue, there are mechanisms in place to identify the responsible parties and take corrective actions promptly. Fairness ensures that AI systems do not introduce bias or discrimination in the surgical process, ensuring equitable healthcare delivery. Lastly, robustness ensures that AI systems can operate effectively in diverse and challenging surgical scenarios, maintaining reliability in critical moments. Trustworthy AI in Telesurgery is pivotal to fostering acceptance and ensuring the safe and effective deployment of AI technologies in healthcare.

The integration of Telesurgery technology brings several potential benefits, yet it also raises critical concerns regarding safety and medicolegal issues [12]. As surgical procedures are conducted remotely, ensuring the reliability and security of data transmission, real-time communication, and equipment functionality becomes paramount.

Any disruption, delay, or data loss could compromise the surgeon's ability to make accurate decisions and execute precise maneuvers. Furthermore, issues surrounding patient consent, liability in case of technical failures, and the attribution of responsibility in the event of adverse outcomes necessitate thorough legal and ethical scrutiny [36].

In addition to these concerns, the teleconsultation of specialists for surgical cases in rural areas is crucial. Surgical procedures extend beyond the operating room, encompassing perioperative care, such as intensive care unit oversight and radiological interventions, which significantly impact patient outcomes. Extensive evidence highlights the substantial reduction in morbidity and mortality when complex surgeries are conducted at high-volume centers specializing in the specific medical condition [12].

Another point is that latency remains insufficiently characterized in preclinical and clinical Telesurgery literature. Subsequent research must acknowledge that signal latency is dynamic, varying throughout a procedure. Existing Telesurgery studies generally focus solely on mean/average signal latency, disregarding both the variance in time delay fluctuations and its potential impact on surgical outcomes [37].

Finally, ensuring patient safety requires a local team capable of autonomously performing the surgery in case of technical failures. This consideration raises concerns regarding the overall advantage of Telesurgery compared to the potential savings from utilizing local resources [38].

## Conclusion

Telesurgery represents a remarkable advancement at the crossroads of medical innovation and technological progress. This transformative approach, enabled by the convergence of robotic systems, telecommunication, and surgical expertise, holds great promise for revolutionizing surgical care. As we journey through the complexities of this new era, we are confronted with opportunities and challenges that extend beyond the limits of traditional operating rooms.

The ethical considerations surrounding Telesurgery are crucial as we expand into this novel territory. From preserving the surgeon-patient relationship to addressing potential dehumanization, patient vulnerability, and conflicts of interest, these ethical details highlight the need for responsible and patient-centered practice. Moreover, the technical complexities of data transmission, cybersecurity, and precision engineering require meticulous attention to ensure the success of Telesurgery. In this scenario, collaboration between medical professionals, technologists, ethicists, and regulatory bodies is mandatory. We believe that finding a balance between innovation and ethical responsibility will define the trajectory and success of Telesurgery's evolution.

## Key points

- We provide a comprehensive overview of the evolution of Telesurgery, tracing its history from its early days to recent technological advancements such as 5G networks.
- It is highlighted the significant humanitarian impact of Telesurgery by improving access to specialized medical care in remote and underserved areas, thereby promoting equity in healthcare delivery.
- The technical aspects of Telesurgery, including the importance of network infrastructure, AI integration, and cybersecurity in ensuring the reliability and safety of remote surgical procedures are discussed.
- Numerous case studies are presented, demonstrating successful Telesurgery applications in various medical specialties, such as urology, neurosurgery, cardiology, and orthopedics.
- We address important ethical challenges associated with Telesurgery, including concerns about patient-surgeon relationships, dehumanization, informed consent, and cybersecurity risks, highlighting the need for responsible and patient-centered practices.

## References

[1] Larkin M. Transatlantic, robot-assisted telesurgery deemed a success. Lancet 2001;358(9287):1074. https://doi.org/10.1016/S0140-6736(01)06240-7.

[2] Marescaux J, Rubino F. Transcontinental robot-assisted remote telesurgery, feasibility and potential applications. In: Teleophthalmology. Springer Berlin Heidelberg; 2006. p. 261–5. https://doi.org/10.1007/3-540-33714-8_31.

[3] Marescaux J, Leroy J, Gagner M, et al. Transatlantic robot-assisted telesurgery. Nature 2001;413(6854):379–80. https://doi.org/10.1038/35096636.

[4] Bonaci T, Herron J, Yusuf T, Yan J, Kohno T, Chizeck HJ. To make a robot secure: an experimental analysis of cyber security threats against teleoperated surgical robots, 2015. Published online April 16 http://arxiv.org/abs/1504.04339.

[5] Fan S, Xu W, Diao Y, et al. Feasibility and safety of dual-console Telesurgery with the KangDuo surgical Robot-01 system using fifth-generation and wired networks: an animal experiment and clinical study. Eur Urol Open Sci 2023;49:6–9. https://doi.org/10.1016/j.euros.2022.12.010.

[6] Li J, Yang X, Chu G, et al. Application of improved robot-assisted laparoscopic telesurgery with 5G technology in urology. Eur Urol 2023;83(1):41–4. https://doi.org/10.1016/j.eururo.2022.06.018.

[7] Ballantyne GH. Robotic surgery, telerobotic surgery, telepresence, and telementoring. Surg Endosc 2002;16(10):1389–402. https://doi.org/10.1007/s00464-001-8283-7.

[8] Akpakwu GA, Silva BJ, Hancke GP, Abu-Mahfouz AM. A survey on 5G networks for the internet of things: communication technologies and challenges. IEEE Access 2017;6:3619–47. https://doi.org/10.1109/ACCESS.2017.2779844.

[9] Hakamada K, Mori M. The changing surgical scene: from the days of Billroth to the upcoming future of artificial intelligence and telerobotic surgery. Ann Gastroenterol Surg 2021;5(3):268–9. https://doi.org/10.1002/ags3.12466.

[10] Nankaku A, Tokunaga M, Yonezawa H, et al. Maximum acceptable communication delay for the realization of telesurgery. PLoS One 2022;17(10). https://doi.org/10.1371/journal.pone.0274328. e0274328.

[11] Nguan CY, Morady R, Wang C, et al. Robotic pyeloplasty using internet protocol and satellite network-based telesurgery. Int J Med Robot Comput Assist Surg 2008;4(1):10–4. https://doi.org/10.1002/rcs.173.

[12] Kirschniak A, Egberts JH, Granderath FA, et al. Augmented reality, cyber-physical systems and robotic surgery: Nice to have or a program with future. Visc Med 2018;34(1):60–5. https://doi.org/10.1159/000487209.

[13] Morohashi H, Hakamada K, Kanno T, et al. Construction of redundant communications to enhance safety against communication interruptions during robotic remote surgery. Sci Rep 2023;13(1):10831. https://doi.org/10.1038/s41598-023-37730-9.

[14] Al Asif MR, Khondoker R. Cyber security threat modeling of a telesurgery system. In: 2020 2nd International conference on sustainable technologies for industry 4.0 (STI); 2020. p. 1–6. https://doi.org/10.1109/STI50764.2020.9350452.

[15] Iqbal S, Farooq S, Shahzad K, Malik AW, Hamayun MM, Hasan O. SecureSurgiNET: a framework for ensuring security in telesurgery. Int J Distrib Sens Netw 2019;15(9), 1550147719873811. https://doi.org/10.1177/1550147719873811.

[16] Kazanzides P, Deguet A, Vagvolgyi B, Chen Z, Taylor RH. Modular interoperability in surgical robotics software. Mech Eng 2015;137(09): S19–22. https://doi.org/10.1115/1.2015-Sep-10.

[17] Zheng J, Wang Y, Zhang J, et al. 5G ultra-remote robot-assisted laparoscopic surgery in China. Surg Endosc 2020;34(11):5172–80. https://doi.org/10.1007/s00464-020-07823-x.

[18] Chu G, Yang X, Luo L, et al. Improved robot-assisted laparoscopic telesurgery: feasibility of network converged communication. Br J Surg 2021;108(11):e377–9. https://doi.org/10.1093/bjs/znab317.

[19] Yang X, Wang Y, Jiao W, et al. Application of 5G technology to conduct tele-surgical robot-assisted laparoscopic radical cystectomy. Int J Med Robot Comput Assist Surg 2022;18(4). https://doi.org/10.1002/rcs.2412.

[20] Tian Z, Lu W, Wang T, Ma B, Zhao Q, Zhang G. Application of a robotic Telemanipulation system in stereotactic surgery. Stereotact Funct Neurosurg 2008;86(1):54–61. https://doi.org/10.1159/000110742.

[21] Patel TM, Shah SC, Pancholy SB. Long distance tele-robotic-assisted percutaneous coronary intervention: a report of first-in-human experience. EClinicalMedicine 2019;14:53–8. https://doi.org/10.1016/j.eclinm.2019.07.017.

[22] Fan Y, Du JP, Liu JJ, et al. Accuracy of pedicle screw placement comparing robot-assisted technology and the free-hand with fluoroscopy-guided method in spine surgery. Medicine 2018;97(22), e10970. https://doi.org/10.1097/MD.0000000000010970.

[23] Adler JR. Remote robotic spine surgery. Neurospine 2020;17(1):121–2. https://doi.org/10.14245/ns.2040088.044.

[24] Ebihara Y, Oki E, Hirano S, et al. Tele-assessment of bandwidth limitation for remote robotics surgery. Surg Today 2022;52(11):1653–9. https://doi.org/10.1007/s00595-022-02497-5.

[25] Morohashi H, Hakamada K, Kanno T, et al. Social implementation of a remote surgery system in Japan: a field experiment using a newly developed surgical robot via a commercial network. Surg Today 2022;52(4):705–14. https://doi.org/10.1007/s00595-021-02384-5.

[26] Nakauchi M, Suda K, Nakamura K, et al. Establishment of a new practical telesurgical platform using the hinotori™ surgical robot system: a preclinical study. Langenbecks Arch Surg 2022;407(8):3783–91. https://doi.org/10.1007/s00423-022-02710-6.

[27] Takahashi Y, Hakamada K, Morohashi H, et al. Reappraisal of telesurgery in the era of high-speed, high-bandwidth, secure communications: Evaluation of surgical performance in local and remote environments. Ann Gastroenterol Surg 2023;7(1):167–74. https://doi.org/10.1002/ags3.12611.

[28] Alafaleq M. Robotics and cybersurgery in ophthalmology: a current perspective. J Robot Surg 2023;17(4):1159–70. https://doi.org/10.1007/s11701-023-01532-y.

[29] Moustris G, Tzafestas C, Konstantinidis K. A long distance telesurgical demonstration on robotic surgery phantoms over 5G. Int J Comput Assist Radiol Surg 2023. https://doi.org/10.1007/s11548-023-02913-2. Published online April 24.

[30] Checcucci E, Cacciamani GE, Amparore D, et al. The Metaverse in urology: ready for prime time. The ESUT, ERUS, EULIS, and ESU perspective. Eur Urol Open Sci 2022;46:96–8. https://doi.org/10.1016/j.euros.2022.10.011.

[31] Lacy AM, Bravo R, Otero-Piñeiro AM, et al. 5G-assisted telementored surgery. Br J Surg 2019;106(12):1576–9. https://doi.org/10.1002/bjs.11364.

[32] Frenkel CH. Telesurgery's Evolution During the Robotic Surgery Renaissance and a Systematic Review of its Ethical Considerations. Surg Innov 2023. https://doi.org/10.1177/15533506231169073. Published online April 11. 155335062311690.

[33] Fuertes-Guiró F, Velasco EV. Ethical aspects involving the use of information technology in new surgical applications: telesurgery and surgical telementoring. Acta Bioeth 2018;24(2):167–79.

[34] Hashimoto DA, Rosman G, Rus D, Meireles OR. Artificial intelligence in surgery: promises and perils. Ann Surg 2018;268(1):70–6. https://doi.org/10.1097/SLA.0000000000002693.

[35] Wickramasinghe CS, Marino DL, Grandio J, Manic M. Trustworthy AI Development Guidelines for Human System Interaction. In: 2020 13th International conference on human system interaction (HSI). IEEE; 2020. p. 130–6. https://doi.org/10.1109/HSI49210.2020.9142644.

[36] Malik MH, Brinjikji W. Feasibility of telesurgery in the modern era. Neuroradiol J 2022;35(4):423–6. https://doi.org/10.1177/19714009221083141.

[37] Barba P, Stramiello J, Funk EK, Richter F, Yip MC, Orosco RK. Remote telesurgery in humans: a systematic review. Surg Endosc 2022;36(5):2771–7. https://doi.org/10.1007/s00464-022-09074-4.

[38] Larcher A, Belladelli F, Capitanio U, Montorsi F. Long-distance robot-assisted teleoperation: every millisecond counts. Eur Urol 2023;83(1):45–7. https://doi.org/10.1016/j.eururo.2022.09.032.

# 16

# Robotic motion planning in surgery

*Yu Tian and Hongliang Ren*

**Department of Electronic Engineering, The Chinese University of Hong Kong, Sha Tin, Hong Kong**
**Shun Hing Institute of Advanced Engineering (SHIAE), The Chinese University of Hong Kong, Sha Tin, Hong Kong**

## Introduction

Humans have an innate ability to make decisions and choose motions effortlessly in their daily lives, rendering planning a seemingly simple task. However, the planning task is not as easy for robots as for humans. Robots need exact digital input to perform actions, while humans mostly plan by intuition [1]. Besides, perception and sensing, which are important parts of planning, are quite challenging for the robot [2]. Recently, robots have been widely

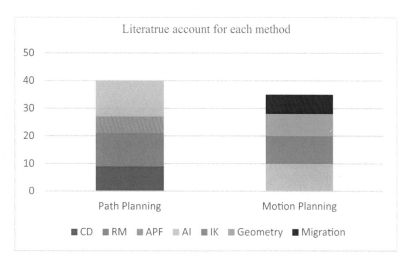

applied in surgery to perform accuracy and high precision. In surgical scenarios [3], complex environments involving deformable structures and extremely low fault tolerance make robotic planning even more difficult. The article's structure is shown in Fig. 1.

## Path planning

In the context of path planning, the robot is treated as a point mass, abstracting away its geometric and structural constraints. Path planning aims to generate a trajectory free of collisions, connecting the initial position to the desired destination with predetermined start and target positions [4]. The main challenges are as follows:

- How to avoid obstacles in a complex environment
- How to find an optical path in the distance cost
- How to improve algorithm efficiency

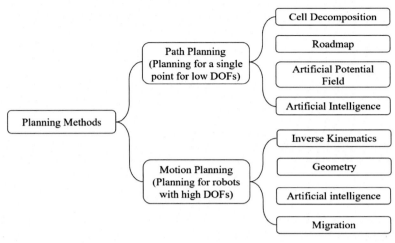

FIG. 1    Chapter structure diagram.

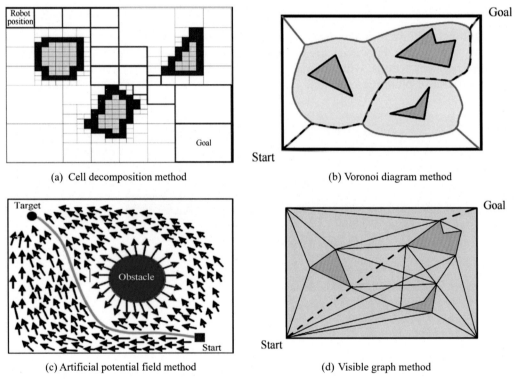

(a)  Cell decomposition method

(b) Voronoi diagram method

(c) Artificial potential field method

(d) Visible graph method

FIG. 2    Diagram of path planning methods [1].

Path planning methods are classified into four parts: cell decomposition (CD) method, roadmap (RM) method, artificial potential field (APF) method, and artificial intelligence (AI) method [5] (Fig. 2).

## CD method

CD method divides space into pure cells (cells without obstacles inside) and corrupted cells and then searches for a feasible way in a pure cell graph to achieve the target cell from the start cell field [6]. In the CD method, the grid method is the most popular decomposition method [7], which uses grids to generate the environment. The main problem with CD methods is how to balance accuracy and efficiency. If a small size of grids is chosen to represent the environment with high accuracy, the exponential rise in memory space and search range could be the result. In addition, in a high-dimension space, the cost of the CD method is also prohibitive. So, the CD method can achieve good results only in a low-dimensional environment.

A-Star algorithm is one of the most popular path-planning methods in grid maps [8,9], where a heuristic distance is introduced to evaluate the cost [10]. Each time, the cells that have been reached with the lowest evaluated value will be set to the next position. The computational complexity of the A-Star algorithm is $O(n^2 \log n)$, where n represents the cell number. A-Star algorithm can find the optical collision-free path in the given gird map but cannot handle a dynamic environment and high-dimension configuration space.

In medical scenarios, the A-Star method is used in some simple grid environments to get a stable and high-quality path. In contrast, path smoothness and algorithm efficiency are potential problems. Mohammadi et al. [11] proposed a modified A-Star algorithm using hexagonal segmentation instead of traditional square grids. They depicted the effectiveness of obstacles and obstacle-avoidance sensitivity to get a smooth trajectory in a deformable environment, which can be employed in typical needle penetration procedures. With the CT image sequence of patients, He et al. [12] constructed a binarized three-dimensional grid map. It used an A-Star algorithm to obtain an endoscopic surgical approach in nasal surgery. Granna et al. [13,14] used a grid-based method to achieve the blood-sucking path planning of robot-assisted cerebral hemorrhage surgery.

## RM method

Roadmap methods represent the environment using a roadmap, which comprises a number of nodes and edges labeled with cost. Edges connect nodes and edges are collision-free. Then, a path from the start node to the target node is detected in the roadmap. Popular RM methods include the visibility graph (VG) algorithm, Voronoi diagram (VD), and sampling-based roadmap methods like probability roadmap (PRM) algorithm and rapid exploring random tree (RRT) algorithm. VG algorithm [15,16] is designed for the 2D environment containing only convex polygonal obstacles. Nodes are set on the vertices of all obstacles and the start and target points, and if two nodes are visible, the line connecting them is an edge in the graph. VD [17] uses equidistant points between two adjacent points on the obstacle boundaries to construct all edges to segment the region into subregions. In sampling-based methods, nodes are randomly generated in the space and then connected with some constraints. In PRM [18], edges are constructed until all nodes are generated, while in RRT [19], once a valid node is generated, the edge linking the new node and the nearest node in the previous nodes will be constructed. After the graph is developed, an optical path can be obtained through graph searching methods, like depth-first search and the Dijkstra algorithm.

In medical scenarios, RRT and its variants [20,21] are the most popular RM method, with its advantages in efficiency and environment adaptability. However, rapidly obtaining a smooth, high-quality path is still an open problem. Zhao et al. [22] combined RRT, greedy-heuristic strategy, and reachability-guided strategy to propose a robot-assisted active, flexible needle algorithm. The algorithm shows advantages in terms of computation speed and robustness of searching ability in dealing with anatomic obstacles and the nonholonomic motion of the needle. Li et al. [23] proposed a planning method for a parallel orthopedic robot based on an improved RRT* method and B-spline curve. An extension strategy was applied to make the explore tree extension more toward the target, which enhances the performance in the situation that the optimal path is not tortuous while weakening the ability to search. Fauser et al. [24] present a k-RRT-Connect that demonstrates the feasibility of preoperative and intra-operative planning of nonlinear access paths in temporal bone surgery, where the operation region is narrow and surrounded by sensitive organs. Guo et al. [25] proposed a local dynamic path planning-based physician training system using RRT to complete the real-time local path planning for vascular interventional surgery. The algorithm demonstrates fast convergence speed and a high-quality path in the experiment. Zhang et al. [26] proposed an RRT* algorithm with an improved sampling strategy, which can find a smoother and shorter path. The algorithm presented in this study demonstrates notable benefits in addressing the nonholonomic constraint associated with flexible needles and effectively managing tissue deformation caused by the needle tip during insertion.

## APF method

In the APF method, an attractive potential field is created by the target point, while obstacles in the space generate a repulsive potential field. The robot's actions are determined using the gradient descent method based on these potential fields [27]. APF method can achieve real-time obstacle-avoidance planning in high-dimension space with low computational cost. However, in obstructed environments, the robot may be stuck at the local minimal position and cannot keep moving toward the target position. Besides, the path obtained by APF is not the optical in the distance. To solve this problem, Lu et al. [28] planned a trajectory by fitting a curve with a goal point and critical oscillating points, which is much smoother than traditional methods. He et al. [29] reconstructed the model and introduced pose threshold gain to overcome linear interference. Liang et al. [30] improved force function with angle factors and adaptive weights.

In medical scenarios, APF shows advantages in obstacle-avoidance ability and smoothness of the path, which greatly increases the safety of surgery. Hao et al. [31] adopted the dynamic gravitational constant and piecewise repulsion function to improve APF to solve the local minimum problem and the problem that the robot cannot reach the targets close to the obstacles. Benefiting from this method, the safety and accuracy of spine surgery robots are improved. Chen et al. [32] improved APF for surgical robot systems with a dual arm. To achieve precise control, the researchers developed a gravitational potential field and a repulsive potential field specifically for the robot joints. Additionally, they employed an enhanced splicing path method to address the challenge of local minimums.

## AI method

Recently, AI has been applied in almost every field, including robot path planning [33]. Among all the AI methods, including Genetic algorithm [34], Firefly algorithm [35] Ant colony algorithm [36], and so on, machine learning (ML) methods are the most popular and show the best result [37–39]. A neural network (NN) is the basis of machine learning. Its functionality resembles the animal neuron system, an interconnected assembly of lower-level processing elements, units, or nodes [37]. Theoretically, if sufficiently trained, neural networks can fit any function. Therefore, they can handle the path planning problem of input map information and output path. However, a trained NN can hardly be generalized to other environments, which is the bottleneck that prevents the progress of AI algorithms.

The great potential of ML brings infinite possibilities, but its uninterpretable nature limits it greatly in the medical environment with very low fault tolerance. Baek et al. [40] combined probabilistic roadmap and reinforcement learning to realize resection automation of cholecystectomy. Milestones were found by PRM, and the Q-learning method was used to find a global path among milestones. Since the Q-learning method is based on a grid map, this method is only suitable for similar and small-scale environments. Nguyen et al. [41] presented a reinforcement learning-based algorithm for pattern-cutting surgery, in which a scissor and a gripper are manipulated concurrently. Unlike former methods, this method used multiple pinch points for the gripper, which shows great superiority when the trajectory is complex. Fauser et al. [42] proposed a deep learning-based automatic segmentation and sampling-based trajectory planning method for temporal bone surgery. This method is task-specific and time-consuming, so it can only be utilized offline. Zhao et al. [43] developed an integrated approach utilizing convolutional neural networks (CNNs) and long short-term memory networks (LSTMs) to create a generative adversarial network (GAN) specifically designed for automated surgical path planning in endovascular surgery. This innovative framework effectively combines the extraction and fusion of spatial features from surgical state images and temporal features from historical tool trajectories, enabling more efficient and accurate path planning in surgical procedures. However, in this method, the deformation of vessels and vascular is not considered. And the sampling efficiency of the dataset needs to be further improved. Zhang et al. [44] used the Q-learning algorithm to realize 3D surgical path planning in anterior spinal surgery. The authors also proposed a curved limit scheme to handle the dimension explosion and improve the efficiency of pathway searching [42]. However, Q-learning can only be applied in discrete gird space. Hence, the application of Q-learning is limited to discrete grid spaces, making it challenging to strike a balance between accuracy and efficiency. The characteristics of the four planning methods introduced in this section are summarized in Tables 1 and 2.

TABLE 1   Characteristics of the four path planning methods.

| Method | Efficiency | Path quality | Environment adaptability | Stability |
|---|---|---|---|---|
| CD | Computation load increase greatly when the number of cell increase | Depending on the size of cells | Suitable for low-dimension environment | Promise to find the path if it exists |
| RM | Can find a solution quickly but takes plenty of time to optimize the result | Asymptotic optimization | Can be applied in high-dimension environment | Promise to find the path if it exists |
| APF | High | Not optical | Can be applied in high-dimension environment | May trapped in a local minimal position |
| AI | Required a larger number of training data and time | Not promise to find an optical path | May fail in an untrained environment | Not promise to find a feasible path |

Unlike path planning, which only focuses on the robot's position, motion planning requires simultaneous planning of multiple robot inputs of robots [45,46]. Generally, each input of the robots is directly coupled and will affect each other; motion planning is more difficult than path planning. Robots usually need to enter a narrow environment to execute tasks in medical surgery. In this case, redundant robots are widely used for their better dexterity brought by redundant degrees of freedom (DOF) [47–50]. For a redundant robot, a specific target end-effect state can correspond to many joint states. Choosing an optical execute position is also a tricky issue. Therefore, motion planning for the redundant robot is more difficult than planning for low-DOF robots. The motion planning method applied to redundant robots can also be applied to low-DOF robots. Hence, the following section focuses on motion planning methods for redundant robots.

## The inverse kinematics (IK) method

The inverse kinematics, based on which many traditional rigid robot planning algorithms are developed, can accurately calculate the joint angles required when the end effector of the robot is in a specific position. However, when it was applied to redundant robots, there were some challenges:

- For redundant robots, an end-effect position could correspond to many different shapes. It is hard to tell which shape is the best one.
- Inverse kinematics mainly focuses on the end effector. When the working environment is narrow and contains multi-obstacles, keeping the rest of the robot in a safe place is challenging.
- Solving the inverse kinematics equation is time-consuming, especially when redundant robots' kinematics are highly complex. Besides, the computational load greatly increases as the DOF of the robot increases. Real-time planning of redundant robots is highly challenging.
- The method based on inverse kinematics can hardly solve the inequality problem. Therefore, these methods cannot adequately handle a multi-obstacles environment.

Marchese et al. [51] proposed an algorithm using a set of constrained optimization problems to resolve locally optimal inverse kinematics and simultaneously considering the entire envelope of the robot in proximity to a confined environment. Based on it, the redundant robots can maneuver through a pipe-like environment. However, the success rate of this algorithm was only around 80% in the simulation, and the speed was rather limited. Chen and Lau [52] applied learning-based support vector regression to fit the inverse kinematics of redundant robots. The prediction accuracy of the proposed method is above 99%. Motahari et al. [53] used the two-by-two searching algorithm to find an inverse kinematics solution to redundant robots. They also developed the removing first collision method to achieve the trajectory tracking problem of redundant robots in a known static obstacle environment. Martin et al. [54] put forward a method based on cyclic coordinate descent to choose coherent and harmonious solutions for the inverse kinematics of redundant robots. In addition, it could allow real-time applications and consider the whole body's movement. Reiter et al. [55] presented prerequisite explicit expressions for the higher-order inverse kinematics to address a time-optimal path for redundant robots. Based on the gradient projection algorithm, Liu et al. [56] raised a weighted additional deviation velocity, which calculated collision avoidance control parameters by the gradient projection algorithm for solving inverse kinematic solutions and solved collision avoidance problems of the redundant robot with a known end-effector path. Liu et al. [57] proposed a simplified kinematics modeling and configuration planning algorithm. To properly address environmental constraints, the researchers introduced an extended virtual joint that facilitates the rapid search for regional solutions and enables equivalent kinematic constraint analysis. By leveraging simplified kinematics and the extended virtual joint, a novel algorithm was developed, allowing redundant robots to navigate around single obstacles and traverse narrow environments. Zhang et al. [58] proposed a "dynamic coordination method" by applying exponential coordinates according to the Lie group theory. The method could provide a more flexible solution for target motion optimization.

In medical scenarios, IK can accurately control the position of robots. The accuracy of IK can greatly increase safety during surgery. However, the weak obstacle-avoidance ability of IK limits its application in a complex or dynamic environment.

## Geometry method

With the advantages of redundancy, redundant robots can easily fit many types of curvatures called backbones. Therefore, if a predetermined backbone is obtained, redundant robots can be fit to the backbone, and a desired configuration can be generated. Therefore, the geometry method can easily plan the overall configuration of redundant robots. However, this type of method faces the following challenges:

- How to generate a proper backbone in a constrained environment cluttered with obstacles or even a dynamic environment?
- How to fit the robot to the backbone?
- How to use a single backbone to guide the whole planning process?
- How to generalize a method for different types of robots?

Chibani et al. [59] used a scheme that assesses the value of kinematics configuration while taking geometric constraints into consideration in a deterministic optimization way. This method simultaneously searches for the best curve depicting the whole kinematic robot configuration and the platform distribution along this curve. Song et al. [60,61] solve the real-time shape estimation and shape estimation with an unknown payload on the end effector based on the Bezier curve. The method can be applied to robots with various kinematics models and offers satisfactory tracking performance and great accuracy of shape reconstruction. Mu et al. [62] segment redundant robots into three geometry parts, namely, shoulder, elbow, and wrist. Then, the kinematics of redundant robots can be separately solved in each segment. This method can avoid joint limits, obstacles, and cross narrow pipelines on the designed robot. Mbakop et al. [63] raised a kinematics model for redundant robots based on Pythagorean Hodograph quintic curve theory and a collision avoidance method using sliding mode theory to ensure a finite time convergence. Bulut et al. [64] presented a simple, robust, and effective geometrical algorithm for the real-time path-planning of redundant manipulators in narrow space with complex obstacles. It can be applied to many redundant robots, even for redundant robots with hundreds of DOFs with different link lengths. Sepehri et al. [65] utilized rapidly exploring trees to determine a suitable backbone, and attractive and repulsive potential fields acted on every trajectory point. Then, they are utilized in a gradient optimization method to generate paths to reach the target position. Tian et al. [66] constructed a virtual guiding pipeline based on the predetermined backbone, and redundant robots were constrained in the pipeline by an improved APF. This method could efficiently guide redundant robots to cross narrow environments with high success rate.

The geometry method can perform well in a known static environment with a specific robot. However, the generalization ability of this category is poor. It cannot be used when the environment is dynamic or unknown, and the method is usually designed only for a specific robot. Besides, most algorithms are designed to obtain the configuration at a single moment but not a series configuration during the process. In this case, completing the whole task could take much computation. Therefore, in medical scenarios,

## AI method

Researchers have put much effort into applying AI methods represented by the Machine Learning method for planning redundant robots.

Srinivasu et al. [67] proposed an approach that utilized reinforcement-based temporal difference and Inverse Kinematics. The dimensionality problem was well-resolved, and multidimensional procedure contexts could be effectively dealt with through that method. However, this CNN-based approach does not fit the dynamic domains well, and it is hard to collect the training data. Wang et al. [68] proposed a demonstration learning method. Models are built using Gaussian mixture models and Gaussian mixture regression to encode the demonstrated trajectory and estimate a suitable path for the manipulators to reproduce a task, such as crossing a hole. Jin et al. [69] construct a dynamic neural network to perform the recurrent computation of manipulability-maximal control actions with redundant robots under strict structure constraints in an inverse-free way. The method has good manipulability in simulations. Li and Hanaford [70] formulated deformable-obstacle avoidance into an optimization problem solved in dual space by Recurrent Neural Networks (RNN). Li et al. [71] formulated a noise-suppression planning as a quadratic problem and used RNN to solve it. The PID information of the target end-effector path is used to construct the equality criterion. Li [72] proposed a zeroing neural network model for cyclical motion planning. Unlike other zeroing neural network models, this method considered the physical limits of redundant robots and was proven to converge in predefined time. Xu et al. [73] suggested a motion-force control method within the theory of projection RNN, which solves the control problem at the velocity level in a quadratic-programming way. Iyengar and Stoyanov [74] developed a control method with no former kinematic model based on a DRL method. To explore curricula use, they attempt to change the target tolerance by training with constant, linear, and exponential decay functions. Joshi et al. [75] tried to find a time-optimal kino-dynamic path by a "Time-Informed Set (TIS)" after finding an initial solution inspired by ideas from reachability analysis.

Learning-based methods have been applied for obstacles avoidance, grab, noise suppression, motion-force control, and other tasks. Furthermore, there are many topics of great interest in the future:

- Efficiency: collecting and training data takes plenty of resources and time.
- Generalization: how to improve the performance in untrained tasks and robots.
- Interpretability: how to explain the result of NN to increase its reliability.

## Migration method

Besides the methods mentioned previously, many methods that originate from other fields have also been migrated to the planning of redundant robots, such as optimization algorithms, graph algorithms, sampling-based methods, and other planning methods for mobile robots.

Guo and Zhang et al. [76] proposed a minimum-acceleration-norm framework for redundancy resolution. They reformulated the framework as a general quadratic program that could be equivalent to linear variational inequality and then converted to a piecewise linear projection equation. Menon et al. [77] present an optimization algorithm, which was derived by using the calculus of variation. The algorithm considered all parts of redundant robots in the vicinity of obstacles and could get natural planning results. Zhang et al. [78] suggested a tricriteria optimization-coordination-motion scheme integrating the minimum velocity norm, repetitive motion planning, and infinity-norm velocity minimization. The scheme is converted to a quadratic programming problem and settled using a simplified-based primal-dual neural network solver. Li et al. [79] raised a fault-tolerant approach with a simultaneous fault-diagnose function for motion planning and controlling redundant robots. It can ensure tracking of the target trajectory even when the fault joints lose their velocity to actuate. Dai et al. [80] presented a sampling-based method with two major components. An initial trajectory was determined by graph search, and then a greedy method was used to optimize the initial trajectory by adaptive fillers in the local regions with large jerks. Using 3D Voronoi and the Dijkstra methods, Gough et al. [81] developed a method to find a trajectory that balances the cost between the robot's shape change and the trajectory's length.

Different migration methods have different characteristics. Most of them suit their specific situation but may not be applied widely. Therefore, if the aforementioned methods cannot satisfy task requirements, the answer might be found in migration methods (Table 2).

TABLE 2  Characteristics of the four motion planning methods.

| Method | Efficiency | Obstacle-avoidance ability | Generalization ability |
|---|---|---|---|
| IK | Computation load increases greatly when DOFs increase | Hard to consider other positions except the end effector and hard to solve the inequality problem | Only need kinematics to be applied to various robots |
| Geometry | Robot needs to be fitted into a curve repetitively each step | Hard to determine the proper backbone in a complex environment | Cannot be applied to robots with different structure |
| AI | Required a larger number of training data and time | Can handle most trained cases | Hard to apply to untrained environments and robots |
| Migration | Depends on the specific method | | |

## Future scenario

Planning in a dynamic obstacle environment needs further exploration for future research. Tasks in which robots must be reactive with the environment, especially deformable environments, need more attention. Besides, solving data efficiency and generalization ability problems with AI methods will be a challenging topic that many researchers are trying to figure out.

## Key points

This chapter discussed popular robotic path and motion planning methods and their variants. In summary, if the planning object is a single-point path, the path planning methods should be chosen, while if all robot inputs need to be planned, motion planning methods are suitable.

In path planning, CD methods are suitable for tasks in low dimensions and small-size environments with high-quality path requirements. RM methods can deal with high-dimension space but cannot obtain an optical path in finite time. APF methods can achieve real-time planning but cannot promise to find a valid solution. AI methods can be widely applied in various environments where a lot of historical data can be obtained, but a lot of training data and time are needed for a new type of task. Besides, the performance is unstable.

As for motion planning, IK methods show superiority in accuracy but are weak in obstacle avoidance. Geometry methods would be the best choice for offline planning when the robot is dexterous. AI methods have the same characteristics as AI path-planning methods. When other methods cannot handle the problem, migration methods might answer.

## Acknowledgment

This work was supported by the Hong Kong Research Grants Council (RGC) Collaborative Research Fund (C4026-21G), General Research Fund (GRF 14211420), and Shenzhen-Hong Kong-Macau Technology Research Programme (Type C) STIC Grant SGDX20210823103535014 (202108233000303).

## References

[1] Patle BK, Babu L G, Pandey A, Parhi DRK, Jagadeesh A. A review: on path planning strategies for navigation of mobile robot. Def Technol 2019;15(4):582–606. https://doi.org/10.1016/j.dt.2019.04.011.

[2] Raja P. Optimal path planning of mobile robots: A review. Int J Phys Sci 2012;7(9). https://doi.org/10.5897/IJPS11.1745.

[3] Diana M, Marescaux J. Robotic surgery. Br J Surg 2015;102(2):e15–28. https://doi.org/10.1002/bjs.9711.

[4] Gasparetto A, Boscariol P, Lanzutti A, Vidoni R. Path planning and trajectory planning algorithms: a general overview. Motion Oper Plan Robot Syst 2015;3–27.

[5] Galceran E, Carreras M. A survey on coverage path planning for robotics. Robot Auton Syst 2013;61(12):1258–76.

[6] Li Y, Chen H, Er MJ, Wang X. Coverage path planning for UAVs based on enhanced exact cellular decomposition method. Mechatronics 2011; 21(5):876–85.

[7] Bailey JP, Nash A, Tovey CA, Koenig S. Path-length analysis for grid-based path planning. Artif Intell 2021;301, 103560.

[8] Ducho☒ F, et al. Path planning with modified a star algorithm for a mobile robot. Procedia Eng 2014;96:59–69.

[9] AlShawi IS, Yan L, Pan W, Luo B. Lifetime enhancement in wireless sensor networks using fuzzy approach and A-star algorithm. IEEE Sens J 2012;12(10):3010–8.

[10] Yao J, Lin C, Xie X, Wang AJ, Hung C-C. Path planning for virtual human motion using improved A* star algorithm. In: 2010 Seventh international conference on information technology: new generations. IEEE; 2010. p. 1154–8.

[11] Mohammadi A, Rahimi M, Suratgar AA. A new path planning and obstacle avoidance algorithm in dynamic environment. In: 2014 22nd Iranian conference on electrical engineering (ICEE). IEEE; 2014. p. 1301–6.

[12] He Y, Zhang P, Qi X, Zhao B, Li S, Hu Y. Endoscopic path planning in robot-assisted endoscopic nasal surgery. IEEE Access 2020;8:17039–48.

[13] Granna J, Guo Y, Weaver KD, Burgner-Kahrs J. Comparison of optimization algorithms for a tubular aspiration robot for maximum coverage in intracerebral hemorrhage evacuation. J Med Robot Res 2017;02(01):1750004. https://doi.org/10.1142/S2424905X17500040.

[14] Granna J, Godage IS, Wirz R, Weaver KD, Webster RJ, Burgner-Kahrs J. A 3-D volume coverage path planning algorithm with application to intracerebral hemorrhage evacuation. IEEE Robot Autom Lett 2016;1(2):876–83. https://doi.org/10.1109/LRA.2016.2528297.

[15] Lacasa L, Luque B, Ballesteros F, Luque J, Nuno JC. From time series to complex networks: the visibility graph. Proc Natl Acad Sci 2008; 105(13):4972–5.

[16] Welzl E. Constructing the visibility graph for n-line segments in O (n2) time. Inf Process Lett 1985;20(4):167–71.

[17] Takahashi O, Schilling RJ. Motion planning in a plane using generalized Voronoi diagrams. IEEE Trans Robot Autom 1989;5(2):143–50.

[18] Kavraki LE, Svestka P, Latombe J-C, Overmars MH. Probabilistic roadmaps for path planning in high-dimensional configuration spaces. IEEE Trans Robot Autom 1996;12(4):566–80. https://doi.org/10.1109/70.508439.

[19] LaValle S. Rapidly-exploring random trees: a new tool for path planning. Research report 1998;9811.

[20] Tahir Z, Qureshi AH, Ayaz Y, Nawaz R. Potentially guided bidirectionalized RRT* for fast optimal path planning in cluttered environments. Robot Auton Syst 2018;108:13–27.

[21] Nasir J, et al. RRT*-SMART: a rapid convergence implementation of RRT. Int J Adv Robot Syst 2013;10(7):299.

[22] Zhao Y-J, et al. 3D motion planning for robot-assisted active flexible needle based on rapidly-exploring random trees. J Autom Control Eng 2015;3(5).

[23] Li J, Cui R, Su P, Ma L, Sun H. A computer-assisted preoperative path planning method for the parallel orthopedic robot. Machines 2022; 10(6):480.

[24] Fauser J, Sakas G, Mukhopadhyay A. Planning nonlinear access paths for temporal bone surgery. Int J Comput Assist Radiol Surg 2018; 13(5):637–46.

[25] Guo J, Sun Y, Guo S. A training system for vascular interventional surgeons based on local path planning. In: 2021 IEEE international conference on mechatronics and automation (ICMA). Takamatsu, Japan: IEEE; 2021. p. 1328–33. https://doi.org/10.1109/ICMA52036.2021.9512808.

[26] Zhang Y, Qi Z, Zhang H. An improved RRT algorithm combining motion constraint and artificial potential field for robot-assisted flexible needle insertion in 3D environment. In: 2021 3rd International conference on industrial artificial intelligence (IAI). Shenyang, China: IEEE; 2021. p. 1–6. https://doi.org/10.1109/IAI53119.2021.9619224.

[27] Khatib O. Real-time obstacle avoidance for manipulators and mobile robots. In: Proceedings. 1985 IEEE international conference on robotics and automation. IEEE; 1985. p. 500–5.

[28] Lu SX, Li E, Guo R. An obstacles avoidance algorithm based on improved artificial potential field. In: 2020 IEEE international conference on mechatronics and automation (ICMA). Beijing, China: IEEE; 2020. p. 425–30. https://doi.org/10.1109/ICMA49215.2020.9233866.

[29] He N, Su Y, Guo J, Fan X, Liu Z, Wang B. Dynamic path planning of mobile robot based on artificial potential field. In: 2020 International conference on intelligent computing and human-computer interaction (ICHCI). Sanya, China: IEEE; 2020. p. 259–64. https://doi.org/10.1109/ICHCI51889.2020.00063.

[30] Liang Q, Zhou H, Xiong W, Zhou L. Improved artificial potential field method for UAV path planning. In: 2022 14th International conference on measuring technology and mechatronics automation (ICMTMA). Changsha, China: IEEE; 2022. p. 657–60. https://doi.org/10.1109/ICMTMA54903.2022.00136.

[31] Hao L, et al. An improved path planning algorithm based on artificial potential field and primal-dual neural network for surgical robot. Comput Methods Programs Biomed 2022;227, 107202. https://doi.org/10.1016/j.cmpb.2022.107202.

[32] Chen Q, Liu Y, Wang P. An Autonomous Obstacle Avoidance Method for Dual-Arm Surgical Robot Based on the Improved Artificial Potential Field Method. In: Liu H, Yin Z, Liu L, Jiang L, Gu G, Wu X, Ren W, editors. Intelligent robotics and applications. Lecture Notes in Computer Science, vol. 13458. Cham: Springer International Publishing; 2022. p. 496–508. https://doi.org/10.1007/978-3-031-13841-6_45.

[33] Puente-Castro A, Rivero D, Pazos A, Fernandez-Blanco E. A review of artificial intelligence applied to path planning in UAV swarms. Neural Comput Applic 2021;1–18.

[34] Mathew TV. Genetic algorithm. Rep. Submitt, IIT Bombay; 2012.

[35] Yang X-S. Nature-inspired metaheuristic algorithms. Luniver Press; 2010.

[36] Dorigo M, Gambardella LM. Ant colony system: a cooperative learning approach to the traveling salesman problem. IEEE Trans Evol Comput 1997;1(1):53–66. https://doi.org/10.1109/4235.585892.

[37] Lv L, Zhang S, Ding D, Wang Y. Path planning via an improved DQN-based learning policy. IEEE Access 2019;7:67319–30.

[38] Zhang B, Mao Z, Liu W, Liu J. Geometric reinforcement learning for path planning of UAVs. J Intell Robot Syst 2015;77(2):391–409.

[39] Wang J, Chi W, Li C, Wang C, Meng MQ-H. Neural RRT*: learning-based optimal path planning. IEEE Trans Autom Sci Eng 2020;17(4):1748–58.

[40] Baek D, Hwang M, Kim H, Kwon D-S. Path planning for automation of surgery robot based on probabilistic roadmap and reinforcement learning. In: 2018 15th International conference on ubiquitous robots (UR). Honolulu, HI: IEEE; 2018. p. 342–7. https://doi.org/10.1109/URAI.2018.8441801.

[41] Nguyen ND, Nguyen T, Nahavandi S, Bhatti A, Guest G. Manipulating soft tissues by deep reinforcement learning for autonomous robotic surgery. In: 2019 IEEE international systems conference (SysCon). IEEE; 2019. p. 1–7.

[42] Fauser J, et al. Toward an automatic preoperative pipeline for image-guided temporal bone surgery. Int J Comput Assist Radiol Surg 2019; 14(6):967–76.

[43] Zhao Y, et al. Surgical GAN: towards real-time path planning for passive flexible tools in endovascular surgeries. Neurocomputing 2022.

[44] Zhang Q, Li M, Qi X, Hu Y, Sun Y, Yu G. 3D path planning for anterior spinal surgery based on CT images and reinforcement learning. In: 2018 IEEE international conference on cyborg and bionic systems (CBS). IEEE; 2018. p. 317–21.

[45] Latombe J-C. Robot motion planning. Vol. 124. Springer Science & Business Media; 2012.

[46] La Valle SM. Motion planning. IEEE Robot Autom Mag 2011;18(2):108–18.

[47] Ikuta K, Hasegawa T, Daifu S. Hyper redundant miniature manipulator" Hyper Finger" for remote minimally invasive surgery in deep area. In: 2003 IEEE International Conference on Robotics and Automation (Cat. No. 03CH37422). IEEE; 2003. p. 1098–102.

[48] Su H, Yang C, Ferrigno G, De Momi E. Improved human–robot collaborative control of redundant robot for teleoperated minimally invasive surgery. IEEE Robot Autom Lett 2019;4(2):1447–53.

[49] Kuo C-H, Dai JS, Dasgupta P. Kinematic design considerations for minimally invasive surgical robots: an overview. Int J Med Robot 2012; 8(2):127–45.

[50] Chikhaoui MT, Burgner-Kahrs J. Control of continuum robots for medical applications: State of the art. In: ACTUATOR 2018; 16th international conference on new actuators. VDE; 2018. p. 1–11.

[51] Agarwal V. Trajectory planning of redundant manipulator using fuzzy clustering method. Int J Adv Manuf Technol 2012;61(5):727–44.

[52] Chen J, Lau HY. Inverse kinematics learning for redundant robot manipulators with blending of support vector regression machines. In: 2016 IEEE workshop on advanced robotics and its social impacts (ARSO). IEEE; 2016. p. 267–72.

[53] Motahari A, Zohoor H, Korayem MH. A new motion planning method for discretely actuated hyper-redundant manipulators. Robotica 2017; 35(1):101–18.

[54] Martin A, Barrientos A, Del Cerro J. The natural-CCD algorithm, a novel method to solve the inverse kinematics of hyper-redundant and soft robots. Soft Robot 2018;5(3):242–57.

[55] Reiter A, Müller A, Gattringer H. On higher order inverse kinematics methods in time-optimal trajectory planning for kinematically redundant manipulators. IEEE Trans Ind Inform 2018;14(4):1681–90.

[56] Liu J, Tong Y, Ju Z, Liu Y. Novel method of obstacle avoidance planning for redundant sliding manipulators. IEEE Access 2020;8:78608–21.

[57] Liu T, Xu W, Yang T, Li Y. A hybrid active and passive cable-driven segmented redundant manipulator: design, kinematics, and planning. IEEEASME Trans Mechatron 2020;26(2):930–42.

[58] Zhang J, Lu K, Yuan J, Di J, He G. Kinematics modeling and motion planning of tendon driven continuum manipulators. J Artif Intell Technol 2020;1(1):28–36. https://doi.org/10.37965/jait.2020.0041.

[59] Chibani A, Mahfoudi C, Chettibi T, Merzouki R, Zaatri A. Generating optimal reference kinematic configurations for hyper-redundant parallel robots. Proc Inst Mech Eng Part J Syst Control Eng 2015;229(9):867–82. https://doi.org/10.1177/0959651815583423.

[60] Song S, Li Z, Meng MQ-H, Yu H, Ren H. Real-time shape estimation for wire-driven flexible robots with multiple bending sections based on quadratic Bézier curves. IEEE Sens J 2015;15(11):6326–34. https://doi.org/10.1109/JSEN.2015.2456181.

[61] Song S, Li Z, Yu H, Ren H. Shape reconstruction for wire-driven flexible robots based on Bézier curve and electromagnetic positioning. Mechatronics 2015;29:28–35. https://doi.org/10.1016/j.mechatronics.2015.05.003.

[62] Mu Z, Yuan H, Xu W, Liu T, Liang B. A segmented geometry method for kinematics and configuration planning of spatial hyper-redundant manipulators. IEEE Trans Syst Man Cybern Syst 2020;50(5):1746–56. https://doi.org/10.1109/TSMC.2017.2784828.

[63] Mbakop S, Tagne G, Lakhal O, Merzouki R, Drakunov SV. Path planning and control of Mobile soft manipulators with obstacle avoidance. In: 2020 3rd IEEE international conference on soft robotics (RoboSoft). New Haven, CT, USA: IEEE; 2020. p. 64–9. https://doi.org/10.1109/RoboSoft48309.2020.9115998.

[64] Bulut Y, Conkur ES. A real-time path-planning algorithm with extremely tight maneuvering capabilities for hyper-redundant manipulators. Eng Sci Technol Int J 2021;24(1):247–58. https://doi.org/10.1016/j.jestch.2020.07.002.

[65] Sepehri A, Moghaddam AM. A motion planning algorithm for redundant manipulators using rapidly exploring randomized trees and artificial potential fields. IEEE Access 2021;9:26059–70. https://doi.org/10.1109/ACCESS.2021.3056397.

[66] Tian Y, Zhu X, Meng D, Wang X, Liang B. An overall configuration planning method of continuum hyper-redundant manipulators based on improved artificial potential field method. IEEE Robot Autom Lett 2021;6(3):4867–74. https://doi.org/10.1109/LRA.2021.3067310.

[67] Srinivasu PN, Bhoi AK, Jhaveri RH, Reddy GT, Bilal M. Probabilistic deep Q network for real-time path planning in censorious robotic procedures using force sensors. J Real-Time Image Process 2021;18(5):1773–85.

[68] Wang H, Chen J, Lau HYK, Ren H. Motion planning based on learning from demonstration for multiple-segment flexible soft robots actuated by electroactive polymers. IEEE Robot Autom Lett 2016;1(1):391–8. https://doi.org/10.1109/LRA.2016.2521384.

[69] Jin L, Li S, La HM, Luo X. Manipulability optimization of redundant manipulators using dynamic neural networks. IEEE Trans Ind Electron 2017;64(6):4710–20. https://doi.org/10.1109/TIE.2017.2674624.

[70] Li Y, Hannaford B. Soft-obstacle avoidance for redundant manipulators with recurrent neural network. In: 2018 IEEE/RSJ international conference on intelligent robots and systems (IROS). Madrid: IEEE; 2018. p. 3022–7. https://doi.org/10.1109/IROS.2018.8594346.

[71] Li Z, Liao B, Xu F, Guo D. A new repetitive motion planning scheme with noise suppression capability for redundant robot manipulators. IEEE Trans Syst Man Cybern Syst 2020;50(12):5244–54. https://doi.org/10.1109/TSMC.2018.2870523.

[72] Li W. Predefined-time convergent neural solution to cyclical motion planning of redundant robots under physical constraints. IEEE Trans Ind Electron 2020;67(12):10732–43. https://doi.org/10.1109/TIE.2019.2960754.

[73] Xu Z, Li S, Zhou X, Cheng T, Guan Y. Dynamic neural networks for motion-force control of redundant manipulators: an optimization perspective. IEEE Trans Ind Electron 2021;68(2):1525–36. https://doi.org/10.1109/TIE.2020.2970635.

[74] Iyengar K, Stoyanov D. Deep reinforcement learning for concentric tube robot control with a goal-based curriculum. In: 2021 IEEE international conference on robotics and automation (ICRA). Xi'an, China: IEEE; 2021. p. 1459–65. https://doi.org/10.1109/ICRA48506.2021.9561620.

[75] Joshi SS, Hutchinson S, Tsiotras P. TIE: time-informed exploration for robot motion planning. IEEE Robot Autom Lett 2021;6(2):3585–91. https://doi.org/10.1109/LRA.2021.3064255.

[76] Guo D, Zhang Y. Acceleration-level inequality-based MAN scheme for obstacle avoidance of redundant robot manipulators. IEEE Trans Ind Electron 2014;61(12):6903–14. https://doi.org/10.1109/TIE.2014.2331036.

[77] Menon MS, Ravi VC, Ghosal A. Trajectory planning and obstacle avoidance for hyper-redundant serial robots. J Mech Robot 2017;9(4), 041010. https://doi.org/10.1115/1.4036571.

[78] Zhang Z, Lin Y, Li S, Li Y, Yu Z, Luo Y. Tricriteria optimization-coordination motion of dual-redundant-robot manipulators for complex path planning. IEEE Trans Control Syst Technol 2018;26(4):1345–57. https://doi.org/10.1109/TCST.2017.2709276.

[79] Li Z, Li C, Li S, Cao X. A fault-tolerant method for motion planning of industrial redundant manipulator. IEEE Trans Ind Inform 2020;16(12):7469–78. https://doi.org/10.1109/TII.2019.2957186.

[80] Dai C, Lefebvre S, Yu K-M, Geraedts JMP, Wang CCL. Planning jerk-optimized trajectory with discrete time constraints for redundant robots. IEEE Trans Autom Sci Eng 2020;17(4):1711–24. https://doi.org/10.1109/TASE.2020.2974771.

[81] Gough E, Conn AT, Rossiter J. Planning for a tight squeeze: navigation of morphing soft robots in congested environments. IEEE Robot Autom Lett 2021;6(3):4752–7. https://doi.org/10.1109/LRA.2021.3067594.

# 17

# Artificial intelligence to understand robotic surgery scenes

*Lalithkumar Seenivasan*[a,b,c] *and Hongliang Ren*[b,c]

[a]Department of Biomedical Engineering, National University of Singapore, Singapore, Singapore [b]Department of Electronic Engineering, The Chinese University of Hong Kong, Sha Tin, Hong Kong [c]Shun Hing Institute of Advanced Engineering (SHIAE), The Chinese University of Hong Kong, Sha Tin, Hong Kong

## Introduction

Robot-Assisted Minimally Invasive Surgery (RAMIS) systems typically consist of a vision arm (endoscopic camera) to provide a high-definition view of the surgical site and surgical instruments that can be controlled by the surgeons through a master–slave system [1]. With close to 1.25 million operations performed worldwide using da Vinci surgical systems alone in 2020 [2], RAMIS is becoming more prominent, with its market expected to reach $15.43 billion by 2029 [3]. While the RAMIS systems currently serve merely as an extension of the surgeon's eyes and arms, allowing them to intuitively control the movements of surgical instruments in confined spaces, RAMIS lacks the intelligence to provide real-time assistance to surgeons [3]. Empowered by artificial intelligence (AI), intelligent RAMIS systems could assist surgeons in real-time in different phases of the surgery (preoperative, intra-operative, and postoperative), improving the quality of surgical procedures and surgical outcomes.

The computer vision domain has witnessed giant leaps of advancements in recent decades, resulting from the enormous progress in Machine Learning (ML) and Deep Learning (DL), subsets of AI. Applications that were once considered difficult to implement due to the need for extensive hand-tailored data processing codes are now being implemented by AI models trained on massive datasets. While the computer vision domain can exploit the advances in AI due to the availability of massive labeled datasets, the use of AI for advancements in the medical domain has been severely limited due to a lack of labeled datasets. A strong emphasis on personal data protection in the medical domain has limited the availability of datasets. Furthermore, surgical procedures were predominantly performed in a person-to-person (surgeon-patient) setting. Generating surgical datasets in such settings often requires additional manpower and hardware setup, which could potentially disrupt the surgical workflow and surgeon's movements. However, with minimally invasive surgeries (MIS) becoming the general mode of surgical intervention in modern times, collecting surgical data has become much easier. Initially introduced to reduce postsurgery trauma, infection rate, and patient recovery time, MIS faced new constraints such as lack of direct vision, lack of tactile feedback, and limited instrument dexterity [1,3]. These constraints were addressed with the introduction of RAMIS systems that employed endoscopic cameras to provide surgeons with a high-definition view of the surgical site and robotic arms for them to control the end-effectors of the surgical instruments [4]. With the inclusion of vision and robotics systems, in addition to addressing the constraints of MIS, the RAMIS systems enable easy acquisition of visual (video), kinematic (joint pose and angle of instruments), and event data [2], without disrupting the surgical workflow and surgeon's movements.

The influx of novel surgical datasets can lead to the development of intelligent AI models for the surgical domain. These novel surgical datasets include preoperative data (medical imaging data, clinical data, and patient history), intra-operative data (visual data, kinematic data, operation time), and postoperative data (surgical outcome, patient-reported outcome measures, morbidity, and mortality) [2]. AI models could be developed and employed (a) to train surgeons, (b) to evaluate surgeon's skills, (c) for preoperative planning, (d) to provide real-time assistance to surgeons during surgery by automating certain tasks, (e) to monitor the progress and predict surgical outcomes, and

*Handbook of Robotic Surgery*
https://doi.org/10.1016/B978-0-443-13271-1.00005-4

(f) for postoperative progress and complications management [2]. However, to perform such highly complex tasks, the AI models should first mimic surgeon-level understanding of the surgical scene, where it can recognize key elements in surgical scenes and infer the different stages and progress of the surgery [3]. Inspired by the works on semantic segmentation, object detection, object-to-object interaction detection, and workflow analysis in the computer vision domain, numerous works have been recently proposed in the surgical domain to enable AI models to perform surgical scene understanding. These include surgical scene segmentation, surgical tool detection (object detection), surgical tool action detection, and surgical phase detection. To assist in surgeon training and surgeon evaluation, works on surgical visual question answering and surgical skill evaluation have also been very recently proposed. Focusing mainly on surgical scene understanding and its applications, in this chapter, the works on AI in robotic surgery are categorized into three main branches: (a) Semantic understanding (object detection and semantic segmentation), (b) surgical process and workflow analysis (action recognition, phase recognition, and real-time monitoring of surgical progress), and (c) training and evaluation (surgical skill evaluation and visual question answering) as shown in Fig. 1. Works on semantic segmentation, action recognition, and visual question answering are discussed in the following sections.

## Semantic segmentation of surgical scenes

Overlaying pre- and intra-medical imaging data over the endoscopic view could significantly help improve the surgeon's perception and overcome the limited field-of-view constraints of RAMIS [5]. Enabling surgeons to visualize underlying critical anatomical structures, blood vessels, and tumors can greatly help them in surgical planning, limit loss of blood, avoid risks, and achieve optimal access to regions of interest. However, intelligent overlay of medical imaging data is vital to avoid obstructing the surgeon's view of the surgical scene with less valuable information [5]. To achieve this, a surgeon-level understanding of the current view of the surgical scene in the endoscope is required to recognize various anatomical structures and surgical instruments. In the computer vision domain, numerous AI models have been introduced to perform pixel-wise semantic segmentation that segments different objects or parts of an object in an image. In pixel-wise semantic segmentation, each pixel in an image is classified from a list of pre-defined object classes. With the recent infusion of novel labeled surgical datasets, pixel-wise segmentation has been introduced into the surgical scene for tissue and instrument segmentation in robotic surgery. Numerous public datasets [5,6] have also been introduced as a part of the EndoVis challenge to promote works on surgical scene segmentation.

Inspired by AI models employed in the computer vision domain, the semantic segmentation of surgical scenes has been predominantly addressed using convolutional neural networks (CNNs) [7–9]. These model architectures usually consist of a feature extractor (encoder) module and a decoder module and rely heavily on local reasoning (neighboring pixels) to classify each pixel. While these models have reported high performance in terms of pixel-wise classification, they lacked spatial consistency, where, despite instruments having parts of fixed shapes, pixels within the instrument region were misclassified into other instrument/tissue classes. To address this and improve spatial consistency, instance-based semantic segmentation has been introduced [10]. Exploiting the temporal optical flow features to

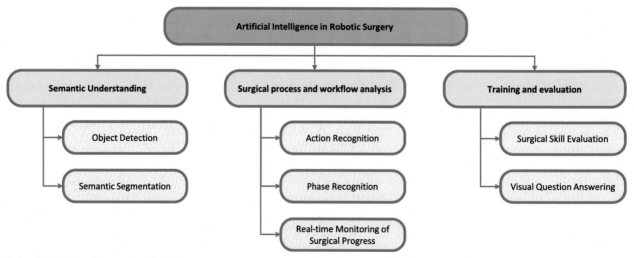

FIG. 1   Artificial intelligence in robotic surgery.

improve semantic segmentation has also been proposed [11,12]. As each instrument moves in the surgical scene, related coherent optical flow features can be observed across consecutive frames (temporal). These temporal optical flow features could be used to group pixels belonging to a moving instrument and distinguish them against the stationary background and other instruments. Incorporated an attention mechanism within its architecture, MF-TAPNet [11] utilized these temporal optical features to perform semantic segmentation with better spatial consistency. Similarly, the ISI-Net [12] exploits the optical flow in temporal frames using a temporal consistency strategy to perform instance-based semantic segmentation.

Relying on the shared-weight convolutional kernels that slide across the image for feature extraction, CNN models reason each pixel based on its neighboring regions and lack reasoning based on global context (distant pixels/overall surgical scene context). For instance, during surgery, a part of the surgical instrument could be covered by blood. Reasoning purely based on local pixels, the blood-covered regions could be wrongly classified as tissue regions. However, if the global context (the remaining part of the instrument) were taken into account, those regions could be classified as the instrument region. Humans often reason based on local relations with the global context. Incorporating global context awareness into the model could greatly improve semantic segmentation [13].

With the global context, the model could award a higher likelihood to specific instruments depending on the surgical scene. To incorporate global context and focus on key regions in the semantic segmentation of the surgical scene, a refined attention-segmentation network called the RASNet [15] was introduced. It incorporated a novel module integrating low-level features with high-level global context features. Alternatively, a graph-based global reasoning module has also been incorporated within a CNN network to perform optimal semantic segmentation [14]. Here, the CNN encoder module performs local neighborhood reasoning using its convolutional kernels, the decoder aggregates semantic segmentation at multiple scales, and the lightweight graph module called GloRe [16] performs global reasoning in latent space to incorporate global context. The performance of CNNs models (ResNet18 [7], LinkNet34 [9]), CNN models enhanced with optimal flow features (MF-TAPNet11 [11]), and CNN model enhanced with global context (Seenivasan et al. [14]) in semantic segmentation of surgical scenes are shown in Fig. 2.

## Surgical action recognition

Mimicking surgeon-level ability in differentiating various anatomical structures and surgical instruments in a surgical scene can be addressed using semantic segmentation. While it helps decide where to overlay pre- and intraoperative medical data, deciding when to overlay (to not hinder the surgeon's view during the crucial phase of the surgery) requires a deeper understanding of a surgical scene, such as surgical actions in it, surgical phase, and current progress of the surgery. With the ability to recognize surgical actions and surgical phases through deep reasoning AI models, novel AI-based methods could also be developed to monitor the progress of the surgery in real-time, predict surgical outcomes and evaluate surgeons' skills. Furthermore, RAMIS often lacks haptic feedback, making it difficult for the surgeon to feel instrument-tissue interactions and gauge the pressure applied to the tissues. Empowered by

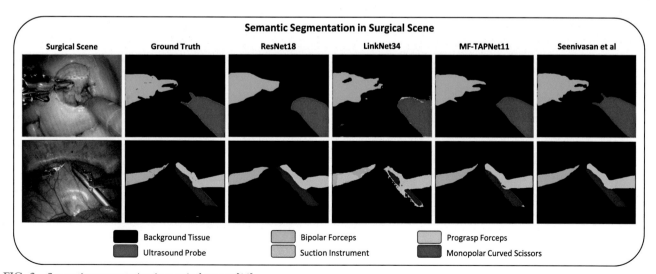

FIG. 2    Semantic segmentation in surgical scenes [14].

deep reasoning AI models, RAMIS can detect instrument-tissue interactions using visual input and could be programmed to simulate haptic feedback to the surgeon's hand, allowing vision-based haptic feedback.

In the AI domain, detecting implicit relations between two objects in an image is termed deep reasoning. However, there are some limitations to recognizing instrument-tissue interaction in a surgical scene. Widely explored conventional AI tasks (classification, regression, and semantic segmentation) are categorized under Euclidean space and are solved using Euclidean space models such as linear layers, CNNs, and transformers [17]. The term Euclidean space refers to a problem statement and model where the model's input and output size remain fixed for all scenarios. Here, the fixed input size corresponds to the fixed size of channels, width, and height of the image and the fixed output size corresponds to the fixed size of classes. However, in an instrument-tissue interaction recognition task, the number of instruments, tissues, and interactions among them varies in each surgical scene. While in most cases, the surgical scenes may contain only two instruments, the scene may sometimes contain up to five instruments. Given a surgical scene, the number ($n$) of objects (instruments and tissues) results in $n^2$ interactions, making the input and/or output sizes vary for each scene. The varying number of inputs and output variables make the tool-tissue interaction recognition (detection) task, a non-Euclidean space problem.

Graph neural networks (GNNs) have been widely explored to address non-Euclidean space problems in the computer vision domain [18,19]. As an alternative to CNNs and Transformer models, GNNs can handle input and output of varied sizes. A GNN typically consists of nodes and edges. The nodes and edges can hold varied features and represent varied objects/relations among objects depending on how the graph is constructed. For instance, in an object-to-object relational reasoning task, each object could be represented as a node, and edges between the nodes could hold relational features or represent a presence of relation (interaction). Alternatively, nodes can represent objects or relations, and edges can represent the presence of a link between object nodes through a relationship node. Based on the graph structure, it can be categorized as a fully connected or partial graph. As shown in Fig. 3, in a fully connected graph, each node is connected to all the other nodes in the graph, whereas, in a partial (sparse) graph, some nodes are not connected to all other nodes. Furthermore, graphs can also be categorized as directed and undirected depending on the direction of the edges that connect the nodes (Fig. 3). In a directed graph, an edge can be unidirectional or bidirectional, which dictates the relationship between the nodes and the information flow (direction) during network propagation. In an undirected graph, all edges are bidirectional and feature information flows in both directions during the network propagation.

Motivated by the works on human-object interaction detection using graph parsing neural network (GPNN) [18] in the computer vision domain, Islam et al. [20] introduced bounding box and scene graph annotation for the EndoVis18 Challenge dataset [5] and employed GPNN to perform instrument-tissue interaction detection in surgical scenes (Fig. 4). Given a scene, the GPNN model constructs a sparse graph, with its nodes representing the instruments and tissue, and edges representing the interaction between the nodes [18]. Here, each node is embedded with visual features extracted using the pretrained feature extraction network from each object region (denoted by a bounding box), and each edge is embedded with features extracted from regions combining both objects (two nodes connected by the edge). Islam et al. [20] further improved the GPNN performance by incorporating GraphSAGE [21] for edge attention and label smoothing [22] for optimal feature extraction.

Alternate to GPNN, which relies solely on visual features of instruments and tissues, the visual-semantic graph attention network (VSGAT) [19] was proposed in the computer vision domain for human-object interaction detection. In addition to the visual features, the models utilize word embedding and semantic relational features to infer instrument-tissue interactions. In the VSGAT model, two small graphs are initially constructed, where the nodes of the first graph are embedded with visual features of the objects (instruments and tissue), and the nodes of the second

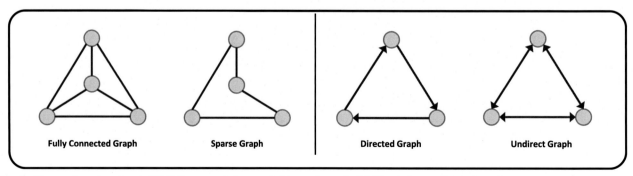

FIG. 3   Types of graph network: (A) Fully connected vs partial (sparse) graph. (B) Directed vs undirected graph.

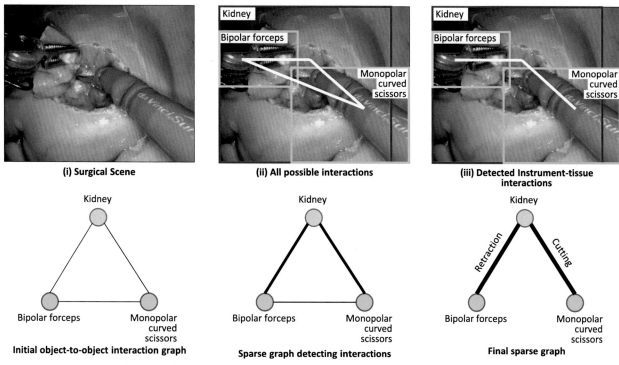

**FIG. 4** Graph parsing neural network for instrument-tissue interaction detection in surgical scenes.

graph are embedded with word embedding of the object names. Upon forward propagation, the nodes of these two graphs are combined to form a final graph, in which the edges are embedded with semantic relation features (features describing the relative spatial location of two objects). The final graph is then propagated and each of its edges is classified to predict the type of interactions. Inspired by this, Seenivasan et al. [14] domain adopted VSGAT for instrument-tissue interaction detection in robotic surgery and introduced global reasoning into the VSGAT to further improve its performance (Fig. 5). Human-level scene analysis requires both local (object-to-object) and global reasoning (whole surgical scene context). Both GPNN and VSGAT models rely on CNN feature extractors to extract visual features for node embedding. As the convolutional kernels in the convolutional layers only reason based on immediate

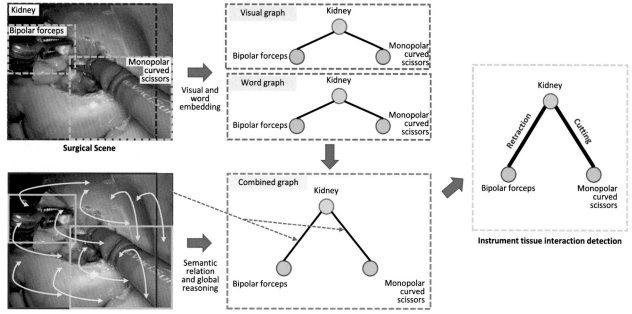

**FIG. 5** Globally reasoned visual-semantic graph attention network for instrument-tissue interaction detection in surgical scenes.

neighboring regions (local), the extracted features only contain features within the object region and lack features from the whole surgical scene. As the graph propagates, the edges allow feature sharing between two nodes, resulting in reasoning based on two object regions. While these object-to-object interaction detection models rely solely on object-to-object local reasoning, incorporating global scene reasoning can significantly improve model performance [23]. For example, if there is no direct contact between an instrument and tissue, the interaction between the instrument and tissue could be inferred as "idle" based on object-to-object local reasoning. Alternatively, based on local reasoning with a global context (global reasoning), it could be inferred as "tool manipulation" if the instrument was used to manipulate another instrument in contact with the tissue. To instill global reasoning into instrument-tissue interaction detection models, Seenivasan et al. [14] incorporated GloRe [16], a graph-based global reasoning network that performs global (nonlocal) relational reasoning in latent space with a minimal computational load, with VSGAT. The visual features extracted from object regions were used for visual node embedding, and together with semantic relational features, globally reasoned (from the whole scene) latent space features are also embedded into the final graph edges. As the graph propagates, the model helps predict interactions based on globally reasoned object-to-object (instrument-tissue) interaction. The performance of globally reasoned VSGAT on instrument-tissue interaction detection in the surgical scene is shown in Fig. 6.

Most graph models, such as GPNN and VSGAT, employed for object-object interaction detection are not end-to-end models. To embed the nodes, they require bounding box annotations and a feature extractor network to extract features from the bounded regions. In a real-time deployment scenario, an object detection network must be employed to detect objects, and a bounding box must be used for the feature extractor to extract features. Any error in the object and bounding box detection will have a ripple effect, affecting the overall object-to-object interaction detection performance. Pang et al. [24] proposed a detection-free gradient-based localized feature extraction approach that can be incorporated with the graph models to make it an end-to-end model. Assuming a scene to have a fixed number of objects and relations, the task of object-to-object interaction detection task was simplified to a Euclidean space problem and addressed using Euclidean models (CNNs and Transformers).

Addressing instrument action detection in non-Euclidean space remains a significant challenge. To further motivate progress on novel instrument-tissue interaction detection, EndoVis Challenge simplified these tasks into Euclidean space problem and introduced two novel public datasets, Cholec Triplet [25] and Petraw [26] dataset. While these have accelerated novel works on action (interaction) detection, another key barrier still limits the use of action detection models in the real world. While the domain shift between the computer vision and surgical domains is prominent, significant domain shifts also exist between different surgeries with the introduction of novel instruments, tissues, and actions. Seenivasan et al. [27] employed an incremental domain generalized instrument-tissue interaction detection model from nephrectomy surgery to transoral robotic surgery. To include novel instruments and tissue in the new domain, it employed incremental learning techniques and to adapt to domain shifts, it employed knowledge distillation learning. Furthermore, it also introduced the Laplace of Gaussian kernel-based curriculum by smoothing to

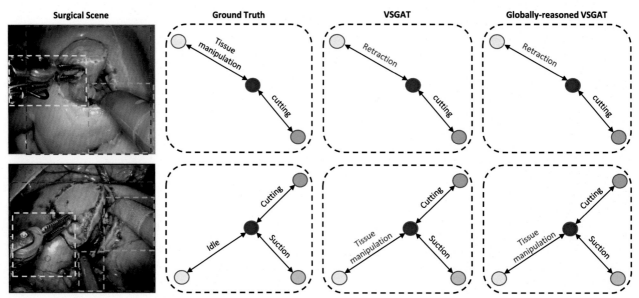

FIG. 6    Instrument-tissue interaction detection using graph models [14].

improve model performance through curriculum learning. Seenivasan et al. [28] also studied class incremental contrastive learning-based domain adaption technique to adapt a network initially trained on nephrectomy surgery to transoral robotic surgery.

## Surgical visual question answering

In recent years, numerous methods have been explored in surgical scene analysis to develop applications to assist in surgeons' training and evaluation in RAMIS. In line with this, visual question and answering (VQA) is also an application where advanced surgical scene analysis could assist in surgeons' training and address the common man's doubts by answering their questions on surgery. Being a highly skilled and specialized domain, the medical domain has left medical students seeking scarce medical professionals' help in answering their doubts and questions about medical diagnosis and surgical procedures [29]. Evolving to be a very highly specialized field, RAMIS is severely short of experts, and existing scarce experts are often overwhelmed with academic and clinical workloads. This makes it hard for medical students to instantly find domain experts to clear their immediate doubts about surgery and surgical procedures. While numerous computer-assisted systems [30,31] and surgical simulators [32,33] have been introduced to help train surgical residents in their surgical skills, they still heavily rely on experts in answering their doubts. A computer-assisted system powered by an AI model trained on the collective knowledge of numerous experts to answer questionnaires on a given surgical scene in RAMIS could significantly help medical students and reduce the workload of experts.

The computer vision domain is witnessing rapid progress in developing multimodality VQA models. As the medical domain includes domain-specific medical terms and visual features that are not common in the computer vision domain, further training is required to domain adapt the computer vision VQA models to the medical domain. Unlike other single-modality scene understanding tasks (semantic segmentation, activity recognition, and phase recognition) that require the model to process a single modality (visual input), the VQA task is a multimodality task that warrants the model to robustly process both visual and natural language inputs, and also establish a relation between the natural language and visual features. While the surgical domain is already challenged by the limited availability of single-modality datasets, the need for coherent multimodality datasets (vision and language) has further complicated the development of Surgical-VQA models. Hindered by the severe shortage of domain-specific multimodality labeled datasets, very few works on VQA tasks exist in the medical domain.

In the medical domain, Sharma et al. [34] first introduced MedFuseNet to answer medical diagnosis questions based on medical images in the MedVQA dataset. They employed attention-based multimodality models to address both categorization and sentence generation VQA tasks. To develop a Surgical-VQA model, Seenivasan et al. [29] utilized the EnoVis18 [5] and Cholec80 [35] to create a novel Surgical-VQA dataset. EndoVis-18-VQA and Cholec80-VQA datasets included question-and-answer pair annotations for classification and sentence-based answering tasks. They then introduced and employed the multimodality VisualBert ResMLP encoder module, where the encoder processes and employs attention to visual and text features. For the classification-based VQA task, they combined the VisualBert ResMLP encoder with a series of linear layers. For the sentence-based VQA task, they combined the VisualBert ResMLP encoder with a transformer decoder model to generate sentence-based answers regressively (word by word). The overview of sentence-based visual question answering proposed by Seenivasan et al. [29] is shown in Fig. 7.

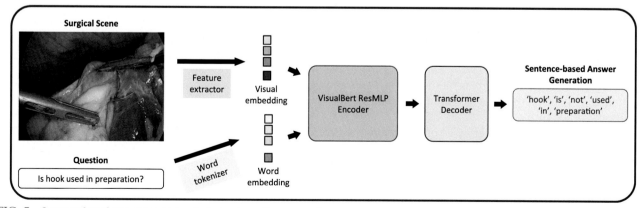

FIG. 7    Sentence-based answering in surgical visual question answering.

The recent advancements in the large language models have showcased the GPT-based model's robust language processing and generation ability. To harness the superior language processing of GPT-based models and improve the performance in Surgical-VQA tasks, Seenivasan et al. [36] also introduced SurgicalGPT. SurgicalGPT is an end-to-end language-vision model that incorporates a vision encoder module and GPT (transformer decoder) module. First, the question sentence is tokenized into word tokens using a standard GPT tokenizer. Second, the vision encoder module converts the surgical scene into vision tokens. Third, the word and vision tokens are further embedded with token type and token pose embedding. Finally, the word and vision tokens are strategically sequenced (word tokens followed by vision tokens) and passed through the GPT model. The final features and the hidden layers features are then used to clarify the answer for the VQA tasks. Apart from large language models (ChatGPT, BARD), large multimodality models are also on the rise and are expected to further improve the robustness of Surgical-VQA tasks and other vision-language tasks such as surgical captioning and surgical report generation.

Robot-assisted minimally invasive surgeries have significantly improved the surgeon's field-of-view and their dexterity in controlling the surgical instruments in confined spaces while reaping the benefits of minimally invasive surgeries. The use of vision systems (endoscopic cameras) and robotic manipulators in RAMIS has also opened a new paradigm where intelligent systems can be developed and employed to assist surgeons in real-time, monitor surgical progress, predict the surgical outcome, automate simple surgical tasks, train surgeons, and evaluate surgical skills. With the recent influx of novel labeled surgical datasets collected from RAMIS systems, numerous AI models have been proposed and employed to perform semantic scene understanding and surgical process and workflow analysis. Fundamentally, these tasks are aimed to mimic the surgeon level understanding of the surgical scenes and procedures and pave the way for developing complex models for tasks, such as surgical skill evaluation and task automation. With the enormous progress currently being made in the surgical AI domain, AI-based surgical task automation, simulators for training, and real-time surgical assistant systems seem like a possible reality soon.

## Future scenario

The rise of large language models and their robust performance in language generation has sparked the development of large vision models, foundation models, and large multimodality models. Trained on enormous datasets and with the option to fine-tune to specific datasets using few examples, these large models are expected to achieve better generalization and performance on most surgical datasets in the near future. These models can be adopted and fine-tuned to perform surgeon-like scene understanding in tasks, such as semantic segmentation, instruction detection, surgical action recognition, surgical phase recognition, report generation, visual question answering, and surgical scene captioning. As the performance of AI models in the surgical scene understanding tasks continues to improve and achieving robust, generalized, and real-time performances seems like a possibility in the foreseeable future, works on employing these scene understanding modules to improve RAMIS are expected to sprout soon. These may include, (a) integrating semantic segmentation and 2D image to 3D point could registration for intelligence overlay of medical images; (b) employing semantic segmentation (blood detection), instrument-tissue interaction detection, and phase recognition for real-time surgery monitoring for surgeon skill evaluation and surgical outcome prediction, (c) employing instrument-tissue interaction detection for vision-based haptic feedback and use of multimodality vision-language models for interactive learning assistant and real-time interactive surgical assistant.

## Key points

- Robot-assisted minimally invasive surgical systems allow easy acquisition of datasets, allowing the development of intelligent systems that can assist surgeons in training, preoperative, intra-operative, and postoperative stages of surgical procedures.
- To allow the intelligent overlay of preoperative medical data, monitor the progress of the surgery, and evaluate surgeons' skills, surgical scene understanding (scene segmentation, surgical tool action detection, and surgical phase detection) is widely being explored in recent times.
- To assist in training surgeons and reduce the workload on domain experts, surgical visual question answering has also been recently proposed.

# Acknowledgment

The work was supported by the Hong Kong Research Grants Council (RGC) Collaborative Research Fund (CRF C4026-21GF), the General Research Fund (GRF #14211420 and GRF #14216022), and the Shenzhen-Hong Kong-Macau Technology Research Programme (Type C) Grant 202108233000303.

# References

[1] Hussain SM, Brunetti A, Lucarelli G, Memeo R, Bevilacqua V, Buongiorno D. Deep learning based image processing for robot assisted surgery: a systematic literature survey. IEEE Access 2022.

[2] Moglia A, Georgiou K, Georgiou E, Satava RM, Cuschieri A. A systematic review on artificial intelligence in robot-assisted surgery. Int J Surg 2021;95, 106151.

[3] Rivas-Blanco I, Pérez-Del-Pulgar CJ, García-Morales I, Muñoz VF. A review on deep learning in minimally invasive surgery. IEEE Access 2021;9:48658–78.

[4] Okamura AM. Haptics in robot-assisted minimally invasive surgery. Encyclopedia Med Robot 2018;317–39.

[5] Allan M, Kondo S, Bodenstedt S, Leger S, Kadkhodamohammadi R, Luengo I, Fuentes F, Flouty E, Mohammed A, Pedersen M, et al. 2018 Robotic scene segmentation challenge. arXiv Preprint; 2020. arXiv:2001.11190.

[6] Luengo I, Grammatikopoulou M, Mohammadi R, Walsh C, Nwoye CI, Alapatt D, Padoy N, Ni Z-L, Fan C-C, Bian G-B, et al. 2020 Cataracts semantic segmentation challenge. arXiv preprint; 2021. arXiv:2110.10965.

[7] He K, Zhang X, Ren S, Sun J. Deep residual learning for image recognition. In: Proceedings of the IEEE conference on computer vision and pattern recognition; 2016. p. 770–8.

[8] García-Peraza-Herrera LC, Li W, Gruijthuijsen C, Devreker A, Attilakos G, Deprest J, Vander Poorten E, Stoyanov D, Vercauteren T, Ourselin S. Real-time segmentation of nonrigid surgical tools based on deep learning and tracking. In: Computer-Assisted and Robotic Endoscopy: Third International Workshop, CARE 2016, Held in Conjunction with MICCAI 2016, Athens, Greece, October 17, 2016, Revised Selected Papers 3. Springer; 2017. p. 84–95.

[9] Chaurasia A, Culurciello E. Linknet: Exploiting encoder representations for efficient semantic segmentation. In: 2017 IEEE visual communications and image processing (VCIP), IEEE; 2017. p. 1–4.

[10] Kletz S, Schoeffmann K, Benois-Pineau J, Husslein H. Identifying surgical instruments in laparoscopy using deep learning instance segmentation. In: 2019 international conference on content-based multimedia indexing (CBMI). IEEE; 2019. p. 1–6.

[11] Jin Y, Cheng K, Dou Q, Heng P-A. Incorporating temporal prior from motion flow for instrument segmentation in minimally invasive surgery video. In: Medical Image Computing and Computer Assisted Intervention–MICCAI 2019: 22nd International Conference, Shenzhen, China, October 13–17, 2019, Proceedings, Part V 22. Springer; 2019. p. 440–8.

[12] González C, Bravo-Sánchez L, Arbelaez P. Isinet: an instance-based approach for surgical instrument segmentation. In: Medical Image Computing and Computer Assisted Intervention– MICCAI 2020: 23rd International Conference, Lima, Peru, October 4–8, 2020, Proceedings, Part III 23. Springer; 2020. p. 595–605.

[13] Zhao H, Shi J, Qi X, Wang X, Jia J. Pyramid scene parsing network. In: Proceedings of the IEEE conference on computer vision and pattern recognition; 2017. p. 2881–90.

[14] Seenivasan L, Mitheran S, Islam M, Ren H. Global-reasoned multi-task learning model for surgical scene understanding. IEEE Robot Autom Lett 2022;7(2):3858–65.

[15] Ni Z-L, Bian G-B, Xie X-L, Hou Z-G, Zhou X-H, Zhou Y-J. Rasnet: Segmentation for tracking surgical instruments in surgical videos using refined attention segmentation network. In: 2019 41st Annual international conference of the IEEE engineering in medicine and biology society (EMBC). IEEE; 2019. p. 5735–8.

[16] Chen Y, Rohrbach M, Yan Z, Shuicheng Y, Feng J, Kalantidis Y. Graph-based global reasoning networks. In: Proceedings of the IEEE/CVF conference on computer vision and pattern recognition; 2019. p. 433–42.

[17] Zhou J, Cui G, Hu S, Zhang Z, Yang C, Liu Z, Wang L, Li C, Sun M. Graph neural networks: a review of methods and applications. AI Open 2020;1:57–81.

[18] Qi S, Wang W, Jia B, Shen J, Zhu S-C. Learning human-object interactions by graph parsing neural networks. In: Proceedings of the European conference on computer vision (ECCV); 2018. p. 401–17.

[19] Liang Z, Liu J, Guan Y, Rojas J. Visual-semantic graph attention networks for human-object interaction detection. In: 2021 IEEE international conference on robotics and biomimetics (ROBIO). IEEE; 2021. p. 1441–7.

[20] Islam M, Seenivasan L, Ming LC, Ren H. Learning and reasoning with the graph structure representation in robotic surgery. In: Medical Image Computing and Computer Assisted Intervention–MICCAI 2020: 23rd International Conference, Lima, Peru, October 4–8, 2020, Proceedings, Part III 23. Springer; 2020. p. 627–36.

[21] Hamilton W, Ying Z, Leskovec J. Inductive representation learning on large graphs. Adv Neural Inf Process Syst 2017;30.

[22] Müller R, Kornblith S, Hinton GE. When does label smoothing help? Adv Neural Inf Process Syst 2019;32.

[23] Wang X, Girshick R, Gupta A, He K. Non-local neural networks. In: Proceedings of the IEEE conference on computer vision and pattern recognition; 2018. p. 7794–803.

[24] Pang W, Islam M, Mitheran S, Seenivasan L, Xu M, Ren H. Rethinking feature extraction: gradient-based localized feature extraction for end-to-end surgical downstream tasks. IEEE Robot Autom Lett 2022;7(4):12623–30.

[25] Nwoye CI, Alapatt D, Yu T, Vardazaryan A, Xia F, Zhao Z, Xia T, Jia F, Yang Y, Wang H, et al. Cholectriplet2021: A benchmark challenge for surgical action triplet recognition. arXiv preprint; 2022. arXiv:2204.04746.

[26] Huaulmé A, Harada K, Nguyen Q-M, Park B, Hong S, Choi M-K, Peven M, Li Y, Long Y, Dou Q, et al. Peg transfer workflow recognition challenge report: Does multi-modal data improve recognition? arXiv preprint; 2022. arXiv:2202.05821.

[27] Seenivasan L, Islam M, Ng C-F, Lim CM, Ren H. Biomimetic incremental domain generalization with a graph network for surgical scene understanding. Biomimetics 2022;7(2):68.

[28] Seenivasan L, Islam M, Xu M, Lim CM, Ren H. Task-aware asynchronous multi-task model with class incremental contrastive learning for surgical scene understanding. Int J Comput Assist Radiol Surg 2023;1–8.

[29] Seenivasan L, Islam M, Krishna AK, Ren H. Surgical-vqa: Visual question answering in surgical scenes using transformer. In: Medical image computing and computer assisted intervention–MICCAI 2022: 25th international conference, Singapore, September 18–22, 2022, Proceedings, Part VII. Springer; 2022. p. 33–43.

[30] Adams L, Krybus W, Meyer-Ebrecht D, Rueger R, Gilsbach JM, Moesges R, Schloendorff G. Computer-assisted surgery. IEEE Comput Graph Appl 1990;10(3):43–51.

[31] Rogers DA, Yeh KA, Howdieshell TR. Computer-assisted learning versus a lecture and feedback seminar for teaching a basic surgical technical skill. Am J Surg 1998;175(6):508–10.

[32] Kneebone R. Simulation in surgical training: educational issues and practical implications. Med Educ 2003;37(3):267–77.

[33] Sarker S, Patel B. Simulation and surgical training. Int J Clin Pract 2007;61(12):2120–5.

[34] Sharma D, Purushotham S, Reddy CK. Medfusenet: an attention-based multimodal deep learning model for visual question answering in the medical domain. Sci Rep 2021;11(1):19826.

[35] Twinanda AP, Shehata S, Mutter D, Marescaux J, De Mathelin M, Padoy N. Endonet: a deep architecture for recognition tasks on laparoscopic videos. IEEE Trans Med Imaging 2016;36(1):86–97.

[36] Seenivasan L, Islam M, Kannan G, Ren H. Surgicalgpt: End-to-end language-vision gpt for visual question answering in surgery. arXiv preprint; 2023. arXiv:2304.09974.

# 18

# Robot assisted prostate biopsy

*Mohammud Shakeel Inder and Taimur Shah*

Imperial College Healthcare NHS Trust, London, United Kingdom

## Introduction

Prostate cancer is the most common malignancy in men and represents a significant health burden and cause of male mortality. Patient at high risk of prostate cancer is initially investigated with a prostate-specific antigen (PSA) blood test and a magnetic resonance imaging (MRI) scan of their prostate gland prior to any invasive intervention.

Conventionally a transrectal (TR) or a transperineal (TP) approach to prostate biopsies has been used for prostate cancer detection. In recent times, the TP approach has superseded the TR approach as we have good evidence showing that the TP approach is safer [1].

Diagnosis of prostate cancer is based on histopathological features of the prostate biopsy specimen. It is therefore of utmost importance to be able to biopsy the target lesions or otherwise high-risk regions of the prostate gland in a precise and safe manner.

The quest to improve on delivery of an accurate diagnosis has led to multiple changes in practice over time:

- Changes in antibiotics prophylaxis to decrease the rate of sepsis
- General anesthesia to local anesthesia to decrease morbidity
- Transrectal to transperineal approach to improve diagnosis and decrease infection rate
- Prebiopsy MRI for better patient selection and to identify target lesions within the prostate gland
- Introduction of different software to aid clinicians to better target areas of interest within the prostate gland

Robot-assisted prostate biopsies have been devised, trialed, and implemented over the last decade, in order to achieve a better diagnostic yield, in a precise and safe manner while eliminating human errors and dynamic changes to the prostate gland while performing prostate biopsies.

In this chapter, we will describe the details of robotic-assisted prostate biopsy, its advantages over conventional biopsy techniques, its limitations, and its future trends.

## Robotic technologies in urology

Robotic technology was introduced in the late 1990s, and it has brought forward significant advantages leading to its use as the gold standard in multiple urological procedures. The implementation of minimally invasive surgery has led to reduced morbidity, early recovery, and decrease hospital stay. From a urological perspective, robots are used for both the treatment of urological malignancies and the management of benign pathologies.

Implementation of robot-assisted prostate biopsies aims to improve prostate cancer detection thus positively impacting the diagnosis and management of patients with prostate cancer. We are indeed at the very start with no randomized controlled trial performed to date to show the efficacy of robot-assisted prostate biopsies over conventional prostate biopsies.

## Robotic-assisted prostate biopsy technique

Multiple robot-assisted prostate biopsy techniques have been described over the years. They differ in terms of the robot hardware, number of robotic arms, number of needles used, software system, and patient positioning. In most of the studies, the procedure is performed under general anesthesia, and the patient lies in the dorsal lithotomy or lateral decubitus position. A transrectal ultrasound (TRUS) probe is inserted into the rectum to visualize the prostate gland, and a three-dimensional image is generated in conjunction with the patient prebiopsy MRI images [2].

Once the prostate gland is visualized, the robotic arm is docked over the patient's pelvis, and the biopsy needles are inserted through the robotic arm. The robotic arms allow for greater precision in needle placement, and more samples can be obtained from different parts of the prostate gland.

Besides, the robotic arm can rotate in all directions (4 degrees of freedom), as described in the literature [3], which further increases the accuracy of needle placement. The robotic arm can also be controlled via a console by the surgeon, who can visualize the prostate gland. With good magnification levels, the surgeon can make real-time adjustments to the needle placement as needed.

The biopsy needles are guided through the prostate gland to obtain tissue samples, which are sent to a laboratory for analysis. The number of samples obtained can vary depending on the size of the prostate gland and suspected cancerous areas. A transperineal approach to prostate biopsy is preferred over a transrectal route [2].

The procedure is done as a day case, and the patient may experience some discomfort or blood in the urine or semen for a few days. Similar advice to having a manually operated TP prostate biopsy is given to patients after the biopsy.

## Evolution and current state of arts for robot-assisted prostate biopsies

Multiple robots have been used to perform prostate biopsies to date, including; MrBot, iSR'obot Mona Lisa Biobot (BioXbot), Soteria Medical BV Robot, Artemis Robot, and PROST Robot amongst others. The way they are operated and function differs as follows:

### MrBot

MrBot is a robotic device made of nonmagnetic and dielectric materials and powered by a pneumatic step motor. The robot uses light sensors and air for actuation. MrBot is a 6 degree-of-freedom robot that is MRI-safe and free of electricity. The device is mounted on the MRI table beside the patient. The nozzle of the needle guide is placed through a 1 cm perineal incision and images are acquired using a 3 Tesla whole-body scanner. Cancer-suspicious regions are defined, and the robot guides the needle to the target with automatic needle depth calculation [4].

### BioXbot

BioXbot is an ultrasound-based and real-time robotic system. It allows MRI-ultrasound fusion that enables the targeting of regions of interest. It utilizes a software-controlled robotic arm, mounted onto the operation table via stabilizer. Using the Mona Lisa intuitive user interface, the operator can create custom biopsy plans with ease and flexibility. UroFusion enables target definition and prostate boundaries while UroBiopsy allows generation of the TRUS-based models for fusion of the MRI and ultrasound images. The robot uses a double dual-cone concept ensuring that the entire prostate can be sampled with two 1 mm perineal skin punctures and defines penetration depth automatically [2,5] (Fig. 1A–D).

### Soteria Medical BV robot

Soteria Medical developed a robotic system consisting of a fully MRI-compatible manipulator made of high-quality plastics. Patients are placed in a 3 Tesla scanner in a prone position and the prostate biopsy is taken transrectally using an MRI-compatible titanium needle. This manipulator is connected to a control unit based in the MRI control room. The software system receives MRI images from the scanner for the planning and positioning of the needle. After registration of the device relative to the patient, images are taken to re-identify the target location prior to target selection. After selecting a target, the robot motion is executed and a quick confirmation scan is taken to confirm that the guide is in the desired position [6].

### Artemis robot

The ARTEMIS device incorporates a semi-robotic mechanical arm with electronic encoders used to scan, digitize, and track the prostate in real-time as well as to precisely navigate biopsy needles. It uses the ProFuse software for

**(A) Biobot Robotic System**

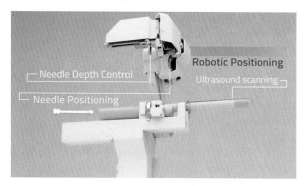

**(B) Patient and Robot Positioning**

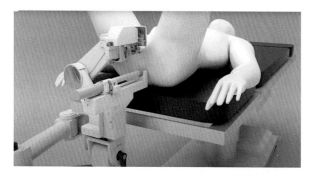

**(C) MRI-Ultrasound Fusion images prior to biopsy**

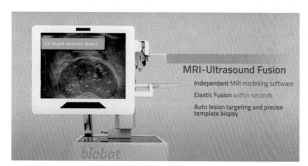

**(D) Needle directions to region of interest**

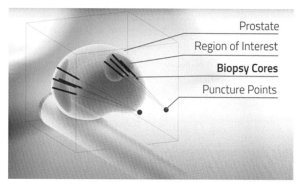

FIG. 1 (A) BioXbot robotic system. (B) Patient and robot positioning. (C) MRI-ultrasound fusion images prior to biopsy. (D) Needle directions to region of interest. *Reproduced with Permission from Biobot Surgical Pte Ltd.*

advanced visualization, computer-assisted detection, and real-time analysis of multiparametric MRI images of the prostate for fusion on the ARTEMIS biopsy system. It uses a TRUS probe and biopsies are taken after image fusion is carried out.

### PROST robot

The PROST robotic system is composed of 2 joint arms that move along parallel planes coupled by an axis passing through the middle that determines needle orientation. The robot has 5 degrees of freedom corresponding to 5 motors. Needle insertion is done manually after the entry point is calculated by the PROST software. The autonomy of the robot is given by the image analysis software, which uses an integrated image fusion algorithm and automatic needle trajectory computation. The robot then directs the needle to the correct depth for each target area.

The robotic systems used to assist in prostate biopsies differ in multiple domains, including lesion targeting (MRI based, MRI-US Fusion based, or US based), ease of use, and autonomy level. They have mainly been used in prospective nonrandomized trials and are FDA approved for clinical use.

The use of robotic-assisted prostate biopsies started in the early 2000s. Among the first trials performed, Ho et al. confirmed that using the BioXbot (under ultrasound guidance) for prostate biopsy had a higher prostate cancer detection rate, a lower complication rate (sepsis and bleeding), and a short procedure time (mean of 18 min). They concluded that the role of the robot-assisted prostate biopsies can be expanded for the management of patients on active surveillance or focal therapy [7].

In 2012, Schouten et al. published their results of using a robotic technique for transrectal MRI-guided prostate biopsies. Prior to 2012, the robotic technique for transperineal seed delivery as well as transperineal access to the prostate gland were described by Stoianovici et al. where an MRI-compatible pneumatic stepper motor was applied to brachytherapy seed placement [8]. This prospective analysis was amongst the first studies done to assess the advantages of robot-assisted prostate biopsies. The sample size was small and showed comparable results to manually operated prostate biopsies in terms of targeting error, accuracy, and procedure time. A comparison of the robotic and manual techniques of prostate biopsy showed that the robotic techniques required extended technical effort but fair better in terms of availability [9]. In the same year, Long et al. presented their initial experience using a 3D ultrasound robotic system for prostate brachytherapy assistance, focal therapy, and prostate biopsies. They showed that the robot was the first system to use intraoperative prostate motion tracking to guide needles into the prostate. Preliminary results showed its ability to reach targets despite prostate motion [10].

A literature review of robotic ultrasound and needle guidance for prostate cancer management was published in 2014 by Kaye et al. who felt that improved needle localization and needle guidance are required. They reported that sampling of the anterior zone may be challenging in patients with an enlarged prostate and complications like urethral injury, injuries to the dorsal venous complex, and neurovascular bundle were related to the technique of fixed needle insertion [11].

With the development of new software to better target prostate lesions, new trials using the MRI-Ultrasound fusion system were launched. Kroenig et al. were amongst the first ones to use a robot-guided transperineal fusion biopsy system. The Biobot, MonaLisa system, used a software-controlled robotic arm and utilized two components; UroFusion and UroBiopsy to aid registration of MRI lesions, fusion of the TRUS, and MRI images and guide biopsy. They showed that the use of a robot system catering to prostate deformation via an elastic fusion algorithm and automatic definition of penetration depth led to higher detection rate of Gleason 8 to 10 in targeted biopsies. Their study also showed that small targets were sampled correctly with an excellent coverage of all areas of the prostate. On the other hand, they had a low overall detection rate on lesions-based analysis (using the PI-RADS version 2 scoring system) that was comparable to other similar studies not employing a robot system [12,13].

In 2017, Ball et al. published their results on the safety and feasibility of the first in human trial of direct MRI-guided prostate biopsy using an MRI-safe robotic device. Although with a very small sample size, they showed that MRI lesions could be accurately targeted with the robot device, with an advantage over fusion-based technology based on needle placement in the cancer suspicious region. The robot used was MRI-safe, powered by a pneumatic motor, using air for actuation and light for the sensors. The main critique was the lengthy procedure time but the technique was considered safe. The most important advantage of a robot system was its targeting accuracy of 2.55 mm [3]. Based on a similar technique, further studies were carried out showing that robot-assisted MRI-guided prostate biopsy had comparable accuracy to the manual template-based approach [2]. Moreira et al. also defined new techniques in order to avoid needle curvature thus enhancing detection rates.

Further studies were carried out in 2019, using transrectal ultrasound-guided prostate biopsy [14] and MRI-US fusion transperineal approach [5]. Lim et al. showed that a transrectal approach enables prostate biopsy with minimal pressure over the prostate and small prostate deformations which in turn helped toward the accuracy of needle targeting. They also showed that the robot-assisted biopsy can be carried out over approximately 13 min. Miah et al. used the

Biobot, adopting a transperineal approach, and demonstrated a high detection of clinically significant prostate cancer when combined with only limited near-field sampling. They also advocate the use of this technique to potentially avoid the morbidity associated with whole gland systematic unguided biopsies.

Maris et al. published their work using the PROST robotic system in 2021 [15]. They used a machine learning algorithm and allowed autonomous positioning of needles aiming to reduce human error. They reported an accuracy of 1 mm regardless of user experience, using a low-cost solution that works independently of the clinician's experience. In the same year, Wetterauer et al. published their work on the assessment of diagnostic accuracy of robotic-assisted transperineal target saturation biopsy of the prostate. They used the MonaLisa Biobot and showed a superior detection of clinically significant cancer, leading to better risk stratification and alteration in treatment decisions [16].

Among the latest studies comparing prostate cancer detection rates of manually operated and robot-assisted in-bore MRI targeted biopsy, Sandahl et al. were not able to prove a significant difference between the two techniques. A systematic review and meta-analysis of robot-assisted MRI-target vs systematic prostate biopsy were carried out by Petov et al. Their results were published in 2023. They showed that robot-assisted targeted prostate biopsy and robot-assisted systematic prostate biopsy were both technically feasible with comparable clinical significance and overall prostate cancer detection rates [17].

Over the last decade, we have seen multiple trials using the robot system to aid in prostate cancer diagnosis. We evolved from trials to clinical implementation, from simple software to state of the art fusion software, using better needle guidance techniques to improve biopsy technique, cancer detection rates, procedure time, and adverse events. Table 1 shows a trend in hardware use, duration, and adverse events recorded.

TABLE 1   Studies conducted with the robot-assisted techniques.

| Author | Year of study/Level of evidence | Prostate volume | MRI lesions (≥ PIRADS 3) | Type of biopsy and hardware | Duration—minutes (Median) | Diagnostic rate of prostate cancer—Overall/clinically significant (%) | Adverse events (%) |
|---|---|---|---|---|---|---|---|
| Sandahl et al. | 2022/Level 2b | 48 (Median) | 379 | Transperineal Fusion Soteria Medical BV | NA (RATP biopsy had 15 min shorter room time) | 73/82 | NA |
| Wetterauer et al. | 2021/Level 2b | 42.7 +/− 17.4 (Mean) | 58 | Transperineal Fusion Biobot (ISR'obot MonaLisa) | NA | NA/78.3 | NA |
| Patel et al | 2020/Level 2b | 53 (Median) | 91 | Transperineal Fusion Biobot (ISR'obot MonaLisa) | 24 | 63.3/53.3 | 1.9 |
| Lee et al. | 2020/Level 2b | 43.2 +/− 18.4 (Mean) | 500 | Transperineal Fusion Biobot (ISR'obot MonaLisa) | NA | 71.6/34.2 | NA |
| Yang et al | 2020/Level 2b | 40 (Median) | NA | Transperineal Fusion Biobot (ISR'obot MonaLisa) | 33 | 63.3/53.3 | 6.7 |
| Miah et al. | 2018/Level 2b | 51.3 +/− 25.24 (Mean | 107 | Transperineal Fusion Biobot (ISR'obot MonaLisa) | NA | NA/51.2 | 1.1 |
| Kaufmann et al. | 2018/Level 3b | NA | NA | Transperineal Fusion Biobot (ISR'obot MonaLisa) | 43 | 61.8/52.7 | 16.3 |
| Kroenig et al. | 2016/Level 2b | 57.6 +/− 26.6 (Mean) | 123 | Transperineal Fusion Biobot (ISR'obot MonaLisa) | NA | 59.6/51.9 | 1.9 |

## Advantages and disadvantages of a robot-assisted prostate biopsies

Robotic-assisted biopsy offers several advantages over traditional biopsy methods. The precision of needle placement can minimize sampling errors and improve accuracy in identifying cancerous tissues. Moreover, it offers the potential for improved patient outcomes and possibly reduced costs. Furthermore, the robotic system allows for real-time imaging and mapping of the prostate gland, which helps the urologist target specific areas for sampling. This can result in improved accuracy of the biopsy, as the urologist can more precisely target suspicious areas, reducing the likelihood of a false negative result.

Robotic-assisted biopsy also allows for shorter procedure times and faster recovery times compared to traditional biopsy methods as evidenced in several studies. The robotic system can quickly and safely perform multiple sampling sites in a single procedure, reducing the need for multiple biopsies.

Other potential advantages of the robot over manual prostate biopsies include better 3D reconstruction of 2D images, automatic alignment of needle to target lesion, and lower prostate deformations.

Additionally, robotic-assisted biopsy can be beneficial for patients who are at high risk for complications with traditional biopsy methods, such as those with bleeding disorders or taking blood-thinning medications. The precision and control of the robotic system can minimize the risk of bleeding or other complications.

Overall, robotic-assisted biopsy is a safe and effective option for patients who require prostate biopsy. It offers improved accuracy, shorter procedure times, and faster recovery times compared to traditional biopsy methods and can reduce the risk of complications for high-risk patients.

## Future scenario

For the foreseeable future, we expect prostate biopsies to be performed by trained physicians or advanced nurse practitioners under local anesthesia using up-to-date fusion imaging technology.

On the other hand, we do believe in the evolution of the diagnostic process towards accepting the use of robotic assistance for an improved patient experience with a higher diagnostic accuracy.

This will indeed be implemented after robust randomized controlled trials approving the benefits of robotic-assisted prostate biopsies. As with the introduction of the robotic system for robotic-assisted urological procedures, we expect the use of the robot for prostate biopsies to be phased in over a time period.

One potential downside of robotic-assisted biopsy is the cost, as the use of the robotic system can add to the overall cost of the procedure. However, for patients with high-risk prostate cancer or those who have had previous negative biopsies, the improved accuracy and lower risk of complications may justify the increased cost. The cost-benefit of the procedure has not been well elaborated in the studies published to date. Some studies have concluded that it is financially beneficial compared to manually operated techniques. Working from the perspective of patients requiring multiple biopsies, imaging (MRI), and outpatient follow-ups to reach a diagnosis indeed adds to the diagnostic costs. This in turn favors having a robot-assisted prostate biopsy where cancer detection rates are higher and thus altering management for this patient cohort and decreasing the diagnostic cost to the minimum. Miah et al. have shown that using a robot-assisted technique can lead to a decrease in the number of cores required for diagnosis, a decrease in costs, an increase in efficiency, and an opportunity for combined prostate biopsy and focal ablation therapy of the prostate in the same session.

Additionally, not all patients are good candidates for robotic-assisted biopsy. Factors such as obesity, or prior abdominal surgeries may make the procedure more difficult and possibly unsafe.

## Conclusion

Overall, robotic-assisted biopsy can be a valuable tool in the diagnosis and management of prostate cancer. It offers improved accuracy and lower risk of complications compared to traditional biopsy methods, but it may not be suitable for all patients.

Important factors like monetary and time expenses have to be taken into consideration. On the other hand, patient satisfaction and tolerability of the procedure need to be taken into account. The challenge remains to run a randomized controlled trial to give us the necessary answers (cost-benefit analysis and patient-reported outcomes) prior to broadly implementing the use of robot-assisted prostate biopsy technique in clinical practice.

# Key points

- Precise prostate biopsies lead to accurate diagnosis and thus guide optimal treatment of patients.
- Use of robot-assisted prostate biopsy is expensive but has been shown to improve accuracy, procedure times, and recovery times compared to traditional prostate biopsy methods currently in place.
- The implementation and use of robot-assisted prostate biopsy does indeed represent a significant financial impact on the diagnosis pathway.
- Further randomized controlled studies are required prior to the recommendation of the same as part of a standard prostate cancer diagnostic pathway.

# References

[1] He J, Guo Z, Huang Y, et al. Comparisons of efficacy and complications between transrectal and transperineal prostate biopsy with or without antibiotic prophylaxis. Urol Oncol 2022;40(5):191.

[2] Kroenig, et al. Diagnostic accuracy of robot-guided, software based Transperineal MRI/TRUS fusion biopsy of the prostate n a high risk population of previously biopsy negative men. Biomed Res Int 2016;2016:6. https://doi.org/10.1155/2016/2384894. 2384894.

[3] Moreira, et al. Evaluation of robot-assisted MRI-guided prostate biopsy: needle path analysis during clinical trials. Phys Med Biol 2019;63(20). https://doi.org/10.1088/1361-6560/aae214.

[4] Ball, et al. Safety and feasibility of direct MRI-guided Transperineal prostate biopsy using a novel MRI-safe robotic device. Urology 2017;109:126–221.

[5] Miah, et al. A prospective analysis of robotic targeted MRI-US fusion prostate biopsy using the centroid targeting approach. J Robot Surg 2020;14:69–74.

[6] Sandahl, et al. Prostate Cancer detection rate of manually operated and robot-assisted in-bore magnetic resonance imaging targeted biopsy. Eur Urol 2022;42:88–94. https://doi.org/10.1016/j.euros.2022.05.002.

[7] Ho, et al. Robotic-assisted Transperineal prostate biopsy with novel device for future prostate interventions: 3-years' clinical experience. J Urol 2010;183:e424–5.

[8] Stoianovici, et al. 'MRI stealth' robot for prostate interventions. Minim Invasive Ther Allied Technol 2007;16(4):241–8.

[9] Schouten, et al. Evaluation of a robotic technique for transrectal MRI-guided prostate biopsies. Eur Radiol 2012;22:476–83.

[10] Long, et al. Development of a novel robot for transperineal needle based interventions: focal therapy, brachytherapy and prostate biopsies. J Urol 2012;188(4):1369–74.

[11] Kaye, et al. Robotic ultrasound and needle guidance for prostate Cancer management: review of the contemporary literature. Curr Opin Urol 2014;24:1. https://doi.org/10.1097/MOU.0000000000000011.

[12] Kesch, et al. Multiparametric MRI and MRI-TRUS fusion biopsy in patients with prior negative prostate biopsy. Der Urologe Ausg 2016;55(8):1071–7.

[13] Cash, et al. The detection of significant prostate cancer is correlated with the prostate imaging reporting and data system (PI-RADS) in MRI/transrectal ultrasound fusion biopsy. World J Urol 2016;34(4):525–32.

[14] Lim, et al. Robotic Transrectal ultrasound-guided prostate biopsy. IEEE Trans Biomed End 2019;66(9):2527–37. https://doi.org/10.1109/TBME.2019.2891240.

[15] Maris, et al. Toward autonomous robotic prostate biopsy: a pilot study. Int J Comput Assist Radiol Surg 2021;16:1393–401.

[16] Wetterauer, et al. Diagnostic accuracy and clinical implications of robotic assisted MRI-US fusion guided target saturation biopsy of the prostate. Sci Rep 2021;11:20250.

[17] Petov, et al. Robot-assisted magnetic resonance imaging-targeted versus systematic prostate biopsy; systematic review and meta-analysis cancers. Cancers 2023;15:118.

# The formation and the training of robotic surgeons

*Pier Paolo Avolio[a,b], Nicolo Maria Buffi[b,c], Paolo Casale[b,c], Maurice Anidjar[a],
Simon Tanguay[a], Wassim Kassouf[a], Armen Aprikian[a],
Giovanni Lughezzani[b,c], and Rafael Sanchez-Salas[a]*

[a]Department of Surgery, Division of Urology, McGill University, Montréal, Quebec, Canada [b]Department of
Biomedical Sciences, Humanitas University, Milan, Italy [c]Department of Urology, IRCCS Humanitas Research Hospital,
Milan, Italy

## Introduction

Robotic surgery has spread rapidly, with a fourfold increase in the number of robot-assisted surgery (RAS) in the last decade [1]. The rapid increase in the use of RAS is due to its numerous advantages compared to open surgery and conventional laparoscopy, such as faster learning curves, elimination of the fulcrum effect, and ability to reflect wrist and hand movements [2,3]. Furthermore, RAS has led to better patient outcomes, such as a significant reduction in blood loss, length of hospital stays and transfusion rate, and lower post-operative pain medication requirements [4]. Therefore, as experienced surgeons switch to robot-assisted surgeries to exploit their advantages, addressing the training needs of current and future surgeons becomes a priority. The robotic surgical platform represents a technological shift from open surgery. Although appearing to be an evolution of laparoscopic surgery, the required skills for naïve robotic surgeons concern console control and maneuvers without haptic feedback, instead of those required for laparoscopic surgery, which as two-dimensional surgery employs instruments with a limited range of motion [5]. A structured curriculum, including the acquirement of basic robotic skills as well as more complex tasks, allows these competencies to be developed safely and gradually in a relatively short learning curve [6,7]. This chapter presents an evidence-based roadmap highlighting the essential elements to be included in a curriculum for basic and advanced robotics training.

## Dry lab training

Dry lab training refers to the employment of non-human, non-animal desktop models for learning and skill acquisition. The materials employed in this setting can be as simple as routine strings/needles/sutures, to sophisticated vascular and bowel models. Most of the exercises are aimed to improve dexterity and control. In robotic surgery courses, these tasks are often presented with virtual reality (VR) simulator exercises. Fried et al. published early and promising results using their McGill Inanimate System for Training and Evaluation of Laparoscopic Skills for laparoscopic training, subsequently corroborated by Sroka et al. [8,9]. These studies proved that inanimate simulator training is beneficial and reproducible in surgical training. The dry lab simulation is convenient and can reliably simulate cutting, suturing, and gripping exercises. Since the user sits on the da Vinci console and uses the robotic instruments to complete the tasks, the accuracy of the simulations is very high [10]. Furthermore, dry training may lead to overcome initial issues with the console, mainly regarding camera, clutch control, as well as hand position [11].

However, the main limitation is the lack of a standardized evaluation model and objective metrics, which is essential in the initial step [12]. Ramos et al. demonstrated that the dry robotic lab activities had face, content, construct, and concurrent validity compared to the corresponding simulator tasks in a prospective study involving naïve and expert surgeons [12]. However, the authors made the comparison using the global evaluative assessment of robotic skills (GEARS) assessment tool [13]. This tool examines seven areas of technical performance; depth perception, bimanual dexterity, efficiency, force sensitivity, autonomy, robotic control, and use of the third arm and assigns each a linear score from 1 to 5. Nevertheless, the GEARS employs a Likert scale and is a subjective score, with individual interpretation and relative variability. Dulan et al. partially addressed the lack of standardization with a training program that consisted of an online training module provided by Intuitive Surgical, followed by a half-day interactive session with a proctor to reinforce the concepts of the online training module [14]. Trainees then moved on to a program of nine exercises, in which each trainee was assessed using the validated Fundamentals of Laparoscopic Surgery approach. The trainees then repeated the exercises over time until they demonstrated two consecutive competence scores. This model is reproducible and its development cost $2200. However, it requires considerable time and effort from the team for the initial set-up, as well as the use of the robotic console and instruments during nonoperational hours or via a dedicated laboratory console [15].

## Wet lab training

Despite dry labs being a valuable learning tool, tissue manipulation, understanding tissue reaction to instrument touch, the use of diathermy and vascular control cannot be acquired in this setting. Wet labs can provide three different types of tool training. Frozen animal parts, frozen human body parts, and live animals, with the cost increasing proportionally [16], Fig. 1. Animal and human body parts are excellent training tools for learning tissue manipulation, dissection, excision, diathermy, and suturing techniques. Although embalmed body parts allow vascular identification and dissection, do not provide a learning field for vascular control. Furthermore, animal models are sometimes not good replicates of the human body, such as radical prostatectomy in the pig model [17]. Nevertheless, the obvious ethical and cost limitations have led to low uptake of anesthetized animal models in the early stages of a training program [18].

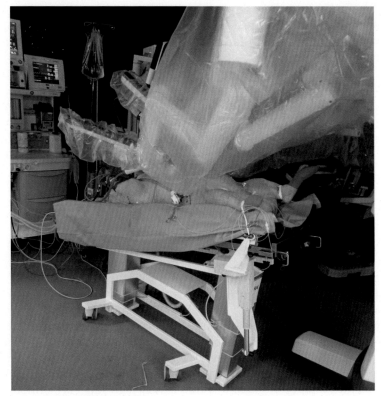

FIG. 1    Showing robotic wet lab training with animal part.

## VR simulator training

VR simulation is increasingly employed in surgical training as well as being considered an important initial step in robotic surgical training [19]. Evaluation and measurement of progress are the main advantages of VR software systems. Furthermore, patients are not exposed to surgeons during the initial learning curve and simulators can be used when operating rooms and robotic platforms are not scheduled for patient care. Five VR simulators are currently available for robotics training. These include the Robotic Surgical Simulator (RoSS; Simulated Surgical Systems, Buffalo, NY), dV-Trainer (Mimic Technologies, Inc., Seattle, WA), SEP Robot (SimSurgeryTM, Norway), the da Vinci Skills Simulator (dVSS Intuitive Surgical, Sunnyvale, CA), and more recently the Robotix mentor (3D systems, formerly Symbionix, Israel) [20–26]. These simulators were assessed for *face validity*, which refers to the realism of the simulator; *content validity*, the skills being tested accurately represent the skills required in robotic surgery; *construct validity*, the simulator's ability to differentiate subjects possessing different levels of skills, except RoSS; *discriminant validity*, the simulator being able to differentiate skill levels within a group with similar experience, except RoSS; *concurrent validity*, indicates the extent to which performance on a new test corresponds to performance on an established test of the same construct; *predictive validity*, performance on the simulator predicts future performance on the robotic platform when used in clinical settings [16,17]. Besides the basic skills simulation, the majority of simulators are now capable of providing complex skills simulations, e.g., Hands-On Surgical Training (HOST) in the RoSS and Maestro AR in the dV-Trainer (http://www.mimicsimulation.com/products/maestro-ar/) [26]. Specifically, the trainee sits at the console grasping the pincer grip, and watches a video of the procedure being performed, while the console arms move in accordance with the video. After recording the movements, the trainee can mimic the surgeon's movements and perform the procedure in real time. Then, the trainee's movement can be tracked and scored by the on-board software. Finally, the latest generation of dV-Trainer (http://www.mimicsimulation.com/products/xperience/) and Robotix mentor provides laparoscopic training for the assistant at the patient's bedside combined with the VR console [27]. Thus, the surgeon at the console can develop advanced procedural skills, while the assistant develops his or her skills at the patient's bedside, promoting greater teamwork. Despite several VR simulators having been developed over the years, few studies validate their deployment as a tool that can improve the console performance of naïve surgeons. The role of HOST training for robotic ureterovesical anastomosis was evaluated by Chowriappa et al. by a randomized controlled trial. They found that participants undergoing HOST training had higher scores in the domains of needle guidance, needle placement and suture placement, respectively 3.0 vs 2.3; $P = 0.042$, 3.0 vs 2.4; $P = 0.033$, 3.4 vs 2.6; $P = 0.014$. Additionally, according to the judgment of 70% of the participants, the HOST training experience was comparable to a real surgical procedure and 75% of the trainees felt that this training could improve confidence in performing a real procedure [28]. Subsequently, Hung et al. in a prospective randomized study demonstrated the validity of three standardized training methods: inanimate, simulator, and in vivo. Furthermore, the authors compared the three different platforms for cross-method training value. They concluded that training on VR simulators significantly improved tissue exercise performance in trainees with low basic robotic skills. However, this improvement was not significant in trainees with high basic robotics skills, suggesting the need to tailor the curriculum individually to each trainee [25]. Similarly, Sangini et al. demonstrated a significant improvement in "completion time" and "economy of movement" for naïve surgeons following training with the dVSS [29]. This program has 30 different exercises to train and test robotic skills. Many exercises test more than one skill domain. Immediate feedback is provided, with an overall score and a breakdown of performance efficiency in time to complete the exercise and economy of movement. Penalties are deducted from this score, including collisions between the endoscope, instruments and environment, use of excessive force, dropping needles/objects, incorrect targeting, and placing instruments out of view. Users can set up recording of their performance to monitor improvements and supervisors can log in to view results. Additional simulation programs for more complex procedures are also available, including one for radical prostatectomy.

## Standard of training for VR simulators

A further area of interest for VR simulation-based training is the evaluation and setting of standards for training. Raison et al., in their longitudinal, observational study, analyzed the results of nine practical VR training courses of the European Association of Urology (EAU) [30]. The aim of the study was to develop objective benchmark competence scores for use during VR robotic surgery training. Competence was set at 75% of the mean expert score. The benchmark scores developed from expert performance provide trainees with challenging scores to achieve during VR simulation training [30]. Objective feedback allows both trainees and trainers to monitor educational progress and ensures that training remains effective. In addition, well-defined goals set by benchmarking provide clear objectives for trainees and allow for a transition to a more efficient competency-based curriculum. Noureldin et al. introduced another valuable application of VR simulation using dVSSS, namely the incorporation of VR simulation into Canadian Objective Structured Clinical Examinations (OSCEs) to assess the basic robotic skills of urology residents (PGTs) [31]. During the OSCE, the station for this segment was only 20 min, allowing only two of the less complex dVSSS skill tests to be performed. Their definition of competence was based on the normo-referenced method, whereby experts performed tasks before trainees, and a passing score was defined as the average of the experts' total scores minus one standard deviation for each task. Although this score passed only one-third of the trainees, the ability of the test to discriminate between less experienced and more experienced trainees should be considered a measure of achievement [31].

## Limitations of VR simulators

Simulators have their limitations. First, there is a lack of consensus on exercises and skills that can be applied to real cases. *Predictive validity* is challenging to demonstrate, as there are many variables in human operation. Also, using the time to perform a procedure as a measure of competence may not be particularly meaningful in the clinical setting, as it may be influenced by many variables, such as case complexity and patient comorbidity; complications during a case may not depend entirely on the operator [17]. Furthermore, cognitive, and human factors constitute a large part of the 'skill' package displayed during a procedure. Variations in anatomy and pathology are perhaps primarily responsible for the challenging and long learning period in all surgical procedures, and simulated tasks and procedures may only provide a very early basis to build clinical experience on [17]. Additionally, it is evident that there is a significant relationship between the set of simulators and the modalities in which they are used in a curriculum. A well-constructed curriculum, both in content and modality, is likely to produce the greatest results from simulation. Another limitation is the lack of standardization of training curricula. Competency-based training is considered the most effective way to acquire the basic skills of robotic surgery. Currently, competence levels often refer to an arbitrary parameter of 70%–80% of the "overall score" produced by the simulation software [32]. The value of a 70%–80% benchmark is unknown because the software metrics were not developed by experienced robotic surgeons and instead represent standard, pre-programmed metrics [33]. Finally, the real cost benefit of robotic simulators remains unclear. Rehman et al. described the indirect cost benefit of using the ROSS at the Roswell Park Cancer Institute. Over a 1-year period, it was determined that approximately $622,784 could be saved if VR simulation replaced dry laboratory training [30]. However, the authors did not investigate whether an hour on the simulator is equivalent to an hour spent on animals or humans operating in terms of training value.

## Training in the operating room

### Patient side training

Similarly to any surgical procedure, the development of robotic skills follows a progression of observation, assistance, supervised execution, and finally independent practice [17,34]. Patient-side training has a two-fold advantage. First, it allows the trainee to assist in the surgical procedure and absorb the skills of the first operator. Second, it also allows the development of the assistant's own skills, such as problem-solving at the patient's side to enable the procedure to run smoothly [16]. Furthermore, the assistant develops an understanding of the ergonomics and access limitations created by the robotic arms. Despite the lack of metrics, such as the number of procedures required or the duration of patient-side training, it is plausible that patient-side skills are acquired relatively quickly and that the acquisition of a basic level of competence allows the transition to the console in a relatively short time [16] (Fig. 1).

### Patient positioning and port placement training

Patient positioning and port placement play a crucial role in the successful outcome of the surgical procedure [35,36]. Adequate patient positioning ensures that each member of the surgical team, such as the patient side assistant, nurse, and anesthetist has appropriate access to the patient, and also allows optimal spatial configuration between the robot arms and the target organ. Similarly, the precise port placement allows access to the target organs, enabling the required triangulation, without any extra- or intracorporeal instrument clashes [16]. These basic skills can be learned in simulated operating rooms by inserting the ports into dummies and testing access and clashes between instruments and emergency release procedures (Fig. 2).

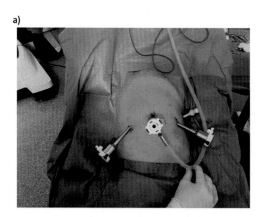

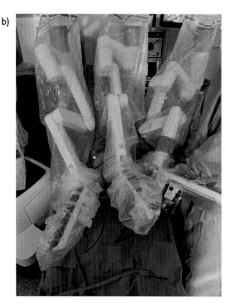

FIG. 2 (A) Showing port placement and (B) robotic arms positioning during robotic ureteral reimplantation after right distal ureteral stricture.

## Basic laparoscopic skills training

The elementary laparoscopic skills required for a robotic surgeon include laparoscopic access as well as the creation of pneumoperitoneum, removal of adhesions hindering port insertion, clip application, suction, and retraction (Fig. 3) [37]. These basic skills can be acquired in a dry lab setting and improved during bedside training. Acquiring basic laparoscopic skills has its own learning curve, but it can help to achieve the robotic learning curve. Angell et al. demonstrated this in a cohort of medical students in a dry lab setting [38]. The investigated task consisted of cutting a spiral structure. The study proved that in a naïve cohort of students, intensive training in basic laparoscopic skills reduced the time required to perform the task robotically and the number of errors [38]. Furthermore, acquiring basic laparoscopic skills may result in the creation of a solid and safe method of positioning and using instruments [39]. Thus, such knowledge can provide a background helping to develop the correct robotic skills on the console, aiding the surgeon in difficult robotic cases [40]. Indeed, the higher degree of movement of the robotic platform allows for the development of many ways to perform a task in a dry lab setting. However, not all methods are necessarily appropriate, and the skills developed in this context may not be generalizable in real-life settings. Since laparoscopic instruments do not

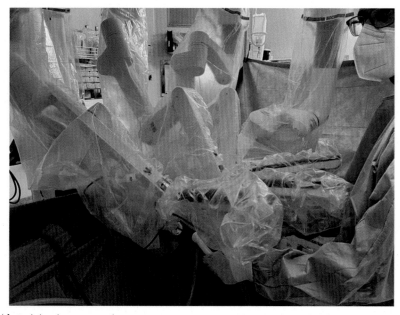

FIG. 3 Showing patient-side training in an operating room.

I. Introduction, technology basics, evolution and next frontiers of surgical robots

allow much freedom, the procedural skills thus developed tend to be the most precise and safe way of performing the task and can be a valuable problem-solving tool for the surgeon in challenging surgical cases. The development of basic laparoscopic skills also improves spatial awareness, the ability to carefully manipulate tissue, and the ability to operate safely in three-dimensional space [41].

## Console training

The Davinci robotic system represents the most widely used robotic approach in the surgical field. Since this is a master–slave system, the surgeon controls the robotic arms during the procedure from a remote console and the robot is merely an instrument [42]. Therefore, knowledge of the console's function is of paramount importance for a surgeon approaching robotic surgery. Thus, online modules are available that introduce the basic concepts of the system currently available (https://www.davincisurgerycommunity.Com/Training?tab1=TR). Attending these online modules is essential before starting any training on the console. The majority of the modules provide descriptions of each component of the system and provide information on problem-solving. Improvement of basic console skills, such as control of the camera, pedal, and finger pressure, can be achieved relatively easily in a dry lab or in a simulated VR setting. Here, each task can be reiterated, refined, and rated. Conversely, advanced skills at the console, such as excision, suturing, and the use of diathermy must be developed in a simulation setting with a tutor, in a VR simulator, in a dry lab or in a wet lab consisting of live animals/cadaveric or human models. Finally, simulation learning provides a platform for the initial development of skills, as well as for their evaluation, while respecting patient safety on the operating table. The console learning should be carried out in a controlled environment, following a modular approach. The role of the tutor is crucial in this process. The modular process begins with the trainee performing the easier part of the procedure, and then progressively addressing progressively more difficult parts, as considered suitable by the tutor. The mentor's transition from preceptor, who intervenes when necessary to proctor, who supervises and allows the trainee sufficient opportunity to operate usually indicates that the trainee is moving forward. Lovegrove et al. developed a safety and evaluation tool to assess the technical skills of surgeons performing robot-assisted radical prostatectomy (RARP) during modular training, based on the Healthcare Failure Mode and Effect Analysis (HFMEA) methodology [43]. Using this methodology, the operative procedure was devised to identify 17 steps and the trainees undergoing modular training were evaluated by their mentors with a score from 1 to 5. A score of 4 or higher indicated the achievement of the objective. Thus, monitoring progress and defining a learning curve for each step of the procedure was possible [43]. The dual console dVSS system, although involving a substantial increase in cost, allows the mentor to immediately intervene and take control without the trainee needing to leave the console, potentially offering the trainee more "operational time" on the console (Fig. 4). Smith et al. demonstrated that this is not significantly different from single-console procedures with regard to operating time and results. Furthermore, dual-console training would be valuable in vascular management [44].

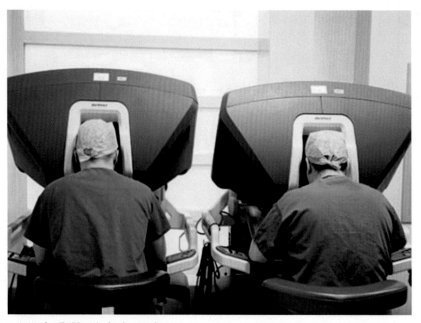

FIG. 4   Showing console training by DaVinci's dual console.

TABLE 1  Showing EAU/ERUS proposed modular scheme for training in robot-assisted radical prostatectomy.

| Robot-assisted radical prostatectomy steps | Required repetitions |
|---|---|
| Bladder detachment | 20 |
| Endopelvic fascia incision | 15 |
| Bladder neck incision | 15 |
| Section of vasa and preparation of seminal vesicles | 15 |
| Dissection of the posterior plane | 10 |
| Dissection of prostatic pedicles | 10 |
| Dissection of neurovascular bundles | 5 |
| Ligation of the Santorini plexus | 10 |
| Apical dissection | 5 |

TABLE 2  Showing EAU/ERUS proposed modular scheme for training in robot-assisted partial nephrectomy.

| Robot-assisted partial nephrectomy steps | Required repetitions |
|---|---|
| Liver/bowel mobilization | 15 |
| Hilar control | 7 |
| Opening gerota's fascia | 10 |
| Tumor demarcation | 10 |
| Vessel clamping and de-clamping | 10 |
| Tumor excision | 5 |
| Inner renorrhaphy | 5 |
| Outer renorrhaphy | 5 |
| Gerota's closure | 20 |

## Modular operative training

A modular approach to robotic surgical training is based on breaking down complete procedures into constitutive steps. These steps follow the order of the surgery, although the trainee may not need to perform the first step before moving to the second step and so forth. Ideally, the different steps require different levels of skill, and the curriculum ensures gradual exposure to progressively more complex steps [6]. The EAU/ERUS educational council has defined a structured modular training framework for robot-assisted radical prostatectomy, including a specific number of times each phase must be performed (see http://urowe-b.org/section/erus/robotic-curriculum/). For instance, robotic radical prostatectomy could be divided into key steps each with associated minimum numbers of repetitions in order to achieve the training (Table 1). A similar scheme has been proposed by the EAU/ERUS for robot-assisted partial nephrectomy, with a required amount of repetitions for each stage, as well as a level of complexity approved by a panel of experts for each stage [45] (Table 2).

## The EAU/ERUS curriculum

To date, the only validated robotics training curriculum in urology that addresses all stages of training from laboratory to expert is the EAU/European Robotic Urology Section (ERUS) curriculum. Volpe et al. published the first validated, standardized robotic training curriculum [46]. A comprehensive training program should start with a basic assessment to confirm the surgeon's current knowledge and skills. Theoretical and safety aspects are covered, usually in e-learning mode. The next steps are VR simulation, dry labs, then cadaveric models, and live animal surgery. As the surgeon approaches in vivo surgery on humans, he or she must gain considerable experience at the patient's bedside as an assistant. This allows for a full understanding of the steps of the procedure and a thorough understanding of patient positioning, setting up the robotic platform, and solving intraoperative problems. Subsequently, the surgeon must be exposed to the operating procedures in a modular system [46]. The curriculum is completed by video performance analysis and marking of a completed RARP procedure. This final video was assessed by experienced, blinded robotic surgeons using generic linear scoring criteria dedicated to each procedure. Dedicated linear scores for each procedural step ranged from 4 to 16, with 10 considered well accepted [46]. A similar program is being developed for robotic partial nephrectomy. Encouragingly, the ERUS curriculum has been adapted and tested in gynecological and thoracic robotic surgery, suggesting its utility and applicability to further specialties [47,48].

## Future scenario

### Non-technical skills development for robotic surgery

Robotics training generally focuses on the acquisition of technical skills, while the teaching of nontechnical skills (NTS) is lacking. The presence of the robot in the operating room poses a unique challenge to team communication and risk management, so the development of non-technical skills is of paramount importance. It has been shown that nontechnical skills, such as teamwork, leadership, situational awareness, and decision-making, have a significant impact on surgical success and can be easily developed in a simulated environment [49]. A recent systematic review by Wood et al. evaluated several nontechnical skills training tools for both the individual surgeon and the team and concluded that the Nontechnical Skills for Surgeons (NOTSS) was the gold standard training tool for training the individual surgeon, while the Oxford Nontechnical Skills Training Tools for the Surgical Team II (NOTECHS) was the most favorable for training the team [50]. These and other NTS assessment tools need further investigation to evaluate their implementation within the training curricula. A major barrier to these assessments is the need for surgical professors to assess the nonsurgical skills of trainees.

### Telementoring

Telementoring is an innovative and exciting technology that allows an expert to provide advice without physically being in the operating theater. The tutor can view the same images as the operating surgeon and provide expert guidance in real time. There are several telementoring initiatives that connect a remote tutor to the robotic operating room [51]. As an educational tool, this has the potential to increase the availability of skills, reduce training costs, and confer higher rates of competence. However, there are important medico-legal issues about shared responsibility that need to be addressed and maybe country specific [51].

### Revalidation and ongoing assessment

Learning and skill acquisition never stop once competence has been achieved; continuous learning is mandatory in surgery. Regular reassessment/revalidation should be considered in robotic surgery. Simulators, dry labs, cadaver lab, video assessment, and telementoring may play a role in the revalidation of experienced surgeons, but there are no published data to suggest their usefulness in the advanced context of reassessment of established robotic surgeons [10,17].

## Conclusions

Well-structured curricula for naïve robotic surgery are becoming increasingly recognized and widespread in the specialist surgical branches. It would be beneficial if these curricula were developed and implemented in a coordinated manner, with the collaboration of professional societies, accredited training centers, and institutions to reach a consensus on standardized training.

## Key points

- Surgical expertise is achieved through well-designed curricula.
- Inanimate simulator training is useful in surgical training.
- Nontechnical skills can be developed in a simulated environment.
- A standardized curriculum in robotic surgery incorporates different aspects of surgical training.
- Patient safety is the ultimate goal of surgical training.

## References

[1] El-Hamamsy D, Geary RS, Gurol-Urganci I, van der Meulen J, Tincello D. Uptake and outcomes of robotic gynaecological surgery in England (2006–2018): an account of hospital episodes statistics (HES). J Robot Surg 2022;16(1):81–8. https://doi.org/10.1007/s11701-021-01197-5.

[2] Buffi NM, Saita A, Lughezzani G, et al. Robot-assisted partial nephrectomy for complex (PADUA score ≥10) tumors: techniques and results from a multicenter experience at four high-volume centers. Eur Urol 2020;77(1):95–100. https://doi.org/10.1016/j.eururo.2019.03.006.

[3] Buffi NM, Lughezzani G, Lazzeri M, et al. The added value of robotic surgery. Urologia 2015;82(Suppl 1):S11–3. https://doi.org/10.5301/uro.5000149.

[4] Egan J, Marhamati S, Carvalho FLF, et al. Retzius-sparing robot-assisted radical prostatectomy leads to durable improvement in urinary function and quality of life versus standard robot-assisted radical prostatectomy without compromise on oncologic efficacy: single-surgeon series and step-by-step guide. Eur Urol 2021;79(6):839–57. https://doi.org/10.1016/j.eururo.2020.05.010.

[5] Joseph JV, Vicente I, Madeb R, Erturk E, Patel HRH. Robot-assisted vs pure laparoscopic radical prostatectomy: are there any differences? BJU Int 2005;96(1):39–42. https://doi.org/10.1111/j.1464-410X.2005.05563.x.

[6] Ahmed K, Khan R, Mottrie A, et al. Development of a standardised training curriculum for robotic surgery: a consensus statement from an international multidisciplinary group of experts. BJU Int 2015;116(1):93–101. https://doi.org/10.1111/bju.12974.

[7] Stegemann AP, Ahmed K, Syed JR, et al. Fundamental skills of robotic surgery: a multi-institutional randomized controlled trial for validation of a simulation-based curriculum. Urology 2013;81(4):767–74. https://doi.org/10.1016/j.urology.2012.12.033.

[8] Fried GM, Feldman LS, Vassiliou MC, et al. Proving the value of simulation in laparoscopic surgery. Ann Surg 2004;240(3):518–25. discussion 525–528. https://doi.org/10.1097/01.sla.0000136941.46529.56.

[9] Sroka G, Feldman LS, Vassiliou MC, Kaneva PA, Fayez R, Fried GM. Fundamentals of laparoscopic surgery simulator training to proficiency improves laparoscopic performance in the operating room-a randomized controlled trial. Am J Surg 2010;199(1):115–20. https://doi.org/10.1016/j.amjsurg.2009.07.035.

[10] Costello DM, Huntington I, Burke G, et al. A review of simulation training and new 3D computer-generated synthetic organs for robotic surgery education. J Robot Surg 2022;16(4):749–63. https://doi.org/10.1007/s11701-021-01302-8.

[11] Bramhe S, Pathak SS. Robotic surgery: a narrative review. Cureus 2022;14(9):e29179. https://doi.org/10.7759/cureus.29179.

[12] Ramos P, Montez J, Tripp A, Ng CK, Gill IS, Hung AJ. Face, content, construct and concurrent validity of dry laboratory exercises for robotic training using a global assessment tool. BJU Int 2014;113(5):836–42. https://doi.org/10.1111/bju.12559.

[13] Goh AC, Goldfarb DW, Sander JC, Miles BJ, Dunkin BJ. Global evaluative assessment of robotic skills: validation of a clinical assessment tool to measure robotic surgical skills. J Urol 2012;187(1):247–52. https://doi.org/10.1016/j.juro.2011.09.032.

[14] Dulan G, Rege RV, Hogg DC, Gilberg-Fisher KK, Tesfay ST, Scott DJ. Content and face validity of a comprehensive robotic skills training program for general surgery, urology, and gynecology. Am J Surg 2012;203(4):535–9. https://doi.org/10.1016/j.amjsurg.2011.09.021.

[15] Carpenter BT, Sundaram CP. Training the next generation of surgeons in robotic surgery. Robot Surg Auckl 2017;4:39–44. https://doi.org/10.2147/RSRR.S70552.

[16] Sridhar AN, Briggs TP, Kelly JD, Nathan S. Training in Robotic surgery-an overview. Curr Urol Rep 2017;18(8):58. https://doi.org/10.1007/s11934-017-0710-y.

[17] Chen R, Rodrigues Armijo P, Krause C, SAGES Robotic Task Force, Siu KC, Oleynikov D. A comprehensive review of robotic surgery curriculum and training for residents, fellows, and postgraduate surgical education. Surg Endosc 2020;34(1):361–7. https://doi.org/10.1007/s00464-019-06775-1.

[18] Lovegrove CE, Elhage O, Khan MS, et al. Training modalities in robot-assisted urologic surgery: a systematic review. Eur Urol Focus 2017;3 (1):102–16. https://doi.org/10.1016/j.euf.2016.01.006.

[19] Piana A, Gallioli A, Amparore D, et al. Three-dimensional augmented reality-guided Robotic-assisted kidney transplantation: breaking the limit of Atheromatic plaques. Eur Urol 2022;82(4):419–26. https://doi.org/10.1016/j.eururo.2022.07.003.

[20] Liss MA, Abdelshehid C, Quach S, et al. Validation, correlation, and comparison of the da Vinci trainer(™) and the daVinci surgical skills simulator(™) using the mimic(™) software for urologic robotic surgical education. J Endourol 2012;26(12):1629–34. https://doi.org/10.1089/end.2012.0328.

[21] Kesavadas T, Stegemann A, Sathyaseelan G, et al. Validation of Robotic surgery simulator (RoSS). Stud Health Technol Inform 2011;163:274–6.

[22] Seixas-Mikelus SA, Kesavadas T, Srimathveeravalli G, Chandrasekhar R, Wilding GE, Guru KA. Face validation of a novel robotic surgical simulator. Urology 2010;76(2):357–60. https://doi.org/10.1016/j.urology.2009.11.069.

[23] Kenney PA, Wszolek MF, Gould JJ, Libertino JA, Moinzadeh A. Face, content, and construct validity of dV-trainer, a novel virtual reality simulator for robotic surgery. Urology 2009;73(6):1288–92. https://doi.org/10.1016/j.urology.2008.12.044.

[24] Sethi AS, Peine WJ, Mohammadi Y, Sundaram CP. Validation of a novel virtual reality robotic simulator. J Endourol 2009;23(3):503–8. https://doi.org/10.1089/end.2008.0250.

[25] Hung AJ, Patil MB, Zehnder P, et al. Concurrent and predictive validation of a novel robotic surgery simulator: a prospective, randomized study. J Urol 2012;187(2):630–7. https://doi.org/10.1016/j.juro.2011.09.154.

[26] Schreuder HWR, Persson JEU, Wolswijk RGH, Ihse I, Schijven MP, Verheijen RHM. Validation of a novel virtual reality simulator for robotic surgery. ScientificWorldJournal 2014;2014:507076. https://doi.org/10.1155/2014/507076.

[27] Ahmad SB, Rice M, Chang C, Zureikat AH, Zeh HJ, Hogg ME. dV-trainer vs. da Vinci simulator: comparison of virtual reality platforms for Robotic surgery. J Surg Res 2021;267:695–704. https://doi.org/10.1016/j.jss.2021.06.036.

[28] Chowriappa A, Raza SJ, Fazili A, et al. Augmented-reality-based skills training for robot-assisted urethrovesical anastomosis: a multi-institutional randomised controlled trial. BJU Int 2015;115(2):336–45. https://doi.org/10.1111/bju.12704.

[29] Sheth SS, Fader AN, Tergas AI, Kushnir CL, Green IC. Virtual reality robotic surgical simulation: an analysis of gynecology trainees. J Surg Educ 2014;71(1):125–32. https://doi.org/10.1016/j.jsurg.2013.06.009.

[30] Raison N, Ahmed K, Fossati N, et al. Competency based training in robotic surgery: benchmark scores for virtual reality robotic simulation. BJU Int 2017;119(5):804–11. https://doi.org/10.1111/bju.13710.

[31] Noureldin YA, Stoica A, Kassouf W, Tanguay S, Bladou F, Andonian S. Incorporation of the da Vinci surgical skills simulator at urology objective structured clinical examinations (OSCEs): a pilot study. Can J Urol 2016;23(1):8160–6.

[32] Leon MG, Carrubba AR, DeStephano CC, Heckman MG, Craver EC, Dinh TA. Impact of robotic single and dual console systems in the training of minimally invasive gynecology surgery (MIGS) fellows. J Robot Surg 2022;16(6):1273–80. https://doi.org/10.1007/s11701-022-01369-x.

[33] Nathan A, Patel S, Georgi M, et al. Virtual classroom proficiency-based progression for robotic surgery training (VROBOT): a randomised, prospective, cross-over, effectiveness study. J Robot Surg 2022;1–7. https://doi.org/10.1007/s11701-022-01467-w. Published online October 17.

[34] Rahimi AO, Ho K, Chang M, et al. A systematic review of robotic surgery curricula using a contemporary educational framework. Surg Endosc 2022. https://doi.org/10.1007/s00464-022-09788-5. Published online December 8.

[35] Covas Moschovas M, Bhat S, Rogers T, Noel J, Reddy S, Patel V. Da Vinci single-port Robotic radical prostatectomy. J Endourol 2021;35(S2):S93–9. https://doi.org/10.1089/end.2020.1090.

[36] Buffi NM, Lughezzani G, Fossati N, et al. Robot-assisted, single-site, dismembered pyeloplasty for ureteropelvic junction obstruction with the new da Vinci platform: a stage 2a study. Eur Urol 2015;67(1):151–6. https://doi.org/10.1016/j.eururo.2014.03.001.

[37] Chauhan R, Ingersol C, Wooden WA, et al. Fundamentals of microsurgery: a novel simulation curriculum based on validated laparoscopic education approaches. J Reconstr Microsurg 2023. https://doi.org/10.1055/a-2003-7425. Published online February 1.

[38] Angell J, Gomez MS, Baig MM, Abaza R. Contribution of laparoscopic training to robotic proficiency. J Endourol 2013;27(8):1027–31. https://doi.org/10.1089/end.2013.0082.

[39] Chahal B, Aydın A, Amin MSA, et al. Transfer of open and laparoscopic skills to robotic surgery: a systematic review. J Robot Surg 2022. https://doi.org/10.1007/s11701-022-01492-9. Published online November 22.

[40] Kowalewski KF, Schmidt MW, Proctor T, et al. Skills in minimally invasive and open surgery show limited transferability to robotic surgery: results from a prospective study. Surg Endosc 2018;32(4):1656–67. https://doi.org/10.1007/s00464-018-6109-0.

[41] Keehner MM, Tendick F, Meng MV, et al. Spatial ability, experience, and skill in laparoscopic surgery. Am J Surg 2004;188(1):71–5. https://doi.org/10.1016/j.amjsurg.2003.12.059.

[42] Moschovas MC, Bhat S, Sandri M, et al. Comparing the approach to radical prostatectomy using the multiport da Vinci Xi and da Vinci SP robots: a propensity score analysis of perioperative outcomes. Eur Urol 2021;79(3):393–404. https://doi.org/10.1016/j.eururo.2020.11.042.

[43] Lovegrove C, Novara G, Mottrie A, et al. Structured and modular training pathway for robot-assisted radical prostatectomy (RARP): validation of the RARP assessment score and learning curve assessment. Eur Urol 2016;69(3):526–35. https://doi.org/10.1016/j.eururo.2015.10.048.

[44] Smith AL, Scott EM, Krivak TC, Olawaiye AB, Chu T, Richard SD. Dual-console robotic surgery: a new teaching paradigm. J Robot Surg 2013;7(2):113–8. https://doi.org/10.1007/s11701-012-0348-1.

[45] Larcher A, De Naeyer G, Turri F, et al. The ERUS curriculum for robot-assisted partial nephrectomy: structure definition and pilot clinical validation. Eur Urol 2019;75(6):1023–31. https://doi.org/10.1016/j.eururo.2019.02.031.

[46] Volpe A, Ahmed K, Dasgupta P, et al. Pilot validation study of the European Association of Urology Robotic training curriculum. Eur Urol 2015;68(2):292–9. https://doi.org/10.1016/j.eururo.2014.10.025.

[47] Rusch P, Kimmig R, Lecuru F, et al. The Society of European Robotic Gynaecological Surgery (SERGS) pilot curriculum for robot assisted gynecological surgery. Arch Gynecol Obstet 2018;297(2):415–20. https://doi.org/10.1007/s00404-017-4612-5.

[48] Veronesi G, Dorn P, Dunning J, et al. Outcomes from the Delphi process of the thoracic Robotic curriculum development committee. Eur J Cardio-Thorac Surg Off J Eur Assoc Cardio-Thorac Surg 2018;53(6):1173–9. https://doi.org/10.1093/ejcts/ezx466.

[49] Nagyné Elek R, Haidegger T. Next in surgical data science: autonomous non-technical skill assessment in minimally invasive surgery training. J Clin Med 2022;11(24):7533. https://doi.org/10.3390/jcm11247533.

[50] Wood TC, Raison N, Haldar S, et al. Training tools for nontechnical skills for surgeons-a systematic review. J Surg Educ 2017;74(4):548–78. https://doi.org/10.1016/j.jsurg.2016.11.017.

[51] Augestad KM, Han H, Paige J, et al. Educational implications for surgical telementoring: a current review with recommendations for future practice, policy, and research. Surg Endosc 2017;31(10):3836–46. https://doi.org/10.1007/s00464-017-5690-y.

# 20

# Surgical training with simulators for robotic surgery

*Vitagliano Gonzalo and Ringa Maximiliano*

Department of Urology, Hospital Alemán, Buenos Aires, Argentina

## Introduction

There is no controversy regarding the need to train before any surgical procedure, especially in urology. The increasing complexity of urological procedures, the sophistication of surgical instruments and robotic platforms along with higher patient expectations and a more litigious working environment make training a must. Different from open surgery, endoscopy, laparoscopy, and robotics share a basic principle. Reality is observed through a camera monitor system; thus, mentoring becomes easy provided the right training elements.

Robotic surgery differentiates from the rest in the absence of haptic feedback. Though some modern platforms are being developed with haptic feedback, most of the mainstream ones lack this feature. In that sense, robotic surgeons must learn how to judge tissue resistance solely by visually monitoring tissue response to traction and compression. A learning curve must be completed before mastering this skill.

In recent years many training modalities have been established for robotic urological surgery. From simple "spikes and rings" to complex hydrogel models, a wide variety of training scenarios have been created to shorten the learning curve and facilitate safer procedures.

## General training environments

A variety of environments has been used in robotic simulation, and each one of them has been analyzed in terms of training and costs (Table 1) [1]. Dry laboratory simulations use classic models for practice with robotic instruments in grasping, cutting, dissecting, and suturing. Unfortunately, this environment has a lack of realism that may affect the skill transfer. In that context, the development of 3D models that mimic the anatomy and tissue elasticity help in training in more realistic scenarios. Alternatively, wet laboratory simulations, using cadaveric and animal models, give the possibility to train tissue responsiveness and hemostasis. The main controversy of both, dry and wet laboratories, is how expensive they are, requiring the availability of the platform for training and in the case of wet laboratory, specifically the material costs, animal maintenance, instruments, and anesthesia. To fill that gap, virtual reality simulators have been developed to offer a substantial number of realistic scenarios with the possibility of replicability and assessment. This is the widely accepted pathway for training while 3D models are in progress. The current Da Vinci training program is based on a combination of these scenarios [2].

## Dry lab

Like laparoscopic training, dry lab utilizes a pelvic trainer adjusted to robotic arms. Using inanimate models for simulation, studies have shown its effectiveness in improving technical skills and reducing the learning curve associated with surgical procedures [3,4]. Furthermore, Da Vinci certification using robotic dry-lab training has shown a decrease in time for real-life RARP steps and an improvement in technical skills.

*Handbook of Robotic Surgery*
https://doi.org/10.1016/B978-0-443-13271-1.00043-1

207

TABLE 1    Training environments.

| Environment | Strengths | Weaknesses |
| --- | --- | --- |
| Dry lab | Same layout and feedback than surgery<br>New 3D printed models with increased realism | Availability of da Vinci Surgical System<br>Current high cost of 3D-printed models<br>Limited metrics and error management<br>Need of mentoring in person |
| Wet lab | Same layout and feedback than surgery<br>Realistic tissue properties, handling, and bleeding | Challenging logistics<br>Ethical and infection control issues<br>Availability and exclusivity of da Vinci Surgical System |
| VR simulation platforms | Safe environment<br>Cost-effective<br>Multiple validated platforms<br>Metrics and curriculums standardized<br>Specific procedural models | High initial investment<br>Different layouts and feedback in most platforms<br>Availability of da Vinci Surgical System console in some platforms (dVSS and ProMIS) |

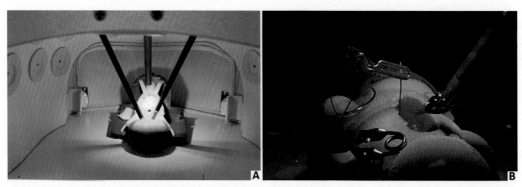

FIG. 1    Two commercially available 3D patient-based models for robotic surgical training. (A) VP Prostate model. (B) VK Kidney model. © Urotrainer, LLC.

Although most frequent exercises were initially related to basic and surgery-deconstructed tasks (e.g., anastomosis), recently a few 3D printed models have been developed with different levels of complexity [5,6].

Advances in 3D-printing technology have facilitated the evolution of synthetic organ models for training robotic surgery in dry lab. The main benefits of these models are based on the realistic layout using both console and robotic arms with tools. Several models have been developed, from early basic prototypes to 3D-printed patient-specific or patient-based models for improving surgical trainee comprehension and localization of renal masses [7–9]. Different materials and combinations were used in order to mimic tissue density and resistance (Fig. 1). Recent synthetic model constructions have focused on integrating them into full procedural simulation and employing methods of validation to confirm their utility as surgical training tools in urology [10,11].

These validated and cost-effective high-fidelity procedural models are currently available for robotic surgery training in urology; however, to enable the widespread adoption of synthetic organ models, it is necessary to engineer scalability for mass production [5]. This would enable a transition in robotic surgical education, where digital and synthetic organ models could replace live animals and cadavers, thus achieving robotic surgery competency. The weaknesses of this environment are expense of the 3D models, as well as the availability of the whole robotic platform.

## Wet lab

Wet laboratory environments utilizing cadaveric and animal models offer an unparalleled opportunity to gain insight into the response of tissue to robotic instruments, develop techniques for gentle tissue handling, and acquire skills like hemostasis that still require more development in dry lab models. Being a higher proportion of animals over cadaveric models, the expense associated with wet laboratory training can create a financial barrier for this pathway, limiting their access to this valuable learning experience [12]. Additionally, ethical concerns surrounding animal welfare may arise with the use of such models. Only a few studies reported face and content validation of porcine models, but no further investigation was done [13,14]. However, wet lab training is considered valuable and effective in improving familiarity and confidence with the robot in real tissues.

# VR simulation

The use of virtual reality training in robotic surgery has emerged as a promising alternative to traditional wet laboratory training. By providing a safe and accessible environment for trainees to practice surgical techniques, VR simulators have the possibility of being easily distributed worldwide where a robotic platform is installed. These simulators utilize feedback and realistic graphics to create a highly immersive experience that closely mimics the tactile and visual cues encountered during real surgical procedures. Multiple studies have shown that VR training can lead to improved technical proficiency establishing it as a valid training tool for robotic surgery [15]. Furthermore, the accessibility and cost-effectiveness of VR training make it an attractive option for trainees who may not have access to expensive dry or wet laboratory facilities, providing an opportunity to level the playing field and enhance surgical education.

Six simulation platforms have been developed for robotic surgery training with different characteristics (Table 2): Simsurgery Educational Platform (SEP), ProMIS, da Vinci Trainer (dv-Trainer), Robotic Surgical Simulator (RoSS), da Vinci Skills Simulator (dVSS), and RobotiX Mentor, the last four being more widely available (Fig. 2 [16]).

The validation literature of these platforms is mainly focused on the first classification of validity which includes face, content, construct, concurrent, and predictive validity. Face validity is concerned with how closely the simulator mimics real-life scenarios. Content validity refers to the extent to which the assessment measures the intended content domain. Construct validity, the objective validation, is related to whether the test is accurately measuring the trait it is intended to assess, including the ability to discriminate between different levels of expertise. Concurrent validity involves how well the results of the simulation correspond with gold standard tests in the same domain and predictive validity assesses the extent to which a simulation can predict future performance. While the ideal curriculum for training is still in discussion and there is no gold standard for training robotic surgery, the skills transfer assessment gained interest in the comparison of different platforms.

## SEP

The SEP Robot is a VR simulator with a console connected to two instruments with a freedom of movement similar to da Vinci. The platform contains a motion tracking device, which detects the controllers and recreates the movements onto a computer screen. Unlike other simulators, the images are not three-dimensional (3D), but the simulator records time, instrument tip trajectory, and error scores for the different tasks.

A trial with 30 participants performing two tasks showed face, content, and construct validation was achieved with 90% realistic score, 87% useful rate, and experts outperforming novices [17]. However, another study failed in showing construct validation with similar times between groups [17,18].

TABLE 2  Simulation platforms characteristics.

| Platform | Manufacturer | Integration of all robotic handling properties[a] | Performance feedback and scoring method | Traditional framework validity | Exercises | Procedure-specific exercises |
|---|---|---|---|---|---|---|
| SEP | Sim Surgery | No | Yes | Face, content, construct | Not specified | No |
| ProMIS | Haptica | No | Yes | Face, content, construct | Not specified | No |
| dV-Trainer | Mimic technologies Inc | Yes | Yes, proficiency system with pass thresholds | Face, content, construct concurrent, predictive | More than 60 | Yes |
| RoSS | Simulated surgical systems LLC | Yes | Yes, point system with pass thresholds | Face, content, construct | More than 50 | Yes |
| dVSS | Intuitive surgical Inc. | Yes | Yes, scaled with pass thresholds | Face, content, construct, concurrent, predictive | Approximately 40 | No |
| RobotiX Mentor | Simbionix USA Inc. | Yes | Yes, proficiency system with pass thresholds | Face, content, construct | More than 50 | Yes |

[a] Endowrist manipulation, camera and fourth arm integration, needle control and driving, energy and dissection.

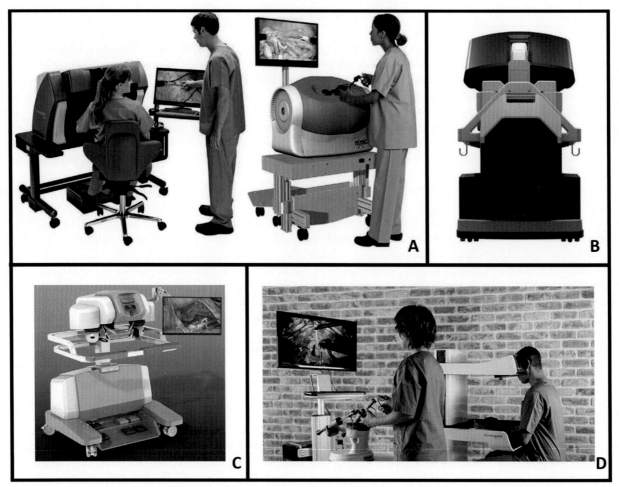

FIG. 2    Four commercially available platforms for robotic surgical training with simulators. (A) dV-Trainer+Xperience Team Trainer © Mimic Technologies, Inc. (B) dVSS console on da Vinci Surgical System © Intuitive Surgical, Inc. (C) RoSS Simulator © Simulated Surgical System, LLC. (D) Simbionix RobotiX Mentor for team training © Surgical Science.

## ProMIS

ProMIS is a surgical simulator initially designed for laparoscopic training and then adapted for robotic platforms. This simulator offers users the ability to interact with both virtual and physical models, providing objective measures of task performance.

A study with eight experienced and 10 novice surgeons showed face, content, and construct validation. The experts rated the platform as accurate for measuring skills, easy to use, and useful for training [19]. Other studies, including a randomized trial, validated the platform for content and construction showing significant differences with a control group without training, improvement after training, and also between novices and experts [20,21].

## dV-Trainer

The dV-Trainer is a portable hardware platform using MSim simulation engine to recreate the look and feel of operating on da Vinci robot. The technology that has been deployed to dVSSs provides access to training outside the OR replicating the vision, hand, and foot controls of the da Vinci robot. The dV-Trainer mimics the hardware of the da Vinci surgical console, the master controllers, and vision cart, and replicates the robotic instrument arms and surgical space through its 3D software. The simulator includes support equipment such as a touch-screen monitor, keyboard, and mouse for instructors to build curricula and manage collected data. The dV-Trainer can simulate the S, Si, and Xi models of the da Vinci robot, and provides independent training outside of the actual surgical system. However, its hardware simulation does not precisely replicate the real robotic controls, with unique controls connected to three cables measuring movement instead of the precise master controllers used in the da Vinci robot.

The system's advantage is its ability to simulate various da Vinci models, offer independent training, and build curricula. On the other hand, its hardware simulation is not as precise as the da Vinci robot, and the controls differ from the master controllers, potentially affecting training quality. Despite the similarity between the dVSS and dV-Trainer exercises, the latter offers new exercises not found in the former, and its visual presentation has been improved in the current version. The dV-Trainer is a large system, smaller than a surgical console with the dVSS attached, and includes an adjustable table for positioning and transporting the simulator device.

In recent years, the dV-Trainer developers have introduced two novel advancements: Mimic's Xperience Team Trainer (XTT) and Flex-VR. XTT is a haptic-enabled laparoscopic trainer that interfaces with the dV-Trainer, enabling team training for both the console surgeon and the first assistant. On the other hand, Flex-VR is a portable robotic surgery training system, offering high mobility, flexibility, and storage.

Multiple studies have examined the face validity of the dv-Trainer, with the majority rating highly in terms of exercise quality, visual and hardware realism, depth perception, precision, and instrument movement [22–27]. Various questionnaires were used to assess these factors, with consistent results across studies [16]. Furthermore, the content of the dv-Trainer has been evaluated by expert surgeons who deemed it effective for training both surgeons and residents [23,25,28–30].

Several studies have investigated the construct validity of the dv-Trainer, revealing some variability in performance between individuals with different levels of previous training. Nonetheless, all studies reported construct validity for the dv-Trainer [22,23,25–31]. Among these studies, previously trained surgeons outperformed control groups in tasks where a comparison was made. In addition, studies of skills transfer for the dv-Trainer have utilized inanimate and animal tissue models [24,32–34]. In two of them where there was a control group, the training group outperformed [24,34].

## RoSS

The RoSS is a self-contained training platform that closely emulates the da Vinci robotic console, similar to the dV-Trainer. Developed by the University of Buffalo and the Roswell Park Cancer Institute, RoSS hardware unit features a 3D monitor, off-the-shelf hand controllers with a limited range of motion comparing da Vinci console, foot pedals, and an instructor monitor. The simulator can simulate both S and Si models of the da Vinci robot. With a curriculum for trainees without prior robotic surgery experience, the simulator offers unique 3D exercises and procedure-specific hands-on surgical training. The exercises incorporate force feedback to guide students through simulated instrument movements and require precise execution before proceeding to the next step.

Multiple studies have assessed the effectiveness of the RoSS system in training surgeons. For face validation, a group of 30 surgeons (24 experienced and 6 without robotic surgery experience) completed basic and complex acquisition and placement tasks. Post-task questionnaires showed that the RoSS platform was a realistic approximation of dVSS for virtual simulation and instrumentation. More than 80% of participants found the pinch device, movement arm, camera movement, and clutch functions to be somewhat or very close to the dVSS platform [35].

To prove content validation, 17 experienced robotic surgeons, 14 intermediate robotic surgeons, and 11 novice surgeons without robotic experience completed modules of progressively increased difficulty, including clutch control, acquisition, precision control and positioning of an object, needle handling, and the use of the fourth arm. The majority of participants rated the modules as "good" or "excellent," and agreed that the platform was useful for training and appropriate for testing residents before operating room access [36].

For construct validation, a group of novice and expert surgeons completed four tasks: ball placement, coordinated tool control, fourth arm control, and needle handling. Expert surgeons demonstrated significantly better performance on all four tasks compared to the novice group [37].

The educational impact and skill transfer were also assessed, with novice participants who trained with RoSS demonstrating a positive impact on performance and time to complete anatomical recognition tasks on the da Vinci platform compared to novices with no prior training [38].

## dVSS

The dVSS is a specialized software package in hardware that connects to the actual surgical console through a single fiber optic network cable. Two dVSS models are currently available, one for each of the da Vinci Surgical Systems on the market, the da Vinci Si, and the da Vinci Xi. The models are not interchangeable and can only be used with the corresponding surgical system. The newer dVSS Xi model is integrated more deeply with the da Vinci Xi Surgical System and offers additional capabilities not available with the original dVSS Si. Both models contain similar psychomotor skill exercises.

The dVSS is considered an embedded trainer because it utilizes the existing equipment of the da Vinci Surgical System. Although the surgical system and simulator are sold separately, purchasing the dVSS may be cost-effective if the da Vinci Surgical System has already been acquired. However, the dVSS is reliant on access to the surgical console for use, which means that it cannot be used during surgical procedures, limiting the availability of trainees. An advantage of the dVSS is that trainees use the actual surgical console that they will use during surgery, allowing the virtual-reality simulator to closely mimic the surgical experience and facilitate higher skill transfer from the training sessions to the actual system.

The simulator includes a set of exercises that allow surgeons to practice various robotic surgical techniques, such as suturing, knot tying, and tissue manipulation. It also provides feedback on the surgeon's performance, including metrics such as accuracy and speed, and allows the surgeon to repeat the exercises until proficiency is achieved.

Numerous face validation studies have reported high scores for the realism of the dVSS, with regards to 3D view, movement, and foot control, among novice and expert surgeons. Despite differences in the Likert scales used, the dVSS was consistently rated highly realistic [39–42]. Additionally, expert robotic surgeons have rated the dVSS as a highly valuable training system for surgical residents, as assessed in content validation studies [39–42].

Construct validation studies have shown that expert or intermediate surgeons significantly outperform novices, although there is variability in the study designs with respect to expertise groups and a number of tasks evaluated [41–45]. Moreover, several studies have evaluated skills transfer for dVSS using inanimate models, animal tissue models, or real patients, and have found that simulator-trained surgeons outperform the control group [46–48]. These findings support the validity of the dVSS and suggest that it may be an effective tool for training novice surgeons.

## RobotiX Mentor

Robotix Mentor is a platform designed to provide comprehensive training including complete robotic clinical procedures.

It is a comprehensive, independent simulator of the da Vinci surgical robot, comparable to the dV-Trainer and RoSS. It consists of two mobile carts that support the replicated surgical console, controls, and vision cart, which are replicated in hardware. The 3D software model reproduces the functions of the robotic arms and the surgical space. The system allows instructors to create custom curricula using available exercises and manage collected data. The innovative free-floating hand controls mimic elements of the da Vinci master controllers and are measured through electromagnetic field tracking with 6 degrees of freedom. The hand controls are tethered to the armrest by a bundled electronic cable, and their attachment and orientation were designed to minimize interference and weighted drag on hand movements. It also allows team-training options with the incorporation of the laparoscopic platform for the surgical assistant to collaborate with the robotic surgeon in practice, working in the same training environment in order to improve communication skills and synchronization.

Two studies with 46 and 37 participants divided into three levels of expertise and performing 9 and 6 tasks respectively, showed face, content, and construct validation [49,50]. A proficiency-based progression curriculum was developed, and skills transfer was assessed in 20 novices for this platform. The completion resulted in improved robotic surgery skills among participants [51].

## VR simulation on urological procedures

The number of robotic surgeries performed each year has been steadily increasing since its approval but at the same time the training for surgeons was an evolving challenge. Since the mentioned platforms with different basic tasks were developed, the initial approach to the robotic interface and controls in novices was resolved. However, the lack of dual consoles in many institutions to train the transition from deconstructed tasks to real surgery steps generated a new gap in robotic surgery training.

In order to improve the robotic learning process, simulated urological procedures were developed in the widely used platforms. In dV-Trainer, a partial nephrectomy module was validated for face, content, and construct [30]. The procedure was rated realistic and effective as a training tool. RoSS has hands-on surgical training, which is an integration of surgical procedure-specific skill exercises with a video from a surgery. In this tool, animated icons are overlaid in the trainer's visual space as instructional guides, indicating specific actions to be taken during the progression of real surgical videos. Unlike separated procedure-specific tasks, RobotiX Mentor provides complete comprehensive surgeries and exercises in a fully simulated anatomical environment.

## Scoring system

The use of proficiency scoring systems in surgical simulators has become increasingly common for evaluating the performance of students in various performance areas. Among the four simulators currently available on the market, each one utilizes a distinct scoring system, with data on student performance collected via the host computer and analyzed through an algorithm to generate a composite score. Acceptable and unsatisfactory scoring levels are established by identifying thresholds, which are determined based on the performance of experienced robotic surgeons. These simulators often utilize a color-based system to convey the quality of a student's performance across multiple metrics.

The dVSS performance scoring system employs a variety of objective metrics to evaluate a student's performance in every exercise. The Classic System provides an overall score represented as a percentage, while the Proficiency-Based System reports the total points earned rather than a percentage. The dVSS does not offer threshold adjustments, which are available in other simulators. The RoSS and dV-Trainer share similarities with the dVSS scoring system, although the metrics collected may differ. In contrast, the scoring metrics for the RobotiX Mentor system vary depending on the simulation module and case, and proficiency benchmarks have yet to be established for all exercises. System administration functions are commonly available in all simulators to create a tailored curriculum and scoring system for lessons, generate new student accounts, and export student scores for evaluation and analysis outside of the simulator device.

## Future scenario

The future of robotic surgery appears highly promising and new tools coupled to previous platforms are currently being developed to enhance surgical training. Innovations in proctoring, like augmented reality tools that facilitate live mentoring during surgical procedures providing real-time feedback or online learning applications equipped with video storage and analytical capabilities, are trying to address the current concerns in the evaluation of surgeon technique [52]. Cutting-edge advancements in artificial intelligence and machine learning are driving the development of innovative tools for the analysis of surgical performance videos. These technologies enable a meticulous evaluation of surgical proficiency, offering precise and individualized real-time feedback to facilitate personalized surgical training programs [53].

Given this perspective, it is anticipated that comprehensive metrics for intricate procedures will be established in the near future. However, there are persisting challenges that demand attention in the training with simulators. The establishment of competency standards for the currently available training tools, the validation of evolving curricula, and the translation of acquired simulation-based skills into tangible outcomes in the operating room and patient care will need attention, particularly during the development of these new technologies.

## Conclusion

Training is an essential component of any surgical procedure, especially in urology due to the ever-increasing complexity of procedures and the need to meet higher patient expectations. Robotic surgery, in particular, requires a specific set of skills due to the absence of haptic feedback. A variety of training environments, including dry labs, wet labs, and virtual reality simulators, have been developed to shorten the learning curve and facilitate safer procedures. Each environment has its advantages and limitations, and the choice of training modality depends on availability, cost-effectiveness, and the trainee's specific needs. Despite their limitations, synthetic organ models and VR simulators have emerged as promising alternatives to traditional wet laboratory training, offering the possibility of being easily distributed worldwide and enhancing surgical education. The improvements of existing training modalities and the adoption of them into a comprehensive curriculum are essential to prepare the next generation of robotic surgeons for the challenges ahead.

## Key points

- The global adoption of robotic surgery has surged dramatically, rendering traditional surgical training models inadequate.
- Robotic urological surgery demands exceptional technical and nontechnical expertise, and is constantly challenging trainers and mentors.
- Numerous training methods have been established, accompanied by diverse scenarios, aiming to accelerate the learning process and enhance procedural safety.
- Various simulation platforms provide realistic environments for learning, incorporating validated assessment tools.
- To meet educational program benchmarks and guarantee patient safety, it is crucial to integrate existing and future robotic surgical training tools into a comprehensive curriculum.

## References

[1] Lovegrove CE, Elhage O, Khan MS, et al. Training modalities in robot-assisted urologic surgery: a systematic review. Eur Urol Focus 2017; 3(1):102–16.

[2] Intuitive learning platform. https://www.intuitive.com/en-us/products-and-services/da-vinci/learning [Accessed 5 April 2023].

[3] Hinata N, Iwamoto H, Morizane S, et al. Dry box training with three-dimensional vision for the assistant surgeon in robot-assisted urological surgery. Int J Urol 2013;20(10):1037–41.

[4] Rashid HH, Leung YYM, Rashid MJ, Oleyourryk G, Valvo JR, Eichel L. Robotic surgical education: a systematic approach to training urology residents to perform robotic-assisted laparoscopic radical prostatectomy. Urology 2006;68(1):75–9.

[5] Costello DM, Huntington I, Burke G, et al. A review of simulation training and new 3D computer-generated synthetic organs for robotic surgery education. J Robot Surg 2022;16(4):749–63.

[6] Parikh N, Sharma P. Three-dimensional printing in urology: history, current applications, and future directions. Urology 2018;121:3–10.

[7] Silberstein JL, Maddox MM, Dorsey P, Feibus A, Thomas R, Lee BR. Physical models of renal malignancies using standard cross-sectional imaging and 3-dimensional printers: a pilot study. Urology 2014;84(2):268–72.

[8] von Rundstedt FC, Scovell JM, Agrawal S, Zaneveld J, Link RE. Utility of patient-specific silicone renal models for planning and rehearsal of complex tumour resections prior to robot-assisted laparoscopic partial nephrectomy. BJU Int 2017;119(4):598–604.

[9] Vitagliano G, Mey L, Rico L, Birkner S, Ringa M, Biancucci M. Construction of a 3D surgical model for minimally invasive partial nephrectomy: the Urotrainer VK-1. Curr Urol Rep 2021;22(9):48.

[10] Ghazi A, Melnyk R, Hung AJ, et al. Multi-institutional validation of a perfused robot-assisted partial nephrectomy procedural simulation platform utilizing clinically relevant objective metrics of simulators (CROMS). BJU Int 2021;127(6):645–53.

[11] Witthaus MW, Farooq S, Melnyk R, et al. Incorporation and validation of clinically relevant performance metrics of simulation (CRPMS) into a novel full-immersion simulation platform for nerve-sparing robot-assisted radical prostatectomy (NS-RARP) utilizing three-dimensional printing and hydrogel casting technology. BJU Int 2020;125(2):322–32.

[12] Moreno Sierra J, Fernández Pérez C, Ortiz Oshiro E, et al. Key areas in the learning curve for robotic urological surgery: a Spanish multicentre survey. Urol Int 2011;87(1):64–9.

[13] Passerotti CC, Passerotti AMAMS, Dall'Oglio MF, et al. Comparing the quality of the suture anastomosis and the learning curves associated with performing open, freehand, and robotic-assisted laparoscopic pyeloplasty in a swine animal model. J Am Coll Surg 2009;208(4):576–86.

[14] Raison N, Poulsen J, Abe T, Aydin A, Ahmed K, Dasgupta P. An evaluation of live porcine simulation training for robotic surgery. J Robot Surg 2021;15(3):429–34.

[15] Moglia A, Ferrari V, Morelli L, Ferrari M, Mosca F, Cuschieri A. A systematic review of virtual reality simulators for robot-assisted surgery. Eur Urol 2016;69(6):1065–80.

[16] These figures were published in Endorobotics, Luigi Manfredi, simulators; 2022. p. 95–113, ISBN:9780128217504. Copyright Elsevier.

[17] Gavazzi A, Bahsoun AN, Van Haute W, et al. Face, content and construct validity of a virtual reality simulator for robotic surgery (SEP robot). Ann R Coll Surg Engl 2011;93(2):152–6.

[18] van der Meijden OAJ, Broeders IAMJ, Schijven MP. The SEP "robot": a valid virtual reality robotic simulator for the Da Vinci surgical system? Surg Technol Int 2010;19:51–8.

[19] McDonough PS, Tausch TJ, Peterson AC, Brand TC. Initial validation of the ProMIS surgical simulator as an objective measure of robotic task performance. J Robot Surg 2011;5(3):195–9.

[20] Feifer A, Al-Ammari A, Kovac E, Delisle J, Carrier S, Anidjar M. Randomized controlled trial of virtual reality and hybrid simulation for robotic surgical training. BJU Int 2011;108(10):1652–6 [discussion 1657].

[21] Jonsson MN, Mahmood M, Askerud T, et al. ProMIS™ can serve as a da Vinci® simulator—a construct validity study. J Endourol 2011;25 (2):345–50.

[22] Lendvay TS, Casale P, Sweet R, Peters C. Initial validation of a virtual-reality robotic simulator. J Robot Surg 2008;2(3):145–9.

[23] Kenney PA, Wszolek MF, Gould JJ, Libertino JA, Moinzadeh A. Face, content, and construct validity of dV-trainer, a novel virtual reality simulator for robotic surgery. Urology 2009;73(6):1288–92.

[24] Korets R, Mues AC, Graversen JA, et al. Validating the use of the mimic dV-trainer for robotic surgery skill acquisition among urology residents. Urology 2011;78(6):1326–30.

[25] Hung AJ, Zehnder P, Patil MB, et al. Face, content and construct validity of a novel robotic surgery simulator. J Urol 2011;186(3):1019–24.

[26] Perrenot C, Perez M, Tran N, et al. The virtual reality simulator dV-trainer(®) is a valid assessment tool for robotic surgical skills. Surg Endosc 2012;26(9):2587–93.

[27] Schreuder HWR, Persson JEU, Wolswijk RGH, Ihse I, Schijven MP, Verheijen RHM. Validation of a novel virtual reality simulator for robotic surgery. ScientificWorldJournal 2014;2014, 507076.

[28] Sethi AS, Peine WJ, Mohammadi Y, Sundaram CP. Validation of a novel virtual reality robotic simulator. J Endourol 2009;23(3):503–8.

[29] Kang SG, Cho S, Kang SH, et al. The tube 3 module designed for practicing vesicourethral anastomosis in a virtual reality robotic simulator: determination of face, content, and construct validity. Urology 2014;84(2):345–50.

[30] Hung AJ, Shah SH, Dalag L, Shin D, Gill IS. Development and validation of a novel robotic procedure specific simulation platform: partial nephrectomy. J Urol 2015;194(2):520–6.

[31] Lee JY, Mucksavage P, Kerbl DC, Huynh VB, Etafy M, McDougall EM. Validation study of a virtual reality robotic simulator—role as an assessment tool? J Urol 2012;187(3):998–1002.

[32] Lerner MA, Ayalew M, Peine WJ, Sundaram CP. Does training on a virtual reality robotic simulator improve performance on the da Vinci surgical system? J Endourol 2010;24(3):467–72.

[33] Whitehurst SV, Lockrow EG, Lendvay TS, et al. Comparison of two simulation systems to support robotic-assisted surgical training: a pilot study (swine model). J Minim Invasive Gynecol 2015;22(3):483–8.

[34] Cho JS, Hahn KY, Kwak JM, et al. Virtual reality training improves da Vinci performance: a prospective trial. J Laparoendosc Adv Surg Tech A 2013;23(12):992–8.

[35] Seixas-Mikelus SA, Kesavadas T, Srimathveeravalli G, Chandrasekhar R, Wilding GE, Guru KA. Face validation of a novel robotic surgical simulator. Urology 2010;76(2):357–60.

[36] Seixas-Mikelus SA, Stegemann AP, Kesavadas T, et al. Content validation of a novel robotic surgical simulator. BJU Int 2011;107(7):1130–5.

[37] Kesavadas T, Kumar A, Srimathveeravalli G, et al. Efficacy of robotic surgery simulator (Ross) for the davinci® surgical system. J Urol 2009;181 (4S):823.

[38] Guru KA, Baheti A, Kesavadas T, Kumar A, Srimathveeravalli G, Butt Z. In-vivo videos enhance cognitive skills for Da Vinci® surgical system. J Urol 2009;181(4S):823.

[39] Alzahrani T, Haddad R, Alkhayal A, et al. Validation of the da Vinci surgical skill simulator across three surgical disciplines: a pilot study. Can Urol Assoc J 2013;7(7–8):E520–9.

[40] Kelly DC, Margules AC, Kundavaram CR, et al. Face, content, and construct validation of the da Vinci skills simulator. Urology 2012;79 (5):1068–72.

[41] Lyons C, Goldfarb D, Jones SL, et al. Which skills really matter? Proving face, content, and construct validity for a commercial robotic simulator. Surg Endosc 2013;27(6):2020–30.

[42] Ramos P, Montez J, Tripp A, Ng CK, Gill IS, Hung AJ. Face, content, construct and concurrent validity of dry laboratory exercises for robotic training using a global assessment tool. BJU Int 2014;113(5):836–42.

[43] Finnegan KT, Meraney AM, Staff I, Shichman SJ. da Vinci skills simulator construct validation study: correlation of prior robotic experience with overall score and time score simulator performance. Urology 2012;80(2):330–5.

[44] Hung AJ, Jayaratna IS, Teruya K, Desai MM, Gill IS, Goh AC. Comparative assessment of three standardized robotic surgery training methods. BJU Int 2013;112(6):864–71.

[45] Connolly M, Seligman J, Kastenmeier A, Goldblatt M, Gould JC. Validation of a virtual reality-based robotic surgical skills curriculum. Surg Endosc 2014;28(5):1691–4.

[46] Vaccaro CM, Crisp CC, Fellner AN, Jackson C, Kleeman SD, Pavelka J. Robotic virtual reality simulation plus standard robotic orientation versus standard robotic orientation alone: a randomized controlled trial. Female Pelvic Med Reconstr Surg 2013;19(5):266–70.

[47] Kiely DJ, Gotlieb WH, Lau S, et al. Virtual reality robotic surgery simulation curriculum to teach robotic suturing: a randomized controlled trial. J Robot Surg 2015;9(3):179–86.

[48] Culligan P, Gurshumov E, Lewis C, Priestley J, Komar J, Salamon C. Predictive validity of a training protocol using a robotic surgery simulator. Female Pelvic Med Reconstr Surg 2014;20(1):48–51.

[49] Whittaker G, Aydin A, Raison N, et al. Validation of the RobotiX Mentor robotic surgery simulator. J Endourol 2016;30(3):338–46.

[50] Alshuaibi M, Perrenot C, Hubert J, Perez M. Concurrent, face, content, and construct validity of the RobotiX Mentor simulator for robotic basic skills. Int J Med Robot 2020;16(3), e2100.

[51] Martin JR, Stefanidis D, Dorin RP, Goh AC, Satava RM, Levy JS. Demonstrating the effectiveness of the fundamentals of robotic surgery (FRS) curriculum on the RobotiX Mentor virtual reality simulation platform. J Robot Surg 2021;15(2):187–93.

[52] Azadi S, Green IC, Arnold A, Truong M, Potts J, Martino MA. Robotic surgery: the impact of simulation and other innovative platforms on performance and training. J Minim Invasive Gynecol 2021;28(3):490–5.

[53] Hung AJ, Chen J, Jarc A, Hatcher D, Djaladat H, Gill IS. Development and validation of objective performance metrics for robot-assisted radical prostatectomy: a pilot study. J Urol 2018;199(1):296–304.

# Education, safety, ethical and administrative aspects for a successful robotic program

# Robotic surgery: Proctoring and teleproctoring

*Alexis Sánchez*[a] *and Omaira Rodríguez González*[b]

[a]Orlando Regional Medical Center, Orlando Health Robotic Surgery Program, Orlando, FL, United States [b]Chief of Surgical Department, Clínicas Caracas Hospital, Faculty of Medicine, Central University of Venezuela, Caracas, Venezuela

## Introduction

Introducing new technologies in the surgical field can bring many benefits, such as improving outcomes, increasing efficiency, and reducing costs. However, it is important to consider the safety implications of new technologies, as they can introduce new risks and challenges. Surgeons and other members of the surgical team should be properly trained and educated on the use of new technologies, to ensure that they are able to use them safely and effectively [1,2].

Robotic surgery has revolutionized the field of minimally invasive surgery, offering improved precision, accuracy, and control to surgeons [3–5]. Mastering the skills required to perform robotic surgery can be a challenging and time-consuming process. Proctoring and teleproctoring have emerged as key strategies for training surgeons in the use of robotic surgical systems [6,7].

This chapter will provide an overview of the concept of proctoring and teleproctoring in robotic surgery, explaining the importance of proctoring in the training process, and the different types of proctoring and teleproctoring that are currently used in the field. It will also explore the relationship between adult learning theories and proctoring, describing the training phases for robotic surgery, and discussing the techniques and technologies that can be used to enhance the proctoring process. Finally, it will examine the safety and regulatory implications of proctoring and teleproctoring, highlighting the technical and legal limitations that should be considered when implementing these strategies.

## Robotic surgery training phases

The training process for robotic surgery is typically divided into several phases, each with its own specific objectives and outcomes. In this section, we will describe the different training phases that are typically used to learn robotic surgery with the da Vinci [8,9] system and provide an overview of the teaching modalities that are used during each phase.

The preclinical or product training component of the training is completed during the first two phases.

**Phase I.** Before beginning hands-on training with the da Vinci system, surgeons typically complete a **pre-training phase that includes online modules and didactic training**. This phase is designed to familiarize trainees with the robotic surgical system. This may include learning about the main steps of specific procedures, as well as the basic principles of robotic surgery, including the use of the da Vinci console, instruments, and patient positioning.

**Phase II.** In this phase, trainees practice using the da Vinci system on **virtual reality simulators**. This allows them to become familiar with the system and develop their dexterity and hand-eye coordination [10]. The main purpose of this phase is to develop trainees' skills in performing more complex procedures and to prepare them for live surgery.

This phase typically includes supervised practice with animal or cadaveric models.

*Handbook of Robotic Surgery*
https://doi.org/10.1016/B978-0-443-13271-1.00035-2

The following phases (Clinical Training) are completed during live surgeries:

**Phase III. Proctored training,** trainees begin to perform surgeries under the direct supervision of an experienced da Vinci surgeon. During this phase, trainees will perform the procedures with the proctor's guidance and feedback, allowing them to develop their skills and confidence.

**Phase IV Unsupervised training,** once the trainee has completed a sufficient number of proctored cases, they may begin to perform surgeries independently, with ongoing feedback and evaluation designed to ensure that surgeons maintain the skills and knowledge required to perform robotic surgery safely and effectively. This phase typically includes continuing education, regular practice, and participation in proctoring or teleproctoring sessions.

It's important to note that the specific training phases and objectives may vary depending on the specific context, the trainee's level of experience, and the surgical system being used [11].

## Proctoring vs mentoring

Proctoring typically refers to the process of having an experienced surgeon supervise or guide a less experienced surgeon during a surgical procedure, usually with an emphasis on technical skills and safety. The proctor's role is to provide real-time guidance, feedback, and support to the trainee surgeon, helping them to improve their skills and perform the procedure safely. The focus of proctoring is on the practical aspects of the procedure and the technical skills involved in using the robotic system (Fig. 1).

Mentoring, on the other hand, is a more holistic and long-term process that focuses on the professional development of the trainee surgeon. A mentor's role is to provide guidance, advice, and support to the trainee in all aspects of their professional life, including technical skills, but also including professional development. A mentor provides feedback and guidance on the trainee's performance but may also provide guidance on broader issues such as communication skills, team dynamics, career advancement, personal growth, and professional ethics (Fig. 2).

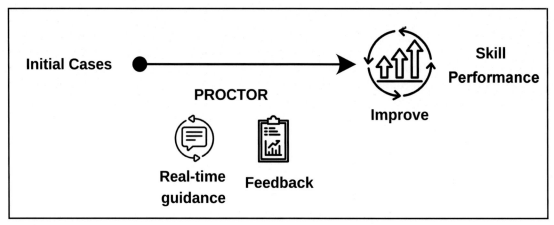

FIG. 1    Proctoring.

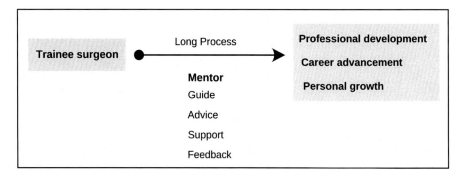

FIG. 2    Mentoring.

While the two concepts are related, they have different goals and objectives. Proctoring focuses on the practical aspects of the procedure and the technical skills involved in using the robotic system, while mentoring is more focused on the long-term professional development of the trainee surgeon. Both proctoring and mentoring can be an important part of the training process and can help to improve the safety and outcomes of the procedure, and the professional development of the trainee [6,7].

## Proctoring and adult learning theories

Proctoring in robotic surgery is aligned with various adult learning theories, it enables transfer of knowledge and skills, enhances patient care quality, and serves as a quality control measure to ensure surgeons meet the necessary standards of care.

**Andragogy** is one adult learning theory that is particularly relevant to proctoring. This theory emphasizes the self-direction and self-motivation of adult learners. In proctoring, the trainee surgeon is able to see a direct connection between the learning and their professional goals, as the proctor guides them to perform the procedure safely and correctly [12].

**Transformative learning theory** also aligns well with proctoring, as proctoring allows trainee surgeon to challenge their existing assumptions and beliefs, leading to new ways of understanding and acting in the surgical world. The proctor can provide guidance, feedback, and support that leads to the trainee's cognitive and emotional development, and this will help them to develop a deeper understanding of the surgical procedures.

**Experiential learning theory**, which emphasizes the importance of hands-on experience and reflection, is also relevant to proctoring. The trainee surgeon is able to actively participate in the surgical procedure, while receiving real-time feedback and guidance from the proctor. This allows them to learn through direct experience, and reflect on their performance to improve their skills.

Additionally, **Kolb's model** of experiential learning is also an important aspect of proctoring, it explains that learning occurs through a cyclical process of experiencing, reflecting, thinking, and acting [13]. Proctoring allows the trainee to experience the procedure, reflect on their performance, think about how to improve and act on the feedback provided by the proctor.

**Self-directed learning theory** also applies to proctoring, as the trainee surgeon is given control over their own learning process and sets their own goals. Proctoring allows the trainee to be self-motivated and self-directed, taking ownership of their learning process, which can lead to better retention.

Finally, the **connectivism theory**, which emphasizes the importance of networks and connections in learning, can also be applied to proctoring. The proctor can provide a network of knowledge and resources for the trainee, allowing them to connect with others and build their skills and understanding of the procedure.

In summary, adult learning theories provide a valuable framework for understanding how adults learn and how to design effective learning experiences [14]. Proctoring aligns well with adult learning theories and can be an effective way to facilitate learning in adults.

## The proctoring process

There is no consensus on how a proctoring session should be done. Different proctors could have totally different proctoring methodologies and styles. The following is the process we have been applying for years of acting as proctors (Fig. 3).

### Before the procedure

The proctoring process should start with **case discussion**. This case discussion before the procedure can help the trainee understand the surgical plan, the key points of the procedures, and the potential risks and benefits, this will help the trainee to be more prepared for the surgery. It also can help the trainee to ask questions and clear up any doubts.

In addition to this, the proctor can use the case discussion to identify areas where the trainee may need additional guidance or training, and to tailor their teaching and mentoring accordingly. It also can help the proctor to understand the trainee's level of knowledge, skills, and experience, and adjust the level of guidance and assistance they provide during the procedure.

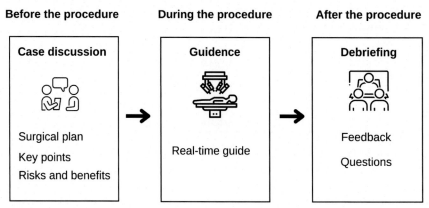

FIG. 3    Proctoring process.

## During the procedure

- The proctor guides the trainee through the use of the robotic surgical system. Starting with patient positioning and portal's location. Depending on the trainee level the proctor would need to focus his attention on different components of the robotic system or the procedure. We would like to emphasize that a proficient surgeon should be familiar with all the components and elements of the process and not only with the console handling.
- The proctor watches the trainee perform the procedure and provides guidance. How active the proctor will be will obviously depend on trainee's experience and skills.
- The proctor may demonstrate specific techniques or maneuvers. For this purpose, the telestration capability of the robotic video tower monitor is a key factor. This function allows the proctor to draw or annotate over the video feed in real time and highlight specific areas of the procedure that need attention.
- The proctor may assist with the actual surgery if necessary. Therefore we consider it important for the proctor to obtain temporary privileges at the trainee facility. Using an internal proctor will help in this matter, since the proctor has privileges to act if necessary.
- The use of a dual console is not indispensable but is of great utility for the proctor. By using the second console, the proctor will also have a three-dimensional view of the operative field and can use the controls to give instructions that will be visible on the active console under the control of the trainee. Additionally, the proctor could take control of the instruments from the console he is using if considered necessary.
- The main goal is to ensure that the trainee becomes proficient in the use of the robotic system and can perform surgeries safely and effectively.

## After the procedure

   **Debriefing** is a process of reviewing and discussing the surgical procedure after it has been completed. This process is considered an essential part of robotic surgery proctoring as it allows the trainee and proctor to reflect on the procedure and identify areas for improvement [13].

   Debriefing can be done immediately after the procedure or at a later time. During debriefing, the trainee and proctor can discuss whether the surgical plan was appropriate, the execution of the procedure, and any challenges or complications that arose. The proctor will provide feedback on the trainee's performance, and the trainee can ask questions and share their own observations.

   Debriefing can be beneficial for a number of reasons:

- It allows the trainee to reflect on their performance and identify areas for improvement
- It provides an opportunity for the proctor to address any misconceptions or errors that the trainee may have
- It allows both the trainee and proctor to learn from the procedure and incorporate that knowledge into future surgeries
- It provides an opportunity for team members to share best practices and learn from other's experiences

# Choosing the right proctor

Proctoring is an art that requires some other features beyond technical skills, and also the proctor himself is an individual in continuous learning. The following are fundamental characteristics of a good Proctor (Fig. 4).

**Technical expertise** is an important aspect of proctoring, as the proctor should have significant experience and expertise in performing the procedures they will be supervising. They should be highly proficient in using the robotic system and have a thorough understanding of the anatomy, physiology, and surgical techniques involved.

**Clinical judgment** is another critical factor in proctoring. The proctor should possess good clinical judgment, be able to identify potential complications and take appropriate measures to mitigate them. They should have the ability to make quick and effective decisions in case of unforeseen events.

**Leadership skills** are also important for a proctor. They should be able to lead and manage a surgical team, set an example of best practices, and create a positive and safe surgical environment. Effective communication, collaboration, and teamwork are essential components of good leadership in the operating room.

**Teaching and mentoring experience** is another important requirement for a proctor. They should possess strong teaching skills and effectively communicate and demonstrate surgical techniques to trainee surgeons. The proctor should be able to provide clear and concise instructions and adjust their teaching style to suit the trainee's learning needs.

**Effective communication** is critical in proctoring. The proctor should have good communication skills, both verbal and non-verbal, in order to provide clear and effective guidance and feedback to the trainee during the procedure.

**Patience** is also an essential characteristic of a proctor. They should have the patience to work with trainees who may be at different levels of experience and proficiency, and be able to provide guidance and feedback in a way that is both supportive and challenging.

**Flexibility**, the proctor should be able to adapt their teaching style to meet the needs of the trainee and be able to modify their approach as necessary to ensure the success of the training process.

**Professionalism** is a key requirement for a proctor. They should demonstrate ethical behavior and adhere to the highest standards of patient care and safety.

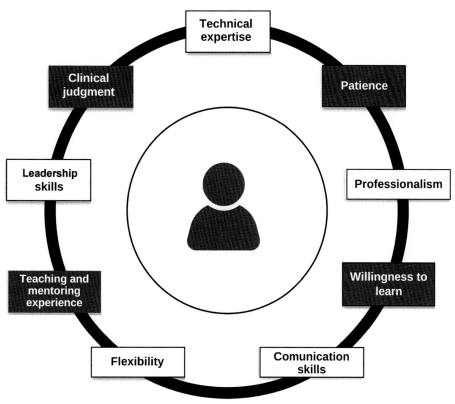

FIG. 4 The right proctor.

A proctor should have a **willingness to learn and adapt** their teaching methods to suit the trainee's learning needs. This includes keeping up-to-date with the latest developments and advancements in the field, as well as seeking feedback and advice from colleagues and other experts.

## Influence of surgeons' style in the proctoring process

The surgeon's style can have a significant impact on the proctoring process, as different surgeons may have different approaches and techniques for performing a procedure. This can affect the trainee's learning experience, as they may be exposed to different techniques and approaches and may have to adapt to the style of the proctor.

For example, some surgeons may have a more hands-off approach, allowing the trainee to take the lead and provide feedback and guidance as needed. This approach can be beneficial for trainees who are more experienced or who have a strong foundation in the procedure. On the other hand, some surgeons may have a more hands-on approach, providing step-by-step instructions and guidance throughout the procedure. This approach can be beneficial for trainees who are less experienced or who need more guidance.

The proctor's style can also affect the trainee's ability to develop their own style and approach to the procedure. Some proctors may encourage trainees to develop their own techniques and approaches, while others may be more focused on teaching a specific technique or approach. This can impact the trainee's ability to become proficient in the procedure and to be able to perform it safely and effectively.

In addition, the surgeon's style can also affect the trainee's ability to apply the skills and knowledge they have learned in other contexts. For example, if the trainee is exposed to a wide variety of techniques and approaches, they may be better able to adapt to different surgical scenarios and to work with different surgeons in the future.

Overall, it is important to consider the impact of the surgeon's style on the proctoring process and to adapt the proctoring approach accordingly. It's also important to expose the trainee to different techniques and approaches, to help them to develop their own style and to be able to adapt to different surgical scenarios.

## Teleproctoring

Teleproctoring in robotic surgery is the use of remote communication technologies, such as videoconferencing or web-based platforms, to provide real-time guidance, supervision, and support to the surgeon operating the robotic system. This allows for expert surgeons to mentor and train less experienced surgeons from a distance, which can be beneficial for surgeons in remote or underserved areas [15].

One of the benefits of teleproctoring is the ability to provide training to surgeons who would otherwise not have access to expert guidance. Additionally, it allows for more efficient use of resources, as it eliminates the need for the proctor to be physically present in the operating room.

High-speed Internet development and new technologies for teleproctoring received a great boost during the pandemic [16]. Limited mobility around the world, and even within the country, led to the development of new platforms to offer remote care.

There is currently no universal consensus on how to provide orders or recommendations during teleproctoring sessions in robotic surgery. Different programs and institutions may have different protocols and procedures in place. However, there are some generally accepted best practices [17] that can help to ensure effective communication and guidance during teleproctoring sessions:

- Clear and concise instructions
- Real-time feedback
- Specific and actionable feedback
- Positive reinforcement
- Respectful communication
- Safety-focused

it's important to note that not all teleproctoring sessions require a defined vocabulary, and it depends on the complexity and specificity of the procedure, the level of experience of the trainee, and the proctor's preferences. For example, a simple and common procedure might not need a defined vocabulary, while a complex and specific procedure would benefit from a defined vocabulary. Defining a common vocabulary before running a teleproctoring session can be beneficial for several reasons: clarity, efficiency, consistency, and standardization.

## Challenges of teleproctoring

Teleproctoring has emerged as a great tool that increases accessibility to qualified support, breaking barriers related to time and mobility. However, it can have some technical and legal limitations that should be taken into account when implementing it in the surgical field. Some of these limitations include.

- **Technical limitations:** Teleproctoring requires a reliable internet connection and the appropriate equipment, such as cameras and microphones, to be in place. These can be a limitation in some locations or settings and can affect the quality of the connection and the training session [18].
  It is important to highlight the importance of performing tests to guarantee the proper functioning of equipment. Additionally, it is necessary to confirm that the appropriate connectivity is in place, and a maximum latency time of between 330 and 450 ms has been suggested [15].
- **Time zone differences:** Teleproctoring sessions may involve trainees and proctors located in different time zones, which can make scheduling training sessions more challenging. Time zone differences can also affect the trainee's ability to stay engaged and focused during the session.
- **Language barriers:** Language barriers can make it difficult for trainees and proctors to communicate effectively during the teleproctoring session. This can make it difficult for the trainee to receive instruction and guidance and for the proctor to provide feedback and guidance.
- **Privacy and security:** Teleproctoring can raise concerns about privacy and security, as sensitive patient information may be shared during the session. It is important to implement appropriate measures, such as encryption and secure data storage, to protect patient information.
- **Legal limitations:** Teleproctoring may be subject to different laws and regulations depending on the location, and it is important to ensure compliance with these regulations [19]. For example, teleproctoring may be subject to different regulations for the protection of personal data, medical malpractice, and telemedicine.
- **Quality control:** Teleproctoring sessions may be subject to quality control measures, such as accreditation, certification, or licensing requirements, to ensure that the training is of high quality and that the trainees are meeting the standards of care.
- **Liability:** Teleproctoring can raise questions about liability, as the proctor may be held responsible for the trainee's actions during the session, and for any complications that may arise. It's important to have a clear understanding of the liability issues associated with teleproctoring and to have appropriate insurance in place.
- **Professionalism:** Teleproctoring can affect the professionalism of the surgeon, as it may limit the direct physical and emotional connection between the proctor and the trainee. It's important to ensure that professionalism is maintained and that the trainee receives the same level of education and guidance as in a direct proctoring session.

## Future scenario

### New technologies and proctoring

In this section, we will provide an overview of the different techniques and technologies that can be used to enhance proctoring in robotic surgery.

- The use of **Objective Performance Indicators** related to the economy of motion and events related to console handling (Clutch, Camera, Swap, etc.) opens a new path in the objective evaluation of skills. This will be crucial for providing optimal feedback to the trainee surgeon and for determining objectively when an adequate level of proficiency has been achieved [20,21]
- **Virtual reality (VR).** VR is one of the most widely used, can be used to simulate the surgical environment, allowing trainees to practice using the robotic surgical system in a safe and controlled environment [22]. VR could also be used for Immersive Robotic OR tours and Remote Case Observation.
- **Distance learning**: VR can be used to provide remote training to surgeons who are located in different parts of the world, making training more accessible, and flexible.

- **Recording and analysis**: VR simulations can be recorded and analyzed, allowing the proctor to review the trainee's performance and provide feedback on specific aspects of the procedure.
- **Augmented Reality (AR):** AR can be used to overlay virtual information and guidance onto the real-world surgical environment [23], allowing the proctor to provide guidance and feedback in real time.
- **Artificial Intelligence (AI):** AI can be used to analyze the trainee's performance based on objective performance indicators and provide feedback, as well as to identify areas for improvement. AI will also be used in the short term to simulate different surgical scenarios and provide case-based learning [24,25].

## Conclusion

Proctoring and teleproctoring have emerged as key strategies for training surgeons in the use of robotic surgical systems. They provide a hands-on and experiential approach that aligns well with adult learning theories and can be an effective way to facilitate learning in adults.

Proctoring and teleproctoring can be an effective way to train surgeons in the use of robotic surgical systems and to ensure that they are able to perform these procedures safely and effectively. However, it's important to consider the technical, legal, and ethical limitations of teleproctoring and to implement appropriate measures to ensure safety and compliance with regulations.

In conclusion, proctoring and teleproctoring are powerful tools that can be used to enhance the training process and improve the quality of care in robotic surgery. As technology advances, it is important to continue to explore new techniques and technologies that can be used to enhance the proctoring process and to ensure that surgeons are able to master the skills required to perform robotic surgery safely and effectively.

## Key points

- Proctoring involves an experienced surgeon guiding a less experienced surgeon during a procedure, focusing on skills and safety.
- Teleproctoring allows experienced surgeons to provide real-time guidance and mentoring to less experienced surgeons from remote location.
- Teleproctoring requires a robust and secure communication infrastructure to ensure uninterrupted real-time video streaming and data transmission. High-speed internet connectivity, reliable video platforms, and encryption protocols are essential to maintain privacy and confidentiality.
- Choosing the right proctor is essential, requiring expertise, judgment, leadership, teaching skills, communication, and adaptability

## References

[1] Sachdeva AK, Russell TR. Safe introduction of new procedures and emerging Technologies in Surgery Education, credentialing, and privileging. Surg Clin North Am 2007;87:853–66.
[2] Stefanidis D, Huffman E, Collins J, Martino M, Satava R, Levy J. Expert consensus recommendations for robotic surgery credentialing. Ann Surg 2022;276:88–93.
[3] Sanchez A, Rodriguez O, Nakhal E, Davila H, Valero R, Sanchez R, Pena R, Visconti M. Robotic-assisted Heller myotomy versus laparoscopic Heller myotomy for the treatment of esophageal achalasia. J Robot Surg 2012;6:213–6.
[4] Peters B, Armijo P, Krause C, Choudhury S, Oleynikov D. Review of emergening surgical robotic technology. Surg Endosc 2018;32:1636–55.
[5] Baig M, Razi SS, Agyabeng-Dadzie K, Stroever S, Muslim Z, Weber J, Herrera L, Bhora F. Robotic-assisted thoracoscopic surgery demonstrate a lower rate of conversion to thoracotomy than video-assisted thoracoscopic surgery for complex lobectomies. Eur J Cardiothorac Surg 2022;62(3).
[6] Zorn K, Gautam G, Shalhay A, Clayman R, Ahlering T, Albala D, et al. Training, credentialing, proctoring and medicolegal risks of robotic urological SurgeryL recommendations of the Society of Urologic Robotic Surgeons. J Urol 2009;182:1126–32.
[7] Lee J, Mucksavage P, Sundaram C, McDougal E. Best practices for robotic surgery training and credentialing. J Urol 2011;185:1191–7.
[8] Green C, Levy J, Martino M, Porterfield J. The current state of surgeon credentialing in the robotic era. Ann Laparosc Endos Surge 2020;5:17–22.
[9] Vanlander A, Mazzon E, Collins J, Mottrie A, Rogiers X, van der Poel H, et al. Orsi consensus meeting on European robotic training (OCERT): results from the first multispecialty consensus meeting on training in robot-assisted surgery. Eur Urol 2020;78:713–6.
[10] Bahler C, Sundaram C. Training in robotic surgery: simulators, surgery, and credentialing. Urol Clin North Am 2014;41:581–9.
[11] Huffman E, Rosen S, Levy J, Martino M, Stefanidis D. Are current credentialing requirements for robotic surgery adequate to ensure surgeon proficiency? Surg Endosc 2021;35:2104–9.
[12] Taylor D, Hamdy H. Adult learning theories: implications for learning and teaching in medical education. AMME guide no. 83. Med Tech 2013;35:1561–72.

[13] Clapper T. Beyond Knowles: what those conducting simulation need to know about adult learning theory. Clin Simul Nurs 2010;6:7–14.

[14] Sanchez A, Rodriguez O, Sanchez R, Inchausti C. Role of simulation in Mnimally Invasiver surgery training. Rev Venez Cir 2022;75:61–9.

[15] Ayoub C, El-Asmar J, Abdulfattah S, El-Hajj A. Telemedicina and Telementoring in urology: a glimpse of the past and a leap into the future. Front Surg 2022;9, 811749.

[16] Musella M, Martines G, Berardi G, Picciariello A, Trigiante G, Vitiello A. Leesons from the Covid-19 pandemic: remote coaching in bariatric surgery. Langenbecks Arch Surg 2022;407:2763–7.

[17] Collins J, Ghazi A, Stoyanov D, Hung A, Coleman M, Cecil T, et al. Utilising an accelerated Delphi process to develop guidance and protocols for telepresence applications in remote robotic surgery training. Eur Urol 2020;22:23–33.

[18] Jin M, Brown M, Patwa D, Nirmalan A, Edwards P. Telemedicine, Telementoring, and Telesurgery for surgical practices. Curr Probl Surg 2021;58, 100986.

[19] Nittari G, Khuman R, Baldoni S, Pallota G, Battineni G, Sirignano A, et al. Telemedicine practice: review of the current ethical and legal challenges. Telemed K E Health 2020;26:1427–37.

[20] Chen J, Oh P, Cheng N, Shah A, Montez J, Jarc A, et al. Use of automated performance metrics to measure surgeons performance during robotic Vesicourethral anastomosis and methodical development of a training tutorial. J Urol 2018;200:895–902.

[21] Chen J, Cheng N, Cacciamani G, Oh P, Lin-Brnade M, Remulla D, et al. Objetive assessment of robotic surgical technical skill: a systematic review. J Urol 2018;201:461–9.

[22] Pirker J, Dengel A. The potential of 360° virtual reality videos and real VR for education-A literature review. IEEE Comput Graph Appl 2021;41:76–89.

[23] Stewart C, Fong A, Payyavula G, DiMaio S, Lafaro K, Tallmon K, et al. Study on augmented reality for robotic surgery bedside assistants. J Robot Surge 2022;16:1019–26.

[24] Bhandari N, Zeffiro T, Reddiboina M. Artificial intelligence and robotic surgeryL current perspective and future directions. Curr Opin Urol 2020;30:48–54.

[25] Chen I, Ghazi A, Sridhar D, Slack M, Kelly K, et al. Evolving robotic surgery training and improving patient safety, with integration of novel technologies. World J Urol 2021;39:2883–93.

# 22

# Mentoring, leadership, and conduction of a robotic surgery team

*José Gadú Campos Salcedo*

Urology Department, Hospital Central Militar, Mexico City, Mexico
Urology Center, Hospital Angeles, Mexico City, Mexico

## Introduction

Robotic surgery has revolutionized the field of surgery and improves patient safety and outcomes of many surgical procedures [1]. Every day robotic surgery gains popularity across different specialties, and the increase in number and availability of robotic systems has become advantageous for surgeons to access these novel minimally invasive procedures [2].

With the arrival of robotic equipment, a change occurred in the organization model of surgery teams, considering the space adaptations of the surgical room where a console will be located, but also the way the surgeon interacts with the rest of the medical and nursing staff. As the development of surgical techniques evolves, surgeons are also obligated to safely and systematically acquire new skills [3]. Thus, surgical programs should be safe, cost-effective, and transferrable to real clinical scenarios. This has led to a greater demand for mentoring of inexperienced surgeons and therefore there has been a need for basic training, assessment, testing, and certification in robotic surgery [2].

As time has passed in the different centers that have a robotic console, the surgical teams have been consolidated thanks to the experience in procedures, in such a way that each center has even developed its own protocols tailored to their needs and resources. The effort made to standardize robotic surgeries has also allowed for the development of highly specialized skills in surgeons, first achieving learning curves to optimize surgery times, perform procedures effectively, and reduce the risk of complications, and also establishing teaching and training programs for other surgical personnel [1,3].

Inevitably the experience of surgeons has resulted in leaders who set the standard in the world of robotic surgery, who innovate further and ensure that the mentoring process and team building are facilitated, which is essential for the development of robotic surgery.

## The importance of mentorship

There are many definitions for mentoring, and it is often confused with training roles or remedial support. Mentoring is described as the process whereby an experienced, highly regarded, empathetic person (the mentor) guides another individual (the mentee) in developing and re-examining their ideas, learning, and personal and professional development [4]. The mentor who often, but not necessarily works in the same organization or field as the mentee, achieves this by listening and talking in confidence to the mentee. Mentoring is not counseling patronage or giving advice. The main purpose of mentoring is to consider the whole surgeon to perform to the best of his/her ability. These may not be clinical skills; mentoring is more likely to address issues related to effective teamwork, leadership, and management skills [5].

Mentoring schemes have proven effective in supporting surgeons' personal and professional development. It can improve their confidence and their ability to deal with difficult situations and challenging communications with patients [4]. The key principles of mentoring are

— The mentoring relationship must be freely entered into and not coerced.
— Discussion between mentor and mentee must take place in confidence (except where patient safety supersedes this).
— The mentor and mentee must agree the boundaries of the relationship and clearly define goals and outcomes.
— A mentoring relationship requires a time commitment on both sides.

Mentoring can be beneficial at any stage in a surgeon's career. Surgeons should seek a mentor to improve their general skills and understanding of their performance and position within new procedures. In robotic surgery, mentoring is an essential practice. The forms of mentoring in robotic surgery include fellowships and proctorships (Table 1) [6]. A formal fellowship is considered one of the most structured forms of proctorship, mentoring, and training program. Most of the programs are modular based and have been proven to have a beneficial educational impact. Its objective is to develop a systematic teaching method, which is reproducible. Amidst longer duration, fellowships provide a greater number of cases, postoperative management of complex cases, and even progression to advanced skills [7,8]. Guided learning has always been part of fellowship wherein trainees perform progressive steps of the operation under the careful guidance of an instructor with constant and immediate feedback. In turn, this translates into a safer and more effective way of surgical education [6].

Proctorship programs have provided efficiency and rapid progression of learning curve among surgeons, they are described as a rotation form of participation in the operation when one takes the role of console surgeon while the other acts like a proctor [6–8]. Provision of proctorship has made trainees achieve an earlier performance of independent robotic cases. This has allowed the conduction of safe surgery with the proctor guiding the trainee throughout the procedure during the early learning curve cases [6].

Some attributes, behaviors, and skills are common to all mentors [4]:

— Taking interest in others and developing others
— Being approachable
— Being confident
— Practicing active listening and observation to the mentee
— Constructively questioning and challenging to help the mentee develop his/her skills

On the other hand, it is the responsibility of the mentee[4]:

— Set objectives, including the aim and purpose of the mentoring process.
— Acknowledge and protect the time required of fellowship or proctorship.
— Define the mentoring agenda.
— Be self-motivated
— Be open to challenge

Mentoring helps surgeons take charge of their own development, release their potential, and achieve results that they value. But mentors require training; every good mentor was once a mentee, it is an evolutionary circle where one of the main objectives, in addition to the acquisition of skills in robotic surgery, is the transmission of learned knowledge to others [6].

There are clear examples of the success of mentoring in the field of robotic surgery, one of the first programs with lectures, standardized laboratory training, analysis of surgery video,s and laparoscopic training in porcine models, was developed by Patel et al. this program benefited surgeons in their transition to robotic surgery [9]. Some urological

TABLE 1  Comparison of robotic surgical learning: mentoring vs proctorship.

| Structure | Duration | Design | Impact (education) | Assessment |
|---|---|---|---|---|
| Fellowship | 12–24 months | Modular, direct, and immediate guidance | Yes, operative cases | Direct proctorship and mentoring by professors |
| Proctored course | 2–5 days (varies) | Lectures, intensive laboratory, and animal or simulation training | Yes, individual private cases | Videotape analysis/mentoring |

robotic surgery programs have shown that in radical prostatectomy, acceptable surgery time and complication rate are achieved after 100 procedures [10]. Thus, each procedure of all specialties that use the robotic approach has its own learning curve [11–16]. Achieving good results allows standardize procedures, establish progressive learning goals, and thereby evaluate adequately the skills and errors of mentees [2].

Additionally, it has been demonstrated in robotic pancreaticoduodenectomy when comparing generations of surgeons who were not trained in robotic surgery, those with curricular training, and surgeons with curricular training plus mentoring, that the sum of a curriculum of skills plus mentoring is the key to reducing the number of cases of the learning curve as well as the complications derived from the procedure [17].

The way of mentoring has also improved with the creation of simulators that resemble clinical scenarios very similar to real cases. Nowadays there are three-dimensional models that not only resemble the anatomy of the sites to be operated on but also the texture of the tissues (Fig. 1) [18]. Simulation procedures have become so dynamic that it is possible to practice possible bleeding and how to act in the event of it [18–20].

On the other hand, the improvement of robotic consoles has also facilitated learning, being simpler with the added value of new instruments, improved ranges of movement, and more intuitive management of the console, which greatly facilitates surgical procedures. And if that were not enough, the ease of having more than one console in the operating room guarantees the learning process where the surgeon in training performs exercises in real time with his mentor and the immediate correction of potential errors if they occur (Fig. 2) [6,7].

The mentoring process will continue to play a fundamental role in the development of robotic surgery, since it is a field of continuous evolution and technological advances will require dissemination and teaching. Currently, distance training has even been carried out where the mentor assesses the evolution of the surgeons in training through virtual reality and at distance, that is basically robotic surgery remote mentoring [21,22]. Although there is no standardized list of requirements worldwide to be a robotic surgery proctor, it is clear that the objective of the surgeon who is going to train others should be to become a mentor so that he gets involved with the personal needs of each mentee.

Finally, it is necessary to emphasize that mentoring programs are not aimed exclusively for surgeons, but also for assistant surgeons, scrub and circulating nurses, as well as biomedical technicians [23,24]. There are continuing education programs for each of the participants in robotic surgery. The entire team has the opportunity to improve and learn from experts who are willing to help them improve their skills and results.

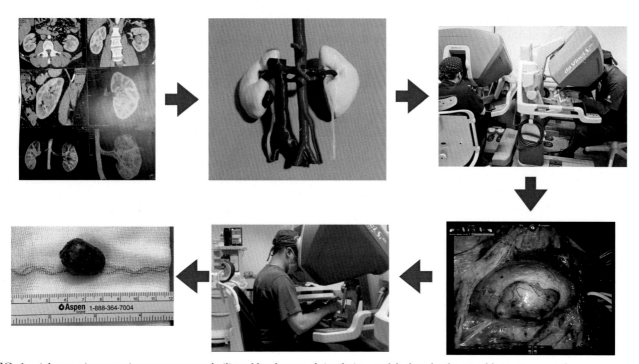

FIG. 1   Advances in mentoring programs are facilitated by the use of simulation models that closely resemble in vivo procedures.

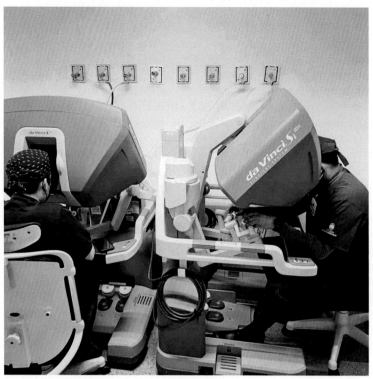

FIG. 2    Mentoring programs through fellowships or proctorships are facilitated when there is more than one console in the operating room since it allows the mentor to evaluate and correct the mentee in real time during surgery.

## Why do we need leaders in robotic surgery?

Leadership involves satisfying team needs and enhancing team effectiveness, it has been defined as the ability to influence a group toward the achievement of a vision or set of goals. Becoming a leader implies the process of influencing others to understand and agree about what needs to be done and how to do it, and the process of facilitating individual and collective efforts to accomplish shared objectives [25].

Having a leader in the robotic surgery team makes it easier to direct the group toward specific goals, organize the team's activities, take charge in solving problems, promote effective communication, and provide feedback on the work of each of the group members [26,27]. Leadership begins from the moment the surgeon arrives at the operating room to keep all staff directed to the proper performance of the procedure, doing it effectively and without complications. The surgeon in his role as leader will direct his assistant surgeons to carry out adequate docking, placement of the instruments he/she will need, mobilization of robotic arms, rearrangement of the camera, and extraction of surgical specimens. In turn, he will command the actions of the nursing staff so that all the supplies required for the surgery are available. With the optimization of procedures, the surgeon's task as a leader is facilitated because everyone has experience of what to do even in the face of possible eventualities [27].

There have been studied different types of leadership. Shared leadership differs from the traditional vertical or top-down approach in that it acknowledges social sources of leadership across a horizontal view of a team (Figs. 3 and 4) [28]. From a shared leadership perspective, leadership exists on a shared group level rather than with a specific individual, it is a dynamic interactive influence process among members of a group for which the objective is to lead one another to the achievement of the team goals. Even though most leadership conducted in the operating room has focused on the surgeon, surgery is characterized by a division of labor in which team members rely on each other's contributions. We cannot consider that there is a more important role in the operating room, the contribution of each of the team members makes the result of the procedure a success or not [29,30].

The model of working in the operating room with a robotic surgery team changes the classic model of hierarchical leadership that puts the surgeon at the top. Because the surgeon is positioned at the console outside the surgical table, there are significant implications and changes to the distribution of power in the team [27,31]. The power structure is

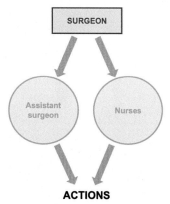

FIG. 3    Traditional and vertical leadership model in the operating room in a top-down way, where the surgeon is the only leader.

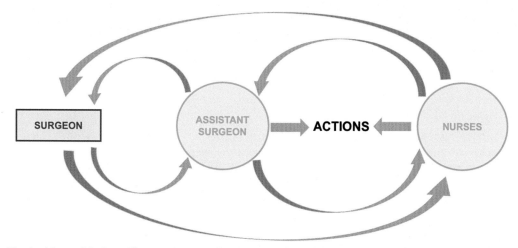

FIG. 4    Shared leadership model where all participants in robotic surgery have established roles that they perform autonomously.

affected because the rest of the team acquires autonomy since the surgeon depends on his tasks to be able to perform the surgery. A proper procedure will not be possible without an attending surgeon and scrub nurse taking control of their roles [32]. Thus, the integration of a robotic console into the operating room and the consequent separation of the surgeon from the team, make the power dynamic more distributed, providing a foundation for shared leadership to occur (Fig. 5) [33].

Leadership has stages that can be divided into two phases: transition and action. During the transition phases, the surgical team will plan and prepare for the procedure, while in the action phase, they will perform the surgery and evaluate the achievement of the goal [26].

*Transition phase*

— *Compose team:* selection of team members and roles that each one will take.
— *Define mission:* determine and communicate the purpose of the team (e.g., expected results of surgery or patient safety)
— *Establish expectations and goals:* identify what the team hopes to achieve, this will depend on each procedure and each patient.
— *Structure and plan:* establish how the work will be done, and specify the activities of each role.
— *Train and develop team:* development of technical and interpersonal skills. All team members in surgery are required to be appropriately trained and qualified.
— *Sensemaking:* identifying critical external events that affect the team and how the event might impact team functioning (e.g., team size or leadership structure).
— *Provide feedback:* reviewing past or current performance to make improvements.

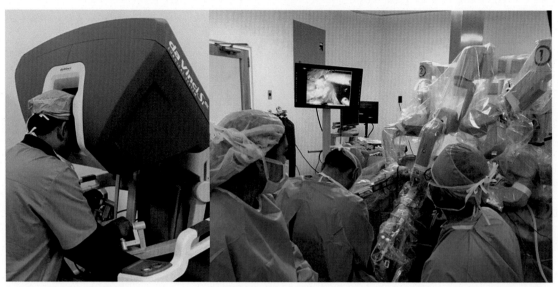

FIG. 5    The separation of the surgeon from the surgical table has contributed to the change of leadership paradigm in robotic surgery operating rooms.

### Action phase

— *Monitor team:* active review of the team while the surgery occurs.
— *Manage team boundaries:* limits between the team and the larger organizational context for logistical reasons.
— *Challenge team:* confront what the team assumes, its methods, and processes to improve.
— *Perform team task:* participate, intervene, and perform the surgery.
— *Problem solver:* ability to diagnose and develop solutions for problems the team faces.
— *Provide resources:* obtaining and provision of financial, material, or informative resources for the team.
— *Encourage team self-management:* always encourage the team to have more autonomy.
— *Support social climate:* maintain a pleasant environment that maintains or increases the production and satisfaction of team members.

It has been proven that when a surgical team has defined their roles as leaders, they can face adversities and challenges more satisfactorily. Particularly, shared leadership has positive effects on team performance in high-risk situations, so in the face of an emergency situation, teams respond with better decision-making and action. The above has been demonstrated not only in the surgical context but also in other scenarios such as flight emergencies, police and fire stations [34,35].

Leadership can also be applied on a larger scale with the work of department heads that further facilitate the organization of robotic surgery teams. Although we have emphasized that the tasks of leadership must be shared, for organizational purposes, it will always be important to have a director who orchestrates all the scaffolding behind the team [36]. It is also important to mention that surgeons can express leadership in academic opinions or as experts in procedures, this contributes to creating teams committed to surgical innovation and research. Being an opinion leader carries an important responsibility because the surgeon becomes a reference for others in search of preparation and improvement of skills.

In conclusion, robotic surgery has revolutionized the leadership model in the operating room, favoring the autonomy of team members and a more effective communication process. This results in the optimization of procedures and obtaining better results that benefit the dynamics of the surgical teams and obviously the prognosis of the patients.

## Robotic surgery teams

As many other important innovation developments throughout medical history, robotic surgery is the result of continuous teamwork. To understand teamwork, first we have to determine what is a team, Dyer defined a team as a unit of at least two people working toward a common goal, where each person has been assigned specific roles or functions to perform, and where completion of the mission requires some form of dependency among the group members [37]. Considering this, it must have been noticed the members of a team interact toward a common and valued goal—objective—mission, and each one has specific roles or functions to perform [36].

Robotic surgery evolution has been the product of individuals working together to combine their efforts to accomplish tasks that require the contributions of multiple members. In the early 2000s Salas and Cannon-Bowers described four reasons why organizations are increasingly dependent upon teams to accomplish goals. First, task complexity and work scope often need contributions from multiple people working together. Second, teams are better prepared for difficult scenarios since their individuals share the responsibility and consequences for their choices. Third, teams used to outperform individuals due to the wisdom of collectives. And fourth, in many organizations as healthcare, teams are "a must" due to the collaborative nature of the work [38].

Teamwork is essential for the proper functioning of robotic surgery equipment [39]. Teamwork is a series of behaviors, cognitions, and attitudes that interact to achieve mutually desired goals and adapt to changes in both internal and external environments. Teamwork consists of knowledge, skills, and attitudes displayed in support of team members, objectives, and mission. The combination of thoughts and actions facilitates adaptive performance and completion of work goals. Compared to individual work, teams are more capable and creative in solving problems, because each individual task and experience is magnified and synergizes with others, achieving a cascade effect [40].

## Creation of robotic surgery teams and subspecialization

The standardization of robotic surgical procedures has allowed a change in the dynamics in which surgical teams operate. Traditionally, due to the lack of training of surgeons, surgeon assistants, and nursery staff, there were few teams in hospitals dedicated exclusively to robotic surgeries. Currently, having the training tools discussed in the previous section and the standardization of procedures since preoperative preparation, it has been possible to train more personnel who are capable of participating with perfection in any procedure, in such a way that mentoring also favors rotation of surgical teams without affecting their performances.

In high-volume centers, the subspecialization of teams is common, particularly with regard to surgeons. The main advantage of this organization model is that they become expert surgeons in certain procedures since they are dedicated exclusively to the treatment of very specific types of pathologies or a particular organ. Although one could assume that this practice would positively impact the results of surgeries, there are no studies at the moment that show a difference in rates of outcomes and complications in centers that do not have these divisions of clinics.

The truth is that the complexity of robotic surgeries requires contributions from a lot of people, therefore teamwork will always be essential. A surgical team with shared responsibility and diverse skill sets is better equipped to get better results and make difficult decisions, it also represents multiple areas of expertise that can outperform any individual experience [40].

## Summary

Throughout this chapter, we have learned that almost two decades since the birth of robotic surgery, there has been improvement not only in the technology of its equipment and consoles or in the improvement of surgical techniques but also in the way of organizing surgical teams, the process of teaching through mentoring programs to perfect skills and results, or simply to get started as robotic surgeons. Additionally, the leadership dynamic within the surgical teams has changed, evolving to greater participation and autonomy of all participants in every procedure.

## Future scenario

It will be interesting to witness the evolution that robotic surgery will have with the advent of new equipment that will reduce costs and possibly favor the expansion of this technology throughout the world. Will the ways of learning and mentoring change? Results of comparative studies between consoles have shown that migrating from one equipment to another to perform robotic surgeries does not represent a challenge for consolidated surgeons with these surgical approaches [41]. It has also been shown that young surgeons have easier understanding and ease in practicing robotic surgery, particularly if they have experience playing video games [42,43].

The future of robotic surgery is promising in terms of technological progress, we are still needed as surgeons to perform procedures but with the gigantic growth of robotics in recent years, it is possible that the dynamics of operating rooms in terms of teamwork and leadership could change even more if we had autonomous consoles that require none or less surgeons' performance.

## Key points

- The ideal way of learning robotic surgery is through mentoring and not proctoring programs.
- Mentors in robotic surgery are proctors who have the added value that personalizing the teaching model to the needs and objectives of each of their mentees.
- Mentoring can be applied to improve the skills of any member of the robotic surgery team.
- Robotic surgery has changed the leadership paradigm of surgeons and the rest of the operating room staff in top-down way and instead has given more autonomy to all members through shared leadership.
- The creation of robotic surgery teams is essential to unite talents, experiences, and attitudes with shared objectives, which could not be achieved through individual work.

## References

[1] Leal GT, Campos CO. 30 years of robotic surgery. World J Surg 2016;40(10):2550–7.
[2] Wang RS, Ambani SN. Robotic surgery training: current trends and future directions. Urol Clin North Am 2021;48(1):137–46.
[3] Alip SL, Jim J, Rha KH, et al. Future platforms of robotic surgery. Urol Clin North Am 2022;49(1):23–38.
[4] Royal College of Surgeons. Mentoring a guide to good practice. 2015 Professional and clinical standards. The Royal College of Surgeons of England.
[5] Steven A, Oxley J, Fleming WG. Mentoring for NHS doctors: perceived benefits across the personal-professional interface. J R Soc Med 2008;101:552–7.
[6] Santok GD, Raheem AA, Kim LHC, et al. Proctorship and mentoring: its backbone and application in robotic surgery. Investig Clin Urol 2016;57(2):S114–20.
[7] Rashid HH, Leung YY, Rashid MJ, et al. Robotic surgical education: a systematic approach to training urology residents to perform robotic-assisted laparoscopic radical prostatectomy. Urology 2006;68:75–9.
[8] Hung AJ, Bottyan T, Clifford TG, et al. Structured learning for robotic surgery utilizing a proficiency score: a pilot study. World J Urol 2017;35(1):27–34.
[9] Patel SR, Hedican SP, Bishoff JT, et al. Skill based mentored laparoscopy course participation leads to laparoscopic practice expansion an assist in transition to robotic surgery. J Urol 2011;186:1997–2000.
[10] Jones A, Eden C, Sullivan ME. Mutual mentoring in laparoscopic urology – $_{6a}$ natural progression from laparoscopic fellowship. Ann R Coll Surg Engl 2007;89:422–5.
[11] Mazzon G, Sridhar A, Busuttil G, et al. Learning curves for robotic surgery: a review of the recent literature. Curr Urol Rep 2017;18(11):89.
[12] Chan KS, Wang ZK, Syn N, et al. Learning curve of laparoscopic and robotic pancreas resections: a systematic review. Surgery 2021;170(1):194–206.
[13] Zheng YL, Feng Q, Yan S, et al. Learning curve of robotic distal and total gastrectomy. Br J Surg 2021;108(9):1126–32.
[14] Wong SW, Crowe P. Factors affecting the learning curve in robotic colorectal surgery. J Robot Surg 2022;16(6):1249–56.
[15] Andolfi C, Umanskiy K. Mastering robotic surgery: where does the learning curve lead us? J Laparoendosc Adv Surg Tech A 2017;27(5):470–4.
[16] Catchpole K, Perkins C, Bresee C, et al. Safety, efficiency and learning curves in robotic surgery: a human factors analysis. Surg Endosc 2016;30(9):3749–61.
[17] Rice MJK, Hodges JC, Bellon J, et al. Learning curve and safety for robotic Pancreaticoduodenectomy. JAMA Surg 2020;155(7):607–15.
[18] Ghazi AE, Teplitz BA. Role of 3D printing in surgical education for robotic urology procedures. Transl Androl Urol 2020;9(2):931–41.
[19] Costello DM, Huntington I, Burke G, et al. A review of simulation training and new 3D computer-generated synthetic organs for robotic surgery education. J Robot Surg 2022;16(4):749–63.
[20] Badash I, Burtt K, Solorzano CA, et al. Innovations in surgery simulation: a review of past, current and future techniques. Ann Transl Med 2016;4(23):453.
[21] Faris H, Harfouche C, Bandle J, et al. Surgical tele-mentoring using a robotic platform: initial experience in a military institution. Surg Endosc 2023. https://doi.org/10.1007/s00464-023-10484-1 [Online ahead of print].
[22] Sinha A, West A, Vasdev N, et al. Current practices and the future of robotic surgical training. Surgeon 2023;21(5):314–22.
[23] Moller L, Hertz P, Grande U, et al. Identifying curriculum content for operating room nurses involved in robotic assisted surgery: a Delphi study. Surg Endosc 2023;37(4):2729–48.
[24] Zamudio J, Woodward J, Kanji FF, et al. Demands on surgical teams in robotic-assisted surgery: an assessment of intraoperative workload within different surgical specialties. Am J Surg 2023;226(3):365–70.
[25] Yukl G. Leadership in organizations. 8th ed. Pearson-Prentice Hall; 2008.
[26] Morgeson FP, Derue DS, Karam EP. Leadership in teams: a functional approach to understanding structures and processes. J Manag 2010;36(1):5–39.
[27] Henrickson PS, Yule S, Flin R, et al. Towards a model of surgeons' leadership in the operating room. BMJ Qual Saf 2011;20(7):570–9.

[28] Pearce CL, Simis HP. Vertical versus shared leadership as predictors of the effectiveness of change management teams: an examination of aversive, directive, transactional, transformational, and empowering leader behaviors. Group Dyn Theory Res Pract 2002;6(2):172–97.

[29] Pearce CL, Simis HP. Shared leadership: toward a multi-level theory of leadership. Team Develop 2000;7:115–39.

[30] Pearce CL, Conger GA. All those years ago: The historical underpinnings of shared leadership. In: Shared leadership: reframig the hows and whys of leaderhip; 2003. p. 1–18.

[31] Stone JL, Aveling EL, Frean M, et al. Effective leadership of surgical teams: a mixed methods study or surgeon behaviors and functions. Ann Thorac Surg 2017;104(2):530–7.

[32] Pasarakonda S, Grote G, Schmutz JB, et al. A strategic core role perspective on team coordination: benefits of centralized leadership for managing task complexity in the operating room. Hum Factors 2020;63(5):910–25.

[33] Pelikan HRM, Cheatle A, Jung MF, et al. Operating at a distance: how a teleoperated surgical robot reconfigures teamwork in the operating room. Proc ACM Human-Comput Interact 2018;2(138).

[34] Bienefeld N, Grote G. Teamwork in an emergency: how distributed leadership improves decision making. Proc Hum Factors Ergon Soc 2011;55 (1):120–2.

[35] Baran BE, Scott CW. Organizing ambiguity: a grounded theory of leadership and sensemaking within dangerous context. Mil Psychol 2010;22 (1):S42–69.

[36] Klein KJ, Ziegart JC, Knight AP, et al. Dynamic delegation: Shared, hierarchical and deindividualized leadership in extreme action teams. Adm Sci Q 2006;51(4):590–621.

[37] Dyer JL. Team reseach and team training: a state-of-the-art review. In: Muckler FA, editor. Human Fact and Ergonomics, vol. 26; 1984. p. 285–323.

[38] Cannon-Bowers JA, Salas E, Converse S. Shared mental moderls in expert team decision making. In: Castellan NJ, editor. Individual an group decision making: Current Issues; 1993. p. 221–46.

[39] Fuji L, Entin E. Robotic surgery and the operating room team. Human Factors Ergo Soc 2005;49(11).

[40] DeChurch LA, Mesmer MJR. The cognitive underpinnings of effective teamwork: a meta-analysis. J Appl Psychol 2010;95(1):32–53.

[41] Bravi CA, Balestrazzi E, De Loof M, et al. Robot-assisted radical prostatectomy performed with different robotic platforms: first comparative evidence between Da Vinci and HUGO robot-assisted surgery robots. Eur Urol Focus 2023;25(23):187.

[42] Moglia A, Perrone V, Ferrari V, et al. Influence of videogames and musical instruments on performances at a simulator for robotic surgery. Minim Invasive Ther Allied Technol 2017;26(3):129–34.

[43] Gupta A, Lawendy B, Goldenberg MG, et al. Can video games enhance surgical skills acquisition for medical students? A systematic review. Surgery 2021;169(4):821–9.

# 23

# Considerations for anesthesiology in robotic-assisted surgeries

*Hon Sen Tan[a] and Ashraf S. Habib[b]*

[a]Department of Women's Anaesthesia, KK Women's and Children's Hospital, Singapore, Singapore [b]Division of Women's Anesthesia, Department of Anesthesiology, Duke University Medical Center, Durham, NC, United States

## Introduction

Robotic-assisted surgery was initially developed to facilitate remote surgery on the battlefield, but has since gained popularity across a wide range of procedures, including general, colorectal, urogynecology, and cardiothoracic surgeries [1]. Within the field of general surgery alone, the use of robotic-assisted surgery has increased from 1.8% in 2012 to 15.1% in 2018 [2].

Robotic-assisted surgery retains the same benefits of laparoscopic surgery compared to open surgery, including smaller incisions, better cosmesis, reduced postoperative pain, lower incidence of wound or respiratory complications, decreased blood loss, and shorter convalescence and hospital stay [3,4]. In addition, robotic-assisted surgery conveys several distinct advantages over laparoscopic surgery, such as increased dexterity and precision due to the greater instrument degrees of freedom, improved surgeon ergonomics, three-dimensional camera view, and stability of the camera platform [3,5]. However, robotic-assisted surgery often involves steep angulation of the operating table, prolonged pneumoperitoneum, and limited patient access which may impact anesthetic management and patient safety [3,6,7]. Hence, the increasing use of robotic-assisted surgery requires anesthesiologists to be familiar with the management of these issues. This article aims to outline the important anesthetic considerations and management of robotic-assisted surgery in order to optimize patient safety and improve perioperative outcomes.

## Overview of the Da Vinci system

At present, the Da Vinci system (Intuitive Surgical Inc., California, USA) is the most commonly encountered commercially available robotic system, comprising of three major components: a surgeon console, patient cart, and vision cart [3,8]. The surgeon console projects a three-dimensional stereoscopic view of the surgical field to the primary surgeon. Hand controls enable the primary surgeon to manipulate the robotic arms and surgical instruments, while selection between the robotic arms, adjustment of the endoscope, and alteration of the instrument settings can be made via the foot pedals. Movements of these controls are translated to electronic signals that can be scaled down and filtered to reduce fine hand tremors, thereby permitting small, precise movements. All electronic signals to and from the surgeon console pass through the vision cart, which acts as a central node integrating power generation, image processing, and information systems across the various Da Vinci system components. The vision cart also displays the live feed of the surgery for the rest of the operative team. Finally, the patient cart receives and translates the electronic signals from the surgeon console into movement of four robotic arms. The first two arms hold and manipulate the surgical instruments, each of which has seven degrees of freedom thus exceeding the capacity of the human hand in open surgery. The third arm contains an endoscope, and an optional fourth arm enables the use of additional instruments or performing tasks like counter-traction. In general, most surgeries are performed by at least two surgeons; the primary surgeon controls the surgical instruments from the surgeon console, while an assistant places the trocars and connects instruments to the robotic arms.

## Physiological effects of pneumoperitoneum and positioning

To obtain appropriate visualization of abdominal and pelvic organs, pneumoperitoneum via carbon dioxide ($CO_2$) insufflation and patient positioning such that noninvolved organs fall away from the surgical site are required. Appreciation of the physiological effects of pneumoperitoneum and steep positioning is required to determine whether the patient can safely tolerate the procedure and to mitigate the potential perioperative issues that may arise.

Insufflation of $CO_2$ into the peritoneal cavity increases its systemic absorption. It was estimated that 38–42 mL. $min^{-1}$ of $CO_2$ are absorbed from the peritoneal cavity during pneumoperitoneum, representing a 30% increase in $CO_2$ load [9]. This excess $CO_2$ is normally eliminated through the lungs, but can lead to hypercarbia and acidosis if intraoperative ventilation is impaired, which in turn increases the risk of cardiac arrhythmias, pulmonary vessel vasoconstriction, and depressed myocardial contractility [10]. It is therefore important to increase minute ventilation to help eliminate the excess $CO_2$ load during pneumoperitoneum, but with care to avoid excessive tidal volumes or barotrauma.

Pneumoperitoneum also creates a state of acutely elevated intra-abdominal pressure, which impacts multiple physiological systems. The effects of increasing intra-abdominal pressure are biphasic; small increases (<10 mmHg) result in increased venous return and cardiac output. However, typical intra-abdominal pressures of 12–15 mmHg during laparoscopic surgery have been associated with decreased stroke volume, while increasing heart rate, systemic vascular resistance, and central venous pressure, all of which culminate in an overall reduction in cardiac output [11–13]. This impairment of cardiac output can lead to ventricular failure in patients with poor baseline cardiac function. The reduction in cardiac function is exacerbated in patients with preoperative hypovolemia, and it is important to ensure a euvolemic status to minimize further cardiac depression resulting from reduced preload.

Pneumoperitoneum exerts direct pressure on renal vasculature and reduces cortical blood flow [14]. In conjunction with the neuroendocrine effects of surgery such as increased vasopressin secretion and activation of the renin-angiotensin-aldosterone system [15], these physiological changes lead to reduced intraoperative urine output [16], although no changes in serum creatinine, blood urea nitrogen, and creatinine clearance were detected after laparoscopic surgery [17,18]. Similarly, pneumoperitoneum exerts pressure on the portal venous system and was associated with hepatic hypoperfusion and acute hepatocyte injury, which may lead to transient increases in transaminase levels that usually resolve within 72 h [19,20].

Increased intra-abdominal pressure shifts the diaphragm cephalad and impedes its movement during respiration. This decreases respiratory compliance and increases airway pressure during ventilation, thereby elevating the risk of atelectasis, barotrauma, and ventilation/perfusion mismatch [21].

Pneumoperitoneum has a direct compressive effect on the inferior vena cava and iliac veins, resulting in up to 57% reduction in lower extremity venous flow in morbidly obese patients undergoing laparoscopic surgery in the reverse Trendelenburg position [22]. Femoral venous flow can be restored by sequential compression devices during laparoscopic surgery, although they were less effective in morbidly obese patients [22,23].

Many of these cardiovascular and respiratory effects associated with pneumoperitoneum are affected by steep patient positioning. For instance, respiratory effects are exacerbated in the Trendelenburg position, but venous return and cardiac output are relatively preserved. Conversely, a head-up position allows better diaphragmatic movement and lower risk of atelectasis, but further reduces venous return, cardiac output, and mean arterial pressure [24]. In addition, the combination of pneumoperitoneum and steep Trendelenburg position can increase intracranial pressure (ICP) and intra-ocular pressure, and may therefore increase the risk of postoperative neurological complications or vision loss in patients with pre-existing intracranial hypertension or ocular disease [25,26]. In rare instances, increased systemic and central venous pressures resulting from the combination of steep Trendelenburg position and pneumoperitoneum can lead to rupture of subcutaneous capillaries within the acoustic meatus, manifesting as bilateral intraoperative otorrhagia [27].

## Perioperative anesthetic considerations

Important perioperative issues and recommended management are summarized in Table 1.

**TABLE 1** Perioperative issues associated with robotic-assisted surgery and recommended management.

| | Anesthetic goals | Recommended management |
|---|---|---|
| Airway and ventilation | • Secure airway to manage limited access, increased ventilatory pressure, and airway edema<br>• Minimize risk of barotrauma, atelectasis, and hypercarbia<br>• Anticipate and assess for airway edema | • Endotracheal intubation<br>• Endotracheal tube adequately secured, position checked before and after patient positioning<br>• Cephalic diaphragmatic displacement can cause endobronchial intubation<br>• Avoid gastric insufflation during airway management, orogastric tube to decompress the stomach<br>• Utilize lung-protective ventilation strategy: tidal volumes 6–8 mL. $kg^{-1}$ predicted body weight and positive end-expiratory pressure (PEEP)<br>• Adequate neuromuscular blockade and pressure-controlled or pressure-controlled volume-guaranteed ventilation<br>• If required for surgery, place patient in steep Trendelenburg position for 2–5 min prior to sterile draping to identify and troubleshoot positioning and ventilation issues<br>• Postoperative recruitment maneuvers, head-up positioning, incentive spirometry, and chest physiotherapy<br>• Cuff leak test to predict postextubation airway obstruction<br>• Identify risk factors for airway edema and poor tolerance for airway obstruction<br>• Close monitoring to rapidly detect and manage airway obstruction postextubation |
| Intravenous fluid management | • Maintain adequate organ perfusion while minimizing risk of fluid overload<br>• Aim for euvolemic status throughout preoperative, intraoperative, and postoperative phases | • Preoperative fasting according to American Society of Anesthesiologists guidelines (clear fluids: 2h, light meal: 6h)<br>• Avoid mechanical bowel preparation if possible, use should be based on surgical and patient indications<br>• Consider providing a clear carbohydrate drink 2h before surgery<br>• Place two peripheral intravenous lines<br>• Replace intraoperative diuresis and insensible losses with balanced crystalloid infusion at 1–3 mL.$kg^{-1}$.$h^{-1}$<br>• Monitor trend of dynamic indices (e.g., stroke volume, stroke volume variation, pulse pressure variation) and their response to small fluid challenges (100–250 mL) to assess fluid responsiveness<br>• Treat hypotension unrelated to hypovolemia with vasopressors, not intravenous fluids<br>• Another assessment and correction of fluid status made toward end of surgery with patient horizontal and without pneumoperitoneum |
| Positioning and access | • Minimize risk of pressure and nerve injuries due to positioning<br>• Avoid dislodgment of lines and patient monitors during positioning<br>• Ensure adequate patient/airway access by surgical and anesthesiology teams<br>• Minimize inadvertent injuries caused by robot arms<br>• Facilitate rapid robot undocking in emergent situations | • Correct positioning of extremities and adequate padding of pressure areas to minimize risk of pressure and nerve injuries<br>• Ensure no patient slippage during steep positioning<br>• Ensure endotracheal tube, intravenous lines, and patient monitors are secure, patent, and with adequate length extensions applied<br>• Take into account footprint of the patient cart, bed position, and body habitus to demarcate access pathways to airway, lines, and patient monitors<br>• Move patient and operating table through full range of possible positions prior to draping to determine safe limits of position and adequate length of lines and monitors<br>• Constant vigilance by surgical and anesthesiology teams to prevent injuries by robot arms<br>• Adequate anesthetic depth and neuromuscular blockade to ensure no patient movement when robotic instruments are docked<br>• Regular training and simulations to facilitate rapid undocking of the patient cart during an emergency |
| Analgesia and prevention of nausea or vomiting | • Adequate postoperative pain relief while minimizing risk of adverse effects<br>• Avoidance of nausea and vomiting | • Multimodal preemptive opioid-sparing analgesia approach, including dexamethasone, acetaminophen, nonsteroidal anti-inflammatory drugs (NSAIDs), and regional anesthesia blocks<br>• Maneuvers to resolve pneumoperitoneum after surgery<br>• Multimodal prophylaxis regimen to prevent nausea or vomiting |
| Thromboembolic prophylaxis | • Minimize risk of perioperative thromboembolic complications | • Mechanical prophylaxis with graduated compression stockings and intermittent pneumatic compression cuffs<br>• Consider pharmacological prophylaxis with low-molecular weight heparin as soon as possible according to institutional protocols |

## Preoperative assessment

In addition to the standard preoperative assessment, anesthesiologists should also focus on identifying comorbidities that may reduce the patient's ability to tolerate robotic-assisted surgery. For instance, prolonged Trendelenburg positioning and pneumoperitoneum may exacerbate difficulties with ventilating patients with morbid obesity or respiratory disease, while those with significant cardiovascular disease or intracranial pathology poorly tolerate the associated changes in preload, afterload, and increased intracranial pressure [24]. Early identification of potential issues with robotic-assisted surgery will help ensure adequate time for multidisciplinary optimization and risk counseling, and if required, the use of an alternative surgical method.

## Airway and ventilation

Standard airway management for robotic-assisted surgery involves endotracheal intubation, given the limited airway access, increased ventilatory pressures, and risk of airway edema and pulmonary aspiration stemming from prolonged Trendelenburg positioning and pneumoperitoneum. The endotracheal tube should be adequately secured, and its position checked before and after patient positioning owing to the risk of dislodgement and limited access to the airway once the patient cart has been docked. It is also important to note that pneumoperitoneum and Trendelenburg position can displace the diaphragm cephalad hence increasing the risk of endobronchial intubation. Gastric decompression via an oro-gastric tube and avoidance of gastric insufflation during mask ventilation are important to avoid obscuring surgical views.

Ventilation may be impaired during pneumoperitoneum and Trendelenburg positioning, with increased risk of barotrauma, atelectasis, and hypercarbia. To a certain extent, hypercarbia can be compensated by increasing minute ventilation, but care must be taken to avoid barotrauma or excessive tidal volumes, especially in patients with underlying respiratory comorbidity such as chronic obstructive pulmonary disease. In general, a lung-protective ventilation strategy is recommended, which includes tidal volumes of $6-8\,mL.kg^{-1}$ predicted body weight, combined with judicious use of positive end-expiratory pressure (PEEP) to optimize lung compliance and reduce the risk of atelectasis. Adequate neuromuscular blockade and pressure-controlled or pressure-controlled volume-guaranteed ventilation may help reduce peak inspiratory pressure. If required for surgery, placing the patient in steep Trendelenburg position for 2–5 min prior to sterile draping and robot docking may facilitate early troubleshooting of positioning and ventilation issues [28]. Postoperative recruitment maneuvers, head-up positioning, incentive spirometry, and chest physiotherapy may help mitigate the risk of atelectasis.

Prolonged Trendelenburg positioning and excessive intravenous fluids increase the risk of airway edema, and it is good practice to assess airway patency prior to extubation. The cuff leak test is commonly used to predict postextubation airway obstruction and need for re-intubation by estimating the adequacy of airflow around the cuff-deflated endotracheal tube. The test can be performed by deflating the endotracheal tube cuff while ventilating the patient in volume-controlled mode. The difference between inspired and expired tidal volumes is measured, with leak volume less than 12%–24% of the delivered tidal volume suggestive of reduced airway patency [29]. A recent meta-analysis reported that the use of cuff leak test in critical care settings predicted airway obstruction and re-intubation with excellent specificity (0.87 and 0.88, respectively) but moderate sensitivity (0.62 and 0.66, respectively), thereby suggesting that patients with reduced cuff leak may require additional management to reduce airway edema prior to extubation, although the presence of normal cuff leak cannot completely exclude airway obstruction [30]. In addition, the decision to extubate should be guided by the presence or absence of risk factors for airway obstruction such as concomitant obesity, obstructive sleep apnea, chronic obstructive pulmonary disease, neuromuscular disease, previous head and neck radiation, Parkinson's disease, rheumatoid arthritis, female sex, and age >70 years [29]. Also, patients with difficult airway, hypoxemia, hypercarbia, acidosis, and cardiovascular instability poorly tolerate the physiological sequelae of airway obstruction and delayed re-intubation. Following extubation, close monitoring to rapidly detect and manage airway obstruction is recommended for all patients with risk factors for airway obstruction.

## Intravenous fluid management

Optimizing intravenous fluid therapy is an important aspect of perioperative management. Hypovolemia from insufficient fluid replacement can exacerbate the surgical stress response, trigger vasopressin release, and increase the risk of inadequate end-organ oxygen delivery [31]. Conversely, excessive fluid therapy may cause interstitial edema and local inflammation, thereby increasing the risk of wound dehiscence, infections, and anastomotic leakage [31]. Therefore, the twin goals of perioperative fluid management are to maintain adequate organ perfusion while minimizing the risk of fluid overload. Fluid management should encompass the continuum through preoperative, intraoperative, and postoperative phases as suboptimal management in one phase can undermine best practice elsewhere.

Ideally, patients should arrive for surgery in a euvolemic state; however, the majority of patients experience a degree of preoperative hypovolemia, which can be exacerbated by excessive fasting and unnecessary bowel preparation [32]. Whenever feasible, preoperative fasting should adhere to the American Society of Anesthesiologists guidelines of clear fluid intake up till 2h and a light meal up till 6h prior to surgery [33] as this was shown by a Cochrane review to have no significant difference in the incidence of aspiration, regurgitation, or postoperative morbidity compared to fasting from midnight to surgery [34]. In addition, the use of mechanical bowel preparation should be individualized based on patient and surgical indications, as routine use contributes to preoperative dehydration and is unpleasant for the patient [35,36]. Of note, although previous meta-analyses have reported no significant clinical benefit of mechanical bowel preparation vs no preparation [37,38], contemporary isosmotic preparations are associated with less physiological consequences compared to older hyperosmotic solutions, and have been shown to reduce surgical site infection and anastomotic leak when combined with oral nonabsorbable antibiotic therapy [39,40]. Finally, ingesting a clear carbohydrate drink 2h before surgery can reduce thirst, hunger, anxiety, and perioperative insulin resistance [41,42].

Intraoperative management should target a net zero fluid balance with individualized replacement of physiologic fluid consumption and surgical losses. Physiologic losses from diuresis and insensible losses can be replaced through an infusion of balanced crystalloid solution of $1-3\,mL.kg^{-1}.h^{-1}$ [43] in combination with additional fluid challenges to replace surgical losses and titrate volume status. However, the major challenge with this strategy lies in accurate assessment of fluid deficit and blood loss. Traditional indicators of hypovolemia, such as decreased central venous pressure, hypotension, and tachycardia, have been shown to be unreliable indicators of volume status and may be influenced by the physiological effects of pneumoperitoneum and steep positioning [44–46]. Urine output is also frequently used as a crude marker of volume status, but oliguria (urine output $<0.5\,mL.kg^{-1}.h^{-1}$) is common during major surgery and may be affected by Trendelenburg positioning and elevated antidiuretic hormone production [47]. There is increasing evidence that intraoperative oliguria does not reflect fluid status or risk of renal failure [47].

In recent years, goal-directed fluid therapy is gaining traction especially within enhanced recovery after surgery (ERAS) protocols [48,49] and has been shown to reduce the incidence of complications following major surgery [50,51]. Goal-directed fluid therapy uses small volume fluid challenges (e.g., 100–250mL) over a short period (e.g., 10min) to identify and treat fluid deficit [52]. An increase in stroke volume (SV) of >10% following this fluid challenge indicates that the patient's cardiovascular status falls within the steep section of the Frank-Starling curve, and is likely to benefit from a further fluid challenge. Conversely, further fluid challenges will not significantly increase SV in a patient on the plateau portion of the Frank-Starling curve [48]. Other than SV, dynamic indices such as stroke volume variation (SVV) or pulse pressure variation (PPV) may be used, with a PPV or SVV of >13% predictive of fluid responsiveness [53]. These indices can be measured using a variety of hemodynamic monitors such as transesophageal Doppler and arterial waveform analysis [49], although anesthesiologists should be aware of their limitations, including the requirement for normal sinus rhythm, normal intrathoracic pressure, and constant tidal volume ventilation [54,55]. Furthermore, elevated intra-abdominal pressure due to pneumoperitoneum can increase SVV and PPV independent of changes in fluid status [56,57], and monitoring the trend of these indices after patient positioning and onset of pneumoperitoneum may be helpful with assessing fluid deficit compared to the use of absolute thresholds. Hypotension unrelated to hypovolemia should be treated with vasopressors instead of intravenous fluids, and another assessment of fluid status should be made toward the end of surgery when the patient is positioned horizontal without pneumoperitoneum [24].

## Patient positioning and access

Patient positioning and robot docking is a dynamic process that demands careful supervision and attention of both surgeon and anesthesiologist. Prior to robot docking, it is important to ensure that the endotracheal tube, intravenous lines, and patient monitors are correctly positioned, secure, patent, and adequate length extensions have been applied. A second intravenous line should be considered as the arms are usually tucked, thus limiting access to the upper extremities. Pressure points should be padded, and adequate measures taken to prevent patient slippage during steep positioning. Patients with arms tucked or those placed in the lithotomy position are at particular risk of ulnar nerve and peroneal nerve injuries.

Proper patient positioning will also allow for safe docking of the patient cart and provide the surgical assistant and anesthesiologist with adequate access to the patient and robot arms. The footprint of the patient cart, bed position, and body habitus of the patient should be taken into account. Access pathways to the airway, lines, and patient monitors should also be demarcated. It may be useful to move the patient and operating table through the full range of possible positions prior to draping to determine the safe limits of patient position, possible pressure points, and to ensure adequate length of lines and monitors. This exercise may also help identify ventilation or hemodynamic issues, especially in obese patients or extremely steep operating positions. Certain surgeries (e.g., head and neck surgery) require the patient's head to be situated away from the anesthesia workstation, and rapid access to the airway is almost impossible.

The robot arms are heavy and bulky, and can inadvertently exert pressure on the patient during surgery. Constant vigilance is needed by both the surgical and anesthesiology teams, especially with changes in operating position or table height. During surgery, movement of the patient while robotic instruments are docked may result in iatrogenic injury with potentially devastating consequences. Hence, adequate anesthetic depth and constant neuromuscular blockade are recommended, and the operating table controls should be secured to prevent inadvertent table movement during surgery. Finally, the limited patient access can pose challenges with assessing intraoperative blood loss and emergent situations. Regular training and simulations should be conducted to facilitate rapid undocking of the patient cart in the event of an emergency. Clear understanding of the roles and communication by all members are needed to ensure safe and prompt emergency undocking, which can be accomplished in under 30 s [58,59].

## Analgesia and prevention of nausea or vomiting

Analgesic management should complement the gains offered by robotic-assisted surgery by achieving adequate postoperative analgesia while minimizing the risk of related adverse effects such as sedation, ileus, and nausea or vomiting. A multimodal, opioid-sparing analgesic approach may reduce the risk of postoperative ileus, nausea, and vomiting [60]. A systematic review of analgesic practices for minimally invasive gynecologic surgery found that preemptive analgesia is a safe and effective method of improving postoperative pain relief as a part of a multimodal analgesic regimen [61]. This includes the use of dexamethasone, acetaminophen, nonsteroidal anti-inflammatory drugs (NSAIDs), and regional anesthesia such as transversus abdominis plane (TAP) blocks [61]. Lidocaine infusion has also been shown to improve analgesia and reduce opioid consumption in patients undergoing abdominal hysterectomy [62]. Similarly, gabapentinoids and ketamine have been associated with reduced postoperative pain and opioid use, but are limited by adverse effects at higher doses and their routine use is not recommended [63,64]. Shoulder pain occurs in up to 80% of patients who undergo laparoscopic gynecologic surgeries, and maneuvers to resolve pneumoperitoneum should be considered [65]. For instance, placing the patient in Trendelenburg position while the anesthesiologist delivered five positive-pressure breaths with the fifth held for 5 s to allow escape of carbon dioxide from the trocar sleeves was associated with a significant reduction in postoperative pain scores after laparoscopic surgery [66].

Postoperative nausea and vomiting are common complications of anesthesia and surgery, and are associated with significant patient dissatisfaction, prolonged hospital admission, and increased healthcare costs [67]. Consequently, the avoidance of postoperative nausea and vomiting is an important goal and a multimodal prophylaxis regimen should be adopted based on the most recent consensus guidelines [67].

## Thromboembolic prophylaxis

Prolonged surgery is an important risk factor for thromboembolism, and prophylaxis using a combination of mechanical (e.g., intermittent pneumatic compression and graduated compression stockings) and pharmacological (e.g., low-molecular weight heparin) should be considered [68,69].

## Future scenario

Although the Da Vinci system is the most commonly used platform at present, several other robotic platforms are available or are at various stages of development. These include Flex (Medrobotics), SPORT—Single Port Orifice Robotic Technology (Titan Medical), and Versius (Cambridge Medical Robotics). Each platform has its own unique components, with varying impact on anesthetic management. Anesthesiologists should be trained and familiarized with the specific platform they will be working with, especially with the aspects pertinent to their anesthetic management.

Advances within the field of robotic-assisted surgery include machines designed to reduce their footprint within the operating room and setup time. Certain platforms such as the Da Vinci SP and SPORT require only a single surgical port, which may improve access to sites that were previously limited, including the nasopharynx and base of skull [70,71]. In addition, miniaturizing robots and robotic instruments facilitate their use in the pediatric population [72]. Newer systems provide haptic and tactile feedback to the surgeon, thereby increasing precision and accuracy [73]. There is also ongoing research in the use of augmented reality and image guidance, which holds promise in aiding identification of critical structures, improving resection, and reducing the operator learning curve [74]. Finally, the use of robots in telesurgery has been discussed for many years, but is limited by issues including the cost required to locate and maintain the robot in a remote location, the need for high speed data transfer, and legal and ethical considerations [75].

# Conclusions

Robotic-assisted surgery is associated with fewer postoperative complications and earlier functional recovery compared to open surgery. However, the physiological effects of steep angulation and pneumoperitoneum, in conjunction with limited patient access and requirement for immobile surgical field require anesthesiologists to be familiar with the management of these issues to optimize patient safety and improve perioperative outcomes.

# Key points

- The use of robotic-assisted surgery is expanding rapidly.
- Robotic-assisted surgery increases dexterity, precision, and surgeon ergonomics.
- Perioperative anesthetic management must account for steep patient positioning, pneumoperitoneum, and limited patient access.
- Development of new robotic platforms requires training and familiarity in their specific issues and impact on perioperative management.

# References

[1] George EI, Brand CTC, Marescaux J. Origins of robotic surgery: from skepticism to standard of care. J Soc Laparoend Surg 2018;22(4).
[2] Sheetz KH, Claflin J, Dimick JB. Trends in the adoption of robotic surgery for common surgical procedures. JAMA Netw Open 2020;3(1): e1918911.
[3] Lee JR. Anesthetic considerations for robotic surgery. Korean J Anesthesiol 2014;66(1):3–11.
[4] Ho C, Tsakonas E, Tran K, et al. Robot-assisted surgery compared with open surgery and laparoscopic surgery. CADTH Technol Overv 2012;2(2).
[5] Giri S, Sarkar DK. Current status of robotic surgery. Indian J Surg 2012;74(3):242–7.
[6] Phong S, Koh L. Anaesthesia for robotic-assisted radical prostatectomy: considerations for laparoscopy in the Trendelenburg position. Anaesth Intensive Care 2007;35(2):281–5.
[7] Tameze Y, Low YH. Outpatient robotic surgery: considerations for the anesthesiologist. Adv Anesth 2022;40(1):15–32.
[8] Intuitive Surgical Inc. Da Vinci Surgical Systems. Accessed 25 Jan 2023 https://www.intuitive.com/en-us/products-and-services/da-vinci/systems.
[9] Nguyen N, Anderson J, Budd M, et al. Effects of pneumoperitoneum on intraoperative pulmonary mechanics and gas exchange during laparoscopic gastric bypass. Surg Endosc Other Interv Tech 2004;18:64–71.
[10] Nguyen NT, Wolfe BM. The physiologic effects of pneumoperitoneum in the morbidly obese. Ann Surg 2005;241(2):219.
[11] Meininger D, Byhahn C, Bueck M, et al. Effects of prolonged pneumoperitoneum on hemodynamics and acid-base balance during totally endoscopic robot-assisted radical prostatectomies. World J Surg 2002;26:1423–7.
[12] Nguyen N, Ho H, Fleming N, et al. Cardiac function during laparoscopic vs open gastric bypass. Surg Endosc 2002;16(1):78.
[13] Perrin M, Fletcher A. Laparoscopic abdominal surgery. Contin Educ Anaesth Crit Care Pain 2004;4(4):107–10.
[14] Are C, Kutka M, Talamini M, et al. Effect of laparoscopic antireflux surgery upon renal blood flow. Am J Surg 2002;183(4):419–23.
[15] Demyttenaere S, Feldman LS, Fried GM. Effect of pneumoperitoneum on renal perfusion and function: a systematic review. Surg Endosc 2007;21:152–60.
[16] Nishio S, Takeda H, Yokoyama M. Changes in urinary output during laparoscopic adrenalectomy. BJU Int 1999;83(9):944–7.
[17] Nguyen NT, Perez RV, Fleming N, Rivers R, Wolfe BM. Effect of prolonged pneumoperitoneum on intraoperative urine output during laparoscopic gastric bypass. J Am Coll Surg 2002;195(4):476–83.
[18] Nguyen NT, Lee SL, Anderson JT, Palmer LS, Canet F, Wolfe BM. Evaluation of intra-abdominal pressure after laparoscopic and open gastric bypass. Obes Surg 2001;11(1):40–5.
[19] Jakimowicz J, Stultiens G, Smulders F. Laparoscopic insufflation of the abdomen reduces portal venous flow. Surg Endosc 1998;12:129–32.

[20] Nguyen NT, Braley S, Fleming NW, Lambourne L, Rivers R, Wolfe BM. Comparison of postoperative hepatic function after laparoscopic versus open gastric bypass. Am J Surg 2003;186(1):40–4.

[21] Strang CM, Hachenberg T, Fredén F, Hedenstierna G. Development of atelectasis and arterial to end-tidal P co2-difference in a porcine model of pneumoperitoneum. Br J Anaesth 2009;103(2):298–303.

[22] Nguyen NT, Cronan M, Braley S, Rivers R, Wolfe BM. Duplex ultrasound assessment of femoral venous flow during laparoscopic and open gastric bypass. Surg Endosc Other Interv Tech 2003;17:285–90.

[23] Schwenk W, Böhm B, Fügener A, Müller J. Intermittent pneumatic sequential compression (ISC) of the lower extremities prevents venous stasis during laparoscopic cholecystectomy: a prospective randomized study. Surg Endosc 1998;12:7–11.

[24] Carey B, Jones C, Fawcett W. Anaesthesia for minimally invasive abdominal and pelvic surgery. BJA Edu 2019;19(8):254–60.

[25] Aceto P, Galletta C, Cambise C, et al. Challenges for anaesthesia for robotic-assisted surgery in the elderly: a narrative review. Eur J Anaesthesiol Intensive Care 2023;2(2), e0019.

[26] Ripa M, Schipa C, Kopsacheilis N, et al. The impact of steep Trendelenburg position on intraocular pressure. J Clin Med 2022;11(10):2844.

[27] Rahe K, Schmidt T, Kohlhoff M, Tonner PH. Intra-operative bilateral otorrhagia during laparoscopic inguinal hernia repair: a case report and review of the literature. Anaesth Cases 2016;4(1):41–4.

[28] Lamvu G, Zolnoun D, Boggess J, Steege JF. Obesity: physiologic changes and challenges during laparoscopy. Am J Obstet Gynecol 2004; 191(2):669–74.

[29] Parotto M, Cooper RM, Behringer EC. Extubation of the challenging or difficult airway. Curr Anesthesiol Rep 2020;10:334–40.

[30] Kuriyama A, Jackson JL, Kamei J. Performance of the cuff leak test in adults in predicting post-extubation airway complications: a systematic review and meta-analysis. Crit Care 2020;24(1):1–11.

[31] Voldby AW, Brandstrup B. Fluid therapy in the perioperative setting—a clinical review. J Intensive Care 2016;4(1):1–12.

[32] Bundgaard-Nielsen M, Jørgensen C, Secher N, Kehlet H. Functional intravascular volume deficit in patients before surgery. Acta Anaesthesiol Scand 2010;54(4):464–9.

[33] American Society of Anesthesiologists. Practice guidelines for preoperative fasting and the use of pharmacologic agents to reduce the risk of pulmonary aspiration: application to healthy patients undergoing elective procedures. Anesthesiology 2017;126:376–93.

[34] Brady MC, Kinn S, Stuart P. Preoperative fasting for adults to prevent perioperative complications. Cochrane Database Syst Rev 2003;4. CD004423.

[35] Jung B, Lannerstad O, Påhlman L, Arodell M, Unosson M, Nilsson E. Preoperative mechanical preparation of the colon: the patient's experience. BMC Surg 2007;7:1–5.

[36] Holte K, Nielsen KG, Madsen JL, Kehlet H. Physiologic effects of bowel preparation. Dis Colon Rectum 2004;47:1397–402.

[37] Rollins KE, Javanmard-Emamghissi H, Lobo DN. Impact of mechanical bowel preparation in elective colorectal surgery: a meta-analysis. World J Gastroenterol 2018;24(4):519.

[38] Leenen JP, Hentzen JE, Ockhuijsen HD. Effectiveness of mechanical bowel preparation versus no preparation on anastomotic leakage in colorectal surgery: a systematic review and meta-analysis. Updates Surg 2019;71:227–36.

[39] Rollins KE, Javanmard-Emamghissi H, Acheson AG, Lobo DN. The role of oral antibiotic preparation in elective colorectal surgery: a meta-analysis. Ann Surg 2019;270(1):43.

[40] Toh JW, Phan K, Hitos K, et al. Association of mechanical bowel preparation and oral antibiotics before elective colorectal surgery with surgical site infection: a network meta-analysis. JAMA Netw Open 2018;1(6):e183226.

[41] Nygren J, Soop M, Thorell A, Efendic S, Nair KS, Ljungqvist O. Preoperative oral carbohydrate administration reduces postoperative insulin resistance. Clin Nutr 1998;17(2):65–71.

[42] Hausel J, Nygren J, Lagerkranser M, et al. A carbohydrate-rich drink reduces preoperative discomfort in elective surgery patients. Anesth Anal 2001;93(5):1344–50.

[43] Chappell D, Jacob M, Hofmann-Kiefer K, Conzen P, Rehm M. A rational approach to perioperative fluid management. J Am Soc Anesthesiol 2008;109(4):723–40.

[44] Pinsky MR. Hemodynamic evaluation and monitoring in the ICU. Chest 2007;132(6):2020–9.

[45] Bundgaard-Nielsen M, Holte K, Secher N, Kehlet H. Monitoring of peri-operative fluid administration by individualized goal-directed therapy. Acta Anaesthesiol Scand 2007;51(3):331–40.

[46] Marik PE, Cavallazzi R. Does the central venous pressure predict fluid responsiveness? An updated meta-analysis and a plea for some common sense. Crit Care Med 2013;41(7):1774–81.

[47] Kheterpal S, Tremper KK, Englesbe MJ, et al. Predictors of postoperative acute renal failure after noncardiac surgery in patients with previously normal renal function. J Am Soc Anesthesiol 2007;107(6):892–902.

[48] Miller TE, Roche AM, Mythen MG. Fluid management and goal-directed therapy as an adjunct to enhanced recovery after surgery (ERAS). Can J Anesth 2016;62(2):158–68.

[49] Kendrick JB, Kaye AD, Tong Y, et al. Goal-directed fluid therapy in the perioperative setting. J Anaesthesiol Clin Pharmacol 2019;35(Suppl 1): S29.

[50] Hamilton MA, Cecconi M, Rhodes A. A systematic review and meta-analysis on the use of preemptive hemodynamic intervention to improve postoperative outcomes in moderate and high-risk surgical patients. Anesth Anal 2011;112(6):1392–402.

[51] Gurgel ST, do Nascimento Jr P. Maintaining tissue perfusion in high-risk surgical patients: a systematic review of randomized clinical trials. Anesth Anal 2011;112(6):1384–91.

[52] Biais M, de Courson H, Lanchon R, et al. Mini-fluid challenge of 100 ml of crystalloid predicts fluid responsiveness in the operating room. Anesthesiology 2017;127(3):450–6.

[53] Marik PE, Cavallazzi R, Vasu T, Hirani A. Dynamic changes in arterial waveform derived variables and fluid responsiveness in mechanically ventilated patients: a systematic review of the literature. Crit Care Med 2009;37(9):2642–7.

[54] Lansdorp B, Lemson J, van Putten MJAM, de Keijzer A, Van Der Hoeven J, Pickkers P. Dynamic indices do not predict volume responsiveness in routine clinical practice. Br J Anaesth 2012;108(3):395–401.

[55] Perel A, Habicher M, Sander M. Bench-to-bedside review: functional hemodynamics during surgery-should it be used for all high-risk cases? Crit Care 2013;17(1):1–8.

[56] Tavernier B, Robin E. Assessment of fluid responsiveness during increased intra-abdominal pressure: keep the indices, but change the thresholds. Crit Care 2011;15:1–2.

[57] Renner J, Gruenewald M, Quaden R, et al. Influence of increased intra-abdominal pressure on fluid responsiveness predicted by pulse pressure variation and stroke volume variation in a porcine model. Crit Care Med 2009;37(2):650–8.

[58] Irvine M, Patil V. Anaesthesia for robot-assisted laparoscopic surgery. Contin Educ Anaesth Crit Care Pain 2009;9(4):125–9.

[59] O'Sullivan O, O'Sullivan S, Hewitt M, O'Reilly B. Da Vinci robot emergency undocking protocol. J Robot Surg 2016;10:251–3.

[60] Richebé P, Brulotte V, Raft J. Pharmacological strategies in multimodal analgesia for adults scheduled for ambulatory surgery. Curr Opin Anesthesiol 2019;32(6):720–6.

[61] Long JB, Bevil K, Giles DL. Preemptive analgesia in minimally invasive gynecologic surgery. J Minim Invasive Gynecol 2019;26(2):198–218.

[62] Xu S-Q, Li Y-H, Wang S-B, Hu S-H, Ju X, Xiao J-B. Effects of intravenous lidocaine, dexmedetomidine and their combination on postoperative pain and bowel function recovery after abdominal hysterectomy. Minerva Anestesiol 2017;83(7):685–94.

[63] Kumar AH, Habib AS. The role of gabapentinoids in acute and chronic pain after surgery. Curr Opin Anesthesiol 2019;32(5):629–34.

[64] Jouguelet-Lacoste J, La Colla L, Schilling D, Chelly JE. The use of intravenous infusion or single dose of low-dose ketamine for postoperative analgesia: a review of the current literature. Pain Med 2015;16(2):383–403.

[65] Kaloo P, Armstrong S, Kaloo C, Jordan V. Interventions to reduce shoulder pain following gynaecological laparoscopic procedures. Cochrane Database Syst Rev 2019;1.

[66] Phelps P, Cakmakkaya OS, Apfel CC, Radke OC. A simple clinical maneuver to reduce laparoscopy-induced shoulder pain: a randomized controlled trial. Obstet Gynecol 2008;111(5):1155–60.

[67] Jin Z, Gan TJ, Bergese SD. Prevention and treatment of postoperative nausea and vomiting (PONV): a review of current recommendations and emerging therapies. Ther Clin Risk Manag 2020;1305–17.

[68] Lyman GH, Carrier M, Ay C, et al. American Society of Hematology 2021 guidelines for management of venous thromboembolism: prevention and treatment in patients with cancer. Blood Adv 2021;5(4):927–74.

[69] Venclauskas L, Maleckas A, Arcelus JI. European guidelines on perioperative venous thromboembolism prophylaxis: surgery in the obese patient. Eur J Anaesthesiol 2018;35(2):147–53.

[70] Tamaki A, Rocco JW, Ozer E. The future of robotic surgery in otolaryngology–head and neck surgery. Oral Oncol 2020;101, 104510.

[71] Sheth KR, Koh CJ. The future of robotic surgery in pediatric urology: upcoming technology and evolution within the field. Front Pediatr 2019;7:259.

[72] Erkul E, Duvvuri U, Mehta D, Aydil U. Transoral robotic surgery for the pediatric head and neck surgeries. Eur Arch Otorhinolaryngol 2017;274:1747–50.

[73] Friedrich DT, Dürselen L, Mayer B, et al. Features of haptic and tactile feedback in TORS-a comparison of available surgical systems. J Robot Surg 2018;12:103–8.

[74] Chan JY, Holsinger FC, Liu S, Sorger JM, Azizian M, Tsang RK. Augmented reality for image guidance in transoral robotic surgery. J Robot Surg 2020;14:579–83.

[75] Hung AJ, Chen J, Shah A, Gill IS. Telementoring and telesurgery for minimally invasive procedures. J Urol 2018;199(2):355–69.

# 24

# Perioperative nursing care and patient positioning in robotic surgery

*Gisele Maia Jünger[a], Camila Mendonça de Moraes[b],*
*and Laís Gonçalves Tolentino[c]*

[a]Institute of Robotic Surgery at Faculty of Medical Sciences of Minas Gerais, Belo Horizonte, Minas Gerais, Brazil [b]Federal University of Rio de Janeiro (UFRJ), Rio de Janeiro, Brazil [c]Vila da Serra Hospital, Nova Lima, Minas Gerais, Brazil

## Introduction

The technology for the development of robot-assisted minimally invasive surgical procedures is constantly growing in the world. To guarantee the safety of the assistance provided to patients, the technical-scientific basis is essential, as well as the improvement of the skills of the nursing team.

The implementation of a Robotic Surgery Program requires the creation of new flows and processes within the institution. Hence, it is necessary to develop care protocols to guide and train the entire team [1].

For the development of specific skills, it is recommended to include a continuing education program, with a theoretical basis, training, and practical simulations for the development of safe assistance actions [2–4].

## Perioperative nursing care

Perioperative nursing assistance in robotic procedures encompasses the entire surgical journey of the patient. This includes the care and guidance of the patient and family in the pre-, intra-, and postoperative period, as well as all the preparation and organization of equipment with the anesthetic-surgical team.

The multidisciplinary team needs technical-scientific knowledge, skills training, and care strategies regarding the handling of the technology, to speed up, facilitate the procedure, and prevent postoperative complications.

### Preoperative

The preoperative period begins from the date surgery is scheduled until the patient enters the surgical center. Care begins with the preoperative nursing visit/interview, where the nurse must collect the patient's data, and information about their current health, and provide preoperative guidance. In addition to collecting specific data from the patient, for personalized assistance, the nurse is essential in active listening, guaranteeing the patient and family members the necessary specific information about the robot and robotic surgery [5,6].

Through the preoperative evaluation (data from the patient, the surgeon, and the surgical procedure to be performed), the nurse must plan which inputs will be used to assemble the room, such as its layout and patient positioning [7].

## Intraoperative

The intraoperative period includes checking and setting up the room, materials, and equipment, patient reception and care, patient identification and monitoring, assistance to the anesthesiologist, trichotomy, urinary catheterization when necessary, patient positioning protocol (analysis of the patient's skin and risk assessment scale), safe surgery checklist and the record of the entire Systematization of Perioperative Nursing Care.

For the prevention of postoperative complications, it is recommended to develop protocols for:

- *Prevention of hypothermia*: use passive and active methods, such as adjusting the room temperature (if it does not interfere with the ideal temperature for the system to work), thermal blanket, heated solutions, heated $CO_2$ gas, and other institutional methods [8].
- *Prevention of Venous Thromboembolism (VTE)*: use mechanical compression methods on the patient's lower limbs, such as elastic compression stockings and pneumatic intermittent compression boots with permanent or disposable leggings, and other measures indicated to the patient [9–11].
- *Prevention of injuries related to the patient's positioning*: it encompasses the assessment of injury risk using the Risk assessment scale for injury related to patient positioning (ELPO) to guide the development of a care plan during patient positioning. The ELPO brings together factors intrinsic to the patient and factors specific to the surgical environment, completing the analysis of seven items that generate a score between 7 and 35 points. Patients with a score greater than 19 are considered at high risk and must take special care to prevent positioning injuries [12].

Robotics technology implies specific activities for the team:

- *Cabling and turning on the system*: connect the cables between the system components and turn them on. Check if the system is working and ready for use. The robotic system must also be onsite, and connected to the internet network so that engineers can provide technical assistance remotely, before and during surgical procedures [13].
- *Robotic system cleaning*: clean the components of the robotic system before draping and after surgeries, with products recommended by the system manufacturer.
- *Room layout (arrangement of the three robotic system components in the operating room)*: perform this activity based on information about the surgical procedure and the medical team, as the layout of each robotic procedure changes according to these two variables.
- *Draping*: placing the drapes (sterile covers to cover the robot arms and maintain the sterility of the system).
- *Completion of robotic surgery forms*: it is recommended to create a checklist of materials for each type of surgical procedure and a form for monitoring specific surgical times. The checklist will help standardize robotic materials and guide nurses in setting up the room. Through the monitoring form, it is possible to assess the performance and learning curve of the multidisciplinary team at each stage of the surgery [1].
- *Robotic material checking*: evaluate permanent and disposable robotics materials. Observe the integrity and shelf-life of the packages, to guarantee the sterility of the materials [6].
- *Patient positioning*: essential to ensure fixation and safety of the procedure, with an explanation below.
- *Surgical safety checklist*: it is suggested to create a specific Surgical Safety checklist, including room layout, *draping*, Si calibration, portal marking, *docking* and its quantity, *undocking*, and the conference of robotic equipment and materials.
- *Errors and failures of the robotic system*: ensure preventive maintenance of the system and be aware of the correct functioning of robotic components during use. In case of error or failure during the procedure, interpret it and take the necessary steps with the engineering staff.
- *Precleaning and transport of robotic materials*:
  - After the surgery, check whether any tweezers have reached the end of their useful life in the inventory on the system's vision tower monitor screen. In X and Xi systems, check the usage indicator on the tweezers housing, which turns red when the number of lives ends [14,15].
  - The precleaning of the materials must be conducted immediately after the surgery [6,16] and its processing must start within 1 h after the end of the same [17].
  - Due to the high cost of robotic materials, and for their control and traceability, a specific form is suggested for sending them to the Central Sterile Supply Department (CSSD).

## Patient positioning in robotic surgery

The perioperative team (nurse, anesthesiologist, surgeon) is responsible for the decision, execution, maintenance, and responsibility for the patient positioning according to the particularities of each surgical procedure and each patient, thus guaranteeing safety and avoiding possible postoperative complications [5,12].

Patient positioning for robotic surgery, in addition to providing the best surgical access, should consider the position and accommodation of the equipment to avoid internal or external collisions of the robot arms, which could cause injuries to the patient [18,19].

The robotic surgical nurse, in addition to having technical-scientific knowledge regarding the anatomy, physiology, and risk factors of the patient that may promote the development of injuries, must be able to develop this activity considering all the particularities of the robotic procedure required, ensuring correct handling of equipment and their proper and safe fitting to the table and the patient [9].

In the preoperative period, the robotic nurse should plan the patient's positioning and assess which materials and devices will be needed, considering information about the surgery (procedure, laterality), the needs of the medical team, and the patient's specificities.

The creation of an institutional positioning protocol for each type of robotic surgical procedure is beneficial to guide the entire team and standardize the care actions provided. The protocol must be built based on evidence and include the patient's risk assessment, guidance on interventions and activities during positioning, support surfaces (SS), and devices necessary to ensure patient safety and comfort [6].

To assess the risk of injury to adult patients, the use of the ELPO scale is recommended to guide nurses in their care conduct [12]. In this sense, it is essential to develop a care protocol that includes the patient's risk score and guidelines related to the use of SSs and protections.

The SSs are all devices used to assist in positioning the patient's body and aim to reduce, relieve, and/or redistribute pressure in certain areas of the body, especially while remaining immobilized on the operating table. It is necessary to know the characteristics of the material that makes up the SS, such as its density and thickness, the ability to distribute pressure, to prevent shearing of the skin and patient sliding on the surgical table, to retain moisture, and to not generate memory [12,18].

Evidence indicates that SSs composed of viscoelastic polymer, as well as foams with a density of 33, have an efficient capacity to reduce tissue interface pressure and, therefore, are indicated for covering armboard, leggings, and the surgical table, for better patient accommodation. However, the choice of single use or permanent material will depend on the particularity of the institution. Cleaning and disinfection after each use must also be considered.

There are currently on the market specific devices for upper limbs for robotic surgery (Fig. 1) and integrated solutions (Figs. 2–4), SS with a thermal blanket, pads, and tube and extension fasteners, which in addition to preventing the slippage of the patient, also avoid shear, skin abrasions, and nerve damage. In addition, it is important to consider that we must organize longer materials and connections, such as tracheas, multiparameter cables, and anesthesia cart connections, as this will facilitate patient positioning according to the type of docking, avoiding traction and accidents during patient movement on the operating table.

Robotic systems are not yet programmed to detect patient slippage during surgeries and change the configuration of arms, trocars, and robotic clamps automatically [20,21]. Thus, it is essential to secure the patient to avoid incisional

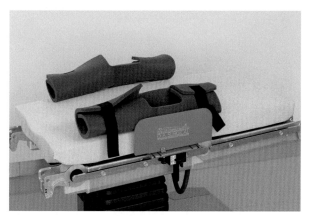

FIG. 1 ArmGuard patient arm protector pads. *Credit: D.A. Surgical, 2020.*

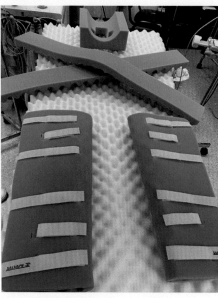

FIG. 2    Device for protection and fixation—Salvapé. *Credit: Authors' archive, 2022.*

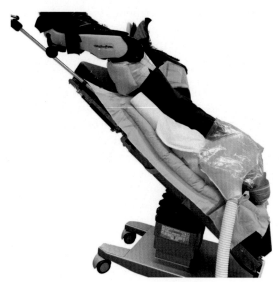

FIG. 3    Opt-shield air Trendelenburg. *Credit: Endocompany, 2023.*

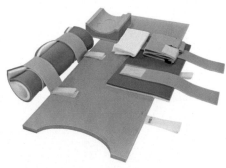

FIG. 4    Pink pad Trendelenburg positioning system. *Credit: Xodus Medical, 2021.*

rupture, prehension of parts of the patient's body in contact with the robot's arms and increased postoperative pain [18].

It is salutary to use nonslip SS, bands, or safety straps to fix the patient on the surgical table to prevent patient movement during the procedure.

The patient return electrode must be disposable and positioned as close as possible to the site to be operated on, in a muscular region with good vascularization, without bony prominences, without hair, with the least possible amount of subcutaneous tissue, and in a path that the current electricity does not pass through the patient's heart and does not interfere with the robotic operative field.

The patient's head (occipital region and face) must also be protected in all robotic procedures. The occipital region must have the pressure relieved, that is, it must not be in contact with the SS and be properly supported on a cushion of the appropriate size for the patient to avoid ischemic necrosis and alopecia. It is recommended to protect the face with a pyramidal foam mattress so that the camera and instruments that can move close to the patient's head during the robotic procedure do not cause injuries to the patient [18].

The eyes must be moistened, closed, and protected so that there is no dryness or injury to the cornea [12,18].

Once the patient cart is coupled to the patient and the surgical table, the patient cannot be moved anymore. Because the arms of the robotic system are connected to the patient's body and the limited access to it, at the end of the patient positioning, the multidisciplinary team should carry out a complete final assessment of the positioning and the patient, and carry out the positioning test, placing the surgical table in the position that it will remain throughout the procedure and then locking it (Example: Trendelenburg, reverse Trendelenburg and lateral), to ensure that the patient will not move during surgery and thus avoid rework and adverse events [22].

It is essential to guarantee safety in all types of patient positioning of robotic procedures for the success of docking and to be able to use the robotic arms in all their range, accurately and without collisions, thus avoiding any damage to the patient [22].

Thus, the robotic nurse should review the height of the positioners and the patient's limbs and areas that need extra protection before *docking*, to avoid collisions between the robot arms and the patient and not to hinder the *docking* performed by the surgeon. When the patient positioning in robotic surgery is not performed with care and care, respiratory, neurological, vascular, or tegumentary complications may occur [23,24], as described in another chapter of this book.

The types of patient positioning performed in robotic surgery are supine, prone, lateral, lithotomy, Trendelenburg, and reverse Trendelenburg position. In most cases, there is an association between them.

Some robotic procedures may have more than one surgical moment and the possibility of several types of positioning at each time, sometimes requiring *undocking* and repositioning of the patient at each surgical time. As an example, we have the robotic esophagectomy, where two surgical stages are usually performed with double docking. The first time, we performed the prone position, and the second time, the supine position [25].

## Supine position

The supine position is used in general robotic surgeries such as colectomy, abdominal wall hernias, pancreatectomy, second-stage esophagectomy, gastroplasty, cholecystectomy, and gastrectomy.

For this positioning (Fig. 5), body alignment must be ensured, and bony prominences should be relieved. Main care recommendations [6,26,27]:

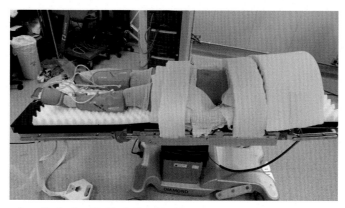

FIG. 5   Supine position with arms along the body. *Credit: Authors' archive.*

- Head: use support that ensures that the occipital region is not in contact with the table;
- Neck: maintain mento-sternal alignment;
- Arms: with one arm open at a maximum of 90° supported by a clamp at the same level as the operating table and with the palm of the hand facing upwards, or with both arms alongside the body with the palms of the hands facing the patient;
- Lower limbs: place a pillow below the knees to better accommodate the sacral region;
- Feet: use SS to keep heels floating.

If the institution opts for the pyramidal foam mattress, it must be complete and cut in accordance with the surgical table along its length. Cut out the head protectors and two straps to secure the patient's thorax and pelvis to the table. In addition to the strap, the patient's pelvis must also be secured with a safety strap to prevent slippage.

If the gel or vacuum mattress is used, also provide the positioners and devices to fix and position the patient on the surgical table.

It is recommended to use adhesive multilayer foam dressings in the sacral region for patients with an ELPO score >19, and in the heels, when it is not possible to keep them floating.

After the patient is positioned dorsally, the surgical table is normally positioned in reverse Trendelenburg, and depending on the robotic procedure, after incline, it is also necessary to perform left or right lateral on the table, with an angulation between 10° and 15°, according to the side of the patient's injury.

Docking can be pelvic or cephalic. In the cephalic docking, the nurse should take the trolley to the surgical table and another person should help her, staying close to the patient's head, signaling the safety distance, so that collisions with the surgical table and damage to the patient do not occur [26].

### *Lateral position*

The lateral position is used mainly in thoracic surgeries and robotic nephrectomies. In this position (Fig. 6), lateralization is on the opposite side to the one to be operated and it is necessary to pay attention to the ear lobe, upper limbs, iliac crest, and lower limbs about the side that is in contact with the operating table [27].

Main care recommendations [6,26–28]:

- Head and neck: support on head support at the height of the patient's shoulder, to guarantee mento-sternal alignment. Pay attention to the pressure on the earlobe in contact with the SS;
- Arms: the lower arm must be stretched at 90° and supported by a clamp and the upper arm raised to shoulder level at most by another clamp or positioner;
- Lower limbs: the patient's lower limb in contact with the table should be flexed and the upper lower limb should be stretched out and on SS a little higher than the bottom, to ensure that the weight of the upper leg does not exert pressure increase on the lower leg;
- Feet: pay attention to the malleolar region and apply adhesive multilayer foam protections when the ELPO score is greater than 19.

It is recommended to use two pads in roll format, one to be placed between the flank and the other in the dorsal region. Place a safety band or strap on the patient's iliac crest region. For kidney surgeries, also place a mattress strip above the patient's chest, to avoid the risk of injury with the robotic arm.

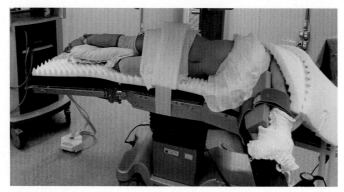

FIG. 6   Lateral position. *Credit: Authors' archive.*

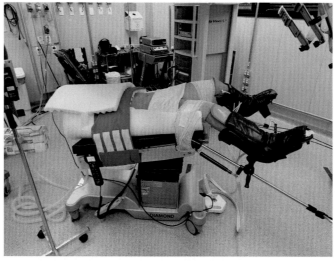

FIG. 7  Positioning of lithotomy associated with Trendelenburg. *Credit: Authors' archive.*

When the operating table is flexed in the Pilet position, one should try to minimize the degree of flexion of the operating table as much as possible [26].

### Lithotomy position

The lithotomy position is used in robotic urological surgeries, such as radical prostatectomy, in gynecological surgeries, such as hysterectomy and endometriosis, and coloproctology procedures, such as robotic rectosigmoidectomy.

Robot-assisted laparoscopic prostatectomy (RALP) is the most performed robotic urological surgery, and for its performance, patients must be positioned in a lithotomy position associated with steep Trendelenburg (Fig. 7), above 20° [9].

Main care recommendations [6,18,20,27,29,30]:

- Head: head support that ensures that the occipital region is not in contact with the table. Apply SS in the form of a hut around the patient's head, to prevent the weight of the fields, cables, and materials used during surgery from causing injury to the patient's face;
- Neck: mento-sternal alignment;
- Arms: positioned along the length of the body, with the palms of the hands facing the body. Protect hands and fingers to avoid pressure injury and collision with system arms;
- Lower limbs: open at the lowest possible angle and supported on stirrups like Allen Yellofins at a uniform height;
- Feet: keep heels floating.

It is recommended to perform the steep Trendelenburg test and check the table angulation [9].

SS should not be used on the shoulders during Trendelenburg in robotic surgeries due to associated neuromuscular injuries, particularly brachial plexus injuries [18,20].

The legs must be simultaneously removed from the stirrups and slowly repositioned in a supine position, immediately after undocking the robot [31].

### Prone position

The prone position is performed in one of the surgical stages of robotic esophagectomy (Fig. 8).

After the anesthetic procedure, place the pneumatic boots on the lower limbs, perform the necessary protections, and then start the rotation to the prone position. The team should act in synchrony, initially placing the patient in lateral decubitus, and then in a modified prone position with the right side of the chest slightly upwards. Pay attention to monitoring cables, probes, endotracheal tubes, and access extensions, avoiding traction, loss of access, and displacement of the tube [32].

Main care recommendations [6,27,33]:

- Head: lean on specific support that ensures pressure relief on the forehead, nostril, and chin. Keep eyes closed, moist, and protected;

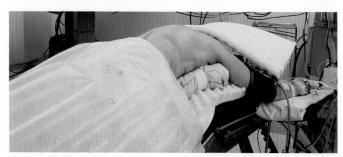

FIG. 8   Modified prone position. *Credit: Fernando Augusto Vasconcelos Santos, 2023.*

- Neck: mento-sternal alignment;
- Arms: positioned in the lower armboard and parallel to the operating table so as not to disturb the docking (the arms must be at a 90° angle with the body and 90° in the elbow region), or with one of the arms closed and close to the body;
- Thorax: place specific SS that guarantees thoracic expansion, and relief of breast pressure in women and male genitalia. When it is not possible to use specific supports, it is recommended to apply a foam roller in the mid-axillary line, to reduce the pressure of the shoulders on the lung, and another two rollers vertically on the abdomen, deviating from the iliac crest, thus forming a type of support. Inverted triangle;
- Lower limbs: bent with support on the thigh and calf to leave the knees free of pressure;
- Feet: support the back of the feet in heel support and let the tips of the fingers float, without contact with the table.

## Postoperative

The nursing team must carry out all the nursing care and interventions provided to patients undergoing minimally invasive surgeries and surgeries using of pneumoperitoneum in the postanesthesia care unit (PACU), inpatient unit, and/or intensive care unit (ICU).

As in any other surgical technique, in robotics, there may also be complications in the preoperative, intraoperative, and postoperative periods [19]. Nurses should be aware of possible harms related to the use of robotic technology: associated damage because of inadequate use of the tool, incorrect positioning, and prolonged surgery time due to the learning curve.

Regarding the inappropriate use of the tool and incorrect positioning, the nurse team should assess the integrity of the patient's skin in the cephalocaudal direction and the presence of bleeding, pain, tingling, weakness, or numbness in any region or limb.

Neuromuscular complications and postoperative pain may be related to inadequate patient positioning and/or the nonuse of adequate positioners [23].

Pain, tingling, and numbness in the lower limbs may be related to the inappropriate use of the surgical table legging associated with prolonged lithotomy. Bad positioning of the leggings in the lower limbs can cause pain, nerve damage, paresthesia, and, in rarer cases, compartment syndrome [31].

Hematomas and skin lesions near surgical incisions may be associated with the incorrect use of robotic trocars. Corneal injuries can result from not using eye solutions or ointments, and/or increased intraocular pressure [24].

Nurses should also be aware of possible complications related to lithotomy associated with accentuated Trendelenburg, a position that is widely used in robotic surgeries and that may lead to cardiovascular alterations, increased intracranial and intraocular pressure, othorragy, respiratory and pulmonary alterations, brachial plexus injury and compartment syndrome [5,6].

Nursing must develop an individualized care plan for each patient, and all nursing care and interventions performed must be recorded and reported to the medical team.

## Future scenario

The introduction of robotic surgery in the surgical center requires nurses to have the technical knowledge and the development of specific skills. Although the robot brings greater complexity to the operating room, it also will bring more safety for the team and the patient, and it is up to the nurse to transform the highly technological environment of the robotic room into a place of reception and humanization of care [2].

# Key points

- Importance of proper patient positioning in robotic surgery: Patient positioning is a critical factor for the success of robotic surgery, as it can affect the visualization of the operative field, access to surgical instruments, and patient physical integrity. The perioperative nurse plays a fundamental role in evaluating and properly positioning the patient before surgery.
- Challenges in nursing care in robotic surgery: Nursing care in robotic surgery presents some challenges, including the need for specialized training for robot operation, protocol creation, and management of multiple high-tech equipment and devices.
- Perioperative nursing care approach in robotic surgery: Perioperative nursing care in robotic surgery should include a thorough patient evaluation before surgery, collaboration with the surgical team to ensure proper patient positioning, and continuous monitoring of the patient during surgery. In addition, the perioperative nurse should be aware of potential risks associated with robotic surgery, such as respiratory and cardiovascular complications, and be prepared to intervene quickly if they occur.

# References

[1] Guimarães GC, Parekh DJ, Távora JEF, Nahar B, Jünger GM. Cirurgia Robótica – Princípios e Fundamentos. 1st ed. Belo Horizonte: Editora Universitária Ciências Médicas; 2022.

[2] Martins RC, Trevilato DD, Jost MT, Caregnato RC. Nursing performance in robotic surgeries: integrative review. Revi Bras Enf 2019;72(3):795–800. https://doi.org/10.1590/0034-7167-2018-0426.

[3] Almeida TFP, Campos MEC, Castro PR, Garcia DPC, Jünger GM, Silva RF, et al. Training in the protocol for robotic undocking for life emergency support (RULES) improves team communication, coordination and reduces the time required to decouple the robotic system from the patient. Int J Med Robot 2022;18(6), e2454. https://doi.org/10.1002/rcs.2454 [Epub 2022 Sep 3] PMID: 35998074.

[4] Cohen TN, Anger JT, Kanji FF, Zamudio J, DeForest E, Lusk C, et al. A novel approach for engagement in team training in high-technology surgery: the robotic-assisted surgery olympics. J Patient Saf 2022;18(6):570–7. https://doi.org/10.1097/PTS.0000000000001056.

[5] Angelo CS, Pachioni CFM, Joaquim EHG, Silva EAL, Santos GG, Bonfim IM, et al. Efetividade do protocolo prevenção de lesões de pele em cirurgias urológicas robóticas. In: Rev SOBECC, São Paulo; 2017. p. 142–60.

[6] Sobecc. Associação Brasileira de Enfermeiros de Centro Cirúrgico, Recuperação Anestésica e Centro de Material e Esterilização. Diretrizes de práticas em enfermagem cirúrgica e processamento de produtos para a saúde. 7th ed. Barueri: Manole; 2017.

[7] Silva FF, Santos PF, Dalto APP, Granandeiro DS, Granandeiro RMA, Melo NGS, et al. Autonomia e gerenciamento do enfermeiro no serviço de cirurgia robótica. Saúde Coletiva (Barueri) 2020;9(51):1954–8. https://doi.org/10.36489/saudecoletiva.2019v9i51p1954-1958. [S. l.]. Available from: http://revistas.mpmcomunicacao.com.br/index.php/saudecoletiva/article/view/182. [Accessed 9 April 2021].

[8] Estes SJ, Goldenberg D, Winder JS, Juza RM, Lyn-Sue JR. Best practices for robotic surgery programs. J Soc Laparoendosc Surg 2017;21(2). https://doi.org/10.4293/JSLS.2016.00102.

[9] Carey RI, Leveillee RJ. Patient positioning for robotic urologic procedures. In: Patel V, editor. Robotic urologic surgery. London: Springer; 2007. p. 61–6.

[10] Hariharan U, Shah SB. Venous thromboembolism and robotic surgery: need for prophylaxis and review of literature. J Hematol Thrombo Dis 2015;3:227. https://doi.org/10.4172/2329-8790.1000227.

[11] Forrest JB, Clemens JQ, Finamore P, Leveillee R, Lippert M, Pisters L, et al. AUA best practice statement for the prevention of deep vein thrombosis in patients undergoing urologic surgery. J Urol 2009;181(3):1170–7. https://doi.org/10.1016/j.juro.2008.12.027 [Epub 2009 Jan 18] PMID: 19152926.

[12] CMM L. Escala de avaliação de risco para o desenvolvimento de lesões decorrentes do posicionamento cirúrgico: construção e validação. 2013. Ribeirão Preto: Universidade de São Paulo; 2013. Tese (Doutorado em Enfermagem).

[13] Intuitive Surgical. Manual do Utilizador Da Vinci Surgical System Si. Sunnyvale; 2009.

[14] Intuitive Surgical. Sistema Manual do Usuário Da Vinci Surgical System X. Sunnyvale; 2019.

[15] Intuitive Surgical. Sistema Manual do Usuário Da Vinci Surgical System Xi. Sunnyvale; 2019.

[16] Anvisa (Agência Nacional de Vigilância Sanitária). Informe técnico nº 01/09. Princípios Básicos para Limpeza de Instrumental Cirúrgico em Serviços de Saúde, 2009. Brasil. Available from: https://www.saude.mg.gov.br/images/documentos/informe_tecnico_1.pdf. [Accessed 29 March 2021].

[17] Intuitive Surgical. Da Vinci S and Si instrument – Reprocessing instructions. Cleaning, disinfection, and sterilization information for reusable instruments that are used with da Vinci S and da Vinci Si systems. Sunnyvale; 2017.

[18] Ali G. Robotics in practice: new angles on safer positioning. Contemporary OB/GYN 2012. October 1.

[19] Jara RD, Guerron AD, Portenier D. Complications of robotic surgery. Surg Clin 2020;100(2):461–8. https://doi.org/10.1016/j.suc.2019.12.008.

[20] Allen D. Patient positioning for robotic surgery - an insider's perspective. OR Today Magazine June 2012.

[21] Hewer CL. The physiology and complications of the Trendelenburg position. Can Med Assoc J 1956;74(4):285–8. PMID: 13293598. PMCID: PMC1824068.

[22] Iqbal H, Gray M, Gowrie-Mohan S. Anestesia para Cirurgia Urológica Auxiliada por Robô. 2019 ABCD Arq Bras Cir Dig 2013;26(3):190–4.

[23] Gezginci E, Ozkaptan O, Yalcin S, Akin Y, Rassweiler J, Gozen AS. Postoperative pain and neuromuscular complications associated with patient positioning after robotic assisted laparoscopic radical prostatectomy: a retrospective non-placebo and non-randomized study. Int Urol Nephrol 2015;47:1635–41.

[24] Gkegkes ID, Karydis A, Tyritzis SI, Iavazzo C. Ocular complications in robotic surgery. Int J Med Rob Comput Assisted Surg 2015;11(3):269–74.

[25] Intuitive Surgical. Da Vinci Esophagectomy - procedure guide. Sunnyvale; 2018.

[26] Sousa CS, Bispo DM, Acunã AA. Criação de um manual para posicionamento cirúrgico: relato de experiência. In: Rev. SOBECC, São Paulo; 2018. p. 169–75.

[27] Aorn. Guidelines for perioperative practice. Denver, CO: Association of Perioperative Registered Nurses (AORN); 2018.

[28] Chang C, Steinberg Z, Shah A, Gundeti MS. Patient positioning and port placement for robot-assisted surgery. J Endourol 2014;28(6):631–8. https://doi.org/10.1089/end.2013.0733 [Epub 2014 Apr 25] PMID: 24548088. PMCID:PMC4382714.

[29] Abdalla RZ, Averbach M, Junior UR, Machado MAC, Filho CRP. Cirurgia Abdominal por Robótica: Experiência Brasileira Inicial. Arq Bras Cir Dig São Paulo 2013;26(3):190–4.

[30] Mangham M. Positioning of the anaesthetised patient during robotically assisted laparoscopic surgery: perioperative staff experiences. J Perioper Pract 2016;26(3):50–2. https://doi.org/10.1177/175045891602600305.

[31] Pridgeon S, Bishop CV, Adshead J. Lower limb compartment syndrome as a complication of robot-assisted radical prostatectomy: the UK experience. BJU Int 2013;112(4):485–8. https://doi.org/10.1111/bju.12201.

[32] Gramont K. Enfermagem Cirúrgica. Brasília: NT Editora; 2016. 182p.

[33] Miranda AB, Fogaça AR, Rizzetto M, Lopes LCC. Posicionamento Cirúrgico: Cuidados de Enfermagem no transoperatório. In: Rev. SOBECC, São Paulo; 2016. p. 52–8.

# Human operator factors affecting robotic surgery

*Luiz Antônio Mondadori, Camila De Souza Hagui,*
*and Eduardo Giroud Joaquim*

Department of Anesthesiology, A.C. Camargo Cancer Center, São Paulo, Brazil

## Concept of human factors

Human factors refer to behaviors, skills, and limitations of individuals that can impact the performance of a task, especially when involving complex and security-critical systems. Human factors include such variables as knowledge, skills, attitudes, motivation, environmental factors, and social interactions, among others.

The acknowledgment and understanding of human factors are important for systems analysis, as they can impact the efficiency, safety, and effectiveness of human performance in several scenarios. The identification and management of human factors can result in major improvements in safety systems, and in performance and efficiency.

Human factors were initially described in the 1950s along with the ongoing development of human sciences and psychology. At the time, practitioners started to focus on human capabilities and limitations, as well as on behavior, and how it influenced effectiveness and safety in complex systems. Initially, it was system security, particularly in aviation, that was the first to address human factors. In the 1980s, human factors began to be described for healthcare due to increasing awareness of the role played by effective communication and teamwork. In 2008, WHO began implementing checklists as part of the patient safety initiative, aimed at improving safety and quality in healthcare worldwide [1].

## Human factors in healthcare

Human factors have a major impact on healthcare, from the development of new technologies to patient care. Some examples include the following [2]:

1. Medical Errors: Human factors such as fatigue, stress, lack of communication, and lack of skills can lead to medical errors and jeopardize patient safety.
2. Communication: Miscommunication among physicians, nurses, and other healthcare professionals can cause errors in patient care and compromise the quality of care.
3. Training: Insufficient or inadequate training can cause errors and endanger patient safety.
4. Psychological Factors: Psychological factors such as stress, anxiety, and depression can impact the performance of healthcare professionals and the quality of patient care.
5. Design of Healthcare Technology: Poorly designed health technologies can lead to errors and jeopardize patient safety.

Human factors are related to human behavior and skills that can impact performance in complex systems, such as healthcare. These factors include fatigue, lack of skills, stress, lack of communication, and time pressure. When such factors are overlooked or improperly managed, they can trigger medical errors. Medical errors are defined as actions or omissions resulting in harm to patients. They can include misdiagnosis, wrong medication, wrong patient, or wrong-site surgery, among others. Medical errors are caused by a series of factors, including human factors. Fatigue and stress are human factors that can trigger medical errors because they affect the attention and concentration of healthcare professionals. Lack of skills and miscommunication among healthcare professionals can also cause medical errors. Furthermore, time pressure and excessive workload can cause medical errors by compelling healthcare professionals to make hasty decisions or to be less careful than required while performing tasks [3].

Miscommunication is one of the most frequent human factors contributing to medical errors. It is associated with lack of precise exchange of information among and between healthcare professionals, patients, and others involved in medical care.

Effective communication is essential to ensure the quality and safety of medical care. Miscommunication can lead to misdiagnosis, medication administered incorrectly, surgery performed on the wrong patient or on the wrong site, among other major medical errors.

Miscommunication can occur at various stages of medical care, including exchange of information between healthcare professionals or between different healthcare institutions, or during healthcare professional-patient communication.

The implementation of clear and reliable communication practices and protocols is crucial for preventing miscommunication. This includes using electronic medical records, clear and accurate records, reviewing important information frequently, and ensuring that all parties involved have access to relevant information. Additionally, training and education are valuable to ensure that healthcare professionals have the communication skills required to assure patient safety [4]. Training is vital for the suitable performance of health professionals and for the safety and quality of healthcare. Human factors, such as knowledge, skills, attitudes, and behaviors, are clearly related to the training and performance of health professionals.

All professionals involved in patient care often need training in nontechnical skills in order to work as a coordinated and objective team. Errors must be seen as opportunities for improvement, as the culture of perfection and infallibility is incompatible with human reality. The prevention of minor or major errors is directly linked to the ability of health professionals to communicate effectively in a teamwork environment.

Proper training helps ensure that healthcare professionals have the required skills and knowledge to perform their roles safely and effectively. This includes training in medical procedures, effective communication, risk management, and problem-solving. Furthermore, training helps shape the attitudes and behaviors of healthcare professionals, which can significantly impact the quality of healthcare. Ethics and value training, for example, can help assure that healthcare professionals are aware of their responsibilities and obligations toward patients and the community.

Continuing education is also important to make sure that healthcare professionals are updated with the latest health discoveries and practices. This can help ensure patients receive the most appropriate and advanced treatments available.

Psychological factors have a significant impact on healthcare, both for healthcare professionals and patients [5].

The mental health and psychological well-being of healthcare professionals can affect their performance and their ability to make appropriate decisions. Excessive stress or burnout, for example, can cause fatigue, affecting the ability of healthcare professionals to pay attention to detail and make appropriate decisions. Furthermore, the mental health of healthcare professionals can interfere with their ability to relate and communicate effectively with patients and team members. This can significantly impact the quality of care and patient satisfaction.

The mental health and psychological well-being of patients are also important for their recovery and for successful outcomes. Conditions such as stress, anxiety, or depression can affect a patient's ability to follow prescribed treatment or adhere to medical recommendations.

Likewise, the mental status of patients can interfere with their ability to communicate effectively with healthcare providers and can lead to misdiagnosis and incorrect treatment.

The design of the technology is an important human factor in healthcare because it shapes how healthcare providers interact with medical technologies and, consequently, how technologies impact patient care [6].

A thoughtful and intuitive design can help healthcare providers to work more efficiently and effectively, minimizing the likelihood of medical errors. Thus, clear, simple user interfaces help providers gain quick access to important data and make educated decisions.

Additionally, technology design can also play an important role in preventing work-related injuries and illnesses, such as repetitive strain injury (RSI) and work-related musculoskeletal disorders (WMSD). An ergonomic keyboard and mouse, for example, can help prevent injuries related to excessive computer use.

The design of the technology is also important for patients. Thoughtfully designed and easy to use medical devices, for example, can help ensure that patients administer their treatments and use medical devices properly at home, without requiring help or supervision.

# Human factors in robotic surgery

Due to their complex architecture, which involves many healthcare professionals and an extensive range of devices in a highly dynamic environment, surgical procedures have a high potential for human errors. Adverse events in robotic surgery can be caused by several factors, such as malfunctioning equipment, miscommunication among surgical team members, suboptimal patient preparation, or procedure planning, among others. Some of the most common adverse events include internal organ injury, infection, excessive bleeding, adverse drug reactions, respiratory disorders, and complications during recovery [7].

Robotic surgery has a unique feature that is absent from other surgical techniques, as the main surgeon is not in the surgical field, but at a console set apart from the other surgical team members, impairing nonverbal communication. The surgeon speaks into a microphone at the console, and it is not always clear to whom the surgeon is directing the command, or there may be distortion in the microphone sound that affects understanding the command. This compels the surgeon to move his/her head away from the console, interfering with the progress of the operation and the surgeon's concentration. Closed-loop communication, a technique employed in simulation scenarios such as cardiac arrest, helps minimize miscommunication and interruptions during surgery [8].

Thus, it is essential to learn from errors and implement safe practices to minimize the occurrence of adverse events in robotic surgery. This may involve establishing strict training programs for surgeons and healthcare staff, performing simulations prior to the actual surgery, in order to improve communication and collaboration among surgical team members, among other measures. It is also important to perform regular evaluations of processes to detect points for continuous improvement and ensure patient safety.

Assessment of the following human factors is key in robotic surgery:

1. **Operator Training and Skills:** Robotic surgery outcomes depend on surgeons' skills and training for handling the robotic system.
2. **Surgical Team Communication:** Clear and effective communication among members of the surgical team is crucial to guarantee patient safety and surgery outcome.
3. **Surgeon Console Ergonomics:** The working position and ergonomics of the surgery console can interfere with a surgeon's ability to control the robotic system and perform the surgery effectively.
4. **Surgeon's Psychological Factors:** Stress, fatigue, and other psychological factors can interfere with the surgeon's ability to control the robotic system and accurately perform the surgery.
5. **Interaction between Robotic System and Patient:** The interaction between the robotic system and the patient is essential to ensure a safe and successful procedure. It is important to consider factors such as patient preparation and proper placement of robotic devices.

Operator training and skills are crucial factors in robotic surgery. The surgeon needs to have a deep understanding of robotic technology, and the skills to control the system efficiently during surgery. Proper training includes simulations, training on animal models, and supervised real-world cases before the surgeon performs surgeries alone [9].

Additionally, it is critical that the surgeon be familiar with the robotic system interface and has appropriate understanding of surgeon console ergonomics to avoid fatigue and guarantee accuracy during surgery.

Clear and effective communication among surgical team members is crucial to ensure patient safety and a successful outcome for the robotic surgery. Intraoperatively, the surgical team must work together to perform the surgery efficiently and safely.

Effective communication includes assigning clear roles to each team member, creating clear protocols for communication, and having an open line of communication intraoperatively. In addition, it is also important that all team members be familiar with the robotic system and equipment, including its capabilities and limitations.

It is also essential that the surgical team be prepared to deal with unforeseen intraoperative incidents. Thus, clear communication among team members is crucial to guarantee a quick and efficient response to any unexpected incident.

Noise and excessive alarms can disrupt robotic surgery in several ways [10] by distraction, confusion, stress, miscommunication, or interruptions. Thus, it is important to keep the robotic surgery environment as quiet and distraction-free as possible, setting alarms appropriately to prevent noise pollution and to keep interruptions at a minimum.

Decking in robotic surgery refers to an intraoperative failure or interruption in robotic system functioning. It can be caused by hardware or software technical glitches, or human failures, such as operating mistakes or inadequate maintenance. Some of the most common problems with decking include power system failures, difficulties with robotic arms, failures in communication between system components, problems with the surgeon's screen or view, and software bugs.

Decking can interrupt the surgery and extend surgical time, which can increase the risk of patient complications. Likewise, decking may cause discomfort or anxiety to the patient and medical staff and may cause loss of confidence in the robotic system. Thus, it is critical that medical and information technology teams work together to minimize the risks of decking in robotic surgeries, through adequate training, regular maintenance, and strict preoperative checking [11].

Regular equipment maintenance and checking are crucial to assure patient safety and the success of robotic surgery. Robotic surgery equipment is complex and requires regular maintenance and checking to ensure it is working properly. Maintenance comprises regular image quality checks, system security checks, and repairs to damaged or worn components. Additionally, preventive maintenance plans in place are essential to prevent intraoperative equipment failure.

Regular equipment checking also includes checking that all system components are present and functioning correctly before surgery. This ensures the system is ready to be used safely during surgery.

The training and skills of the surgeon are critical human factors in robotic surgery. The surgeon is responsible for performing the surgery and making critical intraoperative decisions, thus adequate surgeon training and competence are essential for safe and efficient procedures.

Training includes the surgeon becoming familiar with the robotic system and performing simulated surgeries before doing actual surgeries. Additionally, it is important for the surgeon to have a sound understanding of surgical techniques and human anatomy in order to guarantee a safe and efficient execution of the surgery. Just as the surgeon is trained to perfect the surgical technique, nontechnical skills can also be acquired by training. Simulated reality scenarios, workshops, and e-learning can be used for training.

The surgeon's competence is evaluated through previous experience, adequate training, and ability to carry out surgeries successfully. The outcomes of surgery depend on surgeons' fine motor skills and expertise on how they use surgical instruments. Furthermore, nontechnical surgeon skills, particularly communication and leadership, complement technical skills.

## Nontechnical surgeon skills

The nontechnical skills of a robotic surgery surgeon comprise skills essential for a successful surgery outcome as well as the technical skills specific to the procedure. Some nontechnical skills include the following:

(1) Communication: The ability to communicate effectively with the patient, the healthcare team, and other members of the surgical team is crucial to ensure patient safety and surgery outcome. Clear communication is essential to ensure the safety and outcome of any surgery, even more so in robotic surgery. Clear communication enables the surgical team to share information, solve problems, and coordinate actions effectively.

(2) Leadership: The surgeon performing robotic surgery must be able to lead the surgical team and effectively make clinical decisions intraoperatively.

   Surgeon leadership is decisive during robotic surgery, as the surgeon is responsible for ensuring that the surgical team is aligned and performing efficiently. The surgeon must be able to provide clear direction and make quick decisions intraoperatively. It is also crucial for a surgeon to implement a safety culture, encouraging the team to report errors or failures immediately, and promoting the detection and correction of issues before they cause harm to the patient [12].

   Surgeon leadership is also important to ensure the quality of the work provided by the team, who must be familiar with the techniques and devices used in robotic surgery. In addition, the surgeon must be able to deal with unforeseen scenarios, such as intraoperative incidents, and make evidence-based decisions ensuring the best outcome for the patient.

(3) Teamwork: Robotic surgery is a team effort requiring effective collaboration between the surgeon and other members of the healthcare team.

   Teamwork is a key component of robotic surgery and important for ensuring safe and effective patient outcomes. During robotic surgery, the surgeon controls the surgical robot, but needs other team professionals to help prepare the patient, monitor the patient's clinical condition, and provide additional intraoperative support. Likewise, the surgical team plays a valuable role in preparing and handling robotic technology, including robot and surgical instrument setup before the surgery starts. The surgical team must work together to ensure intraoperative patient safety.

**(4)** Adaptability: The ability to deal with unexpected scenarios and to make quick adjustments is an essential skill for robotic procedure surgeons, as it allows surgeons to respond to unforeseen situations and adapt their technique according to the requirements of each case. This is particularly important in robotic surgery, as it involves complex and highly sophisticated technologies that put high demands on the surgical team. Surgeons must be able to adapt quickly to changes in patient conditions, equipment, or surgical environment. They must also be able to work as a team with other healthcare professionals to guarantee that the patient receives the best possible care.

**(5)** Resilience: Robotic surgeries can be complex and challenging. The surgeon must be able to handle pressure and work effectively even under stressful conditions. Resilience is an important quality for surgeons in robotic surgery, as it is a complex and technologically advanced procedure requiring exceptional skills and abilities. The surgeon needs to be able to manage unexpected challenges, adapt to unanticipated situations, and work under pressure. Here are examples of how resilience is an important attribute for robotic surgery: Technical failures, unpredictability of surgery, and teamwork.

**(6)** Decision Making: The surgeon must be capable of making clinical decisions effectively to ensure patient safety. Proper decision-making is decisive for the successful outcome of robotic surgery. During surgery, the surgeon is responsible for assessing the situation and making quick and accurate decisions that affect surgery outcome, how choice of surgical technique, identification and treatment of complications, continuation, customization, or interruption of Surgery.

In summary, making the right decisions is an important component of robotic surgery and is key to ensure safe and successful outcomes for patients. Thus, all nontechnical skills are just as valuable as technical skills to ensure the success of robotic surgery and preclude adverse events.

## Nontechnical skill training

Training a surgeon's nontechnical skills for robotic surgery can be accomplished in several ways, including [13,14]:

**(1)** Simulation: Simulation is one of the best methods for training nontechnical skills. Surgeons can use virtual simulations to practice tasks and decisions that they would face during an actual surgery. There is no specific ideal number of simulations required for performing robotic surgery, as it depends on several factors, including surgeon skills and experience, complexity of surgery, and training program regulations. Some training programs may require more simulations, while others may accept less. It is important for surgeons to participate in a sufficient number of simulations to feel confident and competent before performing robotic surgery on their own.

Simulation can be used as an ongoing assessment and training tool, so the number of simulations can increase or decrease over time, depending on the surgeon's needs and skills. It is important that surgeons continue to train and improve their nontechnical skills even after having completed initial training.

**(2)** Team Training: Team training can help develop communication, leadership, and teamwork skills. The surgeon may participate in team training sessions with other surgical team members, including nurses, anesthesiologists, and surgical technicians.

**(3)** Peer Feedback: Peer feedback is important to help surgeons improve their nontechnical skills. The surgeon can receive feedback from other surgeons who are already experts in robotic surgery and who can provide advice and tips for improvement.

**(4)** Training with Experienced Instructors: Training with experienced instructors is an excellent way of learning nontechnical skills. The surgeon can participate in courses or training given by experienced robotic surgeons who can provide guidance and feedback.

**(5)** Observation of Real Surgery: Observing actual operations can help the surgeon understand the nontechnical skills required during robotic surgery. The surgeon can observe robotic surgeries firsthand and see how experienced surgeons make decisions and manage complications.

Technological progress in the health sector has shown that technical skills alone do not suffice to guarantee successful surgery outcomes. It is important for the physician to adapt to this new scenario, as mastering technical skills alone is no longer enough to guarantee the success of the surgery. The surgeon must be the team leader, and be able to extract the best from the team, understanding and being clearly understood. It is important to emphasize that nontechnical skills can also be acquired through training, just like technical skills. Training in nontechnical skills can be achieved through a combination of simulations, team training, peer feedback, training with experienced instructors, and observation of actual surgeries. It is important for surgeons to have enough time for training and developing skills before performing robotic surgeries on their own.

Currently, the health system is seen as a service provider, which raises the same expectations and requests as for other types of services. Although it is not possible to guarantee results in the healthcare area, what patients and health operators expect from the healthcare team is that it be prepared to minimize harm, reduce costs, and increase the quality and safety of services offered. This is only possible with a combination of the team's technical efficiency, and a teamwork attitude with an overall view of the process involved in the surgical care of the patient. Therefore, human factors cannot be ignored, as they are crucial for the success of the surgery, especially robotic surgery.

## Future scenario

In the future of robotic surgery, an increase in human-machine collaboration is expected, with surgeons interacting more intuitively with robots. Furthermore, improvements in tactile feedback will allow surgeons to feel the texture of tissues and applied forces during procedures. The miniaturization of surgical instruments will enable less invasive surgeries in hard-to-reach areas, while artificial intelligence and machine learning will play a significant role in the continuous improvement of robotic systems. However, full automation remains uncertain and will depend on technological advancements, regulations, and medical acceptance.

## Key points

- Human factors have been recognized since the 1950s and are important in aviation and healthcare.
- Human factors impact performance in complex systems like robotic surgery.
- Understanding human factors is crucial for efficiency, safety, and effectiveness in healthcare.
- Addressing human factors in robotic surgery improves operator training, communication, ergonomics, and patient interaction.

## References

[1] Parker SH. Human factors science: brief history and applications to healthcare. Curr Probl Pediatr Adolesc Health Care 2015;45(12):390–4.

[2] Keebler JR, Rosen MA, Sittig DF, Thomas E, Salas E. Human factors and ergonomics in healthcare: industry demands and a path forward. Hum Factors 2022;64(1):250–8.

[3] Valentin A, Capuzzo M, Guidet B, Moreno R, Metnitz B, Bauer P, Metnitz P. Research group on quality improvement of the European Society of Intensive Care Medicine (ESICM); sentinel events evaluation (SEE) study investigators. Errors in administration of parenteral drugs in intensive care units: multinational prospective study. BMJ 2009;338:b814.

[4] Catchpole K, Cohen T, Alfred M, Lawton S, Kanji F, Shouhed D, Nemeth L, Anger J. Human factors integration in robotic surgery. Hum Factors 2022:187208211068946.

[5] de Bienassis K, Slawomirski L, Klazinga NS. The economics of patient safety part IV: safety in the workplace: occupational safety as the bedrock of resilient health systems. OECD Health Working Papers; 2021. p. 130. https://EconPapers.repec.org/RePEc:oec:elsaad:130-en.

[6] Dain S. Normal accidents: human error and medical equipment design. Heart Surg Forum 2002;5(3):254–7.

[7] Nik-Ahd F, Souders CP, Houman J, Zhao H, Chughtai B, Anger JT. Robotic urologic surgery: trends in Food and Drug Administration-reported adverse events over the last decade. J Endourol 2019;33(8):649–54.

[8] Yule S, Paterson-Brown S. Surgeons' non-technical skills. Surg Clin North Am 2012;92(1):37–50.

[9] Sharma B, Mishra A, Aggarwal R, Grantcharov TP. Non-technical skills assessment in surgery. Surg Oncol 2011;20(3):169–77.

[10] Siu KC, Suh IH, Mukherjee M, Oleynikov D, Stergiou N. The impact of environmental noise on robot-assisted laparoscopic surgical performance. Surgery 2010;147(1):107–13.

[11] Cofran L, Cohen T, Alfred M, Kanji F, Choi E, Savage S, Anger J, Catchpole K. Barriers to safety and efficiency in robotic surgery docking. Surg Endosc 2022;36(1):206–15.

[12] Parker SH, Yule S, Flin R, McKinley A. Surgeons' leadership in the operating room: an observational study. Am J Surg 2012;204(3):347–54.

[13] Bahler CD, Sundaram CP. Training in robotic surgery: simulators, surgery, and credentialing. Urol Clin North Am 2014;41(4):581–9.

[14] Dixon F, Keeler BD. Robotic surgery: training, competence assessment and credentialing. Bulletin R Coll Surg Engl 2020;102(7):302–6.

# 26

# Organizational level functions aiming the patient safety

*Aline Yuri Chibana*

Department of Anesthesiology, A.C. Camargo Cancer Center, São Paulo, Brazil

## Introduction

Patient safety refers to the prevention of harm or injury to patients during the provision of healthcare services. The World Health Organization (WHO) defines patient safety as "the absence of preventable harm to a patient during the process of healthcare" [1]. The Institute of Medicine (IOM), nos Estados Unidos, further explains that patient safety encompasses the prevention, reduction, reporting, and analysis of adverse events, errors, and near misses that could potentially harm patients [2].

Ensuring patient safety is an essential aspect of healthcare delivery as it promotes trust in the healthcare system, improves clinical outcomes, and reduces healthcare costs associated with medical errors [3]. The implementation of patient safety measures requires a systems-based approach that involves all stakeholders, including healthcare providers, patients, families, and organizations [4].

Healthcare organizations are responsible for ensuring the safety of their patients by providing a safe and effective environment for patients, staff, and visitors. Organizational level functions play a crucial role in achieving this goal by establishing a culture of safety and implementing safety policies and practices. This chapter aims to explore the various organizational level functions aimed at patient safety.

## Background

### Leadership

Effective leadership is crucial in creating a culture of safety in healthcare organizations. Leaders must prioritize patient safety as a key goal and ensure that safety policies and practices are integrated into the organization's operations. Leaders should provide support for safety initiatives and encourage staff to report safety incidents without fear of retaliation. They should also establish a system of accountability for safety and ensure that the organization's goals and objectives are aligned with patient safety. Effective leadership is essential in ensuring that the organization's resources are directed toward improving patient safety.

Additionally, leaders must model the behavior they expect from others, including promoting transparency, accountability, and a commitment to learning from mistakes. Leadership determines organizational priorities and can funnel resources toward important safety initiatives. Ultimately, leaders who promote a positive organizational climate contribute to higher job satisfaction among employees, decreased burnout, fewer medical errors, and an overall improved culture of safety [5].

Effective leaders in patient safety possess several key attributes that enable them to create a culture of safety and continuously improve patient outcomes.

1. Commitment to Safety: Effective leaders in patient safety are committed to creating a culture of safety and making patient safety a top priority. They lead by example, emphasizing the importance of safety in all aspects of care and

ensuring that all team members are accountable for their actions. A study by Pronovost et al. found that leaders who prioritize safety and support a culture of safety are more likely to have improved safety outcomes in their organizations [6].

2. **Communication Skills:** Effective leaders in patient safety are skilled communicators who are able to effectively communicate with team members, patients, and their families. They encourage open and honest communication, listen to feedback, and collaborate with others to improve safety. A study by Leonard et al. found that effective communication is essential for creating a culture of safety and improving patient outcomes [7].

3. **Empathy:** Effective leaders in patient safety are empathetic and understand the experiences of patients and their families. They are able to put themselves in their patients' shoes and recognize the emotional impact that safety incidents can have. A study by Shojania et al. found that empathy is essential for creating a culture of safety and promoting patient-centered care [8].

4. **Systems Thinking:** Effective leaders in patient safety understand that safety is a systems issue and not simply the responsibility of individual team members. They use a systems approach to identify and address root causes of safety incidents, and they encourage their team members to do the same. A study by Dixon-Woods et al. found that leaders who use a systems approach are more likely to successfully implement safety initiatives in their organizations [9].

5. **Continuous Learning:** Effective leaders in patient safety are lifelong learners who continuously seek out opportunities to improve their knowledge and skills. They encourage their team members to do the same and provide resources and support for ongoing education and training. A study by Sexton et al. found that leaders who prioritize continuous learning and improvement are more likely to have improved safety outcomes in their organizations [10].

In conclusion, effective leadership is essential for ensuring patient safety in healthcare organizations. Effective leaders in patient safety possess several key attributes, including a commitment to safety, communication skills, empathy, systems thinking, and a commitment to continuous learning. By cultivating these attributes, leaders can create a culture of safety that leads to improved patient outcomes.

## Governance

Effective governance is critical in ensuring that patient safety is a priority at all levels of the organization. Governance structures should be in place to oversee safety policies and practices, monitor safety performance, and ensure that resources are available to support safety initiatives. This includes the establishment of safety committees and safety performance indicators to monitor the organization's safety performance. Governance structures should also promote transparency and accountability in the organization's safety efforts. An effective governance system can ensure that patient safety is integrated into the organization's strategic goals and objectives.

## High-reliability organization

High-reliability organizations (HROs) are entities that operate in complex, high-risk environments, and consistently achieve near-error-free performance despite the potential for catastrophic failures. These organizations, such as nuclear power plants, air traffic control systems, and healthcare facilities, have developed a set of principles and practices that prioritize safety, communication, and continual learning.

One of the key principles of HROs is a commitment to safety as the highest priority. This means that safety is considered before any other concern, including productivity, profitability, or efficiency. HROs recognize that safety is not just the responsibility of one person or department, but rather the responsibility of everyone within the organization. This commitment to safety is reflected in the culture of the organization, which emphasizes open communication, transparency, and accountability.

Another principle of HROs is a focus on continual learning and improvement. HROs recognize that there is always room for improvement, and they seek to learn from both their successes and their failures. They use tools such as incident reporting systems, root cause analysis, and safety culture surveys to identify areas for improvement and implement changes to prevent future errors. Additionally, HROs prioritize training and education for all employees to ensure that everyone has the knowledge and skills necessary to operate safely.

HROs also prioritize communication and collaboration. They recognize that complex systems require input and coordination from multiple stakeholders, and they work to ensure that all relevant parties are involved in decision-making and problem-solving. This includes not only internal communication within the organization but also communication with external stakeholders, such as regulatory agencies and the public.

Research has shown that organizations that operate in high-risk environments can benefit from adopting HRO principles and practices. For example, a study of healthcare organizations found that those that implemented HRO principles had lower rates of adverse events and better patient outcomes [11]. Similarly, a study of the aviation industry found that HRO principles were associated with improved safety outcomes [12].

Organizations that adopt HRO principles and practices can improve safety outcomes and reduce the risk of catastrophic failures.

## Patient and family involvement

Patient and family involvement in healthcare can improve patient safety by providing valuable feedback on safety issues and contributing to the development of safety policies and practices. Healthcare organizations should encourage patient and family involvement by providing opportunities for engagement and feedback. This includes patient and family advisory councils, surveys, and focus groups. Patient and family involvement can also promote patient-centered care, which can lead to better outcomes and reduced harm.

One way to promote patient and family engagement in patient safety is through open communication. Patients and their families should be encouraged to ask questions, provide feedback, and participate in decision-making processes. Open communication can help identify potential safety concerns and facilitate the development of strategies to mitigate those risks.

Another aspect of patient and family engagement in patient safety is the involvement of patients and their families in care transitions. Care transitions refer to the movement of patients between healthcare providers, settings, or levels of care. Engaging patients and their families in care transitions can help ensure continuity of care and prevent adverse events, such as medication errors or miscommunications between providers.

Several studies have highlighted the importance of patient and family engagement in patient safety. A systematic review by O'Hara et al. found that involving patients and their families in care can improve patient safety by increasing patient participation, improving communication, and reducing adverse events [13].

Another study by Tzeng and Yin highlighted the importance of patient and family engagement in care transitions. The study found that engaging patients and their families in care transitions can improve patient safety by reducing readmissions, improving medication safety, and enhancing patient satisfaction [14].

## Staff training and education

Staff training and education are critical in ensuring that staff have the necessary skills and knowledge to provide safe and effective care. Healthcare organizations should provide ongoing training and education on safety policies and practices, including the use of safety equipment, infection prevention and control, and medication safety. Training should be tailored to meet the specific needs of staff, and the effectiveness of training should be monitored. Continuing education and professional development opportunities can also improve staff knowledge and skills and promote a culture of learning.

It is acknowledged that mistakes are more frequent at the beginning of a surgeon's learning process [15], and there are inherent safety risks for patients when both technology and technique are being learned simultaneously, especially if training is not optimized [16].

The first validated robotic training program was introduced in 2015, and it has since become the gold standard, adopted by multiple societies across various specialties [17,18].

If training is not objectively assessed, benchmarked, and quality assured, weaknesses in an individual's training and subsequent performance may go unnoticed. In contrast, the aviation industry has established international training standards that are benchmarked and quality assured. Pilots are required to demonstrate their proficiency before being permitted to fly with passengers. Unfortunately, such a rigorous approach to surgical training has yet to be implemented [19].

Organizationally, it is necessary to manage the staff shift rosters to ensure team members with sufficient robotic skills are available and those skills are maintained and developed among OR staff [20].

# Quality improvement

Quality improvement initiatives can improve patient safety by identifying and addressing areas of risk and implementing evidence-based practices to prevent harm. Healthcare organizations should prioritize quality improvement initiatives, including the use of root cause analysis and incident reporting systems to identify areas for improvement and monitor progress. Quality improvement efforts should be data-driven, and the effectiveness of interventions should be measured. Regular audits and reviews can also help identify areas for improvement and ensure that the organization's safety efforts are effective.

In a 14-year period, from 2000 to 2013, a study examined safety incidents that occurred during robotic procedures by analyzing all adverse event reports collected from the publicly available FDA MAUDE database across the country. A total of 10,624 events related to robotic systems and instruments were identified, with 98% of the events reported by device manufacturers and distributors, and only 2% voluntarily reported. During the same period, over 1,745,000 robotic procedures were performed in the U.S., resulting in an estimated adverse event rate per procedure of less than 0.6% (95% confidence interval (CI), 0.6–0.62). Among the data analyzed were 1535 (14.4%) adverse events that significantly impacted patients, including 1391 injuries and 144 deaths, as well as 8061 (75.9%) device malfunctions. These findings indicate that healthcare professionals still underreport adverse events, as the vast majority of reported events are related to device malfunctions. Additionally, the lack of notifications related to healthcare professionals' errors suggests that the safety culture needs further strengthening [21].

# Safety culture

Safety culture refers to the shared values, beliefs, attitudes, and practices that determine how an organization manages safety. It involves the commitment of all employees to prioritize safety and make it an integral part of their daily activities. Safety culture has been identified as a critical factor in reducing accidents and improving overall safety performance in various industries such as healthcare, aviation, manufacturing, and transportation.

Safety culture comprises of three main components: organizational culture, safety management, and individual behavior. Organizational culture involves the values and beliefs of the organization, which influence the way employees behave toward safety. Safety management refers to the processes and systems used to manage safety within the organization. Individual behavior refers to the actions and decisions made by individuals that impact safety.

Creating a positive safety culture involves a combination of leadership commitment, employee involvement, communication, and training. Leaders must lead by example and demonstrate their commitment to safety by providing the necessary resources and support. Employees must be encouraged to participate in safety initiatives and share their ideas and concerns. Effective communication is essential to ensure that everyone is aware of the safety goals and objectives of the organization. Regular training and development programs can help employees understand their roles and responsibilities in ensuring safety.

Research has shown that a positive safety culture can lead to improved safety outcomes. For instance, a study by the IOM found that a positive safety culture in healthcare organizations was associated with reduced medical errors and improved patient outcomes [2].

In conclusion, safety culture is a critical aspect of organizational safety. It involves the commitment of all employees to prioritize safety and make it an integral part of their daily activities. Creating a positive safety culture requires leadership commitment, employee involvement, communication, and training. Organizations that prioritize safety culture can achieve improved safety outcomes and a safer working environment for their employees.

# Technology and infrastructure

Technology and infrastructure can play a significant role in improving patient safety. Healthcare organizations should invest in technology that supports safety, such as electronic health records and medication dispensing systems. The physical infrastructure of the organization should also be designed to support safety, such as the use of hand hygiene stations and appropriate lighting. The use of technology can improve communication and collaboration among staff, reducing the risk of errors and improving patient outcomes.

Technical failures in the robotic system can lead to errors during surgery, such as the unintended movement of surgical instruments or the failure of the robotic system to respond to commands.

The physical environment for robotic surgery is critical to ensuring safe and effective operations.

One important aspect of the physical environment for robotic surgery is the design and layout of the operating room. The operating room must be spacious enough to accommodate the robotic surgical equipment and the surgical team. The room should also be designed to minimize the risk of infection and provide adequate lighting and ventilation.

Although ORs' architectural design varies among hospitals, one thing they have in common is that they were mostly constructed in a time when surgical technology was less advanced than it is now. Current surgical technologies, including those utilized in robotic-assisted surgery (RAS), often necessitate more space and require the installation or inclusion of additional equipment, such as monitors, consoles, and a tower, for optimal performance. Furthermore, the incorporation of such advanced surgical technology necessitates adjustments for additional sterile fields, modifications to the room layout, and changes to the overall movement and flow of both equipment and personnel. Moreover, the challenges of accommodating advanced surgical technology can impact various factors, including task flow, patient safety, OR staff safety, instrument storage, sterile protocols, location, and teamwork [22].

Another important consideration is the integration of technology into the operating room. The robotic surgical system should be connected to the hospital's computer network to allow for real-time monitoring and data sharing. The room should also be equipped with imaging and diagnostic tools to assist with the surgical procedure.

The lighting in the operating room is also critical. Adequate lighting is necessary to enable the surgical team to see the surgical field clearly. The lighting should also be adjustable to accommodate the varying needs of different procedures.

The temperature and humidity in the operating room must also be carefully controlled to ensure the comfort and safety of the patient and surgical team. The room should be kept at a comfortable temperature to prevent the patient from becoming too hot or too cold during the procedure. The humidity level should also be maintained at an appropriate level to prevent the accumulation of moisture, which can increase the risk of infection.

## Future scenario

The future of patient safety holds promising trends driven by technological advancements and a deepened understanding of healthcare processes. One key trend is the integration of artificial intelligence (AI) and machine learning (ML) into patient safety protocols. These technologies can analyze vast amounts of data to predict potential risks, identify patterns of errors, and provide real-time alerts to healthcare providers, enabling them to intervene swiftly and prevent adverse events. Additionally, the rise of wearable health devices and remote monitoring tools will enhance patient safety by enabling continuous tracking of vital signs and health indicators. These devices can provide timely insights to both patients and medical professionals, ensuring proactive interventions and reducing the likelihood of complications.

## Conclusion

Robotic surgery is an advanced form of surgery that offers many benefits to patients. However, it also poses unique risks to patient safety that must be addressed to ensure safe and effective care. Healthcare organizations can mitigate these risks by ensuring that surgeons are adequately trained and credentialed in robotic surgery, standardizing procedures, implementing quality control and monitoring programs, and providing patients with information about the risks and benefits of robotic surgery.

## Key points

- Staff training and education are crucial for ensuring safe and effective care in healthcare organizations. Training should be tailored to meet the specific needs of staff, and the effectiveness of training should be monitored.
- Quality improvement initiatives, such as root cause analysis and incident reporting systems, can help identify areas for improvement and prevent harm. These efforts should be data-driven, and the effectiveness of interventions should be measured.
- Safety culture is a critical aspect of organizational safety. It involves the commitment of all employees to prioritize safety and make it an integral part of their daily activities. Creating a positive safety culture requires leadership commitment, employee involvement, communication, and training. Organizations that prioritize safety culture can achieve improved safety outcomes and a safer working environment for their employees.

# References

[1] World Health Organization. Patient safety, 2021. Retrieved from https://www.who.int/health-topics/patient-safety#tab=tab_1.

[2] Institute of Medicine. To err is human: building a safer health system, 2000. Retrieved from https://www.nap.edu/catalog/9728/to-err-is-human-building-a-safer-health-system.

[3] Leape LL, Berwick DM, Bates DW. What practices will most improve safety? Evidence-based medicine meets patient safety. J Gen Intern Med 2009;24(3):364–8.

[4] National Patient Safety Foundation. Free from harm: accelerating patient safety improvement fifteen years after To Err Is Human, 2016. Retrieved from https://www.ihi.org/resources/Pages/Publications/Free-from-Harm-Accelerating-Patient-Safety-Improvement.aspx.

[5] Albright-Trainer B, et al. Effective leadership and patient safety culture, Anesthesia Patient Safety Foundation, 2021. Available at: https://www.apsf.org/article/effective-leadership-and-patient-safety-culture/. [Accessed: March 30, 2023].

[6] Pronovost PJ, Marsteller JA, Goeschel CA. Preventing bloodstream infections: a measurable national success story in quality improvement. Health Aff (Millwood) 2011;30(4):628–34.

[7] Leonard M, Graham S, Bonacum D. The human factor: the critical importance of effective teamwork and communication in providing safe care. Qual Saf Health Care 2004;13(Suppl 1):i85–90.

[8] Drew JR, Pandit M. Why healthcare leadership should embrace quality improvement. BMJ 2020;368, m872. https://doi.org/10.1136/bmj.m872.

[9] Dixon-Woods M, Baker R, Charles K, Dawson J, Jerzembek G, Martin G, McCarthy I, McKee L, Minion J, Ozieranski P, Willars J. Culture and behaviour in the English National Health Service: overview of lessons from a large multimethod study. BMJ Qual Saf 2014;23(2):106–15.

[10] Sexton JB, Berenholtz SM, Goeschel CA, et al. Assessing and improving safety climate in a large cohort of intensive care units. Crit Care Med 2011;39(5):934–9.

[11] Singh H, Petersen LA, Thomas EJ. Understanding and improving patient safety: the psychological safety climate as an essential element. J Patient Saf 2016;12(4):152–8.

[12] Gittell JH, Seidner R, Wimbush J. A relational model of high-reliability organizing: explaining the consistency of safety performance. J Appl Behav Sci 2008;44(2):189–211.

[13] Lawton R, O'Hara JK, Sheard L, et al. Can patient involvement improve patient safety? A cluster randomised control trial of the patient reporting and action for a safe environment (PRASE) intervention. BMJ Qual Saf 2017;26:622–31.

[14] Tzeng HM, Yin CY. Patient engagement in healthcare: A review of literature. In: Healthcare delivery transformation: concepts and practices. Springer; 2017. p. 195–221.

[15] Collins JW, Levy J, Stefanidis D, Gallagher A, Coleman M, Cecil T, et al. Utilising the Delphi process to develop a proficiency-based progression train-the-trainer course for robotic surgery training. Eur Urol 2019;75(5):775–85.

[16] Ahmed K, Khan R, Mottrie A, Lovegrove C, Abaza R, Ahlawat R, et al. Development of a standardised training curriculum for robotic surgery: a consensus statement from an international multidisciplinary group of experts. BJU Int 2015;116(1):93–101.

[17] Volpe A, Ahmed K, Dasgupta P, Ficarra V, Novara G, van der Poel H, et al. Pilot validation study of the European association of urology robotic training curriculum. Eur Urol 2015;68(2):292–9.

[18] Rusch P, Ind T, Kimmig R, Maggioni A, Ponce J, Zanagnolo V, et al. Recommendations for a standardised educational pro- gram in robot assisted gynaecological surgery: consensus from the Society of European Robotic Gynaecological Surgery (SERGS). Facts Views Vis Obgyn 2019;11(1):29–41.

[19] Collins JW, Wisz P. Training in robotic surgery, replicat- ing the airline industry. How far have we come? World J Urol 2019;38(7):1645–51. https://doi.org/10.1007/s00345-019-02976-4.

[20] Catchpole KR, Hallett E, Curtis S, Mirchi T, Souders CP, Anger JT. Diagnosing barriers to safety and efficiency in robotic surgery. Ergonomics 2018;61(1):26–39. https://doi.org/10.1080/00140139.2017.1298845. PMID: 28271956. [Epub 2017 Mar 8. PMCID: PMC6010349].

[21] Alemzadeh H, Raman J, Leveson N, Kalbarczyk Z, Iyer RK. Adverse events in robotic surgery: a retrospective study of 14 years of FDA data. PloS One 2016;11(4), e0151470.

[22] Kanji F, Cohen T, Alfred M, Caron A, Lawton S, Savage S, Shouhed D, Anger JT, Catchpole K. Room size influences flow in robotic-assisted surgery. Int J Environ Res Public Health 2021;18(15):7984. https://doi.org/10.3390/ijerph18157984. PMID: 34360275. PMCID: PMC8345669.

# 27

# Cleaning and sterilization of surgical robotic instruments

*Vanessa de Brito Poveda*[a,b], *Juliana Rizzo Gnatta*[a,b],
*and Giovana Abrahão de Araújo Moriya*[b]

[a]School of Nursing of University of São Paulo, São Paulo, SP, Brazil [b]Brazilian Association of Perioperative Nurses (SOBECC), São Paulo, SP, Brazil

## Introduction

The advent of robotic surgery has led to innovations and unveils new horizons in the improvement of less invasive surgical techniques, with optimized patient recovery. This contributes to patient safety, and also to the lower socio-economic impact by reducing hospitalization time and allowing patients undergoing this procedure to return to work more rapidly.

However, it also creates challenges, like any new technological increment in the area of perioperative care, including the education of the health team to deal with this new perspective in providing healthcare, the maintenance of equipment and, especially, the development of safe protocols for carrying out cleaning, preparation, and sterilization procedures for the medical devices used by the robotic surgery equipment.

Accordingly, the professionals working in the Central Sterile Supply Department (CSSD) play a crucial role in the reprocessing of medical devices and surgical instruments. They are also involved in the production of scientific knowledge to tackle the challenges of efficiently cleaning, preparing, and sterilizing increasingly complex equipment. These challenges include narrow lumens, blind spots, joints that cannot be disassembled, and the inability to directly visualize the presence of soil, as well as limited reprocessing life that needs to be controlled.

## Background

Minimally invasive surgeries or procedures aimed at minimizing invasiveness have been attempted since ancient times. Over the last 500 years, they have developed with the goal of accessing different body cavities for therapeutic purposes. The fundamental principles and equipment used in these procedures have been established since 1850 and are still in use today [1,2].

Despite disagreements about when the first robotic procedure actually took place, it is certain that its expansion began in the 1990s [3], through three main types of robotic systems, namely, active ones, i.e., those that act autonomously, although under the control of the surgeon; semiactive, where elements activated by the surgeon complement a preexisting program; and the most conventional, the Master–Slave system (daVinci and Zeus platforms) that are completely dependent on the surgeons, without any type of prior program installed [4]. Multiarm-Units robots appeared in the 2010s, with these not only having a single station where the robot arms are located to access the patient, but possibly containing up to four, or more, depending on the need (Versius and Hugo platforms), creating a more flexible concept [5].

Along with the growth of robotic surgery systems, the need to prepare the multidisciplinary team to work with the patient at all times during the perioperative period has also increased, with emphasis on perioperative work and on the CSSD to ensure adequate decontamination of the surgical materials used in these devices. The provision of care in this sector must be effective, taking into account the characteristics of robotic instruments and the complex and challenging design of their reprocessing.

Challenges faced by the multidisciplinary team that works with robotic surgery are mainly related to patient safety during the robotic surgical procedure, fear of malfunction, or major equipment errors during the procedure. The need for education and updating considering the technological innovations, the development of nontechnical skills, inter-professional work, and communication are also major concerns [6–8].

Patient safety also depends on the use of controlled-risk medical devices that have been properly reprocessed for use; therefore, it is essential that the work carried out by the CSSD is meticulous, procedural, and well established.

Professionals working in the CSSD face challenges related to the processing of medical devices used in robotic surgery (robotic surgical instruments, endoscopes, and accessories). Cleaning is the most challenging step, as the instruments can, in most cases, be reused many times, depending on the instrument model and manufacturer, cannot be disassembled, and are composed of small parts, with narrow lumens, among other aspects [9].

Research shows the difficulty of cleaning reused robotic instruments through the increased recovery of soil markers from these materials, when compared to those recovered from other types of surgical instruments [10]. Visual inspection seems to be insufficient. Some residual protein detection tests have been recommended such as ATP (an indicator of organic residues and microorganisms by measuring the amount of adenosine triphosphate (ATP), an energy source present in living cells) or proreveal (highly sensitive fluorescence-based protein detection test for checking the presence of residual protein) [11–14].

Guidance regarding the reprocessing of robotic equipment is provided through the instructions for use (IFU) of the manufacturers; however, professionals from various CSSDs have contributed to the standardization and development of operational protocols for the reprocessing of these materials, adapting them to each reality.

In the development of an operational protocol for reprocessing robotic equipment, the IFU provided by the manufacturers must necessarily be taken as a basis, as it is believed that this information has been previously validated by the manufacturers. The IFU must contain data that mention the main characteristics of the instruments, endoscopes, connectors, light cables, electrosurgical cables, and accessories so that professionals can understand the challenges of reprocessing and organize each step adapted to each reality. The flow of the material should start from its exit from the Operating Room which is called point of use (POU), followed by safe transport and ready to be received in CSSD, which will start the unidirectional flow by cleaning and decontamination, inspection, assembly, and preparation practices, sterilization, and concluding within the safe storage. Full validation of all the processes taking into consideration human factors is very important in putting together an IFU along with all standards/guidelines and local standards.

## Challenges in the processing of robotic surgical instruments

The medical devices used in minimally invasive surgeries, especially robotic surgical instruments, are complex and have some particular characteristics that make the processing of these materials challenging, such as

- They have a complex design and conformation, with lumens smaller than 5 mm in diameter, making access difficult for the mechanical removal of soil present on the surface;
- They have joints that favor the accumulation of organic matter;
- Instruments with electric current circulation, such as monopolar/bipolar forceps, monopolar curved scissors, and monopolar cautery, favor the coagulation of biological material on their surfaces, which difficult the process of removing soil;
- They cannot be disassembled, which makes it difficult to access all surfaces of the material, disfavoring the cleaning and decontamination process;
- They present a limited number of reprocessing lives, previously validated by the manufacturer, seeking to guarantee the functionality and safety of the instrument's use.

Due to these characteristics, the processing steps must be meticulous, standardized, and appropriate to the CSSD processes in order to guarantee a medical device that is safe to use. Medical devices should only be handled and reprocessed by trained personnel, as improper processing may result in damage to the device or harm to the patient/professional [15–17].

The necessary common and specific procedures for Master–slave (daVinci) and Multiarm-Units (Versius and Hugo) technologies will be discussed next, considering the cleaning and decontamination (POU, manual, and automated cleaning), preparation, and sterilization steps, based on the respective manufacturer's IFU [15–19]. It should be highlighted that biohazardous measures should be adopted by professionals due to the biological risk present in the materials [20].

## Cleaning and decontamination

### POU cleaning

The cleaning step should start at the "point of use," that is, right after the surgery and still in the operating room. The importance of starting cleaning at the POU is justified by the great complexity of the instruments when compared to conventional surgical instruments, which cannot be disassembled and have irrigation channels with very narrow lumens that need to be flushed to prevent soil drying internally. Neglecting to do so could therefore result in ineffective cleaning.

This step aims to:

- Remove excess soil and keep the residual organic matter moist until the beginning of the cleaning and decontamination carried out in the CSSD;
- Prevent soil from drying on the instrument, which makes its subsequent removal difficult;
- Facilitate the subsequent cleaning steps;
- Improve cleaning efficiency, consequently reducing instrument cleaning time.

While still in the operating room, medical devices must be inspected. If there is any damage or if a defect was reported by the surgical team during use, the material should be reprocessed before being sent for repair, if it is repairable. Excess soil should be removed with a soft, lint-free cloth/pad, including head, shaft, and tip [15–19]. Then, the medical devices must be moistened to avoid drying out the soil and transported hermetically sealed [15–19], following the guidelines presented in Table 1.

To clean any robotic medical device, the following should not be used: saline solution, acid solutions (pH < 7), strongly alkaline solutions (pH > 11), cleaning products based on hydrogen peroxide, cleaning products based on bleach or rinsing aids [15–19].

Upon receiving the medical devices at the CSSD and before starting cleaning, the professional must check that they are intact and whether there are disposable accessories with the material, as is the case of monopolar scissors tip protectors [15–19]. If necessary, disposable materials should be removed and discarded.

The cleaning process in the CSSD includes manual cleaning and automated cleaning. Cleaning is the most important and complex step in medical device processing because, if a device is not properly cleaned, the process can fail. The effectiveness of the cleaning depends on the mechanical removal of dirt and chemical action; therefore, contact time and temperature are factors that must be respected during the process.

### Manual cleaning

In the case of robotic surgical instruments, manual cleaning includes the following steps: soak, flush, spray, brush, and rinse [15–19].

TABLE 1    Recommendations for precleaning materials at the point of use, according to the manufacturers of the daVinci, Versius, and Hugo technologies [15–19].

| | daVinci | Versius | Hugo |
|---|---|---|---|
| Recommendations | Inject with a Luer Tip syringe at least 15 mL of enzymatic solution pH 7–11 or cold water into the primary port (for instruments and Xi endoscope). This step is called prime. Then place the medical device in a bin with solution, or spray to wet all surfaces, and/or wrap in a damp cloth to keep the tip wet.[a] | Inject with a Luer Slip-type syringe 50 mL of a mixture of solution containing pre-mixed pH-neutral (7–11) enzymatic detergent or cold water into the primary port (for instruments); Then, place the medical device in a container, make sure they are fully immersed in a solution containing pre-mixed pH-neutral enzymatic detergent (7–11) or cold water, or wet all surfaces of the medical device with pH-neutral enzymatic foam spray before forwarding to CSSD. | Not cited. |

[a] Predisinfection according to the manufacturer's instructions is optional.

TABLE 2    Minimum times recommended by manufacturers for each manual cleaning step for robotic instruments [17–19].

| Step | daVinci S/Si | daVinci Xi | Versius | Hugo |
|---|---|---|---|---|
| Soak | 30 min | 10 min | 10 min[a] | 10 min |
| Flush | 40 s | 40 s | 2 min | 1–2 min[b] |
| Spray | 30 s | 30 s | 1 min | Not cited |
| Brush | 1 min | 1 min | 4 min | 1–2 min |
| Rinse | 1 min | 1 min | 1.5 min | Not cited |
| Inspect | Not cited | Under magnification 4× | Not cited | |

[a] Unless the manufacturer's instructions recommend a longer soaking period.
[b] If a pressurized water gun is not available, use a syringe to flush each port with 90 mL of water until the water runs clear.

For the soaking step, the instrument must not be disassembled, as the purpose is that the dirt adhered to the material remains wet to facilitate its subsequent removal in the following cleaning steps. The instruments must be immersed in a solution of detergent, recommended, standardized, and validated by the manufacturer prepared inside a suitable container, sink, or recipient according to the manufacturer's recommendation, following the immersion time and temperature guidelines (Table 2). The ports must be irrigated (where applicable) with the aid of a Luer tip or Slip syringe, which is adaptable to the instrument's irrigation orifice, first in port 1 and then in port 2, with this detergent solution. Port 1 irrigates the instrument stem, and the injected liquid exits through the head. Port 2 irrigates the head, and the injected liquid exits through the head itself. For this reason, the irrigation order must be strictly followed in all cleaning steps, port 1, then port 2 [15–19].

After the soaking stage, the accessories indicated by the manufacturer as removable must be disassembled. If disassembly is not done, the process may result in inadequate cleaning, and sterilization will not be effective.

Then, the flushing of the channels with pressurized water begins, also following the port 1–port 2 sequence, respecting the time and pressures recommended by the manufacturer (≤2 bar (30 psi) for daVinci; ≤2.5 bar (36 psi) for Versius; possibility to use a 90-mL syringe to flush for Hugo).

In the spray phase, instruments with open ends must be sprayed with a water gun.

In the brush phase, all surfaces must be brushed with a soft nylon brush. When brushing cautery forceps, care must be taken so that the insulation is not damaged. The rinsing phase must be carried out with potable water (Table 2).

The manufacturers recommend that the temperatures of the solutions and the water during the cleaning step be between 30°C and 42°C for Hugo medical devices, ≤45°C for Versius and that the guidelines of the detergent manufacturers are followed for daVinci.

In the case of endoscopes, recommendations for the cleaning step vary according to the manufacturer [15,18,19] and can be seen in Table 3.

TABLE 3    Cleaning step recommendations for endoscopes used in robotic procedures, according to the manufacturers [15,18,19].

| Step | daVinci S/Si | daVinci Xi[a] | Versius | Hugo* |
|---|---|---|---|---|
| Soak | 15 min | 15 min | 10 min | Not cited |
| Rinse | 1 min | 1 min | 1.5 min | Not cited |
| Flush | Not applicable | 2 min | Not applicable | Not cited |
| Spray | 20 s | 40 s | Not cited | Not cited |
| Bruch | 1 min | 1 min | 1 min | Not cited |
| Rinse | Not cited | Not cited | 1.5 min | Not cited |
| Bruch | Not cited | Not cited | 1 min | Not cited |
| Final Rinse | 1 min | 1 min | 1.5 min | Not cited |

[a] Hugo: the manual states that the instructions of the KARL STORZ – Endoscope Processing Instruction should be consulted: 30-degree scope—26605BA, TIPCAM1 S 3D LAP, 30° 0 degree scope—26605AA, TIPCAM1 S 3D LAP, 0° Fiber Optic Light Cable—Models 495 xx.

TABLE 4 Parameters recommended by manufacturers for automated cleaning in a thermodisinfector [15–19].

| Phase | daVinci Si | | daVinci Xi | | Versius | | Hugo | |
|---|---|---|---|---|---|---|---|---|
| | Time | Temperature | Time | Temperature | Time | Temperature | Time | Temperature |
| Precleaning Cold water | Not cited | Not cited | 1–5 min | Not cited | 2 min | Cold water | 1 min | Cold water |
| Cleaning Hot water | Not cited | Not cited | 10–30 min | Not cited | 2 min | Not cited | 5 min | $\geq 43°C$ |
| Rinsing | Not cited | Not cited | 2–5 min | Not cited | 2 min | Not cited | 15 s | $\geq 43°C$ |
| Thermal disinfection | 1–5 min | 85–93°C | National requirements for $A_0$ | Not cited | National requirements for $A_0$ | Not cited | 1 min | 82°C |
| Drying com hot air | 45 min | Not cited | Not cited | Not cited | 5 min | 115°C | 6 min | 99°C |

## Automated cleaning

Automated cleaning is performed using an ultrasonic washer or automatic washer disinfector (AWD) and can only be performed on equipment validated by the manufacturers and following the IFU of the robotic medical devices. It should be noted that there are different possibilities for automated cleaning (main), favoring the end user who is willing to acquire a robotic system and can think about adapting his CSSD.

The cleaning solutions used in the washers must be validated and recommended by the manufacturer.

It is recommended to inject the detergent solution, with the aid of a syringe that fits into the irrigation holes, before starting the automated cleaning process in an ultrasonic washer, so that the lumens are filled in accordance with the manufacturers' recommendations.

The AWD must be programmed according to each manufacturer's guidelines. Some require special cycles and carriages/racks and irrigation system (daVinci/Hugo), others do not, they can be used in AWD that have carriages/racks with irrigation system (Versius) [15–19]. The hoses must be connected to the main ports of each instrument.

When loading the washer, ensure that the instrument jaws are open and that they do not touch the washer walls or spray arms, that they have sufficient spacing for drainage, and that the irrigation tubes are connected.

Cleaning endoscopes in ultrasonic washers is not recommended for they may suffer irrecoverable damage [21,22].

After automated cleaning, it is recommended that tests be performed to assess cleanliness, according to the manufacturer's recommendations. The final rinse must be performed with purified water to remove some toxic substances such as detergent residues and endotoxins from the cleaning process [15–19,23,24].

Table 4 presents a comparison between the recommended parameters for automated cleaning of different technologies.

The efficiency of the automated cleaning process is directly conditioned to careful loading (appropriate rack or support, depending on the manufacturer's recommendations), type of detergent used, and choice of cleaning cycle.

## Preparation

After automated cleaning, drying must be carried out. In the case of robotic surgical instruments, internal drying can be performed with compressed air (see Table 5 for maximum pressure) and external drying with a soft, lint-free cloth [15–19]. The ports can be dried using this technique, as long as it is possible to control the pressure so that the internal components of the medical devices are not damaged.

After drying, a rigorous inspection, performed with a magnifying lens, should begin of the accessory or instrument for soil, presence of defects, or damage. Endoscopes, for example, should be inspected for defects or damage to the tip

TABLE 5 Maximum pressure for internal drying of robotic surgical instruments, according to the daVinci, Versius, and Hugo technology manufacturers [15–19].

| | daVinci (optional) | Versius | Hugo |
|---|---|---|---|
| Pressure | $\leq 2$ bar (30 psi) | $\leq 5$ bar (73 psi) | Not cited |

TABLE 6   Steam sterilization time at 134°C, according to the daVinci, Versius, and Hugo technology manufacturers [15–19], considering the cycle phase.

| Cycle phase | Exposure time | | |
| --- | --- | --- | --- |
| | daVinci | Versius | Hugo |
| Plateau | 3–18 min | 3–18 min | 3–18 min |
| Dry time (minimum) | 30–50 min | 20 min | 20–40 min |

or lens, light ports, and fiber optic surfaces, as a failure to inspect could result in harm to the patient and/or additional damage to the material itself [15].

Due to the complexity of medical devices, a cleaning efficacy test should be carried out, according to the manufacturer's instructions [15–19].

Then, the articulated parts of the instruments must be lubricated with a lubricant recommended by the manufacturer and categorically in the places indicated by the manufacturers [15–19]. Adequate lubrication is essential to maintain the life of the robotic instrument and ensure its proper functioning during the surgical procedure.

Placing a vapor-permeable protector on the tip of the robotic surgical instrument before putting it in the package is suggested. Robotic instruments can be packed in individual packages, kits, or in boxes/baskets, as validated by the manufacturer [15–19].

Accessories must be packed in boxes, baskets or containers, depending on the manufacturer. The security key must be packaged separately (daVinci).

In the case of endoscopes, they must be packed in boxes suitable for robotic optics, making sure that they are properly positioned and attached to the latches inside the box.

### Sterilization

Sterilization procedures must follow the recommendations of AAMI/ISO 17665-1 or BS EN ISO 17665-1 [24,25]. Accordingly, only equipment with validated parameters and cycles should be used to avoid damage to medical devices or incomplete sterilization cycles [15,18]. In addition to the recommended standards for equipment usage, development, and validation, the sterilization process should be routinely monitored to ensure that sterilization conditions are met.

Robotic instruments must undergo steam sterilization cycles. There are other temperatures validated by the manufacturers, but the focus of this chapter will be on the 134°C temperature that represents most of the international CSSD realities. Other temperatures should be referred to the manufacturer's IFU, which is what is found in most realities.

Table 6 indicates the sterilization time indicated by the manufacturers at a temperature of 134°C.

The endoscopes used in the daVinci and Hugo technologies are thermosensitive and, therefore, have been validated only for low-temperature sterilization methods [15,19].

Unlike other technologies, in the case of the endoscope used in the Versius system, only the steam sterilization process is validated [18].

It is worth pointing out that, it is recommended that only equipment, parameters, and sterilization cycles that have been validated by the manufacturer be used, which must be consulted in the IFU when purchasing the endoscope.

## Future scenario

With the worldwide advance in robotic technology and consequently the growth of Robotic Surgery Programs in institutions, monitoring the performance of these services is becoming increasingly necessary, including the monitoring of the reprocessing of robotic medical devices. This continuous monitoring must take place by measuring data through established metrics that are based on reliable and quality evidence, so that processes are efficiently monitored, and evaluated and improvements implemented.

The metrics allow the CSSD's work to gain visibility within the institution itself, demonstrating the relevance of this sector and the importance of the work of its professionals. The main challenge will be to standardize indicators to measure the reprocessing of robotic medical devices by the CSSD, since the establishment of flows and processes directly impact the success of the procedure and patient safety.

## Conclusion

For the reprocessing of robotic instruments to occur safely, it is essential that the process be based on fully validated instructions. These instructions must be unequivocal, comprehensive, and applicable in the most different realities of the CSSD. Accordingly, the processing of the instruments must fit into the work processes of the CSSD and the professionals who work in the processing must have easy access to the manufacturer's instructions, in addition to receiving specific training in robotic medical devices reprocessing from the manufacturer.

The importance of this early involvement of the client who purchases the robotic system with the manufacturer is highlighted, so that during the technology implementation process there is educational support, through training and monitoring of the process, provided by specialist professionals. The educational support and involvement of manufacturers through feedback from the services that acquire robotic systems are essential for the use of technology to be safe for patients. Furthermore, the importance of the support of managers for the establishment of metrics through indicators to measure the quality of the robotic material processes in the CSSD is highlighted.

## Key points

- The processing of robotic medical devices presents important challenges to professionals working in the Central Sterile Supply Department (CSSD), due to the complexity of materials and the diversity of manufacturers, with their respective specificities.
- It is important to have situational awareness of the CSSD manager's involvement in all decisions from planning to implementation of a robotic system in a healthcare facility.
- The processing of robotic medical devices must be based on the manufacturer's recommendations and previously validated instructions for use.
- There are challenges for professionals to standardize indicators to measure the quality of the processing of robotic instruments in the CSSD and the establishment of continuous education to improve processes and keep up to date with new technologies.

## References

[1] Hargest R. Five thousand years of minimal access surgery: 3000BC to 1850: early instruments for viewing body cavities. J R Soc Med 2020;113 (12):491–6. https://doi.org/10.1177/0141076820967913.

[2] Hargest R. Five thousand years of minimal access surgery: 1850 to 1990: technological developments. J R Soc Med 2021;114(1):19–29. https://doi.org/10.1177/0141076820967918 [A].

[3] Hargest R. Five thousand years of minimal access surgery: 1990-present: organisational issues and the rise of the robots. J R Soc Med 2021;114 (2):69–76. https://doi.org/10.1177/0141076820967907 [B].

[4] Lane T. A short history of robotic surgery. Ann R Coll Surg Engl 2018;100(6_sup):5–7. https://doi.org/10.1308/rcsann.supp1.5.

[5] Morrell ALG, Morrell-Junior AC, Morrell AG, Mendes JMF, Tustumi F, de-Oliveira-E-Silva LG, Morrell A. The history of robotic surgery and its evolution: when illusion becomes reality. Rev Col Bras Cir 2021;18:e20202798.

[6] Kang MJ, De Gagne JC, Kang HS. Perioperative nurses' work experience with robotic surgery: A focus group study. Comput Inform Nurs 2016;34(4):152–8. https://doi.org/10.1097/CIN.0000000000000224.

[7] Schuessler Z, Scott Stiles A, Mancuso P. Perceptions and experiences of perioperative nurses and nurse anesthetists in robotic-assisted surgery. J Clin Nurs 2020;29(1–2):60–74. https://doi.org/10.1111/jocn.15053.

[8] Kanji F, Catchpole K, Choi E, Alfred M, Cohen K, Shouhed D, Anger J, Cohen T. Work-system interventions in robotic-assisted surgery: a systematic review exploring the gap between challenges and solutions. Surg Endosc 2021;35(5):1976–89. https://doi.org/10.1007/s00464-020-08231-x.

[9] Saito Y, Yasuhara H, Murakoshi S, Komatsu T, Fukatsu K, Uetera Y. Novel concept of cleanliness of instruments for robotic surgery. J Hosp Infect 2016;93(4):360–1. https://doi.org/10.1016/j.jhin.2016.04.009.

[10] Saito Y, Yasuhara H, Murakoshi S, Komatsu T, Fukatsu K, Uetera Y. Challenging residual contamination of instruments for robotic surgery in Japan. Infect Control Hosp Epidemiol 2017;38(2):143–6. https://doi.org/10.1017/ice.2016.249.

[11] Chen A, Zou X, Tan Y, Chen Y, Ye X, Hao S. Multicenter comparative study of three "non-destructive" methods of detecting the cleanliness of the da Vinci surgical robotic instrument. Gland Surg 2021;10(12):3305–13. https://doi.org/10.21037/gs-21-814.

[12] Sagourin P, Viallet A, Batista R, Talon D. Validation of the cleaning method for da Vinci XI robotic instruments using protein residue tests. Zentr Steril 2021;29(1):48–53.

[13] Zentr Steril. Guideline compiled by DGKH, DGSV and AKI for the validation and routine monitoring of automated cleaning and thermal disinfection processes for medical devices. 5th edn; 2017.

[14] European Standard (EN). International Organization for Standardization (ISO). Washer-disinfectors—Part 5: Performance requirements and test method criteria for demonstrating cleaning efficacy. EN ISO 15883-5; 2021.

[15] Intuitive Surgical. Reprocessing manual for the *da Vinci S* and *Si* endoscopes. Intuitive Surgical, Inc; 2017.

[16] Intuitive Surgical. Reprocessing manual for *da Vinci S* and Si instruments EndoWrist, Single Site and Stapler. Intuitive Surgical, Inc; 2021.

[17] Intuitive Surgical. Da Vinci Xi instrument reprocessing instructions. Intuitive Surgical, Inc; 2021.

[18] CMR Surgical. Versius summarised reprocessing instructions. CMR Surgical Ltd; 2023.

[19] Medtronic. Instrument and accessory decontamination guide. Medtronic; 2021.

[20] Rutala WA, Weber DJ, Centers for Disease Control, Healthcare Infection Control Practices Advisory Committee (CDC). Guideline for Disinfection and Sterilization in Healthcare Facilities. Atlanta: CDC; 2008. p. 2008.

[21] Association of periOperative Registered Nurses (AORN). Guideline for sterilization. In: Association of periOperative Registered Nurses (AORN), editor. Guidelines for perioperative practice. Denver; 2017.

[22] World Health Organization and Pan American Health Organization (WHO/PAHO). Decontamination and reprocessing of medical devices for health care facilities; 2016. 120 p.

[23] Association for the Advancement of Medical Instrumentation (AAMI). Technical information report 34. TIR34:2014, Water for the reprocessing of medical devices; 2014.

[24] Association for the Advancement of Medical Instrumentation (AAMI). Sterilization of health care products—moist heat—Part 1: requirements for the development, validation, and routine control of a sterilization process for medical devices: AAMI/ISO 17665-1. International Organization for Standardization (ISO); 2006.

[25] European Standard (EN). Sterilization of health care products – moist heat—Part 1: requirements for the development, validation, and routine control of a sterilization process for medical devices: EN ISO 17665-1. International Organization for Standardization (ISO); 2006.

# 28

# The role of the scrub nurse in robotic surgery

*Desrene Gibson and Alexis Sánchez*

Orlando Regional Medical Center, Orlando Health Robotic Surgery Program, Orlando, FL, United States

## Introduction

Robotic surgery is a rapidly advancing field that offers many benefits to patients, including improved precision and accuracy, reduced recovery time, and less pain and scarring [1–3]. However, the success of a robotic surgery procedure depends on many factors, one of which is the role of the scrub nurse or surgical technologist. In this chapter, the role of scrub nurses will be used interchangeably with the surgical technologist. Scrub nurses play a critical role in preparing the robotic surgical instruments before the procedure, maintaining them during the procedure as well as the postprocedure care of the instruments [4]. They are actively involved in designing and updating the preference card, which is the blueprint of the surgical procedure in collaboration with the circulating nurse. The surgical technologist is primarily responsible for ensuring the availability of all necessary instruments and visually checking the robotic instruments for damage or wear and tear during setup. This is vital to the flow and success of the procedure [5,6].

In this chapter, we will explore the responsibilities and tasks of the scrub nurse during a robotic surgery procedure, including preparing the instruments, changing instruments during the procedure, anticipating the surgeon's needs, monitoring the instruments' location, increasing efficiency, and proper care of instruments for transport to sterile processing department (SPD) after the procedure. We will also examine the importance of specialized training and knowledge required for scrub nurses working with robotic instruments, the future of robotic surgery, and the role of the scrub nurse. By understanding the crucial role of the scrub nurse in robotic surgery, we can improve the overall outcomes of these procedures.

## Preparing for the procedure

### Designing and updating the preference card

"Designing and updating the preference card," refers to the process of creating and maintaining a detailed list of the specific instruments, equipment, and supplies that will be needed for a particular surgery. The preference card acts as a blueprint for the surgical procedure, and it is important that it is well designed and updated on a regular basis to ensure that all necessary instruments are available and the room is in the right configuration before the procedure begins [7].

Designing the preference card involves working with the surgical team to identify all the instruments, equipment, and supplies that will be needed for the procedure. This includes not only the standard instruments that are used in most surgeries but also any specialized instruments that may be needed for the specific procedure. The preference card should also include information about the quantity of each item needed, as well as any special instructions or notes that the scrub nurse should be aware of.

Updating the preference card refers to the process of reviewing and revising the preference card as needed, which is important to ensure that it remains accurate and up-to-date. This might include adding new instruments or equipment that have become available, updating information about the quantity of items needed, or removing items that are no longer needed.

*Handbook of Robotic Surgery*
https://doi.org/10.1016/B978-0-443-13271-1.00031-5

It is generally recommended that preference cards be based on evidence and best practices, rather than solely on the preferences of individual surgeons. This is because evidence-based practice and best practices are safe and effective, whereas individual preferences may not have been tested and validated [8].

Having a standardized preference card that is based on evidence and best practices can contribute to ensure that all necessary instruments are available and in the correct configuration before the procedure, which can help to reduce the risk of complications and improve the overall outcome of the procedure. Standardized preference cards can also help to promote consistency and continuity of care by ensuring that all surgeons are using the same instruments and techniques [8,9].

Nonetheless, it is also important to take the individual preferences of the surgeons into account and consider the unique aspects of each case. Surgery is an art and a science, and surgeons have different techniques, therefore, it is important to have a balance between evidence-based practice and best practices and the surgeon's preferences.

## Checking the robotic instruments for damage or wear and tear

Checking the robotic instruments for damage or wear and tear before a surgical procedure is an important step in ensuring that the instruments are in good working condition and can be safely used during the procedure [5].

During this process, the scrub nurse or surgical technologist will visually inspect each instrument to check for any signs of damage, such as bent or broken parts, loose wires, rust, or other visible wear and tear. The instruments should be free of any damage or defects that could affect their function or cause harm to the patient.

The scrub nurse will also check that all the instruments are in proper working order and are properly calibrated. Additionally, the scrub nurse will check that they have been properly cleaned and sterilized by purposefully checking the package or tray indicator to ensure that all the sterilizer parameters have been met before use. This is important to prevent infection and to ensure that the instruments are safe for use [10].

## Specialized training and knowledge required for scrub nurses working with robotic instruments

Working with robotic instruments requires specialized training and knowledge for scrub nurses, as it involves a different set of skills and responsibilities compared to traditional surgery.

Robotic instruments are complex and require a high level of technical proficiency to set up, operate, and maintain. Scrub nurses must be familiar with the specific instruments and equipment used in robotic surgery, as well as their operation, maintenance, and troubleshooting. They need to be able to quickly and efficiently set up the instruments and equipment in accordance with the surgeon's preferences and the procedure's needs [11].

In addition, scrub nurses must have a good understanding of the sterile field, how to maintain it and how to troubleshoot any issues that may arise during the procedure. They must also be able to anticipate the surgeon's needs and have the necessary instruments ready in advance [10].

Additionally, scrub nurses must stay current with new devices and instruments that are being introduced in the OR and be able to learn and quickly adapt to new technology. To acquire this specialized knowledge and skills, scrub nurses should receive specialized training in robotic surgery, which includes both theoretical and practical training. This could include attending workshops, seminars, and training sessions, as well as working alongside experienced robotic surgeons and scrub nurses. Having a basic understanding of the way the da Vinci system works, along with a systematic troubleshooting technique, equips the scrub nurse with the necessary tools to be a valuable asset to the team [5].

## During the procedure

## Assisting with the docking and undocking of the robotic system

Assisting with the docking and undocking of the robotic system refers to the important role that the scrub nurse plays in assisting with the setup and takedown of the da Vinci robotic system before and after the procedure.

The da Vinci robotic system consists of a console where the surgeon sits, a patient cart/robot with multiple arms for instruments, and a vision cart that houses the camera providing a 3D view of the surgical field. The patient cart/robot is positioned next to the operating table and the arms are docked to the cannulas already inserted in the patient (Fig. 1). The scrub nurse is responsible for helping to position the patient-side cart and dock the robotic arms in the correct position [12].

During the docking process, the scrub nurse must ensure that the robotic arms are properly aligned with the surgical site, the instruments are securely attached to the robotic arms, and the self-calibration of the instruments is successful

FIG. 1   Robotic surgery components and operating room configuration.

before proceeding. They must also ensure that the instruments are safely removed before attempting to undock the robotic arms from the patient. The patient-side cart can then be removed from the bedside by the circulator.

In addition, the scrub nurse must also help to troubleshoot any issues that may arise with the robotic system during the procedure, such as instrument malfunctions or system errors. By ensuring that the robotic arms are properly positioned and securely attached, the scrub nurse helps to ensure that the surgery runs smoothly and that the patient's safety is maintained throughout the procedure [11].

## Changing instruments during the procedure

During a robotic surgery procedure, the surgeon may use different instruments or perform slight changes to ports to access different areas of the patient's body or to perform different parts of the procedure. The scrub nurse is responsible for anticipating these needs and providing the surgeon with the necessary instruments in a timely and efficient manner [5].

The scrub nurse must also be vigilant in ensuring that instruments that have become contaminated, such as those that have touched nonsterile surfaces or have come in contact with certain bodily fluids, are replaced promptly to prevent the spread of infection. This involves removing the contaminated instrument and replacing it with a new, sterile instrument [10].

To accomplish this, the scrub nurse needs to have a good understanding of the surgical procedure, be able to anticipate the surgeon's needs, and work efficiently and effectively to provide the necessary instruments. Effective communication with the surgical team is vital to ensure that everyone is on the same page regarding the changing instrument needs during the procedure to ensure patient safety [4].

## Anticipating the surgeon's needs

Anticipating the surgeon's needs refers to the importance of the scrub nurse being aware of the surgeon's needs throughout the procedure and having the necessary instruments and supplies ready for use. The scrub nurse must be aware of the procedure's progression and anticipate the surgeon's needs to have the necessary instruments and supplies ready accordingly.

In addition, the scrub nurse must also be aware of any potential complications that may arise during the procedure and have the necessary instruments and supplies available in case they are needed. They must also be familiar with the emergency undocking procedures should the need arise during the procedure [13].

## Increasing efficiency

## Techniques for streamlining instrument preparation and setup

To increase efficiency during robotic surgery procedures, scrub nurses can utilize many techniques to streamline instrument preparation and setup.

One effective technique is to have a standardized layout and a setup of the instruments and equipment. This involves arranging the instruments in a consistent manner that is easy to remember and is based on the surgical team's preferences. This can save time during the setup process, as the scrub nurse can quickly locate and retrieve the necessary instruments without having to search through the entire instrument tray (Fig. 2) [9].

 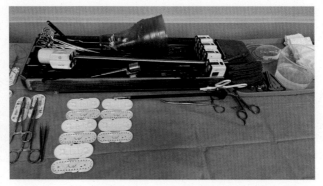

FIG. 2    Table and mayo stand setup.

Another technique is to have a designated area for each type of instrument, which helps to minimize confusion and reduces the risk of picking up the wrong instrument. For example, all scissors could be kept in one area, while all clamps could be kept in another. Organizing the instruments in a logical and efficient manner can also help to increase efficiency during procedures. This can involve arranging the instruments in a consistent order based on the surgical team's preferences or based on the steps of the surgical procedure. The scrub nurse should be familiar with the preferred layout and order of the instruments and be able to retrieve them quickly and efficiently during the procedure.

Finally, effective communication and teamwork between the scrub nurse and the surgical team can also help to increase efficiency. The scrub nurse should be proactive in communicating with the team and anticipating their needs. The surgical team, in turn, should communicate any changes or unexpected needs to the scrub nurse, so that they can be addressed promptly [14].

By utilizing these techniques, scrub nurses can help to streamline the instrument preparation and setup process, which can help to increase efficiency during robotic surgery procedures.

## Strategies for troubleshooting and resolving issues with the instruments during the procedure

Despite careful preparation and maintenance of the instruments, issues may still arise during the robotic surgery procedure. Scrub nurses must be prepared to troubleshoot and resolve any issues that may arise quickly and effectively [5].

One effective strategy is to have a thorough understanding of the instruments and equipment used in the robotic surgery procedure. This includes knowledge of their operation, maintenance, and troubleshooting. The scrub nurse should be familiar with the manufacturer's instructions for use and be able to quickly identify and resolve any issues that may arise.

Another strategy is to have a clear understanding of the surgical procedure and the surgeon's preferences. This can help the scrub nurse to anticipate any issues that may arise and have the necessary backup instruments or supplies available. The scrub nurse should communicate effectively with the surgical team to ensure that everyone is on the same page regarding the troubleshooting procedures and backup plans [15].

In addition, the scrub nurse should be prepared to use problem-solving skills and critical thinking to quickly resolve any issues that may arise. This may involve creative problem-solving skills, such as using alternative instruments or techniques or collaborating with the surgical team to develop a new approach.

Finally, the scrub nurse should have a clear understanding of the emergency procedures and be prepared to implement them quickly and effectively if needed. This requires good communication and teamwork between the scrub nurse and the surgical team, as well as the ability to stay calm and focused in high-pressure situations [13].

By implementing these strategies, scrub nurses can help to increase efficiency and minimize the risk of complications or delays.

## The use of checklists to ensure that all necessary steps are taken before, during, and after the procedure

The use of checklists to ensure that all necessary steps are taken before, during, and after the procedure: One effective technique for increasing efficiency and reducing errors during a robotic surgery procedure is the use of checklists.

Checklists can help to ensure that all necessary steps are taken before, during, and after the procedure, and can help to minimize the risk of errors or oversights. The checklist can include items, such as instrument preparation and setup, patient positioning, and equipment and system checks, backup/standby instruments, or supplies [16,17].

The scrub nurse should work with the surgical team to develop a comprehensive checklist that covers all necessary steps of the procedure. The checklist should be part of the Preference card and be reviewed and updated regularly to reflect any changes in the procedure or equipment.

During the procedure, the scrub nurse should use the checklist to ensure that all necessary steps have been taken and to track the progress of the procedure. This can help to ensure that the procedure is on track and that all necessary instruments and supplies are available.

After the procedure, the scrub nurse should use the checklist to ensure that all necessary postprocedure steps have been taken, such as instrument cleaning and sterilization, equipment shutdown, and patient transport.

In addition, utilizing technology, such as digital checklists or electronic health records, can help to increase efficiency and accuracy during the procedure. These systems can help to automate the checklist process and ensure that all necessary steps are taken and tracked throughout the procedure.

## After the procedure

### Documenting any issues or problems with the instruments

Documenting any issues or problems with the instruments: After the robotic surgery procedure is complete, the scrub nurse should document any issues or problems that were encountered during the procedure.

This can include issues such as instrument malfunctions and system errors that may have occurred. The scrub nurse should document the nature of the issue, the steps that were taken to resolve it, and any other relevant information [18].

This documentation can be used to identify any recurring issues or problems with the instruments or equipment and to develop strategies for addressing them [18,19]. It can also be used to communicate any issues or problems to the manufacturer or service provider, to facilitate repairs or replacement of the equipment.

In addition, the documentation can also be used for quality assurance and process improvement purposes. By reviewing the documentation, the surgical team can identify any areas where improvements can be made in the process or in the use of the instruments or equipment.

### Properly storing instruments

If instruments are processed as peel packs the scrub nurse should inspect the instruments for any signs of wear or damage and communicate with the SPD, which typically cleans and sterilizes them according to the manufacturer's instructions before storage. This can help to ensure that the instruments remain in good working condition and are ready for use in future procedures.

By properly labeling and organizing the instruments, scrub nurses also help to ensure that they are easily accessible and readily available for future use. This may include labeling the instruments with their name, serial number, and expiration date, and storing them in a logical and consistent manner that is easy to remember and locate [5].

By properly storing the instruments after a robotic surgery procedure, scrub nurses can help to ensure that the instruments remain in good condition and help to improve efficiency and reduce the risk of errors or complications in future procedures.

## Conclusions

The role of the scrub nurse in robotic surgery is critical to the success of the procedure and requires specialized knowledge and skills in the preparation, handling, and maintenance of the robotic instruments and equipment.

Overall, the scrub nurse plays a crucial role in ensuring the safety, efficiency, and success of robotic surgery procedures. By following best practices and utilizing effective techniques, scrub nurses can help to optimize the use of the instruments and equipment and improve patient outcomes (Fig. 3).

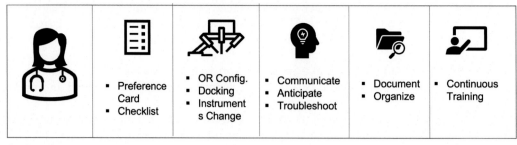

FIG. 3    The role of scrub nurses in robotic surgery.

## Future scenario

### The future of robotic surgery and the role of the scrub nurse

The future of robotic surgery and the role of the scrub nurse or surgical technologist: As the field of robotic surgery continues to evolve and expand, the role of the scrub nurse will continue to be essential to the success of the procedure.

Innovative technologies and techniques are constantly being developed to improve the safety, efficiency, and effectiveness of robotic surgery procedures. As these technologies are introduced, the scrub nurse will need to adapt and expand their knowledge and skills to continue providing high-quality care.

The future of robotic surgery is likely to involve increasing use of artificial intelligence, virtual and augmented reality, and other advanced technologies. The scrub nurse will need to be familiar with these technologies and able to effectively utilize them in the care of their patients [4].

In addition, the role of the scrub nurse may expand to include additional responsibilities, such as monitoring and analyzing data from the robotic instruments and equipment and working with other members of the surgical team to develop new approaches and techniques.

Overall, the future of robotic surgery is promising, and the role of the scrub nurse will continue to be critical to its success. By staying informed and adapting to new technologies and techniques, scrub nurses can continue to provide high-quality care and improve patient outcomes in the field of robotic surgery.

## Key points

- The importance of a well-designed and updated preference card that is based on evidence and best practices, rather than solely on individual surgeon preferences.
- Checking the robotic instruments for damage or wear and tear before, during, and after the procedure, and addressing any issues promptly.
- Changing instruments during the procedure as needed, based on the surgical team's preferences and the steps of the procedure.
- Strategies for troubleshooting and resolving issues with the instruments during the procedure include having a thorough understanding of the instruments and equipment and using critical thinking and problem-solving skills.
- Techniques for streamlining instrument preparation and setup, such as having a standardized layout and setup of the instruments and equipment and utilizing technology to improve efficiency.
- The use of checklists to ensure that all necessary steps are taken before, during, and after the procedure, and to reduce errors and oversights.
- Properly storing instruments after the procedure to ensure that they remain in good condition and are readily available for future use.

## References

[1] Sheetz K, Claflin J, Dimick J. Trends in the adoption of robotic surgery for common surgical procedures. JAMA 2020;3(1), e1918911.
[2] Herrera L, Escalon J, Johnson M, Sanchez A, Sanchez R, Mogollon I. Development of a robot-assisted thoracic surgery (RATS) program. Lessons learned after 2500 cases. J Robot Surg 2023;17(2):405–11.
[3] Teixeira A, Jawad M, Ghanem M, Sanchez A, Inchausti C, Mogollon I, Lind R. Robot-assisted duodenal switch with DaVinci Xi: surgical technique and analysis of a single-institution experience of 661 cases. J Robot Surg 2022; [Online ahead of print].
[4] Venzke E, Segabinazzi L, Treviso P, Zanchi D. Nurse role in robotic surgery: challenges and prospects. Rev SOBECC 2018;23(1):43–51.

[5] Abdel A, Jung H, Don K, Deuk Y, Ho K. Robotic nurse in the urology operative room: 11 years of experience. Asian J Urol 2017;4:116–23.

[6] Thomas CC. Role of the perioperative nurse in robotic surgery. Perioper Nurs Clin 2011;6:227–34.

[7] Shceinker D, Hollingsworth M, Brody A, Phelps C, Bryant W, Pei F, et al. The design and evaluation of a novel algorithm for automated preference card optimization. J Am Med Inform Assoc 2021;28(6):1088–97.

[8] Embick E, Bieri M, Koehler T, Yang A. Cost containment: an experience with surgeon education and universal preference cards at two institutions. Surg Endosc 2020;34(11):5148–52.

[9] Sanchez A, Herrera L, Teixeira A, Mogollon I, Inchausti C, Gibson D, et al. Robotic surgery: financial impact of surgical trays optimization in bariatric and thoracic surgery. J Robot Surg 2023;17(1):163–7.

[10] Wasielewski A. Guideline implementation: minimally invasive surgery, part 1. AORN J 2017;106(1):50–9.

[11] Catchpole K, Hallett E, Curtis S, Mirchi T, Souders C, Anger J. Diagnosis barriers to safety and efficiency in robotic surgery. Ergonomics 2018;61(1):26–39.

[12] Sutton S, Link T, Makic M. A quality improvement project for safe and effective patient positioning during robot-assisted surgery. AORN J 2013;97(4):448–56.

[13] Huser A, Muller D, Brunkhorst V, Kannisto P, Musch M, Kropfl D, et al. Simulated life-threatening emergency during robot-assisted surgery. J Endourol 2014;28:717–21.

[14] Almeras C, Almeras C. Operating room communication in robotic surgery: place, modalities and evolution of a safe system of interaction. J Visc Surg 2019;156(5):397–403.

[15] Schiff L, Tsafrir Z, Aoun J, Taylor A, Theoharis E, Eisenstein D. Quality of communication in robotic surgery and surgical outcomes. JSLS 2016;20(3), e2016.00026.

[16] Russo L, Rodriguez O, Sanchez R, Vegas L, Rosciano J, Jara G, et al. Colecistectomia Laparoscopica: Impacto de la Implementacin de una Lista de Chequeo en la Adecuada Preparacion del Quirofano. Rev Venez Cir 2018;71(1):1–6.

[17] Verdaadonk E, Stassen K, Van der Elst M, Karten T, Dankelman J. Problems with technical equipment during laparoscopic surgery: an observational study. Surg Endosc 2007;2:275–9.

[18] Kaushik D, High R, Clark C, LaGrange C. Malfunction of the DaVinci robotic system during robot-assisted laparoscopic prostatectomy: an international survey. J Endourol 2010;24:571–5.

[19] Lavery H, Thaly R, Albala D, Ahlering T, Shalhav A, Lee D, et al. Robotic equipment malfunction during robotic prostatectomy: a multi-institutional study. J Endourol 2008;22:2165–8.

# 29

# Single-port (SP) robotic surgery: Concept, actual application, and future limits

*Jaya S. Chavali, Nicolas A. Soputro, and Jihad Kaouk*

Glickman Urological and Kidney Institute, Cleveland Clinic, Cleveland, OH, United States

## Introduction

The single-port (SP) purpose-built robot was introduced with the aim to regionalize the surgical dissection to the site of surgery, increase surgeon ergonomics, minimize minimally invasive surgery, and further decrease patient morbidity compared to traditional robotic surgery. The single-arm robot compared to a traditional 3–4 arm robot aims to increase maneuverability in narrow surgical fields like extraperitoneal, perineal, retroperitoneal, oral, anal, and axillary spaces without compromising the surgeon's ergonomics. The focus of the current chapter is on the concept, current applications, and future limits of the purpose-built SP platform.

## History

Single-port surgery has been in practice since the time of laparoscopic surgery, commonly referred to as laparoendoscopic single-site surgery (LESS). The first published description of single-site surgery was retroperitoneal adrenalectomy by Hirano et al. in 2005 [1]. Since the introduction of the robotic platform in 2000 by the U.S. Food and Drug Administration (FDA) for surgical procedures, there has been substantial growth in robotic surgery in the past two decades. There were multiple published reports of different adaptations of LESS single-site surgery using the robotic platform referred to as Robotic-LESS (R-LESS), as first described by Kaouk et al. [2–5]. Most of the current literature involving R-LESS came from urology across various surgeries including prostatectomy, adrenalectomy, partial nephrectomy, and pyeloplasty. Compared to traditional LESS, R-LESS had significant enhancement in ergonomics and optics but had persistent issues with triangulation and instrument clashing. The R-LESS technique had the distinct advantage of shorter postoperative recovery times, less pain, and better cosmetic results. Therefore, numerous efforts were made to improve these difficulties by introducing flexible endoscopes, semi-rigid robotic instruments, curved trocars, and multichannel ports as well as automatic reversal techniques to help with external clashing [6]. There was a steep learning curve and difficulty with ergonomics for the inclusion of R-LESS into common practice until the introduction of a dedicated novel DaVinci SP platform (Fig. 1). The introduction of endowrist technology and double-jointed instruments with the SP robot made the single-port surgery reachable to a wider audience for the first time. DaVinci SP robotic platform, earlier referred to as SP-1098 was first reported clinically in 2012 by Kaouk et al. and studied in various preclinical phases prior to its approval for clinical use by the FDA in 2018 [7–9].

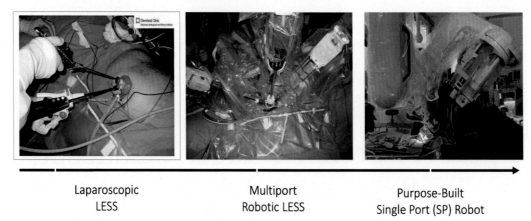

Laparoscopic
LESS

Multiport
Robotic LESS

Purpose-Built
Single Port (SP) Robot

FIG. 1   Evolution of laparoendoscopic single-site surgery (LESS) from multiarm instruments to a dedicated single-port robotic system.

## Single-port robotic platform

### Concept

There are three essential components of the SP surgical system: patient cart, vision cart, and surgeon console (Fig. 2). The vision cart and surgeon console are similar to previous iterations of the DaVinci surgical system. The patient cart, however, is unique and involves a single arm with 360-degree anatomic access. The single-arm design has a smaller footprint and eliminates instrument clashing and allows for various surgical positions and multiquadrant approaches without repositioning the robot (Fig. 3). The SP surgical system empowers the surgeon to operate efficiently in a narrow surgical field using a single incision or a natural orifice.

### Toolbox

The components of the SP toolbox currently in use include a single-port access kit, robotic instrument set, Airseal insufflation system, and Remotely Operated Suction Irrigation (ROSI) system. Spacemaker Kidney Shaped Balloon dissector is utilized in select cases especially if surgery is planned to be done extraperitoneal for the development of the space.

SP access kit consists of a DaVinci SP port, 25 mm multichannel cannula, wound retractor, and laparoscopic assistant port (Fig. 4). The multichannel SP cannula (25-mm entry space) can accommodate three 6-mm double-jointed, articulating instruments and one 8-mm double-jointed, articulating flexible camera. All the instruments are arranged in different quadrants (12,3,6,9 0′ clock) in the cannula and can be independently rotated around the clock (Fig. 5).

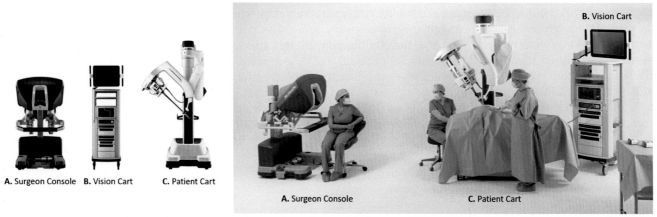

A. Surgeon Console   B. Vision Cart   C. Patient Cart

B. Vision Cart

A. Surgeon Console

C. Patient Cart

FIG. 2   Components of the SP surgical system and animation representing an operation room setup: The components of the SP surgical system include the (A) surgeon console, (B) vision cart, and the (C) patient cart. Images reproduced with permission obtained from Intuitive Surgical.

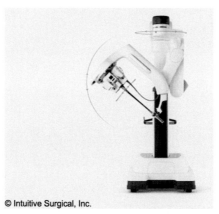

© Intuitive Surgical, Inc.

FIG. 3 Demonstration of Single-Port (SP) robot patient cart: The single arm and 360-degree anatomical rotation access provided by the robot were demonstrated extracorporeally. Images reproduced with permission obtained from Intuitive Surgical.

FIG. 4 Various components of SP access kit: The picture demonstrates the essential parts of the access kit including the bubble port, short entry guide, and wound retractor. The bubble port has two additional ports for the insertion of a laparoscopic trocar and remotely operated suction irrigation (ROSI) system.

The double-jointed instruments and the camera mimic the human elbow and wrist motions. The "Cobra" maneuver of the SP camera, i.e., the unique ability of the endoscope to achieve flexion at 2 separate joints, leads to a S-shaped camera design to maximize the working space away from instruments (Fig. 6).

The single-port DaVinci system includes a wide set of 6-mm instruments for use. The currently available instruments include (1) bipolar (Maryland, fenestrated) forceps, (2) monopolar electrocautery instruments (scissors, hook, etc.), (3) needle driver, (4) Cadière forceps, (5) round tooth retractor, and (6) Hem-o-lok medium-large Weck clip applier. A dedicated SP vessel sealer, stapler, and tenaculum are currently still in the research phase for the current generation SP robotic platform.

The minimum required distance from the tip of the cannula to the target for the activation of the wrist and elbow of instruments include 5 and 10 cm respectively. To increase the working space, all the instruments are introduced using either the originally described floating dock technique or the DaVinci bubble port system with a built-in floating dock.

FIG. 5    SP Multichannel Cannula Components: SP cannula accommodates three 6-mm double-jointed, articulating instruments and one 8-mm double-jointed, articulating flexible camera.

FIG. 6    Different maneuvers of the SP surgical instruments in relation to the camera replicate the camera angles seen with the MP robot. Images reproduced with permission obtained from Intuitive Surgical.

**30° down**              **0°**              **30° up**
© Intuitive Surgical, Inc

The floating dock technique was earlier reported by Kaouk et al. [10] where the instruments were introduced via Alexis wound retractor and Gel point (Applied Medical, CA, USA) system but the Gel Seal Cap and the Alexis are approximately 8 cm off the skin level and Alexis acts as a conduit between trocar and conduit while preserving insufflation. This has increased the working space up to 390% compared to flat surface docking (3) (Fig. 7). The current DaVinci bubble port has an inbuilt working bubble that was developed from the previous idea and is currently in use with proven efficacy to improve working space in narrow spaces, especially when depth is <10 cm.

The ROSI (Remotely Operated Suction Irrigation, Vascular Technology Inc., Nashua, NH) is introduced through a separate port built into the SP access port and directly controlled by the surgeon (Fig. 8). The other options for surgeon-controlled suction include the utilization of a nasogastric tube or foley catheter connected to suction if the ROSI system is unavailable. The laparoscopic assistant trocar used with the SP bubble port is connected to the Air seal insufflation system to maintain constant pressures. There is also an option to connect the trocar to a traditional insufflation. It is, however, advisable to utilize the Air seal system given its ability to maintain constant insufflation pressures even with changes that can occur with suction/smoke evacuation throughout the case given the narrow working space with the SP platform.

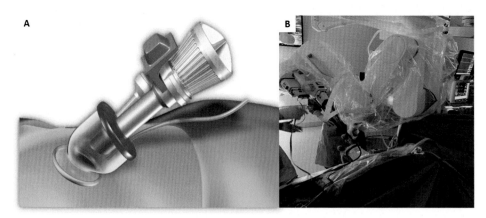

FIG. 7 Floating dock technique: Animation (A) and intraoperative image (B) demonstrating air docking of instruments that were introduced via half-rolled Alexis wound retractor and Gel point to increase the working space of instruments.

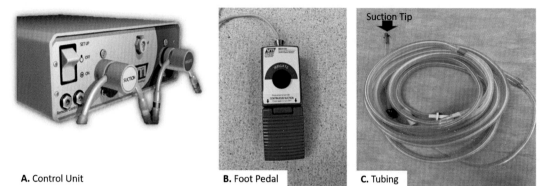

**A.** Control Unit     **B.** Foot Pedal     **C.** Tubing

FIG. 8 Components of Remotely Operated Suction Irrigation (ROSI) System: ROSI system includes (A) Control System, (B) Foot Pedal, and (C) Tubing. *Images reproduced with permission obtained from Vascular Technology Inc. (VTI).*

## Early preclinical experience

Early preclinical literature including safety was studied across various surgical specialties before its widespread FDA approval. The early literature from urology included the utilization of the SP robotic platform to perform various surgeries including prostatectomy, cystectomy, and kidney transplantation, demonstrating safety and feasibility in cadaver models prior to the FDA approval [11–14].

## Current applications

FDA approved the use of a single port platform in 2018. Since its approval, it has been in practice among various surgical specialties including urology, otolaryngology, cardiothoracic, gynecology, and general surgery, for various procedures (Fig. 9). The current chapter focuses on some of its current medical applications and surgical outcomes. It is currently in wide use in the fields of urology and otolaryngology.

### *Urology*

DaVinci SP robot is currently utilized for a wide spectrum of urological surgeries involving the kidney, ureter, bladder, and prostate (Fig. 10). In addition to improvement in cosmesis, the advantages of the SP robotic platform in urology included the ability to facilitate same-day discharge [15], minimizing opioid use [16,17], and early ambulation as compared to traditional robotic surgery.

#### Robotic prostatectomy

Robotic radical prostatectomy (RRP) is a gold standard approach for clinically localized prostate cancer (PCa). Since its advent, the robotic platform has replaced the traditional open radical prostatectomy over the last two decades given the ability to work in a narrow operative field and with improved dexterity in the pelvis, minimal blood loss, and

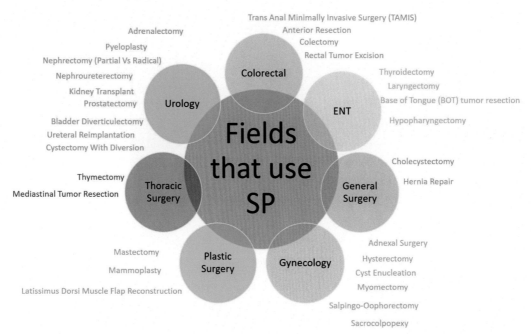

FIG. 9   Current medical applications of the SP surgical system in diverse surgical fields and various surgical procedures attempted with SP robot.

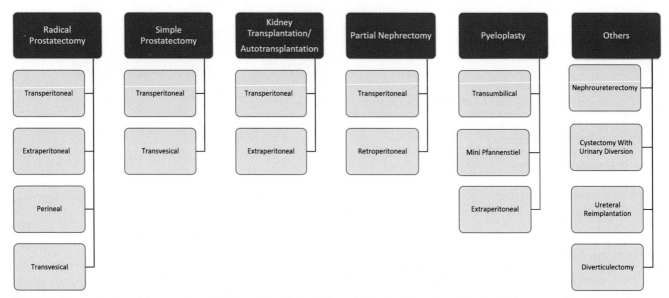

FIG. 10   Demonstration of the current applications of the Davinci SP surgical system in urology including kidney transplantation.

enhanced postoperative. Unlike the open retropubic approach, RRP is usually performed with the patient positioned in a steep Trendelenburg to move the bowels away from the pelvis. The introduction of the current SP platform was able to further regionalize the surgery to the organ of interest without the need for Trendelenburg position. SP robotic radical prostatectomy can now be performed via different approaches including transperitoneal, extraperitoneal, transperineal, and transvesical routes based on the individualized clinical needs, as described below.

*Radical prostatectomy   Transperitoneal*   The current practice of traditional multiport (MP) robotic prostatectomy is largely performed via a transperitoneal approach. Initial single-port surgery tried to replicate the traditional approach during the early learning curve. The initial challenges encountered with the SP robotic platform were similar to the MP experience including the need for the patient to stay in Trendelenburg position and the SP platform did not provide any added benefit.

*Extraperitoneal* After the initial learning curve, the SP surgical platform was utilized to perform prostatectomy using the extraperitoneal approach with minimal Trendelenburg position and avoided peritoneal insufflation. This helped with intraoperative parameters including improved patient ventilation and decreased intraocular pressures. Avoiding the peritoneal cavity also leads to earlier recovery including decreased opioid use [16] and early return of bowel function. After the initial incision of the anterior rectus fascia, the rectus muscles were split in the midline and the extraperitoneal space of retzius was developed using a Spacemaker balloon dissector and SP robot docked into the space (Fig. 11A).

*Transvesical* Further regionalization of surgery to the prostate gland was demonstrated in 2020 by Kaouk et al. with the introduction of the transvesical approach (Fig. 11B) [18]. After the bladder is filled with 200 cc, the robot is percutaneously docked directly into the bladder through a 2 cm vertical cystotomy, and pneumovesicum is achieved. The initial access is similar to performing a suprapubic catheter and the bladder entry is confirmed with a 21-gauge needle with the aspiration of clear urine prior to the cystotomy (Fig. 12).

The single-port robotic transvesical technique is utilized to perform a radical or partial prostatectomy for prostate cancer and simple prostatectomy for benign prostate hypertrophy (BPH). With respect to the transvesical approach, the SP robot is docked directly into the bladder and therefore avoids disruption of both the peritoneum and the space of retzius.

There are multiple advantages to the transvesical approach. Similar to the extraperitoneal approach, the patient is in a minimal Trendelenburg position, which makes it easier to ventilate patients intraoperatively. Second, it has an added advantage in patients with a hostile abdomen that includes a prior history of multiple abdominal and extraperitoneal surgeries including hernia repairs. Third, insufflation is limited to the bladder and has an auxiliary benefit to perform the surgery under epidural analgesia without the need for intubation in select patients [19]. Fourth, it is associated with

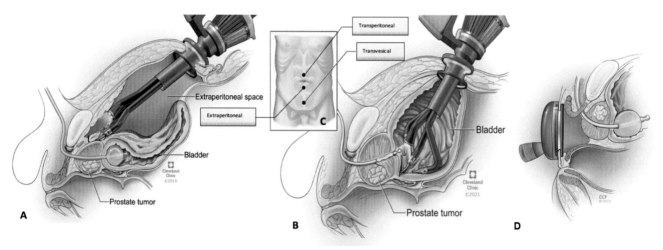

FIG. 11 Illustration of SP extraperitoneal radical prostatectomy options including Extraperitoneal (A), Transvesical (B), and Transperineal (D) radical prostatectomy. (C) Different incision sites of Transperitoneal, Extraperitoneal, and Transvesical radical prostatectomy.

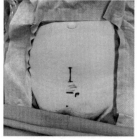

**A 3.5 cm suprapubic incision, 2 fingerbreadths above the pubic symphysis**

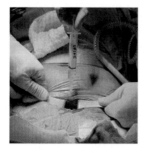

**Bladder Identification using a 21G needle**

**2cm vertical cystotomy**

FIG. 12 Demonstration of steps of the transvesical prostatectomy initial access.

minimal blood loss, minimal to no opioid use, and same-day discharge and early foley removal within 3–4 days post-surgery. [20] Lastly, given the complete sparing of anterior support strictures, the current approach has been associated with early return of urinary control.

*Transperineal*   The transperineal approach was first described as one of the surgical options for prostate cancer by Hugh Hampton Young in 1904. Its practice is currently limited to patients who are not candidates for the traditional retropubic approach given a hostile abdomen (Fig. 11D). It is associated with early postoperative mobility and faster return of urinary continence. The main reason for limited practice is the steep learning curve associated with the surgery and available alternatives including transvesical surgery. Although formerly the surgery of choice for patients with a hostile abdomen, the transperineal approach is currently reserved in hostile abdomen cases with a prior history of bladder surgery or pelvic trauma to the pubic bone and/or the bladder.

*Partial prostatectomy*   The concept of partial prostatectomy using the multiport robot was previously described by Menon et al. as a precision prostatectomy [21]. In the era of single-port surgery and focal therapy for localized low-volume prostate cancer, partial prostatectomy has a unique role among other management modalities. The transvesical approach described above was utilized to perform partial prostatectomy [22,23]. An active transrectal ultrasound probe with fused preoperative magnetic resonance prostate imaging is used intraoperatively for guidance to help achieve negative margins. The advantage of partial prostatectomy compared to radical surgery has been an earlier return of continence and erectile function [24]. Further research regarding long-term oncological outcomes is currently underway.

*Simple prostatectomy*   The single-port robot is also utilized currently for surgical management of BPH. There are multiple available surgical options for the management of BPH including endoscopic and robotic approaches. The surgical options are often limited by the size of the prostate gland. Simple prostatectomy is often preferred when the prostate has an enormous, nonmalignant growth causing obstructive lower urinary tract symptoms. Simple prostatectomy is currently one of the only two management options available for prostate glands greater than 80 g. The initial entry is similar to the aforementioned description of transvesical radical prostatectomy. In a simple prostatectomy, however, only the benign prostatic adenoma is removed leaving the prostate capsule intact.

### Upper tract and renal surgery

Single-port robotic platform is ideal for multiquadrant abdominal surgeries owing to the ability to rotate on its surgical axis through a single incision. One of the ideal platforms for its use in urology is the ability to perform renal auto-transplantation and nephroureterectomy where the procedures involve dual upper tract and pelvic procedures without the need for redocking or use of additional trocars.

*Kidney   Partial nephrectomy*   Partial nephrectomy (PN) is the surgery of choice for small renal masses (<4 cm). The procedure is usually performed via transperitoneal or retroperitoneal approach depending on multiple factors including the location of the tumor, prior abdominal surgeries, and surgeon preference. In SP cases, the patient is usually placed in a lateral, flank position and the procedure is performed via a 3 cm. The incision is either made periumbilical in the transperitoneal approach or over the anterior superior iliac spine in the anterior axillary line in retroperitoneal cases. Recent reports suggested an attempt at renal surgery using low anterior access (formerly known as Pfannenstiel incision) with the patient placed in a supine or relaxed lateral position.

The low anterior retroperitoneal access via a McBurney incision for either transperitoneal or extraperitoneal renal surgery is usually performed with the patient in a relaxed lateral position with the pelvis tilted to be supine for the surgical incision (Fig. 13). The advantage of the low anterior retroperitoneal kidney surgery includes easier access to different renal tumor locations, cosmesis, less postoperative pain, and improved patient post-surgical respiration given evasion of the accessory respiratory muscles with this particular approach.

The SP platform given its low profile and double articulating instruments allows it to work in a narrow retroperitoneal space without significant instrument clashing. It can be utilized to approach different tumors (endophytic versus exophytic) at various locations (upper, mid versus lower pole). However, the SP instruments have a relatively weaker grasp compared to multiport instruments with occasional difficulty observed while performing renal dissection, renorrhaphy compared to the MP robot, and therefore careful patient selection is essential. Current data displayed similar perioperative outcomes with SP robotic PN (RPN) compared to MP RPN for low complex renal lesions [25].

*Nephrectomy*   Radical and donor nephrectomy can be done with SP surgical platform via either a periumbilical incision or a low anterior incision. An additional assistant port is used currently for the vascular stapler given the lack of a dedicated SP stapler. SP donor nephrectomy technique and safety in the early series using a low anterior incision was

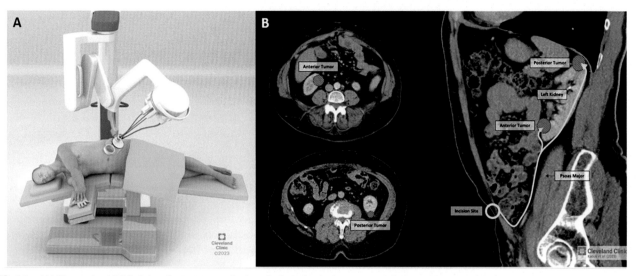

FIG. 13 (A) Illustration highlighting the incision and robot-docking for the low anterior abdominal incision for retroperitoneal upper tract surgery and the (B) demonstration of access to various renal tumor locations using the low incision site.

described by Garden et al. [26] with mean operative time and warm ischemia times including 218 min (SD: 16.3 min) and 3:37 min (SD: 38 s).

*Radical nephroureterectomy* Total nephroureterectomy is recommended in cases of high-grade urothelial carcinoma involving either the renal pelvis or ureter. The surgery involves radical nephrectomy with entire ureter removal with a bladder cuff and cystotomy closure. The SP platform is efficacious for this multiquadrant surgery without the need for redocking for the pelvic portion of the surgery.

*Upper tract reconstruction* The patient is placed in a lateral flank position, a similar 3-cm periumbilical or low anterior incision over the pubic tubercle (Fig. 13A) is performed, and the robot is docked for upper tract reconstruction. It can be performed via transperitoneal versus extraperitoneal approach based on surgeon preference. The approach can be utilized for pyeloplasty [27], ureteroureterostomy, and buccal mucosa graft ureteroplasty for different etiologies leading to stricture or narrowing of the ureter.

The approach is performed via a single incision with minimal pain postoperatively [17] and provides excellent cosmetic results. It is currently in practice for pediatric pyeloplasty as well with comparable outcomes to traditional reconstructive techniques.

*Kidney transplantation and autotransplantation* Robotic kidney transplantation has evolved as a minimally invasive alternative to a conventional open approach given prolonged patient recovery and wound complications with open surgery. The early promising results with the SP platform in prostate and kidney surgery lead to exploring the utilization of robots in the field of transplant surgery. SP kidney transplantation is indicated in patients with end-stage renal disease (ESRD), and SP autotransplantation is indicated in patients with various etiologies including long segment ureteric strictures, prior failed reconstruction, and chronic visceral pain that failed other conservative options.

The patients in these cases are placed in a supine position with a slight Trendelenburg position and a lateral rotation. The surgery is usually performed via a 5-cm periumbilical incision for autotransplant or infra umbilical incision for a kidney transplant. While autotransplantation is performed via a transperitoneal approach, kidney transplantation has moved to an extraperitoneal space for placement of the donor graft. The SP robot with its multiquadrant ability is beneficial in autotransplantation that involves abdominal nephrectomy and later autotransplantation with vascular, ureteric anastomosis in the pelvis without the need for repositioning [28]. The early results from the literature have proven efficacy with similar perioperative outcomes compared to traditional surgery [28,29].

### Adrenalectomy

The SP robotic platform was utilized in robotic adrenalectomy with the patient placed in a lateral position and a 2-cm incision made on the ipsilateral side of umbilicus. An additional assistant port is occasionally used in these cases for dissection and retraction of the liver/spleen for adrenalectomy. The outcomes were comparable to multiport robotic adrenalectomy; however, its application is demonstrated currently in small adrenal masses and further study is required for widespread application [30,31].

### Lower urinary tract surgery

*Radical cystectomy with intracorporeal urinary diversion*    The early series reported the feasibility of radical cystectomy and various urinary diversions including ileal conduit, and neobladder diversions with the SP surgical system. A 3-cm midline incision was made 5 cm above the umbilicus for docking the SP robot. A 12-mm Airseal assistant trocar was used for the assistant port at the premarked stoma site for cases with ileal conduit diversion. The surgical steps replicated the traditional steps of robotic-assisted radical cystectomy (RARC) [32,33].

*Lower urinary tract reconstruction*    The lower urinary tract traditionally comprises the bladder and urethra. Distal ureter strictures that require reimplantation are also included here.

*Distal ureter reconstruction*    The SP platform has been used for lower tract ureter reconstruction including reimplantation in literature. The patients are usually placed in a low lithotomy position or supine for these cases. A 3-cm supra umbilical midline incision is utilized in these cases [34]. An assistant port can often be used for bedside assistance in these cases and later used for drain placement at the end of the case.

*Bladder diverticulectomy*    The bladder diverticulum is usually secondary to an outlet obstruction due to BPH, urethral stricture, etc. The diverticulectomy can be performed via either transperitoneal, extraperitoneal, or transvesical approaches depending on other concomitant procedures being performed.

*Urethral reconstruction*    The posterior urethra is the 1- to 2-in. segment of the urethra that extends from the bladder neck, through the prostate gland and external urinary sphincter and continues distal as anterior urethra. While the anterior urethral reconstruction is approachable via perineal incision for repair, the posterior urethra repairs are often complex and require either perineal, abdominal, or abdominoperineal approaches. SP posterior urethroplasty reconstruction via intraabdominal or transvesical approach with buccal mucosa grafts has been described successfully in literature [35].

### Other urological procedures

The SP platform using a transvesical approach can be used for select cases of vesicovaginal fistula repair. Another use of the SP platform described is the inguinal lymph node dissection performed for select cases of penile cancer [36].

### Otolaryngology, head and neck surgery

The SP robotic platform is utilized in various surgeries done by otolaryngology. Some of the surgical procedures performed via a transoral route that were published include tonsillectomy, resection of the base of tongue tumors or other oropharyngeal tumors, and neck dissection [37].

Subtotal and total thyroidectomy procedures performed via a transoral or a trans-axillary approach with the SP robot were also described in the literature [38].

### Cardiothoracic surgery

The current application of SP robot-assisted thoracic surgery has been limited to case reports including mediastinal tumors and thymectomy. Yang et al. reported the feasibility of resection of superior mediastinal tumors in carefully selected patients [39].

### Plastic and reconstructive surgery

Current clinical reports in plastic surgery include the safety of robot-assisted nipple-sparing mastectomy with immediate reconstruction [40]. Robotic peritoneal flap gender-affirming vaginoplasty surgery (RPGAV) was demonstrated to be feasible using the SP surgical platform with comparable results with respect to vaginal depth and width compared to the traditional Xi approach [41].

### Gynecology and urogynecology

The SP surgical platform used in gynecology has limited case series publications to date. Various gynecological surgeries described include myomectomy, hysterectomy, adnexectomy, and sacrocolpopexy with success without major complications [42,43].

## General surgery

The current utilization of the SP robot in general surgery and colorectal surgery has been limited to early case series publications. Its use in general surgery has been demonstrated in performing cholecystectomy, and inguinal hernia repair. SP Total extraperitoneal (TEP) inguinal hernia repair was described by Lee et al. [44].

With respect to colorectal surgery, its current use has been demonstrated in transrectal surgery. There are initial reports of using an SP robot for transanal minimally invasive surgery (TAMIS) for endoscopic unresectable rectal polyps, early cancers, or low-risk rectal tumors [45] and rectal mucosa graft harvest for reconstructive purposes [46].

## Pediatric surgery

The practice of Hidden incision Endoscopic Surgery (HidES) has been popular among pediatric surgeons. SP robotic innovation has encouraged the adoption of HidES currently in practice. The current data include applications in single-port pyeloplasty via a low anterior incision [27]. There are also reports of the utilization of SP robot for performing simple nephrectomy and mitrofanoff channel creation for neurogenic bladder patients [47].

## Limitations

SP surgical platform has shown great promise in its ability to operate in a small operative space with enhanced optics and ergonomics compared to traditional R-LESS. However, its practice is currently limited to a few educational institutions in the United States and other countries. There is also a limited adaptation due to surgeons being in the early learning curve with the SP robotic system. The potential widespread availability of the robot including a continued adaptation of the platform among other surgical branches can produce high-impact technical data regarding advantages and limitations.

## Future scenario

The single-port robotic application continues to evolve across various surgical specialties in an effort to decrease surgical morbidity. While the SP robot application is increasingly reported in urology, the experience will spread to other surgical branches, in an effort to limit the surgery to the organ of interest. The limited choice of instruments currently available is expected to increase over time with growing adaptation to various specialties.

## Conclusions

The application of single-port robotic platform has been rapidly evolving since its introduction in 2018. The unique features of the SP robot include a low-profile, single arm with 360-degree anatomic access that allows it to work in a limited space, double-jointed camera and instruments, and multiquadrant approachability without often redocking or repositioning the patient.

## Key points

- The single-port (SP) purpose-built robot was introduced with the aim to regionalize the surgical dissection to the site of surgery, increase surgeon ergonomics in limited spaces, and further decrease patient morbidity.
- The unique feature of the SP robot compared to prior surgical platforms includes a single arm with 360-degree anatomic access and double-jointed camera and instruments and multiquadrant approachability.
- It is currently used in the urology and otolaryngology fields with growing adaptation among other surgical branches.

## Appendix: Supplementary material

Supplementary material related to this chapter can be found on the accompanying CD or online at https://doi.org/10.1016/B978-0-443-13271-1.00078-9.

# References

[1] Hirano D, Minei S, Yamaguchi K, Yoshikawa T, Hachiya T, Yoshida T, et al. Retroperitoneoscopic adrenalectomy for adrenal tumors via a single large port. J Endourol 2005;19:788–92.

[2] Kaouk JH, Goel RK, Haber GP, Crouzet S, Stein RJ. Robotic single-port transumbilical surgery in humans: initial report. BJU Int 2009;103:366–9.

[3] Kaouk JH, Goel RK. Single-port laparoscopic and robotic partial nephrectomy. Eur Urol 2009;55:1163–70.

[4] Stein RJ, White WM, Goel RK, Irwin BH, Haber GP, Kaouk JH. Robotic laparoendoscopic single-site surgery using GelPort as the access platform. Eur Urol 2010;57:132–7.

[5] White MA, Haber G-P, Autorino R, Khanna R, Forest S, Yang B, et al. Robotic laparoendoscopic single-site radical prostatectomy: technique and early outcomes. Eur Urol 2010;58:544–50.

[6] Nelson RJ, Chavali JSS, Yerram N, Babbar P, Kaouk JH. Current status of robotic single-port surgery. Urol Ann 2017;9:217.

[7] Garisto JD, Bertolo R, Kaouk J. Technique for docking and port placement using a purpose-built robotic system (SP1098) in human cadaver. Urology 2018;119:91–6.

[8] Kaouk J, Abaza R, Davis J, Eun D, Gettman M, Joseph J, et al. Robotic one access surgery (R-1): initial preclinical experience for urological surgeries. Urology 2019;133(5–10), e1.

[9] Kaouk JH, Khalifeh A, Hillyer S, Haber G-P, Stein RJ, Autorino R. Robot-assisted laparoscopic partial nephrectomy: step-by-step contemporary technique and surgical outcomes at a single high-volume institution. Eur Urol 2012;62:553–61.

[10] Lenfant L, Kim S, Aminsharifi A, Sawczyn G, Kaouk J. Floating docking technique: a simple modification to improve the working space of the instruments during single-port robotic surgery. World J Urol 2021;39:1299–305.

[11] Garisto J, Eltemamy M, Bertolo R, Miller E, Wee A, Kaouk J. Single port robot-assisted transperitoneal kidney transplant using the SP® surgical system in a pre-clinical model. Int Braz J Urol 2020;46:680–1.

[12] Eltemamy M, Garisto J, Miller E, Wee A, Kaouk J. Single port robotic extra-peritoneal dual kidney transplantation: initial preclinical experience and description of the technique. Urology 2019;134:232–6.

[13] Garisto J, Bertolo R, Chan E, Kaouk J. Single-port trans-perineal approach to cystoprostatectomy with intracorporeal ileal conduit urinary diversion and lymph-nodes dissection using a purpose-built robotic system: surgical steps in a preclinical model. Int Braz J Urol 2019;45:854.

[14] Maurice MJ, Kaouk JH. Single-port robot-assisted perineal prostatectomy and pelvic lymphadenectomy: step-by-step technique in a cadaveric model. J Endourol 2018;32. S-93-S-6.

[15] Abaza R, Murphy C, Bsatee A, Brown Jr DH, Martinez O. Single-port robotic surgery allows same-day discharge in majority of cases. Urology 2021;148:159–65.

[16] Sawczyn G, Lenfant L, Aminsharifi A, Kim S, Kaouk J. Predictive factors for opioid-free management after robotic radical prostatectomy: the value of the SP® robotic platform. Minerva Urol Nephrol 2020;73:591–9.

[17] Beksac AT, Wilson CA, Lenfant L, Kim S, Aminsharifi A, Abou Zeinab M, et al. Single-port mini-pfannenstiel robotic pyeloplasty: establishing a non-narcotic pathway along with a same-day discharge protocol. Urology 2022;160:130–5.

[18] Kaouk J, Beksac AT, Abou Zeinab M, Duncan A, Schwen ZR, Eltemamy M. Single port transvesical robotic radical prostatectomy: initial clinical experience and description of technique. Urology 2021;155:130–7.

[19] Kaouk J, Ferguson E, Ramos-Carpinteyro R, Chavali J, Geskin A, Cummings KC, et al. Transvesical percutaneous access allows for epidural anesthesia without mechanical ventilation in single-port robotic radical and simple prostatectomy. Urology 2023.

[20] Ramos-Carpinteyro R, Ferguson EL, Chavali JS, Geskin A, Kaouk J. First 100 cases of transvesical single-port robotic radical prostatectomy. Asian J Urol 2023.

[21] Sood A, Jeong W, Palma-Zamora I, Abdollah F, Butaney M, Corsi N, et al. Description of surgical technique and oncologic and functional outcomes of the precision prostatectomy procedure (IDEAL stage 1–2b study). Eur Urol 2022;81:396–406.

[22] Kaouk JH, Garisto J, Sagalovich D, Dagenais J, Bertolo R, Klein E. Robotic single-port partial prostatectomy for anterior tumors: transvesical approach. Urology 2018;118:242.

[23] Kaouk JH, Sagalovich D, Garisto J. Robot-assisted transvesical partial prostatectomy using a purpose-built single-port robotic system. BJU Int 2018;122:520–4.

[24] Kaouk JH, Ferguson EL, Beksac AT, Abou Zeinab M, Kaviani A, Weight C, et al. Single-port robotic transvesical partial prostatectomy for localized prostate cancer: initial series and description of technique. Eur Urol 2022;82:551–8.

[25] Palacios AR, Morgantini L, Trippel R, Crivellaro S, Abern MR. Comparison of perioperative outcomes between retroperitoneal single-port and multiport robot-assisted partial nephrectomies. J Endourol 2022;36:1545–50.

[26] Garden EB, Al-Alao O, Razdan S, Mullen GR, Florman S, Palese MA. Robotic single-port donor nephrectomy with the da Vinci SP® surgical system. JSLS 2021;25.

[27] Lenfant L, Wilson CA, Sawczyn G, Aminsharifi A, Kim S, Kaouk J. Single-port robot-assisted dismembered pyeloplasty with mini-pfannenstiel or peri-umbilical access: initial experience in a single center. Urology 2020;143:147–52.

[28] Kaouk J, Chavali JS, Ferguson E, Schwen ZR, Beksac AT, Ramos-Carpinteyro R, et al. Single port robotic kidney autotransplantation: initial case series and description of technique. Urology 2023.

[29] Kaouk J, Eltemamy M, Aminsharifi A, Schwen Z, Wilson C, Abou Zeinab M, et al. Initial experience with single-port robotic-assisted kidney transplantation and autotransplantation. Eur Urol 2021;80:366–73.

[30] Fang AM, Fazendin JM, Rais-Bahrami S, Porterfield JR. Comparison of perioperative outcomes between single-port and multi-port robotic adrenalectomy. Am Surg 2022. 00031348221075777.

[31] Lee IA, Kim JK, Kim K, Kang S-W, Lee J, Jeong JJ, et al. Robotic adrenalectomy using the da Vinci SP robotic system: technical feasibility comparison with single-port access using the da Vinci multi-arm robotic system. Ann Surg Oncol 2022;29:3085–92.

[32] Tyson MD, Mi L. Preliminary surgical outcomes after single incision robotic cystectomy (SIRC). Urology 2023;171:127–32.

[33] Kaouk J, Garisto J, Eltemamy M, Bertolo R. Step-by-step technique for single-port robot-assisted radical cystectomy and pelvic lymph nodes dissection using the da Vinci® SP™ surgical system. BJU Int 2019;124:707–12.

[34] Kaouk JH, Garisto J, Eltemamy M, Bertolo R. Robot-assisted surgery for benign distal ureteral strictures: step-by-step technique using the SP® surgical system. BJU Int 2019;123:733–9.

[35] Liu W, Shakir N, Zhao LC. Single-port robotic posterior urethroplasty using buccal mucosa grafts: technique and outcomes. Urology 2022;159:214–21.

[36] Abdullatif VA, Davis J, Cavayero C, Toenniessen A, Nelson RJ. Single-port robotic inguinal lymph node dissection for penile cancer. Urology 2022;161:153–6.

[37] Van Abel KM, Yin LX, Price DL, Janus JR, Kasperbauer JL, Moore EJ. One-year outcomes for da Vinci single port robot for transoral robotic surgery. Head Neck 2020;42:2077–87.

[38] Kang IK, Park J, Bae JS, Kim JS, Kim K. Safety and feasibility of single-port trans-axillary robotic thyroidectomy: experience through consecutive 100 cases. Medicina 2022;58:1486.

[39] Yang B, Chen R, Lin Y, Liu Y. Single-port robotic surgery for mediastinal tumors using the da vinci SP system: initial experience. Front Surgery 2022;9.

[40] Joo OY, Song SY, Park HS, Roh TS. Single-port robot-assisted prosthetic breast reconstruction with the da Vinci SP surgical system: first clinical report. Arch Plast Surg 2021;48:194–8.

[41] Dy GW, Jun MS, Blasdel G, Bluebond-Langner R, Zhao LC. Outcomes of gender affirming peritoneal flap vaginoplasty using the Da Vinci single port versus Xi robotic systems. Eur Urol 2021;79:676–83.

[42] Shin HJ, Yoo HK, Lee JH, Lee SR, Jeong K, Moon H-S. Robotic single-port surgery using the da Vinci SP® surgical system for benign gynecologic disease: a preliminary report. Taiwan J Obstet Gynecol 2020;59:243–7.

[43] Lee JH, Yoo HK, Park SY, Moon HS. Robotic single-port myomectomy using the da Vinci SP surgical system: A pilot study. J Obstet Gynaecol Res 2022;48:200–6.

[44] Kim D, Lee CS. Single-port robotic totally extraperitoneal (TEP) inguinal hernia repair using the da Vinci SP platform: a video vignette. Asian J Surg 2022.

[45] Liu S, Kelley S, Behm K. Single-port robotic transanal minimally invasive surgery (SPR-TAMIS) approach to local excision of rectal tumors. Tech Coloproctol 2021;25:229–34.

[46] Accioly JPE, Zhao H, Ozgur I, Lee GC, Gorgun E, Wood HM. Single-port, robot-assisted Transanal harvest of rectal mucosa grafts for substitution Urethroplasty. Urology 2022;166:1–5.

[47] Parikh N, Findlay B, Boswell T, Granberg C, Gargollo P. Single-port robotic Mitrofanoff in a pediatric patient. J Pediatr Urol 2021;17:424–5.

II. Education, safety, ethical and administrative aspects for a successful robotic program

# 30

# Complications of robotic surgery (related to patient positioning or robot malfunctioning): Causes and solutions

*Camilo Giedelmann[a], Rafael Clavijo[a], Andrés Barrios[a], and René Sotelo[b]*

[a]Urology Department, Hospital de San José, Bogota, Colombia [b]Keck Medical Center at the University of Southern California, Los Angeles, CA, United States

## Introduction

The aim of the robot-assisted laparoscopic surgery is hopefully to reduce surgical complications. Therefore, surgeons must have a precise understanding of the current definition of a successful surgery, which encompasses optimal oncological and functional outcomes and minimal adverse events and complications.

Many articles discuss surgical complications, but few address the issue of complications that may arise due to patient positioning during the procedure. There is no definitive guide on how to prevent or manage these complications, which is concerning since incorrect positioning of the patient, inadequate fixation, or prolonged surgical time are known risk factors for perioperative complications.

The use of robotic technology should not solely rely on its benefits, but also on its drawbacks. Technical problems may include errors in programming, as well as mechanical and communication problems between the robot and the surgical team. Malfunctioning of the robot can result in complications for the patient, such as tissue damage, bleeding, or infection, and may even require conversion to open surgery.

Throughout the course of this chapter, we aim to outline the primary complications associated with patient positioning in robot-assisted laparoscopic surgery, as well as those resulting from equipment malfunction. We will also provide solutions and preventative measures to mitigate these challenges.

## Complications of robotic surgery—Related to patient positioning

Depending on the type of surgical procedure being performed, different patient positions are required to achieve optimal access and improve surgical outcomes. Some of the most commonly used positions in robotic surgery include supine, lithotomy, Trendelenburg, or lateral decubitus.

The Trendelenburg position is commonly employed for surgical interventions involving the pelvic region, owing to its ability to provide a clear and unobstructed surgical field. While the lateral decubitus position is used for most retroperitoneal surgeries [1].

During the perioperative period, it is crucial to consider the patient's comorbidities and medical history to ensure a collaborative approach with the anesthesiology team. This is especially important given the various hemodynamic and ventilatory considerations that may arise due to the patient's position during the procedure.

The surgical team must verify prior to the start of the procedure that all necessary elements, such as pads, gels, and foam, are available to minimize the probability of an incident due to inadequate positioning. Likewise, the operating room can be adapted for the robot equipment, and permanent marking signals (such as colored adhesive tape) can be established on the surgery room floor to ensure proper positioning in relation to the arrangement of the robot arms and anesthesia machine. Proper patient positioning not only helps prevent complications but is also essential for the adequate exposure of the surgical field on the patient's body.

Ocular injuries, neuromuscular injuries, and skin lesions are among the most commonly encountered complications associated with patient positioning in robotic surgery. Therefore, it is essential to taking into consideration and analyzing various risk factors, such as operative time, body mass index, and peripheral vascular disease associated with robotic surgery. This analysis should be conducted before performing any robotic surgical procedure.

Patients with obesity and BMI > 30 are at an increased risk of developing pulmonary complications. The Trendelenburg position may lead to a ventilation-perfusion imbalance with resulting acidosis and hypercapnia due to the increase in thoracic cavity pressure and restriction of the chest wall. Furthermore, the Trendelenburg position can increase cardiac output and vascular tone due to compression of the aorta, resulting in elevated central and intracranial venous and arterial pressure. This, combined with hypercapnia, can lead to the development of cardiac arrhythmias [2,3].

Although the incidence of iatrogenic complications in minimally invasive surgeries in the United States is relatively low, at approximately 1.3%, a majority of these complications have been attributed to patient positioning in robotic surgery. This conclusion was drawn from an analysis of the safety of robotic prostatectomy over time, conducted by Dr. Chughtai, Kaplan et al. [4].

## Skin injuries

Most skin injuries are caused by direct trauma generated by the position of the patient and the support points of the body on the surgical table, which can become pressure points. These injuries are mostly associated with prolonged surgical time and immobilization. However, there may be other skin injuries that are caused by other variables such as the disposition of the incision for surgical piece extraction, inadequate patient fixation, and the positioning of the trocars and robot arms. Scrotal skin injury due to secondary distension from pneumoperitoneum has also been described [5].

## Neuromuscular injuries

Neuromuscular complications are physiopathologically explained by nerve enlargement, which generates compression of the vessels that course intraneurally with a decrease in perfusion inducing ischemia and tissue edema, which could lead to compartment syndrome and nerve tissue damage. This risk of injury increases with every hour of surgery, and depending on the position used, a certain nerve injury will occur. For example, there is a risk of injuring the brachial plexus in the Trendelenburg position, as well as the lesion of the ulnar nerve and femoral nerve in the lateral decubitus and lithotomy position, respectively. During pelvic lymphadenectomy, there may be injury to the obturator nerve, and compression against the surgical table may cause peroneal nerve injury [1,6].

Another complication related to the musculature is rhabdomyolysis, which occurs due to compression of the muscles from the patient's position during surgery and is associated with prolonged surgical time. It is clinically manifested by the presence of postoperative muscle pain and reddish-brown urine, caused by myonecrosis and myoglobinuria. This muscle lytic reaction is triggered in areas of pressure where the blood pressure is less than 10–30 mmHg, inducing ischemia and necrosis in the tissue.

The diagnosis is confirmed by measuring creatine kinase levels, with a positive test result when the values are above 5000 U/L. The main problem with rhabdomyolysis is severe kidney damage, which can even require renal replacement therapy [7], and it is known that there is a directly proportional relationship between the degree of renal injury and the length of the tissue with rhabdomyolysis [8].

## Neurocognitive injuries

Neurocognitive injuries refer to damage to the brain that affects cognitive function, such as memory, attention, perception, and problem-solving abilities.

Although the incidence of postoperative cognitive dysfunction after robotic surgery procedures is low, it is more prevalent in elderly patients due to their increased susceptibility to elevated intracranial pressure resulting from the Trendelenburg position and pneumoperitoneum.

This condition leads to a reduction in cerebral oxygenation and consequent hypoxic deficit.

A rise in intracranial pressure of >20 mmHg is necessary to produce any negative neurological signs. Intraoperative monitoring of intracranial pressure can better manage this variable, and noninvasive techniques such as measuring optic nerve diameter with ultrasonography can be used for this purpose [9]. The anesthesiologist should guide this approach.

## Ocular injuries

Optic neuropathy is a rare complication with a poor prognosis, characterized by blurry vision or loss of vision in the postoperative period. It is caused by increased intraocular pressure resulting from patient positioning and pneumoperitoneum, which leads to vasodilation of the choroid plexus, resulting in congestion, hypoperfusion, and nerve ischemia. However, permanent loss of vision is rare (less than 0.1%) [6,10].

In addition, corneal injuries are described, with an incidence of up to 0.6%, usually associated with ischemic optic neuropathy. Their occurrence is higher in robotic surgery procedures compared to laparoscopic surgery, and the risk increases in relation to the duration of the surgical procedure and patient positioning [1,11].

## Otorrhagia

Otorrhagia is another category of lesions that may manifest during the perioperative period; however, its occurrence is infrequent. Its etiology remains unclear, but a plausible physiopathological explanation is attributed to an elevation in arterial and venous pressure due to pneumoperitoneum and the Trendelenburg position, resulting in capillary rupture and hematoma within the external auditory canal [12].

It typically occurs bilaterally, without any other associated symptoms. Various risk factors that contribute to its development have been described, among others, arterial hypertension, age, and female gender. A postoperative assessment by otolaryngology is advised to rule out other related symptoms such as otalgia, tinnitus, or hearing loss. Its therapy follows a conservative approach using topical corticosteroids, and no auditory sequelae have been reported [13]. Since identification typically occurs upon removal of surgical drapes, no intraoperative management has been proposed.

## Final considerations for the prevention and management of complications related to patient positioning

To prevent complications related to patient positioning, there should be a checklist involving all participants in the surgical event (surgeon, surgical assistant, anesthesiologist, and nurse), ensuring that each one confirms a proper position.

Once the patient has been intubated, proper positioning is performed. Patients undergoing pelvic surgery usually require the Trendelenburg position. For this, the patient is placed in a supine position and then moved caudally until the sacrum is in line with the bottom of the operating table. The legs are adjusted in the stirrups without hyperextending the hips and ensuring that the heels and knees rest comfortably on the stirrups (Figs. 1, 6 and 7). The legs are secured with Velcro, and the arms are adjusted and should be placed against the body and supported by a gel pad, secured with a Velcro strap or a sheet. The hands should be in a neutral position with the thumb up and a foam pad between the fingers (Fig. 2). Finally, shoulder protection is placed using oily liquid on the probable pressure points, followed by the use of pads and support for the shoulders (Fig. 3). Once it is confirmed that all prevention maneuvers related to the patient have been achieved, the operating table is adjusted to the Trendelenburg position, and the surgical procedure begins.

During procedures that require a lateral position, the rotation of the shoulders and hips should be simultaneous to prevent spinal injuries, and the proper positioning of the cervical pillow and axillary roll should be ensured. Special attention should be paid to the flexion of the arm, which should not exceed 90 degrees, and to the placement of supports and/or cushions on pressure points (elbows, knees, and ankles) [6].

Surveillance of the operative time is essential, as the longer this time, the higher the risk of neuromuscular complications. Special attention should be paid in the postoperative period to surgical procedures with operative times greater than 6h, given their increased risk of complications.

As previously described, depending on the position, there is a risk of skin or neuromuscular injuries. Some common injuries include ulnar or radial nerve injury due to hyperextension of the arms, characterized by loss of sensation and strength in the posterior site of the arm and hand. Another common injury is femoral nerve injury, which typically occurs due to hyperextension of the hip in the lithotomy position. Therefore, it is recommended to use a low lithotomy position [1].

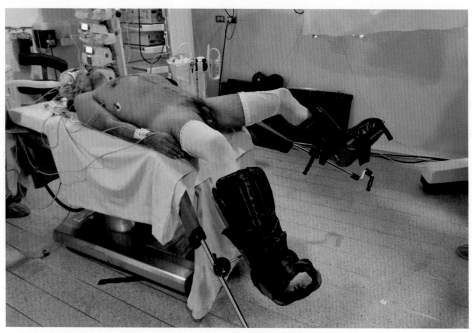

FIG. 1    Safe positioning for legs for surgical interventions involving the pelvic area.

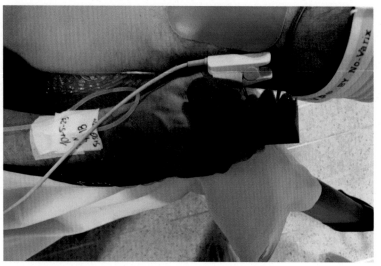

FIG. 2    Neutral position for hands and protection with a foam pad.

If any neurological injury occurs, a neurology and/or physical medicine and rehabilitation evaluation should be sought to determine the need for electromyography to further study the new postoperative condition. Treatment is based on the early initiation of physical therapy and pharmacological therapy with antiinflammatory and antineuropathic medications. Recovery is gradual depending on the demyelination injury and, in the literature, is described a nerve recovery of approximately 1 mm per day after the damage [14,15].

Regarding the management of rhabdomyolysis, the main objective is to limit kidney damage, so the management of intravenous fluids, correction of electrolytes, and correction of metabolic acidosis are aspects that should be prioritized to decrease the risk of kidney injury. Another aspect to take into account is the preoperative monitoring of risk factors (obesity, surgical time, previous renal injury) that should also be done to minimize surgical times if possible, in addition to carefully positioning the patient and having alternatives in case of intraoperative resuscitation [6].

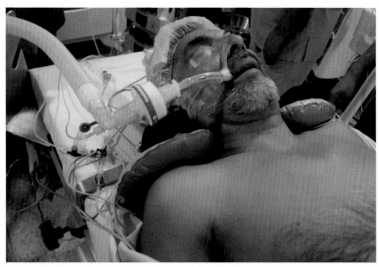

FIG. 3   Protection pads and support for the shoulders.

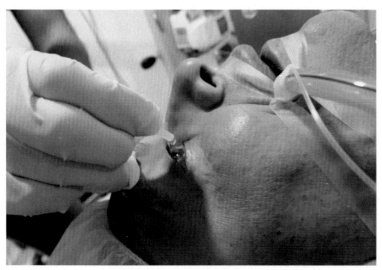

FIG. 4   Use of gel lubricants to maintain eye moisture and ocular patches.

For the management of ocular injuries, the anesthesiology team in the operating room should control the use of gel lubricants to maintain eye moisture and the use of ocular patches to prevent corneal injury, as well as monitor blood pressure levels and surgical time (Figs. 4 and 5). It should be noted that it is possible to pause the surgical procedure to reposition the patient with 5-min intervals in the supine position, and the same applies to the prevention of neurocognitive injuries [16].

The experience of the surgeon and surgical team is crucial in limiting the incidence of complications related to the positioning of the robot; however, if any injury occurs, it should be noted that the majority of them resolve over time, approximately 59.1% within the first month postoperatively, 18.2% between 1 and 6 months, and 22.7% may persist for more than 6 months [17].

## Complications of robotic surgery—Related to robot malfunctioning

### Introduction

With the advancement of technology, new machines and equipment have been developed, which has brought great progress to modern medicine; using this high-tech equipment should look not only at its advantages but also at its

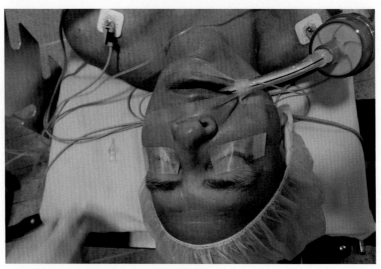

FIG. 5   Use of gel lubricants to maintain eye moisture and ocular patches.

disadvantages, and doctors should be aware of the disadvantages or risk of accident. A glitch in the da Vinci robotic system is one of the flaws, which can lead to different results depending on the severity.

The potential technical advantages of the robotic approach are achieved through complex engineering that is much more complex than laparoscopic instruments, both in terms of hardware and software. In addition, the da Vinci Surgical System (Intuitive Surgical Inc., Sunnyvale, CA) is a complete surgical system solution rather than a set of instruments. Thus, robotic systems are inherently more prone to failure than simpler surgical solutions. As the use of robotics increases, surgeons increasingly rely on computerized systems to function properly to complete cases. Due to reliance on this system, potential failures leading to complications, program aborts, or open transitions are a concern.

According to Intuitive Surgical's 2013 annual report, since 2000, 1.75 million procedures have been performed across specialties in the United States [18]. Surgical robots are able to perform complex minimally invasive procedures with improved visualization, greater precision, and improved skill compared to laparoscopy.

The da Vinci surgical robot is currently the only surgical robot approved by the US Food and Drug Administration (FDA) to perform various types of urological, gynecological, cardiothoracic, and head and neck surgical procedures [19].

## Robotic surgery adverse event facts according to FDA MAUDE

The Manufacturer and User Facility Device Experience ("MAUDE") Database is a public collection published by FDA to document suspected adverse events caused by devices; reporting is mandatory for manufacturers and distributors, but voluntary for healthcare professionals and customers. [20].

As a self-reporting system database, FDA MAUDE suffers from underreporting and inconsistencies [21–23]. However, it provides valuable information about what actually happens during robotic surgery and how it affects patient safety. Data provided by MAUDE on deaths, injuries, and poor device performance can be used as a sample to estimate a lower bound on the incidence of adverse events and to determine their causes and impact on patients and procedures.

Several studies of the FDA-MAUDE database have been conducted over time; the largest was recently completed, and 2.9 million records were analyzed [18]. The correlation between these studies is depicted in Table 1.

## Instrument or robotic device failure—Mortality and injury association

Malfunctions are infrequent, and the necessity of aborting or converting to another technique is rare; the variety and effects of these device malfunctions have varied over time.

Overall, the frequency of equipment malfunctions ranged from 0.5% to 4.6%; this percentage varied depending on the experience of the surgeons and the number of cases they had performed. [31–33]. In the review panel at the University of Illinois [18], five primary causes of malfunction were identified, either that they negatively impacted patients, either by causing injuries or complications, or by disrupting the procedure or prolonging the time required

TABLE 1   Some studies conducted over time with the FDA-MAUDE database.

| Study | No. reports (years) | System under study | Surgical specialties |
|---|---|---|---|
| Murphy et al. [24] | 38 system failures, 78 adverse events (2006–2007) | da Vinci system | N/A |
| Andonian et al. [25] | 189 (2000–2007) | ZEUS and da Vinci | N/A |
| Lucas et al. [26] | 1914 (2003–2009) | da Vinci system models dV and dVs | N/A |
| Fuller et al. [27] | 605 (2001–2011) | da Vinci system | N/A |
| Friedman et al. [28] | 565 (2009–2010) | da Vinci instruments | N/A |
| Gupta et al. [29] | 741 (2009–2010) | da Vinci system | Urology, gynecology |
| Manoucheri et al. [30] | 50 injuries/deaths (2006–2012) | da Vinci system | Gynecology |
| Alemzadeh et al. [18] | 10,624 (2000–2013) | Da Vinci robotic systems and instruments | Gynecology, urology, general, colorectal, cardiothoracic, and head and neck surgery |

FDA MAUDE is a self-reporting system database about robotic surgery adverse event facts.

for surgery. This division also has the greatest influence in the analysis of robot failures, as it encompasses the greatest number of cases studied (2.9 million records) and the highest number of adverse events reported (10,624).

As the da Vinci robotic system comprises multiple software and hardware components, the downfall is that each component can lead to an unintended failure.

- **Errors in the system and video/audio glitches**
  - These are the most common, accounting for 7.4% of adverse events, and were the leading cause of surgical interruptions, including the number of reboots in the system, the conversion of proceedings to a nonrobotic approach (59.2% of all conversions), and the need to reschedule or abort procedures (81.8% of all cases).
  - System failures have increased as a result of the existing security mechanisms in place because of troubleshooting that cannot be corrected automatically; the majority of the time this is remedied with a manual system reboot (recoverable error); however, there are instances when the robotic procedure must be halted (unrecoverable error).
  - The most common cause of failure was a malfunction in the robotic arm and joint system (71.4%) [33]. The arm and optical systems were the primary culprits of inoperability [34].
- **Falling pieces or burnings in the patient's body**
  - They accounted for approximately 1557 (14.7%) of the adverse events. In almost every instance, the procedure is halted, and the surgical team spends time searching for the missing pieces in order to recover them from the patient (in 119 instances, an injury to the patient was reported, and in 1 instance, a death was reported).
- **Devices that generate electricity via arcs, flames, or burning**
  - Regarding burns or holes on tips covers, they accounted for 1111 incidents (10.5%) of the reports, which resulted in around 193 injuries, such as burn tissue.
  - Instrumental failures involve accessories to the insulation tip, such as the monopolar scissors. Studies showed 25% to 33% of insulation failure accessories after a single surgical use. Single use of each insulation device is recommended [35].
- **Unintended operation of instruments.**
  - Unbridled movements and spontaneous switch On or Off occurred in 1078 adverse events (10.1%), including 52 injuries and 2 deaths.
- **Malfunctions that could not be classified into another criteria, such as breaking of instrument's string.**

Despite a fairly high number of reports, the vast majority of procedures were successful, and no problems were present. In the analysis with most reports, the number of injuries or death events per procedure has remained fairly stable since 2007. Still, the total number of failures reported per procedure (0.46%) was six times below the average number of malfunction by the procedure (3%). Also, the total number of injuries and deaths reported by the procedure (0.08%) was nearly the same as that prognosticated for robotic surgery complications [36], but in a lesser magnitude than the lower rate of complications reported for robotic surgery in former studies (2%) [37].

## Precautionary or recuperative system measures

In practice, the use of a robotic platform interface is a sophisticated machine with surgeons in an area of patient care and safety. From a technological point of view, the use of security practices and controls, mainly bettered in its design, operation safety, and confirmation of robotic surgical systems, could help situations of failure and its consequences.

## Some recommendations to minimize patient's risk

- Human-machine interfaces and improved surgical simulators that train surgical brigades to handle specialized problems [38,39] and estimate their conduct in real time during surgery.
- Report real-time information to the surgeon on the anatomical paths and safe actions that can be taken.
- By having previous knowledge of the anatomical limits that help robotics tools to advance further than anticipated, leading to dangerous situations for the case by entering certain areas of the workspace during surgery [40], this is achieved grounded in case specific anatomical models which are programmed depending on the surgery to be performed and tracking surgeon's surgical movements at the robotic console while using simulators and live cases performed preliminarily [41].
- Take into account new security machines for monitoring procedures (including the surgeon, the case, and device status) and give complete information to the surgical platoon about events and troubleshooting procedure, therefore precluding dislocations.
- All health professional working in the area of robotic surgery should know the process for the enrollment and announcement of educated incidents during the proceedings, to be more precise in security information, effectiveness on surgical systems and by learning from situations formerly endured by other surgeons.
- The learning curve might be an implicit factor in these malfunctions and might be related to misplaced port planning, docking fashion, and improper movement of the arms or misunderstanding toward the limited range of motion. Unfamiliarity might result in a total shutdown of the robotic system. Having the knowledge helps one to take protective actions.
- When the robotic system shuts off constantly, the arms, the robot position, and ports must be repositioned precisely. These malfunctions in which repeated stops are reported, are common during the first procedures; once the learning curve is overcome, the reasons for malfunctions are mainly mechanical dysfunction or overuse of instruments. Maintenance and regular updates are essential to avoid these problems.
- Discussion with patients and their families regarding pitfalls of mechanical failure and indispensable surgeries is important before surgery [42].
- It is recommended that at the time of patient's admission to the operating room, the robotic system is powered up and has proven its proper functioning. The day before any surgery, the machine should be tested. However, the technician should be communicated immediately and expressed of the problem for results, if a malfunction occurs any time.
- If a critical error occurs, conversion to open or laparoscopic surgery is an alternative. However, rescheduling is another option, if the failure occurs before induction of anesthesia. Eventually, if finances allow, an alternate robot da Vinci is another option.

## Final considerations on complications of robotic surgery related to robot malfunctioning

The robotic surgical systems have been successfully espoused in numerous surgical specialties. It is an extremely safe and dependable system for surgery in multiple specialties. It should be noted that the malfunction is rare and the threat of critical failure is veritably low. The total number of injuries and death events per procedure has been fairly constant over the times.

Knowledge of adverse situations that can happen is critical since it will allow the necessary measures to help and avoid that those events occur, indeed more when it is known that the malfunction of bias and instruments have affected thousands of cases and surgical teams causing complications and prolonged surgery times. As surgical robotic systems continue to evolve with new technologies, standard and uniformed disciplined habits in training surgical equipment, more advanced human–machine interfaces, bettered accident disquisition, reporting mechanisms and design ways interfaces grounded on security, incident rates should be reduced in the future.

Robot-assisted surgery has brought new implicit specialized problems for the surgeon, but most of these problems can be corrected or temporarily overwhelmed to complete the operation. Robotic surgery provides a safe way of minimally invasive treatment.

## The future scenario

### Robotic surgery and prevention of surgical complications: A urological perspective

Robotic surgery has emerged as an innovative paradigm in the field of urology, fostering significant advancements in the precision and safety of surgical procedures. Attention has been directed particularly toward the prevention of surgical complications and the optimization of patient positioning, which has led to the exploration of novel methodologies and technologies to further enhance clinical outcomes.

In relation to the innovative approach of the extended reality that has a potential for the optimization of patient positioning in urological robotic surgeries, the aim is to harness the capability of robotic systems to automatically adjust patient positioning according to the robot's rotational workspace. By integrating extended reality, greater precision in patient alignment is achieved, thereby minimizing risks associated with poor positioning and establishing a robust foundation for future research in this domain.

As robotic surgery advances toward automation, a critical question arises regarding the ongoing role of healthcare professionals. In the urological realm, collaboration between surgeons and robotic systems is a cornerstone. While automation can enhance accuracy and efficiency, clinical decision-making and adaptability to unforeseen situations remain inherently human attributes. The symbiotic interaction between robots and surgeons will enable safer and more effective urological surgery. Additionally, concerning complications due to patient positioning, the use of sensors at pressure points will help prevent injuries that were previously considered inadvertent [43].

In summary, the future of robotic surgery in urology appears as a synergy between technological automation and human clinical expertise. These advancements promise to propel the ongoing evolution of urological robotic surgery, enhancing outcomes and the quality of life for patients with urological conditions [44].

Key points on complications of robotic surgery—Related to patient positioning

- Doing a checklist involving all surgical team members is essential for proper patient positioning and preventing complications.
- Trendelenburg position is commonly used for pelvic surgeries; keep in mind patient positioning involves supine placement, and secured with a Velcro strap or a sheet and the entire body needs to rest comfortably on the stirrups.
- Lateral position requires synchronized shoulder and hip rotation, correct cervical pillow, and axillary roll placement.
- Prolonged operative times increase the risk of neuromuscular complications, particularly in surgeries exceeding 6h.
- Skin and neuromuscular injuries are associated with different positions; femoral and ulnar/radial nerve injuries are common.
- Neurological injuries should be evaluated by neurology or physical medicine specialists, and treatment includes early physical therapy and medication.
- Rhabdomyolysis management prioritizes fluid balance, electrolyte correction, and metabolic acidosis treatment to mitigate kidney damage.
- Prevention of ocular injuries includes maintaining eye moisture and using patches, alongside monitoring of blood pressure and surgical time.
- Experience of the surgeon and the team is crucial to minimize positioning-related complications; most injuries resolve over time, with varying incidence rates.

Key points on complications of robotic surgery—Related to robot malfunctioning

- Robotic surgical systems are widely used across various specialties, offering safety and reliability.
- Awareness of potential adverse situations of robot malfunction is crucial for taking necessary measures to address complications and prolonged surgery times.

- Evolution of surgical systems, including advanced technology, training, human-machine interfaces, and reporting mechanisms, should reduce incident rates of adverse events.
- The learning curve may be an underlying factor of robotic failures. Increased experience and knowledge gives the opportunity to take protective actions.
- If a critical error occurs, conversion to open or laparoscopic surgery is an alternative. However, before induction of anesthesia, rescheduling is another option.

# Appendix

See Figs. 6 and 7.

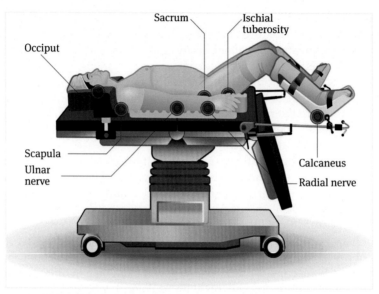

FIG. 6    Pressure points for protection in Trendelenburg position.

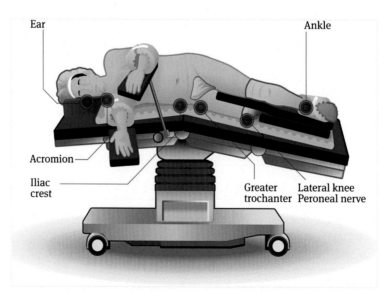

FIG. 7    Pressure points for protection in lateral decubitus position.

# References

[1] Tourinho-Barbosa RR, et al. Complications in robotic urological surgeries and how to avoid them: a systematic review. Arab J Urol 2018;16 (3):285–92.

[2] Bauman TM, Potretzke AM, Vettter JM, et al. Pulmonary disease increase risk of complications with robotic partial nephrectomy. J Endourol 2016;30(3):293e9. https://doi.org/10.1089/end.2015.0534 [Epub 2015 Dec 17].

[3] Lim PC, Kang E. How to prepare the patient for robotic surgery: before and during the operation. Best Pract Res Clin Obstet Gynaecol 2017. https://doi.org/10.1016/j.bpobgyn.2017.04.008.

[4] Chughtai B, Isaacs AJ, Mao J, Lee R, Te A, Kaplan S, et al. Safety of robotic prostatectomy over time: a national study of inhospital injury. J Endourol 2015;29:181–5.

[5] Sotelo RJ, Haese A, Machuca V, Medina L, Nuñez L, Santinelli F, et al. Safer surgery by learning from complications: a focus on robotic prostate surgery. Eur Urol 2016;69(2).

[6] Raed A, Azhar an M. Elkoushy. Complicaciones of positioning. Chapter 9. In: Complications in robotic urologic surgery. Springer; 2018.

[7] Reisiger KE, Landman J, Kibel A, Clayman RV. Laparoscopic renal surgery and the risk of rhabdomyolysis: diagnosis and treatment. Urology 2005;66(5 Suppl):29–35. https://doi.org/10.1016/j.urology.2005.06.009. PMID: 16194704.

[8] Shaikh S, Nabi G, McClinton S. Risk factors and prevention of rhabdomyolysis: after laparoscopic nephrectomy. BJU Int 2006;98(5):960–2.

[9] Kim MS, Bai SJ, Lee J-R, Choi YD, Kim YJ, Choi SH. Increase in intracranial pressure during carbon dioxide pneumoperitoneum with steep trendelenburg positioning proven by ultrasonographic measurement of optic nerve sheath diameter. J Endourol 2014;28:801–6.

[10] Zillioux JM, Krupski TL. Patient positioning during minimally invasive surgery: what is current best practice? Robot Surg 2017;4:69–76. https://doi.org/10.2147/RSRR.S115239. PMID: 30697565. PMCID: PMC6193419.

[11] Weber ED, Colyer MH, Lesser RL, Subramanian PS. Posterior ischemic optic neuropathy after minimally invasive prostatectomy. J Neuroophthalmol 2007;27(4):285–7. https://doi.org/10.1097/WNO.0b013e31815b9f67. PMID: 18090562.

[12] Aloisi A, Pesce JE, Paraghamian SE. Bilateral otorrhagia after robotically assisted gynecologic surgery in the setting of a reduced Trendelenburg position and low-pressure pneumoperitoneum: a case report and review of the literature. J Minim Invasive Gynecol 2017;24(7):1229–33. https://doi.org/10.1016/j.jmig.2017.04.009.

[13] Jones WS, Klafta JM. Bilateral bloody otorrhagia after robotic-assisted laparoscopic prostatectomy. A A Case Rep 2015;5(6):91–2. https://doi.org/10.1213/xaa.0000000000000.

[14] Winnfree CJ, Kline DG. Intraoperative positioning nerve injures. Surg Neurol 2005;63:5–18.

[15] Li J-R, Cheng C-L, Weng W, Hung S, Yang C-R. Acute renal failure after prolonged pneumoperitoneum in robot-assisted prostatectomy: a rare complication report. J Robot Sur 2008;1(4):313–4. https://doi.org/10.1007/s11701-007-0060-8.

[16] Gkegkes ID, Karydis A, Tyritzis SI, Iavazzo C. Ocular complications in robotic surgery. Int J Med Robot Comput 2015;11:269–74.

[17] Mills JT, Burris MD. Warburton dj et at. Positioning injuries associated with robotic assisted urological surgery. J Urol 2013;190:580–4.

[18] Alemzadeh H, Raman J, Leveson N, Kalbarczyk Z, Iyer RK. Adverse events in robotic surgery: a retrospective study of 14 years of FDA data. PloS One 2016;11(4), e0151470. https://doi.org/10.1371/journal.pone.0151470.

[19] The da Vinci surgical system, Intuitive Surgical Inc.; http://www.intuitivesurgical.com/products/davinci_surgical_system/.

[20] MAUDE: Manufacturer and user facility device experience, U.S. Food and Drug Administration. http://www.accessdata.fda.gov/scripts/cdrh/cfdocs/cfMAUDE/search.CFM.

[21] Adverse event reporting of medical devices. U.S. Department of Health and Human Services, Office of Inspector General; 2009, October [OEI-01-08-00110] https://oig.hhs.gov/oei/reports/oei-01-08-00110.

[22] Hauser RG, et al. Deaths and cardiovascular injuries due to device-assisted implantable cardioverter-defibrillator and pacemaker lead extraction. Europace 2010;12(3):395–401. https://doi.org/10.1093/europace/eup375. PMID: 19946113.

[23] Cooper MA, et al. Underreporting of robotic surgery complications. J Healthc Qual 2013.

[24] Murphy D, et al. Complications in robotic urological surgery. Minerva urologica e nefrologica = Ital J Urol Nephrol 2007;59(2):191–8. PMID: 17571055.

[25] Sero A, et al. Device failures associated with patient injuries during robot-assisted laparoscopic surgeries: a comprehensive review of FDA MAUDE database. Can J Urol 2008;15(1):3912.

[26] Lucas SM, Pattison EA, Sundaram Chandru P. Global robotic experience and the type of surgical system impact the types of robotic malfunctions and their clinical consequences: an FDA MAUDE review. BJU Int 2012;109(8):1222–7. https://doi.org/10.1111/j.1464-410X. 2011.10692.x. PMID: 22044556.

[27] Andrew F, Vilos GA, Pautler SE. Electrosurgical injuries during robot assisted surgery: insights from the FDA MAUDE database. SPIE BiOS 2012;8207:820714.

[28] Friedman Diana CW, Lendvay TS, Blake H. Instrument failures for the da Vinci surgical system: a Food and Drug Administration MAUDE database study. Surg Endosc 2013;27(5):1503–8. https://doi.org/10.1007/s00464-012-2659-8. PMID: 23242487.

[29] Priyanka G, et al. 855 adverse events associated with the davinci surgical system as reported in the fda maude database. J Urol 2013;189(4), e351.

[30] Elmira M, et al. MAUDE-analysis of robotic-assisted gynecologic surgery. J Minim Invasive Gynecol 2014;21(4):592–5. https://doi.org/10.1016/j.jmig.2013.12.122. PMID: 24486535.

[31] Zorn KC, Gofrit ON, Orvieto MA, et al. Da Vinci robot error and failure rates: single institution experience on a single three-arm robot unit of more than 700 consecutive robot-assisted laparoscopic radical prostatectomies. J Endourol 2007;21:1341–4.

[32] Borden Jr LS, Kozlowski PM, Porter CR, Corman JM. Mechanical failure rate of da Vinci robotic system. Can J Urol 2007;14:3499–501.

[33] Chen C-C, Yen-Chuan O, Yang C-K, Chiu K-Y, Wang S-S, Chung-Kuang S, Ho H-C, Cheng C-L, Chen C-S, Lee J-R, Chen W-M. Malfunction of the da Vinci robotic system in urology. Int J Urol 2012;19:736–40.

[34] Lavery HJ, Thaly R, Albala D, et al. Robotic equipment malfunction during robotic prostatectomy: a multi-institutional study. J Endourol 2008;22. 2165–8.24.

[35] Engebretsen SR, Huang GO, Wallner CL, Anderson KM, Schlaifer AE, Arnold II DC, Olgin G, Baldwin DD. A prospective analysis of robotic tip cover accessory failure. J Endourol 2013;27(7):914–7.

[36] Clare R, et al. Relative effectiveness of robot? Assisted and standard laparoscopic prostatectomy as alternatives to open radical prostatectomy for treatment of localized prostate cancer: a systematic review and mixed treatment comparison metaanalysis. BJU Int 2013;112(6):798–812. https://doi.org/10.1111/bju.12247. PMID: 23890416.

[37] Stefan B, et al. Robotic-assisted versus laparoscopic cholecystectomy: outcome and cost analyses of a case-matched control study. Ann Surg 2008;247(6):987–93. https://doi.org/10.1097/SLA.0b013e318172501f. PMID: 18520226.

[38] Homa A, et al. A software framework for simulation of safety hazards in robotic surgical systems. SIGBED Review 2015. (Special Issue on Medical Cyber Physical Systems Workshop) 12.4.

[39] Alemzadeh H, et al. Simulation-based training for safety incidents: lessons from analysis of adverse events in robotic surgical systems. In: American College of Surgeons' 8th Annual Meeting of the Consortium of ACS-accredited Education Institutes; 2015, March.

[40] Taylor RH, et al. Medical robotics and computer-integrated surgery. In: Springer handbook of robotics. Berlin Heidelberg: Springer; 2008. p. 1199–222.

[41] Lin HC, Shafran I, Yuh D, Hager GD. Towards automatic skill evaluation: detection and segmentation of robot-assisted surgical motions. Comput Aided Surg 2006;11(5):220–30. PMID: 17127647.

[42] Kaushik D, High R, Clark CJ, LaGrange CA. Malfunction of the da Vinci robotic system during robot-assisted laparoscopic prostatectomy: an international survey. J Endourol 2010;24:571–5.

[43] Fiorini L. Advancement on human-robot interaction: perception, cognitive architecture and field tests. Int J Soc Robot 2023;15:369–70.

[44] Żelechowski M, Faludi B, Karnam M, et al. Automatic patient positioning based on robot rotational workspace for extended reality. Int J CARS 2023. https://doi.org/10.1007/s11548-023-02967-2.

# 31

# Troubleshooting in robotic surgery and complications prevention

*Valeria Celis[a], Aref S. Sayegh[a], Jaime Poncel[a], Luis G. Medina[a], Omaira Rodriguez[b], and René Sotelo[a]*

[a]Catherine and Joseph Aresty Department of Urology, USC Institute of Urology, Keck School of Medicine, University of Southern California, Los Angeles, CA, United States [b]Robotic Surgery Program, University Hospital of Caracas, Medicine Faculty, Central University of Venezuela, Caracas, Venezuela

## Introduction

With advances in technology, new machines and devices have been developed, and robotic surgery certainly represents the state-of-the-art technological innovation in the surgical field over the past two decades, which has led to the implementation and great advances of this technology worldwide [1,2]. The number of robotic-assisted procedures is exponentially increasing in all surgical specialties in the last 20 years [3,4]. The use of robotic surgery has increased significantly, providing many benefits over traditional surgical methods. These benefits include increased precision, less invasiveness, reduced blood loss, and shorter hospital stays. However, the use of such a high-technology platform should not only be based on its advantages but also on its shortcomings, and surgeons should be aware of the risks of malfunction or unexpected events [2]. Therefore, it is crucial to be well-versed in how to approach and effectively manage the problems that can arise when using a robotic surgical system.

Robotic equipment malfunction can occur in 0.5%–4.6% of cases [5]. Despite having the Manufacturer and User Facility Device Experience (MAUDE) database in which adverse events derived from the robotic platform are recorded and reported, the actual incidence still suffers from underreporting and inconsistencies. The documentation of these events will aid in the creation of guidelines for identifying, preventing, and managing similar incidents in the future.

Troubleshooting involves finding the real cause of a problem instead of just addressing the symptoms and then developing a solution to resolve the root cause. Robot malfunction can drastically reduce the efficiency and safety of the workspace.

This chapter provides a comprehensive review of the various troubleshooting techniques used in robotic surgery. Also, how to identify the most common sites and types of failures and technical issues that may arise during the robotic procedure, including hardware and software issues with the da Vinci Surgical System.

## System errors

The da Vinci Surgical System (Intuitive Surgical Inc., Sunnyvale, CA) performs self-monitoring to detect technical and mechanical errors. When a fault is detected, the system will determine if it is recoverable or nonrecoverable, alert the surgical team by sounding a series of error beeps, displaying a notification describing the error on the monitors and flashing color lights (usually yellow for recoverable or red for nonrecoverable errors) on all arms or just in one if it is an arm-specific fault. Additionally, it will lock all the arms making the instrument harder to move, therefore preventing intraoperative lesions.

Nonrecoverable errors tend to be related to some type of electric or software malfunction; examples are power outages, power supply voltage out of range, and a processor not completing a step during system startup within the allotted time. If a fault is nonrecoverable, a system restart will be required to resolve the issue. A message will be displayed on the screen monitors. For example: "Non-recoverable fault: XXXX Restart System to continue" (Fig. 1). The system can be restarted without having to remove any of the instruments from the patient. First, press the Power button on any system component to power off the system; this will take several seconds. When complete, all system Power buttons will be lit yellow (Fig. 2), indicating standby mode, and readiness for restart. At this point, the Power button should be pressed once more to restart the system. It is important to keep in mind that during the system restart, the video component at the Surgeon Console viewer and touchscreen monitor is temporarily unavailable [6].

To override recoverable faults, all that is needed is to tap the "Recover Fault" option on the touchpads or touchscreen. Doing this will silence the alarms and have the system recover after a few seconds. If the fault condition remains, the system will immediately raise the alarms again. Although this troubleshooting procedure is very simple and effective, it should never be carried out before fully understanding the cause that triggered the fault. Overriding a fault without understanding its cause may result in uncontrolled movement of the arms up to 2 cm, or uncontrolled motion of the master hand controls up to 5 cm [6].

FIG. 1    Nonrecoverable fault message. Message displayed on the screen monitors when a nonrecoverable fault occurs.

FIG. 2    Power button lit yellow. Indicates that the robot is in standby mode and is ready to be restarted.

**A**                              **B**

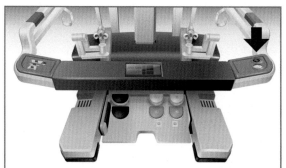

FIG. 3   Emergency stop button. *Arrow: location of the* Emergency Stop button on the surgeon's console (A) and patient cart (B).

Some recoverable errors that can present during robotic-assisted surgery (RAS) comprise robotic arm malfunctions, detection of power fluctuation, low battery backup, and on/off failure. Of all the robotic system components, the robotic arm and joint system is the one that most commonly malfunctions [7,8]. A collision or clashing between surgical arms, difficulty in determining the position of an instrument arm, extreme angulation causing unusual twisting on the instrument, or excessive force on the surgical arm can all trigger an error message that renders the specific mechanism immobile until the error is acknowledged. Keeping a proper distance between surgical instruments and robot arms and maintaining appropriate approach angles is crucial to prevent errors of this type. In general, the da Vinci S and Si ports should be separated by a distance of 8–10 cm, while for the da Vinci Xi, a distance of 6–10 cm between ports is enough [6,9]

If an arm-specific error occurs and cannot be corrected, the system displays a disabled arm option on the touchscreen or touchpad. This feature lets the surgeon complete a procedure with the remaining arms. Once the arm has been disabled, the arm cannot be reenabled until the next power cycle, but the arm clutch and port clutch buttons can still be used to move the arm out of the way [6].

Another malfunction that may arise is a nonresponsive system. In these circumstances, first, you should check the screen for messages to see if the system is performing an action (for example, downloading instrument data or performing automatic calibration of the endoscope). If this is not the case, proceed to press the Emergency Stop button on the patient cart or surgeon console (Fig. 3). A "Resume Use" option will appear on the touchpads or touchscreen, tap it, and confirm that the system is functioning properly. In cases where the problem persists, the next step is performing a system restart in the same fashion as described for nonrecoverable errors. If the system fails to restart, you will need to perform a "hard-cycle power on". To accomplish this, you will need to Remove all alternant current (AC) power from the Vision Cart and Surgeon Console by switching each AC power switch to off (indicated by "O" near each switch) (Fig. 4). Then press the emergency power off (EPO) button on the Patient Cart (it will remain partially pressed in) to

## AC power switches

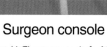

Surgeon console          Patient cart          Vision cart (at base)

FIG. 4   Alternant current (AC) power switch. *Yellow Circle: location of the* AC power switch on the surgeon console, patient cart, and vision cart, respectively.

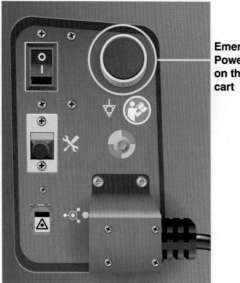

FIG. 5    Emergency power off (EPO) Button. *Yellow Circle*: location of the EPO button on the patient cart.

remove all power (Fig. 5). Wait 2 s and then press the EPO button again to reset it (it will rebound to its fully extended ready position). Make sure the Vision Cart, Surgeon Console, and Patient Cart AC power switches are in the "on" position. After 30 s, all three system components should return to default standby mode (power buttons will be yellow) and the robot can be powered on normally by pressing the power button [6].

In the rare case you need to abort the robotic approach, you should remove all the instruments including the camera, then disconnect the ports from the robotic arms, and retract them from the patient, followed by removal of the robot from the area. All conversions should be reported to Intuitive's da Vinci Surgery Technical Assistance Team (dVSTAT).

## Video/Image problems

There are mainly two video/image problems, either no image being shown or the image quality is suboptimal. The first is usually due to an electrical/cable fault. Ensure the vision cart is plugged in and all its components (endoscope controller, video processor, core) power switches are in the "on" position. Check that fiber cables (blue) are correctly connected between the vision cart's core, the surgical cart, and the surgeon console. If none of this seems to be the cause, there may be a problem with the operation of the internal power supply, in which case you will need to contact the dVSTAT.

When it comes to problems in the quality of the image, the most common one is blurriness that usually comes from smudges or condensation on the surface of the lens. Fogging can occur due to a difference in temperatures between the endoscope tip and the pneumoperitoneum. Intraoperative cleaning of the camera can help resolve these issues; this should be done with a soft cloth. Additionally, briefly submerging the tip of the endoscope in warm water (<131°F or 55°C) and using heated insufflation can help avoid fogging of the camera lens.

Less commonly, the reason behind a low-quality image may be incorrect image settings. In the settings option on the surgeon console or vision cart (Fig. 6), you can adjust the brightness (with the brightness slider), the digital zoom (by selecting ×1), and the white balance (restore to factory defaults). If all else fails, another endoscope should be tried.

Apart from the troubleshooting for the specific situations mentioned above, in order for the Vinci Surgical System (Intuitive Surgical Inc., Sunnyvale, CA) to operate optimally, it is necessary to perform hardware components maintenance and software system updates routinely.

## Instrument malfunction

Any inherent defect in a robotic instrument that restricts its normal operation is called instrument malfunction. Correctly assessing these events' incidence can be challenging because, in many opportunities, they can go unnoticed and

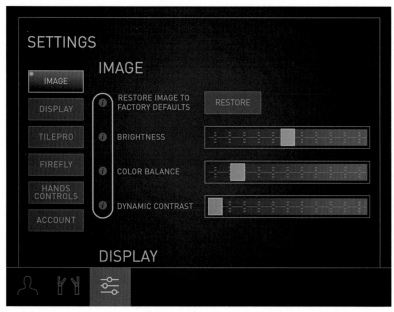

FIG. 6    Image settings display. Display in the surgeon console or vision cart where you can adjust the brightness and the white balance.

do not result in patient injury or complications [10,11]. Additionally, even if they are identified, the reporting is not mandatory, leading to underreporting [10,12]. Reported rates of instrument malfunction are between 0.25% and 1.1% [7,11].

Proper instrument installation is indispensable to avoid triggering errors and possible instrument damage. The endoscope/camera should be placed first so that instrument installation can be done under direct visualization. Instruments must be inspected thoroughly for any damage prior to their insertion. An experienced robotic surgical team with knowledge about instrument functions and proper instrument handling is an invaluable asset in the identification of damages and malfunctions before surgery.

During the loading process, instruments need to be adequately placed by inserting the closed tip into the port and sliding the instrument housing into the sterile adapter. Insertion should be stopped immediately if resistance is met. If any errors occur, the system will make them known by showing an error message on the screen and triggering visual and audible alarms. In these cases, remove the instrument, confirm the instrument hub is positioned properly, and then reload the instrument.

Instrument malfunctions can be classified into electrical or mechanical.

## Electrical malfunctions

Instrument electrical malfunctions represent around 10% of all robotic equipment malfunctions reported in the MAUDE [5], 90% of the time they are manifested as an electric arcing [10], a form of electric discharge that occurs when electricity flows through the air from one conductive point to another due to defects in insulation. Temperatures 700°C and higher can be generated by a single spark. This can result in severe thermal tissue damage, especially in cases of hollow organ involvement, as the mucosa is particularly sensitive, and even one single arc can damage it, producing a perforation instantly or in the 3–15 days to follow due to damage of the vascular supply [13].

The most likely times when the instrument's insulation can be damaged are during instrument handling (installation on the robotic arm and reprocessing) or secondary to prolonged use of coagulation current. Regarding location, the two sites where insulation defects can present are the shaft and the tip cover accessory (TCA) [14].

The TCA is a single-use sleeve that provides insulation to the metallic joint of monopolar curved scissors preventing, in this way, electric arcing from the wrist and alternate site burns [15]. Problems with this accessory are mainly due to wrongful colocation or damage by a member of the surgical team. In our experience, human error leading to the incorrect placement of TCAs on robotic instruments is the main cause of TCA-related electrical malfunctions. This can occur due to a variety of factors, such as lack of training, distraction, fatigue, or miscommunication among surgical team members. In some cases, the TCA may not be properly aligned with the instrument's tip, or it may be attached incorrectly overriding part of the shaft of the instrument due to a misinterpretation of the instructions.

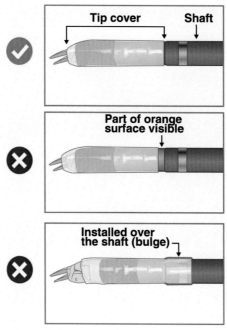

FIG. 7   Tip Cover Accessory (TCA) placement. *Green Checkmark*: illustrates the correct placement of a TCA. *Red Cross*: 2 most common mistakes when placing a TCA.

Surgical teams should be trained on the proper use and placement of tip cover accessories, including how to attach and remove them. To correctly place a TCA with the help of the included applicator first, the scissor blades must be closed, and the instrument wrist straightened. Then the TCA, with the help of the installation tool, should be slid onto the distal end of the instrument. When correctly placed, the TCA should completely cover (without going beyond) the orange surface on the distal end and the wrist of the instrument (Fig. 7) [15]. Incorrect placement can cause damage to the instrument. The accessory may interfere with the instrument's movement, causing it to malfunction or break. Additionally, if the accessory is not properly attached, it can come loose during the procedure, potentially causing injury to the patient.

Another thing to keep in mind to help prevent damage to the insulating capacity of the TCA is to maintain electrocautery settings within the tolerated range. Intuitive Surgical Inc. (Sunnyvale, CA). recommends settings that do not exceed 3 kV [15]. Even when doing this, the possibility of damage is not zero. Using the lowest power setting for the shortest duration possible while still yielding the desired effect is recommended.

Sometimes it can be difficult to determine where the spark is coming from, and therefore, it may not be possible to establish at first glance where the insulation defect is located. The correct approach in this instance is to remove the instrument from the patient and do a thorough inspection looking for visible insulation defects. If none are found, both the instrument and the TCA should be replaced.

It is imperative to make an early diagnosis and effectively manage instruments' electrical malfunctions, as this hampers tissue damage and the potential for complications. Thermal injuries must be carefully evaluated, treated, and followed over time. It is important to remember that the full extent of the damage may not be evident initially. Thermal injuries can damage the vasculature, which supplies a larger area than the one that was initially burned. Ischemia and necrosis take days to weeks to show their full extent [13].

## Mechanical malfunctions

This type of robotic instrument malfunction comprises any physical defects that compromise the normal function and range of motion of the accessory. This in turn hinders the surgeon's dexterity and ability to carry out specific movements during RAS. Mechanical malfunction can affect any part of the instrument, but the most commonly reported sites of malfunction are the wrist and end effector [10].

Wrist malfunctions mainly affect the range of motion. When it comes to end effector malfunctions, as expected, what it affected is the specific function carried out by the damaged instrument. The instruments that constitute the majority of reported malfunctions are the Prograsp forceps (16.3%), monopolar curved scissors (15.6%), Maryland bipolar (12.3%), and fenestrated bipolar (11.7%). The lowest percentages of malfunctions are seen in the crocodile grasper (0.8%) and tenaculum forceps (0.5%) [3]. In general, any instruments with an articulated jaw can be more susceptible to malfunctions of the end effector and presentation can vary greatly.

Another situation that may present secondary to end effector mechanical malfunction is the dislodging of broken instrument pieces into the patient's body. It has been reported that this constitutes about 14.7% of all the adverse events in RAS [5] and represents the cause of 16.8% of reported injuries [3]. Nevertheless, is hard to say if this is an accurate representation of its incidence or if they are simply more frequently reported because they are more easily noticed.

In most cases, it is possible to identify the dislodged fragment and extract it. However, depending on the size of the broken piece, retrieval can be difficult. The surgeon must immediately stop the surgical procedure to locate the broken piece and remove it from the patient's body because any additional manipulation could accidentally push the fragment deeper [16], complicating its retrieval and increasing the possibility of injury [17].

When examining the surgical field to localize the broken piece, it should be started in the last region that was manipulated, followed by a systematic evaluation of the rest of the areas by quadrants, all this while avoiding the mobilization of any structures [18].

If localization of the broken piece is unsuccessful with the above-mentioned method, intraoperative image modalities can be used such as fluoroscopic images [19]. Taking fluoroscopic images in both the anterior–posterior and lateral planes can help locate the fragment. Patient repositioning and the need for an available radiologist are disadvantages of using intraoperative imaging studies. It may be necessary to convert to an open procedure if the surgeon cannot find the broken piece using fluoroscopic imaging or if fluoroscopic imaging is unavailable [16].

Although not common, shaft malfunctions can arise. Some reports state around 13.5% of instrument malfunctions occur on the shaft [10]. Collision/clashing of the instrument with the robotic ports or arms can cause the shaft to bend, peel, or break.

Whenever an instrument malfunction is noticed, no matter the type, location, or size, the instrument should be immediately removed and discarded. Attempting to unbend, reattach, or fix the instrument in any way should not be considered. For this reason, a backup set of all instruments needed for surgery should be procured prior to the start of any RAS in case replacement is needed.

Malfunctions can happen at any time during surgery. Even though there is no way to completely eliminate them, taking measures to prevent or minimize their incidence is the best approach.

As mentioned before, a surgical team with extensive knowledge and hands-on experience in normal instrument function and proper handling of robotic instruments is indispensable for the preoperative identification of faulty instruments as well as the prevention of future damage. Prior to surgery, a meticulous inspection of all instruments should be performed by the surgical team looking for cracked, bent, worn, or broken components. During surgery, to avoid damaging the instrument, always keep the wrist straight when passing it through the robotic port. Before engaging the instrument, the bedside assistant must straighten the wrist by rotating the disc on the instrument housing (the wrist should not be manipulated directly). At the time of disengagement, the surgeon should straighten the wrist using the master controllers (Fig. 8).

**Flexed**     **Straight**

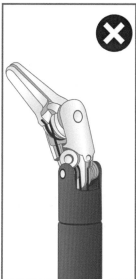 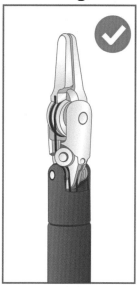

FIG. 8   Correct instrument wrist positioning for robotic port passage. *Arrow:* Dissected tubular structure entering the bladder.

II. Education, safety, ethical and administrative aspects for a successful robotic program

## Future Scenario

The field of robotic surgery is expected to undergo significant advancements in the years to come. Troubleshooting will become more efficient with the integration of technologies like augmented reality and artificial intelligence. Implementation of such technologies has the potential to greatly improve the process of identification and resolution of technical issues during complex procedures. Real-time imaging and augmented reality can provide surgeons with enhanced visualization, allowing them to identify and address unexpected challenges promptly. Additionally, machine learning algorithms will enable robotic systems to detect and respond to technical errors proactively, minimizing complications and improving patient outcomes. Overall, these advancements promise to revolutionize robotic surgery by making it safer and more effective.

## Conclusion

One critical aspect of troubleshooting in robotic surgery is the need for continuous monitoring during the procedure. The surgical team must be vigilant and able to detect technical issues early to prevent complications that could potentially harm the patient. Therefore, training and education programs are essential to ensure that surgeons and surgical teams are adequately equipped to handle technical issues that may arise during robotic surgery.

In conclusion, robotic surgery has revolutionized the field of surgery, offering numerous benefits over traditional surgical methods. However, troubleshooting in robotic surgery is a critical component to ensure patient safety and the overall success of the surgery. By being familiar with common technical issues and troubleshooting techniques, surgical teams can prevent technical errors and maintain the highest standards of patient care. Continued research and development in this field will further enhance the safety and efficacy of robotic surgery and pave the way for future advancements in the field of surgical robotics.

## Key points

- Robotic surgery offers increased precision, reduced invasiveness, decreased blood loss, and shorter hospital stays, but surgeons must be aware of the risks associated with equipment malfunctions.
- Troubleshooting techniques are crucial for effectively managing problems during robotic procedures and maintaining efficiency and safety.
- System errors in the da Vinci Surgical System can be recoverable or nonrecoverable, requiring different solutions for resolution.
- Video/image problems can arise from electrical/cable faults or image quality issues, which can be resolved through proper troubleshooting steps.
- Instrument malfunction, including electrical and mechanical issues, can impede surgical operations, emphasizing the importance of instrument handling, maintenance, and prompt diagnosis for prevention and resolution.

## References

[1] Cacciamani G, Desai M, Siemens DR, Gill IS. Robotic urologic oncologic surgery: ever-widening horizons. J Urol 2022;208(1):8–9.

[2] Cuevas CAG, Rodriguez RAC. Robot failure Complicat. Robot Urol Surg 2018;23–8.

[3] Nik-Ahd F, Souders CP, Houman J, Zhao H, Chughtai B, Anger JT. Robotic urologic surgery: trends in Food and Drug Administration-reported adverse events over the last decade. J Endourol 2019;33(8):649–54.

[4] Sheetz KH, Claflin J, Dimick JB. Trends in the adoption of robotic surgery for common surgical procedures. JAMA Netw Open 2020;3(1), e1918911.

[5] Alemzadeh H, Raman J, Leveson N, Kalbarczyk Z, Iyer RK. Adverse events in robotic surgery: a retrospective study of 14 years of FDA data. Lee HS, editor, PLoS One 2016;11(4). e0151470.

[6] Intuitive Surgical Inc. daVinci X System: User Manual. Intuitive Surgical Inc.; 2021. Available from: https://manuals.intuitivesurgical.com/c/document_library/get_file?uuid=dcac42ee-c041-0c16-727c-802b1c11f826&groupId=73750789.

[7] Chen CC, Ou YC, Yang CK, Chiu KY, Wang SS, Su CK, et al. Malfunction of the da Vinci robotic system in urology: malfunction of the da Vinci robotic system. Int J Urol 2012;19(8):736–40.

[8] Lavery HJ, Thaly R, Albala D, Ahlering T, Shalhav A, Lee D, et al. Robotic equipment malfunction during robotic prostatectomy: a multi-institutional study. J Endourol 2008;22(9):2165–8.

[9] Intuitive surgical Inc. daVinci Si system: User manual. Intuitive surgical Inc.; 2012.

[10] Friedman DCW, Lendvay TS, Hannaford B. Instrument failures for the da Vinci surgical system: a Food and Drug Administration MAUDE database study. Surg Endosc 2013;27(5):1503–8.

[11] Kim WT, Ham WS, Jeong W, Song HJ, Rha KH, Choi YD. Failure and malfunction of da Vinci surgical systems during various robotic surgeries: experience from six departments at a single institute. Urology 2009;74(6):1234–7.

[12] Lucas SM, Pattison EA, Sundaram CP. Global robotic experience and the type of surgical system impact the types of robotic malfunctions and their clinical consequences: an FDA MAUDE review: surgical system and time impact robot malfunctions. BJU Int 2012;109(8):1222–7.

[13] Mendez-Probst CE, Vilos G, Fuller A, Fernandez A, Borg P, Galloway D, et al. Stray electrical currents in laparoscopic instruments used in da Vinci ® robot-assisted surgery: an *in vitro* study. J Endourol 2011;25(9):1513–7.

[14] Mues AC, Box GN, Abaza R. Robotic instrument insulation failure: initial report of a potential source of patient injury. Urology 2011;77(1):104–7.

[15] Intuitive Surgical Inc. daVinci Xi and daVinci X instruments and accessories user manual. Intuitive Surgical Inc.; 2022. Available from: https://manuals.intuitivesurgical.com/c/document_library/get_file?uuid=51bc6c20-fa53-1cb8-6eca-d1f1592c20df&groupId=73750789.

[16] Park SY, Cho KS, Lee SW, Soh BH, Rha KH. Intraoperative breakage of needle driver jaw during robotic-assisted laparoscopic radical prostatectomy. Urology 2008;71(1):168–e5.

[17] Sotelo RJ, Haese A, Machuca V, Medina L, Nuñez L, Santinelli F, et al. Safer surgery by learning from complications: a focus on robotic prostate surgery. Eur Urol 2016;69(2):334–44.

[18] Medina LG, Martin O, Cacciamani GE, Ahmadi N, Castro JC, Sotelo R. Needle lost in minimally invasive surgery: management proposal and literature review. J Robot Surg 2018;12:391–5.

[19] Lee Z, Eun D. Instrument Malfunction. In: Sotelo R, Arriaga J, Aron M, editors. Complications in robotic urologic surgery. Cham: Springer International Publishing; 2018. p. 29–37. Available from: http://link.springer.com/10.1007/978-3-319-62277-4.

# 32

# Identifying patient-related factors for surgical complications

*Andrew Kerner[a],\*, Leanne Iorio[a],\*, Romina Bianchi[b], Joshua Karas[a], and Antonio Caycedo-Marulanda[a,b,c]*

[a]Orlando Health Colon and Rectal Institute, Orlando, FL, United States [b]Queens University, Kingston, Ontario, Canada [c]University of Central Florida, Orlando, FL, United States

## Introduction

Robotic-assisted surgery has significantly changed the landscape of minimally invasive surgery in many of the same ways that laparoscopy did; however, robotic surgery has been demonstrated to carry a shorter learning curve and a decreased incidence of conversion to open surgery. As more surgeons in training are learning robotic surgery in residency and established surgeons are incorporating the robotic platform into their practices, more patients will have undergone robotic-assisted surgeries than ever before. Therefore, it is imperative to recognize and understand patient risk factors that may lead to postoperative complications to mitigate and prevent unnecessary morbidity.

## Comorbidities

Two major prerequisites for successful robotic surgery include occasional steep patient positioning and using insufflation to provide adequate exposure in the body cavity where the intended target is found. Patient positioning utilizes gravity to make the target more isolated and accessible, but gravity is a force that significantly tests the hemodynamics of the patient over the course of the operation. Carbon dioxide is the gas of choice for insufflation given its cost-effectiveness, solubility, and lack of combustibility. However, like patient positioning, insufflation is not without risk, and these must be kept in mind with certain patient populations with unique risk factors and comorbidities that can lead to surgical complications.

Surgery in the abdomen and pelvis requires pneumoperitoneum, which results in an increase in intra-abdominal pressure from baseline. This increase in pressure alters the physiology of different hemodynamic systems. For instance, pneumoperitoneum results in decreased renal perfusion, which can compromise renal function [1,2]. Although the clinical significance may not always be realized in healthy patients, it may become more problematic in patients with baseline kidney disease, relative hypovolemia (i.e., dehydration, peripheral edema, sepsis), and in operations that anticipate steep patient positioning. Mitigating risks to the renal system include achieving a euvolemic state prior to incision, stopping medications that will suppress the renin-angiotensin-aldosterone axis in the preoperative period, and using lower insufflation pressures as able [3]. Although not the primary focus of this chapter, another incentive of lower insufflation pressure includes decreased postoperative pain and subsequent decrease in analgesic requirement due to less stretching of the diaphragm and/or phrenic nerve irritation [4].

The hemodynamic effects of abdominal insufflation impact and transcend the diaphragm. As intra-abdominal pressure increases, the chance of bradycardia due to vagal stimulation also increases, putting patients with arrhythmias at

---

\* These authors are equal contributors as first authors.

a potential higher risk for robotic surgery. It is therefore recommended to insufflate gradually to a desired pressure, desufflate immediately as the first step to reverse profound bradycardia, and not exceed an intra-abdominal pressure of 15 mmHg [5]. When intra-abdominal pressures exceed 15 mmHg, cardiac output begins to suffer due to a proportional decrease in preload due to IVC compression and increased intrathoracic pressure from lifting the diaphragm. As the body compensates by increasing the afterload (or systemic vascular resistance), cardiac output is maintained by a reflexive increase in heart rate. Patients with underlying cardiac dysfunction are vulnerable because of a compromised ability to compensate. Patients with diastolic dysfunction are more vulnerable to myocardial ischemia in this scenario due to hastened filling times, with diastolic dysfunction being shown as an independent risk factor for adverse cardiovascular events after noncardiac surgery [6]. In patients with congestive heart failure, the combination of elevated intra-abdominal pressure, prolonged steep positioning, and general anesthesia can cause the cardiac index to be reduced as much as 50% [7], an important consideration in counseling patients with decreased cardiopulmonary reserve about robotic surgery. However, laparoscopy has been demonstrated to be a safe alternative in appropriately selected patients with congestive heart failure [8], and this gives reason to extrapolate the data from laparoscopy in patients with cardiac concern into the robotic frontier.

Patients with hypercoagulable disorders may be at a relatively higher risk for deep venous thrombosis or venous thromboembolism complications when undergoing robotic surgery. Abdominal insufflation and reverse Trendelenburg positioning increase the venous stasis in the lower extremities, with significant changes measured in femoral vein diameter, cross-sectional area, peak velocity, and volume flow [9]. This venous stasis puts patients with hypercoagulable disorders (such as Factor V Leiden and cancer) and patients with a history of DVT at even higher risk of developing a DVT. Risk-stratifying patients with assessment models such as the Caprini score is recommended [10].

Diabetes is another patient risk factor for surgical complications in robotic surgery as diabetics are more prone to surgical site infections, which prolong the length of stay, increase rehospitalization rates, and significantly increase healthcare cost [11]. Although not fully discerned, the association of diabetes with hyperglycemia has been proposed as a causative factor. This hypothesis gained further traction after data revealed postoperative hyperglycemia was associated with increased morbidity and mortality, even in nondiabetic patients [12]. In patients with significant peripheral vascular disease secondary to diabetes, atherosclerosis, or arteriosclerosis, a diminished circulatory response to catecholamine secretion secreted in association with abdominal insufflation or steep positioning can lead to low-flow states that promote thrombosis [13].

Patients with glaucoma are also at risk for complications in robotic surgery requiring steep Trendelenburg position (i.e., operating in the pelvis) as this positioning increases the intraocular pressure in the eyes. Prolonged duration with increased intraocular pressure can lead to perioperative vision loss [14] due to a low perfusion state leading to retinal cell ganglion dysfunction. This can possibly be mitigated with interval supine positioning of patients during longer cases in steep Trendelenburg [15].

## Body habitus

Another patient-related factor that may contribute to robotic surgery complications is the patient's body habitus. It is well documented that the obese population (BMI > 30) has a higher risk of complications due to their associated comorbidities. Laparoscopic surgery in these patients has been associated with higher conversion to laparotomy, longer operative times, greater blood loss, increased postoperative complications, and longer length of hospital stays. The higher rate of conversions and perioperative complications have been associated with excessive adipose tissue, decreased operating space within the peritoneal cavity, and a thickened omentum and mesentery making dissection more challenging. Consequently, several studies have documented a higher risk of ileus, wound infections, perioperative morbidity, and anastomotic leak rates in patients depending on BMI [16]

Specifically, obesity can lead to pulmonary dysfunction secondary to pro-inflammatory cytokines, decreased expiratory reserve volume, and decreased residual capacity [17–19]. Obese patients have an increased work of breathing, decreased $PaO_2$, and increased $PaCO_2$ [20]. Insufflation of the obese abdomen changes respiratory physiology and can further exacerbate the underlying hypercarbia and hypoxemia. As the abdomen is insufflated, airway pressures are increased and pulmonary compliance is reduced by the displacement of the diaphragm cephalad and the increased respiratory rate necessary to exhale $CO_2$. As a result, V/Q mismatching can occur resulting in hypoxemia, while the increased $CO_2$ tension that is not exhaled can dissolve into the blood causing hypercarbia [21]. This can lead to hypercapnia with postoperative respiratory failure, heart failure, and prolonged intubation.

These phenomena are exacerbated in patients with baseline respiratory compromise such as patients with chronic obstructive pulmonary disease (COPD) or obesity specifically when placing the patient in steep Trendelenburg while performing robotic pelvic surgery. This position along with the pneumoperitoneum can increase peak inspiratory pressure and intra-abdominal pressure while decreasing tidal volume resulting in pulmonary complications. Interestingly though, the benefit of using the robot rather than straight stick instruments may allow precise pelvic surgery without the need for very steep Trendelenburg position [22,23].

In addition to the physiological parameters, the anatomical consequences of obesity can complicate robotic surgery as well. A thicker abdominal wall and increased intraabdominal visceral fat can decrease the working space within the abdomen. Intraoperative time can increase with these limitations. Depending on the length of the trocar, freedom of motion may also be limited. However, BMI alone can be misleading as it talks about overall body composition and not components or distribution of body habitus. Waist-to-hip ratio and waist circumference better identify those with central obesity and may be more important when considering robotic surgery risks [24]. These values have been shown to be predictive of adverse outcomes in surgery [25,26]. Generally speaking, cutoff points for waist-to-hip ratio associated with increased metabolic risk have been reported as $\geq 0.90$ cm in men and $\geq 0.85$ cm in women and cut-off points for waist circumference have been reported as $>94$ cm in men and $>80$ cm in women. Caution must be taken in using these specific numeric values, as studies from various countries show not only sex, but ethnicity plays a role in adipose distribution and therefore these values cannot be applied to all [27].

Although obesity is a risk factor in robotic surgery, there is some evidence there are fewer complications with robotic surgery as compared to open and laparoscopic surgery. A meta-analysis published in the Journal of Robotic Surgery in 2018 compared patients who underwent abdominal robotic versus laparoscopic surgery from 2012 to 2014 identified in the American College of Surgery-National Surgical Quality Improvement Program (ACS-NSQIP) database. The data revealed there were fewer conversions to laparotomy, shorter length of stay, and decreased postoperative ileus in the robotic subset of patients [16]. Another multicenter retrospective review of 1264 patients undergoing robotic anatomical pulmonary resections from 2002 to 2018 found that BMI did not have a significant impact on complications [28] as opposed to open thoracotomy [29].

As in any surgery, optimizing chronic conditions may help avoid unwanted complications. Patients may undergo prehabilitation that refers to a variety of methods to improve a patient's condition prior to surgery including aerobic and deep breathing exercises [30]. There is evidence that respiratory physiotherapy such as preoperative inspiratory training specifically in the obese population can lead to decreased postoperative pulmonary complications by resulting in an increased $PaO_2/FiO_2$ ratio. [31,32] This may be a consideration in optimizing obese patients prior to robotic surgery, although there is no set standard of who should be screened with preoperative function tests and who should be prehabilitated. [33]

On the contrary to obesity, underweight patients with a paucity of fat are at risk for hypothermia. Hypothermia tends to occur during the first hour of an operation before active warming strategies have time to counteract the drop in temperature. Hypothermia is typically defined as a core temperature below 36 degrees centigrade and affects up to 50% of surgical patients. Perioperative hypothermia can lead to increased blood loss and transfusion requirements, length of stay, and surgical site infection [34]. Active warming options to mitigate perioperative hypothermia include warmed intravenous fluid preoperatively forced-air warming and warmed insufflation. [35]

## Prior surgeries

Another important patient risk factor that increases the risk of robotic surgery complications is the severity of adhesions formed from prior surgeries. Previous abdominal surgery is an independent risk factor associated with the highest postoperative morbidity [36]. The main culprit of this morbidity is the formation of adhesions after surgery, with laparoscopic surgery generating fewer adhesions than those seen in open surgery [37]. Intraperitoneal adhesions can lead to chronic pain, obstruction, and secondary female infertility. During surgery, these adhesions increase the risk of inadvertent enterotomies and blood loss [38].

This inadvertent enterotomy during the reopening of the abdomen or subsequent adhesion dissection is a feared complication of surgery after a previous laparotomy. The incidence can be as high as 20% in open surgery and between 1% and 100% in laparoscopy depending on the underlying disease. Delayed postoperative detection of enterotomy is a particular feature of laparoscopy associated with significant morbidity and mortality. Adhesions to the ventral abdominal wall are responsible for the majority of trocar injuries. Both trocar injuries and inadvertent enterotomies result in conversion from laparoscopy to laparotomy in almost 100% of cases.

When adhesions are suspected prior to robotic surgery, it is important to consider safe ways of entering the abdomen to avoid an initial injury. The open entry technique (Hasson) results in fewer failed attempts at entry; however, it does not necessarily have a decreased rate of visceral or vasculature injuries. When periumbilical adhesions or umbilical hernia is suspected, an entry in the left upper quadrant at Palmer's point should be considered. When using the closed technique with the Veress needle, low intraabdominal pressure (<10 mmHg) is considered reliable for correct intraperitoneal insertion, and other safety checks are not recommended. Additionally, the angle of insertion should be 45 degrees and should be adjusted to 90 degrees in obese patients [39,40].

Postoperative morbidity and mortality of patients who need adhesiolysis are higher than those who do not. The necessity to dissect adhesions is associated with increased hospital stay and is considered a main reason for conversion from laparoscopy to laparotomy. Various studies confirmed that previous abdominal surgery is an independent risk factor for complications, even if we use the benefits provided by multiquadrant new robotic platforms. [41,42]

## Patient age

The global demographic landscape is shifting in the proportion of individuals aged 65 years and above. This demographic has far-reaching implications across various sectors, most notably health care. As the elderly population requiring surgical interventions continues to grow, the proportion of individuals aged >65 years in the total world population has increased to 7.6% in 2010 from 5.2% in 1950, and it is expected to increase to 17.6% in 2060. It is therefore imperative to appreciate age as a risk factor as age is linked to a decline in functional reserve capacity [43], making the elderly less tolerant of surgical stress. As the body ages, there is a decline in the normal function of most organ systems, which can become clinically significant when subjected to the physiological stress of robotic surgery. While age itself may not predict mortality, age-related comorbidities play a significant role in perioperative complications [44]. By recognizing age-related risks and implementing tailored approaches, surgeons can optimize outcomes for older patients undergoing robotic surgery [45,46].

Perhaps more important than age as a number however is biological age as aging does not impact everyone the same way. Two important traits to look for in elderly patients are frailty and sarcopenia. Frailty is defined as a condition of increased vulnerability to developing negative health-related events when exposed to exogenous or endogenous stressors [47]. Frailty can manifest as unintentional weight loss, slowed walking speed, low physical activity, low energy, and low grip strength. Any patient that shows the aforementioned signs of frailty should undergo a comprehensive geriatric assessment preoperatively. Not only does a comprehensive geriatric assessment increase the likelihood of patients being discharged home rather than a rehabilitation facility but also increases the likelihood of being alive in their own home at 3-month and 12-month follow-up [48]. Sarcopenia can be characterized as a loss of muscle mass, which is a direct correlate with frailty. This can be objectively measured on a computed tomography scan looking at the skeletal mass index at the third lumbar level to evaluate the psoas, lumbar, and rectus abdominus muscles [49]. Quantitative loss of muscle mass is associated with risks of postoperative complications and decreased overall survival. When sarcopenia or frailty is identified, preoperative rehabilitation programs should be considered.

Preoperative rehabilitation programs aim to improve a patient's resilience before elective surgery. These programs typically include physical exercise, focus on nutrition, psychosocial recovery, and smoking cessation. Although preoperative rehabilitation has not been shown to reduce in-hospital mortality, ICU length of stay, or readmission rates, research has shown benefits in postoperative functional status, lower complication rates, shorter overall length of stays, and increase in home discharges [50]. One published example of this was a statewide rollout of a clinical prehabilitation program among 21 hospitals within the Michigan Surgical Quality Collaborative with a goal preoperative duration of 4 weeks but included patients having procedures up to 7 days prior. This included individualized instruction in physical exercise, nutrition, and stress reduction techniques. Patients enrolled in this initiative were found to have significantly reduced length of stays and increased discharges to home [51].

## Medications

The preoperative period is an opportune time to review a patient's medication prescriptions. Failure to do so can lead to intraoperative and postoperative complications. This is especially true in patients with frailty and polypharmacy, with the former having a diminished capacity to compensate for periods of stress such as surgery and the latter being more prone to delirium. Postoperative delirium is a significant risk factor for 5-year mortality [52], underscoring the importance of adjudicating when to hold and resume certain home medications. Although not an exhaustive list of medication classes, common classes of medications to give attention to include antihypertensives, hypoglycemics, antiplatelets, and anticoagulants. Included in the antihypertensive category are ACE inhibitors, angiotensin receptor blockers, beta-blockers, calcium channel blockers, and diuretics. It is reasonable to withhold ACE inhibitors and angiotensin receptor blockers in the perioperative period as these blunt the sympathetic response, which can lead to dramatic change in vascular tone with a patient receiving general anesthesia [53]. Prescribed beta-blockers should be continued during the operative period unless the patient develops bradycardia or hypotension, but starting a new beta blocker postoperative is discouraged due to the risk of masking insidious tachycardia and a higher risk of stroke [54]. Diuretics should be suspended on the day of surgery and restarted when the intravascular volume is stable, and the patient is tolerating a diet [55]. The perioperative management of hypoglycemics that can be summarized with hyperglycemia is generally safer than hypoglycemia. Patients undergoing surgery are ordered nothing per mouth hours before their surgery. Therefore, in most instances, it is recommended that hypoglycemics such as metformin should be held on the day of major surgery to avoid postoperative hypoglycemia and resumed 48h later [56]. Patients prescribed antiplatelets such as Aspirin and Plavix should elevate the surgeon's awareness of underlying cardiovascular burden, such as history of stroke or cardiac stents. In most instances, Aspirin should be continued through the perioperative period while Plavix should be held 5 days prior to elective surgery to reduce the incidence of postoperative bleeding [57] However, patients with cardiac stents deserve special attention as suspending antiplatelet therapy significantly increases the risk of stent thrombosis, which purports a high risk of mortality. Therefore, the decision to suspend antiplatelet therapy should be discussed with the patient's cardiologist as it may be advised to delay elective surgery up to 6 months from stent placement when able [58]. Patients prescribed anticoagulants are another special group that deserves special attention. Weighing the risk of intraoperative and postoperative bleeding versus thrombus formation is a multidisciplinary discussion beyond the scope of this book chapter, as the decision-making must take into account high thrombotic risk factors such as recent stroke, prosthetic valves, and hypercoagulable disorders as well as the bleeding risk of the surgery itself [59]. As with patients on antiplatelets, a multidisciplinary approach to suspending and resuming anticoagulants is advised.

## Future scenario

In view of the above discussed, in the next few years, the main patients-related risk factors for complications in robotic surgery will be similarly present and perhaps will become more frequent due to the worldwide population aging. To perform a holistic evaluation of each individualized patient, foreseeing and taking into account their comorbidities, biological age, body habitus, previous surgeries, medications, and preferences will be still fundamental for the best postoperative outcomes of robotic-assisted surgery.

## Conclusion

The patient risk factors for surgical complications in robotic surgery are no different than the risk factors seen in laparoscopy or open surgery. However, the combination of insufflation and steep positioning that are commonly seen in robotic surgery may not be well tolerated by all patients, which emphasizes the importance of patient selection and informed consent. Prolonged robotic surgeries affect the systemic physiology of the body, so knowing how to identify patients with higher risks is important to avoid or mitigate morbidity after surgery. While minimally invasive surgery has demonstrated immense value in decreasing postoperative pain, length of stay, wound infection, and incisional hernias, in select cases open surgery may be safer in some patients with some of the risk factors described in this chapter. Nevertheless, the very same risk factors for complications after robotic surgery may be even higher in open surgery, making the decision to proceed with a robotic-assisted procedure an individualized process.

## Key points

- There are several patient-related risk factors that can lead to complications in robotic surgery including comorbidities, body habitus, prior surgeries, age, and medication consumption.
- Recognizing and mitigating these patient-related risk factors is important in preventing postoperative complications and improving patient outcomes in robotic surgery.
- Patient-related risk factors are present regardless of surgical approach (open versus minimally invasive).
- As robotic surgery continues to grow, our understanding of risk factors that were traditionally labeled as contraindications for minimally invasive surgery will continue to be challenged.

# References

[1] Demyttenaere S, Feldman LS, Fried GM. Effect of pneumoperitoneum on renal perfusion and function: a systematic review. Surg Endosc 2007;21(2):152–60. https://doi.org/10.1007/s00464-006-0250-x.

[2] Umano GR, Delehaye G, Noviello C, Papparella A. The "Dark Side" of Pneumoperitoneum and Laparoscopy. Wang PH, ed. Minim Invasive Surg 2021;2021. https://doi.org/10.1155/2021/5564745. 5564745.

[3] de Seigneux S, Klopfenstein CE, Iselin C, Martin PY. The risk of acute kidney injury following laparoscopic surgery in a chronic kidney disease patient. NDT Plus 2011;4(5):339–41. https://doi.org/10.1093/ndtplus/sfr071.

[4] Ortenzi M, Montori G, Sartori A, et al. Low-pressure versus standard-pressure pneumoperitoneum in laparoscopic cholecystectomy: a systematic review and meta-analysis of randomized controlled trials. Surg Endosc 2022;36(10):7092–113. https://doi.org/10.1007/s00464-022-09201-1.

[5] Sood J. Advancing frontiers in anaesthesiology with laparoscopy. World J Gastroenterol 2014;20(39):14308–14. https://doi.org/10.3748/wjg.v20.i39.14308.

[6] Fayad A, Ansari MT, Yang H, Ruddy T, Wells GA. Perioperative diastolic dysfunction in patients undergoing noncardiac surgery is an independent risk factor for cardiovascular events: a systematic review and Meta-analysis. Anesthesiology 2016;125(1):72–91. https://doi.org/10.1097/ALN.0000000000001132.

[7] Joris JL, Noirot DP, Legrand MJ, Jacquet NJ, Lamy ML. Hemodynamic changes during laparoscopic cholecystectomy. Anesth Analg 1993;76 (5):1067–71. https://doi.org/10.1213/00000539-199305000-00027.

[8] Speicher PJ, Ganapathi AM, Englum BR, Vaslef SN. Laparoscopy is safe among patients with congestive heart failure undergoing general surgery procedures. Surgery 2014;156(2):371–8. https://doi.org/10.1016/j.surg.2014.03.003.

[9] Lu M, Bi J, Li Y, Hao Z. Effect of obesity on venous hemodynamics of the lower limbs during laparoscopic cholecystectomy. Phlebology 2022;37 (5):381–5. https://doi.org/10.1177/02683555221081634.

[10] Bahl V, Hu HM, Henke PK, Wakefield TW, Campbell DA, Caprini JA. A validation study of a retrospective venous thromboembolism risk scoring method. Ann Surg 2010;251(2):344–50. https://doi.org/10.1097/SLA.0b013e3181b7fca6.

[11] Dimick JB, Chen SL, Taheri PA, Henderson WG, Khuri SF, Campbell DA. Hospital costs associated with surgical complications: a report from the private-sector National Surgical Quality Improvement Program. J Am Coll Surg 2004;199(4):531–7. https://doi.org/10.1016/j.jamcollsurg.2004.05.276.

[12] Ata A, Lee J, Bestle SL, Desemone J, Stain SC. Postoperative hyperglycemia and surgical site infection in general surgery patients. Arch Surg Chic Ill 1960 2010;145(9):858–64. https://doi.org/10.1001/archsurg.2010.179.

[13] Richmond BK, Thalheimer L. Laparoscopy associated mesenteric vascular complications. Am Surg 2010;76(11):1177–84.

[14] Moriyama Y, Miwa K, Yamada T, Sawaki A, Nishino Y, Kitagawa Y. Intraocular pressure change during laparoscopic sacral colpopexy in patients with normal tension glaucoma. Int Urogynecol J 2019;30(11):1933–8. https://doi.org/10.1007/s00192-018-03866-w.

[15] Molloy B, Watson C. A comparative assessment of intraocular pressure in prolonged steep Trendelenburg position versus level supine position intervention. J Anesthesiol Clin Sci 2012;1(1). https://doi.org/10.7243/2049-9752-1-9.

[16] Harr JN, Haskins IN, Amdur RL, Agarwal S, Obias V. The effect of obesity on laparoscopic and robotic-assisted colorectal surgery outcomes: an ACS-NSQIP database analysis. J Robot Surg 2018;12(2):317–23. https://doi.org/10.1007/s11701-017-0736-7.

[17] Hegewald MJ. Impact of obesity on pulmonary function: current understanding and knowledge gaps. Curr Opin Pulm Med 2021;27(2):132–40. https://doi.org/10.1097/MCP.0000000000000754.

[18] Kaw R, Wong J, Mokhlesi B. Obesity and obesity hypoventilation, sleep hypoventilation, and postoperative respiratory failure. Anesth Analg 2021;132(5):1265–73. https://doi.org/10.1213/ANE.0000000000005352.

[19] Melo LC, Silva MAMd, Calles ACdN. Obesity and lung function: a systematic review. Einstein Sao Paulo Braz 2014;12(1):120–5. https://doi.org/10.1590/s1679-45082014rw2691.

[20] Gabrielsen AM, Lund MB, Kongerud J, Viken KE, Røislien J, Hjelmesæth J. The relationship between anthropometric measures, blood gases, and lung function in morbidly obese white subjects. Obes Surg 2011;21(4):485–91. https://doi.org/10.1007/s11695-010-0306-9.

[21] Atkinson TM, Giraud GD, Togioka BM, Jones DB, Cigarroa JE. Cardiovascular and Ventilatory consequences of laparoscopic surgery. Circulation 2017;135(7):700–10. https://doi.org/10.1161/CIRCULATIONAHA.116.023262.

[22] Blecha S, Harth M, Zeman F, et al. The impact of obesity on pulmonary deterioration in patients undergoing robotic-assisted laparoscopic prostatectomy. J Clin Monit Comput 2019;33(1):133–43. https://doi.org/10.1007/s10877-018-0142-3.

[23] Wysham WZ, Kim KH, Roberts JM, et al. Obesity and perioperative pulmonary complications in robotic gynecologic surgery. Am J Obstet Gynecol 2015;213(1):33.e1–7. https://doi.org/10.1016/j.ajog.2015.01.033.

[24] Zorn KC. Robotic surgery techniques for obese patients. Can Urol Assoc J 2010;4(4):255–6.

[25] Balentine CJ, Robinson CN, Marshall CR, et al. Waist circumference predicts increased complications in rectal Cancer surgery. J Gastrointest Surg 2010;14(11):1669–79. https://doi.org/10.1007/s11605-010-1343-3.

[26] Kartheuser AH, Leonard DF, Penninckx F, et al. Waist circumference and waist/hip ratio are better predictive risk factors for mortality and morbidity after colorectal surgery than body mass index and body surface area. Ann Surg 2013;258(5):722–30. https://doi.org/10.1097/SLA.0b013e3182a6605a.

[27] World Health Organization. Waist circumference and waist-hip ratio : Report of a WHO expert consultation, Geneva, 8-11 December 2008. Published online 2011. [Accessed 13 September 2023] https://apps.who.int/iris/handle/10665/44583.

[28] Cao C, Louie BE, Melfi F, et al. Outcomes of major complications after robotic anatomic pulmonary resection. J Thorac Cardiovasc Surg 2020;159 (2):681–6. https://doi.org/10.1016/j.jtcvs.2019.08.057.

[29] Agostini P, Cieslik H, Rathinam S, et al. Postoperative pulmonary complications following thoracic surgery: are there any modifiable risk factors? Thorax 2010;65(9):815–8. https://doi.org/10.1136/thx.2009.123083.

[30] Hughes MJ, Hackney RJ, Lamb PJ, Wigmore SJ, Christopher Deans DA, Skipworth RJE. Prehabilitation before major abdominal surgery: a systematic review and Meta-analysis. World J Surg 2019;43(7):1661–8. https://doi.org/10.1007/s00268-019-04950-y.

[31] Tenório LHS, de Lima AMJ, Brasileiro-Santos MdS. The role of respiratory physiotherapy in the lung function of obese patients undergoing bariatric surgery. A review. Rev Port Pneumol 2010;16(2):307–14. https://doi.org/10.1016/s0873-2159(15)30028-3.

[32] Lloréns J, Rovira L, Ballester M, et al. Preoperative inspiratory muscular training to prevent postoperative hypoxemia in morbidly obese patients undergoing laparoscopic bariatric surgery. A randomized clinical trial. Obes Surg 2015;25(6):1003–9. https://doi.org/10.1007/s11695-014-1487-4.

[33] Pouwels S, Smeenk FWJM, Manschot L, et al. Perioperative respiratory care in obese patients undergoing bariatric surgery: implications for clinical practice. Respir Med 2016;117:73–80. https://doi.org/10.1016/j.rmed.2016.06.004.

[34] Sun Z, Honar H, Sessler DI, et al. Intraoperative Core temperature patterns, transfusion requirement, and hospital duration in patients warmed with forced air. Anesthesiology 2015;122(2):276–85. https://doi.org/10.1097/ALN.0000000000000551.

[35] Shaw CA, Steelman VM, DeBerg J, Schweizer ML. Effectiveness of active and passive warming for the prevention of inadvertent hypothermia in patients receiving neuraxial anesthesia: a systematic review and meta-analysis of randomized controlled trials. J Clin Anesth 2017;38:93–104. https://doi.org/10.1016/j.jclinane.2017.01.005.

[36] Buchs NC, Addeo P, Bianco FM, et al. Perioperative risk assessment in robotic general surgery: lessons learned from 884 cases at a single institution. Arch Surg Chic Ill 1960 2012;147(8):701–8. https://doi.org/10.1001/archsurg.2012.496.

[37] van Goor H. Consequences and complications of peritoneal adhesions. Colorectal Dis Off J Assoc Coloproctology G B Irel 2007;9(Suppl 2):25–34. https://doi.org/10.1111/j.1463-1318.2007.01358.x.

[38] Van Der Krabben AA, Dijkstra FR, Nieuwenhuijzen M, Reijnen MM, Schaapveld M, Van Goor H. Morbidity and mortality of inadvertent enterotomy during adhesiotomy. Br J Surg 2000;87(4):467–71. https://doi.org/10.1046/j.1365-2168.2000.01394.x.

[39] Vilos GA, Ternamian A, Dempster J, Laberge PY, Clinical Practice Gynaecology Committee. Laparoscopic entry: a review of techniques, technologies, and complications. J Obstet Gynaecol Can JOGC 2007;29(5):433–47. https://doi.org/10.1016/S1701-2163(16)35496-2.

[40] Thepsuwan J, Huang KG, Wilamarta M, Adlan AS, Manvelyan V, Lee CL. Principles of safe abdominal entry in laparoscopic gynecologic surgery. Gynecol Minim Invasive Ther 2013;2(4):105–9. https://doi.org/10.1016/j.gmit.2013.07.003.

[41] Fantola G, Brunaud L, Nguyen-Thi PL, Germain A, Ayav A, Bresler L. Risk factors for postoperative complications in robotic general surgery. Updates Surg 2017;69(1):45–54. https://doi.org/10.1007/s13304-016-0398-4.

[42] ten Broek RPG, Strik C, Issa Y, Bleichrodt RP, van Goor H. Adhesiolysis-related morbidity in abdominal surgery. Ann Surg 2013;258(1):98–106. https://doi.org/10.1097/SLA.0b013e31826f4969.

[43] Seishima R, Okabayashi K, Hasegawa H, et al. Is laparoscopic colorectal surgery beneficial for elderly patients? A systematic review and meta-analysis. J Gastrointest Surg Off J Soc Surg Aliment Tract 2015;19(4):756–65. https://doi.org/10.1007/s11605-015-2748-9.

[44] Dagrosa LM, Ingimarsson JP, Gorlov IP, Higgins JH, Hyams ES. Is age an independent risk factor for medical complications following minimally invasive radical prostatectomy? An evaluation of contemporary American College of Surgeons National Surgical Quality Improvement (ACS-NSQIP) data. J Robot Surg 2016;10(4):343–6. https://doi.org/10.1007/s11701-016-0605-9.

[45] Madden N, Frey MK, Joo L, et al. Safety of robotic-assisted gynecologic surgery and early hospital discharge in elderly patients. Am J Obstet Gynecol 2019;220(3):253.e1–7. https://doi.org/10.1016/j.ajog.2018.12.014.

[46] PACE Participants, Audisio RA, Pope D, et al. Shall we operate? Preoperative assessment in elderly cancer patients (PACE) can help. A SIOG surgical task force prospective study. Crit Rev Oncol Hematol 2008;65(2):156–63. https://doi.org/10.1016/j.critrevonc.2007.11.001.

[47] Proietti M, Cesari M. Frailty: what is it? Adv Exp Med Biol 2020;1216:1–7. https://doi.org/10.1007/978-3-030-33330-0_1.

[48] Ellis G, Gardner M, Tsiachristas A, et al. Comprehensive geriatric assessment for older adults admitted to hospital. Cochrane Effective Practice and Organisation of Care Group, ed Cochrane Database Syst Rev 2017;2017(9). https://doi.org/10.1002/14651858.CD006211.pub3.

[49] Wang S, Xie H, Gong Y, et al. The value of L3 skeletal muscle index in evaluating preoperative nutritional risk and long-term prognosis in colorectal cancer patients. Sci Rep 2020;10(1):8153. https://doi.org/10.1038/s41598-020-65091-0.

[50] Barberan-Garcia A, Ubré M, Roca J, et al. Personalised Prehabilitation in high-risk patients undergoing elective major abdominal surgery: a randomized blinded controlled trial. Ann Surg 2018;267(1):50–6. https://doi.org/10.1097/SLA.0000000000002293.

[51] Mouch CA, Kenney BC, Lorch S, et al. Statewide Prehabilitation program and episode payment in Medicare beneficiaries. J Am Coll Surg 2020;230(3):306–313e6. https://doi.org/10.1016/j.jamcollsurg.2019.10.014.

[52] Moskowitz EE, Overbey DM, Jones TS, et al. Post-operative delirium is associated with increased 5-year mortality. Am J Surg 2017;214(6):1036–8. https://doi.org/10.1016/j.amjsurg.2017.08.034.

[53] Hollmann C, Fernandes NL, Biccard BM. A systematic review of outcomes associated with withholding or continuing angiotensin-converting enzyme inhibitors and angiotensin receptor blockers before noncardiac surgery. Anesth Analg 2018;127(3):678–87. https://doi.org/10.1213/ANE.0000000000002837.

[54] POISE Study Group, Devereaux PJ, Yang H, et al. Effects of extended-release metoprolol succinate in patients undergoing non-cardiac surgery (POISE trial): a randomised controlled trial. Lancet Lond Engl 2008;371(9627):1839–47. https://doi.org/10.1016/S0140-6736(08)60601-7.

[55] Howell SJ. Preoperative hypertension. Curr Anesthesiol Rep 2018;8(1):25–31. https://doi.org/10.1007/s40140-018-0248-7.

[56] Membership of the Working Party, Barker P, Creasey PE, et al. Peri-operative management of the surgical patient with diabetes 2015: Association of Anaesthetists of Great Britain and Ireland. Anaesthesia 2015;70(12):1427–40. https://doi.org/10.1111/anae.13233.

[57] Mikhail MA, Mohabbat AB, Ghosh AK. Perioperative cardiovascular medication Management in Noncardiac Surgery: common questions. Am Fam Physician 2017;95(10):645–50.

[58] Banerjee S, Angiolillo DJ, Boden WE, et al. Use of antiplatelet therapy/DAPT for post-PCI patients undergoing noncardiac surgery. J Am Coll Cardiol 2017;69(14):1861–70. https://doi.org/10.1016/j.jacc.2017.02.012.

[59] Johnson S, Haywood C. Perioperative medication management for older people. J Pharm Pract Res 2022;52(5):391–401. https://doi.org/10.1002/jppr.1834.

# Robotic surgery after previous abdominal surgeries

*Paulo Roberto Stevanato Filho*[a,b]

[a]Colorectal Cancer Reference Center, A.C. Camargo Cancer Center, São Paulo, SP, Brazil [b]Brazilian Society of Surgical Oncology–SBCO, Rio de Janeiro, Brazil

Previous abdominal surgeries are associated with an increased risk of perioperative complications in subsequent surgical procedures. The main contributing factor to this increased risk is the presence of intra-abdominal adhesions, which hinder intracavitary access and the creation of pneumoperitoneum during laparoscopic surgeries. Trocar site changes, robotic docking positioning, anatomical distortion, and visual field limitation are also associated with increased risks of perioperative complications, including injuries to intraperitoneal visceral and vascular structures [1,2].

Compared with robotic surgeries, laparoscopic surgeries have a higher conversion rate to open surgery and longer operating times in patients with intra-abdominal adhesions. However, advances in laparoscopic instrumentation and increased experience have improved the use and safety of laparoscopic surgery in most patients with a history of previous abdominal surgery.

One of the advantages of robotic surgery is its lower rate of unplanned conversions compared to traditional laparoscopic procedures. However, it is worth noting that certain specialties have historically faced technical difficulties with laparoscopy and have reported such challenges. For example, prospective studies have demonstrated that robotic-assisted endopelvic access for total mesorectal excision is associated with lower conversion rates compared to laparoscopy. This advantage is particularly prominent in obese patients and men, as robotic instrumental skills allow for better maneuverability in regions with limited space [3].

It is hypothesized that robotic surgery could have a lower conversion rate compared to laparoscopy in patients with adhesions. This is justified by the potential technical advantages of robotic surgery, such as enlarged 3D vision with a better operative field, preservation of the natural eye-hand-instrument alignment, and precision-controlled EndoWrist instruments with better ergonomics and reduced physiological tremor. The XI robot generation, which permits the insertion of the endoscope at 0- or 30-degree angles through all 8mm trocars, offers a substantial advantage when performing surgeries in challenging or hostile abdominal cavities. This feature allows surgeons to achieve a more comprehensive view of the surgical field, thereby improving their capacity to navigate through adhesions and assess the issue's extent. This results in increased flexibility and precision during minimally invasive procedures.

These advantages are further highlighted in cases of distorted abdominal anatomy related to adhesions, which could hinder visualization and make the surgical procedure more difficult [4,5].

A meta-analysis [6] included a total of 14,329 patients who underwent various surgical procedures including colorectal, esophagogastric, hepatobiliary, pancreatic, endocrine, urological, and gynecological surgeries. Among these patients, 6472 underwent robotic surgery and 7857 underwent laparoscopic surgery. Overall, the robotic approach was associated with fewer conversions compared to the laparoscopic approach (OR 1.53; 95%CI, 1.12–2.10, $P = 0.007$). Furthermore, an analysis of procedures performed by "expert surgeons" showed a statistically significant difference with a lower conversion rate with robotic surgery (OR 1.48; 95%CI, 1.03–2.12, $P = 0.03$); however, there was no difference between novice robotic surgeons.

A reduction in conversion rates specifically associated with adhesions has been identified in patients who underwent colorectal cancer robotic surgery (OR 2.62; 95%CI, 1.20–5.72, $P = 0.02$). The robotic approach may be a valid surgical option in patients with abdominal adhesions, especially when performed by experienced surgeons and in the subgroup of patients undergoing colorectal cancer resection [6]. A possible explanation for these findings is that colorectal cancer surgery often requires access to several abdominal quadrants, especially the supra- and inframesocolic spaces. Therefore, the presence of adhesions could significantly affect the procedure to a greater extent than it does other specialties or procedures that are limited to an abdominal or pelvic compartment [6].

On the other hand, to safely and efficiently proceed with the intended operation, the surgical plan must include the handling of abdominal adhesions. However, although many colorectal surgeons are comfortable with gastrointestinal manipulation, there is greater controversy among specialists from other areas regarding it. Although robotic surgery has many technical advantages, it lacks tactile feedback, which may difficult intracavitary access for less experienced surgeons.

Some studies [7] have evaluated the impact of liver resection after a previous surgical procedure. The results showed that there was no significant increase in surgical time in patients with a history of previous surgeries, including the subgroups of patients who had previously undergone upper abdominal surgery, open surgery, or major organ resections. The subgroup of patients with a history of liver resection had a longer surgical duration compared with the duration of patients without a previous history of surgical procedures. There were also no changes in morbidity rates in patients with previous abdominal surgery. The study concluded that patients with a history of abdominal surgery do not have an increased risk of postoperative complications after robotic liver resection; furthermore, the surgery duration was not affected by previous surgeries. These findings corroborate those of a previous study on laparoscopic liver surgery after previous abdominal surgeries [8] and support the safety and feasibility of robotic liver surgery.

Several studies on robotic prostatectomy [9] that compared patients who had undergone abdominal or inguinal surgeries reported no significant differences in blood loss and total operative time compared with robot-assisted procedures. Surgeons specialized in the minimally invasive extraperitoneal approach have argued that their dissections secure access to the prostate, while reducing the risks of intestinal injury and complications. On the other hand, the transperitoneal approach has a higher incidence of adhesions compared with the extraperitoneal approach, mainly in patients who have undergone colectomy. However, previous abdominal or inguinal surgery is not a contraindication for transperitoneal robotic prostatectomy. Moreover, some studies have concluded that the procedure is safe and does not increase the risk of complications for these patients.

Robot-assisted radical cystectomy has been increasingly used as a minimally invasive surgical option for patients with locally advanced bladder cancer. Despite prior abdominal surgeries and adhesions being more common in older patients with bladder cancer, some studies [10] have reported that previous operations seem not to affect the likelihood of a safe robotic operation.

Regarding gynecological procedures, several studies [11] have compared the performance of robotic surgery and laparoscopy in total hysterectomy. The robotic approach reportedly showed significantly reduced operating time and decreased blood loss in the group with severe adhesions, which implies that robotic surgery is potentially superior to laparoscopy for complicated hysterectomies with severe adhesions that contraindicate the use of conventional laparoscopy. Furthermore, the robotic group had lower pain scores and needed fewer opioids than did the laparoscopic group regardless of the adherence score, which demonstrates that postoperative pain was affected only by the type of surgery. The results of several series have also demonstrated that robotic cancer surgeries are associated with favorable short-term surgical outcomes during hysterectomies in patients with severe adhesions, including in the treatment of endometrial and cervical cancer [12,13]. One possible explanation for this observation is that the robotic approach provides a more accurate dissection of pelvic adherence; therefore, it reduces operating time and tissue damage.

The management of pelvic adhesion can be time-consuming and technically difficult during surgery. Some authors [11] suggest that with appropriate training, the robotic approach is a better alternative for the treatment of patients with severe adhesions, with experienced surgeons further reducing the operating time and length of hospital stay. The robotic EndoWrist instrument has been a useful tool in dealing with these situations (Fig. 1). Certain strategies are useful in managing adhesions between the uterus and the intestines. For example, by bending the bipolar forceps at a 90° angle, surgeons can push the adherent uterus into an anterior position using the bipolar arm. Extra forceps can also be used to grasp the intestines through the accessory port to help separate the tissues. Counteraction is then created between the attached tissues to reveal the adhesive interface. This configuration helps dissect the adhesions more efficiently without damaging the intestinal serosa. The EndoWrist instrument also facilitates dissections at difficult angles in the pelvic cavity, thus reducing surgical difficulties in cases of patients with severe adhesions. These resources can also help to decrease the rate of intraoperative conversion to laparotomy in cases of severe adhesions.

A systematic review of 14 studies (38,057 patients) analyzed the surgical outcomes of patients with small bowel obstruction who underwent laparoscopic and exploratory laparotomy for adhesion lysis [14]. The study showed that

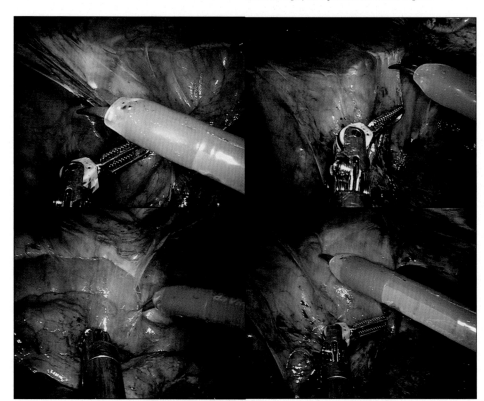

FIG. 1    Lysis of intra-abdominal and pelvic adhesions.

the laparoscopic approach had a significant advantage compared to the conventional open approach, including reduced risk of hospitalization, mortality, postoperative complications, and the incidence of bowel resection and surgical site infection [14]. The study's results also show that the presence of adherences was not a contraindication for minimally invasive surgeries, which may be a better choice for these patients.

In most studies [6] on esophagogastric surgery (including gastric cancer and bariatric procedures) as well as robotic and laparoscopic pancreatic surgery, no heterogeneity has been observed among the included studies. These studies demonstrated that there were no statistically significant differences in adhesion-related conversion rates, even when specifically analyzing the group of surgeries performed by experienced surgeons. However, it is important to highlight that these procedures have a limited number of cases compared with procedures involving patients with a history of previous abdominal surgeries.

Partial nephrectomies and other procedures have been evaluated at multiple centers [15,16] and have been described as safe procedures in patients with previous abdominal surgery. There was reportedly no significantly increased total surgical time, console time, conversion rate, renal function change, surgical margin, blood transfusion, or infection rates.

Some shared strategies are part of the same time as traditional laparoscopy, such as the creation of pneumoperitoneum for trocar positioning in abdominal surgeries before robotic docking. Several associated complications are related to the initial access to the peritoneal cavity [17], mainly with the use of a Veress needle that is blindly inserted into the abdomen before insufflation, which is described as the closed technique. The classic Veress needle puncture site is at the midline of the abdomen, close to the umbilicus; however, some studies refer to a lower risk of iatrogenic injury when the first puncture is at a site away from the midline, in an area that has not been previously manipulated [17–19]. Median incisions present the highest risk of adhesions around the umbilical scar, and many incisions also cause adherence in the umbilical region, despite being performed in another location. Thus, Palmer [18] described an alternative Veress needle puncture in the left hypochondrium, a few centimeters from the costal margin, on the midclavicular line. This site has been used by some surgeons in obese patients and in patients with a history of previous laparotomy (Fig. 2).

The presence of abdominal adhesions resulting from previous open abdominal surgery should no longer be considered an absolute contraindication. Several retrospective studies have shown that patient history and the finding of multiple adhesions is often a huge challenge; however, technical advances and the development and improvement of instrumentation, growing surgical experience, and increasing laparoscopic skills over time have made laparoscopic

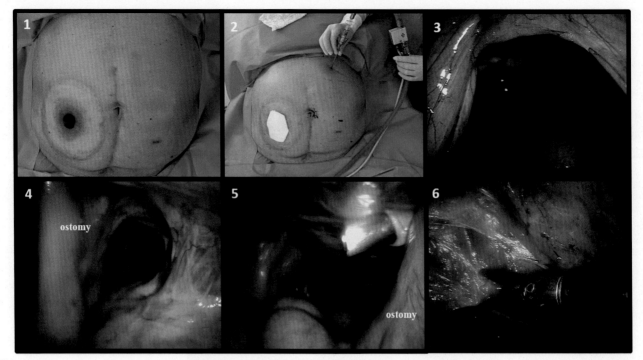

FIG. 2    The creation of pneumoperitoneum for trocar positioning in abdominal surgeries before robotic docking. (1) Patient with midline incision and ostomy; (2) and (3) Veress needle puncture in the left hypochondrium; (4) the trocars are passed under direct visualization; (5) precise lower rectal transection and endorectal plug exposure; (6) Lysis of adhesions for trocar positioning before robotic docking.

surgery the method of choice for several complexes and technically demanding procedures in some centers. Similarly, robotic surgeries have improved, especially in terms of a smaller learning curve in most procedures.

According to some studies, the incidence of adhesions caused by anterior laparotomy is up to 93%; moreover, up to 10% of cases of adhesions are not related to prior surgeries [14,20]. Abdominal cavity adhesions can be the result of factors other than open surgery and previous laparoscopic surgery, as intra-abdominal and idiopathic inflammation and infections are risk factors. Laparoscopic surgery has a much lower incidence of adhesions (33%) and is less likely to require more extensive dissection. However, most surgeons agree that adhesions can complicate procedures; therefore, they require greater care and attention. Each specialty should be individually analyzed and the surgeon's expertise in highly complex procedures, such as the handling of intestinal loop adhesions, must be considered before safely and efficiently proceeding with the intended operation.

## Future scenario

- The future of surgical practice, particularly in the context of patients with a history of abdominal surgeries and adhesions, is likely to witness several advancements and trends. One notable trend is the continued evolution and integration of robotic-assisted surgery. Robotic platforms are expected to become even more sophisticated and versatile. These robots offer advantages like 3D vision, precise instrument control, and ergonomic benefits, which can significantly aid in managing adhesions and challenging abdominal conditions. The insertion of the endoscope in all trocars is an additional advantage in hostile abdominal cavities, enhancing visualization and surgical maneuverability. As robotic technology continues to advance, it is anticipated that more surgeons will adopt and master these tools, leading to better outcomes for patients with abdominal adhesions.
- The field of minimally invasive surgery is likely to see the development of more specialized instruments and techniques for managing adhesions. These innovations will aim to reduce conversion rates and complications, making surgery safer and more efficient for patients with a history of previous abdominal procedures. Additionally, the integration of augmented reality, artificial intelligence, and telemedicine may further enhance the surgical process by providing real-time guidance and decision support to surgeons. As surgical teams gain access to more comprehensive data and technological support, the management of adhesions and the overall safety of surgeries in patients with abdominal adhesions are expected to improve in the coming years.

## Key points

- Robotic surgery is a viable option for most patients with a history of abdominal surgery, with advancements in instruments, like EndoWrist, aiding in dissection and reducing conversion rates.
- Previous abdominal surgeries increase the risk of perioperative complications, primarily due to intra-abdominal adhesions. Robotic surgery exhibits a lower conversion rate compared to laparoscopy, especially in complex cases.
- Robotic surgery provides technical benefits, such as improved 3D vision and precision-controlled instruments for adhesion management.
- Specific strategies for trocar placement, pneumoperitoneum creation, and adhesion management in robotic surgery help mitigate complications.

## Competing interests

The author declares no conflicts of interest.

## Appendix: Supplementary material

Supplementary material related to this chapter can be found on the accompanying CD or online at https://doi.org/10.1016/B978-0-443-13271-1.00044-3.

## References

[1] Figueiredo MN, Campos FG, D'Albuquerque LA, Nahas SC, Cecconello I, Panis Y. Short-term outcomes after laparoscopic colorectal surgery in patients with previous abdominal surgery: a systematic review. World J Gastrointest Surg 2016;8:533–40.
[2] Zeng WG, Liu MJ, Zhou ZX, Hou HR, Liang JW, Wang Z, et al. Impact of previous abdominal surgery on the outcome of laparoscopic resection for colorectal cancer: a case-control study in 756 patients. J Surg Res 2015;199:345–50.
[3] Jayne D, Pigazzi A, Marshall H, et al. Effect of robotic-assisted vs conventional laparoscopic surgery on risk of conversion to open laparotomy among patients undergoing resection for rectal cancer: the ROLARR randomized clinical trial. JAMA 2017;318(16):1569–80. https://doi.org/10.1001/jama.2017.7219.
[4] Park S, Kang J, Park EJ, Baik SH, Lee KY. Laparoscopic and robotic surgeries for patients with colorectal cancer who have had a previous abdominal surgery. Ann Coloproctol 2017;33:184–91.
[5] Hu M, Miao C, Wang X, Ma Y. Robotic surgeries for patients with colorectal cancer who have undergone abdominal procedures: protocol for meta-analysis. Medicine (Baltimore) 2018;97, e0396.
[6] Milone M, Manigrasso M, Anoldo P, D'Amore A, Elmore U, Giglio MC, Rompianesi G, Vertaldi S, Troisi RI, Francis NK, et al. The role of robotic visceral surgery in patients with adhesions: a systematic review and meta-analysis. J Pers Med 2022;12:307. https://doi.org/10.3390/jpm12020307.
[7] Feldbrügge L, Ortiz S, Frisch O, Benzing C, Krenzien F, Riddermann A, Kästner A, Nevermann N, Malinka T, Schoening W, Pratschke J, Schmelzle M. Safety and feasibility of robotic liver resection after previous abdominal surgeries. Surg Endosc 2022;36. https://doi.org/10.1007/s00464-021-08572-1.
[8] Feldbrügge L, Wabitsch S, Benzing C, Krenzien F, Kästner A, Haber PK, Atanasov G, Andreou A, Öllinger R, Pratschke J, Schmelzle M. Safety and feasibility of laparoscopic liver resection in patients with a history of abdominal surgeries. HPB 2019;22(8):1191–6.
[9] Siddiqui SA, Krane LS, Bhandari A, Patel MN, Rogers CG, Stricker H, Peabody JO, Menon M. The impact of previous inguinal or abdominal surgery on outcomes after robotic radical prostatectomy. Urology 2010;75(5):1079–82. ISSN 0090-4295 https://doi.org/10.1016/j.urology.2009.09.004.
[10] Yuh BE, Ciccone J, Chandrasekhar R, Butt ZM, Wilding GE, Kim HL, Mohler JL, Guru KA. Impact of previous abdominal surgery on robot-assisted radical cystectomy. JSLS 2009;13(3):398–405 [PMID: 19793483; PMCID: PMC3015968].
[11] Chiu L-H, Chen C-H, Tu P-C, Chang C-W, Yen Y-K, Liu W-M. Comparison of robotic surgery and laparoscopy to perform total hysterectomy with pelvic adhesions or large uterus. J Minim Access Surg 2015;11(1):87–93. https://doi.org/10.4103/0972-9941.147718.
[12] El Hachem L, Acholonu Jr UC, Nezhat FR. Postoperative pain and recovery after conventional laparoscopy compared with robotically assisted laparoscopy. Obstet Gynecol 2013;121:547–53.
[13] Soliman PT, Langley G, Munsell MF, Vaniya HA, Frumovitz M, Ramirez PT. Analgesic and antiemetic requirements after minimally invasive surgery for early cervical cancer: A comparison between laparoscopy and robotic surgery. Ann Surg Oncol 2013;20:1355–9.
[14] Sajid MS, Khawaja AH, Sains P, Singh KK, Baig MK. A systematic review comparing laparoscopic vs open adhesiolysis in patients with adhesional small bowel obstruction. Am J Surg 2016;212:138–50.
[15] Zargar H, Isac W, Autorino R, Khalifeh A, Nemer O, Akca O, et al. Robot-assisted laparoscopic partial nephrectomy in patients with previous abdominal surgery: single center experience. Int J Med Robot 2015;11:389–94.
[16] Abdullah N, Rahbar H, Barod R, Dalela D, Larson J, Johnson M, et al. Multicentre outcomes of robot-assisted partial nephrectomy after major open abdominal surgery. BJU Int 2016;118:298–301. https://doi.org/10.1111/bju.13408 [PMID: 27417163].

[17] Rohatgi A, Widdison AL. Left subcostal closed (Veress needle) approach is a safe method for creating a pneumoperitoneum. J Laparoendosc Adv Surg Tech A 2004;14(5):278–80.

[18] Palmer R. Safety in laparoscopy. J Reprod Med 1974;13(1):1–5.

[19] Schwartz ML, Drew RL, Andersen JN. Induction of pneumoperitoneum in morbidly obese patients. Obes Surg 2003;13(4):601–4.

[20] Seetahal S, Obirieze A, Cornwell EE, Fullum T, Tran D. Open abdominal surgery: A risk factor for future laparoscopic surgery? Vol. 209. Society of Black Academic Surgeons; 2015. p. 623–6.

# 34

# Basics of economic analysis of robotic surgery in its third decade

*Gustavo Cardoso Guimarães*[a,b,c] *and Renato Almeida Rosa de Oliveira*[b,c]

ᵃHospital Israelita Albert Einstein, São Paulo, SP, Brazil ᵇUrology Department, Beneficência Portuguesa de São Paulo, São Paulo, SP, Brazil ᶜInstituto de Urologia, Oncologia e Cirurgia Robótica, São Paulo, SP, Brazil

## Introduction

Significant improvements have been made to robotic surgery since its introduction in the late 20th century, as it has become a rapidly growing technique that has transformed the way surgical procedures are performed. Robotic surgery utilizes advanced computer technology and robotics to assist surgeons in performing complex procedures with greater precision, accuracy, and control. It has led to reduced complications, faster recovery times, and better outcomes for patients. However, with these technology enhancements, the costs of robotic surgery have become a significant concern for hospitals and patients. This chapter aims to provide a comprehensive overview of the basics of economic analysis in robotic surgery, shedding light on the financial aspects and the value that robotic surgery brings to healthcare systems. Related costs will be discussed as well as the initial investment, maintenance, and procedural costs [1].

## Advantages of robotic surgery

### Improved precision

Robotic surgery allows surgeons to perform intricate procedures with unparalleled precision and accuracy. The robotic arms, controlled by the surgeon, offer enhanced dexterity and the ability to perform complex maneuvers. This precision reduces the occurrence of complications, leading to improved patient outcomes [2].

### Shorter recovery times

The minimally invasive nature of robotic surgery contributes to shorter recovery times for patients. Smaller incisions result in less postoperative pain, a reduced risk of infection, and a quicker return to daily activities [2]. This faster recovery leads to a decreased hospital length of stay, reducing healthcare costs [3].

### Improved patient outcomes

Studies have shown that robotic surgery can lead to improved patient outcomes in terms of reducing blood loss, lowering the need for blood transfusions, and decreasing postoperative complications rates [4]. These positive outcomes result in cost savings by reducing the need for additional interventions and treatments.

## Economic evaluation of robotic surgery

Economic evaluation is crucial in determining the cost effectiveness and value delivered to patients through new medical technologies such as robotic surgery. It involves an assessment of the potential benefits, costs, and risks associated with these innovations. Several primary methods are commonly used for economic analysis in healthcare, including cost-effectiveness analysis (CEA), cost-utility analysis (CUA), and cost-benefit analysis (CBA) [5].

### Cost-effectiveness analysis (CEA)

Cost-effectiveness analysis is a widely used economic evaluation tool that compares the costs of healthcare interventions to their outcomes in terms of health improvements. The key parameter in CEA is the unitary cost of health gain, such as cost per life-year gained, cost per quality-adjusted life-year (QALY) gained, or cost per successful surgical outcome. Robotic outcomes must be compared to the outcomes of open and laparoscopic surgeries. New technologies tend to be more expensive and more efficient. For this reason, the calculated ratio is analyzed by comparing thresholds for each selected dimension of gains. In the robotic surgery context, CEA is a valuable tool to access the economic impact and advantages over conventional surgical methods [5].

### Cost-utility analysis (CUA)

Cost-utility analysis complements CEA by incorporating quality of life into the assessment of healthcare interventions. It measures the cost per QALY gained, considering the patient's health-related quality of life along with the associated costs. CUA provides a more comprehensive evaluation of the economic value of robotic surgery by capturing both clinical outcomes and the patient's perspective. The calculation is performed the same way as CEA [5].

### Cost-benefit analysis (CBA)

Cost-benefit analysis goes beyond CEA and CUA by quantifying the monetary value of the health outcomes compared to the costs involved. CBA assesses the return on investment (ROI) of robotic surgery, considering both the direct and indirect economic benefits generated by using the technology. This analysis aids decision makers in understanding broader economic implications and the long-term sustainability of implementing robotic surgery [5].

## Factors affecting the economics of robotic surgery

Several factors influence the economic analysis of robotic surgery, including:

### Initial investment and maintenance costs

The acquisition and maintenance costs of robotic surgical systems are substantially high. These costs encompass the acquisition price, system maintenance, upgrades, and instrument costs. Economic analysis should consider the lifespan of the system and the utilization rate to calculate the ROI accurately.

#### *Initial investment costs*

Robotic systems are very expensive and require a significant initial investment from hospitals. The world's most used technology is the Da Vinci Surgical System, which is manufactured by Intuitive Surgical Inc., and has costs between $1.5 and $2.5 million, depending on the model and the number of instruments included.

Additionally, hospitals must also invest in training programs for surgeons and staff, which can consume several weeks and can cost between $20,000 and $50,000 per surgeon, including both onsite training and travel expenses for the surgeon and team.

#### *Maintenance costs*

Robotic surgery systems require regular maintenance to ensure that they are functioning properly. Maintenance costs include the cost of replacing worn or damaged instruments as well as the cost of software updates and system upgrades. These costs can be significant, with some hospitals reporting annual maintenance costs of between $150,000 and $250,000.

Moreover, hospitals must also consider the cost of repairing eventual system malfunctions or breakdowns, which can vary significantly depending on the extent of the damage and the device life cycle. Some repairs can be relatively inexpensive while others can cost tens of thousands of dollars.

## Procedure time and efficiency

Robotic surgery has the potential to reduce operation times, leading to a shorter hospital length of stay and quicker patient recovery. Reduced procedure times can result in cost savings for hospitals and healthcare systems. Economic evaluation should examine the impact of robotic surgery on operating room utilization and associated cost savings [3,6].

## Procedural costs

The costs related to robotic surgery procedures can vary significantly depending on the type of procedure, the hospital, and the team's expertise. In general, robotic surgery procedures are more expensive than traditional open or laparoscopic surgeries. This is primarily due to the higher cost of the robotic surgery system, supplies, and the additional training required for surgeons and staff.

The cost of a robotic surgery procedure can also be influenced by the surgical length and the number of instruments used. Longer procedures or those that require greater numbers of instruments can be more expensive. Furthermore, some procedures may require the use of additional equipment, such as image-guided or disposable devices, which may increase the overall costs.

## Reimbursement and insurance coverage

The availability and amount of reimbursement or insurance coverage greatly influence the economic viability of robotic surgery. Understanding the reimbursement policies and potential changes in payment models is crucial for conducting a comprehensive economic analysis.

### Insurance coverage

In general, insurance providers are more likely to cover robotic surgery procedures that are considered medically necessary and have been proven to be effective. However, coverage may be limited or denied for procedures that are considered experimental or have not yet been proven to be effective.

Coverage policies may vary among different companies and countries and in some cases, patients may be required to partially pay costs out of pocket. In developing countries, coverage for robotic procedures by health insurers can be a great challenge.

### Reimbursement policies

Robotic surgery's economic viability is closely related to reimbursement policies established by insurance payers and healthcare providers. As the technology evolves, reimbursement models must adapt to ensure access to patients and fair compensation for hospitals, surgeons, and anesthesiologists.

### Key considerations for reimbursement policies include

Current reimbursement landscape: Reimbursement rates for robotic surgery vary across regions and healthcare systems. Some countries have specific reimbursement models for robotic procedures while others compose them with traditional laparoscopic or open surgery payments. The continuous evaluation and adjustment of such policies are necessary to align with the evolving landscape of robotic surgery.

### Value-based reimbursement

Value-based reimbursement models, which emphasize quality outcomes, risk sharing, and cost effectiveness, are gaining prominence. The better outcomes of robotic surgery and reduced healthcare system use can guide the development of value-based reimbursement frameworks.

## Patient outcomes and complication rates

An economic analysis must consider the potential cost savings resulting from the improved outcomes. Reduced complication rates and readmissions, reduced blood loss, lower transfusion rates, shorter hospital stays and bed occupancy, and lower postoperative pain [6] may positively impact costs and should be considered as potential drivers of economic benefits. Furthermore, it is important to highlight the reduction in the need for analgesics, especially opioids and their related abuse problems [3,6,7].

## Learning curve and surgeon training

The learning curve associated with robotic surgery affects procedure length and outcomes. Economic analysis should address the training costs for surgeons and the surgical team as well as the impact of expertise on cost effectiveness [2,8].

### Understanding the learning curve

The robotic learning curve refers to the process of acquiring the necessary skills and proficiency in performing robotic-assisted procedures. Surgeons must familiarize themselves with the robotic platform, including the console interface, instrumentation, and robotic arm control. They must also exhibit proficiency in maneuvering and operating the surgical robot for precise and efficient surgical interventions [9–12].

### Challenges in robotic surgery learning curve

Technical complexity: Robotic systems are complex, comprising multiple components and advanced technologies. Surgeons must undergo comprehensive training to understand the system's intricacies and gain proficiency in using it effectively [10–12].

### Psychomotor skills acquisition

Robotic surgery demands the development of specialized psychomotor skills, including hand-eye coordination, depth perception, and instrument manipulation. These skills often require time and practice to develop, contributing to the learning curve [11,12].

### Surgical decision making

As surgeons master the technical aspects of robotic surgery, they must also learn to adapt their decision-making process to the unique environment and capabilities of the robotic system. Developing the ability to navigate through complex surgical scenarios is an essential aspect of the learning curve.

### Importance of experience and training

Several studies have emphasized the significance of experience and training in optimizing the learning curve and enhancing surgical outcomes in robotic surgery.

The British Association of Urologic Surgeons considers that the learning curve for a robotic radical prostatectomy can take between 50 and 250 cases, with most authors finding that the learning curve plateaued after 80–100 robotic-assisted surgeries for prostate cancer, indicating that surgeon experience strongly influences procedural success and patient safety [13]. Another study focused on robotic-assisted gynecologic surgeries, concluded that surgeons who completed fellowship training and underwent additional mentorship had shorter learning curves and better outcomes [14,15].

The American College of Surgeons and the Society of American Gastrointestinal and Endoscopic Surgeons have recognized the importance of training programs and simulation-based training for surgeons venturing into robotic surgery [16,17].

## Market competition and technological advances

The competitiveness of the market and the pace of technological advancements influence the price of robotic surgical systems and instruments. Economic analysis should consider the impact of market dynamics and future innovations on cost-related issues.

Despite the high costs associated with robotic surgery, there is evidence to suggest that it may be cost effective in some cases by comparing costs to clinical benefits and outcomes provided. Studies have shown that robotic surgery can lead to reduced complication rates, shorter hospital stays, and faster recovery times, which can ultimately lead to cost savings for hospitals and patients. CEA plays a crucial role in evaluating the economic impact of this technology [6].

However, robotic cost effectiveness depends on several factors, including the type of procedure, the patient population, and the hospital's overall financial situation. Hospitals must carefully consider the costs and benefits of robotic surgery before making a significant investment in the technology.

Several key factors contribute to CEA.

## Direct costs

Robotic surgery involves significant upfront investments in acquisition, maintenance, and instrument costs. Although the initial investment is substantial, studies suggest that over time, unitary costs can be comparable to or even lower than traditional surgical approaches due to reduced hospital stays, decreased complications, and faster patient recovery.

## Indirect costs

Indirect costs encompass various factors, such as increased operative time and the learning curve associated with adopting robotic technology. Initially, surgeons and operating room staff may require additional training and experience to maximize the benefits of robotic surgery, which can impact procedural efficiency and increase costs. However, as experience grows, these indirect costs tend to diminish.

## Long-term financial implications

Beyond the immediate cost effectiveness and reimbursement considerations, understanding the long-term financial implications of robotic surgery is essential. These implications extend to various stakeholders, including hospitals, surgeons, patients, and healthcare systems as a whole.

### Hospital economics

Hospitals must evaluate the ROI associated with robotic surgery. While the initial capital expenditure may be significant, some factors are directly and indirectly implicated in financial incomes: increased surgery volume, increased services portfolio, reduced complications, and improved patient satisfaction can contribute to long-term financial benefits. Robotic surgery also allows hospital positioning as reference centers, increasing volume and profits. Strategic planning, accurate volume projections, and efficient resource allocation are crucial for optimizing the financial impact of robotic surgery on hospitals.

### Surgeon economics

Surgeons considering the adoption of robotic surgery should assess the potential impact on their practice and financial well-being. Factors such as surgery volume, patient demand, and reimbursement rates directly influence the financial viability of incorporating robotic surgery into their practice. Analyzing the surgeon's current caseload and estimating the growth in robotic procedures can inform the financial decision-making process.

### Patient affordability

As robotic surgery becomes more frequent, ensuring patient affordability and access to this technology is of paramount importance. Balancing the benefits of robotic surgery with the potential financial burden on patients is a critical consideration. Collaboration among healthcare providers, insurance companies, and policymakers is necessary to address issues related to patient affordability and equitable access to robotic surgery.

### Market forces and technological advancements

As robotic surgery enters its third decade, it is encountering a landscape shaped by market forces and rapid technological advancements. Robotic surgical system manufacturers face competition from both established players and emerging start-ups, driving innovation and continual improvement. This competition has led to advancements such as smaller robotic platforms, improved imaging capabilities, and enhanced haptic feedback, making robotic surgery more accessible and appealing to a wider range of surgical specialties.

## Balancing the costs and benefits

While the economic analysis of robotic surgery showcases its remarkable potential, it is crucial to achieve a balance between costs and benefits. The high upfront costs associated with robotic systems must be weighed against the potential long-term savings resulting from reduced postoperative complications, shorter hospital stays, and improved patient outcomes. Considering this, healthcare providers and policymakers need to judiciously evaluate the cost effectiveness and long-term financial implications of adopting robotic surgery.

## Future scenario

Despite the proven benefits, robotic surgery faces numerous challenges for broader adoption. One significant barrier is the high initial investment required for robotic systems. Consequently, smaller healthcare facilities may face financial constraints limiting their access to this technology. Efforts are under way to develop cost-effective robotic systems that could alleviate this challenge [18].

Additionally, the role of health technology assessments, including value-based assessments, in evaluating robotic surgery's economic impact cannot be understated. Continued research and long-term studies should focus on patient outcomes, socioeconomic factors, and regional variations of costs and benefits, ensuring a comprehensive assessment of the economics of robotic surgery [19–21].

## Conclusion

As with other branches of industry, the acquisition of technological innovations in medicine is certain, as is the concern with its economic sustainability. Robotic surgery needs to be approached with a comprehensive understanding of its unique characteristics, the benefits delivered to patients, the viability of investments, the market forces that govern it, and the perpetuity of the entire project.

## Key points

- Robotic surgery has become the new standard in many procedures despite its higher costs.
- Control of inventory, supplies, and surgical time are essential to ensure the economic viability of the robotic surgery program.
- Dedicated program management is fundamental to ensure the economic viability of the robotic surgery program.

## References

[1] Probst P. A review of the role of robotics in surgery: to DaVinci and beyond! Mo Med 2023;120(5):389–96. PMID: 37841561. PMCID: PMC10569391.
[2] Morris B. Robotic surgery: applications, limitations, and impact on surgical education. MedGenMed 2005;7(3):72. PMID: 16369298. PMCID: PMC1681689.
[3] Guimarães GC, Oliveira RAR, Santana TBM, Favaretto RL, Mourão TC, Rocha MM, Campos RM, Zequi SC. Comparative analysis of functional outcomes between two different techniques after 1088 robotic-assisted radical prostatectomies in a high-volume cancer center: a clipless approach. J Endourol 2019;33(12):1017–24. https://doi.org/10.1089/end.2019.0361. Epub 2019 Nov 7.
[4] Ahlering TE, Skarecky D, Lee D, Clayman RV. Successful transfer of open surgical skills to a laparoscopic environment using a robotic interface: initial experience with laparoscopic radical prostatectomy. J Urol 2003;170(5):1738–41.
[5] Kotsis SV, Chung KC. Fundamental principles of conducting a surgery economic analysis study. Plast Reconstr Surg 2010;125(2):727–35. https://doi.org/10.1097/PRS.0b013e3181c91501. PMID: 19910842. PMCID: PMC4414117.
[6] de Oliveira RAR, Guimarães GC, Mourão TC, de Lima FR, Santana TBM, Lopes A, de Cassio ZS. Cost-effectiveness analysis of robotic assisted versus retropubic radical prostatectomy: a single cancer center experience. J Robot Surg 2021;15(6):859–68. https://doi.org/10.1007/s11701-020-01179-z. Epub 2021 Jan 8 PMID: 33417155.
[7] Trinh QD, Sammon J, Sun M, Ravi P, Ghani KR, Bianchi M, Jeong W, Shariat SF, Hansen J, Schmitges J, Jeldres C, Rogers CG, Peabody JO, Montorsi F, Menon M, Karakiewicz PI. Perioperative outcomes of robot-assisted radical prostatectomy compared with open radical prostatectomy: results from the nationwide inpatient sample. Eur Urol 2012;61(4):679–85. https://doi.org/10.1016/j.eururo.2011.12.027. Epub 2011 Dec 22 PMID: 22206800.
[8] Moorthy K, Munz Y, Sarker SK, Darzi A. Objective assessment of technical skills in surgery. BMJ 2003;327(7422):1032–7. https://doi.org/10.1136/bmj.327.7422.1032. PMID: 14593041. PMCID: PMC261663.

[9] Collins JW, Levy J, Stefanidis D, Gallagher A, Coleman M, Cecil T, Ericsson A, Mottrie A, Wiklund P, Ahmed K, Pratschke J, Casali G, Ghazi A, Gomez M, Hung A, Arnold A, Dunning J, Martino M, Vaz C, Friedman E, Baste JM, Bergamaschi R, Feins R, Earle D, Pusic M, Montgomery O, Pugh C, Satava RM. Utilising the Delphi process to develop a proficiency-based progression train-the-trainer course for robotic surgery training. Eur Urol 2019;75(5):775–85. https://doi.org/10.1016/j.eururo.2018.12.044. Epub 2019 Jan 19 PMID: 30665812.

[10] Burke JR, Fleming CA, King M, El-Sayed C, Bolton WS, Munsch C, Harji D, Bach SP, Collins JW. Utilising an accelerated Delphi process to develop consensus on the requirement and components of a pre-procedural core robotic surgery curriculum. J Robot Surg 2023;17(4):1443–55. https://doi.org/10.1007/s11701-022-01518-2. Epub 2023 Feb 9 PMID: 36757562. PMCID:PMC9909133.

[11] Brook NR, Dell'Oglio P, Barod R, Collins J, Mottrie A. Comprehensive training in robotic surgery. Curr Opin Urol 2019;29(1):1–9. https://doi.org/10.1097/MOU.0000000000000566. PMID: 30394945.

[12] Sridhar AN, Briggs TP, Kelly JD, Nathan S. Training in robotic surgery—an overview. Curr Urol Rep 2017;18(8):58. https://doi.org/10.1007/s11934-017-0710-y. PMID: 28647793. PMCID: PMC5486586.

[13] BAUS. Robotic surgery curriculum—guidelines for training, https://www.baus.org.uk/_userfiles/pages/files/Publications/Robotic%20Surgery%20Curriculum.pdf.

[14] Altok M, Achim MF, Matin SF, Pettaway CA, Chapin BF, Davis JW. A decade of robot-assisted radical prostatectomy training: time-based metrics and qualitative grading for fellows and residents. Urol Oncol 2018;36(1):13.e19–25. https://doi.org/10.1016/j.urolonc.2017.08.028. Epub 2017 Sep 28 PMID: 28964658.

[15] Soliman PT, Iglesias D, Munsell MF, Frumovitz M, Westin SN, Nick AM, Schmeler KM, Ramirez PT. Successful incorporation of robotic surgery into gynecologic oncology fellowship training. Gynecol Oncol 2013;131(3):730–3. https://doi.org/10.1016/j.ygyno.2013.08.039. Epub 2013 Sep 19 PMID: 24055616. PMCID: PMC3856555.

[16] Satava RM, Stefanidis D, Levy JS, Smith R, Martin JR, Monfared S, Timsina LR, Darzi AW, Moglia A, Brand TC, Dorin RP, Dumon KR, Francone TD, Georgiou E, Goh AC, Marcet JE, Martino MA, Sudan R, Vale J, Gallagher AG. Proving the effectiveness of the fundamentals of robotic surgery (FRS) skills curriculum: a single-blinded, multispecialty, multi-institutional randomized control trial. Ann Surg 2020;272(2):384–92. https://doi.org/10.1097/SLA.0000000000003220. PMID: 32675553.

[17] Chen R, Rodrigues Armijo P, Krause C, SAGES Robotic Task Force, Siu KC, Oleynikov D. A comprehensive review of robotic surgery curriculum and training for residents, fellows, and postgraduate surgical education. Surg Endosc 2020;34(1):361–7. https://doi.org/10.1007/s00464-019-06775-1. Epub 2019 Apr 5 PMID: 30953199.

[18] Shademan A, Decker RS, Opfermann JD, Leonard S, Krieger A, Kim PC. Supervised autonomous robotic soft tissue surgery. Sci Transl Med 2016;8(337):337ra64. https://doi.org/10.1126/scitranslmed.aad9398. PMID: 27147588.

[19] Nabi J, Friedlander DF, Chen X, Cole AP, Hu JC, Kibel AS, Dasgupta P, Trinh QD. Assessment of out-of-pocket costs for robotic cancer surgery in US adults. JAMA Netw Open 2020;3(1), e1919185. https://doi.org/10.1001/jamanetworkopen.2019.19185.

[20] Salkowski M, Checcucci E, Chow AK, Rogers CC, Adbollah F, Liatsikos E, Dasgupta P, Guimaraes GC, Rassweiler J, Mottrie A, Breda A, Crivellaro S, Kaouk J, Porpiglia F, Autorino R. New multiport robotic surgical systems: a comprehensive literature review of clinical outcomes in urology. Ther Adv Urol 2023;15, 17562872231177781. https://doi.org/10.1177/17562872231177781. PMID: 37325289. PMCID: PMC10265325.

[21] Guimarães GC. New platforms in robotic surgery. In: Robotic surgery devices in surgical specialties. Cham: Springer International Publishing; 2023. p. 225–32.

# 35

# The role of a clinical engineer in a robotic program: Conceptualization, devices choice, hospital facilities preparation, technical assistance

*Lucio Flavio de Magalhães Brito*

Pontifical Catholic University of São Paulo, Exact and Technological Sciences Faculty—Biomedical Engineering, São Paulo, Brazil

## Introduction

What is clinical engineering? Several definitions have been used to define this profession; however, the analysis of the etymology of the word clinical facilitates the understanding and helps in the perception of the scope of this area of activity of engineering professionals in the health area. A query in the online etymology dictionary on the word clinical [1] found the following definition:

> clinic (n.)
> 1620s, "bedridden person, one confined to his bed by sickness," from French clinique (17c.), from Latin clinicus "physician that visits patients in their beds," from Greek klinike (techne) "(practice) at the sickbed," from klinikos "of the bed," from kline "bed, couch, that on which one lies," from suffixed form of PIE root *klei- "to lean."

Over time, several authors and organizations have created definitions that help in understanding the possibilities of action and in the dissemination of the benefits that clinical engineering professionals have been causing in health systems.

In 1991, Bauld defined the clinical engineer as follows:

> "a professional who supports and advances patient care by applying engineering and managerial skills to health care technology," [2].

In 1991, Pacela suggested a definition for the biomedical engineer, with the aim of differentiating this professional from the clinical engineer.

> "applies a wide spectrum of engineering level knowledge and principles to the understanding, modification or control of human or animal biological systems" [3].

Founded in 1990, the American College of Clinical Engieering, defined clinical engineer in 1992, with the following statement:

> "a Clinical Engineer is a professional who supports and advances patient care by applying engineering and managerial skills to healthcare technology [4]."

In 1992, in the book Management of Medical Technology edited by Bronzino, the following definition is described:

*"an engineer who has graduated from an accredited academic program in engineering or who is licensed as a professional engineer or engineer-in-training and is engaged in the application of scientific and technological knowledge developed through engineering education and subsequent professional experience within the health care environment n support of clinical activities [5]."*

In 2017, the World Health Organization presented the term clinical engineering as a synonym for biomedical engineering, defining it as follows:

*"Biomedical engineering" includes equivalent or similar disciplines, whose names might be different, such as medical engineering, electromedicine, bioengineering, medical and biological engineering, and clinical engineering [6].*

Finally, in 2019, the Clinical Engineering Division (CED) of the International Federation for Medical and Biological Engineering (IFMBE) defined clinical engineers with the following statement:

*"Professionals who are qualified by education and/or registration to practice engineering in the health care environment where technology is created, deployed, taught, regulated, managed or maintained related to health services" [7].*

All definitions, in a way, place the performance of this professional close to the patient and, together with other professionals who work in health systems, such as doctors, nurses, physiotherapists, psychologists, dentists, administrators, economists, etc.

In practice, there are numerous activities, areas of knowledge, and challenges for the development and consolidation of these professionals' activities. Examples of activities in actual practice include, but are not limited to, those mentioned as follows:

- Supervision of a hospital clinical engineering department that includes clinical engineers and biomedical equipment technicians (BMETs).
- **Prepurchase evaluation and planning for new medical technology.**
- Design, modification, or repair of sophisticated medical instruments or systems.
- **Cost-effective management of a medical equipment calibration and repair service.**
- Safety and performance testing of medical equipment by BMETs.
- **Inspection of all incoming equipment (new and returning repairs).**
- Establishment of performance benchmarks for all equipment.
- Medical equipment inventory control.
- **Coordination of outside services and vendors.**
- **Training of medical personnel for the safe and effective use of medical devices and systems.**
- Clinical application engineering, such as custom modification of medical devices for clinical research or evaluation of new noninvasive monitoring systems.
- **Biomedical computer support.**
- **Input to the design of clinical facilities where medical technology is used [e.g., operating rooms (ORs) or intensive care units].**
- Development and implementation of documentation protocols required by external accreditation and licensing agencies [5].

Among the activities aforementioned, those shown in bold are those that are most in line with the content of this chapter. These activities are developed by the clinical engineer, when in the process of adopting a new technology, which will be used in the hospital environment.

However, before detailing these activities in practice, it is important that the clinical engineer and physicians, hospital administrators, and health professionals keep in mind the concepts and definitions associated with the process of healthcare technology assessment.

Figure 1, adapted by Panerai and Mhor [8] from Banta et al. [9], shows a way of visualizing the healthcare technology life cycle.

Innovation is the phase where the equipment, device, or work process is created, where money is usually spent on research and development activities. It ends with the first practical use.

Initial diffusion occurs when new technology is presented at conferences, scientific events and in the mass media. In this phase, knowledge of the new technology generates strong expectations in all interested parties, such as patients, physicians, hospital administrators, and other health professionals. For patients, the new technology can bring benefits, alleviate or eliminate their health problems; for physicians, it may allow obtaining additional income, professional prestige, or the belief about the tangible benefits it can offer to patients; for administrators, the new technology can attract better medical teams, competitive advantages over competing hospitals, and finally, the new technology can lead to the need for professional updating, given the impact it will have on work routines.

**Technology Life-Cycle**

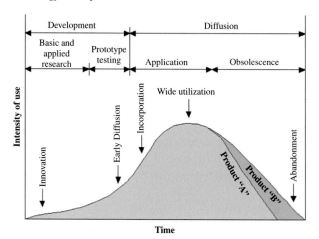

FIG. 1 Presents a schematic design of the life cycle of a health technology and its main phases in relation to the intensity of use over time.

Incorporation, in turn, occurs when the technology is already established and when there are those who are willing to pay for it. Be it health plans that decide to reimburse patients or subsidize the new technology due to the set of health benefits or the quality of care that are associated with it. This is the phase where new benefits and risks are introduced in hospitals.

Wide utilization is the most important phase as this is where health benefits are accrued, critical resources are wasted, and new security risks are detected. Despite more than 20 years of use, there are studies that show the negative or dark side of this technological resource, which must be known by clinical engineering professionals before adoption. In this sense, it is convenient for the clinical engineer to verify alerts, recalls, and adverse events, verify their existence in the regulatory agencies of the country where the equipment will be installed or in the systems of the countries in which the equipment offered is manufactured. The following addresses may be useful when carrying out this activity: FDA (Food and Drug Administration) https://www.accessdata.fda.gov/scripts/cdrh/cfdocs/cfres/res.cfmin the United States and in the United Kingdon MHRA (Medicines and Healthcare products Regulatory Agency) in the following link https://www.gov.uk/drug-device-alerts. The detailed evaluation of the existing information, if any, allows us to know about past problems, whether they were solved and what to do based on the data obtained, to improve the hospital's technosurveillance actions.

Abandonment is the phase in which one technology is replaced by another. New technology may offer faster results, be more accurate, safer, or more cost-effective. There may also be resistance to change, either due to the lack of recovery of time or economic resources previously invested in the previous technology. An example of abandonment currently taking place is the replacement of X-ray film developing machines by digital systems such as CR (computed radiography) or DR (direct radiography).

Other example of abandonment was the use of LED (light-emitting diode) technology to replace fluorescent lamps, used in the treatment of hyperbilirubinemia in newborn patients. LED technology, in addition to a much longer useful life, eliminated the risk of exposure to mercury present in the previous technology and ensured greater stability of the power offered ($W/cm^2$) in the phototherapy equipment used for this purpose.

Having conceptualized clinical engineering and the life cycle of a health technology, it is convenient, for practical purposes, to list a set of desirable actions by the clinical engineering professional during the process of adopting a robotic surgery system.

## Prepurchase evaluation and planning for new medical technology

For planning purposes, the clinical engineer and interested parties should keep in mind the following macro steps of the planning process for the adoption of a health technology, proposed by the ECRI (Emergency Care Research Institute) in its publication, Medical Equipment Planning: A Guide to Getting it Right the First Time:

- Implement a clear process for making medical equipment decisions.
- Stay mindful of the "bigger picture" for clinical services.
- Don't stop at choosing the right equipment. Also consider your total cost of ownership (TCO).
- Dive into the details, too.
- Partner with unbiased experts who give evidence-based guidance [10].

## Instrument for obtaining supplier market information and comparing offers

During the daily routine, the clinical engineering service is asked to participate in acquisition processes of new technological resources. Due to the set of theoretical and practical engineering knowledge that they possess, they work with other professionals with the objective of acquiring technological resources in health, in a rational, sustainable, and safe way.

In these situations, it is reasonable for the clinical engineer to understand to what extent the clinical and administrative strategies of the organization's survival impact on the technological strategy of what is desired to be incorporated as a health resource [11].

A practical way, used by many professionals in the field, is to prepare a set of questions to be answered by each of the proponents. Questions are sent to bidders under the title of request for information (RFI), and based on the answers, the hospital has a greater set of information about the offer and can better elaborate the specification of what it wants, in relation to what is being offered.

An example of RFI that can be used or adapted for each situation is mentioned in Table 1, containing the following main parts: general information, patient safety, medical application, robotic surgery system, peripherals, technology management, financial equalization, and training program.

In the set of general information, it is convenient to know all the ways to contact the manufacturer during the planning and use process and to know some aspects related to regulatory aspects that can impact the robotic surgery system offered.

TABLE 1    Presents the set of information to be requested from the supplier market.

| General information | Proposal | 1.0 | Technical and commercial proposal number. |
|---|---|---|---|
| | | 1.1 | Date of submission of the technical and commercial proposal. |
| | | 1.2 | Expiration date of the submitted proposal. |
| | | 1.3 | Do the values presented already include the value of taxes? (Yes/No) |
| | | 1.4 | Has all information related to the technology offered been sent? (Yes/No) |
| | Equipment | 2.0 | Equipment name. |
| | | 2.1 | Equipment model. |
| | | 2.2 | Inform the version/generation of the equipment. |
| | | 2.3 | Year of first equipment sold. |
| | | 2.4 | Year of first equipment sold in the country where the SCR will be installed. |
| | | 2.5 | Number of equipment installed in the world. |
| | | 2.6 | Number of equipment installed in the country where the SCR will be installed. |
| | | 2.7 | Number of procedures performed registered in the world. |
| | | 2.8 | Number of procedures performed registered in the country where the SCR will be installed. |
| | | 2.9 | Number of scientific publications registered in the world. |
| | | 2.10 | Number of scientific publications registered in the country where the SCR will be installed. |
| | | 2.11 | Number of tutors trained in technology in the world. |
| | | 2.12 | Number of tutors trained in technology in the country where the SCR will be installed. |
| | | 2.13 | Inform the name of three or organizations that make use of the proposed technology. |
| | | 2.14 | Inform the contact name in the organizations that make use of the proposed technology. |
| | | 2.15 | Inform the telephone number of the contacts in the organizations that make use of the proposed technology. |

TABLE 1 Presents the set of information to be requested from the supplier market—cont'd

| | | |
|---|---|---|
| Manufacturer | 3.0 | Manufacturer's name. |
| | 3.1 | Manufacturer homepage. |
| | 3.2 | Manufacturer address. |
| | 3.3 | It informs the telephone number of contact with the manufacturer, in the country where the equipment is manufactured. |
| | 3.4 | Name of the sales manager for the region where the country is part. |
| | 3.5 | Email address of the sales manager of the region in which the interested country takes part. |
| | 3.6 | Cell phone number of the sales manager in the region the interested country is a part of. |
| Representative | 4.0 | Representative's name |
| | 4.1 | Representative home page |
| | 4.2 | Representative's address |
| | 4.3 | Name of the sales manager for the region the interested country is part of. |
| | 4.4 | Email address of the sales manager of the region in which the interested country takes part. |
| | 4.5 | Cell phone number of the sales manager in the region in which the interested country takes part. |
| National manager of sales | 5.0 | Name of the national sales manager. |
| | 5.1 | National Sales Manager email address. |
| | 5.2 | Cell phone number of the national sales manager. |
| | 5.3 | Name of Regional Sales Manager. |
| | 5.4 | Regional Sales Manager email address. |
| | 5.5 | Cell phone number of the regional sales manager. |
| Engineering | 6.0 | Name of service manager. |
| | 6.1 | Help desk manager email address. |
| | 6.2 | Cell phone number of the national sales manager. |
| Clinical applications | 7.0 | Name of person responsible for supporting clinical applications. |
| | 7.1 | Email address for clinical application support. |
| | 7.2 | Telephone number of the person responsible for supporting clinical applications. |
| Regulatory issues | 8.0 | The equipment is FDA (Food and Drug Administration) regulatory compliant. (year/no) |
| | 8.1 | The equipment is regulatory compliant with the requirements of the European Community (CE). (year/no) |
| | 8.2 | The equipment is in regulatory compliance with the Sanitary Surveillance Agency of the country where the SCR will be installed. (year/no) |
| | 8.3 | Inform the registration number of the equipment at the Sanitary Surveillance Agency of the country where the SCR will be installed. |
| | 8.4 | Expiration date of registration at the Sanitary Surveillance Agency of the country where the SCR will be installed. |
| | 8.5 | Inform if there is any recall, alert, or field service notification in the FDA for the equipment offered in the last 2 years. (Yes/No) |
| | 8.6 | If the answer is yes, inform the reference number(s) for the respective reports. |
| | 8.7 | Inform if there is any recall, alert, or field service notification in the MHRA (Medicines and Healthcare products Regulatory Agency—United Kingdom) for the equipment offered in the last 2 years. (Yes/No) |

*Continued*

II. Education, safety, ethical and administrative aspects for a successful robotic program

TABLE 1    Presents the set of information to be requested from the supplier market—cont'd

| | | | |
|---|---|---|---|
| | | 8.8 | If the answer is affirmative, inform the reference numbers for the respective reports. |
| | | 8.9 | Inform if there is any recall, alert, or field service notification in the Sanitary Surveillance Agency of the country where the SCR will be installed for the equipment offered in the last 2 years. (Yes/No) |
| | | 8.10 | If the answer is affirmative, inform the reference numbers for the respective reports. |
| Patient safety | Medical application | 9.0 | Is the equipment intended for general surgery specialty? (yes/no/in clinical research) |
| | | 9.1 | Is the equipment intended for application in bariatric surgery? (yes/no/in clinical research) |
| | | 9.2 | Is the equipment intended for application in gastrectomy surgery? (yes/no/in clinical research) |
| | | 9.3 | Is the equipment intended for application in hernioplasty surgery? (yes/no/in clinical research) |
| | | 9.4 | Is the equipment intended for application in rectosigmoidectomy surgery? (yes/no/in clinical research) |
| | | 9.5 | Is the equipment intended for application in total/partial colectomy surgery? (yes/no/in clinical research) |
| | | 9.6 | Is the equipment intended for application in pancreatectomy surgery? (yes/no/in clinical research) |
| | | 9.7 | Is the equipment intended for application in whipple surgery? (yes/no/in clinical research) |
| | | 9.8 | Is the equipment intended for the specialty of thoracic surgery? (yes/no/in clinical research) |
| | | 9.9 | Is the equipment intended for application in lobectomy surgery? (yes/no/in clinical research) |
| | | 9.10 | Is the equipment intended for use in mediastinal tumor resection surgery? (yes/no/in clinical research) |
| | | 9.11 | Is the equipment intended for application in esophagectomy surgery? (yes/no/in clinical research) |
| | | 9.12 | Is the equipment intended for application in thymectomy surgery? (yes/no/in clinical research) |
| | | 9.13 | Is the equipment intended for head and neck surgery specialty? (yes/no/in clinical research) |
| | | 9.14 | Is the equipment intended for application in thyroidectomy surgery? (yes/no/in clinical research) |
| | | 9.15 | Is the equipment intended for application in tongue base tumor resection surgery? (yes/no/in clinical research) |
| | | 9.16 | Is the equipment intended for application in laryngectomy surgery? (yes/no/in clinical research) |
| | | 9.17 | Is the equipment intended for the specialty of urological surgery? (yes/no/in clinical research) |
| | | 9.18 | Is the equipment intended for application in radical prostatectomy surgery? (yes/no/in clinical research) |
| | | 9.19 | Is the equipment intended for application in total/partial nephrectomy surgery? (yes/no/in clinical research) |
| | | 9.20 | Is the equipment intended for application in radical cystectomy surgery? (yes/no/in clinical research) |
| | | 9.21 | Is the equipment intended for application in pelvic lymphadenectomy surgery? (yes/no/in clinical research) |

TABLE 1    Presents the set of information to be requested from the supplier market—cont'd

| | | |
|---|---|---|
| | 9.22 | Is the equipment intended for application in pyeloplasty surgery? (yes/no/in clinical research) |
| | 9.23 | Is the equipment intended for application in ureteral reimplantation surgery? (yes/no/in clinical research) |
| | 9.24 | Is the equipment intended for the specialty of gynecological surgery? (yes/no/in clinical research) |
| | 9.25 | Is the equipment intended for application in benign/malignant hysterectomy surgery? (yes/no/in clinical research) |
| | 9.26 | Is the equipment intended for application in endometriosis surgery? (yes/no/in clinical research) |
| | 9.27 | Is the equipment intended for application in myomectomy surgery? (yes/no/in clinical research) |
| | 9.28 | Is the equipment intended for application in oophorectomy surgery? (yes/no/in clinical research) |
| | 9.29 | Inform other possible applications of the equipment in addition to those informed above. |
| Good practices in cyber security | 10.0 | Are there regulatory or legal requirements that need to be considered when connecting CLINICAL equipment data to the cloud? (Yes/No) |
| | 10.1 | If you answered "yes" to the previous question, specify which ones. |
| | 10.2 | Are there specific cybersecurity practices that need to be considered when connecting CLINICAL data to the cloud? (Yes/No) |
| | 10.3 | If you answered "yes" to the previous question, specify which ones. |
| | 10.4 | Are there infrastructure requirements to consider when connecting CLINICAL data to the cloud? (Yes/No) |
| | 10.5 | If you answered "yes" to the previous question, specify which ones. |
| | 10.6 | Are there regulatory or legal requirements that need to be considered when connecting TECHNICAL equipment data to the cloud? (Yes/No) |
| | 10.7 | If you answered "yes" to the previous question, specify which ones. |
| | 10.8 | Are there specific cybersecurity practices that need to be considered when connecting TECHNICAL data to the cloud? |
| | 10.9 | If you answered "yes" to the previous question, specify which ones. |
| | 10.10 | Are there infrastructure requirements to consider when connecting TECHNICAL data to the cloud? (Yes/No) |
| | 10.11 | If you answered "yes" to the previous question, specify which ones. |
| Connectivity and cyber security | 11.0 | The equipment has the capacity to connect your TECHNICAL data to the cloud using a 4G cell phone. (Yes/No) |
| | 11.1 | The equipment can connect your TECHNICAL data to the cloud using a 5G mobile device. (Yes/No) |
| | 11.2 | The equipment can connect your TECHNICAL data to the cloud using Wi-Fi. (Yes/No) |
| | 11.3 | The equipment can connect your TECHNICAL data to the cloud using a wired Ethernet network. (Yes/No) |
| | 11.4 | The equipment has the capacity to connect your CLINICAL data to the cloud using a 4G cell phone. (Yes/No) |
| | 11.5 | The equipment has the capacity to connect your CLINICAL data to the cloud using a 5G mobile device. (Yes/No) |
| | 11.6 | The equipment can connect your CLINICAL data to the cloud using Wi-Fi. (Yes/No) |

*Continued*

TABLE 1    Presents the set of information to be requested from the supplier market—cont'd

|  |  |  |  |
|---|---|---|---|
|  |  | 11.7 | The equipment has the capacity to connect your CLINICAL data to the cloud using a wired Ethernet network. (Yes/No) |
|  |  | 11.8 | Does the equipment and its devices require authentication certified by some standard such as IEEE 802.1× to transmit patient data? (Yes/No) |
|  |  | 11.9 | If the answer is yes, inform the required standard. |
|  |  | 11.10 | How often can the equipment be connected to a wired ETHERNET network without interrupting normal use of the equipment? (ongoing, daily, weekly, monthly, ad hoc, on demand, other (mention) |
|  |  | 11.11 | How often can the equipment be connected to a Wi-Fi or Cellular network without interrupting normal use of the equipment? (ongoing, daily, weekly, monthly, ad hoc, on demand, other) (mention) |
|  |  | 11.12 | The equipment has some CLINICAL data transmission indicator. (Yes/No) |
|  |  | 11.13 | The equipment has some TECHNICAL data transmission indicator. (Yes/No) |
| Medical application | Ergonomics | 12.1 | Inform the surgeon's position during the surgical procedure. (standing/sitting) |
|  |  | 12.2 | Considering that the surgery position is sitting, is the chair to be used by this professional part of the offer? (Yes/No) |
|  |  | 12.3 | If the answer is "yes" to the previous question. Inform if the seat available has adjustable casters, seat height, backrest height, backrest inclination? (Yes/No) |
|  |  | 12.4 | Regarding the surgeon's field of view, it can be enlarged (optical or approximation zoom) in how many times. (number of times) |
|  |  | 12.5 | Regarding the surgeon's field of view, it can be enlarged (electronic zoom) how many times. (number of times) |
|  |  | 12.6 | The surgeon's field of view is 3D capable. (Yes/No) |
|  |  | 12.7 | The equipment can eliminate any involuntary movements. (Yes/No) |
|  |  | 12.8 | The equipment offers tactile sensation to the surgeon. (Yes/No) |
|  |  | 12.9 | Length of the articulated segment of the surgical instrument. (mm) |
|  | Safety, learning and management | 13.0 | Inform the number of hours that the equipment is tested before being made available for sale. (hours) |
|  |  | 13.1 | The SCR offered allows the active collection of data from procedures for auditing, research, and standardization of procedures. (Yes/No) |
|  |  | 13.2 | Regarding productivity, inform the average number of procedures that can be performed per day. (number of procedures) |
|  |  | 13.3 | Inform the time required for a surgeon experienced in conventional procedures to learn to fully operate the SCR. (hours) |
|  |  | 13.4 | Considering the historical database of the technology offered, inform data on the learning curve of medical professionals. (number of procedures/hours of surgery) |
|  |  | 13.5 | Inform if, during the SCR implementation process, the availability of a "representative" nursing professional able to share technology knowledge is an integral part. (Yes/No) |
|  |  | 13.6 | If you answered "yes" to the previous question, which professional will work under (permanent/temporary) regime? |
|  |  | 13.7 | Inform if the asset management software robotic surgery system is included. (Yes/No) |
|  |  | 13.8 | If you answered "yes" to the previous question, the software reports the productivity and work characteristics of the robotic surgery system. (Yes/No) |
|  |  | 13.9 | If you answered "yes" to the previous question, inform the commercial name of the software. (Yes/No) |

TABLE 1  Presents the set of information to be requested from the supplier market—cont'd

| Robotic surgery system | Infrastructure of the operating room | 14.0 | The equipment offered can be used in any typical operating room, requiring no adaptations. (Yes/No) |
|---|---|---|---|
| | | 14.1 | Should the operating room be exclusive for the use of the equipment? (Yes/No) |
| | | 14.2 | Inform the minimum area required for the operating room for the proper use of the technology. ($m^2$) |
| | | 14.3 | Inform the minimum free height necessary for the operating room intended for the application of the technology. (m) |
| | | 14.4 | Inform the minimum height of the ceiling void that is required for full operation of the robotic system. (m) |
| | | 14.5 | Inform the ideal ambient temperature range for the operation of the robotic surgery system. (°C) |
| | | 14.6 | Inform the ideal relative humidity range for the operation of the robotic surgery system. (%) |
| | | 14.7 | Inform the ideal atmospheric pressure range for the operation of the robotic surgery system. (mmHg) |
| | | 14.8 | Inform if the equipment requires an exclusive electric power supply system. (Yes/No) |
| | | 14.9 | If the answer is affirmative, inform how many independent subcircuits are necessary for the SCR to work. (number of circuits) |
| | | 14.10 | Inform how many subcircuits, in addition to those exclusive to the independent SCR, are needed to function the SCR. (number of circuits) |
| | | 14.11 | Inform if it is necessary to use a system/circuit linked to an uninterruptible power source, type "no-break". (Yes/No) |
| | | 14.12 | Inform if the equipment offered requires the use of a voltage stabilizer. (Yes/No) |
| | | 14.13 | If you answered "yes" to the previous question, inform the power of the stabilizer. (kVA) |
| | | 14.14 | If voltage stabilizer is not required, inform the percentage of voltage variation that the equipment admits in the power supply network (+ − %). |
| | | 14.15 | Regarding the protection elements (circuit breakers), inform if there are any special requirements. (Yes/No) |
| | | 14.16 | If the previous answer was affirmative, inform the type of curve that the circuit breaker must have. (curve type) |
| | Cleaning, disinfection and sterilization | 15.0 | Inform if the SCR is supplied together with a cleaning and disinfection kit for the instruments that are part of the equipment. (Yes/No) |
| | | 15.1 | Inform the "part number" (order code) of the kit for cleaning and disinfecting SCR instruments. (#PN) |
| | | 15.2 | If so, inform the cost of the kit mentioned above. ($) |
| | | 15.3 | Inform how long after the purchase has been signed, the "packing list" of the products that make up the kit will be sent to the hospital. (days) |
| | | 15.4 | If the cleaning and disinfection agents are typical of the market, inform the name of this agent(s). |
| | Area required for SCR repair and costs maintenance | 16.0 | Inform how much time will be needed to open SCR boxes and packages. (h) |
| | | 16.1 | Inform the minimum dimensions required for unpacking the SCR. (L × W × H) (m) |
| | | 16.2 | Regarding the displacement of the SCR inside the hospital, inform the minimum width (L) required. (m) |
| | | 16.3 | Regarding the displacement of the SCR inside the hospital, inform the minimum height (A) required. (m) |

*Continued*

II. Education, safety, ethical and administrative aspects for a successful robotic program

TABLE 1    Presents the set of information to be requested from the supplier market—cont'd

| | | |
|---|---|---|
| | 16.4 | Inform the minimum load capacity of the elevator intended for the vertical transport of the SCR around the hospital. (kg) |
| | 16.5 | Inform the minimum area of the elevator intended for the vertical transport of the SCR around the hospital. (LxW) (m) |
| | 16.6 | Inform the minimum width of the elevator door required to move the equipment. (m) |
| | 16.7 | Inform the minimum area required for the storage and repair room for the robotic surgery system. (m$^2$) |
| | 16.8 | Inform the required ambient temperature range to the robotic surgery system storage and repair room. (°C) |
| | 16.9 | Inform the relative humidity range required for the storage and repair room of the robotic surgery system. (%) |
| | 16.10 | Inform the required atmospheric pressure range to the storage and repair room for the robotic surgery system. (Pa) |
| | 16.11 | Inform whether the robotic surgery system storage and repair room requires electrical power supply. (Yes/No) |
| | 16.12 | Inform whether the robotic surgery system storage and repair room requires data network provision. (Yes/No) |
| | 16.13 | Inform the need for cabinets/shelves for the storage of instruments and consumables applied to the robotic surgery system. (Yes/No) |
| | 16.14 | If you answered "yes" to the previous question, please specify the minimum required dimensions of the storage item. (L × W × H) (cm) |
| | 16.15 | Inform the annual value of the preventive and corrective maintenance contract for the equipment, including only the labor service. ($) |
| | 16.16 | Inform the annual value of the equipment preventive maintenance contract including preventive parts ($) |
| | 16.17 | Inform the annual value of the preventive maintenance contract for the equipment, including preventive parts and corrective maintenance service. ($) |
| | 16.18 | Inform the annual value of the equipment preventive maintenance contract including preventive parts and corrective maintenance service including corrective parts. ($) |
| Surgeon's console | 17.0 | Inform how many simultaneous outputs for video images are available for use by different devices. (outputs) |
| | 17.1 | Inform how many simultaneous inputs for video images are available for use by different devices. (Inputs) |
| | 17.2 | Inform if the technology has the possibility for upgrades. (Yes/No) |
| | 17.3 | Inform if the technology offered is compatible for more than one console. (Yes/No) |
| | 17.4 | Inform if the technology offered has reflective movement. If so, characterize this function in the remarks field. (Yes/No) |
| | 17.5 | If you answered "yes" to the previous question. Inform the commercial name of this application. |
| | 17.6 | Inform the dimensions of the equipment in rest condition. (L × W × H) (cm) |
| | 17.7 | Inform the dimensions of the equipment in surgical procedure condition. (L × W × H) (cm) |
| | 17.8 | Inform the weight of the equipment. (kg) |
| | 17.9 | Inform the electrical power required by the equipment (kVA) |
| | 17.10 | Inform the operating voltage. (V) |
| | 17.11 | Inform if the equipment offered requires the use of a voltage stabilizer. (Yes/No) |

TABLE 1    Presents the set of information to be requested from the supplier market—cont'd

|  |  |  |
|---|---|---|
|  | 17.12 | If you answered "yes" to the previous question, inform the power of the stabilizer. (kVA) |
|  | 17.13 | If voltage stabilizer is not required, inform the percentage of voltage variation that the equipment admits in the power supply network $(+ - \%)$. |
| Patient car | 18.0 | Inform how many simultaneous outputs for video images are available for use by different devices. (outputs) |
|  | 18.1 | Inform how many simultaneous inputs for video images are available for use by different devices. (inputs) |
|  | 18.2 | Inform the dimensions of the equipment in rest condition. $(L \times W \times H)$ (cm) |
|  | 18.3 | Inform the dimensions of the equipment in surgical procedure condition. $(L \times W \times H)$ (cm) |
|  | 18.4 | Inform the electrical power required by the equipment (kVA) |
|  | 18.5 | Inform the operating voltage. (V) |
|  | 18.6 | Inform the weight of the equipment. (kg) |
|  | 18.7 | The robotic surgery system is able to act in synchronicity with the operating table. (Yes/No) |
| Vision car | 19.0 | Inform how many simultaneous outputs for video images are available for use by different devices. (outputs) |
|  | 19.1 | Inform how many simultaneous inputs for video images are available for use by different devices. (inputs) |
|  | 19.2 | Inform the minimum distance that the vision car must be from the operative field. (m) |
|  | 19.3 | Inform the dimensions of the equipment in rest condition. $(L \times W \times H)$ (cm) |
|  | 19.4 | Inform the dimensions of the equipment in surgical procedure condition. $(L \times W \times H)$ (cm) |
|  | 19.5 | Inform the electrical power required by the equipment (kVA) |
|  | 19.6 | Inform the operating voltage. (V) |
|  | 19.7 | Inform the weight of the equipment. (kg) |
| Accessories | 20.0 | Inform if the camera comes with the robotic surgery system. (Yes/No) |
|  | 20.1 | If you answered "yes" to the previous question, inform how many cameras come with the offer. |
|  | 20.2 | Inform if the fiber optic cable comes with the robotic surgery system. (Yes/No) |
|  | 20.3 | If you answered "yes" to the previous question, inform how many cables are included in the offer. |
|  | 20.4 | Inform if the light source comes with the robotic surgery system. (Yes/No) |
|  | 20.5 | If you answered "yes" to the previous question, inform how many light sources come with the offer. |
|  | 20.6 | Inform if the 0° endoscope comes with the robotic surgery system. (Yes/No) |
|  | 20.7 | If you answered "yes" to the previous question, inform how many 0° endoscopes are included in the offer. |
|  | 20.8 | Inform if the 30° endoscope comes with the robotic surgery system. (Yes/No) |
|  | 20.9 | If you answered "yes" to the previous question, inform how many 30° endoscopes are included in the offer. |
|  | 20.10 | Inform if the offered SCR requires a dedicated stapler. (yes/no/optional) |
|  | 20.11 | If the answer is yes to the previous question, inform the value of the stapler. ($) |
|  | 20.12 | Inform if the stapler comes with the robotic surgery system (Yes/No) |

*Continued*

TABLE 1  Presents the set of information to be requested from the supplier market—cont'd

| | |
|---|---|
| 20.13 | Enter how many staples make up a recommended stapler load. |
| 20.14 | Inform if the 8 mm blade without obturator (disposable) comes with the robotic surgery system. (Yes/No) |
| 20.15 | If you answered "yes" to the previous question, inform how many 08 mm blades without obturator (disposable) are included in the offer. |
| 20.16 | Inform if the 8 mm blunt obturator blade comes with the robotic surgery system. (Yes/No) |
| 20.17 | If you answered "yes" to the previous question, inform how many 08 mm shutter blades are included in the offer. |
| 20.18 | Inform if the 5 mm 8 mm universal joint (disposable) comes with the robotic surgery system. (Yes/No) |
| 20.19 | If you answered "yes" to the previous question, inform how many universal gaskets 5 mm 8 mm (disposable) are included in the offer. |
| 20.20 | Inform if the 8 mm cannula (disposable) comes with the robotic surgery system. (Yes/No) |
| 20.21 | If you answered "yes" to the previous question, inform how many 8 mm cannulas are included in the offer. |
| 20.22 | Inform if the monopolar cable comes with the robotic surgery system. (Yes/No) |
| 20.23 | If you answered "yes" to the previous question, inform the length of the monopolar cable that comes with the offer. (m) |
| 20.24 | If you answered "yes" to the previous question, inform how many monopolar cables are included in the offer. |
| 20.25 | Inform if the bipolar cable comes with the robotic surgery system. (Yes/No) |
| 20.26 | If you answered "yes" to the previous question, inform the length of the bipolar cable that comes with the offer. (m) |
| 20.27 | If you answered "yes" to the previous question, inform how many bipolar cables are included in the offer. |
| 20.28 | Inform if the calibrating pin comes with the robotic surgery system. (Yes/No) |
| 20.29 | If you answered "yes" to the previous question, inform how many calibrator pins are included in the offer. |
| 20.30 | Inform if the introducer instrument comes with the robotic surgery system. (Yes/No) |
| 20.31 | If you answered "yes" to the previous question, inform how many introducer instruments come with the offer. |
| 20.32 | Inform whether the cover for the monopolar curved scissor holder (disposable) is included with the robotic surgery system. (Yes/No) |
| 20.33 | If you answered "yes" to the previous question, inform how many caps for monopolar curved scissor tips (disposable) are included in the offer. |
| 20.34 | Inform if the spine cover comes with the robotic surgery system. (Yes/No) |
| 20.35 | If you answered "yes" to the previous question, inform how many columns covers come with the offer. |
| 20.36 | Inform if the cover for the robotic arm comes with the robotic surgery system. (Yes/No) |
| 20.37 | If you answered "yes" to the previous question, inform how many covers for the robotic arm come with the offer. |
| 20.38 | Inform if it is necessary to use exclusive 3D glasses for the operation of the SCR. (Yes/No) |
| 20.39 | Inform whether 3D glasses come with the robotic surgery system. (Yes/No) |

TABLE 1    Presents the set of information to be requested from the supplier market—cont'd

|  |  |  |
|---|---|---|
|  | 20.40 | If you answered "yes" to the previous question, please inform how many 3D glasses come with the offer. |
|  | 20.41 | Inform if the technology offered has simulators of activities and procedures. (Yes/No) |
|  | 20.42 | If the previous answer was "yes", inform if the console is attached to the surgeon's console (Yes/No) |
|  | 20.43 | If the previous answer was "yes", inform whether it is possible to perform a surgical procedure with the simulator attached. (Yes/No) |
|  | 20.44 | Inform and list other accessories not mentioned above, discriminating them in type and quantities that are part of the offer of the robotic surgery system. |
| Surgical instruments | 21.0 | Inform how many types of tweezers are available with the SCR offered. |
|  | 21.1 | Inform the number of incisions conventionally required in a standard procedure with SCR offered. |
|  | 21.2 | Inform the diameter of the incisions conventionally needed in a standard procedure with SCR offered. (mm) |
|  | 21.3 | Inform the length of the tweezers available with the SCR offered. (mm) |
|  | 21.4 | Inform the diameter of the tweezers available with the SCR offered. (mm) |
|  | 21.5 | Inform the length of the optic provided with the SCR offered. (mm) |
|  | 21.6 | Inform the diameter of the optics available with the SCR offered. (mm) |
|  | 21.7 | Inform the position of the optics chip. (distal/proximal) |
|  | 21.8 | Inform the characteristic of use of the instruments (disposable/consumable) |
|  | 21.9 | Inform if the large conductor needle comes with the robotic surgery system. (Yes/No) |
|  | 21.10 | If you answered "yes" to the previous question, please inform how many large conductor needles are included in the offer. |
|  | 21.11 | Inform whether the "Prograsp" type grasping forceps comes with the robotic surgery system. (Yes/No) |
|  | 21.12 | If you answered "yes" to the previous question, inform how many "Prograsp" type grasping forceps come with the offer. |
|  | 21.13 | Inform whether the "Maryland" type bipolar forceps is included with the robotic surgery system. (Yes/No) |
|  | 21.14 | If you answered "yes" to the previous question, please inform how many "Maryland" type bipolar clamps are included in the offer. |
|  | 21.15 | Inform if the monopolar curved scissors come with the robotic surgery system. (Yes/No) |
|  | 21.16 | If you answered "yes" to the previous question, please inform how many monopolar curved shears are included in the offer. |
|  | 21.17 | Inform if the folding fenestrated grasper accompanies the robotic surgery system. (Yes/No) |
|  | 21.18 | If you answered "yes" to the previous question, inform how many folding fenestrated graspers come with the offer. |
|  | 21.19 | Inform if the conductive needle type "suture cut" comes with the robotic surgery system. (Yes/No) |
|  | 21.20 | If you answered "yes" to the previous question, inform how many "suture cut" conductive needles are included in the offer. |

*Continued*

II. Education, safety, ethical and administrative aspects for a successful robotic program

TABLE 1    Presents the set of information to be requested from the supplier market—cont'd

| Peripherals | Surgical table | 22.0 | The robotic surgery system requires a dedicated operating table for operation. (Yes/No) |
|---|---|---|---|
| | | 22.1 | If so, please inform the manufacturer of the operating table. |
| | | 22.2 | Inform the model of the surgical table. |
| | | 22.3 | Inform the number of ports on the surgical table. (doors) |
| | | 22.4 | Inform the external dimensions of the surgical table. ($L \times W \times H$) (cm) |
| | | 22.5 | Inform the weight of the surgical table. (kg) |
| | | 22.6 | Inform whether the electrical supply system is single-phase or three-phase. (single-phase/three-phase) |
| | | 22.7 | Inform the operating voltage. (V) |
| | | 22.8 | Inform the electrical frequency of operation. (Hz) |
| | | 22.9 | Inform the construction material of the surgical table in accordance with the AISI standard. |
| | | 22.10 | Inform if the surgical table can be activated by remote control via cable. (Yes/No) |
| | | 22.11 | Inform if the operating table has a drive system (control panel) installed in the body of the equipment. (Yes/No) |
| | | 22.12 | Inform if the equipment has a rack for the use of anesthesia accessories. (baskets) |
| | | 22.13 | Enter the dimensions of each basket. ($L \times W \times H$) (cm) |
| | | 22.14 | Inform the number of carts to support the racks that are part of the offer. (car) |
| | | 22.15 | Inform the annual value of the preventive and corrective maintenance contract for the equipment, including only the labor service. ($) |
| | | 22.16 | Inform the annual value of the equipment preventive maintenance contract including preventive parts ($) |
| | | 22.17 | Inform the annual value of the preventive maintenance contract for the equipment, including preventive parts and corrective maintenance service. ($) |
| | | 22.18 | Inform the annual value of the equipment preventive maintenance contract including preventive parts and corrective maintenance service including corrective parts. ($) |
| | Electrosurgical unit | 23.0 | The robotic surgery system requires an exclusive surgical electrosurgical unit for its operation. (Yes/No) |
| | | 23.0 | If the answer is yes, please inform the manufacturer of the surgical number. |
| | | 23.1 | If the electrosurgical unit is not exclusive, inform the compatible equipment. |
| | | 23.2 | Inform the model of the surgical scalpel. |
| | | 23.3 | Inform the external dimensions of the surgical scalpel. ($L \times W \times H$) (cm) |
| | | 23.4 | Inform the weight of the equipment. (kg) |
| | | 23.5 | Inform whether the electrical supply system is single-phase or three-phase. (single-phase/three-phase) |
| | | 23.6 | Inform the operating voltage. (V) |
| | | 23.7 | Inform the electrical frequency of operation. (Hz) |
| | | 23.8 | Inform the construction material of the surgical scalpel in accordance with the AISI standard. |
| | | 23.9 | Inform the maximum operating power by the surgical scalpel. (kVA) |
| | | 23.10 | Inform if the equipment has an alarm for disconnecting the patient board. (Yes/No) |
| | | 23.11 | Inform the annual value of the preventive and corrective maintenance contract for the equipment, including only the labor service. ($) |

II. Education, safety, ethical and administrative aspects for a successful robotic program

TABLE 1   Presents the set of information to be requested from the supplier market—cont'd

|  |  |  |
|---|---|---|
|  | 23.12 | Inform the annual value of the equipment preventive maintenance contract including preventive parts ($) |
|  | 23.13 | Inform the annual value of the preventive maintenance contract for the equipment, including preventive parts and corrective maintenance service. ($) |
|  | 23.14 | Inform the annual value of the equipment preventive maintenance contract including preventive parts and corrective maintenance service including corrective parts. ($) |
| Ultrasonic surgical unitco | 24.0 | O sistema de cirurgia robótica requer uma unidade cirúrgica ultrassônica exclusivo para o seu funcionamento. (sim/não) |
|  | 24.1 | If so, please inform the manufacturer of the ultrasonic surgical unit. |
|  | 24.2 | Inform the model of the ultrasonic surgical unit. |
|  | 24.3 | Inform the ultrasound frequency used by the equipment. (Hz) |
|  | 24.4 | Inform the external dimensions of the ultrasonic surgical unit. (L × W × H) (cm) |
|  | 24.5 | Inform the weight of the equipment. (kg) |
|  | 24.6 | Inform whether the electrical supply system is single-phase or three-phase. (single-phase/three-phase) |
|  | 24.7 | Inform the operating voltage. (V) |
|  | 24.8 | Inform the electrical frequency of operation. (Hz) |
|  | 24.9 | Inform the constructive material of the ultrasonic surgical unit in accordance with the AISI standard. |
|  | 24.10 | Inform the maximum operating power by the ultrasonic number. (kVA) |
|  | 24.11 | Inform the annual value of the preventive and corrective maintenance contract for the equipment, including only the labor service. ($) |
|  | 24.12 | Inform the annual value of the equipment preventive maintenance contract including preventive parts ($) |
|  | 24.13 | Inform the annual value of the preventive maintenance contract for the equipment, including preventive parts and corrective maintenance service. ($) |
|  | 24.14 | Inform the annual value of the equipment preventive maintenance contract including preventive parts and corrective maintenance service including corrective parts. ($) |
| Surgical insuflator | 25.0 | The robotic surgery system requires a dedicated surgical insufflator for its operation. (Yes/No) |
|  | 25.1 | If so, please inform the manufacturer of the surgical insufflator. |
|  | 25.2 | Inform the model of the surgical insufflator. |
|  | 25.3 | Inform the external dimensions of the surgical insufflator. (L × W × H) (cm) |
|  | 25.4 | Inform the weight of the equipment. (kg) |
|  | 25.5 | Inform whether the electrical supply system is single-phase. (Yes/No) |
|  | 25.6 | Inform the operating voltage. (V) |
|  | 25.7 | Inform the electrical frequency of operation. (Hz) |
|  | 25.8 | Inform the construction material of the surgical insufflator in accordance with the AISI standard. |
|  | 25.9 | Inform the maximum flow offered by the surgical insufflator. (L/min) |
|  | 25.10 | Inform the maximum pressure of the surgical insufflator. (mm Hg) |
|  | 25.11 | Inform if the equipment has a rotameter to quantify the volume of gas used? |
|  | 25.12 | Inform if the equipment has an alarm for leak detection. (Yes/No) |
|  | 25.13 | Inform the annual value of the preventive and corrective maintenance contract for the equipment, including only the labor service. ($) |

*Continued*

II. Education, safety, ethical and administrative aspects for a successful robotic program

TABLE 1    Presents the set of information to be requested from the supplier market—cont'd

| | | |
|---|---|---|
| | 25.14 | Inform the annual value of the equipment preventive maintenance contract including preventive parts ($) |
| | 25.15 | Inform the annual value of the preventive maintenance contract for the equipment, including preventive parts and corrective maintenance service. ($) |
| | 25.16 | Inform the annual value of the equipment preventive maintenance contract including preventive parts and corrective maintenance service including corrective parts. ($) |
| Ultrasonic washer | 26.0 | The robotic surgery system requires an ultrasonic instrument washer, exclusive for its operation. (Yes/No) |
| | 26.1 | If so, please inform the manufacturer of the ultrasonic cleaner. |
| | 26.2 | Inform the model of the ultrasonic washer. |
| | 26.3 | Inform the external dimensions of the ultrasonic washer. (L × W × H) (cm) |
| | 26.4 | Inform the ultrasound frequency used by the equipment. (Hz) |
| | 26.5 | Inform the dimensions of the internal chamber (vat) of the ultrasonic washer. (L × W × H) (cm) |
| | 26.6 | Inform the weight of the equipment. (kg) |
| | 26.7 | Inform whether the electrical supply system is single-phase or three-phase. (single-phase/three-phase) |
| | 26.8 | Inform the operating voltage. (V) |
| | 26.9 | Inform the electrical frequency of operation. (Hz) |
| | 26.10 | Inform the constructive material of the ultrasonic washer in accordance with the AISI standard. |
| | 26.11 | Inform the number of ultrasonic cleaning programs available on the equipment. (Software) |
| | 26.12 | Inform if the equipment has a command display with a touch screen. (Yes/No) |
| | 26.13 | Inform the dimension of the display. (inches) |
| | 26.14 | Inform if the equipment has an integrated printer. (Yes/No) |
| | 26.15 | Inform the annual value of the preventive and corrective maintenance contract for the equipment, including only the labor service. ($) |
| | 26.16 | Inform the annual value of the equipment preventive maintenance contract including preventive parts ($) |
| | 26.17 | Inform the annual value of the preventive maintenance contract for the equipment, including preventive parts and corrective maintenance service. ($) |
| | 26.18 | Inform the annual value of the equipment preventive maintenance contract including preventive parts and corrective maintenance service including corrective parts. ($) |
| Thermal disinfector washer | 27.0 | The robotic surgery system requires an exclusive thermo-disinfection system for its operation. (Yes/No) |
| | 27.1 | If the answer is yes, please inform the manufacturer of the thermodisinfector. |
| | 27.2 | Inform the thermodisinfector model. |
| | 27.3 | Inform the number of ports on the thermodisinfector. (doors) |
| | 27.4 | Inform the external dimensions of the thermodisinfector. (L × W × H) (cm) |
| | 27.5 | Inform the dimensions of the internal chamber of the thermodisinfector. (L × W × H) (cm) |
| | 27.6 | Inform the weight of the equipment. (kg) |
| | 27.7 | Inform the capacity of DIN standard baskets that can be used in the equipment. (baskets) |
| | 27.8 | Inform whether the electrical supply system is single-phase or three-phase. (single-phase/three-phase) |

TABLE 1  Presents the set of information to be requested from the supplier market—cont'd

| | | |
|---|---|---|
| | 27.9 | Inform the operating voltage. (V) |
| | 27.10 | Inform the electrical frequency of operation. (Hz) |
| | 27.11 | Inform the frequency value(s) used to clean the instruments (Hz) |
| | 27.12 | Inform the construction material of the thermodisinfector in accordance with the AISI standard. |
| | 27.13 | Inform the number of thermodisinfection programs available on the equipment. (Software) |
| | 27.14 | Inform if the equipment has a command display with a touch screen. (Yes/No) |
| | 27.15 | Inform the dimension of the display. (inches) |
| | 27.16 | Inform if the equipment has an integrated printer. (Yes/No) |
| | 27.17 | Inform if the equipment has a rack for general use. If the answer is yes, inform the number of baskets it holds. (baskets) |
| | 27.18 | Inform if the equipment has a rack for the use of anesthesia accessories. (baskets) |
| | 27.19 | Enter the dimensions of each basket. (C × XH) (cm) |
| | 27.20 | Inform the number of carts to support the racks that are part of the offer. (car) |
| | 27.21 | Inform the annual value of the preventive and corrective maintenance contract for the equipment, including only the labor service. ($) |
| | 27.22 | Inform the annual value of the equipment preventive maintenance contract including preventive parts ($) |
| | 27.23 | Inform the annual value of the preventive maintenance contract for the equipment, including preventive parts and corrective maintenance service. ($) |
| | 27.24 | Inform the annual value of the equipment preventive maintenance contract including preventive parts and corrective maintenance service including corrective parts. ($) |
| Low temperature sterilizer | 28.0 | The robotic surgery system requires a unique low temperature sterilization system for its operation. (Yes/No) |
| | 28.1 | Inform the technology used in the low temperature sterilization process. |
| | 28.2 | If so, please inform the manufacturer of the low temperature sterilizer. |
| | 28.3 | Inform the model of the low temperature sterilizer. |
| | 28.4 | Inform the number of ports of the low temperature sterilizer. (doors) |
| | 28.5 | Inform the external dimensions of the low temperature sterilizer. (L × W × H) (cm) |
| | 28.6 | Inform the dimensions of the internal chamber of the low temperature sterilizer. (L × W × H) (cm) |
| | 28.7 | Inform the weight of the equipment. (kg) |
| | 28.8 | Inform the capacity of DIN standard baskets that can be used in the equipment. (baskets) |
| | 28.9 | Inform whether the electrical supply system is single-phase or three-phase. (single-phase/three-phase) |
| | 28.10 | Inform the operating voltage. (V) |
| | 28.11 | Inform the electrical frequency of operation. (Hz) |
| | 28.12 | Inform the constructive material of the low temperature sterilizer in accordance with the AISI standard. |
| | 28.13 | Inform the number of thermodisinfection programs available on the equipment. (Software) |
| | 28.14 | Inform if the equipment has a command display with a touch screen. (Yes/No) |
| | 28.15 | Inform the dimension of the display. (inches) |
| | 28.16 | Inform if the equipment has an integrated printer. (Yes/No) |

*Continued*

II. Education, safety, ethical and administrative aspects for a successful robotic program

TABLE 1    Presents the set of information to be requested from the supplier market—cont'd

| | 28.17 | Inform if the equipment has a basket holder rack for general use. If the answer is yes, inform the number of baskets it holds. (baskets) |
|---|---|---|
| | 28.18 | Enter the dimensions of each basket. (L × W × H) (cm) |
| | 28.19 | Inform the number of carts to support the racks that are part of the offer. (car) |
| | 28.20 | Inform the value of the bottle with the sterilizing agent. ($) |
| | 28.21 | Inform the volume of sterilizing agent contained in the bottle. (mL) |
| | 28.22 | Inform the number of sterilization cycles that can be performed with a bottle. (sterilization cycles) |
| | 28.23 | Inform the annual value of the preventive and corrective maintenance contract for the equipment, including only the labor service. ($) |
| | 28.24 | Inform the annual value of the equipment preventive maintenance contract including preventive parts ($) |
| | 28.25 | Inform the annual value of the preventive maintenance contract for the equipment, including preventive parts and corrective maintenance service. ($) |
| | 28.26 | Inform the annual value of the equipment preventive maintenance contract including preventive parts and corrective maintenance service including corrective parts. ($) |
| Irrigator - secretion aspirator | 29.0 | The robotic surgery system requires a unique long-type irrigation/aspiration system for its operation. (Yes/No) |
| | 29.1 | If yes, please inform the manufacturer of the long irrigator/aspirator |
| | 29.2 | Inform the model of the irrigator/vacuum cleaner. |
| | 29.3 | Inform the minimum required length of instruments used with the irrigator/aspirator. (cm) |
| | 29.4 | Inform the type of use of the instrument applied with the irrigator/aspirator. (permanent/disposable) |
| | 29.5 | Inform the external dimensions of the surgical irrigator/aspirator. (L × W × H) (cm) |
| | 29.6 | Inform the weight of the surgical irrigator/aspirator. (kg) |
| | 29.7 | Inform the operating voltage. (V) |
| | 29.8 | Inform the electrical frequency of operation. (Hz) |
| | 29.9 | Inform the constructive material of the surgical irrigator/aspirator in accordance with the AISI standard. (Yes/No) |
| | 29.10 | Inform the maximum flow offered by the surgical irrigator/aspirator. (L/min) |
| | 29.11 | Inform the maximum pressure of the surgical irrigator/aspirator. (mm Hg) |
| | 29.12 | Inform the annual value of the preventive and corrective maintenance contract for the equipment, including only the labor service. ($) |
| | 29.13 | Inform the annual value of the equipment preventive maintenance contract including preventive parts ($) |
| | 29.14 | Inform the annual value of the preventive maintenance contract for the equipment, including preventive parts and corrective maintenance service. ($) |
| | 29.15 | Inform the annual value of the equipment preventive maintenance contract including preventive parts and corrective maintenance service including corrective parts. ($) |
| Surgical retractor | 30.0 | The robotic surgery system requires an exclusive upper digestive surgery retractor for its operation. (Yes/No) |
| | 30.1 | If so, please inform the manufacturer of the upper digestive surgery retractor. |
| | 30.2 | Inform the model of the upper digestive surgery retractor. |
| | 30.3 | Inform the dimensions of the upper digestive surgery retractor. (L × W × H) (cm) |
| | 30.4 | Inform the weight of the upper digestive surgery retractor. (kg) |

TABLE 1 Presents the set of information to be requested from the supplier market—cont'd

| | Video recorder | 31.0 | The robotic surgery system requires an exclusive video recorder for its operation. (Yes/No) |
|---|---|---|---|
| | | 31.1 | If so, please inform the VCR manufacturer. |
| | | 31.2 | Inform the model of the video recorder. |
| | | 31.3 | Inform the external dimensions of the video recorder. (L × W × H) (cm) |
| | | 31.4 | Inform the weight of the video recorder. (kg) |
| | | 31.5 | Inform the operating voltage. (V) |
| | | 31.6 | Inform the electrical frequency of operation. (Hz) |
| Technology management | Maintenance manuals and software | 32.0 | The operating manuals of the offered equipment are an integral part of the equipment. (Yes/No) |
| | | 32.1 | The maintenance manuals for the equipment offered are an integral part of the equipment. (Yes/No) |
| | | 32.2 | The manuals (operation and maintenance) of the equipment will be updated by the manufacturer or its representative, whenever changes in their content are necessary to guarantee the safety of their operation. (Yes/No) |
| | | 32.3 | The supplier undertakes to install all diagnostic software available for maintenance and troubleshooting or faults with the equipment. (Yes/No) |
| | Technical support | 33.0 | Telephone support with qualified technical personnel should be provided while the equipment is in use in the hospital. This support includes access to the hospital's technical engineering staff to discuss configuration issues, error codes, and identify possible solutions to be implemented, etc. This service will be provided at no additional cost to the customer. |
| | | 33.1 | The representative's telephone number and the manufacturer's identification number must be displayed in a visible manner, facilitating communication whenever necessary. |
| | | 33.2 | The bidder undertakes to provide technical support and service coordination through the engineering service, during the warranty period. (Yes/No) |
| | | 33.3 | The proponent undertakes to provide a technical professional based in the hospital in full, during the period of implementation of the SCR. (Yes/No) |
| | | 33.4 | The proponent undertakes to make a technical professional available in the hospital in full, during the post-implementation period of the SCR and in the follow-up of surgical procedures. (Yes/No) |
| | Technical training | 36.0 | The manufacturer undertakes to provide additional training for the hospital's engineering staff, at additional costs. If so, mention the number of hours of training recommended for carrying out the first service, with greater quality and correctness (Yes/No). |
| | | 36.1 | Complete maintenance training will be applied to all equipment for at least one engineering member during the warranty period. This training will be provided at no additional cost to the customer and will have the general objective of better preparing the clinical engineering of the hospital to perform the first service, before contacting the manufacturer, thus facilitating greater access by the manufacturer to the problems that the equipment presents. (Yes/No) |
| | Guarantee | 37.0 | The manufacturer/representative undertakes to correct any failure of the offered equipment that appears during the acceptance period. If they are not able to resolve the issue properly and quickly, the representative will provide other equipment and remove the damaged equipment from the customer. (Yes/No) |
| | | 38.1 | The bidder agrees to start the warranty period after the date on which the equipment is accepted. (Yes/No) |
| | | 38.2 | The representative undertakes to carry out periodic inspections and maintenance activities on the equipment within the warranty period, at no additional cost to the customer. (Yes/No) |

*Continued*

II. Education, safety, ethical and administrative aspects for a successful robotic program

TABLE 1    Presents the set of information to be requested from the supplier market—cont'd

| | | | |
|---|---|---|---|
| | | 38.3 | A direct and exclusive line of communication will be available with the supplier's technical teams during the warranty period or in case of a maintenance contract, at no additional cost. (Yes/No) |
| | Preventive and corrective maintenance | 39.0 | The bidder undertakes to present documentation containing the cleaning, calibration, periodic replacement of parts and functionality verification procedures. These services will receive oversight from the hospital's engineering service. Technicians authorized by the hospital's engineering department will be able to carry out maintenance procedures during this period, under the guidance of the technical representative, without reducing or eliminating the warranty period. (Yes/No) |
| | | 39.1 | During all maintenance procedures, the representative will hand over a copy of the service report. (Yes/No) |
| | | 39.2 | The service report will contain the description of the problem, the description of the solution, identification of the numbers of the parts and materials used, the cost of the same, travel time, time and the identification of the equipment recognized by the SEC software. (Yes/No) |
| | | 39.3 | The supplier undertakes to provide an annual summary of service reports performed on the equipment, 1 month before the expiration date of the equipment warranty period. This summary will be reviewed in conjunction with the engineering service so that any trends or defect patterns, failures, or user errors will be identified, addressed, and resolved. (Yes/No) |
| | | 39.4 | The supplier shall provide maintenance training on a recycling basis. Inform the value in USD and the number of hours. |
| | | 39.5 | Number of years that the maintenance service provider company has been operating in the market. |
| | | 39.6 | What is the location of the technical staff (City)? |
| | | 39.7 | Inform the time in hours between a maintenance call and the actual presence of the professional at the Hospital? (hours) |
| | | 39.8 | Are maintenance services offered directly by the manufacturer or through a distributor/representative? |
| | | 39.9 | Number of people located in the technical office or workshop who are accredited to carry out maintenance procedures on the equipment in question and what is their experience and training? |
| | | 39.10 | Location of the back office (city)? |
| | | 39.11 | For the purposes of this supply, consider that the telephone response time guarantee must be 01:30h. (Yes/No) |
| | | 39.12 | Inform for the equipment offered, the percentage of parts and spare parts available in the manufacturer's warehouse in the national territory, for the typical defects that the equipment presented considering its history in the country where equipment will be installed. (%) |
| Financial equalization | Administration | 40.0 | Inform the delivery time after sending the SWIFT. (Days) |
| | | 40.1 | Inform the total cost of the equipment offered (SCR—robotic surgery system). (USD) |
| | | 40.2 | Inform the estimated costs related to the logistics involved in obtaining the offered robotic system. (USD) |
| | | 40.3 | Inform the origin of shipment/dispatch of the equipment. (Country) |
| | | 40.4 | Inform the percentage value of import taxes that will be levied on the price of the offered robotic system. (%) |
| | | 40.5 | Inform the estimate of total taxes that must be paid for the equipment to be available for use in the country where the SCR will be installed. (USD) |
| | | 40.5 | Number of robotic procedures foreseen for cost estimation. (procedures/day) |
| | | 40.7 | Inform, for planning purposes, the accounting depreciation rate, practiced for the technology offered (%/year). |

TABLE 1 Presents the set of information to be requested from the supplier market—cont'd

| | | |
|---|---|---|
| | 40.8 | If the equipment has the EOL (End Of Life) declared, please inform how many years after purchase. (years) |
| | 40.9 | Inform the manufacturer's recommendation for determining the life cycle for the equipment offered. (years) |
| | 40.10 | Inform time in months for the warranty regarding manufacturing defects. (months) |
| | 40.11 | Inform the annual value imposed on the equipment for the acquisition of an extended warranty against manufacturing defects. (additional USD/year) |
| | 40.12 | Inform the time for offering the updates required to keep the equipment operating in safe conditions? We consider that up dates are free. (weeks) |
| | 40.13 | Inform the time to offer the necessary upgrades to improve the performance of the equipment operating in safe and current conditions, from the identification of your need. (months) |
| | 40.14 | Inform the value of the software that allows asset management, informing the productivity and work characteristics of the equipment offered. (USD) |
| Costs of peripherals and accessories | 41.0 | Inform the cost of the SCR. (USD) |
| | 41.1 | Inform the cost of the recommended surgical table. (USD) |
| | 41.2 | Inform the cost of the recommended molding mattress, chest strap and hand protector. (USD) |
| | 41.3 | Inform the cost of the recommended upper digestive surgery retractor. (USD) |
| | 41.4 | Enter the cost of the recommended stapler kit. (USD) |
| | 41.5 | Inform the cost of the recommended 3D glasses. (USD) |
| | 41.6 | Inform the cost of the recommended surgical scalpel. (USD) |
| | 41.7 | Inform the cost of the recommended ultrasonic scalpel. (USD) |
| | 41.8 | Inform the cost of the recommended blower. (USD) |
| | 41.9 | Inform the cost of the sterilizer at the recommended low temperature. (USD) |
| | 41.10 | Inform the cost of the recommended ultrasonic washer. (USD) |
| | 41.11 | Inform the cost of the recommended thermodisinfector washer. (USD) |
| | 41.12 | Inform the cost of the recommended long irrigator/vacuum. (USD) |
| | 41.13 | Inform the cost of the recommended laparoscopic tray. (USD) |
| | 41.14 | Enter the cost of the recommended video recorder. (USD) |
| | 41.15 | Inform the estimated operating expense per surgery. (USD/surgery) |
| | 41.16 | Inform the cost of the camera for patient observation. (USD) |
| | 41.17 | Calibration frequency required by the SCR and recommended by the manufacturer? (annual, semi-annual, daily, automatic, etc.) |
| | 41.18 | Calibration value required by the SCR. (USD) |
| | 41.19 | The manufacturer undertakes to provide training in the operation and use of the equipment to its users at any time, after the warranty period, similar in terms of hours to the initial training. Inform the number of class hours and the value in USD of said training if requested in the future. |
| | 41.20 | The supplier must provide a price list of commonly replaced parts, in accordance with the maintenance history of this type of equipment, held by the manufacturer or maintenance representative. Provide the list as an attachment, including price (USD) and part number. Inform at least 5 types of parts. |
| | 41.21 | Value of the calibration report and SCR performance certificate after the warranty period (performance) in USD. |
| | 41.22 | Inform the value of the software that allows asset management, informing the productivity and work characteristics of the equipment offered. (USD) |

*Continued*

II. Education, safety, ethical and administrative aspects for a successful robotic program

TABLE 1 Presents the set of information to be requested from the supplier market—cont'd

| Training program | Medical trainning | 42.0 | Inform the training location of the surgical team professionals (doctors). |
|---|---|---|---|
| | | 42.1 | Inform how many medical professionals trained in the city where the SCR will be installed. (number of doctors) |
| | | 42.2 | Inform whether the proponent will provide complete operational training on the operation and use of the robotic surgery system, including the certification and preceptorship of surgeons. (Yes/No) |
| | | 42.3 | If the answer to the previous item is affirmative, inform the value of the training. ($) |
| | | 42.4 | Inform if the training fee is included in the proposal. (Yes/No) |
| | | 42.5 | If the value of the training is not included in the proposal, inform the value of the separate training per team. ($/surgical team) |
| | | 42.6 | If the value of the training is not included in the proposal, inform the value of the separate training per participant. ($/surgeon) |
| | | 42.7 | Inform if the SCR comes with a "kit" of instruments for training. (Yes/No) |
| | | 42.8 | If the answer to the previous item is affirmative, inform how many training sessions the referred "kit" can be used. ($) |
| | | 42.9 | Inform the selling price of this kit, for future needs. ($) |
| | Nursing | 43.0 | Inform if the proponent will offer a nursing professional dedicated exclusively to the hospital. (Yes/No) |
| | | 43.1 | If there is a need for a nursing professional dedicated exclusively to the hospital, inform the monthly cost. ($) |
| | | 43.2 | Inform the training location of the surgical team professionals (nurses). |
| | | 43.3 | Inform how many professional nurses trained in the city of where equipment will be installed. (number of nurses) |
| | | 43.4 | Inform whether the proponent will provide complete operational training on the preparation, installation, performance of pre-operational tests, hygiene and cleaning procedures and pre-operational care of the robotic surgery system, including the certification of the nursing professionals involved. (Yes/No) |
| | | 43.5 | If the answer to the previous item is affirmative, inform the value of the training. ($) |
| | | 43.6 | Inform if the training fee is included in the proposal. (Yes/No) |
| | | 43.7 | If the value of the training is not included in the proposal, inform the value of the separate training per team. ($/nursing team) |
| | | 43.8 | If the value of the training is not included in the proposal, inform the value of the separate training per participant. ($/professional) |
| | Engineering | 44.0 | Supplier undertakes to provide additional refresher training to at least one member of the Engineering Department during the warranty period. This training will be offered by the supplier at no additional cost. (Yes/No) |
| | | 44.1 | The manufacturer undertakes to provide training in the operation and use of the diagnostic software included in the service training package, free of charge to the customer. This training aims to increase the capacity of the current technical team to perform the first assistance at the hospital, before calling the manufacturer, if necessary. (Yes/No) |
| | | 44.2 | The manufacturer undertakes to provide training in the operation and use of the equipment to its users before and during the first use. Enter the number of class hours. (hours) |
| | | 44.3 | The syllabus of each of the required training courses (technical and operational/clinical) must be sent (15 days after formalizing the purchase) to the hospital's engineering department for evaluation and formalization purposes with users and the training program. Continuing education of the organization. The professionals assigned to this activity by the supplier must have proven experience in the hospital's specialty. |
| | | 44.4 | During the warranty period and through a maintenance contract, the bidder must offer specialized clinical support to the operators, both by telephone and via the WEB, including, if necessary, remote access to the equipment offered. (Yes/No) |

*In the table, USD is the monetary value expressed in American dollars, and $ is the value of the local currency, used in the country where the equipment will be sold.*

In the subject of patient safety, it is important to know what the product can be used for and to know aspects of cybersecurity.

Regarding medical applications, it is possible to highlight ergonomic issues and aspects related to the support offered for professional training in the safe use of the offered system.

In a robotic surgery system, we seek to obtain more data on the infrastructure of the operating room where the equipment will be used, aspects of cleaning, disinfection and sterilization related to the equipment, area required for system repair and maintenance costs, the parts that make up the equipment and its accessories and instruments.

Concerning of peripherals, we sought to obtain more information about the interaction of the SCR with equipment such as the operating table, electric scalpel, insufflator, irrigator, etc. At this stage, it is possible to assess the extent to which equipment that already exists in the hospital can be used in conjunction with the proposed equipment.

In terms of healthcare technology management, we can learn more about the operational and technical documentation of the equipment, how technical support will be offered, the depth of technical training for the hospital staff, warranty, and maintenance issues.

In financial equalization, we seek to obtain information of interest to management such as costs. This information should allow an analysis of the break-even point of the operation in theoretical and practical terms. They should be enough to determine the fixed cost that the SCR will impose on the hospital, and the variable unit cost, which is the one that each new surgery will add to the fixed cost of the operation. In addition, it is convenient to know the costs of peripherals compatible with each SCR, which the hospital may have to buy to assemble the complete robotic surgery system.

Finally, in the training program, we developed questions related to medical, nursing, and engineering training.

For the purpose of objectively comparing answers and accelerating the decision process, it is important to guide each proponent on the following points: use the same units of measurement constant in each question and not create new lines between the existing ones that, if necessary, add them at the end of the table.

Armed with this information and having fully consolidated what it wants to obtain, the hospital can generate, in the same format, a formal request for a proposal to the market, which can be called a request for proposal (RFP). Based on this set of information, the final evaluation is carried out and the decision to buy or not to buy can be taken with greater confidence.

## Summary of clinical engineering activities in the planning and management of robotic surgery systems

There are several important activities in which clinical engineers can make a contribution. The scope and detail of each depend on the support received from the hospital administration. It is still common in most hospitals to use the term maintenance instead of engineering in the organization chart. As maintenance and repair belong to engineering, it is convenient for hospital administration to evaluate the advantages and disadvantages of using the term engineering in the organization chart, which is more in line with the following list of activities necessary for the safe adoption of new technologies.

The **needs assessment** is crucial for the engineering team. It is convenient that this team participates from the beginning in all meetings involving administration, medical, and nursing staff.

The **assessment of the facilities** allows clinical engineers to know the impact that the new technology will cause in the facilities and the verification of the needs to renovate, adapt, or expand the existing spaces.

The preparation of a **budget** involving all the costs that the new technology will impose on the hospital. The data obtained, preferably, should reflect the total cost of ownership.

The **preparation of the specification** of what will be purchased, not only the main equipment but also its accessories and peripherals.

The preparation of the **reference term** that will be sent to each bidder, containing, on the part of the hospital and from the engineering point of view, the terms and conditions of supply.

Receipt and final **evaluation of proposals**, about technical aspects related to the robotic surgery system.

**Receipt** of all purchased equipment and verification that they meet the terms of reference.

**Monitoring the installation** of the equipment involving the following parts: confirmation that it was installed as it should (installation qualification), confirmation that it operates as it should (operation qualification), and the monitoring of **acceptance and performance tests** (performance qualification).

**Monitoring and recording** of technical and operational training at the level and depth negotiated during the acquisition process.

Receiving training on operator maintenance, which defines the responsibility of the care team in relation to the robotic surgery system.

Carrying out or monitoring **inspection** activities, **preventive and corrective maintenance**.

**Inventory** of each component part of the system, to manage all processes related to the SCR through the **computerized maintenance management system** (CMMS) used by the hospital.

Prepare **management reports** throughout the SCR life cycle within the organization.

## Future scenario

Currently, the scope of clinical engineering services depends on the hospital where it operates. It depends on the support of the administration and the results that it is currently obtaining, in the planning processes and technological management required by each hospital.

The clinical engineering service will be as developed as the ways in which hospital administrators and health professionals work. The most important aspect is the change from a view that engineering lends itself solely to maintenance activities, to a broader view, which is related to the capacity of hospitals to plan and manage their technological resources with the support of an engineering service, complete with prepared people, instruments, and work processes.

In addition, it is convenient to reinforce that the future scenario of clinical engineering will depend on how the current service contributes to hospital administration, in facing the pressures caused by technological forces, economic forces, demographic and cultural forces, and the regulatory forces that shape the new healthcare models [12].

## Key points

- The implantation of the robotic surgery system in hospitals can generate good or bad impact on its operation, depending on the system acquired and the acquisition process.
- Clinical engineering professionals can help from the beginning of the acquisition planning process, to the management of the SCR itself, where activities such as inventory, control of requests for preventive and corrective maintenance services, calibration, compliance with laws and regulations, management of risks, costing of the operation, monitoring of recalls and alerts related to the product, creation of performance indicators, etc., which will comprehensively guarantee a rational, sustainable, and safe adoption process for the patient and interested parties.

## References

[1] Iadanza E, editor. Clinical engineering. Clinical engineering handbook. Academic Press; 2020. p. 3–6, ISBN:978-0-12-813467-2. https://doi.org/10.1016/B978-0-12-813467-2.00001-8.

[2] Bauld TJ. The definition of a clinical engineer. J Clin Eng 1991;16:403–5. https://doi.org/10.1097/00004669-199109000-00011.

[3] Pacela A. Career "fact sheets" for clinical engineering and biomedical technology. J Clin Eng 1991;16:407–16.

[4] American College of Clinical Engineering., 2023. Retrieved from: https://accenet.org/about/Pages/ClinicalEngineer.aspx.

[5] Bronzino JD. Clinical engineering: Evolution of a discipline. In: Dyro JF, editor. Biomedical engineering, clinical engineering handbook. Academic Press; 2004. p. 3–7, ISBN:9780122265709. https://doi.org/10.1016/B978-012226570-9/50003-X. http://www.sciencedirect.com/science/article/pii/B9780122265709500003X.

[6] Human resources for medical devices, the role of biomedical engineers. Geneva: World Health Organization; 2017. Retrieved from: https://www.who.int/publications/i/item/9789241565479. License: CC BY-NC-SA 3.0 IGO.

[7] IFMBE/CED CE-HTM Definitions., 2023. Retrieved from: https://ced.ifmbe.

[8] Panerai RB, Mhor JP. Health technology assessment methodologies for developing countries. Pan American Health Organization; 1989, ISBN:92-75-12023-4.

[9] Banta HD, C.J. Behney y J.S. Willems. Toward rational Technology in Medicine. New York: Springer Publ. Co.; 1981.

[10] Emergency Care Research Institute (ECRI). A Guide to Getting it Right the First Time, 2023. Retrieved from: https://www.ecri.org/a-guide-to-getting-it-right-the-first-time.

[11] Radice L. Challenge in surgical robot development. In: Iadanza E, editor. Clinical engineering handbook. 2nd ed. Academic Press; 2020. p. 469–72, ISBN:978-012-813467-2. https://doi.org/10.1016/B978-0-12-813467-2.00071-7.

[12] Grimes SL. The future of clinical engineering: the challenge of change. In: Dyro JF, editor. Biomedical engineering. Clinical Engineering Handbook: Academic Press; 2004. p. 623–7, ISBN:9780122265709.

# 36

# Credentialing and training in robotic surgery in middle-income countries

*Alex Barbosa[a] and Héber Salvador de Castro Ribeiro[b,c]*

[a]Department of Surgical Oncology, Santa Casa de Maceió, Alagoas, Brazil [b]Department of Abdominal Surgery, A.C. Camargo Cancer, São Paulo, Brazil [c]Reference Center on Upper GI & HPB Oncology, A.C. Camargo Center, São Paulo, Brazil

## Introduction

Robotic surgery has been one of the most promising areas of medicine in recent years. Since its introduction to the market in the 2000s, its utilization has rapidly grown worldwide. Robotic surgery is being increasingly adopted in various fields such as urology, gynecology, colorectal surgery, and even in less complex procedures, despite being associated with higher costs compared to other approaches like laparoscopy and open surgery [1].

Initially, robotic surgery was seen as an evolution of laparoscopic surgery. However, the techniques have different learning curves, and surgical time in robotics is faster than laparoscopy when the surgeon is inexperienced in both platforms [2]. Over the past 20 years, the robotic platform has evolved with improvements, offering articulated wristed instruments, eliminating surgeon tremors, and providing 3D visualization, which enhances depth perception and enables access to narrow spaces and more complex procedures [3].

In Latin America, like in other developing countries, robotic surgery is increasingly being utilized. The technology has been used to perform complex surgeries with greater precision, reducing recovery time, lower conversion rates to open surgery, lower complication rates, and minimizing risks associated with traditional surgery [4,5].

Despite the numerous benefits of the technology, the lack of proper regulation can lead to inappropriate and potentially dangerous practices. Many Latin American countries have expressed concerns about patient safety as robotic technology rapidly advances and spreads. Regulatory societies have struggled to keep up with the rapid pace of technological advancements and often lag behind the regulations of developed countries.

## Credentialing and certification

Robotic surgery is already recognized as an important therapeutic option by various regulatory authorities such as the Food and Drug Administration (FDA) in the United States in 2000, the National Health Surveillance Agency (ANVISA) in Brazil in 2008, and the National Institute for Health and Care Excellence (NICE) in United Kingdon in 2015. However, to be considered safe and effective, it must be used appropriately and with comprehensive training of the surgical team in hospitals that have the technical and physical capabilities to handle high-complexity cases [6].

Credentialing for robotic surgery is a crucial process to ensure patient safety and the qualification of professionals involved in this surgical technique. The purpose of this process is to evaluate and certify the surgeon's ability to successfully perform robotic surgeries, ensuring they have undergone adequate training and are updated on the best practices and guidelines for this type of procedure.

Currently, there are many robotic platforms available, the most common are Da Vinci (Intuitive Surgical-Sunivaley CA-US), Versius (CMR Surgical-Cambridge-UK), Senhance (TransEnterix-Durham-NC-US), HUGO (Medtronic-Minneapolis-MT-US), among others. This diversity, combined with the heterogeneity among regulatory societies, leads to a lack of standardization and uniformity in training and credentialing guidelines, which can hinder the training of qualified professionals.

Among Latin American countries, Brazil has stood out in the deployment and diffusion of the robotic system throughout the country. According to data from Strattner, the representative of the Da Vinci System in Brazil, the pioneering in the country, over 2500 robotic surgeons have been trained and more than 100,000 robotic procedures have been performed using the Da Vinci platform in Brazil [7].

Considering the need to guide the certification process for robotic surgery in Brazil, the Brazilian Medical Association (AMB), in conjunction with affiliated specialty societies, has based its guidelines on Resolution No. 2.311 of the Federal Medical Council (CFM) dated March 28, 2022. These guidelines may undergo specific modifications as deemed necessary by the affiliated specialty societies.

Other well-established regulations are used in robotic training centers, such as those of the pioneering Da Vinci robotic system (Intuitive Surgical) and The Fundamentals of Robotic Surgery (FRS).

## Federal Council of Medicine resolution

Robotic surgery can only be performed by physicians who are required to have a Specialist Qualification Registration (RQE) in the Regional Medical Council (CRM) in the surgical area related to the procedure.

These surgeons must undergo specific training in robotic surgery during Medical Residency or through specific qualifications after completing two training stages. The first stage consists of theoretical and practical basic training, including knowledge about the robotic equipment and the functioning of the robot, provided by the manufacturer; online training in robotic surgery fundamentals or a similar platform.

### Stage 1—Basic training

This stage involves the initial theoretical and specific practical training for each available robotic platform, focusing on adapting to the robotic system through simulation and aiming to develop psychomotor skills.

(a) Theoretical knowledge about the robotic equipment and the functioning of the robot, provided by the manufacturer;
(b) Online training in robotic surgery fundamentals or a similar platform;
(c) Watching edited or unedited videos of robotic surgeries in a virtual environment;
(d) Participation of the surgeon in a minimum of 10 robotic surgeries as an assistant surgeon at the patient's bedside in any surgical area, with at least three of them in the specific surgical specialty in which they wish to practice. This step is necessary before the candidate starts performing the 10 surgeries as primary/principal surgeon.
(e) Training on a validated robotic simulator for this purpose. The minimum required time for these simulation exercises is 20h;
(f) In-service training in which the surgeon must simulate movements and procedures to be used during the actual surgery on the robot console, using simulation molds, for a minimum of 2h.

### Stage 2—Advanced training

This phase involves the surgeon performing robotic surgery as the primary surgeon under the supervision of a robotic surgery instructor, who will guide the technical management of the robot (console and instruments).

(a) This phase must include a minimum number of surgeries with the evaluation and approval of the robotic surgery instructor, who will attest to the competence of the primary surgeon to perform robotic surgery.
(b) In this phase, it will be necessary to participate as the primary surgeon in a minimum of 10 robotic surgeries in the specialty of practice under the supervision of a robotic surgery instructor.
(c) After completing all training stages and the minimum number of surgeries, the primary surgeon will undergo an evaluation with a robotic surgery instructor, who will attest to their competence in robotic surgery if the primary surgeon is approved.

The surgeon in the training phase, after completing the basic training stage, can only perform robotic surgery under the supervision and guidance of a robotic surgery instructor.

The surgeon will have the autonomy to perform robotic surgery without the participation of the robotic surgery instructor after completing and being approved in the training with the instructor, having performed a minimum of 10 robotic surgeries.

The robotic surgery instructor will be responsible for guiding the robot handling and evaluating the competence of the primary surgeon but will not directly participate in patient care.

The robotic surgery instructor has the autonomy to interrupt the robot-assisted modality if deemed necessary for the benefit of the patient. To act as a robotic surgery proctor, the surgeon must have performed a minimum of 50 robotic surgeries as the primary surgeon.

## Da Vinci training program

Intuitive suggests four training phases in the da Vinci technology for surgeons. However, on their website, they make it clear that they are not responsible for training on the product but rather training on the use of the da Vinci Surgical System. The information provided during Intuitive Surgical's training is not intended to replace formal medical training or certification. Intuitive Surgical is in no way responsible for surgical credentialing or training in surgical procedures or techniques, nor do the training programs provided by Intuitive Surgical substitute for hospital credentialing requirements [8,9].

Phase 1: da Vinci training corresponds to an introduction to the technology, involving activities that familiarize the surgeon with the da Vinci system. Contact with the technology allows understanding of its features and capabilities, and observation of videos and cases helps to understand clinical applications and techniques.

Phase 2: The surgeon has access to online training to better understand the features of the da Vinci system. Following that, the surgeon undergoes hands-on training on the functions of the da Vinci system and develops skills with the help of a simulator and the actual robotic system. The visit to the training center allows for a practical day with various activities, including exercises on organic tissue.

Phase 3: This phase involves the integration of the da Vinci technology and surgical practice, with the start of procedures guided by an instructor surgeon. Evaluation of the results and constant alignment of the team enable continuous improvement of clinical performance.

Phase 4: Continuous development on the platform.

## Certificate of da Vinci system training for residents and fellows

Intuitive Surgical offers a training program for residents and fellows in accredited academic centers, which has been well-received in centers of excellence worldwide. It is important to include robotic surgery certification in the training of residents, as it is already a reality in medical training centers, requiring residents to keep up with the advancements in surgical technology [10].

In Brazil, the Brazilian Society of Oncological Surgery became the first specialty society accredited by the Brazilian Medical Association (AMB) to introduce a comprehensive robotic surgery training program for residents as an integral part of its curriculum. Nowadays the majority of medical specialties societies have their specific robotic surgery training programs. This program involves an additional year of training (fourth year) dedicated to advanced video-assisted surgery and comprehensive training and certification in robotic surgery. The training curriculum is based on the program developed by Intuitive Surgical, a leader in robotic surgical systems. By incorporating this specialized training, residents gain extensive knowledge and hands-on experience performing robotic surgery, ensuring they are well-equipped to provide high-quality care in this advanced surgical field.

On the other hand, there are certain specialties that facilitate certification programs in robotic surgery for fellows, ensuring they gain hands-on experience in the field's technology. These programs aim to provide comprehensive training and expertise in robotic surgical procedures within specific medical specialties. By participating in these certification programs, fellows have an opportunity to acquire the necessary skills and proficiency in using robotic systems for surgical interventions.

There are six criteria for the Certificate of da Vinci System Training for Residents and Fellows [11]:

Criteria 1: Completion of da Vinci System Online Training for Surgeons, Residents, and Fellows. This includes video-based online training covering system component overviews, instructions for using system components, instruments, accessories, and advanced technologies.

Criteria 2: Completion of da Vinci System Overview In-Service Training. This includes hands-on training covering system component overview, docking, port placement, da Vinci instrument overview, Surgeon Console overview, emergency procedures, and system shutdown.

Criteria 3: Completion of da Vinci System Online Assessment. The above training activities are prerequisites for the da Vinci System Online Assessment. An online training certificate can be printed upon completion of the online assessment.

Criteria 4: Performance of da Vinci Procedures in the Primary Role of Console Surgeon (20 recommended) and Primary Role of Bedside Assistant (10 recommended).

Criteria 5: Obtain a Letter of Verification of da Vinci System Training and Completed Procedures from the Chief of Surgery or Program Director.

Criteria 6: Submit a Copy of the Online Training Certificate, Case Log, and Letter of Verification to an Intuitive Representative.

## The fundamentals of robotic surgery (FRS)

The FRS is a multispecialty curriculum based on proficiency in basic technical skills to train and evaluate surgeons in the safe and efficient performance of robotic-assisted surgery. It was developed by over 80 national/international experts in robotic surgery, behavioral psychologists, medical educators, statisticians, and psychometricians. Clinical robotic surgery subject matter experts represented all major surgical specialties in the United States currently performing robotic-assisted surgical procedures.

The methodology consists of four modules: introduction to robotic systems, didactic introductions to robotic surgery, psychomotor skills curriculum, and team training and communication skills.

The Institute for Surgical Excellence (ISE) managed the international validation testing of the proficiency-based progression training program, which included 12 leading American College of Surgeons' Accredited Education Institute (ACS-AEI) centers. The results were published in the Annals of Surgery in August 2020, demonstrating better performance by those trained following FRS compared to controls in a transfer test [12].

This training program has already been validated and provided to several countries with adapted language translation when linked to registered healthcare institutions.

## Conclusion

Robotic surgery has become a promising field in medicine, with its increasing use worldwide. While it offers significant benefits, the lack of proper regulation can lead to inadequate and dangerous practices, especially in middle-income countries. In these countries, such as those in Latin America, robotic surgery is being increasingly used, but regulatory bodies struggle to keep up with the rapid advancement of technology.

Accreditation and certification are fundamental aspects to ensure patient safety and the qualification of professionals involved in robotic surgery. Several regulatory bodies recognize the importance of this surgical modality by establishing specific guidelines and requirements. However, the lack of standardization and uniformity in training and accreditation guidelines can hinder the consistent formation of qualified professionals.

In Brazil, for example, the Brazilian Medical Association (AMB), along with affiliated specialty societies, has established a certification process for robotic surgery qualification based on the resolution of the Federal Medical Council (CFM). This process involves two stages of training (basic and advanced) and the supervision of a robotic surgery instructor surgeon. This model has been mirrored by other societies with specific modifications as needed, and we believe it yields positive results in professional qualification.

Furthermore, companies that develop robotic systems, such as Intuitive Surgical, provide specific training programs for surgeons. These programs include online training, practical training on simulators and the actual robotic system, and integration of the da Vinci technology with surgical practice. However, it is important to emphasize that these programs do not replace hospital accreditation requirements and formal medical certification.

Another important initiative is the Fundamentals of Robotic Surgery (FRS), a multispecialty curriculum based on basic technical skills. Developed by experts in robotic surgery, the FRS aims to train and evaluate surgeons to perform robotic-assisted surgery safely and efficiently.

In summary, accreditation and certification are crucial processes to ensure the safety and quality of robotic surgery. While there are available guidelines and training programs, continuous efforts are necessary to promote standardization and constant updates of practices and regulations, especially in middle-income countries, in order to keep pace with the rapid advancement of this technology and ensure the best care for patients.

## Future scenario

The future of training and certification in robotic surgery points toward greater integration of these skills into surgeons' residency and fellowship programs. With several robotic platforms in development, it is anticipated that specific curricula in robotic surgery will be established, covering everything from fundamentals to advanced procedures, equipping surgeons in training for a well-founded clinical practice.

Additionally, it is believed that quality standards, advanced simulation, and continuous education will play crucial roles in ensuring that surgeons acquire high-level skills and are prepared to navigate the ongoing evolution of technology.

In order for patient safety and risk minimization to remain priorities, there is a need for regulatory and oversight agencies to ensure that professionals are proficient in using robotic surgery effectively and safely in their clinical practices.

## Key points

- Accreditation and certification are crucial processes to ensure the safety and quality of robotic surgery.
- Developing or middle-income countries need to establish rules for certification and credentialing of robotic surgery, at the same velocity as this surgical modality has been expanded in these countries.
- Continuous efforts are necessary to promote standardization and constant updates of practices and regulations, especially in middle-income countries,
- Fundamentals of Robotic Surgery (FRS), a multispecialty curriculum, aims to train and evaluate surgeons to perform robotic-assisted surgery safely and efficiently.

## References

[1] Sheetz KH, Claflin J, Dimick JB. Trends in the adoption of robotic surgery for common Surgical procedures. JAMA Netw Open 2020;3(1), e1918911. https://doi.org/10.1001/jamanetworkopen.2019.18911 [published Online First: 20200103].

[2] Flynn J, et al. The learning curve in robotic colorectal surgery compared with laparoscopic colorectal surgery: a systematic review. Colorectal Dis 2021;23(11):2806–20.

[3] Spinoglio G, Summa M, Priora F, Quarati R, Testa S. Robotic laparoscopic surgery with the da Vinci® system: an early experience. Surg Technol Int 2009;18:70–4.

[4] Gross JL. Directions for robotic surgery in the treatment of thoracic diseases in Brazil. J Bras Pneumol 2020;46(1), e20190427.

[5] Feng Q, et al. Robotic versus laparoscopic surgery for middle and low rectal cancer (REAL): short-term outcomes of a multicentre randomised controlled trial. Lancet Gastroenterol Hepatol 2022;7(11):991–1004.

[6] Regulation of robotic surgery in Brazil. CFM Resolution No. 2.311/2022; 2022 [Published in the D.O.U. of March 28, 2022, Section I, p. 234].

[7] Treinamentos Cirurgia Robótica da Vinci., 2023. Disponível em: https://www.strattner.com.br/cirurgia-robotica/treinamentos-em-cirurgia-robotica. Acessado em: 05 mai.

[8] Intuitive Surgical. Product training disclaimer., 2023, https://www.intuitive.com/en-us/about-us/company/legal/product-training-disclaimer. Acessado em 9 Mai.

[9] Da Vinci Training Passport, Supporting surgeons on a path to success. By Intuitive 10/2018.

[10] Carpenter BT, Sundaram CP. Training the next generation of surgeons in robotic surgery. Robot Surg 2017;21(4):39–44.

[11] Surgical. Certificate of da Vinci System Training for Residents and Fellows. Pdf acessado em 10 Mai; 2023.

[12] Satava RM, et al. Proving the effectiveness of the fundamentals of robotic surgery (FRS) skills curriculum: a single-blinded, multispecialty, multi-institutional randomized control trial. Ann Surg 2020;272(2):384–92.

# 37

# Robotic surgery advantages and disadvantages

*Fernando Korkes*[a,b,c], *Diego Abreu*[d], *and Alexandre Kyoshi Hidaka*[a]

[a]Urology Division, Centro Universitario FMABC, Santo André, SP, Brazil [b]Hospital Municipal da Vila Santa Catarina, São Paulo, SP, Brazil [c]Hospital Israelita Albert Einstein, São Paulo, SP, Brazil [d]Department of Urology, Hospital Pasteur, Montevideo, Uruguay

## Introduction

The past decades have seen exponential technological growth. The information turnover is almost unpredictable as it spreads worldwide. In a matter of a couple of years, all devices become "old" and need to be replaced to be "updated" with the current information.

In the health industry, information turnover might be slower than in other industries, but yet it is also inexorable.

Nearly four decades ago, since the development of laparoscopy in France, a worldwide increment in techniques and surgeons performing these techniques occurred. After several decades of development, benefits of robotic surgery have also emerged as the ultimate revolution in minimally invasive surgery.

Since the first robotic prostatectomy was performed in the United States in 2000, the use of RS in urology has increased dramatically. Robotic radical prostatectomy procedures grew from 13.6% to 72.6% in frequency between 2003 and 2012 [1]. Robotic surgery massive adoption was much faster than the adoption of laparoscopic surgery.

We will thoroughly examine the benefits and drawbacks of RS in this chapter.

## Looking outside the box

The virtual world is ruled by codes. These codes not only precisely mimic real-world actions but can actually be enhanced. The current robotic systems use consolidated technology from other fields in RS, such as airplane systems, artificial intelligence (AI), and even weapons systems. This integration can also help improve the cost effectiveness and efficiency. Any ineffective move generates a cost to the institution and a bill to the patient. The search to solve this problem demands an ultraspecialized team and an economic plan to apply this kind of procedure.

### RS is a business

The costs for RS include buying the robotic system and the associated materials. The variable costs involve the replacement of these instruments as well as staff recruitment and/or staff training. Indeed, the success of the program is determined by the procedure volume. A Chicago group's experience showed the need to perform at least five procedures per week in the beginning to continue program development. After 5 years, they improved their mark to as many as 350 cases per year [2].

A business plan must be developed together with a market screening. To offer a new product, it's necessary to track those who offer the same product as well as how much they charge and when they deliver, as with any start-up business.

The availability of RS is another drawback. Some patients might not be able to receive RS, even though they might benefit from it, because not all hospitals or surgical centers have access to the technology. This can restrict patient access to the technology and result in less-than-ideal results.

## Robotic surgeon journey

It takes a person several years to graduate from medical school, several more long years to become a surgeon, and then a few more to learn to perform RS. In the end, it will take more than a decade to become a surgeon who uses RS. The learning curve associated with RS is another drawback. To operate the robot, surgeons must go through specific training, which can be difficult and time consuming. This implies that it might require more time for surgeons to master the technology, which might result in longer operation times and possibly higher complication rates [3].

## Training

The surgeon console, patient cart, and vision cart as well as the robotic accessories, cables, and connectors must all be understood for a robotic training curriculum to be effective. Before performing their first robotic surgery, trainees should be comfortable with the different customizable console settings. These include programmable features for the camera, motion scaling, digital zoom, energy control, secondary console control, and communication, among others. Knowing how to resolve common issues such as expired instruments or instrument collisions is also beneficial.

Cimen and colleagues [4] showed that trainees with expertise with robotic bedside devices performed better on the console than learners without that experience. These results were ascribed to an improvement in self-assurance and problem-solving skills. Compared to surgeons without expertise, those with bedside experience may be able to handle difficult cases earlier. A minimal number of cases to acquire the essential bedside assistant skills is unclear and probably varies depending on the situation because there is no learning curve for successful bedside experience. Every RS resident should spend some time working as a bedside assistant throughout all common robotic procedures.

In terms of training and certification, there is a lack of consistency because RS is still a relatively new technique. This means that depending on the surgeon and the facility, the quality of RS can vary greatly. The absence of standards may result in lower-quality outcomes and higher patient risk [5].

## Operating room

The current generation of robotic platforms is robust and large, requiring a lot of space in the operating room (OR). A robotic platform basic composition includes: One surgeon console, one patient cart, one vision cart and several accessories, cables, and connectors. To run the surgery, it's necessary to have a specialized robotic operator nurse, a supervisor nurse, one circulating nurse, one engineer, an anesthesiologist, and one surgeon. That means at least six people are needed to manage an RS OR (Fig. 1).

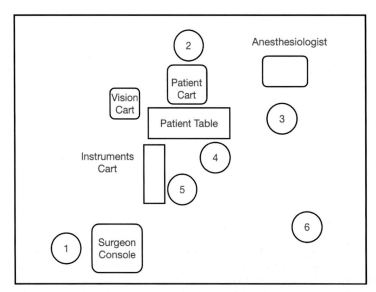

FIG. 1   Configuration of the operation room for robotic-assisted radical cystectomy: (1) surgeon, (2) nurse, (3) anesthesiologist, (4) assistant, (5) surgical technician, and (6) nurse assistant.

## Ergonomics

For surgeons, RS may be more pleasant because the surgeon sits at a console and uses a joystick and foot pedals to control the robotic arms rather than spending hours on end standing over the patient. This can make the surgeon feel less worn out and uncomfortable, which will ultimately benefit the patient.

Hislop et al. in a meta-analysis demonstrated that RS is associated with reduced surgeons' discomfort. More appropriate back, elbow, wrists and hand positions were reported in the study. These findings suggest a protective effect of RS, increasing the potential for a surgeon's professional longevity [6].

## The incisions

Smaller incisions than in conventional surgery are needed. Current RS ports needs at least an 8 mm incision. As a result, patients have a faster recovery time and less discomfort and scarring as well as less risk of infection.

According to a systematic review of colorectal cancer surgery, laparoscopic patients had the lowest rate of wound infection, and there was a statistically significant difference between them and open surgery patients (OR = 0.65; 95% CI, 0.49 and 0.82). However, there was no statistically significant difference between them and patients who underwent RS (OR = 1.09; 95% CI, 0.11 and 8.45) [6].

## Robotic hands

Limited sensory feedback is another drawback of RS. To complete the treatment, surgeons rely on visual cues and feedback from the robot, which can impair their capacity to sense tissue tension and other crucial tactile sensations. This may make it more challenging to evaluate the tissue's quality.

The ability of RS to access more difficult-to-reach body parts is another benefit. In RS, surgeons can access places that may be challenging to reach using conventional surgical techniques because the robotic arms can be maneuvered more precisely than a human hand.

Greater precision during treatment is one of the most important benefits of RS. Because robotic arms are significantly more accurate than human hands, surgeons may carry out intricate surgeries with more precision. In delicate treatments, this precision can be very significant.

## Intraoperative complications

Moreover, there is less blood loss during robotic surgery. This is because there is less stress to the tissues because the robot utilizes smaller devices, a bipolar grasper, and magnified vision. In procedures where blood loss is a concern, such as surgery on the liver or spleen, this can be particularly crucial. The decreased bleeding and transfusion rates in the robot-assisted group may be due to RS, which causes high abdominal pressure during surgery by pneumoperitoneum.

According to the findings of a study by Moran, compared to open surgery, a robot-assisted radical prostatectomy was significantly associated with a lower predicted blood loss and transfusion rate. The difference in blood loss between the two methods was linked with a significant level of heterogeneity (I2 = 98%), which reflected the various ways this was measured in the included trials. More consistency was seen in the transfusion rate outcomes (RR 0.23, 95% CI 0.18–0.29, P 0.001, I2 = 17%) [7]. Novara reports of the robotic approach demonstrated a transfusion rate of less than 2% [8].

Equipment that is extremely sophisticated and precise is required for RS. Significant problems or even death could result from a mechanical breakdown or malfunction of the robot during operation. Although this risk is minimal, surgeons should be aware of it nonetheless.

## Postoperative

Patients often recover faster from robotic surgery because it is less invasive than traditional surgery. This translates into reduced downtime and an earlier return to their regular activities. For patients who lead busy lives or who must rapidly return to work or school, this may be especially crucial.

In urologic surgery, one of the major concerns following radical prostatectomy is urinary and sexual recovery. RS has better outcomes than open surgery in terms of sexual function at 12 months (RR 1.60, 95% CI 1.33–1.93, P = 0.001), urinary function at 12 months (RR 1.06, 95% CI 1.02–1.11, P = 0.009), and the percentage of pT2 tumors with positive surgical margins (RR 0.63, 95% CI 0.49–00.81, P 0.001). Differences in sexual function and favorable surgical margins might suggest that patients undergoing RS stand to benefit, even though the difference in urinary function is of questionable clinical significance given the modest degree of benefit shown and the poor methodological quality of the evidence on which it is based [7].

Patients experience reduced pain and discomfort because of robotic surgery. As there is less tissue trauma due to the smaller incisions, there is less pain and suffering during the healing process. Furthermore, RS often has fewer complications, which can lessen the pain and discomfort for patients.

## Future scenario

Based on technology developments we are seeing in other fields far from the health and medical areas, we expect that the patient cart, vision cart, and surgeon cart will be minimized in size while more precise movements will guide the surgeon during surgery. With the rising use of AI, the intraoperative use of image fusion guidance and/or intracorporeal devices may enhance the surgeon's ability to find endophytic tumors in solid organs such as the liver or kidney. Machine learning techniques would be used to extract recommendations from thousands of prior examples, which would then be stored in the cloud and made available as needed. Predictive movement models may also be used by AI to improve this process and allow the surgeon-console signal adequate time to travel over very long distances [9].

Nowadays, with the end of Intuitive's Da Vinci system patent, several robotic platforms may enter the market. They differ in cost effectiveness with several patterns of use. The market is going to be ruled by supply and demand, and some platforms may rise as an alternative to today's standard of care. Some cheaper options may be utilized in simpler surgeries, such as a cholecystectomy or abdominal hernia repair. Smaller and less-complex surgeries are going to be in the RS scope in the next few years.

## Key points

- Robotic surgery is a worldwide rising market with remarkable growth potential.
- Future advancements in automation may further lower the learning curve for RS.
- With new products and businesses entering the market, the 10 ten years will see significant development in this industry.

## References

[1] Hu JC, O'Malley P, Chughtai B, et al. Comparative effectiveness of cancer control and survival after robot-assisted versus open radical prostatectomy. J Urol 2017;197(1):115–21.

[2] Palmer KJ, Lowe GJ, Coughlin GD, Patil N, Patel VR. Launching a successful robotic surgery program. J Endourol 2008;22(4):819–24. https://doi.org/10.1089/end.2007.9824. PMID: 18419223.

[3] Wang RS, Ambani SN. Robotic surgery training: current trends and future directions. Urol Clin North Am 2021;48(1):137–46. https://doi.org/10.1016/j.ucl.2020.09.014. Epub 2020 Nov 5 PMID: 33218588.

[4] Cimen HI, Atik YT, Gul D, et al. Serving as a bedside surgeon before performing robotic radical prosta-tectomy improves surgical outcomes. Int Braz J Urol 2019;45(6):1122–8.

[5] Schönburg S, Anheuser P, Kranz J, Fornara P, Oubaid V. Cognitive training for robotic surgery: a chance to optimize surgical training? A pilot study. J Robot Surg 2021;15(5):761–7. https://doi.org/10.1007/s11701-020-01167-3. Epub 2020 Nov 13 PMID: 33185847. PMCID:PMC8423692.

[6] Hislop J, Hensman C, Isaksson M, Tirosh O, McCormick J. Self-reported prevalence of injury and discomfort experienced by surgeons performing traditional and robot-assisted laparoscopic surgery: a meta-analysis demonstrating the value of RALS for surgeons. Surg Endosc 2020;34 (11):4741–53. https://doi.org/10.1007/s00464-020-07810-2. Epub 2020 Jul 24 PMID: 32710214.

[7] Moran PS, O'Neill M, Teljeur C, Flattery M, Murphy LA, Smyth G, Ryan M. Robot-assisted radical prostatectomy compared with open and laparoscopic approaches: a systematic review and meta-analysis. Int J Urol 2013;20(3):312–21. https://doi.org/10.1111/iju.12070. Epub 2013 Jan 14 PMID: 23311943.

[8] Novara G, Ficarra V, Rosen RC, Artibani W, Costello A, Eastham JA, Graefen M, Guazzoni G, Shariat SF, Stolzenburg JU, Van Poppel H, Zattoni F, Montorsi F, Mottrie A, Wilson TG. Systematic review and meta-analysis of perioperative outcomes and complications after robot-assisted radical prostatectomy. Eur Urol 2012;62(3):431–52. https://doi.org/10.1016/j.eururo.2012.05.044. Epub 2012 Jun 2 PMID: 22749853.

[9] Kim SSY, Dohler M, Dasgupta P. The internet of skills: use of fifth-generation telecommunications, haptics and artificial intelligence in robotic surgery. BJU Int 2018;122(3):356–8.

# 38

# The nursing role in ethics, safety, privacy, and legal aspects of robotic surgery

*Rita de Cássia Burgos de Oliveira[a], Cecilia da Silva Angelo[b], and Yasmin Russo de Toledo[b]*

[a]University of São Paulo, School of Nursing, São Paulo, Brazil [b]Surgical Center Nursing, A.C. Camargo Cancer Center, São Paulo, Brazil

## The role of nursing ethics, safety, privacy, and legal aspects of robotic surgery

Ethics is the part of philosophy responsible for investigating the principles that motivate, distort, discipline, or guide human behavior, majorly with reflections on the essence of norms, values, prescriptions, and exhortations present in any social reality, that is, it is a set of moral and judgmental rules and precepts of an individual, a social group, or a society.

After many years studying ethics, especially with real-life experience as a nurse in the Surgical Center Unit and teacher of that same discipline for critical and acute patients, I understand that ethics is nothing more than being fully aware of what one is doing, and of the risks and consequences, especially complications. In this space, I refer to human ethics in a professional approach that establishes the set of standards that should govern the individual's professional behavior, thus maintaining an adequate posture and technical adjustment according to the work environment. In this professional environment, multiple elements are anticipated: professionals and patients, norms, techniques, discernment, respect, and responsibility [1].

The role of the nursing staff in preserving patient privacy during robotic surgery is essential to ensure respect for patient rights and compliance with ethical and legal standards. The ethical principles and confidentiality guidelines concerning patient information must be adhered to. All personal and medical patient information must be treated with utmost confidentiality.

The nurse must provide clear and accurate information to patients about the surgical procedure, including aspects related to privacy, such as who will have access to the operating room, the use of cameras and other devices, and how their information will be protected. The terms of consent must be applied voluntarily to the patient undergoing robotic surgery. Only the team involved in the procedure must be allowed access to the operating room, and the patient's positioning must be done in a way that preserves their privacy as much as possible during the entire procedure, only exposing areas that are necessary for the surgery. It is mandatory that all monitoring equipment such as cameras is configured in a way that preserves the patient's privacy. Nurses must also be empathetic and respectful toward the patient's privacy during robotic surgery and be available to answer questions and reassure the patient as needed.

Protecting patient privacy in surgical procedures is critical to ensuring the integrity of medical records and patient trust. Ensuring compliance with security, ethics, privacy, and all legal aspects is a shared responsibility between the professionals and institutions involved.

Surgeries continue to evolve at fast pace, especially since the adoption of less invasive procedures and instruments [1]. Such procedures began in the late 1980s in the United States, and that period in the history of surgery has become known as the "laparoscopic revolution" [1]. Surgeons and perioperative nurses rigorously observed the physiopathology of surgical trauma, seeking better therapeutic results through minimal intrusion and fewer occurrences of trauma [2].

Minimal surgical trauma then was redefined as a period of fast resumption of organic functions, less pain, better respiratory function, better humoral response, and less time spent by the patient in the hospital. These are the main arguments that determine and justify the use of minimal incisions, even taking into consideration that reduced operating surfaces may cause techniques to become more complex [2].

Technology notwithstanding, patient and staff safety must be ensured at all times. In 2004, with the aim of preserving surgical patient safety, the World Health Organization (WHO) created the World Alliance for Patient Safety. The objectives of this program (which was renamed the Patient Safety Program) were, among others, to organize the concepts and definitions of patient safety and propose measures to reduce risks and mitigate adverse events [3]. A safe surgery checklist has been outlined, to ensure that all the necessary steps for safe surgery are executed properly. The safe surgery checklist must be verified three times: before the anesthesia (patient's entry into the operating room/sign in), before the surgical incision (time in and time out), and before the patient leaves the operating room (sign out). That is performed by the nursing team and involves the entire interprofessional team, i.e., the surgeons, the anesthesiologist, and the surgical instrument nurse (scrub nurse). The tool includes multiple verification items such as full name (safety goal number 1), identification number (health institution registration), surgical technique, surgery site, control of materials and instruments, and other items. Patient safety is the top priority in surgical procedures. All processes and protocols in Perioperative Care are specifically designed to reduce risks and mitigate AEs (Adverse Events). No amount of cautiousness applied during the process is enough to eliminate completely the risk of preventable errors, so practices and protocols must be put in place to ensure patient safety [4–7].

Ensuring the safety and efficacy of the process requires that patients undergoing robotic surgery be qualified for the procedure and can benefit from this technique.

The patient must be adequately prepared before robotic surgery, which includes explaining the procedure and identified risks, providing clear instructions on preoperative preparation such as fasting and use of medication, explaining critical surgical positioning in this surgical modality and precautions taken through diagnosis and nurse assessment, and ensuring that the patient is physically and psychologically fit for the surgery.

The patient is evaluated through a process with multiple stages that comprise the analysis of the patient's history before, during, and after the surgery. This process is known as Systematization of Perioperative Nursing Care (SPNC) [6].

Based on the patient's clinical history and particularities, the nurse plans all the assistance to be provided throughout the perioperative period, by prescribing nursing care and evaluation of admission to the Surgical Center.

When the nurse admits the patient, they perform an evaluation to determine the patient's clinic history, medication being used, allergies, surgical history, and current condition [4,6]. The nurse instructs the patient on the surgical procedure and postoperative care.

Nurses must be alert to the emotional condition of patients, which is often frail in individuals with high levels of anxiety, helplessness, and fear. This procedure must take place in a calm and peaceful environment so the patient can understand the information given and assimilate the context of the surgical procedure and the environment to which they will be exposed. Therapeutic techniques in this scenario can reduce stress and anxiety and contribute to the patient's psychological health (music therapy, color therapy, and others). The nurse must ensure that the patient is in optimal condition to ensure the success of the procedure and a speedy recovery [8].

The nurse is also responsible for ensuring patient safety since the preoperative period, planning in advance all care measures for the surgical day including patient risk assessment, evaluation of associated disease conditions, such as diabetes, hypertension, and other problems such as kidney disease, heart disease, or related to the musculoskeletal system or integumentary system.

This entire assessment is directly linked to patient safety in the intraoperative period and to the prevention of unwanted events or adverse events.

The immediate postoperative period is the last period corresponding to the perioperative period.

## Members of the operating room in robotic surgery

The robotic procedure requires adjustments to the layout of the operating room (interference map) such that the robot, controlled by the surgeon, can perform its function without interference. The nurse is in charge of that process, based on the protocols and the procedure to be executed according to the health institution. Taking into account the time length of the surgery and the patient's particularities, the nurse manages the best arrangement for the operating room according to the surgical procedure to be performed. It is a controlled and organized environment where a team cooperates to ensure a successful procedure. Each member of the team has specific and fundamental roles in the adequate execution of the surgical process [6,9].

The members of a robotic surgery team are the surgeon, the assistant surgeon, the anesthesiologist, the nurse, the nurse technician, and the surgical scrub nurse. The surgeon is the leader of the surgical team, tasked with performing the procedure and making the necessary decisions during the surgery. He controls the surgical robot through the surgeon's console and the "manual commands," supported by his/her team, which help in carrying out the surgical procedures and in the manipulation of instruments attached to robotic arms and specific instruments and equipment. [6]

The anesthesiologist is responsible for ensuring the patient's hemodynamics. A robotic procedure requires thorough monitoring of the patient by means of electrocardiography, oximetry, blood pressure, temperature, anesthesia depth monitoring (Bispectral Index), and neuromuscular blockade monitoring. Total general anesthesia, balanced anesthesia, and spinal or epidural anesthesia are appropriate for robotic procedures. Anesthetic techniques vary according to the patient's particular condition and the need for the procedure, determined by the careful evaluation of the anesthesiologist and surgeon [5,6].

Nurses in robotic surgery have two roles: the coordinator of the robotic program and the reference nurse who provides the assistance to the robotic surgery patient. The coordinating nurse is responsible for regularly training and developing the team, designing and applying protocols in accordance with guidelines and recommended good practices, prioritizing and ensuring the safety of the patient and everyone involved, planning and enforcing the robotic surgery map in accordance with the surgery schedule, participating in the entire planning and strategy for the procurement of supplies and robotic materials along the supply chain, actively collaborating with the medical coordinator of the program in the execution and statistical survey of patients and training of doctors as "qualified doctor in the program and proctor" according to society and the CFM (Federal Medicine Council, e.g.), observing and assessing the use of robotic tweezers by the surgeons for later evaluation of purchases and specific commercial packages for the program areas and medical coordination, contributing to statistical analyses of the surgical volume of the teams and/or medical departments, performing analysis of the infection rate in the surgical site for the creation of indicators, organizing and holding meetings with the reference nurse and robotics nurse team for assertive communication of the processes and organization [5,6].

The perioperative care reference nurse of the robotic procedures is responsible for checking personal data, patient clinical history, and preoperative exams. Based on the robotic surgery map, he/she plans the arrangement of the operating room, outlines and checks materials, checks instruments with the Material and Sterilization Center, supplies and equipment, and provides these materials. The nurse is also responsible for assembling the entire robotic system, calibrating the optics in the SI system, and for positioning the patient adequately, safely, and effectively. During the entire procedure, the nurse is in charge of ensuring proper functioning of the equipment and the robot, enforces and feeds the robot's checklist daily and at every period, evaluates the entire performance and connections of the robotic system with the clinical engineering team, monitors the statistics provided by the program coordinator and analyzes the operational performance of the robotic nursing team involved in the operation, analyzes the performance of the team as a whole and contributes to the reports made to the coordination in periodic meetings, collaborates with the Material and Sterilization Center in analyzing the robotic kits according to specialization, and inserts the computerized document where reports are generated that will be evaluated by the program coordinators (nursing and physician), providing statistical data and performance evaluation of the surgeons (SPNC Robotics document) [5,6].

Both the coordinating and the care reference nurse accompany the patient in all periods, from the surgical proposal to the postoperative phase, which is in compliance with the Robotic Systematization of Perioperative Nursing Care (Robotic SPNC). Robotic SPNC carries out the relevant planning and interventions in order to ensure a safe procedure and adequate recovery without preventable adverse events.

The Nurse at the Material and Sterilization Center is responsible for processing and managing robotic materials. Robotic instruments are considered critical and expensive materials. The responsible nurse and everyone involved in this process must be trained to carry out all the stages of processing, from receiving the material at the Material and Sterilization Center to storage and distribution, including every inspection stage, cleaning, drying, final review and testing, preparation, and sterilization. The scrub nurse must ensure efficiency in processing in collaboration with the robotics coordinator nurse, and both nurses must establish routine protocols, ensuring proper functioning of the materials and efficient inventory control. A computerized system is also necessary to guarantee the traceability of materials, with photographic identification of the reference in the system of the robotic tweezers used in each surgery [6].

The circulating nursing technician is responsible for ensuring that all necessary equipment, materials, and supplies are available and accessible to team members, overseeing proper patient positioning with the team, identifying and submitting surgical pathology samples for analysis, and assisting the robotics reference nurse in all his technology-related activities in the operating room.

The surgical instrumentation technician (scrub nurse) is responsible for preparing and organizing the surgical instruments on the surgical table, assisting the nurse in placing the drapes on the robotic arms of the robotics platforms assisting the team with their attire, providing instruments to the surgeon and medical team throughout the entire surgical procedure, assembling and disassembling the surgical table, and collaborating with the assistant surgeon who remains present donning proper attire [6].

The nurse must ensure that the robot is connected to the mains "on and charging" and with the connections attached and connected, functioning correctly before the procedure is initiated. The nurse is responsible for assembling, disassembling, and cleaning the robot, also checking the equipment connections, testing the instruments, and ensuring that the software is updated and the robot remains "online," i.e., connected to the system's specific platform network. It is also important to check for compatibility issues between the instruments and the robot, to avoid complications during surgery. The nurse and the surgeon must be aware of every movement of the robotic arms and continuously monitor the progress of the surgery [6].

It is important to prevent unforeseen delays during the procedure by ensuring that the patient is correctly positioned on the surgical table. Positioning must be carefully planned to avoid injuries caused by prolonged periods in the same position. Safety devices must be confirmed to be in full working order for that.

## Surgical positioning in robotic surgery

The choice of the surgical position depends on the type of surgery to be performed, patient's anatomy, procedure technique, and the patient's health condition. Positioning is a fundamental aspect of preoperative preparation, for the success of the surgery largely depends on the surgeon's ability to properly access and visualize the area of interest. The type of positioning used will depend on the region of the body to be operated on, the specific needs of the procedure, and the clinical condition of the patient. Positioning is performed after anesthesia with the patient already sedated and intubated [5,7].

Surgical robots can adapt to different angles and positions, allowing less conventional and more personalized positioning in order to maximize the precision and effectiveness of the surgery. Positioning in robotic procedures is more laborious in comparison with other surgical techniques because the nurse must position the patient in the best possible way, with the specific concern of access to surgery and prevention of injuries and complications, also providing adequate access to the robotic arms [9].

Correct positioning requires knowledge of anatomical and physiological implications of each surgical position inside the patient's body, associated with the type of surgery, duration of the surgical procedure, as well as safety and protection measures available to ensure adequate positioning and avoid postoperative complications [5,7].

Positioning of the patient is carried out by the nurse, the medical team (surgeon and anesthesiologist), and the nursing technician. This entire interdisciplinary team must be aware of the risks associated with improper positioning and know how to mitigate and avoid them to ensure patient safety during the surgery. The entire team must have adequate training in positioning techniques in robotic procedures, technical and scientific knowledge of anatomy, prior knowledge of the patient's particularities and risks of complications and development of injuries, as well as the use of modern equipment for fixation of the patient on the operating table, considering that most robotic procedures use positions with a high degree of inclination. After positioning the patient, before starting the procedure, it is necessary to test the positioning by adjusting the table at the angle at which the procedure will take place to verify that there is no risk of the patient moving from the surgical table and that all regions and limbs are protected [5,6].

## Future scenario

1. Future of robotic surgery, what are the prospects? Food for thought! Will robotic surgery at some point be accessible to everyone who needs this technology?

   Technology progresses at fast pace and robotic surgical procedure emerges as a promising technique with significant impact on the world of medicine. The technology is expected to become more advanced as systems allow surgeons to perform more complex procedures with greater safety and efficiency. Professionals increasingly seek the improvement of technologies and practices based on guidelines and scientific evidence because it is an innovative modality in the market. Inevitable additions to minimally invasive technologies in surgical care such as AI (artificial intelligence) and 5G technology (with remote surgeries) are not too far in the future as shown by their exponential evolution.

Currently, robotic surgery accounts for about 5% of minimally invasive and laparoscopic surgeries in the world. However, in countries such as Brazil, it is still rarely available to patients who depend on the Brazilian Unified Health System (Sistema Unico de Saúde-SUS).

**2.** Is there a risk of overreliance on surgical robots, which would affect surgeons' ability to perform manual procedures?

Mastering robotic surgery and operating the robot efficiently pose a significant learning curve to surgeons. However, once they become proficient in using the robotic system, they may become less adept at performing procedures manually. It is imperative that surgeons maintain continuous training, alternating between traditional and robotic techniques, thus preserving their proficiency in both approaches and developing their aptitude in areas that are complementary to robotic surgery, such as laparoscopic techniques, which also require advanced manual skills. Even though robotic procedures are minimally invasive procedures with a low risk of complications, the surgeon must be prepared and capable of transitioning to a manual approach in the event of technical difficulties or unforeseen hurdles during the surgery.

## Key points

- Robotic surgeries—the revolution of surgical approaches and how they require a competent multidisciplinary team.
- The role of the nurses in ensuring patient safety, from the preparations to the surgery, to the postprocedure recovery stage, including assessment, guidance, anxiety management, and ensuring ideal conditions for the procedure;
- Compliance with security, ethics, and privacy requirements and legal aspects and shared responsibility between the professionals and institutions involved.

## References

[1] LEITE, Cássia Burgos de Oliveira Rd. DANTAS, Stela Manual de Normas Rotinas do Centro Cirúrgico do Hospital Aristides Maltez. Salvador; 1996.
[2] Martin RF. Robotic Surgery. Surg Clin North Am 2020;100(2):xiii–xiv. https://doi.org/10.1016/j.suc.2020.02.001. PMID: 32169191.
[3] Brasil. Ministério da Saúde, Fundação Oswaldo Cruz, Agência Nacional de Vigilância Sanitária. Documento de referência para o Programa Nacional de Segurança do Paciente. Brasília: Ministério da Saúde; 2014 [accessed on 05 May 2023]. https://bit.ly/2P04MwB.
[4] Rabêlo PPC, Prazeres PN, Cunha Bezerra T, Leite Cruz dos Santos DdJ, Venção de Moura NA, D'Eça Júnior A. Enfermagem e a aplicação da lista de cirurgia segura: uma revisão integrativa. Rev SOBECC. February 08, 2023. https://revista.sobecc.org.br/sobecc/article/view/856.
[5] Diretrizes de práticas em Enfermagem Perioperatória em Processamento de Produtos de Saúde. 8ª Edição. Brasil: Sociedade Brasileira de Enfermeiros de Centro Cirúrgico Recuperação Anestésica e Centro de Material e Esterilização – SOBECC; 2021.
[6] Cirurgia Robotica Principios E Fundamentos. 1ª edição. Brasil: Fundação Educacional Lucas Mac; 2022.
[7] Morais RM, Oliveira IKM, Marques KMAP. Cuidados de enfermagem para a prevenção de complicações anestésico-cirúrgicas no pós-operatório imediato. Sanare 2022;21(2):53–60.
[8] Joseph A, Khoshkenar A, Taaffe KM. RIPCHD.OR study group, et al minor flow disruptions, traffic-related factors and their effect on major flow disruptions in the operating room. BMJ Qual Saf 2019;28:276–83.
[9] Ângelo CdS, Silva Érica ALd, Souza Ad, Bonfim IM, Joaquim EHG, Apezzato MLdP. Posicionamento cirúrgico em cirurgia robótica pediátrica: relato de experiência. Rev SOBECC 2020;25(2):120–3. https://revista.sobecc.org.br/sobecc/article/view/581.

# The current and future clinical applications of robotic surgery among medical specialties

# Urologic robotic surgery for kidney and upper urinary tract

*José Ignacio Nolazco[a,b], Leonardo O. Reis[c], and Steven Lee Chang[a,d]*

[a]Department of Urology, Brigham and Women's Hospital, Harvard Medical School, Boston, MA, United States [b]Servicio de Urología, Hospital Universitario Austral, Universidad Austral, Pilar, Argentina [c]UroScience, State University of Campinas, Unicamp, and Urologic Oncology Department, School of Life Sciences, Pontifical Catholic University of Campinas, Campinas, São Paulo, Brazil [d]Lank Center for Genitourinary Oncology, Dana-Farber Cancer Institute, Boston, MA, United States

## Introduction to urologic robotic surgery for kidney and upper tract surgery

Over the last two decades, there has been a clear trend toward using robotic approaches in a wide range of surgical fields, especially Urology [1]. This shift in surgical approach is attributed to the shorter learning curve associated with the robotic surgical platform compared with conventional laparoscopic procedures. The robotic platform offers multiple distinct advantages. Surgical precision is enhanced by three-dimensional, high-definition surgical field visualization. Additionally, robotic surgical systems provide an unprecedented degree of freedom in instrument movement, granting surgeons improved maneuverability. Features like tremor filtering and superior suturing precision contribute to increased surgical dexterity, a crucial benefit, particularly in complex surgeries [2]. Although radical prostatectomy has emerged as the most common procedure employing robot-assisted technology in urologic surgery [3,4], there is arguably no more suitable application of robotic surgery than for kidney and upper tract surgery, where accuracy, precision, and economy of motion are paramount. Consequently, the robotic approach has gained great significance in other areas of urologic surgery, notably in complex partial nephrectomies, which is associated with less postoperative pain, reduced bleeding, shorter hospital stays, and superior cosmetic results without compromising oncologic outcomes. Herein we will explore important facets of robotic surgery in urology and specifically review the application of robotic surgery in various urological procedures for the upper tract urinary system.

## Robotic preoperative planning

The decision to perform a robotic-assisted surgery requires meticulous patient selection. Similar to any surgical intervention, a comprehensive assessment of the patient's overall health status, the existence of comorbidities, and suitability for anesthesia and surgery is imperative [5,6]. Despite the many advantages of robotic surgery, certain conditions may present relative contraindications. Patients with a history of extensive abdominal surgery, severe obesity, or significant cardiopulmonary diseases may not be ideal candidates due to the potential for increased risk or procedural complexity [7,8].

In preparation for surgery, a thorough preoperative medical evaluation of the patient is essential. Given that the upper urinary tract cannot be readily assessed via physical examination, a detailed review of recent cross-sectional imaging, such as CT or MRI scans, should complement routine laboratory tests. Such images are crucial for accurately characterizing kidney lesions and appreciating the unique anatomical considerations of each patient. Before surgery, patients should be informed of the potential risks and complications associated with the procedure, underlining the importance of informed consent. Due to the proximity of kidneys and ureters to the aorta, inferior vena cava, and branches of these great vessels, the risk for bleeding must be discussed in the preoperative setting.

The complexity associated with a robotic procedure requires effective team coordination and communication. Preoperative briefing aids in ensuring all surgical team members are synchronized with the procedural plan [9]. Considerations such as using pneumatic or mechanical compression devices for thrombosis prevention, or placing a Foley catheter for bladder drainage, should be discussed and agreed upon. In particular, patients undergoing minimally invasive upper urinary tract surgery are frequently maintained in a lateral decubitus position and special attention must be paid to unique pressure points including the ear, acromion process, iliac crest, greater trochanter, lateral aspect of the knee, and the malleolus.

## Patient positioning for robotic renal surgery

The patient's position during surgery is vital, typically involving a modified flank position. Care must be taken to avoid nerve injuries resulting from improper positioning. Attention should also be given to potential pressure points to prevent perioperative injuries. The transperitoneal route is most commonly employed for robotic surgeries when it comes to the surgical approach [10]. While this approach is typically associated with a wide field of view and access to multiple organs, there are disadvantages such as a need for bowel mobilization that may prolong postoperative ileus as well as the creation of bowel adhesions potentially leading to internal hernia and an elevated risk for complications for future surgeries. To avoid these issues, the retroperitoneal approach is ideal as it allows for direct access to the kidneys, thus eliminating the concern for postoperative adhesions and theoretically minimizing the issue of ileus. However, this retroperitoneal approach is generally considered more technically challenging due to the limited visibility and unfamiliar anatomy. Thus, for retroperitoneal surgery, the robotic surgical platform truly shines with enhanced precision and dexterity of the robotic arms and wristed instruments in the limited working space as well as the three-dimensional visualization to better appreciate the subtle landmarks of the retroperitoneum [11].

## Port-placement for robotic renal surgery

Following the meticulous patient positioning, strategic planning for port placement becomes indispensable for the procedure's effectiveness. The specific site selection for trocar insertion in renal surgeries depends on many considerations, including the patient's unique anatomical features, tumor location, the surgeon's predilections, and the specific robotic system. Initial entrance into the peritoneal cavity can be achieved through the Veress needle technique or the Hassan approach, performed under direct visualization as in laparoscopic techniques. Once access to the cavity is established and the pneumoperitoneum is insufflated, subsequent trocars are placed under direct visual guidance. Optimally positioned trocars ensure direct access to the kidney, minimize the risk of instrument clashing, and optimize the instrument's range of motion.

## Robotic radical nephrectomy

### Oncological outcomes

In the realm of kidney cancer surgery, three primary hierarchical objectives are pursued: first and foremost, the achievement of optimal oncological outcomes, followed by functional outcomes, and lastly, the mitigation of complications. A systematic review showed no differences in local recurrence rates nor in all-cause cancer-specific mortality between robotic and laparoscopic radical nephrectomy (LRN) for localized kidney cancer [12]. Conversely, a large multicenter study compared robotic radical nephrectomy (RRN) and LRN and found an association between the robotic approach and a higher positive margin rate and greater risk of recurrence or metastasis than LRN ($P < 0.01$). However, patients who underwent RRN presented with more advanced or progressive disease, characterized by higher tumor grades and pathological stages. There was no difference in overall survival between groups in this study [13]. Golombos et al. delved into the SEER-Medicare linked database, discovering no significant distinctions between the two techniques concerning overall and cancer-specific survival. Nevertheless, they noted a considerably higher cost associated with RRN [14].

## Functional outcomes

A systematic review and meta-analysis, including seven studies comprised of 1832 patients, showed no difference between robotic and LRN regarding perioperative outcomes, surgical time, blood loss, conversion rate, and complications [15]. Even for higher stage kidney cancers (nonmetastatic RCC >10 cm in size [pT2b]), RRN demonstrated comparable results compared with LRN and open radical nephrectomy [16].

## RRN in advanced kidney cancer: Renal vein and IVC thrombus

In advanced venous thrombi cases, RRN has emerged as a minimally invasive alternative to open surgery when technically feasible. The increased dexterity and precision, coupled with the improved range of motion afforded by the robotic approach, is uniquely advantageous over the laparoscopic approach for these complex cases. A propensity-score matched retrospective cohort study including 324 patients comparing open, laparoscopic, and robotic approaches for radical nephrectomy with inferior vena cava thrombectomy found that RRN had the lowest operative time, blood loss, and transfusion rates, while the open approach presented the highest complication rate and longest postoperative hospital stay [17].

## Robotic partial nephrectomy (RPN)

Robotic surgery offers several advantages, specifically for partial nephrectomy, which is a complex procedure in which optimal precision and efficiency are keys to success.

## Hilar clamping and ischemia management

The *first-assistant sparing technique* (FAST) aims to minimize warm ischemia time (WIT) and total operative duration. Before clamping, all sutures designated for renorrhaphy are strategically positioned intracorporeally, along with the bulldog clamps and the ultrasound probe. This arrangement grants the surgeon improved control over the subsequent critical steps following clamping [18].

Numerous options exist for managing hilar clamping. Early unclamping has effectively reduced WIT compared with conventional techniques without a significant increase in blood loss or complications in laparoscopic partial nephrectomy (LPN) [19]. This was also true when applied to RPN surgeries [20].

Selective clamping has been described as an alternative to minimize healthy parenchymal ischemia during partial nephrectomy. By clamping, only the branches supplying blood to the tumor, perfusion to the remaining parenchyma is maintained, potentially minimizing ischemia's impact on postoperative function. Nevertheless, some studies have not identified a significant advantage associated with this technique [21–23].

Furthermore, off-clamping techniques were shown to have a significant impact on postoperative renal function [24,25]. However, to establish the clinically significant effect of this approach, a positive impact must be demonstrated over the long term. A meta-analysis comparing off-clamp and conventional RPN outcomes revealed no significant differences in renal function 3 months postsurgery [26]. Nonetheless, the authors acknowledged unmeasured confounding factors in their analysis and concluded there is no evidence to recommend one ischemia technique over the other. A recent randomized controlled trial (RCT) involving 80 patients showed that off-clamping RPN yielded comparable functional outcomes to the on-clamp technique, indicating no discernible advantage in preserving patients' renal function [27].

## Tumor excision

In partial nephrectomy, three primary resection techniques are employed for tumor removal. First, enucleation involves the tumor excision without any surrounding healthy parenchyma, utilizing the tumor's pseudocapsule as the limit to detach the tumor from the kidney. Second, enucleoresection comprises the removal of the tumor along with a thin rim of peritumoral parenchyma. Lastly, resection refers to removing the tumor in addition to a "consistent" rim of healthy parenchyma (approximately 1 cm) [28].

A prospective multicenter study evaluating partial nephrectomy resection techniques (enucleation, enucleoresection, and resection) determined that the resection technique was a significant predictor of positive surgical margins. Maintaining a wide parenchymal margin distant from the tumor limit can protect against positive surgical margins, particularly with imperfectly spherical tumors. The enucleoresection technique had a significantly higher risk of positive margins than enucleation (OR 2.42, 95% CI 1.08–5.43) and resection (OR 5.73, 95% CI 1.26–26.17). This association persisted even after excluding the surgical approach from the analysis [29]. However, a recent systematic review, including 20 studies, concluded that positive margins are unrelated to the resection technique in robotic-assisted partial nephrectomy [30]. There are likely additional unmeasured factors that may explain the earlier conflicting data including the location of the renal masses and proximity to surrounding structures as well as possibly the degree of trauma to the renal mass during resection. Additional studies are necessary to clarify if there is an indisputable optimal surgical technique for tumor excision.

## Renal reconstruction

Following tumor excision, various renorrhaphy techniques can be employed. Traditionally, the primary objectives of renorrhaphy have been to minimize complications by ensuring adequate hemostasis and preventing urinary leaks. However, reducing the ischemic impact on renal parenchyma has recently emerged as a crucial goal in addition to the classic objectives [31].

A systematic review comparing single-layer renorrhaphy (typically a more superficial cortical layer excluding the collecting system) and double-layer renorrhaphy (involving both a deeper layer potentially including the collecting system as well as a superficial cortical closure) demonstrated a reduction in operative time and WIT for single-layer closure. Importantly, this advantage did not result in an increase in urinary leak or other complication rates [32]. Similar to the challenges of evaluating resection techniques, the successes and failures of renorrhaphy techniques are linked to the unique aspects of the tumor itself. More superficial tumors that do not involve the collecting system are likely well managed with a single-layer renorrhaphy. In contrast, for deep tumors that are likely to be associated with a longer operative time and WIT, incorporating the collecting system and transected segmental vessels in the deeper layer may be necessary to avoid postoperative urinary leak and pseudoaneurysm, respectively.

## Oncological outcomes

Partial nephrectomy provides oncological results noninferior to those of radical nephrectomy. Van Poppel et al. conducted a RCT comparing the overall survival of 541 patients with small ($\leq$5 cm) solitary renal tumors (T1–T2 N0 M0). This investigation showed that both surgical approaches exhibited similar oncological outcomes, thereby establishing partial nephrectomy as a noninferior alternative to radical nephrectomy (HR: 2.06 [95% CI 0.62–6.84]) [33]. Also, partial nephrectomy is safe for larger tumors ($\geq$7 cm), offering similar oncological and functional outcomes [34,35]. In experienced hands, RPN for T1–T2 Renal tumors has comparable perioperative short-term oncological and functional results as compared with the LPN approach [36]. In a retrospective propensity score-matched study comparing open, laparoscopic, and robotic partial nephrectomy, no differences were observed in local recurrence, distant metastasis, or cancer-related deaths at the 5-year follow-up [37].

## Functional outcomes

After prioritizing adequate oncological outcomes, preserving renal function remains crucial. Kidney cancer patients often face comorbidities like hypertension, obesity, and diabetes, further impacting renal function and potentially leading to chronic kidney disease (CKD) [38,39]. Nephron-sparing surgery emerged as a preferable alternative to traditional radical nephrectomy. Solid evidence highlights partial nephrectomy's superior renal function preservation compared with radical nephrectomy [40,41]. Concerning approach, a meta-analysis reveals RPN's comparable or marginally better functional outcomes vs LPN, potentially due to RPN's shorter WIT association [42].

A meta-analysis found that RPN was associated with a lower conversion rate, shorter ischemia time, and shorter length of stay in the hospital than LPN [43]. In concordance with these findings, a recent meta-analysis further substantiated these results, demonstrating that RPN exhibited shorter warm ischemic times, superior postoperative renal function, and lower conversion to open and fewer intraoperative complication rates for complex renal tumors characterized by renal nephrometry score of $\geq$7 [44].

# Pentafecta

The gold standard for the long-term results of nephron-sparing surgery is the "pentafecta," achieving (i) negative surgical margins, (ii) WIT ≤25 min, (iii) no major complications, (iv) >90% preservation of baseline estimated glomerular filtration rate (eGFR), and (v) no upgrading of CKD stage.

# Complications

## Bleeding

Bleeding is the most common complication in RPN and can occur during or after surgery [45]. Therefore, meticulous preoperative evaluation of tumor location and its relationship with the vasculature is paramount. For preoperative planning, 3D impressions (or virtual reality reconstructions) of the tumor can assist the surgeon in orienting, strategizing ahead, and reviewing the surgical steps before surgery [46]. Before tumor resection, the complete mobilization of the kidney with adequate exposure of the renal hilum and identification of all hilum elements is crucial [47]. If pulsatile bleeding is observed after clamping the renal artery, it may be due to an unrecognized supernumerary renal artery. Also, postoperative bleeding in partial nephrectomy can occur days after surgery. Late bleeding is most commonly due to the formation of pseudoaneurysms. This complication requires urgent angiography to identify and control with selective embolization in the same setting [48,49]. When suspicion is high, this approach is preferred over a CT scan or duplex ultrasound.

## Urine leak

The collecting system might be opened during partial nephrectomy, necessitating surgical repair. Ensuring adequate closure of the collecting system is vital to prevent urinary extravasation. A large multicenter analysis involving five different institutions and 1791 patients who underwent partial nephrectomy showed a urine leak rate of 0.78% for patients who underwent robotic partial nephrectomy. When measured against open or laparoscopic methods, this comparatively low rate of urine leakage was attributed to the enhanced visualization and suturing techniques accompanying the robotic approach [50]. Some risk factors are associated with this complication: tumor proximity to the collecting system, the surgeon's early operative experience, and preoperative moderate or severe CKD [51]. The intrarenal pelvis is associated with a higher risk of urine leak following partial nephrectomy [52]. The renal pelvic anatomy, measured by the renal pelvic score (RPS), can predict urine leaks following open and robotic partial nephrectomy [53]. Most of these leaks resolve spontaneously without requiring surgical intervention. Management strategies may include the placement of a urinary catheter to decompress the urinary system [54]. Another management method involves the application of a tissue glue agent, such as cyanoacrylate glue, to promote wound healing [55,56].

## Acute renal failure

Preoperative identification of patients more susceptible to developing acute kidney injury (AKI) after nephrectomy is paramount [57]. Pre-existing conditions such as CKD, hypertension, diabetes, and obesity are among some risk factors for developing AKI following nephrectomy [58–60]. Although not routinely used, some biomarkers, including Cystatin C and neutrophil gelatinase-associated lipocalin (NGAL), have been identified as potentially useful predictive tools for AKI following partial and radical nephrectomy [61]. A comprehensive preoperative evaluation, which includes assessing all factors that increase the risk of AKI, can inform more effective patient management strategies, potentially reducing the incidence of postoperative AKI.

# Technological advances in partial nephrectomy

## Image-guided navigation

Intraoperative imaging technologies, most commonly ultrasound, are routinely utilized to locate and target the tumor for removal, especially for highly endophytic tumors. Real-time imaging provides crucial information about the tumor's location and nearby blood vessels, aiding in careful tissue removal. Furthermore, the assessment for a lack renal blood flow following clamping with the Doppler function is a useful and easy method to minimize the risk of significant bleeding during tumor resection and renal renorrhaphy [62].

Another emerging technology that may enhance intraoperative and postoperative outcomes in patients undergoing PN is fluorescence imaging with indocyanine green (ICG). This technique has improved real-time visualization of the tumor anatomy and identified feeding arterial vessels, allowing for super selective clamping of tumor-specific branches of the renal artery, reducing ischemia time, and preserving normal renal function with acceptable oncological results [63]. A retrospective study found that patients undergoing ICG-guided RPN had superior renal functional preservation at 3 months compared with the standard technique. ICG was also associated with reduced blood loss in malignant kidney tumors [64]. While more extensive comparative trials are still needed, this novel technology shows promise in improving outcomes and progressing partial nephrectomy techniques.

### Advanced imaging and 3D printing

Preoperative imaging, such as MRI and CT scans, can help create detailed 3D models of the patient's kidney anatomy and tumor. Surgeons can use these models to plan the surgery and simulate the procedure before entering the operating room, improving surgical precision, and reducing complications.

### Surgical planning, simulation, and training

Virtual reality (VR) and augmented reality (AR) technologies can aid in surgical planning, rehearsal, and training. Surgeons can practice complex procedures in a virtual environment before performing them on actual patients [65].

## Robotic nephroureterectomy

In patients with localized upper tract urothelial carcinoma (UTUC), radical nephroureterectomy (RNU) with bladder cuff excision is the gold standard and offers durable control and cancer-specific survival [66]. The first robot RNU was reported by Rose et al. [67]. Although the robotic approach has been progressively adopted in the United States [68], there is no strong evidence indicating superiority over both laparoscopic and open RNU. A systematic review and meta-analysis revealed no clear advantage of one minimally invasive technique over another in treating UTUC via nephroureterectomy [69].

Various excision techniques exist to manage the bladder cuff, including the transvesical, extravesical, and endoscopic approaches. Notably, the endoscopic approach was associated with a higher risk of intravesical recurrence rates [70].

Template-based lymph node dissection (LND) has been shown to improve cancer-specific survival in patients with muscle-invasive disease and reduce the risk of local recurrence [71,72]. Clinical guidelines recommend offering template-based LND to all patients undergoing RNU for high-risk, nonmetastatic UTUC [73]. Intriguingly, Pearce et al. found that patients undergoing RNU were more likely to undergo LND (27%) compared with those having open (15%) and laparoscopic (10%) procedures [74]. This might be attributable to the augmented dexterity provided by the robotic technology facilitating this complex surgical maneuver. Another study corroborating this trend reported LND rates of 41%, 35%, and 27% for robotic, open, and laparoscopic approaches, respectively. Furthermore, RNU was associated with the highest lymph node yield, having the greatest number of LNs removed. Importantly, this study found an association between increased LN yield and improved overall survival in patients with pN0 disease [75].

Regarding oncological outcomes, robotic-assisted surgery has demonstrated equivalent results to laparoscopic and open nephroureterectomy for organ-confined UTUC [66,76,77]. However, minimally invasive approaches have demonstrated superior perioperative outcomes, such as significantly reduced bleeding and shorter hospital stays, as well as decreased need for postoperative analgesics during the recovery period when compared with the open approach [78]. A recent multicenter propensity score study comparing laparoscopic vs robotic approaches revealed that robot-assisted nephroureterectomy was associated with fewer postoperative complications (primarily relating to low-grade complications) and a shorter hospital stay [79].

## Robotic pyeloplasty

Ureteropelvic junction obstruction (UPJO) is characterized by the anterograde obstruction of urine flow from the renal pelvis to the proximal ureter. This obstruction can lead to multiple complications, including hydronephrosis, renal failure, renal atrophy, lithiasis, and recurrent urinary tract infections [80–82].

In 1949, Anderson and Hynes described the first open dismembered pyeloplasty [83]. Traditionally, open pyeloplasty has been the gold standard for treating UPJO. Over the years, minimally invasive approaches have been employed with increasing frequency, demonstrating comparable success rates and lower associated morbidity [84]. Minimally invasive approaches are becoming more widespread due to their effectiveness and associated recovery [85,86]. In particular, robot-assisted pyeloplasty is a feasible and safe approach, boasting excellent success in relieving obstruction and low perioperative and postoperative morbidity [87]. In particular, the wristed instrumentation of the robotic platform allows for relatively straightforward suturing of the anastomosis, which is arguably the most challenging part of the procedure. Robot-assisted ureteroplasty has demonstrated long-term resolution of obstruction for the majority of patients [88].

A systematic review and meta-analysis comparing single-port robotic-assisted pyeloplasty with multiport robotic-assisted pyeloplasty demonstrated that the single-port approach had comparable effectiveness and safety, with superior outcomes and less postoperative pain compared with the multiport approach [89].

## Future scenario

Genetic and molecular profiling advances will translate into more personalized treatment plans, tailoring the surgical approach and treatment strategies to the specific characteristics of each patient's tumor [90]. Bioengineered tissue for reconstruction or tissue repair is expected to improve results, mainly in nephron-sparing techniques. Artificial intelligence (AI) algorithms might analyze real-time data to predict potential complications, suggest optimal strategies for safer surgeries, and assist surgeons in decision-making, tissue identification, and precise instrument control [91]. Also, AR and VR may enhance accuracy and visualization of robotic surgery [92].

## Conclusions

To conclude, data suggest that robotic-assisted urologic surgeries for the upper urinary tract have equivalency, if not superiority, to traditional open and laparoscopic approaches in specific contexts. This is reflected in similar oncological outcomes coupled with often superior perioperative and postoperative metrics, including reduced bleeding, lower reliance on postoperative analgesics, decreased length of hospital stays, and a lesser incidence of postoperative complications. However, patient selection is a crucial aspect of these procedures. Identifying the most appropriate surgical approach necessitates comprehensively evaluating various essential factors. These encompass the distinctive characteristics of the disease, the patient's explicit preferences and health status, and the surgeon's expertise, among other considerations. While the initial acquisition and ongoing maintenance costs of robotic equipment are significant, it is crucial to consider the broader advantages of these systems. The positive impact on patient outcomes and the overall recovery trajectory may counterbalance these financial burdens in the longer term. The robotic-assisted urologic surgical field is rapidly evolving, highlighting an urgent need for high-quality, randomized, controlled trials. This would support the existing evidence base and provide a more profound understanding of these procedures' cost-efficacy and long-term implications.

## Key points

- Robotic surgery is increasingly utilized for urologic procedures involving the kidney and upper urinary tract due to advantages including improved visualization and enhanced dexterity compared with open and laparoscopic techniques, respectively.
- Robotic surgery has enabled more minimally invasive approaches for complex renal procedures by enhancing surgical precision beyond the limits of conventional laparoscopic techniques.
- High-quality investigations are still needed to determine the equivalency and/or superiority of the robotic approach over conventional methods for many oncological and functional endpoints.
- Robotic surgery faces barriers to widespread global adoption, including high equipment acquisition and maintenance costs.
- Further research, along with technological advances like AI-assisted navigation, virtual reality simulation, and advanced 3D imaging and printing, will likely continue to optimize the precision, safety, and accessibility of robotic platforms for complex urologic surgery involving the kidney and upper tract.

# References

[1] McGuinness LA, Prasad RB. Robotics in urology. Ann R Coll Surg Engl 2018;100(6_sup):38–44.

[2] Hussain A, Malik A, Halim MU, Ali AM. The use of robotics in surgery: a review. Int J Clin Pract 2014;68(11):1376–82.

[3] Moretti TBC, Magna LA, Reis LO. Radical prostatectomy technique dispute: analyzing over 1.35 million surgeries in 20 years of history. Clin Genitourin Cancer 2023;21(4):e271–8.e42.

[4] Trinh QD, Sammon J, Sun M, Ravi P, Ghani KR, Bianchi M, et al. Perioperative outcomes of robot-assisted radical prostatectomy compared with open radical prostatectomy: results from the nationwide inpatient sample. Eur Urol 2012;61(4):679–85.

[5] Lee JR. Anesthetic considerations for robotic surgery. Korean J Anesthesiol 2014;66(1):3–11.

[6] Vasdev N, Poon ASK, Gowrie-Mohan S, Lane T, Boustead G, Hanbury D, et al. The physiologic and anesthetic considerations in elderly patients undergoing robotic renal surgery. Rev Urol 2014;16(1):1–9.

[7] Nazemi T, Galich A, Smith L, Balaji KC. Robotic urological surgery in patients with prior abdominal operations is not associated with increased complications. Int J Urol 2006;13(3):248–51.

[8] Thompson J. Myocardial infarction and subsequent death in a patient undergoing robotic prostatectomy. AANA J 2009;77(5):365–71.

[9] Gill A, Randell R. Robotic surgery and its impact on teamwork in the operating theatre. J Perioper Pract 2016;26(3):42–5.

[10] Zhou J, Liu ZH, Cao DH, Peng ZF, Song P, Yang L, et al. Retroperitoneal or transperitoneal approach in robot-assisted partial nephrectomy, which one is better? Cancer Med 2021;10(10):3299–308.

[11] Marconi L, Challacombe B. Robotic Partial Nephrectomy for posterior Renal Tumours: retro or Transperitoneal approach? Eur Urol Focus 2018;4(5):632–5.

[12] Asimakopoulos AD, Miano R, Annino F, Micali S, Spera E, Iorio B, et al. Robotic radical nephrectomy for renal cell carcinoma: a systematic review. BMC Urol 2014;14:75.

[13] Anele UA, Marchioni M, Yang B, Simone G, Uzzo RG, Lau C, et al. Robotic versus laparoscopic radical nephrectomy: a large multi-institutional analysis (ROSULA collaborative group). World J Urol 2019;37(11):2439–50.

[14] Golombos DM, Chughtai B, Trinh QD, Mao J, Te A, O'Malley P, et al. Adoption of technology and its impact on Nephrectomy Outcomes, a U.S. population-based Analysis (2008-2012). J Endourol 2017;31(1):91–9.

[15] Li J, Peng L, Cao D, Cheng B, Gou H, Li Y, et al. Comparison of Perioperative Outcomes of Robot-Assisted vs. laparoscopic radical Nephrectomy: a systematic review and Meta-Analysis. Front. Oncologia 2020;10, 551052.

[16] Pahouja G, Sweigert SE, Sweigert PJ, Gorbonos A, Patel HD, Gupta GN. Does size matter? Comparing robotic versus open radical nephrectomy for very large renal masses. Urol Oncol 2022;40(10):456.e1–7.

[17] Zhang Y, Bi H, Yan Y, Liu Z, Wang G, Song Y, et al. Comparative analysis of surgical and oncologic outcomes of robotic, laparoscopic and open radical nephrectomy with venous thrombectomy: a propensity-matched cohort study. Int J Clin Oncol 2023;28(1):145–54.

[18] Berg WT, Rich CR, Badalato GM, Deibert CM, Wambi CO, Landman J, et al. The first assistant sparing technique robot-assisted partial nephrectomy decreases warm ischemia time while maintaining good perioperative outcomes. J Endourol 2012;26(11):1448–53.

[19] Nguyen MM, Gill IS. Halving ischemia time during laparoscopic partial nephrectomy. J Urol 2008;179(2):627–32. discussion 632.

[20] San Francisco IF, Sweeney MC, Wagner AA. Robot-assisted partial nephrectomy: early unclamping technique. J Endourol 2011;25(2):305–8.

[21] Takahara K, Kusaka M, Nukaya T, Takenaka M, Zennami K, Ichino M, et al. Functional outcomes after selective clamping in robot-assisted partial nephrectomy. J Clin Med Res 2022;11(19). Available from: https://doi.org/10.3390/jcm11195648.

[22] Paulucci DJ, Rosen DC, Sfakianos JP, Whalen MJ, Abaza R, Eun DD, et al. Selective arterial clamping does not improve outcomes in robot-assisted partial nephrectomy: a propensity-score analysis of patients without impaired renal function. BJU Int 2017;119(3):430–5.

[23] Badani KK, Kothari PD, Okhawere KE, Eun D, Hemal A, Abaza R, et al. Selective clamping during robot-assisted partial nephrectomy in patients with a solitary kidney: is it safe and does it help? BJU Int 2020;125(6):893–7.

[24] Kaczmarek BF, Tanagho YS, Hillyer SP, Mullins JK, Diaz M, Trinh QD, et al. Off-clamp robot-assisted partial nephrectomy preserves renal function: a multi-institutional propensity score analysis. Eur Urol 2013;64(6):988–93.

[25] Brassetti A, Cacciamani GE, Mari A, Garisto JD, Bertolo R, Sundaram CP, et al. On-clamp vs. Off-clamp robot-assisted partial nephrectomy for cT2 renal tumors: retrospective propensity-score-matched multicenter outcome analysis. Cancer 2022;14(18). Available from: https://doi.org/10.3390/cancers14184431.

[26] Greco F, Autorino R, Altieri V, Campbell S, Ficarra V, Gill I, et al. Ischemia techniques in nephron-sparing surgery: a systematic review and Meta-Analysis of surgical, oncological, and Functional Outcomes. Eur Urol 2019;75(3):477–91.

[27] Anderson BG, Potretzke AM, Du K, Vetter JM, Bergeron K, Paradis AG, et al. Comparing Off-clamp and On-clamp Robot-assisted Partial Nephrectomy: a prospective randomized trial. Urology 2019;126:102–9.

[28] Daugherty M, Bratslavsky G. Surgical techniques in the Management of Small Renal Masses. Urol Clin North Am 2017;44(2):233–42.

[29] Minervini A, Campi R, Lane BR, De Cobelli O, Sanguedolce F, Hatzichristodoulou G, et al. Impact of resection technique on Perioperative Outcomes and surgical margins after Partial Nephrectomy for localized Renal masses: a prospective Multicenter study. J Urol 2020;203(3):496–504.

[30] Bertolo R, Pecoraro A, Carbonara U, Amparore D, Diana P, Muselaers S, et al. Resection techniques during Robotic Partial Nephrectomy: a systematic review. Eur Urol Open Sci 2023;52:7–21.

[31] Porpiglia F, Bertolo R, Amparore D, Fiori C. Nephron-sparing suture of Renal parenchyma after Partial Nephrectomy: which technique to go for? Some Best Practices Eur Urol Focus 2019;5(4):600–3.

[32] Bertolo R, Campi R, Mir MC, Klatte T, Kriegmair MC, Salagierski M, et al. Systematic review and pooled Analysis of the impact of Renorrhaphy techniques on Renal Functional Outcome after Partial Nephrectomy. Eur Urol Oncol 2019;2(5):572–5.

[33] Van Poppel H, Da Pozzo L, Albrecht W, Matveev V, Bono A, Borkowski A, et al. A prospective, randomised EORTC intergroup phase 3 study comparing the oncologic outcome of elective nephron-sparing surgery and radical nephrectomy for low-stage renal cell carcinoma. Eur Urol 2011;59(4):543–52.

[34] Long CJ, Canter DJ, Kutikov A, Li T, Simhan J, Smaldone M, et al. Partial nephrectomy for renal masses $\geq$ 7 cm: technical, oncological and functional outcomes. BJU Int 2012;109(10):1450–6.

[35] Bratslavsky G. Argument in favor of performing partial nephrectomy for tumors greater than 7 cm: the metastatic prescription has already been written. Urol Oncol 2011;29(6):829–32.

[36] Alimi Q, Peyronnet B, Sebe P, Cote JF, Kammerer-Jacquet SF, Khene ZE, et al. Comparison of short-term Functional, oncological, and perioperative outcomes between laparoscopic and Robotic Partial Nephrectomy beyond the learning curve. J Laparoendosc Adv Surg Tech A 2018;28 (9):1047–52.

[37] Chang KD, Abdel Raheem A, Kim KH, Oh CK, Park SY, Kim YS, et al. Functional and oncological outcomes of open, laparoscopic and robot-assisted partial nephrectomy: a multicentre comparative matched-pair analyses with a median of 5 years' follow-up. BJU Int 2018;122(4):618–26.

[38] Saly DL, Eswarappa MS, Street SE, Deshpande P. Renal cell Cancer and chronic kidney disease. Adv Chronic Kidney Dis 2021;28(5). 460–8.e1.

[39] Chang A, Finelli A, Berns JS, Rosner M. Chronic kidney disease in patients with renal cell carcinoma. Adv Chronic Kidney Dis 2014;21(1):91–5.

[40] Lane BR, Fergany AF, Weight CJ, Campbell SC. Renal functional outcomes after partial nephrectomy with extended ischemic intervals are better than after radical nephrectomy. J Urol 2010;184(4):1286–90.

[41] Wood AM, Benidir T, Campbell RA, Rathi N, Abouassaly R, Weight CJ, et al. Long-term Renal function following renal cancer surgery: historical perspectives, current status, and future considerations. Urol Clin North Am 2023;50(2):239–59.

[42] Leow JJ, Heah NH, Chang SL, Chong YL, Png KS. Outcomes of Robotic versus laparoscopic Partial Nephrectomy: an updated Meta-Analysis of 4,919 patients. J Urol 2016;196(5):1371–7.

[43] Choi JE, You JH, Kim DK, Rha KH, Lee SH. Comparison of perioperative outcomes between robotic and laparoscopic partial nephrectomy: a systematic review and meta-analysis. Eur Urol 2015;67(5):891–901.

[44] Jiang YL, Yu DD, Xu Y, Zhang MH, Peng FS, Li P. Comparison of perioperative outcomes of robotic vs. laparoscopic partial nephrectomy for renal tumors with a RENAL nephrometry score ≥7: a meta-analysis. Front Surg 2023;10:1138974.

[45] Tanagho YS, Kaouk JH, Allaf ME, Rogers CG, Stifelman MD, Kaczmarek BF, et al. Perioperative complications of robot-assisted partial nephrectomy: analysis of 886 patients at 5 United States centers. Urology 2013;81(3):573–9.

[46] Kyung YS, Kim N, Jeong IG, Hong JH, Kim CS. Application of 3-D printed kidney model in Partial Nephrectomy for predicting surgical Outcomes: a feasibility study. Clin Genitourin Cancer 2019;17(5):e878–84.

[47] Arora S, Rogers C. Partial Nephrectomy in central renal tumors. J Endourol 2018;32(S1):S63–7.

[48] Ghoneim TP, Thornton RH, Solomon SB, Adamy A, Favaretto RL, Russo P. Selective arterial embolization for pseudoaneurysms and arteriovenous fistula of renal artery branches following partial nephrectomy. J Urol 2011;185(6):2061–5.

[49] Sutherland DE, Williams SB, Rice D, Jarrett TW, Engel JD. Vascular pseudoaneurysms in urology: clinical characteristics and management. J Endourol 2010;24(6):915–21.

[50] Potretzke AM, Knight BA, Zargar H, Kaouk JH, Barod R, Rogers CG, et al. Urinary fistula after robot-assisted partial nephrectomy: a multicentre analysis of 1 791 patients. BJU Int 2016;117(1):131–7.

[51] Zargar H, Khalifeh A, Autorino R, Akca O, Brandão LF, Laydner H, et al. Urine leak in minimally invasive partial nephrectomy: analysis of risk factors and role of intraoperative ureteral catheterization. Int Braz J Urol 2014;40(6):763–71.

[52] Tomaszewski JJ, Cung B, Smaldone MC, Mehrazin R, Kutikov A, Viterbo R, et al. Renal pelvic anatomy is associated with incidence, grade, and need for intervention for urine leak following partial nephrectomy. Eur Urol 2014;66(5):949–55.

[53] Tomaszewski JJ, Smaldone MC, Cung B, Li T, Mehrazin R, Kutikov A, et al. Internal validation of the renal pelvic score: a novel marker of renal pelvic anatomy that predicts urine leak after partial nephrectomy. Urology 2014;84(2):351–7.

[54] Meeks JJ, Zhao LC, Navai N, Perry Jr KT, Nadler RB, Smith ND. Risk factors and management of urine leaks after partial nephrectomy. J Urol 2008;180(6):2375–8.

[55] Seo IY, Lee YH, Rim JS. Case report: percutaneous fibrin glue injection for urine leakage in laparoscopic partial nephrectomy. J Endourol 2008;22 (5):959–62.

[56] Goel R, Nayak B, Singh P, Gamanagatti S, Yadav R. Percutaneous Management of Persistent Urine Leak after Partial Nephrectomy: sealing the leak site with glue. J Endourol Case Rep 2020;6(4):472–5.

[57] Schmid M, Krishna N, Ravi P, Meyer CP, Becker A, Dalela D, et al. Trends of acute kidney injury after radical or partial nephrectomy for renal cell carcinoma. Urol Oncol 2016;34(7):293.e1–293.e10.

[58] Schmid M, Abd-El-Barr AER, Gandaglia G, Sood A, Olugbade Jr K, Ruhotina N, et al. Predictors of 30-day acute kidney injury following radical and partial nephrectomy for renal cell carcinoma. Urol Oncol 2014;32(8):1259–66.

[59] Wenzel M, Kleimaker A, Uhlig A, Würnschimmel C, Becker A, Yu H, et al. Impact of comorbidities on acute kidney injury and renal function impairment after partial and radical tumor nephrectomy. Scand J Urol 2021;55(5):377–82.

[60] Rosen DC, Kannappan M, Kim Y, Paulucci DJ, Beksac AT, Abaza R, et al. The impact of obesity in patients undergoing Robotic Partial Nephrectomy. J Endourol 2019;33(6):431–7.

[61] Antonelli A, Allinovi M, Cocci A, Russo GI, Schiavina R, Rocco B, et al. The predictive role of biomarkers for the detection of acute kidney injury after Partial or radical Nephrectomy: a systematic review of the literature. Eur Urol Focus 2020;6(2):344–53.

[62] Mues AC, Okhunov Z, Badani K, Gupta M, Landman J. Intraoperative evaluation of renal blood flow during laparoscopic partial nephrectomy with a novel Doppler system. J Endourol 2010;24(12):1953–6.

[63] Gadus L, Kocarek J, Chmelik F, Matejkova M, Heracek J. Robotic Partial Nephrectomy with Indocyanine green fluorescence navigation. Contrast Media Mol Imaging 2020;(2020):1287530.

[64] Yang YK, Hsieh ML, Chen SY, Liu CY, Lin PH, Kan HC, et al. Clinical benefits of indocyanine green fluorescence in robot-assisted partial nephrectomy. Cancer 2022;14(12). Available from: https://doi.org/10.3390/cancers14123032.

[65] Thakker PU, O'Rourke Jr TK, Hemal AK. Technologic advances in robot-assisted nephron sparing surgery: a narrative review. Transl Androl Urol 2023;12(7):1184–98.

[66] Margulis V, Shariat SF, Matin SF, Kamat AM, Zigeuner R, Kikuchi E, et al. Outcomes of radical nephroureterectomy: a series from the upper tract urothelial carcinoma collaboration. Cancer 2009;115(6):1224–33.

[67] Rose K, Khan S, Godbole H, Olsburgh J, Dasgupta P, Guy's and St. Thomas' Robotics Group. Robotic assisted retroperitoneoscopic nephroureterectomy - - first experience and the hybrid port technique. Int J Clin Pract 2006;60(1):12–4.

[68] Tinay I, Gelpi-Hammerschmidt F, Leow JJ, Allard CB, Rodriguez D, Wang Y, et al. Trends in utilisation, perioperative outcomes, and costs of nephroureterectomies in the management of upper tract urothelial carcinoma: a 10-year population-based analysis. BJU Int 2016;117(6):954–60.

[69] O'Sullivan NJ, Naughton A, Temperley HC, Casey RG. Robotic-assisted versus laparoscopic nephroureterectomy; a systematic review and meta-analysis. BJUI Compass 2023;4(3):246–55.

III. The current and future clinical applications of robotic surgery among medical specialties

[70] Xylinas E, Rink M, Cha EK, Clozel T, Lee RK, Fajkovic H, et al. Impact of distal ureter management on oncologic outcomes following radical nephroureterectomy for upper tract urothelial carcinoma. Eur Urol 2014;65(1):210–7.

[71] Dominguez-Escrig JL, Peyronnet B, Seisen T, Bruins HM, Yuan CY, Babjuk M, et al. Potential benefit of lymph node dissection during radical Nephroureterectomy for upper tract urothelial carcinoma: a systematic review by the European Association of Urology guidelines panel on non-muscle-invasive bladder Cancer. Eur Urol Focus 2019;5(2):224–41.

[72] Dong F, Xu T, Wang X, Shen Y, Zhang X, Chen S, et al. Lymph node dissection could bring survival benefits to patients diagnosed with clinically node-negative upper urinary tract urothelial cancer: a population-based, propensity score-matched study. Int J Clin Oncol 2019;24(3):296–305.

[73] Uroweb - European Association of Urology. Upper urinary tract urothelial cell carcinoma, 2023. Available from: https://uroweb.org/guidelines/upper-urinary-tract-urothelial-cell-carcinoma/chapter/disease-management.

[74] Pearce SM, Pariser JJ, Patel SG, Steinberg GD, Shalhav AL, Smith ND. The effect of surgical approach on performance of lymphadenectomy and perioperative morbidity for radical nephroureterectomy. Urol Oncol 2016;34(3). 121.e15–21.

[75] Lenis AT, Donin NM, Faiena I, Salmasi A, Johnson DC, Drakaki A, et al. Role of surgical approach on lymph node dissection yield and survival in patients with upper tract urothelial carcinoma. Urol Oncol 2018;36(1):9.e1–9.

[76] Rodriguez JF, Packiam VT, Boysen WR, Johnson SC, Smith ZL, Smith ND, et al. Utilization and Outcomes of Nephroureterectomy for upper tract urothelial carcinoma by surgical approach. J Endourol 2017;31(7):661–5.

[77] Ribal MJ, Huguet J, Alcaraz A. Oncologic outcomes obtained after laparoscopic, robotic and/or single port nephroureterectomy for upper urinary tract tumours. World J Urol 2013;31(1):93–107.

[78] Lee H, Kim HJ, Lee SE, Hong SK, Byun SS. Comparison of oncological and perioperative outcomes of open, laparoscopic, and robotic nephroureterectomy approaches in patients with non-metastatic upper-tract urothelial carcinoma. PloS One 2019;14(1), e0210401.

[79] Veccia A, Carbonara U, Djaladat H, Mehazin R, Eun DD, Reese AC, et al. Robotic laparoscopic Nephroureterectomy for upper tract urothelial carcinoma: a Multicenter Propensity-Score Matched pair "tetrafecta" Analysis (ROBUUST collaborative group). J Endourol 2022;36(6):752–9.

[80] Williams B, Tareen B, Resnick MI. Pathophysiology and treatment of ureteropelvic junction obstruction. Curr Urol Rep 2007;8(2):111–7.

[81] Krajewski W, Wojciechowska J, Dembowski J, Zdrojowy R, Szydełko T. Hydronephrosis in the course of ureteropelvic junction obstruction: an underestimated problem? Current opinions on the pathogenesis, diagnosis and treatment. Adv Clin Exp Med 2017;26(5):857–64.

[82] Wang QF, Zeng G, Zhong L, Li QL, Che XY, Jiang T, et al. Giant hydronephrosis due to ureteropelvic junction obstruction: a rare case report, and a review of the literature. Mol Clin Oncol 2016;5(1):19–22.

[83] Anderson JC, Hynes W. Retrocaval ureter; a case diagnosed pre-operatively and treated successfully by a plastic operation. Br J Urol 1949;21 (3):209–14.

[84] Patel VR, Patil NN, Coughlin G, Dangle PP, Palmer K. Robot-assisted laparoscopic pyeloplasty: a review of minimally invasive treatment options for ureteropelvic junction obstruction. J Robot Surg 2008;1(4):247–52.

[85] Elaarag M, Alashi H, Aldeeb M, Khalil I, Al-Qudimat AR, Mansour A, et al. Salvage minimally invasive robotic and laparoscopic pyeloplasty in adults: a systematic review. Arab J Urol 2022;20(4):204–11.

[86] Varda BK, Wang Y, Chung BI, Lee RS, Kurtz MP, Nelson CP, et al. Has the robot caught up? National trends in utilization, perioperative outcomes, and cost for open, laparoscopic, and robotic pediatric pyeloplasty in the United States from 2003 to 2015. J Pediatr Urol 2018;14(4):336. e1–8.

[87] Moretto S, Gandi C, Bientinesi R, Totaro A, Marino F, Gavi F, et al. Robotic versus Open Pyeloplasty: Perioperative and Functional Outcomes. J Clin Med Res 2023;12(7). Available from: https://doi.org/10.3390/jcm12072538.

[88] Hopf HL, Bahler CD, Sundaram CP. Long-term Outcomes of Robot-assisted laparoscopic Pyeloplasty for Ureteropelvic junction obstruction. Urology 2016;90:106–10.

[89] Gu L, Li Y, Li X, Liu W. Single port versus multiple port robot-assisted laparoscopic pyeloplasty for the treatment of ureteropelvic junction obstruction (UPJO): a systematic review and meta-analysis. J Endourol 2023. Available from: https://doi.org/10.1089/end.2023.0064.

[90] Kotecha RR, Motzer RJ, Voss MH. Towards individualized therapy for metastatic renal cell carcinoma. Nat Rev Clin Oncol 2019;16(10):621–33.

[91] Dagli MM, Rajesh A, Asaad M, Butler CE. The use of artificial intelligence and machine learning in surgery: a comprehensive literature review. Am Surg 2023;89(5):1980–8.

[92] Ghani KR, Trinh QD, Sammon J, Jeong W, Dabaja A, Menon M. Robot-assisted urological surgery: current status and future perspectives. Arab J Urol 2012;10(1):17–22.

# 40

# Urology robotic prostate surgery

*Asher Mandel, Adriana M. Pedraza, Manish Kumar Choudhary,*
*Dhruti M. Patel, Ziyan Cao, Vinayak Wagaskar, and Ashutosh Tewari*

The Milton and Carroll Petrie Department of Urology at The Icahn School of Medicine at Mount Sinai, New York, NY, United States

## Robotic prostate surgery indications

## BPH

Benign prostatic hyperplasia (BPH) is a condition that commonly develops in men as they age, leading to a constellation of symptoms known as lower urinary tract symptoms (LUTS). Various treatment options, including medical and surgical interventions, are available to alleviate these symptoms. For most men who experience bothersome LUTS, the first step in management is medical monotherapy or combination regimens. Three classes of medications are available: alpha-blockers (e.g., tamsulosin), 5-alpha reductase inhibitors (e.g., finasteride), and PDE 5 inhibitors (e.g., Cialis). These medications play a significant role in symptom relief and improving overall quality of life for individuals with BPH.

The evaluation of BPH involves several diagnostic procedures to assess the condition and guide treatment decisions. Cystoscopy is employed to visualize the outlet obstruction and gain insights into the patient's anatomy, specifically considering factors such as the presence of a large median lobe, intravesical protrusion, or lateral lobe dominant subtypes. Additionally, the diagnostic workup involves a volume assessment through transrectal ultrasound or magnetic resonance imaging (MRI) of the prostate. It is crucial to emphasize that prostate cancer screening with prostate-specific antigen (PSA) and possible digital rectal exam (DRE) should be conducted before considering surgical intervention for BPH to ensure the absence of concurrent malignancy and to guide appropriate management decisions.

Patients who experience LUTS refractory to medical management or are unable to tolerate the medication side effects are offered surgical treatment options. Minimally invasive endoscopic procedures like transurethral resection of the prostate (TURP), photo vaporization of the prostate (PVP), Urolift, and Rezum are usually effective for small- to medium-sized prostate glands, or those between 30 and 80 g. However, for individuals with prostate glands larger than 80–100 g, the optimal treatment approach is enucleation of the prostate. Enucleation can be achieved through utilization of the holmium laser (HoLEP) or by robotic-assisted simple prostatectomy (RASP). These surgical interventions have proven to be successful in addressing LUTS in patients with larger prostate glands, providing symptom relief and improving overall quality of life.

## Prostate cancer (PCa)

PCa is a commonly diagnosed malignancy in men, ranking second only to skin cancer. In the United States alone, over 250,000 cases are diagnosed each year [1]. The screening process for PCa typically commences between the ages of 40 and 50, depending on individual risk factors, and involves a blood test known as PSA testing. Clinicians may recommend earlier screening based on certain risk factors, such as a family history of prostate cancer or a history of BRCA-related cancers, including ovarian or breast cancer. Additional risk factors include being of Black race or having known germline mutations. The decision regarding the appropriate time to initiate PCa screening is arrived at through shared decision-making between the patient and the physician, following thorough discussions about the potential benefits and risks associated with screening.

When a DRE shows abnormalities or there are two consecutive elevated PSAs, a physician may opt to perform a biopsy. The decision to conduct the biopsy may involve additional factors, such as PSA density, percentage of free PSA, and PSA doubling time (PSA kinetics). Some newer products that have shown promise for diagnostic utility include 4K-PSA, Select mdx, and ExosomeDx. If patients are determined to need a prostate biopsy based on these factors, they may undergo an imaging study like multiparametric MRI, followed by transrectal or transperineal prostate biopsy. Commonly, MRI imaging is fused with ultrasound imaging at the time of biopsy to ensure accurate sampling of abnormal lesions of the prostate. The prostate is sampled in 12 different areas according to the standard template, as well as any MRI lesions as targeted biopsies.

After pathological confirmation of PCa, patients undergo staging with a bone scan or PSMA-PET scan and risk stratification. Risk stratification involves assessing several factors, including Gleason grade (a measure of cancer aggressiveness), clinical stage, PSA, PSA density, and the extent of tissue positivity. This risk stratification plays a significant role in determining the appropriate treatment options for patients with localized disease.

Localized PCa can be effectively managed with radical prostatectomy, radiation therapy, or focal therapies like high-intensity focused ultrasound (HIFU) or cryoablation. Radical prostatectomy can be performed by either open, laparoscopic, or robotic-assisted approach (RARP), and in areas where robotic platforms are available, RARP is the preferred approach. For locally advanced prostate cancers, a multimodal treatment approach is often employed. This typically involves a combination of RARP or radiotherapy along with hormonal therapy, which suppresses testosterone production that fuels the growth of prostate cancer cells. In cases of locally recurrent disease following prior treatments like radiotherapy or HIFU, salvage therapy options are considered. RARP can be performed in these salvage settings to address the recurrent disease.

RARP for primary treatment is recommended for patients with clinically localized intermediate, high-risk, or very high-risk prostate cancer, provided they have a life expectancy of at least 10 years. The choice of treatment modality depends on various factors, including the patient's overall health, tumor characteristics, and individual preferences. The decision-making process involves a comprehensive evaluation and consultation with a multidisciplinary team of healthcare professionals, including urologists, radiation oncologists, and medical oncologists, to determine the most appropriate treatment strategy for each patient. The radical prostatectomy procedure involves the complete removal of the prostate gland and seminal vesicles. Pelvic lymph node dissection may be indicated according to the Briganti or Memorial Sloan Kettering nomograms [2]. During the surgery, some hospitals utilize pathologists to review the frozen surgical margins to ensure all cancer has been removed intraoperatively.

To further enhance treatment decision-making, significant advancements have been made in utilizing genetic testing. Genetic testing plays a valuable role in distinguishing which intermediate prostate cancers are more likely to exhibit aggressive behavior, enabling earlier or more aggressive interventions. This includes both germline testing, which examines inherited genetic mutations, and sequencing of the DNA of the tumor itself (oncotype sequencing) using pathologic tissue obtained from biopsies or surgical specimens. Commercial products such as OncotypeDx and Decipher have emerged to aid in analyzing and gaining insights into the genetic profile of the tumor. This understanding can help in identifying biomarkers that indicate aggressiveness of cancer, or even which therapies are likely to be effective. This information helps in tailoring treatment strategies and determining the most appropriate interventions for individual patients, ensuring personalized and effective care.

## Evolution of surgical technique

Robotic surgery has revolutionized surgical techniques by providing surgeons with enhanced precision and dexterity in performing procedures, especially in intricate anatomical areas. The most common platform used is the Intuitive Surgical da Vinci robot. The use of the robot allows for meticulous dissection in tight and complex anatomical zones. One of the key advantages of robotic surgery is the magnification and stereoscopic visualization provided at the surgeon's console, offering a three-dimensional binocular view through the robotic laparoscopic camera.

In oncologic surgery, a crucial balance is in cancer control by achieving negative margins, while preserving the integrity of vital surrounding structures. During radical prostatectomy, special attention is given to avoid damage to the muscles, ligaments, and peripheral autonomic nerves which are responsible for continence and erections. Maintaining the integrity and function of these structures is crucial for optimizing postsurgical functional outcomes in patients. By leveraging the precision and fine control offered by the robotic system, surgeons can carefully navigate these critical structures, minimizing the risk of damage and aiming for optimal functional preservation.

The development of robotic prostatectomy has its roots in the open surgical technique. Open prostatectomy involves a retrograde approach, where the transection of the urethra is the initial step. In contrast, robotic prostatectomy is typically performed in an anterograde fashion, beginning with the incision of the bladder neck. Various approaches have been established to gain access to the prostate during robotic prostatectomy. The following sections will provide an overview of these approaches: traditional, retzius-sparing, transvesical, and extraperitoneal. A summary of the pros and cons of each approach is provided in Table 1. A simple prostatectomy will be discussed subsequently. Finally, special techniques in nerve-sparing and continence preservation will be discussed as well.

## Traditional approach

The patient is placed on the operating room table in a supine position. After intubation and prepping, access to the abdomen is achieved with the open (Hasson) or closed (Veress Needle) techniques. After insufflation with carbon dioxide, the *laparoscopic* ports are placed. These include one 12mm camera port and three 8mm robotic arm instrument ports. Surgeon preference guides the number and size of the assistant ports. Lysis of adhesions is sometimes needed during the process of port placement, typically in patients with a history of prior intraabdominal surgery. To take advantage of gravity, the patient is strapped down to the table and placed in steep Trendelenburg. This allows the intestines to fall superiorly toward the dependent peritoneal regions away from the pelvis. Once the robot is docked, the surgeon begins the operation.

By way of general orientation—to remove the prostate, the attachments to the surrounding structures must be dissected off the prostate. The important structures surrounding the prostate include the bladder (anterior and superior), the pubic bone (anterior, with the attached puboprostatic ligament), the seminal vesicles (SVs) (posterior and superior, with vas deferens), and the rectum (posterior). Other important structures to note include the endopelvic fascia, the lateral pelvic fascia, and Denonvillier's fascia, between the rectum and prostate.

The procedure begins with the incision of the anterior peritoneum, lateral to the medial umbilical folds. The inverted U-incision is followed by tracing the attachments of the bladder to allow its descent (known as the "bladder drop"). These attachments include the remnant of the urachus, or the median umbilical ligament. The plane is developed between the peritoneum and the transversalis fascia. Once these attachments are cut, the bladder will fall posteriorly, permitting visualization of the prostate - inferior to the bladder.

Once the prostate is visualized and de-fatted, a proximal cut is made in the bladder neck. Preserving a larger bladder neck has been shown to improve continence after surgery. This cut is made with the monopolar curved scissors and is followed posteriorly until the Denonvillier's fascia is encountered. At this point, a retrotrigonal incision is made medially and followed inferiorly until the SVs and vas deferens are encountered. The vas runs into the internal ring of the inguinal canal and is therefore cut to free the SV from that attachment. The SVs are dissected gently and carefully to free them from surrounding connective tissue attachments, as the proximal neurovascular plate lies close to these structures.

The theme of the ensuing dissection revolves around the idea that the medial sagittal plane is avascular, while the lateral planes contain the pedicles to the SV and prostate with penetrating arteries as well. Thus, the medial avascular plane is developed posteriorly from the base of the prostate almost to the apex. This is accomplished with sharp and blunt dissection with minimal or no need for cautery. As this plane is carried inferiorly, the lateral aspects need to be addressed as well.

TABLE 1   A summary of pros and cons of the various surgical approaches.

| Approaches | Pros | Cons |
|---|---|---|
| Traditional (Intraperitoneal) | – Most studied<br>– Widely practiced | – Less return of continence at 1 week |
| Retzius-sparing | – Access to prostate without disrupting anterior structures<br>– Can decrease time from surgery to return of continence [3] | – Higher rate of positive surgical margins<br>– Difficult learning curve<br>– Lack of sexual function recovery |
| Transvesical | – Clear visualization of the prostate<br>– Precise dissection and removal of gland | – Bleeding<br>– Temporary incontinence |
| Extraperitoneal | – Reduce risks of intraperitoneal complications | – Robotic arms mobility limitation with less working room |

To travel laterally, control of the vessels that supply the prostate is required. Two sets of pedicles (one for the SVs and the other for the prostate itself) are ligated with either clips or electrocautery at the posterolateral edges of the prostate. The main vessels supplying the prostate arise from the inferior vesicular artery, which is a branch of the internal iliac artery.

The next step involves freeing the prostate from its surrounding connective tissue attachments. The surgeon divides the puboprostatic fascia, which overlies the dorsal venous complex (DVC). DVC bleeding can be controlled in a number of ways including using a stapler but is commonly ligated using sutures as well. In addition, the distal prostatic urethra is cut. At this stage, posterior reconstruction and urethrovesical anastomosis are conducted to reconnect the distal urethra to the bladder. The anastomosis is accomplished with several sutures to bring the two sides together.

Finally, if metastatic disease or lymph node invasion is suspected preoperatively, a lymph node dissection is conducted bilaterally. The lymph nodes taken may include the internal and external iliac and the obturator nodes. The general margins for this dissection are the bifurcation of the iliac artery superiorly, Cooper's ligament inferiorly (superior pubic ramus), and the genitofemoral nerve laterally. Some use the obturator nerve as the inferior landmark.

## Retzius-sparing approach

The retzius-sparing approach, also known as the posterior approach, aims to preserve the Retzius space, which is located between the pubic bone and bladder (retropubic space). By sparing this space, the surgeon can access the prostate without traversing the peritoneum, potentially minimizing postoperative complications, particularly leading to earlier return of continence in some studies.

## Transvesical approach

The transvesical approach involves making an incision in the bladder to gain access to the prostate. This approach allows for excellent visualization of the prostate and facilitates precise dissection and removal of the gland.

## Extraperitnoeal approach

The extraperitoneal approach involves accessing the prostate by creating a working space between the pelvic organs and the peritoneum, without entering the peritoneal cavity. This approach can be advantageous in terms of reducing the risk of intraperitoneal complications.

## Simple prostatectomy

In addition, the technique of RASP can be explored. RASP involves the enucleation or removal of the prostate tissue in patients with enlarged prostates. It is typically performed using a robotic-assisted approach, allowing for precise and controlled removal of the prostate tissue while minimizing damage to surrounding structures.

## Nerve sparing (NS) technique

Despite using stereoscopic visualization, identifying the nerves around the prostate intraoperatively is difficult due to their small, net-like structure, anatomic variation, scar tissue from prior inflammation, and distortion from the tumor. Accidental damage to these delicate structures can be difficult to avoid. Indeed, review studies have found very wide variability in post-RARP recovery of erectile function, between 15% and 87 % [4]. Other series honing in on complete sexual recovery show that only 7% of patients regain complete erectile function within 1 year of RARP and 16% within 2 years of RARP [5]. The resulting erectile dysfunction can have severe ramifications for the quality of life and mental health of these patients [6,7].

Through cadaveric dissection, great efforts have been made to define the anatomy of the hammock-like distribution of cavernous nerves [8]. The spinal nuclei enable erections to form from the S2–S4 levels. Their axons travel ventrally, joining with other fibers responsible for bladder and rectal innervation, forming the sacral visceral efferents. Sympathetic nerves join them to form the pelvic plexus, which is a subsection of the inferior hypogastric plexus. The pelvic plexus is on the anteriolateral rectal wall, and the inferior portion is the prostatic plexus. The midpoint of the pelvic plexus is approximately at the tip of the SV. The neurovascular bundle (NVB) arises from there and is enveloped between two fascial layers. The predominant NVB is typically found posterolateral to the prostate: at the level of

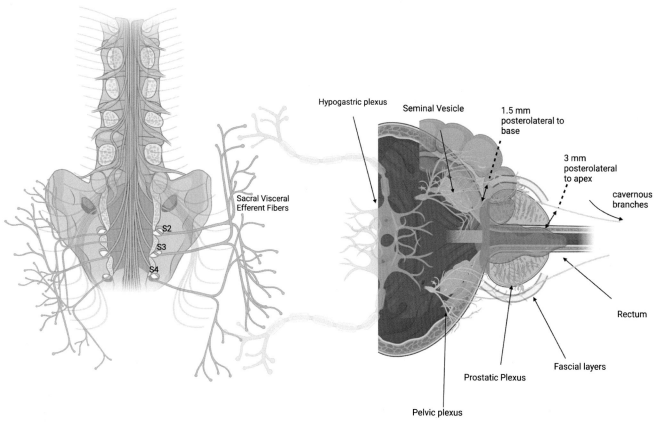

FIG. 1    Anatomic relationships of nerves and the prostate.

the base—1.5 mm, and at the level of the apex—3 mm. The cavernous branches extend toward the base of the penis anterior to the rectum. Understanding these anatomic relationships is paramount to avoiding injury to the nerves. These anatomical relationships are depicted in Fig. 1.

Following posterior peritoneotomy, the SVs are encountered. Particularly near the SV, it is advisable to use blunt dissection in an avascular plane due to the proximity of the nerves [9]. The SVs are enveloped by fascia, with nerves and vasculature traversing laterally. Consequently, the logical site for dissection is medial. Subsequently, the SVs are elevated, and access is gained to Denonvillier's Fascia at the midline to develop the posterior plane. This plane is extended distally to expose the underside of the prostate-urethral junction. Further, the plane is extended laterally ipsilateral to the radiologically predicted side with less aggressive cancer. For grade 1 NS, a plane is developed between the periprostatic venous plexus and the lateral aspect of the prostatic capsule. This is particularly feasible in the distal region, as there are fewer perforating vessels in that zone. Any encountered vessels are sharply severed and may necessitate the application of clips. A similar plane is developed contralaterally, thereby releasing the entire posterior aspect of the prostate from the hammock. At this juncture, the NVB remains connected to the prostate solely at the base. There, the prostatic pedicle enters the prostate and SVs. The hammock also retains its attachment at the anterolateral aspect, where fascial compartments merge with the endopelvic fascia [9].

The graded NS RARP technique was developed by Tewari et al. [10] NS is graded on a scale from 1 to 4 with 1 being the highest grade, corresponding to a resection plane very close to the prostatic capsule. Meanwhile grade 4, on the other end of the spectrum, entails a wide excision without attempting NS. They described three zones of potential nerve damage—proximal, posterolateral, and apical. The decision for what grade is possible for each individual patient is based on the likelihood of ipsilateral extraprostatic extension (Martini nomogram) [11]. The Tewari method helps risk stratify patients according to the schema in Fig. 2. The variable dissection planes are described in Fig. 3, along with the visual representation in Fig. 4. In one retrospective study, >2000 patients underwent NS technique of various grades, and erectile function was followed postoperatively [10]. They found that grade 1 NS had the highest rate (90%) of return to baseline sexual function with a gradual decline in rate as the NS grade went up. Grade 2 had ~80%, grade 3 had ~70%, and grade 4 had ~60% recovery. Positive surgical margins were insignificantly different between cohorts.

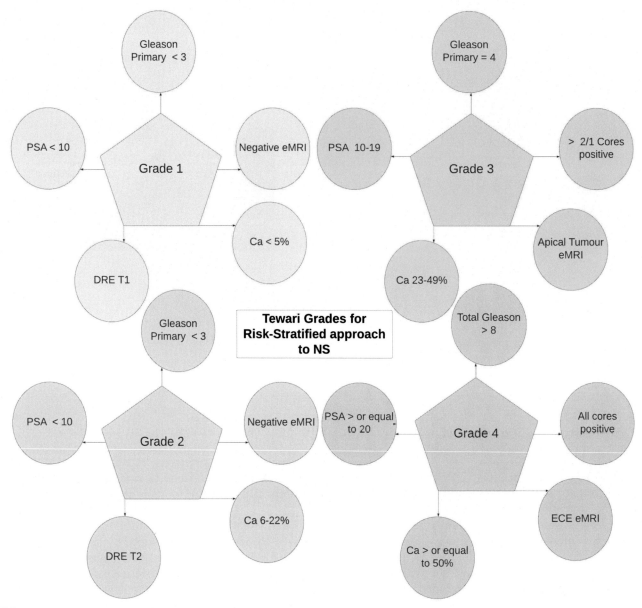

FIG. 2    Tewari Algorithm for risk-stratified approach to NS. PSA levels in ng/mL, maximum % cancer in any biopsy core.

The primary challenge of NS RARP is the visualization of delicate cavernous nerves to avoid accidental injury. To decrease rates of nerve injury and subsequent postoperative erectile dysfunction, several suggestions have been proposed. These include preoperative and intraoperative imaging, utilizing MRI, PET-CT, and PSMA-PET, to improve nerve visualization [12]. There are also a few reports of using an intraoperative robotically manipulated transrectal ultrasound, so-called tandem RARP or T-RARP [13,14]. Although no single approach has been shown to be consistently superior, there is growing interest in the use of intraoperative ultrasound, fluorescence-guided surgery, and neuro-monitoring to assist in the identification and preservation of the nerves. While these techniques show promise, further research is needed to determine their effectiveness in improving NS RARP outcomes.

## Return of continence

In addition to the preservation of the motor neurons that innervate the sphincter muscles, anatomical reconstruction also has an influence on recovery of continence after RARP. Reconstruction involves several steps including the posterior aspect, which entails suturing Denonvillier's fascia to the posterior bladder neck. The urethrovesical anastomosis

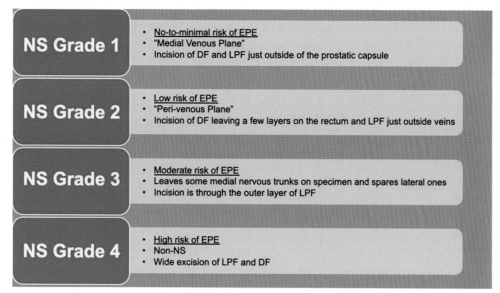

FIG. 3   Surgical approach for risk-stratified NS. *EPE*, extraprostatic extension; *DF*, Denonvillier's Fascia; *LPF*, Lateral Pelvic Fascia.

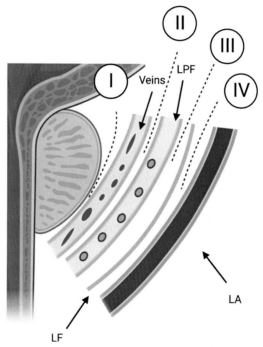

FIG. 4   Visual representation for risk-stratified NS. *LPF*, Lateral Pelvic Fascia; *LF*, Lateral layer of lateral pelvic Fascia; *LA*, Levator Ani.

connects the bladder neck to the distal remnant of the urethra. Other anatomic structures that contribute to continence are the anterior structures, consisting of the detrusor apron, arcus tendinosus, and puboprostatic ligament [15]. These structures may be preserved using an approach known as the hood technique [16].

## Cancer control

Cancer control may be achieved by choosing a dissection plane based on the likelihood of the ipsilateral extraprostatic extension with a more conservative approach taken with a higher likelihood of advanced disease. Intraoperative frozen margins, particularly common at the apex, can guide a surgeon about whether more tissue needs resection.

Positive surgical margins, particularly with longer length of margin positivity and higher Gleason score at the margin, increase the risk for biochemical recurrence (defined as a rising PSA after RARP) [17].

Conventional apical transection after DVC ligation is suboptimal because visualization is distorted and inadvertent capsulotomy can occur, leading to apical margin positivity. Other approaches with circum-apical or retro-apical synchronous urethral transections have been found to have a lower rate of margin positivity (1.4% compared to 4.4%.) [18]. The challenge is in distinguishing the apical prostate from the sphincter structures of the membranous urethra and DVC. Another concern is damaging the sphincter leading to delayed return of continence. The posterior aspect of the apical prostate is covered only by Denonvillier's fascia and the rectourethralis muscle, making it a clearer approach to this area. Challenges of the apical transection include the imperceptible transition of apex to urethra, the fact that apex lacks a truly defined capsule unlike other parts of the prostate, the anterior aspect of the apex is covered with large veins intermingled with fibromuscular stroma, and at the apex, the peripheral zone wraps anteriorly bringing cancer potentially closer to the capsule.

## Postoperative course

Following RARP, patients are monitored for potential complications, such as hematoma, lymphocele, and thromboembolism. These complications are carefully assessed to ensure timely intervention, if necessary. Additionally, long-term recovery from surgery focuses on regaining sexual function and urinary continence.

One notable advantage of robotic surgery compared to open surgery is the significant reduction in postoperative pain experienced by patients. This improved pain control allows for more effective management and can contribute to a more comfortable recovery process. As a result, patients often experience shorter hospital stays compared to traditional open surgery.

In recent times, during the COVID-19 pandemic, there has been increased interest in optimizing healthcare resources. Same-day discharge, also known as outpatient or ambulatory surgery, has gained attention to alleviate strain on the healthcare system. Analysis of the NSQUIP (National Surgical Quality Improvement Program) database has demonstrated that same-day discharge after robotic prostatectomy does not result in an increased risk of complications or mortality [16]. This finding supports the feasibility and safety of same-day discharge, particularly in well-selected patients who meet specific criteria for early recovery and adequate social support.

## Current robotic platforms

The robot was initially developed through several iterations of the Da Vinci system, developed by Intuitive. The two primary types of robotic surgery are the multiport and single port systems. Currently, Da Vinci offers the multiport in both X and Xi versions. The single port device came on the market in 2018 and has been challenging surgeons to innovate for a future when robotic surgery patients all have only one incision.

Recently two other companies, Medtronic and CMR have come out with new robotic surgery platforms that are enrolling in trials for safety and efficacy analysis. Medtronic is working on a prototype known as Hugo and CMR is developing Versius. Both not only aim to deliver the same quality surgical precision as Intuitive but also improve in terms of cost, versatility, and space requirements.

## Future scenario

Continued research and development of new surgical techniques will likely lead to further improvements in outcomes for patients undergoing RARP. Specific innovations involving nerve sparing are an active area of research for improved erectile function following RARP.

Genetic analysis of germline and tumor-based oncotype sequencing are adding new clinical information at a rapid pace. Tools utilizing these data help guide patient management decisions. Particularly for patients with intermediate risk, this tool may help add granularity to their risk stratification to clarify who needs surgery and who can be observed. Whereas germline analysis can be done with any of the patient's cells such as from mucosal swab or blood test, the tumor sequencing requires pathology tissue from the biopsy. Additional management decisions, such as whether a patient needs adjuvant radiation, may be aided by further genetic analysis of the final prostatectomy tissue.

Fluorescent image-guided nerve-sparing RARP is an exciting frontier with imminent translational potential. Probes (such as Human Neural Peptide-401) that target nerves with high signal-to-background ratios can deliver fluorophores (such as oxazine) with favorable emission properties. To use these designed probes, optical coherence tomography, and confocal laser endomicroscopy are both still viable options to facilitate visualization with fast-paced innovation occurring, particularly with dual-modality capabilities for zoom and larger perspective options. Larger human trials are needed to truly evaluate the efficacy and feasibility of these technologies for use in nerve-sparing RARP.

## Key points

- Prostate diseases such as BPH and prostate cancer are commonly managed robotically. Robotic radical prostatectomy and robotic simple prostatectomy are commonly used for prostate cancer and BPH, respectively.
- During robotic radical prostatectomy for prostate cancer, care is taken to maintain the integrity of continence and erectile functions, for which several techniques have been developed.
- The robotic platform has decreased hospital length of stay postoperatively for patients undergoing prostate surgery with an average length of stay of 1 day [19].

# References

[1] Siegel RL, Miller KD, Fuchs HE, Jemal A. Cancer statistics, 2022. CA Cancer J Clin 2022;72:7–33.
[2] Di Pierro GB, et al. Comparison of four validated nomograms (memorial Sloan Kettering Cancer center, Briganti 2012, 2017, and 2019) predicting lymph node invasion in patients with high-risk prostate Cancer candidates for radical prostatectomy and extended pelvic lymph node dissection: clinical experience and review of the literature. Cancer 2023;15.
[3] Chang L-W, Hung S-C, Hu J-C, Chiu K-Y. Retzius-sparing robotic-assisted radical prostatectomy associated with less bladder neck descent and better early continence outcome. Anticancer Res 2018;38:345–51.
[4] Ficarra V, et al. Retropubic, laparoscopic, and robot-assisted radical prostatectomy: a systematic review and cumulative analysis of comparative studies. Eur Urol 2009;55:1037–63.
[5] Koehler N, et al. Erectile dysfunction after radical prostatectomy: the impact of nerve-sparing status and surgical approach. Int J Impot Res 2012;24:155–60.
[6] Nelson CJ, Scardino PT, Eastham JA, Mulhall JP. Back to baseline: erectile function recovery after radical prostatectomy from the patients' perspective. J Sex Med 2013;10:1636–43.
[7] Penson DF, et al. General quality of life 2 years following treatment for prostate cancer: what influences outcomes? Results from the prostate cancer outcomes study. J Clin Oncol 2003;21:1147–54.
[8] Kirschner-Hermanns R, Jakse G. Quality of life following radical prostatectomy. Crit Rev Oncol Hematol 2002;43:141–51.
[9] Tewari A, et al. An operative and anatomic study to help in nerve sparing during laparoscopic and robotic radical prostatectomy. Eur Urol 2003;43:444–54.
[10] Tewari AK, Srivastava A, Huang MW. Anatomical grades of nerve sparing: a risk-stratified approach to neural-hammock sparing during robot-assisted radical prostatectomy (RARP). BJU Int 2011.
[11] Martini A, Gupta A, Lewis SC, et al. Development and internal validation of a side-specific, multiparametric magnetic resonance imaging-based nomogram for the prediction of extracapsular extension of prostate cancer. BJU Int 2018;122(6):1025–33. https://doi.org/10.1111/bju.14353.
[12] Bahler CD, et al. Assessing extra-prostatic extension for surgical guidance in prostate cancer: comparing two PSMA-PET tracers with the standard-of-care. Urol Oncol 2023;41(48):e1–48.e9.
[13] Hung AJ, et al. Robotic transrectal ultrasonography during robot-assisted radical prostatectomy. Eur Urol 2012;62:341–8.
[14] Han M, et al. Tandem-robot assisted laparoscopic radical prostatectomy to improve the neurovascular bundle visualization: a feasibility study. Urology 2011;77:502–6.
[15] Arroyo C, Martini A, Wang J, Tewari AK. Anatomical, surgical and technical factors influencing continence after radical prostatectomy. Ther Adv Urol 2019;11. 1756287218813787.
[16] Wagaskar VG, Mittal A, Sobotka S, et al. Hood technique for robotic radical prostatectomy-preserving Periurethral anatomical structures in the space of Retzius and sparing the pouch of Douglas, enabling early return of continence without compromising surgical margin rates. Eur Urol 2021;80(2):213–21.
[17] Lee W, Lim B, Kyung YS, Kim CS. Impact of positive surgical margin on biochemical recurrence in localized prostate cancer. Prostate Int 2021;9(3):151–6.
[18] Tewari AK, et al. Anatomical retro-apical technique of synchronous (posterior and anterior) urethral transection: a novel approach for ameliorating apical margin positivity during robotic radical prostatectomy. BJU Int 2010;106:1364–73.
[19] Hill GT, Jeyanthi M, Coomer W, et al. Same-day discharge robot-assisted laparoscopic prostatectomy: feasibility, safety and patient experience. BJU Int 2023;132(1):92–9. https://doi.org/10.1111/bju.16002.

# 41

# Urology: Robotic bladder surgery

*Walter Henriques da Costa, Maurício Murce Rocha, and Rafael Ribeiro Zanotti*

Urology Division, A.C. Camargo Cancer Center, São Paulo, São Paulo, Brazil

## Introduction

Bladder cancer (BC) is a prevalent and aggressive disease. More than 550,000 new cases are diagnosed yearly worldwide [1]. Such tumors are characterized to be one of the most expensive human cancers to manage, with a significant impact on patients' long-term quality of life [2]. The vast majority of BC are urothelial carcinoma (UC) in histological subtype and are best stratified into nonmuscle-invasive (NMI) and muscle-invasive (MI) diseases. The natural history of NMI and MI cancers differs widely, so treatments reflect risks of progression and metastases [3].

As oncological surgery moves toward personalized and tailored approaches, we enter an era of precision treatment for the management of genitourinary cancers. This transformation is possible due to the implementation of new technologies, among which robotic surgery stands out for a significant impact [4]. BC treatment is multidisciplinary and varies enormously depending on characteristics associated with the tumor aggressiveness, therapeutic response to conservative options, and patient's performance status. Patients with low- and intermediate-risk NMI disease are best treated with endoscopic procedures and intravesical therapy. Radical cystectomy (RC) is indicated in patients with MI bladder cancer (MIBC) and treatment-refractory non-MIBC (NMIBC).

Robotic-assisted radical cystectomy (RARC) for BC was first described by Menon et al. in 2003 [5] and has been widely discussed due to the advantages and limitations of this surgical technique compared to conventional procedures. One of the major advantages for surgeons using robotic platforms is the possibility of achieving sufficient skill in a shorter time compared with laparoscopic surgery. The development of specific and increasingly technologically advanced simulators has certainly shortened the learning curve and increased surgeons' skills, consequently improving results [6]. Such advances make it possible to perform complex surgical procedures while maintaining oncological results and add the advantages of minimally invasive procedures, such as lower rates of perioperative complications. We will discuss this topic further in the course of this chapter.

BC is the second most common malignant neoplasm of the urinary tract, with an estimated global incidence of 573,278 new cases of the disease in 2020 [7]. In the same year, there were 4595 deaths in Brazil, while 16,710 deaths related to BC are expected in the USA in 2023 [8,9].

UC represents the most common histological type, and approximately 25% of cases are diagnosed as muscle-invasive disease (MIBC) [10]. In patients with nonmuscle-invasive disease (NMIBC), 11 to 21% will progress to MIBC during treatment, and depending on the risk factors present, the progression rate can reach 45% [11,12].

RC with pelvic lymphadenectomy and urinary diversion represents the gold standard surgical treatment for patients with MIBC and for those with high-risk NMIBC refractory to intravesical BCG treatment, offering excellent cancer control and low incidence of local pelvic recurrence [13,14]. However, it is associated with considerable morbidity, and approximately 30% of patients experience complications during the first 90 days postsurgery. Perioperative mortality is about 3% [14].

Radical laparoscopic cystectomy emerged as an option and an attempt to reduce the morbidity associated with the procedure. However, its steep learning curve made the procedure as challenging or more so than ORC, making it less commonly adopted. Robot-assisted radical cystectomy (RARC) emerged as an option to pure laparoscopy, allowing the reproduction of the advantages of robotic surgery known in procedures for the treatment of prostatic and renal disease [15].

## RARC: perioperative care and surgical technique

Preoperative preparation of the patient should involve assessment with a stoma nurse in those patients who will undergo ileal conduit, in order to evaluate the best location of the stoma, as well as to receive guidance on its functioning and changes related to body image [16]. In patients whose ileal segment will be used for urinary reconstruction, there is no indication for bowel preparation, as this does not result in the reduction of surgical complications and may be related to *Clostridium difficile* infections [17].

The use of an enhanced recovery after surgery protocol is associated with better postoperative recovery, reducing time to bowel function initiation and length of hospital stay (LOS). The following measures can be adopted in the preoperative and postoperative period: carbohydrate fluid intake up to 2h before surgery, use of alvimopam (μ-opioid antagonist), use of unfractionated or low-molecular-weight heparin associated with compression elastic stockings and pneumatic compressors to prevent venous thromboembolism; prophylactic antibiotic starting up to 1h before surgical incision, which can be maintained for up to 48h in certain cases; avoiding routine use of nasogastric tube; early initiation of oral diet after surgery; use of chewing gum in the immediate postoperative period; early mobilization; respiratory and motor physiotherapies; and avoiding excessive fluid intake [16–18].

In our service, RARC is performed according to previous publications by other authors with minor modifications [15,16,19], as described below.

After anesthesia, the patient is positioned (for da Vinci Si and X Plataforms: in lithotomy position, allowing the insertion of the patient cart between the legs. For da Vinci Xi, patients may be in dorsal decubit with the patient cart inserted laterally. Fixed to the surgical table to prevent positioning injuries and inadvertent movements during surgery). Pneumoperitoneum is performed with a Veress needle or, in patients with previous abdominal surgeries, by the Hasson technique. When using the Da Vinci Si platform (Intuitive Surgical Inc., Sunnyvale, CA, USA), a 12-mm port is placed 4cm above the umbilical scar. In the case of the Da Vinci Xi platform (Intuitive Surgical Inc., Sunnyvale, CA, USA), a robotic 8-mm port is used. Other ports are introduced under vision: 3 robotic 8mm ports, two 12mm ports, and one 5mm port. Note that one of the 12mm ports will be inserted at the site of the fourth arm of the robot on the left side of the patient and another 8mm robotic port is introduced through it (Fig. 1). During intracorporeal reconstruction, the fourth arm is temporarily uncoupled for the use of an intestinal stapler through the 12mm port. Xi platform is possible to use specifically dedicated staplers (Endowrist) coupled with the robotic arms, under the guidance of the console surgeon.

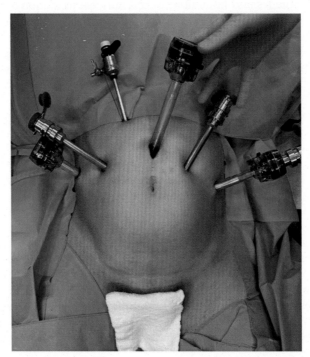

FIG. 1   Placement of trocars for the Si platform. The 8-mm trocar on the left flank can be removed for the assistant to pass the stapler through the 12-mm trocar during reconstruction.

After the introduction of the ports, the patient is positioned in Trendelenburg with an inclination of about 30 degrees and the robot docking is performed. At the time of intracorporeal urinary diversion, the inclination is reduced to 15 degrees.

Initially, the posterior peritoneum is opened over the common iliac vessels and the ureter is identified and dissected caudally to the bladder. Then, the lateral peritoneum is opened lateral over the obliterated umbilical artery and, through blunt dissection, the space lateral to the bladder is created up to the endopelvic fascia, which is opened. The vas deferens is identified, ligated with polymer clips, and sectioned. At this point, we perform pelvic lymphadenectomy from the aortic bifurcation, including common, external, and internal iliac vessels, obturator fossa, and presacral region. The previous steps are then performed on the left side.

We continue with the opening of the peritoneum in the posterior cul-de-sac between the ureters and the creation of the plane between the seminal vesicles, bladder, prostate, anterior, and posterior structures: rectum and Denonvillier's fascia.

Ureters are then clipped and sectioned, and margins are sent for frozen section analysis. The lateral pedicles are controlled using polymer clips. If a nerve-sparing surgery is intended, dissection is initiated near the base of the seminal vesicles with antegrade release of the neurovascular bundle, adjacent to the prostate up to the prostatic apex.

Anterior bladder release is performed, and Retzius space is dissected. The dorsal venous plexus is sectioned using monopolar scissors and later controlled with hemostatic sutures. Urethra is then clipped and sectioned, and the distal margin is sent for frozen section.

If an ileal conduit is planned, a segment of approximately 15 cm of the distal ileum is isolated about 20 cm from the ileocecal valve using a linear stapler, which is also used for reconstruction of the intestinal transit with a side-to-side anastomosis. Vascularization of the distal ureter can be verified using fluorescence imaging. Uretero-ileal anastomosis is performed using the Bricker technique, and the ileal conduit is exteriorized and matured to the skin at the site previously marked for the stoma [20].

If an orthotopic diversion is planned, an ileal neobladder is constructed like the open Studer modified technique. The main surgical steps to form the intracorporeal orthotopic ileal neobladder are isolation of 65 cm of small bowel; anastomosis of small bowel; detubularization of the bowel; suturing of the posterior wall of the neobladder; neobladder-urethral anastomosis and cross-folding of the pouch; and uretero-enteral anastomosis [21,22].

## RC in female patients

In women, the uterus should be anteriorly tractioned with the assistance of the fourth arm of the robot or through trans-fixing sutures through the abdominal wall. The Douglas pouch is opened just below the junction of the uterus with the posterior wall of the vagina. The ureters are identified at the crossing point with the iliac vessels and dissected caudally until the bladder, where they will be clipped and cut. The ovarian pedicles are identified, ligated and cut, and the cardinal and uterosacral ligaments are divided bilaterally. At this moment, a bilateral pelvic lymphadenectomy is performed.

The umbilical artery and the lateral vesical pedicle are identified, clipped, and cut. Dissection continues until the endopelvic fascia, with smaller vessels being controlled using bipolar energy if necessary.

With the assistance of a sponge stick device placed inside the vagina, it is opened at its posterior junction with the uterus, and the vaginal incision is directed laterally toward the urethra to ensure removal of a segment of the anterior vaginal wall. To prevent air leakage through the vagina, it can be filled with a moist compress. Control of the anterior blood vessels to the bladder, vagina, and uterine cervix is performed using bipolar or harmonic devices.

The Retzius space is accessed anteriorly, and the bladder is dissected from the anterior wall. The dorsal venous complex is then cut and controlled posteriorly, and the urethra is identified. If urinary diversion with an ileal conduit is planned, the urethra is resected en bloc with the bladder. In the case of orthotopic ileal neobladder diversion, the urethra is cut along with the bladder, and a specimen is sent for frozen biopsy.

The specimen is placed in an endocatch bag and removed through the vagina, which is then closed using continuous sutures with a barbed suture.

## Analysis of the evidence

In 2003, Menon et al. published the first series of 17 patients undergoing RARC (14 men and 3 women), with urinary diversion performed extracorporeally in all cases. The mean time to perform cystectomy, ileal conduit construction, and orthotopic neobladder formation was 140, 120, and 168 min, respectively. The mean blood loss was less than 150 mL, and there were no cases with positive surgical margins [15].

Just like robotic radical prostatectomy, RARC has been increasingly adopted as the method of choice for the surgical treatment of bladder cancer. Evaluation of national cancer databases in the United States showed an increase in the use of RARC from 0.6% in 2004 to 23% in 2012 [23]. Brassetti et al. [24] reported that the use of RARC in a tertiary referral center in the United Kingdom increased from 10% in 2006 to 100% in 2013.

Even though initial case series of RARC showed advantages related to blood loss, transfusion rate, and hospital stay [25], there were still doubts about possible disadvantages such as the lack of tactile sensation allowing for a higher rate of positive surgical margins in advanced stages, longer operative time related to robotic radical cystectomy plus urinary diversion, and, especially, the possibility of atypical recurrences such as peritoneal carcinomatosis and implants at the port sites related to pneumoperitoneum [26,27].

Recently, a series of cases with a larger number of patients, systematic reviews, and randomized prospective studies have provided better evidence about the use of RARC and its advantages. Analysis of the database of the International Robotic Cystectomy Consortium evaluated 939 patients who underwent RARC at about 20 participating institutions [28]. Four hundred and fifty (48%) patients presented complications within 90 days of postoperative follow-up (modified Clavien system). Two hundred and seventy-three (29%) patients had grade 1–2 (low-grade) complications, and 19% (175) patients had grade 3–5 (high-grade) complications. Gastrointestinal, infectious, and genitourinary complications were more common. Fifty-three patients underwent reoperation within the first 30 days, and the 90-day mortality rate was 4.2%. A retrospective study evaluated 481 patients undergoing RC over 10 years at Mayo Clinic. Two hundred and three patients underwent RARC and 278 underwent ORC. RARC was associated with a lower incidence of complications, mainly due to a lower blood transfusion rate (33.5% vs 55.8%, $P < 0.001$) [29].

Two randomized clinical trials evaluated perioperative outcomes related to RARC. A study conducted by Bochner et al. did not show a significant difference between complication rates within 90 days after surgery between 60 patients in the RARC group and 58 patients in the ORC group (62% vs 66%, $P = 0.7$) even when considering only high-grade complications. However, patients in the RARC group had significantly lower estimated blood loss and longer mean operative time compared to the ORC group. LOS was on average 8 days for both groups [26]. The RAZOR Trial was a randomized clinical trial with 15 participating centers in the USA that evaluated 176 patients in the RARC group and 174 in the ORC group. There was no significant difference in overall complication rates between the groups or in the rates of more severe complications. Patients undergoing RARC had significantly lower rates of estimated blood loss, intraoperative and postoperative blood transfusions, and LOS. Once again, the average operating time was significantly longer for patients undergoing RARC [30].

The oncological outcomes of RARC do not seem to differ much from the results of the ORC series. A systematic review showed a median number of removed lymph nodes of 19 (range 3–55) and a positive lymph node rate of 22%. The average rate of positive surgical margins among studies was 5.6% (range 0–26%). Overall and cancer-specific survivals at 5 years varied from 66% to 80% and from 39% to 66%, respectively [31]. Results of randomized studies also did not show significant differences between the groups regarding the number of lymph nodes removed, rates of positive surgical margins, overall occurrence of local recurrence, or atypical recurrences. RARC is not inferior to ORC regarding recurrence-free survival at 2 years [26,30].

RC is known for its influence on the patient's quality of life. Patients undergoing RARC with intracorporeal urinary diversion, when compared to ORC, have better quality of life scores up to 12 weeks after surgery. The degree of disability after ORC tends to be higher up to 12 weeks after the surgical procedure; however, this difference does not persist after 26 weeks [32]. In addition, in this randomized trial, patients undergone RARC presented fewer wound infections and a quarter of thromboembolic events in comparison with those treated by open surgery.

Functional outcomes are difficult to assess and compare because they depend on the type and technique of urinary diversion used and on the baseline condition of each patient. In patients undergoing RARC and orthotopic ileal neobladder with nerve-sparing surgery, daytime continence in men is about 88.2%, while nighttime continence is 73.5%. At 12 months, in women, these values are 66.7% for both daytime and nighttime. Approximately 81.2% of men were potent with or without the use of 5 phosphodiesterase at 12 months [33].

Materials involved in RARC make the procedure up to 19% more expensive compared to ORC; however, RARC may be more cost-effective when considering patients undergoing ileal conduit, if the operative time does not exceed 360 min and when LOS is less than 6.6 days [34,35].

## Robotic-assisted partial cystectomy

RC is the gold standard for the surgical treatment of MIBC; however, it is associated with a high complication rate and negatively affects patient's quality of life [36]. For highly selected patients, bladder-sparing surgery may provide similar outcomes to RC [37].

While it is clear that partial cystectomy (PC) offers a less morbid procedure than RC, it is unclear if robotic-assisted partial cystectomy (RAPC) offers additional morbidity benefits over open partial cystectomy (OPC).

PC can be considered a bladder-sparing alternative to radical RC. The success of PC for patients diagnosed with MIBC depends on careful patient selection, surgery technique, and follow-up. Robotic assistance is being increasingly used to perform complex urologic surgeries.

RAPC can be a reasonable alternative for some situations and has become an increasingly used approach for cases in which PC is indicated [38]. In highly selected cases, this approach can offer comparable oncologic outcomes while avoiding the morbidity of radical surgery and urinary diversion. A retrospective review showed a recurrence-free survival of 80% for highly selected patients diagnosed with MIBC who underwent RAPC [39].

Some series identified appropriate candidates for PC. These include patients with a solitary tumor in a location suitable for resection with a 2-cm margin of normal bladder, no associated carcinoma in situ, and no history of prior bladder tumors. Adequate bladder capacity and function should be maintained with the 2 cm margin of resection depending on initial bladder characteristics. Factors such as positive lymphnodes, positive surgical margins, tumor multifocality, and carcinoma in situ are related to increased risk of local recurrence in MIBC following PC [40]. Before these criteria were established, relapse rates were as high as 78% [41]. Tumors located at the bladder neck or trigone may not be suitable for PC as achieving a suitable negative margin may be difficult.

Strict selection criteria for PC are only applicable to 3%–10% of patients with MIBC, who have tumors that allow excision with free margins leaving a functional bladder [40].

Achieve negative surgical margins and adequately select patients are cornerstone points to achieve success in PC. Using strict criteria for performing PC resulted in the 5-year overall survival rate of 67%–70%, recurrence-free survival rates of 39%–62%, and cancer-specific survival rates of 84%–87%, as reported by contemporary series [40,42,43].

In a retrospective analysis of 35 patients with at least 12 months of follow-up, authors found a 40% rate of complications with 11.4% being grade 3 or higher. About 35% of the patients had a recurrence during follow-up. They found a rate of positive surgical margin of 9%. In this series, authors used gemcitabine to reduce high-grade bladder cancer recurrence after tumor excision [44].

The study of Bailey et al. showed longer operative time for RAPC but shorter postoperative hospital stays. This study directly compared OPC vs RAPC. However, there were benign and malign indications for PC [45].

Study of Owyong and col. showed no statistically significant difference in the rate of positive surgical margins (PSM) between OPC, LPC, and RAPC. Their result suggests that PC using the laparoscopic or robotic-assisted approach does not carry a higher risk of PSMs and that tumor location and biology, rather than the surgical approach, may be the important determinants of PSMs. There was no significant difference in the rate of PSMs between patients who underwent OPC (19.6%), LPC (18.2%), and RPC (21.6%). Tumors in a dome/urachal location had lower odds of PSMs than those in nondome/urachal locations (OR 0.67; $P = 0,022$). Another interesting finding was that RAPC was more often performed on tumors <3 cm (20.2%) than on tumors ≥3 cm (15.2%), while OPC was more often performed on tumors ≥3 cm (66.7%) than on tumors <3 cm (60.3%); however, this association between tumor size and surgical approach was not statistically significant ($P = 0.051$). Nonetheless, this represents a potential bias toward the selection of smaller tumors for minimally invasive surgery [46].

RAPC is also reported for benign conditions like endometriosis and mesh removal [47,48].

## Robot-assisted bladder augmentation

Bladder augmentation is one of the possible treatments for neurogenic detrusor overactivity and poor bladder compliance [49,50]. Minimally invasive approaches offer the potential benefits of decreased recovery time and wound and bowel complications [51,52].

Flum et al. analyzed 22 patients who underwent Robot-Assisted Augmentation Enterocystoplasty in a total intracorporeal fashion. They found a mean operative time of 365 (220–778) minutes; a mean time for return of bowel function of 5 (2–17) days and median length of hospitalization of 6 days (4–44). [53] Potential benefits that robotic assistance can provide are cosmesis, decreased time to return bowel function, and shorter duration of postoperative hospitalization.

Robotic technology provides three-dimensional observation, improving the degrees and freedom of movement, improved ergonomics, stabilization of surgical movements, and ease of intracorporeal suturing. This can be especially useful for PC and bladder augmentation surgery, where these characteristics can help delineate the tumors and suture the bladder.

Literature on this subject is scarce. The series is composed of a few patients and with a short follow-up time [43]. Robotic surgery costs are potentially the main limitation for adopting this way of approach to perform PC and robotic-assisted bladder augmentation.

## Future scenario

With the wide dissemination of surgeries with assistance from robotic platforms, we imagine that more and more patients will have access to this technology. As the limits are pushed, an increasing number of surgeons will use robotic platforms to perform surgery, with the entire procedure being performed intracorporeal. The application of new platforms for cystectomy is still in its infancy; however, over time, it will be more frequent to see different types of platforms and strategies for performing the procedure.

## Key points

- Radical cystectomy is a major procedure. The benefits of robotic surgery are getting more noticeable.
- The use of the robotic platform makes it possible to maintain the advantages of open surgery, with better ergonomics and range of motion; in addition, to maintaining the benefits of minimally invasive surgery.

## Appendix: Supplementary material

Supplementary material related to this chapter can be found on the accompanying CD or online at https://doi.org/10.1016/B978-0-443-13271-1.00054-6.

## References

[1] Cumberbatch MGK, Jubber I, Black PC, Esperto F, Figueroa JD, Kamat AM, et al. Epidemiology of bladder cancer: a systematic review and contemporary update of risk factors in 2018. Eur Urol 2018;74(6):784–95.

[2] Catto JWF, Downing A, Mason S, Wright P, Absolom K, Bottomley S, et al. Quality of life after bladder cancer: a cross-sectional survey of patient reported outcomes. Eur Urol 2021;79(5):621 32.

[3] Rouprêt M, Babjuk M, Burger M, Capoun O, Cohen D, Compérat EM, et al. European Association of Urology guidelines on upper urinary tract urothelial carcinoma: 2020 update. Eur Urol 2021;79(1):62–79.

[4] Autorino R, Porpiglia F, Dasgupta P, Rassweiler J, Catto JW, Hampton LJ, et al. Precision surgery and genitourinary cancers. Eur J Surg Oncol 2017;43(5):893–908.

[5] Menon M, Hemal AK, Tewari A, Shrivastava A, Shoma AM, El-Tabey NA, et al. Nerve-sparing robot-assisted radical cystoprostatectomy and urinary diversion. BJU Int 2003;92(3):232–6.

[6] Dal Moro F. How robotic surgery is changing our understanding of anatomy. Arab J Urol 2018;16(3):297–301.

[7] The Global Cancer Observatory., 2023. Available from: https://gco.iarc.fr/today/data/factsheets/cancers/30-Bladder-fact-sheet.pdf.

[8] Siegel RL, Miller KD, Wagle NS, Jemal A. Cancer statistics, 2023. CA Cancer J Clin 2023;73(1):17–48.

[9] INCA. Estimativa 2023: incidência de câncer no Brasil [Internet], 2022. [cited 2023 Mar 24]. Available from: https://www.inca.gov.br/sites/ufu.sti.inca.local/files//media/document//estimativa-2023.pdf.

[10] Burger M, Catto JWF, Dalbagni G, Grossman HB, Herr H, Karakiewicz P, et al. Epidemiology and risk factors of urothelial bladder cancer. Eur Urol 2013;63(2):234–41.

[11] Van Den Bosch S, Witjes JA. Long-term cancer-specific survival in patients with high-risk, non-muscle-invasive bladder cancer and tumour progression: a systematic review. Eur Urol 2011;60(3):493–500.

[12] Sylvester RJ, Van Der Meijden APM, Oosterlinck W, Witjes JA, Bouffioux C, Denis L, et al. Predicting recurrence and progression in individual patients with stage ta T1 bladder cancer using EORTC risk tables: a combined analysis of 2596 patients from seven EORTC trials. Eur Urol 2006;49(3):466–77.

[13] Chang SS, Bochner BH, Chou R, Dreicer R, Kamat AM, Lerner SP, et al. Treatment of non-metastatic muscle-invasive bladder Cancer: AUA/ASCO/ASTRO/SUO guideline. J Urol 2017;198(3):552–9.

[14] Stein JP, Lieskovsky G, Cote R, Groshen S, Feng AC, Boyd S, et al. Radical cystectomy in the treatment of invasive bladder Cancer: long-term results in 1,054 patients. J Clin Oncol 2001;19(3):666–75.

[15] Menon M, Hemal AK, Tewari A, Shrivastava A, Shoma AM, El-Tabey NA, et al. Nerve-sparing robot-assisted radical cystoprostatectomy and urinary diversion. BJU Int 2003;92(3):232–6.

[16] Ahmed YE, Hussein AA, Kozlowski J, Guru KA. Robot-assisted radical cystectomy in men: technique of spaces. J Endourol 2018;32(May):S44–8.

[17] Wilson TG, Guru K, Rosen RC, Wiklund P, Annerstedt M, Bochner BH, et al. Best practices in robot-assisted radical cystectomy and urinary reconstruction: recommendations of the Pasadena consensus panel. Eur Urol 2015;67(3):363–75.

[18] Pak JS, Lee JJ, Bilal K, Finkelstein M, Palese MA. Utilization trends and short-term outcomes of robotic versus open radical cystectomy for bladder Cancer. Urology 2017;103:117–23.

[19] Buffi N, Mottrie A, Lughezzani G, Koliakos N, Schatteman P, Carpentier P, et al. Surgery illustrated - surgical atlas robotic radical cystectomy in the male. BJU Int 2009;104(5):726–45.

[20] Medina LG, Baccaglini W, Hernández A, Rajarubendra N, Ashrafi AN, Tafuri A, et al. Robotic intracorporeal ileal conduit: technical aspects. Arch Esp Urol 2019;72(3):299–308.

[21] Chopra S, de Castro Abreu AL, Berger AK, Sehgal S, Gill I, Aron M, et al. Evolution of robot-assisted orthotopic ileal neobladder formation: a step-by-step update to the University of Southern California (USC) technique. BJU Int 2017;119(1):185–91.

[22] Goh AC, Gill IS, Lee DJ, De Castro Abreu AL, Fairey AS, Leslie S, et al. Robotic intracorporeal orthotopic ileal neobladder: replicating open surgical principles. Eur Urol 2012;62(5):891–901.

[23] Matulewicz RS, DeLancey JOL, Manjunath A, Tse J, Kundu SD, Meeks JJ. National comparison of oncologic quality indicators between open and robotic-assisted radical cystectomy. Urol Oncol 2016;34(10):431.e9–431.e15.

[24] Brassetti A, Möller A, Laurin O, Höijer J, Adding C, Miyakawa A, et al. Evolution of cystectomy care over an 11-year period in a high-volume tertiary referral centre. BJU Int 2018;121(5):752–7.

[25] Pruthi RS, Nielsen ME, Nix J, Smith A, Schultz H, Wallen EM. Robotic radical cystectomy for bladder cancer: surgical and pathological outcomes in 100 consecutive cases. J Urol 2010;183(2):510–5.

[26] Bochner BH, Dalbagni G, Sjoberg DD, Silberstein J, Keren Paz GE, Donat SMH, et al. Comparing open radical cystectomy and robot-assisted laparoscopic radical cystectomy: a randomized clinical trial. Eur Urol 2015;67(6):1042–50.

[27] El-Tabey NA, Shoma AM. Port site metastases after robot-assisted laparoscopic radical cystectomy. Urology 2005;66(5):1110.e1–3.

[28] Johar RS, Hayn MH, Stegemann AP, Ahmed K, Agarwal P, Balbay MD, et al. Complications after robot-assisted radical cystectomy: results from the international robotic cystectomy consortium. Eur Urol 2013;64(1):52–7.

[29] Faraj KS, Abdul-Muhsin HM, Rose KM, Navaratnam AK, Patton MW, Eversman S, et al. Robot assisted radical cystectomy vs open radical cystectomy: over 10 years of the Mayo clinic experience. Urol Oncol 2019;37(12):862–9.

[30] Parekh DJ, Reis IM, Castle EP, Gonzalgo ML, Woods ME, Svatek RS, et al. Robot-assisted radical cystectomy versus open radical cystectomy in patients with bladder cancer (RAZOR): an open-label, randomised, phase 3, non-inferiority trial. Lancet 2018;391(23):2525–36.

[31] Novara G, Catto JWF, Wilson T, Annerstedt M, Chan K, Murphy DG, et al. Systematic review and cumulative analysis of perioperative outcomes and complications after robot-assisted radical cystectomy. Eur Urol 2015;67(3):376–401.

[32] Catto JWF, Khetrapal P, Ricciardi F, Ambler G, Williams NR, Al-Hammouri T, et al. Effect of robot-assisted radical cystectomy with Intracorporeal urinary diversion vs open radical cystectomy on 90-day morbidity and mortality among patients with bladder Cancer: a randomized clinical trial. JAMA 2022;327(21):2092–103.

[33] Tyritzis SI, Hosseini A, Collins J, Nyberg T, Jonsson MN, Laurin O, et al. Oncologic, functional, and complications outcomes of robot-assisted radical cystectomy with totally intracorporeal neobladder diversion. Eur Urol 2013;64(5):734–41.

[34] Fujimura T. Current status and future perspective of robot-assisted radical cystectomy for invasive bladder cancer. Int J Urol 2019;26(11):1033–42.

[35] Cai PY, Khan AI, Scherr DS, Shoag JE. Robotic radical cystectomy in the contemporary Management of Bladder Cancer. Urol Clin North Am 2021;48(1):45–50.

[36] Chang SS, Bochner BH, Chou R, Dreicer R, Kamat AM, Lerner SP, et al. Treatment of non-metastatic muscle-invasive bladder cancer: AUA/ASCO/ASTRO/SUO guideline. J Urol 2017;198(3):552–9.

[37] Biagioli MC, Fernandez DC, Spiess PE, Wilder RB. Primary bladder preservation treatment for urothelial bladder cancer. Cancer Control 2013;20(3):188–99.

[38] Cha EK, Donahue TF, Bochner BH. Radical transurethral resection alone, robotic or partial cystectomy, or extended lymphadenectomy. Urol Clin North Am 2015;42(2):189–99.

[39] Alanee S, El-Zawahry A. Robotic-assisted partial cystectomy for muscle invasive bladder cancer: contemporary experience. Int J Med Robot Comput Assist Surg 2022;18(4).

[40] Holzbeierlein JM, Lopez-Corona E, Bochner BH, Herr HW, Donat SM, Russo P, et al. Partial cystectomy: a contemporary review of the memorial sloan-kettering cancer center experience and recommendations for patient selection. J Urol 2004;172(3):878–81.

[41] Faysal MH, Freiha FS. Evaluation of partial cystectomy for carcinoma of bladder. Urology 1979;14(4):352–6.

[42] Kassouf W, Swanson D, Kamat AM, Leibovici D, Siefker-Radtke A, Munsell MF, et al. Partial cystectomy for muscle invasive urothelial carcinoma of the bladder: a contemporary review of the M. D. Anderson Cancer center experience. J Urol 2006;175(6):2058–62.

[43] Smaldone MC, Jacobs BL, Smaldone AM, Hrebinko RL. Long-term results of selective partial cystectomy for invasive urothelial bladder carcinoma. Urology 2008;72(3):613–6.

[44] Alanee S, El-Zawahry A. Robotic-assisted partial cystectomy for muscle invasive bladder cancer: contemporary experience. Int J Med Robot Comput Assist Surg 2022;18(4).

[45] Bailey GC, Frank I, Tollefson MK, Gettman MT, Knoedler JJ. Perioperative outcomes of robot-assisted laparoscopic partial cystectomy. J Robot Surg 2018;12(2):223–8.

[46] Owyong M, Koru-Sengul T, Miao F, Razdan S, Moore KJ, Alameddine M, et al. Impact of surgical technique on surgical margin status following partial cystectomy. Urol Oncol 2019;37(12):870–6.

[47] Ramesmayer C, Lusuardi L, Griessner H, Gruber R, Oberhammer L. Robotic partial cystectomy with excision of mesh after inguinal hernia repair: a case report. BMC Urol 2023;23(1):27.

[48] Bahadur A, Mundhra R, Sherwani P, Kumar S. Robot-assisted partial cystectomy for bladder endometriosis: dual approach involving cystoscopy and robotic surgery. BMJ Case Rep 2021;14(8).

[49] Kalkan S, Jaffe WI, Simma-Chiang V, Li ESW, Blaivas JG. Long term results of augmentation cystoplasty and urinary diversion in multiple sclerosis. Can J Urol 2019;26(3):9774–80.

[50] Blaivas JG, Weiss JP, Desai P, Flisser AJ, Stember DS, Stahl PJ. Long-term followup of augmentation enterocystoplasty and continent diversion in patients with benign disease. J Urol 2005;173(5):1631–4.

[51] Gill IS, Rackley RR, Meraney AM, Marcello PW, Sung GT. Laparoscopic enterocystoplasty. Urology 2000;55(2):178–81.

[52] Elliott SP, Meng MV, Anwar HP, Stoller ML. Complete laparoscopic ileal cystoplasty. Urology 2002;59(6):939–43.

[53] Flum AS, Zhao LC, Kielb SJ, Wilson EB, Shu T, Hairston JC. Completely Intracorporeal robotic-assisted laparoscopic augmentation Enterocystoplasty with Continent Catheterizable Channel. Urology 2014;84(6):1314–8.

III. The current and future clinical applications of robotic surgery among medical specialties

# Robotic urinary diversion

*Tarek Ajami, Ankur Malpani, Archan Khandekar, and Bruno Nahar*

Desai Sethi Urology Institute, Miller School of Medicine, University of Miami, Miami, FL, United States

## Introduction

The decision regarding the choice of urinary diversion after radical cystectomy (RC) is a collaborative process that considers various factors such as patient characteristics, medical conditions, and surgical considerations [1]. Patient preferences, oncologic factors, comorbidities, baseline conditions, and reported quality of life with each form of diversion are examples of patient factors [2]. The available options for urine diversion presented to patients are largely determined by the surgeon's knowledge, experience, and personal preferences. Most urine diversions were carried out extracorporeal after robotic-assisted radical cystectomy (RARC) in the 2000s. More recently, there has been a trend toward more use of robotic intracorporeal urinary diversions (ICUDs). This chapter focuses on the surgical technique description of ICUD, mainly ileal conduit (IC), continent diversion, and orthotopic neobladders (ONBs).

## IC urinary diversion

The IC remains the most frequent urinary diversion. It is universally recognized as being the most generalizable, cost-effective, and reliable solution in the long term [3]. Patients with certain conditions are advised to consider an IC because these conditions are relative contraindications to a continent orthotopic or continent cutaneous urinary diversion. These conditions include physical conditions limiting clean intermittent catheterization, neurologic disorders like dementia, intellectual impairment, urethral stricture disease, chronic inflammatory bowel disease, and hepatic or renal impairment [4,5]. The ICUD-EAU International Consultation on Bladder Cancer stated that patients with a serum creatinine of 150 mol/L should be advised against a continent diversion in favor of an IC [5].

### Step-by-step technique

#### Patient positioning and port placement

The patient undergoing surgery with the Xi robot (Intuitive Surgical, Sunnyvale, California, EUA USA) is placed in supine position. Secure positioning and protection of pressure points are important to avoid injuries, using foam pads at elbows and wrists. Similarly, to fix the head and shoulders and ensure the stability of the patient when the Trendelenburg position is done, TrenGuard (D.A. Surgical, Newbury, OH, USA) could be used. Pneumoperitoneum is established using the Veress needle or Hasson technique according to the surgeon's preference. Once the pneumoperitoneum is established, the patient is given a steep Trendelenburg position (which will be reduced to 15–17 degrees for the urinary diversion). A seven-port transperitoneal approach is used with a camera, three robotic, and three assistant ports, as shown in Fig. 1. The camera port is placed 20 cm above the pubic symphysis, and four ports are placed in a horizontal line 2 fingerbreadths above the umbilicus. Two robotic ports are placed on the patient's right and one robotic port and a 15-mm assistant port are placed on the patient's left. A 12-mm AirSeal port (ConMed, Utica, NY, USA) is placed in the right upper quadrant, triangulating between the camera port and the right robotic port. Similarly, a 5-mm port is placed between the camera and the robotic port in the left upper quadrant. The 15-mm robotic

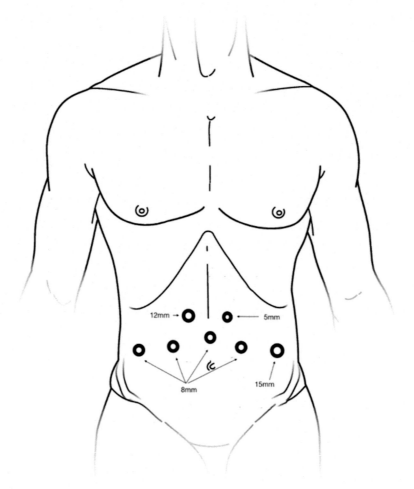

FIG. 1    The seven-port placement we prefer for the procedure. 15-mm left-sided port is for stapler use.

port allows the use of handheld bowel staplers. The procedure is carried out throughout in 0-degree lens scope. A 30-degree scope may be used in certain cases with urethral involvement or in cases of a narrow pelvis. We use Maryland forceps, Robotic scissors, Cadiere forceps, Vessel Sealer, and needle drivers for the surgery.

### Isolation of bowel segment

To begin with, we have two ureters clipped and divided. We place two Hem-o-Lok (Teleflex, Morrisville, NC, USA) clips at the distal ureter where it enters the bladder and divide it in between. The proximal clip has a Vicryl suture tied to it. On the left side ureter, we use a dyed suture, while on the right side, we use an undyed suture. This allows easy identification of the laterality of both ureters. The left ureter is transposed behind the sigmoid mesentery. Then, we use a thread attached to the Hem-o-Lok clip being used for clipping the ureter to allow easy manipulation of the ureter. It is necessary to make sure the retro-mesenteric window is wide enough to allow transposition of the left ureter to the opposite side and allow tension-free anastomosis and no ureteral kinks.

The first step is to identify the ileocecal valve, where we measure around 20 cm from the same proximally in the terminal ileum (Fig. 2). This point marks the distal end of the conduit loop which we will create. Proximal marking is done around 15 cm from this point. The segment of the small bowel used for the conduit is typically 15–20 cm long.

When using the ECHELON FLEX stapler (Ethicon, USA), we use the 60-mm blue cartridge (Fig. 3). It is important to position the stapler perpendicular to the bowel segment and mesentery to minimize the risk of devascularization.

The fourth arm is used to elevate the distal end of the bowel segment and straighten the loop to ensure the conduit is of adequate length before dividing the bowel. The proximal segment of the IC is then divided with the stapler, and depending on the mobility of the isolated segment, the vessel sealer can be used to deepen the mesenteric division.

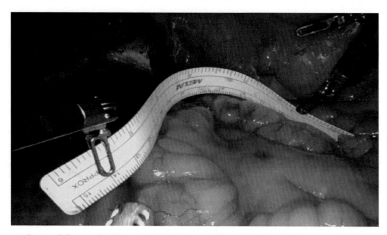

FIG. 2 Marking of bowel loop to be used for conduit. Typically, 15–20 cm of terminal ileum 20 cm proximal to the IC junction is used for conduit.

FIG. 3 (A) We use an ECHELON FLEX 60-mm stapler with a *blue cartridge* to divide the intestinal loop. Care has to be taken to avoid encasing mesenteric vessel loops in the staple line. (B) Small enterotomy is made in proximal and distal ileal segments, and side-to-side anastomosis is done using a stapler. (C) After ensuring adequate side-to-side anastomosis, a single staple below the enterotomies is made, and the redundant bowel segment is removed.

### Bowel anastomosis

Using the table motion feature, Trendelenburg is reversed to 15–17 degrees to prevent the displacement of bowel into the upper abdomen and to aid in the manipulation of the bowel. We must ensure adequate separation and mobility of the IC from the bowel anastomosis that is to follow. A side-to-side bowel anastomosis is then performed (Fig. 3). It is important to ensure that the conduit is placed below the small bowel anastomosis with the correct orientation. Placing stay sutures on the antimesenteric aspect of segment edges can help in two ways. First, it allows easy manipulation of the bowel, and second, it allows maintaining the orientation of the bowel. Two small enterotomies are made using electrocautery at the antimesenteric corners of the two bowel stumps marked by the stay sutures. A stapler is introduced from the left lateral port, and the two bowel segments are placed individually on each blade of the stapler through their respective enterotomies. It is helpful to use the stay sutures to advance each bowel segment to the jaw of the stapler blades before closing the stapler. Care is taken to avoid inclusion of small bowel mesentery into the staple line before the stapler is fired. A second stapler load is introduced from the lateral port, passed simultaneously into the proximal and distal bowel segments, and stapled to ensure a widely patent side-to-side anastomosis. Finally, the stump of the side-to-side anastomosis is stapled off at the top using the robotic stapler.

### Uretero-Ileal anastomosis

Our preference is to perform uretero-ileal anastomosis using the Bricker technique. Basic principles of ureteric anastomosis are followed here too. We avoid kinking or twisting of the ureter and avoid its devascularization. Performing left-side anastomosis first is preferable, though any side can be performed first [6–8]. A small enterotomy is made at the appropriate location on the conduit, close to the proximal end. The ureter is incised at an appropriate location, maintaining adequate length and vascularity. Leaving the distal ureteral tail intact allows us to have a place to hold the ureter while doing the anastomosis (Fig. 4). Spatulation is done according to ureter width. The uretero-ileal anastomosis is completed using a 4-0 Vicryl suture in an interrupted manner using a fine reverse cutting needle (Fig. 5).

FIG. 4  Terminal redundant portion of ureter should not be disconnected until completion of uretero-ileal anastomosis. Allows tension-free handling of the ureter.

FIG. 5  We place two Vicryl 4-0 sutures at the apex and use each open to complete one wall in an interrupted fashion.

The main objective of this step is to achieve a spatulated, tension-free mucosa-to-mucosa anastomosis. The first suture is placed at the apex of the spatulation from outside to inside the ureter and then inside to outside on the bowel. The posterior wall of the anastomosis is completed. A second suture is placed at the apex, and the anterior wall of the anastomosis is commenced. Now a 7 Fr close-ended Single J ureteral stent is introduced. The stent is placed in the ureter and passed to the collecting system using both robotic arms. The guidewire is removed, and the distal end of the stent is passed through the distal end of the conduit. The distal ureteral tail is cut, and the uretero-ileal anastomosis is completed. This procedure is repeated for the right ureter.

**Wallace technique**

The steps for enterotomy are the same. The difference lies in the number of uretero-ileal anastomoses. Here both ureters are split open and then the posterior wall is sutured to result in a single lumen which is then anastomosed to the bowel (Fig. 6). Specific indication where this method should be considered is a case where the ureteric lumen is too narrow and future uretero-ileal anastomotic stricture is expected. While strictures are less common with the Wallace technique, if it happens, it will affect both kidneys.

*Stoma creation*

We prefer creating the stoma before undocking the robot. We remove the robotic port placed at the site of the stoma, which is marked before the start of the procedure. A circular skin incision is made over the stoma site, subcutaneous tissue is cut using cautery, and fascia is reached. We use robotic scissors to facilitate the dissection at the site of the stoma from inside the abdominal cavity. The posterior sheath is divided, and the rectus muscles are separated. We divide the rectus sheath in a cruciate fashion making sure that at least two fingers can be accommodated. Under

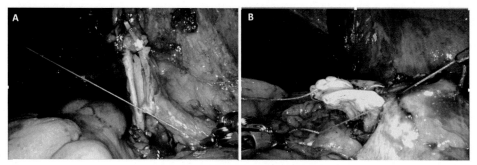

FIG. 6    (A) Wallace method of anastomosis: the posterior wall of both ureters anastomosed to allow forming a single lumen. The principles of anastomosis remain essentially the same as Bricker's method. (B) Completion of the posterior layer of uretero-ileal anastomosis. The lumen should look adequate after one wall anastomosis before placement of a stent.

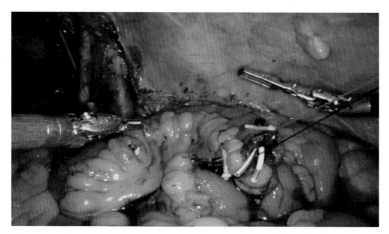

FIG. 7    Use of Lapra-Ty clips (Ethicon LLC, Raritan, NJ, USA) with dyed and undyed sutures to fix the stents to the terminal conduit end and allow identification of laterality. This method is used to ensure stents are not pulled out while delivering the conduit through the stomal incision.

camera guidance, the distal conduit loop is pulled through the abdominal wall using the stay sutures. It is crucial at this point to make sure the bowel loop is not twisted while bringing it out; otherwise, it could compromise vascularity. The technique of using tagged Vicryl sutures at ureter ends delivers benefits in ensuring this principle (Fig. 7). Using Vicryl 2-0, the base of the stoma is fixed to the fascia and skin through the same suture making sure that a rosebud stoma is created. We avoid taking sutures at the mesenteric sides of the bowel. Everted mucosa ensures perfect maturation of the stoma.

The abdomen is now de-sufflated, and the specimen is extracted through a lower midline incision in a male patient. In female patients, we remove the specimen through the vagina and create a vaginal flap to close the vaginal lumen using Vicryl 1-0/Vicryl -0 SH needle.

## ONBs

Although IC is the most common diversion after RC, ONB has become the procedure of choice due to the restoration of body image and facilitating normal voiding, and therefore, improving the quality of life [9]. However, the complexity of this procedure requires surgical dexterity as well as active patient participation for proper maintenance. Nowadays, ICUD with robotic neobladder is increasingly the modality of interest for urinary diversions in high-volume robotic centers.

### Patient selection

Appropriate patient selection is critical to ensure the success of ONB. Disease extent and anatomic considerations can limit reconstructive options [10]. Prior abdominal surgeries, radiation therapy, and inadequate renal and hepatic function are unfavorable factors. Other relative contraindications include lack of patient motivation or psycho-social and cognitive impairment which hinder the patient's management of ONB.

- *Oncological considerations*: Disease at the bladder neck and/or the vaginal wall and the presence of tumor in the prostatic urethra and stroma increase the risk of urethral recurrence. Disease extent into the pelvis should be evaluated on a case-by-case basis given the low cancer-specific survival in patients with nodal disease. Therefore, the time to mature the neobladder into an effective reservoir should be balanced with time to recurrence [9].
- *Anatomic and metabolic factors*: Patients with urethral strictures are not suitable for neobladder reconstruction. Compromised renal and hepatic function may cause metabolic alterations (acidosis, hepatic encephalopathy, etc.).

With the advent of minimally invasive robot-assisted surgeries, the techniques of ONB have shown an important advancement, although the design and configuration of the continent reservoir remain almost the same. Despite the technical complexity, patients should still be offered the option for ONB regardless of the surgical approach.

## Surgical technique

There are different types of neobladder according to the selected intestinal segment. The most used are the ileal neobladders. Most published series have adopted open techniques in robotic surgeries. The most described technique is the standard and modified Studer neobladder which was described by the Karolinska Institute group [11,12]. Other techniques include the modified Hautmann, the Vescica Ileale Padovana neobladder, Y technique, and the Florence technique.

### Techniques of bowel configuration

- *The Studer ileal neobladder* [13–17]: This type of neobladder is created by using a 55-cm segment of ileum and brought to the pelvis as in U shape and anastomosed to the urethra (Fig. 8). The right limb has 15 cm while the left one has 40 cm. The ileum is detubularized except for the proximal 10-cm segment of the left limb to create the chimney. The posterior wall is constructed by bringing the medial aspects of both limbs and then folding the anterior wall, bringing the top right segment to the apex of the left limb, creating a spherical reservoir. The staple line of the afferent reservoir loop is excised to perform uretero-ileal anastomosis using the Wallace technique.

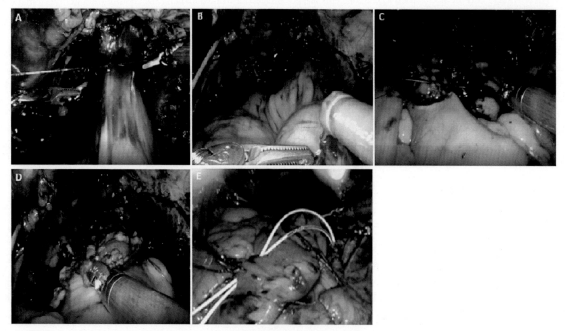

FIG. 8  Development of modified Studer neobladder. (A) Anastomosis of the dependent portion of the small bowel to the urethra. (B) Detubularization of the ileum, 15 cm of the right limb and 30 cm of the left limb leaving 10 cm of the proximal end for the chimney. (C) Anastomosis of the posterior wall of the neobladder. (D) Development of the anterior wall after folding the right limb to the apex of the left one. (E) Uretero-ileal anastomosis to the chimney of the neobladder.

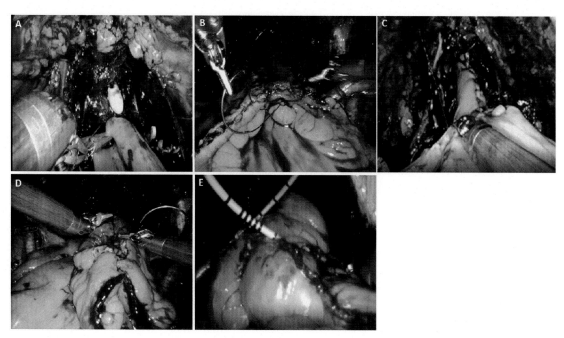

FIG. 9   Development of Y-shaped neobladder. (A) Anastomosis of the dependent portion of the small bowel to the urethra. (B) Detubularization of the ileum, both limbs being symmetric and anastomosis of the posterior wall of the neobladder. (C) Development of the neck of the neobladder with direct anastomosis of the dependent portion. (D) Folding of the posterior plate anteriorly. (E) Bilateral uretero-ileal anastomosis.

- *The W-configuration ileal neobladder* [17]: For this type of neobladder, 50-cm ileum is used, and stay sutures are placed to fix the W configuration (on the dependent part of the right limb, at the end of the right chimney, and at 10 cm proximally on the ascending loop of the right limb. The same is done for the left part). Then, detubularization is performed including both sides of chimneys, and the mesenteric borders are sutured continuously. The posterior plate of the neobladder is anastomosed to the posterior urethral plate. Afterward, the anterior wall is closed in a T-shaped manner.
- *The Y-configuration ileal neobladder* [18–20]: This type of neobladder is similar to the Studer neobladder, but the ileal segment is folded in a "Y" shape to increase its capacity (Fig. 9). The urethral anastomosis is done first, and then both limbs are detubularized. The posterior plate is created, and the anterior reconstruction is completed following a folding of the posterior plate.
- *The robotic Padua ileal neobladder* [21,22]: A 42-cm segment of the most dependent portion of the ileum is isolated, and the neobladder is constructed according to the following measures:
  - 8 cm for the right plate
  - 10 cm for the neck configuration
  - 8 cm for the left plate
  - 16 cm folded in a "U" configuration to create an 8-cm dome.

The optimal point to perform the urethra-ileal anastomosis is approximately 13 cm proximal to the distal margin of the ileal segment. A 10-cm inverted U-shaped neobladder neck is created. The neobladder neck is created with an Endo-GIA (Medtronic Parkway, Minneapolis, MN, USA) stapler or using running sutures from the two branches of the inverted U (5 cm of most dependent part of each limb). The first 8 cm of the distal ileum (right limb) is now detubularized along the antimesenteric border. The remaining 24 cm of the ileum (left limb) is then detubularized, and the neobladder is then folded to create a triangular shape with 8 cm sides and the vertex at the inverted U-shaped neobladder neck. The anterior wall is then closed.

### Ureteric anastomoses

In case of intracorporeal neobladders, ureteral anastomoses vary according to the neobladder configuration, either unilaterally (transposition of the left ureter) to the neobladder chimney as in the case of Studer neobladders and its variants, or bilaterally to the lateral horns of the reservoir. Ureteral catheters (simple-J) are advanced using a Seldinger technique from an inferior abdominal incision, and then through a small window at the anterior wall of the neobladder and then placed in both ureters (Fig. 10). Stents could also be internalized after being placed percutaneously [13].

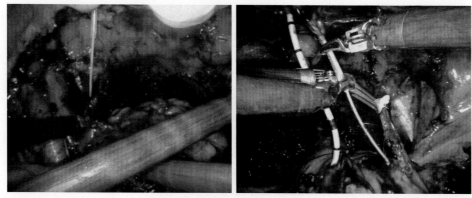

FIG. 10   Percutaneous catheterization of ureters using the Seldinger technique by external insertion of a guidewire and then the ureteral stents.

## Challenges in Intracorporeal ONB

Many technical challenges could arise during the reconstruction of ONBs: in some cases, it could lead to either conversion to open surgery or conversion to an IC. Major factors described are body mass index, short and fat mesentery, adhesions, low-volume surgeon, and oncologic factors (positive surgical margins, nodal disease, etc.).

Urethro-ileal anastomosis is one of the major steps where problems could be encountered while bringing the dependent part of the ileum to the pelvis. Tension-free anastomosis is critical to prevent leak or stricture [23]. For this reason, most of the described surgical robotic techniques start with maneuvering the intestinal segment to the urethra, or even completing the urethra-ileal anastomosis before bowel resection to achieve a tension-free anastomosis.

Many maneuvers have been described, for instance, preservation of maximal urethra length (Fig. 11), careful selection of ileal loop with long mesentery, reduction of peritoneum and taking the patient out of Trendelenburg position, and parallel incisions in the mesentery peritoneum to allow better mobilization of the ileum.

## Nonorthotopic continent diversions

There is a paucity of data on robotic continent diversions, mainly singular case series and case reviews for patients having a robotically constructed continent pouch after a RC. Goh et al. reported the first case of robot-assisted ICIP (Intracorporeal Continent Ileal Pouch) with an operating time (OT) of 180 min and no perioperative complications after 12 months of follow-up [24]. Desai et al. reported a clinical series of 10 patients with a median age of 68 years, of which 80% were males. The median total OT (skin-to-skin) was 369 min, and the median OT time was 210 min. Early minor/ major and late minor/major complication rates were 40%/0% and 0%/20%, respectively. All patients were able to self-catheterize postoperatively. However, one case elected diversion to an IC due to noncompliance with catheterization. The main short-term complications were prolonged urine leakage, alkalosis, and subcutaneous emphysema (all Clavien III), while long-term complications were uretero-enteric stricture (Clavien IIIb) [25].

FIG. 11   Dissection of full prostatic urethra to gain more urethral length for uretero-ileal anastomosis.

## Surgical technique

The technique is described by Aron et al. [26]; following the extirpative phase of the procedure, the left ureter is repositioned to the right side of the sigmoid. Using a 60-mm linear laparoscopic stapler with a blue or white load (60 mm Endo-GIA stapler), the terminal ileum is divided about 10 cm proximal to the ileocecal valve via the assistant port in the left upper quadrant. The cecum and right ascending colon are then mobilized cephalad toward the transverse colon. The colon is divided around 25 cm distal to the ileocecal valve using a 60-mm laparoscopic stapler with a blue or white through the assistant port. Dyed and undyed sutures are used to mark the proximal and distal ends of the bowel segment intended for pouch creation, respectively.

The restoration of bowel continuity follows the previously described method in the IC. The surgical procedure involves an appendectomy and the creation of a cecostomy via the appendiceal stump for the purpose of washing out the isolated colonic segment. A 24 Ch catheter is introduced through the appendiceal stump to vigorously cleanse the segment using a 60-mL catheter tip syringe until clear. The distal staple line on the colonic segment is then excised, and the bowel is opened along the antimesenteric surface, with care taken to preserve the cecal cap intact.

Once the posterior plate is formed, the next step is to perform ureteral anastomosis. Two full-thickness incisions are created in the posterior wall of the ascending colon, and each ureter is brought into the pouch lumen through the incision. The end of each ureter is then spatulated for at least 1 cm and anastomosed to the colonic wall from within the pouch using a 4-0 Vicryl running or interrupted suture. To avoid potential ureteral disruption, all efforts must be made to suture the ureteral wall to the full thickness of the colonic wall. The left ureter is anastomosed first and closer to the ileocecal valve before the right ureteral anastomosis. Once the caudal part of the anastomosis is completed, a 4.8 Ch double J stent is placed through the assistant port. This process is repeated for the right ureter.

The colonic plate is folded into an inverted U shape, and the adjacent edges are sutured together. The proximal stapled end of the ileal segment is then exteriorized and delivered to the skin surface, where the staple line is excised, and a 12F IFT is inserted through the ileocecal valve into the colonic pouch and secured. A 24 Ch catheter is also placed through the appendiceal stump cecostomy into the pouch and brought out through the right 12-mm anterior axillary port and secured with a purse-string suture. The efferent limb is tapered along its antimesenteric border using a stapler with a blue/white load, and the ileocecal valve is buttressed with several interrupted sutures. The efferent limb is then exteriorized with an Allis forceps through the umbilical incision, and the distal end is sutured to the umbilical skin with interrupted sutures to create a catheterizable stoma.

## Future scenario

Throughout this chapter, we have described multiple techniques and fashions of intracorporeal diversions. Almost all the published data are based on single-center experience with unique model for developing the ICUD. Randomized studies comparing outcomes from different types of neobladder, the performance in an intra- or extracorporeal manner, or the performance of an IC neobladder versus IC-IC are lacking in the literature. The best diversion is what the surgeon can offer, with lower rates of complications, better patient acceptance and adaptation, and following physiologic principles of good functioning urinary conduit or reservoir.

The implementation of robotic techniques continues to expand, with robot-assisted ICUD gaining increasing acceptance as a technique for adult patients undergoing RC. One major concern is whether the learning curve for intracorporeal techniques, particularly continent diversions, may limit their utilization, potentially resulting in fewer patients being offered neobladders. To address this, it is reasonable to select ECUD during the learning curve of RARC and introduce ICUD after overcoming the learning curve. Therefore, the establishment of proctoring programs for RARC and urinary diversions is of utmost importance to ensure the safe adoption and implementation of this technique.

Throughout this chapter, we have discussed various techniques and approaches to ICUD. The existing literature mainly comprises single-center experiences with unique models for performing ICUD. However, there is a lack of randomized head-to-head studies comparing outcomes between different types of neobladders and intra- versus extracorporeal approaches. Ongoing trials comparing intra to extracorporeal diversions, including one at our institution, are highly anticipated and will provide valuable insights.

## Key points

- Robotic RC with intracorporeal diversion has become increasingly utilized for the management of bladder cancer.
- Continent and noncontinent ICUD are feasible in experienced robotic centers with good patient selection.
- Technique standardization with a stepwise approach is paramount to obtain good surgical outcomes and help shorten the learning curve.

# References

[1] Hautmann RE. Urinary diversion: ileal conduit to neobladder. J Urol 2003;169(3):834–42.

[2] Lee RK, Abol-Enein H, Artibani W, Bochner B, Dalbagni G, Daneshmand S, Fradet Y, Hautmann RE, Lee CT, Lerner SP, Pycha A. Urinary diversion after radical cystectomy for bladder cancer: options, patient selection, and outcomes. BJU Int 2014;113(1):11–23.

[3] Colombo R, Naspro R. Ileal conduit as the standard for urinary diversion after radical cystectomy for bladder cancer. Eur Urol Suppl 2010;9 (10):736–44.

[4] Hussein AA, May PR, Jing Z, Ahmed YE, Wijburg CJ, Canda AE, Dasgupta P, Khan MS, Menon M, Peabody JO, Hosseini A. Outcomes of intracorporeal urinary diversion after robot-assisted radical cystectomy: results from the International Robotic Cystectomy Consortium. J Urol 2018;199(5):1302–11.

[5] Hautmann RE, Abol-Enein H, Davidsson T, Gudjonsson S, Hautmann SH, Holm HV, Lee CT, Liedberg F, Madersbacher S, Manoharan M, Mansson W. ICUD-EAU international consultation on bladder cancer 2012: urinary diversion. Eur Urol 2013;63(1):67–80.

[6] Rehman J, Sangalli MN, Guru K, de Naeyer G, Schatteman P, Carpentier P, Mottrie A. Total intracorporeal robot-assisted laparoscopic ileal conduit (Bricker) urinary diversion: technique and outcomes. Can J Urol 2011;18(1):5548–56.

[7] Goh AC, Gill IS, Lee DJ, de Castro Abreu AL, Fairey AS, Leslie S, Berger AK, Daneshmand S, Sotelo R, Gill KS, Xie HW. Robotic intracorporeal orthotopic ileal neobladder: replicating open surgical principles. Eur Urol 2012;62(5):891–901.

[8] Pruthi RS, Nix J, McRackan D, Hickerson A, Nielsen ME, Raynor M, Wallen EM. Robotic-assisted laparoscopic intracorporeal urinary diversion. Eur Urol 2010;57(6):1013–21.

[9] Tan WS, Lamb BW, Kelly JD. Evolution of the neobladder: a critical review of open and intracorporeal neobladder reconstruction techniques. Scand J Urol 2016;50:95–103.

[10] Tyritzis SI, Hosseini A, Collins J, Nyberg T, Jonsson MN, Laurin O, et al. Oncologic, functional, and complications outcomes of robot-assisted radical cystectomy with totally intracorporeal neobladder diversion. Eur Urol 2013;64(5):734–41.

[11] Maqboul F, Thinagaran JKR, Dovey Z, Wiklund P. The contemporary status of robotic intracorporeal neobladder. Mini-invasive. Surgery 2021.

[12] Wiklund NP, Poulakis V. Surgery illustrated - surgical atlas robotic neobladder. BJU Int 2011;107(9):1514–37.

[13] Otaola-Arca H, Seetharam Bhat KR, Patel VR, Moschovas MC, Orvieto M. Totally intracorporeal robot-assisted urinary diversion for bladder cancer (part 2). Review and detailed characterization of the existing intracorporeal orthotopic ileal neobladder. Asian J Urol Editorial Off Asian J Urol 2021;8:63–80.

[14] Bhattu AS, Ritch CR, Jahromi M, Banerjee I, Gonzalgo ML. Robotic intracorporeal orthotopic neobladder in the supine Trendelenburg position: a stepwise approach (Internet). Can J Urol 2021;28. Available from: www.canjurol.com.

[15] Collins JW, Hosseini A, Sooriakumaran P, Nyberg T, Sanchez-Salas R, Adding C, et al. Tips and tricks for intracorporeal robot-assisted urinary diversion. Curr Urol Rep 2014;15.

[16] Lavallee E, Sfakianos J, Mehrazin R, Wiklund P. Detailed description of the Karolinska technique for intracorporeal Studer neobladder reconstruction. J Endourol 2022;36:S67–72.

[17] Nouhaud FX, Williams M, Yaxley W, Cho J, Perera M, Thangasamy I, et al. Robot-assisted orthotopic "W" ileal neobladder in male patients: step-by-step video-illustrated technique and preliminary outcomes. J Robot Surg 2020;14(5):739–44.

[18] Gaston R, Ramírez P, Ramírez Rodríguez-Bermejo P. Current status of robotic surgery in urology intracorporeal neobladder @ correspondence.

[19] Asimakopoulos AD, Gubbiotti M, Agrò EF, Morini E, Giommoni V, Piechaud T, et al. "Bordeaux Neobladder": first evaluation of the urodynamic outcomes. Eur Urol Open Sci 2023;47:102–9.

[20] Asimakopoulos AD, Campagna A, Gakis G, Corona Montes VE, Piechaud T, Hoepffner JL, et al. Nerve sparing, robot-assisted radical cystectomy with Intracorporeal bladder substitution in the male. J Urol 2016;196(5):1549–57.

[21] Cacciamani GE, de Marco V, Sebben M, Rizzetto R, Cerruto MA, Porcaro AB, et al. Robot-assisted Vescica Ileale Padovana: a new technique for intracorporeal bladder replacement reproducing open surgical principles. Eur Urol 2019;76(3):381–90.

[22] Simone G, Papalia R, Misuraca L, Tuderti G, Minisola F, Ferriero M, et al. Robotic Intracorporeal Padua Ileal bladder: surgical technique, perioperative. Oncol Funct Outcom Eur Urol 2018;73(6):934–40.

[23] Almassi N, Zargar H, Ganesan V, Fergany A, Haber GP. Management of challenging urethro-ileal anastomosis during robotic assisted radical cystectomy with intracorporeal neobladder formation. Eur Urol 2016;69(4):704–9.

[24] Goh AC, Aghazadeh MA, Krasnow RE, Pastuszak AW, Stewart JN, Miles BJ. Robotic intracorporeal continent cutaneous urinary diversion: primary description. J Endourol 2015;29(11):1217–20.

[25] Desai MM, de Abreu ALC, Goh AC, Fairey A, Berger A, Leslie S, et al. Robotic intracorporeal urinary diversion: technical details to improve time efficiency. J Endourol 2014;28(11):1320–7.

[26] Aron M, Chopra S, Desai MM. Robotic continent cutaneous diversion: Indiana pouch in chapter 59. Robotic Urinary Diversion. Hinman's Atlas of Urologic Surgery; 2016. p. 59.

# Urology: Pelvic lymphadenectomy

*Rafael Ribeiro Meduna[a], Éder Silveira Brazão, Jr.[a], and Stênio de Cássio Zequi[b]*

[a]Urology Division, A.C. Camargo Cancer Center, São Paulo, São Paulo, Brazil [b]Urology Division, Referral Center for Urological Tumors and Graduate School, A.C. Camargo Cancer Center, São Paulo, São Paulo, Brazil

## Abbreviations

| | |
|---|---|
| **BC** | bladder cancer |
| **BRFS** | biochemical recurrence-free survival |
| **CSS** | cancer-specific survival |
| **e-PLND** | extended pelvic lymph node dissection |
| **EAU** | European Association of Urology |
| **MIS** | minimally invasive surgery |
| **OS** | overall survival |
| **PSA** | prostate-specific antigen |
| **PeC** | penile cancer |
| **PLND** | pelvic lymph node dissection |
| **PC** | prostate cancer |
| **RP** | radical prostatectomy |
| **R-PLND** | robot-assisted pelvic lymph node dissection |
| **UC** | urethral cancer |
| **UUTUC** | upper urinary tract urothelial carcinoma |

## Introduction

Pelvic lymph node dissection (PLND) represents one of the steps in the treatment of urological cancers, as prostate, bladder, upper urinary tract, penile, and urethral cancers (UCs) [1,2].

The main purpose of lymph node dissection in patients with any neoplasia is to identify nodal metastasis [1,2]. The PLND has several roles in uro-oncology: enhance locoregional staging, stratify the indication for adjuvant therapies, and ultimately, promote better oncological outcomes [3]. However, this is a controversial topic, especially regarding the long-term oncological benefits [1–3].

Over the past three decades, the introduction of minimally invasive surgery (MIS) has revolutionized the uro-oncology [4]. In addition, the robot-assisted procedures made the surgeon's life easier, providing the perioperative advantages of an MIS approach, as well as oncological outcomes comparable to open surgery [1,3].

Our objective is to review the indications, surgical techniques, outcomes, and complications of robot-assisted pelvic lymph node dissection (R-PLND) for the main urological cancers.

## PLND for prostate cancer

The role of PLND in the surgical treatment of prostate cancer (PC) is a controverse [5]. Despite recent advances in imaging technology and analysis, the PLND remains the gold standard method for nodal staging [5,6]. However, there is still debate about the ideal extension of lymphadenectomy and its therapeutic benefit [7].

The indication for performing lymphadenectomy in PC is related to risk group [8]. In low-risk PC, PLND is not recommended [9]. In intermediate-risk disease, the indication of lymphadenectomy is based on the risk of lymph node involvement [5]. Some nomograms can be used to guide treatment [10,11]. If the risk of lymph node involvement is >5%–7%, depending on the nomogram used, PLND is indicated in cases of intermediate risk [10–13]. In high-risk disease, PLND is always indicated.

The templates of PLND for PC can be limited or extended [6]. Limited template includes lymph nodes in the topography of the obturator fossa [14]. Extended PLND (e-PLND) includes the removal of obturator, internal, and external iliac nodes [8]. American Urological Association and European Association of Urology (EUA) guidelines support an e-PLND when there is an indication to perform PLND [8,9]. e-PLND produces more efficient lymph node staging [10]. Despite this, oncological benefit of e-PLND has not yet been proven [5]. Two prospective randomized studies failed to show oncological benefit of extended versus limited lymphadenectomy [7,14]. Subgroup analysis showed a biochemical recurrence-free survival (BRFS) benefit for e-PLND in the International Society of Urological Pathology prognostic groups ⊇ 3. [14]

Salvage PLND may be considered in some patients with exclusive pelvic nodal recurrence [15]. In general, the main goal of this procedure is to delay systemic treatments [16]. In a systematic review, the five-year BRFS and cancer-specific survival (CSS) ranged from 0% to 56.1% and 71.1% to 100%, respectively. The complete response rate, defined as: prostate-specific antigen (PSA) serum levels <0.2 ng/mL within 2 months of the procedure) ranged from 0% to 82.3%. This variability is associated with the heterogeneity among studies [17]. In this sense, Fossati et al. evaluated the ideal candidate for salvage PLND in a multi-institutional series. At multivariable analysis, Gleason grade group 5, time from radical prostatectomy (RP) to PSA rising, hormonal therapy administration at PSA rising after RP, retroperitoneal uptake at PET/CT scan, three or more positive spots at PET/CT scan, and PSA level at salvage PLND were significant predictors of clinical recurrence [18]. However, the evidence supporting salvage PLND in this scenario is still unclear [17,18]. Thus, while we await robust evidence, individualized patient selection and tumor board meetings may be the best way to achieve good results [16,17].

Salvage PLND template after radical prostatectomy has yet to be defined [19]. No comparative studies evaluating the optimal template (bilateral, unilateral, image-guided) have been published [20–22]. In most studies, the salvage PLND template includes obturator, internal, external, and common iliac nodes [20,22,23]. Presacral sampling depends on the research center [23]. Few studies have limited salvage lymphadenectomy exclusively to positive lymph nodes on PET-CT, under penalty of incomplete resection of microscopic lymph node disease not detected by imaging [20].

## Pelvic lymphadenectomy for bladder cancer (BC)

The standard treatment for muscle-invasive BC and for some patients with high-risk non-muscle-invasive BC is radical cystectomy associated with bilateral PLND [24]. Currently, clinical staging and prognostic criteria do not allow for selective or risk-adapted PLND [25]. Approximately 25% of patients undergoing radical cystectomy for muscle-invasive BC have positive lymph nodes [26]. Therefore, PLND is an essential part of the surgical treatment of BC [24,25].

The extent of lymphadenectomy in the treatment of BC can be classified in different ways [27,28]. Currently, the most commonly used classification was developed by a panel of experts from the EUA Working Group on Muscle-Invasive Bladder Cancer, which determined the boundaries of dissection [27]. The template was defined as follows: limited (lymph node resection is limited to the obturator fossa), standard (removal of lymph nodes from the limited template associated with the topography of the internal iliac artery, external iliac artery, presacral area), extended (includes lymph nodes from the standard template associated with the topography of the common iliac artery up to the bifurcation of the aorta), and superextended (resection of lymph nodes from the extended template with cranial extension up to the inferior mesenteric artery) [27,29]. Another older classification was developed by Leissner et al. [28]. In this system, three levels of PLND were defined: levels I, II, and III (corresponding to the limited and standard template, to the extended template, and to the superextended template, respectively) [28,30].

The oncological control is a debated topic regarding the benefits of performing PLND [31]. A systematic literature review compared the performance of lymphadenectomy with nonperformance; all seven evaluation studies reported a better oncological outcome for the group that underwent lymphadenectomy [27]. However, this association can be a result of the Will Rogers' phenomenon, where a better staging of the disease leads to the transfer of a patient from a healthy group to a diseased one: patients who would be classified as pN0 in a limited lymphadenectomy could be restaged as pN1 if an e-PLND is performed [24].

The oncological benefit of removing micrometastases through lymphadenectomy is described in the literature [32,33]. However, there is a significant discussion regarding the ideal anatomical extent of PLND [34,35]. A more extensive lymphadenectomy is associated with a greater number of resected lymph nodes, increasing the number of positive lymph nodes, and may be associated with better survival and a lower rate of pelvic recurrence [36,37]. A meta-analysis compared the performance of the e-PLND with nonextended lymphadenectomy, identifying better recurrence-free survival in patients with negative lymph nodes, positive lymph nodes, and stage pT3 and pT4 in the group that underwent the e-PLND [30]. In 2018, the first prospective randomized phase III study comparing the extended and nonextended templates did not show a statistical difference between the two groups in the primary endpoint (recurrence-free survival) and secondary endpoints (CSS and overall survival (OS)) [35]. Another phase III randomized study evaluating these two templates is currently being conducted by the southwest oncology group (S-1011) [24,25,31]. The results of this study are expected in the coming years and will provide high-level evidence on the utility of e-PLND in the BC treatment. The superextended template was compared to the e-PLND in two retrospective studies, achieving similar results between the groups [38,39].

There are studies that suggest that the greater the number of lymph nodes resected during cystectomy, the better the patient's oncological outcome [40]. Some retrospective series recommend a minimum of 9–16 lymph nodes to be removed [36,41–43]. However, there are no randomized studies evaluating the minimum number of lymph nodes that should be removed to optimize the result [40].

## PLND for penile cancer (PeC)

Approximately 20%–30% of patients with PeC and positive inguinal lymph nodes will present disease in the pelvic lymph nodes [44,45]. The presence of positive pelvic lymph nodes is associated with lower five-year survival rates [46]. Thus, lymphadenectomy plays an important role in staging. The EUA guidelines indicate performing PLND in patients with two or more positive inguinal lymph nodes on one side [47]. The National Comprehensive Cancer Network indicates resection of the pelvic lymph nodes if three or more positive inguinal lymph nodes are present. Both guidelines also indicate PLND in the presence of lymph nodes with extracapsular extension [44]. In the absence of these risk factors, the chance of pelvic lymph node involvement is <5%, on retrospective studies [45,46,48,49]. Currently, prospective randomized studies evaluating this topic are lacking [46]. The oncological role of PLND in the treatment of PeC is a debated topic [46,48]. Retrospective studies suggest a benefit of PLND in this scenario [49,50]. However, it is described that only 0%–33% of patients with pelvic disease can be cured [50].

The extent of PLND in the treatment of PeC presents controversies [47,49]. Some authors describe the resection of the chains of the external and internal iliac arteries, and obturator fossa [44]. Others, in an e-PLND, include the region of the common iliac artery [48]. The results achieved with both of these approaches are not comparable [44,47]. Another discussion is related to the performance of ipsilateral or bilateral PLND in patients with unilateral positive inguinal node [46]. Involvement of pelvic lymph nodes always arises from positive ipsilateral inguinal lymph nodes. Thus, there would be no involvement of pelvic lymph nodes without inguinal involvement, or pelvic lymph node involvement originating from contralateral positive inguinal lymph node. The data support the performance of ipsilateral lymphadenectomy. Zargar-Shoshtari et al. suggested that the number of positive inguinal lymph nodes should guide the performance of unilateral or bilateral PLND. Bilateral approach should be considered in the case of four or more positive inguinal lymph nodes [51]. PLND can be performed concomitantly with inguinal lymphadenectomy or in a second stage [45,48].

## PLND for upper urinary tract urothelial carcinoma (UUTUC)

Upper urinary tract urothelial carcinoma (UUTUC) represents 5%–10% of urothelial carcinomas [52]. Pielocaliceal tumors are approximately twice as common as ureteral tumors [53]. At radical nephroureterectomy, 20%–30% of patients have lymph node involvement [54]. The presence of lymph node metastasis is an independent predictor of poor survival [53,54]. Tumor grade and stage are two main risk factors for the presence of lymph node metastasis in UUTUC [55]. Lymphadenectomy is indicated in all cases of radical nephroureterectomy with nonmetastatic high-risk tumors [55,56]. For cTa/cT1 tumors, lymphadenectomy should not be performed routinously [54,56].

The template of lymphadenectomy in UUTUC of dissection is guided by the location of the primary tumor [57–59]. PLND is indicated only in distal ureter tumors, from the crossing of the common iliac artery to the ureteral meatus [58]. Also, resection of the obturator fossa and ipsilateral common, external, and internal iliac arteries chains is indicated [56].

Lymphadenectomy is fundamental in the staging of UUTUC [53]. However, their oncological benefits are controverse: some studies show benefit, while others show no benefit [54,59–62]. Some benefit was suggested, with the resection of at least eight lymph nodes [62]. There is no prospective randomized studies, and the majority of these series are retrospective, reporting nonstandardized templates of dissection [53,56,63].

## PLND for urethral neoplasia

Primary UC corresponds to less than 1% of genitourinary neoplasms [64]. There is a paucity of data to guide its treatments [65]. UC is the predominant histological type, followed by squamous cell carcinoma and adenocarcinoma [64]. The presence of enlarged lymph nodes in UC represents metastatic disease in 84% of patients [66]. Lymphatic drainage from the anterior urethra in men occurs to superficial and deep inguinal lymph nodes and subsequently to pelvic lymph nodes [67]. The posterior urethra drains to pelvic chains [66]. In women, lymphatic drainage from the proximal urethra occurs to pelvic lymph nodes, while the distal two-thirds drain to superficial and deep inguinal lymph nodes [64].

Lymphadenectomy, both inguinal and/or pelvic, is indicated for clinically enlarged inguinal/pelvic lymph nodes or in invasive tumors [68]. In this scenario, some studies associate OS benefit with the performance of lymphadenectomy [65,67–69]. There is no evidence supporting the prophylactic lymphadenectomy in patients with UC [64]. The extent and location of lymphadenectomy should be guided by the location of the tumor in the urethra due to different lymphatic drainage patterns [68,69]. PLND is indicated in cases of proximal UC [65]. The lymphadenectomy template in these cases is similar to the BC [66]. Further studies are needed to evaluate the ideal extent and the benefits of lymphadenectomy in primary UC [64–66].

## Minimally invasive and robotic PLND

Robot-assisted oncological treatment has been increasingly adopted as a standard procedure based on similar outcomes compared to open surgery [3]. However, PLND has not been evaluated as a primary outcome in the studies [24]. The RAZOR study was the first randomized phase III trial evaluating robotic and open cystectomy for BC treatment. This evaluation showed the noninferiority of robotic surgery compared to open surgery in radical cystectomy regarding two-year progression-free survival. The surgical method did not influence the choice of the extent of lymphadenectomy, the number of lymph nodes removed, and the pathological stage. Additionally, no significant difference was identified in the assessment of lymphadenectomy-related complications between the groups [70]. Similarly, in the treatment of PC, PLND can be performed using open, laparoscopic, or robotic techniques with similar complication rates and similar lymph node retrieval number [71]. The literature does not present studies evaluating R-PLND in ureter, penis, and urethral tumors, as they are uncommon. Nevertheless, since the templates for these tumors are similar to those used in the treatment of PC and BC, the efficacy results of the robotic technique could be extrapolated to the treatment of this rare malignancies.

Thus, the literature supports the data that R-PLND provides equivalent results to open surgery in oncologic urologic surgery [25]. Experienced surgeons can achieve, using robotic surgery, lymph node resection and a similar complication rate as with the open procedure [24].

## R-PLND-patient positioning and operating room layout RPLND

The procedure is performed under general anesthesia, and a single prophylactic dose of antibiotics is administered. Elastic stockings and pneumatic compressors on the patient's lower limbs, along with the use of low molecular weight heparin, are employed to minimize the risk of deep vein thrombosis.

The patient is placed on a gel mattress in the lithotomy position using leg stirrups. Alternatively, the patient can be placed in a horizontal dorsal decubitus position if using the da Vinci XI robotic platform. Attention is then given to patient fixation and protection. The arms are carefully protected and placed alongside the patient's body in a neutral anatomical position with the thumbs turned upward to prevent nerve injuries. Ensure that the upper and lower limbs are not under traction or pressure. The operating room is set up with the robotic patient cart positioned between the patient's legs. If using the da Vinci XI robot, the robotic cart can be perpendicular to the patient. The surgeon will control the robotic console, and the assisting surgeon is positioned on the patient's bedside where there is only one robotic arm.

For the treatment of distal ureteral tumors requiring nephroureterectomy and PLND, the patient is positioned in a lateral decubitus position (75–90 degrees) with the operative side facing upward on a gel mattress. As previously described, attention is given to patient fixation and protection. The robotic cart is positioned perpendicular to the patient. Similarly, the surgeon will control the robotic console, and the assisting surgeon is positioned in front of the patient.

# Port placement and docking for prostate, bladder, penis, and UC

The procedure is performed using a transperitoneal approach with six ports (four robotic ports and two assistant ports). A supraumbilical incision is made, followed by the creation of pneumoperitoneum using the Hasson technique or a Veress needle, maintaining a pressure of 12–14 mmHg throughout the procedure. In cases of radical cystectomy and e-PLND, a more cranial incision can be made, approximately 2–4 cm above the umbilical scar, to provide a larger surgical field. Through the supraumbilical incision, a 12-mm trocar (Si platform) or an 8-mm robotic trocar (Xi platform) is inserted into the peritoneal cavity for camera placement. A 0-degree or 30-degree downward-facing camera can be used, depending on the surgeon's preference. The arrangement of the trocars is similar in the Si and Xi platforms. Two 8-mm robotic trocars are inserted in the right and left pararectal lines, approximately 8–10 cm lateral to the camera port. The third 8-mm robotic trocar is placed 8–10 cm laterally to the left pararectal robotic port, near the left anterior iliac crest. Finally, a 12-mm assistant trocar is inserted near the right iliac crest (8–10 cm lateral to the right robotic trocar), and a permanent 5-mm trocar is placed between the camera port and the right robotic port (Fig. 1). It is important to maintain a minimum distance of 2 cm between the trocars and bony prominences. Alternatively, the mirrored configuration of the ports described above can be used if the surgeon prefers to have two robotic arms on the right side of the patient (Fig. 2).

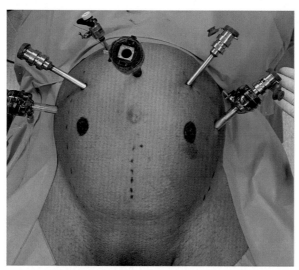

FIG. 1    Trocars placement, for R-PLND in the SI-da Vinci platform.

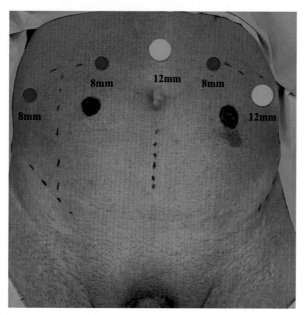

FIG. 2    The mirrored configuration of the ports, described in Fig. 1, can be used if the surgeon prefers to have two robotic arms on the right side of the patient.

III. The current and future clinical applications of robotic surgery among medical specialties

After the trocars are inserted, the patient is placed in a Trendelenburg position (25–28 degrees). This allows the intestines to be positioned outside the pelvic region, facilitating access to the R-PLND. Next, the docking is performed by positioning the surgical system (Si or Xi) between the patient's legs, which is the more traditional approach. In cases where patients have difficulty with hip abduction, lateral docking can be performed. In these cases, when using the Si platform, docking is done on the left or right side of the patient, forming an angle of approximately 45 degrees with the main axis of the surgical table. In the Xi platform, the robot is positioned on the patient's side, forming a 90-degree angle with the patient's main axis, as this device has a rotational boom that allows the arms to rotate in the desired direction. The surgical instruments used are monopolar scissors, bipolar Maryland or fenestrated forceps, and ProGrasp forceps.

## Port placement and docking for UUTUC

Radical nephroureterectomy associated with PLND is performed using a transperitoneal approach. Pneumoperitoneum is created using the Hasson technique or Veress needle while maintaining a pressure of 12–14 mmHg throughout the procedure. In the da Vinci Xi platform, the four robotic trocars are placed in a straight line oblique to the lateral abdominal rectus muscle, with a distance of 6–8 cm between them. Initially, the camera port is placed just above and lateral to the umbilical scar. The other trocars are inserted with the assistance of a 30-degree camera. The most cranial robotic port is located 2 cm from the costal margin, and the most caudal port is placed near the suprapubic midline. The last robotic trocar is positioned on the same line between the camera and the most caudal robotic port. Additionally, a 12-mm auxiliary trocar is inserted in the midline just above the umbilical scar. On the right side, a second 5-mm auxiliary trocar may be necessary in the midline subxiphoid region for liver retraction. Next, the docking is performed by positioning the Xi robotic system perpendicular to the table. The surgical instruments used are monopolar scissors, bipolar fenestrated forceps, and ProGrasp forceps.

Some maneuvers can be performed to optimize visualization during distal ureterectomy and PLND if it is not adequate. The camera can be switched to the caudal robotic trocar, adjacent to the optical port. Additionally, the camera target can be changed to the region of the ureterovesical junction. These maneuvers allow for a broader access in the pelvis to complete the resection of the bladder cuff without the repositioning the robotic system.

## Description of the robot-assisted pelvic lymphadenectomy technique

Next, a step-by-step description of the extended R-PLND performed in the treatment of BC will be provided. It is aiming to serve as a guide for performing less extensive PLND according to the treatment of different neoplasms. The lymphadenectomy can be performed as either the first or last step of the surgical treatment, depending on the surgeon's preference.

1. Identification of important anatomical landmarks such as the medial umbilical ligament, external iliac vessels, vas deferens, and internal inguinal ring. This will facilitate lymph node dissection and prevent complications (Fig. 3).

FIG. 3    Identification of important anatomical landmarks such as the medial umbilical ligament, external iliac vessels, vas deferens, and internal inguinal ring. This will facilitate lymph node dissection and prevent complications.

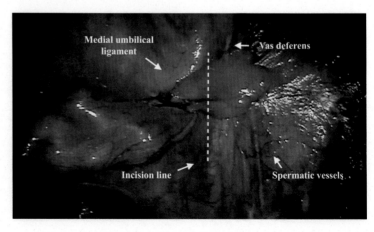

2. Occasionally, there are sigmoid adhesions in left pelvic topography. This may facilitate exposure of the lymph nodes located to the right of the pelvis. For this reason, procedure is usually started from right side.

3. After locating iliac vessels, the first step is proximal incision of peritoneum in topography of common iliac artery, over external iliac artery and distally to vas deferens. Then, a peritoneal incision is made laterally to the median umbilical ligament and medially to the internal inguinal ring. The next step is identifying the external iliac vein and common and external iliac artery. Lymphatic tissue dissection is performed with caution to avoid ureteral injury (Fig. 4).

4. Initially, all fibrous fat tissue is removed along the region of common and external iliac vessels. Boundaries in region of the common iliac artery are aortic bifurcation, bifurcation of common iliac vessels, lateral wall of artery, and medial border of common iliac vein. Boundaries in external iliac artery region are bifurcation of common iliac vessels, circumflex iliac vein, lateral wall of artery, and medial border of external iliac vein.

5. All tissue within obturator fossa is removed, leaving obturator nerve skeletonized. Dissection starts at the angle between the external iliac vein and the pubic bone. In this way, lymph node bundle of external iliac vein is released, following it to the lateral wall of the pelvis. Dissection is performed posteriorly until identification of obturator nerve and vessels. Next, lymph node bundle is dissected with distal extension to the level of circumflex iliac vein or Cloquet's lymph node, and cranially to the confluence with hypogastric artery. Occasionally, we can find small accessory veins (<5 mm in diameter) present in the distal portion of obturator fossa that drain into the posterior/medial walls of the external iliac vein (Fig. 5).

6. Lymphadenectomy of the lateral and medial region of the internal iliac artery is performed. Dissection limit is represented by bifurcation of common iliac vessels, pelvic floor, bladder wall, and obturator nerve.

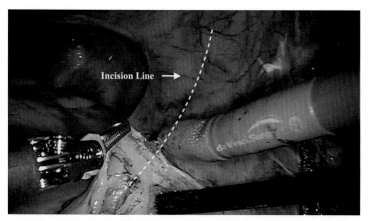

FIG. 4 After locating iliac vessels, the first step is proximal incision of peritoneum in topography of common iliac artery, over external iliac artery and distally to vas deferens. Then, a peritoneal incision is made laterally to the median umbilical ligament and medially to the internal inguinal ring. The next step is identifying the external iliac vein and common and external iliac artery. Lymphatic tissue dissection is performed with caution to avoid ureteral injury; in this way, the fourth robotic arm can perform a slight traction to medial and to the promontory of the peritoneal flap medially to the ureter (Fig. 4).

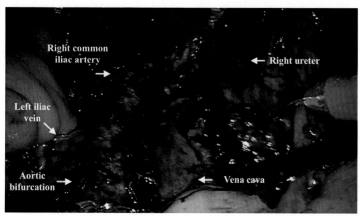

FIG. 5 A panoramic view at the level of aortic bifurcation, during R-PLND.

III. The current and future clinical applications of robotic surgery among medical specialties

FIG. 6 Lymphadenectomy of the lateral and medial region of the internal iliac artery is performed. Dissection limit is represented by bifurcation of common iliac vessels, pelvic floor, bladder wall, and obturator nerve. Presacral lymph nodes are dissected, represented by tissue next to the sacral bone.

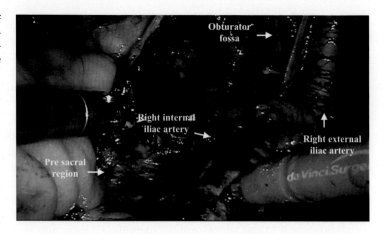

7. Presacral lymph nodes are dissected, represented by tissue next to the sacral bone. During this lymph node dissection, attention should be paid to presacral venous branches, as they are a frequent cause of nuisance bleeding (Fig. 6).
8. After right R-PLND, attention is given to left R-PLND, which is performed similarly as described above.
9. Resected lymph nodes can be removed at the end of each template, or through an extractor bag at the end of procedure.

## Complications

Intraoperative complications in this procedure include injuries to vessels, nerves, and adjacent organs. Vascular injuries include injury to common, external, internal, circumflex, and epigastric iliac vessels. During PLND, injury to genitofemoral or obturator nerves may occur. In addition, ureteral, bladder, and intestinal injury may occur, also. These injuries must be identified and promptly repaired intraoperatively, avoiding complications from a late diagnosis.

Postoperative complications include lymphocele, thromboembolic events, neuropraxia, hematoma, ileus, surgical wound infection, chronic lymphedema, and voiding dysfunctions.

## Future scenario

In the future, molecular staging imaging methods, fluorescence-guided surgery, augmented reality, and real-time three-dimensional reconstruction techniques with artificial intelligence support could enhance the accurate identification of the extension of affected lymph nodes and to guide the more complete resection of them. Newer energy sources and vessel sealers may decrease the lymphatic complications. Single-port surgeries may reduce surgical trauma and the need for intraoperative repositioning of robots.

## Conclusion

PLND is part of the surgical treatment of main urological neoplasms. Lymphadenectomy remains the best lymph nodal staging method. Thus, it can help individualized adjuvant treatments for these patients. The oncological role of PLND are controversial. R-PLND has facilitated the performance of this procedure, being a minimally invasive option with all its benefits, while maintaining a similar complication rate and efficacy outcome to open surgery.

## Key points

- Lymphadenectomy plays an important role in lymph node staging.
- Oncological benefits proportionated by the PLND are controversial in urologic neoplasms.
- R-PLND is a well-established treatment.

# Appendix: Supplementary material

Complementary data: The video: Extended Robotic Pelvic Lymphadenectomy-A.C. Camargo Cancer Center Technique, associated with this chapter, can be found online, at https://doi.org/10.1016/B978-0-443-13271-1.00017-0.

# References

[1] Pini G, Matin SF, Suardi N, et al. Robot assisted lymphadenectomy in urology: pelvic, retroperitoneal and inguinal. Minerva Urol Nefrol 2017;69 (1):38–55. https://doi.org/10.23736/S0393-2249.16.02823-X.

[2] Heidenreich A. Still unanswered: the role of extended pelvic lymphadenectomy in improving oncological outcomes in prostate cancer. Eur Urol 2021;79(5):605–6. https://doi.org/10.1016/j.eururo.2021.01.033.

[3] Vanni AJ, Moinzadeh A. Lymphadenectomy in minimally invasive urologic oncology. Curr Opin Urol 2008;18(2):163–6. https://doi.org/10.1097/MOU.0b013e3282f4f00b.

[4] Loughlin KR. Robotic urology: remember the future. Urol Clin North Am 2021;48(1):xiii–xiv. https://doi.org/10.1016/j.ucl.2020.10.001.

[5] Fossati N, Willemse PPM, Van den Broeck T, et al. The benefits and harms of different extents of lymph node dissection during radical prostatectomy for prostate cancer: a systematic review. Eur Urol 2017;72(1):84–109. https://doi.org/10.1016/j.eururo.2016.12.003.

[6] Mattei A, Fuechsel FG, Bhatta Dhar N, et al. The template of the primary lymphatic landing sites of the prostate should be revisited: results of a multimodality mapping study. Eur Urol 2008;53(1):118–25. https://doi.org/10.1016/j.eururo.2007.07.035.

[7] Touijer KA, Sjoberg DD, Benfante N, et al. Limited versus extended pelvic lymph node dissection for prostate cancer: a randomized clinical trial. Eur Urol Oncol 2021;4(4):532–9. https://doi.org/10.1016/j.euo.2021.03.006.

[8] Mottet N, Bellmunt J, Bolla M, et al. EAU-ESTRO-SIOG guidelines on prostate cancer. Part 1: screening, diagnosis, and local treatment with curative intent. Eur Urol 2017;71(4):618–29. https://doi.org/10.1016/j.eururo.2016.08.003.

[9] Sanda MG, Cadeddu JA, Kirkby E, et al. Clinically localized prostate cancer: AUA/ASTRO/SUO guideline. Part II: recommended approaches and details of specific care options. J Urol 2018;199(4):990–7. https://doi.org/10.1016/j.juro.2018.01.002.

[10] Gandaglia G, Ploussard G, Valerio M, et al. A novel nomogram to identify candidates for extended pelvic lymph node dissection among patients with clinically localized prostate cancer diagnosed with magnetic resonance imaging-targeted and systematic biopsies. Eur Urol 2019;75 (3):506–14. https://doi.org/10.1016/j.eururo.2018.10.012.

[11] Cimino S, Reale G, Castelli T, et al. Comparison between Briganti, Partin and MSKCC tools in predicting positive lymph nodes in prostate cancer: a systematic review and meta-analysis. Scand J Urol 2017;51(5):345–50. https://doi.org/10.1080/21681805.2017.1332680.

[12] Briganti A, Larcher A, Abdollah F, et al. Updated nomogram predicting lymph node invasion in patients with prostate cancer undergoing extended pelvic lymph node dissection: the essential importance of percentage of positive cores. Eur Urol 2012;61(3):480–7. https://doi.org/10.1016/j.eururo.2011.10.044.

[13] Gandaglia G, Fossati N, Zaffuto E, et al. Development and internal validation of a novel model to identify the candidates for extended pelvic lymph node dissection in prostate cancer. Eur Urol 2017;72(4):632–40. https://doi.org/10.1016/j.eururo.2017.03.049.

[14] Lestingi JFP, Guglielmetti GB, Trinh QD, et al. Extended versus limited pelvic lymph node dissection during radical prostatectomy for intermediate- and high-risk prostate cancer: early oncological outcomes from a randomized phase 3 trial. Eur Urol 2021;79(5):595–604. https://doi.org/10.1016/j.eururo.2020.11.040.

[15] Steuber T, Jilg C, Tennstedt P, et al. Standard of care versus metastases-directed therapy for PET-detected nodal oligorecurrent prostate cancer following multimodality treatment: a multi-institutional case-control study. Eur Urol Focus 2019;5(6):1007–13. https://doi.org/10.1016/j.euf.2018.02.015.

[16] Aluwini SS, Mehra N, Lolkema MP, et al. Oligometastatic prostate cancer: results of a Dutch multidisciplinary consensus meeting. Eur Urol Oncol 2020;3(2):231–8. https://doi.org/10.1016/j.euo.2019.07.010.

[17] Fantin JPP, Furst MCB, Tobias-Machado M, et al. Role of salvage lymph node dissection in patients previously treated for prostate cancer: systematic review. Int Braz J Urol 2021;47(3):484–94. https://doi.org/10.1590/S1677-5538.IBJU.2020.0051.

[18] Fossati N, Suardi N, Gandaglia G, et al. Identifying the optimal candidate for salvage lymph node dissection for nodal recurrence of prostate cancer: results from a large, multi-institutional analysis. Eur Urol 2019;75(1):176–83. https://doi.org/10.1016/j.eururo.2018.09.009.

[19] Bandini M, Fossati N, Briganti A. Salvage surgery for nodal recurrent prostate cancer. Curr Opin Urol 2017;27(6):604–11. https://doi.org/10.1097/MOU.0000000000000437.

[20] Abreu A, Fay C, Park D, Quinn D, Dorff T, Carpten J, Kuhn P, Gill P, Almeida F, Gill I. Robotic salvage retroperitoneal and pelvic lymph node dissection for 'node-only' recurrent prostate cancer: technique and initial series. BJU Int 2022;120(3):401–8. https://doi.org/10.1111/bju.13741.

[21] Bravi CA, Fossati N, Gandaglia G, et al. Assessing the best surgical template at salvage pelvic lymph node dissection for nodal recurrence of prostate cancer after radical prostatectomy: when can bilateral dissection be omitted? Results from a multi-institutional series. Eur Urol 2020;78 (6):779–82. https://doi.org/10.1016/j.eururo.2020.06.047.

[22] Linxweiler J, Saar M, Al-Kailani Z, et al. Robotic salvage lymph node dissection for nodal-only recurrences after radical prostatectomy: perioperative and early oncological outcomes. Surg Oncol 2018;27(2):138–45. https://doi.org/10.1016/j.suronc.2018.02.010.

[23] Kolontarev K, Govorov A, Kasyan G, Rasner P, Vasiliev A, Pushkar D. Extended robotic salvage lymphadenectomy in patients with 'node-only' prostate cancer recurrence: initial experience. Cent Eur J Urol 2018;71(2):162–7. https://doi.org/10.5173/ceju.2018.1478.

[24] Małkiewicz B, Kiełb P, Gurwin A, et al. The usefulness of lymphadenectomy in bladder cancer-current status. Medicina 2021;57(5):1–16. https://doi.org/10.3390/medicina57050415.

[25] Sung HH, Lerner SP. Utility of lymphadenectomy in bladder cancer: where do we stand? Curr Opin Urol 2020;30(3):407–14. https://doi.org/10.1097/MOU.0000000000000750.

[26] Packiam VT, Tsivian M, Boorjian SA. The evolving role of lymphadenectomy for bladder cancer: why, when, and how. Transl Androl Urol 2020;9(6):3082–93. https://doi.org/10.21037/tau.2019.06.01.

[27] Bruins HM, Veskimae E, Hernandez V, et al. The impact of the extent of lymphadenectomy on oncologic outcomes in patients undergoing radical cystectomy for bladder cancer: a systematic review. Eur Urol 2014;66(6):1065–77. https://doi.org/10.1016/j.eururo.2014.05.031.

[28] Leissner J, Ghoneim MA, Abol-Enein H, et al. Extended radical lymphadenectomy in patients with urothelial bladder cancer: results of a prospective multicenter study. J Urol 2004;171(1):139–44. https://doi.org/10.1097/01.ju.0000102302.26806.fb.

[29] Dorin RP, Daneshmand S, Eisenberg MS, et al. Lymph node dissection technique is more important than lymph node count in identifying nodal metastases in radical cystectomy patients: a comparative mapping study. Eur Urol 2011;60(5):946–52. https://doi.org/10.1016/j.eururo.2011.07.012.

[30] Bi L, Huang H, Fan X, et al. Extended vs non-extended pelvic lymph node dissection and their influence on recurrence-free survival in patients undergoing radical cystectomy for bladder cancer: a systematic review and meta-analysis of comparative studies. BJU Int 2014;113(5 B):39–48. https://doi.org/10.1111/bju.12371.

[31] Wang YC, Wu J, Dai B, et al. Extended versus non-extended lymphadenectomy during radical cystectomy for patients with bladder cancer: a meta-analysis of the effect on long-term and short-term outcomes. World J Surg Oncol 2019;17(1):1–9. https://doi.org/10.1186/s12957-019-1759-5.

[32] May M, Herrmann E, Bolenz C, et al. Association between the number of dissected lymph nodes during pelvic lymphadenectomy and cancer-specific survival in patients with lymph node-negative urothelial carcinoma of the bladder undergoing radical cystectomy. Ann Surg Oncol 2011;18(7):2018–25. https://doi.org/10.1245/s10434-010-1538-6.

[33] Brössner C, Pycha A, Toth A, Mian C, Kuber W. Does extended lymphadenectomy increase the morbidity of radical cystectomy? BJU Int 2004;93(1):64–6. https://doi.org/10.1111/j.1464-410X.2004.04557.x.

[34] Mandel P, Tilki D, Eslick GD. Extent of lymph node dissection and recurrence-free survival after radical cystectomy: a meta-analysis. Urol Oncol Semin Orig Investig 2014;32(8):1184–90. https://doi.org/10.1016/j.urolonc.2014.01.017.

[35] Gschwend JE, Heck MM, Lehmann J, et al. Extended versus limited lymph node dissection in bladder cancer patients undergoing radical cystectomy: survival results from a prospective, randomized trial (figure presented). Eur Urol 2019;75(4):604–11. https://doi.org/10.1016/j.eururo.2018.09.047.

[36] Wright JL, Lin DW, Porter MP. The association between extent of lymphadenectomy and survival among patients with lymph node metastases undergoing radical cystectomy. Cancer 2008;112(11):2401–8. https://doi.org/10.1002/cncr.23474.

[37] Herr HW, Faulkner JR, Grossman HB, et al. Surgical factors influence bladder cancer outcomes: a cooperative group report. J Clin Oncol 2004;22(14):2781–9. https://doi.org/10.1200/JCO.2004.11.024.

[38] Simone G, Abol Enein H, Ferriero M, et al. 1755 extended versus super-extended Plnd during radical cystectomy: comparison of two prospective series. J Urol 2012;187(4S), e708. https://doi.org/10.1016/j.juro.2012.02.1771.

[39] Zehnder P, Studer UE, Skinner EC, et al. Super extended versus extended pelvic lymph node dissection in patients undergoing radical cystectomy for bladder cancer: a comparative study. J Urol 2011;186(4):1261–8. https://doi.org/10.1016/j.juro.2011.06.004.

[40] Koppie TM, Vickers AJ, Vora K, Dalbagni G, Bochner BH. Standardization of pelvic lymphadenectomy performed at radical cystectomy: can we establish a minimum number of lymph nodes that should be removed? Cancer 2006;107(10):2368–74. https://doi.org/10.1002/cncr.22250.

[41] Leissner J, Hohenfellner R, Thüroff JW, Wolf HK. Lymphadenectomy in patients with transitional cell carcinoma of the urinary bladder; significance for staging and prognosis. BJU Int 2000;85(7):817–23. https://doi.org/10.1046/j.1464-410X.2000.00614.x.

[42] Konety BR, Joslyn SA, O'Donnell MA. Extent of pelvic lymphadenectomy and its impact on outcome in patients diagnosed with bladder cancer: analysis of data from the surveillance, epidemiology and end results program data base. J Urol 2003;169(3):946–50. https://doi.org/10.1097/01.ju.0000052721.61645.a3.

[43] Herr HW, Bochner BH, Dalbagni G, Donat SM, Reuter VE, Bajorin DF. Impact of the number of lymph nodes retrieved on outcome in patients with muscle invasive bladder cancer. J Urol 2002;167(3):1295–8. https://doi.org/10.1016/S0022-5347(05)65284-6.

[44] Clark PE, Spiess PE, Agarwal N, et al. Penile cancer. J Natl Compr Canc Netw 2013;11(5):594–615. https://doi.org/10.6004/jnccn.2013.0075.

[45] Lughezzani G, Catanzaro M, Torelli T, et al. The relationship between characteristics of inguinal lymph nodes and pelvic lymph node involvement in penile squamous cell carcinoma: a single institution experience. J Urol 2014;191(4):977–82. https://doi.org/10.1016/j.juro.2013.10.140.

[46] O'Brien JS, Perera M, Manning T, et al. Penile cancer: contemporary lymph node management. J Urol 2017;197(6):1387–95. https://doi.org/10.1016/j.juro.2017.01.059.

[47] Hakenberg OW, Compérat EM, Minhas S, Necchi A, Protzel C, Watkin N. EAU guidelines on penile cancer: 2014 update. Eur Urol 2015;67(1):142–50. https://doi.org/10.1016/j.eururo.2014.10.017.

[48] Lont AP, Kroon BK, Gallee MPW, van Tinteren H, Moonen LMF, Horenblas S. Pelvic lymph node dissection for penile carcinoma: extent of inguinal lymph node involvement as an indicator for pelvic lymph node involvement and survival. J Urol 2007;177(3):947–52. https://doi.org/10.1016/j.juro.2006.10.060.

[49] Zargar-Shoshtari K, Sharma P, Djajadiningrat R, et al. Extent of pelvic lymph node dissection in penile cancer may impact survival. World J Urol 2016;34(3):353–9. https://doi.org/10.1007/s00345-015-1593-5.

[50] Li ZS, Deng CZ, Yao K, et al. Bilateral pelvic lymph node dissection for Chinese patients with penile cancer: a multicenter collaboration study. J Cancer Res Clin Oncol 2017;143(2):329–35. https://doi.org/10.1007/s00432-016-2292-3.

[51] Zargar-Shoshtari K, Djajadiningrat R, Sharma P, et al. Establishing criteria for bilateral pelvic lymph node dissection in the management of penile cancer: lessons learned from an international multicenter collaboration. J Urol 2015;194(3):696–702. https://doi.org/10.1016/j.juro.2015.03.090.

[52] Lenis AT, Donin NM, Faiena I, et al. Role of surgical approach on lymph node dissection yield and survival in patients with upper tract urothelial carcinoma. Urol Oncol Semin Orig Investig 2018;36(1):9.e1–9. https://doi.org/10.1016/j.urolonc.2017.09.001.

[53] Dominguez-Escrig JL, Peyronnet B, Seisen T, et al. Potential benefit of lymph node dissection during radical nephroureterectomy for upper tract urothelial carcinoma: a systematic review by the European Association of Urology guidelines panel on non-muscle-invasive bladder cancer. Eur Urol Focus 2019;5(2):224–41. https://doi.org/10.1016/j.euf.2017.09.015.

[54] Lughezzani G, Jeldres C, Isbarn H, et al. A critical appraisal of the value of lymph node dissection at nephroureterectomy for upper tract urothelial carcinoma. Urology 2010;75(1):118–24. https://doi.org/10.1016/j.urology.2009.07.1296.

[55] Dong F, Xu T, Wang X, et al. Lymph node dissection could bring survival benefits to patients diagnosed with clinically node-negative upper urinary tract urothelial cancer: a population-based, propensity score-matched study. Int J Clin Oncol 2019;24(3):296–305. https://doi.org/10.1007/s10147-018-1356-6.

[56] Xia HR, Li SG, Zhai XQ, Liu M, Guo XX, Wang JY. The value of lymph node dissection in patients with node-positive upper urinary tract urothelial cancer: a retrospective cohort study. Front Oncol 2022;12(June):1–8. https://doi.org/10.3389/fonc.2022.889144.

[57] Matin SF, Sfakianos JP, Espiritu PN, Coleman JA, Spiess PE. Patterns of lymphatic metastases in upper tract urothelial carcinoma and proposed dissection templates. J Urol 2015;194(6):1567–74. https://doi.org/10.1016/j.juro.2015.06.077.

[58] Kondo T, Nakazawa H, Ito F, Hashimoto Y, Toma H, Tanabe K. Primary site and incidence of lymph node metastases in urothelial carcinoma of upper urinary tract. Urology 2007;69(2):265–9. https://doi.org/10.1016/j.urology.2006.10.014.

[59] Kondo T, Hashimoto Y, Kobayashi H, et al. Template-based lymphadenectomy in urothelial carcinoma of the upper urinary tract: impact on patient survival. Int J Urol 2010;17(10):848–54. https://doi.org/10.1111/j.1442-2042.2010.02610.x.

[60] Cho KS, Choi HM, Koo K, et al. Clinical significance of lymph node dissection in patients with muscle-invasive upper urinary tract transitional cell carcinoma treated with nephroureterectomy. J Korean Med Sci 2009;24(4):674–8. https://doi.org/10.3346/jkms.2009.24.4.674.

[61] Brausi MA, Gavioli M, De Luca G, et al. Retroperitoneal lymph node dissection (RPLD) in conjunction with nephroureterectomy in the treatment of infiltrative transitional cell carcinoma (TCC) of the upper urinary tract: impact on survival{a figure is presented}. Eur Urol 2007;52(5):1414–20. https://doi.org/10.1016/j.eururo.2007.04.070.

[62] Roscigno M, Shariat SF, Margulis V, et al. Impact of lymph node dissection on cancer specific survival in patients with upper tract urothelial carcinoma treated with radical nephroureterectomy. J Urol 2009;181(6):2482–9. https://doi.org/10.1016/j.juro.2009.02.021.

[63] Duquesne I, Ouzaid I, Loriot Y, Moschini M, Xylinas E. Lymphadenectomy for upper tract urothelial carcinoma: a systematic review. J Clin Med 2019;8(8). https://doi.org/10.3390/jcm8081190.

[64] Gakis G, Bruins HM, Cathomas R, et al. European Association of Urology guidelines on primary urethral carcinoma-2020 update. Eur Urol Oncol 2020;3(4):424–32. https://doi.org/10.1016/j.euo.2020.06.003.

[65] Wu J, Su HC, Shou JZ. The role of regional lymph node dissection in men with primary urethral carcinoma. World J Urol 2022;40(5):1247–9. https://doi.org/10.1007/s00345-021-03835-x.

[66] Janisch F, Abufaraj M, Fajkovic H, et al. Current disease management of primary urethral carcinoma. Eur Urol Focus 2019;5(5):722–34. https://doi.org/10.1016/j.euf.2019.07.001.

[67] Calderón Cortez JF, Territo A, Fontana M, et al. Primary urethral carcinoma: results from a single center experience. Actas Urol Esp (English Ed) 2022;46(2):70–7. https://doi.org/10.1016/j.acuroe.2020.10.017.

[68] Werntz RP, Riedinger CB, Fantus RJ, et al. The role of inguinal lymph node dissection in men with urethral squamous cell carcinoma. Urol Oncol Semin Orig Investig 2018;36(12):526.e1–6. https://doi.org/10.1016/j.urolonc.2018.09.014.

[69] Karnes RJ, Breau RH, Lightner DJ. Surgery for urethral cancer. Urol Clin North Am 2010;37(3):445–57. https://doi.org/10.1016/j.ucl.2010.04.011.

[70] Parekh DJ, Reis IM, Castle EP, et al. Robot-assisted radical cystectomy versus open radical cystectomy in patients with bladder cancer (RAZOR): an open-label, randomised, phase 3, non-inferiority trial. Lancet 2018;391(10139):2525–36. https://doi.org/10.1016/S0140-6736(18)30996-6.

[71] Mottrie A, Volpe A. The robotic approach does not change the current paradigms of pelvic lymph node dissection for prostate cancer. Eur Urol 2014;65(1):17–9. https://doi.org/10.1016/j.eururo.2013.05.017.

# 44

# Urology: Robotic retroperitoneal lymphadenectomy

*Rafael Ribeiro Zanotti*[a], *Rodrigo Sousa Madeira Campos*[a], *and Stênio de Cássio Zequi*[b]

[a]Urology Division, A.C. Camargo Cancer Center, São Paulo, São Paulo, Brazil [b]Urology Division, Referral Center for Urological Tumors and Graduate School, A.C. Camargo Cancer Center, São Paulo, São Paulo, Brazil

## Abbreviations

| | |
|---|---|
| **GCTC** | germ cell testis cancer |
| **MI-RPLND** | minimally invasive retroperitoneal lymph node dissection |
| **NSGCT** | nonseminoma germ cell tumor |
| **O-RPLND** | open retroperitoneal lymph node dissection |
| **RPLND** | retroperitoneal lymph node dissection |
| **R-RPLND** | robotic retroperitoneal lymph node dissection |

## Introduction

The management of testicular cancer involves an initial radical orchidectomy, followed by observation, retroperitoneal lymph node dissection (RPLND), radiotherapy, or cisplatin-based chemotherapy, depending on the stage and risk classification [1].

The development of cisplatin-based chemotherapy has significantly improved the survival rates of patients with testicular cancer, from 5%–10% to 80%–90% [2]. As a result, the focus of treatment has shifted from solely improving survival to also reducing treatment-related side effects. RPLND has been increasingly used for this purpose, as its side effects are mostly limited to the immediate postoperative period.

RPLND plays a crucial role in the management of testicular cancer. While open RPLND (O-RPLND) has been the standard approach for staging patients with germ cell testis cancer (GCT) and has therapeutic benefits for some patients, it is associated with significant perioperative morbidity and extended recovery times [3]. To overcome these challenges, minimally invasive RPLND (MI-RPLND) was developed with the aim of achieving comparable oncologic efficacy to O-RPLND while providing the advantages of a minimally invasive approach.

The concept of MI-RPLND was first described in 1992, utilizing a laparoscopic approach. However, the significant learning curve and concerns regarding the ability to achieve adequate dissection posterior to the great vessels and control bleeding have limited the progress of this approach. Laparoscopic RPLND is technically challenging and requires extensive training, which has hindered its widespread adoption. The introduction of robotic RPLND (R-RPLND) has provided additional resources that have made this challenging surgery more feasible. Robotic surgery offers several potential advantages over laparoscopic surgery.

The first description of R-RPLND was provided by Davol et al. in 2006 [4]. The robotic approach offers several advantages, including a shortened learning curve, improved visualization, and better ergonomics. These benefits allow surgeons to potentially harness the advantages of minimally invasive surgery while overcoming the technical challenges associated with laparoscopic procedures [1].

Similarly to O-RPLND, R-RPLND should adhere to fundamental surgical principles, such as obtaining adequate exposure, performing meticulous dissection, and sparing nerves. The robotic platform has expanded the scope of minimally invasive urologic procedures and provides significant advantages over conventional laparoscopy. The enhanced vision with 3D optics and magnification, improved dexterity with greater instrument precision, and ergonomic advantages for the surgeon all contribute to the application of robotic techniques in various surgical procedures. These advantages generally enable a more controlled and delicate dissection compared to laparoscopy. In the context of R-RPLND, specific advantages include the ability to perform circumferential dissection around structures due to the flexibility of wristed instruments and improved control of bleeding from the lumbar vessels. The robotic platform allows the reproduction of the oncologic principles of open RPLND in a minimally invasive setting, providing access to nodal tissue posterior to the great vessels, which can be challenging with conventional laparoscopy [5].

Despite the recent expansion in the applicability of robotic surgery, the literature on robotic retroperitoneal lymphadenectomy remains limited. Most research on R-RPLND has a low level of evidence. Therefore, R-RPLND is still considered an emerging treatment option.

## Oncological safety and morbidity

The most recent systematic review in this field provides valuable insights into the outcomes of primary and postchemotherapy R-RPLND. In primary R-RPLND, the majority of patients (88.9%) had nonseminoma germ cell tumor (NSGCT), and approximately 31% underwent bilateral dissection. The mean operative time ranged from 175 to 540 min, with a median estimated blood loss ranging from 5 to 450 mL. The conversion rate to open surgery was relatively low at 2.2%, with poor exposure, failure to progress, and vascular injury being the main reasons for conversion. The overall complication rate was reported as 16.7%, with 4.1% of complications categorized as Clavien Dindo ≥3. The median lymph node yield varied between 11 and 58.5, and the median hospital stay ranged from 1 to 4 days [6].

For postchemotherapy R-RPLND, the mean operative time ranged from 134 to 550 min, and the median estimated blood loss ranged from 43 to 2300 mL. The overall conversion rate to open surgery was 9%, with vascular injury, poor exposure, failure to progress, tumor spillage, robotic malfunction, and ventilation problems being the main reasons for conversion. The median hospital stays ranged from 1 to 8 days. In the pooled analysis of primary R-RPLND data, the disease recurrence rate was reported as 6.7% (18 out of 267 patients) over a median follow-up period ranging from 3 to 47 months. For postchemotherapy R-RPLND, the disease recurrence rate was 2.2% (4 out of 183 patients) during a median follow-up period of 4–49 months [6].

Regarding functional outcomes, the majority of patients who underwent primary R-RPLND (69%) received a modified template dissection, and 71.1% of them preserved antegrade ejaculation in the postoperative period. In postchemotherapy R-RPLND, 29.3% of patients received a modified template dissection, and 52.3% maintained preserved antegrade ejaculation [6].

Comparative studies between R-RPLND and O-RPLND are limited, but the available evidence suggests several advantages of R-RPLND. The operative time is similar between the two approaches, but R-RPLND has been associated with significantly lower median estimated blood loss compared to O-RPLND (0.9% vs 14.5%). The overall complication rate is lower for R-RPLND, although major complications appear to be similar. Lymph node yield is comparable between the two approaches. However, the length of hospital stay is significantly shorter for patients undergoing R-RPLND [6].

Recurrence rates between O-RPLND and R-RPLND are similar, as observed in studies by Li et al. and Grenabo Bergdahl et al. These studies reported disease recurrence rates of 19% and 10% for O-RPLND, and 8.6% and 6.8% for R-RPLND, respectively. Another systematic review found that R-RPLND had a lower transfusion rate and fewer overall complications compared to O-RPLND. However, there is a notable rate of conversion to open surgery in the postchemotherapy setting (9%). The length of follow-up in most series is limited to less than 2 years, and the number of harvested lymph nodes varies widely [6–8].

Retrospective studies comparing R-RPLND and O-RPLND have generally shown favorable outcomes for R-RPLND, including decreased intraoperative blood loss, shorter hospital stays, comparable complication rates, and similar short-term oncologic outcomes. A retrospective review also found that operative time decreased and overall complications decreased with increasing experience in performing R-RPLND. Predictive factors for conversion or high-grade complications have not been consistently identified. Garg et al. demonstrated that R-RPLND is associated with less blood loss, a lower blood transfusion rate, fewer overall complications, and a shorter hospital stay compared to O-RPLND. The median lymph node yield and early oncological outcomes were similar between the two groups [9–11].

One important consideration in RPLND is the completeness of dissection. Most comparative studies have reported similar lymph node yields between R-RPLND and O-RPLND [8,11,12].

# Patient positioning, trocars placement, instruments, and exposure

Most urologists are familiar with retroperitoneal anatomy through the lateral approach, which has been commonly used for RPLND. However, for bilateral dissection in the flank position, the robot often needs to be redocked to access the contralateral lymph node packet. This can result in limited exposure for contralateral lymph node dissection. To address this concern, a transition from the flank position to the supine Trendelenburg position has been implemented. Stepanian et al. have described the evolution of the robotic approach for RPLND in detail. Both the lateral and supine approaches have their own limitations, and currently, there is insufficient evidence to establish the superiority of one approach over the other. However, the da Vinci Xi surgical platform has been instrumental in overcoming many of the challenges associated with earlier platforms, particularly in terms of docking. The ability to position the robot laterally to the patient has been beneficial, and the da Vinci Xi platform is the most used for R-RPLND in contemporary series [13–15].

## Lateral approach docking

This is a five-port (Fig. 1) technique with the 30° lens. The patient lies in a contralateral position to the template that will be dissected. Similarly to a nepherectomy docking, the robot is positioned posterior to the patient (for da Vinci Si and Xi). This technique is appropriate for unilateral template dissection and allows resection of the inguinal cord.

## Supine approach docking

The supine Trendelenburg position offers excellent exposure of the retroperitoneal mass located near the aorta and inferior vena cava (IVC) while minimizing bowel mobilization. This position allows for a complete bilateral dissection without the need to reposition the patient. By placing the patient in Trendelenburg position, the bowel falls toward the head, providing better visualization of the iliac vessels and cecum. Robotic assistance can be used to retract the bowel, further enhancing exposure. With the da Vinci Xi surgical platform, the robot is docked alongside the patient, simplifying the room set-up. This positioning facilitates the supine approach for R-RPLND [15] (Fig. 2).

The previous limitations associated with the Si platform for R-RPLND, such as docking over the patient's head and the need to redock for excising the spermatic cord, have been overcome with the da Vinci Xi platform. The Xi system allows for side docking, enabling the surgeon to perform a complete bilateral template dissection and excise the remaining spermatic cord without the need for redocking. This improvement in the robotic system enhances the efficiency and ease of performing R-RPLND.

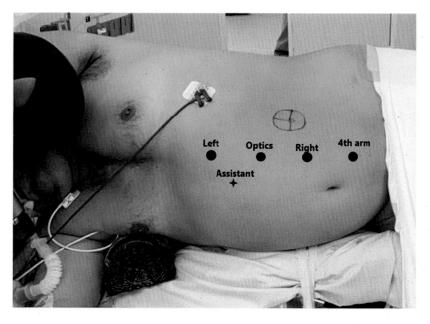

FIG. 1 Suggested positioning if the intention is to dissect using a modified template. In this decubitus position, for paraaortic template. Patient in right lateral decubitus position, suitably fixed and with protections on bony prominences. The table is broken in order to open the flank, allowing more space for work. In the image, the suggested placement of the trocars was designed according to the da Vinci Xi platform.

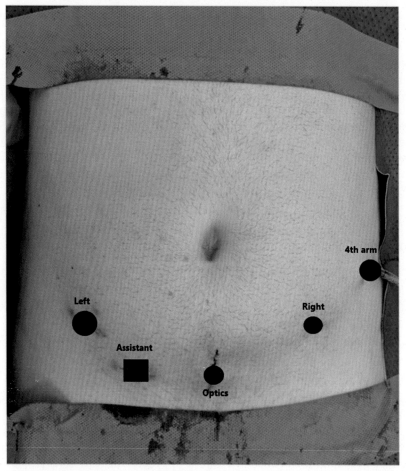

FIG. 2    This demo for organizing portals can be used for both the da Vinci Si and Xi platforms. However, using Si, it would be necessary to perform redocking to achieve dissection of the inguinal cord.

## Instruments

During the R-RPLND procedure, specific instruments and techniques are employed. A monopolar scissor is placed in the right robotic port, a fenestrated bipolar forceps in the left robotic port, and a Prograsp forceps in the fourth robotic port. Initially, a 0° lens is used for visualization, and once the retroperitoneum is exposed, it is switched to a 30° down-angled lens for better visibility.

To gain access to the retroperitoneum, the cecum and ileum are mobilized. This is achieved by making an incision in the posterior peritoneum medial to the cecum and extending it upward toward the ligament of Treitz. To elevate the cecum and small bowel and expose the retroperitoneum, a monofilament suture with a straight needle is passed through the anterior abdominal wall, threaded through the cut edge of the posterior peritoneum, and then passed back through the anterior abdominal wall. This procedure is performed on both sides of the cut posterior peritoneum.

It is important to note that there is currently no direct comparison of outcomes between the supine and flank approaches, and the decision regarding patient positioning (supine or flank) is based on the surgeon's discretion.

## Surgical principles and recurrences

Recently, authors from a high-volume center skilled for decades in performing open RPLND for GCTC, reported their initial experience in managing five patients who developed early (<270 days) and unusual intraabdominal tumor recurrences after undergoing robotic RPLND in other institutions. They reiterate the recommendations to follow the basic surgical technical and oncologic principles of open surgery in minimally invasive robotic approaches, which have been increasingly utilized in recent years [16]. In a related editorial, there were mentioned reports of local recurrences after historical series of as in open, as in videolaparoscopic approaches for RPLND, it is not possible to conclude on all factors involved in the reported cases earlier [17]. Nowadays, proceeding cautiously with robotic RPLND and encouraging the production of large series by combining critically searched data are more than recommended actions for fostering the outcomes of R-RPLND for patients with testicular cancer [17].

## Future scenario

In the future, perhaps fluorescence-guided surgery, augmented reality, real-time three-dimensional reconstruction techniques with artificial intelligence support could enhance the accurate identification of the extension of affected retroperitoneal lymph nodes and guide the more complete resection of them, associated with enhanced visual identification of retroperitoneal autonomic nervous system fibers to be preserved. Actual molecular image methods are not recommended for GCTC, we do not know if future modalities of these image tests could, for example, to avoid the nontherapeutic RPLND (precisely identifying the presence of fibrosis or necrosis in residual mass). Newer energy sources and vessel sealers may decrease the lymphatic complications. We are looking forward to new platforms that optimize surgical field exposition. Large prospective series are awaited.

## Conclusions

Indeed, R-RPLND (Robotic Retroperitoneal Lymph Node Dissection) holds promise as a less morbid procedure for selected patients with testicular cancer. Preliminary evidence suggests that it can provide satisfactory surgical and oncological outcomes. As advancements continue to be made in this approach, its boundaries are being pushed further, contributing to its progress.

However, it is important to note that the optimal patient selection criteria for R-RPLND are still being elucidated. Further research and studies are needed to refine and establish the criteria for selecting patients who would benefit the most from this approach. As more data becomes available and experience with R-RPLND accumulates, a clearer understanding of its applicability and effectiveness will emerge.

## Key points

- Knowledge and expertise about robotic RPLND have increased significantly in recent years, both for patients in the initial stages and for patients submitted to resections of residual masses after chemotherapy.
- Robotic RPLND should be performed in high-volume referral centers and by experienced surgeons.
- Robotic RPLND has reproduced results similar to open surgery, with reduced bleeding rates and shorter hospitalization time.
- Robotic RPLND requires a smaller learning curve and allows us to overcome technical difficulties inherent to the videolaparoscopic approach.
- Performing robotic RPLND requires maintaining the basic oncological precepts and surgical techniques historically used in the open surgeries.
- Large multi-institutional series are expected.

## Appendix: Supplementary material

Supplementary material related to this chapter can be found on the accompanying CD or online at https://doi.org/10.1016/B978-0-443-13271-1.00026-1.

## References

[1] Yang H, Obiora D, Tomaszewski JJ. Outcomes and expanding indications for robotic retroperitoneal lymph node dissection for testicular cancer. Transl Androl Urol 2021;10:2188–94. AME Publishing Company.
[2] Stephenson AJ, Gilligan TD. Neoplasms of the testis. In: Partin AW, Dmochowski RR, Kavoussi LR, Peters CA, editors. Campbell-Walsh-Wein-Urology. 12 ed. Philadelphia, PA: Elsevier; 2021. p. 1680–710.

[3] Baniel J, Sella A. Complications of retroperitoneal lymph node dissection in testicular cancer: primary and post-chemotherapy. Semin Surg Oncol 1999;17(4):263–7.

[4] Davol P, Sumfest J, Rukstalis D. Robotic-assisted laparoscopic retroperitoneal lymph node dissection. Urology 2006;67(1):199.e7–8.

[5] Abdel-Aziz KF, Anderson JK, Svatek R, Margulis V, Sagalowsky AI, Cadeddu JA. Laparoscopic and open retroperitoneal lymph-node dissection for clinical stage I Nonseminomatous germ-cell testis tumors. J Endourol 2006;20(9):627–31.

[6] Garg H, Mansour AM, Psutka SP, Kim SP, Porter J, Gaspard CS, et al. Robot-assisted retroperitoneal lymph node dissection: a systematic review of perioperative outcomes. BJU Int 2023.

[7] Ray S, Pierorazio PM, Allaf ME. Primary and post-chemotherapy robotic retroperitoneal lymph node dissection for testicular cancer: A review. Transl Androl Urol 2020;9:949–58. AME Publishing Company.

[8] Grenabo Bergdahl A, Månsson M, Holmberg G, Fovaeus M. Robotic retroperitoneal lymph node dissection for testicular cancer at a national referral Centre. BJUI Compass 2022;3(5):363–70.

[9] Schwen ZR, Gupta M, Pierorazio PM. A review of outcomes and technique for the robotic-assisted laparoscopic retroperitoneal lymph node dissection for testicular cancer. Adv Urol 2018;2018:1–7.

[10] Schermerhorn SMV, Christman MS, Rocco NR, Abdul-Muhsin H, L'Esperance JO, Castle EP, et al. Learning curve for robotic-assisted laparoscopic retroperitoneal lymph node dissection. J Endourol 2021;35(10):1483–9.

[11] Li R, Duplisea JJ, Petros FG, González GMN, Tu SM, Karam JA, et al. Robotic Postchemotherapy retroperitoneal lymph node dissection for testicular Cancer. Eur Urol Oncol 2021;4(4):651–8.

[12] Lloyd P, Hong A, Furrer MA, Lee EWY, Dev HS, Coret MH, et al. A comparative study of peri-operative outcomes for 100 consecutive post-chemotherapy and primary robot-assisted and open retroperitoneal lymph node dissections. World J Urol 2022;40(1):119–26.

[13] Tselos A, Moris D, Tsilimigras DI, Fragkiadis E, Mpaili E, Sakarellos P, et al. Robot-assisted retroperitoneal lymphadenectomy in testicular Cancer treatment: a systematic review. J Laparoendosc Adv Surg Tech 2018;28(6):682–9.

[14] Cheney SM, Andrews PE, Leibovich BC, Castle EP. Robot-assisted retroperitoneal lymph node dissection: technique and initial case series of 18 patients. BJU Int 2015;115(1):114–20.

[15] Stepanian S, Patel M, Porter J. Robot-assisted laparoscopic retroperitoneal lymph node dissection for testicular Cancer: evolution of the technique. Eur Urol 2016;70(4):661–7.

[16] Calaway A, Einhorn L, Masterson T, et al. Adverse surgical outcomes associated with robotic retroperitoneal lymph node dissection among patients with testicular cancer. Eur Urol 2019;76:607–9.

[17] Porter J, Eggener S, Castle E, Pierorazio P. Recurrence after RoboticRetroperitoneal lymph node dissection raises more questions than answers. Eur Urol 2019;76(5):610–1.

# Urology robotic inguinal lymphadenectomy

*Marcos Tobias-Machado*[a,b], *Idvaldo Salazar-M-Messias*[a,c], *René Sotelo*[c], *and Luis G. Medina*[c]
*Penile Cancer Collaborative Coalition Latin-America*
[a]Department of Urology, Dr Arnaldo Vieira de Carvalho Cancer Institute, São Paulo, SP, Brazil [b]ABC Medical School, Santo André, SP, Brazil [c]Catherine and Joseph Aresty Department of Urology, USC Institute of Urology, Keck School of Medicine, University of Southern California, Los Angeles, CA, United States

## Overview of penile cancer (PC)

PC is rare in developed countries but more prevalent in less developed regions such as Africa, Asia, and South America where it can account for 10%–20% of all male malignancies [1,2]. Typically, it affects men with an average age of 55–58 years [3]. Nevertheless, around 19% of patients are below the age of 40, and 7% are below the age of 30 [4]. Greater rates have been seen in underdeveloped countries, such as Uganda (2.8/100,000) and areas of Brazil (1.5–3.7/100,000). Studies have describe. Maranhão, both within Brazil and globally, records the highest rates of PC. The tumors are typically in an advanced stage at the time of diagnosis, and there is a concerning prevalence among young individuals. The patients' low socioeconomic status poses challenges in completing treatment and accessing necessary follow-up care [5].

Risk factors for PC include genital warts, penile tears, chronic penile rash, lichen sclerosis (also known as balanitis xerotica obliterans), penile injuries, and phimosis. Human papillomavirus is identified in 50% of cases, while human immunodeficiency virus increases the risk by 4–8 times. Tobacco exposure makes it three times more likely and is also a risk factor [6]. A recent study confirmed that zoophilia is a risk factor for PC. Men who engage in sexual activities with animals are found to have twice the likelihood of developing this type of cancer. The researchers suggest that micro-traumas to the penis and contact with animal secretions may play a role in increasing the risk of this type of cancer [7].

PC is known for it early spread, initially, through the superficial nodes followed by profound spread to the pelvic lymphatic system; it is widely acknowledged that the presence of metastatic disease in the inguinal or pelvic lymph nodes is indicative of a poor prognosis. As a result, surgical staging and resection of the inguinal and pelvic areas play a crucial role in the multidisciplinary management of this condition [8]. Despite satisfactory survival rates, the open surgery is associated with high rates of surgical complications such as skin necrosis, wound dehiscence, lymphorrhea, infection, and lymphedema [9]. Designed to minimize complications, particularly incision-related complications, R-VEIL was introduced with the objective of decreasing local complications while preserving oncological outcomes. R-VEIL has shown to reduce morbidity, shorten hospital stays [10], and minimize blood loss when compared to open inguinal lymphadenectomy [11].

## PC and inguinal spread

PC typically spreads to the inguinal lymph nodes, with a low incidence of distant metastases (1%–2%). Approximately 20% of patients with nonpalpable inguinal nodes may have hidden metastases. The first group of draining lymph nodes is in the superficial and deep inguinal region, while the second group extends to the ipsilateral pelvic area and retroperitoneum [3].

Patients with low-risk PC (Tis, Ta, T1G1, No LVI) have a lower metastatic rate to lymph nodes. Intermediate-risk patients (T1G2) have a higher rate of 17%–50%, and high-risk patients (T1G3, >T2, LVI) have a metastatic rate of 68%–73% [12].

Radical resection of inguinal metastases is the standard treatment, and the extent of lymph node involvement is a significant predictor of survival in penile squamous cell carcinoma [3]. The number of affected nodes impacts the five-year survival rate, with men who have one involved node (N1) having a 100% three-year disease-specific survival, while those with over five positive nodes have 0% survival rates in 5 years [3].

Conventional open radical inguinal lymphadenectomy has a high morbidity rate with complications ranging between 50% and 90%.

## Recommendations for inguinal lymphadenectomy in PC

The management of regional lymph nodes is crucial in PC. Radical lymphadenectomy is preferred and can cure the disease confined to regional nodes. However, the decision to perform inguinal lymphadenectomy should be based on selective indications, considering the patient's risk group and surgical morbidity [13,14].

### Patients with nonpalpable or visible inguinal lymph nodes (cN0)

In clinically negative cases, up to 25% of patients may have micro metastatic disease, requiring invasive lymph node staging. Inguinal lymphadenectomy is recommended for intermediate- and high-risk (pT2 or higher) patients. Superficial inguinal lymphadenectomy is an option, when these lymph nodes are negative at frozen section biopsies, and radical lymphadenectomy should be considered if one pathological lymph node is found without extranodal extension [15].

### Patients with clinically positive inguinal lymph nodes (cN1/cN2)

Patients with palpable inguinal nodes should undergo prompt lymph node surgery. In these cases, a radical inguinal lymphadenectomy is the preferred treatment approach. If the primary lesion is of high risk, a complete inguinal lymphadenectomy with a contralateral lymphadenectomy is recommended. In patients with two or more inguinal nodes positive and greater than or equal to 2 cm, or when one inguinal node is positive with extra nodal extension and greater than or equal to 1 cm, an ipsilateral pelvic lymph node dissection is recommended [16].

### Patients with fixed inguinal nodes

In cases of fixed inguinal nodes, a combination of chemotherapy and surgery is often recommended. Radical inguinal lymphadenectomy may be necessary for metastatic disease in mobile lymph nodes that are ≥4 cm, while positive lymph nodes that are ≥2 cm or have extranodal extension may require pelvic lymphadenectomy and adjuvant chemotherapy. For lymph nodes that are ≥4 cm and fixed or mobile bilateral, neoadjuvant chemotherapy may be used, preferentially after a positive fine needle aspiration result. If there is a good response to chemotherapy, inguinal and pelvic lymphadenectomy may follow [17].

### Additional applications aside from PC

Inguinal lymphadenectomy (IL) has also been indicated in vulvar cancer, anal cancer, melanoma, and other skin neoplasms, where R-VEIL has been reported to be safe and effective [14].

## The advancement of minimally invasive techniques for inguinal lymph node dissection (ILND)

In 2002, Ian M. Thompson and Jay T. Bishop proposed an endoscopic and subcutaneous approach using laparoscopic techniques for this procedure. They described the endoscopic subcutaneous modified inguinal lymphadenectomy procedure in a cadaveric model, which identified anatomical structures and demonstrated overall feasibility [18].

In 2006, Tobias-Machado et al. reported the first successful VEIL procedure in humans. The group went on to conduct a study in 2007, comparing VEIL and open surgery in 10 patients (one side for each technique). The study demonstrated reduced complications (20% vs. 70%), shorter hospital stays, favorable cosmetic outcomes, and satisfactory short-term oncological results. In the same year, Sotelo et al. also reported their initial experience with VEIL, achieving similar results [11,19,20].

A meta-analysis of 10 comparative studies found that VEIL has clear advantages over other techniques in terms of skin events, lymphedema, and severe complications. Further prospective evaluation is required to establish robotics efficacy compared to traditional endoscopic procedures. This chapter describes the surgical technique for R-VEIL.

## Endoscopic visualization of the anatomy of the femoral triangle region

Video endoscopic surgery is an effective option for treating inguinal disease, with the retrograde technique being the most commonly used approach. The initial landmarks to identify during video endoscopic inguinal surgery include the boundaries of the skin, Camper's fascia, and Scarpa's fascia. The $CO_2$ gas insufflation on a 12 mm port located at the apex of the femoral triangle and a 0 degree optic are inserted, and blunt optic dissection is utilized to separate the skin from the underlying lymphatic and vascular structures. The sartorius muscle forms the lateral limit of the femoral triangle, the adductor longus muscle the medial limit, and the inguinal ligament serves as the superior boundary [19,20].

Both Camper's fascia and Scarpa's fascia are easily identifiable through external palpation. After crossing the medial boundary, the saphenous vein and great saphenous vein become visible. In patients without palpable lymph nodes, the saphenous vein can often be spared. It is crucial to perform dissection beneath Scarpa's fascia, preserving subcutaneous adipose tissue and minimizing energy dissipation to prevent skin devascularization. As dissection progresses, lymph nodes can be identified and separated from the skin for inclusion in the surgical specimen (Fig. 1).

Performing proximal dissection will eventually reveal the fossa ovalis, along with the accessory saphenous vein and other saphenous tributaries. Under normal conditions, seven tributaries drain into the femoral vein in the fossa ovalis (see Fig. 2 for schematic and Fig. 3 for an endoscopic view). The junction between the femoral and saphenous veins can be located externally about two fingerbreadths lateral and two fingerbreadths inferior to the pubic tubercle. The saphenous travels from the superficial inguinal region to the fossa ovalis. The inguinal lymph nodes are separated into superficial and deep groups by the fascia lata. The superficial inguinal lymph nodes are located in the deep

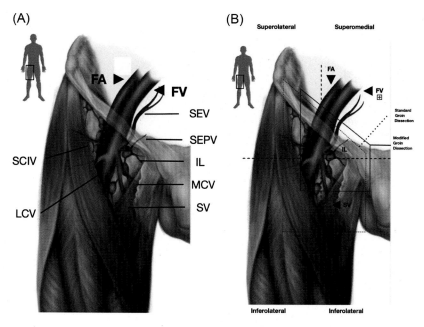

FIG. 1 Inguinal anatomy. (A) and (B): *FA*, femoral artery; *FT*, femoral triangle; *IL*, inguinal ligament; *LCV*, lateral cutaneous vein; *MCV*, medial cutaneous vein; *SCIV*, superior circumflex inferior vein; *SEPV*, superior epigastric posterior vein; *SEV*, superior epigastric vein; *SV*, saphenous vein.

FIG. 2   (A) Retrograde dissection from the femoral triangle vertex until the inguinal ligament preserving fat under the skin. (B) Removing all lymph nodes inside the template.

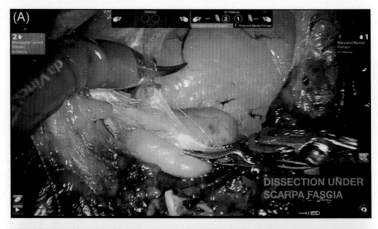

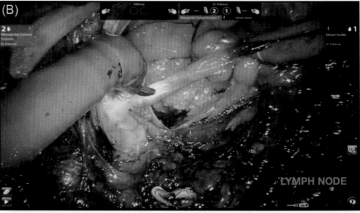

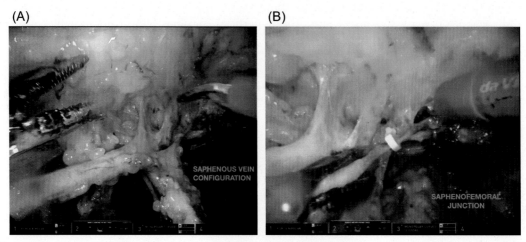

FIG. 3   (A) Endoscopic view of tributaries of the saphena cross. (B) Saphenofemoral junction.

membranous layer of the superficial fascia of the thigh, commonly consisting of 4–25 nodes. These superficial inguinal nodes have been classified into five anatomical groups by Daseler [3,21] (Fig. 3):

1. Group I: Superomedial nodes located around the superficial external pudendal and superficial epigastric veins.
2. Group II: Superolateral nodes located around the superficial circumflex vein.
3. Group III: Inferolateral nodes located around the lateral femoral cutaneous and superficial circumflex veins.
4. Group IV: Inferomedial nodes located around the greater saphenous vein.
5. Group V: Central nodes located around the saphenofemoral junction.

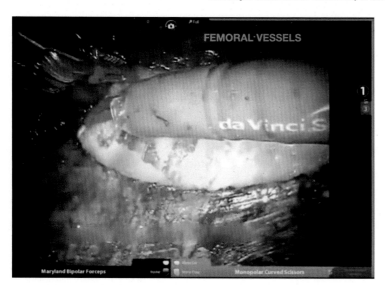

FIG. 4 The fascia lata is opened exactly in the area that covers the femoral artery to gain access to deep lymph nodes.

The majority of lymph nodes are positioned above the fascia lata, particularly on the medial side of the saphenous-femoral junction and can be identified by their coloration, which is usually brown or green. It is crucial to ensure that all the areolar tissues between the skin, inguinal cord, and saphenous vein are included during dissection. Preoperative ultrasound guidance and palpation can be advantageous in marking the skin above the most prominent nodes. The structures located under the fascia lata within the femoral triangle, from lateral to medial, include the femoral nerve, femoral artery, femoral vein, and deep inguinal lymph nodes, with the latter three situated inside the femoral sheath [3]. To access the deep node compartment, where the deep inguinal nodes are located, it is necessary to section the fascia lata over the femoral artery pulse (Fig. 4). This involves resecting all nodal and areolar tissues medial to the femoral vein and lateral to the adductor longus muscle until Cloquet's node, located inside the femoral channel, is identified. Identification of Cloquet's node signifies completion of this resection and identification of the crucial structures within the femoral triangle.

According to Daseler, the standard or full template for ILND involves the removal of both superficial and deep inguinal nodes from the femoral triangle. The dissection is limited to the lateral side of the femoral artery to avoid damage to the femoral nerve. Additionally, this template includes the ligation of the saphenous vein as it emerges from the femoral vein [21].

In a modified template differs from the standard or full template in that it involves the superficial dissection of the 1, a portion of 2 and 5 lymph node groups. The inferior group nodes are left undisturbed, while the deep dissection remains the same. This modified approach has been shown to reduce the risk of complications, which can also be further minimized by using a minimally invasive approach [22] (Fig. 5).

(A)  (B)  (C)

FIG. 5 Important structures preserved after resection of inguinal lymph nodes: (A) inguinal ligament, (B) femoral vein, and (C) femoral artery.

## Preoperative

As mentioned earlier, palpable inguinal lymph nodes may be associated with inflammatory conditions in 30%–50% of cases. To reduce the risk of postoperative inguinal infections, a four-week course of antibiotics is typically prescribed after penectomy, along with intraoperative prophylactic antibiotics. The most commonly used broad-spectrum antibiotics include ampicillin with aminoglycoside or ciprofloxacin. Ideally, the time interval between penectomy and ILND is 6 weeks.

Due to the high risk of deep venous thrombosis in PC patients, we recommend the routine use of low-molecular-weight heparin, starting the night before surgery and continuing while the patient is in bed. We also strongly encourage early ambulation. However, the use of heparin may increase the risk of wound hematoma and serous wound drainage due to the ongoing extravasation of the lymph [23].

## R-VEIL surgical steps

### Patient positioning

The patient is positioned in a low lithotomy position. To perform a right-sided dissection, the assistant stands lateral to the right leg, while for the left side, they stand between the legs. To guide trocar placement and identify the extent of dissection, bony and soft tissue landmarks are marked on the skin surface. These markings create an inverted triangle with the base forming a line connecting the anterior superior iliac spine to the pubic tubercle, following the course of the inguinal ligament. The lateral boundary is the sartorius muscle angling toward the apex, while the medial limit is the adductor longus muscle, extending toward the apex.

Initially, we positioned the robot on the right side of the table as it was an SI system, without the need for repositioning, only instrument re-docking. Next, a 3-cm medial incision was made inside the region of the sartorius muscle joint with the adductor longus in the femoral triangle.

The white subcutaneous layer corresponding to Scarpa's fascia is identified, and finger dissection is performed to create skin flaps at the apex of the triangle in both directions, allowing for additional ports to be placed. Digital dissection is performed toward the inguinal ligament, with a combination of blunt finger dissection and lateral movements toward the upper limit. This technique is used to dissect the potential space beneath Scarpa's fascia for the superficial plane of lymph nodes (Fig. 6).

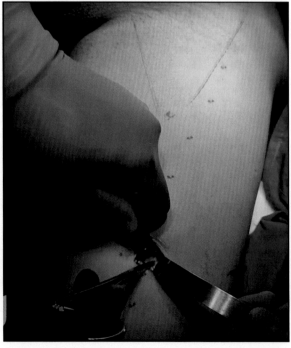

FIG. 6   Initial skin incision under the Scarpa's fascia distal for vertex of femoral triangle and finger dissection to create working space.

## Trocar placement and docking

We create space by performing three additional incisions, followed by placement of two robotic 8-mm ports in a triangular fashion using finger-guided techniques, one laterally and one medially. Additionally, an optional 5- to 12-mm port may be placed between the scope and medial robotic port for the assistant. The scope is used to sweep the fat and extend the subcutaneous workspace under the Scarpa's fascia, creating a superficial subcutaneous flap.

Simultaneously, both legs are prepared for the robot docking as the description proceeds. The da Vinci X or Xi system allows for smaller incisions and quick camera switching to another port, resulting in a faster and more optimal position in the center of the patient's abducted legs. In all da Vinci systems, we use a 0-degree lens, and after the ports are placed, we maintain a $CO_2$ insufflation pressure of 10 mmHg (as shown in Fig. 7).

For the first procedure, the robot is positioned at a 45-degree angle opposite to the affected side (left) and on the same side (right) for the second procedure. The assistant stands beside the patient's knees (refer to Fig. 8 for visualization).

## Anterior, posterior, and lateral dissection

After the docking is complete, the 0-degree lens scope is positioned, and all instruments are inserted while being visually guided. Various instruments such as the bipolar Maryland or forceps can be utilized in the left robotic arm, while monopolar scissors are used in the right arm to dissect the lymphatic and membranous tissues located beneath the Camper's fascia. The objective is to meticulously create an anterior working space along the inguinal ligament. At the end of this process, the superior limit of the dissection is typically identified as a transverse structure with white fibers, signifying the inguinal ligament.

Using a robotic vessel sealer is a contemporary approach for dissecting and sealing vessels up to 8 mm. The packet of superficial nodes is located over the fascia lata by following the anatomical landmarks of the femoral triangle (Fig. 4). An additional proposal to detect and mark the lymph nodes is through the utilization of fluorescence. Typically, we administer an injection of 5–10 mL indocyanine near the tumor site at the onset of the procedure, and a minimally invasive resection can assist in the extraction of any dubious nodes using near infrared fluorescence during R-VEIL procedure (Fig. 9) [24].

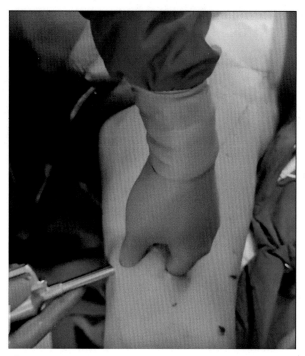

FIG. 7  Trocar placement and initial working space creation. The skin is marked with femoral triangle boundaries. Digital and optical dissections are done to create the space. Transillumination helps the surgeon to maintain dissection limits.

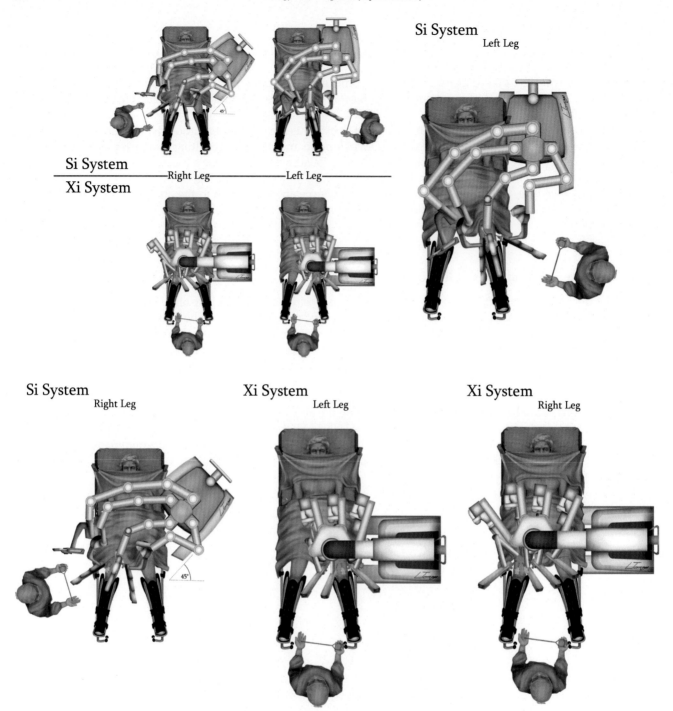

FIG. 8    *SI-and-XI-combine*: Position of assistant and comparative docking for Si and Xi. *SI_L*: Position of assistant and docking for the left leg in the Si. *SI_R*: Position of assistant and docking for the right leg in the Si. *Xi_L*: Position of assistant and docking for the left leg in the Xi. *Xi_R*: Position of assistant and docking for the right leg in the Xi.

The nodal tissue can be gently rolled inward on both sides using blunt dissection. This technique is then continued inferiorly and on both sides until the inferior apex of the nodal packet is defined. The saphenous vein is identified at the internal border of the dissection near the apex of the femoral triangle, and the surgeon can follow it to the saphenous arch, where it joins the superficial femoral vein at the fossa ovalis. Sharp and blunt dissection are used to dissect the packet away from the fascia lata as the dissection continues superiorly.

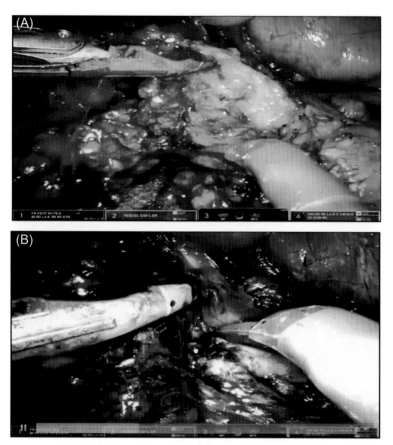

FIG. 9   (A) Superficial nodal package observed without fluorescence. (B) With fluorescence, nodes and lymphatic channels are easy to observe versus normal vision.

The nodal packet is typically lifted by the nondominant hand while the dominant hand uses monopolar scissors to advance the dissection. Once the fossa ovalis is encountered, the packet is dissected away at its superolateral and superomedial limits, narrowing it and pulling it away from the inguinal ligament. This is where the superficial and deep planes of dissection join and separate the package from the inguinal ligament. At this point, venous tributaries are clipped as the nodal packet is circumferentially dissected, except for its attachments to the saphenous arch.

The pulsations of the femoral artery in close proximity serve as a distinctive reference point. If feasible, efforts will be made to separate the packet from the saphenous vein. However, if this is not achievable, the vein can be ligated with Hem-o-Lok clips at the saphenous arch. Whenever possible, it is important to preserve the saphenous vein to minimize the chances of lymphedema developing after the operation.

For deep node dissection, we cut through the fascia lata using trocars placed laterally to the adductor longus and sartorius muscles, allowing us to work beneath the fascia. $CO_2$ insufflation provides a working space, and we use the middle of the muscles to locate the femoral nerve and vessels before removing the femoral nodes. Inferomedial dissection is performed around the femoral vein to enable the resection of the deep inguinal nodes, continuing up to the level of the femoral canal until the pectineus muscle is visible to ensure complete nodal retrieval. Subsequently, we remove the trocars (without redocking) to remain in the superficial space.

We begin by lifting and tracking the superficial nodal packet toward the inguinal ligament, following the course of the saphenous vein and continuing up to the saphenofemoral junction to complete the superficial node dissection. Control of the saphenofemoral junction can be achieved using either metallic or 10 mm Hem-o-Lok clips. The nodal packet is then placed in a bag and removed through the initial incision or the main port under camera visualization. Finally, a drain is inserted into each leg through the 5-mm incision trocar.

## Alternative methods

Two variations of the VEIL technique have been reported. The first is the lateral leg access (L-VEIL), which was described in only one comparative study and did not show any clear advantages over the standard technique [25]. The other variation is the hypogastric subcutaneous approach (H-VEIL), which offers the advantage of using fewer trocars to perform bilateral VEIL and the same trocars to operate pelvic nodes in the same procedure, if necessary. However, a simultaneous inguinal approach is not possible with this technique, and the area of skin dissection is more extensive [26].

## Postoperative management following inguinal lymphadenectomy

Patients who undergo ILND may experience high drainage of lymphatic fluid and are advised to maintain proper hydration. Although patients are recommended to restrict saturated fat intake, there are no other dietary restrictions. Additionally, we encourage early or immediate deambulation after the procedure as use of fitted elastic stockings (6 months) and sequential compression devices.

To prevent lymphorrhea and seroma formation, postoperative suction drainage should be employed, and the drains should be removed once the 24-h output is less than 50 mL, which usually occurs between 3 and 15 days after surgery. Prophylactic antibiotics should be continued for a week after surgery or until all drains have been removed to avoid bacterial colonization and spread along the drains, which may increase the risk of infection.

## Our data and literature analysis

Between 2015 and 2020, we performed 18 cases (36 limbs) of robotic ILND; the median age was 56 years old. All patients had a penile squamous cell carcinoma diagnosis confirmed by biopsy and partial or total penectomy. The TNM for the patients was T1-3N0M0G1-3. The mean operative time was 120 min for both sides. The mean blood loss was 40 mL. None of the patients were converted to an open approach.

The total mean number of dissected lymph nodes was nine for each limb. The presence of positive lymph nodes disease was 33%. Mean hospital stay was 2 days. The mean duration of the drainage was 7 days, with three cases prolonged to 3 weeks. The most frequent complication was lymphocele treated without hospitalization (Table 1).

TABLE 1   Intra and postoperative outcomes.

| | |
|---|---|
| No of patients | 18 (36 legs) |
| Age of patient (years) | (40–74) 56 |
| Histological type | Penile SCC |
| Mean operative time (min) | (80–230) 120 |
| Mean blood loss (mL) | (15–85) 40 |
| Conversion to open | N (0) |
| Nr of dissected lymph nodes positive nodes | 6–15 (9) 6/18 (33%) |
| Mean hospital stay (in days) | 1–4 (2) |
| Main duration of drainage (in days) | 3–21 (7) |
| Overall postoperative complications (%) | 25% |

## Discussion

Compared to traditional open procedures, this minimally invasive approach has shown to result in fewer complications, particularly with regard to skin necrosis, while maintaining oncologic control. The utilization of robotic technology in R-VEIL allows for improved visibility, identification of anatomical landmarks, and greater dexterity and ergonomics. However, the implementation of R-VEIL is limited by factors such as high costs, technology availability, and the learning curve required to approach the femoral triangle and anatomical variations.

Josephson et al. reported the first case of a non-simultaneous bilateral robotic-assisted inguinal lymphadenectomy (RAIL) using the da Vinci Si robotic system [27].

In 2013, Matin et al. conducted a prospective study to evaluate the efficacy of R-VEIL for staging penile squamous cell carcinoma. Ten patients, between T1 and T3 stage with cN0-cN1, underwent the procedure [28]. The authors reported a median age of 62 years, a mean of nine nodes per limb, mean blood loss of 100 mL, and few complications. The surgeon assessed the dissection field through a small inguinal incision [29].

Several years later, Ahlawat and colleagues presented a study on the reproducibility of different techniques of simultaneous bilateral RAIL without relocating the robotic system. This technique enables bilateral staging in a single surgical time, making it a valuable option for certain patients [30].

Russell et al. reported in 2017 on the minimally invasive approach to robotic VEIL, comparing the complications between R-VEIL and VEIL techniques in a small group of patients (14 in total). They found that the R-VEIL group had a lower rate of complications (10%) than the VEIL group (40%). The most common complications were lymphocele, wound infection, and flap necrosis, which were consistent with the findings of other studies [31].

In 2015, Corona-Montes and colleagues reported the first successful bilateral RAIL procedure in Mexico. The patient was 73 years old with penile squamous cell carcinoma (T3N0M0G1) and a previous radical penectomy 4 weeks earlier. The surgery was performed using da Vinci Si, and both saphenous veins were preserved. Additionally, they collaborated with Machado et al. to publish a study of 18 patients (36 limbs) who received REIL. The patients had a mean age of 56 years old, the operative time was 120 min (80–230), and the blood loss was 40 mL. There were no conversions to open surgery, and the node dissection resulted in 6/18 positive cases of the disease. The mean hospitalization was 2 days (1–4), and the most common complication was lymphocele [1,10].

In one of the most recent studies on R-VEIL, Sign et al. compared the outcomes of robotic and open inguinal approaches [32] (Table 2). The study found a similar incidence of lymphocele but fewer cases of lymphedema, as well as higher rates of skin preservation due to smaller incisions and less traumatic tissue manipulation provided by the minimally invasive procedure. The use of clips and the three-dimensional view provided by the robotic technology may also help in avoiding lymphoceles and lymphedemas by ensuring proper closure of lymphatic vessels.

The majority of groups that have conducted robotic ILND have reported longer operative times, similar to those of VEIL, despite the quicker set-up time of the endoscopic procedure. Furthermore, most of these publications were authored by groups that had prior experience with VEIL.

Hu and colleagues conducted a comparison of 10 studies on the VEIL technique and found several benefits, including reduced intraoperative blood loss, shorter hospital stays, shorter drain usage, and a lower incidence of wound infection, skin necrosis, and lymphedema [33].

While VEIL and R-VEIL offer perioperative benefits over open lymphadenectomy, there is still ongoing debate about their oncological outcomes compared to the open approach. More well-designed studies with long-term follow-up are needed to further evaluate this aspect. However, current evidence suggests that RAIL can result in good oncological outcomes with reduced morbidity, less blood loss, and shorter hospitalization time [34–36].

## Future scenario

As the technology spreads to developing countries, where the incidence of this disease is higher, there is a growing understanding of the long-term oncological outcomes of this minimally invasive procedure.

## Conclusions

Radical resection of PC inguinal metastases is the primary treatment and the most significant predictor of survival in penile squamous cell carcinoma. Open inguinal lymphadenectomy, while effective, is associated with increased morbidity and complication rates such as flap skin necrosis, longer hospitalization, and infections. Recent retrospective studies have shown that patients who underwent robotic-assisted video endoscopic inguinal lymphadenectomy had fewer morbidity rates.

TABLE 2  Literature reports of R-VEIL.

| Author | Year | Case report/ case series | Number of patients (no. of limbs) | Age (mean years) | Penile cancer (histologic) | T stage | Pre-LND cN stage | Lymph nodes dissected (n)—mean | Operative time (min) | Blood loss (mL) | Complications |
|---|---|---|---|---|---|---|---|---|---|---|---|
| Josephson et al. | 2009 | Case report | 1 (2) | 37 | SCC[b] | T3 | cN2 | 10/9[a] | 120/130[a] | 100/50[a] | None |
| Matin et al. | 2013 | Case series | 10 (20) | 62 (58–69) | SCC[b] | T1–T3 | cN0–cN1 | Left, 9; right, 9 | 180–240 | 100 (mean) | Cellulitis (1/10), wound breakdown (1/10), skin necrosis (1/10) |
| Sotelo et al. | 2013 | Case report | 1 (2) | 64 | SCC[b] | T3 | cN0 | 33 | 360 | 100 (10–200) | Lymphocele |
| Ahlawat et al. | 2016 | Case series | 3 (6) | 56 | SCC[b] | T2–T3 | cN1–cN2 | Left, 18; right,14 | 453 | 147 (mean) | Lymphocele (1/3) |
| Russel et al. | 2017 | Case series | 14 (27) | 72 (62–76) | SCC[b] | T1–T3 | cN0–cN2 | 8 | 136.8 | 50 (15–50) | (3/14) Lymphocele, wound infection, flap necrosis |
| Corona-Montes et al. | 2018 | Case series | 12 (24) | 58 | SCC[b] | T2–T3 | cN1–cN2 | 12 | 110 | 59 (mean) | Lymphocele (2/12) |
| Singh et al. | 2018 | Case series | 51 (102) | 58 (50–68) | SCC[b] | T1–T3 | cN0–cN2 | 12 | 75/per limb | 75 (65–80) | Edge necrosis, flap necrosis (2%), lymphocele (23%) |

[a] R-VEIL performed in two separate procedures (one OR time per limb).
[b] Squamous cell carcinoma of the penis.

# Key points

- Radical resection of PC inguinal lymph node metastases is the primary locoregional treatment and the most significant predictor of survival in penile squamous cell carcinoma.
- PC is not confined to one side and can metastasize to either groin regardless of the primary lesion's location. However, PC does not spread to the pelvic nodes without prior involvement of the inguinal nodes.
- R-VEIL may have limitations, such as high costs, technology availability, and the learning curve to approach the femoral triangle and anatomical variations.
- R-VEIL leads to shorter recovery time, improved disease-free survival, and faster return to normal activities.

# References

[1] Ferlay J, Ervik M, Lam F, Colombet M, Mery L, Piñeros M, et al. Global cancer observatory: cancer today. vol. 3(20). Lyon, France: International Agency for Research on Cancer; 2018. p. 2019.

[2] Corona-Montes V, Moyo-Martínez E, Almazán-Treviño L, Ríos-Dávila V, Santiago-Hernández Y, Mendoza-Rojas E. Linfadenectomía inguinal robot asistida (LIRA) para cáncer de pene. Rev Mex Urol 2015;75(5):292–6.

[3] Partin AW, Wein AJ, Kavoussi LR, Peters CA. Campbell-Walsh urology E-book. Elsevier Health Sciences; 2015.

[4] Favorito LA, Nardi AC, Ronalsa M, Zequi SC, Sampaio FJ, Glina S. Epidemiologic study on penile cancer in Brazil. Int Braz J Urol 2008;34:587–93.

[5] Coelho RWP, Pinho JD, Moreno JS, do Nascimento AMT, Larges JS, Calixto JRR, et al. Penile cancer in Maranhão, Northeast Brazil: the highest incidence globally? BMC Urol 2018;18(1):1–7.

[6] Daling JR, Madeleine MM, Johnson LG, Schwartz SM, Shera KA, Wurscher MA, et al. Penile cancer: importance of circumcision, human papillomavirus and smoking in in situ and invasive disease. Int J Cancer 2005;116(4):606–16.

[7] Zequi SC, Guimarães GC, da Fonseca FP, Ferreira U, de Matheus WE, Reis LO, et al. Sex with animals (SWA): behavioral characteristics and possible association with penile cancer. A multicenter study. J Sex Med 2012;9(7):1860–7.

[8] Ficarra V, Akduman B, Bouchot O, Palou J, Tobias-Machado M. Prognostic factors in penile cancer. Urology 2010;76(2):S66–73.

[9] Stuiver MM, Djajadiningrat RS, Graafland NM, Vincent AD, Lucas C, Horenblas S. Early wound complications after inguinal lymphadenectomy in penile cancer: a historical cohort study and risk-factor analysis. Eur Urol 2013;64(3):486–92.

[10] Corona-Montes VE, Gonzalez-Cuenca E, Tobias-Machado M. Robotic-assisted inguinal lymphadenectomy (RAIL). Medical robotics-new achievements. IntechOpen; 2019.

[11] Tobias-Machado M, Tavares A, Ornellas AA, Molina Jr WR, Juliano RV, Wroclawski ER. Video endoscopic inguinal lymphadenectomy: a new minimally invasive procedure for radical management of inguinal nodes in patients with penile squamous cell carcinoma. J Urol 2007;177(3):953–8.

[12] Hegarty P, Dinney C, Pettaway C. Controversies in ilioinguinal lymphadenectomy. Urol Clin 2010;37(3):421–34.

[13] OW H. Penile cancer – European Association of Urology Guidelines 2018. Available from: https://uroweb.org/guideline/penilecancer/.

[14] Thomas A, Necchi A, Asif M, Tobias-Machado M, Huyen TAT, Spiess PE, et al. Penile cancer (Primer). Nat Rev Dis Primers 2021;7(1).

[15] Kulkarni JN, Kamat MR. Prophylactic bilateral groin node dissection versus prophylactic radiotherapy and surveillance in patients with N (0) and N (1-2A) carcinoma of the penis. Eur Urol 1994;26(2):123–8.

[16] Lughezzani G, Catanzaro M, Torelli T, Piva L, Biasoni D, Stagni S, et al. Relationship between lymph node ratio and cancer-specific survival in a contemporary series of patients with penile cancer and lymph node metastases. BJU Int 2015;116(5):727–33.

[17] Pizzocaro G, Nicolai N, Milani A. Taxanes in combination with cisplatin and fluorouracil for advanced penile cancer: preliminary results. Eur Urol 2009;55(3):546–51.

[18] Bishoff JT. Endoscopic subcutaneous modified inguinal lymph node dissection for squamous cell carcinoma of the penis. In: Smith Arthur D, editor. Smith's textbook of endourology. Blackwell Publishing Ltd; 2012. p. 917–23. https://doi.org/10.1002/9781444345148.ch78.

[19] Tobias-Machado M, Tavares A, Silva MNR, Molina J, Rica W, Forseto PH, Juliano RV, et al. Can video endoscopic inguinal lymphadenectomy achieve a lower morbidity than open lymph node dissection in penile cancer patients? J Endourol 2008;22(8):1687–92.

[20] Sotelo R, Sánchez-Salas R, Carmona O, Garcia A, Mariano M, Neiva G, et al. Endoscopic lymphadenectomy for penile carcinoma. J Endourol 2007;21(4):364–7.

[21] Daseler E, Anson B, Reimann A. Radical excision of the inguinal and iliac lymph glands; a study based upon 450 anatomical dissections and upon supportive clinical observations. Surg Gynecol Obstet 1948;87(6):679–94.

[22] Levi D'Ancona CA, Goncalves de Lucena R, de Oliveira A, Querne F, Tavares Martins MH, Denardi F, Rodrigues Netto NJ. Long-term followup of penile carcinoma treated with penectomy and bilateral modified inguinal lymphadenectomy. Commentarie: J Urol 2004;172(2):498–501.

[23] Clark PE, Spiess PE, Agarwal N, Biagioli MC, Eisenberger MA, Greenberg RE, et al. Penile cancer. J Natl Compr Canc Netw 2013;11(5):594–615.

[24] Sávio LF, Barboza MP, Alameddine M, Ahdoot M, Alonzo D, Ritch CR. Combined partial penectomy with bilateral robotic inguinal lymphadenectomy using near-infrared fluorescence guidance. Urology 2018;113:251.

[25] Nayak SP, Pokharkar H, Gurawalia J, Dev K, Chanduri S, Vijayakumar M. Efficacy and safety of lateral approach-video endoscopic inguinal lymphadenectomy (L-VEIL) over open inguinal block dissection: a retrospective study. Indian J Surg Oncol 2019;10:555–62.

[26] Wang H, Li L, Yao D, Li F, Zhang J, Yang Z. Preliminary experience of performing a video endoscopic inguinal lymphadenectomy using a hypogastric subcutaneous approach in patients with vulvar cancer. Oncol Lett 2015;9(2):752–6.

[27] Josephson DY, Jacobsohn KM, Link BA, Wilson TG. Robotic-assisted endoscopic inguinal lymphadenectomy. Urology 2009;73(1):167–70.

[28] Matin SF, Cormier JN, Ward JF, Pisters LL, Wood CG, Dinney CP, et al. Phase 1 prospective evaluation of the oncological adequacy of robotic assisted video-endoscopic inguinal lymphadenectomy in patients with penile carcinoma. BJU Int 2013;111(7):1068–74.

[29] Sotelo R, Cabrera M, Carmona O, de Andrade R, Martin O, Fernandez G. Robotic bilateral inguinal lymphadenectomy in penile cancer, development of a technique without robot repositioning: a case report. Ecancermedicalscience 2013;7.

[30] Ahlawat R, Khera R, Gautam G, Kumar A. Robot-assisted simultaneous bilateral radical inguinal lymphadenectomy along with robotic bilateral pelvic lymphadenectomy: a feasibility study. J Laparoendosc Adv Surg Techn 2016;26(11):845–9.

[31] Russell CM, Salami SS, Niemann A, Weizer AZ, Tomlins SA, Morgan TM, et al. Minimally invasive inguinal lymphadenectomy in the management of penile carcinoma. Urology 2017;106:113–8.

[32] Singh A, Jaipuria J, Goel A, Shah S, Bhardwaj R, Baidya S, et al. Comparing outcomes of robotic and open inguinal lymph node dissection in patients with carcinoma of the penis. J Urol 2018;199(6):1518–25.

[33] Hu J, Li H, Cui Y, Liu P, Zhou X, Liu L, et al. Comparison of clinical feasibility and oncological outcomes between video endoscopic and open inguinal lymphadenectomy for penile cancer: a systematic review and meta-analysis. Medicine 2019;98(22).

[34] Tobias-Machado M, Moschovas MC, Ornellas AA. History of minimally invasive inguinal lymphadenectomy. In: Delman K, Master V, editors. Malignancies of the groin: surgical and anatomic considerations. Springer, Cham; 2018. p. 1–7. https://doi.org/10.1007/978-3-319-60858-7_1.

[35] Tobias-Machado M, Almeida-Carrera RJ. Video endoscopic inguinal lymphadenectomy. In: Muneer A, Horenblas S, editors. Textbook of penile cancer. Cham: Springer; 2016. p. 207–19. https://doi.org/10.1007/978-3-319-33220-8_15.

[36] Tobias-Machado M, Moschovas MC. Inguinal lymphadenectomy. In: Sotelo R, Arriaga J, Aron M, editors. Complications in robotic urologic surgery. Cham: Springer; 2018. p. 305–10. https://doi.org/10.1007/978-3-319-62277-4_32.

CHAPTER

# 46

# Urology: Female robotic reconstructive surgery
## Surgical technique

*Paul J. Oh and Howard Goldman*

Glickman Urological and Kidney Institute, Cleveland Clinic, Cleveland, OH, United States

## Robotic abdominal sacrocolpopexy

The robotic sacrocolpopexy is typically performed with the patient in a modified dorsolithotomy position. When using a *da Vinci Si* robotic system (Intuitive Surgical, Inc., Sunnyvale, CA), a standard inverted "W" port placement pattern is utilized (Fig. 1). When using a *da Vinci Xi* system, all ports are in a transverse line about 2 cm above the umbilicus. We utilize 8 mm ports throughout for the robotic arms. With newer systems available in some locales, port placement may differ. After port placement, the patient is placed in steep Trendelenburg (Fig. 2).

In patients with a uterus, our preference is to perform a concomitant supracervical hysterectomy and bilateral salpingectomy. A total hysterectomy (robotic, laparoscopic, or vaginal approach) can also be performed as data suggests that there are no differences in intraoperative complications, overall recurrent prolapse rates, or patient-reported symptom improvement [1,2]. Some authors suggest a higher incidence of anterior prolapse recurrence with supracervical hysterectomy [1,2]. However, performing a total hysterectomy must be balanced against the potential increase in mesh extrusion risk at the vaginal cuff in light of conflicting reports regarding mesh erosion rates [1–4]. If a patient prefers to retain her uterus, we perform a sacrohysteropexy.

We begin the robotic sacrocolpopexy by placing a grasping forcep (EndoWrist ProGrasp Forceps, Intuitive Surgical) in the far left arm (third arm), a bipolar forceps (EndoWrist Cautery Fenestrated Bipolar Forceps, Intuitive Surgical) in the left arm, and a monopolar scissor (EndoWrist Cautery Hot Shears, Intuitive Surgical) in the right arm. A 12 mm assistant port is utilized as well with port placement lateral to the right arm.

The small bowel is gently moved out of the pelvis cephalad and the sigmoid is retracted to the left. We identify the site where the fifth lumbar body ends and drops down to the sacral promontory. The peritoneum overlying this is incised between and caudad to the iliac vessels. The anterior longitudinal ligament is then carefully identified, and an area is cleared off for the upcoming mesh attachment.

We then carefully dissect the bladder off of the vagina to the level of the intertrigonal ridge anteriorly and dissect posteriorly as well (Fig. 3). A vaginal retractor such as an exam under anesthesia metal sound or flat retractor blade can be used by an assistant to displace the anterior and posterior fornices. This can aid the dissection of the vagina from the bladder and rectum. If we are performing a concomitant supracervical hysterectomy, we first separate the fallopian tubes and the uterus from their lateral attachments using cautery. Staying close to the uterus during this step is crucial to avoid trauma to the nearby ureters. We then use the uterus for retraction to perform the anterior (and sometimes the posterior) dissection of the vaginal wall before amputating across the cervix. If a total hysterectomy is performed, it is critical to perform a high-quality multilayer closure of the vaginal cuff prior to attaching the mesh.

Once the dissection is complete, we tailor a lightweight, synthetic, wide pore polypropylene "Y" mesh (some use two separate strips of mesh) such that the vagina will be suspended without tension. Some have started to use autologous fascia for this. We prefer to attach the mesh to the vagina with delayed absorbable sutures, but nonabsorbable monofilament suture may also be used. There are conflicting reports of higher mesh and suture exposure noted with nonabsorbable monofilament [5–8]. Some surgeons prefer to utilize a barbed, absorbable monofilament suture (such as Medtronic V-Loc, Minneapolis, MN) to sew continuously in a Z pattern to avoid utilizing multiple sutures. The proximal end of the mesh is sutured to the anterior longitudinal ligament with a permanent suture (Fig. 4). For this step, a

*Handbook of Robotic Surgery*
https://doi.org/10.1016/B978-0-443-13271-1.00011-X

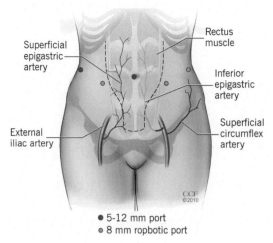

FIG. 1    The camera port is placed about 2 cm cephalad to the umbilicus. The left and right arm ports are positioned about 8–9 cm inferolateral to the camera port. The assistant port is placed the same distance, superolateral to the right arm port at the level of the camera port. An optional "third arm" is placed superolateral to the left arm port at the level of the camera.

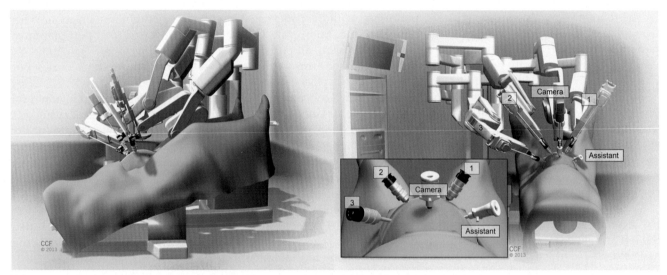

FIG. 2    After port placement has been completed, the patient should be placed in Trendelenburg position to displace bowel cephalad and aid the proximal dissection of the anterior longitudinal ligament. Then, the robot is docked and instruments loaded.

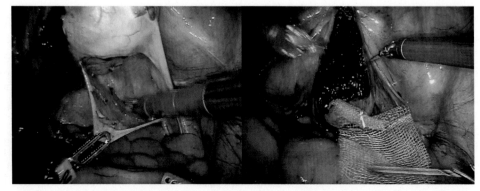

FIG. 3    *Left*—the posterior vaginal plane is developed bluntly after initial incision of the posterior vaginal wall peritoneum. *Right*—the anterior vaginal plane has been developed and the Y-mesh is introduced into the surgical field for subsequent attachment to the vaginal cuff. *Surgeon: Howard Goldman MD.*

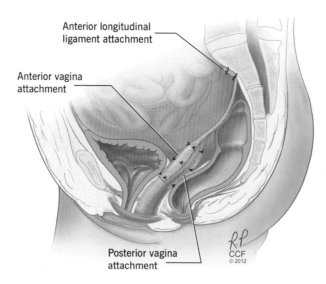

FIG. 4 The sacrocolpopexy mesh has been attached to the anterior and posterior vagina distally. Proximally, the mesh has been fixed to the anterior longitudinal ligament. Adequate tension of this mesh results in prolapse reduction.

FIG. 5 Peritoneal closure over the mesh is performed to decrease the risk of bowel obstruction, adhesion formation, and mesh erosion. *Surgeon: Howard Goldman MD.*

delayed absorbable suture has also been shown to be effective [9]. It is critical not to overtension the mesh. The peritoneum is closed over the mesh (Fig. 5). If a hysterectomy was performed, the specimen is removed in an EndoCatch bag (Medtronic, Minneapolis, MN) through the camera port, and the procedure is completed. Our practice is to additionally perform a concomitant rectocele repair/perineorrhaphy when the genital hiatus is larger than 4 cm. This may help address and/or prevent posterior prolapse, as there is evidence that the majority of recurrence reoperations require a rectocele repair [10].

Both objective and subjective outcomes of the robotic sacrocolpopexy have been compared with the open abdominal sacrocolpopexy and the laparoscopic sacrocolpopexy. Improvements in prolapse cure rates and pelvic floor function are comparable between the open abdominal and robotic sacrocolpopexy [11,12]. Additionally, there are no differences in intraoperative blood loss, postoperative complications, postoperative stress incontinence, mesh exposure, and prolapse recurrence rates among the open abdominal, laparoscopic, and robotic sacrocolpopexy [11,12]. Similarly, there are minimal differences in outcomes between laparoscopic and robotic sacrocolpopexy. Robotic sacrocolpopexy may be associated with lower blood loss, intraoperative complications, and conversion rate but longer operating time than laparoscopic sacrocolpopexy [13]. Of note, minimally invasive approaches may allow for a shorter hospital length of stay compared to the open approach. We found that same-day discharge is both feasible and safe after robotic-assisted pelvic floor reconstruction surgeries (predominantly robotic sacrocolpopexy) while maintaining equivalent patient satisfaction compared to patients who were hospitalized after surgery [14]).

Robotic sacrocolpopexy specifically has high rates of medium-term and long-term anatomic success with success rates approaching 99% at over 2 years of follow-up. Reoperation rates for persistent or recurrent prolapse are low at about 6% [10]. In the case of rare apical prolapse recurrence, sacrospinous ligament fixation or total vaginal mesh repair may be considered. Mesh erosion rates range from 4% to 8% and can sometimes be managed with local topical estrogen therapy [10–12]. However, these erosions may sometimes require transvaginal or transabdominal mesh removal. The most common intraoperative complication is cystotomy, which occurs in about 3% of cases [10,15]. Intraoperative blood loss typically ranges about 50–150 mL but is often minimal [11,16]. Additionally, with robotic sacrocolpopexy, large improvements are seen in quality of life while improving dyspareunia and maintaining sexual activity after postoperative recovery.

Data comparing the Da Vinci single port SP robotic system with the multiport robot for robotic sacrocolpopexy show no significant differences in intraoperative blood loss, postoperative pain, or improvements in short-term anatomic and quality of life outcomes [17]. Additionally, there may be evidence that there are advantages to utilizing the single port robotic system with quicker robot docking and faster cervix suturing in addition to a more favorable incision compared to the multiport robot [18]. Therefore, the single port robotic approach may become an increasingly popular option to perform the robotic sacrocolpopexy if long-term outcomes are sustained.

## Sacrohysteropexy

If a sacrohysteropexy is to be performed to preserve the uterus, multiple techniques may be used. One option is to dissect the vagina and attach an anterior and posterior mesh to the vagina. The anterior mesh is then brought through an opening in the broad ligament on the right side of the uterus, and the anterior and posterior meshes are then fixed to the anterior longitudinal ligament [19]. Alternative options which avoid mesh placement to the vagina and instead attach it to the uterus also exist. One technique is to suture a strip of mesh to the cervical isthmus and back of the uterus and transfix that to the anterior longitudinal ligament. A second option is to attach one or two strips of mesh anteriorly to the uterus, bringing them through windows in the broad ligament and then attaching these strips, along with a posterior mesh (with or without Y-formation), to the ligament [20] (Fig. 6).

The decision to keep or remove the uterus at the time of pelvic organ prolapse treatment largely is the patient's decision. Given the rising trend in women who desire keeping their uterus, understanding the sacrohysteropexy technique is beneficial in these cases [21,22]. Studies have shown that robotic sacrohysteropexy may have a relatively steep learning curve but once comfortable, intraoperative blood loss and robotic console time are similar to a robotic sacrocolpopexy [19,20,23]. Additionally, patients report improvements in health, wellbeing, and sexual function while also reporting less nervousness, shame, embarrassment, and frustration. They also report a large decrease in sensation of vaginal bulge or pressure. Anatomically, sacrohysteropexy is quite durable with reported success rates of >90% [19,20]. One study with long-term outcomes data reported a recurrence rate of 20% at 5 years postoperatively [23].

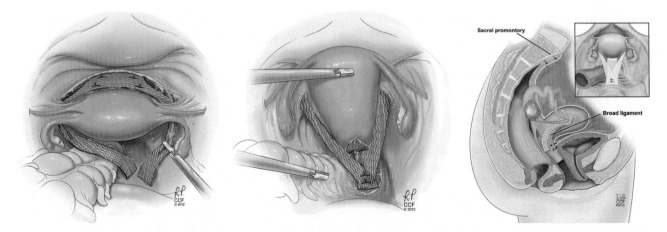

FIG. 6  *Left*—the mesh has been attached to the anterior surface of the uterus with the mesh wings placed through channels created in the broad ligament. *Middle*—the mesh is then attached, along with the posterior mesh wing, to the anterior longitudinal ligament. *Right*—sagittal view of completed sacrohysteropexy.

# Repair of mesh complication

## Painful, tight, or eroded ASC mesh

Patients who are having localized pain which clearly began after RASC may benefit from mesh removal. If there is generalized pain, it may be best to remove all the mesh. If the mesh is thought to be too tight, the patient may benefit from having the mesh cut and an interposition placed. These procedures should only be done after an extensive evaluation as to the cause of the discomfort, a trial of conservative management, and a detailed discussion with the patient. Specifically, how much mesh to be removed and from what location should be tailored to the individual patient and problem.

Surgically it is easiest to identify the mesh between its attachment to the ligament and the vagina – somewhere along the base of the "Y." Once that has been identified, one can carefully dissect the overlying and underlying tissues off of the mesh. At the level of the vagina, the bladder and/or rectum are carefully dissected off the mesh caudally. To dissect the mesh off of the vagina, it is typically easier to cut the mesh proximally just beyond where the "Y" splits and then continue the dissection distally. If the mesh has extruded, close the vaginal wall where the extrusion occurred once the mesh is removed. If the mesh is found to be intravesical, a multilayer closure of the bladder should be done with no mesh in proximity to the repair.

If it is felt that the mesh attached to the ligament should be removed, one should start distally where the mesh can be easily identified and then work proximally to the ligament. Great care should be taken to stay as close to the mesh as possible due to multiple blood vessels, nerves, and bowel that may be adherent to the mesh. This dissection should be done in a slow and meticulous fashion.

## Mid-urethral sling causing retropubic pain or passing through the bladder

The exact amount and sites of mesh removal must be individualized for each patient based on their pathology and wishes. After robotic port access has been obtained, the bladder is taken down from the abdominal wall. Blunt dissection can be performed to a point in-between the pubis and anterior vaginal wall although it is not always necessary to dissect this distally. At that level, one should be able to find both arms of a retropubic sling. With careful blunt and sharp dissection, the sling arm can be dissected distally under the pubis to a point in between the pubis and the anterior vaginal wall. If the vaginal portion of the mesh has been previously removed, the dissection can be carried to where the cut edge of the sling is encountered. Toward the abdominal wall, the mesh can be carefully dissected free of the fascia and in some cases teased all the way to the end of the mesh in the subcutaneous fat. In other cases, it may be necessary to make small prepubic incisions to reach the ends of the mesh from the skin. If the aim is not to remove all of the mesh, then a more limited dissection can be performed.

If mesh is within the bladder, one can amputate the mesh above the bladder and then simply follow the mesh into the bladder and then back out distally where it can be amputated. The bladder is then closed in a multilayer fashion after mesh removal.

## Robotic vesicovaginal fistula repair

Vesicovaginal fistulas may be repaired through multiple approaches depending on the location, size, and complexity of the fistula. Our usual preference is a transvaginal fistula repair with an interposition labial Martius or peritoneal flap as needed. However, a robotic fistula repair may be advantageous in certain cases. These include nulliparous patients or those with a small introitus, which hinder vaginal exposure and dissection. Also, more cephalad fistulas, large and complex fistulas with significant induration, recurrent or irradiated fistulas, and fistulas with concomitant distal ureteral injuries may benefit from a robotic approach [24–26].

A robotic vesicovaginal repair begins with cystoscopy in the lithotomy position to confirm the location and characteristics of the fistula. Ureteral stents may be placed at this time if desired. A guidewire or 5-French catheter is then placed through the fistula and into the vagina. If this is unsuccessful, a colposcopy or vaginal speculum can be used and the catheter or guidewire attempted through the vaginal aspect of the fistula instead. After cannulating the fistulous tract, the robot may be docked. Port placement mirrors the robotic sacrocolpopexy described earlier, and the patient is placed in Trendelenburg position. Of note, an AirSeal system (ConMed, Utica, NY) may be useful to maintain pneumoperitoneum once the vagina and bladder are later opened. The posterior bladder is then dissected to the level of the preplaced wire or catheter highlighting the tract. Anterior vaginal manipulation with a vaginal retractor by an assistant can aid this dissection. We typically separate the bladder from the vagina for at least 1 cm distal to the

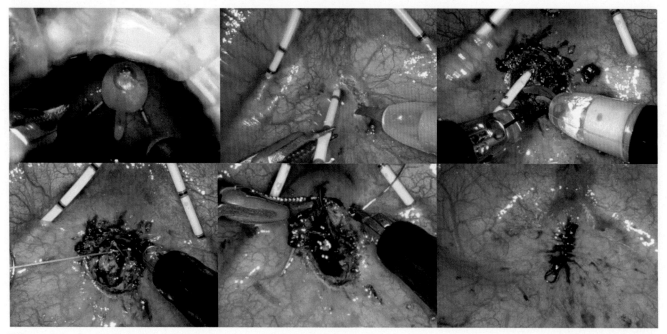

FIG. 7    *Top*—The bladder is entered after transvesical robotic port placement. 5-French ureteral catheters are placed in each ureteral orifice as well as through the fistula. The fistulous tract is dissected, and any surrounding fibrous tissue is excised. *Bottom*—the deeper vaginal layer is closed first followed by the closure of the bladder defect. *Surgeon: Mohamed Eltemamy MD.*

fistulous tract. Next, a multilayer closure with absorbable suture is performed on the vagina and bladder, typically in a perpendicular fashion to prevent overlap of suture lines. It is crucial to mobilize the posterior bladder circumferentially around the fistula to ensure a tension-free closure of only healthy, well-vascularized tissue. At this time, the authors prefer performing a greater omentum interposition flap between the repaired tissues. Alternatively, a peritoneal, buccal mucosal, rectus myofascial, or sigmoid appendix epiploic flap may be used [27–29]. The bladder is then filled with 120–180 cc of saline to confirm a water-tight closure before undocking the robot. The bladder catheter is removed about 2–3 weeks postoperatively, either with or without a cystogram performed prior.

A transvesical approach has also been described for robotic vesicovaginal fistula repair. Utilizing a multiport robot, the dome of the bladder is opened, and fistula dissection is initiated at the healthy bladder mucosa of the fistulous tract. This is then advanced posteriorly, and the fibrous fistulous tract is excised. At our institution, we have performed single-port robotic vesicovaginal fistula repair. This approach allows us to dock the robot directly into the bladder to then perform an inside-out fistula repair (Fig. 7). There are some advantages to this approach. These include omitting cystoscopy at the beginning of the case, as a 5-French catheter or guidewire can be placed through both the fistula and ureteral orifices through the robotic port. Also, a transvesical single-port approach avoids transabdominal entry and may decrease the risk of bowel injury, particularly in patients who have had previous abdominal surgeries. Lastly, the pneumoperitoneum of the abdomen can be avoided as insufflation is limited only to the bladder and may be beneficial to patients with cardiopulmonary pathologies. One difficulty with this approach is the lack of surrounding tissue that can be utilized for coverage of the repair, although reports have shown high success rates even without a flap interposition [30,31].

## Repair of distal ureteral structure—robotic ureteral reimplant

The patient is first positioned in dorsolithotomy if utilizing the *da Vinci Si* robot or supine if utilizing the *da Vinci Xi* robot. Then, the patient is placed in steep Trendelenburg position. Port placement is similar to the previously described robotic sacrocolpopexy but with ports placed more cephalad to allow for bladder manipulation. If using the *da Vinci Si* robot, the 8 mm left and right arm ports are placed inferolateral to the camera port in a W configuration. The assistant port is placed lateral to the right arm port at the level of the camera port. If using the *da Vinci Xi* robot, the assistant port and right/left arm ports are placed in a straight line at the level of the camera port. Note that for a more proximal ureteral stricture, optimal port placement differs, with ports generally placed in a vertical orientation along the mid-clavicular line.

Prior to docking the robot, additional steps may be performed to help identify the ureter and stricture. One option is to perform a flexible cystoscopy to inject indocyamine green (25 mg in 10 mL water) into the ureter through an open-ended catheter [32,33]. This will then aid the identification of the ureter when utilizing the *da Vinci* Fluorescence Imaging Vision System (Firefly). Additionally, the ureteral stricture is demarcated by a decrease in fluorescence relative to the surrounding, well-vascularized, and healthy ureter [34]. Alternatively, cystoscopy may be performed to place a guidewire into the ureter followed by a flexible ureteroscope, which can be brought to the level of the stricture to help identify its position when performing the ureteral dissection robotically.

After docking the robot, the ureter is identified anterior to the iliac vessels at the level of the iliac bifurcation. The ureter is traced proximally and distally with care not to crush the ureter or strip the periureteral tissue. A vessel loop may be helpful to retract the ureter atraumatically during this step. Sufficient mobilization is necessary to help ensure a tension-free anastomosis. The distal ureter is transected just proximal to the strictured area. If there are any oncologic concerns, the area distal to the stricture can be clipped prior to transection. The segment of diseased ureter may be resected and sent for pathology.

Next, the bladder is mobilized from anterior bladder attachments to develop the space of Retzius. This allows the bladder to drop posteriorly to further allow for a tension-free anastomosis with the remaining ureter. The bladder is filled with about 200 mL of saline or water. If there is any oncologic suspicion for the ureteral stricture, then the distal stump of the ureter and bladder cuff surrounding the ureteral orifice is resected. The defect is closed in two layers with an absorbable suture.

The reimplantation of the ureter to the bladder is then performed in an area of healthy bladder usually just lateral to the posterior dome of the bladder. A cystotomy is performed robotically, and the anastomosis is performed using absorbable suture after sufficient spatulation of the ureter. These steps leading up to the ureteral anastomosis are shown in Fig. 8. A ureteral stent is then usually placed retrograde with the proximal loop in the renal pelvis and the distal loop in the bladder. If a nephrostomy catheter was in place prior to surgery, it is typically removed at this time while the stent is grasped robotically to ensure it is not dislodged. This stent typically remains for 4–6 weeks prior to removal cystoscopically. Upon completion of the anastomosis, the bladder is filled with 120–180 cc of saline or water to ensure the anastomosis is water-tight. Some surgeons opt to place a closed suction surgical drain near the anastomosis, which may be removed when outputs decrease sufficiently. A urethral catheter is left in place postoperatively for about 2 weeks. Some surgeons opt to perform a cystogram to assess for leak prior to catheter removal.

If there is insufficient length of ureter to the bladder to perform a tension-free anastomosis, a psoas hitch and/or Boari flap may be performed robotically. The psoas hitch is performed by exposing the ipsilateral psoas tendon.

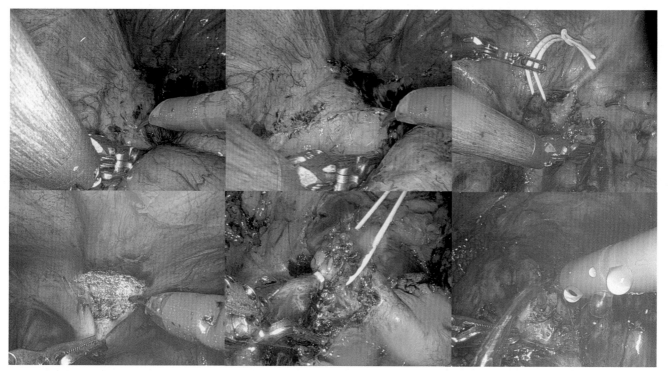

FIG. 8  *Top row*—Incision made overlying ureter at the level of iliac artery bifurcation, followed by identification and careful dissection of the ureter. A vessel loop may be useful for atraumatic ureteral retraction. *Bottom row*—release of bladder from anterior attachments followed by ureteral transection above level of stricture and cystotomy at dome of bladder to minimize distance for vesicoureteral anastomosis. *Surgeon: Mohamed Eltemamy MD.*

FIG. 9  *Left*—Bladder has been scored with cautery in a trapezoid configuration with a wide base at the dome of the bladder. *Middle*—Initial incision made at distal edge of Boari flap, revealing the tip of the Foley catheter. *Right*—Boari flap has been completely mobilized to provide length for a tension-free anastomosis with the ureter. *Surgeon: Mohamed Eltemamy MD.*

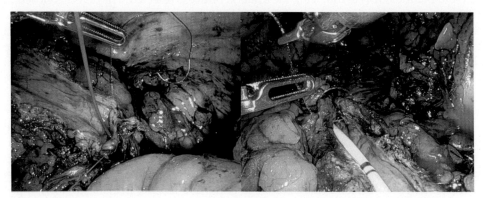

FIG. 10  *Left*—Completed back-wall anastomosis of the ureter to Boari flap with retrograde guidewire placed for stent placement. *Right*—tubularization of Boari flap after completion of vesicoureteral anastomosis. *Surgeon: Mohamed Eltemamy MD.*

The posterior bladder's detrusor and serosal layers are then fixed to the tendon with absorbable suture placed in a longitudinal orientation to avoid harming the genitofemoral nerve. If additional mobilization is needed, the contralateral bladder pedicle can also be ligated and divided. A Boari flap may additionally be performed to provide additional mobility. A wide-based trapezoid-shaped strip of bladder is dissected from the anterior bladder longitudinally with the thinner apex about 3 cm proximal to the bladder neck and the thicker base closer to the bladder dome (Fig. 9). A length-to-width ratio of 2:1 is ideal to maintain adequate vascularity [35]. The flap is then fixed to the psoas and tubularized. The spatulated ureter is then anastomosed to the tubularized Boari flap with absorbable suture (Fig. 10). A flap of omentum or peritoneum can be considered to wrap around the anastomosis to provide additional vascularity to the reconstruction.

Postoperatively, a renal ultrasound is performed 2–3 months after stent removal. A nuclear medicine renal scan can be considered if there are signs of persistent ureteral obstruction. There are few comparative studies between surgical approaches for a distal ureteral reimplantation. Based on current literature, when compared to open distal ureteral reimplant, the robotic approach has been shown to have less blood loss and a shorter hospitalization duration while maintaining success and recurrence rates [36,37]. When also compared with the laparoscopic reimplantation, the robotic approach maintained superiority in estimated blood loss and hospitalization duration [38].

## Future scenarios

Robotic surgery for female reconstructive urologic surgeries has gained significant popularity and has revolutionized surgeries as seen with the sacrocolpopexy. With the expected increase in players in the realm of surgical robotics (i.e., Moon Surgical, Medtronic Hugo, Senhance Surgical, Vicarous Surgical, Inc.) as well as the increased adoption of the Da Vinci SP robot, we expect this trend to continue. For example, descriptions of robotic artificial urinary sphincter placement at the bladder neck for stress urinary incontinence have recently been described [39]. Additionally, platforms have recently been developed for transvaginal robotic surgery, with the Hominis Surgical System obtaining FDA approval for hysterectomy in 2021. Novel transvaginal robotic techniques may further revolutionize female reconstructive surgeries such as the sacrospinous ligament fixation and uterosacral suspension.

## Key points

- Robotic sacrocolpopexy and hysteropexy are effective in curing pelvic organ prolapse while decreasing hospitalization duration.
- Complications from mesh including mid-urethral slings can be managed robotically.
- The robotic approach is effective at treating complex and cephalad vesicovaginal fistulas. A transvesical single-port approach for those not requiring flap interposition may further decrease morbidity.
- Robotic reconstructive ureteral surgery highlighted in this chapter is important to understand for the management of iatrogenic injuries.

## Appendix: Supplementary material

Supplementary material related to this chapter can be found on the accompanying CD or online at https://doi.org/10.1016/B978-0-443-13271-1.00011-X.

## Acknowledgments

The authors would like to acknowledge and thank Dr. Mohamed Eltemamy for providing images and video from his female robotic reconstructive cases.

## References

[1] Davidson ERW, Thomas TN, Lampert EJ, Paraiso MFR, Ferrando CA. Route of hysterectomy during minimally invasive sacrocolpopexy does not affect postoperative outcomes. Int Urogynecol J 2019;30(4):649–55. https://doi.org/10.1007/s00192-018-3790-4.

[2] Gupta A, Ton J, Maheshwari D, et al. Route of hysterectomy at the time of Sacrocolpopexy: a multicenter retrospective cohort study. Female Pelvic Med Reconstr Surg 2022;28(2). https://doi.org/10.1097/SPV.

[3] Dallas K, Taich L, Kuhlmann P, et al. Supracervical hysterectomy is protective against mesh complications after minimally invasive abdominal Sacrocolpopexy: a population-based cohort study of 12,189 patients. J Urol 2022;207(3):669–75. https://doi.org/10.1097/JU.0000000000002262.

[4] Das D, Carroll A, Mueller M, et al. Mesh complications after total vs supracervical laparoscopic hysterectomy at time of minimally invasive sacrocolpopexy. Int Urogynecol J 2022;33(9):2507–14. https://doi.org/10.1007/s00192-022-05251-0.

[5] Powell CR, Tachibana I, Eckrich B, Rothenberg J, Hathaway J. Securing mesh with delayed absorbable suture does not increase risk of prolapse recurrence after robotic sacral Colpopexy. J Endourol 2021;35(6):944–9. https://doi.org/10.1089/end.2018.0029.

[6] Matthews CA, Geller EJ, Henley BR, et al. Permanent compared with absorbable suture for vaginal mesh fixation during total hysterectomy and sacrocolpopexy: a randomized controlled trial. Obstet Gynecol 2020;136(2):355–64. https://doi.org/10.1097/AOG.0000000000003884.

[7] Bretschneider CE, Kenton K, Geller EJ, Wu JM, Matthews CA. Pain after permanent versus delayed absorbable monofilament suture for vaginal graft attachment during minimally invasive total hysterectomy and sacrocolpopexy. Int Urogynecol J 2020;31:2035–41. https://doi.org/10.1007/s00192-020-04471-6/Published.

[8] Tan-Kim J, Menefee SA, Lippmann Q, Lukacz ES, Luber KM, Nager CW. A pilot study comparing anatomic failure after sacrocolpopexy with absorbable or permanent sutures for vaginal mesh attachment. Perm J 2014;18(4):40–4. https://doi.org/10.7812/TPP/14-022.

[9] Linder BJ, Anand M, Klingele CJ, Trabuco EC, Gebhart JB, Occhino JA. Outcomes of robotic Sacrocolpopexy using only absorbable suture for mesh fixation. Female Pelvic Med Reconstr Surg 2017;23(1):13–6. https://doi.org/10.1097/SPV.0000000000000326.

[10] Hudson CO, Northington GM, Lyles RH, Karp DR. Outcomes of robotic sacrocolpopexy: a systematic review and meta-analysis. Female Pelvic Med Reconstr Surg 2014;20(5):252–60. https://doi.org/10.1097/SPV.0000000000000070.

[11] Geller EJ, Parnell BA, Dunivan GC. Robotic vs abdominal sacrocolpopexy: 44-month pelvic floor outcomes. Urology 2012;79(3):532–6. https://doi.org/10.1016/j.urology.2011.11.025.

[12] Pushkar DY, Kasyan GR, Popov AA. Robotic sacrocolpopexy in pelvic organ prolapse: a review of current literature. Curr Opin Urol 2021;31(6):531–6. https://doi.org/10.1097/MOU.0000000000000932.

[13] Kenton K, Mueller ER, Tarney C, Bresee C, Anger JT. One-year outcomes after minimally invasive Sacrocolpopexy. Female Pelvic Med Reconstr Surg 2016;22(5):382–4. https://doi.org/10.1097/SPV.0000000000000300.

[14] Lloyd JC, Guzman-Negron J, Goldman HB. Feasibility of same day discharge after robotic assisted pelvic floor reconstruction. Can J Urol 2018;25(3):9307–12.

[15] Moreno Sierra J, Ortiz Oshiro E, Fernandez Pérez C, et al. Long-term outcomes after robotic sacrocolpopexy in pelvic organ prolapse: prospective analysis. Urol Int 2011;86(4):414–8. https://doi.org/10.1159/000323862.

[16] Jong K, Klein T, Zimmern PE. Long-term outcomes of robotic mesh sacrocolpopexy. J Robot Surg 2018;12(3):455–60. https://doi.org/10.1007/s11701-017-0757-2.

[17] Matanes E, Boulus S, Lauterbach R, Amit A, Weiner Z, Lowenstein L. Robotic laparoendoscopic single-site compared with robotic multi-port sacrocolpopexy for apical compartment prolapse. Am J Obstet Gynecol 2020;222(4):358.e1–358.e11. https://doi.org/10.1016/j.ajog.2019.09.048.

[18] Lee SR, Roh AM, Jeong K, Kim SH, Chae HD, Sung MH. First report comparing the two types of single-incision robotic sacrocolpopexy: single site using the da Vinci Xi or Si system and single port using the da Vinci SP system. Taiwan J Obstet Gynecol 2021;60(1):60–5. https://doi.org/10.1016/j.tjog.2020.10.007.

[19] Mourik SL, Martens JE, Aktas M. Uterine preservation in pelvic organ prolapse using robot assisted laparoscopic sacrohysteropexy: quality of life and technique. Eur J Obstet Gynecol Reprod Biol 2012;165(1):122–7. https://doi.org/10.1016/j.ejogrb.2012.07.025.

[20] Lee T, Rosenblum N, Nitti V, Brucker BM. Uterine sparing robotic-assisted laparoscopic sacrohysteropexy for pelvic organ prolapse: safety and feasibility. J Endourol 2013;27(9):1131–6. https://doi.org/10.1089/end.2013.0171.

[21] Kow N, Goldman HB, Ridgeway B. Management options for women with uterine prolapse interested in uterine preservation. Curr Urol Rep 2013;14(5):395–402. https://doi.org/10.1007/s11934-013-0336-7.

[22] Zucchi A, Lazzeri M, Porena M, Mearini L, Costantini E. Uterus preservation in pelvic organ prolapse surgery. Nat Rev Urol 2010;7(11):626–33. https://doi.org/10.1038/nrurol.2010.164.

[23] Grimminck K, Mourik SL, Tjin-Asjoe F, Martens J, Aktas M. Long-term follow-up and quality of life after robot assisted sacrohysteropexy. Eur J Obstet Gynecol Reprod Biol 2016;206:27–31. https://doi.org/10.1016/j.ejogrb.2016.06.027.

[24] Blaivas JG, Heritz DM, Romanzi LJ. Early versus late repair of vesicovaginal fistulas: vaginal and abdominal approaches. J Urol 1995;153 (1110–1113):1110–3.

[25] Kumar S, Kekre NS, Gopalakrishnan G. Vesicovaginal fistula- An update. Indian J Urol 2007;(April–June):187–91.

[26] McKay E, Watts K, Abraham N. Abdominal approach to Vesicovaginal fistula. Urol Clin North Am 2019;46(1):135–46. https://doi.org/10.1016/j.ucl.2018.08.011.

[27] Randazzo M, Lengauer L, Rochat CH, et al. Best practices in robotic-assisted repair of Vesicovaginal fistula: a consensus report from the European Association of Urology robotic urology section scientific working Group for Reconstructive Urology. Eur Urol 2020;78(3):432–42. https://doi.org/10.1016/j.eururo.2020.06.029.

[28] Sandhu RS, Cheung F. Robotic-assisted surgery—a highly effective modality for Vesicovaginal fistula repairs. Curr Urol Rep 2023. https://doi.org/10.1007/s11934-022-01140-7.

[29] Schimpf MO, Morgenstern JH, Tulikangas PK, Wagner JR. Vesicovaginal fistula repair without intentional Cystotomy using the laparoscopic robotic approach: a case. Report 2007.

[30] Miklos JR, Moore RD. Laparoscopic extravesical vesicovaginal fistula repair: our technique and 15-year experience. Int Urogynecol J 2015;26 (3):441–6. https://doi.org/10.1007/s00192-014-2458-y.

[31] Martini A, Dattolo E, Frizzi J, Villari D, Paoletti MC. Robotic vesico-vaginal fistula repair with no omental flap interposition. Int Urogynecol J 2016;27(8):1277–8. https://doi.org/10.1007/s00192-016-2989-5.

[32] Lee Z, Moore B, Giusto L, Eun DD. Use of indocyanine green during robot-assisted ureteral reconstructions. Eur Urol 2015;67(2):291–8. https://doi.org/10.1016/j.eururo.2014.08.057.

[33] van Manen L, Handgraaf HJM, Diana M, et al. A practical guide for the use of indocyanine green and methylene blue in fluorescence-guided abdominal surgery. J Surg Oncol 2018;118(2):283–300. https://doi.org/10.1002/jso.25105.

[34] Lee Z, Simhan J, Parker DC, et al. Novel use of indocyanine green for intraoperative, real-time localization of ureteral stenosis during robot-assisted ureteroureterostomy. Urology 2013;82(3):729–33. https://doi.org/10.1016/j.urology.2013.05.032.

[35] Fuller TW, Daily AM, Buckley JC. Robotic ureteral reconstruction. In: Urologic clinics of North America, Vol. 49. W.B. Saunders; 2022. p. 495–505. https://doi.org/10.1016/j.ucl.2022.05.002.

[36] Kozinn SI, Canes D, Sorcini A, Moinzadeh A. Robotic versus open distal ureteral reconstruction and reimplantation for benign stricture disease. J Endourol 2012;26(2):147–51. https://doi.org/10.1089/end.2011.0234.

[37] Musch M, Hohenhorst L, Pailliart A, Loewen H, Davoudi Y, Kroepfl D. Robot-assisted reconstructive surgery of the distal ureter: single institution experience in 16 patients. BJU Int 2013;111(5):773–83. https://doi.org/10.1111/j.1464-410X.2012.11673.x.

[38] Elsamra SE, Theckumparampil N, Garden B, et al. Open, laparoscopic, and robotic ureteroneocystotomy for benign and malignant ureteral lesions: a comparison of over 100 minimally invasive cases. J Endourol 2014;28(12):1455–9. https://doi.org/10.1089/end.2014.0243.

[39] Peyronnet B, Gray G, Capon G, Cornu JN, Van Der AF. Robot-assisted artificial urinary sphincter implantation. Curr Opin Urol 2021;31(1):2–10. https://doi.org/10.1097/MOU.0000000000000837.

# Robotics in kidney stone treatment

*Cian L. Jacob, Lucas B. Vergamini, Wilson R. Molina, and Bristol B. Whiles*
Department of Urology, University of Kansas, Kansas City, KS, United States

## Introduction

Kidney stone disease, or nephrolithiasis, is a common and costly condition. Affecting 1 in 11 people, nephrolithiasis results in more than $2 billion in healthcare spending annually in the United States alone [1,2]. Additionally, the recurrence of nephrolithiasis is high, seen in approximately 50% of patients within 10 years [3]. Its high incidence, recurrence, and cost of care make urolithiasis treatment highly relevant. Nephrolithiasis treatment itself has significantly changed since its inception, as our technology and tools have adapted from rigid-only instruments to now include flexible scopes and robotic platforms. This now allows for customized treatment plans with management recommendations that are unique for each patient and dependent upon stone size, stone location, and other patient-specific factors (AUA Surgical Stone Management Guideline).

The first ureteroscopy procedure was performed by HH Young in 1912 when he used a cystoscope to access an infant's dilated ureter [4]. Almost 40 years later, Dr. Hirschowitz developed the first fiberoptic endoscope in 1957 for gastroenterologic procedures [5]. These discoveries then paved the way for Dr. Marshall to introduce the first flexible ureteroscope in Urology. This newfound flexibility enabled urologists to better visualize the urinary tract and allowed them to perform the first flexible ureteroscopy (fURS) in 1964 [6]. However, Marshall's ureteroscope lacked a working channel and a deflection mechanism. These deficits in the first flexible scopes then led Dr. Bagley to design the first flexible scope with a working channel in 1987 [7]. Bagley's improved design enabled urologists to use the ureteroscope for both diagnostic and interventional purposes, forever changing the landscape of endourologic procedures including stone treatment.

Despite further advancements in scope technology over the coming decades, our flexible fiberoptic ureteroscopes continued to have issues with poor durability and high repair costs. Researchers began to research, evaluate, and put into use modified versions of the flexible ureteroscope. Digital ureteroscopes were first introduced in 2006, and they greatly improved the clarity and visualization of the upper urinary tract [8]. A few years later, single-use, disposable ureteroscopes hit the market. The full impact of the disposable ureteroscopes is yet to be seen, but they were initially postulated to have the potential for lowering the overall costs to the healthcare system since expensive repairs for individual scopes are not required [7].

The next great advancement in endoscopic urologic procedures was more recent, with the introduction of robotic platforms for utilization in kidney stone surgery. The goal of many of these robotic platforms is to reduce the physical burden on surgeons when performing nephrolithiasis procedures, ensuring continued high patient safety and good clinical outcomes. Here, we review the various robotic systems that have been developed and discuss their potential use for ureteroscopic or percutaneous stone surgery as well as for percutaneous renal access. Additionally, the advantages and potential pitfalls of the various robotic platforms for kidney stone treatment will be discussed (Table 1).

## Sensei robot

The first robotic flexible ureteroscopy was performed in 2008 by the Sensei–Magellan system, designed by Hansen Medical [9]. Initially designed for cardiac procedures, the Sensei robot was modified to perform urologic procedures.

TABLE 1    Robotic kidney stone treatment and renal access platforms.

|  | Sensei robot | Roboflex Avicenna | ILY | easyUretero (Zaminex R) | Monarch | ANT-X |
|---|---|---|---|---|---|---|
| Company | Hansen Medical | ELMED | STERLAB | ROEN Surgical Inc. | Ethicon, Johnson & Johnson | NDR Medical Technology |
| Location | USA | Turkey | France | Korea | USA | Singapore |
| Platform type | Robotic telemanipulation system | Robotic telemanipulation system | Robotic telemanipulation system | Robotic telemanipulation system | Robotic telemanipulation system and Electromagnetic-guided percutaneous access system | Robotic needle targeting system |
| Surgeon control type | Joysticks on surgeon console | Joysticks on surgeon console | Wireless remote controller | Handle on surgeon console | Wired remote controller | Monitor +controller |
| Surgery use | fURS | fURS ECIRS | fURS | fURS | fURS, Mini-PCNL, Percutaneous access | Percutaneous access |
| Benefits | First robotic ureteroscopy system | Allows laser fiber firing control | First device with remote controller for surgeon | Can control working instruments Able to fire the laser fiber Measure stones compared to ureteral sheath size | Capable of controlling laser fiber and basket Great versatility | Automatically aligns percutaneous needle using targeting system |
| Limitations | Only allows passive defection of the scope | Requires manual manipulation of working instructions such as stone retrieval basket | Requires manual manipulation of all instruments in the working channel including stone retrieval basket Unable to fire laser | Limited types of ureteroscopes compatible with system | Implementation in clinical practice yet to be seen | Surgeon must determine depth of insertion. Remainder of procedure is done manually |

*fURS*, flexible ureteroscopy; *ECIRS*, endoscopic combined intrarenal surgery; *Mini-PCNL*, mini-percutaneous nephrolitotripsy.

The robot is composed of four components: (1) a flexible catheter system, (2) a remote catheter manipulation system, (3) a surgeon's console, and (4) an electronic rack containing the system's hardware and video system. The Sensei system has an outer access sheath (14/12 Fr) with an inner guide (12/10 Fr) that allows for the insertion of a 7.5 Fr fiberoptic flexible ureteroscope for use in this robotic ureteroscopy platform [10]. In a 2008 study by Desai et al., the feasibility and accuracy of the Sensei robot in performing retrograde ureteroscopy were assessed in 10 porcine renal units. A 14Fr catheter system was used to passively manipulate a fiberscope that was mounted on the catheter manipulator. This revealed that the ureteroscope could be manipulated by the Sensei robot into 83 out of 85 (98%) renal calyces. The catheter system was successfully and primarily introduced into 8 out of 10 ureters, while balloon dilation was required for the other 2 ureters to accommodate the sheath. The robot system was also rated an 8 out of 10 on the visual analog scale for the reproducibility of accessing calyces and a 10 out of 10 for the stability of the instrument [9]. These initial findings for the Sensei robot were promising for ergonomics, increased range of motion, and the stability of the sheath.

Although the Sensei robot was successful in the initial animal study, additional research was needed to further corroborate the findings in humans. This same research group then went on to conduct the first-in-human clinical study with the Sensei robot in 18 patients with nephrolithiasis. The robotic system was manually introduced into the patient's collecting system, and then a ureteroscope was placed through the access sheath guide. Stone fragmentation and relocation were then performed from the robotic system [11]. Unfortunately, since the Sensei robot is an attachable actuation unit, it could only passively manipulate the ureteroscope. This ultimately led to the project's discontinuation after 18 clinical trial subjects [12]. In addition, both limited maneuverability and a limited workspace also contributed to the lack of adoption of this first robotic system in nephrolithiasis treatment [13,14]. Nevertheless, the Sensei robot was a pioneer in robotic ureteroscopy, and its limitations shaped the path for future robotic systems in nephrolithiasis treatment.

## Roboflex Avicenna

The next major development in robotic ureteroscopy was the creation of the Roboflex Avicenna by ELMED Ankara, Turkey. The Roboflex Avicenna robot is a robotic telemanipulation system composed of a surgeon console and a bedside robotic manipulator arm that stabilizes and adjusts an attached ureteroscope [15]. During its development, various advancements and improvements were made to the joystick design, to its maneuverability, to the system's scope deflection capabilities, to the screen attached to the surgeon's console. By using joysticks to control the deflection and orientation of the scope manipulator, the touch screen monitor to adjust irrigation and the laser fiber, and the foot pedals to activate the laser, the Roboflex allowed urologists to perform the robotic laser lithotripsy procedure while seated at the surgeon console [15]. One other benefit is that the system allows for surgeons to robotically rotate 440°, which reduces the risk of endoscopic torsion [16]. Roboflex Avicenna also has a force-controlled deflection mechanism for the scope, straight laser positioning, and an overall reduced risk of scope damage [15]. These features may help to minimize human error and injury during laser lithotripsy.

The first clinical study with Roboflex was in 2014 and assessed the outcomes of fURS as well as the robot's effects of ergonomics and safety [17]. Of the 81 patients with renal stones treated in this study, Roboflex successfully treated stones in 79 patients (97.5%). Ergonomics were evaluated via a questionnaire to assess the surgeon's discomfort during the procedure. The Roboflex-assisted fURS had a significantly lower discomfort score compared to classic fURS (5.6 vs 31.3, $P < 0.01$), illustrating that robotic fURS reduced surgeons' physical burden [17].

Next, in 2016, a study compared the efficacy of Roboflex fURS with classic fURS and also evaluated treatment and laser lithotripsy times [18]. In their prospective randomized trial, they recruited 132 patients who had renal calculi and randomly assigned them to either the classic fURS group or the Roboflex-assisted group. They found similar treatment (50 min vs 51 min) and fragmentation times (39 min vs 37 min) for classic and robotic fURS, respectively ($P > 0.05$). There was one case of intraoperative bleeding in the standard fURS group, but zero intraoperative complications in the robotic fURS group. Finally, the robotic fURS group had significantly lower retreatment rates (9.1%) compared to the classic fURS group (15%). These findings illustrate the potential benefits of robotic fURS with the Roboflex Avicenna platform.

Klein et al. also investigated the outcomes of Roboflex-assisted fURS treatment. Their study was the largest human study to date that implemented robotic ureteroscopy. This was completed after Roboflex Avicenna obtained the CE certification for use in the European Union and Turkey. Their prospective study demonstrated a 90% stone-free rate (SFR) in 240 patients with renal stones in the study [19].

This robotic platform has also been utilized in percutaneous stone treatment procedures. In a study conducted by Tokatli et al. Roboflex was efficacious for robot-assisted endoscopic combined intrarenal surgery (ECIRS), a combination of mini-percutaneous nephrolithotomy (mini-PCNL) and fURS with laser lithotripsy [20]. A total of 42 patients were analyzed retrospectively, and stones were successfully removed utilizing a single percutaneous tract in 38 of 44 renal units with an SFR of 95.5%. This promising data shows that robot-assisted mini-ECIRS may be a highly efficacious procedure. Future studies are needed to further evaluate the SFR and complication rates when compared to standard mini-ECIRS procedures.

The Roboflex system has many potential benefits and uses. The spatial orientation and scope stability are similar between surgeons, regardless of their ureteroscopy training, ultimately illustrating the platform's minimal learning curve for use [21]. However, there are some limitations with the Roboflex Avicenna that should be highlighted. These limitations include its high cost and lack of tactile feedback. Although the robotic platforms help with firing of the laser fiber, other instruments utilized in the working channel of the ureteroscope must be manual manipulated such as stone retrieval baskets. Overall, the Roboflex Avicenna fURS system has numerous potential advantages and has emerged as a suitable option for robotic ureteroscopy.

## ILY

The ILY robotic system was developed after Roboflex by STERLAB in Vallauris, France. Similar to the Roboflex, ILY is a robotic telemanipulation system. The robot itself sits docked near the patient but the surgeon controls the device with a wireless remote controller. This design is different than the Roboflex system, which utilized a joystick configuration on the main console for ureteroscope manipulation. Furthermore, the ILY is more versatile in its ability to accommodate different ureteroscopes of variable sizes such as reusable, flexible, and single-use devices [15].

The typical protocol for the ILY robot requires that the surgeon manually place a guidewire into the renal pelvis followed by a ureteral access sheath. This then allows for the sheath type and size to be selected and modified based on patient factors. After placement, the ureteral access sheath is then fixed onto the ureteroscope holder. The ureteroscope is introduced by the surgeon, and once it reaches the patient's renal pelvis, it too is then fixed onto the ILY robot. The entire setup process for the robot takes approximately 5 min with an additional 3 min to install the scope.

The remote control is activated by the surgeon who can now manipulate the ureteroscope. The ILY robot enables the surgeon to easily explore the kidney, locate stones, fragment, or dust the stone with the assistance of a laser [15]. The robot is also able to rotate 360° each direction, and its increased mobility is an added benefit for its application in clinical practice.

However, it should be noted that the ILY robot differs from other robot systems because it cannot replicate the exact movements urologists use in manual fURS. Instead, surgeons need to learn the movements of the scope based on the remote controller to safely perform their procedures [15]. Additionally, the ILY robot cannot adjust the laser fiber nor can it control the laser by itself. As a result, laser fiber adjustment and stone basket use must still be done manually, and these disadvantages make it difficult to implement the ILY robot in clinical practice by many surgeons [10].

## easyUretero (Zaminex R)

The next robotic stone platform to emerge was the easyUretero system. This system received approval from the Korean Ministry of Food and Drug Safety in 2022 and was developed and is marketed by Roen Surgical Inc., which recently renamed and rebranded the device as Zaminex R (Fig. 1). EasyUretero can control flexible ureteroscopes, laser fibers, and stone baskets. The robot is divided into a master console and a slave console. The master console consists of a touch screen, clutch, armrest, and handle. The slave manipulator is the component of the robotic platform that shifts the ureteroscope back and forth, adjusts laser positioning, and controls the stone basket [22]. The slave manipulator performs these functions without assistance at the bedside, ultimately decreasing the burden on the surgeon and their assistants. Additionally, the movements of the robot can be easily repeated, which aids in the efficient process of stone retrieval during fURS [10]. EasyUretero is also able to measure and compare stone size relative to the size of the ureteral access sheath, helping to prevent injury during the retrieval of large stones [10,15].

Park et al. assessed the ergonomics and feasibility of the easyUretero system in removing renal calculi in both in vitro and in vivo models. For the in vitro design, renal stone removal was performed four times using an artificial kidney–ureter-bladder model. Both manual fURS and robotic fURS with stone basket retrieval were performed and compared regarding completion time and ergonomics. The times for robotic fURS and manual fURS completed by

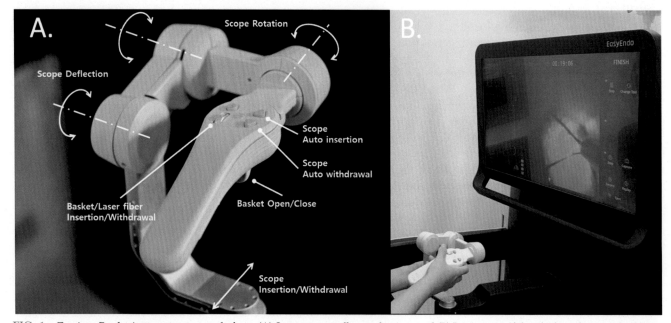

FIG. 1    Zaminex R robotic stone treatment platform. (A) Surgeon controller mechanism, and (B) In vivo use of the platform for stone basketing.

experts were 282.6 ± 92.4 vs 73.6 ± 43.3 s, and for residents 281.3 ± 111.0 vs 188.6 ± 138.6 s, respectively (both with $P < 0.001$). The completion times for experts, fellows, and residents to complete the fURS with the easyUretero system were significantly longer than manual fURS at all skill levels. Despite these longer procedural times, the ergonomics were rated better when participants used easyUretero for robotic fURS than manual fURS. Additionally, there were no safety issues presented during in vivo testing, which shows potential for easyUretero [22].

The first clinical trial in humans to evaluate the easyUretero robotic fURS system was conducted in April 2022 at Seoul National University Hospital and Severance Hospital (Yonsei University College of Medicine, Seoul, Korea). The purpose of the study was to assess the safety and efficacy of easyUretero. A total of 47 patients undergoing retrograde intrarenal surgery (RIRS) for the treatment of renal calculi were enrolled in this multicenter, prospective, single-arm study. The results from this trial are still pending but will hopefully shed light on easyUretero's use in human patients as well as future applications in urologic procedures [15].

Although easyUretero is a promising addition to robotic ureteroscopy, the system has a few limitations. The slave console can only mount to a limited number of flexible ureteroscopes [15]. Depending on the type of ureteroscope available, some urologists may be unable to utilize this robotic system in their routine clinical practice. Further advancements are needed to allow for easyUretero to accommodate all types of ureteroscopes. Additionally, there is limited information to date about the use of this platform in human patients. The results from the clinical trial at Yonsei University College of Medicine as well as future studies are needed for practitioners to better understand how the easyUretero system may be incorporated into their routine practice.

## Monarch

The Monarch platform was first developed by Auris (now acquired by Ethicon, Johnson & Johnson) for flexible bronchoscopy [10]. It was then utilized for the first time in a urologic procedure in 2019, when 10 cases were completed in India [23]. And more recently, in the Spring 2022, the US Food and Drug Administration approved 501(k) clearance for the Monarch platform use in endourologic procedures.

The Monarch platform is composed of three separate components: (1) the three-arm robotic patient cart, (2) a fluid management system for irrigation and suction, and (3) a screen cart with connected surgeon controller. For the robotic patient cart, two of the robotic arms are deployed under the patient's leg while in the dorsal lithotomy position and can be utilized for cystoscopy or fURS. After placement of a retrograde ureteral access sheath, the sheath is docked to one of the robotic arms and a robotic ureteroscope is inserted and docked to the other arm. The location of the tip of the access sheath is denoted, which allows the robot to move quickly when within the sheath itself. The device contains pressor sensors to ensure the Monarch does not employ too much force during its movements within the patient. A robotic laser fiber as well as a basket can also be utilized via the robotic ureteroscope.

During mini-ECIRS, the third robotic arm is placed about 1 handsbreadth away from the planned site of percutaneous access and is first utilized to help obtain renal access. This arm's electromagnetic field generator, when used in combination with the robotic ureteroscope and access needle, helps create a graphic interface on the surgeon's screen. The surgeon selects the desired calyx and "marks" it with the ureteroscope using the electromagnetic field generated by the robotic arm. The Monarch is then able to determine the location of the desired calyx and the necessary access trajectory with its knowledge of the exact location of the ureteroscope and needle, creating this target-like screen graphic to help guide the surgeon during needle insertion. After gaining percutaneous access into the desired calyx, the tract is dilated, and a percutaneous sheath is placed and docked to the robotic arm. This allows for insertion of the irrigation and aspiration catheter, which enables the surgeon to maneuver the catheter to the stone within the collecting system. The irrigation and aspiration catheter are used to keep the stone in place as well as suction out stone dust and small fragments that are generated during laser lithotripsy via the retrograde placed robotic ureteroscope's laser fiber [16,24] (Fig. 2).

The major benefits of this robotic platform stem from its compact robotic arm design as well as the handheld gaming-type controller. This leads to improved stability and easier navigation of robotic instruments. The Monarch platform may also be able to reduce procedural fatigue and enable urologists to complete more complex procedures. It should also be highlighted that this versatile system can be used to perform both fURS and/or PCNL procedures [15].

One potential benefit of the Monarch system that has been studied is its ability to separate the surgeon from the radiation source, helping to limit the surgeon's exposure. Radiation usage and exposure during PCNL access obtained using traditional fluoroscopy were compared to the Monarch platform [25]. "Novice" surgeons, or those who use IR for access, have gained access less than 30 times after training, or gain percutaneous access less than 24 times per year, were included in the study. The radiation time when using Monarch was 0.11 ± 0.01 min, while the radiation time when using fluoroscopy was 1.30 ± 0.33 min ($P < 0.05$). Additionally, the radiation dosage was 1.00 ± 0.15 mGy

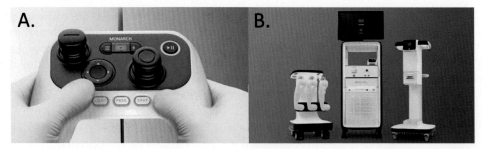

FIG. 2   Monarch robotic stone treatment platform. (A) Surgeon remote controller, and (B) Robotic components including patient console and video tower.

($P < 0.01$) for Monarch and $5.43 \pm 0.88$ mGy for the fluoroscopy-guided group. Therefore, the Monarch platform decreased the radiation exposure to the patient and therefore also to the surgeon, when compared with traditional fluoroscopic access in less experienced urologists [25].

Chi et al. also explored the feasibility and safety of Monarch and compared it to standard devices in URS and PCNL procedures using a porcine model. A total of 24 renal units from 12 pigs were used in their study, evenly divided into a PCNL group and a URS group. Safety was evaluated via intraoperative complication assessment. A retrograde pyelogram was performed at the conclusion of each procedure and demonstrated no safety events occurred when using Monarch for either PCNL or URS procedures. In this porcine model, Monarch was deemed to have comparable usability and safety compared to standard procedures [26].

In February 2023, the Monarch platform was utilized for the first time in the United States to perform a robotically assisted electromagnetic guided percutaneous access and mini-ECIRS procedure. This pioneering procedure was completed by Dr. Jamie Landman at the University of California Irvine, where he and his team have an ongoing clinical trial utilizing the Monarch robotic platform. The implementation and utilization of the Monarch device in routine clinical practice is yet to be seen and results from their clinical trial will be important findings regarding the clinical utility of this device.

## ANT-X

One additional robotic platform that is important to discuss for percutaneous renal stone access is ANT-X (Automated Needle Targeting with X-ray). ANT-X is a robotic, automated needle targeting system that utilizes artificial intelligence image registration software. Using a single C-arm X-ray image, it provides closed loop, autoalignment feedback for the percutaneous needle using a bullseye technique. This allows the surgeon to focus solely on controlling the depth of needle insertion when obtaining percutaneous access, since the ANT-X robotically aligns the needle to the desired calyx. It recently received 510(k) clearance via the US Food and Drug Administration in June 2023.

A few studies have evaluated use of the ANT-X system. A randomized control trial with 71 patients with renal stones was conducted to assess the success rate of ANT-X [27]. Use of the ANT-X device for robotic-assisted fluoroscopic guided renal access demonstrated comparable single puncture success rates to ultrasound-guided access (50% vs 34.3%, $P = 0.2$). However, there were significantly fewer needle punctures (1.82 vs 2.51 times, $P = 0.025$) and shorter time required for percutaneous access (5.5 vs 8 min, $P = 0.049$) compared to ultrasound guided access in PCNL procedures performed by novice surgeons. There were also no instances in which the resident could not obtain access in the ANT-X group (0% compared vs 14.3% in the standard access group, $P = 0.025$) [27,28]. They also noted no differences in postoperative complications between the ANT-X and traditional care groups [27]. The ANT-X technology is promising in that it has the potential to allow for increased performance of percutaneous access by urologic surgeons with limited experience performing traditional fluoroscopic or ultrasound percutaneous access.

## Other developments in nephrolithiasis treatment via artificial intelligence

Robotic platforms for kidney stone surgery are rapidly evolving. However, our discussion of these systems would not be complete without a brief overview of other developments in AI. Use of AI for both the diagnosis, treatment, and medical management of kidney stones is likely to be part of our future.

AI, deep learning, and machine learning models are also being used to predict surgical outcomes as well as to determine stone composition. For example, an artificial neural network (ANN) algorithm by Aminsharifi et al. has successfully predicted PCNL outcomes in 81%–98.2% of 454 patients [29]. Recently, various prediction-based models and convolutional neural networks have also been developed to detect the composition of stones. A machine learning model by Hokamp et al. incorporates dual-energy CT to create images of renal stones [30]. Two hundred stones were used to train the model, and it accurately identified the composition in 90% of stones. Smartphone microscopy has also been incorporated into the development of some of these stone prediction models. A novel recognition system based on smartphone microscopy has obtained a weighted accuracy of 87% and an overall accuracy of 88% in 37 total stones [31]. Further studies are needed to evaluate the role of these models in the routine care of patients.

Extended reality is a form of augmented reality, mixed reality, and virtual imaging. This has been utilized to aid in percutaneous needle access during PCNL. The hyper-accuracy 3D reconstruction (HA3D) model generates a three-dimensional PDF of holograms that can overlap with the patient. This mixed reality experience can then be used to identify certain anatomical regions, ultimately pinpointing the entry site for the percutaneous needle. The visualization of mixed reality images coupled with real-time needle guidance is associated with an increased rate of successful, first-attempt punctures when compared to the control group (100% vs 50%, $P = 0.032$) [32].

The Hisense computer-assisted system (CAS system) can also be used to visualize and identify important anatomical structures. A few studies to date have evaluated its use. First, when compared to standard punctures, the CAS system had an improved single puncture success rate (90% vs 67%, $P = 0.028$) [33]. Additionally, the CAS system resulted in an improved initial stone-free rate when compared to the standard procedure (87% vs 63%, $P = 0.037$). Tan et al. confirmed these results for first-time success rate of puncture, and also determined that the CAS system had higher rates than the standard procedure (87.5% vs 47.8%, $P = 0.03$) [34]. However, it should be noted that a lack of real-time imaging and generalizability is a limitation of these models.

Finally, 3D printing has been useful for training and improving PCNL skills and has emerged as a tool for preoperative planning [35]. One study compared the effect of 3D printing combined with PCNL versus conventional PCNL in the treatment of complex renal calculi. The test group would have their CT scan extracted and processed by a specific software able to translate the image to a 3D printed form. This 3D model would be used as preoperative planning. The puncture point, depth, and angle were drawn before printing. With the model in hand, the surgeon would cognitively translate what was represented to the patient in question using predefined anatomical landmarks. This study concluded that the use of the 3D printing model was associated with a lower operation time (103 vs 126 min, $P < 0.001$), a better stone-free rate (96% vs 80%, $P = 0.03$), a lower complication rate (6.67% vs 22.22%, $P = 0.02$), and a lower puncture time (7.2 vs 9.7 min, $P < 0.001$) [36]. In a prospective study by Huang et al., 120 patients with complex kidney stones were recruited. The researchers implemented a 3D reconstruction model to determine the exact position of the collecting systems and their stones. The preoperative 3D model led to lower operation times (85.57 vs 92.42 min, $P = 0.006$) and the mean number of percutaneous sticks necessary for a successful puncture was lower in the 3D reconstruction model when compared with a standard puncture (2 vs 3, $P < 0.001$). A lower intraoperative blood loss was also observed in the 3D reconstruction model (63.8 vs 109.8 mL, $P < 0.001$) [37]. Advances in these different technologies may in turn advance the success and use of robotics in kidney stone treatment.

## Conclusions

Nephrolithiasis treatment has experienced tremendous advancements, specifically in the type of surgical treatment options available to patients through technological advancements in our tools. Although the shift from rigid-only ureteroscopes to digital and flexible ureteroscopes has greatly improved patient treatment options and outcomes, concerns regarding surgeon ergonomics and comfort have arisen as well as the desire for improved stone-free rates. Numerous robotic platforms have been developed in the past two decades that could better accommodate the surgeon and potentially improve stone clearance during nephrolithiasis treatment. The Sensei robot was a pioneer in this field, yet it could only passively manipulate ureteroscopes, which prevented its widespread use. Nevertheless, it inspired other robotic advancements in ureteroscopy. The Roboflex, ILY, easyUretero, and Monarch platforms are all robotic telemanipulation systems that allow for robotic stone treatment. There are significant differences in how these robotic systems function and how they are controlled, which may lead to different surgeon preferences for a particular robotic system. For example, the joysticks on the Roboflex are located on the main console, while the ILY, easyUretero, and Monarch systems rely on a remote controller. The Roboflex does not have a built-in feature to remove stones, so urologists may prefer one of the other robotic platforms that has the capability to assist with laser lithotripsy and stone basketing. There has also been emergence of robotic-assisted percutaneous access devices such as that within the

Monarch platform as well as the ANT-X system. Although the use of these access systems may not be necessary for urologists who are already facile at obtaining their own percutaneous renal access, they provide new tools that may add to the armamentarium for urologists who do not currently obtain their own access. Robotics in nephrolithiasis treatment is promising and continued advancements to these systems are anticipated. Although additional human trials are warranted, these robotic platforms will likely change the way we treat stones in the future.

## Future scenario

There is continued opportunity for the ongoing development of robotics use in kidney stone treatment, ushering in a new era of innovation within urology. As this technology becomes more refined, we anticipate the integration of advanced imaging techniques, such as real-time 3D visualization, allowing unparalleled precision and efficiency for nephrolithiasis treatment. Furthermore, the use of AI-driven software could enhance decision-making both before and during procedures while accounting for variables, including stone dimension, composition, and patient positioning. As miniaturization and remote-control capabilities improve, these robotic tools will likely become even more versatile, granting access and navigation in the most challenging ureters and collecting systems while also reducing surgeons' physical strain. Although robotic platforms are currently expensive to utilize, we anticipate their costs to decrease over time. The convergence of robotics, imaging, and artificial intelligence will likely revolutionize our standard of care practices in kidney stone surgery.

## Key points

- The recurring nature of nephrolithiasis poses a threat to both affected individuals and the U.S. healthcare system.
- There have been numerous advancements in nephrolithiasis treatment. Currently, the field is experiencing a boom in robotic platforms, which have the potential to improve patient outcomes while also reducing the physical burden on surgeons.
- Currently, available robotic platforms vary from attachable actuation systems, such as the Sensei robot, to robotic telemanipulation systems, such as the Roboflex and Monarch platforms. The various robotic platforms discussed have their own unique advantages and disadvantages, and urologists may choose one platform over another based on these.
- AI and machine learning are also effective for percutaneous access during urologic procedures, and the intersection of AI and robotic technologies for nephrolithiasis treatment will be an important area of continued research and expansion in the future.

## References

[1] Scales CD, Smith AC, Hanley JM, Saigal CS, Project UDiA. Prevalence of kidney stones in the United States. Eur Urol 2012;62(1):160–5. https://doi.org/10.1016/j.eururo.2012.03.052.
[2] San Juan J, Hou H, Ghani KR, Dupree JM, Hollingsworth JM. Variation in spending around surgical episodes of urinary stone disease: findings from Michigan. J Urol 2018;199(5):1277–82. https://doi.org/10.1016/j.juro.2017.11.075.
[3] Uribarri J, Oh MS, Carroll HJ. The first kidney stone. Ann Intern Med 1989;111(12):1006–9. https://doi.org/10.7326/0003-4819-111-12-1006.
[4] Dhinakar L. A retrospective study of ureteroscopy performed at the sultan qaboos hospital, Salalah from august 2001–august 2006. Oman Med J 2007;22(3):24–32.
[5] Monga M. Ureteroscopy: Indications, Instrumentation & Technique. 1 ed. Current Clinical Urology, Humana Totowa: SpringerLink; 2023.
[6] Marshall VF. Fiber optics in urology. J Urol 1964;91:110–4. https://doi.org/10.1016/S0022-5347(17)64066-7.
[7] Mazzucchi E, Marchini GS, Berto FCG, et al. Single-use flexible ureteroscopes: update and perspective in developing countries. A narrative review. Int Braz J Urol 2022;48(3):456–67. https://doi.org/10.1590/S1677-5538.IBJU.2021.0475.
[8] Gridley CM, Knudsen BE. Digital ureteroscopes: technology update. Res Rep Urol 2017;9:19–25. https://doi.org/10.2147/RRU.S104229.
[9] Desai MM, Aron M, Gill IS, et al. Flexible robotic retrograde renoscopy: description of novel robotic device and preliminary laboratory experience. Urology 2008;72(1):42–6. https://doi.org/10.1016/j.urology.2008.01.076.
[10] Lee JY, Jeon SH. Robotic flexible ureteroscopy: a new challenge in endourology. Investig Clin Urol 2022;63(5):483–5. https://doi.org/10.4111/icu.20220256.
[11] Desai MM, Grover R, Aron M, et al. Robotic flexible ureteroscopy for renal calculi: initial clinical experience. J Urol 2011;186(2):563–8. https://doi.org/10.1016/j.juro.2011.03.128.
[12] Rassweiler J, Fiedler M, Charalampogiannis N, Kabakci AS, Saglam R, Klein JT. Robot-assisted flexible ureteroscopy: an update. Urolithiasis 2018;46(1):69–77. https://doi.org/10.1007/s00240-017-1024-8.

[13] Zhao J, Li J, Cui L, Shi C, Wei G. Design and performance investigation of a robot-assisted flexible Ureteroscopy system. Appl Bionics Biomech 2021;2021:6911202. https://doi.org/10.1155/2021/6911202.

[14] Sinha MM, Gauhar V, Tzelves L, et al. Technical aspects and clinical outcomes of robotic Ureteroscopy: is it ready for primetime? Curr Urol Rep 2023;01:1–10. https://doi.org/10.1007/s11934-023-01167-4.

[15] Gauhar V, Traxer O, Cho SY, et al. Robotic retrograde intrarenal surgery: a journey from "Back to the future". J Clin Med 2022;11(18). https://doi.org/10.3390/jcm11185488.

[16] Humphreys M. AUA2021 State-Of-The-Art Lecture. AUA2021 State-Of-The-Art Lecture: Robotic Ureteroscopy. Accessed 07/24/2023. https://soundcloud.com/auauniversity/aua2021-state-of-the-art-lecture-robotic-ureteroscopy.

[17] Saglam R, Muslumanoglu AY, Tokatlı Z, et al. A new robot for flexible ureteroscopy: development and early clinical results (IDEAL stage 1-2b). Eur Urol 2014;66(6):1092–100. https://doi.org/10.1016/j.eururo.2014.06.047.

[18] Geavlete P, Saglam R, Georgescu D, et al. Robotic flexible Ureteroscopy versus classic flexible Ureteroscopy in renal stones: the initial Romanian experience. Chirurgia (Bucur) 2016;111(4):326–9.

[19] Klein J, Charalampogiannis N, Fiedler M, Wakileh G, Gözen A, Rassweiler J. Analysis of performance factors in 240 consecutive cases of robot-assisted flexible ureteroscopic stone treatment. J Robot Surg 2021;15(2):265–74. https://doi.org/10.1007/s11701-020-01103-5.

[20] Tokatli Z, Ibis MA, Sarica K. Robot-assisted Mini-endoscopic combined intrarenal surgery for complex and multiple calculi: what are the real advantages? J Laparoendosc Adv Surg Tech A 2022;32(8):890–5. https://doi.org/10.1089/lap.2022.0124.

[21] Proietti S, Dragos L, Emiliani E, et al. Ureteroscopic skills with and without Roboflex Avicenna in the K-box. Cent European J Urol 2017;70(1):76–80. https://doi.org/10.5173/ceju.2017.1180.

[22] Park J, Gwak CH, Kim D, et al. The usefulness and ergonomics of a new robotic system for flexible ureteroscopy and laser lithotripsy for treating renal stones. Investig Clin Urol 2022;63(6):647–55. https://doi.org/10.4111/icu.20220237.

[23] Desai M, Singh A, Ganpule A, Sabnis R, Desai M, Landman J. PD41-05 initial clinical experience of mini-percutaneous nephrolithotomy (PCNL) WITH ureteroscopic (URS) lithotripsy and synchronous percutaneous evacuation of fragments using a steerable suction device. Abstract. 2021-September; 2021. https://doi.org/10.1097/JU.0000000000002051.05.

[24] Landman J. Development and initial clinical experience of a novel endoscopic robotic platform: monarch PCNL presentation - Jaime Landman. UroToday 2023. Accessed 07/24/2023 https://www.urotoday.com/video-lectures/aua-2023/video/3349-development-and-initial-clinical-experience-of-a-novel-endoscopic-robotic-platform-monarch-pcnl-jaime-landman.html.

[25] Humphreys M, Wymer K, Chew B, et al. PD40-12 robotic-assisted electromagnetic guidance minimizes radiation exposure in gaining percutaneous access for nephrolithotomy: a cadaveric study with novices. Abstract. 2022-May; 2022. https://doi.org/10.1097/JU.0000000000002601.12.

[26] Chi T, Hathaway L, Chok R, Stoller M. Abstracts of the 39th world congress of Endourology: WCE 2022. J Endourol 2022;36(S1):A1–A315. https://doi.org/10.1089/end.2022.36001.abstracts.

[27] Taguchi K, Hamamoto S, Okada A, et al. A randomized, single-blind clinical trial comparing robotic-assisted fluoroscopic-guided with ultrasound-guided renal access for percutaneous nephrolithotomy. J Urol 2022;208(3):684–94. https://doi.org/10.1097/JU.0000000000002749.

[28] Gauhar V, Giulioni C, Gadzhiev N, et al. An update of in vivo application of artificial intelligence and robotics for percutaneous Nephrolithotripsy: results from a systematic review. Curr Urol Rep 2023;24(6):271–80. https://doi.org/10.1007/s11934-023-01155-8.

[29] Aminsharifi A, Irani D, Tayebi S, Jafari Kafash T, Shabanian T, Parsaei H. Predicting the postoperative outcome of percutaneous Nephrolithotomy with machine learning system: software validation and comparative analysis with Guy's stone score and the CROES nomogram. J Endourol 2020;34(6):692–9. https://doi.org/10.1089/end.2019.0475.

[30] Große Hokamp N, Lennartz S, Salem J, et al. Dose independent characterization of renal stones by means of dual energy computed tomography and machine learning: an ex-vivo study. Eur Radiol 2020;30(3):1397–404. https://doi.org/10.1007/s00330-019-06455-7.

[31] Onal EG, Tekgul H. Assessing kidney stone composition using smartphone microscopy and deep neural networks. BJUI Compass 2022;3(4):310–5. https://doi.org/10.1002/bco2.137.

[32] Porpiglia F, Checcucci E, Amparore D, et al. Percutaneous kidney puncture with three-dimensional mixed-reality hologram guidance: from preoperative planning to intraoperative navigation. Eur Urol 2022;81(6):588–97. https://doi.org/10.1016/j.eururo.2021.10.023.

[33] Qin F, Sun Y-F, Wang X-N, et al. Application of a novel computer-assisted surgery system in percutaneous nephrolithotomy: a controlled study. World J Clin Cases 2022;10(18):60039–6049. https://doi.org/10.12998/wjcc.v10.i18.6039. In press.

[34] Tan H, Xie Y, Zhang X, Wang W, Yuan H, Lin C. Assessment of three-dimensional reconstruction in percutaneous Nephrolithotomy for complex renal calculi treatment. Front Surg 2021;8, 701207. https://doi.org/10.3389/fsurg.2021.701207.

[35] Hameed BMZ, Shah M, Pietropaolo A, et al. The technological future of percutaneous nephrolithotomy: a Young Academic Urologists Endourology and Urolithiasis Working Group update. Curr Opin Urol 2023;33(2):90–4. https://doi.org/10.1097/MOU.0000000000001070.

[36] Cui D, Yan F, Yi J, et al. Efficacy and safety of 3D printing-assisted percutaneous nephrolithotomy in complex renal calculi. Sci Rep 2022;12(1):417. https://doi.org/10.1038/s41598-021-03851-2.

[37] Huang YS, Zhu XS, Wan GY, Zhu ZW, Huang HP. Application of simulated puncture in percutaneous nephrolithotomy. Eur Rev Med Pharmacol Sci 2021;25(1):190–7. https://doi.org/10.26355/eurrev_202101_24384.

# 48

# Head and neck and transoral robotic surgery

*José Guilherme Vartanian*[a], *Renan Bezerra Lira*[a,b], *and Luiz Paulo Kowalski*[a,c]

[a]Department of Head and Neck Surgery and Otorhinolaryngology, A.C. Camargo Cancer Center, São Paulo, São Paulo, Brazil [b]Robotic Surgery Program, Hospital Israelita Albert Einstein, São Paulo, Brazil [c]Head and Neck Surgery Department and LIM 28, University of São Paulo Medical School, São Paulo, Brazil

## Introduction

Transoral robotic surgery (TORS) is a cutting-edge technique that allows a minimally invasive approach, which facilitates access to more complex areas that would be more difficult to access, or that would be more morbid when performed in the conventional approach. After its approval by the American FDA in 2009, robotic surgery has been widely used in head and neck surgery, mainly for the treatment of oropharyngeal tumors [1].

In addition to the oropharyngeal site, robotic surgery in the head and neck has been used to treat surgical diseases of the oral cavity, larynx, hypopharynx, parapharyngeal space, and as a remote alternative access for the treatment of cervical surgical pathologies, such as neck dissection, thyroidectomies, salivary gland, benign nodules, and others [2,3].

The pioneering and most widely used robotic system in head and neck surgery is the *da Vinci* system, developed by Intuitive Surgical Inc. in the USA [4]. This system allows less invasive access, magnification of the image of the surgical field in high-definition 3D view, facilitating the identification of important structures and providing safer surgeries. The robotic surgical arms are controlled by the surgeon through a console, reproducing their movements accurately, without tremors, thus providing greater dexterity and efficiency in tissue dissection.

Initially, this transoral robotic technique was used to treat benign and early-stage malignant lesions of the oropharynx, but with the development of the technique and the gain of experience of the surgical teams, the indications were expanded, being indicated for selected cases of surgical pathologies of the larynx (supraglottis), hypopharynx, metastatic lymph node metastases with unknown primary tumors, parapharyngeal space tumors, sleep apnea, and thyroid nodules [3,5,6].

In addition to the *da Vinci* system, several other robotic systems are under development to optimize flexibility and access to the lower pharynx and larynx [4].

## TORS advantages

Several studies have demonstrated the benefits of robotic transoral surgery, especially in the treatment of oropharyngeal tumors. The transoral approach avoids the need for external incisions, which could decrease the risk of infections, bleeding, and patients' dissatisfaction with aesthetic results. Furthermore, less manipulation of surrounding tissues allows for faster recovery, with less pain, decreasing hospital stay, and promoting an early return to patients' usual activities [1,7]. Most patients undergoing TORS do not require tracheostomies or alternative feeding tubes, as they experience earlier swallowing rehabilitation [1,8–10].

In a retrospective survey of 2015 TORS procedures performed by 45 surgeons, fewer than 6% required tracheotomy, with most patients (62%) initiating oral intake on the first postoperative day [10].

## TORS limitations

One of the limitations for the widespread use of TORS is the difficult access for some regions, mainly middle and low-income countries due to its high cost [11,12]. Additionally, the learning curve may also be a factor to consider (between 20 and 30 cases), requiring specialized training and experience to achieve the best results [13].

From a clinical applicability perspective, limiting factors would be a difficulty for mouth opening and oropharyngeal exposure (trismus, cervical spine osteoarthrosis, smaller interincisor gap, and larger neck circumference) and a retropharyngeal carotid artery [14]. Another limitation of the method would be the loss of palpation capacity during surgical resection, which may be important in some cases, helping to obtain appropriate deep margins.

## TORS for oropharyngeal tumors

The most common indication for TORS is the treatment of HPV-positive oropharyngeal cancer. Historically, the treatment was based on surgical resection, usually through a split mandible access, or based on radiotherapy or chemoradiation. Such treatment strategies carry their own morbidity and potential functional complications [9,15,16]. HPV-related disease carries a better prognosis than HPV-negative oropharyngeal cancer, and most patients are younger, usually with less comorbid conditions, present low risk of second primary tumors, and probably will survive for a long time. In this scenario, the morbidity with chemoradiation could be considerable, and strategies for toxicity mitigation and optimizing quality of life (QOL) have become an important issue [16,17].

The main advantages of TORS in this group of patients are the less invasive approach, decreasing the need of radiotherapy (or at least decreasing the radiation dose), early recovery, and the avoidance of tracheostomy and a feeding tube [1,7–10].

Several studies have addressed the oncological outcomes of TORS in the treatment of oropharyngeal tumors. In an NCDB population-based study, Mahmoud et al. reported similar 3-year survival between TORS versus primary radiotherapy (95% vs 91%) in the HPV-positive patients, but in the HPV-negative group, TORS presented better survival rates than patients treated by radiotherapy (84% vs 66%) [18]. A large randomized multicenter study has shown that for patients with intermediate risk, which is based on the pathological findings after TORS (mainly surgical margins, and lymph node status), it is possible to decrease the postoperative radiation dose and overleap chemotherapy with preservation of high survival rate and good QOL [19].

Regarding the functional results after TORS, several publications reported good functional and QOL outcomes, emphasizing the low rate of the need for feeding tubes and tracheostomy [8,9]. In the systematic review by Hutcheson et al., the crude gastrostomy rate ranged from 18% to 39% in the TORS-based treatment and from 29% to 60% in the radiotherapy-based group [9]. In a more recent publication from MD Anderson, Barbon et al. reported no significant group differences were found in modified barium swallow studies, dysphagia prevalence as well as in the mean MDADI scores among groups at subacute and late time points. This finding can suggest that patients submitted to TORS (usually followed by adjuvant therapy) could present similar dysphagia rate and degree compared with patients receiving primary chemoradiotherapy [20]. However, in this single-center study, feeding tube placement and duration significantly differed among primary treatment groups. Most patients in the TORS group have placed a feeding tube (89%) with a median use of 3 days, and 37% of patients in the bilateral radiotherapy group have placed a feeding tube, however with a median use of 104 days [20].

A systematic review including 20 papers addressing the post-treatment QOL and swallow function of patients treated with TORS, most of them retrospective series, reported superior QOL and swallowing outcomes when compared with primary chemoradiation or open approaches. However, the outcomes were dependent on baseline function, T stage, and adjuvant treatment status [21].

However, the first randomized trial with 68 patients allocated to chemoradiotherapy or TORS showed that swallowing-related QOL 1 year after treatment is better in the chemoradiotherapy group, but that the difference (not significant) is reduced over 3 years [22]. Regarding this publication, some considerations should be analyzed, as most patients were submitted to a protective tracheostomy (after one enrolled patient died at the beginning of the study), which could be associated with worse swallowing scores, and a higher death rate (3%) compared to other publications. In a retrospective survey of 2015 TORS procedures performed by 45 surgeons (cited earlier), Chia et al. reported a death rate of 0.3% [10].

Patients undergoing TORS for oropharyngeal tumors generally experience less postoperative pain, shorter recovery time, and better functional and QOL outcomes compared to open surgery, and at least similar functional outcomes compared to chemoradiation. The outcomes are directly related to the patient selection and the surgeon's experience with the transoral robotic technique [22,23].

## Larynx/hypopharynx tumors

TORS for larynx/hypopharynx tumors usually follows the same principles of transoral laser microsurgery (TLM), with its advantages over open procedures and radiotherapy in this scenario, promoting better functional results with similar oncological outcomes. TORS procedures for larynx tumors could be supraglottic and glottic (cordectomy) resections for early-stage tumors (T1–2) and highly selected T3 lesions, and some authors reported selected cases suitable for total laryngectomy in more advanced stages [23,24].

The potential advantages of TORS compared to TLM include better 3D magnified high-definition imaging, a greater range of movement due to robotic endowrists, which can provide greater dexterity, eliminate the surgeon's possible tremors, and also make it possible to perform sutures. The limitations to using TORS for larynx/hypopharynx lesions are the costs, robotic system availability, increased procedure time, and a "crowded" operative field due to robotic arms and cameras. This last limitation could be solved in the near future with the dissemination of the single port system (da Vinci SP) [23].

A systematic review with 422 patients treated with TORS supraglottic resections was reported by LeChien et al. Most tumors were T1–T2 (83%) and the location was epiglottis and aryepiglottic fold in 86% of patients. The 2-year local control rates ranged from 94.3% to 100%. The 2-year and 5-year overall survival rates ranged from 66.7% to 88.0% and 78.7% to 80.2%, respectively. In this review, feeding tubes were used in 62% of patients and a percutaneous endoscopic gastrostomy in 9%, and the return to oral diet varied from 1 day to 5 weeks [25].

In another review by Turner et al., including 503 patients submitted to a supraglottic TORS, the mean hospital stay was 9.5 days (3.9–15.1), positive surgical margins rate was 5%, mean operative time for tumor resection varied from 25.3 to 124.0 min, bleed rate of 3.7%, 33.9% were treated with elective tracheostomy, the 2-year local control rates ranged from 94.3% to 100%, and 5-year overall survival rates ranged from 78.7% to 80.2% [26].

TORS for glottic cancers is more challenging because it requires more exposure as it is more inferior/distal in the larynx. For glottic cases, patient selection is of paramount importance for the right exposure. The current evidence for TORS in the treatment of glottic tumors is limited. A review including 114 patients with a T1–T2 glottic tumors reported that the exposure was inadequate in 4% of cases, leading to conversion for TLM. Other findings included positive resection margins in 4.5% of patients, local recurrence in 10.7% of patients, with a wide range of feeding tube use (15%) and tracheostomy (23%). The mean duration of surgery and hospital stay ranged from 30 to 82 min and 2 to 7 days, respectively. Complications reported rate was 12% in this review, including dyspnea, bleeding, granuloma, synechia, tongue hematoma, and tongue dysesthesia. The authors' conclusion was that the current robotic system does not appear adequate for TORS cordectomy. Probably with the advance of robotic systems, such as da Vinci SP, robotic cordectomy could be more feasible.

Total laryngectomy using TORS is not well established, with just a few series reporting the feasibility of the procedure. The rationale could be an attempt to decrease de morbidity of the open procedures, preserving more mucosa, decreasing the healing time, and decreasing the pharyngocutaneous fistula rate, and also it should be performed in cases in which there is no need for neck dissection (mainly bilateral neck dissection), such as salvage procedures after chemoradiation without imaging evidence of lymph node disease, histologys with no need of elective neck dissection as salivary gland tumors, chondrosarcomas and in cases of functional laryngectomy for the severe dysfunctional larynx [23]. The largest series that reported the outcomes of 10 patients submitted to TORS total laryngectomy. In this series, the mean robotic operative time was 278 min, and the mean hospital stay was 13.9 days (8–28 days). A patient with laryngeal chondrosarcoma presented local recurrence 3 years after surgery. No other local recurrence reported [27]. Currently, due to the limited number of studies and patients, the benefits of TORS total laryngectomy are still debatable.

## Parapharyngeal space tumors

This technological development of TORS expanded its indications, including the resection of parapharyngeal tumors. The refinement of the robotic system, with better 3D magnified vision and articulated instruments, led to a better exposure of parapharyngeal tumors, a reduction of tumor rupture risk, decreased the risk of incomplete removal, and improved the management of potential complications during the surgeries [27].

A systematic review from Italy, including 22 studies with 113 patients reported that all tumors were successfully resected. Most parapharyngeal space tumors had a benign histology (90%), with pleomorphic adenoma the most common (58%), followed by schwannomas (10%). The median tumor size was 4.8 cm, and a combined open approach (transcervical or transparotid) was necessary in 15% of patients. Tumor fragmentation occurred in 10% of patients.

Most patients returned to an oral diet on the first postoperative day (68%), with a median hospital stay of 3 days. Complications reported were dysphagia (4.5%), hematoma (3.6%), Horner's syndrome (2.7%), pharyngeal dehiscence (1.8%), trismus (1.8%), first bite syndrome (1.8%), vocal cord palsy/laryngeal paralysis (1.8%), phlegmon (0.9%%), and cervical emphysema (0.9%) [27].

## Metastatic lymph nodes from occult primary tumors

TORS can also be employed in cases of unknown primary tumors (UPT). The diagnostic work-up for UPT usually consists of a physical exam with an endoscopic evaluation of the upper aerodigestive tract (UADT), CT and/or MRI, and a PET-CT scan, followed by examination under general anesthesia, with biopsies guided by the imaging findings. If no primary tumor is identified, it can be performed an ipsilateral tonsillectomy, and/or random biopsies from UADT. In this scenario, TORS could be a final step in this investigation, which could enable a bilateral diagnostic resection of the base of the tongue mucosa (lingual tonsillectomy), increasing the primary identification rate, up to 72%–100% of cases [28].

In a systematic review addressing the role of TORS, TLM, and lingual tonsillectomy in the identification of head and neck squamous cell carcinoma of unknown primary origin, the authors reported that TORS/TLM identified the primary tumor in 80% of patients overall. The base of tongue mucosal resection (lingual tonsillectomy) identified the primary tumor in 18/25 (72%%) patients with no clinical findings.

Other series also reported that pharyngeal and lingual tonsillectomy using TORS can improve the identification rate of occult primary tumors, directing treatment to the primary tumor subsite (surgical +/− radiotherapy), potentially decreasing the treatment morbidity [29].

## Obstructive sleep apnea

TORS was introduced for the treatment of obstructive sleep apnea (OSA) in 2010. Selected patients with OSA could benefit by a surgical intervention, usually when the OSA is due to hypertrophic tonsils, which is described in most patients [30]. Compared with other techniques, TORS improves the surgical view, and facilitates the instrumentation and surgical dexterity of the surgeon.

A review study addressing the effect of TORS base of tongue reduction on sleep-related outcomes in patients with OSA, including 353 patients, reported significant improvement in the clinical and patient-reported outcomes measures in about 70% of patients. Most patients reported improved outcomes in the apnea-hypopnea index (AHI), Epworth Sleepiness Scale, lowest oxygen saturation, and snoring visual analog scale. As demonstrated by the authors, the surgical success rate (defined as a >50% reduction of AHI with a postoperative AHI < 20) was observed in 68.4% of patients, while the cure rate (defined as a postoperative AHI <5) was achieved by 23.8% of patients [30]. In this study, the most frequent complications were transitory taste alteration, tongue numbness, tongue soreness, bleeding, edema, and dysphagia. In only three patients taste alteration lasted more than 1 year for resolution. Four patients presented postoperative bleeding requiring additional surgery, and one patient presented postoperative pharyngeal stenosis treated by a surgical intervention.

## Thyroidectomies

Since the beginning of the 2000s, several alternative techniques have been developed for the surgical treatment of the thyroid, with the aim of reducing morbidity, mainly related to cosmetic issues. Initial experience with minimally invasive video-assisted techniques (MIVAT), followed by transcervical video approaches, and finally remote access through axillar, breast, postauricular, and transoral routes. The latter, the transoral endoscopic thyroidectomy vestibular approach (TOETVA) was the only technique without visible scars on the skin, developed in Thailand by Anuwong, which was the most widespread technique in other countries [31–33].

From 2007, robot-assisted thyroid surgery has been described, but the transoral approach was introduced a few years later. The transoral technique compared with the other robotic approaches, provides a midline exposure with access to both thyroid lobes and central compartment and has a more familiar view of the operative field by the surgeon, which carries an advantage over the other lateral robotic access (axillary, breast, or retroauricular routes) [34].

Several series reported similar oncological and functional results between transoral and conventional open techniques, with the cosmetic advantages of transoral access [34,35]. However, the transoral approach is a technique that requires training, reserved for patients with low-volume thyroid surgical pathologies and for patients who are motivated to avoid a visible cervical scar.

When we compare the TOETVA with the robotic transoral technique (TORTVA), some studies suggest that the robotic technique provides a better magnified 3D image and the endowrist of the robotic arms would facilitate the procedure. A systematic review comparing TORTVA with other robotic techniques and with TOETVA demonstrated greater cosmetic satisfaction with TORTVA; however, it showed a longer operative time, longer hospitalization period, and more pain in the early postoperative period than the TOETVA technique [36].

## Neck dissection

In patients with oropharyngeal and supraglottic malignant tumors, due to the high risk of lymph node dissemination, a neck dissection is necessary. Such neck dissection can be performed simultaneously with resection of the primary tumor, or in a sequential staged manner. It is recommended that ligation of ascending pharyngeal artery, lingual artery, and facial artery should be done to decrease the bleeding rate.

In selected cases, the neck dissection could be done by a remote access (retroauricular) also using the robotic system. This technique is discussed in other section of this book.

## TORS technique

The technique for TORS includes the setup of a self-retaining mouth gag with a tongue retractor (FK-WO Olympus, Crowe-Davis, or Dingman) to maintain the working space, and dock of the robotic system, with 3 arms. A 0-degree or a 30-degree camera is usually inserted in the central robotic arm and a bipolar Maryland forceps and a monopolar scissors (or spatula tip monopolar cautery) in the lateral arms (Fig. 1). The surgical bedside assistant should be at the head of the patient to perform suction, retraction, or to place hemostatic clips (Fig. 2).

For a radical tonsillectomy, a mucosal incision in the soft palate through the superior constrictor muscle, the deep margin being the parapharyngeal space fat, the superior margin the soft palate, the posterior oropharyngeal wall posteriorly, the retromolar trigone anteriorly and the base of the tongue as the inferior limit (Fig. 3).

For the tongue base tumors, we try to plan a compartmental resection that is limited by the oral tongue anteriorly, palatine tonsil laterally, midline medially, and vallecula posteriorly. The deep margin at the tongue muscle will be

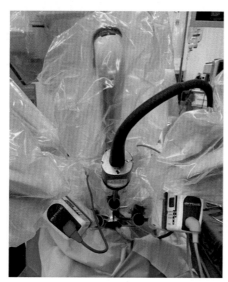 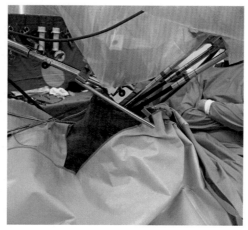

FIG. 1   Dock of the robotic system with 3 arms.

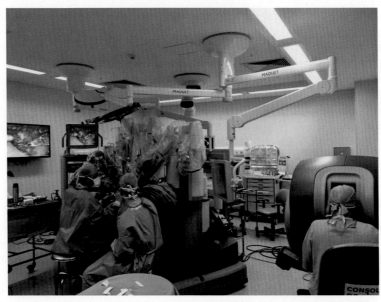

FIG. 2    Surgical team disposition. Surgeon at the robotic console and the surgical bedside assistant at the head of the patient.

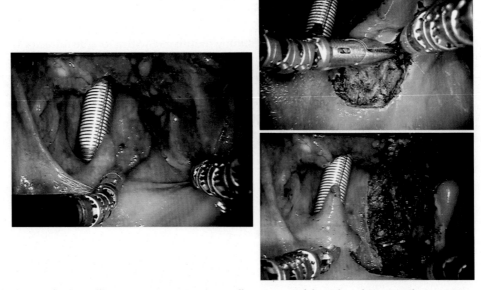

FIG. 3    Transoral robotic radical tonsillectomy to treat a squamous cell carcinoma of the right palatine tonsil.

defined by the depth of invasion of the tumor. In most cases, the lingual artery must be clipped and sectioned. For cases of unknown primaries, the base of tongue resection is restricted to the ipsilateral lingual tonsil and mucosal layer, sparing the muscles.

For supraglottic larynx or hypopharynx resection, after proper exposure and docking, the surgical resection is planned in a way that proper margins can be obtained. Besides, the limitations imposed by the narrow working space, on-bloc resections can be performed in the majority of cases.

For resection of parapharyngeal space tumors, a 3–6 cm incision should be done at the anterior tonsil pillars, through the superior constrictor muscle to access the parapharyngeal space. After tumor resection, a primary closure of the mucosal defect should be performed (Fig. 4).

For OSA, the most frequent procedure is similar to the resection of lingual tonsil performed for patients with neck metastasis of UPT, but the resection is more prominent at the central part of tongue base, including a superficial part of the muscle layer, but avoiding the lateral part due to the risk of lingual artery bleeding.

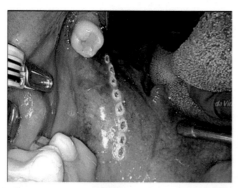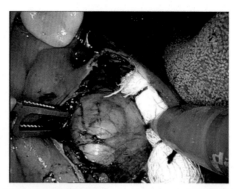
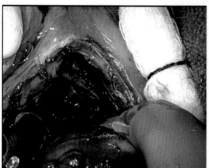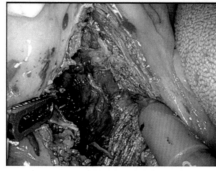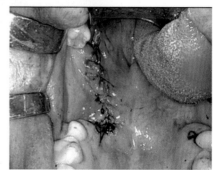

FIG. 4    Transoral robotic resection of a left parapharyngeal tumor (pleomorphic adenoma).

## Future scenario

Some institutions have reported the use of neoadjuvant chemotherapy followed by TORS resection in patients with HPV-related oropharyngeal tumors. This combination can potentially decrease the use of radiotherapy, as the main treatment option as well as in the adjuvant setting, mitigating the acute and long-term morbidity of radiotherapy. Preliminary reports have shown encouraging results [37].

Another perspective in TORS is the optically guided surgery. The combination of TORS with new intraoperative imaging of fluorescent molecules can proportionate the visualization of tissues of interest to guide the surgery, such as lymph nodes or to guide surgical margins. Most studies addressing optically guided surgery have utilized Indocyanine green. The *da Vinci* robotic system has the option to adopt a built-in optical camera system [38].

It is important to highlight that TORS is constantly evolving. The new *da Vinci* SP (single port) provides 360° of anatomical access with a single arm designed to eliminate the possibility of external collisions and clashing of arms. Currently, its use is restricted for just a few centers, with interesting results [39]. However, with the dissemination of its use worldwide, we will likely see the expansion of the use of TORS in the field of head and neck surgery. Additionally, continued advances in robotic technology and surgical training may further improve outcomes and expand indications for TORS in the future.

## Key points

* Transoral robotic surgery is constantly evolving.
* The most frequent indication is in the treatment of HPV-related oropharyngeal carcinoma as an option to decrease the long-term morbidity of radiation/chemoradiation as well as the morbidity associated with the conventional open approach.
* Other indications of TORS include the treatment of supraglottic larynx tumors, highly selected patients with glottic tumors, parapharyngeal space tumors, and also as another option to detect an unknown primary tumor with base of tongue mucosal resection.
* TORS can be useful in the treatment of selected patients with obstructive sleep apnea and in highly selected and motivated patients with thyroid tumors.

# References

[1] Weinstein GS, O'Malley BW, Snyder W, et al. Transoral robotic surgery: radical tonsillectomy. Arch Otolaryngol Head Neck Surg 2007; 133(12):1220–6.

[2] Byeon HK, Koh YW. The new era of robotic neck surgery: the universal application of the retroauricular approach. J Surg Oncol 2015; 112(7):707–16.

[3] Lira RB, Kowalski LP. Robotic head and neck surgery: beyond TORS. Curr Oncol Rep 2020;22(9):880.

[4] Poon H, Li C, Gao W, et al. Evolution of robotic systems for transoral head and neck surgery. Oral Oncol 2018;87:82–8.

[5] Dziegielewski PT, Kang SY, Ozer E. Transoral robotic surgery (TORS) for laryngeal and hypopharyngeal cancers. J Surg Oncol 2015; 112(7):702–6.

[6] Miller SC, Nguyen SA, Ong AA, et al. Transoral robotic base of tongue reduction for obstructive sleep apnea: a systematic review and meta-analysis. Laryngoscope 2017;127(1):258–65.

[7] Richmon JD, Quon H, Gourin CG. The effect of transoral robotic surgery on short-term outcomes and cost of care after oropharyngeal cancer surgery. Laryngoscope 2014;124(1):165–71.

[8] Albergotti WG, Jordan J, Anthony K, Abberbock S, Wasserman-Wincko T, Kim S, Ferris RL, Duvvuri U. A prospective evaluation of short-term dysphagia after transoral robotic surgery for squamous cell carcinoma of the oropharynx. Cancer 2017;123(16):3132–40.

[9] Hutcheson KA, Holsinger FC, Kupferman ME, Lewin JS. Functional outcomes after TORS for oropharyngeal cancer: a systematic review. Eur Arch Otorhinolaryngol 2015;272(2):463–71.

[10] Chia SH, Gross ND, Richmon JD. Surgeon experience and complications with Transoral robotic surgery (TORS). Otolaryngol Head Neck Surg 2013;149(6):885–92.

[11] Dombrée M, Crott R, Lawson G, et al. Cost comparison of open approach, transoral laser microsurgery and transoral robotic surgery for partial and total laryngectomies. Eur Arch Otorhinolaryngol 2014;271:2825–34.

[12] Krishnan G, Mintz J, Foreman A, Hodge JC, Krishnan S. The acceptance and adoption of transoral robotic surgery in Australia and New Zealand. J Robot Surg 2019;13(2):301–7.

[13] Albergotti WG, Gooding WE, Kubik MW, Geltzeiler M, Kim S, Duvvuri U, Ferris RL. Assessment of surgical learning curves in Transoral robotic surgery for squamous cell carcinoma of the oropharynx. JAMA Otolaryngol Head Neck Surg 2017;143(6):542–8.

[14] Gaino F, Gorphe P, Vander Poorten V, Holsinger FC, Lira RB, Duvvuri U, Garrel R, Van Der Vorst S, Cristalli G, Ferreli F, De Virgilio A, Giannitto C, Morenghi E, Colombo G, Malvezzi L, Spriano G, Mercante G. Preoperative predictors of difficult oropharyngeal exposure for transoral robotic surgery: the Pharyngoscore. Head Neck 2021;43(10):3010–21.

[15] Pauloski BR. Rehabilitation of dysphagia following head and neck cancer. Phys Med Rehabil Clin N Am 2008;19(4):889–928.

[16] Lechner M, Liu J, Masterson L, Fenton TR. HPV-associated oropharyngeal cancer: epidemiology, molecular biology and clinical management. Nat Rev Clin Oncol 2022;19(5):306–27.

[17] Ang KK, Harris J, Wheeler R, Weber R, Rosenthal DI, Nguyen-Tân PF, Westra WH, Chung CH, Jordan RC, Lu C, Kim H, Axelrod R, Silverman CC, Redmond KP, Gillison ML. Human papillomavirus and survival of patients with oropharyngeal cancer. N Engl J Med 2010;363(1):24–35.

[18] Mahmoud O, Sung K, Civantos FJ, Thomas GR, Samuels MA. Transoral robotic surgery for oropharyngeal squamous cell carcinoma in the era of human papillomavirus. Head Neck 2018;40(4):710–21.

[19] Ferris RL, Flamand Y, Weinstein GS, et al. Phase II randomized trial of transoral surgery and low-dose intensity modulated radiation therapy in resectable p16+ locally advanced oropharynx cancer: an ECOG-ACRIN Cancer research group trial (E3311). J Clin Oncol 2022;40(2):138–49.

[20] Barbon CEA, Yao CMK, Peterson CB, Moreno AC, Goepfert RP, Johnson FM, Chronowski GM, Fuller CD, Gross ND, Hutcheson KA. Swallowing after primary TORS and unilateral or bilateral radiation for low- to intermediate-risk tonsil Cancer. Otolaryngol Head Neck Surg 2022; 167(3):484–93.

[21] Castellano A, Sharma A. Systematic review of validated quality of life and swallow outcomes after Transoral robotic surgery. Otolaryngol Head Neck Surg 2019;161(4):561–7.

[22] Nichols AC, Theurer J, Prisman E, et al. Randomized trial of radiotherapy versus transoral robotic surgery for oropharyngeal squamous cell carcinoma: long-term results of the ORATOR trial. J Clin Oncol 2022;40(8):866–75.

[23] Jia W, King E. The role of robotic surgery in laryngeal Cancer. Otolaryngol Clin North Am 2023;56(2):313–22.

[24] Lawson G, Mendelsohn A, Fakhoury R, et al. Transoral robotic surgery total laryngectomy. ORL J Otorhinolaryngol Relat Spec 2018;80:171–7.

[25] Lechien JR, Fakhry N, Saussez S, et al. Surgical, clinical and functional outcomes of transoral robotic surgery for supraglottic laryngeal cancers: a systematic review. Oral Oncol 2020;109, 104848.

[26] Turner MT, Stokes WA, Stokes CM, et al. Airway and bleeding complications of transoral robotic supraglottic laryngectomy (TORS-SGL): a systematic review and meta-analysis. Oral Oncol 2021;118, 105301.

[27] De Virgilio A, Costantino A, Mercante G, Di Maio P, Iocca O, Spriano G. Trans-oral robotic surgery in the management of parapharyngeal space tumors: a systematic review. Oral Oncol 2020;103, 104581.

[28] Fu TS, Foreman A, Goldstein DP, de Almeida JR. The role of transoral robotic surgery, transoral laser microsurgery, and lingual tonsillectomy in the identification of head and neck squamous cell carcinoma of unknown primary origin: a systematic review. J Otolaryngol Head Neck Surg 2016;45(1):28.

[29] Larsen MHH, Channir HI, von Buchwald C. Human papillomavirus and squamous cell carcinoma of unknown primary in the head and neck region: a comprehensive review on clinical implications. Viruses 2021;13(7):1297.

[30] Miller SC, Nguyen SA, Ong AA, Gillespie MB. Transoral robotic base of tongue reduction for obstructive sleep apnea: a systematic review and meta-analysis. Laryngoscope 2017;127(1):258–65.

[31] Miccoli P, Berti P, Conte M, Bendinelli C, Marcocci C. Minimally invasive surgery for thyroid small nodules: preliminary report. J Endocrinol Invest 1999;22:849–51.

[32] Lee KE, Rao J, Youn YK. Endoscopic thyroidectomy with the da Vinci robot system using the bilateral axillary breast approach (BABA) technique: our initial experience. Surg Laparosc Endosc Percutan Tech 2009;19:e71–5.

[33] Anuwong A. Transoral endoscopic thyroidectomy vestibular approach: a series of the first 60 human cases. World J Surg 2016;40:491–7.

[34] Chai YJ, Kim HY, Kim HK, Jun SH, Dionigi G, Anuwong A, Richmon JD, Tufano RP. Comparative analysis of 2 robotic thyroidectomy procedures: Transoral versus bilateral axillo-breast approach. Head Neck 2018;40:886–92.

[35] You JY, Kim HY, Chai YJ, Kim HK, Anuwong A, Tufano RP, Dionigi G. Transoral robotic thyroidectomy versus conventional open thyroidectomy: comparative analysis of surgical outcomes in thyroid malignancies. J Laparoendosc Adv Surg Tech A 2019;29:796–800.

[36] Kang YJ, Cho JH, Stybayeva G, Hwang SH. Safety and efficacy of Transoral robotic thyroidectomy for thyroid tumor: a systematic review and Meta-analysis. Cancers (Basel) 2022;14(17):4230.

[37] Sadeghi N, Mascarella MA, Khalife S, Ramanakumar AV, Richardson K, Joshi AS, Taheri R, Fuson A, Bouganim N, Siegel R. Neoadjuvant chemotherapy followed by surgery for HPV-associated locoregionally advanced oropharynx cancer. Head Neck 2020;42(8):2145–54.

[38] Lee YJ, Krishnan G, Nishio N, et al. Intraoperative fluorescence-guided surgery in head and neck squamous cell carcinoma. Laryngoscope 2021;131(3):529–34.

[39] Costantino A, Sampieri C, Meliante PG, De Virgilio A, Kim SH. Transoral robotic surgery in oropharyngeal squamous cell carcinoma: a comparative study between da Vinci single-port and da Vinci xi systems. Oral Oncol 2023;148, 106629.

III. The current and future clinical applications of robotic surgery among medical specialties

CHAPTER

# 49

# Robotic neck dissection

*Renan Bezerra Lira[a,b], José Guilherme Vartanian[b], and Luiz Paulo Kowalski[b,c]*

[a]Robotic Surgery Program, Hospital Israelita Albert Einstein, São Paulo, Brazil [b]Department of Head and Neck Surgery and Otorhinolaryngology, A.C. Camargo Cancer Center, São Paulo, São Paulo, Brazil [c]Head and Neck Surgery Department and LIM 28, University of São Paulo Medical School, São Paulo, Brazil

## Introduction

Neck dissection remains as a key part of the surgical treatment of head and neck squamous cell carcinoma. Thyroid carcinoma lymph node metastasis is treated surgically as well, in the great majority of cases. These procedures are usually performed using large anterior neck incisions and associated with significant fibrosis and local sensorial changes, resulting in different degrees of potentially avoidable aesthetic and functional sequelae, that might be associated with psychosocial repercussions [1]. Therefore, the number of patients that need neck dissection and would benefit from a less morbid approach is very significant.

Alongside the establishment of transoral robotic surgery (TORS) as a standard procedure in oropharyngeal carcinoma treatment, head and neck surgeons around the world tried to find beneficial applications of this new technological tool. Aiming to avoid visible neck scars reducing aesthetic and psychological consequences, especially for young patients, some remote access approaches to the neck were described in the last decades [2–11]. These new approaches gained enthusiasm especially in Asia, where cultural aspects motivate patients to avoid a visible neck scar [12], and expanded the potential roles of robot-assisted surgery in the head and neck. In this scenario, a retroauricular approach was reported as a viable and versatile option for different neck procedures, including selective and comprehensive robotic or endoscopic neck dissections [2,6–8,13,14]. This technique led not only better cosmetic results, but also it reduced the risk of postoperative complications and was associated with similar oncologic outcomes in several case series [2,6–8,13–17]. To date, the retroauricular approach remains as the stablished technique for robotic neck dissection. However, due to various concerns of exposure and visualization, many head and neck surgeons remain hesitant to routinely adopt robotic neck dissection, despite these encouraging initial safety and oncological outcomes [18–25], sustaining significant controversy around the effectiveness of such procedures when compared to conventional well-stablished techniques [26,27].

## Operative technique

Under general anesthesia, the patient is positioned over the surgical table in the supine position with arms alongside the body and contralateral head rotation with minimal neck extension.

The incision is planned from the inferior limit of the ear implantation ascending in the retroauricular area, gently curving downward and then following the hairline about 5 mm inside the scalp (Fig. 1A). Using conventional instrumentation and a head light, skin flap dissection is performed just superficial of the sternocleidomastoid (SCM) muscle and fascia, exposing and preserving the great auricular nerve and external jugular vein. The posterior border of the platysma muscle is then identified in that area, and the subplatysmal plan is followed until the midline and superiorly and inferiorly enough to expose the neck levels that will be dissected. During this part of the dissection, it is helpful to identify and dissect the accessory nerve from the posterior border of the SCM muscle to the posterior belly of the

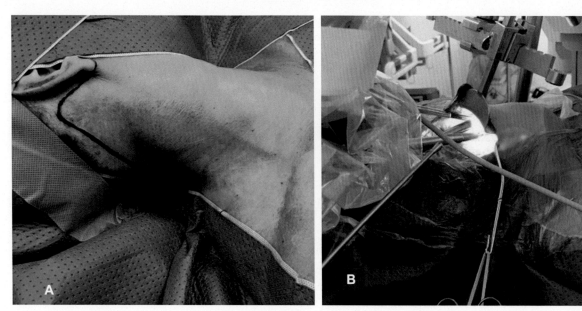

FIG. 1    Robotic neck dissection retroauricular approach (*right side*). (A) Incision planning. (B) Docking of robotic platform.

digastric muscle, removing the level IIb. The next step is to setup a self-retaining retractor (Bookwalter or Thompson) to maintain the working space, and dock of the robot system (Fig. 1B).

The robotic platform is placed at the contralateral side, with its central column about 30 cm lateral to the patient shoulder in a 45–60-degree angle with the surgical table. A 30-degree downward camera is inserted in the central robotic arm and bipolar Maryland forceps and monopolar scissors in the lateral arms.

The procedure continues with the robotic dissection of levels II–III (and, if indicated, level IV) from superior to inferior, dissecting and preserving the internal jugular vein, carotid arteries, as well as the vagus and hypoglossal nerves. The jugular tributary veins can be preserved or removed. Hem-o-lock clips are applied when needed for hemostasis and for lymphatic ducts sealing. Surgical specimen containing levels II–IV nodes is then removed (Fig. 2A–C). Level V also can be easily reached and dissected through the retroauricular approach, preserving the accessory nerve and cervical transverse artery.

For level Ib dissection, initially we identify, dissect, and preserve the marginal branch of the facial nerve. Then, level Ib is approached from posterior to anterior, detaching it from the posterior belly of the digastric muscle and parotid inferior pole. The facial artery is exposed, dissected and clipped with a Hem-o-lock just above the posterior belly of the digastric muscle. The dissection continues along the inferior mandibular edge, sealing and cutting the distal facial vessels while preserving the marginal branch. The mylohyoid muscle is then exposed and the lingual and hypoglossal nerves are identified and preserved, while dividing the lingual nerve branch into the submandibular gland and the submandibular duct. Level Ib dissection is completed removing all the fibroadipose tissue from mylohyoid muscle, reaching the anterior ipsilateral digastric belly (Fig. 2D). Level Ia is then dissected from the hyoid bone to the mandible between the anterior digastric muscle bellies, following the underlying mylohyoid muscle. After the specimen is removed, hemostasis is secured, and suction drain placed. Closure is then carried out in 2 layers with absorbable sutures and Dermabond.

## Learning curve

For head and neck surgeons, a comprehensive anatomy understanding of landmarks and precise location of lymph node chains, muscles, vessels and nerves is mandatory for conventional neck surgery and even more for robotic neck dissection. Ideally, only well-experienced surgeons in conventional neck dissection should step into the learning curve of robotic neck surgery. Although the anatomy is the same, surgeons must get used to new surgical devices and, most importantly, to operate in different attack angles, from lateral and superior, to medial and inferior.

In a prospective study, Kim et al. [18] assessed the learning curve for robotic neck dissection using the retroauricular approach. There were 50 modified radical neck dissections and 40 selective neck dissections performed by a single

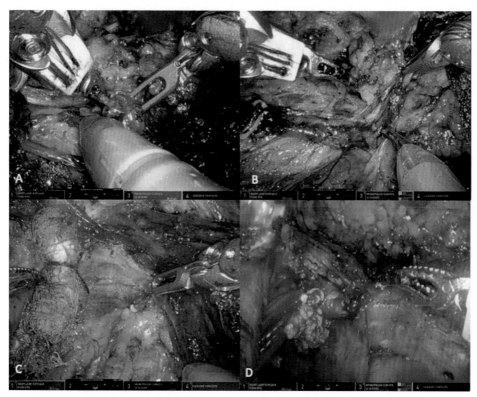

FIG. 2  Surgical field of robotic neck dissection using retroauricular approach (*right side*). (A) Level II. (B) Level III. (C) Level IV. (D) Level Ib.

surgeon. There was a significant reduction in the average operating time in both groups. The average operating time for the modified radical neck dissection group decreased by 29% during the course of the study (initial subgroup, 298.1 min; last subgroup, 212.4 min). In the subgroup of selective neck dissections, the mean operating time decreased by 53% (initial subgroup, 226.5 min; last subgroup, 106.1 min). To date, we do not have a threshold number for the proper learning curve in robotic neck dissection, but approximately 30 cases would be sufficient for an experienced head and neck surgeon to become proficient.

## Surgical and oncological results

Even though robotic neck dissection has benefits, the need for specific training and a stipe learning curve limited a massive routine adoption of this technique by most head and neck departments. Nevertheless, a few reference centers around the world are consistently performing robotic neck dissection using the retroauricular approach, with encouraging results, reaffirming its surgical and oncological safety in several case series, that have shown favorable outcomes without any major complications or surgery-related deaths. A systematic review and metaanalysis including 11 studies and more than 200 robotic neck dissection cases reported similar results regarding hospital stay, lymph node yield and recurrence rate when compared to patients submitted to conventional neck dissection. The only disadvantage was longer operative time. All included studies that assessed cosmetic outcomes reported higher patient satisfaction in those submitted to robotic neck dissection when compared to the conventional approach [28,29]. In some papers that looked at functional outcomes (edema, fibrosis, movement, and sensory loss following neck dissection), robotic neck dissection was associated with lesser postoperative neck edema and sensory loss [29].

Our group has gathered a significant experience using retroauricular endoscopic and robotic neck surgery. We have performed more than 400 robotic and endoscopic neck dissections so far, with comparable oncological and surgical outcomes compared to our historic series of conventional surgery. In our opinion, there has been a significant aesthetic and functional advantage in these patients that are being submitted to endoscopic and robotic neck dissection in our center (Fig. 3) [14,22–24,30,31].

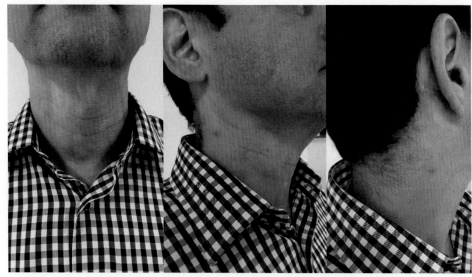

FIG. 3    Aesthetic outcome of robotic neck dissection using retroauricular approach (*right side*—levels I–III).

## Future scenario

The evolution of robotic surgery has been intensified in last few years, with several companies trying to develop different robotic platforms, more adapted to specific procedures, with different mechanical and surgical features. Besides that, the simultaneous growth of artificial intelligence and augmented reality is also very promising in the future of digital surgery. It is difficult to precisely preview the next paradigm shift in this field; however, it is certain that this rising evolution in robotic and digital surgery will be reshaping the way that we treat our patients in the years to come.

## Key points

- The applications of robotic-assisted procedures are on the rise in the head and neck surgery.
- Retroauricular approach combined with robotic surgery gives great access to the ipsilateral neck, allowing oncologically sound robotic neck dissections with better cosmetic and functional outcomes.
- The use of the robotic platform is already bringing technical advantages to robotic neck dissection, such as magnified 3D view and movement filter, making it a more "precise" procedure.
- In the future, with more adapted robotic systems and increased integration with artificial intelligence and augmented reality, robotics will have a greater hole in both neck and transoral/transnasal surgery.

## References

[1] Weymuller EA, Yueh B, Deleyiannis FW, Kuntz AL, Alsarraf R, Coltrera MD. Quality of life in patients with head and neck cancer: lessons learned from 549 prospectively evaluated patients. Arch Otolaryngol Head Neck Surg 2000;126(3):329–35. discussion 335–336.
[2] Lee HS, Kim WS, Hong HJ, Ban MJ, Lee D, Koh YW, et al. Robot-assisted supraomohyoid neck dissection via a modified face-lift or retroauricular approach in early-stage cN0 squamous cell carcinoma of the oral cavity: a comparative study with conventional technique. Ann Surg Oncol 2012;19(12):3871–8.
[3] Terris DJ, Singer MC. Robotic facelift thyroidectomy: Facilitating remote access surgery. Head Neck 2012;34(5):746–7.
[4] Lee HS, Lee D, Koo YC, Shin HA, Koh YW, Choi EC. Endoscopic resection of upper neck masses via retroauricular approach is feasible with excellent cosmetic outcomes. J Oral Maxillofac Surg 2013;71(3):520–7.
[5] Lee J, Chung WY. Robotic thyroidectomy and neck dissection: past, present, and future. Cancer J 2013;19(2):151–61.
[6] Park YM, Holsinger FC, Kim WS, Park SC, Lee EJ, Choi EC, et al. Robot-assisted selective neck dissection of levels II to V via a modified facelift or retroauricular approach. Otolaryngol-Head Neck Surg 2013;148(5):778–85.
[7] Tae K, Ji YB, Song CM, Min HJ, Kim KR, Park CW. Robotic selective neck dissection using a gasless postauricular facelift approach for early head and neck cancer: technical feasibility and safety. J Laparoendosc Adv Surg Tech A 2013;23(3):240–5.
[8] Byeon HK, Holsinger FC, Koh YW, Ban MJ, Ha JG, Park JJ, et al. Endoscopic supraomohyoid neck dissection via a retroauricular or modified facelift approach: preliminary results: endoscopic supraomohyoid neck dissection. Head Neck 2014;36(3):425–30.

[9] Mohamed SE, Noureldine SI, Kandil E. Alternate incision-site thyroidectomy. Curr Opin Oncol 2014;26(1):22–30.

[10] Lee HS, Kim D, Lee SY, Byeon HK, Kim WS, Hong HJ, et al. Robot-assisted versus endoscopic submandibular gland resection via retroauricular approach: a prospective nonrandomized study. Br J Oral Maxillofac Surg 2014;52(2):179–84.

[11] Byeon HK, Holsinger FC, Kim DH, Kim JW, Park JH, Koh YW, et al. Feasibility of robot-assisted neck dissection followed by transoral robotic surgery. Br J Oral Maxillofac Surg 2015;53(1):68–73.

[12] Mohamed SE, Noureldine SI, Kandil E. Alternate incision-site thyroidectomy. Curr Opin Oncol 2014;26(1):22–30.

[13] Greer Albergotti W, Kenneth Byrd J, De Almeida JR, Kim S, Duvvuri U. Robot-assisted level II-IV neck dissection through a modified facelift incision: initial North American experience: robot-assisted neck dissection. Int J Med Robot 2014;10(4):391–6.

[14] Chulam TC, Lira RB, Kowalski LP. Robotic-assisted modified retroauricular cervical approach: initial experience in Latin America. Rev Col Bras Cir 2016;43(4):289–91.

[15] Tae K, Ji YB, Song CM, Jeong JH, Cho SH, Lee SH. Robotic selective neck dissection by a postauricular facelift approach: comparison with conventional neck dissection. Otolaryngol-Head Neck Surg 2014;150(3):394–400.

[16] Lira RB, Chulam TC, de Carvalho GB, Schreuder WH, Koh YW, Choi EC, et al. Retroauricular endoscopic and robotic versus conventional neck dissection for oral cancer. J Robot Surg 2017. Available from: http://link.springer.com/10.1007/s11701-017-0706-0.

[17] Lira RB, Chulam TC, Kowalski LP. Variations and results of retroauricular robotic thyroid surgery associated or not with neck dissection. Gland Surg 2018;7(S1):S42–52.

[18] Byeon HK, Holsinger FC, Tufano RP, Chung HJ, Kim WS, Koh YW, et al. Robotic total thyroidectomy with modified radical neck dissection via unilateral retroauricular approach. Ann Surg Oncol 2014;21(12):3872–5.

[19] Byeon HK, Holsinger FC, Kim DH, Kim JW, Park JH, Koh YW, et al. Feasibility of robot-assisted neck dissection followed by transoral robotic surgery. Br J Oral Maxillofac Surg 2015;53(1):68–73.

[20] Anuwong A, Ketwong K, Jitpratoom P, Sasanakietkul T, Duh QY. Safety and outcomes of the transoral endoscopic thyroidectomy vestibular approach. JAMA Surg 2017. Available from: http://archsurg.jamanetwork.com/article.aspx?doi=10.1001/jamasurg.2017.3366.

[21] Kim MJ, Nam KH, Lee SG, Choi JB, Kim TH, Lee CR, et al. Yonsei experience of 5000 gasless transaxillary robotic thyroidectomies. World J Surg 2017. Available from: http://link.springer.com/10.1007/s00268-017-4209-y.

[22] Lira RB, Chulam TC, de Carvalho GB, Schreuder WH, Koh YW, Choi EC, et al. Retroauricular endoscopic and robotic versus conventional neck dissection for oral cancer. J Robot Surg 2017. Available from: http://link.springer.com/10.1007/s11701-017-0706-0.

[23] Lira RB, Chulam TC, Kowalski LP. Safe implementation of retroauricular robotic and endoscopic neck surgery in South America. Gland Surg 2017;6(3):258–66.

[24] Lira RB, Chulam TC, Kowalski LP. Variations and results of retroauricular robotic thyroid surgery associated or not with neck dissection. Gland Surg 2018;7(S1):S42–52.

[25] Kim HK, Chai YJ, Dionigi G, Berber E, Tufano RP, Kim HY. Transoral robotic thyroidectomy for papillary thyroid carcinoma: perioperative outcomes of 100 consecutive patients. World J Surg 2019;43(4):1038–46.

[26] Berber E, Bernet V, Fahey TJ, Kebebew E, Shaha A, Stack BC, et al. American thyroid association statement on remote-access thyroid surgery. Thyroid Off J Am Thyroid Assoc 2016;26(3):331–7.

[27] Hinson AM, Kandil E, O'Brien S, Spencer HJ, Bodenner DL, Hohmann SF, et al. Trends in robotic thyroid surgery in the United States from 2009 through 2013. Thyroid 2015;25(8):919–26.

[28] Sukato DC, Ballard DP, Abramowitz JM, Rosenfeld RM, Mlot S. Robotic versus conventional neck dissection: a systematic review and meta-analysis. Laryngoscope 2019;129(7):1587–96.

[29] Ji YB, Song CM, Bang HS, Park HJ, Lee JY, Tae K. Functional and cosmetic outcomes of robot-assisted neck dissection by a postauricular facelift approach for head and neck cancer. Oral Oncol 2017;70:51–9.

[30] Lira RB, Kowalski LP. Robotic neck dissection: state of affairs. Curr Opin Otolaryngol Head Neck Surg 2020;28(2):96–9.

[31] Kowalski LP, Lira RB. Anatomy, technique, and results of robotic retroauricular approach to neck dissection. Anat Rec Hoboken 2021;304 (6):1235–41.

# 50

# Otolaryngology: Sleep apnea and benign diseases robotic surgery

*Wei Li Neo[a], Chwee Ming Lim[a,b], and Song Tar Toh[a,c]*

[a]Department of Otorhinolaryngology—Head and Neck Surgery, Singapore General Hospital, Singapore, Singapore
[b]Surgery Academic Clinical Programme, Duke-NUS Graduate Medical School, Singapore, Singapore [c]SingHealth Duke-NUS Sleep Centre, Singapore, Singapore

## Introduction

The introduction of the da Vinci robotic system since the early 2000s has made a significant impact on the management of otolaryngology patients. Broadly speaking, surgeons have adopted its use in various areas including the oropharynx, hypopharynx, parapharyngeal space, and even thyroidectomies and neck dissections.

Advantages of a transoral robotic approach include superior visualization due to improved magnification and use of angled scopes in a narrow field and improved precision in tissue handling [1]. Furthermore, this minimally invasive approach allows access to previously difficult-to-reach areas without the need for the morbidity of an open transcervical approach.

In this chapter, we will discuss the use of transoral robotic surgery in sleep apnea surgery and for resection of benign oropharyngeal and parapharyngeal tumors.

## Sleep apnea surgery

Obstructive sleep apnea (OSA) is a prevalent problem, with an estimated prevalence of 30.5% in some adult populations [2]. It is associated with significant cardiovascular morbidity [3], cerebrovascular accidents [4], and increased road traffic accidents. Multilevel collapse is often seen in OSA. Although positive airway pressure (PAP) therapy is the gold standard for the treatment of OSA, compliance continues to be a problem with nonadherence rates of 29%–83% [5]. In patients who are intolerant of PAP therapy, surgery may be considered. Transoral robotic surgery (TORS) may be considered for tongue base reduction and partial epiglottectomy as part of multilevel surgery in OSA.

## Preoperative management

### Patient selection

Preoperative management should include taking a comprehensive history, clinical examination, and overnight sleep study. Patients with diagnosed OSA should then undergo a trial of PAP therapy. For patients who are intolerant of PAP therapy, they should undergo a drug-induced sleep endoscopy (DISE) to determine the levels of upper airway collapse.

Indications for transoral robotic surgery include the following [1,6]:

1. Base of tongue obstruction
2. Lingual tonsillar hypertrophy (Friedman type 3 or 4)
3. Epiglottic collapse
4. Supraglottic collapse

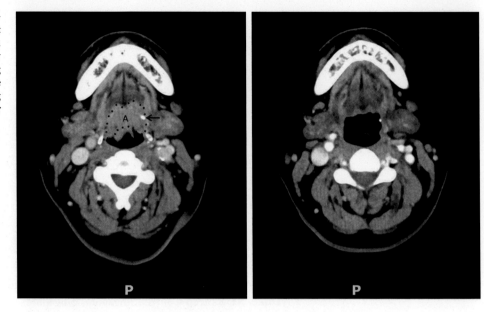

Contraindications for transoral robotic surgery include the following [1,7]:

1. Severe cardiac or pulmonary disease
2. Contraindications to general anesthesia
3. Severe trismus or inadequate mouth opening (less than 2 cm)
4. Unrealistic expectations of surgery
5. Uncontrolled psychiatric disorders

Preoperative imaging such as a computed tomography (CT) scan with intravenous contrast may be helpful but it is not mandated prior to surgery in our institution. Fig. 1 shows a preoperative CT scan of a patient with significant lymphoid hyperplasia.

## Intraoperative management

### *Relevant anatomy*

Lingual artery hemorrhage is a feared complication of TORS. As such, in-depth knowledge of the anatomy of the lingual artery is essential.

The deep lingual artery is the terminal branch of the lingual artery. Li et al. described the use of the computed tomographic angiogram (CTA) for identification. The average depth of the lingual arteries is $29.27 \pm 5.39$ mm [8]. At the foramen cecum, the average distance between the two lingual arteries was $27.78 \pm 6.57$ mm [8]. Based on the average distance and 95% confidence intervals (CI), a surgical functional zone was identified, in which 14.9 mm was the distance between the two lingual arteries, and 18.7 mm was the distance to the tongue surface [8].

Other methods of determining the course of the lingual artery included ultrasonography [9,10]. In an ultrasonographic study by Liu et al., the depth of the lingual arteries was $2.2 \pm 0.21$ cm and the distance between the two lingual arteries was found to be $1.61 \pm 0.27$ cm at the level of the foramen caecum [9]. The depth of the lingual arteries was significantly deeper in the OSA group compared to the control group ($P < 0.01$). Similarly, the distance between bilateral lingual arteries was wider in the OSA group compared to the control group ($P < 0.01$) [9].

Additionally, surgeons should be mindful that the position of the hypoglossal nerve - lingual artery neurovascular bundle is different in a resting position compared to its position during surgery, i.e., the tongue is protruded and elevated. A cadaveric dissection study by Cohen et al. showed that the average distance from the foramen caecum to the origin of the dorsal lingual artery was $2.0 \pm 0.75$ cm and $1.34 \pm 0.29$ cm in resting and suspended positions, respectively [11].

## Setup

The patient is placed in a supine position with the patient's neck slightly extended. The da Vinci robotic system is placed on the right side of the patient's bed at 30°–45° angle to the bed. The anesthetist and the anesthetic machine should be on the opposite side of the patient's bed. A surgical assistant is seated at the head of the patient's bed with a good view of the vision cart. A schematic diagram of the setup is shown in Fig. 2.

The authors preferred nasal intubation as the tube would be away from the surgical field, with adequate padding to prevent alar necrosis. Other institutions favor an orotracheal intubation as it may make performing a combined nasal or palatal surgery more convenient with reduced risk of pressure necrosis of the nostril [7]. The eyes of the patient should be protected with eye covers.

## Surgical technique

A silk 2/0 suture is placed on the anterior third of the tongue. This acts as an anchor to retract the tongue anteriorly, allowing better visualization and improved access particularly where the tongue is bulky.

A Crowe-Davis gag with Davis-Meyer tongue blades of varying sizes should be prepared. The gag is introduced and suspended using draffin bipods, ensuring an adequate view of lingual tonsils (Fig. 3). The da Vinci robotic system is docked into position. The robotic arms are then positioned, and the 12 mm 30° endoscope is mounted. A 8.5 mm 30° endoscope is an option for patients with a smaller mouth opening.

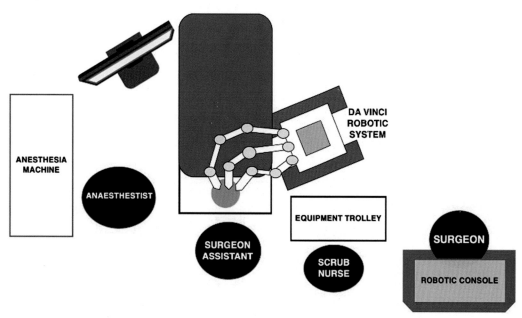

FIG. 2   Schematic diagram showing suggested layout of the operating room.

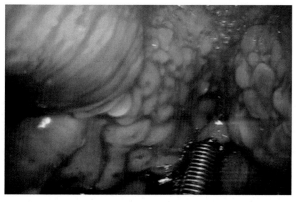

FIG. 3   Intraoperative photo using the da Vinci system showing bilateral lingual tonsil hypertrophy causing significant upper airway narrowing prior to resection.

III. The current and future clinical applications of robotic surgery among medical specialties

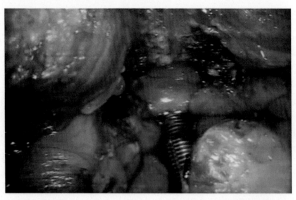

FIG. 4    Intraoperative view postlingual tonsil resection.

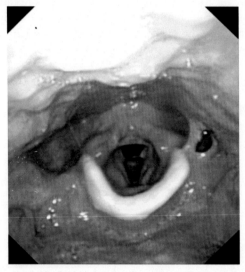

FIG. 5    Flexible nasoendoscopy photo of a patient who had undergone lingual tonsillectomy showing a significant reduction in the size of the lingual tonsils.

The da Vinci 5 mm Endowrist Maryland dissector and monopolar bovie diathermy are then introduced. The surgical assistant should be seated at the head of the patient and may assist in providing counter traction. The monopolar diathermy is used to create a 1 cm deep incision beginning at the foramen cecum, in the midline, posteriorly toward the glossoepiglottic ligament. The Maryland forceps is then used to retract the lingual tissues laterally, and the incision may then be deepened further. Dissection proceeds from medially to laterally and should be in a plane superficial to the muscles. It is crucial to avoid the lingual artery and hypoglossal nerve neurovascular bundles laterally as described in the earlier section. The right and left hemibase of tongue is delivered separately. Hemostasis is secured and checked with a Valsalva maneuver. The volume of the resected tissue is then measured using the water displacement technique. A postresection photo is shown in Fig. 4, and a postoperative photo taken using flexible nasoendoscopy is shown in Fig. 5.

For patients with significant inspiratory collapse of the epiglottis during DISE, a transoral robotic approach may be used for resection of the upper 1/3 of the epiglottis. The Maryland dissector is used to grasp the epiglottis, and a midline cut is made about 1/3 the length of the epiglottis. The incision is then carried laterally, and each half of the epiglottis is delivered separately. Hemostasis is secured with monopolar diathermy.

Postprocedure, a nasopharyngeal airway is placed and kept overnight. Most patients in our center are then extubated and transferred to the high dependency unit for continuous oxygen saturation monitoring. If significant airway edema is present postresection, the patient may be kept intubated and monitored in the intensive care unit.

## Postoperative management

Patients should be nursed in a 30° head up position, with continuous oxygen saturation monitoring. In terms of postoperative analgesia, acetaminophen and nonsteroidal antiinflammatory drugs (NSAIDs) are recommended with proton pump inhibitors. In addition, benzydamine hydrochloride lozenge and throat sprays may be helpful for postoperative pain management.

In addition, 24 h of intravenous antibiotics (e.g., coamoxiclav) should be given and the patient should be discharged with oral antibiotics for 7 days postsurgery. A short course of low-dose steroids is recommended to reduce postoperative inflammation and swelling.

Patients are recommended to be on fluids to soft diet for the first 7–14 days postsurgery.

The patient is usually discharged 1–2 days postsurgery. They are then reviewed in the outpatient clinic 7–10 days postsurgery.

## Outcomes

TORS tongue base reduction is frequently employed as one component of multilevel airway surgery. As an isolated procedure for OSA, Lin et al. reported a series of 27 patients treated with TORS tongue base reduction without any other concomitant upper airway procedures. There was a significant reduction of the apnea-hypopnea index (AHI) from $43.9 \pm 41.1$ to $17.6 \pm 16.2$ with an average AHI reduction of 56.2% [12]. Additionally, there was a significant reduction in the Epworth sleepiness scale (ESS), snoring intensity, and daytime somnolence levels [12]. In a retrospective study of patients with moderate to severe OSA treated with TORS tongue base surgery, partial epiglottectomy, and palatal surgery, the overall surgical cure rate was 35% with a surgical success rate of 55%, as defined by Sher's criteria [13].

Two metaanalyses reporting the use of TORS as part of multilevel surgery for OSA showed that the summary estimate of reduction of AHI of 24.0–24.25 [14,15]. The summary estimate of ESS reduction was 7.2–7.92 [14,15]. The estimated surgical success was 48.2%–69% [14,15].

The pooled data from 902 patients in the metaanalysis by Lechien et al. reported that the mean hospital stay was 3.65 days [15].

Few studies discuss predictive factors affecting the outcomes of TORS. First, BMI is a significant predictor of success. Hoff et al. reported that patients with a BMI $< 30 \, \text{kg/m}^2$ had higher surgical success (69.4% vs 41.7%, $P = 0.004$) in patients with moderate to severe OSA compared to those with BMI $> 30 \, \text{kg/m}^2$. Furthermore, patients with a lower BMI $< 25 \, \text{kg/m}^2$ had a 78.3% success rate [16]. Similarly, Lin et al. discussed that patients with BMI $< 30 \, \text{kg/m}^2$ had significantly higher success rates of 88.2% compared to their counterparts with a BMI 30–40 $\text{kg/m}^2$ or $> 40 \, \text{kg/m}^2$, where the success rates were only 31.3% and 16.7%, respectively [17].

While no AHI cut-offs have been established for patient selection, Turhan et al. reported that surgical responders had a significantly lower AHI than the nonresponders (38.8/h vs 50.5 events/h) [18]. Lin et al. similarly reported that patients with AHI $\geq 60$ /h had a significantly lower surgical success compared to those with AHI $< 60$ /h (18.2% vs 67.9%, $P = 0.011$) [17].

Lin et al. also reported that patients with lateral velopharyngeal collapse had significantly reduced surgical success compared to their counterparts without (25% vs 66.7%, $P = 0.035$) [17]. Similarly, Meraj et al. showed that patients without lateral pharyngeal collapse were more likely to improve postsurgery, though this did not affect the rate of success or cure [19]. In this study, the degree of collapse during DISE did not improve success or cure rates amongst patients who had TORS as part of multilevel surgery [19].

There is no consensus as to the ideal volume of tissue resection from the tongue base. Hoff et al. reported that the volume of resected tissue can predict improvement in AHI, and a significantly smaller volume of tissue was resected in patients with poorer surgical outcome (6.07 vs 8.24 mL, $P = 0.03$) [16]. Vicini et al. recommended at least 7 mL of tissue resection to relieve the tongue base obstruction [20]. Eesa et al. described a higher rate of success in patients who had 10–20 cc of tissue removed [21]. Folk et al. also reported that a larger volume of tissue removed (10.3 vs 8.6 mL, $P = 0.02$) was predictive of surgical success [22]. However, the authors advocate that there is no one size fits all, and a larger patient would typically have more redundant tissue. As such, these volumes should only be used as a guide. Two metaanalyses have reported an overall complication rate of 17.8%–22.3% [14,23].

Varying opinions exist regarding the need for tracheotomy in patients undergoing TORS tongue base resection. Some authors advocate that a tracheotomy may be useful in airway protection in event of hemorrhage [24,25] or if there are anticipated difficulties with reintubation (Cormack Lehane grade 3–4) [25]. In a large series of 166 patients by Glazer et al., none of the patients required a tracheostomy [26], suggesting that a routine tracheotomy may be avoided in most circumstances.

The most common complication was transient dysphagia. A nasogastric tube may be considered, though it is not routinely necessary. One study reported return to oral diet within 5 h, but most report return to oral diet around 24 h [14]. Notably, Eesa et al. described that there was no correlation between the amount of tissue resected and risk of aspiration [21]. There was also no significant difference in subjective swallowing outcomes based on MD Anderson dysphagia inventory (MDADI) [21]. However, about 2–4 weeks may still be required to return to normal diet [14].

Other complications included bleeding. The rate of postoperative bleeding requiring management in the operating room was not more than 5.7% [27]. Other complications included pain, dysgeusia, aspiration, and oropharyngeal stenosis or synechiae [14].

# Benign parapharyngeal diseases

## Introduction

The traditional approach for the removal of benign prestyloid parapharyngeal space tumor is via a transcervical transparotid route. This approach requires a complete dissection of the facial nerve and its five peripheral branches. The superior parotid gland is reflected laterally, thereby exposing the deep parotid lobe and the prestyloid compartment. When the tumor extends superiorly towards the medial pterygoid muscle/pterygoid plate, freeing this portion from the transcervical route may be challenging due to poor visualization and the need to maneuver the intervening facial nerve. In some instances, a combined mandibulotomy approach may be required.

The role of a transoral approach to prestyloid PPS tumor was previously neglected due to unfamiliarity of the transoral anatomy and the potential risk of tumor spillage. Additionally, the limited view and therefore limited ability to control the great vessels made this approach unfavorable. [29] The improved appreciation of the transoral anatomy [30] and magnified visualization with the robotic system has equipped surgeons to remove these tumors safely without the morbidity of a transcervical approach and/or mandibulotomy.

### Advantages and disadvantages of transoral approach

The advantages and disadvantages are summarized in Table 1. [31]

## Preoperative considerations

Appropriate transoral access is essential for TORS resection. Contraindications for a TORS procedure include severe trismus (less than 3 fingerbreaths width), limited cervical spine extension, and a high modified mallampati score [29].

A contrast-enhanced magnetic resonance imaging (MRI) (Figs. 6 and 7) is often favored over a CT scan due to better delineation of the tissue planes [31]. Scans should be carefully reviewed to assess if the tumor is well encapsulated with no osseous involvement or mucosal involvement, as this may necessitate reconstruction after resection [29]. The course of the internal carotid artery (ICA) should be examined carefully, in particular, the relation of the tumor to the ICA. Displacement of the ICA by tumor laterally is not an absolute contraindication to TORS excision. Factors that may preclude a TORS approach include if the carotid is displaced anteromedially or if there is no clear fat plane between the carotid artery and the tumor [31,32].

Fine needle aspiration (FNA) of the parapharyngeal mass is useful in elucidating the nature of the lesion and may be performed in most lesions which are not vascular in nature [31]. The general consensus is that TORS should not be attempted in suspected malignant parapharyngeal space tumors, though it has been suggested that it is a viable approach in selected malignant lesions where neck dissection is not required, [33] or for diagnostic purposes [33].

Additionally, combined TORS and transcervical or transparotid approaches have been described for larger lesions or lesions that may be at risk of tumor fragmentation. [34] If a combined approach is anticipated, the patient should be appropriately counseled.

TABLE 1   Advantages and disadvantages of transoral robotic excision of parapharyngeal space tumors.

| Advantages | Disadvantages |
| --- | --- |
| — Lack of external neck scar | — More limited exposure |
| — Avoidance of neck numbnesss | — Possibility of tumor spillage from lack of haptic feedback |
| — Reduced first bite syndrome | — Possibility of trismus |
| — Reduced risk to facial nerve | |
| — Avoids Frey's syndrome | |
| — Avoids sialocele | |

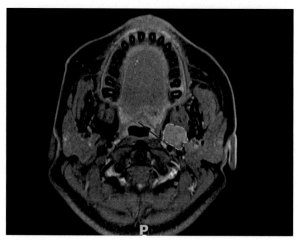

FIG. 6    T1-weighted axial images of magnetic resonance imaging (MRI) scan showing a well-encapsulated left parapharyngeal space lesion (*blue dotted line*). The *red arrow* indicates the plane between the carotid sheath and the lesion.

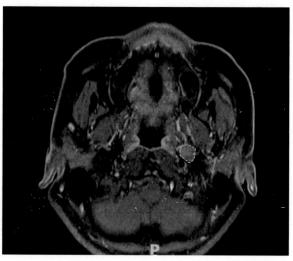

FIG. 7    T1-weighted axial images of magnetic resonance imaging (MRI) scan showing a well-encapsulated left parapharyngeal space (*blue dotted line*). Arrow estimates the position of the facial nerve (between styloid process and mastoid) which is usually some distance away from the tumor capsule.

## Surgical technique

A similar operating room setup is adopted as shown in Fig. 2. Nasotracheal intubation is favored as the endotracheal tube would be out of the field of surgery. Adequate protection of the ala with a soft sponge is required to avoid pressure necrosis. A facial nerve monitor may be considered if the lateral aspect of the tumor is close to the deep lobe of the parotid.

Exposure may be achieved using a Crowe-Davis gag with the Davis-Meyer blade. This is suspended with draffin rods. In contrast to the access for tongue base surgery discussed in the earlier part of the chapter, a midline retraction suture is not required.

Incision is marked from a paramedian position extending obliquely to the pterygomandibular raphe (Fig. 8). Incision is made using monopolar diathermy and is carried through the soft palate and palatoglossus and superior constrictors into the parapharyngeal fat. Retracting sutures may be placed on either side of the incision to provide further retraction. Once the tumor is identified, blunt dissection should be carried out in an extracapsular plane circumferentially (Fig. 9), until the lesion can be delivered. This may be assisted with surgical patties and retraction by the surgical assistant. A intraoperative photo of the surgical bed is shown in Fig. 10 after the tumor is removed, and the specimen is shown in Fig. 11.

The incision may then be closed via the conventional transoral route with a headlight.

III. The current and future clinical applications of robotic surgery among medical specialties

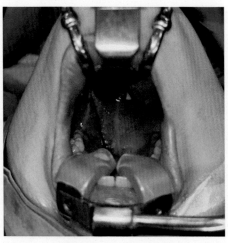

FIG. 8    Operative photo showing the eventual exposure of oral cavity using Crowe-Davis gag with the planned incision marked out.

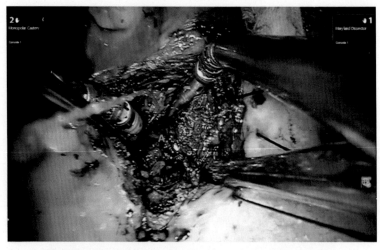

FIG. 9    Intraoperative photo showing blunt dissection of the tumor (A) with surgical patties.

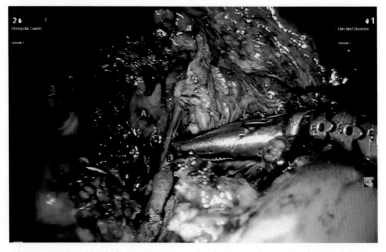

FIG. 10    Intraoperative photo showing surgical bed after tumor removal, where A indicates the fascia of the carotid sheath, B indicates the pterygoid venous plexus, C is the parapharyngeal fat.

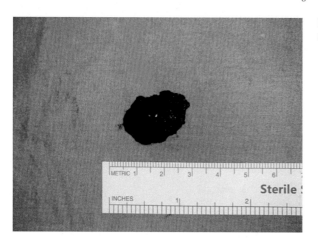

FIG. 11 Parapharyngeal space tumor excised via a transoral robotic approach. Histology revealed oncocytic neoplasm.

A tracheotomy may be considered if there is a high risk of airway compromise or hemorrhage and in patients with poor cardiopulmonary reserve who may not tolerate a secondary bleed or to protect his/her airway.

## Clinical outcomes

In a systematic review performed by de Virgilio et al. 90.3% of tumors excised were benign, with the predominant histology being pleomorphic adenoma (58.4%) [33]. Of the malignant tumors, the most common subtype was carcinoma ex-pleomorphic adenoma. Other histological subtypes included adenoid cystic carcinoma, mucoepidermoid carcinoma, and medullary thyroid carcinoma [33]. The median tumor size of parapharyngeal space tumors removed using the TORS approach was 5.4 cm [33]. However, the removal of parapharyngeal space tumors up to 8 cm in size via a TORS approach has been described [35].

De Virgilio et al. further described that capsule disruption occurred in 14.5% (11 cases) and tumor fragmentation was observed in 10.3% (7 cases) [33]. The median intraoperative blood loss was 30.4 mL, and the median TORS procedure time was 102.5 min, with a total surgery time of 147.3 min [32]. Median hospitalization time was 3 days [33].

Porcuna et al. compared the use of lateral (transcervical) approaches with various medial (TORS or combined) approaches for prestyloid tumors. Tumor length, volume, and recurrence rates were similar in both groups [36]. In their series of 28 patients, the patients who had a medial approach generally had a shorter hospital stay (5.8 days vs 8.9 days) [36]. Additionally, the mean duration of surgery was longer for the lateral approach rather than medial approaches (221.4 vs 133.3 min, $P = 0.02$). There were also more postsurgical complications in the lateral approaches [36].

In De Virgilio et al. systematic review of 111 patients, the most commonly described complication was postoperative dysphagia (4.5%). However, 68% of patients were able to tolerate oral diet from postoperative day 1 [33]. The use of nasogastric tubes was described in two studies in the immediate postoperative period, with a mean time of 7.6 days [29] and 10 days to oral feeding [37]. Additionally, mucosal dehiscence requiring nasogastric tube feeding for 3 weeks was described in 1.8% of patients [38]. There was eventual complete healing of the surgical wound with conservative management. Other complications described included hematoma formation (3.6%), trismus (1.8%), and first bite syndrome (1.8%).

## Benign oropharyngeal diseases

TORS may also be applied in benign oropharyngeal diseases such as the excision of benign oropharyngeal cysts such as vallecula cysts [39], and rarer entities such as lingual thyroids [40,41] (Fig. 12) and salivary fistulas [42].

A similar setup may be adopted with a Crowe-Davis retractor or Feyh-Kastenbauer (FK) retractor as in a TORS lingual tonsillectomy setup. An anchoring suture is placed in the anterior third of the tongue as in tongue base resection and is used to retract the tongue anteriorly to ensure adequate visualization. The surgical approach to excision of a lingual thyroid was previously described by Teo et al. [40] The incision should be made anterior to the mass, allowing a cuff of muscle to be taken together with the mass, so as not to tear the thyroid tissues. The lateral incisions are then

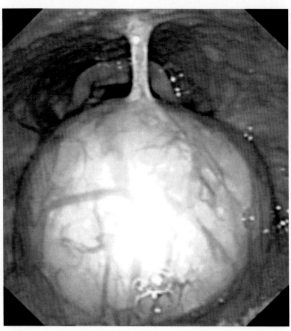

FIG. 12    Preoperative nasoendsocopy showing a large lingual thyroid, partially obscuring the view of the larynx.

performed, and dissection was carried from anteriorly to laterally in a broad plane with careful attention near the lingual arteries. Hemostasis was then secured with monopolar diathermy. The larynx was also inspected for edema prior to extubation. Other authors have described a similar transoral robotic approach with the use of a LISA laser [43].

This minimally invasive approach negates the risks of a traditional transcervical approach such as swallowing impairment requiring a nasogastric tube, risk of pharyngocutaneous fistula, prolonged hospitalization, and external scar [40]. Most patients are able to commence oral feeding on postoperative day 1 [40,43].

## Future scenarios

There have been continual advancements in TORS. Newer robotic systems include the da Vinci single port systems. The da Vinci single-port robotic system offers the ability to have a third working instrument arm and improves access with all the three individually articulating instruments passing through a single port [44]. The results thus far have been encouraging when used for resection of oropharyngeal tumors with a low rate of conversion to open and few complications described [44,45]. Perhaps, this may also be adopted in sleep apnea surgery. Other robotic systems which have previously been used for abdominal surgeries such as the Versius (CMR surgical) are in their initial phases of being adopted for TORS. Additionally, with future miniaturization of instruments and the working arms as well as the use of flexible systems, we envision that robotic systems may be considered for use for laryngeal work.

## Conclusion

The use of robotics for transoral surgery has expanded our capabilities in addressing previously difficult-to-reach areas such as the oropharynx and parapharynx with greater precision, superior visualization, and maneuverability. The expanded applications have improved our ability to address multilevel airway obstruction in sleep apnea surgery with a good safety profile and superior outcomes compared to traditional approaches. However, appropriate patient selection and counseling remains crucial to optimize outcomes.

## Key points

• In appropriately selected patients, transoral robotic tongue base reduction and partial epiglottectomy may be used as part of multilevel surgery for the treatment of OSA patients with a good safety profile and favorable results.

- Familiarity with the anatomy of the oropharynx and parapharynx is crucial, in particular, with critical structures like the lingual artery and the hypoglossal nerve.
- Careful review of preoperative imaging and preoperative counseling is crucial for transoral robotic excision of parapharyngeal space lesions.

# References

[1] Toh ST, Hsu PP. Robotic obstructive sleep apnea surgery. Robotic obstructive sleep apnea surgery. Adv Otorhinolaryngol 2017;80:125–35. https://doi.org/10.1159/000470882. Epub 2017 Jul 17.

[2] Tan A, Cheung YY, Yin J, Lim W, Tan LWL, Lee CH. Prevalence of sleep-disordered breathing in a multiethnic Asian population in Singapore: a community-based study. Respirology 2016;21(5):943–50.

[3] Gottlieb DJ, Yenokyan G, Newman AB, O'Connor GT, Punjabi NM, Quan SF, Redline S, Resnick HE, Tong EK, Diener-West M, Shahar E. Prospective study of obstructive sleep apnea and incident coronary heart disease and heart failure: the sleep heart health study. Circulation 2010;122 (4):352–60.

[4] Yaggi HK, Concato J, Kernan WN, Lichtman JH, Brass LM, Mohsenin V. Obstructive sleep apnea as a risk factor for stroke and death. N Engl J Med 2005;353(19):2034–41.

[5] Sunwoo BY, Light M, Malhotra A. Strategies to augment adherence in the management of sleep-disordered breathing. Respirology 2020;25 (4):363–71.

[6] Baptista PM, Diaz Zufiaurre N, Garaycochea O, Alcalde Navarrete JM, Moffa A, Giorgi L, Casale M, O'Connor-Reina C, Plaza G. TORS as part of multilevel surgery in OSA: the importance of careful patient selection and outcomes. J Clin Med 2022;11(4):990. https://doi.org/10.3390/jcm11040990. PMID: 35207264. PMCID: PMC8878188.

[7] Lin HC, Friedman M. Transoral robotic OSA surgery. Auris Nasus Larynx 2021;48(3):339–46. https://doi.org/10.1016/j.anl.2020.08.025. Epub 2020 Sep 8 PMID: 32917413.

[8] Li S, Shi H. Lingual artery CTA-guided midline partial glossectomy for treatment of obstructive sleep apnea hypopnea syndrome. Acta Otolaryngol 2013;133(7):749–54.

[9] Liu C, Qin J, Xing D, Lu H, Yue R, Li S, Wu D. Ultrasonic measurement of lingual artery and its application for midline Glossectomy. Ann Otol Rhinol Laryngol 2020;129(9):856–62. https://doi.org/10.1177/0003489420913581. Epub 2020 Apr 21 PMID: 32316740.

[10] Zheng C, Shi L, Xing D, Qin J, Ji P, Li S, Wu D. Measurement of lingual artery using ultrasound versus computed tomography angiography for midline Glossectomy in patients with obstructive sleep apnea. Ann Otol Rhinol Laryngol 2022;131(11):1210–6. https://doi.org/10.1177/00034894211062697. Epub 2021 Dec 1 PMID: 34852648.

[11] Cohen DS, Low GMI, Melkane AE, Mutchnick SA, Waxman JA, Patel S, Shkoukani MA, Lin HS. Establishing a danger zone: an anatomic study of the lingual artery in base of tongue surgery. Laryngoscope 2017;127:110–5.

[12] Lin HS, Rowley JA, Badr MS, Folbe AJ, Yoo GH, Victor L, Mathog RH, Chen W. Transoral robotic surgery for treatment of obstructive sleep apnea-hypopnea syndrome. Laryngoscope 2013;123(7):1811–6. https://doi.org/10.1002/lary.23913.

[13] Toh ST, Han HJ, Tay HN, Kiong KL. Transoral robotic surgery for obstructive sleep apnea in Asian patients. A Singapore sleep Centre experience. JAMA Otolaryngol Head Neck Surg 2014;140(7):624–9.

[14] Justin GA, Chang ET, Camacho M, Brietzke SE. Transoral robotic surgery for obstructive sleep apnea: a systematic review and Meta-analysis. Otolaryngol Head Neck Surg 2016;154(5):835–46. https://doi.org/10.1177/0194599816630962. Epub 2016 Mar 1 PMID: 26932967.

[15] Lechien JR, Chiesa-Estomba CM, Fakhry N, Saussez S, Badr I, Ayad T, Chekkoury-Idrissi Y, Melkane AE, Bahgat A, Crevier-Buchman L, Blumen M, Cammaroto G, Vicini C, Hans S. Surgical, clinical, and functional outcomes of transoral robotic surgery used in sleep surgery for obstructive sleep apnea syndrome: a systematic review and meta-analysis. Head Neck 2021;43(7):2216–39. https://doi.org/10.1002/hed.26702. Epub 2021 Apr 16 PMID: 33860981.

[16] Hoff PT, Glazer TA, Spector ME. Body mass index predicts success in patients undergoing transoral robotic surgery for obstructive sleep apnea. ORL 2014;76:266–72.

[17] Lin HS, Rowley JA, Folbe AJ, Yoo GH, Badr MS, Chen W. Transoral robotic surgery for treatment of obstructive sleep apnea: factors predicting surgical response. Laryngoscope 2015;125(4):1013–20. https://doi.org/10.1002/lary.24970. Epub 2014 Oct 24 PMID: 25346038.

[18] Turhan M, Bostanci A. Robotic tongue-base resection combined with Tongue-Base suspension for obstructive sleep apnea. Laryngoscope 2020;130(9):2285–91. https://doi.org/10.1002/lary.28443. Epub 2019 Nov 29 PMID: 31782809.

[19] Meraj TS, Muenz DG, Glazer TA, Harvey RS, Spector ME, Hoff PT. Does drug-induced sleep endoscopy predict surgical success in transoral robotic multilevel surgery in obstructive sleep apnea? Laryngoscope 2017;127(4):971–6. https://doi.org/10.1002/lary.26255. Epub 2016 Oct 31 PMID: 27796047.

[20] Vicini C, Montevecchi F, Scott Magnuson J. Robotic surgery for obstructive sleep apnea. Curr Otorhinolaryngol Rep 2013;1:130–6.

[21] Eesa M, Montevecchi F, Hendawy E, et al. Swallowing outcome after TORS for sleep apnea: short- and long-term evaluation. Eur Arch Otorhinolaryngol 2015;272:1537–41.

[22] Folk D, D'Agostino M. Transoral robotic surgery vs. endoscopic partial midline glossectomy for obstructive sleep apnea. World J Otorhinolaryngol Head Neck Surg 2017;3(2):101–5. https://doi.org/10.1016/j.wjorl.2017.05.004. PMID: 29204587. PMCID: PMC5683621.

[23] Meccariello G, Cammaroto G, Montevecchi F, Hoff PT, Spector ME, Negm H, Shams M, Bellini C, Zeccardo E, Vicini C. Transoral robotic surgery for the management of obstructive sleep apnea: a systematic review and meta-analysis. Eur Arch Otorhinolaryngol 2017;274(2):647–53. https://doi.org/10.1007/s00405-016-4113-3. Epub 2016 May 24. PMID: 27221389.

[24] Vicini C, Dallan I, Canzi P, Frassineti S, Nacci A, Seccia V, Panicucci E. Grazia, La Pietra M, Montevecchi F, Tschabitscher M: Transoral robotic surgery of the tongue base in obstructive sleep apnea-hypopnea syndrome: anatomic considerations and clinical experience. Head Neck 2012;34:15–22.

[25] de Bonnecaze G, Vairel B, Dupret-Bories A, Serrano E, Vergez S. Transoral robotic surgery of the tongue base for obstructive sleep apnea: preliminary results. Eur Ann Otorhinolaryngol Head Neck Dis 2018;135(6):411–5. https://doi.org/10.1016/j.anorl.2018.09.001. Epub 2018 Oct 24 PMID: 30430999.

[26] Glazer TA, Hoff PT, Spector ME. Transoral robotic surgery for obstructive sleep apnea: perioperative management and postoperative complications. JAMA Otolaryngol Head Neck Surg 2014;140(12):1207–12. https://doi.org/10.1001/jamaoto.2014.2299. PMID: 25275670.

[27] Thaler ER, Rassekh CH, Lee JM, et al. Outcomes for multilevel surgery for sleep apnea: obstructive sleep apnea, transoral robotic surgery, and uvulopalatopharyngoplasty. Laryngoscope 2016;126(1):266–9.

[29] Panda S, Sikka K, Thakar A, et al. Transoral robotic surgery for the parapharyngeal space: expanding the transoral corridor. J Robotic Surg 2020;14:61–7.

[30] Lim CM, Mehta V, Chai R, Pinheiro CN, Rath T, Snyderman C, Duvvuri U. Transoral anatomy of the tonsillar fossa and lateral pharyngeal wall: anatomic dissection with radiographic and clinical correlation. Laryngoscope 2013;123(12):3021–5.

[31] Larson AR, Ryan WR. Transoral excision of Parapharyngeal space tumors. Otolaryngol Clin North Am 2021;54:531–41.

[32] Boyce BJ, Curry JM, Luginbuhl A. Transoral robotic approach to Parapharyngeal space tumors: case series and technical limitations. Laryngoscope 2016;2016(126):1776–82.

[33] De Virgilio A, Costantino A, Mercante G, Di Maio P, Iocca O, Spriano G. Trans-oral robotic surgery in the management of parapharyngeal space tumors: a systematic review. Oral Oncol 2020;103, 104581. https://doi.org/10.1016/j.oraloncology.2020.104581. Epub 2020 Feb 12 PMID: 32058293.

[34] Chu F, De Berardinis R, Tagliabue M, Zorzi S, Bandi F, Ansarin M. The role of transoral robotic surgery for Parapharyngeal space: experience of a tertiary center. J Craniofac Surg 2020;31(1):117–20. https://doi.org/10.1097/SCS.0000000000005912. PMID: 31634316.

[35] Hussain A, Ah-See KW, Shakeel M. Trans-oral resection of large parapharyngeal space tumours. Eur Arch Otorhinolaryngol 2014;271(3):575–82. https://doi.org/10.1007/s00405-013-2550-9. Epub 2013 May 10 PMID: 23661062.

[36] Porcuna DV, Munoz LP, Soria CV, Nicastro V, Viarnes MP, Guisasola CP. A retrospective analysis of surgery in prestyloid parapharyngeal tumors: lateral approaches vs transoral robotic surgery. Laryngoscope Investig Otolaryngol 2021;6(5):1062–7.

[37] De Virgilio A, Park YM, Kim WS, Byeon HK, Lee SY, Kim SH. Transoral robotic surgery for the resection of parapharyngeal tumour: our experience in ten patients. Clin Otolaryngol 2012;37(6):483–8. https://doi.org/10.1111/j.1749-4486.2012.02525.x. PMID: 23253343.

[38] O'Malley BW, Quon H, Leonhardt FD, Chalian AA, Weinstein GS. Transoral robotic surgery for parapharyngeal space tumors. ORL J Otorhinolaryngol Relat Spec 2010;72:332–6.

[39] McLeod IK, Melder PC. Da Vinci robot-assisted excision of a vallecular cyst: a case report. Ear Nose Throat J 2005;84(3):170–2.

[40] Teo EH, Toh ST, Tay HN, Han HJ. Transoral robotic resection of lingual thyroid: case report. J Laryngol Otol 2013;127(10):1034–7. https://doi.org/10.1017/S0022215113002156. Epub 2013 Oct 14 PMID: 24125062.

[41] Pellini R, Mercante G, Ruscito P, Cristalli G, Spriano G. Ectopic lingual goiter treated by transoral robotic surgery. Acta Otorhinolaryngol Ital 2013;33(5):343–6.

[42] Rassekh CH, Kazahaya K, Livolsi VA, Loevner LA, Cowan AT, Weinstein GS. Transoral robotic surgery-assisted excision of a congenital cervical salivary duct fistula presenting as a branchial cleft fistula. Head Neck 2016;38(2):E49–53. https://doi.org/10.1002/hed.24123. Epub 2015 Jul 15 PMID: 25974185.

[43] Prisman E, Patsias A, Genden EM. Transoral robotic excision of ectopic lingual thyroid: case series and literature review. Head Neck 2015;37(8): E88–91. https://doi.org/10.1002/hed.23757. Epub 2015 May 26 PMID: 24816912.

[44] Van Abel KM, Yin LX, Price DL, Janus JR, Kasperbauer JL, Moore EJ. One-year outcomes for da Vinci single port robot for transoral robotic surgery. Head Neck 2020;42(8):2077 87. https://doi.org/10.1002/hed.26143. Epub 2020 Mar 19 PMID: 32190942.

[45] Park YM, Choi EC, Kim SH, et al. Recent progress of robotic head and neck surgery using a flexible single port robotic system. J Robotic Surg 2022;16:353–60. https://doi-org.libproxy1.nus.edu.sg/10.1007/s11701-021-01221-8.

# 51

# Robotic cardiac surgery: Advancements, applications, and future perspectives

*Burak Ersoy and Burak Onan*

Department of Cardiovascular Surgery, University of Health Sciences, Istanbul Mehmet Akif Ersoy Thoracic and Cardiovascular Surgery Training and Research Hospital, Istanbul, Turkey

## Introduction

### Cardiovascular diseases: A global health burden

Cardiovascular diseases account for more than 17 million deaths worldwide each year[p1] [1]. Surgical interventions play a vital role in managing complex cardiovascular conditions [2]. With the rising prevalence of cardiovascular diseases, there is a growing need for innovative surgical techniques that can offer improved patient outcomes.

### Advancements in surgical techniques: Robotic cardiac surgery

Robotic cardiac surgery has emerged as an advanced surgical technique that combines the precision of robotics with the expertise of cardiac surgeons. It offers innovative solutions for improved patient outcomes [3]. By leveraging robotic technology, surgeons can achieve enhanced precision, improved visualization, and reduced invasiveness compared to traditional methods [4]. These advancements have the potential to revolutionize cardiac surgery and improve patient care.

### Objectives and structure of the chapter

This chapter aims to explore the historical development and evolution of robotic cardiac surgery. It investigates the applications, clinical outcomes, and limitations of robotically assisted procedures. Furthermore, it discusses future prospects, including technological advancements and challenges. By providing a comprehensive overview of robotic cardiac surgery, this chapter aims to shed light on the potential of this field and its implications for patient outcomes and healthcare systems.

## Development of robotic cardiac surgery

### Historical overview

The field of robotic cardiac surgery traces back to the 1990s when telemanipulation and telerobotic experiments in cardiac surgery were first conducted [5]. These early experiments laid the foundation for the development of robotic-assisted cardiac procedures. Dr. Marco A. Zenati's pioneering work in robotic-assisted coronary artery bypass grafting (CABG) marked a significant milestone in the development of robotic cardiac surgery [5]. Since then, there have been remarkable advancements in technology and techniques, driving the progress of this field.

## Technological advancements

Advancements in robotic cardiac surgery have been driven by the continuous improvement of robotic platforms and instruments. The miniaturization of instruments such as robotic arms, endoscopes, and robotic suturing devices has enabled more precise and minimally invasive procedures [5]. Advanced robotic platforms, such as the da Vinci Surgical System and the RoboCABG system, have been developed and utilized in robotic cardiac surgeries. These platforms offer enhanced control and visualization, facilitating complex cardiac procedures. The integration of 3D high-definition vision systems has further provided surgeons with enhanced depth perception, improving their ability to perform intricate tasks.

## Robotic systems in cardiac surgery

Several robotic systems have been employed in cardiac surgery, with the da Vinci Surgical System being one of the most widely utilized systems [6]. Known for its modular design and intuitive controls, the da Vinci Surgical System allows surgeons to perform precise and complex procedures with enhanced dexterity. Other robotic systems, including the Titan Medical SPORT Surgical System and the Medtronic Hugo Surgical System, have also been utilized in cardiac surgeries [7]. These robotic systems offer advantages such as increased precision, tremor filtration, and improved ergonomics, which can contribute to better surgical outcomes [3].

## Robotically assisted mitral valve surgery

### Importance of mitral valve surgery

Mitral valve diseases are prevalent and often require surgical intervention for improved outcomes [8]. The mitral valve is a crucial component of the heart that regulates blood flow between the left atrium and the left ventricle. Conventional approaches to mitral valve surgery, such as sternotomy (a surgical incision through the sternum), pose challenges in complex mitral valve repair [8]. Robotic-assisted mitral valve surgery has emerged as a valuable technique for overcoming these challenges. Surgeon can visualize mitral valve in a 3D fashion with 4K definition at console. Surgeons can perform challenging mitral valve repair procedures with the help of both better visualization of camera and more precise movements of robotic arms. Robotic arms are more delicate than conventional surgical instruments and hence improve the movement ability of surgeon in tackling positions during mitral valve procedures.

### Benefits of robotic-assisted mitral valve surgery

Robotic-assisted mitral valve surgery offers several benefits compared to traditional approaches. One key advantage is enhanced visualization and magnification, which enables surgeons to perform intricate repairs with greater precision [9]. The use of robotic instruments allows for precise movements and fine suturing, reducing trauma to surrounding tissues. This minimally invasive approach also results in reduced blood loss and postoperative complications compared to traditional methods [10]. These benefits contribute to improved patient outcomes and faster recovery. Robotically assisted surgery requires smaller incision compared to conventional surgical approaches and postoperative infection such as mediastinitis and sternal osteomyelitis are less commonly observed after robotically assisted surgical procedures.

### Techniques and innovations in robotically assisted mitral valve surgery

Totally endoscopic approaches utilizing robotic arms and endoscopic vision systems have been developed for mitral valve surgery [3]. Surgical steps are;

1. Incision and Port Placement: The surgeon creates several small incisions (approximately 1–2 cm in length) on the patient's chest. These incisions serve as access points for the robotic instruments and camera.
2. Robotic System Setup: The surgical team sets up the robotic-assisted surgical system, which typically consists of robotic arms equipped with specialized instruments and a high-definition camera. The surgeon controls the robotic arms from a console in the operating room.
3. Camera Insertion: The camera port is inserted through one of the incisions, providing a clear and magnified 3D view of the heart structures.

4. Trocar Insertion: Trocars (thin, tube-like instruments) are inserted through the other incisions to allow the robotic instruments to access the heart.
5. Pericardium Opening: The pericardium, the protective sac surrounding the heart, is carefully opened to expose the mitral valve.
6. Mitral Valve Inspection: The surgeon inspects the mitral valve to assess the severity of the condition and decide whether repair or replacement is necessary.
7. Mitral Valve Repair: If repair is feasible, the surgeon uses the robotic instruments to perform the necessary repairs. Techniques may include resizing, reshaping, or repositioning the valve leaflets and annuloplasty to stabilize the valve ring.
8. Mitral Valve Replacement: In cases where repair is not possible, the surgeon removes the damaged mitral valve and replaces it with a bioprosthetic or mechanical valve. The choice of the valve type depends on the patient's condition and other factors.
9. Closure and Recovery: After completing the surgical procedure, the surgeon removes the robotic instruments and trocars, closes the incisions, and allows the heart to resume its normal function.

## Clinical outcomes and studies

Clinical studies have shown promising outcomes with robotic-assisted mitral valve surgery. Comparative studies have demonstrated improved patient outcomes, including lower mortality rates, reduced hospital stays, and improved quality of life [11]. Long-term follow-up studies have shown durability and functional improvements in patients who undergo robotic-assisted mitral valve surgery [12]. These findings highlight the potential of this technique to significantly impact patient outcomes.

## Limitations and challenges

While robotic-assisted mitral valve surgery offers several advantages, there are limitations and challenges that need to be addressed. Surgeon experience and training play a crucial role in the successful implementation of robotic-assisted procedures [3]. The initial setup time and equipment costs may also limit the widespread adoption of robotic systems [13]. Complex pathologies and reoperations require careful patient selection and planning to ensure optimal outcomes [3]. Overcoming these challenges will be essential to fully leverage the potential of robotic-assisted mitral valve surgery.

## Robotically assisted coronary revascularization

### Challenges in coronary artery disease treatment

Coronary artery disease (CAD) is a leading cause of mortality and morbidity globally [14]. It occurs when the coronary arteries, which supply blood to the heart, become narrowed or blocked due to the buildup of plaque. Traditional coronary revascularization techniques, such as on-pump CABG, have been the standard of care for treating CAD surgically. However, these techniques have limitations, and robotic-assisted coronary revascularization has emerged as an alternative approach.

### Robotic-assisted CABG

Robotic CABG enables precise grafting with reduced tissue trauma and blood loss [14]. This technique utilizes robotic arms to assist the surgeon in harvesting grafts from other blood vessels, such as the internal mammary artery or radial artery, and then suturing them onto the blocked coronary arteries. Robotic-assisted CABG can be performed off-pump, which means that it does not require the use of a heart-lung machine [3]. Off-pump techniques reduce the need for cardiopulmonary bypass and associated complications, further improving patient outcomes. Robotic-assisted CABG can also be performed with the help of cardiopulmonary bypass; heart can be taken into an asystole status by cross clamping ascending aorta and using cardioplegia or preload is minimized with the help of cardiopulmonary bypass and so heart can be easily manipulated if there is need for complex CABG procedures. The totally endoscopic coronary artery bypass (TECAB) procedure can also be applied in selected patients, and there is no need for extra incision for anastomosis. TECAB can be applied to a very small portion of patients who undergoes robotically assisted CABG due to its limitations in coronary anatomy and revascularization sites.

## Techniques in robotically assisted CABG

Surgical steps are as follows:

1. Incision and Port Placement: The surgeon creates several small incisions (about 1–2 cm in length) on the patient's chest. These incisions serve as access points for the robotic instruments and camera.
2. Robotic System Setup: The surgical team sets up the robotic-assisted surgical system, which includes robotic arms equipped with specialized instruments and a high-definition camera. The surgeon controls the robotic arms from a console in the operating room.
3. Camera Insertion: The camera port is inserted through one of the incisions, providing a magnified 3D view of the heart and coronary arteries.
4. Trocar Insertion: Trocars (thin, tube-like instruments) are inserted through the other incisions to allow the robotic instruments to access the heart.
5. Heart Stabilization: To facilitate precise movements during surgery, the heart is stabilized using a special mechanical stabilizer or stabilizing device.
6. Artery and/or Vein Harvesting: The surgeon identifies suitable arteries or veins (often from the patient's leg or chest) to use as grafts. The robotic instruments may be employed to harvest these vessels.
7. Graft Preparation: The harvested artery or vein is prepared for grafting. For arterial grafts, the robotic system allows for meticulous dissection and preparation.
8. Graft Attachment: The surgeon performs the bypass by attaching one end of the graft to the aorta and the other end to the coronary artery downstream of the blockage. This restores blood flow to the heart muscle beyond the blockage.
9. Completion of Grafts: Multiple grafts may be performed during a single procedure, depending on the number of blockages and the complexity of the CAD.
10. Robotic Instrument and Trocar Removal: Once all grafts are completed, the robotic instruments and trocars are removed, and the heart is allowed to resume its normal function.
11. Closure and Recovery: The small incisions are closed, and the patient is carefully monitored during the recovery process.

## Clinical outcomes and studies

Clinical studies have demonstrated equivalent or improved short-term outcomes with robotic-assisted CABG compared to traditional on-pump and off-pump CABG. Reduced postoperative pain, shorter hospital stays, and improved graft patency have been reported [5]. Long-term studies demonstrate similar survival rates and freedom from major adverse cardiac events [15]. These findings suggest that robotic-assisted CABG can be a safe and effective alternative for treating CAD.

## Limitations and future directions

While robotic-assisted coronary revascularization has shown promise, there are limitations and areas for further development. Future directions include expanding the applications of robotic-assisted techniques to complex multi-vessel and high-risk patients [16]. A continued improvement of anastomotic techniques is crucial for achieving optimal graft patency and long-term outcomes [17]. Additionally, the integration of robotic assistance with percutaneous coronary interventions for hybrid approaches is an area of ongoing research, aiming to provide personalized treatment options for patients with complex CAD [18]. The heart team plays a critical role at planning hybrid approaches; some cardiac centers have the ability to perform percutaneous coronary interventions just after undocking of the robotic system from the patient in a hybrid operating room.

## Robotically assisted other cardiac surgeries

### Atrial septal defect repair

In addition to mitral valve surgery and coronary revascularization, robotic-assisted approaches have been applied to other cardiac surgeries. Atrial septal defect (ASD) repair, which involves closing a hole in the wall between the heart's atrial chambers, can benefit from robotic assistance. Robotic-assisted approaches provide enhanced visualization and precise defect closure [19]. These minimally invasive techniques result in shorter hospital stays and faster recovery compared to conventional open-heart surgery. Comparative studies have demonstrated comparable outcomes with conventional approaches, supporting the use of robotic-assisted techniques in ASD repair [20]. Partial anomalous pulmonary venous return (PAPVR) disease can also be corrected with the aid of robotic surgery. PAPVR can be a challenging operation even with median sternotomy because of disease nature; however, the better visualization of interatrial septum and pulmonary vein connection with the robotic system provides a good approach and good postoperative results [21].

## Ventricular septal defect repair

Robotic assistance has also been utilized in ventricular septal defect (VSD) repair, which involves closing a defect in the ventricular septum. Robotic technology allows for a precise closure of complex VSDs, minimizing morbidity and improving patient outcomes [22]. Studies have shown reduced morbidity, shorter ventilation times, and improved quality of life in patients who undergo robotic-assisted VSD repair [23]. Long-term follow-up studies have demonstrated durable repairs and improved quality of life, further supporting the effectiveness of robotic-assisted approaches in VSD repair [24].

## Atrial fibrillation ablation

Robotic systems have also been used for minimally invasive ablation procedures in the treatment of atrial fibrillation (AF), a common heart rhythm disorder. Robotic-assisted ablation procedures enable precise mapping and ablation of the abnormal electrical pathways responsible for AF [25]. Advanced mapping technologies and robotic instruments improve procedural success rates and reduce the risk of complications [25]. Comparative studies have demonstrated comparable outcomes to traditional ablation techniques, making robotic-assisted ablation a viable option for patients with AF [26].

## Cardiac tumor resection

Robotic technology has also found application in the resection of cardiac tumors. Cardiac tumor resection requires precise removal of the tumor while preserving the surrounding healthy tissue. Robotic-assisted approaches facilitate this precise resection with minimal invasiveness [27]. Improved visualization and maneuverability provided by robotic systems aid in complete tumor removal. Case studies have shown successful outcomes with minimal complications and shorter hospital stays, supporting the use of robotic-assisted techniques in cardiac tumor resection [28].

## Beyond: Emerging applications and innovations

The applications of robotic cardiac surgery extend beyond the procedures discussed above. Robotic-assisted tricuspid valve surgery, interventions for congenital heart diseases, and transcatheter robotics are areas of ongoing research and innovation [29]. These emerging applications hold promise for expanding the scope of robotic cardiac surgery and providing innovative solutions for complex cardiac conditions. Ongoing clinical trials and registries are being conducted to assess outcomes and refine techniques, ensuring the safety and efficacy of these emerging approaches [29].

## Future scenario

### Technological advancements and innovations

The future of robotic cardiac surgery lies in the continual refinement of robotic platforms and instruments. Technological advancements aim to increase precision, versatility, and integration with other technologies. For example, the integration of advanced imaging modalities, such as intracardiac echocardiography and fusion imaging, can enhance surgical planning and guidance, enabling surgeons to perform procedures with greater accuracy. Real-time navigation systems and intraoperative decision-support tools based on artificial intelligence (AI) can further improve procedural guidance, enhancing surgical outcomes [30]. Continued research and innovation in these areas will shape the future of robotic cardiac surgery.

### Artificial intelligence and machine learning applications

AI and machine learning (ML) are rapidly advancing fields that hold great potential for improving robotic cardiac surgery. AI algorithms are being developed for surgical planning, risk stratification, and personalized treatment algorithms. These algorithms can analyze patient data, imaging studies, and surgical outcomes to assist surgeons in making informed decisions. ML-based surgical skill assessment, training, and performance optimization are also being explored [30]. By analyzing surgical data and providing feedback, ML algorithms can help surgeons enhance their skills and improve patient outcomes. Furthermore, the development of adaptive and autonomous robotic systems using AI techniques is an area of active research, aiming to create robotic systems that can adapt to the surgeon's needs and perform certain tasks independently. These AI and ML applications have the potential to enhance the precision, efficiency, and safety of robotic cardiac surgery.

## Enhanced surgical training and education

Surgical training and education are essential for the successful adoption and advancement of robotic cardiac surgery. Simulation-based training programs incorporating virtual reality and haptic simulators are being utilized to enhance surgical training. These training programs provide a realistic environment for surgeons to practice and refine their skills before performing procedures on patients. Collaboration between surgeons, engineers, and educators is crucial for designing comprehensive training curricula that address the specific challenges and nuances of robotic cardiac surgery [31]. Global training initiatives are being established to ensure widespread adoption and skill transfer, enabling surgeons from different regions to benefit from robotic cardiac surgery advancements [32].

## Cost-effectiveness and wide adoption

The cost-effectiveness and widespread adoption of robotic cardiac surgery are important considerations for healthcare systems. Health economic evaluations are necessary to assess the cost-effectiveness and long-term sustainability of robotic cardiac surgery [33]. These evaluations take into account factors such as the cost of robotic systems, operating room utilization, length of hospital stays, and patient outcomes. Strategies to optimize resource utilization, reduce equipment costs, and streamline workflow are being explored to enhance the cost-effectiveness of robotic cardiac surgery. Collaboration between healthcare systems, industry, and regulatory bodies is crucial for ensuring the affordability and accessibility of robotic systems. By addressing cost-related challenges, robotic cardiac surgery can become more widely available, benefiting a larger number of patients.

## Conclusion

Robotic cardiac surgery has revolutionized the field of cardiovascular surgery, offering numerous advantages over traditional approaches. The applications of robotically assisted mitral valve surgery, robotically assisted coronary revascularization, and other cardiac surgeries have shown promising outcomes, demonstrating the potential of this technology. However, several challenges and limitations must be addressed to fully realize the future potential of robotic cardiac surgery. Continued research, technological advancements, interdisciplinary collaborations, and robust training programs will shape the future of this field. By embracing these opportunities, robotic cardiac surgery has the potential to transform the management of cardiovascular diseases and improve patient outcomes on a global scale.

## Key points

- Robotic cardiac surgery offers improved precision, enhanced visualization, and reduced invasiveness.
- Robotic-assisted mitral valve surgery demonstrates improved patient outcomes and reduced complications.
- Robotic-assisted coronary revascularization enables precise grafting and reduces postoperative morbidity.
- Other cardiac surgeries, such as atrial septal defect repair and cardiac tumor resection, benefit from robotic assistance.
- Future prospects include advancements in technology, integration with imaging and navigation systems, AI applications, enhanced training, and cost-effectiveness.

## References

[1] Cardiovascular diseases (CVDs). https://www.who.int/news-room/fact-sheets/detail/cardiovascular-diseases-(cvds). [Accessed 11 September 2023].

[2] Sepehripour AH, Garas G, Athanasiou T, Casula R. Robotics in cardiac surgery. Ann R Coll Surg Engl 2018;100(Suppl 7):22–33. https://doi.org/10.1308/RCSANN.SUPP2.22.

[3] Chapter 44. Minimally Invasive and Robotic Mitral Valve Surgery, Cardiac Surgery in the Adult, 4e, AccessSurgery, McGraw Hill Medical. [Accessed 11 September 2023]. https://accesssurgery.mhmedical.com/content.aspx?bookid=476&sectionid=39679059.

[4] Moss E, Murphy DA, Halkos ME. Robotic cardiac surgery: current status and future directions. Robot Surg: Res Rev 2014;27. https://doi.org/10.2147/RSRR.S35929. Published online October.

[5] Bonatti J, Wallner S, Crailsheim I, Grabenwöger M, Winkler B. Minimally invasive and robotic coronary artery bypass grafting-a 25-year review. J Thorac Dis 2021;13(3):1923–44. https://doi.org/10.21037/JTD-20-1535.

[6] Nishimura RA, Otto CM, Bonow RO, et al. 2017 AHA/ACC focused update of the 2014 AHA/ACC guideline for the Management of Patients with Valvular Heart Disease: a report of the American College of Cardiology/American Heart Association task force on clinical practice guidelines. J Am Coll Cardiol 2017;70(2):252–89. https://doi.org/10.1016/J.JACC.2017.03.011.

[7] Whellan DJ, McCarey MM, Taylor BS, et al. Trends in robotic-assisted coronary artery bypass grafts: a study of the Society of Thoracic Surgeons adult cardiac surgery database, 2006 to 2012. Ann Thorac Surg 2016;102(1):140–6. https://doi.org/10.1016/J.ATHORACSUR.2015.12.059.

[8] Cerny S, Oosterlinck W, Onan B, et al. Robotic cardiac surgery in Europe: status 2020. Front Cardiovasc Med 2022;8. https://doi.org/10.3389/FCVM.2021.827515/FULL.

[9] Sepehripour AH, Garas G, Athanasiou T, Casula R. Robotics in cardiac surgery. Ann R Coll Surg Engl 2018;100(Suppl 7):22–33. https://doi.org/10.1308/RCSANN.SUPP2.22.

[10] Fatehi Hassanabad A, Nagase FNI, Basha AM, et al. A systematic review and meta-analysis of robot-assisted mitral valve repair. Innovations (Phila) 2022;17(6):471–81. https://doi.org/10.1177/15569845221141488.

[11] Hawkins RB, Mehaffey JH, Mullen MM, et al. A propensity matched analysis of robotic, minimally invasive, and conventional mitral valve surgery. Heart 2018;104(23):1970. https://doi.org/10.1136/HEARTJNL-2018-313129.

[12] Virani SS, Alonso A, Benjamin EJ, et al. Heart disease and stroke Statistics-2020 update: a report from the American Heart Association. Circulation 2020;141(9):E139–596. https://doi.org/10.1161/CIR.0000000000000757.

[13] Williams ML, Hwang B, Huang L, et al. Robotic versus conventional sternotomy mitral valve surgery: a systematic review and meta-analysis. Ann Cardiothorac Surg 2022;11(5):490–503. https://doi.org/10.21037/ACS-2022-RMVS-21.

[14] Cao C, Indraratna P, Doyle M, et al. A systematic review on robotic coronary artery bypass graft surgery. Ann Cardiothorac Surg 2016;5(6):530. https://doi.org/10.21037/ACS.2016.11.08.

[15] Cavallaro P, Rhee AJ, Chiang Y, Itagaki S, Seigerman M, Chikwe J. In-hospital mortality and morbidity after robotic coronary artery surgery. J Cardiothorac Vasc Anesth 2015;29(1):27–31. https://doi.org/10.1053/J.JVCA.2014.03.009.

[16] Gofus J, Cerny S, Shahin Y, et al. Robot-assisted vs. conventional MIDCAB: a propensity-matched analysis. Front Cardiovasc Med 2022;9. https://doi.org/10.3389/FCVM.2022.943076.

[17] Halkos ME, Liberman HA, Devireddy C, et al. Early clinical and angiographic outcomes after robotic-assisted coronary artery bypass surgery. J Thorac Cardiovasc Surg 2014;147(1):179–85. https://doi.org/10.1016/J.JTCVS.2013.09.010.

[18] de Jong AR, Gianoli M, Namba HF, et al. A Nationwide study of clinical outcomes after robot-assisted coronary artery bypass surgery and hybrid revascularization in the Netherlands. Innovations (Phila) 2023;18(1):73–9. https://doi.org/10.1177/15569845231154046.

[19] Kim JE, Jung SH, Kim GS, et al. Surgical outcomes of congenital atrial septal defect using da VinciTM surgical robot system. Korean J Thorac Cardiovasc Surg 2013;46(2):93–7. https://doi.org/10.5090/KJTCS.2013.46.2.93.

[20] Xiao C, Gao C, Yang M, et al. Totally robotic atrial septal defect closure: 7-year single-institution experience and follow-up. Interact Cardiovasc Thorac Surg 2014;19(6):933–7. https://doi.org/10.1093/ICVTS/IVU263.

[21] Onan B, Aydin U, Kadirogullari E, Onan IS, Sen O, Kahraman Z. Robotic repair of partial anomalous pulmonary venous connection: the initial experience and technical details. J Robot Surg 2020;14(1):101–7. https://doi.org/10.1007/S11701-019-00943-0.

[22] Mirzai S, Hibino N, Torregrossa G, Balkhy HH. Adult ventricular septal defect repair using a robotic totally endoscopic approach: a case report. Innovations (Phila) 2020;15(4):372–5. https://doi.org/10.1177/1556984520922978.

[23] Gao C, Yang M, Wang G, Wang J, Xiao C, Zhao Y. Totally endoscopic robotic ventricular septal defect repair. Innovations (Phila) 2010;5 (4):278–80. https://doi.org/10.1097/IMI.0B013E3181EE94CB.

[24] Balkhy HH, Nisivaco S, Torregrossa G, et al. Multi-spectrum robotic cardiac surgery: early outcomes. JTCVS Tech 2022;13:74–82. https://doi.org/10.1016/J.XJTC.2021.12.018.

[25] Roberts HG, Wei LM, Dhamija A, Cook CC, Badhwar V. Robotic assisted cryothermic biatrial cox-maze. J Cardiovasc Electrophysiol 2021;32 (10):2879–83. https://doi.org/10.1111/jce.15075.

[26] Almousa A, Mehaffey JH, Wei LM, et al. Robotic-assisted cryothermic cox maze for persistent atrial fibrillation: longitudinal follow-up. J Thorac Cardiovasc Surg 2023;165(5):1828–1836.e1. https://doi.org/10.1016/J.JTCVS.2022.05.012.

[27] Liu Y, Liu Z, Li X, et al. A comparison of total thoracoscopic versus robotic approach for cardiac myxoma resection: a single-center retrospective study. J Robot Surg 2023;17(4). https://doi.org/10.1007/S11701-023-01531-Z.

[28] Yang M, Yao M, Wang G, et al. Comparison of postoperative quality of life for patients who undergo atrial myxoma excision with robotically assisted versus conventional surgery. J Thorac Cardiovasc Surg 2015;150(1):152–7. https://doi.org/10.1016/J.JTCVS.2015.01.056.

[29] Rivero-Moreno Y, Echevarria S, Vidal-Valderrama C, et al. Robotic surgery: a comprehensive review of the literature and current trends. Cureus 2023;15(7). https://doi.org/10.7759/cureus.42370. e42370.

[30] Tan S, Lopuszko A, Bashir M. Artificial intelligence in cardiac surgery. In: Intelligence-based cardiology and cardiac surgery; 2024. p. 243–6. https://doi.org/10.1016/B978-0-323-90534-3.00023-8. Published online January 1.

[31] Bakhuis W, Max SA, Maat APWM, Bogers AJJC, Mahtab EAF, Sadeghi AH. Preparing for the future of cardiothoracic surgery with virtual reality simulation and surgical planning: a narrative review. *Shanghai* Chest 2023;7(0). https://doi.org/10.21037/SHC-22-63/COIF.

[32] Badhwar V, Wei LM, Geirsson A, et al. Contemporary robotic cardiac surgical training. J Thorac Cardiovasc Surg 2023;165(2):779–83. https://doi.org/10.1016/j.jtcvs.2021.11.005.

[33] Moss E, Halkos ME. Cost effectiveness of robotic mitral valve surgery. Ann Cardiothorac Surg 2017;6(1):33–7. https://doi.org/10.21037/ACS.2017.01.03.

III. The current and future clinical applications of robotic surgery among medical specialties

# 52

# Robotic thoracic surgery

*Jennifer Pan[a], Ammara Watkins[b], and Elliot Servais[b]*

[a]Department of General Surgery, Beth Israel Deaconess Medical Center, Boston, MA, United States [b]Division of Thoracic and Cardiovascular Surgery, Lahey Hospital & Medical Center, Burlington, MA, United States

## Introduction

Robotic-assisted thoracic surgical volume has grown rapidly since it was adopted in the early 2000s [1–4]. The first robotic thoracic case series was conducted in 2001 by Melfi et al., demonstrating its feasibility, advantages, and hurdles within the operating system. Robotic-assisted lobectomies, tumor enucleations, and bullae closure were described by this group for a variety of thoracic disease [2]. Since then, many have demonstrated the significant advantages of minimally invasive surgery (MIS) compared to open thoracic surgery. Minimally invasive thoracic surgery has now become the standard of care in the majority of thoracic oncologic operations [5]. Robotic-associated thoracic surgery (RATS) has now surpassed video-assisted thoracoscopic surgery (VATS) for anatomic lung resection [3] (Fig. 1).

The da Vinci system (Intuitive Surgical, Sunnyvale, CA) is the most utilized platform. It has several advantages when compared to VATS. The advantages are improved maneuverability with wristed instruments, dexterity, elimination of physiologic tremor, surgeon control of the camera and retraction arms, three-dimensional vision, and magnification [1]. With the robotic approach, the surgeon uses carbon dioxide insufflation of the thorax for improved visualization. Although capnothorax can also be utilized in VATS surgery, it is less frequently employed. These properties result in improved visualization of the surgical field and easier access to traditionally difficult areas, including the mediastinum and thoracic outlet. The da Vinci Xi system also has an intraoperative near-infrared fluorescence, which can be used with indocyanine green (ICG) for the identification of vasculature, lung planes, and tumor marking [6].

## Robotic lung resection

The advantages of RATS have largely been described in robotic lung resections. Multiple studies have found that RATS lung resection yields either equivalent or improved outcomes to traditional open or VATS lung resection.

### Robotic compared to open lung resection

Compared to thoracotomy, RATS has demonstrated clear advantages including significantly lower postoperative complication rates, length of stay (LOS), and mortality despite longer operative times [7–14]. Multiple database studies have found improved outcomes with robotic lung resections. A multicenter study comparing RATS lobectomy to thoracotomy outcomes from the Society of Thoracic Surgeons (STS) database found that robotic-assisted lobectomy had significantly lower postoperative blood transfusion rates, air-leak rates, chest tube duration, and LOS [15]. Rajaram et al. found no difference in margin positivity, 30-day readmission rates or death at 30- or 90-days between RATS or open lobectomies for stage I-IIIA nonsmall cell lung cancer (NSCLC) but those who underwent robotic-assisted lobectomies had a significantly shorter LOS [16]. Robotic-assisted lung resection for malignancy has a decreased rate of perioperative complications and a shorter hospital stay.

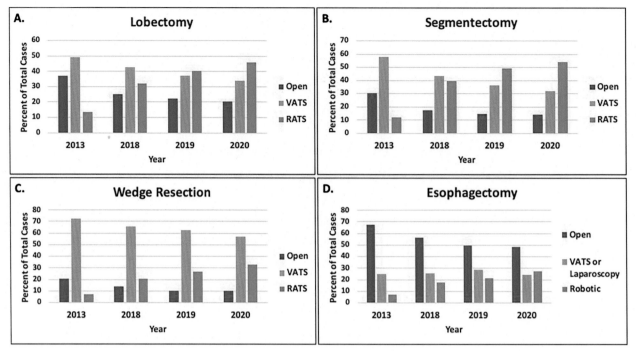

FIG. 1    Proportion of cases performed by modality over time. Modalities were thoracotomy (*dark blue* bars), VATS (*gray* bars), and RATS (*blue* bars). *Reprinted from E. Servais et al. The society of thoracic surgeons general thoracic surgery database: 2022 update on outcomes and research. Ann Thorac Surg 2023; 115: 43–50. Copyright 2023 by the Society of Thoracic Surgeons.*

Patients also experience less postoperative pain, following a minimally invasive lobectomy. Kwon et al. found a significant decrease in acute pain and chronic numbness following RATS or VATS compared to open resection [17]. Patients who undergo RATS lobectomy had faster postoperative pulmonary recovery than those who underwent thoracotomy [18].

## Robotic compared to VATS lung resection

Perioperative outcomes between RATS and VATS lung resection are similar. VATS and RATS lung resection have comparable rates of postoperative complications, intraoperative blood loss, in-hospital mortality, and LOS [7,10,13,14,16,19–23]. An advantage of RATS when compared to VATS is a lower rate of conversion to open thoracotomy as shown by Servais et al. [24] In a retrospective review of lobectomies from 21 centers, the rate of conversion to open, for all clinical stages, was lower in RATS lobectomies compared to VATS [8]. In a review of the Premier Healthcare database, RATS lobectomies had a significantly lower conversion to open rate and lower 30-day complication rate [25]. There is also an advantage of robotic surgery in obese patients. Seder et al. found that VATS had a five times greater conversion rate to thoracotomy than RATS lung resection in patients with BMI > 24 [26].

## Lymphadenectomy

The extent of lymph node dissection is an important aspect in oncologic care, particularly for NSCLC. The robotic platform has improved visualization, dexterity, and accurate application of bipolar energy which can facilitate a thorough lymph node dissection. Robotic lung resection has excellent quality outcomes in the lymph node resection [21,22,27,28]. In a multiinstitutional retrospective review, Wilson et al. found that robotic surgery had similar rates of nodal upstaging to thoracotomy and higher rates of upstaging to VATS [29]. In a retrospective review of 1053 patients, there was no difference between RATS and open lymph node upstaging, but both were better than VATS [30]. Robotic lymph node sampling may result in higher rates of upstaging than VATS; this is an advantage that should be considered when performing an oncologic operation and warrants further investigation.

## Oncologic outcomes

In terms of long-term outcomes, robotic lobectomy for NSCLC has appropriate overall survival (OS) and disease-free survival (DFS) [22,28,31–33]. In a propensity-matched analysis of RATS and VATS lobectomy for stage I-IV NSCLC, patients had no difference in OS or cancer-specific mortality at 3 years [34]. This was consistent with a large metaanalysis of lobectomy outcomes of stage I-IV NSCLC where no differences in 5-year OS were seen between RATS, VATS, or open lobectomy [8,13]. In the recent PORTaL survival analysis of IA-IIIA NSCLC, 5-year OS was significantly higher for open and RATS lobectomies compared to VATS [35].

## Costs

One of the major critiques of robotic surgery is that the cost of surgery is higher than VATS or open [10,36,37]. In a retrospective review of lung resection costs, robotic cases were determined to be more expensive due to robotic-specific supplies and depreciation [38]. In a database study of the Premier Healthcare database of lobectomies, the upfront cost of robotic surgery was more expensive due to OR and supply costs. However, this became cost neutral in hospitals with >25 annual robotic cases due to shorter hospital stays [9]. More recently, Kneuetz et al. found no difference in cost across all three lobectomy approaches [39].

## Segmentectomy

In select patients with NSCLC who have tumors <2 cm or significant comorbidities, segmentectomy is an option to preserve pulmonary function [40]. Since the initial trial comparing lobectomy and pulmonary limited resection [40], there have been two randomized control trials (JCOG0802 and CALGB 140503) comparing sublobar resection to lobectomy for small peripheral nonsmall cell tumors. These have demonstrated that for patients with tumors <2 cm and node-negative disease, segmentectomy has similar survival outcomes to lobectomy, albeit increased recurrence rates [41,42].

Minimally invasive segmentectomies have become more common [3]. VATS segmentectomy has been shown to be a safe procedure with similar oncologic outcomes, fewer complications, and shorter hospital stay than open segmentectomy [43,44]. Robotic-assisted segmentectomy has also been shown to be safe for both benign and malignant lesions [27]. When compared to open segmentectomy, the robotic approach was found to have similar mortality, shorter hospital stays, and fewer complications including chest tube duration and prolonged air leak [45]. Compared to VATS, robotic-assisted segmentectomy has similar morbidity and mortality rates with shorter hospital stay and lower conversion to open rates [45-51]. RATS segmentectomy was also found to yield a greater number of lymph nodes compared to VATS or open [49,52,53].

## Indications

Eligibility criteria for RATS lung resection is typically similar to the VATS approach. In experienced hands, studies have shown appropriate oncologic and perioperative outcomes even for complex resections compared to thoracotomy [54-59]. The preoperative evaluation for patients undergoing RATS, VATS, or thoracotomy is similar. Patients undergo pulmonary evaluation including standard preoperative pulmonary function tests (PFTs). Tumors are staged with preoperative bronchoscopy, CT chest, PET/CT, and brain MRI, if indicated, in addition to lymph node evaluation. Patients should also undergo cardiopulmonary risk stratification prior to surgery.

A few techniques have been described—including a three-arm technique and four-arm technique with or without a utility incision [60-62]. Similar techniques have been described for sublobar resections [63-65]. Segmentectomy is technically challenging, requiring dissection to the segmental structures. Three-dimensional renditions of the pulmonary anatomy including location of lesion can aid in preoperative planning [57]. ICG can be used to identify the pulmonary nodule, draining lymph nodes, and intersegmental planes [66,67]. Herein, we describe our four-arm, completely portal technique. This allows for less reliance on the bedside assistant and a near autonomous operation by the console surgeon. While our general approach to the operations is described in the following section, it is not uncommon to have slight variance due to intraoperative factors such as patient anatomy.

## Patient and robot positioning

Single-lung ventilation is used through a double-lumen endotracheal tube. Positioning of the endotracheal tube is confirmed on bronchoscopy. Patients are positioned in a lateral decubitus position with the operative side up.

The set-up at our institution consists of four robotic ports (two 8 mm and two 12 mm) with an assistant port (12 mm). All robotic ports are placed along the eighth intercostal space to minimize postoperative pain. The initial port (8 mm) is the camera port and placed at the highest point of chest convexity to optimize visualization. The 12-mm ports are placed on either side of the initial port. This allows the stapler to be used from the anterior or posterior approach. The final 8-mm port is placed in the posterior-most position. The ports are generally one hand width apart. The assistant port is placed anteriorly and inferiorly, usually in the tenth or eleventh intercostal space, triangulating between the two anterior most ports (Fig. 2). The assistant port uses the AirSeal (ConMed, NY, USA). The robot is then docked. We do not perform "targeting" but do place the laser crosshairs over the camera port prior to docking. Docking is optimized when the camera arm is parallel to the highest point and long axis of the patient. The subsequent arms are spaced approximately two hand width apart. The clearance of the arms is dropped maximally (Fig. 3).

### *Instruments*

Typically, a zero-degree camera is used. For retraction and visualization, commercially available rolled-up gauzes are used for atraumatic handling of the tissue. Vessel loops are used to encircle critical structures for retraction when passing staplers for transection of critical structures. We use the tip-up fenestrated grasper, long bipolar forceps, and

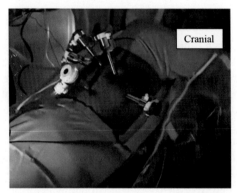

FIG. 2   Right-sided positioning for a lobectomy. Robotic ports are in the same intercostal space and the assistant port is the inferior-most part of pleural cavity.

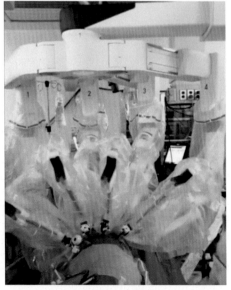

FIG. 3   Robot docked with the clearance of the arms dropped down maximally.

fenestrated bipolar for lung resections. The tip-up fenestrated bipolar and fenestrated bipolar forceps are used for retraction. The fenestrated bipolar and bipolar forceps are used to divide tissue and small vessels in addition to moving the lung tissue. A vessel sealer is occasionally utilized to divide smaller blood vessels not amenable to stapling. The endoscopic stapler (SureForm, Intuitive Surgical, CA) is used for dividing larger structures. The assistant port is used for exchanging gauzes and removal of specimens.

## Lung resection

We typically begin by dividing the inferior pulmonary ligament and proceed with the posterior hilar dissection. The posterior pleura is opened to expose the posterior hilum, which is dissected to expose the bronchus, pulmonary artery (PA) and branches, and pulmonary veins. We proceed to complete the fissure if necessary. We aim to perform a lymphadenectomy prior to vascular division for both oncologic benefit and to facilitate easier passage of the stapler. The pulmonary vein is typically transected first followed by the PA or branches. We always perform a clamp test to ensure ventilation to remaining lung parenchyma prior to dividing the airway. Intraoperative ICG can be used during sublobar resections to identify the extent of resection.

## Mediastinum

The mediastinum is the thoracic space between the pleural cavities. There a number of ways to divide the space; most divide it into three or four compartments [68]. We will use the method described by the International Thymic Malignancy Interest Group (ITMIG). The three compartments constitute prevascular, visceral, and paravertebral [68,69].

Historically, mediastinal surgery involved a sternotomy or thoracotomy to obtain the appropriate exposure. MIS has gained popularity in mediastinal dissections [70–74]. Numerous studies have demonstrated the feasibility of robotic-assisted mediastinal surgery for thymic tumors, germ cell tumors, thyroid and parathyroid tumors, lymphoma, pericardial cysts, bronchial cysts, and neuroendocrine tumors [73,75–85].

## Prevascular compartment

Robotic surgery in this compartment has primarily been studied in thymus pathology and, less commonly, in ectopic parathyroid or thyroid goiter excision [78,86–88]. The most common indications for a thymectomy are in myasthenia gravis and thymoma. In patients with myasthenia gravis, a thymectomy may be recommended even in the absence of a thymoma as patients have better long-term clinical outcomes following surgery [89,90].

Robotic surgery for thymoma is feasible and safe [75,80,91,92]. Compared to open surgery, robotic-assisted thymectomy decreases postoperative complications, 30-day readmission rates, and mortality [74,93]. Others have demonstrated decreased blood loss, shorter chest tube duration, shorter hospital LOS, and improved quality of life with a robotic approach [94–101]. Compared to VATS, RATS has shown equivalent or decreased intraoperative blood loss, chest tube duration, LOS, and fewer postoperative complications [93,102,103]. Ruckert et al. compared VATS and RATS thymectomies for myasthenia gravis. Both groups had similar rates of conversion to open but the robotic group had a higher remission rate [104].

Multiple studies have demonstrated comparable rates of R0 resection, recurrence rates of tumors, reduction of myasthenia gravis symptoms, and long-term survival [17,74,91,95,96,98–100,105–107]. Regarding oncologic outcomes, Kang et al. compared RATS thymectomy to transsternal thymectomy, and all patients had an R0 resection. There was no statistically significant difference in three-year OS rates or recurrence rates between RATS and open thymectomy. Furthermore, both groups had a reduction in their myasthenic symptoms [105]. Yang et al. analyzed postthymectomy patients with stage I to III thymoma from the National Cancer Database and found that MIS thymectomies had similar margin positivity and 5-year OS when compared to sternotomy. There was no difference in outcomes between VATS and RATS thymectomies [71].

Early studies have shown feasibility of robotic-associated thymectomy in tumors <4 cm in size [97]. It was previously thought that larger thymomas should be resected in an open fashion for oncologic outcomes. Studies have demonstrated the feasibility and safety of a robotic approach for thymomas >4 cm [73,95,96,108]. Soder et al. had a median tumor size of 4.9 cm with the largest tumor being 13 cm. Size was not an absolute contraindication for robotic approach but rather, adjacent organ invasion influenced the approach more [95].

FIG. 4    Robotic port placement for thymectomy with three 8-mm ports and an assistant port. Occasionally, a right-sided laparoscopic port is placed for visualization of the right phrenic nerve.

Mediastinal lesions can be approached robotically from right, left, subxiphoid, or a combination of these approaches. There are limited studies comparing the different robotic approaches to the mediastinum. We generally prefer a left-sided approach for robotic-associated thymectomy (Fig. 4) as this allows for straight forward identification of the left phrenic and innominate vein [109]. Thymic tissue has been shown to extend further to the left, and this provides better visualization of the left phrenic nerve [110]. We use a vessel sealer to remove thymic tissue. In the setting of a predominantly right-sided anterior mediastinal pathology, a right-sided approach may be preferable.

## Visceral and paravertebral compartment

There is a lack of data comparing RATs, VATS, and open resection in visceral and paravertebral compartments. To date, studies have shown the feasibility and safety of robot-assisted surgery to access visceral and paravertebral lesions which are difficult areas to reach [79,82]. The robotic approach has been used to resect bronchogenic or esophageal cysts, lymphoma, metastasis, teratomas, and nervous system tumors (schwannoma, paraganglioma, and ganglioneuroma) [79–85,111,112]. In a retrospective study of 130 patients who underwent RATS or VATS resection of posterior mediastinal tumors, patients who underwent RATS were found to have similar chest tube duration and conversion to open rates but had decreased blood loss and LOS when compared to VATS. There was no significant difference in the postoperative complication rate [84].

There are several advantages in using a robotic-assisted approach for mediastinal masses. The robotic platform provides improved visualization and access to restricted areas such as the mediastinal compartments. The degree of articulation of the robotic arms, filtration of hand tremors, three-dimensional images, and use of insufflation all aid to the exposure and maneuverability [77,80,92,100]. Broussard et al. and Cerfolio et al. describe different approaches to access various locations with ease [79,81]. There is no current standardized approach. We approach our posterior mediastinal masses in a similar fashion as a RATS lobectomy. We typically use cadiere forceps, fenestrated bipolar, and bipolar forceps for dissection. For ganglioneuromas, the nerves are isolated, clipped, and ligated with a vessel sealer.

## Complex airway surgery

### Pneumonectomy and sleeve lobectomy

A limited number of series and case studies investigated robotic-assisted pneumonectomy [113–116]. Patton et al. reviewed their RATS pneumonectomy data and found that the robotic group had a shorter operative time, decreased blood loss, and more sampled lymph nodes than the converted to open group [117]. In a review of the National Cancer Database, Hennon et al. found that there was no difference in perioperative mortality between RATS, VATS, or open pneumonectomy. The MIS group resected a higher number of lymph nodes than the open group [118].

Sleeve lobectomy is an alternative to pneumonectomy. Sleeve lobectomy involves the removal of the lobar bronchus and surrounding parenchyma. It serves as a lung parenchyma preserving option for diseases that would otherwise require a pneumonectomy. A number of series have compared sleeve lobectomy to pneumonectomy. Patients who underwent sleeve lobectomy had improved long-term survival rates, preserved pulmonary function, and lower operative mortality [119,120].

Minimally invasive sleeve lobectomies have been shown to be safe [58,59,121–124]. Jiao et al. found that preoperative comorbidity, elderly, and early surgeon experience were risk factors for postoperative complications [125]. Qui et al. did not identify any significant complication rate difference between open, VATS, or robotic [126]. In their series, Geraci et al. found that all patients had an R0 resection with a median LOS of 3 days without 30- or 90-day mortality [127]. When robotic-assisted sleeve lobectomy was compared to VATS and thoracotomy, there was no difference in 90-day mortality or morbidity. The RATS group had decreased blood loss, shorter operative time, and chest tube duration compared to open and VATS. All patients in the RATS group had negative margins. There was no difference in DFS among the three groups [126]. Liu et al. studied the long-term outcomes of robotic-assisted sleeve lobectomy. There was no endobronchial or perianastomotic recurrence, and 5-year DFS and OS rates were comparable to open and VATS [128].

Patients with advanced malignant lung disease are considered for pneumonectomy. Other indications for pneumonectomy included severely damaged lung from chronic infections [129]. If a tumor is in the lobar bronchus or invading the main bronchus or there is a benign bronchial stricture, patients can be considered for a sleeve resection. Patients should complete a cardiopulmonary assessment and complete oncologic staging with mediastinal lymph node evaluation. For both, we position and place our ports in the same fashion as lobectomies. Infrared imaging is used to visualize the extent of bronchial resection in a sleeve lobectomy [59].

## Tracheobronchoplasty

Tracheobronchomalacia (TBM) and excessive dynamic airway collapse (EDAC) are two pathologies under excessive central airway collapse (ECAC). This is an underdiagnosed disease, often resulting in significant morbidity. The collapse of the airway lumen results in a persistent cough, dyspnea, and respiratory infections. Patients with severe ECAC are evaluated for tracheobronchoplasty (TBP), which involves stabilization of the central airways. TBP has been traditionally performed with an open posterolateral thoracotomy, as described by Gangadharan et al. [130] Following TBP, patients have often had improvement in health-related quality of life questionnaires (HRQOL) and improvement in their six-minute walk test [131–135]. However, improvements are not always seen in the rest of the PFTs [136].

While there are limited studies published on robotic-assisted TBP, the robotic approach has been demonstrated [137]. Patients were found to have a shorter LOS, improvement on their postoperative HRQOL questionnaires, and their PFTs [138]. Lazzaro et al. published on their robotic-assisted TBP method which begins with the left mainstem bronchus (LMSB), followed by the trachea, and then, the right mainstem. The mesh is cut into three pieces to better align with the posterior airway [137]. Seasteadt et al. noted a similar order of repair. However, if the exposure of the LMSB is challenging, then the tracheoplasty is performed first to enable better access to the LMSB [139].

## Other thoracic surgery

### First rib resection

Thoracic outlet syndrome (TOS) describes a constellation of symptoms which result from compression of the brachial plexus, subclavian vein, or artery. The brachial plexus, subclavian artery, and vein run between the clavicle and first rib. TOS occurs with prolonged compression of these structures. Causes include a cervical rib, prolonged transverse process, and persistent muscular bands between the scalene muscles or trauma.

Neurologic TOS is the most common type of TOS and due to irritation or compression of the brachial plexus. Patients experience pain and upper extremity paresthesias. Venous TOS is known as Paget-Schroetter syndrome and results from compression of the subclavian vein. This is typically with repetitive motions causing muscle enlargement. Patients typically have upper extremity swelling and heaviness. Arterial TOS usually results from bone abnormality that causes compression of brachial artery. Patients can develop poststenotic dilation or subclavian artery aneurysms and distal emboli [129,140].

A few open approaches are utilized but all face challenges with adequate exposure. The advantage of the robotic system is its superior ability to visualize the anatomy and instrument articulation. Early reports by Kocher et al. and Gharagozloo et al. demonstrate promising results of a robotic approach to first rib resections without any intraoperative or postoperative complications [141,142]. Similar studies have demonstrated the safety of this approach with the resolution of symptoms [143–146]. Compared to the supraclavicular approach, the robotic approach had similar operative times and hospital stay but found to have less pain, fewer overall complications, including brachial plexus palsy [147,148].

Patients who are candidates for open surgery are candidates for robotic intervention. Patients with vascular TOS or persistent neurogenic TOS after nonoperative management are considered for surgery. They undergo a thorough physical examination and MRI to evaluate the anatomy. Patients with vascular TOS also should undergo noninvasive Doppler evaluation, arteriography, or venography for diagnostic and therapeutic intervention. Burt et al. describes their 12 steps of a standard first rib resection [149].

## Diaphragm plication

The diaphragm muscle plays an important role in respiratory effort. During inspiration, the diaphragm contracts to increase lung volume. The diaphragm is innervated by the phrenic nerve (C3–5) bilaterally and controlled by the brain stem. Any injury or damage along the pathway can result in diaphragm paralysis. Unilateral diaphragm paralysis can occur after thoracic or cardiac surgery, malignancy, or injury to the nervous system. With diaphragm paralysis, the diaphragm flattens and moves paradoxically during respiration. This leads to a decrease in vital capacity by 20% with unilateral paralysis [129,150].

For symptomatic unilateral diaphragm paralysis, open diaphragm plication has been well described. Following transthoracic diaphragm plication, patients had subjective and objective improvement in symptoms and PFTs up to 14 years postoperatively [151–154]. Laparoscopic plication has demonstrated significant improvements in quality of life and PFTs [155,156]. The VATS approach had shorter LOS and complication rates compared to open [157]. Robotic-assisted diaphragm plication has been shown to be safe and effective with shorter LOS compared to open plication [158–160]. Asaf et al. found significant improvement in PFTs after robotic plication via transabdominal or transthoracic approach [161].

Symptomatic patients with hemidiaphragm paralysis should undergo PFTs and evaluation of other causes of their respiratory symptoms. A fluoroscopic sniff test can identify lack of movement or paradoxical movement of the paralyzed diaphragm [162]. In setting up for a RATS plication, the robot boom is rotated such that the instruments point inferiorly towards the diaphragm. The line of plication is sutured with barbed suture and pledgets (Fig. 5).

## Conclusion

Since the robotic series published by Melfi et al [2], the platform has become widely adopted for thoracic surgeries including malignant and benign pathologies. Overall, the robotic approach results in decreased postoperative pain and shorter hospital stay but with higher operative costs. In malignancy, robotic-assisted thoracic surgery demonstrates similar oncologic outcomes with higher lymph node sampling rates.

## Future directions

As the robotic platform expands, more thoracic surgeons will incorporate this modality into their practice. While robotic surgery has multiple advantages over other modalities, its major disadvantage is the lack of haptic feedback. This remains an active area of innovation research to create sensors that can transmit this information. Another future direction is single-site surgery through which the operation is completed via one small port. This allows for less pain and even faster recovery [6].

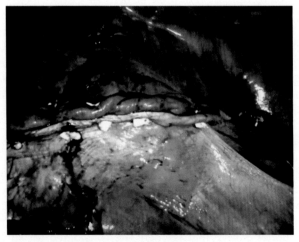

FIG. 5   Plicated diaphragm with barbed suture and pledgets along each stitch.

## Key points

- The robotic platform has been demonstrated to be feasible and safe in many thoracic options.
- There are clear benefits in using the robotic platform for thoracic operations, largely due to optimal visualization and access.
- The robotic approach has demonstrated similar oncologic outcomes and possible improved lymph node sampling for nonsmall cell lung cancer resections.
- Robotic-assisted thymectomies for thymoma had similar oncologic outcomes as other approaches.

# References

[1] Zirafa CC, Romano G, Key TH, Davini F, Melfi F. The evolution of robotic thoracic surgery. Ann Cardiothorac Surg 2019;8(2):210–7. https://doi.org/10.21037/acs.2019.03.03.

[2] Melfi FMA, Menconi GF, Mariani AM, Angeletti CA. Early experience with robotic technology for thoracoscopic surgery. Thorac Surg 2002;21(5):864–8.

[3] Servais EL, Blasberg JD, Brown LM, et al. The Society of Thoracic Surgeons general thoracic surgery database: 2022 update on outcomes and research. Ann Thorac Surg 2023;115(1):43–9. https://doi.org/10.1016/j.athoracsur.2022.10.025.

[4] Abdelfatah E, Jordan S, Dexter EU, Nwogu C. Robotic thoracic and esophageal surgery: a critical review of comparative outcomes. Ann Laparosc Endosc Surg 2021;6:10. https://doi.org/10.21037/ales.2020.04.03.

[5] Demmy TL, Yendamuri S, D'Amico TA, Burfeind WR. Oncologic equivalence of minimally invasive lobectomy: the scientific and practical arguments. Ann Thorac Surg 2018;106(2):609–17. https://doi.org/10.1016/j.athoracsur.2018.02.089.

[6] Lazar JF, Hwalek AE. A review of robotic thoracic surgery adoption and future innovations. Thorac Surg Clin 2023;33(1):1–10. https://doi.org/10.1016/j.thorsurg.2022.07.010.

[7] Kent M, Wang T, Whyte R, Curran T, Flores R, Gangadharan S. Open, video-assisted thoracic surgery, and robotic lobectomy: review of a National Database. Ann Thorac Surg 2014;97(1):236–44. https://doi.org/10.1016/j.athoracsur.2013.07.117.

[8] Kent MS, Hartwig MG, Vallières E, et al. Pulmonary open, Robotic and thoracoscopic lobectomy (PORTaL) study: an analysis of 5,721 cases. Ann Surg 2021. https://doi.org/10.1097/SLA.0000000000005115. Publish Ahead of Print.

[9] Nguyen DM, Sarkaria IS, Song C, et al. Clinical and economic comparative effectiveness of robotic-assisted, video-assisted thoracoscopic, and open lobectomy. J Thorac Dis 2020;12(3):296–306. https://doi.org/10.21037/jtd.2020.01.40.

[10] Subramanian MP, Liu J, Chapman WC, et al. Utilization trends, outcomes, and cost in minimally invasive lobectomy. Ann Thorac Surg 2019;108(6):1648–55. https://doi.org/10.1016/j.athoracsur.2019.06.049.

[11] Cao C, Manganas C, Ang SC, Yan TD. A systematic review and meta-analysis on pulmonary resections by robotic video-assisted thoracic surgery. Ann Cardiothorac Surg 2012;1(1).

[12] Zhang O, Alzul R, Carelli M, Melfi F, Tian D, Cao C. Complications of robotic video-assisted Thoracoscopic surgery compared to open thoracotomy for Resectable non-small cell lung cancer. J Pers Med 2022;12(8):1311. https://doi.org/10.3390/jpm12081311.

[13] Aiolfi A, Nosotti M, Micheletto G, et al. Pulmonary lobectomy for cancer: systematic review and network meta-analysis comparing open, video-assisted thoracic surgery, and robotic approach. Surgery 2021;169(2):436–46. https://doi.org/10.1016/j.surg.2020.09.010.

[14] Agzarian J, Fahim C, Shargall Y, Yasufuku K, Waddell TK, Hanna WC. The use of robotic-assisted thoracic surgery for lung resection: a comprehensive systematic review. Semin Thorac Cardiovasc Surg 2016;28(1):182–92. https://doi.org/10.1053/j.semtcvs.2016.01.004.

[15] Adams RD, Bolton WD, Stephenson JE, Henry G, Robbins ET, Sommers E. Initial multicenter community robotic lobectomy experience: comparisons to a national database. Ann Thorac Surg 2014;97(6):1893–900. https://doi.org/10.1016/j.athoracsur.2014.02.043.

[16] Rajaram R, Mohanty S, Bentrem DJ, et al. Nationwide assessment of robotic lobectomy for non-small cell lung Cancer. Ann Thorac Surg 2017;103(4):1092–100. https://doi.org/10.1016/j.athoracsur.2016.09.108.

[17] Kang CH, Na KJ, Park S, Park IK, Kim YT. Long-term outcomes of robotic Thymectomy in patients with Thymic epithelial tumors. Ann Thorac Surg 2021;112(2):430–5. https://doi.org/10.1016/j.athoracsur.2020.09.018.

[18] Lacroix V, Kahn D, Matte P, et al. Robotic-assisted lobectomy favors early lung recovery versus limited thoracotomy. Thorac Cardiovasc Surg 2021;69(06):557–63. https://doi.org/10.1055/s-0040-1715598.

[19] Ye X, Xie L, Chen G, Tang JM, Ben XS. Robotic thoracic surgery versus video-assisted surgery for lung cancer: a meta-analysis. Interact Cardiovasc Thorac Surg 2015;21(4):409–14. https://doi.org/10.1093/icvts/ivv155.

[20] Louie BE, Farivar AS, Aye RW, Vallières E. Early experience with robotic lung resection results in similar operative outcomes and morbidity when compared with matched video-assisted thoracoscopic surgery cases. Ann Thorac Surg 2012;93(5):1598–605. https://doi.org/10.1016/j.athoracsur.2012.01.067.

[21] Liang H, Liang W, Zhao L, et al. Robotic versus video-assisted lobectomy/segmentectomy for lung cancer: a meta-analysis. Ann Surg 2018;268(2):254–9. https://doi.org/10.1097/SLA.0000000000002346.

[22] Li C, Hu Y, Huang J, et al. Comparison of robotic-assisted lobectomy with video-assisted thoracic surgery for stage IIB–IIIA non-small cell lung cancer. Transl Lung Cancer Res 2019;8(6):820–8. https://doi.org/10.21037/tlcr.2019.10.15.

[23] Louie BE, Wilson JL, Kim S, et al. Comparison of video-assisted thoracoscopic surgery and robotic approaches for clinical stage I and stage II non-small cell lung cancer using the Society of Thoracic Surgeons database. Ann Thorac Surg 2016;102(3):917–24. https://doi.org/10.1016/j.athoracsur.2016.03.032.

[24] Servais EL, Miller DL, Thibault D, et al. Conversion to thoracotomy during thoracoscopic vs robotic lobectomy: predictors and outcomes. Ann Thorac Surg 2022;114(2):409–17. https://doi.org/10.1016/j.athoracsur.2021.10.067.

[25] Reddy RM, Gorrepati ML, Oh DS, Mehendale S, Reed MF. Robotic-assisted versus thoracoscopic lobectomy outcomes from high-volume thoracic surgeons. Ann Thorac Surg 2018;106(3):902–8. https://doi.org/10.1016/j.athoracsur.2018.03.048.

[26] Seder CW, Farrokhyar F, Nayak R, et al. Robotic vs thoracoscopic anatomic lung resection in obese patients: a propensity-adjusted analysis. Ann Thorac Surg 2022;114(5):1879–85. https://doi.org/10.1016/j.athoracsur.2021.09.061.

[27] Toker A, Özyurtkan MO, Demirhan Ö, Ayalp K, Kaba E, Uyumaz E. Lymph node dissection in surgery for lung cancer: comparison of open vs. video-assisted vs. robotic-assisted approaches. Ann Thorac Cardiovasc Surg 2016;22(5):284–90. https://doi.org/10.5761/atcs.oa.16-00087.

[28] Yang HX, Woo KM, Sima CS, et al. Long-term survival based on the surgical approach to lobectomy for clinical stage I nonsmall cell lung cancer: comparison of robotic, video-assisted thoracic surgery, and thoracotomy lobectomy. Ann Surg 2017;265(2):431–7. https://doi.org/10.1097/SLA.0000000000001708.

[29] Wilson JL, Louie BE, Cerfolio RJ, et al. The prevalence of nodal upstaging during robotic lung resection in early stage non-small cell lung cancer. Ann Thorac Surg 2014;97(6):1901–7. https://doi.org/10.1016/j.athoracsur.2014.01.064.

[30] Kneuertz PJ, Cheufou DH, D'Souza DM, et al. Propensity-score adjusted comparison of pathologic nodal upstaging by robotic, video-assisted thoracoscopic, and open lobectomy for non–small cell lung cancer. J Thorac Cardiovasc Surg 2019;158(5):1457–1466.e2. https://doi.org/10.1016/j.jtcvs.2019.06.113.

[31] Cerfolio RJ, Ghanim AF, Dylewski M, Veronesi G, Spaggiari L, Park BJ. The long-term survival of robotic lobectomy for non–small cell lung cancer: a multi-institutional study. J Thorac Cardiovasc Surg 2018;155(2):778–86. https://doi.org/10.1016/j.jtcvs.2017.09.016.

[32] Kneuertz PJ, D'Souza DM, Richardson M, Abdel-Rasoul M, Moffatt-Bruce SD, Merritt RE. Long-term oncologic outcomes after robotic lobectomy for early-stage non–small-cell lung cancer versus video-assisted thoracoscopic and open thoracotomy approach. Clin Lung Cancer 2020;21(3):214–224.e2. https://doi.org/10.1016/j.cllc.2019.10.004.

[33] Park BJ, Melfi F, Mussi A, et al. Robotic lobectomy for non–small cell lung cancer (NSCLC): long-term oncologic results. J Thorac Cardiovasc Surg 2012;143(2):383–9. https://doi.org/10.1016/j.jtcvs.2011.10.055.

[34] Sesti J, Langan RC, Bell J, et al. A comparative analysis of long-term survival of robotic versus thoracoscopic lobectomy. Ann Thorac Surg 2020;110(4):1139–46. https://doi.org/10.1016/j.athoracsur.2020.03.085.

[35] Kent MS, Hartwig MG, Vallières E, et al. Pulmonary open, robotic and thoracoscopic lobectomy (PORTaL) study: survival analysis of 6,646 cases. Ann Surg 2023. https://doi.org/10.1097/SLA.0000000000005820. Publish Ahead of Print.

[36] Swanson SJ, Miller DL, McKenna RJ, et al. Comparing robot-assisted thoracic surgical lobectomy with conventional video-assisted thoracic surgical lobectomy and wedge resection: results from a multihospital database (premier). J Thorac Cardiovasc Surg 2014;147(3):929–37. https://doi.org/10.1016/j.jtcvs.2013.09.046.

[37] Nasir BS, Bryant AS, Minnich DJ, Wei B, Cerfolio RJ. Performing robotic lobectomy and segmentectomy: cost, profitability, and outcomes. Ann Thorac Surg 2014;98(1):203–9. https://doi.org/10.1016/j.athoracsur.2014.02.051.

[38] Deen SA, Wilson JL, Wilshire CL, et al. Defining the cost of care for lobectomy and segmentectomy: a comparison of open, video-assisted Thoracoscopic, and robotic approaches. Ann Thorac Surg 2014;97(3):1000–7. https://doi.org/10.1016/j.athoracsur.2013.11.021.

[39] Kneuertz PJ, Singer E, D'Souza DM, Abdel-Rasoul M, Moffatt-Bruce SD, Merritt RE. Hospital cost and clinical effectiveness of robotic-assisted versus video-assisted thoracoscopic and open lobectomy: a propensity score–weighted comparison. J Thorac Cardiovasc Surg 2019;157(5):2018–2026.e2. https://doi.org/10.1016/j.jtcvs.2018.12.101.

[40] Rubinstein LV. Randomized trial of lobectomy versus limited resection for T1 N0 Non - Small cell lung cancer. Ann Thorac Surg 1995;615–23.

[41] Saji H, Okada M, Tsuboi M, et al. Segmentectomy versus lobectomy in small-sized peripheral non-small-cell lung cancer (JCOG0802/WJOG4607L): a multicentre, open-label, phase 3, randomised, controlled, non-inferiority trial. Lancet 2022;399(10335):1607–17. https://doi.org/10.1016/S0140-6736(21)02333-3.

[42] Altorki N, Wang X, Kozono D, et al. Lobar or sublobar resection for peripheral stage IA non–small-cell lung cancer. N Engl J Med 2023;388(6):489–98. https://doi.org/10.1056/NEJMoa2212083.

[43] Leshnower BG, Miller DL, Fernandez FG, Pickens A, Force SD. Video-assisted thoracoscopic surgery segmentectomy: a safe and effective procedure. Ann Thorac Surg 2010;89(5):1571–6. https://doi.org/10.1016/j.athoracsur.2010.01.061.

[44] Ghaly G, Kamel M, Nasar A, et al. Video-assisted Thoracoscopic surgery is a safe and effective alternative to thoracotomy for anatomical Segmentectomy in patients with clinical stage I non-small cell lung Cancer. Ann Thorac Surg 2016;101(2):465–72. https://doi.org/10.1016/j.athoracsur.2015.06.112.

[45] Zhang Y, Chen C, Hu J, et al. Early outcomes of robotic versus thoracoscopic segmentectomy for early-stage lung cancer: a multi-institutional propensity score-matched analysis. J Thorac Cardiovasc Surg 2020;160(5):1363–72. https://doi.org/10.1016/j.jtcvs.2019.12.112.

[46] Demir A, Ayalp K, Ozkan B, Kaba E, Toker A. Robotic and video-assisted thoracic surgery lung segmentectomy for malignant and benign lesions†. Interact Cardiovasc Thorac Surg 2015;20(3):304–9. https://doi.org/10.1093/icvts/ivu399.

[47] Zhou N, Corsini EM, Antonoff MB, et al. Robotic surgery and anatomic segmentectomy: an analysis of trends, patient selection, and outcomes. Ann Thorac Surg 2022;113(3):975–83. https://doi.org/10.1016/j.athoracsur.2021.03.068.

[48] Alwatari Y, Khoraki J, Wolfe LG, et al. Trends of utilization and perioperative outcomes of robotic and video-assisted thoracoscopic surgery in patients with lung cancer undergoing minimally invasive resection in the United States. JTCVS Open 2022;12:385–98. https://doi.org/10.1016/j.xjon.2022.07.014.

[49] Kodia K, Razi SS, Alnajar A, Nguyen DM, Villamizar N. Comparative analysis of robotic Segmentectomy for non-small cell lung cancer: a national cancer database study. Innov Technol Tech Cardiothorac Vasc Surg 2021;16(3):280–7. https://doi.org/10.1177/1556984521997805.

[50] Cerfolio RJ, Watson C, Minnich DJ, Calloway S, Wei B. One hundred planned robotic segmentectomies: early results, technical details, and preferred port placement. Ann Thorac Surg 2016;101(3):1089–96. https://doi.org/10.1016/j.athoracsur.2015.08.092.

[51] Kneuertz PJ, Zhao J, D'Souza DM, Abdel-Rasoul M, Merritt RE. National trends and outcomes of segmentectomy in the Society of Thoracic Surgeons database. Ann Thorac Surg 2022;113(4):1361–9. https://doi.org/10.1016/j.athoracsur.2021.07.056.

[52] Ma J, Li X, Zhao S, Wang J, Zhang W, Sun G. Robot-assisted thoracic surgery versus video-assisted thoracic surgery for lung lobectomy or segmentectomy in patients with non-small cell lung cancer: a meta-analysis. BMC Cancer 2021;21(1):498. https://doi.org/10.1186/s12885-021-08241-5.

[53] Kneuertz PJ, Abdel-Rasoul M, D'Souza DM, Zhao J, Merritt RE. Segmentectomy for clinical stage I non–small cell lung cancer: national benchmarks for nodal staging and outcomes by operative approach. Cancer 2022;128(7):1483–92. https://doi.org/10.1002/cncr.34071.

[54] Amirkhosravi F, Kim MP. Complex robotic lung resection. Thorac Surg Clin 2023;33(1):51–60. https://doi.org/10.1016/j.thorsurg.2022.08.006.

III. The current and future clinical applications of robotic surgery among medical specialties

[55] Cerfolio RJ, Bryant AS, Minnich DJ. Minimally invasive chest wall resection: sparing the overlying, uninvolved extrathoracic musculature of the chest. Ann Thorac Surg 2012;94(5):1744–7. https://doi.org/10.1016/j.athoracsur.2012.05.132.

[56] Mariolo AV, Casiraghi M, Galetta D, Spaggiari L. Robotic hybrid approach for an anterior pancoast tumor in a severely obese patient. Ann Thorac Surg 2018;106(3):e115–6. https://doi.org/10.1016/j.athoracsur.2018.03.013.

[57] Muriana P, Perroni G, Novellis P, Veronesi G. Robotic surgery for locally advanced non-small cell lung cancer. J Visc Surg 2021;7:26. https://doi.org/10.21037/jovs-20-114.

[58] Pan X, Gu C, Wang R, Zhao H, Shi J, Chen H. Initial experience of robotic sleeve resection for lung cancer patients. Ann Thorac Surg 2016;102(6):1892–7. https://doi.org/10.1016/j.athoracsur.2016.06.054.

[59] Watkins AA, Quadri SM, Servais EL. Robotic-assisted complex pulmonary resection: sleeve lobectomy for cancer. Innov Technol Tech Cardiothorac Vasc Surg 2021;16(2):132–5. https://doi.org/10.1177/1556984521992384.

[60] Veronesi G, Galetta D, Maisonneuve P, et al. Four-arm robotic lobectomy for the treatment of early-stage lung cancer. J Thorac Cardiovasc Surg 2010;140(1):19–25. https://doi.org/10.1016/j.jtcvs.2009.10.025.

[61] Veronesi G. Robotic lobectomy and segmentectomy for lung cancer: results and operating technique. J Thorac Dis 2015;7.

[62] Ramadan OI, Wei B, Cerfolio RJ. Robotic surgery for lung resections—total port approach: advantages and disadvantages. J Visc Surg 2017;3:22. https://doi.org/10.21037/jovs.2017.01.06.

[63] Perroni G, Veronesi G. Robotic segmentectomy: indication and technique. J Thorac Dis 2020;12(6):3404–10. https://doi.org/10.21037/jtd.2020.02.53.

[64] Wei B, Cerfolio R. Technique of robotic segmentectomy. J Visc Surg 2017;3:140. https://doi.org/10.21037/jovs.2017.08.13.

[65] Pardolesi A, Park B, Petrella F, Borri A, Gasparri R, Veronesi G. Robotic anatomic Segmentectomy of the lung: technical aspects and initial results. Ann Thorac Surg 2012;94(3):929–34. https://doi.org/10.1016/j.athoracsur.2012.04.086.

[66] Geraci TC, Ferrari-Light D, Kent A, et al. Technique, outcomes with navigational bronchoscopy using Indocyanine green for robotic Segmentectomy. Ann Thorac Surg 2019;108(2):363–9. https://doi.org/10.1016/j.athoracsur.2019.03.032.

[67] Pardolesi A, Veronesi G, Solli P, Spaggiari L. Use of indocyanine green to facilitate intersegmental plane identification during robotic anatomic segmentectomy. J Thorac Cardiovasc Surg 2014;148(2):737–8. https://doi.org/10.1016/j.jtcvs.2014.03.001.

[68] Carter BW, Tomiyama N, Bhora FY, et al. A modern definition of mediastinal compartments. J Thorac Oncol 2014;9(9):S97–S101. https://doi.org/10.1097/JTO.0000000000000292.

[69] Carter BW, Benveniste MF, Madan R, et al. ITMIG classification of mediastinal compartments and multidisciplinary approach to mediastinal masses. Radiographics 2017;37(2):413–36. https://doi.org/10.1148/rg.2017160095.

[70] Rieger R, Schrenk P, Woisetschliger R, Wayand W. Videothoracoscopy for the management of mediastinal mass lesions. Surg Endosc 1996;10(7):715–7.

[71] Yang CFJ, Hurd J, Shah SA, et al. A national analysis of open versus minimally invasive thymectomy for stage I to III thymoma. J Thorac Cardiovasc Surg 2020;160(2):555–567.e15. https://doi.org/10.1016/j.jtcvs.2019.11.114.

[72] Yoshino I, Hashizume M, Shimada M, et al. Thoracoscopic thymomectomy with the da Vinci computer-enhanced surgical system. J Thorac Cardiovasc Surg 2001;122(4):783–5. https://doi.org/10.1067/mtc.2001.115231.

[73] Alvarado CE, Worrell SG, Bachman KC, et al. Robotic approach has improved outcomes for minimally invasive resection of mediastinal tumors. Ann Thorac Surg 2022;113(6):1853–8. https://doi.org/10.1016/j.athoracsur.2021.05.090.

[74] Hurd J, Haridas C, Potter A, et al. A national analysis of open versus minimally invasive thymectomy for stage I–III thymic carcinoma. Eur J Cardiothorac Surg 2022;62(3):ezac159. https://doi.org/10.1093/ejcts/ezac159.

[75] Bodner J, Wykypiel H, Wetscher G, Schmid T. First experiences with the da Vinci™ operating robot in thoracic surgery☆. Eur J Cardiothorac Surg 2004;25(5):844–51. https://doi.org/10.1016/j.ejcts.2004.02.001.

[76] DeRose JJ, Swistel DG, Safavi A, Connery CP, Ashton RC. Mediastinal mass evaluation using advanced robotic techniques. Ann Thorac Surg 2003;75(2):571–3. https://doi.org/10.1016/S0003-4975(02)04295-9.

[77] Savitt MA, Gao G, Furnary AP, Swanson J, Gately HL, Handy JR. Application of robotic-assisted techniques to the surgical evaluation and treatment of the anterior mediastinum. Ann Thorac Surg 2005;79(2):450–5. https://doi.org/10.1016/j.athoracsur.2004.07.022.

[78] Ismail M, Maza S, Swierzy M, et al. Resection of ectopic mediastinal parathyroid glands with the da Vinci® robotic system. Br J Surg 2010;97(3):337–43. https://doi.org/10.1002/bjs.6905.

[79] Cerfolio RJ, Bryant AS, Minnich DJ. Operative techniques in robotic thoracic surgery for inferior or posterior mediastinal pathology. J Thorac Cardiovasc Surg 2012;143(5):1138–43. https://doi.org/10.1016/j.jtcvs.2011.12.021.

[80] Chen K, Zhang X, Jin R, et al. Robot-assisted thoracoscopic surgery for mediastinal masses: a single-institution experience. J Thorac Dis 2020;12(2):105–13. https://doi.org/10.21037/jtd.2019.08.105.

[81] Broussard BL, Wei B, Cerfolio RJ. Robotic surgery for posterior mediastinal pathology. Ann Cardiothorac Surg 2016;5(1):62–4.

[82] Hong Z, Gou W, Cui B, et al. Analysis of the efficacy of the da Vinci robot in surgery for posterior mediastinal neurogenic tumors. BMC Surg 2022;22(1):413. https://doi.org/10.1186/s12893-022-01855-x.

[83] Hsu DS, Banks KC, Velotta JB. Surgical approaches to mediastinal cysts: clinical practice review. Mediastinum 2022;6:32. https://doi.org/10.21037/med-22-20.

[84] Li XK, Cong ZZ, Xu Y, et al. Clinical efficacy of robot-assisted thoracoscopic surgery for posterior mediastinal neurogenic tumors. J Thorac Dis 2020;12(6):3065–72. https://doi.org/10.21037/jtd-20-286.

[85] Li H, Li J, Huang J, Yang Y, Luo Q. Robotic-assisted mediastinal surgery: the first Chinese series of 167 consecutive cases. J Thorac Dis 2018;10(5):2876–80. https://doi.org/10.21037/jtd.2018.04.138.

[86] Profanter C, Schmid T, Prommegger R, Bale R, Sauper T, Bodner J. Robot-assisted mediastinal parathyroidectomy. Surg Endosc Interv Tech 2004;18(5):868–70. https://doi.org/10.1007/s00464-003-4272-3.

[87] Karagkounis G, Uzun DD, Mason DP, Murthy SC, Berber E. Robotic surgery for primary hyperparathyroidism. Surg Endosc 2014;28(9):2702–7. https://doi.org/10.1007/s00464-014-3531-9.

[88] Amore D, Cicalese M, Scaramuzzi R, Di Natale D, Curcio C. Antero mediastinal retrosternal goiter: surgical excision by combined cervical and hybrid robot-assisted approach. J Thorac Dis 2018;10(3):E199–202. https://doi.org/10.21037/jtd.2018.01.169.

[89] Davenport E, Malthaner RA. The role of surgery in the management of thymoma: a systematic review. Ann Thorac Surg 2008;86(2):673–84. https://doi.org/10.1016/j.athoracsur.2008.03.055.

III. The current and future clinical applications of robotic surgery among medical specialties

[90] Wolfe GI, Kaminski HJ, Aban IB, et al. Randomized trial of thymectomy in myasthenia gravis. N Engl J Med 2016;375(6):511–22. https://doi.org/10.1056/NEJMoa1602489.

[91] Geraci TC, Ferrari-Light D, Pozzi N, Cerfolio RJ. Midterm results for robotic thymectomy for malignant disease. Ann Thorac Surg 2021;111(5):1675–81. https://doi.org/10.1016/j.athoracsur.2020.06.111.

[92] Rea F, Marulli G, Bortolotti L, Feltracco P, Zuin A, Sartori F. Experience with the "Da Vinci" robotic system for Thymectomy in patients with myasthenia gravis: report of 33 cases. Ann Thorac Surg 2006;81(2):455–9. https://doi.org/10.1016/j.athoracsur.2005.08.030.

[93] Li R, Ma Z, Qu C, et al. Comparison of perioperative outcomes between robotic-assisted and video-assisted thoracoscopic surgery for mediastinal masses in patients with different body mass index ranges: a population-based study. Front Surg 2022;9:963335. https://doi.org/10.3389/fsurg.2022.963335.

[94] Jurado J, Javidfar J, Newmark A, et al. Minimally invasive Thymectomy and open Thymectomy: outcome analysis of 263 patients. Ann Thorac Surg 2012;94(3):974–82. https://doi.org/10.1016/j.athoracsur.2012.04.097.

[95] Soder SA, Pollock C, Ferraro P, et al. Post-operative outcomes associated with open versus robotic thymectomy: a propensity matched analysis. Semin Thorac Cardiovasc Surg 2021. https://doi.org/10.1053/j.semtcvs.2021.11.011. Published online November. S1043067921004846.

[96] Marulli G, Maessen J, Melfi F, et al. Multi-institutional European experience of robotic thymectomy for thymoma. Ann Cardiothorac Surg 2016;5(1).

[97] Marulli G, Rea F, Melfi F, et al. Robot-aided thoracoscopic thymectomy for early-stage thymoma: a multicenter European study. J Thorac Cardiovasc Surg 2012;144(5):1125–32. https://doi.org/10.1016/j.jtcvs.2012.07.082.

[98] Seong YW, Kang CH, Choi JW, et al. Early clinical outcomes of robot-assisted surgery for anterior mediastinal mass: its superiority over a conventional sternotomy approach evaluated by propensity score matching†. Eur J Cardiothorac Surg 2014;45(3):e68–73. https://doi.org/10.1093/ejcts/ezt557.

[99] Marulli G, Comacchio GM, Schiavon M, et al. Comparing robotic and trans-sternal thymectomy for early-stage thymoma: a propensity score-matching study. Eur J Cardiothorac Surg 2018;54(3):579–84. https://doi.org/10.1093/ejcts/ezy075.

[100] Hess NR, Sarkaria IS, Pennathur A, Levy RM, Christie NA. Minimally invasive versus open thymectomy: a systematic review of surgical techniques, patient demographics, and perioperative outcomes. Ann Cardiothorac Surg 2016;5(1).

[101] Balduyck B, Hendriks JM, Lauwers P, Mercelis R, Ten Broecke P, Van Schil P. Quality of life after anterior mediastinal mass resection: a prospective study comparing open with robotic-assisted thoracoscopic resection☆. Eur J Cardiothorac Surg 2011;39(4):543–8. https://doi.org/10.1016/j.ejcts.2010.08.009.

[102] Shen C, Li J, Li J, Che G. Robot-assisted thoracic surgery versus video-assisted thoracic surgery for treatment of patients with thymoma: a systematic review and meta-analysis. Thorac Cancer 2022;13(2):151–61. https://doi.org/10.1111/1759-7714.14234.

[103] O'Sullivan KE, Kreaden US, Hebert AE, Eaton D, Redmond KC. A systematic review of robotic versus open and video assisted thoracoscopic surgery (VATS) approaches for thymectomy. Ann Cardiothorac Surg 2019;8(2):174–93. https://doi.org/10.21037/acs.2019.02.04.

[104] Rückert JC, Swierzy M, Ismail M. Comparison of robotic and nonrobotic thoracoscopic thymectomy: a cohort study. J Thorac Cardiovasc Surg 2011;141(3):673–7. https://doi.org/10.1016/j.jtcvs.2010.11.042.

[105] Kang CH, Hwang Y, Lee HJ, Park IK, Kim YT. Robotic Thymectomy in anterior mediastinal mass: propensity score matching study with Trans-sternal Thymectomy. Ann Thorac Surg 2016;102(3):895–901. https://doi.org/10.1016/j.athoracsur.2016.03.084.

[106] Melfi F, Fanucchi O, Davini F, et al. Ten-year experience of mediastinal robotic surgery in a single referral Centre. Eur J Cardiothorac Surg 2012;41(4):847–51. https://doi.org/10.1093/ejcts/ezr112.

[107] Marulli G, Schiavon M, Perissinotto E, et al. Surgical and neurologic outcomes after robotic thymectomy in 100 consecutive patients with myasthenia gravis. J Thorac Cardiovasc Surg 2013;145(3):730–6. https://doi.org/10.1016/j.jtcvs.2012.12.031.

[108] Kneuertz PJ, Kamel MK, Stiles BM, et al. Robotic Thymectomy is feasible for large Thymomas: a propensity-matched comparison. Ann Thorac Surg 2017;104(5):1673–8. https://doi.org/10.1016/j.athoracsur.2017.05.074.

[109] Seastedt KP, Watkins AA, Kent MS, Stock CT. Robotic mediastinal surgery. Thorac Surg Clin 2023;33(1):89–97. https://doi.org/10.1016/j.thorsurg.2022.08.007.

[110] Ismail M, Swierzy M, Rückert RI, Rückert JC. Robotic Thymectomy for myasthenia gravis. Thorac Surg Clin 2014;24(2):189–95. https://doi.org/10.1016/j.thorsurg.2014.02.012.

[111] Lacquet M, Moons J, Ceulemans LJ, De Leyn P, Van Raemdonck D. Surgery for mediastinal neurogenic tumours: a 25-year single-Centre retrospective study. Interact Cardiovasc Thorac Surg 2021;32(5):737–43. https://doi.org/10.1093/icvts/ivab002.

[112] Shidei H, Maeda H, Isaka T, et al. Mediastinal paraganglioma successfully resected by robot-assisted thoracoscopic surgery with en bloc chest wall resection: a case report. BMC Surg 2020;20(1):45. https://doi.org/10.1186/s12893-020-00701-2.

[113] Giulianotti PC, Buchs NC, Caravaglios G, Bianco FM. Robot-assisted lung resection: outcomes and technical details. Interact Cardiovasc Thorac Surg 2010;11(4):388–92. https://doi.org/10.1510/icvts.2010.239541.

[114] Spaggiari L, Galetta D. Pneumonectomy for lung Cancer: a further step in minimally invasive surgery. Ann Thorac Surg 2011;91(3):e45–7. https://doi.org/10.1016/j.athoracsur.2010.12.008.

[115] Emmert A, Straube C, Buentzel J, Roever C. Robotic versus thoracoscopic lung resection: a systematic review and meta-analysis. Medicine (Baltimore) 2017;96(35), e7633. https://doi.org/10.1097/MD.0000000000007633.

[116] Khan N, Fikfak V, Chan E, Kim M. "Five on a dice" port placement allows for successful robot-assisted left pneumonectomy. Thorac Cardiovasc Surg Rep 2017;06(1):e42–4. https://doi.org/10.1055/s-0037-1613714.

[117] Patton BD, Zarif D, Bahroloomi DM, Sarmiento IC, Lee PC, Lazzaro RS. Robotic pneumonectomy for lung Cancer: perioperative outcomes and factors leading to conversion to thoracotomy. Innov Technol Tech Cardiothorac Vasc Surg 2021;16(2):136–41. https://doi.org/10.1177/1556984520978227.

[118] Hennon MW, Kumar A, Devisetty H, et al. Minimally invasive approaches do not compromise outcomes for pneumonectomy: a comparison using the National Cancer Database. J Thorac Oncol 2019;14(1):107–14. https://doi.org/10.1016/j.jtho.2018.09.024.

[119] Deslauriers J, Grégoire J, Jacques LF, Piraux M, Guojin L, Lacasse Y. Sleeve lobectomy versus pneumonectomy for lung cancer: a comparative analysis of survival and sites or recurrences. Ann Thorac Surg 2004;77(4):1152–6. https://doi.org/10.1016/j.athoracsur.2003.07.040.

[120] Gaissert HA, Mathisen DJ, Moncure AC, Hilgenberg AD, Grillo HC, Wain JC. Survival and function after sleeve lobectomy for lung cancer. J Thorac Cardiovasc Surg 1996;111(5):948–53. https://doi.org/10.1016/S0022-5223(96)70369-0.

[121] Caso R, Watson TJ, Khaitan PG, Marshall MB. Outcomes of minimally invasive sleeve resection. J Thorac Dis 2018;10(12):6653–9. https://doi.org/10.21037/jtd.2018.10.97.

[122] Lin MW, Kuo SW, Yang SM, Lee JM. Robotic-assisted thoracoscopic sleeve lobectomy for locally advanced lung cancer. J Thorac Dis 2016;8 (7):1747–52. https://doi.org/10.21037/jtd.2016.06.14.

[123] Pan X, Chen Y, Shi J, Zhao H, Chen H. Robotic assisted extended sleeve lobectomy after neoadjuvant chemotherapy. Ann Thorac Surg 2015;100 (6):e129–31. https://doi.org/10.1016/j.athoracsur.2015.08.084.

[124] Cerfolio RJ. Robotic sleeve lobectomy: technical details and early results. J Thorac Dis 2016;8(Suppl 2):S223–6.

[125] Jiao W, Zhao Y, Qiu T, et al. Robotic bronchial sleeve lobectomy for central lung tumors: technique and outcome. Ann Thorac Surg 2019;108 (1):211–8. https://doi.org/10.1016/j.athoracsur.2019.02.028.

[126] Qiu T, Zhao Y, Xuan Y, et al. Robotic sleeve lobectomy for centrally located non–small cell lung cancer: A propensity score–weighted comparison with thoracoscopic and open surgery. J Thorac Cardiovasc Surg 2020;160(3):838–846.e2. https://doi.org/10.1016/j.jtcvs.2019.10.158.

[127] Geraci TC, Ferrari-Light D, Wang S, et al. Robotic sleeve resection of the airway: outcomes and technical conduct using video vignettes. Ann Thorac Surg 2020;110(1):236–40. https://doi.org/10.1016/j.athoracsur.2020.01.077.

[128] Liu A, Zhao Y, Qiu T, et al. The long-term oncologic outcomes of robot-assisted bronchial single sleeve lobectomy for 104 consecutive patients with centrally located non-small cell lung cancer. Transl Lung Cancer Res 2022;11(5):869–79. https://doi.org/10.21037/tlcr-22-298.

[129] LoCicero J, Feins RH, Colson YL, Rocco G. Shields' General Thoracic Surgery. 8th ed. Wolters Kluwer; 2019.

[130] Gangadharan SP, Bakhos CT, Majid A, et al. Technical aspects and outcomes of tracheobronchoplasty for severe Tracheobronchomalacia. Ann Thorac Surg 2011;91(5):1574–81. https://doi.org/10.1016/j.athoracsur.2011.01.009.

[131] Majid A, Guerrero J, Gangadharan S, et al. Tracheobronchoplasty for severe Tracheobronchomalacia. Chest 2008;134(4):801–7. https://doi.org/10.1378/chest.08-0728.

[132] Bezuidenhout AF, Boiselle PM, Heidinger BH, et al. Longitudinal follow-up of patients with Tracheobronchomalacia after undergoing Tracheobronchoplasty: computed tomography findings and clinical correlation. J Thorac Imaging 2019;34(4):278–83. https://doi.org/10.1097/RTI.0000000000000339.

[133] McGinn J, Herbert B, Maloney A, Patton B, Lazzaro R. Quality of life outcomes in tracheobronchomalacia surgery. J Thorac Dis 2020;12 (11):6925–30. https://doi.org/10.21037/jtd.2020.03.08.

[134] Digesu CS, Ospina-Delgado D, Ascanio J, et al. Obese patients undergoing Tracheobronchoplasty have excellent outcomes. Ann Thorac Surg 2022;114(3):926–32. https://doi.org/10.1016/j.athoracsur.2021.07.018.

[135] Buitrago DH, Majid A, Wilson JL, et al. Tracheobronchoplasty yields long-term anatomy, function, and quality of life improvement for patients with severe excessive central airway collapse. J Thorac Cardiovasc Surg 2023;165(2):518–25. https://doi.org/10.1016/j.jtcvs.2022.05.037.

[136] Wilson JL, Wallace JS. Tracheobronchoplasty outcomes: a narrative review. J Visc Surg 2022;8:16. https://doi.org/10.21037/jovs-21-10.

[137] Lazzaro RS, Bahroloomi D, Wasserman GA, Patton BD. Robotic Tracheobronchoplasty: technique. Oper Tech Thorac Cardiovasc Surg 2022;27 (2):218–26. https://doi.org/10.1053/j.optechstcvs.2021.06.017.

[138] Lazzaro R, Patton B, Lee P, et al. First series of minimally invasive, robot-assisted tracheobronchoplasty with mesh for severe tracheobronchomalacia. J Thorac Cardiovasc Surg 2019;157(2):791–800. https://doi.org/10.1016/j.jtcvs.2018.07.118.

[139] Seastedt KP, Wilson JL, Gangadharan SP. Robotic surgery for Tracheobronchomalacia. Thorac Surg Clin 2023;33(1):61–9. https://doi.org/10.1016/j.thorsurg.2022.09.003.

[140] Gharagozloo F, Atiquzzaman N, Meyer M, Tempesta B, Werden S. Robotic first rib resection for thoracic outlet syndrome. J Thorac Dis 2021;13 (10):6141–54. https://doi.org/10.21037/jtd-2019-rts-04.

[141] Kocher GJ, Zehnder A, Lutz JA, Schmidli J, Schmid RA. First rib resection for thoracic outlet syndrome: the robotic approach. World J Surg 2018;42(10):3250–5. https://doi.org/10.1007/s00268-018-4636-4.

[142] Gharagozloo F, Meyer M, Tempesta B, Gruessner S. Robotic transthoracic first-rib resection for Paget–Schroetter syndrome. Eur J Cardiothorac Surg 2019;55(3):434–9. https://doi.org/10.1093/ejcts/ezy275.

[143] Pupovac SS, Lee PC, Zeltsman D, Jurado J, Hyman K, Singh V. Robotic-assisted first rib resection: our experience and review of the literature. Semin Thorac Cardiovasc Surg 2020;32(4):1115–20. https://doi.org/10.1053/j.semtcvs.2020.04.016.

[144] Zehnder A, Lutz J, Dorn P, et al. Robotic-assisted Thoracoscopic resection of the first rib for vascular thoracic outlet syndrome: the new gold standard of treatment? J Clin Med 2021;10(17):3952. https://doi.org/10.3390/jcm10173952.

[145] Gkikas A, Lampridis S, Patrini D, et al. Thoracic outlet syndrome: single center experience on robotic assisted first rib resection and literature review. Front Surg 2022;9:848972. https://doi.org/10.3389/fsurg.2022.848972.

[146] Azenha LF, Kocher GJ, Kestenholz PB, Gioutsos K, Minervini F. Thoracic outlet syndrome: a retrospective analysis of robotic assisted first rib resections. J Robot Surg 2022. https://doi.org/10.1007/s11701-022-01486-7. Published online November 3.

[147] Burt BM, Palivela N, Cekmecelioglu D, et al. Safety of robotic first rib resection for thoracic outlet syndrome. J Thorac Cardiovasc Surg 2021;162 (4):1297–1305.e1. https://doi.org/10.1016/j.jtcvs.2020.08.107.

[148] Palivela N, Lee HS, Jang HJ, et al. Improvement of disability in neurogenic thoracic outlet syndrome by robotic first rib resection. Ann Thorac Surg 2022;114(3):919–25. https://doi.org/10.1016/j.athoracsur.2021.07.052.

[149] Burt BM, Palivela N, Karimian A, Goodman MB. Transthoracic robotic first rib resection: twelve steps. JTCVS Tech 2020;1:104–9. https://doi.org/10.1016/j.xjtc.2020.01.005.

[150] Ricoy J, Rodríguez-Núñez N, Álvarez-Dobaño JM, Toubes ME, Riveiro V, Valdés L. Diaphragmatic dysfunction. Pulmonology 2019;25 (4):223–35. https://doi.org/10.1016/j.pulmoe.2018.10.008.

[151] Graham DR, Kaplan D, Evans CC, Hind CRK, Donnelly RJ. Diaphragmatic plication for unilateral diaphragmatic paralysis: a 10-year experience. Ann Thorac Surg 1990;49(2):248–52. https://doi.org/10.1016/0003-4975(90)90146-W.

[152] Higgs SM, Hussain A, Jackson M, Donnelly RJ, Berrisford RG. Long term results of diaphragmatic plication for unilateral diaphragm paralysisq. Thorac Surg 2002. Published online.

[153] Freeman RK, Van Woerkom J, Vyverberg A, Ascioti AJ. Long-term follow-up of the functional and physiologic results of diaphragm plication in adults with unilateral diaphragm paralysis. Ann Thorac Surg 2009;88(4):1112–7. https://doi.org/10.1016/j.athoracsur.2009.05.027.

[154] Celik S, Celik M, Aydemir B, Tunckaya C, Okay T, Dogusoy I. Long-term results of diaphragmatic plication in adults with unilateral diaphragm paralysis. J Cardiothorac Surg 2010;5(1):111. https://doi.org/10.1186/1749-8090-5-111.

[155] Groth SS, Rueth NM, Kast T, et al. Laparoscopic diaphragmatic plication for diaphragmatic paralysis and eventration: an objective evaluation of short-term and midterm results. J Thorac Cardiovasc Surg 2010;139(6):1452–6. https://doi.org/10.1016/j.jtcvs.2009.10.020.

[156] Gritsiuta AI, Gordon M, Bakhos CT, Abbas AE, Petrov RV. Minimally invasive diaphragm plication for acquired unilateral diaphragm paralysis: a systematic review. Innov Technol Tech Cardiothorac Vasc Surg 2022;17(3):180–90. https://doi.org/10.1177/15569845221097761.

[157] Gazala S, Hunt I, Bedard ELR. Diaphragmatic plication offers functional improvement in dyspnoea and better pulmonary function with low morbidity. Interact Cardiovasc Thorac Surg 2012;15(3):505–8. https://doi.org/10.1093/icvts/ivs238.

[158] Schumacher L, Zhao D. Outcomes and technique of robotic diaphragm plication. J Thorac Dis 2021;13(10):6113–5. https://doi.org/10.21037/jtd-2019-rts-01.

[159] Kwak T, Lazzaro R, Pournik H, Ciaburri D, Tortolani A, Gulkarov I. Robotic thoracoscopic plication for symptomatic diaphragm paralysis. J Robot Surg 2012;6(4):345–8. https://doi.org/10.1007/s11701-011-0328-x.

[160] Nardini M, Jayakumar S, Migliore M, Nosotti M, Paul I, Dunning J. Minimally invasive plication of the diaphragm: a single-center prospective study. Innov Technol Tech Cardiothorac Vasc Surg 2021;16(4):343–9. https://doi.org/10.1177/15569845211011583.

[161] Bin Asaf B, Kodaganur Gopinath S, Kumar A, Puri HV, Pulle MV, Bishnoi S. Robotic diaphragmatic plication for eventration: a retrospective analysis of efficacy, safety, and feasibility. Asian J Endosc Surg 2021;14(1):70–6. https://doi.org/10.1111/ases.12833.

[162] Groth SS, Andrade RS. Diaphragm plication for eventration or paralysis: a review of the literature. Ann Thorac Surg 2010;89(6):S2146–50. https://doi.org/10.1016/j.athoracsur.2010.03.021.

# 53

# Esophagus/foregut and pancreatic robotic surgery

*Felipe J.F. Coimbra, Rebeca Hara Nahime, Silvio Melo Torres, and Igor Correia Farias*

Department of Abdominal Surgery, Reference Center on Upper GI & HPB Oncology, São Paulo, Brazil

## Introduction

In recent years, robotic surgery has revolutionized the field of esophagus/foregut and pancreatic surgery, ushering in a new era of precision and innovation. This groundbreaking approach has rapidly gained popularity among surgeons and patients due to its potential for improved outcomes and enhanced surgical capabilities. This chapter aims to provide a comprehensive overview of the generic concepts and innovations in robotic surgery for esophageal, foregut, and pancreatic disorders, while also offering practical insights into medical practitioners and surgeons.

Robotic surgery offers numerous advantages over traditional open or laparoscopic techniques, including superior visualization, increased dexterity and enhanced ergonomic control. These benefits translate into superior surgical precision and reduced tissue trauma, contributing to faster patient recovery time and minimized postoperative complications.

As we delve into the specific applications of robotic technology in esophagus/foregut and pancreatic surgery, we will explore a wide range of procedures, such as robotic esophagectomy, robotic fundoplication for gastroesophageal reflux disease (GERD), robotic Heller myotomy for achalasia, robotic-assisted Whipple procedure for pancreatic tumors and many more. Some of the most important procedures will be examined in more detail, highlighting the potential benefits and limitations of the robotic approach.

In addition to the technical aspects, this chapter will also shed light on the main surgical procedures and indications. Advances in robotic technology, including multijointed robotic arms, high-definition 3D visualization and haptic feedback, will significantly contribute to further improving surgical outcomes and expanding the repertoire of feasible robotic procedures.

Ultimately, this chapter aims to serve as a valuable resource for medical professionals seeking to harness the full potential of robotic surgery in esophagus/foregut and pancreatic procedures. By understanding the generic concepts, exploring the latest innovations and gaining practical insights, surgeons can confidently embrace robotic technology as a transformative tool in advancing patient care and surgical outcomes.

## Background/history

The history of robotic surgery for esophagus, stomach, duodenum and pancreas has evolved significantly over the past few decades, marked by numerous milestones that have shaped its current state. In 1997, the first robotic cholecystectomy (RC) was performed, marking the first use of the da Vinci Surgical System [1]. In the early 2000s, pioneering studies like the one done by Cadière et al. [2] demonstrated the feasibility of robotic-assisted laparoscopic surgery for various gastrointestinal procedures, laying the groundwork for future advancements. Subsequently, innovative surgeons such as Marescaux et al. [3] showcased the first robotic-assisted Whipple procedure, marking a crucial milestone in pancreatic surgery. As technology continued to improve, robotic systems, such as the da Vinci Surgical System, gained popularity for their enhanced dexterity and precision. Recently, robot-assisted surgery has allowed for the possibility of fully automated surgical operations. The Smart Tissue Autonomous Robot (STAR), designed at John Hopkins University, performed the first autonomous intestinal anastomosis in 2022 on porcupines over a one-week period [4]. The results indicated that the automated system outperformed expert surgeons and robot-assisted surgery in terms of both consistency and accuracy, demonstrating the intricacy of robotics and the potential future of robotic surgery.

With advancements in instrumentation and imaging, robotic esophagectomy for esophageal cancer emerged as a promising option, as evidenced by studies like that by Cerfolio et al. [5]. Over the years, this field has witnessed the growth of robotic-assisted fundoplication for GERD, myotomy for achalasia, and minimally invasive approaches for duodenal diseases, among others. Recent studies, including those by Zureikat et al. [6], have reported favorable outcomes in robotic pancreatic surgery, solidifying its position as a viable option for complex pancreatic disorders. Today, robotic surgery continues to evolve, with ongoing research and clinical trials aimed at expanding its applications and refining its techniques, promising a bright future for patients with esophagus, foregut, and pancreatic pathologies.

## Potential advantages of robotic surgery in esophagus/foregut and pancreatic surgery

Enhanced visualization: Robotic systems provide high-definition 3D visualization and magnification, offering surgeons an improved view of the surgical field [5]. This enhanced visualization aids in precise dissection and reduces the risk of injury to vital structures, especially during careful lymph node dissection and organ mobilization for cancer.

Increased dexterity: The robotic instruments offer seven degrees of freedom, enabling precise and delicate movements within the confined spaces of the thoracic and abdominal cavities [3]. This increased dexterity allows for meticulous dissection and suturing, critical in complex procedures, such as lymph node dissection, gastrointestinal anastomosis and bowel suturing, which has great value for oncological operations.

Reduced tissue trauma: Robotic instruments allow for finer and more controlled movements, leading to reduced tissue trauma and less postoperative pain [5]. This benefit can facilitate a faster recovery and shorter hospital stay for patients. This is particularly interesting, considering the long operations necessary for esophagus, gastric, and pancreas resections.

Ergonomic advantages: The ergonomic design of robotic consoles minimizes surgeon fatigue and discomfort during lengthy procedures, promoting better surgical outcomes [5].

Minimized scarring: The use of smaller incisions in robotic surgery leads to less visible scarring, resulting in improved cosmesis and patient satisfaction [2].

## Disadvantages of robotic surgery in esophagus/foregut and pancreatic surgery

### Cost

Robotic surgery systems are expensive to acquire and maintain, making it a significant financial investment for healthcare institutions [5]. The high costs may limit its availability and accessibility for some patients. For some specific procedures, cost may overcome health systems possibilities to allow its feasibility in large scale. Nowadays, this is a great challenge for developing countries.

### Learning curve

Mastering robotic surgical techniques requires extensive training and a learning curve for surgeons [5]. Adequate experience is essential to maximize the benefits and minimize the risks of robotic surgery. Particularly for esophagus and pancreas surgery, this may be crucial, since it has shown the impact of the learning curve and volume for patients outcomes [7].

### Setup time

Robotic systems need time for setup and docking, which may prolong the overall operative time [5]. This may impact scheduling and resource allocation in busy surgical units. But, as we and many have noticed, the procedure starts faster as the team gets used to the routine.

## Lack of haptic feedback

Current robotic systems lack haptic feedback, which can compromise the surgeon's tactile sense during delicate procedures [3]. Although the visual and auditory cues compensate to some extent, haptic feedback remains an area for improvement. It may be particularly important for upper gastrointestinal surgery, such as, for soft pancreas anastomosis, positioning the small bowel and other situations.

## Limited availability

The widespread adoption of robotic surgery may be limited due to its availability only in specialized centers and high-resource settings, making it less accessible in remote or underprivileged regions [2]. For example, even in a country like Brazil, most of the cholecystectomies are still performed with an open access in approximately 60% of the procedures [8].

In conclusion, robotic surgery in esophagus/foregut and pancreatic procedures offers significant advantages, including enhanced visualization, increased dexterity, reduced tissue trauma and improved ergonomics. However, it is crucial to acknowledge the disadvantages, such as high cost, learning curve, setup time, lack of haptic feedback, and limited availability. With proper training and selection of appropriate cases, robotic surgery can be a valuable tool in the armamentarium of surgeons for selected complex cases. Further research, cost-effectiveness studies, and technological advancements are needed to optimize the role of robotic surgery in esophagus/foregut and pancreatic procedures.

## Main robotic procedures in esophagus, foregut, and pancreatic surgery

### Robotic esophagectomy

Robotic-assisted esophagectomy has gained acceptance as a minimally invasive approach for esophageal cancer. This procedure involves resecting the affected part of the esophagus and reconstructing it using the stomach or a segment of the colon. Studies have demonstrated equivalent oncological outcomes and reduced postoperative morbidity with robotic esophagectomy when compared to open surgery [9]. The robotic platform allows for precise dissection and suturing in the narrow mediastinal space, minimizing trauma to surrounding structures and improving patient recovery (Fig. 1).

Robotic fundoplication for GERD: Robotic-assisted fundoplication has become a well-established treatment for GERD. This procedure aims to reinforce the lower esophageal sphincter to prevent acid reflux into the esophagus. The robotic approach offers enhanced visualization and dexterity, facilitating precise suturing of the fundoplication

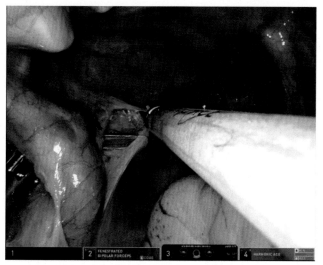

FIG. 1   Azygos vein dissection at the thoracic phase of subtotal esophagectomy.

III. The current and future clinical applications of robotic surgery among medical specialties

[10]. Studies have reported excellent outcomes in terms of symptom relief and patient satisfaction, with comparable efficacy to the traditional laparoscopic fundoplication.

Robotic Heller myotomy: Robotic-assisted Heller myotomy is a well-established treatment for achalasia, a motility disorder of the esophagus. This procedure involves cutting the muscle fibers of the lower esophageal sphincter to relieve obstruction and improve esophageal conduction, and thus swallowing. The robotic system's wristed instruments allow for precise dissection and suturing in the confined space of the esophagus site [11]. Studies have shown favorable outcomes in terms of symptom relief and quality of life, with reduced rates of postoperative reflux compared to laparoscopic myotomy.

## Robotic-assisted Whipple procedure

The Whipple procedure, or pancreaticoduodenectomy, is a complex surgery for pancreatic tumors involving the head of the pancreas. The robotic-assisted Whipple procedure has emerged as a promising option, offering improved visualization and dexterity during the intricate dissection of the pancreaticoduodenal region [12]. While still relatively new, initial studies have reported comparable safety and feasibility with laparoscopic or open approaches (Fig. 2).

## Robotic gastrectomy

Robotic-assisted gastrectomy is increasingly being used for the treatment of gastric cancer. This procedure involves partial or complete removal of the stomach, along with nearby lymph nodes (Fig. 3). Robotic surgery allows meticulous lymph node dissection and precise reconstruction of the gastrointestinal tract [13]. Studies have shown comparable oncological outcomes and improved short-term postoperative outcomes, such as reduced blood loss and shorter hospital stays, compared to conventional laparoscopic gastrectomy [14].

FIG. 2    Dissecting the portal vein tunnel.

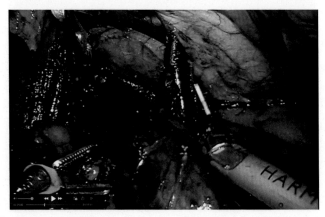

FIG. 3    Lymphadenectomy of chains 19, 20, 110, and 111 for Siewert 2 esophagogastric junction tumor.

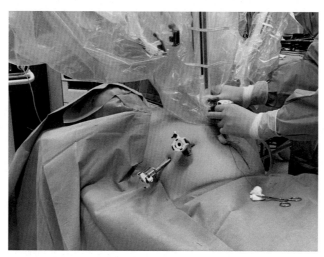

FIG. 4 Portal distribution for gastrectomy and pancreaticoduodenectomy.

In conclusion, robotic surgery has demonstrated significant advancements in the treatment of esophagus/foregut and pancreatic disorders. Procedures such as robotic esophagectomy, robotic fundoplication for GERD, robotic Heller myotomy, robotic-assisted Whipple procedure, and robotic gastrectomy have shown promising results in terms of safety, efficacy, and patient outcomes (Fig. 4). As the field continues to evolve, ongoing research and long-term follow-up are crucial to further establish the role of robotic surgery in esophagus/foregut and pancreatic procedures.

## Critical analysis on the evidence and outcomes on robotic upper gastrointestinal surgery

Robotic-assisted upper GI surgery is a rapidly advancing field, due to benefits including providing a high degree of instrument freedom and stabilizing the surgeon's tremor. However, the current evidence base does not yet fully support its widespread use or justify the associated expense [15].

Robotic-assisted MI esophagectomy (RAMIE) was introduced in 2003 as a safe and viable option for esophagectomy. The ROBOT RCT [16] showed that RAMIE yielded comparable oncologic outcomes to open esophagectomy, with lower rates of surgically related postoperative complications, lower median blood loss, improved functional recovery at postoperative day 14, and better quality of life at discharge and at 6 weeks postdischarge. Long-term survival analysis showed that overall and disease-free survival was comparable, supporting the use of robotic surgery in esophageal cancer [17]. Additionally, Yang et al. [18] showed that RAMIE yielded shorter operation time with improved lymph node dissection compared to MIE, with no difference in complications including vocal cord paralysis, anastomotic leak, pulmonary complications, blood loss and conversion rate. Long-term survival data from this trial is currently awaited. Furthermore, a systematic review supports the use of RAMIE, showing comparable mortality and reduced morbidity rates; however, operative time was found to be longer in patients receiving RAMIE compared to MIE [19].

The first robotic gastrectomy was described by Hashizume et al. [20]. Since then, several studies have analyzed perioperative outcomes such as blood loss, number of resected lymph nodes, anastomotic fistula rate, length of hospital stay, surgical site infection, major complications (Clavien-Dindo ≥3), time to first flatus, daily liquid intake, and readmission rate. Tao et al. [14], in a metaanalysis including eight studies and a total of 2763 patients undergoing robotic and laparoscopic subtotal gastrectomy, concluded that robotic surgery had lower blood loss, higher rate of resected lymph nodes, and earlier liquid intake compared to the laparoscopic approach. Special care should be taken during the learning curve and ideally the first cases be accompanied by a proctor.

When we focus specifically in pancreatic robotic surgery, we are dealing with technically demanding procedures including distal pancreatectomy, pancreaticoduodenectomies, enucleations, and others. For better outcomes, it is suggested that these surgeries be performed only in specialized high-volume centers such as in open and in minimally invasive (laparoscopic and robotic) procedures. Even so, learning the training curve outcomes can help establish the training curriculum for specialists in such centers and other developing institutions. Some studies have shown that robotic pancreaticoduodenectomy has a more challenging learning period than a distal pancreatectomy (36.5 vs 23.5 cases) [21]. Robotic approaches in both procedures also require more cases performed to achieve the learning curve.

This can be attributed to the technically demanding nature of the procedure and robotics. Patient selection and surgeon expertise when implementing robotic techniques must be highlighted [22]. Overall, the literature shows no significant difference between the number of cases required to surmount the learning curve for laparoscopic pancreatoduodenectomy vs robotic pancreatoduodenectomy (laparoscopic pancreatoduodenectomy 34.1 [95% confidence interval 30.7–37.7] vs robotic pancreatoduodenectomy 36.7 [95% confidence interval 32.9–41.0]; $P = 0.8241$) and laparoscopic distal pancreatectomy vs robotic distal pancreatectomy (laparoscopic distal pancreatectomy 25.3 [95% confidence interval 22.5–28.3] vs robotic distal pancreatectomy 20.7 [95% confidence interval 15.8–26.5]; $P = 0.5997$) [23].

This Miami International Evidence-based Guidelines on Minimally Invasive Pancreas Resection were published in 2020 and are the result of a comprehensive review of the available literature and aim to provide evidence-based recommendations for the use of minimally invasive techniques in pancreas resection surgeries. The recommendations emphasize the importance of surgeon expertise, appropriate patient selection, and multidisciplinary collaboration in achieving successful outcomes with minimally invasive approaches [24].

This scientific article, authored by Hesse et al. [25], examines intraoperative conversion and complications in robotic-assisted primary and redo gastric bypass surgery. The study investigates the incidence of conversion from robotic to open surgery during the procedure and analyzes associated complications. The findings suggest that the rate of intraoperative conversion in robotic-assisted gastric bypass surgery is relatively low. The study also identifies certain factors that may increase the likelihood of conversion, such as patient characteristics and anatomical considerations. Additionally, the authors discuss the types of complications that can occur during the conversion process. The article highlights the importance of surgeon experience, patient selection, and appropriate preoperative planning to minimize the risk of conversion and complications in robotic-assisted gastric bypass surgery. Overall, this study provides insights into the challenges and considerations associated with intraoperative conversion and complications in this surgical procedure [25].

## Technique example

### Subtotal gastrectomy

Subtotal gastrectomy with D2 lymphadenectomy begins with access to the lesser sac, performing transection of the omentum 3 cm far from the gastroepiploic arcade, followed by omentectomy up to the N4sb chain and ligation of the left gastroepiploic vessels. In the opposite direction, we go to the right gastroepiploic vessels to ligate them and resect the lymph nodes of the N6 chain. The next step is the duodenotomy, followed by ligation of the right gastric artery and removal of the supraduodenal lymph nodes from the N5 chain, dissection of the N12a, N8a, N9, N7, and N11p lymph node chains. After completing the lymphadenectomy, we proceed to the gastrotomy and removal of the surgical specimen. Gastroenteroanastomosis is performed latero-laterally using a stapler, as well as the entero-entero anastomosis.

### Training

As in any other technological advance in surgical procedures, it is important to create an educational environment that allows an easier, cheaper, and safer technique development and means to estimate and measure this development. Thus, many aspects of this scenario have been discussed. Nevertheless, the determination of the learning curve required for the secure and optimal use of this new technology has been studied.

Primarily, the learning curve, as to say the time required to master a new technique, is usually composed of three phases: the first phase is the starting point, the period of a rapid increase in learning in which the operative time and estimated blood loss are higher; the second phase is the plateau period, in which the operative time and the estimated blood loss are almost the same as the average level, indicating that the surgical procedure has become more stable; and the final phase is the steady improvement period, in which the operative time and the estimated blood loss decrease below the average level, which indicates that further modification of the procedure will contribute to better outcomes. Only this final phase is stable and may objectively reflect the advantages and disadvantages of a novel technique. Furthermore, functional and clinical results must be considered, and the confounders should be minimized to draw reliable conclusions [26].

This worthy metaanalysis, authored by Moekotte et al. [27], presents a systematic review focused on the safe implementation of minimally invasive pancreas resection, and reviewed assesses to various aspects of safe implementation, including patient selection, surgical techniques, perioperative outcomes, and long-term results. The findings suggest that minimally invasive pancreas resection, as an example of complex surgery, can be safely performed in selected patients, with favorable perioperative outcomes and comparable long-term survival rates to open surgery. The authors emphasize the importance of surgeon expertise, appropriate patient selection, and a multidisciplinary approach in achieving successful outcomes with minimally invasive techniques [27].

The information pertaining to the learning curve in minimally invasive surgery can not only be used as a guide for training surgeons to gauge their own performance but also be used in the decision-making of adopting newer innovative techniques especially in lower volume hospitals. Cost-effectiveness is a central and unavoidable metric which should be considered in the adoption of new procedures, and the learning curve plays a major role in determining cost-effectiveness. Previous studies have shown that shortening the learning curve is critical in managing the costs of minimally invasive surgery [21].

Finally, training has several factors that can influence on the outcomes such as the nature of the procedure, the surgical workload, surgical instruments and technology used, the training program, innate ability of the surgeon, patient factors (type of lesion and anatomy), and the ability of decision-making (intraoperatively and under pressure). However, at last stance, well-established training methodologies, surgical teams, institutions, and medical associations should create a baseline curriculum for robotic training along with laparoscopy for future surgeons and those who intend to apply the technology [21].

## Future scenario

We envision a transformative era where robotic technology continues to revolutionize surgical approaches and patient outcomes, especially in the field of robotic surgery in esophagus, stomach, duodenum, and pancreas.

With ongoing advancements in robotic platforms, we anticipate the development of next-generation robots specifically designed for the intricate and complex anatomy of these structures. These robots will offer enhanced dexterity, precision, ergonomics, and haptics, empowering surgeons to perform even the most challenging procedures with improved outcomes.

In the future, we also expect robotic surgery to be seamlessly integrated with advanced imaging technologies, artificial intelligence, and predictive modeling. Surgeons will have access to sophisticated preoperative planning tools that enable personalized surgical approaches tailored to each patient's unique anatomy and disease characteristics. This precision will enhance surgical outcomes and minimize potential complications.

Augmented reality and virtual reality technologies will likely play a significant role in the future of robotic surgery. Surgeons will benefit from enhanced visualization and navigation capabilities, allowing for more accurate tumor localization, precise tissue dissection, and improved preservation of vital structures. This integration of advanced visualization tools will further enhance safety and precision in complex surgeries.

Teamwork and telemedicine will foster greater collaboration among surgeons worldwide. Through telemedicine and remote surgical mentorship, experts will be able to provide guidance and support to less-experienced surgeons in real time, expanding access to high-quality surgical care in underserved areas. This collaborative approach will promote continuous learning and drive further advancements in the field.

As we envision this future scenario, it is crucial to recognize that these projections are speculative but grounded in the rapid advancements we have witnessed in recent years. The integration of robotic technology, personalized planning, enhanced visualization, and global collaboration holds immense promise for improving patient outcomes, revolutionizing surgical practices, and shaping the future of healthcare.

## Conclusions

Minimally invasive (MI) surgery has become the revolutionized surgery, becoming the standard of care in many countries around the globe. Demonstrated benefits over traditional open surgery include reduced pain, shorter hospital stay, and decreased recovery time. Gastrointestinal surgery was an early adaptor to both laparoscopic and robotic surgery. Robotic surgery introduced three-dimensional vision output, instrumentation with a significantly higher degree of movement freedom compared to laparoscopic instruments. Still, this came hand-to-hand with increased cost and use of rather sizable pieces of equipment.

# Key points

- Robotic surgery has revolutionized the field of esophagus/foregut and pancreatic surgery, bringing superior visualization, increased dexterity, and enhanced ergonomic control.
- Nowadays, procedures such as robotic esophagectomy, gastrectomy, fundoplication for GERD, and pancreatectomies are well accepted and have benefits such as less tissue trauma, faster recovery, and smaller hospital stay.
- Throughout the training of esophagus/foregut and pancreatic robotic surgery, the outcomes should be assessed systematically in order to improve results and cost-effectiveness.
- In the future, artificial intelligence, predictive modeling, and augmented reality incorporated into the robotic surgical field will allow better preoperative planning, enabling personalized surgical approaches tailored to each patient.

# References

[1] Himpens J, Leman G, Cadiere GB. Telesurgical laparoscopic cholecystectomy. Surg Endosc 1998;12(8):1091. https://doi.org/10.1007/s004649900788.

[2] Cadière GB, Himpens J, Vertruyen M, et al. Evaluation of telesurgical (robotic) NISSEN fundoplication. Surg Endosc 2001;15(9):918.

[3] Marescaux J, Leroy J, Gagner M, et al. Transatlantic robot-assisted telesurgery. Nature 2001;413(6854):379–80.

[4] Saeidi H, Opfermann JD, Kam M, Wei S, Leonard S, Hsieh MH, et al. Autonomous robotic laparoscopic surgery for intestinal anastomosis. Sci Robot 2022;7(62). https://doi.org/10.1126/scirobotics.abj2908. eabj2908.

[5] Cerfolio RJ, Bryant AS, Minnich DJ. Starting a robotic program in general thoracic surgery: why, how, and lessons learned. Ann Thorac Surg 2011;91(6):1729–36.

[6] Zureikat AH, Postlewait LM, Liu Y, et al. A multi-institutional comparison of perioperative outcomes of robotic and open Pancreaticoduodenectomy. Ann Surg 2016;264(4):640–9.

[7] Nickel F, Wise P, Müller PC, Kuemmerli C, Cizmic A, Salg G, Steinle V, Niessen A, Mayer P, Mehrabi A, Loos M, Müller-Stich BP, Kulu Y, Büchler MW, Hackert T. Short-term outcomes of robotic versus open Pancreatoduodenectomy - propensity score-matched analysis. Ann Surg 2023. https://doi.org/10.1097/SLA.0000000000005981 [Epub ahead of print. PMID. 37389886].

[8] Olijnyk JG, Valandro IG, Rodrigues M, Czepielewski MA, Cavazzola LT. Cohort cholecystectomies in the Brazilian public system: is access to laparoscopy universal after three decades? Rev Col Bras Cir 2022;(49), e20223180. English, Portuguese https://doi.org/10.1590/0100-6991e-20223180-en, PMID: 35858035.

[9] Park S, Lee H, Lee HJ, et al. Robotic-assisted minimally invasive esophagectomy versus thoracoscopic minimally invasive esophagectomy for resectable esophageal cancer: a propensity score-matched analysis. Ann Surg 2020;272(1):87–93.

[10] Garg S, Nayar S, Bapaye A, et al. Robotic-assisted laparoscopic vs open Nissen fundoplication for gastroesophageal reflux disease: a systematic review and meta-analysis. World J Gastrointest Surg 2016;8(11):670–81.

[11] Andolfi C, Fisichella PM. Robot-assisted Heller myotomy for the treatment of achalasia: a comprehensive review. Surgery 2018;164(4):695–702.

[12] Wang SE, Shyr BU, Lin PW, et al. Robotic pancreaticoduodenectomy in patients with a history of abdominal surgery: current status and future perspectives. Int J Med Robot 2021;17(1), e2177.

[13] Hu Y, Huang C, Sun Y, et al. Morbidity and mortality of laparoscopic versus open D2 distal gastrectomy for advanced gastric cancer: a randomized controlled trial. J Clin Oncol 2017;35(3):277–84.

[14] Sun T, Wang Y, Liu Y, Wang Z. Perioperative outcomes of robotic versus laparoscopic distal gastrectomy for gastric cancer: a meta-analysis of propensity score-matched studies and randomized controlled trials. BMC Surg 2022;22(1):427. https://doi.org/10.1186/s12893-022-01881-9. PMID: 36517776. PMCID: PMC9749346.

[15] Kinross JMMS, Mylonas G, Darzi A. Next-generation robotics in gastrointestinal surgery. Nat Rev Gastroenterol Hepatol 2020;17:430–40. https://doi.org/10.1038/s41575-020-0290-z.

[16] van der Sluis PC, van der Horst S, May AM, Schippers C, Brosens LAA, Joore HCA, et al. Robot-assisted minimally invasive thoracolaparoscopic esophagectomy versus open transthoracic esophagectomy for resectable esophageal cancer: a randomized controlled trial. Ann Surg 2019;269(4):621–30. https://doi.org/10.1097/SLA.0000000000003031.

[17] de Groot EM, van der Horst S, Kingma BF, Goense L, van der Sluis PC, Ruurda JP, et al. Robot-assisted minimally invasive thoracolaparoscopic esophagectomy versus open esophagectomy: long-term follow-up of a randomized clinical trial. Dis Esophagus 2020;33(Supplement_2). https://doi.org/10.1093/dote/doaa079. doaa079.

[18] Yang Y, Zhang X, Li B, Li Z, Sun Y, Mao T, et al. Robot-assisted esophagectomy (RAE) versus conventional minimally invasive esophagectomy (MIE) for resectable esophageal squamous cell carcinoma: protocol for a multicenter prospective randomized controlled trial (RAMIE trial, robot-assisted minimally invasive esophagectomy). BMC Cancer 2019;19(1):608.

[19] Angeramo CA, Bras Harriott C, Casas MA, Schlottmann F. Minimally invasive ivor Lewis esophagectomy: robot-assisted versus laparoscopic-thoracoscopic technique. Systematic review and meta-analysis. Surgery 2021;170(6):1692–701. https://doi.org/10.1016/j.surg.2021.07.013.

[20] Hashizume M, Shimada M, Tomikawa M, et al. Early experiences of endoscopic procedures in general surgery assisted by a computer-enhanced surgical system. Surg Endosc 2002;16:1187–91. https://doi.org/10.1007/s004640080154.

[21] Fung G, Sha M, Kunduzi B, Froghi F, Rehman S, Froghi S. Learning curves in minimally invasive pancreatic surgery: a systematic review. Langenbecks Arch Surg 2022;407(6):2217–32. https://doi.org/10.1007/s00423-022-02470-3. PMID: 35278112. [Epub 2022 Mar 12. PMCID: PMC9467952].

[22] Kulu Y, Büchler MW, Hackert T. Evidenz für die. Evidence for robotics in oncological pancreatic surgery. Chirurg 2021;92(2):102–6. https://doi.org/10.1007/s00104-020-01299-0. PMID: 33064158.

[23] Chan KS, Wang ZK, Syn N, Goh BKP. Learning curve of laparoscopic and robotic pancreas resections: a systematic review. Surgery 2021;170(1):194–206. https://doi.org/10.1016/j.surg.2020.11.046. Epub 2021 Feb 2 PMID: 33541746.

[24] Asbun HJ, Moekotte AL, Vissers FL, Kunzler F, Cipriani F, Alseidi A, D'Angelica MI, Balduzzi A, Bassi C, Björnsson B, Boggi U, Callery MP, Del Chiaro M, Coimbra FJ, Conrad C, Cook A, Coppola A, Dervenis C, Dokmak S, Edil BH, Edwin B, Giulianotti PC, Han HS, Hansen PD, van der Heijde N, van Hilst J, Hester CA, Hogg ME, Jarufe N, Jeyarajah DR, Keck T, Kim SC, Khatkov IE, Kokudo N, Kooby DA, Korrel M, de Leon FJ, Lluis N, Lof S, Machado MA, Demartines N, Martinie JB, Merchant NB, Molenaar IQ, Moravek C, Mou YP, Nakamura M, Nealon WH, Palanivelu C, Pessaux P, Pitt HA, Polanco PM, Primrose JN, Rawashdeh A, Sanford DE, Senthilnathan P, Shrikhande SV, Stauffer JA, Takaori K, Talamonti MS, Tang CN, Vollmer CM, Wakabayashi G, Walsh RM, Wang SE, Zinner MJ, Wolfgang CL, Zureikat AH, Zwart MJ, Conlon KC, Kendrick ML, Zeh HJ, Hilal MA, Besselink MG. International study group on minimally invasive pancreas surgery (I-MIPS). The miami international evidence-based guidelines on minimally invasive pancreas resection. Ann Surg 2020;271(1):1–14. https://doi.org/10.1097/SLA.0000000000003590. PMID: 31567509.

[25] Hesse UJ, Lenz J, Dubecz A, Stein HJ. Intraoperative conversion and complications in robotic assisted primary and redo gastric bypass surgery. J Robot Surg 2022;16(1):235–9. https://doi.org/10.1007/s11701-021-01212-9.

[26] Shen D, Wang H, Wang C, Huang Q, Li S, Wu S, Xuan Y, Gong H, Li H, Ma X, Wang B, Zhang X. Cumulative sum analysis of the operator learning curve for robot-assisted Mayo Clinic level I-IV inferior vena cava Thrombectomy associated with renal carcinoma: a study of 120 cases at a single center. Med Sci Monit 2020;26, e922987. https://doi.org/10.12659/MSM.922987. PMID: 32107362. PMCID: PMC7063847.

[27] Moekotte AL, Rawashdeh A, Asbun HJ, Coimbra FJ, Edil BH, Jarufe N, Jeyarajah DR, Kendrick ML, Pessaux P, Zeh HJ, Besselink MG, Abu Hilal M, Hogg ME. International evidence-based guidelines on minimally invasive pancreas resection (IG-MIPR). Safe implementation of minimally invasive pancreas resection: a systematic review. HPB (Oxford) 2020;22(5):637–48. https://doi.org/10.1016/j.hpb.2019.11.005 [Epub 2019 Dec 10. PMID: 31836284].

III. The current and future clinical applications of robotic surgery among medical specialties

# 54

# Bariatric robotic surgery

*Francisco Guerra, Jr.[a], Shinil K. Shah[a,b],*
*Erik B. Wilson[a], and Melissa M. Felinski[a]*

[a]Division of Minimally Invasive and Elective General Surgery, Department of Surgery, McGovern Medical School at UT Health, Houston, TX, United States [b]Michael E. DeBakey Institute for Comparative Cardiovascular Science and Biomedical Devices, Texas A&M University, College Station, TX, United States

## Introduction

Bariatric surgery (BS) remains the most effective long-term treatment for the chronic disease of morbid obesity and its related medical comorbidities. The number of bariatric surgical procedures has increased significantly and continues to rise. BS is indicated for patients with a body mass index (BMI) greater than $40 \, kg/m^2$, or a BMI between 35 and $40 \, kg/m^2$ with an associated medical comorbidity, including hypertension, diabetes, obstructive sleep apnea, and cardiopulmonary disease. Guidelines are changing to recognize the benefit of BS in patients with a BMI $\geq 35 \, kg/m^2$ regardless of the presence of associated medical comorbidities [1] (Ref according to the rules). The trend in bariatric cases has continued to shift in the last decade, with a high prevalence of sleeve gastrectomy currently (likely due to perceived relative ease) and a decrease in the number of bypasses and duodenal switches performed [2] This is expected to change with the increase in revisional cases as well as the increased popularity of single anastomosis variants of the traditionally described operation.

The majority of BS procedures have been performed with laparoscopic techniques, dating back to 1994, with Wittgrove and Clark performing the first laparoscopic Roux-en-Y gastric bypass (RYGB) [3]. Robotic approaches have aimed to eliminate the inherent limitations of laparoscopy, especially in revisional cases and in patients with higher BMI, by affording increased dexterity, which in turn improves dissection and suturing, visualization, and ergonomics for the surgeon [4]. Since the first robotic BS was performed in 1999 by Himpens and Cadiere [5], the application of robotic surgery in BS has continued to grow. The focus of this chapter is to help those surgeons transitioning into robotic BS by summarizing the most commonly performed BS procedures with robotic assistance, including setup, steps, and common pitfalls.

## Robotic limitations

Intra-abdominal adhesions remain a possible problem in any patient with previous surgeries, including those undergoing revisional surgery. We approach this by obtaining safe access to the peritoneal cavity and performing laparoscopic lysis of adhesions as required. In a RYGB, this is especially important to ensure adequate bowel mobilization for anastomoses. Depending on a surgeon's preference for length of the Roux and biliopancreatic limbs, this often means ensuring 200–300 cm of small bowel from the ligament of Treitz is mobilized. Similarly, when performing biliopancreatic diversion/duodenal switch or single anastomosis duodenal ileal bypass (SADI), it is important to ensure enough free distal bowel to allow for a single or Roux-en-Y anastomosis.

## Entry, ports, and patient positioning

In this chapter, we focus on sleeve gastrectomy, RYGB, and SADI. Port placement remains relatively consistent throughout all primary and revisional bariatric procedures. The only general variation that exists is when an assist port is used for stapling. This section will apply to nearly every robotic bariatric surgical procedure performed.

The patient is positioned supine with a footboard in place, as they will be placed in reverse Trendelenburg. It is important to adequately secure the legs and pad the foot and ankle to prevent hyperflexion of the foot. The arms are extended on arm boards, and all remaining pressure points are padded appropriately. The calibration tube of choice is advanced into the stomach. The patient is then placed in reverse Trendelenburg, typically 15–20 degrees. A 5 mm optical trocar is used to enter the abdomen in the right upper quadrant. Depending on the patient's individual operative history, a left upper quadrant entry may be preferred. The peritoneal cavity is insufflated. A diagnostic laparoscopy is performed to assess for adhesions; if present, subsequent trocars will generally allow for adequate lysis of adhesions. We then place our first 8-mm trocar just above and left of the umbilicus. This allows for placement of the port through the rectus sheath to minimize the risk of future incisional hernias. Additionally, it helps to avoid interference with visualization by the falciform ligament while obtaining a global view, as this will be the camera port. All additional trocars are generally placed in line with the camera port. A left lateral 8 mm trocar is placed, and an additional 8 mm trocar is placed equidistant between the left lateral and camera port. If a robotic stapler is being utilized, a 12 mm robotic trocar is placed about 5–7 cm to the right of the camera port. If one chooses to have an assist port for stapling, then we exchange our right upper quadrant optical entry port for an 8 mm robotic port. In place of a 12 mm robotic trocar, we use a regular 12–15 mm trocar (depending on the stapler and cartridge preference of the surgeon). If no bedside stapling is performed, the optical entry trocar can then be kept or switched out for the surgeon's preferred insufflation or assist port. Liver retraction can be performed via internal liver retractors or via an external (e.g., Nathanson) liver retractor. We generally dock the robotic patient side cart to the trocars from the patient's left side. Port placement is depicted in Fig. 1.

## Roux-en-Y gastric bypass

### Instruments

- Tip-up fenestrated grasper × 2 (arms 1 and 4)
- Monopolar shears (arm 3)
- Suction irrigator (bedside or robotic)
- Robotic stapler (arm 1) (depending on stapling preference)
- Large needle driver (arms 1 and 3)
- Cadiere grasper (this can be used in place of a tip-up in arm 1)

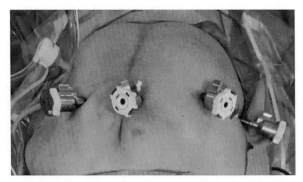

FIG. 1    Port placement for bariatric case when using robotic stapler.

## Operative steps

### Creation of the gastric pouch

We first dissect the peritoneum off the angle of His to expose the superior and inferior portions of the left crus. We then create a tunnel posterior to the stomach using a pars flaccida approach (Fig. 2A) using a combination of fenestrated graspers (arms 1 and 4) and monopolar shears (arm 3). We connect this tunnel to the superior dissection at the angle of His. The preference of our group is to utilize a vascular load stapler with absorbable staple line reinforcement to secure the perigastric mesentery, making sure to remain inferior enough to preserve the left gastric artery. Certainly, a perigastric approach may be utilized. In revisional cases, a perigastric approach may be preferred. Generally, 2–3 appropriately sized 45–60 mm staple loads (with or without absorbable staple line reinforcement, as per the surgeon's choice) are used to create the gastric pouch. The first staple line is horizontal and below the level of the left gastric artery (Fig. 2B), followed by lateral staple lines parallel to a bougie/sizing tube toward the gastroesophageal junction (Fig. 2C). When using a robotic stapler, all staple fires are done through arm 1. Some surgeons may also elect to use a 12 mm robotic trocar at the left mid abdominal trocar (arm 3) to facilitate the parallel staple fires.

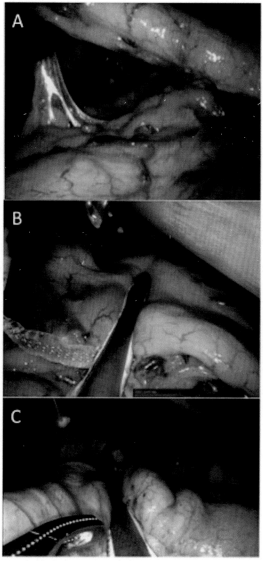

FIG. 2    Gastric pouch creation. (A) Creation of the posterior gastric tunnel via a pars flaccida (lesser curvature) approach. (B) After completion of the first staple fire to secure the lesser curvature mesentery below the level of the left gastric artery, a horizontal staple fire starts the gastric pouch creation. (C) Sequential horizontal staple fires complete the pouch creation. The bougie or sizing tube is utilized as a guide.

## Omental division

Depending on the intra-abdominal anatomy, the surgeon may elect to divide the omentum to facilitate tension-free placement of the Roux limb. If the surgeon chooses to divide the omentum, it is retracted superiorly using fenestrated graspers in arms 1 and 4. The omentum is then divided from the transverse colon to its free end using the surgeon's preferred energy device.

## Creation of omega loop and gastrojejunal anastomosis

The omega loop is created initially by retracting the transverse colon superiorly and locating the ligament of Treitz. We typically run the jejunum from the ligament of Treitz between 50 and 100 cm (depending on the patient's BMI, medical comorbidities, and the surgeon's preference for the length of the biliopancreatic limb). Recent randomized data suggests longer biliopancreatic limb length may improve weight loss [6]. This is done using fenestrated graspers in arms 1 and 4. This loop is then brought up to the gastric pouch and held in place with a fenestrated grasper in arm 4; arms 1 and 3 are replaced with needle drivers (Fig. 3A). The orientation of the omega loop is exceedingly important to prevent Roux-en-O anatomy. The biliopancreatic (afferent) limb should be on the patient's left side, and the Roux limb (efferent) limb should be on the patient's right side. We have previously published technical details on creating a truly isoperistaltic gastrojejunostomy [7].

We then suture the antimesenteric portion of the small bowel (serosa) to the staple line of the pouch (Fig. 3B); this can be performed using an absorbable braided or monofilament suture of the surgeon's choice or a 2–0 double-armed, absorbable, and barbed suture. We then create our gastrotomy and enterotomy using monopolar shears; this is done

FIG. 3 Gastrojejunotomy. (A) The jejunum is ran distally starting at the ligament of Trietz to allow for an appropriately sized biliopancreatic limb; it is then brought to the gastric pouch to create an omega loop configuration. In this configuration, the biliopancreatic limb (BP) (afferent) is on the patient's left side (image *right*) and the Roux limb (efferent) is on the patient's right side (image *left*). (B) The posterior row of the gastrojejunostomy is created by approximating the horizontal staple line of the pouch (full thickness bites) to the serosa of the jejunum. (C) The inner row of the gastrojejunostomy is completed. After completion of the posterior outer and posterior inner row, the bougie/sizing tube is passed through the anastomosis into the Roux limb. (D) Appearance of the gastrojejunostomy during suturing of the anterior outer Lembert row of sutures. (E) The small bowel is transected to create a biliopancreatic limb (image *right*, patient's *left side*) and Roux limb (image *left*, patient's *right side*).

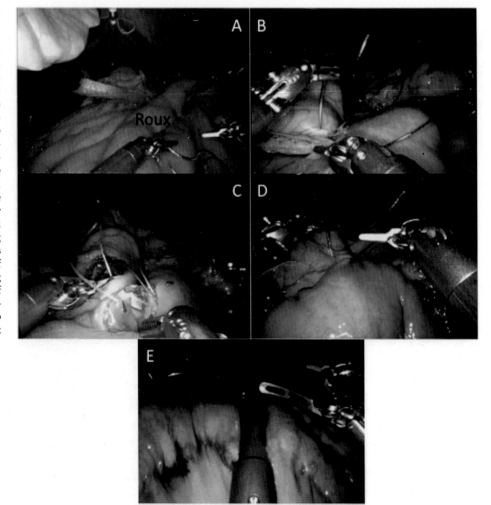

at about the width of the bougie. We then use a double-armed 2–0 braided absorbable suture to perform the inner layer of the anastomosis (Fig. 3C). Once the posterior inner row is completed, the bougie is passed through the anastomosis and into the Roux limb. The inner and anterior rows of the suture are then completed (Fig. 3D). After the anastomosis is completed, the bougie is removed.

## Creation of the Roux and biliopancreatic limbs

The small bowel is then divided just proximal to the pouch (patient's left) (Fig. 3E). We first create a window in the mesentery using blunt dissection with our two needle drivers (arms 1 and 3), while the fenestrated grasper in arm 4 provides retraction. The small bowel is then divided using a stapler of the surgeon's choice. At this point, the Roux limb should be on the patient's right side, and the biliopancreatic limb should be on the patient's left side.

## Jejunojejunostomy

Starting at the gastrojejunostomy, we run the Roux limb between 100 and 150 cm using fenestrated graspers in arms 1 and 4. The length of the Roux limb is based on the surgeon's preference. Making the Roux limb longer than the bilio-pancreatic limb may help prevent inadvertent Roux-en-O anatomy. We avoid a very short Roux limb (i.e., less than 75–100 cm) to avoid bile reflux. We then align the Roux limb to the blind end of the biliopancreatic (BP) limb; this can be secured with a proximal and/or distal stay stitch. Placing a distal stay stitch can allow the surgeon to retract the jeju-nojejunostomy toward the spleen using a fenestrated grasper in arm 4, allowing for good exposure. Monopolar shears are placed in arm 3, and an enterotomy is made on the antimesenteric portion of each limb, 1 cm proximal to the staple line on the biliopancreatic limb. If a robotic stapler is being utilized, the fenestrated grasper in arm 1 is then moved to arm 3, and a stapler is brought into arm 1. The jejunojejunostomy is then created (Fig. 4A). The stapler is removed, and needle drivers are introduced into arms 1 and 3. The common channel enterotomy is closed using a single layer of running 2–0 braided absorbable suture (Fig. 4B). Care should be taken to avoid large bites on the Roux limb, as this may be a potential cause of an early mechanical small bowel obstruction at the jejunojejunostomy. We then close the mesenteric defect with a permanent braided 2–0 suture (Fig. 4C). We generally do not close Petersen's space in our antecolic, antegastric RYGB. Barbed sutures can be used for all steps, as per the surgeon's preference. We perform a modified, traditional antiobstruction Brolin stitch with the mesenteric closure suture [8].

## Case conclusion

We routinely perform an upper endoscopy with an intra-operative leak test by submerging the gastrojejunostomy under saline and insufflating; a fenestrated grasper is used to occlude the Roux limb. We also examine for patency of the gastrojejunal anastomosis as well as to ensure no intra-luminal bleeding. We then ensure that all external staple lines are hemostatic. We close the facial defect of any 12–15 mm port.

## Revisions and common pitfalls

Performing a RYGB in a patient who has had a prior sleeve gastrectomy is typically straightforward, as creating the posterior gastric tunnel to create the gastric pouch is generally less tedious given the prior mobilization and resection of the fundus.

Patients who have significant scarring or inflammation in the lesser sac can pose a particular challenge to pouch creation. This can result from prior fixed or adjustable gastric band placement, vertical banded gastroplasty, prior fun-doplication, or from previous inflammatory diseases, such as pancreatitis, or in patients who have had prior distal pancreatectomy or splenectomy. This can pose difficulty when creating the posterior gastric tunnel. In these cases, it is often helpful to mobilize the greater curvature (similar to a sleeve gastrectomy or during a fundoplication). This allows for a lateral to medial dissection, which can often facilitate appropriate identification of key anatomy, including the left gastric artery. If greater curvature mobilization is performed, after completion of mobilization and gastric pouch creation, the perfusion of the superior gastric remnants should be confirmed. In certain cases, the fundus may require resection. If prior staple lines are present, such as in the case of a prior vertical banded gastroplasty or nondivided gastric bypass, they should all be resected. The pouch should be created superior and medial to all fixed bands and staple lines. In general, after the pouch has been created, a partial remnant gastrectomy will need to be

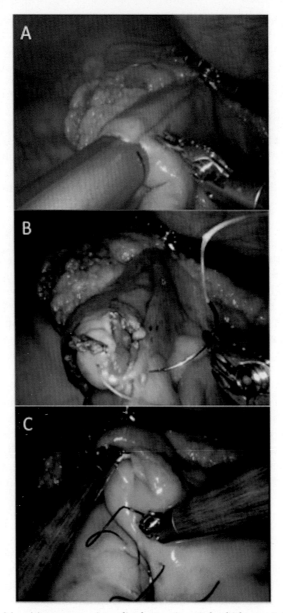

FIG. 4    Jejunostomy. (A) Creation of the jejunojejunostomy using a distal retraction stitch which retracts the anastomosis to the left upper quadrant. (B) Closure of the common channel enterotomy using a single layered running braided absorbable suture. (C) Closure of the mesenteric defect at the jejunojejunostomy.

performed that includes both the fixed band (if present) and the staple line of the previous gastroplasty. Failure to resect a lateral staple line may lead to a persistent gastrogastric fistula or gastrocele.

As mentioned, omental division can mitigate tension in a gastrojejunostomy. Sometimes, gastric pouch mobilization by a formal hiatal dissection can be helpful in rare cases. While retrocolic placement of the Roux limb to decrease tension is an option, this is generally not necessary.

Accidental Roux-en-O reconstruction is a rare but potentially serious complication. Keeping the Roux limb longer than the biliopancreatic limb generally will facilitate identification of a potential Roux-en-O configuration intra-operatively.

Placing ports too high may cause issues with visualization of the ligament of Treitz, identification and manipulation of the jejunum, as well as difficulty creating the jejunojejunostomy. Our guidelines for port placement are described previously.

Prior to stapling of the gastric pouch or any bariatric surgical procedure, it is important that there are no other tubes within the stomach besides the surgeon's preferred sizing tube/bougie. A stapling "time-out" should be performed to ensure that there are no additional nasogastric/orogastric tubes or temperature probes in the nose or mouth.

# Sleeve gastrectomy

## Instruments

- Tip-up fenestrated grasper × 2 (arms 1 and 4)
- Vessel Sealer (arm 3)
- Suction irrigator (bedside or robotic)
- Robotic stapler (arm 1) (depending on the stapling preference)
- Cadiere (this can be used in place of a tip-up in arm 1, if the surgeon prefers)

## Operative steps

### Greater curve mobilization

Prior to starting our dissection, we advance our sizing tube (bougie) and suction the air out of the stomach; the tube is then kept off suction. We generally use a 36–40 Fr sizing tube. We first identify the pylorus and identify a point on the greater curvature 4–5 cm proximal to it. The stomach is elevated and retracted medially, and the omentum is retracted laterally with fenestrated graspers in arms 1 and 4. Using a vessel sealing (energy) device, we begin the mesenteric division at the midpoint of the greater curve. Once a window is obtained in the lesser sac (Fig. 5A), we mobilize the greater curvature distally until we reach our target 4–5 cm proximal to the pylorus (Fig. 5B). We then take down any posterior adhesions; this method will allow better retraction of the stomach when dissecting more proximally. The proximal greater curve is then mobilized to the angle of His securing the short gastric vessels. The left crus is completely exposed. During mobilization of the greater curvature, care is taken to avoid injury to the right gastroepiploic artery. Of note, we approach this proximal dissection (i.e., from the superior pole of the spleen to the left crus) in a posterior/medial to lateral fashion (Fig. 5C); we retract the stomach superiorly and caudally to takedown any medial posterior adhesions, making sure not to injure the vasculature of the lesser curvature of the stomach (specifically the left gastric artery). This approach also minimizes the risk of splenic capsule tears and bleeding. Once the greater curve is mobilized, the bougie is advanced to the pylorus.

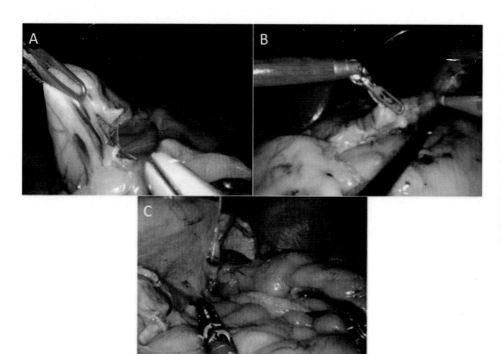

FIG. 5 Sleeve gastrectomy—mesenteric dissection. (A) Entry into the lesser sac in accomplished at the mid greater curvature of the stomach. (B) The stomach is mobilized distally toward a point 4–5 cm proximal to the pylorus. (C) Once the greater curvature has been mobilized to the superior pole of the spleen, the stomach is retracted superiorly and inferiorly to allow for a posterior/medial approach to secure the remaining short gastric vessels. Dissection takes place toward the base of the left crus, and the remaining short gastric vessels are then secured in a medial to lateral fashion.

## Vertical sleeve gastrectomy

Stapling is started 4–5 cm from the pylorus with a robotic or bedside stapling device. Staple cartridges appropriate for thick tissues are generally used for the first 1–2 staple fires, with progressively smaller staples as the surgeon proceeds toward the angle of His. Particular attention should be paid to the first two fires, as these two should be sufficiently away from the incisura to avoid distal narrowing at this site (Fig. 6A and B). Prior to each fire, we ensure there are no loose residual staples (or staple line reinforcement, if utilized) at the most proximal portion of the staple line, as well as examine for adequate staple formation. After completion, the staple line is assessed for bleeding and adequate hemostasis (Fig. 6C). An upper endoscopy is performed to assess sleeve anatomy, narrowing at the incisura, as well as intraluminal bleeding. A leak test is performed as described earlier. The specimen is then removed, and fascial defects for all 12–15 mm ports are closed.

## Common pitfalls

Placing ports too high (specifically the stapling port) may cause issues with the first two staple fires. Generally, the internal aspect of the stapling ports should be approximately 5 cm below the site of the first staple fire on the antrum.

Narrowing of the incisura may contribute to chronic nausea, reflux, as well as leaks after sleeve gastrectomy from increased intraluminal pressure. Using an appropriately sized bougie can help avoid this. Care should be taken to ensure that the first two staple fires do not cause incisural narrowing. When approaching the proximal staple fires at the fundus, it is very important to obtain adequate lateral and posterior/inferior retraction to avoid retaining a fundus or a significant "dog ear" at the GE junction (Fig. 6D).

## Revisions

Generally, patients with prior fixed gastric bands or vertical banded gastroplasties should not be revised to sleeve gastrectomy. The most common revisional procedure encountered by bariatric surgeons is the revision of a prior adjustable gastric band to sleeve gastrectomy. It is important to ensure that the pseudocapsule (cicatrix) created by the band is dissected and removed anteriorly, laterally, and posteriorly to ensure adequate staple line formation. In addition, it is important to ensure that any gastropexy or plication is completely taken down to restore normal anatomy prior to the start of a stapling. When revising patients who have had a prior endolumenal plication

FIG. 6 Sleeve gastrectomy—stapling. (A) The first staple line starts 4–5 cm proximal to the pylorus. (B) It is important to ensure that the first and second staple line (depicted in this image) does not cause narrowing at the incisura. All stapling occurs around a sizing tube/bougie. In this case, a 40 Fr bougie is being utilized. (C) After completion of the sleeve gastrectomy, the staple line is carefully examined for bleeding. (D) Posterior and lateral retraction of the fundus during the last staple fire helps to ensure no large "dog ear" or significant retained fundus in the superior aspect of the sleeve.

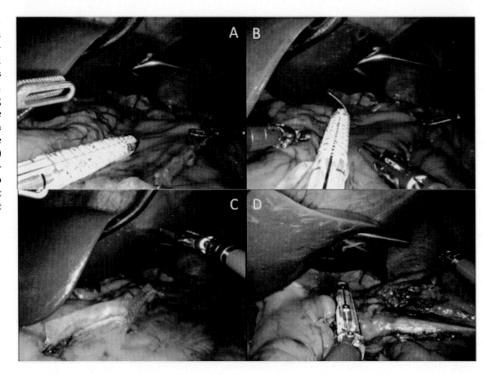

(endolumenal sleeve gastroplasty with sutures or fasteners), preoperative endoscopic examination is mandatory to ensure adequate distance from the lesser curvature to completely resect all prior sutures/fasteners, as well as to ensure that sutures/fasteners will not be in the staple line. Particular attention should be paid to the incisura on preoperative upper endoscopy. These points regarding the revision of adjustable gastric band to stapled operations apply also when revising the RYGB.

## Single anastomosis duodenal ileal bypass

### Instruments

- Tip-up fenestrated grasper × 2 (arms 1 and 4)
- Monopolar shears (arm 3)
- Suction irrigator (bedside or robotic)
- Robotic stapler (arm 1) (depending on the stapling preference)
- Large needle driver (arms 1 and 3)
- Vessel sealer (arm 3)
- Cadiere (this can be used in place of a tip-up in arm 1, if the surgeon prefers)

### Gastric and duodenal dissection

This procedure begins, as with all cases, with a diagnostic laparoscopy to ensure that there are no adhesions precluding adequate small bowel length for our common channel. We then proceed with mobilization of the greater curvature and duodenum, paying particular focus on removing any posterior gastric adhesions. Generally, two methods can be utilized, including carrying the greater curvature dissection distally onto the duodenal bulb to allow about 2.5 cm of duodenal mobilization. Alternatively, the most distal mesentery to the antrum and prepyloric area can be preserved, and a tunnel made under the duodenal bulb for transection. This posterior tunnel creation is performed with the combination of a vessel sealing device, fenestrated graspers, and/or monopolar shears. Once the dissection is complete, we place our fenestrated grasper in arm 4 through the duodenal tunnel to help facilitate the lateral/superior duodenal mobilization. A vessel loop or Penrose drain is placed through the tunnel and left in place to help facilitate duodenal transection. It is important to avoid injury to the common bile duct or head of the pancreas, as well as the gastroduodenal artery and portal vein. Preservation of the lesser curvature blood supply, particularly the right gastric artery, helps to ensure adequate perfusion.

### Sleeve gastrectomy

Our focus then turns to completing the sleeve gastrectomy in the same manner described previously.

### Duodenal division

Once we have completed our sleeve gastrectomy, we proceed with our duodenal transection about 2–2.5 cm distal to the pylorus using a bedside or robotic stapler, as per the surgeon's preference (Fig. 7A).

### Duodeno-ileostomy

After the duodenum is divided, we identify the ileocecal valve and run the small intestine proximally to the desired length of the common channel (generally 250–300 cm). The small intestine is manipulated with fenestrated graspers in arms 1 and 4 and brought up to the duodenum; the intestine is then held in place with arm 4. In the same fashion we create a gastrojejunostomy in bypasses, we start by creating the outer posterior row with double-armed, 2–0 barbed monofilament absorbable suture (needle drivers in arms 1 and 3). The duodenotomy and enterotomy are made using monopolar shears in arm 3. We suture our inner layer similar to our technique for RYGB (double armed 2–0 braided absorbable suture) (Fig. 7B). Prior to starting the anterior row, the sizing tube/bougie is advanced past the anastomosis into the ileum. After completion of the anterior, inner, and outer rows of suture (Fig. 7C), we perform an upper endoscopy to assess for sleeve anatomy, incisural narrowing, intra-luminal bleeding, and an anastomotic leak test is

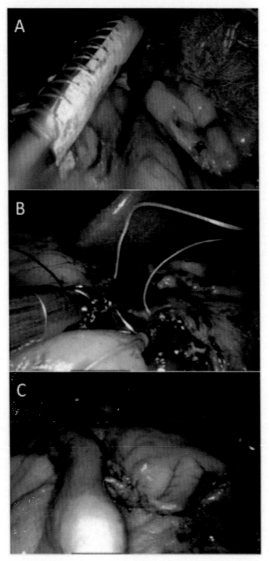

FIG. 7   Single Anastomosis Duodenal-ileal bypass. (A) Duodenal transection is performed about 2–2.5 cm distal to the pylorus using a bedside or robotic stapler. (B) Double layered duodeno-ileostomy, inner row. (C) Completed duodeno-ileostomy.

completed. The resected stomach is removed through the 12- or 15-mm port. As per previous descriptions, we routinely close the fascial defect for any 12- or 15-mm trocar.

## Common pitfalls

Challenges regarding sleeve gastrectomy are discussed previously. Generally, the most challenging portion of the SADI is the duodenal mobilization, specifically when creating the posterior tunnel. This can be avoided by approaching the dissection inferiorly and posteriorly. Initial superior dissection by surgeons without significant experience may result in difficulty to control bleeding. It is important to avoid overuse of blunt dissection, utilizing cautery appropriately, in addition to not forcing the tip-ups through the periduodenal fat. Severe bile reflux after a single anastomosis procedure may require conversion to a traditional Roux-en-Y reconstruction as described with biliopancreatic diversion/duodenal switch. Converting a SADI to a Roux-en-Y duodenal switch can be performed by using an identical technique as utilized for RYGB after creation of the omega loop gastrojejunostomy.

## Future scenario

Continued evolution of robotic assisted techniques in bariatric and metabolic surgery is expected with the introduction of new robotic platforms, including endolumenal robotic platforms, as well as decision-making augmentation.

As more surgeons begin to adopt surgical platforms in any field, it will lead to the development of new robotic platforms, which could lead to further improvements in robotic BS.

## Conclusion

Robotic assisted laparoscopic BS has potential advantages over traditional laparoscopy, especially in patients with higher BMI, anastomotic procedures, and revisional cases. Multiple large studies have demonstrated the safety of this procedure.

## Key points

- We aim to provide a concise surgical atlas for approaching common bariatric procedures.
- The approach to any bariatric surgery, especially port placement, is key to an efficient and smooth surgery.
- This chapter provides examples of the benefits and pitfalls of robotic surgery, and what the surgeon can do to mitigate them.

## References

[1] Eisenberg D, Shikora SA, Aarts E, Aminian A, Angrisani L, Cohen RV, de Luca M, Faria SL, Goodpaster KPS, Haddad A, Himpens JM, Kow L, Kurian M, Loi K, Mahawar K, Nimeri A, O'Kane M, Papasavas PK, Ponce J, Pratt JSA, Rogers AM, Steele KE, Suter M, Kothari SN. 2022 American Society of Metabolic and Bariatric Surgery (ASMBS) and International Federation for the Surgery of Obesity and Metabolic Disorders (IFSO) indications for metabolic and bariatric surgery. Obes Surg 2023;33(1):3–14. https://doi.org/10.1007/s11695-022-06332-1.

[2] ASMBS. Estimate of bariatric surgery numbers, 2011–2020, 2022. https://asmbs.org/resources/estimate-of-bariatric-surgery-numbers. [Accessed 16 April 2023].

[3] Wittgrove AC, Clark GW, Tremblay LJ. Laparoscopic gastric bypass, roux-en-Y: preliminary report of five cases. Obes Surg 1994;4(4):353–7. https://doi.org/10.1381/096089294765558331.

[4] Iranmanesh P, Fam J, Nguyen T, Talarico D, Chandwani KD, Bajwa KS, Felinski MM, Katz LV, Mehta SS, Myers SR, Snyder BE, Walker PA, Wilson TD, Rivera AR, Klein CL, Shah SK, Wilson EB. Outcomes of primary versus revisional robotically assisted laparoscopic roux-en-Y gastric bypass: a multicenter analysis of ten-year experience. Surg Endosc 2021;35(10):5766–73. https://doi.org/10.1007/s00464-020-08061-x.

[5] Cadiere GB, Himpens J, Vertruyen M, Favretti F. The world's first obesity surgery performed by a surgeon at a distance. Obes Surg 1999;9(2):206–9. https://doi.org/10.1381/096089299765553539.

[6] Zerrweck C, Herrera A, Sepulveda EM, Rodriguez FM, Guilbert L. Long versus short biliopancreatic limb in roux-en-Y gastric bypass: short-term results of a randomized clinical trial. Surg Obes Relat Dis 2021;17(8):1425–30. https://doi.org/10.1016/j.soard.2021.03.030.

[7] Chamely EA, Hoang B, Jafri NS, Felinski MM, Bajwa KS, Walker PA, Barge J, Wilson EB, Cen P, Shah SK. Palliative endoscopic salvage of a functionally obstructed gastrojejunostomy—report of technique. CRSLS 2022;9(1). https://doi.org/10.4293/CRSLS.2021.00094.

[8] Brolin RE. The antiobstruction stitch in stapled roux-en-Y enteroenterostomy. Am J Surg 1995;169(3):355–7. https://doi.org/10.1016/S0002-9610(99)80175-5.

# Robotic surgery for perihilar biliary tract cancer

*Trenton Lippert[a], Sharona Ross[b],*
*Alexander Rosemurgy[b], and Iswanto Sucandy[b]*

[a]University of South Florida Morsani College of Medicine Tampa, FL, United States [b]Digestive Health Institute, AdventHealth Tampa, Tampa, FL, United States

## Introduction

The introduction of the robotic surgery technique has improved clinical outcomes in hepatobiliary surgery while maintaining the indication and standard operative technique in liver surgery, including perihilar biliary tumor resection [1]. This minimally invasive platform has only changed the technical approach without inventing a new oncological operation, which is important to maintain long-term cancer surgery outcomes. Utilization of the robotic technique and implementation of minimally invasive surgery tools provide an opportunity to minimize the physiological stress that patients suffer while undergoing curative surgeries [2–4]. The avoidance of rib retraction and morbid right upper quadrant incisions in liver surgery has significantly lowered the incidence of wound complications, postoperative pulmonary issues, and improved postoperative pain management [5].

Perihilar cholangiocarcinoma is a rare, aggressive malignancy diagnosed in approximately 9000 people in the United States each year [6]. Over the past 30 years, the five-year survival rate has double to tripled with the advancement of diagnostic and surgical techniques that improve diagnosis, surgical planning, and postoperative outcomes [7–10]. Although the gain is significant, the five-year survival rate for patients undergoing resection for extrahepatic cholangiocarcinoma remains below 50%, supporting the aggressive nature of the tumor to invade nearby vital structures [11]. Notably, the inflow via the portal vein and hepatic artery are the most commonly affected anatomical structures, requiring meticulous, aggressive dissection and often necessitating vascular resection with reconstruction. In this chapter, we describe each operative step in detail. Application of an enhanced recovery after surgery (ERAS) protocol aids our goal to reduce postoperative length of stay with minimal impact on readmission rates.

Liver transplantation had been described as an option for early-stage and limited perihilar cholangiocarcinoma; however, currently it is not a widely accepted curative option for perihilar cholangiocarcinoma, as in its intrahepatic counterpart. Complete surgical resection to gain R-0 resection margins, followed by biliary reconstruction remains the main curative option for patients [12]. Indications for adjuvant and neoadjuvant treatment are dictated by the patient's condition, preoperative staging, response to preoperative treatment, operative clinical outcomes, and postoperative final staging via pathology [12]. The technique of choice for preoperative biliary drainage is a controversial topic that must be determined by each institution based on the availability of experienced endoscopists vs interventional radiologists with the skill to manage the possible complications associated with biliary decompression in these patients.

Traditionally, patients with perihilar cholangiocarcinoma undergo resection via the open approach; however, the application of the robotic approach under the guidance of an experienced robotic hepatobiliary surgeon is safe and feasible in carefully selected patients [1]. Undertaking this operation via the robotic platform provides improved dexterity compared to laparoscopy, and superior portal lymphadenectomy yield vs the open and laparoscopic approaches [13]. We have found and believe that perioperative outcomes in patients undergoing robotic resection are superior to that in other approaches [1].

## Indications

Diagnosis of the tumor is the first step, even though positive cancer diagnosis is only achieved in about two-thirds of the patients preoperatively. The preoperative suspicion for extrahepatic and perihilar cholangiocarcinoma is determined by clinical presentation and high-resolution imaging findings. We employ and recommend a strict selection criteria by only offering Klatskin types 1–3 based on preoperative Bismuth Corlette classification to be resected using the robotic approach. At this point, we do not feel we have adequate solid clinical data for Klatskin type 4 to be resected via the minimally invasive approach.

Most patients present with jaundice and dark urine. Preoperative tissue diagnosis is attempted via biliary endoscopy with endoluminal biopsies or brushing for most patients [12]. Due to the paucicellular nature of perihilar cholangiocarcinoma, brushing is only associated with a 30% positivity rate [14,15]. Patients with locally advanced or suspicious metastatic lesions undergo neoadjuvant chemotherapy or chemoradiation, followed by restaging evaluation. Guidelines vary; however, resectable criteria generally include the lack of retropancreatic or paraceliac lymph node involvement and lack of local organ invasion beyond the hepatic parenchyma. Traditionally, contralateral portal vein and hepatic artery involvement are contraindications to resection, but at high volume institutions with experienced hepatobiliary surgeons, en bloc resection can be performed with vascular reconstruction [16]. Delays in undertaking the operation may be associated with reigning in other uncontrolled medical conditions, including diabetes, cardiopulmonary comorbidities, necessary weight loss, or additional preoperative testing and procedures (e.g., biliary decompression, cardiac clearance, and coronary angiography) to optimize the anticipated perioperative course and recovery.

## Preoperative steps

High-quality imaging is obtained by 1-mm cut computed tomography (CT) or magnetic resonance (MRI) imaging scan. The CT scan is typically the preferred imaging due to its lower cost, better inflow/outflow vascular details, and surgeon's familiarly with the CT scan. However, MRI is a useful imaging modality as well to detect small intrahepatic lesions in a case of suspicious occult hepatic metastasis. Magnetic resonance cholangiopancreatography (MRCP) is useful to road map the perihilar and intrahepatic biliary branching pattern and to rule out any unrecognized biliary anatomy variations, which can complicate the biliary reconstructive plan. For example, a separate right posterior hepatic duct insertion into the left hepatic duct is important to recognize preoperatively in the case of a Klatskin type 3B resection where a total anatomical left hepatectomy needs to be undertaken. Failure to recognize such a variation can cause prolonged bile leak, sepsis, and even necessitate a return to the operation room.

A 3D liver volumetric reconstruction using an advanced software imaging is crucial when a major hepatectomy is planned or is likely, such as in Type 3 and 4 Klatskin tumor resections, in addition to the biliary resection/reconstruction. A formal right hepatectomy or trisectionectomy can lead to postoperative hepatic insufficiency in patients with future liver remnant volume <30% in the setting of underlying hepatic parenchymal dysfunction. Patients with perihilar cholangiocarcinoma often present with a history of liver cirrhosis, hepatic steatosis, steatohepatitis, chemotherapy-induced parenchymal damage, and cholestatic liver disease. In these types of circumstances, the future liver remnant volume needs to be generally >45%. To reduce the risk of mortality from posthepatectomy liver failure, it is important to recognize in advance, patients who need induction of future liver remnant hypertrophy via either a preoperative portal vein embolization, associating liver partition and portal vein ligation for staged hepatectomy (ALPPS), or hepatic venous deprivation [11].

## Operative planning

Preoperatively, we always obtain high-resolution CT imaging with triple phase 1-mm cuts of the chest, abdomen,-pelvis, and abdominal MRI with MRCP. The high-resolution imaging allows for detailed assessments of the inflow and outflow vasculatures and to detect any possible aberrant vasculatures or interval vascular invasion of the cancer. MRCP evaluates the degree of intrahepatic biliary disruption secondary to the extrahepatic tumor and provides the preoperative plan for biliary reconstruction. CT scan of the chest is important to exclude any extrahepatic metastases that preclude surgical resection. A positron emission tomography (PET) scan can be useful to further exclude any evidence of metastatic disease; however, in general, PET scan is an expensive modality, which can only be obtained when a tissue diagnosis of malignancy had been achieved. Most of the insurance carriers in the United States employ a high scrutiny for the use of PET scan by clinicians before the operation.

## Patient optimization

Patients with hyperbilirubinemia >3 mg/dL, cholangitis, or those who were to undergo a major hepatectomy with anticipated FLR volume of ≤35% should be referred for preoperative biliary drainage. Hyperbilirubinemia and cholestatic liver disease are the known risk factors for posthepatectomy liver failure, and both increase the risk of mortality after perihilar cholangiocarcinoma resection [14]. Both endoscopic retrograde cholangiopancreatography (ERCP) by a gastrointestinal endoscopist, and percutaneous transhepatic biliary drainage (PTBC) by an interventional radiologist are valid options to decompress the biliary system preoperatively. The preference is varied among institutions, and referral patterns may affect the use of these two modalities.

In our institution, ERCP with the placement of an endobiliary plastic stent is the primary approach to biliary drainage, which allows for brushings or endoluminal biopsies to be collected if no tissue diagnosis has been previously obtained. Percutaneous transhepatic cholangiography (PTC) can be utilized in patients with persistent hyperbilirubinemia following ERCP or after unsuccessful ERCP. Once the biliary drainage is obtained, we prefer to wait until the total serum bilirubin level is as low as ≤3 mg/dL before taking the patients to the operating room. Hyperbilirubinemia of 3–10 mg/dL however, is not an absolute contraindication to the operation, even though this is not an ideal circumstance in perihilar cholangiocarcinoma resection.

## Patient preparation

An extensive informed consent from the patients must be obtained, which preferably involves their family members in the decision making. This includes explaining the reason and steps of the planned operative intervention in terms that they can understand. We routinely draw diagrams of the operative steps and list all of the common complications they could experience. The most substantial risk to the patient is death, and it is important to convey this information in a manner that the patient can understand without inducing excessive fear.

An important point for patients to realize is that the minimally invasive approach to this operation is clearly advantageous to their postoperative recovery; however, the surgical approach does not change the perihilar cholangiocarcinoma resection into a minor operation. Robotic perihilar cholangiocarcinoma is a major operation that is still associated with risks of postoperative complications known in the open method. The operative duration typically ranges from 6 to 9 h. The length of hospital stay is approximately 5–6 days, and the length of postsurgical recovery at home is approximately 4 weeks. A home care nursing team may also be needed for patients who are discharged home with their abdominal drain(s). Older and deconditioned patients may need to go to a rehabilitation center for 2–3 weeks.

Prior to leaving their surgical consultation, patients receive our detailed ERAS protocol that aims to minimize complications related to preoperative nutritional deficits and physical deconditioning. It begins with the patients drinking immune enhancing fluids for 5 days prior to their operation. An exercise program at home is also a part of this protocol in addition to cessation of smoking and alcohol drinking. Additionally, this protocol emphasizes aggressive postoperative early ambulation, appropriate perioperative pain control through multimodal analgesia and minimization of opioid use, early nutritional intake, and intensive physical therapy.

## Operating room set-up and patient positioning

Patients are taken to the operating room and laid supine on the operating table with their arms at 90° abducted on arm boards, followed by induction of general anesthesia. We do not feel the absolute advantage of using the French position. A central arterial line and central venous catheter are placed by our anesthesia colleagues for optimal hemodynamic monitoring during the operation. An orogastric tube is inserted for gastric decompression along with a Foley catheter for urinary bladder decompression. The abdomen is prepped using an alcohol-based solution followed by the application of a betadine-based plastic drape after a 3-min dry time.

We begin the operation with the patient in 15° reverse Trendelenburg and 5° right side up. The da Vinci Xi robotic surgical system (Intuitive Surgical, Sunnyvale, CA, USA) dual console system is located about 10 ft away from the patient's bed to allow for appropriate space for the bedside surgeon, scrub technician, and anesthesia team to work unincumbered by the console system. The robotic system is docked from the patient's right shoulder area. The bedside surgeon stands to the right of the patient and the scrub technician stands to the left of the patient in the opposite position (Fig. 1).

# OR Set-up

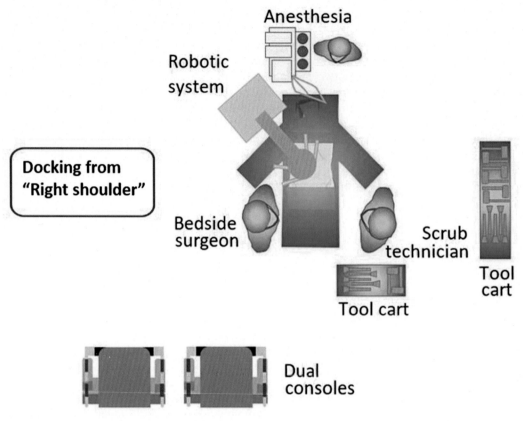

FIG. 1    Operating room set-up.

## Key operative steps

### Port placement and OR set-up

Local anesthesia is placed prior to each skin incision. The first incision is made at the periumbilical area, where an 8-mm robotic blunt robotic port is placed. The umbilical incision should not be beyond the 8-mm necessary to accommodate the port to avoid an air leak from the peritoneal cavity upon insufflation. The pneumoperitoneum is established at 15 mmHg and maintained. The robotic 30-degree camera is placed in the umbilical port, and it is first utilized to perform diagnostic laparoscopy to rule out any possible peritoneal metastasis. Any suspicious nodule is biopsied and sent for a frozen section pathological examination.

Once peritoneal metastases are excluded, additional ports can then be placed; additional 8-mm ports at the right midclavicular line and left anterior axillary line, a 12-mm port is also placed along the left midclavicular line at the level of the umbilicus for the use of a robotic stapler. The final port incision is 3–5-cm in length for an Access Gelport (Applied Medical, Rancho Santa Margarita, California, USA) between the right midclavicular line and the umbilicus, slightly inferior to the umbilicus (Fig. 2). The Gelport provides access for the bedside surgeon or the first assist to work throughout the operation for suctioning, providing exposure, and finally specimen extraction. An additional 5-mm incision is made in the right upper quadrant for the 5-mm AirSeal (ConMed, Largo, Florida, USA) port, which provides reliable smoke evacuation, rapid insufflation, and insertion of a laparoscopic liver retractor if necessary (Fig. 2). The da Vinci Xi robotic surgical system is paired to the operating table and docked over the right shoulder of the patient, allowing for intraoperative positional changes (Fig. 1).

Organization and implementation of each robotic tool is essential to avoid clashes of the instruments and associated frustration during the operation. A robotic fenestrated bipolar forceps instrument is placed in arm one, the robotic

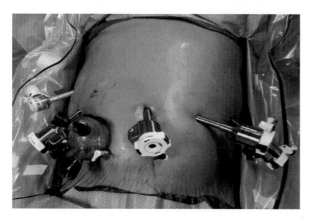

FIG. 2    Robotic port placement.

scope is controlled by arm two, the robotic hook cautery or monopolar curved scissors is placed in arm three, and atraumatic bowel grasper is utilized in arm four. In the Gelport, the bedside surgeon employs a laparoscopic suctioning device and laparoscopic atraumatic bowel grasping tool. We begin by dividing the falciform ligament to expose the anterior surface of the liver and evaluate for potential subcapsular occult hepatic metastases. The dissection is carried cephalad to expose the hepatocaval confluence. The root of hepatic veins is exposed in cases of planned hemihepatectomy, such as in Klatskin types 3A and 3B.

The liver retractor inserted via the AirSeal port retracts the liver cephalad as needed to create an adequate exposure to the porta hepatis. In small livers, the retraction can be simply achieved by placing an atraumatic bowel grasper in robotic arm four and gently retract the segment 4B cephalad. This retraction can be easily adjusted throughout the operation as needed to expose the hilar plate.

## Systematic radical portal lymphadenectomy

We employ a meticulous dissection within the porta hepatis and the surrounding regions to collect periportal lymph nodes while skeletonizing the hepatic artery, portal vein, and common bile duct (CBD). This portion of the operation is completed mainly with the robotic fenestrated bipolar and hook cautery or scissor cautery. The bedside surgeon helps provide traction to improve exposure using the laparoscopic suction device.

Our lymphadenectomy begins by opening the gastrohepatic ligament and exposing the caudate lobe of the liver. The common hepatic artery origin at the celiac trunk is exposed, and all the lymph nodes along the hepatic artery trajectory are removed, preserving the left portal vein (Fig. 3). The superior border of the head, neck, and proximal body of the pancreas is exposed. As we move the cephalad toward the hilar plate, all the lymphatic bearing tissues ventral to the hepatoduodenal ligaments are removed, leaving the bile duct and the right and left hepatic arteries well exposed. Mobilization of the duodenum is necessary to provide access to the retroduodenal lymph nodes for harvesting. We achieve this via the Kocher maneuver, beginning with identifying the vena cava through the foramen of Winslow. The duodenum is routinely kocherized to help expose the distal CBD and remove retroduodenal/ retropancreatic/retroportal lymph nodes. The ligament of Treitz can be left intact.

## Transection of distal CBD

After the completion of retroduodenal lymphadenectomy, the next step is transection of the distal CBD as low as possible at the level of head of pancreas. The distal transection of the CBD needs to be performed low but without injuring the pancreatic duct junction (Fig. 4). At this point, the distal CBD specimen is sent for a frozen section analysis to evaluate distal margins. While awaiting intraoperative pathology results, we use intraoperative cholangioscopy to endoluminally assess proximal tumor extent in common hepatic duct (CHD), and biliary confluence. The location of the proximal bile duct transection and the type of Klatskin tumor is now determined. If cholecystectomy has not been performed, the gallbladder is then removed en bloc with the extrahepatic bile duct. In the case of Klatskin type 3 tumor, the ipsilateral hemihepatectomy is now planned based on the tumor extension into the secondary and tertiary biliary radicles.

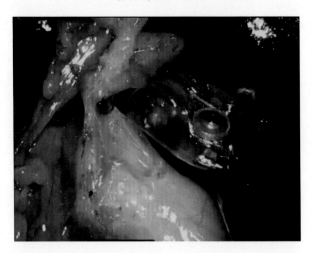

FIG. 3    Portal lymphadenectomy.

FIG. 4    Distal common bile duct transection.

Once the distal CBD margins are confirmed to be tumor free, the opening of the distal CBD at the entrance of the pancreatic head is closed with a running 4–0 polypropylene suture to prevent retrograde pancreatic leak thorough the distal CBD lumen.

## Transection of proximal hepatic duct

After transection of the distal CBD, a temporary suture is placed on the proximal CBD segment to stop the ongoing bile leak and to provide cephalad traction to the CHD. The main portal vein is now widely exposed, and further retroportal lymphadenectomy can be completed, leaving the Foramen of Winslow clean of any nodal tissues.

The dorsal aspect of the CHD is dissected off the right hepatic artery which is typically crossing in a posterior manner to the CHD as it makes its way into the right hepatic lobe Glissonean pedicle. The surgeon must be ready to undertake an arteriography using 5–0 or 6–0 polypropylene sutures since this dissection is often difficult. Once circumferentially dissected free from surrounding vascular structures, the bile duct bifurcation is then prepared for the proximal hepatic duct transection.

Upon reaching the hilar plate, the Laennec's capsule at the inferior aspect of segment IVB was exposed by lowering the hilar plate. This maneuver allows for a further higher dissection of the left and right hepatic duct to obtain negative proximal margins. The level of proximal bile duct transection is dependent on the Bismuth-Corlette classification and intraoperative cholangioscopy. We advance to the secondary biliary radicles before accepting R-1 resection margins. Both right and left hepatic duct margins are sent separately for frozen section analysis.

## Ipsilateral hepatectomy

Concomitant hepatectomy with en bloc caudate resection is undertaken for Bismuth classifications, three and greater. Our approach to robotic left and right hepatectomy has been previously described in detail [17,18]. Since

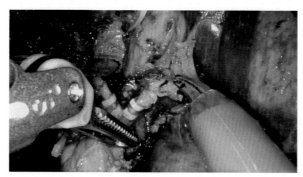

FIG. 5   Left hepatic artery transection.

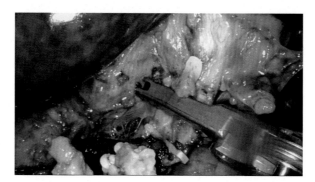

FIG. 6   Left portal vein ligation.

the porta hepatis was previously dissected and cleared of lymphatic nodule tissue, the inflow vessels into the right or left hemiliver can be readily isolated and ligated (Figs. 5, 6). Once the inflow control is secured, the liver parenchyma is split along the demarcation line using a monopolar scissor cautery for the superficial (first 1- to 2-cm) liver parenchyma (Fig. 7). For deeper liver parenchyma, we mainly utilize the crush and clamp technique using robotic fenestrated bipolar forceps and a vessel sealer. Crossing vessels larger than 7 mm are divided between locking clips. Finally, the hepatic vein is transected at its origin flush to the IVC using a liner stapler (Fig. 8).

For the purpose of robotic type 3A and 3B perihilar cholangiocarcinoma resection, we routinely include caudate lobe resection as part of the liver resection since caudate lobe is a frequent site for later tumor recurrence. Caudate lobe bile duct branches are known to drain directly into the biliary bifurcation, serving as a means for cholangiocarcinoma spread/invasion. Dissection and division of the short hepatic veins separating the caudate lobe from the inferior vena cava (IVC) are greatly facilitated by the robotic system (Fig. 9). Minor bleeding from the IVC can be easily handled with robotic suturing.

Once the specimen has been freed from future liver remnants, it is placed in a laparoscopic extraction bag and removed through the right lower quadrant incision that previously housed the Gelport. In cases of type 3B Klatskin tumor resection, the right anterior and right posterior sectoral bile ducts are joined together using ductoplasty techniques, similar to the technique utilized in liver transplantation. The ductoplasty can be easily undertaken using 5–0 or 6–0 absorbable monofilament sutures to enable a classical single end-to-side hepaticojejunostomy anastomosis, as long as a tension-free approximation can be achieved (Fig. 10).

## Roux-en-Y hepaticojejunostomy biliary reconstruction

Once the resection phase of the operation is completed and negative resection margins are confirmed, we now move toward the biliary reconstruction. We create a mesenteric window approximately 35-cm distal to the ligament of Treitz, and the proximal jejunum is divided with a 45-mm robotic blue load stapler (Fig. 11). The roux limb is then measured to 60-cm in length and a standard stapled side-to-side jejunojejunostomy is created next. The jejunal mesentery is divided with robotic vessel sealers as necessary. The common enterotomy is closed with running 3–0 barbed sutures in a watertight fashion.

III. The current and future clinical applications of robotic surgery among medical specialties

FIG. 7    Liver transection, initial.

FIG. 8    Liver transection, complete.

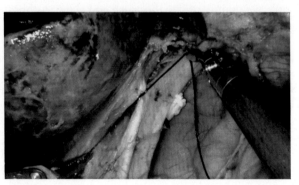

FIG. 9    Dissection of caudate lobe from inferior vena cava.

The roux limb is then transposed toward the hilar plate, where we begin to create the hepaticojejunostomy. Our technique of choice is a running single layer hepaticojejunostomy using two 9-in. 4–0 absorbable barbed sutures. The index suture is placed at the 9 o'clock position, and the posterior layer of the hepaticojejunostomy is done first (Fig. 12). Once the posterior layer is completed, the anterior layer is created next from the 9 to 3 o'clock position (Fig. 13). The robotic system facilitates fine suturing for this anastomosis and allows for ambidextrous hand movements. The posterior and the anterior sutures are tied at the 3 o'clock position, completing the anastomosis (Fig. 14). The abdomen is liberally irrigated with one to two liters of saline solution. A closed suction drain is placed near the anastomosis and brought out to the abdominal wall through the robotic arm one skin incision. A nasogastric tube (NGT) is placed prior to leaving the operating room and transferred to the surgical floor.

FIG. 10    Ductoplasty, right anterior and posterior intrahepatic ducts.

FIG. 11    Side-to-side jejunojejunostomy.

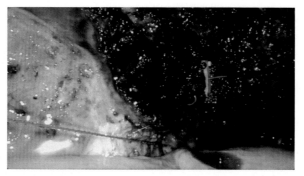

FIG. 12    Hepaticojejunostomy, suture posterior aspect.

FIG. 13    Hepaticojejunostomy, suture anterior aspect.

III. The current and future clinical applications of robotic surgery among medical specialties

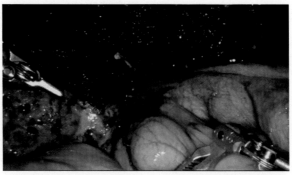

FIG. 14   Hepaticojejunostomy, complete.

## Postoperative instructions and outcomes

Postoperative care continues with the previously described ERAS protocol. Patients remain nil per os for the first 24 h postoperatively while they begin to ambulate. If no concerning output is seen from the NGT during post-op day (POD) one, it is removed along with the central line and the Foley catheter after a voiding trial. On POD two, patients are graduated to a clear liquid diet and receive inpatient physical therapy at least twice per day. If they can tolerate a clear liquid diet without difficulties, the diet then advances to a full liquid on POD three. Discharge to home is expected to occur between POD days four and six after adequate ambulation and pain control have been accomplished. We prefer to remove the closed suction drain prior to discharge from the hospital.

Patients should be followed closely with early postoperative follow-up in the clinic within 1–2 weeks of hospital discharge. A follow-up phone call is made by our nurse practitioners to ensure the patient's well-being within 2–3 days of discharge.

## Future scenario

We have undertaken 27 robotic perihilar cholangiocarcinoma resections since 2019 after completing more than 200 robotic major hepatectomies for liver tumors. We have also had the technical experience of robotic pancreatic resection since 2016. This way, we ensured completion of our learning curve in robotic major hepatectomy prior to embarking on this complex hepatobiliary cancer resection. We concluded that the robotic approach for perihilar cholangiocarcinoma resection is safe and effective in carefully selected patients in the hands of experienced robotic hepatobiliary surgeons [1]. As experience with the robotic approach continues to grow, we believe that it may become the preferred method due to the benefits carried by minimally invasive surgery approaches. Ongoing long-term oncologic outcome research is being performed to ensure no oncologic sacrifices are being made for the short-term benefits achieved by this minimally invasive approach.

## Key points

- Patient optimization with extensive efforts to treat preoperative hyperbilirubinemia is essential to minimizing postoperative complications.
- Preoperative preparation with comprehensive imaging is key to the success of perihilar cholangiocarcinoma resection.
- Robotic resection of perihilar cholangiocarcinoma is safe and feasible in the hands of an experienced minimally invasive surgeon.

## References

[1] Sucandy I, Shapera E, Jacob K, Luberice K, Crespo K, Syblis C, Ross SB, Rosemurgy AS. Robotic resection of extrahepatic cholangiocarcinoma: institutional outcomes of bile duct cancer surgery using a minimally invasive technique. J Surg Oncol 2022;125(2):161–7.
[2] Yang HY, Rho SY, Han DH, Choi JS, Choi GH. Robotic major liver resections: surgical outcomes compared with open major liver resections. Ann Hepatobiliary Pancreat Surg 2021;25(1):8–17.

[3] Zheng H, Huang SG, Qin SM, Xiang F. Comparison of laparoscopic versus open liver resection for lesions located in posterosuperior segments: a meta-analysis of short-term and oncological outcomes. Surg Endosc 2019;33(12):3910–8.

[4] Sotiropoulos GC, Prodromidou A, Kostakis ID, Machairas N. Meta-analysis of laparoscopic vs open liver resection for hepatocellular carcinoma. Updates Surg 2017;69(3):291–311.

[5] Aragon RJ, Solomon NL. Techniques of hepatic resection. J Gastrointest Oncol 2012;3(1):28–40.

[6] Abou-Alfa G. Cholangiocarcinoma, https://rarediseases.org/rare-diseases/cholangiocarcinoma/. [accessed 4/2/2023].

[7] Boerma EJ. Research into the results of resection of hilar bile duct cancer. Surgery 1990;108(3):572–80.

[8] Dondossola D, Ghidini M, Grossi F, Rossi G, Foschi D. Practical review for diagnosis and clinical management of perihilar cholangiocarcinoma. World J Gastroenterol 2020;26(25):3542–61.

[9] Soares KC, Jarnagin WR. The landmark series: hilar cholangiocarcinoma. Ann Surg Oncol 2021;28(8):4158–70.

[10] Yan Y, Lu N, Tian W, Liu T. Evolution of surgery for Klatskin tumor demonstrates improved outcome: a single center analysis. Tumori 2014;100(6):e250–6.

[11] Jang JY, Kim SW, Park DJ, Ahn YJ, Yoon YS, Choi MG, Suh KS, Lee KU, Park YH. Actual long-term outcome of extrahepatic bile duct cancer after surgical resection. Ann Surg 2005;241(1):77–84.

[12] Vogel A, Bridgewater J, Edeline J, Kelley RK, Klumpen HJ, Malka D, Primrose JN, Rimassa L, Stenzinger A, Valle JW, Ducreux M, ESMO Guidelines Committee. Electronic address: clinicalguidelines@esmo.org. Biliary tract cancer: ESMO clinical practice guideline for diagnosis, treatment and follow-up. Ann Oncol 2023;34(2):127–40.

[13] Ratti F, Cipriani F, Ingallinella S, Tudisco A, Catena M, Aldrighetti L. Robotic approach for lymphadenectomy in biliary tumours: The missing ring between the benefits of laparoscopic and reproducibility of open approach? Ann Surg 2022.

[14] Trikudanathan G, Navaneethan U, Njei B, Vargo JJ, Parsi MA. Diagnostic yield of bile duct brushings for cholangiocarcinoma in primary sclerosing cholangitis: a systematic review and meta-analysis. Gastrointest Endosc 2014;79(5):783–9.

[15] Barr Fritcher EG, Voss JS, Brankley SM, Campion MB, Jenkins SM, Keeney ME, Henry MR, Kerr SM, Chaiteerakij R, Pestova EV, Clayton AC, Zhang J, Roberts LR, Gores GJ, Halling KC, Kipp BR. An optimized set of fluorescence in situ hybridization probes for detection of Pancreatobiliary tract cancer in cytology brush samples. Gastroenterology 2015;149(7):1813–24 [e1].

[16] Hewitt DB, Brown ZJ, Pawlik TM. Surgical management of cholangiocarcinoma. Hepatoma Res 2021;7:75.

[17] Gravetz A, Sucandy I, Wilfong C, Patel N, Spence J, Ross S, Rosemurgy A. Single-institution early experience and learning curve with robotic liver resections. Am Surg 2019;85(1):115–9.

[18] Sucandy I, Gravetz A, Ross S, Rosemurgy A. Technique of robotic left hepatectomy: how we approach it. J Robot Surg 2019;13(2):201–7.

# 56

# Lato sensu graduate program in robotic surgery for developing countries

*Saulo Borborema Teles[a], Bianca Bianco[a,b], and Arie Carneiro[a]*

[a]Department of Urology, Hospital Israelita Albert Einstein, São Paulo, SP, Brazil [b]Discipline of Sexual and Reproductive Health and Population Genetics, Department of Collective Health, Faculdade de Medicina Do ABC, São Paulo, SP, Brazil

In the past decades, the robotic surgery has revolutionized urological procedures, offering enhanced precision, shorter recovery times, less bleeding, and less transfusion rates, improving patient outcomes. As the demand for robotic urological procedures continues to rise, it becomes imperative to develop a well-designed postgraduation program in robotic surgery.

As technology continues to evolve, new robotic platforms are emerging in the field of urology. Incorporating these new platforms into a postgraduation program is essential to expose urologists to the latest advancements and equip them with the skills needed to operate these robotic systems effectively.

Before initiating the development of a postgraduation program, it is essential to design a solid curriculum that encompasses the core competencies required for robotic surgery in urology. The curriculum should include simulation exercises, hands-on training, clinical cases debate, and proctored surgery. It should also incorporate progressive learning, allowing participants to advance from basic to complex procedures as they gain proficiency [1].

The postgraduate student should learn not only the robotic skills, but also the operation room setting, patient positioning, trocar position, docking, bedside assistance, postoperatory evaluation, and complication management. The urologist interested in the postgraduate course must have contact with the necessary skills to perform radical prostatectomy, radical and partial nephrectomy, radical cystectomy, simple prostatectomy, use of laparoscopic ultrasound, and use of robotic contrasts as indocyanine, among others.

For implementation of the program, it is crucial to establish the necessary infrastructure and acquire the required resources. This includes dedicated robotic surgical consoles, robotic simulators, adequate operating room, and other equipment necessary for training. Adequate physical space, including simulation centers, is essential for conducting training sessions and workshops.

Simulated training provides a safe environment for trainees to practice and refine their skills, reducing the learning curve associated with complex robotic procedures. Utilizing validated training modules and proficiency metrics can help ensure trainees attain the necessary competence before performing procedures on patients. With the guidance of a more experienced surgeon, the proctor, it is possible to reduce the student's learning curve, ensuring standardization of the surgical technique and approach by a higher experienced surgeon and management in more critical moments.

Robotic surgery presents unique challenges in emergency situations, such as intraoperative bleeding or complications and device malfunction. It is crucial to prepare trainees for these scenarios to ensure they can respond appropriately and manage emergencies effectively [2]. Incorporating training modules that simulate emergency situations, including bleeding control, conversion to open surgery, or alternative approaches, allows trainees to develop the necessary decision-making skills and technical proficiency to handle unforeseen complications. Simulated emergency scenarios can be created using robotic simulators, anatomical models, and high-fidelity patient simulators.

It is also extremely important to develop a training program to the nursing team within the postgraduate course. The team must be prepared for a correct layout of the operating room, effective performance of the scrub nurse, management of surgical instruments, and action in eventual intercurrences with the device and clinical complications.

---

**561**

The standardization of the surgical procedure as well as the operating room preparation and patient care is essential for good results [3].

The postgraduate program at our institution (Hospital Albert Einstein—São Paulo/Brazil) is grounded in a curriculum that encompasses both theoretical knowledge and practical hands-on experience. It entails a rigorous 360 hours theoretical component, featuring in-depth discussions and lectures covering various aspects of robotic surgical techniques. Complementing this theoretical foundation, the practical aspect of the program comprises 40 hours of training on surgical simulators, allowing participants to develop dexterity and familiarity with robotic systems. Furthermore, trainees gain valuable insights into observing 10 surgeries as active observers and providing assistance in three surgical procedures. To ensure proficiency in the use of the robotic system, participants are required to complete the online console training provided by the Intuitive for the Si and Xi platforms. This comprehensive approach equips urology professionals with the skills and knowledge necessary in the field of robotic surgery.

Until recently, certification for performing robotic surgery was in the hands of the company that developed the robotic platform. Currently, this has changed to postgraduate courses, which allow greater access to different technologies, presenting all available robotic platforms, guaranteeing teaching in large specialized centers and with highly specialized and experienced surgeons. With this, access to education becomes increasingly democratic and with probably better quality.

The main barriers found for the development of a quality graduate program in robotic surgery in urology are (1) difficulty in building an adequate curriculum, involving theoretical teaching, robotic simulators, simulation of adverse situations, hands-on models, and practice on real patients. (2) Financial resources to ensure the contact with the newest technologies available and keep equipment up to date. (3) Specialized and experienced human resources with robotic technology. This involves both the medical team and the nursing team. (4) Follow-up after postgraduation to build a quality continuing education to keep participants continuously involved with technology and evolve together with the devices.

## Future scenario

The main future prospects involve the increasing availability of robotic platforms in developing countries and the growing interest in specialization for the use of these technologies. It is imperative to maintain the quality of graduate programs in order to maintain quality teaching and good results. In this way, teaching will be increasingly widespread and will lead to the availability of robotic technology to more patients.

## Key points

- A robust curriculum with simulations, hands-on practice, and progressive learning is essential.
- A successful postgraduation program requires dedicated infrastructure, including robotic consoles, simulators, and proper operating rooms, to ensure safe and effective training.
- Beyond surgeons, nursing teams must also be trained to manage operating room setup, surgical instruments, and patient care, contributing to standardized and high-quality outcomes.

## References

[1] Stockheim J, Perrakis A, Sabel BA, et al. RoCS: robotic curriculum for young surgeons. J Robot Surg 2023;17(2):495–507.
[2] Almeida TFP, Campos MEC, Castro PR, et al. Training in the protocol for robotic undocking for life emergency support (RULES) improves team communication, coordination and reduces the time required to decouple the robotic system from the patient. Int J Med Robot 2022;18(6):e2454.
[3] Sun X, Okamoto J, Masamune K, et al. Robotic technology in operating rooms: a review. Curr Robot Rep 2021;2(3):333–41.

# General abdominal robotic surgery: Indications and contraindications

*Flavio Daniel Saavedra Tomasich*[a,b,c,d,e], *Luiz Carlos Von Bahten*[a,b,f], *Marcos Gómez Ruiz*[b,g,h], *Paulo Roberto Corsi*[b,i,j], *and Sergio Roll*[i,k,l]

[a]Department of Surgery, Federal University of Paraná, Curitiba, Paraná, Brazil [b]National Directory, Brazilian College of Surgeons (CBC), Rio de Janeiro, Brazil [c]Committee on Minimally Invasive Surgery and Robotic Surgery, Brazilian College of Surgeons (CBC), Rio de Janeiro, Brazil [d]Brazilian Society of Oncologic Surgery (SBCO), Rosemont, IL, United States [e]Department of Surgery, Erasto Gaertner Cancer Center, Curitiba, Paraná, Brazil [f]Department of Surgery, Pontificia Universidade Católica do Paraná (PUC-PR), Curitiba, Brazil [g]Marques de Valdecilla University Hospital, Santander, Spain [h]Valdecilla Biomedical Research Institute (IDIVAL), Santander, Spain [i]Department of Surgery, Faculty of Medical Sciences of Santa Casa de São Paulo (FCMSCSP), São Paulo, Brazil [j]Brazilian Chapter of the American College of Surgeons, Rio de Janeiro, Brazil [k]Abdominal Wall Group of the Department of Surgery at the Faculty of Medical Sciences of Santa Casa de São Paulo—(FCMSCSP), São Paulo, Brazil [l]Hernia Surgery Center at German Hospital Oswaldo Cruz, São Paulo, Brazil

## Introduction

In recent decades, technology has both changed and significantly boosted surgical procedures, bringing along the necessity for the development of new techniques and their respective indications and contraindications. The introduction of technological advances in the operating room improves the quality of the surgical outcome and benefits patients around the world, with "Robotic Surgery" being one of the major advancements [1].

Robotic surgery is a minimally invasive surgery modality that uses advanced technology to help surgeons perform complex procedures with greater precision, safety, and resolution. In the area of abdominal surgery, its use is increasing, based on the advantages it presents over traditional surgical techniques.

The most complex intracavitary operations are the ones that most benefit from the use of surgical robots, the modality has become the gold standard for several procedures, in the search for better clinical results.

In general surgery, almost all procedures are feasible with the use of robotic platforms: hernias, esophageal surgery, gastric surgery, and intestinal and colorectal malignancies. However, reservations must be made regarding cost and prolonged surgical time [2–5].

A very important factor to be considered is the training of the surgeon to operate this equipment, establishing a new reality about professional training worldwide [6]. It is changed by the computerized interface of surgical robots: training becomes more complete, with smaller learning curves and reduces morbidity and mortality.

There are also disadvantages of current robotic platforms, such as the high cost of installation and maintenance, the suboptimal use in multiquadrant operations (which has enhanced with new robot generations), and the lack of tactile sensitivity requiring greater compensatory visual attention from the surgeon [7–11].

Economic costs remain one of our biggest concerns. The acquisition, implementation, and maintenance values of a robotic surgery program should not be evaluated detachedly, and the benefits in the postoperative period and in the quality of life of patients must be considered. So far, there are no large-scale studies that have evaluated this aspect from a global perspective.

In this chapter, we present an overview of robotic surgery currently available, highlighting its main indications in the field of general surgery.

However, the concept that we would like to record very clearly is that robotic systems are nothing more than very sophisticated tools, nevertheless, tools at the service of the surgeon. The professional commitment must be with patient safety. To ensure this premise, we must know the indications and fundamentally know when not to use the platform.

## Robotic surgery in esophagus disorders

The use of the robotic platform has shown advantages over laparoscopic surgery only in selected situations in esophageal surgery.

The first robotic fundoplication was reported by Cadiére et al., in 1999 [12]. This procedure is the most common in the esophagus-gastric transition, with its performance on the robotic platform proved to be feasible, but it does not provide objective benefits for the patients [13,14]. However, some surgical maneuvers are facilitated with the use of the robotic platform, such as posterior esophageal dissection, mobilization of the greater gastric curvature, and performing sutures [15]. As a disadvantage, robotic surgery has a longer operating time and higher cost. The most recommended indication is in paraoesophageal hernias, bulky hernias with an enlarged hiatus, and in reoperations in the esophageal hiatus region [16].

Heller's cardiomyotomy associated with laparoscopic fundoplication is considered the gold standard for the treatment of achalasia. Melvin et al., in 2001, reported the first robot-assisted Heller surgery, currently considered as safe as the laparoscopic approach to treat achalasia [17]. Comparative studies have shown similar results in dysphagia relief, although the rate of esophageal perforations in robotic surgery has been significantly lower. Long-term results in alleviating dysphagia and the need for reoperations are better with robotic surgery [17,18].

Esophagectomy for the treatment of esophageal cancer plays an important role in the therapeutic options for this disease, but it carries postoperative complications and may result in mortality. Minimally invasive surgery has revolutionized esophageal resection due to far superior results when compared to open surgery.

The robotic system, initially used only in the thoracic phase, showed better results in lymph node dissection with a longer surgical time. Currently, the thoracic and abdominal times have been performed with the robot and only the cervical anastomosis, when indicated, performed with cervicotomy. All esophagectomy techniques can be performed by robot: in three fields, with intrathoracic or trans hiatal anastomosis [19,20].

In our understanding, robotic esophagectomy should be restricted to reference centers in the treatment of esophageal cancer.

Robotic surgery has also been used for the treatment of less prevalent conditions such as esophageal leiomyomas, considering the meticulous dissection of the mucosa requires delicate movements and high-definition visualization. Resections of supradiaphragmatic esophageal diverticula also have technical benefits from the use of the robot due to the three-dimensional vision and freedom of movement [15].

## Robotic surgery in stomach affections

The use of a robotic platform is a viable option for many gastric surgical procedures, although there are situations where robotic surgery still does not bring benefits to the surgical patient.

Gastric cancer is the fifth most common malignancy and the fourth leading cause of cancer death worldwide. Its main treatment is radical gastrectomy with lymphadenectomy. There is insufficient evidence to evaluate the long-term outcomes of robotic radical gastrectomy. When comparing long-term studies of robotic radical gastrectomy and laparoscopic radical gastrectomy, it was not possible to establish statistical difference in overall survival and recurrence [21–24]. Short-term analyses demonstrate that the operative time of robotic surgery was longer, intraoperative blood loss was lower [21–23], earlier ambulation, and faster postoperative ileus recovery. Surgeons felt more confident in early refeeding of their patients, interpreting this evidence as faster immediate postoperative recovery [21–24]. A recent study showed a greater number of recovered lymph nodes [23].

A comparison of the robotic platform and the laparotomic approach for gastrectomy with D2 lymphadenectomy was performed in groups with similar patients, resulting in similar average number of lymph nodes harvested, and a longer operative time for the robotic procedure. As a benefit, there was a greater than 50% reduction in intraoperative bleeding [25].

General surgeons adequately trained by robotic surgery specialists can perform robotic distal gastrectomy safely.

Metabolic and bariatric surgeries use a minimally invasive approach 90% of the time. With the increasing prevalence of robotic platform use for metabolic and bariatric surgery, there is no consensus on the superiority of either laparoscopic or robotic approaches, especially in revisional procedures (vertical gastrectomy to Roux-en-Y gastric bypass conversion). Intraoperatively, no significant difference was found in total morbidity; however, operative times were significantly longer in robotic procedures. Postoperatively, no significant differences were found in discharge day, 30-day readmission, reoperation rate, additional intervention rate, or 30-day mortality [26].

In our understanding, obese patients with preexisting cardiac or renal comorbidities, venous thromboembolism, and limited functional status, as well as patients who have undergone previous metabolic or bariatric surgery, should not be operated by surgeons in the early stages of their learning curve with the robotic platform.

There is a worldwide increase in the use of the robotic platform approach in bariatric surgery. Concerns exist regarding the potential increase in healthcare costs associated with this technique. More studies are needed to establish key performance indicators, costs, training guidelines, and use of this approach.

## Robotic colorectal surgery

Minimally invasive colorectal surgery is not yet fully implemented in the Western world. Despite its growth over the last 10 years, it represents less than 50% of the procedures, which is probably due to technical reasons [27].

As a more relevant example, rectal cancer surgery, in the minimally invasive approach, helps to overcome anatomical difficulties due to its versatility in improving the vision in the pelvis. Despite this, minimally invasive approach is not universal; the laparotomic approach still is common in some centers, justified by the difficulty of obtaining adequate laparoscopic exposure or pelvises with less elastic tissues (typical of men, obese or patients who have previously received radiotherapy) [28,29].

Another significant benefit is intracorporeal anastomosis after minimally invasive colon cancer resection. Several recently published prospective multicenter studies have demonstrated its benefit for the patient [30,31], although its implementation is scarce. This is likely due to the limited training of colorectal or general surgeons in intracorporeal suturing and the inherent technical difficulty of performing this anastomosis with laparoscopic instruments.

Robotic surgery offers an alternative approach to these difficulties of the minimally invasive approach in colorectal conditions. Increasing in the number of robotic surgical system installations from different manufacturers (mainly Intuitive Surgical, -Sunnyvale, CA-US and also Medtronic-Minneapolis-MT-US, Cambridge Medical Robotics-Cambridge-UK, and others) will hopefully lead to expanding the access that colorectal surgeons have worldwide to these instruments.

In colorectal surgery, it is difficult to define the application of robotic surgery. It may be limited if we understand that the most complex procedures are those that benefit the most from the robotic approach. However, there are many procedures in which the surgeon gains from using the robotic approach, as it provides greater resolution.

We highlight three indications for using the robotic platform:

**1.** High anterior resection for rectal cancer

This is a good indication for patients with distal sigmoid or upper rectum neoplasm. Robotic platforms facilitate central vascular ligation, lowering of the splenic flexure, and partial excision of the mesorectum with greater precision. Robotic platforms can be especially useful in obese patients or those with locally advanced tumors at risk of tumor rupture during dissection.

**2.** Sigmoidectomy for complicated diverticular disease

Diverticular disease is often treated by general surgeons. The conversion rate to open surgery increases in cases of microperforation of the sigmoid, presence of an inflammatory mass with intestinal loops, abscesses, or bladder fistula [32]. Robotic surgical systems allow for more efficient dissection with less risk of perforation or conversion due to three-dimensional vision and articulated instruments. Bladder suturing after resection of a colovesical fistula may also be more feasible for less experienced surgeons due to the same attributes.

**3.** Intracorporeal suturing after segmental resection of the right colon

There is growing evidence of the benefits of intracorporeal anastomosis in minimally invasive bowel resections (lower overall rate of complications, shorter postoperative stay) [30,31]. Despite this, extracorporeal anastomoses continue to be the most frequently performed procedure. In this context, the lack of training and experience of many general surgeons in performing this type of anastomosis can be overcome by using robotic surgical systems. The simplest anastomosis in these cases is the mechanical side-by-side enterotomy, although similar results have been obtained with the fully manual side-by-side anastomosis [33].

## Robotic surgery in hepato-pancreato-biliary disorders

The use of robotic platforms in hepato-pancreato-biliary surgeries is being analyzed regarding indications and postoperative results. Noninferiority and financial aspects are the main focuses of studies [34].

Laparoscopic cholecystectomy is the current gold standard of treatment for benign gallbladder disease. It is one of the most commonly performed surgical procedures worldwide, with ample clinical evidence supporting its application. Cholecystectomy with the use of a robotic platform has become frequent; however, its exact role remains undefined.

Multiple evidence demonstrates that cholecystectomy via the robotic platform is not inferior to that performed via laparoscopy. It allows the surgeon to perform the procedure with greater dexterity and better visualization. Robotic technology facilitates minimally invasive surgery in a way that conventional laparoscopy cannot achieve [35,36].

The indications are similar for both approaches and include symptomatic cholelithiasis, complicated cholelithiasis (acute cholecystitis, choledocholithiasis, and biliary pancreatitis), biliary dyskinesia, biliary sludge, gallbladder polyps, and cholangitis. If the surgeon responsible for operating on the patient is adequately trained in both methods, there is no clinical contraindication documented that justifies choosing one method over the other [35].

However, experts say that the laparoscopic approach remains appropriate for most patients requiring cholecystectomy. It is the most cost-effective treatment modality for gallbladder disease [35,36].

Currently, robotic cholecystectomy is not able to improve clinical outcomes enough to justify its additional costs. Although it has inherent technical advantages and facilitates the learning curve or using the robotic platform, it is imagined that in the future, the use of the robotic platform may be strongly indicated in more complex cases or involving interventions associated with the bile duct.

Bile duct approaches require greater expertise from the surgeon. Most patients with benign or malignant biliary obstruction require surgical treatment with biliary-enteric anastomosis. In these cases, fine dissection and advanced suturing are necessary, and robotic surgery can overcome some limitations of conventional laparoscopic surgery. The precise role of robotic biliary surgery has not yet been defined.

Patients requiring complex bile duct surgery may benefit from using the robotic platform for a variety of benign and malignant indications. Acceptable intraoperative performance has been demonstrated regarding blood loss, conversion to open surgery, and intraoperative complications. This approach has been proven efficient and safe in experienced hands. Referral of these patients to a specialized high-volume center is strongly recommended [37].

At the beginning of the robotic surgery era, authors prophesied the pros and cons of using this technology, with much skepticism regarding its use in hepato-pancreato-biliary surgery [38,39]. Some limitations were overcome, and others persist, such as questioning the actual oncological results. Extensive literature review has examined with particular attention the oncological aspects of robotic surgery and its possible impact on treatment outcomes.

Robotic surgery for oncological indications shows promise, although high-quality studies are lacking. Future experience should consider the oncological benefit and new indications in the rapidly changing field of antineoplastic regimens [34].

The use of minimally invasive surgery in the liver is still limited. The main technical challenges include the difficult access to the inferior vena cava and major hepatic veins, the precision required for hepatic hilum dissection, and the propensity for liver bleeding. There is a learning curve that hampers its practice outside of high-volume centers.

Indications for robotic hepatectomy are like those for laparoscopic hepatectomy. Both benign and malignant tumors can be resected robotically. Patients must have physiological reserve to tolerate general anesthesia and prolonged pneumoperitoneum, their clinically controlled coagulopathy disorders [38].

In the current literature, robotic liver surgery has mostly been associated with noninferior outcomes compared to laparoscopy, although it is suggested that the robotic approach has a shorter learning curve, lower conversion rates, and less intraoperative blood loss. Robotic surgical systems offer a more realistic image with integrated three-dimensional systems. Furthermore, the improved dexterity offered by robotic surgical systems may lead to better intra- and postoperative outcomes [39].

Current results demonstrate that robotic hepatectomy is a safe and effective procedure that will surely continue to increase in utilization. It will likely continue to evolve in parallel with technological developments that enhance robots' abilities. However, it will face important obstacles for broader implementation in daily clinical practice, including associated costs, technical difficulties, and limited amount of evidence. Despite all its promise, until the benefits are more clearly defined, robotic liver surgery will likely be practiced by a select group of surgeons in high-volume centers [38,39].

A very similar situation is experienced regarding robotic surgery in the pancreas. The approach to benign and malignant lesions of the pancreas using this technology has become well accepted and the number of centers using it is rapidly expanding. The most studied robotic pancreatic surgeries are pancreatoduodenectomy and distal pancreatectomy. Most studies report that it is a safe and feasible procedure. Convincing data on the costs of robotics versus conventional techniques is lacking [40].

Robotic pancreas surgery is still in its early stages. It promises to become the new surgical standard for pancreatic resections; however, more research is still needed to establish its safety and effectiveness [40,41]. Currently, there is no gold standard for evaluating a learning curve. It is necessary to establish a standardization in the assessment of the learning curve centered on patient safety [41].

## Robotic-assisted abdominal wall surgery

Laparoscopic treatment of inguinal hernia is justified by the shorter recovery time, less postoperative pain, and equivalent long-term recurrence rates when compared to traditional open mesh repair [42].

More recently, the description of the use of a robotic platform to perform preperitoneal transabdominal inguinal hernia repair (robotic-assisted transabdominal preperitoneal hernia repair/re-TAPP) has the benefits of improved ergonomics, three-dimensional visualization, and greater dexterity for dissection and suture of the mesh [43].

New minimally invasive surgery techniques are emerging for the treatment of complex abdominal wall hernias, and robotic access is obtaining promising results for these patients.

Historically, over the past 20 years, open complex hernia surgery has had a significant complication rate [44]. The robotic platform allows surgeons to perform technically difficult surgeries in a minimally invasive way due to better visualization and a great ability to move in reduced spaces. Along with technique development, advancements in mesh biomaterials have also contributed to better outcomes [45].

Despite the advantages offered by minimally invasive surgery, open surgery is still commonly performed for ventral and inguinal hernia repairs due to the degree of technical difficulty [46]. The learning curve for minimally invasive hernia repair is long, and this is an important factor that contributes to the low rate of adherence to the technique. With the advent of the robotic platform, the learning curve was shortened, with an increase in all minimally invasive techniques [45,46].

The first reports of robotic inguinal hernia repair were published by urologists while performing radical prostatectomy. Since then, numerous articles on robotic inguinal hernia repair have appeared in the literature. Initially, robotic surgery had relative contraindications such as: previous laparoscopic repair, ascites, peritoneal dialysis, large hernias, multiple previous abdominal surgeries, and an inability to tolerate pneumoperitoneum. Some persist, but after the learning curve and the advent of new instruments, several of these relative contraindications were overcome [47].

Nevertheless, international guidelines suggest alternating the approach for the recurrent hernia treatment, and there are encouraging results to maintain the minimally invasive treatment in reoperations [46,48]. There are some advantages of the robotic platform in the repair of recurrent hernias after a posterior approach. Enhanced three-dimensional visualization allows for more accurate dissection of tissues, especially around the bladder and iliac vessels. Better ergonomics and surgical dexterity make it possible to shorten the operative time in complex dissections (mean time 168.9 min, including couplings and unilateral and bilateral repairs) [44,49].

Patients who undergo laparoscopic hernia repair and develop recurrence can be approached using the robotic platforms, by experienced surgeons with safety and efficacy. In these cases, it is necessary to point out that this is a technically demanding approach, for which not everyone has sufficient training [47–49].

The treatment of complex incisional hernias gained prominence with the advent of the robotic platform, rescuing cases that until then were performed using the open technique. An example of this is posterior component separation surgery (transversus abdominis release—TAR), technically complex and with a long learning curve, the use of robotic platforms has been frequent and effective in repairing these defects. TAR in the open approach results in reconstitution of the linea alba without division of the neurovascular bundles. Performing the medial union of the rectus abdominis muscles is possible through the previous division of the transversus abdominis muscle (TA) along its entire length. In the robotic approach, however, the posterior separation of the components has not presented complications such as cutaneous ischemia, necrosis, and seroma. A large preperitoneal space can be created due to extensive myofascial release, thus allowing reinforcement with meshes. The advantages of robotic TAR include shorter hospital stays, less wound complications, less postoperative pain, and a quicker return to work and daily activities. In addition, populations considered at high risk, such as diabetic patients, body mass index (BMI) > 30, and smokers, showed encouraging results [44,47,50].

Robotic repair of ventral and incisional hernias is comparable to open and laparoscopic approaches, considering postoperative complications. Robot-assisted ventral repair significantly shortens hospital stay compared to the open approach, although operative time is still longer. For ventral hernias that would normally require an open procedure, robotic surgery may be a good option.

For inguinal hernias, the benefit of robotic surgery is still questionable, as the literature has not yet demonstrated its superiority compared to the laparoscopic approach.

## Robotic operations in emergency care

The scenario of emergency surgeries has long been dominated by the laparoscopic video approach. Is it possible to use the robotic platform in emergency surgeries?

Currently, there are few data on robotic-assisted operations in emergency surgery. However, operations for acute appendicitis, cholecystitis, hernias, and acute gastrointestinal conditions have been described. The most common operations are acute cholecystectomies and incarcerated hernias. The feasibility of robotic operations has been demonstrated for all indications, with no reported higher rates of complications.

Another clinical situation that the general surgeons encounter is prior abdominal surgeries. This situation can make the minimally invasive approach difficult. In these cases, due to the inherent characteristics of robotic surgery, its application can facilitate the release of adhesions, reduce the conversion rate, improve postoperative recovery, and reduce the rates of incisional hernias.

Several urgent operations in general surgery can be performed robotically without increased risk. The available data does not allow for a final evidence-based assessment [51].

## Future scenario

Robotic surgery is undergoing transformation toward its complete clinical applicability. We can say that the practice of robotic surgery will soon be different from what we experience today. It is easy to predict that robotic systems will be smaller and more functional, and the popularization of this surgical technology will be a consequence of the emergence of new platforms from different competing companies.

These new machines must incorporate technology in a sequential manner, which will enhance their virtues and facilitate their use. Chapters like this will probably see an increase in the number of indications in the future and a drastic reduction in contraindications. We believe that technological advances, biological knowledge, and artificial intelligence will be factors of change in robotic surgery of the future.

## Conclusions

Robotic surgery represents a qualitative leap in surgical instrumentation, as well as the possibility of offering a minimally invasive approach to our patients, even though the procedures to be performed are technically complex or in anatomical locations that are "uncomfortable" for the surgeon.

The most complex intracavitary operations are the ones that most benefit from the use of surgical robots. The cost of implementation and maintenance has limited their utilization.

However, we can say that the true potential of this surgical approach has not yet been reached. With certainty, soon the practice of robotic surgery will be different from what we experience today.

Robotic systems will become smaller, and as competition emerges, they will also become more affordable. The digital transformation and incorporation of artificial intelligence are emerging factors that transform the current scenario.

Technological advances will continue to provide innovations that, combined with biological knowledge, will raise the current level of robotic surgery.

We look forward to these future innovations, which will continue to drive the advancement of robotic surgery.

# Key points

- Robotic surgery is a major advance approach, incorporating cutting-edge technological knowledge, improving the quality of surgical results.
- The robotic platform helps the surgeon to perform complex procedures with greater precision, safety, and resolution.
- The most complex intracavitary operations benefit most from the use of surgical robots.
- The surgeon who operates with the help of the robotics platform must know the main indications, with a strong professional commitment to patient safety.

# References

[1] Satava RM. Future directions in robotic surgery. In: Rosen J, Hannaford B, Satava RM, editors. Surgical robotics: systems applications and visions; 2011. p. 791–824. https://doi.org/10.1007/978-1-4419-1126-1_33 [chapter 1].

[2] Ojima T, Nakamura M, Hayata K, Kitadini J, Katsuda M, Takeuchi A, et al. Short-term outcomes of robotic gastrectomy vs. laparoscopic gastrectomy for patients with gastric cancer: a randomized clinical trial. JAMA Surg 2021;156:954–63.

[3] Jayne D, Pigazzi A, Marshall H, Croft J, Corrigan N, Copeland J, Phil Q, et al. Effect of robotic-assisted vs conventional laparoscopic surgery on risk of conversion to open laparotomy among patients undergoing resection for rectal cancer: the ROLARR randomized clinical trial. JAMA Surg 2017;318:1569–80.

[4] Lee L, de Lacy B, Ruiz MG, Liberman AS, Albert MR, Monson J, et al. A multicenter matched comparison of transanal and robotic total mesorectal excision for mid and low-rectal adenocarcinoma. Ann Surg 2019;270:1110–6.

[5] Chaar ME, King K, Salem JF, Arishi A, Galvez A, Stoltzfus J. Robotic surgery results in better outcomes following Roux-en-Y gastric bypass: metabolic and Bariatric Surgery Accreditation and Quality Improvement Program analysis for the years 2015–2018. Surg Obes Relat Dis 2021;17:694–700.

[6] Nacul MP, Melani AGF, Zilberstein B, Benevenuto DS, Cavazzola LT, Araujo RLC, et al. Educational note: teaching and training in robotic surgery. An opinion of the Minimally Invasive and Robotic Surgery Committee of the Brazilian College of Surgeons. Rev Col Bras Cir 2020;47(1). https://doi.org/10.1590/0100-6991. e2020681.

[7] Araujo RLC, Benevenuto DS, Zilberstein B, Sallum RA, Aguiar S, Cavazzola LT, et al. Overview and perspectives about the robotic surgical certification process in Brazil: the new statement and a national web-survey. Rev Col Bras Cir 2020;47(1), e20202714. https://doi.org/10.1590/0100-6991.

[8] MacFarlane M, Rosen J, Hannaford B, Pellegrini C, Sinanan M. Force feedback grasper helps restore the sense of touch in minimal invasive surgery. J Gastrointest Surg 1999;3:278–85.

[9] Lanfranco AR, Castellanos AE, Desai JP, Meyers WC. Robotic surgery: a current perspective. Ann Surg 2004;239(1):14–21. https://doi.org/10.1097/01.sla.0000103020.19595.7d.

[10] Stylopoulos N, Rattner D. Robotics and ergonomics. Surg Clin N Am 2003;83(6):1321–37. https://doi.org/10.1016/S0039-6109(03)00161-0.

[11] Panait L, Rafiq A, Mohammed A, Mora F, Merrell RC. Robotic assistant for laparoscopy. J Laparoendosc Adv Surg Tech 2006;16(2):88–93. https://doi.org/10.1089/lap.2006.16.88.

[12] Cadiére GB, Himpens J, Vertruyen M, Bruyns J, Fourtanier G. Nissen fundoplication done by remotely controlled robotic technique. Ann Chir 1999;53(2):137–41.

[13] Draaisma WA, Ruurda JP, Scheffer RCH, Simmermacher RKJ, Gooszen HG, Rijnhart-de Jong HG, Buskens E, Broeders IAMJ. Randomized clinical trial of standard laparoscopic versus robot-assisted laparoscopic Nissen fundoplication for gastro-oesophageal reflux disease. Br J Surg 2006;93:1351–9. https://doi.org/10.1002/bjs.5535.

[14] Huttman MM, Robertson HF, Smith AN, Biggs SE, Dewi F, Dixon LK, et al. Blencowe on behalf of RoboSurg collaborative group. A systematic review of robot-assisted anti-reflux surgery to examine reporting standards. J Robot Surg 2023;17:313–24.

[15] Straughan DM, Azoury SC, Bennett RD, Pimiento JM, Fontaine JP, Toloza EM. Robotic-assisted esophageal surgery. Cancer Control 2015;22 (3):335–9.

[16] Kastenmeier A, Gonzales H, Gould JC. Robotic applications in the treatment of diseases of the esophagus. Surg Laparosc Endosc Percutan Tech 2012;22:304–9.

[17] Melvin WS, Needleman BJ, Krause KR, Wolf RK, Michler RE, Ellison EC. Computer-assisted robotic heller myotomy: initial case report. J Laparoendosc Adv Surg Tech A 2004;11(4):251–3. https://doi.org/10.1089/109264201750539790.

[18] Iqbal A, Haider M, Desai K, Garg N, Kavan J, Mittal S, et al. Technique and follow-up of minimally invasive Heller myotomy for achalasia. Surg Endosc 2006;20(3):394–401. https://doi.org/10.1007/s00464-005-0069-x.

[19] Esagian SM, Ziogas IA, Skarentzos K, Katsaros I, Tsoulfas G, Molena D, et al. Robot-assisted minimally invasive Esophagectomy versus open esophagectomy for esophageal cancer: a systematic review and meta-analysis. Cancer 2022;14:3177. https://doi.org/10.3390/cancers14133177.

[20] Banks KC, Hsu DS, Velotta JB. Outcome of minimally invasive and robot-assisted Esophagectomy for esophageal cancer. Cancer 2022;14 (15):3667. https://doi.org/10.3390/cancers14153667.

[21] Li JT, Lin JX, Wang FH, Wang JB, Lu J, Chen QY, et al. Comparison of long-term outcomes after robotic versus laparoscopic radical gastrectomy: a propensity score-matching study. Surg Endosc Other Interv Tech 2022;36(11):8047–59.

[22] Beyer K. Surgery matters: progress in surgical management of gastric cancer. Curr Treat Options Oncol 2023;24(2):108–29.

[23] Sun T, Wang Y, Liu Y, Wang Z. Perioperative outcomes of robotic versus laparoscopic distal gastrectomy for gastric cancer: a meta-analysis of propensity score-matched studies and randomized controlled trials. BMC Surg 2022;22(12):427.

[24] Ong CT, Schwarz JL, Roggin KK. Surgical considerations and outcomes of minimally invasive approaches for gastric cancer resection. Cancer 2022;128(22):3910.

[25] Ribeiro U, Dias AR, Ramos MFKP, Yagi OK, Oliveira RJ, Pereira MA, et al. Short-term surgical outcomes of robotic gastrectomy compared to open gastrectomy for patients with gastric cancer: a randomized trial. J Gastrointest Surg 2022;26(12):2477.

[26] Seton T, Mahan M, Dove J, Villanueva H, Obradovic V, Falvo A, et al. Is robotic revisional bariatric surgery justified? An MBSAQIP analysis. Obes Surg 2022;32(12):3863–8.

[27] Yeo H, Niland J, Milne D, ter Veer A, Bekaii-Saab T, Farma JM, et al. Incidence of minimally invasive colorectal cancer surgery at National Comprehensive Cancer Network Centers. J Natl Cancer Inst 2015;107(1):362.

[28] Gomez FM. From Miles' procedure to robotic transanal proctectomy. Cir Esp 2014;92(8):507–9. https://doi.org/10.1016/j.ciresp.2014.01.001.

[29] Rullier E, Sa Cunha A, Couderc P, Rullier A, Gontier R, Saric J. Laparoscopic intersphincteric resection with coloplasty and coloanal anastomosis for mid and low rectal cancer. Br J Surg 2003;90:445–51.

[30] Cleary RK, Silviera M, Reidy TJ, McCormick J, Johnson CS, Sylla P, at al. Intracorporeal and extracorporeal anastomosis for robotic-assisted and laparoscopic right colectomy: short-term outcomes of a multi-center prospective trial. Surg Endosc 2022;36(6):4349–58. https://doi.org/10.1007/s00464-021-08780-9.

[31] Gómez Ruiz M, Espin-Basany E, Spinelli A, Cagigas Fernández C, Bollo Rodriguez J, María Enriquez Navascués J, et al, Mircast Study Group. Early outcomes from the minimally invasive right colectomy anastomosis study (MIRCAST). Br J Surg 2023;1–8. https://doi.org/10.1093/bjs/znad077.

[32] Larkins K, Mohan H, Apte SS, Chen V, Rajkomar A, Larach JT, et al. A systematic review and meta-analysis of robotic resections for diverticular disease. Colorectal Dis 2022;24(10):1105–16. https://doi.org/10.1111/codi.16227.

[33] Harji D, Rouanet P, Cotte E, Dubois A, Rullier E, Pezet D, et al. A multicentre, prospective cohort study of handsewn versus stapled intracorporeal anastomosis for robotic hemicolectomy. Colorectal Dis 2022;24(7):862–7. https://doi.org/10.1111/codi.16096.

[34] Bahra M, Saidy RRO. Current status of robotic surgery for hepato-pancreato-biliary malignancies. Expert Rev Anticancer Ther 2022;22(9):939–46. https://doi.org/10.1080/14737140.2022.210521.

[35] Chandhok S, Chao P, Koea J, Srinivasa S. Robotic-assisted cholecystectomy: current status and future application. Lap Endosc Robot Surg 2022;5(3):85–91. https://doi.org/10.1016/j.lers.2022.06.002.

[36] Singh A, Panse NS, Prasath V, Arjani S, Ravi J, Chokshi RJ. Cost-effectiveness analysis of robotic cholecystectomy in the treatment of benign gallbladder disease. Surgery 2023;173(6):1323–8. https://doi.org/10.1016/j.surg.2023.01.017.

[37] D'Hondt M, Wicherts DA. Robotic biliary surgery for benign and malignant bile duct obstruction: a case series. J Robot Surg 2023;17:55–62.

[38] Leung U, Fong Y. Robotic liver surgery. Hepatobiliary Surg Nutr 2014;3(5):288–94. https://doi.org/10.3978/j.issn.2304-3881.2014.09.02.

[39] Bozkurt E, Sijberden JP, Hilal MA. What is the current role and what are the prospects of the robotic approach in liver surgery? Cancer 2022;14(17):4268. https://doi.org/10.3390/cancers14174268.

[40] Khachfe HH, Habib JR, Al Harthi S, Suhool A, Hallal AH, Jamali FR. Robotic pancreas surgery: an overview of history and update on technique, outcomes, and financials. J Robot Surg 2022;16:483–94.

[41] Fung G, Sha M, Kunduzi B, Frogghi F, Rehman S, Froghi S. Learning curves in minimally invasive pancreatic surgery: a systematic review. Langenbeck's Arch Surg 2022;407:2217–32.

[42] Miserez M, Peeters E, Aufenacker T, Bouillot JL, Campanelli G, Conze J, et al. Update with level 1 studies of the European hernia society guidelines on the treatment of inguinal hernia in adult patients. Hernia 2014;18(2):151–63. https://doi.org/10.1007/s10029-014-1236-6.

[43] Escobar Dominguez JE, Gonzalez A, Donkor C. Robotic inguinal hernia repair. J Surg Oncol 2015;112(3):310–4. https://doi.org/10.1002/jso.23905.

[44] Amaral MVF, Guimaraes JR, Volpe P, Oliveira FMM, Domene CE, Roll S, Cavazzola LT. Robotic transversus abdominis release (TAR): is it possible to offer minimally invasive surgery for abdominal wall complex defects? Rev Col Bras Cir 2017;44(02):216–9. https://doi.org/10.1590/0100-69912017002009.

[45] Telem DA. Is robotic surgery the future for abdominal wall hernia repair? Not so fast. Ann Surg 2018;267(2):218–9. https://doi.org/10.1097/SLA.0000000000002336.

[46] Claus CMP, Oliveira FMM, Furtado ML, Azevedo MA, Roll S, Soares G, Nacul MP, et al. Guidelines of the Brazilian hernia society (BHS) for the management of inguinocrural hernias in adults. Rev Col Bras Cir 2019;46(4), e20192226. https://doi.org/10.1590/0100-6991e-20192226.

[47] Henriksen NA, Jensen KK, Muysoms F. Robot-assisted abdominal wall surgery: a systematic review of the literature and meta-analysis. Hernia 2019;23:17–27.

[48] Van Den Heuvel BJ, Wijsmuller AR, Fitzgibbons RJ. Indications-treatment options for symptomatic and asymptomatic patients. Hernia surge group. International guidelines for groin hernia management. Hernia 2018;22(1):1–165. https://doi.org/10.1007/s10029-017-1668-x.

[49] Amaral PHF, Pivetta LGA, Dias ERM, Carvalho JPV, Furtado M, Malheiros CA, Roll S. Robotic re-TAPP: a minimally invasive alternative for the failed posterior repair / re-TAPP robótico: uma alternativa minimamente invasiva Para falha da via posterior. Rev Col Bras Cir 2022;49, e20223063. https://doi.org/10.1590/0100-6991e-20223063.

[50] Nguyen T, Kunes K, Crigler C, Ballecer C. Robotic transversus abdominis release for ventral hernia repairs. Int J Abdom Wall Hernia Surg 2022;5(3):103–9.

[51] Reinisch A, Liese J, Padberg W, Ulrich F. Robotic operations in urgent general surgery: a systematic review. J Robot Surg 2023;17:275–90.

# 58

# Robotic surgery in abdominal wall hernias

## A. Betancourt[a], C. Hartmann[b], and E. Parra Davila[b]

[a]General Surgery Department, Good Samaritan Medical Center, West Palm Beach, FL, United States [b]Department of Hernia and Abdominal Wall Reconstruction, Good Samaritan Medical Center-TENET Health, West Palm Beach, FL, United States

## Introduction

Since the introduction of robot-assisted laparoscopic surgery, the number of procedures in which this approach has been adopted has increased, and in recent years also included abdominal wall surgery. While some surgeons have embraced the approach, others remain skeptical as to whether or not its use is justified [1]. Ventral hernias have traditionally been a challenge for the surgeon due to the high rate of recurrence [2]. Using the robot facilitates the dissection of the elements, the defect closing, and the optimal fixation of the mesh.

The robotic repair of ventral hernias was first described in 2002 by GH Ballantyne [3]. Boasting the virtues of improved visualization, tremor-less precision, and superior ergonomics forms the basis for the emergence of robotic techniques in the hernia space.

The robotic platform enables the exploration of the individual layers of the abdominal wall. Virtually, any well-established surgical plane of the abdominal wall can be exploited and dissected for the subsequent placement of mesh in a preperitoneal, retromuscular, and even onlay position, effectively protected from the visceral cavity by the body's autologous tissue. While this approach has been demonstrated with conventional laparoscopy, it remains technically challenging [4].

## Advantages of robotic approach

The robotic approach offers numerous advantages when compared to laparoscopy, including several degrees of motion, three-dimensional (3D) imaging, and superior ergonomics that enable easy and precise intracorporeal suturing. Other reports have demonstrated the ease of intracorporeal suturing of the mesh to the abdominal wall [5]. Thus, this device is an ideal tool for intracorporeal suturing of mesh to the posterior fascia of the anterior abdominal wall for ventral hernia repair.

## Limitations

Conversely, disadvantages may include high financial costs, longer operative times, the need for a table side surgeon, and conflicting evidence that outcomes are improved.

## Preoperative considerations

Obtaining a thorough history and physical is mandatory to coordinate an operative plan. Specifically, co-morbidities such as diabetes, obesity, smoking, and collagen vascular disease may critically affect the operative plan.

A CT scan of the abdomen and pelvis is critical to preoperative planning and remains the gold standard imaging test. This imaging modality can delineate the size and location of the hernia defect, the content of the hernia sac, and possibly the position of previously placed mesh.

A complete medical history along with imaging offers the opportunity for surgeons to construct a risk/benefit ratio. This scale may then be presented to the patients so they could make an informed decision to proceed or not with surgery and what kind of repair fits better for each specific hernia.

# Techniques

Hernia repair techniques amenable to the robotic approach include:

- IPOM Intraperitoneal onlay mesh bridge.
- IPOM Intraperitoneal onlay mesh after primary closure of the defect.
- Preperitoneal placement of mesh.
- Placement of retromuscular mesh with or without posterior components separation.

These individual techniques are chosen based on the location of the hernia defect, size of defect, and perhaps most importantly, surgeon experience. This chapter will provide detailed instructions on each technique along with author insight where applicable.

## Intraperitoneal onlay mesh after primary closure of the defect

### Patient positioning, trocar placement and docking

For the majority of patients with defects in the midline, supine positioning with the arms tucked is preferred, unless trocar access to the lateral abdomen is obscured by this position. In this situation, the arm is placed on a board set at 90° from the trunk. For mid-abdominal hernias, the trocars should be placed at the most extreme lateral, cranial, and caudal positions possible. The most lateral position of the camera and two instrument arms will allow for a full range of motion which facilitates dissection and suturing on the anterior abdominal wall.

Gaining safe intra-abdominal access remains the first step in minimally invasive surgery. This can be made difficult in the multiply operated abdomen. Sites of previous operative intervention will certainly influence the strategy to gain initial access. Optical entry with a 5-mm trocar with or without initial Veress needle insufflation in the left upper quadrant is generally safe. A 12- or 8-mm trocar for the camera is placed as far lateral to the ipsilateral edge of the defect. This in most cases obviates the need to place trocars on the contralateral abdomen when securing the mesh to the ipsilateral abdominal wall. An 8-mm dV trocar is placed in the lower lateral abdomen, and the initial 5-mm optical trocar is then replaced with an 8-mm dV trocar or by the camera trocar (Fig. 1).

Another consideration is the accessory port. The accessory port is used to aid with the mesh introduction and orientation, suture introduction and removal, and suture cutting. We found that using the accessory trocar for the larger mesh introduction under direct visualization was safer and more efficient than introducing the mesh and sutures through the 12-mm camera port. The accessory port is less useful for the repair of smaller ventral hernias, where the orientation of the mesh and the retraction of the mesh for exposure in suture placement are less cumbersome.

The accessory port location must also be determined in relationship to the three da Vinci arms. The optimal position is located opposite the defect between one instrument arm and the camera arm trocar and also at the subxiphoid or suprapubic area and that way may serve both sides if needed. It is crucial to place the accessory port as far from the defect as possible to allow for an increased range of motion and effectiveness (Fig. 2).

Generally, for mid-abdominal hernias a neutral supine position is sufficient. Any patient position manipulation required, however, must be performed before docking the robot. The robotic cart is driven directly over the abdomen and in line with the trocar sites.

### Instrumentation

For right-handed surgeons, a dV prograsp (or fenestrated bipolar) is placed in arm #2, a 12-mm 30° up camera in the camera port, and the dV monopolar scissors are placed in arm #1.

The dV suture cut needle driver is used to primarily close the hernia defect as well as fixate the mesh to the abdominal wall. You can use the fenestrated bipolar instead of the prograsp if the surgeon prefers (Fig. 3).

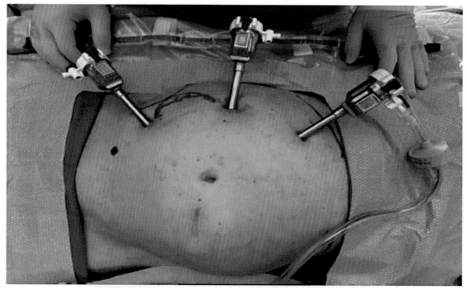

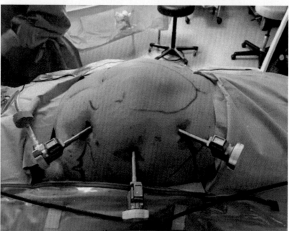

FIG. 1   Trocars setting.

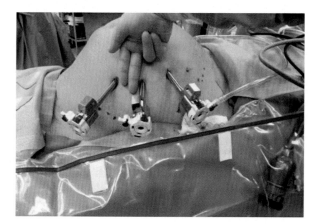

FIG. 2   Robotic arms separation.

## Essential steps

### *Adhesiolysis*

The essential steps of robotic hernia repair are analogous to that of conventional laparoscopic repair. Adhesiolysis of the abdominal wall to isolate the hernia defect must be performed meticulously to avoid iatrogenic injury to the abdominal viscera. The dV platform facilitates adhesiolysis through its 3D visualization, extended range of motion,

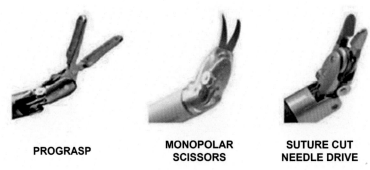

**PROGRASP**          **MONOPOLAR**          **SUTURE CUT**
                      **SCISSORS**            **NEEDLE DRIVE**

FIG. 3    Robotic instrumentation.

tremor-less precision, and superior ergonomics. For direct bowel handling, the dV fenestrated bipolar grasper is less traumatic to bowel serosa. It is important to emphasize the loss of haptic feedback when performing robotic surgery. This shortcoming is overcome by the improved ability to visualize individual stretched fibers. Special attention is therefore required to prevent iatrogenic bowel injury and excessive bleeding by way of atraumatic handling and judicious use of cautery. Complete adhesiolysis is mandatory to insure a complete evaluation of the abdominal wall. If necessary, the falciform ligament is taken down to allow for the flush placement of mesh against the abdominal wall. In the setting of dense adhesions, the robotic harmonic scalpel or dV vessel sealer may facilitate hemostasis.

### Primary closure of the defect

Successful primary closure of the defect is facilitated by the use of the barbed V-loc suture (Covidien). The ability to primarily close defects without component separation is based on the principles of Ramirez regarding the width and location of the hernia defect [6]. Of course, this is based on open technique and not working against the forces of pneumoperitoneum. As a general rule, a less than 10-cm wide defect is amenable to primary closure but also depends on body habitus, age, and abdominal wall compliance. Desufflating the abdominal cavity to 6–8-mmHg pneumoperitoneum may be necessary. The suture is introduced into the intra-abdominal cavity through the 8-mm dV trocar or the accessory port. Opening and bending the needle slightly to facilitate both the introduction and subsequent removal of the suture will accomplish this.

### Mesh placement and fixation

A tissue-separating mesh is utilized when placed in the intraperitoneal position. The size of the mesh upholds the principle of maintaining at least a 5-cm overlap in all directions. For larger defects primarily closed under some degree of tension, a wider mesh is used. Depending on the size of the prosthetic, it can be introduced through the 8-mm dV trocar, camera port, or accessory 10–15-mm port.

There are a myriad of options and permutations of technique to secure the mesh to the abdominal wall, including reproducing the standard LVHRP lease, writing for the first time the meaning of the VLHR abbreviate technique with a combination of tacks and transfascial sutures, or securing the mesh to the abdominal wall with circumferential suture fixation.

With the mesh positioned on the abdominal wall by using a scrolling technique or the self-expanding mesh device (Echo mesh, Bard), a full-length nonabsorbable suture (00 or 0 prolene Ethicon) is introduced into the intra-abdominal cavity through the trocar of the needle holder. The external end of the suture situated outside the trocar is secured with a hemostat. This technique avoids excessive suture in the intra-abdominal cavity thereby facilitating fixation. In a running fashion, the suture is then placed around the circumference of the mesh. It may be necessary to use one or two sutures for larger prosthetics.

Upon completion of mesh fixation, the robot is undocked. Only the 10–12-mm trocar fascial sites are closed with a suture passer under direct laparoscopic vision.

IPOM Intraperitoneal onlay mesh bridge has the same steps described above but without the closure of the defect but is mostly abandoned due to failure rates.

## Robotic TAPP ventral hernia repair

Exploiting the layers of the abdominal wall is made possible by the precision the dV robot affords. While possible to do with conventional laparoscopy, working high on the anterior abdominal wall remains technically demanding and ergonomically challenging. Placing mesh in the preperitoneal space obviates the need for a more costly tissue-separating mesh, allows the mesh to incorporate directly on fascia thereby decreasing the need for sutures or tack fixation which cause postoperative pain, and avoids complications inherent with leaving mesh in the intraperitoneal position, i.e., bowel erosion, fistula, or severe adhesions.

## Reperitonealization of mesh

### Essential steps

Patient positioning, trocar placement, docking, and instrumentation are analogous to the above-described procedure.

#### *Developing a preperitoneal plane*

The peritoneum is incised at least 5-cm proximal to the hernia defect (Fig. 4). A preperitoneal plane is then developed widely in a cephalad to caudad direction with a combination of blunt and sharp technique. Care is taken to avoid disrupting the posterior fascia. In the event the posterior fascia is breached and the rectus muscle is visible, it is subsequently closed with a suture. The hernia sac is reduced, and dissection continues distal to the defect thereby allowing for the placement of an adequately sized mesh. Wide distal dissection allows for the creation of a large flap in which to completely reperitonealize the mesh.

#### *Primary closure of the defect*

The hernia defect is closed with 0 or 1 V-lock running barbed permanent or long-term absorbable suture (Covidien). Desufflation of the abdominal cavity may need to be employed to facilitate the closure of the hernia defect (Fig. 5).

#### *Mesh placement, fixation, and reperitonealization*

The mesh is introduced into the intra-abdominal cavity and placed flat on the abdominal wall. Large overlap of the closed defect (5-cm minimum in all directions) is insured. The mesh is secured to the abdominal wall with four absorbable tacks (AbsorbaTack, COVIDIEN) placed at the cardinal points of the mesh or with sutures as per the surgeon's preference. Once adequate fixation and hemostasis are achieved, the peritoneal flap is re-approximated to cover the mesh with a continuous, absorbable 3–0 V-Loc suture (Ethicon) running suture or tacks (Fig. 6).

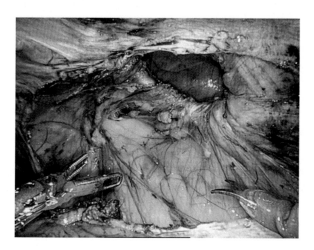

FIG. 4   Peritoneum incised.

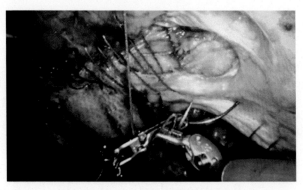

FIG. 5    Primary closure of the defect.

FIG. 6    Mesh placement and reperitonealization.

## Subxiphoid hernias

Traditionally, subxiphoidal hernias have been difficult to repair laparoscopically because of the difficulty in reliably securing the mesh to the lower thoracic outlet. The preperitoneal technique obviates the need for full-thickness transfascial sutures because the mesh is effectively sandwiched between the abdominal wall and peritoneum. The technique itself is analogous to that of the TAPP ventral hernia for mid-abdominal defects involving dissecting a large preperitoneal plane, exposing the defect, primary closure of the defect, mesh placement, and reperitonealization. The falciform ligament and associated peritoneum can be exploited to ensure full coverage of the mesh. If the preperitoneal space is not accessible an IPOM can be easily achieved and the mesh is secured by suturing the mesh to the posterior wall or diaphragm carefully avoiding the cardiac area.

The patient is placed in a supine position with the arm tucked. The strategy again is to place the camera trocar at least 15–20 cm from the caudal aspect of the defect. Depending on body habitus and torso length, an infraumbilical incision for initial access generally works well. Two or three dV 8-mm trocars are placed in line with the 12-mm trocar with at least 6–10 cm of space between trocars. Patient positioning must be completed prior to docking of the robot. The robot is then docked over the right or left shoulder.

# Suprapubic hernias

The challenges of laparoscopic suprapubic hernia repair include the requisite mobilization of the bladder, creating a pelvic dissection within the space of Retzius, and fixating the mesh along the pelvic rim. Robotic preperitoneal repair facilitates the bladder mobilization, visualization of the pelvic rim, and creation of a large preperitoneal space to accommodate overlapping mesh that is especially difficult in the setting of recurrent hernias or in patients with previous open prostatectomy (Fig. 7).

The patient is placed in a supine lithotomy position. A three-way Foley catheter is placed which is used to distend the bladder for proper identification. The patient is positioned in a slight Trendelenburg position.

A 12-mm camera trocar is placed in a supraumbilical location for initial access. The camera port must be at least 15–20 cm from the superior aspect of the hernia defect. Two or three dv 8-mm trocars are placed in line with the camera trocar and the robot is docked in between the legs.

## Essential steps

A preperitoneal plane has incised a minimum of 5-cm cephalad to the superior aspect of the hernia defect. A wide plane of dissection is necessary to accommodate a large sheet of overlapping mesh. The hernia defect is reduced. The superior dome of the bladder may occupy the hernia sac, and therefore, great care and meticulous dissection are performed to avoid bladder injury. This is facilitated by instilling 300 cc of sterile saline into the bladder for easy identification. The retro inguinal space (space of Bogros) is developed bilaterally to expose Cooper's ligament. Caudal mobilization of the bladder reveals the space of Retzius (Fig. 8). This space can be dissected inferiorly to insure adequate overlap of mesh inferior to the caudal aspect of the hernia defect.

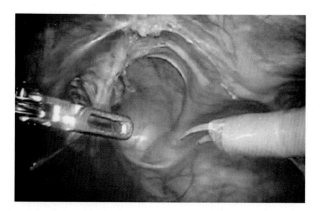

FIG. 7   Suprapubic hernia.

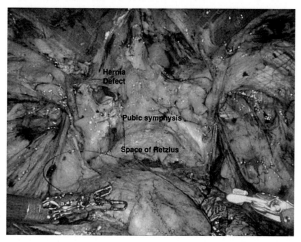

FIG. 8   Dissection on suprapubic space.

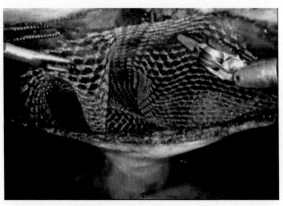

FIG. 9    Sized light or medium weight polypropylene mesh is introduced.

The hernia defect is primarily closed with 0 or 1 V-loc barbed suture (Covidien) as described previously. Partial desufflation of the abdominal cavity may be required to adequately close the defect. The dome of the defect may also be incorporated into the closure to obliterate the dead space, thereby reducing the risk of seroma formation.

An adequately sized light- or medium-weight polypropylene mesh is introduced into the abdominal cavity (Fig. 9). Absorbable tacks or sutures are placed to secure the mesh to the abdominal wall. 00 or 0 prolene suture is used to secure the mesh to Cooper's ligament bilaterally as well as the symphysis pubis. Upon completion of mesh fixation, the mesh is reperitonealized with 3–0 absorbable-barbed suture.

## Robotic Rives repair

The retromuscular hernia repair as described by Rives is considered by many to be the standard by which all hernia repairs are judged [8–10]. The posterior component separation (PCS) technique allows for the closure of large hernia defects with wide prosthetic mesh overlap. These two techniques performed in tandem have traditionally been exclusive to open hernia repair.

The retromuscular repair as described by Rives uses the natural myofascial planes of the abdominal wall while preserving the integrity of the subcutaneous tissue [12]. In this technique, the mesh is secured in the retrorectus position, sandwiched by the closure of the anterior fascia above and by the posterior fascia below. With recurrence rates reported to be in the range of 0%–4%, many consider this technique of open ventral hernia repair as the gold standard for all hernia repairs [8,9]. The limitation of the Rives-Stoppa repair is that the maximal transverse diameter of the mesh is confined to the lateral edge (linea semilunaris) of the rectus muscles.

For the majority of patients with large defects in the midline, supine positioning with the arms tucked is preferred, unless trocar access to the lateral abdomen is obscured. Trocars are placed in the lateral abdomen similar to conventional laparoscopic repair. The optical trocar technique preferably in a location remote to previous surgical intervention is used to gain initial access. An 8–12-mm trocar is placed in the lateral abdomen and then two 8-mm trocars follow on each side of this trocar (Fig. 10).

### Essential steps

#### *Posterior sheath incision*

The anterior abdominal wall is cleared of all adhesions to adequately define and size the hernia defect. The retromuscular space is accessed by incision and subsequent mobilization of the posterior sheath. Below the arcuate line, the peritoneum and transversalis fascia is mobilized in a similar fashion. The degree of cranial-caudal dissection is based on the size of the defect, assuring a minimum of 5-cm overlap (Fig. 11).

#### *Closure of the anterior sheath, mesh placement, and posterior sheath closure*

Closure of the anterior sheath is accomplished utilizing a 0 V-loc (Covidien) suture in a running fashion. The subcutaneous tissue and hernia sac is incorporated into the closure to obliterate the anterior dead space. This step restores the linea alba and the rectus abdominis muscle in its anatomical and physiologic correct position (Fig. 12).

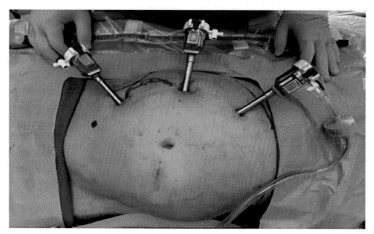

FIG. 10   Patient positioning and trocar placement.

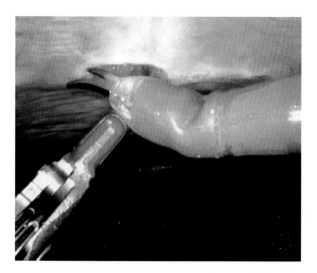

FIG. 11   Posterior sheath incision.

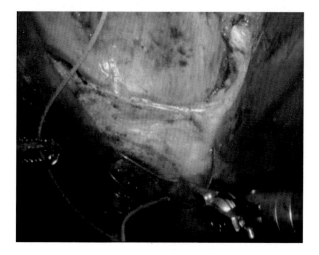

FIG. 12   Closure of defect.

III. The current and future clinical applications of robotic surgery among medical specialties

FIG. 13    Placement of mesh after closure of defect.

FIG. 14    Posterior sheath closure.

The extent of dissection is measured in cranial-caudal and axial dimensions to choose an appropriately sized mesh. A single central transfascial suture in the retromuscular positions light or mid-weight polypropylene mesh. Circumferential fixation is accomplished with an absorbable tacker or suture (Fig. 13).

The posterior sheath is then reapproximated using 0 V-loc suture (Coviden). It is often helpful to incorporate a bite of mesh to elevate the two leaves of the posterior sheath away from the intra-abdominal viscera. The peritoneum is re-approximated below the arcuate line (Fig. 14).

### Drain placement

Secondary to pneumoperitoneum, the retromuscular space represents a large potential space for seroma formation. Trocars are withdrawn from the intraperitoneal cavity into the retrorectus space. In this position, adequate hemostasis can be confirmed and two 19F drains are placed.

## Robotic inguinal hernia repair

For the robotic platform, the transabdominal preperitoneal (TAPP) approach is the preferred method given the space limitations and arm position issues encountered. Many of the recommendations listed below are from the International Endohernia Society (IEHS) guidelines regarding TAPP and personal experience with the robotic approach.

# Preoperative evaluation

## Selection

If the clinician is unsure whether a hernia exists or an additional contralateral hernia is present, an ultrasound examination with valsalva or a noncontrast CT of the pelvis can be informed. Likewise, a TAPP approach allows for a rapid assessment of the contralateral side. The patient should be counseled preoperatively on the possibility of bilateral hernia repair.

## Contraindications

- Inability to tolerate pneumoperitoneum.
- Ascites.
- Gross contamination.
- Previous lower abdominal surgery (relative).

# Robotic TAPP for inguinal hernia repair

## Positioning

Place the patient supine on the operating table. Tuck both arms.

Prior to docking the robotic arms, place the patient in a 15-degree Trendelenburg. If it is a unilateral hernia, you can tilt the bed 15 degrees away from the side of the hernia.

## Perioperative management

Have the patient void immediately prior to entering the operating room to avoid the need for a Foley catheter. Foley can be used alternatively, particularly with patients who may have sliding hernias or advanced age with incomplete voiding. If difficult dissection is anticipated or a hernia has a scrotal component, a Foley catheter can be placed. It is important to have a decompressed bladder during the case to avoid the risk of bladder injury. Restricting the *peri* and postoperative quantity of intravenous fluids can reduce the need for Foley catheter and reduce postoperative urinary retention. Ideally, this would be kept to <500 cc IVF total.

# Trocar placement and docking

Initial entry is using a Veress needle in the left upper quadrant (Palmer's point). In our practice, we use three 8-mm trocars in the upper, epigastric, right, and left upper quadrant. An extra fourth trocar can be used in case of a large inguinoscrotal hernia (Fig. 15).

Prior to docking the robot place the patient in 15-degree Trendelenburg position. We routinely use the Xi platform. Instruments are placed under direct vision. We begin the case with a force bipolar grasper in the left port and a scissor with monopolar cautery on the right.

## Essential steps

### Opening of peritoneum

#### Dissection/creation of the peritoneal flap

*Mesh insertion*   Mesh is introduced through an 8-mm trocar. Medium polypropylene mesh (Bard © 3Dmax) is our preference. The shape is helpful with orientation, and placement, and reduces the need for anchoring. Large ($10.8 \times 16$ cm) is generally utilized. X-Large ($12.4 \times 17 \times 3$ cm) may be preferable with a large hernia defect or in large patients.

*Mesh fixation*   An advantage to the robotic platform is the ability to place sutures (rather than utilizing a tacking device). We generally place three 2–0 Vicryl simple air knots sutures medially in the rectus muscle, 1 laterally to the epigastric vessels, into the transversus musculature, and 1 at the Coopers ligament.

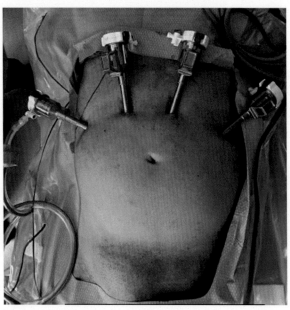

FIG. 15    Port placement for large inguinoscrotal hernia.

*Peritoneal closure*    An advantage of the robotic over laparoscopy is the ease of closing with a running suture. The closure is performed with an absorbable, barbed suture.

## Cost considerations during robotic inguinal hernia repair

There are several advantages that the robotic platform allows over pure laparoscopic that can decrease cost below even laparoscopic approaches in some situations.

— Specific instrument selection minimizes the instruments that are opened. For a standard hernia repair, there shouldn't be a need for anything other than a Cadiere or fenestrated bipolar, scissors, and needle driver.
— Do not use a tacker device. If choosing to secure the mesh use 3–0 Vicryl (or equivalent) and for peritoneal closure use absorbable barbed suture (V-loc at our institution). In some circumstances, securing the mesh may not be necessary.
— Not using the balloon dissector as commonly used for TEP leads to lower overall costs.
— Robotic TAPP allows for reusable trocars whereas many practices use disposable ones for the laparoscopic approach. A Hassan may still be used at the umbilicus.

## Future scenario

Robotic surgical systems are evolving to include specific features and improvements of the bedside cart and effector's arms (*lightweight, smaller size, mounted on operating table or on separate carts, single arm with a variety of instruments inside*), instruments (*tactile feedback, micro- motors*), console (*open, closed, semi-open*) or without a console, and 3D HD video technology (*polarized glasses, oculars, mirror technology*). Currently, there are several companies on the quest for better robotic devices for abdominal surgeries, here are some of the names:

— Intuitive, Google/ Johnson&Johnson,Medtronic,Medrobotics,CMR,Vicarious surgical Transenetirx,Verb Surgical, Virtual incision,Memic Med,Accuray,EndoControl,Freehand,Endomaster,Titan Medical,Meere Co.,Avatera Revo-I and Raven.

## Digitalization of robotic surgery

Innovation in robotic surgery will continue to parallel advancements in technology; especially with the considerable progress in computer science and artificial intelligence (AI). Novel distinct features, such as haptic gloves, cellular image guidance, or even autonomy might be the next step in the evolution of next-generation devices. Shademan et al. have described in vivo supervised autonomous soft tissue surgery in an open surgical setting, enabled by a plenoptic 3D and near-infrared fluorescent imaging system that supports an autonomous suturing algorithm. A computer program generates a plan to complete complex surgical tasks on deformable soft tissue, such as suturing an intestinal anastomosis based on expert human surgical practices [13].

## Key points

- Robotic abdominal wall and hernia repair procedures have been performed at increasing rate in both the United States and Europe.
- The range of motion of the robotic instruments and the improvement of surgeons' ergonomics has allowed for high-performance suturing of the abdominal wall making it possible to perform component separation in a minimal invasive manner.
- High financial costs, longer operative times, the need for a table side surgeon, are disadvantages of robotic abdominal hernia repairs.
- Novel surgical modalities with new technology, perhaps will decrease costs, and smaller size devices will become available.

## References

[1] Telem DA. Is robotic surgery the future for abdominal wall hernia repair? Not so fast. Ann Surg 2018;267(2):218–9. https://doi.org/10.1097/SLA.0000000000002336.
[2] Novitsky YW, Elliott HL, Orenstein SB, et al. Transversus abdominis muscle release: a novel approach to posterior component separation during complex abdominal wall reconstruction. Am J Surg 2012;204:709–16.
[3] Ballantyne GH. Robotic surgery, telerobotic surgery, telepresence, and telementoring: review of early clinical results. Surg Endosc 2002;16:1389–402.
[4] Prasad P, Tantia O, Patle NM, Khanna S, Sen B. Laparoscopic transabdominal preperitoneal repair of ventral hernia: a step towards physiological repair. Indian J Surg 2011;73:403–8.
[5] Ballantyne GH, Hourmont K, Wasielewski A. Telerobotic laparoscopic repair of incisional ventral hernias using intraperitoneal prosthetic mesh. JSLS 2003;7:7–14.
[6] Ramirez OM, Ruas E, Dellon AL. "Components separation" method for closure of abdominal-wall defects: An anatomic and clinical study. Plast Reconstr Surg 1990;86:519–26.
[8] Jin J, Rosen MJ. Laparoscopic versus open ventral hernia repair. Surg Clin North Am 2008;88:1083–100.
[9] Novitsky YW, Elliott HL, Orenstein SB, et al. Transversus abdominis muscle release: a novel approach to posterior component separation during complex abdominal wall reconstruction. Am J Surg 2012;204:709–16.
[10] Rives J, Pire JC, Flament JB, et al. Treatment of large eventrations. New therapeutic Indications apropos of 322 cases. Chirurgie 1985;111:215–25.
[12] Heniford BT, Park A, Ramshaw BJ, et al. Laparoscopic ventral and incisional hernia repair in 407 patients. J Am Coll Surg 2000;190:645–50.
[13] Cha J, Shademan A, Le HN, Decker R, et al. Multispectral tissue characterization for intestinal anastomosis optimization. J Biomed Opt 2015;20, 106001.

# 59

# Colorectal robotic surgery

*Samuel Aguiar, Jr.*[a] *and Tomas Mansur Duarte de Miranda Marques*[b]

[a]Colorectal Cancer Reference Center, A.C. Camargo Cancer Center, São Paulo, SP, Brazil [b]Department of Colorectal Surgery, A.C. Camargo Cancer Center, São Paulo, Brazil

## Introduction

Colorectal surgery, as we know it today, began in the eighteenth century with a significant evolution in the approach to intestinal diseases. It transitioned from small procedures for trauma treatment to more complex surgeries involving resections and intestinal diversions. Advances in the knowledge of human anatomy, improvement of antiseptic techniques, and the development of anesthesia were major milestones in this evolution and the success of surgical procedures [1].

The first colon resection on record dates to 1823 when Reybard performed a sigmoid tumor resection. In the late nineteenth and early twentieth centuries, there were significant advancements in colorectal surgical techniques, such as the Miles procedure described in 1908. In the 1940s, anterior resection of the rectum with colorectal anastomosis and organ preservation gained prominence. Later, in the 1980s, the concept of total mesorectal excision (TME) was introduced by surgeon Richard Heald, making a significant impact [1,2].

Minimally invasive colorectal surgery took its first steps in the 1990s when a series of 20 colon cases operated by laparoscopy were published in 1991 [3]. Since then, there has been a significant advancement in technologies and surgical techniques that aim to maintain the surgical and oncological principles of open surgery while improving aesthetic results, surgical recovery, and early return to daily routines. The main advantages of minimally invasive surgery include reduced blood loss, smaller incisions, improved intestinal function, earlier hospital discharge, and a lower inflammatory response. However, one of the main limitations of laparoscopic surgery is the restricted movement of the surgical instruments and visualization in small cavities, such as the male pelvis. Robotic surgery emerged as an alternative to overcome these limitations, offering enhanced mobility and freedom of movement, impressive tremor-free precision, and immersive 3D vision. Robotic surgery has been gaining increasing prominence in colorectal surgery and has a shorter learning curve compared to laparoscopy [4,5].

The first reported robotic colorectal surgery was performed in 2002, a robotic-assisted sigmoidectomy for diverticular disease. The first robotic oncological surgeries were performed in 2003 in Japan, including right colectomy, left colectomy, and sigmoidectomy. Since then, robotic surgery has become an integral part of colorectal surgery modalities, but its safe application and benefits for cancer treatment needed to be evaluated [6,7].

Colorectal cancer ranks third in global incidence and is the third most common cancer in men and the second most common in women. Although there has been a decline in incidence among the elderly, its occurrence is increasing among people younger than 50. In terms of mortality, colorectal cancer ranks second globally, according to the World Health Organization (WHO). Approximately 73% of diagnoses occur without the presence of metastatic lesions. Despite the range of therapeutic possibilities for colorectal tumors, surgery remains one of the few alternatives that can achieve a curative outcome. In recent years, several studies have been conducted to evaluate the safety of robotic surgery in the treatment of colon and rectal cancer, as well as to compare its outcomes with other modalities [8].

Currently, it can be stated that robotic surgery is safe and non-inferior for the treatment of colon cancer. In rectal cancer, the superiority of robotic surgery over other modalities has been confirmed in terms of short-term outcomes such as quality of resection and better postoperative recovery.

## Colon robotic surgery

Minimally invasive approaches account for approximately 50% of colon surgeries performed. The technique is well-established and safe, demonstrating superiority in terms of hospital stay, time to resumption of regular diet, and pain compared to open surgery. In addition to demonstrating non-inferiority in the treatment of colon cancer, robotic surgery offers an alternative to conventional laparoscopy by adding superior robot characteristics. A meta-analysis of 9 studies involving over 4000 patients showed that in left colectomy for diverticular disease, the conversion rate was lower in robotic surgery compared to laparoscopy (7.4%, vs 12.5% $P < 0.001$), with shorter hospital stays, but at the cost of longer surgical time [9]. Another study evaluating the role of robotic surgery in left colectomy also demonstrated lower rates of anastomotic fistula, surgical site infection, and general complications when the procedure was performed by robot assisted [10].

One of the main advantages of robotics over laparoscopy is the greater range of motion, flexibility, and precision of movement and surgical vision. The application of these characteristics in colectomy surgery increases the accuracy of intracorporeal anastomosis. A meta-analysis of 37 studies involving over 23,000 patients was conducted to compare robotic and laparoscopic right colectomy, as well as the type of anastomosis (intracorporeal or extracorporeal). This study identified that, overall, robotic surgery had shorter hospital stays, lower conversion rates, faster bowel function recovery, and lower complication rates. However, in cases of extracorporeal anastomosis, there was no difference in hospital stay duration. Perhaps the greatest benefit of robotic surgery over laparoscopy is the increased performance of intracavitary anastomosis [11]. Zhang et al. analyzed that incision is smaller ($P < 0.000001$) and confirmed the relationship between robotic intracorporeal anastomosis and a lower rate of adynamic ileus [12].

Complete mesocolic excision (CME) is still a controversial topic in the literature. It involves the resection of the colon while respecting the entire embryological component, including central vessel ligation with lymphadenectomy extending to the superior mesenteric vein up to the Henle trunk. This approach increases the median number of resected lymph nodes and, in the long term, achieves better overall survival and disease-free survival rates [13]. The performance of robotic CME has been shown to be safe and effective, yielding better surgical outcomes with more resected lymph nodes and lower conversion rates [14].

## Rectal robotic surgery

Rectal surgery is always a major challenge for surgeons. The narrow anatomy of the pelvis and the intimate relationship between organs pose an additional difficulty to this procedure. In the 1980s, Heald established what he called the Holy plane, which is a precise anatomical plane between the mesorectal fascia and presacral fascia, avascular and crucial for complete mesorectal excision with quality. In addition to its oncological significance, preserving this plane protects the hypogastric plexus and, thus, preserves urinary and sexual function. The male pelvis is even more challenging, especially in obese patients, as it is narrower, making visualization of planes and structures more difficult [2].

Due to the characteristics of the pelvis, robotic surgery has emerged as a promising approach for performing these procedures. Its 3D vision, combined with greater depth perception and camera stability, allows for a more precise and comprehensive view of the small pelvis. Another notable feature is the wide mobility of the robotic instrument in small spaces, enabling precise maneuvers and dissections. Because of these qualities, robotic-assisted pelvic surgery has gained strength and prominence in recent years.

In the early 2000s, Weber et al. described the first experience of robotic colorectal surgery, performing a sigmoidectomy and right colectomy for diverticular disease, demonstrating the feasibility of the procedure. From that point on, numerous studies were conducted to assess the benefits of robotic surgery compared to laparoscopic and conventional approaches. The initial studies demonstrated that robotic surgery was feasible and safe for the treatment of both benign and malignant rectal diseases. Subsequently, several other studies were conducted to directly compare the different surgical approaches and their short-term surgical outcomes. Conversion rate, blood loss, mortality, anastomotic fistula, and the quality of surgery assessed by margin evaluation, number of resected lymph nodes, and functional preservation were evaluated and compared mainly with laparoscopic surgery. A systematic review with meta-analysis conducted by Wang et al., which included over 5000 patients, demonstrated that in robotic surgery, the conversion rate to open surgery was lower (OR 0.55, 95% CI 0.44–0.69), hospital stay was shorter (OR −0.15, 95% CI −0.30 to 0.00), intestinal function recovery was faster (OR −0.38, 95% CI −0.74 to 0.02), and the incidence of postoperative complications was lower (OR 0.79, 95% CI 0.65–0.97) [15]. Another study conducted by Ng, with over 160,000 patients, identified less intraoperative blood loss ($q = 0.01$, I2 = 88%; REM: MD −18.05, 95% CI −32.24 to −3.85; mL) and fewer wound infections ($P < 0.001$, $I_2 = 0$%; FEM: OR 1.24, 95% CI 1.11–1.39) [16].

Regarding the preservation of urinary and sexual function, robotic surgery also showed higher preservation rates. On the other hand, there was no difference between the surgical approaches in terms of mortality and anastomotic fistula. Due to the difficulty of accessing the pelvic space, the conversion rate from laparoscopic surgery to open surgery is approximately 10.1%. A multicenter randomized study with 471 patients compared the conversion rate between robotic and laparoscopic surgery for rectal cancer treatment. Although the conversion rate was lower for robotic surgery (8.1% vs 12.2%), the study did not reach statistical significance ($P = 0.16$). However, subgroup analysis showed a higher conversion rate in male patients ($P = 0.04$) and obese patients ($P = 0.001$) [17].

In oncological surgery, the use of robotics has been demonstrating better oncological outcomes in terms of the quality of surgery [15]. In rectal cancer, the involvement of lymph nodes and tumor extension beyond the mesorectal fascia increases the risk of incomplete resection, compromised margins, local recurrence, and worse oncological prognosis. Therefore, surgical manipulation must always be performed with great care to avoid tumor rupture, compromised margins, and incomplete mesorectal resection. Laparoscopic surgery has never been able to demonstrate superiority in terms of the quality of resection compared to open surgery [18–20]. A prospective randomized study evaluated the short-term outcomes of robotic surgery compared to laparoscopic surgery for rectal cancer treatment. Resection with a clear circumferential margin was superior in robotic-assisted surgeries (4% vs 7.2%, $P = 0.023$). The circumferential margin is a risk factor for local recurrence and worse survival in rectal cancer. The fact that robotic surgery reduces the risk of margin involvement may indicate better oncological outcomes, but long-term results are still awaited [21]. Preserving the hypogastric plexus is crucial for adequate preservation of genitourinary function. A meta-analysis that evaluated 1286 patients from 10 studies showed that urinary function in men was significantly better in the robotic surgery group compared to the laparoscopic surgery group at 6 months (mean difference [MD] $-1.36$, 95% CI $-2.31$ to $-0.40$; $P = 0.005$) and at 12 months (MD $-1.08$, 95% CI $-1.85$ to $-0.30$; $P = 0.007$)[22]. Regarding annal sphincter function, a meta-analysis of 32 publications, including 5565 patients, showed that anorectal function recovered significantly better 1 year after robotic TME (3.8 [95% CI $-9.709$ to 17.309]) compared to laparoscopic TME (26.4 [95% CI 19.524 to 33.286], $P = 0.006$), open TME (26.0 [95% CI 24.338 to 29.702], $P = 0.002$), and transanal TME (27.9 [95% CI 22.127 to 33.669], $P = 0.003$) [23].

Another advantage of the robot is the shorter learning curve. The possibility of using a robotic simulator with predefined exercises improve the learning curve. Additionally, surgeons already bring experience from laparoscopy when they start performing robotic surgeries. Some studies demonstrate that the learning phase ranges from 20 to 40 cases, and consolidation occurs after approximately 40–50 cases. Right colectomy via robotic approach has been used as a procedure for the learning curve, as it encompasses all the steps of robot handling and its functions, besides being less complex compared to rectal surgeries [24].

On the other hand, the major disadvantage of robotic surgery is the cost. The implementation and use of robotic technology are expensive compared to other techniques. Despite many studies demonstrating shorter hospital stays, the cost remains higher. The dissemination of technology and the opening of the market to new companies and models will allow for a cost reduction. Another negative point worth mentioning is the surgical time. Robotic procedures require patient preparation, docking time, and the procedure itself takes longer than conventional laparoscopic approach. However, with increased experience, the time required for these steps decreases. Furthermore, the gains achieved through robotic approaches outweigh these disadvantages [25].

## Surgery technique (Da Vinci system, S/X and Xi platforms)

Robotic surgery is a minimally invasive approach. That is, small incisions are made through which ports (trocars) are inserted for the passage of instruments and a camera used for the surgical procedure. Carbon dioxide gas is used to create a pneumoperitoneum, creating an intracavitary space for surgery.

Incisions of 8 mm are made for the passage of ports, distributed with a spacing of 5–10 cm between them, depending on the Si/X or Xi system. Additionally, a 12 mm trocar is inserted as a port for the camera (Si/X platform), and another one is used for the assistant. Generally, in the Si/X series, all three arms of the robot are used with four ports, so that arm two initiates the surgery in a position for the release of the splenic flexure of the colon and mobilization of the descending colon to the pelvis. Subsequently, this arm is switched to the lower position for manipulation of the pelvic area. In right colectomy surgery, contralateral passage of trocars is adopted. In this case, only two arms of the robot are used. A 12 mm port is also inserted for the camera, and another one for the assistant. In the Xi platform, all arms are set in a linear formation and the camera can be used in any position on both sides. A 12 mm port is inserted for the assistant (Fig. 1).

FIG. 1    (A) Ports placement with SI/X system (*green*) to a TME surgery. In this technique three arms and four ports are used. When operating on the splenic flexure of the colon, the cranial port is utilized, and to operate on the pelvis, arm 2 is switched to the lower port. (B) Ports placement with Xi system (green) to a TME surgery, in a linear formation and one assistant port (*yellow*).

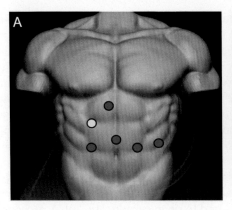
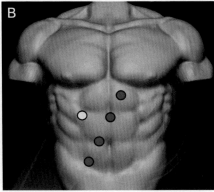

## Future scenario

The main barrier for the dissemination of robotic colorectal surgery is still the costs, especially in developing countries. The use of robotic platforms for colon surgery, for example, is far to be cost-effective. However, new players from the industry are already on market, and others are coming. The diversity among different and concurrent robotic platforms will naturally decrease costs, and the use of robotic surgery even in colon surgery seems to be unreversible.

On the technology view, two advances are expected. One is the incorporation of robotic arms in endoscopic devices. This advance can allow the resection of huge adenomas endoscopically, as well as turn the trans-anal rectal surgery much easier and safer. The other expected advance is the incorporation of image navigation on the robotic surgical platform, which can improve the quality of rectal cancer surgery, by improving radial margins in more advanced cases, or turning intersphincteric dissections more precise, with better functional outcomes.

## Key points

- Robotic surgery can already be considered standard of care for total mesorectal excision among men, overweighted patients, or low rectal tumors, especially for short-term outcomes.
- For long-term outcomes, data are still expected for demonstrating superiority of robotic over laparoscopic surgery.
- For colon cancer surgeries, the use of robots is not cost-effective yet, but it will probably change soon.

## References

[1] Tebala GD. History of colorectal surgery: A comprehensive historical review from the ancient Egyptians to the surgical robot. Int J Colorectal Dis 2015;30:723–48. Springer Verlag.

[2] Heald RJ. The "holy plane" of rectal surgery. J R Soc Med 1988;81(9):503–8.

[3] Jacobs M, Verdeja JC, Goldstein HS. Minimally invasive colon resection (laparoscopic colectomy). Surg Laparosc Endosc 1991;1(3):144–50.

[4] Yamauchi S, Matsuyama T, Tokunaga M, Kinugasa Y. Minimally invasive surgery for colorectal cancer. JMA J 2021;4:1.

[5] Sivathondan PC, Jayne DG. The role of robotics in colorectal surgery. Ann R Coll Surg Engl 2018;100:42–53.

[6] Weber PA, Merola S, Wasielewski A, Ballantyne GH. Telerobotic-assisted laparoscopic right and sigmoid colectomies for benign disease. Dis Colon Rectum 2002;45(12):1689–94. discussion 1695–6.

[7] Hashizume M, Shimada M, Tomikawa M, Ikeda Y, Takahashi I, Abe R, et al. Early experiences of endoscopic procedures in general surgery assisted by a computer-enhanced surgical system. Surg Endosc 2002;16(8):1187–91.

[8] Siegel RL, Miller KD, Wagle NS, Jemal A. Cancer statistics, 2023. CA Cancer J Clin 2023;73(1):17–48.

[9] Giuliani G, Guerra F, Coletta D, Giuliani A, Salvischiani L, Tribuzi A, et al. Robotic versus conventional laparoscopic technique for the treatment of left-sided colonic diverticular disease: a systematic review with meta-analysis. Int J Colorectal Dis 2022;37(1):101–9.

[10] Solaini L, Bocchino A, Avanzolini A, Annunziata D, Cavaliere D, Ercolani G. Robotic versus laparoscopic left colectomy: a systematic review and meta-analysis. Int J Colorectal Dis 2022;37(7):1497–507.

[11] Genova P, Pantuso G, Cipolla C, Latteri MA, Abdalla S, Paquet JC, et al. Laparoscopic versus robotic right colectomy with extra-corporeal or intra-corporeal anastomosis: a systematic review and meta-analysis. Langenbecks Arch Surg 2021;406(5):1317–39.

[12] Zhang T, Sun Y, Mao W. Meta-analysis of randomized controlled trials comparing intracorporeal versus extracorporeal anastomosis in minimally invasive right hemicolectomy: upgrading the level of evidence. Int J Colorectal Dis 2023;38(1):147.

[13] Díaz-Vico T, Fernández-Hevia M, Suárez-Sánchez A, García-Gutiérrez C, Mihic-Góngora L, Fernández-Martínez D, et al. Complete Mesocolic excision and D3 lymphadenectomy versus conventional colectomy for Colon Cancer: a systematic review and meta-analysis. Ann Surg Oncol 2021;28(13):8823–37. https://doi.org/10.1245/s10434-021-10186-9.

[14] Oweira H, Reissfelder C, Elhadedy H, Rahbari N, Mehrabi A, Fattal W, et al. Robotic colectomy with CME versus laparoscopic colon resection with or without CME for colon cancer: a systematic review and meta-analysis. Ann R Coll Surg Engl 2022;105(2):113–25. https://doi.org/10.1308/rcsann.2022.0051.

[15] Wang X, Cao G, Mao W, Lao W, He C. Robot-assisted versus laparoscopic surgery for rectal cancer: a systematic review and meta-analysis. J Cancer Res Ther 2020;16(5):979–89.

[16] Ng KT, Tsia AKV, Chong VYL. Robotic versus conventional laparoscopic surgery for colorectal cancer: a systematic review and Meta-analysis with trial sequential analysis. World J Surg 2019;43(4):1146–61.

[17] Jayne D, Pigazzi A, Marshall H, Croft J, Corrigan N, Copeland J, et al. Effect of robotic-assisted vs conventional laparoscopic surgery on risk of conversion to open laparotomy among patients undergoing resection for rectal cancer the rolarr randomized clinical trial. JAMA 2017;318(16):1569–80.

[18] van der Pas MH, Haglind E, Cuesta MA, Fürst A, Lacy AM, Hop WC, et al. Laparoscopic versus open surgery for rectal cancer (COLOR II): short-term outcomes of a randomised, phase 3 trial. Lancet Oncol 2013;14(3):210–8.

[19] Stevenson ARL, Solomon MJ, Lumley JW, Hewett P, Clouston AD, Gebski VJ, et al. Effect of laparoscopic-assisted resection vs open resection on pathological outcomes in rectal Cancer: the ALaCaRT randomized clinical trial. JAMA 2015;314(13):1356–63.

[20] Fleshman J, Branda M, Sargent DJ, Boller AM, George V, Abbas M, et al. Effect of laparoscopic-assisted resection vs open resection of stage II or III rectal Cancer on pathologic outcomes: the ACOSOG Z6051 randomized clinical trial. JAMA 2015;314(13):1346–55.

[21] Feng Q, Yuan W, Li T, Tang B, Jia B, Zhou Y, et al. Robotic versus laparoscopic surgery for middle and low rectal cancer (REAL): short-term outcomes of a multicentre randomised controlled trial. Lancet Gastroenterol Hepatol 2022;7(11):991–1004.

[22] Fleming CA, Cullinane C, Lynch N, Killeen S, Coffey JC, Peirce CB. Urogenital function following robotic and laparoscopic rectal cancer surgery: meta-analysis. Br J Surg 2021;108(2):128–37.

[23] Grass JK, Chen CC, Melling N, Lingala B, Kemper M, Scognamiglio P, et al. Robotic rectal resection preserves anorectal function: systematic review and meta-analysis. Int J Med Robot 2021;17(6), e2329.

[24] Raimondi P, Marchegiani F, Cieri M, Cichella A, Cotellese R, Innocenti P. Is right colectomy a complete learning procedure for a robotic surgical program? J Robot Surg 2018;12(1):147–55.

[25] Morelli L, Guadagni S, Lorenzoni V, Di Franco G, Cobuccio L, Palmeri M, et al. Robot-assisted versus laparoscopic rectal resection for cancer in a single surgeon's experience: a cost analysis covering the initial 50 robotic cases with the da Vinci Si. Int J Colorectal Dis 2016;31(9):1639–48.

# 60

# Robotics in gynecologic surgery

## Glauco Baiocchi[a] and Mario M. Leitao, Jr.[b]

[a]Department of Gynecologic Oncology, AC Camargo Cancer Center, São Paulo, Brazil [b]Gynecology Service, Department of Surgery, Memorial Sloan-Kettering Cancer Center and Department of Obstetrics and Gynecology, Weill Cornell Medical College New York, New York, NY, United States

## Introduction

Minimally invasive surgery (MIS) has been established as an essential tool in surgeons´ armamentarium. The major advantages are reduced postoperative pain, fewer infection events, less blood loss, less surgical adhesion, fewer hospital stay length, and faster return to work and normal activities [1]. MIS has been performed with standard laparoscopic instruments that have limited range of motions. The uptake of conventional laparoscopy has been challenging and usually a time-consuming task [2], mainly due to the limitations of imaging systems that render only 2-dimensional images that negatively impact movement precision and depth perception.

Additionally, the instruments are counterintuitive and, as mentioned, have restricted degrees of movement. The advent of robotics-assisted surgical platforms should be considered as a technological evolution in laparoscopy, which can mitigate or eliminate some of the conventional laparoscopic disadvantages [3,4]. It is important to note that these "robotic" platforms are truly not "robotic" in the sense that the current platforms do not perform automated procedures. Computer-assisted surgical platform is likely a more appropriate way to view these systems. However, we will refer to "robotic" for ease throughout this chapter.

Robotic surgery is well known to overcome many of the limitations of standard laparoscopic instrumentation. It also tends to mimic many open surgery characteristics. The most widely used platform is the da Vinci Surgical System (Intuitive Surgical Inc., Sunnyvale, CA) that received approval from the United States Food and Drug Administration (FDA) for gynecologic surgery in 2005. Some advantages are three-dimensional high-definition imaging, magnification of up to 10 times, surgeon's hand movements can be scaled, physiological tremor reduction, the instruments provide seven degrees of freedom, and an ergonomically surgeon sitting position [5]. The learning curve for robotic surgery has been noted to be less steep compared to laparoscopy, yet some disadvantages should be noted. The surgeon still needs extensive training and a dedicated theater (operating room) team that is familiar with the device setup and solutions for problems [3–5]. It is important to have a dedicated team, especially, when first introducing robotics platforms and in institutions with lower case volumes. However, "dedicated" team structures will evolve as programs grow and become more efficient with higher volumes.

Other putative disadvantages are that trocars are often placed higher with aesthetical impact, and less direct access to patients with concern so that the time to convert to laparotomy may be delayed in case of an emergency. Trocar placement is not set and can be modified and reduced in number for certain procedures as surgeons gain experience. A further disadvantage is the absence of tactile feedback, however it may be solved after extensive training and partly compensated by the three-dimensional vision system [5–7].

## Learning curve

Assessing the learning of surgical procedures and interventions can be challenging and methodologies vary. Learning curve may be objectively measured as procedure-based (operating time, hospital stay length) and/or as patient-based (blood loss, complications). A dedicated theater team plays a key factor in reducing time and costs. Therefore, the whole team, including surgeons, anesthesiologists, nurses, and technicians must receive a focused training. Generally, three time points are usually measured when assessing operative times. First, the setup time (robotic preparation), second the docking time (positioning and installing of the robotic for surgery), and third the console time (surgeon's robotic procedure) [4].

The procedure-based learning curve improvement is certainly faster for a committed group in a high-volume setting, where 20–50 cases may be needed for robotic assembly as timing drops from 45 to 35 min, and 50 robotic hysterectomy cases are needed for operating time improvement at 95 min [8].

Others have reported various cutoffs for learning curves with the operative time improving after 20 cases for hysterectomies and myomectomies (benign conditions) [9], and 20 cases for endometrial cancer staging (hysterectomy and lymphadenectomy) [10]. Lim et al. suggested a shorter learning curve for robotic surgery compared to conventional laparoscopy (24 vs 49 cases) for hysterectomy and lymph node dissection in endometrial cancer [11].

It is suggested that skills developed after repetition and previous experience in conventional laparoscopy may play an important role for faster expertise in robotic complex surgeries [4,12]. While some controversies still exist, technological advancements will continue to impact surgical skill development and robotics−/computer-assisted instruments certainly promote more efficient learning and higher adoption of MIS compared to conventional laparoscopic or vaginal techniques.

## Costs

In 2017, gynecology was ranked the highest among specialties in robotic procedure volume with 252,000 performed cases in the United States [13]. A constant discussion regarding robotic surgery is about the costs of the platforms and instruments, especially, when compared to conventional laparoscopy and open surgery. Cost-effectiveness analyses are challenging to conduct properly. It is critical that the correct endpoints and impacts are analyzed when assessing costs. In regard to MIS, these may include direct hospital costs (including or not the robotic device acquisition amortization) and societal costs (including lost wages, employer costs) [14–17]. Many analyses focus on the costs of the immediate perioperative period comparing robotics assisted cases to nonrobotic cases.

Multiple early studies reported an overall significant increase of costs and operating room times for robotics-assisted surgery when compared to conventional laparoscopy [14,15,18,19]. A large database study ($n = 36,188$) reported that robotic hysterectomy has a higher cost compared to laparoscopic hysterectomy for both inpatient (US\$9640 vs US\$6973 in 2007 to 2008; $P < 0.01$) and outpatient settings (US\$7920 vs \$5949; $P < 0.01$) [20]. It is important to note that these early studies were confounded by the early learning curves of surgeons using the robotics platforms, which will lead to longer operative times and possible costs. Also, none of these studies assessed how incorporating the robotics platforms impacted overall MIS rates and therefore possibly having implications for cost assessments.

A cost-comparison among laparoscopic, robotic, and open hysterectomy for endometrial cancer found societal costs of US\$10,128, US\$11,476, and US\$12,847, respectively, being conventional laparoscopy the least expensive regarding to hospital costs, independently of the initial investment [14]. Additionally, a study of a high-volume cancer center suggested an increase in cost of US\$3157 higher for robotic surgery compared to conventional laparoscopy (US\$23,646 vs US\$20,489; $P < 0.01$) if the amortized costs of the capital purchase was included [16]. However, the non-amortized costs were the same for robotics-assisted and standard laparoscopy in the same study. This means that using the robotics platform once a robotics platform has been purchased and is available does not lead to continued increase in cost. Another study of the same center, regarding morbidly obese patients demonstrated a large cost savings by incorporating the robotics platform in its practice [21]. The rate of MIS in this morbidly obese cohort went from 6% to 57% with the introduction of the robotics platforms. This led to an overall reduction in length of stays and complications. There was a nearly \$9000 cost reduction per patient across the cohort at 6 months. Therefore, the costs of robotics need to assess how MIS rates are impacted and also look beyond the immediate perioperative event.

Despite costs being an important argument against robotics, the use of laparoscopy for hysterectomy in the United States before the era of robotics was poor. Of 518,828 hysterectomies performed in 2015, with open approach, vaginal techniques, and laparoscopy being performed in 64%, 22%, and 14% of cases, respectively [22]. The adoption of robotics-assisted surgery compared to conventional laparoscopy directly correlates to decrease in laparotomy rates, mainly for complex cases and in institutions with low previous expertise in laparoscopy [23]. For endometrial cancer, an ACS-NSQIP evaluation reported a trend of MIS increase over the study years from 16.7% in 2006 to 48% in 2010, and every 10% of MIS increase leads to 41 less postoperative complications and 600 days/year of hospitalization [24].

In summary, the decrease in laparotomy rate and MIS increase after robotics will likely neutralize the costs of robotics platforms and instrumentation and in many cohorts actually lead to a cost reduction overall [16,25]. Despite the difference in perioperative costs between conventional laparoscopy and robotics, the overall enhancement in MIS rates may finally contribute to major health care savings, mainly due to less perioperative complications and reduced hospital stays in favor of MIS compared to laparotomy [26].

## Surgical indications

## General gynecology

### *Hysterectomy for benign diseases*

Hysterectomy for benign conditions with robotics approach has been addressed by some studies. A large database retrospective analysis ($n = 804,551$) evaluated robotics-assisted and standard laparoscopic hysterectomies for benign diseases and showed overall similar patient complication outcomes. The robotics cases had lower blood transfusions (2.1% vs 3.1%; $P < 0.001$), however with higher likelihood of postoperative pneumonia (RR 2.2, 95% CI 1.24–3.78; $P = 0.005$) [27]. However, a 2015 Cochrane review of hysterectomy techniques reported better outcomes in favor of vaginal approach, recording the fewest intraoperative complications, quickest return to baseline activities, and the fewest number of urinary/bowel dysfunction and dyspareunia. No significant differences in morbidity were reported between conventional laparoscopy and robotics [28]. A more recent Cochrane review reported that the effectiveness and safety of robotics compared with conventional laparoscopy were still uncertain for benign gynecology diseases, but suggested a comparable morbidity profile [6].

A multicentric study analyzed 32,118 hysterectomy cases and found that, while robotic cases had a higher rate of adhesive diseases, larger uterus, and morbid obese patients compared to laparoscopy and vaginal approaches, robotics cases had the shortest hospital length of stay. Moreover, no difference was found in intraoperative complications between the MIS approaches [29].

Some randomized trials have been published on this topic. Sarlos et al. reported a higher surgical time length for robotics compared to conventional laparoscopy (106 vs 75 min; $P < 0.001$) and better QoL postoperative scores [30]. Paraiso et al. found no difference in QoL scores when robotics was compared to laparoscopy for up to 6 months [31]. Lönnerfors et al. investigated the hospital cost and short-term outcome of conventional vs robotic hysterectomy and found similar cost when the robotics was a previous investment and robotics was related to less blood loss (50 vs 100 mL; $P < 0.05$) and fewer postoperative complications [32]. All of these trials have significant limitations. Primarily, they were done during the early learning curve of robotic surgery among surgeons who had an extensive nonrobotics MIS experience. Also, none of them addressed whether the rates of MIS were improved or not by introducing the robotics platforms. A retrospective analysis noted that robotics approaches by experienced robotic users led to lower length of stay, rates of transfusions, and complications compared to all other approaches including vaginal techniques [29].

In summary, while studies suggest an overall comparable outcome between robotics and laparoscopy, MIS should be offered as the substitute to the open approach. Surgeon preferences and experience as well as hospital resources finally impact the surgical route decision.

### *Myomectomy*

MIS approaches are also feasible for women undergoing myomectomy. About 25% of women may have significant symptoms related to leiomyoma that requires a surgical intervention [33]. Advincula et al. retrospectively compared robotics myomectomy to open myomectomy and noted a lower blood loss (195.7 vs 354.7 mL; $P < 0.05$) and longer surgical time (231 vs 154 min; $P < 0.05$) for robotics approach. Moreover, robotics and conventional laparoscopy had no difference in blood loss, hospitalization time, number and size of myomas, and postoperative complications [34]. A metaanalysis included 2852 women from 20 studies and demonstrated that robotic myomectomy was associated with fewer complications, such as lower blood loss, fewer conversions, and postoperative bleeding when compared to conventional laparoscopy [35].

With regards of fertility outcomes, studies suggested no difference between robotics and laparoscopy [36,37]. Pitter et al. retrospectively evaluated 872 women submitted to robotics-assisted myomectomy, and 127 pregnancies were achieved (60.6% spontaneous), with a 21% pregnancy loss rate in up to 14 weeks [38]. In summary, the available data suggest a benefit in favor of robotics-assisted approach compared to conventional and open approach for myomectomy.

## *Sacrocolpopexy*

Pelvic organ prolapse has as higher prevalence in women between 70 and 74 years and is related to parity [39]. Sacrocolpopexy has been described as an efficient treatment of pelvic organ prolapse and MIS has emerged as the best approach. Geller et al. compared 73 cases that underwent robotic sacrocolpopexy to 105 open approach and showed less estimated blood loss (103 vs 255 mL; $P < 0.001$), longer operative times (328 vs 225 min), and shorter hospital stay length (1.3 vs 2.7 days; $P < 0.001$). Similar Pelvic Organ Prolapse Quantification System (POP-Q) scores were reported [40].

Moreover, a small, randomized trial compared conventional laparoscopy ($n = 38$) to robotics ($n = 40$) and revealed longer operative times, higher costs, and worse pain scores for the robotics-assisted group. Notably, surgeons involved in the study were highly skilled in both techniques and POP-Q scores up to 1 year exhibited no significant differences [41].

In summary, MIS sacrocolpopexy has many benefits compared to the open approach. Yet, as in other surgical procedures, surgical skills and training are the key factors for choosing the best MIS approach.

## Gynecologic oncology

### *Endometrial cancer*

Laparoscopy has emerged as the standard approach in endometrial cancer staging. Laparoscopy has the same oncologic outcomes compared to laparotomy and with the MIS benefit of less complication rates. The largest published randomized trial was GOG LAP2, where 2181 women were randomized to undergo either laparoscopy or laparotomy. Robotics platforms were not widely available during the trial study period and all cases were done via standard laparoscopy. Patients randomized to laparoscopy had lower complication rates (21% vs 14%; $P < 0.001$) and less hospital stay length compared to laparotomy [42]. There was a 25% conversion to laparotomy rate in the laparoscopic cohort, which is quite high compared to currently reported rates of conversion in laparoscopy with or without robotics assistance. Importantly, the 3-year recurrence rates did not differ between groups (11.3% vs 10.2%) [43]. The LAP2 trial findings are supported by other six trials [44–49] that were pooled in a Cochrane metaanalysis [50]. Similar oncologic outcomes between laparoscopy and laparotomy are recorded even for high-risk histologies. A posthoc analysis of GOG LAP2 trial showed no difference in patterns of recurrences in surgical approach [51] and a recent metaanalysis of nine observational studies underscored this statement [52]. A recent single center review noted similar oncologic outcomes for uterine serous carcinomas between MIS and open approaches [53].

Despite the benefits of MIS in endometrial cancer and it being considered the "standard" approach, the rate of MIS in endometrial cancer has been poor using standard laparoscopic instrumentation. The introduction and adoption of the robotics platforms has led to significant increases in the rate of MIS for endometrial cancer. Fader et al. retrospectively evaluated the US Nationwide Inpatient Sample database ($n = 32,560$) of women who underwent surgery for endometrial cancer and found an increase from 22% to 50.8% in MIS hysterectomy between 2007 and 2011, mainly due to increasing of robotic surgery. Yet, women with Medicaid compared to private insurance, black or hispanic and low volume hospitals are less likely have MIS [54]. Another interesting database review (NSQIP; $n = 12,283$) reported a significant increase in MIS (24.1% to 71.4%) from 2008 to 2014, with a comprehensive decrease in 30 days complications rates [55]. A metaanalysis published in 2010 compared robotics ($n = 589$) to conventional laparoscopy ($n = 396$) and laparotomy ($n = 606$). The study demonstrated no difference of perioperative outcomes between robotics and conventional laparoscopy, and both recorded better outcomes compared to laparotomy [56].

Recent metaanalyses have been published on this topic. First by Fu et al. analyzed the oncologic outcomes with regard of surgical approach. Robotics recorded a long-term oncologic outcome similar to conventional laparoscopy, however with an unexpected better overall survival (HR = 0.68, 95% CI: 0.57–0.80), and recurrence free survival (HR = 0.79, 95% CI: 0.65–0.96) compared to laparotomy [57]. Second, the RECOURSE study by Leitao et al. performed a systematic review and metaanalysis of long-term outcomes associated with robotically assisted surgery for endometrial, cervical, colorectal, lung and prostate cancer. For endometrial cancer, robotic surgery recorded better overall survival compared to open surgery (HR = 0.77, 95% CI: 0.71–0.83; $P < 0.001$) [58]. It is not evident why robotics-assisted MIS would lead to better survival. However, at a minimum, we can say that the use of robotics platforms is safe from an oncologic standpoint. There is unlikely to ever be a randomized trial comparing robotic vs standard laparoscopy for various reasons including, both approaches that are laparoscopic and surgeon's experience is a confounder that cannot be overcome any longer in this area.

A noted potential benefit of robotics in endometrial cancer is for morbidly obese patients. Leitao et al. analyzed 426 women with BMI $> 40$ mg/m$^2$ and found an MIS approach during laparoscopic period of 6% (1993–2007) and increased to 57% in a subsequent robotics period (2008–2012). MIS was related to lower hospital stay length, lower complication rates (15% vs 36%; $P < 0.001$), and lower cost [21]. Similar results were noted by Tang et al. for women with BMI $> 30$ kg/m$^2$, recording a conversion rate of only 12% [59].

In summary, MIS has been established as the standard surgical approach of endometrial cancer, recording lower morbidity and similar oncological outcomes. The goal of surgeons treating endometrial cancer should be a minimum MIS rate of 80% [60]. Despite the discussion on costs, robotics approach may ultimately decrease the laparotomy rates and related complications. Yet, more women are having the benefit of MIS because of robotics-assisted laparoscopy.

## Cervical cancer

Results from endometrial cancer led surgeons to develop techniques for more complex surgical approaches in cervical cancer either by conventional laparoscopy or robotics-assisted. The MIS approach and the advantages of robotics-assisted surgery may help in the visualization and precise dissection of lymph nodes, ureter dissection, and radical parametrial resection in radical hysterectomy. However, in 2018 data from the Laparoscopic Approach to Cervical Cancer (LACC) trial [61] were published and did not support the theoretical advantages in better quality of life and lower morbidity for MIS. Additionally, women who were randomized to MIS had almost four times the risk of recurrence or death than those who underwent laparotomy.

Subsequently, several retrospective reports, including well-designed observational studies, were published, mostly confirming the LACC trial data [62–64], whereas others did not generate the same results [65–70]. Similar conflicting results were found in a systematic review and metaanalysis. A metaanalysis that included 15 studies ($n = 9499$) recorded 71% and 56% higher risks of recurrence and death for women who underwent MIS [71]. On the other hand, the recent RECOURSE Study, also a metaanalysis, suggested similar overall and disease-free survival between robotics and conventional laparoscopy or open (HR 1.18, 95%CI: 0.99–1.41; $P = 0.06$) surgery [58].

We should take care that the LACC trial is considered as the definitive study. There are other randomized noninferiority trials ongoing [72,73], two of them with robotics-assisted laparoscopy. All studies add surgical technique modifications such as exclusion of uterine manipulator and tumor containment methods.

## Ovarian carcinoma

MIS in ovarian carcinoma is still debatable. In contrast to endometrial and cervical cancer, ovarian carcinoma is mostly diagnosed as an advanced disease with stages III and IV in 75% of cases. While peritoneal dissemination is nearly always found, the surgical aim is cytoreduction with no visible residual disease. Surgery also has an important role in comprehensive staging that includes not only hysterectomy and bilateral salpingo-oophorectomy but also omentectomy, peritoneal washings, as well as pelvic and para-aortic lymphadenectomy up to the renal vessels.

For the early-stage disease, Bogani et al. published in 2017 a metaanalysis that included 1450 patients who underwent laparoscopic staging and 1615 by laparotomy. They found that survival was not influenced by surgical route and laparoscopy was superior in perioperative complications [74].

Nevertheless, another recent metaanalysis evaluated current evidence of MIS in ovarian carcinoma [75]. Ten studies addressed staging and included 1509 patients who underwent MIS and 1787 patents who underwent laparotomy, when no difference of recurrence-free or overall survival was reported. Moreover, for interval cytoreduction after neoadjuvant chemotherapy, four studies ($n = 543$ MIS and $n = 2801$ laparotomy) showed no difference in survival outcomes [75]. Yet, Melamed et al. analyzed the National Cancer Database ($n = 3071$) and found no negative impact of MIS compared to laparotomy in interval debulking [76].

Notably, data on the value of robotics-assisted surgery compared to conventional laparoscopy are lacking. Results from National Cancer Database ($n = 1901$) addressed MIS approaches. Compared to conventional laparoscopy, robotics-assisted laparoscopy had less conversion rate (7.2% vs 17.9%; $P < 0.001$) and there were no significant differences in survival between robotics-assisted laparoscopy and conventional laparoscopy [77].

In summary, though most published studies in ovarian carcinoma are retrospective and with inherent risk of bias, MIS is not associated with higher risk of recurrence and death compared to open approach for staging and in interval cytoreduction. Data on the role of robotics are still scarce.

## Future scenario

Robotics-assisted laparoscopy is undoubtedly a definitive technological evolution of MIS and new robotics platforms have been already developed and launched. Despite the debate on higher costs in a single case basis compared to conventional laparoscopy, a great number of patients will receive MIS and all the well-known benefits of MIS instead of open surgery due to the robotics implementation.

## Conclusions

The advantages and disadvantages of each MIS approach must be weighed with regard of surgeon expertise and each institution's characteristics. Robotic surgery can help surgeons with low expertise in complex laparoscopic procedures in providing MIS to their patients. Yet, for experienced surgeons, robotics can add to the inherent advantages and facilities of surgical techniques.

## Key points

- The advantages and disadvantages of each approach must be weighed against the surgeon's expertise and each institution's resources.
- Robotics-assisted surgery provides the benefits of minimally invasive surgery to a high number of patients who would have otherwise undergone laparotomy ("open").
- The costs of incorporating robotics platforms is debatable but requires thoughtful discussion and a broader global surgical care perspectives.

## References

[1] Johnson N, Lethaby A, Farquhar C, Garry R, Barlow D. Surgical approaches to hysterectomy for benign gynaecological disease. In: Cochrane database Syst. Rev. Chichester, UK: John Wiley & Sons, Ltd; 2002. https://doi.org/10.1002/14651858.CD003677.

[2] Mayooran Z, Rombauts L, Brown TIH, Tsaltas J, Fraser K, Healy DL. Reliability and validity of an objective assessment instrument of laparoscopic skill. Fertil Steril 2004;82:976–8. https://doi.org/10.1016/j.fertnstert.2004.05.067.

[3] De Wilde RL, Herrmann A. Robotic surgery – advance or gimmick? Best Pract Res Clin Obstet Gynaecol 2013;27:457–69. https://doi.org/10.1016/j.bpobgyn.2012.12.005.

[4] Varghese A, Doglioli M, Fader AN. Updates and controversies of robotic-assisted surgery in gynecologic surgery. Clin Obstet Gynecol 2019;62:733–48. https://doi.org/10.1097/GRF.0000000000000489.

[5] Lanfranco AR, Castellanos AE, Desai JP, Meyers WC. Robotic surgery. Ann Surg 2004;239:14–21. https://doi.org/10.1097/01.sla.0000103020.19595.7d.

[6] Lawrie TA, Liu H, Lu D, Dowswell T, Song H, Wang L, Shi G. Robot-assisted surgery in gynaecology. Cochrane Database Syst Rev 2019;2019. https://doi.org/10.1002/14651858.CD011422.pub2.

[7] van Dam P, Hauspy J, Verkinderen L, Trinh XB, van Dam P-J, Van Looy L, Dirix L. Are costs of robot-assisted surgery warranted for gynecological procedures? Obstet Gynecol Int 2011;2011, 973830. https://doi.org/10.1155/2011/973830.

[8] Lenihan JP, Kovanda C, Seshadri-Kreaden U. What is the learning curve for robotic assisted gynecologic surgery? J Minim Invasive Gynecol 2008;15:589–94. https://doi.org/10.1016/j.jmig.2008.06.015.

[9] Pitter MC, Anderson P, Blissett A, Pemberton N. Robotic-assisted gynaecological surgery—establishing training criteria; minimizing operative time and blood loss. Int J Med Robot Comput Assist Surg 2008;4:114–20. https://doi.org/10.1002/rcs.183.

[10] Seamon LG, Fowler JM, Richardson DL, Carlson MJ, Valmadre S, Phillips GS, Cohn DE. A detailed analysis of the learning curve: robotic hysterectomy and pelvic-aortic lymphadenectomy for endometrial cancer. Gynecol Oncol 2009;114:162–7. https://doi.org/10.1016/j.ygyno.2009.04.017.

[11] Lim PC, Kang E, Park DH. A comparative detail analysis of the learning curve and surgical outcome for robotic hysterectomy with lymphadenectomy versus laparoscopic hysterectomy with lymphadenectomy in treatment of endometrial cancer: a case-matched controlled study of the first o. Gynecol Oncol 2011;120:413–8. https://doi.org/10.1016/j.ygyno.2010.11.034.

[12] Cook JA, Ramsaya CR, Fayers P. Statistical evaluation of learning curve effects in surgical trials. Clin Trials 2004;1:421–7. https://doi.org/10.1191/1740774504cn042oa.

[13] Childers CP, Maggard-Gibbons M. Estimation of the acquisition and operating costs for robotic surgery. JAMA 2018;320:835. https://doi.org/10.1001/jama.2018.9219.

[14] Barnett JC, Judd JP, Wu JM, Scales CD, Myers ER, Havrilesky LJ. Cost comparison among robotic, laparoscopic, and open hysterectomy for endometrial Cancer. Obstet Gynecol 2010;116:685–93. https://doi.org/10.1097/AOG.0b013e3181ee6e4d.

[15] Behera MA, Likes CE, Judd JP, Barnett JC, Havrilesky LJ, Wu JM. Cost analysis of abdominal, laparoscopic, and robotic-assisted myomectomies. J Minim Invasive Gynecol 2012;19:52–7. https://doi.org/10.1016/j.jmig.2011.09.007.

[16] Leitao MM, Bartashnik A, Wagner I, Lee SJ, Caroline A, Hoskins WJ, Thaler HT, Abu-Rustum NR, Sonoda Y, Brown CL, Jewell EL, Barakat RR, Gardner GJ. Cost-effectiveness analysis of robotically assisted laparoscopy for newly diagnosed uterine cancers. Obstet Gynecol 2014;123:1031–7. https://doi.org/10.1097/AOG.0000000000000223.

[17] Desille-Gbaguidi H, Hebert T, Paternotte-Villemagne J, Gaborit C, Rush E, Body G. Overall care cost comparison between robotic and laparoscopic surgery for endometrial and cervical cancer. Eur J Obstet Gynecol Reprod Biol 2013;171:348–52. https://doi.org/10.1016/j.ejogrb.2013.09.025.

[18] Wright KN, Jonsdottir GM, Jorgensen S, Shah N, Einarsson JI. Costs and outcomes of abdominal, vaginal, laparoscopic and robotic hysterectomies. JSLS J Soc Laparoendosc Surg 2012;16:519–24. https://doi.org/10.4293/108680812X13462882736736.

[19] Judd JP, Siddiqui NY, Barnett JC, Visco AG, Havrilesky LJ, Wu JM. Cost-minimization analysis of robotic-assisted, laparoscopic, and abdominal Sacrocolpopexy. J Minim Invasive Gynecol 2010;17:493–9. https://doi.org/10.1016/j.jmig.2010.03.011.

[20] Pasic RP, Rizzo JA, Fang H, Ross S, Moore M, Gunnarsson C. Comparing robot-assisted with conventional laparoscopic hysterectomy: impact on cost and clinical outcomes. J Minim Invasive Gynecol 2010;17:730–8. https://doi.org/10.1016/j.jmig.2010.06.009.

[21] Leitao MM, Narain WR, Boccamazzo D, Sioulas V, Cassella D, Ducie JA, Eriksson AGZ, Sonoda Y, Chi DS, Brown CL, Levine DA, Jewell EL, Zivanovic O, Barakat RR, Abu-Rustum NR, Gardner GJ. Impact of robotic platforms on surgical approach and costs in the Management of Morbidly Obese Patients with newly diagnosed uterine Cancer. Ann Surg Oncol 2016;23:2192–8. https://doi.org/10.1245/s10434-015-5062-6.

[22] Jacoby VL, Autry A, Jacobson G. Nationwide use of laparoscopic hysterectomy compared with abdominal. Obstet Gynecol 2009;114:1041–8.

[23] Peiretti M, Zanagnolo V, Bocciolone L, Landoni F, Colombo N, Minig L, Sanguineti F, Maggioni A. Robotic surgery: changing the surgical approach for endometrial cancer in a referral cancer center. J Minim Invasive Gynecol 2005;16:427–31. https://doi.org/10.1016/j.jmig.2009.03.013.

[24] Scalici J, Laughlin BB, Finan MA, Wang B, Rocconi RP. The trend towards minimally invasive surgery (MIS) for endometrial cancer: an ACS-NSQIP evaluation of surgical outcomes. Gynecol Oncol 2015;136:512–5. https://doi.org/10.1016/j.ygyno.2014.11.014.

[25] Lau S, Vaknin Z, Ramana-kumar AV. Outcomes and cost comparisons after introducing a robotics program for endometrial cancer surgery. Obstet Gynecol 2012;119:717–24. https://doi.org/10.1097/AOG.0b013e31824c0956.

[26] Xu T, Hutfless SM, Cooper MA, Zhou M, Massie AB, Makary MA. Hospital cost implications of increased use of minimally invasive surgery. JAMA Surg 2015;150:489. https://doi.org/10.1001/jamasurg.2014.4052.

[27] Rosero EB, Kho KA, Joshi GP, Giesecke M, Schaffer JI. Comparison of robotic and laparoscopic hysterectomy for benign gynecologic disease. Obstet Gynecol 2013;122:778–86. https://doi.org/10.1097/AOG.0b013e3182a4ee4d.

[28] Aarts JW, Nieboer TE, Johnson N, Tavender E, Garry R, Mol BWJ, Kluivers KB. Surgical approach to hysterectomy for benign gynaecological disease. Cochrane Database Syst Rev 2015;2015. https://doi.org/10.1002/14651858.CD003677.pub5.

[29] Lim PC, Crane JT, English EJ, Farnam RW, Garza DM, Winter ML, Rozeboom JL. Multicenter analysis comparing robotic, open, laparoscopic, and vaginal hysterectomies performed by high-volume surgeons for benign indications. Int J Gynecol Obstet 2016;133:359–64. https://doi.org/10.1016/j.ijgo.2015.11.010.

[30] Sarlos D, Kots L, Stevanovic N, von Felten S, Schär G. Robotic compared with conventional laparoscopic hysterectomy: a randomized controlled trial. Obstet Gynecol 2012;120:604–11. https://doi.org/10.1097/AOG.0b013e318265b61a.

[31] Paraiso MFR, Ridgeway B, Park AJ, Jelovsek JE, Barber MD, Falcone T, Einarsson JI. A randomized trial comparing conventional and robotically assisted total laparoscopic hysterectomy. Am J Obstet Gynecol 2013;208(368):e1–7. https://doi.org/10.1016/j.ajog.2013.02.008.

[32] Lönnerfors C, Reynisson P, Persson J. A randomized trial comparing vaginal and laparoscopic hysterectomy vs robot-assisted hysterectomy. J Minim Invasive Gynecol 2015;22:78–86. https://doi.org/10.1016/j.jmig.2014.07.010.

[33] Stewart E, Cookson C, Gandolfo R, Schulze-Rath R. Epidemiology of uterine fibroids: a systematic review. BJOG An Int J Obstet Gynaecol 2017;124:1501–12. https://doi.org/10.1111/1471-0528.14640.

[34] Advincula AP, Xu X, Goudeau S, Ransom SB. Robot-assisted laparoscopic myomectomy versus abdominal myomectomy: a comparison of short-term surgical outcomes and immediate costs. J Minim Invasive Gynecol 2007;14:698–705. https://doi.org/10.1016/j.jmig.2007.06.008.

[35] Wang T, Tang H, Xie Z, Deng S. Robotic-assisted vs. laparoscopic and abdominal myomectomy for treatment of uterine fibroids: a meta-analysis. Minim Invasive Ther Allied Technol 2018;27:249–64. https://doi.org/10.1080/13645706.2018.1442349.

[36] Flyckt R, Soto E, Nutter B, Falcone T. Comparison of Long-term fertility and bleeding outcomes after robotic-assisted, laparoscopic, and abdominal myomectomy. Obstet Gynecol Int 2016;2016:1–8. https://doi.org/10.1155/2016/2789201.

[37] Bedient CE, Magrina JF, Noble BN, Kho RM. Comparison of robotic and laparoscopic myomectomy. Am J Obstet Gynecol 2009;201(566): e1–566.e5. https://doi.org/10.1016/j.ajog.2009.05.049.

[38] Pitter MC, Gargiulo AR, Bonaventura LM, Lehman JS, Srouji SS. Pregnancy outcomes following robot-assisted myomectomy. Hum Reprod 2013;28:99–108. https://doi.org/10.1093/humrep/des365.

[39] MacLennan AH, Taylor AW, Wilson DH, Wilson D. The prevalence of pelvic floor disorders and their relationship to gender, age, parity and mode of delivery. BJOG An Int J Obstet Gynaecol 2000;107:1460–70. https://doi.org/10.1111/j.1471-0528.2000.tb11669.x.

[40] Geller EJ, Siddiqui NY, Wu JM, Visco AG. Short-term outcomes of robotic Sacrocolpopexy compared with abdominal Sacrocolpopexy. Obstet Gynecol 2008;112:1201–6. https://doi.org/10.1097/AOG.0b013e31818ce394.

[41] Paraiso MFR, Jelovsek JE, Frick A, Chen CCG, Barber MD. Laparoscopic compared with robotic Sacrocolpopexy for vaginal prolapse. Obstet Gynecol 2011;118:1005–13. https://doi.org/10.1097/AOG.0b013e318231537c.

[42] Walker JL, Piedmonte MR, Spirtos NM, Eisenkop SM, Schlaerth JB, Mannel RS, Spiegel G, Barakat R, Pearl ML, Sharma SK. Laparoscopy compared with laparotomy for comprehensive surgical staging of uterine cancer: Gynecologic oncology group study LAP2. J Clin Oncol 2009;27:5331–6. https://doi.org/10.1200/JCO.2009.22.3248.

[43] Walker JL, Piedmonte MR, Spirtos NM, Eisenkop SM, Schlaerth JB, Mannel RS, Barakat R, Pearl ML, Sharma SK. Recurrence and survival after random assignment to laparoscopy versus laparotomy for comprehensive surgical staging of uterine Cancer: Gynecologic oncology group LAP2 study. J Clin Oncol 2012;1–7. https://doi.org/10.1200/JCO/2011/405506.

[44] Zullo F, Palomba S, Falbo A, Russo T, Mocciaro R, Tartaglia E, Tagliaferri P, Mastrantonio P. Laparoscopic surgery vs laparotomy for early stage endometrial cancer: long-term data of a randomized controlled trial. Am J Obstet Gynecol 2009;200(296):e1–9. https://doi.org/10.1016/j.ajog.2008.10.056.

[45] Janda M, Gebski V, Davies LC, Forder P, Brand A, Hogg R, Jobling TW, Land R, Manolitsas T, Nascimento M, Neesham D, Nicklin JL, Oehler MK, Otton G, Perrin L, Salfinger S, Hammond I, Leung Y, Sykes P, Ngan H, Garrett A, Laney M, Ng TY, Tam K, Chan K, Wrede CD, Pather S, Simcock B, Farrell R, Robertson G, Walker G, Armfield NR, Graves N, McCartney AJ, Obermair A. Effect of total laparoscopic hysterectomy vs total abdominal hysterectomy on disease-free survival among women with stage i endometrial cancer: a randomized clinical trial. JAMA - J Am Med Assoc 2017;317:1224–33. https://doi.org/10.1001/jama.2017.2068.

[46] Lu Q, Liu H, Liu C, Wang S, Li S, Guo S, Lu J, Zhang Z. Comparison of laparoscopy and laparotomy for management of endometrial carcinoma: a prospective randomized study with 11-year experience. J Cancer Res Clin Oncol 2013;139:1853–9. https://doi.org/10.1007/s00432-013-1504-3.

[47] Malzoni M, Tinelli R, Cosentino F, Perone C, Rasile M, Iuzzolino D, Malzoni C, Reich H. Total laparoscopic hysterectomy versus abdominal hysterectomy with lymphadenectomy for early-stage endometrial cancer: a prospective randomized study. Gynecol Oncol 2009;112:126–33. https://doi.org/10.1016/j.ygyno.2008.08.019.

[48] Tozzi R, Malur S, Koehler C, Schneider A. Laparoscopy versus laparotomy in endometrial cancer: first analysis of survival of a randomized prospective study. J Minim Invasive Gynecol 2005;12:130–6. https://doi.org/10.1016/j.jmig.2005.01.021.

[49] Mourits MJ, Bijen CB, Arts HJ, ter Brugge HG, van der Sijde R, Paulsen L, Wijma J, Bongers MY, Post WJ, van der Zee AG, de Bock GH. Safety of laparoscopy versus laparotomy in early-stage endometrial cancer: a randomised trial. Lancet Oncol 2010;11:763–71. https://doi.org/10.1016/S1470-2045(10)70143-1.

III. The current and future clinical applications of robotic surgery among medical specialties

[50] Galaal K, Donkers H, Bryant A, Lopes AD. Laparoscopy versus laparotomy for the management of early stage endometrial cancer. Cochrane Database Syst Rev 2018;2018. https://doi.org/10.1002/14651858.CD006655.pub3.

[51] Fader AN, Java J, Tenney M, Ricci S, Gunderson CC, Temkin SM, Spirtos N, Kushnir CL, Pearl ML, Zivanovic O, Tewari KS, O'Malley D, Hartenbach EM, Hamilton CA, Gould NS, Mannel RS, Rodgers W, Walker JL. Impact of histology and surgical approach on survival among women with early-stage, high-grade uterine cancer: an NRG oncology/gynecologic oncology group ancillary analysis. Gynecol Oncol 2016;143:460–5. https://doi.org/10.1016/j.ygyno.2016.10.016.

[52] Kim NR, Lee AJ, Yang EJ, So KA, Lee SJ, Kim TJ, Shim S-H. Minimally invasive surgery versus open surgery in high-risk histologic endometrial cancer patients: a meta-analysis. Gynecol Oncol 2022;166:236–44. https://doi.org/10.1016/j.ygyno.2022.06.004.

[53] Sia TY, Basaran D, Dagher C, Sassine D, Brandt B, Rosalik K, Mueller JJ, Broach V, Makker V, Soslow RA, Abu-Rustum NR, Leitao MM. Laparoscopy with or without robotic assistance does not negatively impact long-term oncologic outcomes in patients with uterine serous carcinoma. Gynecol Oncol 2023;175:8–14. https://doi.org/10.1016/j.ygyno.2023.05.064.

[54] Fader AN, Weise RM, Sinno AK, Tanner EJ, Borah BJ, Moriarty JP, Bristow RE, Makary MA, Pronovost PJ, Hutfless S, Dowdy SC. Utilization of minimally invasive surgery in endometrial cancer care. Obstet Gynecol 2016;127:91–100. https://doi.org/10.1097/AOG.0000000000001180.

[55] Casarin J, Multinu F, Ubl DS, Dowdy SC, Cliby WA, Glaser GE, Butler KA, Ghezzi F, Habermann EB, Mariani A. Adoption of minimally invasive surgery and decrease in surgical morbidity for endometrial cancer treatment in the United States. Obstet Gynecol 2018;131:304–11. https://doi.org/10.1097/AOG.0000000000002428.

[56] Gaia G, Holloway RW, Santoro L, Ahmad S, Di Silverio E, Spinillo A. Robotic-assisted hysterectomy for endometrial Cancer compared with traditional laparoscopic and laparotomy approaches. Obstet Gynecol 2010;116:1422–31. https://doi.org/10.1097/AOG.0b013e3181f74153.

[57] Fu H, Zhang J, Zhao S, He N. Survival outcomes of robotic-assisted laparoscopy versus conventional laparoscopy and laparotomy for endometrial cancer: a systematic review and meta-analysis. Gynecol Oncol 2023;174:55–67. https://doi.org/10.1016/j.ygyno.2023.04.026.

[58] Leitao MM, Kreaden US, Laudone V, Park BJ, Pappou EP, Davis JW, Rice DC, Chang GJ, Rossi EC, Hebert AE, Slee A, Gonen M. The RECOURSE study: Long-term oncologic outcomes associated with robotically assisted minimally invasive procedures for endometrial, cervical, colorectal, lung, or prostate Cancer: a systematic review and Meta-analysis. Ann Surg 2023;277:387–96. https://doi.org/10.1097/SLA.0000000000005698.

[59] Tang KY, Gardiner SK, Gould C, Osmundsen B, Collins M, Winter WE. Robotic surgical staging for obese patients with endometrial cancer. Am J Obstet Gynecol 2012;206(513):e1–6. https://doi.org/10.1016/j.ajog.2012.01.002.

[60] Bergstrom J, Aloisi A, Armbruster S, Yen T-T, Casarin J, Leitao MM, Tanner EJ, Matsuno R, Machado KK, Dowdy SC, Soliman PT, Wethington SL, Stone RL, Levinson KL, Fader AN. Minimally invasive hysterectomy surgery rates for endometrial cancer performed at National Comprehensive Cancer Network (NCCN) centers. Gynecol Oncol 2018;148:480–4. https://doi.org/10.1016/j.ygyno.2018.01.002.

[61] Ramirez PT, Frumovitz M, Pareja R, Lopez A, Vieira M, Ribeiro R, Buda A, Yan X, Shuzhong Y, Chetty N, Isla D, Tamura M, Zhu T, Robledo KP, Gebski V, Asher R, Behan V, Nicklin JL, Coleman RL, Obermair A. Minimally invasive versus abdominal radical hysterectomy for cervical cancer. N Engl J Med 2018;379:1895–904. https://doi.org/10.1056/NEJMoa1806395.

[62] Rodriguez J, Rauh-Hain JA, Saenz J, Isla DO, Rendon Pereira GJ, Odetto D, Martinelli F, Villoslada V, Zapardiel I, Trujillo LM, Perez M, Hernandez M, Saadi JM, Raspagliesi F, Valdivia H, Siegrist J, Fu S, Hernandez Nava M, Echeverry L, Noll F, Ditto A, Lopez A, Hernandez A, Pareja R. Oncological outcomes of laparoscopic radical hysterectomy versus radical abdominal hysterectomy in patients with early-stage cervical cancer: a multicenter analysis. Int J Gynecol Cancer 2021;31:504–11. https://doi.org/10.1136/ijgc-2020-002086.

[63] Uppal S, Gehrig PA, Peng K, Bixel KL, Matsuo K, Vetter MH, Davidson BA, Cisa MP, Lees BF, Brunette LL, Tucker K, Staley AS, Gotlieb WH, Holloway RW, Essel KG, Holman LL, Goldfeld E, Olawaiye A, Rose SL. Recurrence rates in patients with cervical cancer treated with abdominal versus minimally invasive radical hysterectomy: a multi-institutional retrospective review study. J Clin Oncol 2020;38:1030–40. https://doi.org/10.1200/JCO.19.03012.

[64] Chiva L, Zanagnolo V, Querleu D, Martin-Calvo N, Arévalo-Serrano J, Căpîlna ME, Fagotti A, Kucukmetin A, Mom C, Chakalova G, Aliyev S, Malzoni M, Narducci F, Arencibia O, Raspagliesi F, Toptas T, Cibula D, Kaidarova D, Meydanli MM, Tavares M, Golub D, Perrone AM, Poka R, Tsolakidis D, Vujić G, Jedryka MA, Zusterzeel PLM, Beltman JJ, Goffin F, Haidopoulos D, Haller H, Jach R, Yezhova I, Berlev I, Bernardino M, Bharathan R, Lanner M, Maenpaa MM, Sukhin V, Feron JG, Fruscio R, Kukk K, Ponce J, Minguez JA, Vázquez-Vicente D, Castellanos T, Chacon E, Alcazar JL. SUCCOR study: an international European cohort observational study comparing minimally invasive surgery versus open abdominal radical hysterectomy in patients with stage IB1 cervical cancer. Int J Gynecol Cancer 2020;30:1269–77. https://doi.org/10.1136/ijgc-2020-001506.

[65] Li P, Chen L, Ni Y, Liu J, Li D, Guo J, Liu Z, Jin S, Xu Y, Li Z, Wang L, Bin X, Lang J, Liu P, Chen C. Comparison between laparoscopic and abdominal radical hysterectomy for stage IB1 and tumor size <2 cm cervical cancer with visible or invisible tumors: a multicentre retrospective study. J Gynecol Oncol 2021;32. https://doi.org/10.3802/jgo.2021.32.e17.

[66] Kim SI, Cho JH, Seol A, Kim YI, Lee M, Kim HS, Chung HH, Kim J-W, Park NH, Song Y-S. Comparison of survival outcomes between minimally invasive surgery and conventional open surgery for radical hysterectomy as primary treatment in patients with stage IB1–IIA2 cervical cancer. Gynecol Oncol 2019;153:3–12. https://doi.org/10.1016/j.ygyno.2019.01.008.

[67] Kwon BS, Roh HJ, Lee S, Yang J, Song YJ, Lee SH, Kim KH, Suh DS. Comparison of long-term survival of total abdominal radical hysterectomy and laparoscopy-assisted radical vaginal hysterectomy in patients with early cervical cancer: Korean multicenter, retrospective analysis. Gynecol Oncol 2020;1–7. https://doi.org/10.1016/j.ygyno.2020.09.035.

[68] Jensen PT, Schnack TH, Frøding LP, Bjørn SF, Lajer H, Markauskas A, Jochumsen KM, Fuglsang K, Dinesen J, Søgaard CH, Søgaard-Andersen E, Jensen MM, Knudsen A, Øster LH, Høgdall C. Survival after a nationwide adoption of robotic minimally invasive surgery for early-stage cervical cancer – a population-based study. Eur J Cancer 2020;128:47–56. https://doi.org/10.1016/j.ejca.2019.12.020.

[69] Brandt B, Sioulas V, Basaran D, Kuhn T, LaVigne K, Gardner GJ, Sonoda Y, Chi DS, Long Roche KC, Mueller JJ, Jewell EL, Broach VA, Zivanovic O, Abu-Rustum NR, Leitao MM. Minimally invasive surgery versus laparotomy for radical hysterectomy in the management of early-stage cervical cancer: survival outcomes. Gynecol Oncol 2020;156:591–7. https://doi.org/10.1016/j.ygyno.2019.12.038.

[70] Baiocchi G, Ribeiro R, Dos Reis R, Falcao DF, Lopes A, Costa RLR, Pinto GLS, Vieira M, Kumagai LY, Faloppa CC, Mantoan H, Badiglian-Filho L, Tsunoda AT, Foiato TF, Andrade CEMC, Palmeira LO, Gonçalves BT, Zanvettor PH. Open versus minimally invasive radical hysterectomy in cervical cancer: the CIRCOL group study. Ann Surg Oncol 2022;29:1151–60. https://doi.org/10.1245/s10434-021-10813-5.

[71] Nitecki R, Ramirez PT, Frumovitz M, Krause KJ, Tergas AI, Wright JD, Rauh-Hain JA, Melamed A. Survival after minimally invasive vs open radical hysterectomy for early-stage cervical cancer: a systematic review and meta-analysis. JAMA Oncol 2020;6:1019–27. https://doi.org/10.1001/jamaoncol.2020.1694.

[72] Chao X, Li L, Wu M, Ma S, Tan X, Zhong S, Lang J, Cheng A, Li W. Efficacy of different surgical approaches in the clinical and survival outcomes of patients with early-stage cervical cancer: protocol of a phase III multicentre randomised controlled trial in China. BMJ Open 2019;9, c029055. https://doi.org/10.1136/bmjopen-2019-029055.

[73] Falconer H, Palsdottir K, Stalberg K, Dahm-Kähler P, Ottander U, Lundin ES, Wijk L, Kimmig R, Jensen PT, Zahl Eriksson AG, Mäenpää J, Persson J, Salehi S. Robot-assisted approach to cervical cancer (RACC): an international multi-center, open-label randomized controlled trial. Int J Gynecol Cancer 2019;29:1072–6. https://doi.org/10.1136/ijgc-2019-000558.

[74] Bogani G, Borghi C, Leone Roberti Maggiore U, Ditto A, Signorelli M, Martinelli F, Chiappa V, Lopez C, Sabatucci I, Scaffa C, Indini A, Ferrero S, Lorusso D, Raspagliesi F. Minimally invasive surgical staging in early-stage ovarian carcinoma: A Systematic Review and Meta-analysis. J Minim Invasive Gynecol 2017;24:552–62. https://doi.org/10.1016/j.jmig.2017.02.013.

[75] Knisely A, Gamble CR, St CM, Clair JY, Hou F, Khoury-Collado AA, Gockley JD, Wright AM. The role of minimally invasive surgery in the care of Women with ovarian cancer: a systematic review and meta-analysis. J Minim Invasive Gynecol 2021;28:537–43. https://doi.org/10.1016/j.jmig.2020.11.007.

[76] Melamed A, Nitecki R, Boruta DM, del Carmen MG, Clark RM, Growdon WB, Goodman A, Schorge JO, Rauh-Hain JA. Laparoscopy compared with laparotomy for Debulking ovarian cancer after neoadjuvant chemotherapy. Obstet Gynecol 2017;129:861–9. https://doi.org/10.1097/AOG.0000000000001851.

[77] Facer B, Wang F, Grijalva CG, Alvarez RD, Shu X-O. Survival outcomes for robotic-assisted laparoscopy versus traditional laparoscopy in clinical stage I epithelial ovarian cancer. Am J Obstet Gynecol 2020;222(474):e1–474.e12. https://doi.org/10.1016/j.ajog.2019.10.104.

# 61

# Endometriosis and robotic surgery

*Renato Moretti-Marques[a], Mariana Costa Rossette[b], Gil Kamergorodsky[c], Vanessa Alvarenga-Bezerra[a], and Sérgio Podgaec[d]*

[a]Department of Gynecological Oncology, Hospital Israelita Albert Einstein, São Paulo, SP, Brazil [b]Gynecologic Division, Universidade Federal de São Paulo, Escola Paulista de Medicina, São Paulo, SP, Brazil [c]Department of Gynecology, Federal University of São Paulo (Dr. Kamergorodsky), São Paulo, Brazil [d]Department of Obstetrics and Gynecology, Medical School Hospital, University of São Paulo, São Paulo, Brazil

## Introduction

Endometriosis is a chronic, estrogen-dependent, and inflammatory disease characterized by endometrial-like tissue outside the uterus, often accompanied by fibrosis [1]. The lesions are typically pelvic but can affect other locations such as the bowel, diaphragm, umbilicus, and pleural cavity. It is a common disease affecting approximately 10% of women during their reproductive years, representing around 190 million women with the disease in the world [2].

Endometriosis is usually associated with chronic pelvic pain, severe dysmenorrhea, dyspareunia, and infertility, although asymptomatic cases arise [1]. The painful, debilitating symptoms significantly impact patients' quality of life, with negative consequences on daily activities, relationships, sexual function, and productivity at work [3]. The economic burden of the disease is enormous, considering both healthcare costs related to treatment and work loss. Like other chronic conditions, such as Crohn's disease, rheumatoid arthritis, and diabetes, a woman with endometriosis lose, on average, 11 h of work per week [4]. Considering these consequences, endometriosis should be regarded as a public health issue of significant importance.

The disease is heterogeneous, with three well-described presentations: superficial peritoneal lesions, ovarian endometriomas, and deep infiltrating endometriosis. Superficial lesions are considered those that occur on the peritoneum surface with less than 5 mm invasion. Ovarian endometriomas are cystic masses that grow within the ovary and can have various dimensions. The most severe phenotype is the deep disease when lesions penetrate deeper than 5 mm under the peritoneal surface or infiltrate the muscularis propria of other organs close to the uterus, such as the bladder, rectosigmoid, and ureter. Moreover, up to 12% of patients can experience extrapelvic locations, including the gastrointestinal and urinary tract, diaphragm surface, lungs, pleura, skin, and nervous system—field [1,2,5].

Because of its chronic nature, endometriosis requires lifelong management of patients. Current treatments include surgical removal of lesions, drugs that suppress ovarian hormone production, and menstruation, besides assisted reproductive therapies. The choice of the ideal approach should be individualized based on the clinical presentation, symptom severity, patients' priorities, disease extent, and location [6]. Medical therapy represents the cornerstone of endometriosis management since the pharmacological treatment for endometriosis related-pain may be necessary for decades during the reproductive years, aiming to minimize multiple surgical procedures in the life course of women [1,2,6]. On the other hand, surgery should be considered in some specific situations, such as persistent pain even after hormonal therapy, lesion progression, bowel occlusion, ureteral involvement, large endometriomas (>6.0 cm), suspicious adnexal mass, and some cases of infertility (Table 1). Disadvantages of surgery include risk of injury (especially the bowel and bladder), possible reduction of ovarian reserve, fibrosis, and adhesion formation [2,6]. Laparoscopy surgery has been the technique of choice for treating endometriosis because it shows good outcomes with the advantages of minimally invasive surgery (MIS), including shorter hospital stay, better visualization and dissection of tissues, faster recovery, and better cosmetic results [7,8].

TABLE 1   Surgical indications for endometriosis.

Persistent pain despite medical therapy
Lesion progression
Contraindications to or refusal of medical therapy
Large endometriomas (>6 cm) or need of exclusion of malignancy in an adnexal mass
Obstruction of the bowel or urinary tract
Infertility (desire for spontaneous pregnancy or after failed in vitro fertilization)

There are limited data about robotic surgery (RS) for endometriosis and comparing robot-assisted surgery with standard laparoscopy (SL). The available evidence shows that a robotic approach is feasible, effective, and safe, delivering similar surgical outcomes to conventional laparoscopy [9–17]. However, uncertainties on cost-effectiveness remain. This chapter presents the techniques, strategical use, and possible advantages of robot-assisted surgery in the treatment of endometriosis.

## Robotic-assisted surgery vs laparoscopy for endometriosis

Since the da Vinci surgical system FDA approval in 2005, RS has been introduced to improve surgical performance. Increased dexterity, better vision and depth perception, and a wide range of motion are some advantages of robotic-assisted techniques. Its limitations are mainly related to the lack of tactile feedback and increased cost (Table 2) [18,19].

RS in gynecology covers a broad spectrum of uses and is growing fast. The robotic system may assist the gynecological surgeon in treating benign conditions such as fibroids, endometriosis, pelvic organ prolapse, and adnexal masses or for malignancies such as endometrial, cervical, or ovarian cancer [18].

The surgical treatment of deep endometriosis remains one of the most challenging operations because of the disease's adhesions, inflammation, and anatomical distortion. Therefore, expert dissection skills are required to perform the proper surgical excision of the disease with the preservation of nerves and normal tissue of the surroundings [2,6]. Considering this, it is rational to suppose that using a tool that improves better precision could favor a more conservative and less extensive surgery to preserve an organ's function. Araujo et al., 2016 reported a case in which the use of RS favored the realization of a conservative surgery to treat extensive bowel endometriosis with rectal preservation field [9]. Mosbrucker et al., 2018, in a randomized control trial of 98 patients, compared the visualization of histologically confirmed endometriosis lesions with a robotic scope versus a standard laparoscope. The study found that more lesions were detected using robotic visualization with odds 2.36 times greater than SL [20]. Similar to these findings; many studies show that robot-assisted surgery is a feasible and safe option to treat deep endometriosis [17,21].

It is well established that MIS is the gold standard for endometriosis surgical treatment. During the last decades, SL has been accepted as the technique of choice because of its better outcomes and cosmetic results when compared to laparotomy [7,8,19]. More recently, robotic-assisted surgery has been proposed as a good alternative to overcome the limitations of SL, especially regarding complex and multiorgan deep endometriosis cases [12–14]. However, specialists' opinions about the benefits of RS for endometriosis still conflict mainly because of the lack of good evidence regarding long-term pain relief and pregnancy rates that could justify its higher costs [19]. Majority of the data comparing laparoscopy to RS come from retrospective studies with many limitations, especially due to the lack of randomization and heterogenous stages of the diseases treated using both approaches.

TABLE 2   Advantages and disadvantages of the robotic platform for the surgical treatment of endometriosis.

| Advantages | Disadvantages |
| --- | --- |
| The ergonomic position of the surgeon reduces fatigue and improves focus | Lack of haptic feedback |
| Reduced tremor interference, precision, and finesse to the surgical procedure | Cost-effectiveness |
| Seven degrees of freedom of the instruments, better access to challenging anatomical spaces | |
| Three-dimensional image, high-definition image, and image magnification, better identification of lesions | |
| Ability to control four surgical instruments, including the camera | |
| Firefly technology, an infrared light source, helps to identify structures and evaluate vascularization when stained with indocyanine green | |
| Better teaching and training possibilities, reducing learning curve (simulation exercises, drawing touch screen, dual console system | |

Berlanda et al., 2017 in a systematic review, concluded that RS did not show perioperative advantages compared with SL, whereas it was associated with a longer operating time in some studies [19]. Nevertheless, the quality of the studies included was low. On the other hand, Soto et al., 2017, in a multicenter randomized controlled trial comparing robotics with laparoscopy for the treatment of endometriosis, did not observe differences in operative time between groups. Additionally, there were no significant clinical differences observed over the 6-month follow-up period regarding complications and quality of life [11]. However, long-term results were not accessed in this study. To better answer questions about the possible superiority of RS compared to SL for the treatment of deep endometriosis, a randomized controlled trial, ROBEndo, is being conducted. The protocol will include the investigation of short-term perioperative and postoperative outcomes as well as long-term quality-of-life outcomes up to 2 years after surgery [14]. Therefore, at this time, the available data are limited and heterogeneous, making it difficult to demonstrate concrete conclusions.

## Surgical planning

### Preoperative assessment and preparation for surgery

Care of the endometriosis patient can be complex and difficult. Once medical management has been exhausted and symptoms persist, patients should be offered surgical treatment [6].

Endometriosis surgical planning begins with a good evaluation of the patient's symptoms, priorities, and desired surgical outcomes. It is essential to discuss the extent and approach of the desired surgery [2,6,8,22]. Concerning radicality, the surgical management of endometriosis can be, by definition, either conservative or definitive. Conservative surgery involves the excision or ablation of endometriotic lesions with the preservation of the uterus and ovarian tissue. It is the treatment of choice for most women being less morbid and providing good results concerning pain reduction and better fertility outcomes. The disadvantage of conservative treatment is the higher possibility of pain and disease recurrence. Definitive surgery includes a hysterectomy with or without oophorectomy. It is usually offered to women that have completed childbearing with debilitating symptoms. The main disadvantages of definitive surgery are the surgical risks, loss of fertility, and surgical menopause [6,22].

When there is multiorgan involvement, it is mandatory to have a multidisciplinary approach and preoperative consultation with other specialists that may be involved in the surgical treatment, such as a coloproctologist, urologist, or thoracic surgeon, for example. This is important in order to best inform the patient of possible risks, outcomes, and expectations of postoperative recovery. Moreover, informed consent covering the typical risks, benefits, alternatives, complications, and outcomes of the surgical intervention should be given and explained to the patient preoperatively [23].

### Imaging

Surgery with histologic confirmation is historically considered the gold standard for the diagnosis of endometriosis; however, laparoscopic for diagnostic reasons is no longer recommended. In women with persistent pelvic pain with or without suggestive clinical examination findings, the use of transvaginal sonography and magnetic resonance imaging under specific protocols has been shown to accurately diagnose ovarian and deep infiltrating endometriosis.

MRI for deep endometriosis has a sensitivity of 94% and specificity of 77%, while the US has shown in a review and meta-analysis a pooled sensitivity of 79% (range 69%–89% and specificity of 94% (range 88%–100%) [24]. Although the accuracy of either method will vary depending on the skills and expertise of imagers professionals, in advanced centers, the sensitivity and specificity for both US and MRI are above 80%–90% with no substantial differences between methods [25].

Deep endometriosis is a complex disease that usually requires complex surgical treatment. It can affect multiple pelvic sites, including retro cervical/uterosacral ligament, rectosigmoid, ovaries, bladder, and ureter, besides extra pelvic locations. For this reason, adequate surgical planning can be well defined when there is proper preoperative mapping by high-quality image studies [26]. The definition of the extension of the disease and its correlation to the clinical complaints of the patients is essential for preoperative counseling about surgical strategies, the need for a multidisciplinary team, outcomes, and even possible complications.

## Anesthesia and ERAS protocol applied to gynecology minimally invasive procedures

The anesthesia management in urologic and gynecologic surgery for robotic procedures requires very close collaboration between anesthesiologists and surgeons. The anesthesiology team should be aware of the procedure extension and complexity, in particular, changes in physiology brought about by Trendelenburg positioning (around 30 degrees) and $CO_2$ pneumoperitoneum. Especially when combined with a pneumoperitoneum, a "steep" Trendelenburg can lead to a 10%–30% reduction in cardiac output due to decreased venous return from caval compression and dependent venous pooling [27–29]. Furthermore, the prolonged Trendelenburg position can result in rare but relevant complications in the recovery period that must be known by the anesthesiologist, such as laryngeal edema, postoperative confusion, and delirium presumably secondary to cerebral edema and inadequate clearance of $CO_2$ [27,30].

Before inducing general anesthesia, the patient must be properly monitored. This applies to caliber access, pulse oximetry, capnography, electrocardiography, blood pressure monitoring (invasive if indicated), temperature, level of consciousness, and neuromuscular controls. Orogastric tube placement decompresses the stomach and avoids gastric trauma with port placement. Anesthesia management intraoperatively should be of constant awareness in order to identify and correct critical issues. Intraoperative fluid overload should be avoided [27].

Enhanced recovery after surgery (ERAS) programs have been implemented in several surgical specialties, mainly for complex surgeries. It is a compilation of evidence-based best practices applied across the perioperative period in order to reduce the physiologic stress response for surgery and promote better recovery. In agreement with most of the international protocols, it is being largely adopted in our institution for RSs with the focus on optimizing patient education and perioperative expectations, decreasing the perioperative fasting period, maintaining euvolemia and normothermia, increasing mobilization, providing multimodal pain relief, providing multimodal nausea and vomiting prophylaxis, and decreasing unnecessary or prolonged use of catheters and drains [28,29,31].

Postoperative pain relief is usually achieved through a multimodal analgesia technique. Intravenous dipyrone and nonsteroidal antiinflammatory drugs are commonly used at our institution in an attempt to reduce the use of narcotic medication. The use of transversus abdominis plane blocks and wound infiltration with local anesthesia has also been described as preemptive anesthesia concepts. The neuraxial blockade is generally not required for postoperative pain relief, and thus, it is rarely used. Nausea and vomiting may persist in the postoperative, principally due to ileus, and antiemetic medication should be prophylactically prescribed [31].

## Patient positioning

Optimal patient positioning is likely to be the essential step of RS as it enables the surgeon to properly perform the surgery, decreases compression injury rate, and maximizes movements of the robotic arms. Inappropriate patient positioning is associated with inadequate exposure to the operative field as well as detrimental complications that may lead to long-term side effects. These issues can be reduced with the use of proper or strategic positioning techniques [30], such as:

- Arms should be tucked at the patient's sides.
- The armpit angle should be less than 90 degrees, ideally 60 degrees if the arms are not tucked.
- The head should be positioned in the midline of the operating room (OR) table and should not flex laterally.
- The shoulder should be kept in a position that prevents its posterior displacement.
- Egg crates or a foam mattress, instead of shoulder braces, should be used to prevent shifting.
- The thigh-trunk angle should be approximately 170 degrees.
- Allen stirrups should be used to bring the patient to a modified lithotomy position.
- The hip's flexion, abduction, and external rotation should be restrained.
- The knees should be padded, and the knees flexion should be restricted to less than 90 degrees.
- The weight of the legs should be on the heels.
- After the OR table is brought to Trendelenburg, the positions of the hands, legs, and feet should be rechecked against their displacement.

In pelvic gynecological procedures, attention to the patient's positioning is essential in order to ensure safety. The patient should be placed in a dorsal lithotomy position with adequate sacral protection and support, legs placed in stirrups under compression stockings and pneumatic compression devices, and both arms should be padded with foam and tucked alongside the body in an anatomically neutral position with bedsheets to prevent injury to the brachial plexus and peripheral nerves [30]. To avoid cephalad movement in a steep Trendelenburg position, the operating

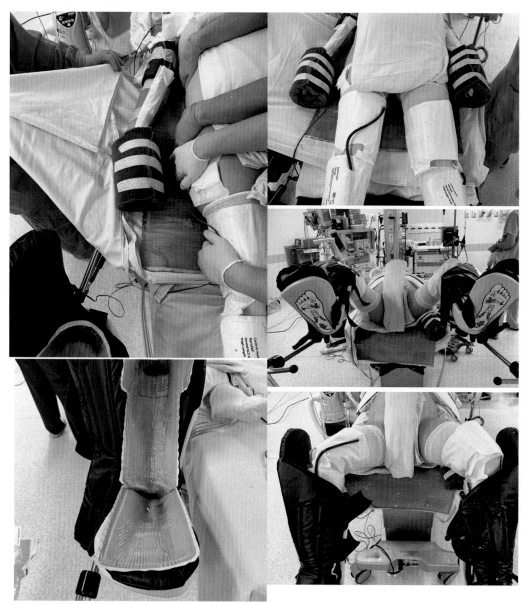

FIG. 1  Patient position before draping. *Photo: courtesy of Dr. Moretti-Marques's personal archive.*

table should have a gel pad to allow friction and molding to the patient's body. When surgery estimating time is over 2 h, it is also recommended to protect bone projections with foam dressings to avoid pressure injuries [23,27,30] (Fig. 1).

## Port placement

Port planning and placement are essential to a successful surgery. Port placement aims to prevent arm collisions and improve the range of motion of instruments and endoscopes. Surgical procedure and physical patient characteristics should be considered when deciding on trocar insertion; moreover, it should be thoroughly discussed with a senior surgeon to avoid inadequate port placement and difficulties in carrying out the surgery.

In general, the camera port is placed in the umbilical scar and should be at a minimum of 10–20 cm from the targeted surgical field. Two to three additional 8-mm robotic ports are then placed for the robotic instruments. Each of these ports needs to be approximately 10 cm apart to prevent collision. Ports should be placed perpendicularly to the abdominal wall tissue with the remote center at the level of the abdominal muscle. The remote center represents a fixed point around which the surgical arm will move. If the trocar is not correctly placed, unnecessary torque may occur on the abdominal wall causing a potential risk of lesions [23]. Some instructions are found in Strattner User's Guide, Intuitive Surgical Inc., as follows (Table 3).

Portals are generally placed in an arch or "W" in da Vinci Si and in line in da Vinci Xi (Figs. 2 and 3).

It is essential to follow the safety concepts proposed by the manufacturer and individualize the cases according to the characteristics of the patients, maintenance of surgical technical feasibility, and functional and aesthetic outcomes. For example, the patient below has short stature and previous abdominoplasty surgery, dislocating his umbilical scar caudally. In this case, the camera port is not at the umbilical scar but still respects the manufacturer's principles of a distance of 20 cm from the targeted organ (Fig. 4).

TABLE 3    Recommendations for port placement for robot-assisted surgery.

For a successful procedure, determine port placement based on a pattern of parallel lines
Perform the first trocar placement and inflate pneumoperitoneum before measuring the position of other trocars
Use the camera location as the center point (keep it 10–20 cm from the target anatomy)
Draw parallel lines **8–10 cm** (in da Vinci Si) or **6–10 cm** (in da Vinci Xi) apart based on the line from the target anatomy to the chamber door
Place da Vinci doors along the lines, keeping a distance of **10–20 cm** from the target anatomy and a distance of 8–10 cm for Si and 6–10 cm for Xi from each other ports
Triangulate da Vinci ports so they are closer to or farther from the target anatomy as needed for the procedure
In the case of accessory ports, keep a distance of at least **5 cm** (in da Vinci Si) and **7 cm** (in da Vinci Xi) from other ports, with a clear trajectory to the target anatomy
Reattach the patient cart (in da Vinci Si) or set a new target (in da Vinci Xi) if you are working on more than two quadrants

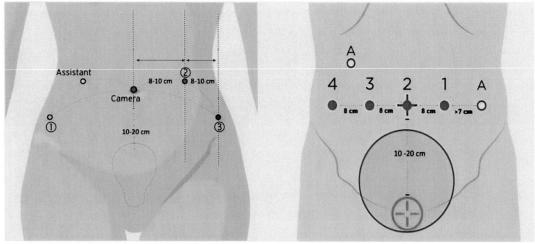

FIG. 2    Port placement illustration. Da Vinci Si on the left, where ports are placed over parallel lines 8–10 cm apart from one another, forming an arch. Da Vinci Xi on the right, where ports are placed in line at the umbilical scar height, at 6–10 cm from one another, an average of 8 cm in the picture. *Modified photo from Strattner User's Guide, Intuitive Surgical Inc.*

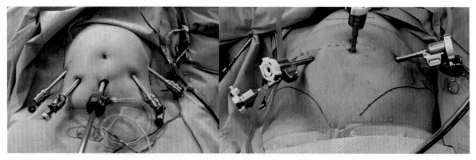

FIG. 3    Examples of port placement in da Vinci Si (left) and Xi (right). *Photo: courtesy of Dr. Renato Moretti-Marques's personal archive.*

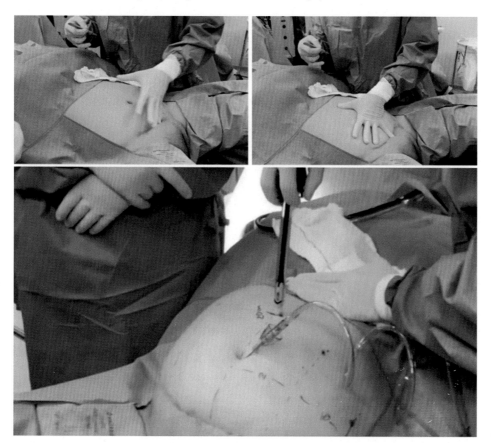

FIG. 4  Camera port placed cephalad to umbilical scar, respecting 20 cm distance from the targeted organ. *Photo: courtesy of Dr. Renato Moretti-Marques's personal archive.*

From an aesthetic point of view, there is a discussion about robotics being worse than laparoscopy due to the size and location of the incisions. In some cases, we can perform side or lower incisions that promote better cosmetic satisfaction (Fig. 5). However, each case should be evaluated individually.

Since endometriosis can have a disseminated pattern, sometimes a multiquadrant surgery is necessary to access the pelvis and the superior abdomen, such as in diaphragmatic endometriosis (DE). In these cases, special attention should be made to trocar planning to make the procedure easier and the docking changes faster. The use of assistant trocars is essential to optimize exposure, localization can vary, and more than one trocar can be necessary (Fig. 6).

Concerning the docking change, in the Si platform, it is necessary to undock all arms, move the patient's cart, and reposition it to the patient's shoulder (Fig. 7).

On the Xi platform, the patient's cart is positioned to the patient's left, and the target anatomy can be selected on display. For multiquadrant surgeries, it is not necessary to reposition the cart. Arms should be uncoupled, the boom moved, and new target set (Figs. 8 and 9).

## The complexity of surgeries and benefits of robotic surgery for endometriosis

### Distal ureteral reanastomosis

Of women diagnosed with endometriosis, approximately 1% have urinary tract endometriosis. The most frequently involved site is the urinary bladder (85%), followed by the ureter (10%), kidney (4%), and urethra (2%) [32]. The left distal ureter is the most commonly affected site. Ureteral endometriosis is rare, with an estimated prevalence of 0.1%. It can be classified as extrinsic or intrinsic and is usually unilateral [33]. Extrinsic ureteral endometriosis is the most usual finding. It is characterized by the infiltration of the peritoneum, the uterosacral ligament, the ureteral adventitia, and the surrounding connective tissue, determining extrinsic compression of the ureteral wall. Intrinsic ureteral

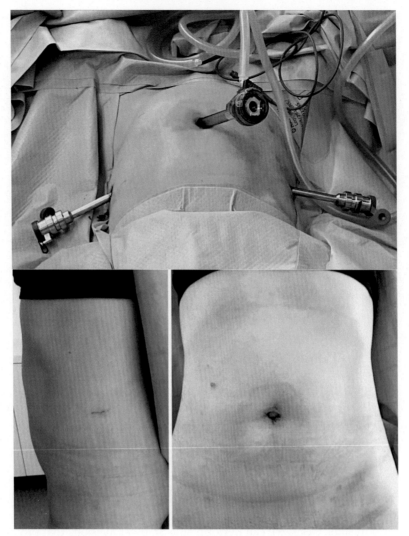

FIG. 5   Robotic treatment of posterior compartment endometriosis using da Vinci Si platform in a side docking. Ports were placed on both abdominal flanks, and the patient had a satisfactory aesthetic result. *Photo: courtesy of Dr. Renato Moretti-Marques's personal archive.*

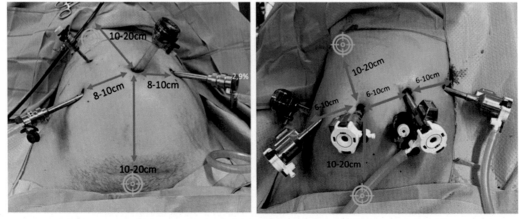

FIG. 6   Examples of port placement in da Vinci Si (left) and Xi (right) for a multiquadrant surgery for the treatment of deep and diaphragmatic endometriosis. *Photo: courtesy of Dr. Renato Moretti-Marques's personal archive.*

III. The current and future clinical applications of robotic surgery among medical specialties

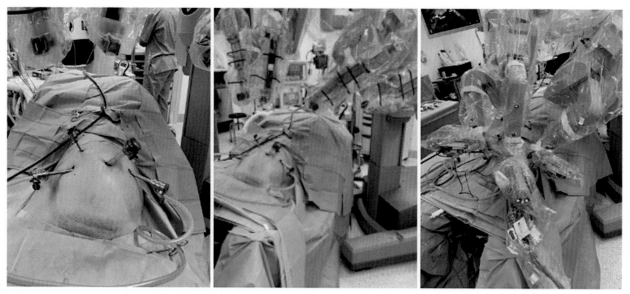

FIG. 7 Port placement and side docking for diaphragmatic endometriosis approach using the Si da Vinci platform. The patient is tilted to the left side, robotic ports are obliquely placed targeting the diaphragm, and the patient cart is side docked for the upper abdomen. *Photo: courtesy of Dr. Renato Moretti-Marques's personal archive.*

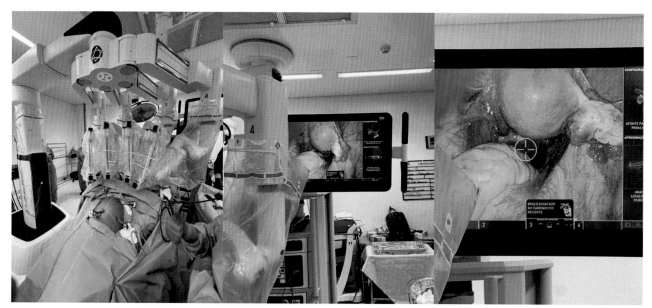

FIG. 8 The patient cart is placed on the patient's left side, the boom is rotated to the pelvis, and the target is set on the uterus. *Photo: courtesy of Dr. Renato Moretti-Marques's personal archive.*

endometriosis is less common and involves the mucosa or muscularis propria [12]. At the same time, up to 50% of women with ureteral endometriosis are asymptomatic, and around 25% are present with lower back pain, hematuria, and recurrent urinary tract infections [32].

Surgery is usually necessary to treat ureteral endometriosis in order to address urinary obstruction and preserve renal function. The treatment options may vary based on the lesion's size, location, and consequences. The general aim is to isolate and free the ureter from the endometriotic nodules, removing all endometriotic foci in the vicinity to prevent disease recurrence or progression and future ureteral stenosis. The surgical management of ureteral endometriosis includes conservative ureteral lysis or radical approaches such as urethrectomy with end-to-end anastomosis and ureteroneocystostomy [34]. Ureterolysis is the first step of all ureteral procedures, as it removes the endometriotic fibrotic tissue involving the ureter, mobilizes the ureter, and relieves the obstruction. If resection of the stenotic

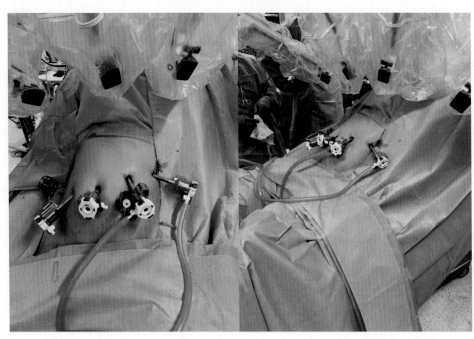

FIG. 9   The patient cart is placed on the patient's left side, the boom is rotated to the upper abdomen, and the target is set on the liver. *Photo: courtesy of Dr. Renato Moretti-Marques's personal archive.*

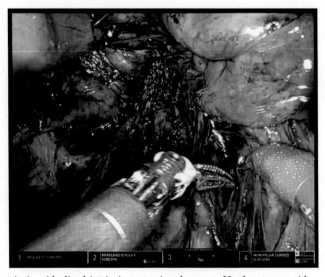

FIG. 10   Parametrial deep endometriosis with distal intrinsic ureter involvement. Urethrectomy with psoas hitch ureteroneocystostomy was performed. *Photo: courtesy of Dr. Renato Moretti-Marques's personal archive.*

segment is required in the middle or upper third, ureteroureteral anastomosis usually is performed. When the ureteral lesion is in the distal third of the ureter, ureteral resection and reimplantation (ureteroneocystostomy) are indicated [35]. Since most ureteral endometriotic lesions are located in the distal ureter, 3–4 cm above the vesicoureteral junction, ureteral resection with ureteroneocystostomy is typically performed to remove the stenotic segment and safely reimplant the ureter. In this technique, the ureter is incised proximal to the stricture and is reimplanted into the bladder, typically at the dome. A bladder-psoas hitch can be required to achieve a tension-free anastomosis (Fig. 10). An antireflux plasty of the bladder valve is performed to avoid ascending infections.

In the last decades, the literature confirms that ureterolysis can be adequately achieved by a laparoscopic approach with acceptable complication rates and a low risk of recurrences. Bosev et al., 2009 published the largest series on the laparoscopic approach to ureteral endometriosis, observing 1% of ureteral complications [36]. However, these results largely depend on the operator's dexterity and experience, together with the complexity of the disease. Moreover, the majority of the studies about the treatment of ureteral endometriosis by laparoscopy have only focused on ureterolysis. On the other hand, laparoscopic treatment for ureteral implantation is technically difficult, requires high suture skills, and has higher overall complication rates compared to ureterolysis in the treatment of endometriosis [37]. With advances in science and technology, robot-assisted laparoscopic reconstructive surgery of the ureter has been shown to be safe and feasible. Robot-assisted surgery provides three-dimensional magnified images and increases wrist flexibility with a reduction in tremors, and allows for more complex surgical procedures to be performed through a laparoscopic approach. However, robot-assisted laparoscopic ureteral reconstruction for ureteral endometriosis has rarely been reported. The literature available consists basically of retrospective case series with small sample sizes, and case reports [12,33,38]. In our experience, robotic-assisted surgery for the treatment of urinary endometriosis, especially involving the distal ureter, provides an unmeasurable benefit for the realization and success of the procedure, being the first choice for most of the urologists of our institution.

## Diaphragmatic endometriosis

The diaphragm and the thoracic cavity are among the most common localization of extrapelvic endometriosis. The incidence of the disease in the general population is not known. However, it has been reported in <1% of women undergoing pelvic surgery for suspected or known pelvic endometriosis [39].

The diagnosis and management of DE still represent a challenge, especially due to the scanty data reported in the literature, the poor confidence of the gynecologist with upper-abdominal surgical procedures, and the difficulty of preoperative diagnosis [5]. The majority of patients (70%) are asymptomatic, but some women can present nonspecific symptoms, such as catamenial shoulder, right upper quadrant, and arm or chest pain.

The pathogenesis is not completely understood, but diaphragmatic implants can be attributed to Sampson's theory of retrograde menstruation and to the physiological circulation of peritoneal fluid toward the subphrenic regions, allowing cells to infiltrate the diaphragm or to migrate to the pleural cavity via diaphragmatic fenestrations.

There are just case reports and a few case series about the treatment and management of DE in the literature. Most of the cases were treated by SL [5,13,16,39,40]. However, robotic-assisted surgery has shown promising results because of the better access to the upper abdomen with the same trocar ports, strategic exposure of this challenging region, and broad dissection possibilities using its articulated arms (Fig. 11).

Ceccaroni et al., 2021 classify diaphragmatic lesions in three different presentations [5]:

1. "Foci": bidimensional superficial lesions, generally ≤1 cm diameter (Fig. 1A).
2. "Nodules": solid, tridimensional implants associated with partial or full-thickness infiltration of underneath muscle layers and generally >1 cm diameter.
3. "Plaques": fibrotic bidimensional lesions, superficially infiltrating and thickening the diaphragmatic peritoneum, often resulting from the confluence of different foci and responsible for dense adhesions between diaphragm and liver, > 1 cm diameter.

The literature describes different approaches and techniques to treat DE [40].

Argon beam coagulator can be used to vaporize small superficial lesions (i.e., without infiltration of muscle layers). When the laser is not available, superficial lesions can be treated with diathermocoagulation. Superficial lesions with a diameter >1 cm are usually excised by "peritoneal stripping." When the disease infiltrates the muscle superficially, nodulectomy can be performed by excising only a small amount of the involved fibers without opening the thoracic cavity. In case of full-thickness infiltration of the diaphragm or dense adhesions to the central tendon of the muscle, entry into the pleural cavity is usually unavoidable. In these cases, after informing the anesthesiologist, the surgeon should remove the lesion with a "full-thickness resection" of the diaphragm [13].

## Bowel sparing

Rectovaginal and bowel endometriosis are forms of deep infiltrating endometriosis that invades at least to the level of the bowel muscularis. Most bowel lesions do not infiltrate the full thickness of the bowel wall. Among women with endometriosis, the reported prevalence of rectovaginal or bowel involvement ranges widely, from 5% to 25%, with the bowel being the most common site for extragenital endometriosis [41]. Patients are usually present with the classic symptoms of endometriosis (dysmenorrhea, dyspareunia, and infertility) and/or with gastrointestinal symptoms

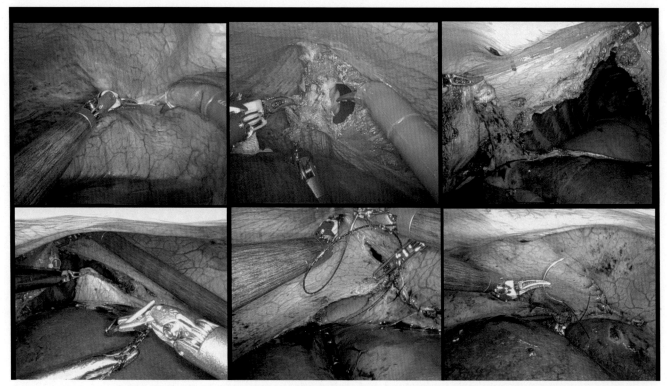

FIG. 11  Endometriosis affecting the diaphragmatic dome. Full-thickness excision and diaphragm closure using the da Vinci Xi platform. The pneumothorax was evacuated before closing the final stitch, after adequate expansion of the lung, obtained by positive pressure ventilation and putting the aspirator into the pleural cavity. *Photo: courtesy from Dr. Vanessa Alvarenga-Bezerra's and Dr. Renato Moretti-Marques's personal archive.*

(painful defecation, dyschezia, rectal bleeding, constipation, and/or bloating). The rectovaginal disease often presents with localized symptoms (deep dyspareunia and dyschezia) [23,38,41].

Medical management is often recommended as an initial therapy, it is not curative, and surgery is often required as an adjunct for the management of symptoms [23]. MIS is the standard of care, providing less blood loss, quicker recoveries, and better outcomes compared to laparotomy [42]. The surgical management of advanced-stage bowel endometriosis can be technically difficult, and SL can offer limitations even in the hands of an experienced surgeon. RS allows for important depth perception with 3D imaging, improved range of motion, tremor filtering, the scale of motion, and improved ergonomics [9,38]. However, again currently available evidence regarding the use of RS for the treatment of bowel endometriosis does not show its superiority compared to laparoscopy.

Bowel endometriosis surgical treatment consists of the excision of all the diseases compromising the viscera. It can be more conservative with a shaving or disc colectomy, or it may require a segmental resection depending on the depth of invasion and extension of the disease. In some case reports and in our experience, RS can improve rectal-sparing procedures because of its precision benefits (Fig. 12) [9]. The literature, however, does not allow us to make this affirmation.

## Parametrial endometriosis

Lateral parametrium (LP) is defined as the retroperitoneal connective areolar tissue that extends from the uterus to the pelvic side wall, surrounding uterine vessels and enveloping lymphatic structures and nerves. The ureter passes through the LP dividing it into the cranial and caudal region, called para cervix, in which the deep uterine vein represents the main anatomic landmark for the pelvic autonomic nerves [43].

LP endometriosis has been associated with more severe disease, ureteral stenosis, voiding dysfunctions, and hypogastric nerve involvement. Its surgical treatment with complete eradication of endometriosis can cause important morbidity related to urinary and rectal voiding dysfunctions that should be highly discussed with the patient preoperatively [44,45].

Access to this region should be done cautiously because of the numerous important structures in the surroundings that should be preserved whenever possible [45]. There are case reports of the use of RS for the treatment of this condition successfully [46]. Fig. 13 shows a case of parametrial and sacral plexus endometriosis that was treated with

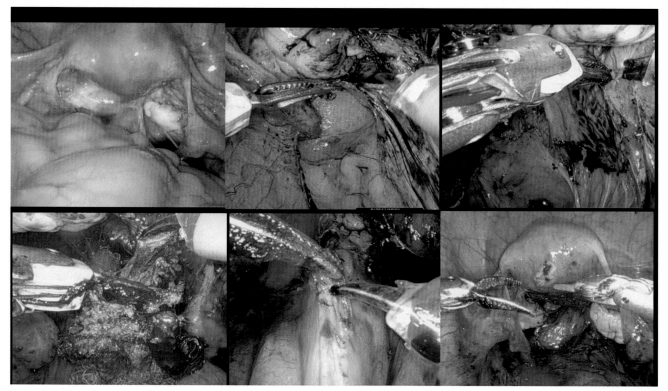

FIG. 12 Cul-de-sac obliteration due to deep endometriosis, compromising uterine torus, lateral parametria, anterior rectum, and rectovaginal septum. A posterior peritonectomy and rectal shaving was performed with complete excision of the disease. *Photo: courtesy of Drs. Mariana Rossette and Renato Moretti-Marques's personal archives.*

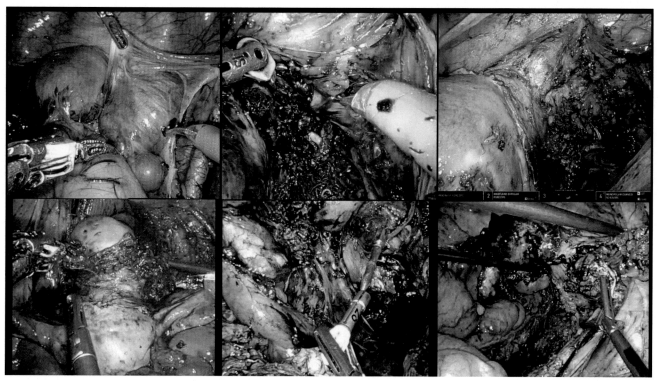

FIG. 13 Parametrial and sacral plexus deep endometriosis causing right ureter obstruction and sensorial and motor dysfunction due to sacral nerve roots and sciatic nerve involvement. *Photo: courtesy of Renato Moretti-Marques's personal archive.*

complete removal of deep endometriosis lesions of the right parametrium, ureter reimplantation, segmental rectosigmoid resection, and somatic nerves decompression. The patient evolved exceptionally well after surgery, substantially improving chronic pain, quality of life, and motor dysfunction.

## Future scenario

In our experience, the use of RS for the treatment of endometriosis results in better preservation of normal tissues and nerves with excellent outcomes. However, the literature is conflicting about its superiority when compared to SL. Until we do not have enough data to support the benefits of RS in the surgical treatment of endometriosis, we can affirm at least that its shorter learning curve can allow more surgeons to perform the surgical treatment of endometriosis. On the other hand, we recognize that access to this technology patients should be democratized by reducing its costs.

New robotic platforms are being incorporated with the promise of cost reduction such as Versius and Hugo Ras; however, there aren't publications about their use in the endometriosis surgery field [47,48]. Our institution has recently acquired the Hugo Ras robot, and our group has successfully performed a few endometriosis surgeries (Fig. 14). In our perception, the high-quality images provided by the new Medtronic robot platform may be a benefit to increase surgical precision. Fig. 14 shows a case of a patient with deep endometriosis and the diagnosis of cervical cancer stage IB1 that was submitted to radical hysterectomy with salpingo-oophorectomy using the Hugo Ras platform. Colpectomy and vaginal closure were performed vaginally, respecting oncologic precepts.

## Conclusions

Endometriosis is a challenging heterogeneous disease that can cause great anatomical distortion and involve multiple organs. Surgical treatment of endometriosis has been shown to improve patients' symptoms and fertility. Moreover, it may be indicated in some dangerous conditions, such as deep infiltrative bowel disease and ureter and thoracic endometriosis.

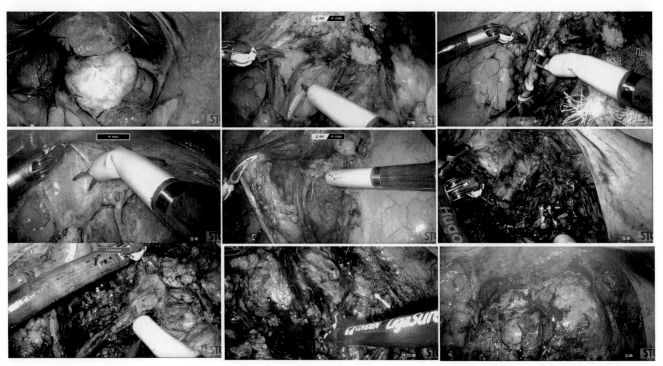

FIG. 14    Radical hysterectomy and bilateral salpingo-oophorectomy performed in a patient with deep endometriosis. *Photo: courtesy of Dr. Renato Moretti-Marques's archive.*

MIS is the standard of surgical care for deep endometriosis showing numerous advantages when compared to laparotomy; however, laparoscopy can have limitations when very complex surgeries are required, even for very experienced surgeons. RS is a surgical modality that demonstrates promising benefits in the treatment of deep endometriosis, especially for advanced diseases. Ergonomics, the precision of movements, articulated movements, and efficiency of monopolar and bipolar energy provide surgical efficiency and safer access to difficult anatomic regions.

## Key points

- Surgical resection of endometriosis lesions has consistently improved pain symptoms, fertility, and patient's quality of life.
- RS is a surgical modality that demonstrates promising benefits in the treatment of deep endometriosis, especially for advanced diseases.
- The available evidence shows that a robotic approach is feasible, effective, and safe, delivering similar surgical outcomes compared to conventional laparoscopy.
- The superiority of robotics in terms of surgical outcomes compared to laparoscopy is still to be proven and uncertainties on cost-effectiveness remain.

## References

[1] Horne AW, Missmer SA. Pathophysiology, diagnosis, and management of endometriosis. BMJ 2022;379, e070750.

[2] Zondervan KT, Becker CM, Missmer SA. Endometriosis. N Engl J Med 2020;382:1244–56.

[3] Missmer SA, Tu FF, Agarwal SK, et al. Impact of endometriosis on life-course potential: a narrative review. Int J Gen Med 2021;14:9–25.

[4] Simoens S, Hummelshoj L, D'Hooghe T. Endometriosis: cost estimates and methodological perspective. Hum Reprod Update 2007;13:395–404.

[5] Ceccaroni M, Roviglione G, Farulla A, Bertoglio P, Clarizia R, Viti A, Mautone D, Ceccarello M, Stepniewska A, Terzi AC. Minimally invasive treatment of diaphragmatic endometriosis: a 15-year single referral center's experience on 215 patients. Surg Endosc 2021;35:6807–17.

[6] Becker CM, Bokor A, Heikinheimo O, et al. ESHRE guideline: endometriosis. Hum Reprod Open 2022;2022, hoac009.

[7] Vercellini P, Aimi G, Busacca M, Apolone G, Uglietti A, Crosignani PG. Laparoscopic uterosacral ligament resection for dysmenorrhea associated with endometriosis: results of a randomized, controlled trial. Fertil Steril 2003;80:310–9.

[8] Duffy JMN, Arambage K, Correa FJS, Olive D, Farquhar C, Garry R, Barlow DH, Jacobson TZ. Laparoscopic surgery for endometriosis. Cochrane Database Syst Rev 2014; CD011031.

[9] Araujo SEA, Seid VE, Marques RM, Gomes MTV. Advantages of the robotic approach to deep infiltrating rectal endometriosis: because less is more. J Robot Surg 2016;10:165–9.

[10] Capozzi VA, Scarpelli E, Armano G, et al. Update of robotic surgery in benign gynecological pathology: systematic review. Medicina 2022. https://doi.org/10.3390/medicina58040552.

[11] Soto E, Luu TH, Liu X, Magrina JF, Wasson MN, Einarsson JI, Cohen SL, Falcone T. Laparoscopy vs. robotic surgery for endometriosis (LAROSE): a multicenter, randomized, controlled trial. Fertil Steril 2017;107. 996–1002.e3.

[12] Giannini A, Pisaneschi S, Malacarne E, Cela V, Melfi F, Perutelli A, Simoncini T. Robotic approach to ureteral endometriosis: surgical features and perioperative outcomes. Front Surg 2018;5:51.

[13] Delara R, Suárez-Salvador E, Magrina J, Magtibay P. Robotic excision of full-thickness diaphragmatic endometriosis. J Minim Invasive Gynecol 2020;27:815.

[14] Terho AM, Mäkelä-Kaikkonen J, Ohtonen P, Uimari O, Puhto T, Rautio T, Koivurova S. Robotic versus laparoscopic surgery for severe deep endometriosis: protocol for a randomised controlled trial (ROBEndo trial). BMJ Open 2022;12, e063572.

[15] Kanno K, Andou M, Aiko K, Yoshino Y, Sawada M, Sakate S, Yanai S. Robot-assisted nerve plane-sparing eradication of deep endometriosis with double-bipolar method. J Minim Invasive Gynecol 2021;28:757–8.

[16] Roman H, Dennis T, Grigoriadis G, Merlot B. Robotic management of diaphragmatic endometriosis in 10 steps. J Minim Invasive Gynecol 2022;29:707–8.

[17] Saget E, Peschot C, Bonin L, et al. Robot-assisted laparoscopy for deep infiltrating endometriosis: a retrospective French multicentric study (2008-2019) using the Society of European Robotic Gynecological Surgery endometriosis database. Arch Gynecol Obstet 2022;305:1105–13.

[18] Moon AS, Garofalo J, Koirala P, Vu M-LT, Chuang L. Robotic surgery in gynecology. Surg Clin North Am 2020;100:445–60.

[19] Berlanda N, Frattaruolo MP, Aimi G, Farella M, Barbara G, Buggio L, Vercellini P. "Money for nothing". The role of robotic-assisted laparoscopy for the treatment of endometriosis. Reprod Biomed Online 2017;35:435–44.

[20] Mosbrucker C, Somani A, Dulemba J. Visualization of endometriosis: comparative study of 3-dimensional robotic and 2-dimensional laparoscopic endoscopes. J Robot Surg 2018;12:59–66.

[21] Restaino S, Mereu L, Finelli A, et al. Robotic surgery vs laparoscopic surgery in patients with diagnosis of endometriosis: a systematic review and meta-analysis. J Robot Surg 2020;14:687–94.

[22] Practice Committee of the American Society for Reproductive Medicine. Treatment of pelvic pain associated with endometriosis: a committee opinion. Fertil Steril 2014;101:927–35.

[23] Hur C, Falcone T. Robotic treatment of bowel endometriosis. Best Pract Res Clin Obstet Gynaecol 2021;71:129–43.

[24] Nisenblat V, Bossuyt PMM, Farquhar C, Johnson N, Hull ML. Imaging modalities for the non-invasive diagnosis of endometriosis. Cochrane Database Syst Rev 2016;2, CD009591.

[25] Mattos LA, Goncalves MO, Andres MP, Young SW, Feldman M, Abrão MS, Kho RM. Structured ultrasound and magnetic resonance imaging reports for patients with suspected endometriosis: guide for imagers and clinicians. J Minim Invasive Gynecol 2019;26:1016–25.

[26] Trippia CH, Zomer MT, Terazaki CRT, Martin RLS, Ribeiro R, Kondo W. Relevance of imaging examinations in the surgical planning of patients with bowel endometriosis. Clin Med Insights Reprod Health 2016;10:1–8.

[27] Aceto P, Beretta L, Cariello C, et al. Joint consensus on anesthesia in urologic and gynecologic robotic surgery: specific issues in management from a task force of the SIAARTI, SIGO, and SIU. Minerva Anestesiol 2019;85:871–85.

[28] Scheib SA, Thomassee M, Kenner JL. Enhanced recovery after surgery in gynecology: a review of the literature. J Minim Invasive Gynecol 2019;26:327–43.

[29] Stone R, Carey E, Fader AN, et al. Enhanced recovery and surgical optimization protocol for minimally invasive gynecologic surgery: an AAGL white paper. J Minim Invasive Gynecol 2021;28:179–203.

[30] Takmaz O, Asoglu MR, Gungor M. Patient positioning for robot-assisted laparoscopic benign gynecologic surgery: a review. Eur J Obstet Gynecol Reprod Biol 2018;223:8–13.

[31] Alvarenga-Bezerra V, Moretti-Marques R, Barbosa MG, Rios GM, Mengai ACS, Assir FF, Pimenta ECS, Podgaec S. Short-term outcomes after an enhanced recovery after surgery protocol in abdominal hysterectomies for leiomyomas in a teaching hospital. Int J Gynaecol Obstet 2022. https://doi.org/10.1002/ijgo.14594.

[32] Schneider A, Touloupidis S, Papatsoris AG, Triantafyllidis A, Kollias A, Schweppe K-W. Endometriosis of the urinary tract in women of reproductive age. Int J Urol 2006;13:902–4.

[33] Hung Z-C, Hsu T-H, Jiang L-Y, Chao W-T, Wang P-H, Chen W-J, Chen Y-J, Lin ATL. Robot-assisted laparoscopic ureteral reconstruction for ureter endometriosis: case series and literature review. J Chin Med Assoc 2020;83(3):288–94. https://doi.org/10.1097/JCMA.0000000000000249.

[34] Di Maida F, Mari A, Morselli S, et al. Robotic treatment for urinary tract endometriosis: preliminary results and surgical details in a high-volume single-institutional cohort study. Surg Endosc 2020;34:3236–42.

[35] Drain A, Jun MS, Zhao LC. Robotic ureteral reconstruction. Urol Clin North Am 2021;48:91–101.

[36] Bosev D, Nicoll LM, Bhagan L, Lemyre M, Payne CK, Gill H, Nezhat C. Laparoscopic management of ureteral endometriosis: the Stanford University hospital experience with 96 consecutive cases. J Urol 2009;182:2748–52.

[37] Mereu L, Gagliardi ML, Clarizia R, Mainardi P, Landi S, Minelli L. Laparoscopic management of ureteral endometriosis in case of moderate-severe hydroureteronephrosis. Fertil Steril 2010;93:46–51.

[38] Nezhat C, Hajhosseini B, King LP. Robotic-assisted laparoscopic treatment of bowel, bladder, and ureteral endometriosis. JSLS 2011;15:387–92.

[39] Nezhat C, Lindheim SR, Backhus L, Vu M, Vang N, Nezhat A, Nezhat C. Thoracic endometriosis syndrome: a review of diagnosis and management. JSLS 2019. https://doi.org/10.4293/JSLS.2019.00029.

[40] Moawad G, Youssef Y, Ayoubi JM, Feki A, De Ziegler D, Roman H. Diaphragmatic endometriosis: robotic approaches and techniques. Fertil Steril 2022;118:1194–5.

[41] Nezhat C, Li A, Falik R, et al. Bowel endometriosis: diagnosis and management. Am J Obstet Gynecol 2018;218:549–62.

[42] Siesto G, Ieda N, Rosati R, Vitobello D. Robotic surgery for deep endometriosis: a paradigm shift. Int J Med Robot 2014;10:140–6.

[43] Mabrouk M, Raimondo D, Arena A, Iodice R, Altieri M, Sutherland N, Salucci P, Moro E, Seracchioli R. Parametrial endometriosis: the occult condition that makes the hard harder. J Minim Invasive Gynecol 2019;26:871–6.

[44] Lemos N, D'Amico N, Marques R, Kamergorodsky G, Schor E, Girão MJBC. Recognition and treatment of endometriosis involving the sacral nerve roots. Int Urogynecol J 2016;27:147–50.

[45] Kanno K, Yanai S, Sawada M, Sakate S, Andou M. Nerve-sparing surgery for deep lateral parametrial endometriosis. Fertil Steril 2022;118:992–4.

[46] Alboni C, Farulla A, Facchinetti F, Ercoli A. Robot-assisted nerve-sparing resection of bilateral parametrial deep infiltrating endometriosis. J Minim Invasive Gynecol 2021;28:18–9.

[47] Gueli Alletti S, Chiantera V, Arcuri G, Gioè A, Oliva R, Monterossi G, Fanfani F, Fagotti A, Scambia G. Introducing the new surgical robot HUGO™ RAS: system description and docking settings for gynecological surgery. Front Oncol 2022;12, 898060.

[48] Monterossi G, Pedone Anchora L, Gueli Alletti S, Fagotti A, Fanfani F, Scambia G. The first European gynaecological procedure with the new surgical robot Hugo™ RAS. A total hysterectomy and salpingo-oophorectomy in a woman affected by BRCA-1 mutation. Facts Views Vis Obgyn 2022;14:91–4.

# Breast robotic surgery

*Marina Sonagli[a], Antonio Toesca[b], Giada Pozzi[b],*
*Guglielmo Gazzetta[b], and Fabiana Baroni Alves Makdissi[a]*

[a]Department of Mastology, A.C. Camargo Cancer Center, São Paulo, Brazil [b]Division of Breast Surgery, European
Institute of Oncology, Istituto di Ricovero e Cura a Carattere Scientifico (IRCCS), Milan, Italy

## Introduction

Robotic breast surgery was first described by Toesca et al. in 2015, when three cases of robot-assisted Nipple-sparing mastectomies were reported. These initial cases comprised BRCA mutation carrier patients with a previous history of breast cancer surgery who decided to a delayed contralateral prophylactic NSM and Immediate breast reconstruction [1,2]. Before the development of robot-assisted surgery, endoscopic NSM was attempted to achieve better cosmetic outcomes and greater patient satisfaction [3,4]. Endoscopic technique was indicated to patients with small breasts in whom Breast conserving therapy could result in poor cosmesis [5]. However, complication rates and cosmesis satisfaction of endoscopic NSM were not statistically different when compared to open NSM [3]. In addition, the 2-dimensional endoscopic camera associated with rigid instruments turns this technique more difficult, contributing to the lack of spread of endoscopic breast surgery worldwide [3]. As an evolution of endoscopic NSM, better feasibility of robot-assisted NSM promoted its diffusion and acceptance worldwide, with many studies demonstrating the advantages of robotic NSM since Toesca's report [6].

da Vinci robotic surgery system has also been used for breast reconstruction. The first robot-assisted plastic surgery was reported in 2011 by Patel et al. as a latissimus dorsi flap for shoulder reconstruction for sarcoma resection [7]. In the next year, Selber et al. reported a case series of robot-assisted latissimus dorsi flap reconstruction, in which five cases were for breast reconstruction [8]. Since then, many studies have shown feasibility of immediate robotic breast reconstructive surgery with autologous flap or breast implants [2,9–13].

## Background

Robotic Nipple-sparing mastectomy with immediate breast reconstruction is a relatively new technique developed to seek for less complication rates and better cosmetic outcomes compared to the conventional technique, C-NSM. One of the most feared complications that R-NSM tries to avoid is nipple necrosis. The three-dimensional camera of da Vinci can offer a tenfold magnified view of the operating field, facilitating the identification of skin and nipple vascularity and avoiding their damages. Also, the intense lighting attached to the camera enables a better differentiation of blood vessels, lymphatics, adipose lobules, ligaments, the mammary gland, skin, and nipple through translucence, contributing to less nipple and skin necrosis [13,14].

Rates of nipple ischemia related to R-NSM were reported as varying from 5.6%–13%, although, not every nipple ischemia evolved to nipple necrosis [15–17]. The highest rates of nipple necrosis related to R-NSM ranged from 1.1%–2.2% [17–20]. Other complications described for R-NSM comprise skin ischemia (6.1%) [15], infection (2.1%–7.3%) [14,18,19], implant loss (1.6%–5%) [14,18,19,21], hematoma (1.9%–10%) [16,19,21], and seroma (5.3%) [19].

Studies have compared R-NSM and C-NSM. Overall complication rates reported ranged from 21.8%–58.5% for R-NSM and 27.5%–50% for C-NSM [16,18,20–22]. None of these studies showed statistical differences between the two techniques. Some studies have described more nipple necrosis, skin necrosis, implant loss, and hematoma for C-NSM, even if not statistically relevant [16,20,21,23]. A metaanalysis that evaluated 49 studies and included 13,886 cases of R-NSM and C-NSM reported overall complication rates of 3.9% for R-NSM and 7% for C-NSM, however, this difference was not statistically significant ($P = 0.07$). This metaanalysis also showed slightly higher rates of implant removal (4.1% for R-NSM vs 3.2% for C-NSM), hematoma (4.3% for R-NSM vs 2% for C-NSM), and infection (8.3% for R-NSM vs 4% for C-NSM), although, without statistically significant differences [6].

Despite the absence of statistical differences shown in other studies, Park et al. reported a 62% reduction on the risk of nipple necrosis ($P = 0.044$, HR 0.48; and CI 95% 0.15–0.98) and a 38% reduction in the risk of grade 3 postoperative complications ($P = 0.045$, HR 0.62; IC 95% 0.39–0.99) related to R-NSM when compared to C-NSM [20]. Similar results were described by Lai et al.. In this study, grade 3 complications were reported in 13.8% of C-NSM and only in 2.6% of R-NSM ($P = 0.045$) [23].

A few studies evaluated postoperative pain related to R-NSM [23–25]. Lai et al. reported a higher postoperative pain in C-NSM when compared to R-NSM ($P = 0.02$) [23]. Moon et al. found a significantly higher pain score in patients who underwent R-NSM on the first 6 postoperative hours ($P = 0.007$), however, no differences were observed at other periods ($P = 0.338$) and no need for additional analgesics during the first 48 postoperative hours [24]. Nevertheless, to minimize pain after R-NSM, an intraoperative pectoralis nerve II block under ultrasound guidance might lead to a lower pain score within the first 2 h ($P < 0.05$) and less consumption of fentanyl within the first 24 postoperative hours [25].

Another objective of R-NSM is to improve cosmetic outcomes impacting on quality of life and patients' self-esteem. Not only the wound length, but also Nipple-areola complex sensitivity and NAC positioning contribute to a better body image after surgery. Toesca et al. reported NAC sensitivity preservation on 31.6% of R-NSM versus loss of NAC sensitivity of C-NSM ($P = 0.0002$) and a higher satisfaction with NAC positioning. In this study, 88.9% of patients who underwent R-NSM reported being satisfied and very satisfied compared to 55.6% of patients who had C-NSM ($P = 0.0005$) [21]. Greater overall satisfaction rate for R-NSM (92% for R-NSM vs 75.6% for C-NSM, $P = 0.046$) and higher satisfaction with NAC positioning with R-NSM were also described by Lai et al. [16]. Other studies, however, did not demonstrate significant differences in patients' satisfaction [22,23], physical well-being [20,23], sexual well-being [20,23], and psychosocial well-being [23] between R-NSM and C-NSM.

Advantages of robotic surgery have been described not only for patients, but also for surgeon. In some cases, Nipple-sparing mastectomy might lead to a physical demand on the surgeon, such as neck, back pain, and fatigue. The use of a console, with ergonomic adjustment, can prevent these complaints [26].

Robot-assisted Nipple-sparing mastectomy was reported for prophylactic and therapeutic purposes. This technique is indicated for patients with small breasts (cup size A, B, or C) and without accentuated ptosis. For breast cancer treatment, R-NSM is suitable for patients with clinical tumor size of up to 5 cm, with a tumor-skin distance of at least 3 mm and without clinical NAC involvement. Therapeutic R-NSM is not recommended to patients with inflammatory breast cancer or with tumors involving skin, pectoralis of major muscle or chest wall. Neoadjuvant treatment is not a contraindication for R-NSM [27]. Sentinel node biopsy or axillary dissection can be performed through the same incision of R-NSM, however, patients with low disease burden in the axilla are more suitable for R-NSM [27,28].

Due to the recent use of R-NSM, a few studies could evaluate oncologic outcomes. After a 18-month follow-up, Park et al. did not find significant differences in Disease-free survival and Overall survival between R-NSM and C-NSM [20]. Also, no recurrences were observed by Toesca et al. after 19 months of follow-up in 56 therapeutic R-NSM [19]. The longer follow-up of R-NSM reported was a median of 27 months. In this study, 2.7% of local recurrence events in C-NSM group and 1.3% in R-NSM group were detected and no mortality was reported [23]. An updated analysis of a previous RCT conducted by Toesca et al. [21] did not show significant differences in OS ($P = 0.970$) and DFS ($P = 0.760$) between R-NSM and C-NSM after a 42-month follow-up. There were no nipple recurrences in the R-NSM group [29]. Nevertheless, time to evaluate oncological safety of R-NSM is still too short, requiring studies with a higher level of evidence. There are several ongoing studies to assess the 5-year oncologic outcomes as primary or secondary endpoints [30–33].

Costs, operating time, and learning curve are disadvantages that hinder the use of R-NSM worldwide. According to Houvenaeghel et al. costs of R-NSM with immediate breast reconstruction with implants were 34.7% higher than C-NSM and 30% higher when R-NSM was associated with reconstruction with autologous flap [22]. Higher costs related to R-NSM were also reported by Lai et al. ($P < 0.01$) [16,23]. Many variables may contribute to increased cost of a procedure and operating time might be one of them. Studies demonstrated a longer time for R-NSM (ranging from 182 to 224 min, on average) when compared to C-NSM (190–197 min) ($P < 0.01$) [16,18,23].

As any surgical technique, a learning curve to be proficient is necessary. As expected, simpler procedures require a faster learning curve, while innovative and more complex procedures require longer training. Learning curve varies among studies. Toesca et al. reported a gradual reduction in operating time: the total length of the first robotic surgery (positioning patient, docking, mastectomy, and reconstruction) dropped from 7–3 h after 21 cases [2]. Loh et al., in turn, described a significant improvement in time for R-NSM, reconstruction and total procedure after the 26th surgery [17]. Learning curve observed by Lai et al. had 10–12 cases to achieve an operating time close to C-NSM [34]. After analyzing a larger number of cases, Lai et al. reported a significant drop in the operative time after 14 cases: docking time dropped from 13 to 9 min ($P = 0.024$), R-NSM plus immediate reconstructive time dropped from 121 to 89 min ($P = 0.018$), and total procedure time dropped from 287 to 235 min ($P = 0.019$) [35].

Because of difficulties for breast surgeons to gain experience using the robotic surgical system alone, Lee et al. developed a training program using simulators of the basic robotic surgical system, cadaveric and animal models. In this training program, 12 procedures were performed (10 using cadaveric model and 2 using porcine model). Trainees' feedback revealed that the program met the satisfactory skills development [36]. Due to the higher operating time reported in the literature, especially, during the initial learning curve, R-NSM is not recommended for patients with a poor performance and multiple comorbidities in the beginning of the learning curve [27].

## Technique

The use of da Vinci Si (Intuitive Surgical, Sunnyvale, CA) [13] or da Vinci Xi (Intuitive Surgical, Sunnyvale, CA) [37] for R-NSM has been described in the literature. The surgical technique is the same in both risk-reducing and therapeutic NSMs.

### Positioning the patient

The first challenge of R-NSM rests on positioning the patient. The space required for robot docking can create blind spots to the camera, collision of the device's arms and patient's arm and lead to inadvertent neuropraxia [14]. The patient is placed in a comfort supine position with the ipsilateral arm abducted 90 degrees [17,35,37]. Other possibility is to place the ipsilateral upper limb slightly anteflexed above the head [2,38]. More importantly, all experts agreed that the ideal limb position was the one that can avoid interference with the robotic arms [27]. Patient's position is shown in Fig. 1.

### Mastectomy technique

After positioning the patient, a 3–5 cm lateral-thoracic incision, in the midaxillary line in the axillary fossa promotes a scar hidden by patient's arm [2,14,17]. The subcutaneous flap is dissected under directed vision in a 3 cm area. Through this incision it is possible to perform axillary lymph node surgery or sentinel lymph node biopsy and to introduce the single port (Fig. 2) [2,17]. A physiological saline solution of 1000 mL containing 0.05% xylocaine 20 mL, epinephrine 1 mL, and $NaHCO_3$ 20 mL or 0.05% lidocaine and epinephrine 1:1,000,000 can be injected into

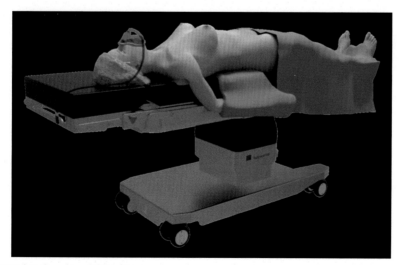

FIG. 1    Right thorax positioned at the edge of OR table.

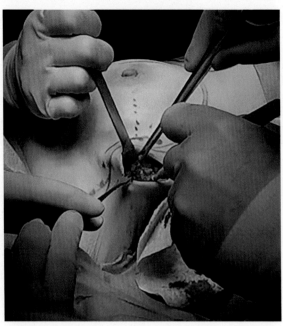

FIG. 2   Skin flap dissection around the axillary incision to create a subcutaneous cavity.

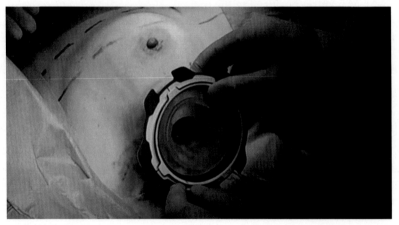

FIG. 3   Monoport insertion through the incision.

the subcutaneous layer to minimize perioperative bleeding [17,35]. Before introducing the single port, a subcutaneous dissection is performed as far as possible with scissors [14].

Some single port devices are reported to be used: Nelis glove port (Glove Port, Nelis, Gyeonggido, Korea) with three 5–12 mm channels [17], GelPoint mono-trocar [38,39], and a laparoscopic adapted single-port (Access Transformer OCTO; Seoul, Korea) consisting of four 5–12 mm channels and a silicone gas pipe (Fig. 3) [2]. After placement of the single port, carbon dioxide is inflated to create a space for mastectomy, and the pressure is constantly maintained at 8 mmHg [2,14,35]. Then the robot is docked with the cart on contralateral side of the operation and the robotic arms nearly parallel with the floor, to avoid conflicts during dissection (Fig. 4) [2]. Robot can also be docked behind the patient with the two robotic arms and the endoscope extending over the patient [17,35].

After completing docking, the surgeon moves to the console. A 30° 12-mm diameter camera (Intuitive Surgical, Sunnyvale, CA) is used in the upper port to enable a central view [13,14,17,35]. Dissection can be performed using a 5 mm monopolar cautery with a cautery spatula tip (Intuitive Surgical, Sunnyvale, CA) [2] or an 8 mm monopolar scissor (Intuitive Surgical, Sunnyvale, CA) [17,35] on the right robotic arm. To maintain an excellent exposure and stretch-out of the tissue by traction and counter-traction, an 8 mm ProGrasp forceps (Intuitive Surgical, Sunnyvale, CA) is used on the left robotic arm [2,17,35].

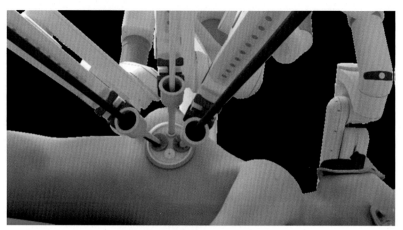

FIG. 4   da Vinci Xi docking from a patient's right for left nipple-sparing mastectomy.

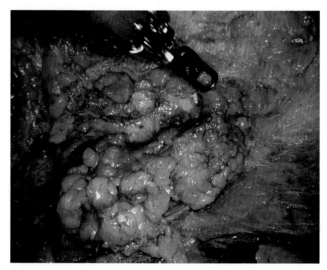

FIG. 5   da Vinci Xi endoscopic glandular dissection.

The dissection starts with from the superficial skin flaps, and moving from the axillary toward the NAC (Fig. 5) [2]. When reaching the NAC, subareolar biopsy can be performed and then continued below the NAC up to the breast fold along the lateral, inferior, and internal margins. During flap dissection, the assistant of the operating table can check, through the transillumination of the skin flap, the position of the instruments, cooperating with the surgeon to avoid skin injury. The operation proceeds with subfascial dissection, lifting the breast of the pectoralis major muscle. When the gland is fully mobilized, it can be extracted through the axillary incision [2,17].

Breast reconstruction can be performed using prosthetic implant [2,16,37] or autologous muscle flap [8,40,41]. For reconstructions with implants, after extraction of the mammary gland, the single port is repositioned to dissect the submuscular pocket. To create the submuscular pocket, the pectoralis major muscle is completely released from the thorax wall, medially to the sternum and inferiorly to the inframammary fold [2,17]. After the completion of submuscular pocket, the implant is inserted manually, and drains are placed in both submuscular and subcutaneous planes [2]. For reconstructions with latissimus dorsi flap, after removal of breast through the axillary wound, the thoracodorsal pedicle is identified and marked with a vessel loop. Then, the patient is turned to a prone position, and the port is docked to the robotic arms. Insufflation of carbon dioxide gas is maintained at a pressure of 10–12 mmHg. The latissimus dorsi muscle is dissected from the inferoposterior border and the scapular border to be pulled out from the axilla. After that, the robotic arms are undocked and the latissimus dorsi flap transferred to the mastectomy site. The patient is shifted to supine position again and the flap is allocated and fixed [40].

## Future scenario

Most studies used da Vinci Si or Xi (Intuitive Surgical, Sunnyvale, CA) to perform R-NSM. Positioning the patient, however, is a challenge and can cause an undesirable brachial neuropraxia [42]. To minimize this issue, da Vinci Single Port (Intuitive Surgical, Sunnyvale, CA) robot system is under testing. The SP system consists of a single port through which four instruments emerge and enabling the use of two robotic arms through one incision [43,44].

The use of HUGO RAS robotic system to perform breast surgeries has not been described in the literature yet. Results of a Korean prospective trial using a single port robot to perform therapeutic and prophylactic R-NSMs are expected. This multiinstitutional study aims to evaluate complication rates, recurrence-free survival, cancer incidence rate, patient and surgeon satisfaction, and cost-effectiveness of the Robot Endoscopic Surgery (MARRES) in prophylactic and therapeutic R-NSMs [15].

## Conclusions

Robotic Nipple-sparing mastectomy is a promising breast surgery technique. Some studies reported less postoperative complications, including the feared nipple necrosis. Another advantage comprises better aesthetic outcomes with a greater patients' satisfaction with the preservation of NAC sensitivity and position and the small and hidden scar. However, costs, operating time, and learning curve are still barriers for robotics' use. Besides, more studies are required to determine oncologic safety of robotic Nipple-sparing mastectomy.

## Key points

- High-definition and power lighting cameras allow better identification of breast structures and surgical plans.
- Robotic Nipple-sparing mastectomy allows breast removal through a small and hidden wound.
- Less grade 3 complications are reported in robotic Nipple-sparing mastectomy when compared to open techniques.
- Rates of nipple and skin necrosis of robotic Nipple-sparing mastectomy are lower than conventional surgery.
- Better cosmetic outcomes, patients' satisfaction, and nipple sensation and position are reported in robotic Nipple-sparing mastectomy.
- Robotic Nipple-sparing mastectomy can be indicated for prophylactic or therapeutic purposes. Lymph node biopsy or axillary dissection can be performed through the same incision of mastectomy.
- Higher costs, operating time, and learning curves are some disadvantages related to robotic Nipple-sparing mastectomy.
- Positioning the patient is one of the major steps of Nipple-sparing mastectomy. The objective is to place the patient's arm in a way to avoid collision with robotic arms and brachial neuropraxia.
- Breast reconstruction can be performed using the robotic surgery system. Immediate breast reconstruction with breast implants or latissimus dorsi flap are reported.

## References

[1] Toesca A, Manconi A, Peradze N, Loschi P, Panzeri R, Granata M, et al. 1931 preliminary report of robotic nipple-sparing mastectomy and immediate breast reconstruction with implant. Eur J Cancer 2015;51:S309. Available from: https://linkinghub.elsevier.com/retrieve/pii/S0959804916308802.
[2] Toesca A, Peradze N, Manconi A, Galimberti V, Intra M, Colleoni M, et al. Robotic nipple-sparing mastectomy for the treatment of breast cancer: feasibility and safety study. Breast 2017.
[3] Leff DR, Vashisht R, Yongue G, Keshtgar M, Yang G-Z, Darzi A. Endoscopic breast surgery: where are we now and what might the future hold for video-assisted breast surgery? Breast Cancer Res Treat 2011;125(3):607–25. Available from: http://link.springer.com/10.1007/s10549-010-1258-4.
[4] Sakamoto N, Fukuma E, Higa K, Ozaki S, Sakamoto M, Abe S, et al. Early results of an endoscopic nipple-sparing mastectomy for breast Cancer. Indian J Surg Oncol 2010;1(3):232–9. Available from: http://link.springer.com/10.1007/s13193-011-0057-7.
[5] Keshtgar MR, Fukuma E. Endoscopic mastectomy: what does the future hold? Women Health 2009;5(2):107–9. Available from: http://journals.sagepub.com/doi/10.2217/17455057.5.2.107.
[6] Filipe MD, de Bock E, Postma EL, Bastian OW, Schellekens PPA, Vriens MR, et al. Robotic nipple-sparing mastectomy complication rate compared to traditional nipple-sparing mastectomy: a systematic review and meta-analysis. J Robot Surg 2022;16(2):265–72. Available from: https://link.springer.com/10.1007/s11701-021-01265-w.

[7] Patel NP, Van Meeteren J, Pedersen J. A new dimension: robotic reconstruction in plastic surgery. J Robot Surg 2012;6(1):77–80. Available from: http://link.springer.com/10.1007/s11701-011-0300-9.

[8] Selber JC, Baumann DP, Holsinger FC. Robotic latissimus Dorsi muscle harvest. Plast Reconstr Surg 2012;129(6):1305–12. Available from: http://journals.lww.com/00006534-201206000-00014.

[9] Houvenaeghel G, Bannier M, Rua S, Barrou J, Heinemann M, Knight S, et al. Robotic breast and reconstructive surgery: 100 procedures in 2-years for 80 patients. Surg Oncol 2019;31:38–45. Available from: https://linkinghub.elsevier.com/retrieve/pii/S0960740418304067.

[10] Bishop SN, Selber JC. Minimally invasive robotic breast reconstruction surgery. Gland Surg 2021;10(1):469–78. Available from: https://gs.amegroups.com/article/view/51285/html.

[11] Egan KG, Selber JC. Modern innovations in breast surgery: robotic breast surgery and robotic breast reconstruction. Clin Plast Surg 2023;50 (2):357–66. Available from: https://linkinghub.elsevier.com/retrieve/pii/S0094129822001134.

[12] Jeon DN, Kim J, Ko BS, Lee SB, Kim EK, Eom JS, et al. Robot-assisted breast reconstruction using the prepectoral anterior tenting method. J Plast Reconstr Aesthet Surg 2021;74(11):2906–15. Available from: https://linkinghub.elsevier.com/retrieve/pii/S1748681521002242.

[13] Toesca A, Peradze N, Galimberti V, Manconi A, Intra M, Gentilini O, et al. Robotic nipple-sparing mastectomy and immediate breast reconstruction with implant. Ann Surg 2017;266(2):e28–30. Available from: http://insights.ovid.com/crossref?an=00000658-201708000-00038.

[14] Sarfati B, Struk S, Leymarie N, Honart JF, Alkhashnam H, Tran de Fremicourt K, et al. Robotic prophylactic nipple-sparing mastectomy with immediate prosthetic breast reconstruction: a prospective study. Ann Surg Oncol 2018.

[15] Ryu JM, Kim JY, Choi HJ, Ko B, Kim J, Cho J, et al. Robot-assisted nipple-sparing mastectomy with immediate breast reconstruction. Ann Surg 2022;275(5):985–91. Available from: https://journals.lww.com/10.1097/SLA.0000000000004492.

[16] Lai H-W, Chen S-T, Mok CW, Lin Y-J, Wu H-K, Lin S-L, et al. Robotic versus conventional nipple sparing mastectomy and immediate gel implant breast reconstruction in the management of breast cancer- a case control comparison study with analysis of clinical outcome, medical cost, and patient-reported cosmetic result. J Plast Reconstr Aesthet Surg 2020;73(8):1514–25. Available from: https://linkinghub.elsevier.com/retrieve/pii/S1748681520300802.

[17] Loh Z-J, Wu T-Y, Cheng FT-F. Evaluation of the learning curve in robotic nipple-sparing mastectomy for breast Cancer. Clin Breast Cancer 2021;21(3):e279–84. Available from: https://linkinghub.elsevier.com/retrieve/pii/S1526820920302573.

[18] Lee J, Park HS, Lee H, Lee DW, Song SY, Lew DH, et al. Post-operative complications and nipple necrosis rates between conventional and robotic nipple-sparing mastectomy. Front Oncol 2021;10. Available from: https://www.frontiersin.org/articles/10.3389/fonc.2020.594388/full.

[19] Toesca A, Invento A, Massari G, Girardi A, Peradze N, Lissidini G, et al. Update on the feasibility and Progress on robotic breast surgery. Ann Surg Oncol 2019;26(10):3046–51. Available from: http://link.springer.com/10.1245/s10434-019-07590-7.

[20] Park HS, Lee J, Lai H-W, Park JM, Ryu JM, Lee JE, et al. Surgical and oncologic outcomes of robotic and conventional nipple-sparing mastectomy with immediate reconstruction: international multicenter pooled data analysis. Ann Surg Oncol 2022;29(11):6646–57. Available from: https://link.springer.com/10.1245/s10434-022-11865-x.

[21] Toesca A, Sangalli C, Maisonneuve P, Massari G, Girardi A, Baker JL, et al. A randomized trial of robotic mastectomy versus open surgery in women with breast Cancer or BrCA mutation. Ann Surg 2022;276(1):11–9. Available from: http://www.ncbi.nlm.nih.gov/pubmed/34597010.

[22] Houvenaeghel G, Barrou J, Jauffret C, Rua S, Sabiani L, Van Troy A, et al. Robotic versus conventional nipple-sparing mastectomy with immediate breast reconstruction. Front Oncol 2021;11. Available from: https://www.frontiersin.org/articles/10.3389/fonc.2021.637049/full.

[23] Lai H-W, Chen D-R, Liu L-C, Chen S-T, Kuo Y-L, Lin S-L, et al. Robotic versus conventional or endoscopic assisted nipple sparing mastectomy and immediate prothesis breast reconstruction in the management of breast cancer—a prospectively designed multicenter trial comparing clinical outcomes, medical cost, and patient. Ann Surg 2023;25. Available from: https://journals.lww.com/10.1097/SLA.0000000000005924.

[24] Moon J, Lee J, Lee DW, Lee HS, Nam DJ, Kim MJ, et al. Postoperative pain assessment of robotic nipple-sparing mastectomy with immediate prepectoral prosthesis breast reconstruction: a comparison with conventional nipple-sparing mastectomy. Int J Med Sci 2021;18(11):2409–16. Available from: https://www.medsci.org/v18p2409.htm.

[25] Moon J, Park HS, Kim JY, Lee HS, Jeon S, Lee D, et al. Analgesic efficacies of intraoperative pectoralis nerve II block under direct vision in patients undergoing robotic nipple-sparing mastectomy with immediate breast reconstruction: a prospective, randomized controlled study. J Pers Med 2022;12(8):1309. Available from: https://www.mdpi.com/2075-4426/12/8/1309.

[26] Stewart C, Raoof M, Fong Y, Dellinger T, Warner S. Who is hurting? A prospective study of surgeon ergonomics. Surg Endosc 2022;36(1):292–9. Available from: https://link.springer.com/10.1007/s00464-020-08274-0.

[27] Lai H-W, Toesca A, Sarfati B, Park HS, Houvenaeghel G, Selber JC, et al. Consensus statement on robotic mastectomy—expert panel from international endoscopic and robotic breast surgery symposium (IERBS) 2019. Ann Surg 2020;271(6):1005–12. Available from: https://journals.lww.com/10.1097/SLA.0000000000003789.

[28] Chen K, Beeraka MN, Zhang J, Reshetov IV, Nikolenko VN, Sinelnikov MY, et al. Efficacy of da Vinci robot-assisted lymph node surgery than conventional axillary lymph node dissection in breast cancer – A comparative study. Int J Med Robot Comput Assist Surg 2021;17(6). Available from: https://onlinelibrary.wiley.com/doi/10.1002/rcs.2307.

[29] Toesca A, Park HS, Ryu JM, Kim YJ, Lee J, Sangalli C, et al. Robot-assisted mastectomy: next major advance in breast cancer surgery. Br J Surg 2023;110(4):502–3. Available from: https://academic.oup.com/bjs/article/110/4/502/7008352.

[30] Park HS. Prospective study of mastectomy with reconstruction including robot endoscopic surgery (MARRES) - NCT04585074, 2023. Available from: https://classic.clinicaltrials.gov/ct2/show/NCT04585074.

[31] Oncology EI of. Robotic Nipple-Sparing Mastectomy Vs Conventional Open Technique - NCT03440398., 2023. Available from: https://classic.clinicaltrials.gov/ct2/show/NCT03440398?term=NCT03440398&draw=2&rank=1.

[32] Park HS. Surgical and oncologic outcomes after robotic nipple sparing mastectomy and immediate reconstruction (SORI) - NCT04108117, 2023. Available from: https://classic.clinicaltrials.gov/ct2/show/NCT04108117?term=NCT04108117&draw=2&rank=1.

[33] Yonsei University. Robot-assisted vs. Open Nipple-sparing Mastectomy With Immediate Breast Reconstruction (ROM) - NCT05490433, 2023. Available from: https://classic.clinicaltrials.gov/ct2/show/NCT05490433?term=NCT05490433&draw=2&rank=1.

[34] Lai HW, Chen ST, Lin SL, Chen CJ, Lin YL, Pai SH, et al. Robotic nipple-sparing mastectomy and immediate breast reconstruction with gel implant: Technique, Preliminary Results and Patient-Reported Cosmetic Outcome. Ann Surg Oncol 2019.

III. The current and future clinical applications of robotic surgery among medical specialties

[35] Lai H-W, Wang C-C, Lai Y-C, Chen C-J, Lin S-L, Chen S-T, et al. The learning curve of robotic nipple sparing mastectomy for breast cancer: an analysis of consecutive 39 procedures with cumulative sum plot. Eur J Surg Oncol 2019;45(2):125–33. Available from: https://linkinghub.elsevier.com/retrieve/pii/S0748798318314409.

[36] Lee J, Park HS, Lee DW, Song SY, Yu J, Ryu JM, et al. From cadaveric and animal studies to the clinical reality of robotic mastectomy: a feasibility report of training program. Sci Rep 2021;11(1):21032. Available from: https://www.nature.com/articles/s41598-021-00278-7.

[37] Sarfati B, Struk S, Leymarie N, Honart J-F, Alkhashnam H, Kolb F, et al. Robotic nipple-sparing mastectomy with immediate prosthetic breast reconstruction: surgical technique. Plast Reconstr Surg 2018;142(3):624–7. Available from: https://journals.lww.com/00006534-201809000-00009.

[38] Houvenaeghel G, Bannier M, Rua S, Barrou J, Heinemann M, Van Troy A, et al. Breast cancer robotic nipple sparing mastectomy: evaluation of several surgical procedures and learning curve. World J Surg Oncol 2019;17(1):27. Available from: https://wjso.biomedcentral.com/articles/10.1186/s12957-019-1567-y.

[39] Chen K, Zhang J, Beeraka NM, Sinelnikov MY, Zhang X, Cao Y, et al. Robot-assisted minimally invasive breast surgery: recent evidence with comparative clinical outcomes. J Clin Med 2022;11(7):1827. Available from: https://www.mdpi.com/2077-0383/11/7/1827.

[40] Lai H-W, Lin S-L, Chen S-T, Lin Y-L, Chen D-R, Pai S-S, et al. Robotic nipple sparing mastectomy and immediate breast reconstruction with robotic latissimus dorsi flap harvest - technique and preliminary results. J Plast Reconstr Aesthet Surg 2018;71(10):e59–61. Available from: http://www.ncbi.nlm.nih.gov/pubmed/30122600.

[41] Houvenaeghel G, Cohen M, Ribeiro SR, Barrou J, Heinemann M, Frayret C, et al. Robotic nipple-sparing mastectomy and immediate breast reconstruction with robotic latissimus Dorsi flap harvest: technique and results. Surg Innov 2020;27(5):481–91. Available from: http://www.ncbi.nlm.nih.gov/pubmed/32418492.

[42] Park KU, Cha C, Pozzi G, Kang Y-J, Gregorc V, Sapino A, et al. Robot-assisted nipple sparing mastectomy: recent advancements and ongoing controversies. Curr Breast Cancer Rep 2023;15(2):127–34. Available from: https://link.springer.com/10.1007/s12609-023-00487-1.

[43] Go J, Ahn JH, Park JM, Choi SB, Lee J, Kim JY, et al. Analysis of robot-assisted nipple-sparing mastectomy using the da Vinci SP system. J Surg Oncol 2022;126(3):417–24. Available from: https://onlinelibrary.wiley.com/doi/10.1002/jso.26915.

[44] Park HS, Lee J, Lee H, Lee K, Song SY, Toesca A. Development of robotic mastectomy using a single-port surgical robot system. J Breast Cancer 2020;23(1):107. Available from: https://ejbc.kr/DOIx.php?id=10.4048/jbc.2020.23.e3.

# Pediatric robotic surgery for benign diseases

*Kentaro Mizuno[a], Yutaro Hayashi[a], Hidenori Nishio[a], Junya Hata[b], Yuichi Sato[b], and Yoshiyuki Kojima[b]*

[a]Department of Pediatric Urology, Nagoya City University Graduate School of Medical Sciences, Nagoya, Japan [b]Department of Urology, Fukushima Medical University School of Medicine, Fukushima, Japan

## Abbreviations

| | |
|---|---|
| APV | appendicovesicostomy |
| AR | augmented reality |
| DVSS | da Vinci surgical system |
| HIdES | hidden incision endoscopic surgery |
| ICG | indocyanine green |
| LP | laparoscopic pyeloplasty |
| NIRF | near-infrared fluorescence |
| OR | odds ratio |
| RABD | robot-assisted bladder diverticulectomy |
| RALHN | robot-assisted laparoscopic heminephrectomy |
| RALIUU | robot-assisted laparoscopic ipsilateral ureteroureterostomy |
| RAL-O | robot-assisted laparoscopic orchiopexy |
| RAL-P | robot-assisted laparoscopic pyeloplasty |
| RALS | robot-assisted laparoscopic surgery |
| RALS-UL | robot-assisted laparoscopic surgery for urolithiasis |
| RALUR | robot-assisted laparoscopic ureteral reimplantation |
| RCT | randomized control study |
| UPJO | ureteropelvic junction obstruction |
| UTI | urinary tract infection |
| VUR | vesicoureteral reflux |

## Introduction

According to the progress of several robotic platforms, as typified by the da Vinci Surgical System (DVSS) (Intuitive Surgical Inc., Sunnyvale, CA, USA) in 1999, robot-assisted laparoscopic surgeries (RALS) are widely performed among adult and pediatric patients [1]. Although laparoscopic surgeries are indicated in addition to open surgeries in the field of pediatric surgery, several challenges have been observed with pediatric patients, such as being time-consuming [2], the burden on surgeons [3], prolonged learning curve [4,5], and so on. One of the reasons for these challenges is the large volume of plastic and reconstructive surgeries in laparoscopic surgeries for pediatric patients. The robotic platform has the advantages of a stable magnified 3-D view, tremor filtering, robotic arms with high flexibility, and motion scaling, allowing for precise intracorporeal exposure and suturing [6].

Initially, fundoplication was performed as RALS for pediatric patients [7]. Many reports on RALS have been documented after that, especially in the field of pediatric urology [1,8–22]. This trend is not independent of rapid application in adult urology, especially robot-assisted laparoscopic prostatectomy. In this chapter, we review the current status and future scenario of RALS with a focus on pediatric urology.

## Pyeloplasty

Hydronephrosis is a pathological condition in which the renal pelvis and calyces dilate due to stagnation or reflux of urine. Antenatal hydronephrosis is found approximately 1%–5% in all pregnancies, with the ureteropelvic junction obstruction (UPJO) variant being the most common, with a prevalence of 10%–30% [23]. Pyeloplasty has been the gold standard treatment for UPJO since the era of open surgery [24]. The obstruction of the urinary tract can cause a variety of symptoms, including flank pain, abdominal mass, gross hematuria, febrile urinary tract infection (UTI), and urolithiasis; however, some cases are asymptomatic. Urinary tract reconstruction surgery is performed to improve subjective symptoms and avoid renal parenchymal damage due to urinary tract obstruction. Patients of all ages, including infants, older children, and adults, can undergo this procedure. Robot-assisted laparoscopic pyeloplasty (RAL-P) for children using DVSS was initially introduced in 2002 [25], and many studies have been reported [8,9,26–35]. According to an investigation in North America, the number of RAL-P for pediatric patients increased compared to the decrease in open surgery; in 2015, above 80% of adolescents (13–18 years old) underwent RAL-P [10].

While open pyeloplasty is generally performed through the retroperitoneal approach, RAL-P is mostly performed through the transperitoneal approach. The robotic working space is small, especially for infant or toddler patients, and the retroperitoneal approach is limited [6]. Chua et al. reported a meta-analysis on the laparoscopic pyeloplasty (LP) approach for UPJO and found no significant differences in success rate, operative time, or complications between transperitoneal and retroperitoneal approaches [26]. They also demonstrated that conversion rates are higher, diet occurs faster, and drain duration is shorter with the retroperitoneal approach than with the transperitoneal approach. More recently, Blanc et al. performed a multicenter, comparative, and prospective study between intra- and retroperitoneal RAL-P [27]. While the treatment efficacy and complication rate were similar, set-up time, anastomotic time, and console time were significantly shorter with the transperitoneal approach. Conversely, the median hospital stay was longer after transperitoneal than retroperitoneal RAL-P. Thus, through the retroperitoneal approach, RAL-P can avoid the injury of abdominal organs and urine leakage into the intraperitoneal cavity. However, using DVSS, we are comfortable with the transperitoneal approach because the working space is not limited [8]. The transmesenteric approach is a better option for left transperitoneal RAL-P with little or no bowel manipulation. It may be more appropriate for children and applicable to RAL-P [36] (Fig. 1).

Although dismembered pyeloplasty is considered the standard in RAL-P [6], nondismembered procedures, such as Y-V [8], flap [28], and bypass [29], are also applicable. While the nondismembered procedure is technically easier because the ureter is not completely transected from the renal pelvis, the main advantage of the dismembered procedure is the complete excision of ureteropelvic junction stenotic segments. The selection of these surgical procedures

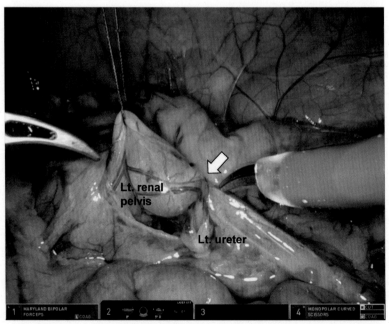

FIG. 1   Robot-assisted laparoscopic pyeloplasty with transmesenteric approach in a 3-year-old boy. Left renal pelvis is dilated and ureter is bended at ureteropelvic junction (arrow).

depends on intraoperative findings, adhesions, and surgeons' preferences. Although the definition of surgical success in RAL-P remains unclear, the vast majority of reports included the following criteria: reduction of hydronephrosis on ultrasonography, preservation of split renal function, improvement in the drainage curve on diuretic renogram, and symptom resolution postoperatively [8].

Several reports have indicated that RAL-P achieved a similar success rate, shorter hospital stay, and lower narcotic usage compared to open pyeloplasty [6]. Meanwhile, disadvantages such as longer operative time and higher costs have also been reported [30]. Compared to LP, there is one randomized control study (RCT) [9] and several systematic reviews [31,32,35]. There were no significant differences in success rates in RCT (92.6% vs 100%) [9], systematic reviews in infants (97.5% vs 94.8%) [35], or case–control studies (96.05% vs 97.86%) between LP and RAL-P, respectively [33]. However, two systematic reviews revealed significantly higher success rates for RAL-P compared to LP (odds ratio [OR] = 0.51, 95% confidence interval [95% CI]: 0.31–0.84, $P = 0.008$ [31]; OR = 2.51, 95% CI: 1.08–5.83, $P = 0.03$ [32]). In this context, conflicting results exist concerning the advantages of RAL-P regarding success rates, and several reasons, including indirectness or risk of bias, such as variability of patients' age and lack of coherence in the definition of surgical success, account for this conflict.

With regard to the complication rate, there were no significant differences between RAL-P and LP in RCT (7.6% vs 7.5%) [9], two case–control studies (18% vs 13% [34] and 1.3% vs 3.6% [33]), and two systematic reviews [31,32], respectively. However, the complication rate of RAL-P was significantly higher than that of LP (16% vs 9%, $P = 0.03$) for infants, especially the Clavien grade 3–4 (5.7% vs. 1.5%, $P = 0.01$) [35]. The authors pointed out that the use of larger robotic ports and instruments in the small space of the infant abdomen might be responsible for higher complications in RAL-P, including a significantly larger number of port-site hernias [35].

Recently, the feasibility of RAL-P as a secondary procedure has increased [37–39]. Even in RAL-P, a redo pyeloplasty remains challenging because of the extensive scarring and fibrosis from the prior procedure [6]. Chandrasekharam et al. concluded that LP and RAL-P seem to be good alternatives to open pyeloplasty for redo cases, with comparable success and complication rates, from the result of their systematic review and meta-analysis [39].

## Ureteral reimplantation

Primary vesicoureteral reflux (VUR) is a congenital anomaly characterized by the retrograde flow of urine accumulated in the bladder back to the kidney because of anatomical or functional abnormalities of the preventive mechanism against reflux [40]. VUR causes repeated febrile UTIs and subsequently deteriorates renal function. The rate of spontaneous resolution of VUR in infants within 1–4 years was 50%, and the resolution rate was higher when the VUR grade was lower [40]. Interventional approaches range from endoscopic injection at the ureterovesical junction to open, laparoscopic, and robotic surgical correction [41]. Although laparoscopic surgery has been shown to be a feasible treatment for VUR with intra- or extravesical approaches [42] as robot-assisted laparoscopic ureteral reimplantation (RALUR), the number of extravesical cases is increasing [11]. First described by Peters in 2004 [43], the intravesical RALUR approach is technically challenging and associated with difficulty maintaining pneumovesicum. The procedure cannot be performed in young children because of their small bladder capacities (less than 130 mL) [12]. Meanwhile, a systematic review and meta-analysis comparing intra- and extravesical open surgeries for VUR revealed no significant differences in VUR persistence, and the extravesical approach had shorter operative time and hospital stay than the intravesical approach [44]. Besides, bilateral procedures with an extravesical approach had a higher risk of postoperative acute urinary retention than the intravesical approach (OR = 4.40, 95% CI: 1.33–14.58, $P = 0.02$) [44]. In this context, we reported the first surgical outcome of extravesical RALUR in Japan, with a comparable success rate (93.3%) to previous studies [11] (Fig. 2).

Unlike the intravesical cross-trigonal procedure, extravesical RALUR allows the child to retain normal anatomy and is associated with decreased morbidity, including lower frequencies of postoperative hematuria and bladder spasms [11]. According to a recent systematic review of RALUR in pediatric patients, the overall success rate was 92%, and intra- and postoperative complications were 1.5% and 10.7%, respectively [45]. They also reported mean rates of 4.4% postoperative urinary retention and 3.9% reoperation. Previous studies on VUR resolution rates after RALUR reported a range of 77%–100%, and this variability may have been affected by case selection and surgeon learning curves [6]. Although Herz et al. reported laterality, age, VUR grade, and preoperative bladder and bowel dysfunction as risk factors for surgical failure [46], Esposito et al. explicitly mentioned no correlation between the success rates and complex anatomy or previous procedures [45].

On the other hand, to optimize surgical outcomes, Gundeti et al. described RALUR technique modifications [47], particularly LUAA, which represents the length of the detrusor tunnel (L), the use of a U stitch (U), the placement of

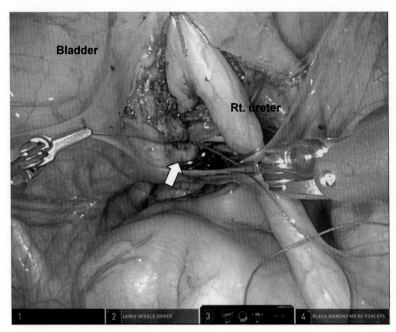

FIG. 2    Robot-assisted laparoscopic ureteral reimplantation in a 2-year-old boy. By the extravesical approach, the bladder muscularis is closed over right ureter (arrow).

permanent ureteral alignment suture (A), and the inclusion of ureteral adventitia (A) in detrusorrhaphy. We also previously reported the usefulness of ureteral advancement sutures for VUR resolution, which are used to advance the ureter underneath the self of the bladder muscle towards the bladder neck [42]. It is thought that the results of RALUR are gradually improving due to these innovative techniques. Indeed, a more recent systematic review showed greater success (92.8%) and lower failure (5.2%) rates compared to earlier papers (90.9% and 9.2%, respectively) [45]. In addition, fewer postoperative complications, Clavien 1–2 and 3 grade, and reoperations were reported lately compared with early papers ($P = 0.001$) [45]. Although the major complications include urinary retention and ureteral injury with ureteral obstructions or leakages, urinary retention is the most common in bilateral cases [6]. Several branches of the pelvic plexus were observed to travel to the ureterovesical junction and surround the distal ureter as a fine network [48]. Bilateral dissection of the posterior bladder may disrupt the pelvic nerve plexus, resulting in postoperative urinary retention. As previously reported, nerve-sparing dissection has been proposed to reduce this complication. Since unconsidered dissection of this area can increase the possibility of bleeding and an unclear field of view, careful maneuvering is required. Additionally, roughly blunt dissections or the heavy use of electrocautery in this area could cause nerve injuries [6].

While the number of RAL-P for UPJO is rapidly increasing, RALUR remains relatively uncommon in many pediatric centers [13]. This may be related to less frequency of diagnosis due to new recommendations on the evaluation of the first UTI by the American Academy of Pediatrics in 2011 [14], the necessary learning curve [45], and the increase of other interventions such as endoscopic injection [49].

Although the treatment options for VUR differ by surgeon's preferences in part, previous studies have revealed that RALUR has comparative surgical outcomes compared to open or laparoscopic surgeries for VUR. An improved learning curve may enable the RALUR application to catch up with established surgical corrections for VUR [1].

## Ureteroureterostomy

Ureterocele is a congenital cystic dilatation of the intravesical ureter, while ectopic ureter is any ureter that does not enter the trigonal area of the bladder. Ureterocele and ectopic ureter are often associated with ureteral duplication and are involved in recurrent UTIs, a decline in renal function of the upper moiety, and urinary incontinence. Ureterocele has associated anatomic and pathophysiologic features, including intravesical ureteral obstruction, dysplasia or obstructive nephropathy of the ureterocele-associated moiety (40%–70%), and VUR to the ipsilateral inferior moiety (50%) or contralateral renal unit (25%) [15].

The optimal management of ureterocele and ectopic ureter with duplication may include an upper or lower tract surgical approach. Upper tract approaches have traditionally included pyelopyelostomy, ureteropyelostomy, and heminephrectomy of the superior moiety. The first two procedures are seldom used due to no advantage and higher morbidity compared to heminephrectomy [15]. As lower tract surgical approaches, endoscopic ureterocele incision or ipsilateral ureteroureterostomy are used for patients whose obstructed moiety has significant functionality. As RALS, robot-assisted laparoscopic heminephrectomy (RALHN) or robot-assisted laparoscopic ipsilateral ureteroureterostomy (RALIUU) have been described. Malik et al. reported 16 cases of RALHN and demonstrated the advantages of robotic procedures, including allowing easy and efficient identification of the ureters, limiting traction on the vasculature, and reducing the risk of resultant ischemic damage. They also compared success outcomes and complications to open or laparoscopic procedures [50]. Conversely, Ellison et al. reviewed 47 cases of RALIUU and reported that older children might benefit more from minimally invasive surgery, and the robotic platforms offer surgeons improved management of the distal ureteral stump. They also pointed out a lack of intermediate or long-term outcomes of RALIUU [16].

Although some debate exists about which procedures are optimal, Sahadev et al. recently performed a retrospective review of consecutive children undergoing RALIUU and RALHN among two hospitals [51]. Surgical success rates were 100% across cohorts, and operative time favored RALHN. They concluded that RALIUU and RALHN are definitive surgical interventions in children with complex duplex moieties, delivering satisfactory surgical outcomes with a low complication profile and marginal differences in postoperative patient outcomes.

## Bladder diverticulectomy

Bladder diverticula can develop as a consequence of bladder outlet obstruction, neurogenic bladder, or VUR. Diverticula that retain urine causing incontinence and UTIs should be removed. After the initial case of robot-assisted bladder diverticulectomy (RABD) described in 2006, approximately 100 robotic cases have been reported in the literature up to 2022 [17]. The surgical techniques for RABD are not yet standardized due to the limited number of cases. Some authors prefer an extravesical approach, while others prefer the transvesical or transdiverticular approach [52]. While some authors have proposed that bladder diverticula can be identified via injection of saline solution through the indwelling catheters, others have suggested using cystoscopy. Although we have reported laparoscopic bladder diverticulectomy using the indigo carmine solution to identify the diverticulum wall [53], Vedovo et al. reported four RABD cases to detect bladder diverticulum using Firefly® technology [54]. Validation of standardized procedures and a large number of case series in multicenter settings with long-term follow-up is warranted [17].

## Appendicovesicostomy

Bladder reconstruction with bowel is the treatment of choice for patients with a neurogenic bladder that failed medical management. The congenital neurogenic bladder is caused by several abnormalities, including posterior urethral valve, spina bifida, tethered cord syndrome, and sacral agenesis [55]. Impaired bladder function leads to decreased quality of life, urinary incontinence, and deteriorating renal function.

In the vast majority of cases, an augmentation ileocystoplasty is accompanied by creating a continent catheterizable channel to overcome this problem. In proportion to the widely performed minimally invasive surgeries, an increasing number of publications have advocated for the excellent results achieved by performing the procedure laparoscopically compared to the original open approach [18]. In addition, robot-assisted ileocystoplasty with Mitrofanoff appendicovesicostomy (APV) [19] and bladder neck reconstruction has been reported [56]. Howe et al. reviewed several reports concerning robot-assisted APV and found a 67%–100% success rate [22]. Grimsby et al. reported large case series comparing 39 robotic APVs and 28 open APVs. Although there were no significant differences in complication rates (26%–29%), the robotic group showed more severe complications (Clavien grade 3) requiring secondary surgery [57]. Meanwhile, Gundeti et al. also reported a multiinstitutional study of robotic-assisted laparoscopic Mitrofanoff APV surgical outcomes, and continence rates were comparable to previous open series with low complication rates [20]. A recent systematic review of the literature comparing open vs robotic APV revealed that both approaches performed equally well regarding overall postoperative complications, surgical reintervention, and stomal stenosis. Besides, the authors found a significantly shorter postoperative length of stay in the robotic group ($P = 0.001$) in their single center [18]. Robot-assisted APVs are seldom performed in pediatric urology centers and are considered rare cases in most institutions. Thus, a surgical training program on robot-assisted APV conducted by dedicated pediatric urologists produced favorable surgical outcomes [18].

## Miscellaneous procedures

The prostatic utricle and Mullerian duct cysts are uncommon congenital anomalies, generally thought to be derived from Mullerian duct remnants. The prostatic utricle is typically accompanied by hypospadias, cryptorchidism, and disorders of sex development [58]. The prevalence of the prostatic utricle is reported to be 1%–5%, and most cases are asymptomatic. However, an enlarged and symptomatic prostatic utricle is sometimes found with complaints of recurrent UTIs or epididymo-orchitis. Since the prostatic utricle is located in the deep pelvis and confluence of the posterior urethra, the approach is limited by visibility, and excision of the prostatic utricle remains challenging [59]. Recently, a robotic platform has been applied to excise symptomatic prostatic utricles [21,22,59–66]. Lima et al. reported three pediatric cases who underwent robot-assisted excision for prostatic utricle and excellent visualization of the retrovesical structures, lowering the risk of injury to the vas deferens, ureters, rectum, and bladder [60]. Bayne et al. also reported four cases in which patients underwent the retrovesical approach and demonstrated the usefulness of concomitant cystoscopy in defining the location and limit of dissection [59]. Although there is limited literature reporting prostatic utricles treated using a robotic platform, the robot-assisted excision allowed for better visualization during this procedure, allowing us to explore the delicate local anatomy [61].

Cryptorchidism is one of the most common congenital urological anomalies, including the risk of future infertility and malignancy. While laparoscopic procedures have been central in treating abdominal undescended testis, in recent years, robot-assisted laparoscopic orchiopexy (RAL-O) has gradually been reported in small case series [62,63]. However, RAL-O has not been shown to offer advantages over the conventional laparoscopic approach [21].

Urachal anomalies occur when the urachus fails to obliterate during normal embryologic development. Symptomatic urachal remnants representing abdominal pain or umbilical discharge should be removed. While laparoscopic excision of the urachal remnant is well described, there are limited data for the robot-assisted approach in children [22]. Rivera et al. reported 11 case series (eight adults and three children) who underwent robot-assisted excision for urachal remnants. The median operative time was 85 min (90 min for the children), and the median length of stay was 1 day [64]. Osumah et al. reported 14 robotic cases performed by a transperitoneal approach with hidden incision endoscopic surgery (HIdES). They performed partial cystectomy in 10/14 (71.4%) cases and reported that median operative time was significantly longer compared to the traditional open approach (136 vs 33 min, $P < 0.01$). Complete excision can be achieved using HIdES port configuration, allowing for excellent cosmetic outcomes [65].

While most patients with urolithiasis are treated via ureteroscopy, retrograde intrarenal surgery, percutaneous nephrolithotomy, or extracorporeal shock wave lithotripsy, the literature on the robotic approach for urolithiasis is limited. Robot-assisted laparoscopic surgery for urolithiasis (RALS-UL) includes robot-assisted pyelolithotomy, ureterolithotomy, and flexible ureteroscopy [66]. RALS-UL is often performed concomitantly with RAL-P for UPJO and has been demonstrated to be safe, effective, and with high stone-free rates in some specific cases [66]. There are no studies comparing the RALS-UL to other minimally invasive surgeries.

In addition, other robot-assisted surgeries for pediatric patients including Malone antegrade continence enema [67], posterior urethral diverticula [68], seminal vesicle cyst [69], varicocelectomy [70], and gynecologic surgeries [71–73] have been reported. However, the number of these robotic procedures remains few.

## Future scenario

Two decades have passed since the first pediatric robot-assisted surgery, and the literature for patients under 1-year-old or 10 kg is increasing. Cascini et al. found that RAL-P for UPJO in infants is feasible and safe compared to open surgery. Although there were no significant differences in surgical failure (4.2% vs 5.2%) and postoperative complication (10.9% vs 10.0%) rates, RAL-P presented a longer operative time (144.0 min vs 129.4 min) and shorter length of hospital stay (2.2 days vs 3.2 days) compared to the open approach [74]. The authors explicitly described the limitation of using RAL-P for UPJO in infants depending on the technical aspects, such as limited working space. They suggest limiting RAL-P in infants to only those high-volume centers with experienced surgeons [74].

Lombardo et al. pointed out that children have unique physiologic and anatomic differences, such as faster gastric emptying times, shifted bladders, and increased abdominal wall laxity; they suggested that surgeons should learn these characteristics of infants or toddlers, as well as the limitations of working space, trocar placement, and surgical instruments, to achieve safe and effective robot-assisted surgeries [75].

With the advancement of DVSS (S/Si/Xi), these devices have acquired improved vision, easier docking, and narrower robotic arms. While Lei et al. reported that the Xi system has better perioperative outcomes for radical prostatectomy compared with the Si system [76], there are few reports about the differences of the robotic system in the field of pediatric urology and surgery. The narrower robotic arms in the Xi system may reduce the collision even in infants. It is assumed that the increase in the literature for infants in recent years is largely due to advances in equipment as well as surgeon experience. Nevertheless, because of the collision due to the fourth arm of DVSS, only three arms are generally used for surgeries in the pediatric patients. Barashi et al. reported that in children taller than 5 ft., it is possible to use the fourth robotic arm, which will help in traction and counter traction during some crucial surgical steps [55].

Contrastingly, several new technologies have been introduced for pediatric patients in the robot-assisted surgery era. First, near-infrared fluorescence (NIRF) imaging with indocyanine green (ICG) has been adopted to improve intraoperative visualization of anatomic structures and facilitate surgery [77]. ICG-guided NIRF imaging has been utilized with the Firefly system of DVSS. Although there is very scarce evidence in robotic surgery for pediatric patients, Esposito et al. reported varicocelectomy, nephrectomy, partial nephrectomy, and renal cyst deroofing [77]. A recent review also concluded that ICG-guided NIRF imaging is gaining popularity due to the excellent results from previous reports [78].

Second, regarding instrument size, 5 mm and 8 mm instruments are often unsuitable for small pediatric patients [1]. To date, the Senhance (Asensus surgical Inc., Durham, NC, USA) is the only robotic system providing an adequate portfolio of 3 mm instruments. Although smaller instruments are crucial for surgery in small pediatric patients, the 3 mm instruments that are currently available cannot be angulated. Brownlee et al. stated that the port size and the size of the robotic system are significant factors for pediatric robotic surgery [79]. Sheth et al. overviewed a novel robotic platform for pediatric patients and described the need for more delicate tissue handling and adaptation to a smaller operative working space posed a further challenge in minimally invasive surgery [80]. More recently, Granberg et al. reported case series utilizing the SP robotic platform in children. A single, 2.5 cm incision allows for placement of a port that admits a fully wristed camera as well as three fully-wristed instruments, all controlled by the surgeon at the console [81]. In Japan, as a novel robotic platform, the Hinotori Surgical Robot System was newly developed and has already been introduced into routine clinical practice [82]. However, there are still no reports of the surgical outcomes for pediatric patients and further experiences are needed.

Finally, augmented reality (AR) and intelligence might represent the next steps in robot-assisted surgery for pediatric patients. For example, surgical guidance aims to use AR to provide information that is difficult to access intraoperatively or that may be inaccessible to the surgical team [1]. Furthermore, in the future, the combination of virtual reality technology and robotic surgery may lead to a completely new era of surgery, including autonomous robotic surgery [80].

Despite several restrictions, such as high cost, small market, larger instrument size, and the lack of haptic feedback [80,83], robust evidence about robot-assisted surgery for pediatric patients, especially in pediatric urology, has been accumulated. The accumulation of evidence and technological innovations would gain more popularity with safety and efficacy, which would significantly benefit pediatric patients.

## Key points

- A large number of pediatric robotic surgeries for benign diseases are reconstructive procedures in the field of pediatric urology.
- While the number of RAL-P for UPJO is rapidly increasing, RALUR remains relatively uncommon because of less frequency of diagnosis and increase of other interventions.
- Along with widespread of the robotic platform, the literature of surgical outcomes for younger children, infants, is increasing.
- By further technological innovation, the robotic platform suitable for pediatric patients is eagerly awaited.

## References

[1] Krebs TF, Schnorr I, Heye P, Häcker F-M. Robotically assisted surgery in children-a perspective. Children 2022;9(6). Available from: https://doi.org/10.3390/children9060839.

[2] Piaggio LA, Franc-Guimond J, Noh PH, Wehry M, Figueroa TE, Barthold J, et al. Transperitoneal laparoscopic pyeloplasty for primary repair of ureteropelvic junction obstruction in infants and children: comparison with open surgery. J Urol 2007;178(4 Pt 2):1579–83.

[3] Stucky C-CH, Cromwell KD, Voss RK, Chiang Y-J, Woodman K, Lee JE, et al. Surgeon symptoms, strain, and selections: systematic review and meta-analysis of surgical ergonomics. Ann Med Surg (Lond) 2018;27:1–8.

[4] Passerotti C, Peters CA. Pediatric robotic-assisted laparoscopy: a description of the principle procedures. ScientificWorldJournal 2006;6:2581–8.

[5] Woo R, Le D, Krummel TM, Albanese C. Robot-assisted pediatric surgery. Am J Surg 2004;188(4A Suppl):27S–37S.

[6] Mizuno K, Kojima Y, Nishio H, Hoshi S, Sato Y, Hayashi Y. Robotic surgery in pediatric urology: current status. Asian J Endosc Surg 2018;11(4):308–17.

[7] Gutt CN, Markus B, Kim ZG, Meininger D, Brinkmann L, Heller K. Early experiences of robotic surgery in children. Surg Endosc 2002;16(7):1083–6.

[8] Mizuno K, Kojima Y, Kurokawa S, Kamisawa H, Nishio H, Moritoki Y, et al. Robot-assisted laparoscopic pyeloplasty for ureteropelvic junction obstruction: comparison between pediatric and adult patients-Japanese series. J Robot Surg 2017;11(2):151–7.

[9] Silay MS, Danacioglu O, Ozel K, Karaman MI, Caskurlu T. Laparoscopy versus robotic-assisted pyeloplasty in children: preliminary results of a pilot prospective randomized controlled trial. World J Urol 2020;38(8):1841–8.

[10] Varda BK, Wang Y, Chung BI, Lee RS, Kurtz MP, Nelson CP, et al. Has the robot caught up? National trends in utilization, perioperative outcomes, and cost for open, laparoscopic, and robotic pediatric pyeloplasty in the United States from 2003 to 2015. J Pediatr Urol 2018;14 (4):336.e1–8.

[11] Hayashi Y, Mizuno K, Kurokawa S, Nakane A, Kamisawa H, Nishio H, et al. Extravesical robot-assisted laparoscopic ureteral reimplantation for vesicoureteral reflux: initial experience in Japan with the ureteral advancement technique. Int J Urol 2014;21(10):1016–21.

[12] Gundeti MS, Kojima Y, Haga N, Kiriluk K. Robotic-assisted laparoscopic reconstructive surgery in the lower urinary tract. Curr Urol Rep 2013;14(4):333–41.

[13] Baek M, Koh CJ. Lessons learned over a decade of pediatric robotic ureteral reimplantation. Investig Clin Urol 2017;58(1):3–11.

[14] Kurtz MP, Leow JJ, Varda BK, Logvinenko T, McQuaid JW, Yu RN, et al. The decline of the open ureteral Reimplant in the United States: National Data from 2003 to 2013. Urology 2017;100:193–7.

[15] Timberlake MD, Corbett ST. Minimally invasive techniques for management of the ureterocele and ectopic ureter: upper tract versus lower tract approach. Urol Clin North Am 2015;42(1):61–76.

[16] Ellison JS, Lendvay TS. Robot-assisted ureteroureterostomy in pediatric patients: current perspectives. Robot Surg 2017;(4):45–55.

[17] Giannarini G, Rossanese M, Macchione L, Mucciardi G, Crestani A, Ficarra V. Robot-assisted bladder diverticulectomy using a transperitoneal extravesical approach. Eur Urol Open Sci 2022;44:162–8.

[18] Juul N, Persad E, Willacy O, Thorup J, Fossum M, Reinhardt S, Robot-Assisted vs. Open appendicovesicostomy in pediatric urology: a systematic review and single-center case series. Front Pediatr 2022;10:908554.

[19] Gundeti MS, Acharya SS, Zagaja GP, Shalhav AL. Paediatric robotic-assisted laparoscopic augmentation ileocystoplasty and Mitrofanoff appendicovesicostomy (RALIMA): feasibility of and initial experience with the University of Chicago technique. BJU Int 2011;107(6):962–9.

[20] Gundeti MS, Petravick ME, Pariser JJ, Pearce SM, Anderson BB, Grimsby GM, et al. A multi-institutional study of perioperative and functional outcomes for pediatric robotic-assisted laparoscopic Mitrofanoff appendicovesicostomy. J Pediatr Urol 2016;12(6):386.e1–5.

[21] Savio LF, Nguyen HT. Robot-assisted laparoscopic urological surgery in children. Nat Rev Urol 2013;10(11):632–9.

[22] Howe A, Kozel Z, Palmer L. Robotic surgery in pediatric urology. Asian J Urol 2017;4(1):55–67.

[23] Kohno M, Ogawa T, Kojima Y, Sakoda A, Johnin K, Sugita Y, et al. Pediatric congenital hydronephrosis (ureteropelvic junction obstruction): medical management guide. Int J Urol 2020;27(5):369–76.

[24] Persky L, Krause JR, Boltuch RL. Initial complications and late results in dismembered pyeloplasty. J Urol 1977;118(1 Pt 2):162–5.

[25] Gettman MT, Peschel R, Neururer R, Bartsch G. A comparison of laparoscopic pyeloplasty performed with the daVinci robotic system versus standard laparoscopic techniques: initial clinical results. Eur Urol 2002;42(5):453–7; discussion 457–8.

[26] Chua ME, Ming JM, Kim JK, Milford KL, Silangcruz JM, Ren L, et al. Meta-analysis of retroperitoneal vs transperitoneal laparoscopic and robot-assisted pyeloplasty for the management of pelvi-ureteric junction obstruction. BJU Int 2021;127(6):687–702.

[27] Blanc T, Abbo O, Vatta F, Grosman J, Marquant F, Elie C, et al. Transperitoneal versus retroperitoneal robotic-assisted laparoscopic pyeloplasty for ureteropelvic junction obstruction in children. A multicentre, prospective study. Eur Urol Open Sci 2022;41:134–40.

[28] Hopf HL, Bahler CD, Sundaram CP. Long-term outcomes of Robot-assisted laparoscopic pyeloplasty for ureteropelvic junction obstruction. Urology 2016;90:106–10.

[29] Haga N, Sato Y, Ogawa S, Yabe M, Akaihata H, Hata J, et al. Laparoscopic modified bypass pyeloplasty: a simple procedure for straightforward ureteral spatulation and intracorporeal suturing. Int Urol Nephrol 2015;47(12):1933–8.

[30] Chang S-J, Hsu C-K, Hsieh C-H, Yang SS-D. Comparing the efficacy and safety between robotic-assisted versus open pyeloplasty in children: a systemic review and meta-analysis. World J Urol 2015;33(11):1855–65.

[31] Uhlig A, Uhlig J, Trojan L, Hinterthaner M, von Hammerstein-Equord A, Strauss A. Surgical approaches for treatment of ureteropelvic junction obstruction—a systematic review and network meta-analysis. BMC Urol 2019;19(1):112.

[32] Taktak S, Llewellyn O, Aboelsoud M, Hajibandeh S, Hajibandeh S. Robot-assisted laparoscopic pyeloplasty versus laparoscopic pyeloplasty for pelvi-ureteric junction obstruction in the paediatric population: a systematic review and meta-analysis. Ther Adv Urol 2019;11:1756287219835704.

[33] Hong P, Ding G, Zhu D, Yang K, Pan J, Li X, et al. Head-to-head comparison of modified laparoscopic pyeloplasty and robot-assisted pyeloplasty for ureteropelvic junction obstruction in China. Urol Int 2018;101(3):337–44.

[34] Wong YS, Pang KKY, Tam YH. Comparing Robot-Assisted laparoscopic Pyeloplasty vs. laparoscopic Pyeloplasty in infants aged 12 months or less. Front Pediatr 2021;9:647139.

[35] Chandrasekharam VVS, Babu R. A systematic review and meta-analysis of conventional laparoscopic versus robot-assisted laparoscopic pyeloplasty in infants. J Pediatr Urol 2021;17(4):502–10.

[36] Kojima Y, Umemoto Y, Mizuno K, Tozawa K, Kohri K, Hayashi Y. Comparison of laparoscopic pyeloplasty for ureteropelvic junction obstruction in adults and children: lessons learned. J Urol 2011;185(4):1461–7.

[37] Lindgren BW, Hagerty J, Meyer T, Cheng EY. Robot-assisted laparoscopic reoperative repair for failed pyeloplasty in children: a safe and highly effective treatment option. J Urol 2012;188(3):932–7.

[38] Asensio M, Gander R, Royo GF, Lloret J. Failed pyeloplasty in children: is robot-assisted laparoscopic reoperative repair feasible? J Pediatr Urol 2015;11(2):69 e1–6.

[39] Chandrasekharam VVS, Babu R. A systematic review and metaanalysis of open, conventional laparoscopic and robot-assisted laparoscopic techniques for re-do pyeloplasty for recurrent uretero pelvic junction obstruction in children. J Pediatr Urol 2022;18(5):642–9.

[40] Miyakita H, Hayashi Y, Mitsui T, Okawada M, Kinoshita Y, Kimata T, et al. Guidelines for the medical management of pediatric vesicoureteral reflux. Int J Urol 2020;27(6):480–90.

[41] Schober MS, Jayanthi VR. Vesicoscopic ureteral Reimplant: is there a role in the age of robotics? Urol Clin North Am 2015;42(1):53–9.

[42] Kojima Y, Mizuno K, Umemoto Y, Yasui T, Hayashi Y, Kohri K. Ureteral advancement in patients undergoing laparoscopic extravesical ureteral reimplantation for treatment of vesicoureteral reflux. J Urol 2012;188(2):582–7.

[43] Peters CA. Robotically assisted surgery in pediatric urology. Urol Clin North Am 2004;31(4):743–52.

[44] Law ZW, Ong CCP, Yap T-L, Loh AHP, Joseph U, Sim SW, et al. Extravesical vs. intravesical ureteric reimplantation for primary vesicoureteral reflux: a systematic review and meta-analysis. Front Pediatr 2022;10:935082.

[45] Esposito C, Castagnetti M, Autorino G, Coppola V, Cerulo M, Esposito G, et al. Robot-assisted laparoscopic extra-vesical ureteral reimplantation (Ralur/Revur) for pediatric vesicoureteral reflux: a systematic review of literature. Urology 2021;156:e1–11.

[46] Herz D, Fuchs M, Todd A, McLeod D, Smith J. Robot-assisted laparoscopic extravesical ureteral reimplant: a critical look at surgical outcomes. J Pediatr Urol 2016;12(6):402.e1–9.

[47] Gundeti MS, Boysen WR, Shah A. Robot-assisted laparoscopic extravesical ureteral reimplantation: technique modifications contribute to optimized outcomes. Eur Urol 2016;70(5):818–23.

[48] Mizuno K, Hayashi Y, Kojima Y. Bladder, bladder neck, and ureteral anatomy and innervations. In: Gundeti MS, editor. Surgical Techniques in Pediatric and Adolescent Urology. Jaypee Brothers Medical Publishers Ltd.; 2020. p. 655–64.

[49] Esposito C, Yamataka A, Varlet F, Castagnetti M, Scalabre A, Fourcade L, et al. Current trends in 2021 in surgical management of vesico-ureteral reflux in pediatric patients: results of a multicenter international survey on 552 patients. Minerva Urol Nephrol 2021. https://doi.org/10.23736/S2724-6051.21.04430-X. Available from:.

[50] Malik RD, Pariser JJ, Gundeti MS. Outcomes in pediatric robot-Assisted laparoscopic heminephrectomy compared with contemporary open and laparoscopic series. J Endourol 2015;29(12):1346–52.

[51] Sahadev R, Rodriguez MV, Kawal T, Barashi N, Srinivasan AK, Gundeti M, et al. Upper or lower tract approach for duplex anomalies? A bi-institutional comparative analysis of robot-assisted approaches. J Robot Surg 2022;16(6):1321–8.

[52] Liu S, Pathak RA, Hemal AK. Robot-assisted laparoscopic bladder diverticulectomy: adaptation of techniques for a variety of clinical presentations. Urology 2021;147:311–6.

[53] Nishio H, Mizuno K, Kamisawa H, Nakagawa M, Yasui T, Hayashi Y. Detailed presurgical evaluation of a case of congenital bladder diverticulum. Urol Case Rep 2019;27:100905.

[54] Vedovo F, de Concilio B, Zeccolini G, Silvestri T, Celia A. New technologies for old procedures: when firefly improves robotic bladder diverticulectomy. Int Braz J Urol 2019;45(5):1080.

[55] Barashi NS, Rodriguez MV, Packiam VT, Gundeti MS. Bladder reconstruction with bowel: robot-assisted laparoscopic ileocystoplasty with mitrofanoff appendicovesicostomy in pediatric patients. J Endourol 2018;32(S1):S119–26.

[56] Bagrodia A, Gargollo P. Robot-assisted bladder neck reconstruction, bladder neck sling, and appendicovesicostomy in children: description of technique and initial results. J Endourol 2011;25(8):1299–305.

[57] Grimsby GM, Jacobs MA, Gargollo PC. Comparison of complications of robot-assisted laparoscopic and open appendicovesicostomy in children. J Urol 2015;194(3):772–6.

[58] Dai L-N, He R, Wu S-F, Zhao H-T, Sun J. Surgical treatment for prostatic utricle cyst in children: a single-center report of 15 patients. Int J Urol 2021;28(6):689–94.

[59] Bayne AP, Austin JC, Seideman CA. Robotic assisted retrovesical approach to prostatic utricle excision and other complex pelvic pathology in children is safe and feasible. J Pediatr Urol 2021;17(5):710–5.

[60] Lima M, Maffi M, Di Salvo N, Ruggeri G, Libri M, Gargano T, et al. Robotic removal of Müllerian duct remnants in pediatric patients: our experience and a review of the literature. Pediatr Med Chir 2018;40(1). https://doi.org/10.4081/pmc.2018.182. Available from.

[61] Able C, Srinivasan A, Alzweri L. Robotic management of a large mullerian duct cyst: a case report and review of surgical options. Transl Androl Urol 2022;11(9):1354–60.

[62] Shumaker A, Neheman A. Robot-assisted modified one-stage Orchiopexy: description of a surgical Technique. Urology 2021;153:355–7.

[63] Higganbotham C, Cook G, Rensing A. Bilateral robot-assisted laparoscopic orchiopexy for undescended testes. Urology 2021;148:314.

[64] Rivera M, Granberg CF, Tollefson MK. Robotic-assisted laparoscopic surgery of urachal anomalies: a single-center experience. J Laparoendosc Adv Surg Tech A 2015;25(4):291–4.

[65] Osumah TS, Granberg CF, Butaney M, Gearman DJ, Ahmed M, Gargollo PC. Robot-assisted laparoscopic urachal excision using hidden incision endoscopic surgery technique in pediatric patients. J Endourol 2020. Available from: https://doi.org/10.1089/end.2019.0525.

[66] Ballesteros N, Snow ZA, Moscardi PRM, Ransford GA, Gomez P, Castellan M. Robotic management of urolithiasis in the pediatric population. Front Pediatr 2019;7:351.

[67] Thakre AA, Bailly Y, Sun LW, Van Meer F, Yeung CK. Is smaller workspace a limitation for robot performance in laparoscopy? J Urol 2008;179(3):1138–42:discussion 1142–3.

[68] Alsowayan O, Almodhen F, Alshammari A. Minimally invasive surgical approach to treat posterior urethral diverticulum. Urol Ann 2015;7(2):273–6.

[69] Moore CD, Erhard MJ, Dahm P. Robot-assisted excision of seminal vesicle cyst associated with ipsilateral renal agenesis. J Endourol 2007;21(7):776–9.

[70] Hidalgo-Tamola J, Sorensen MD, Bice JB, Lendvay TS. Pediatric robot-assisted laparoscopic varicocelectomy. J Endourol 2009;23(8):1297–300.

[71] Benson AD, Kramer BA, McKenna PH, Schwartz BF. Robot-assisted laparoscopic sacrouteropexy for pelvic organ prolapse in classical bladder exstrophy. J Endourol 2010;24(4):515–9.

[72] Nakib G, Calcaterra V, Scorletti F, Romano P, Goruppi I, Mencherini S, et al. Robotic assisted surgery in pediatric gynecology: promising innovation in mini invasive surgical procedures. J Pediatr Adolesc Gynecol 2013;26(1):e5–7.

[73] Pelizzo G, Nakib G, Calcaterra V. Pediatric and adolescent gynecology: treatment perspectives in minimally invasive surgery. Pediatr Rep 2019;11(4):8029.

[74] Cascini V, Lauriti G, Di Renzo D, Miscia ME, Lisi G. Ureteropelvic junction obstruction in infants: open or minimally invasive surgery? A systematic review and meta-analysis. Front Pediatr 2022;10:1052440.

[75] Lombardo AM, Gundeti MS. Review of robot-assisted laparoscopic surgery in management of infant congenital urology: advances and limitations in utilization and learning. Int J Urol 2022. Available from: https://doi.org/10.1111/iju.15105.

[76] Lei KY, Xie WJ, Fu SQ, Ma M, Sun T. A comparison of the da Vinci Xi vs. da Vinci Si surgical systems for radical prostatectomy. BMC Surg 2021;21(1):409.

[77] Esposito C, Coppola V, Del Conte F, Cerulo M, Esposito G, Farina A, et al. Near-infrared fluorescence imaging using indocyanine green (ICG): emerging applications in pediatric urology. J Pediatr Urol 2020;16(5):700–7.

[78] Esposito C, Borgogni R, Autorino G, Cerulo M, Carulli R, Esposito G, et al. Applications of Indocyanine green-guided near-infrared fluorescence imaging in Pediatric minimally invasive surgery urology: a narrative review. J Laparoendosc Adv Surg Tech A 2022;32(12):1280–7.

[79] Brownlee EM, Slack M. The role of the versius surgical robotic system in the paediatric population. Children 2022;9(6). Available from: https://doi.org/10.3390/children9060805.

[80] Sheth KR, Koh CJ. The future of robotic surgery in Pediatric urology: upcoming technology and evolution within the field. Front Pediatr 2019;7:259.

[81] Granberg C, Parikh N, Gargollo P. And then there was one … incision. First single-port pediatric robotic case series. J Pediatr Urol 2023. https://doi.org/10.1016/j.jpurol.2023.03.038.

[82] Hinata N, Yamaguchi R, Kusuhara Y, Kanayama H, Kohjimoto Y, Hara I, Fujisawa M. Hinotori surgical robot system, a novel robot-assisted surgical platform: preclinical and clinical evaluation. Int J Urol 2022;29(10):1213–20.

[83] Shen LT, Tou J. Application and prospects of robotic surgery in children: a scoping review. World J Pediatr Congenit Heart Surg 2022;5(4): e000482.

# 64

# Robotic surgery in pediatric oncology

*Daniel DaJusta and Molly Fuchs*

Nationwide Children's Hospital Division of Pediatric Urology, The Ohio State University
Department of Urology, Columbus, OH, United States

## Introduction

Robotic surgery has brought several advantages in adult oncology over open and standard laparoscopic techniques without sacrificing oncological outcomes. The benefits of smaller incisions, reduced blood loss, and recovery over open techniques have been demonstrated in various oncological surgical procedures, such as prostate, bladder, kidney, and colon cancers. On the other hand, benefits over conventional laparoscopy are more challenging to demonstrate as standard laparoscopic surgery does offer similar benefits as described above. Nevertheless, the ability to work in places of the body with limited space, the freedom of movement provided by the robotic *endo*-wrist, and the ergonomic benefits for the surgeon is among the main advantages over standard laparoscopy.

In the field of pediatric surgery, both pediatric surgeons and urologists have made use of the robotic platform to further their Minimally invasive surgery (MIS) techniques. While standard laparoscopy has been a long-standing feature of both specialties, its use has been limited to more straightforward cases such as orchiopexy and pyloromyotomy, with only a select few surgeons venturing to attempting complex cases. Yet, the introduction of the robotic platform led to an immediate popularization of the MIS technique in children in more complex cases. An excellent example of this increase in popularity was the changes in pyeloplasty techniques in the US over time [1]. In 2003, despite being described for over 10 years, standard laparoscopy only comprised a small portion of the overall pyeloplasties done in the US. Robotic pyeloplasty was introduced in 2004, and 10 years later, it has become the more commonly performed technique, a trend that has only increased with time. Similar trends are currently being observed in other procedures where the robotic approach has been introduced.

Nevertheless, the use of robotic surgery in children to resect tumors has lagged far behind its utilization in adults. There is likely not a single specific factor for the underutilization of the robot in pediatric cancer. But slowly, there has been more and more information starting with the utilization of standard MIS techniques, which usually is followed by the utilization of the robotic platform. In this chapter, we will discuss what has been done and explore possible future trends. We plan to do so by breaking down the discussion based on specific cancer types for which the robotic or at least MIS technique has been attempted.

## Renal and retroperitoneal tumors

Utilization of the MIS technique for nephrectomy has become the standard of practice in the adult world when managing renal tumors. Robotic nephrectomy and partial nephrectomies are considered the standard procedures in managing Renal cell carcinoma (RCC) in adults, one of the most common renal tumors in this population [2]. However, the same cannot be said for pediatric renal tumors, despite the widespread utilization of MIS in children for various benign renal diseases requiring either total or partial nephrectomy. Open surgery continues to be the mainstream approach for managing retroperitoneal and renal tumors in children, such as neuroblastoma and Wilms, the two most common malignant tumors in this region.

A recent multicenter study in France showed their experience with the first 100 robotic cancer resection cases in children [3]. This was the first of a kind of reports as it looked for robotics utilization in children for a variety of different tumors. Overall 93 procedures were performed to remove 100 tumors that vary in location (abdomen, 67%; thorax, 17%; pelvis, 10%; and retroperitoneum, 6%). There was an 8% rate of conversion to open. The two largest groups of tumors comprised 31 neuroblastic tumors (including 19 neuroblastomas) and 24 renal tumors (including 20 Wilms tumors). No tumor spillage was reported during robotic dissection. The median age of the cohort was 8.2 years, with the youngest patient being 5 months. Five of the seven conversions to open occurred in the Wilms tumor population. One additional conversion occurred due to uncontrollable bleeding. At a median follow-up of 2.4 years, there were only two recurrences, both in Wilms tumors patients. Overall, despite the widespread indication and the heterogenicity of application, the manuscript showed the feasibility and safety of the technique. It also is one of the largest series showing the use of robotic MIS technique for neuroblastomas.

Next, we plan to discuss further the utilization of robotic and MIS techniques in specific renal tumor modalities:

## Wilms tumors

Treatment of Wilms tumor is done through standardized protocols following either the SIOP (International Society of Pediatric Oncology) or the COG (Children's Oncology Group) guideline. While nephrectomy is part of both protocols, there are general differences related to the timing of the surgery, which is beyond the scope of this chapter. Yet, both protocols have advocated open nephrectomy for unilateral tumors with node sampling. This recommendation stands even though most tumors are extremely chemotherapy-sensitive, which could make this disease a good model for MIS surgery and consideration for kidney-sparing resection. It is easy to envision using MIS techniques to resect these tumors. Yet especially, in the US, where most resections are done upfront, before chemotherapy by pediatric surgeons, MIS techniques have been almost inexistent. A study showed that in the US, from 1994 to 2014, a total of 9259 patients underwent unilateral nephrectomy. Of these patients, only 1.3% had it done using an MIS technique [4].

Yet, when the SIOP protocol is followed, there appears to be growing interest in using MIS techniques to remove the kidney. Warmman et al. in 2014 evaluated a subset of patients from the SIOP 2001, who had undergone a laparoscopic nephrectomy following neoadjuvant chemotherapy [5]. Of the 4220 patients in the original study protocol, 24 underwent a laparoscopic nephrectomy. The overall oncological outcome was similar to the other patients; this cohort had 100% survival. The average age of this population was 40 months, with an average tumor volume of 117 mL. Unfortunately, no recovery or blood loss information was provided to compare to the primary cohort. But this indicated that the MIS technique could be safely used in Wilms tumor without compromising the oncological outcome. Lymph nodes were able to be sampled in the majority of patients. Additional studies, from Bouty et al., with a total of 50 patients, and Gavens et al., with 14 patients, showed similar results [6,7].

Bouty et al. evaluated the risk of local recurrence following MIS nephrectomy for Wilms tumor. In this manuscript, where 114 laparoscopic nephrectomies were performed for Wilms tumor, they recorded a local recurrence rate of 3.9% [8]. This rate was noted to be below the rate of open resection. However, the authors pointed out that the tumors using the MIS technique were smaller despite the average volume of 227 mL.

Finally, could partial nephrectomy be an option in this population, as it has become quite popular for treating small renal tumors? Again, given that this tumor has an excellent response to chemotherapy and patients endure excellent survival rates, it would be natural to think partial nephrectomy should be considered, at least for some patients. Yet the highly successful current management makes implementing less proven techniques difficult. McKay et al. evaluated MIS nephrectomy and partial nephrectomy in the resection of Wilms tumors following neoadjuvant chemotherapy [9]. Despite the small number of patients in the MIS group, the manuscript did show similar oncological outcomes. It also showed reduced use of narcotics and hospital stays, which is one of the main touted benefits of the MIS approach.

## Renal cell carcinoma

The mainstream treatment for RCC continues to be surgical resection, as this disease does not respond well to either chemotherapy or radiation. Thus, complete surgical resection becomes essential for patient survival outcomes. For a while now, MIS techniques to perform nephrectomy and partial nephrectomy have become well accepted in managing RCC for adult patients. However, RCC is a much rare condition in the pediatric population. Thus, the management has been based on the adult literature.

The robotics technique in adults for resection of RCC has shown equivalent cancer outcomes. Furthermore, it has lower rates of estimated blood loss, shorter length of hospital stays, and fewer complications compared to open nephrectomy. While it may be challenging to obtain a reasonable number of pediatric patients to prove similar results in this population, given that most RCC in children occurs in older patients, similar results are to be expected.

Given the similar oncological outcome, open, laparoscopic, and robotics techniques are viewed as comparable approaches. Both laparoscopic and robotics techniques may provide similar benefits of faster recovery, less intraoperative blood loss, and less postoperative pain. The main advantage of the robotics technique may be in the partial nephrectomy, which can be a demanding procedure done using a straight laparoscopic approach. Yet, the advent of the robotic platform that makes intracorporeal suturing and other complex laparoscopic maneuvers easier to perform has allowed for more widespread use of the MIS technique for partial nephrectomy. It has also led surgeons to expand the indication for larger and centrally located tumors. This can be done by combining several specific techniques, such as vascular mapping with infrared fluorescence imaging, on-demand ischemia, early unclamping, enucleoresection techniques, and intracorporeal hypothermia to achieve complete resection and minimize warm ischemia and renal parenchyma damage. Again, given that pediatric cancer patient has even longer expected life span, preserving renal tissue when possible should be one of the primary goals without compromising the oncological outcome.

## Other renal tumors

Ultimately the choice of an MIS technique to remove a renal mass of unknown etiology in the pediatric population at this should be a combined decision between the surgeon and family. Luckily in the pediatric population, most renal lesions that require resection outside the abovementioned condition are rare and tend to be benign. Most of which can be managed with nephrectomy. Therefore, robotic assisted nephrectomy is an excellent option when available and can be performed even in smaller patients, such as infants as early as 6 weeks and at least 5 k in weight [10].

## Adrenal tumors

Robotic surgery techniques have been utilized to perform adrenalectomies for both benign and malignant lesions. Both transperitoneal and retroperitoneal approaches have been successfully used. However, the published pediatric experience is small and restricted to a few case series. In 2020, Mitra et al. described their initial experience with robotic adrenalectomy in children [11]. They used a transperitoneal approach with all the ports positioned in the midline. A total of 3 patients were described. Tumor pathology was ganglioneuroblastoma in 2 patients and pheochromocytoma in one. Median operative time and blood loss were 244 min and 100 mL, respectively. Hospital length of stay day averaged 2 days. At a median follow-up of 19 months, all patients were noted to be disease free.

Blanc et al. performed 26 robotic adrenal resections described as part of an extensive series of 100 pediatric tumor resections using the robotics technique [3]. All adrenal cases were complete without conversion and with only one reported complication in a bilateral case that required percutaneous drainage of a retroperitoneal collection. Final pathology results showed several pheochromocytomas, a few adrenocortical adenomas, two bilateral Carney complex, and two bilateral McCune–Albright. All patients had no evidence of disease on follow-up. Thus, these two series show the robotics technique's feasibility and safety for children's adrenal tumors.

## Bladder and prostate cancer (Rhabdomyosarcoma)

Despite the current widespread use of the robotics technique in adult prostate cancer and, to a lesser stent, in bladder cancer, the same has not been the case for malignancies of the prostate or bladder in children requiring resection. This issue is directly related to the fact that in the pediatric population, the most common tumor affecting these organs is exceptionally aggressive and does not respond well to resection alone. The most common malignant tumor affecting the bladder and prostate in children is rhabdomyosarcoma. Current treatment protocols for this condition are based on chemotherapy and radiation therapy upfront, and surgery is reserved only for relapse and no responder as a last resort. It is not infrequently that a complete pelvic exenteration is required when surgery is needed. To date, there are just a few case reports of using robotics techniques to resect this aggressive tumor. Agarwal et al. describe a single case report for prostatic resection mimicking an adult robotic prostatectomy technique to resect a prostatic rhabdomyosarcoma [12]. The patient was 7 years of age when he underwent the procedure and had a hospital stay of 3 days. Early postoperative evaluation at 3 months showed a continent patient with no evidence of disease. This case illustrates the safety and feasibility of this approach and its potential benefits.

## Retroperitoneal lymph node dissection for testicular tumors

Orchiectomy is the usual first step in managing testicular tumors, and the robotics technique has no role in this particular procedure. Nevertheless, these patients will often go on to require a Retroperitoneal lymph node dissection (RPLND). This procedure is well known to be an operation with significant morbidity when done open, requiring a large midline incision. In addition, it can lead to a lengthy recovery period with a considerable need for postoperative narcotics and prolonged hospital stay. Thus, it is easy to see why the MIS technique could be an alternative if it could provide the well-known benefits associated with this approach.

Nevertheless, despite these potential benefits, the utilization of the robotics technique for RPLND in patients with germ cell tumors is still considered controversial given the need for complete adequate resection and concern for poor oncological outcomes if this is not accomplished. Currently, only high-volume centers with significant experience should consider it for patients with germ cell tumors. This may be, especially, true for patients being seen in pediatric hospital settings. Brown et al. evaluated RPLND procedures done in pediatric hospitals across the US from 2008 to 2014 [13]. Only 90 cases were performed in the various pediatric hospitals across the US, which pale when compared to the number of procedures performed in adult centers. Additionally, of those procedures, only four were done using the robotics technique.

Yet, in older patients with rhabdomyosarcoma where RPLND is used for staging and to guide further therapy, Robotics-assisted RPLND should be considered as it can offer less postprocedure pain and narcotic use coupled with faster recovery and early chemotherapy initiation. Due to the low volume of these cases, currently, there are only a few case series publications on the subject. Cost et al. presented two patients with an average lymph node count of 20, which is comparable to open procedures series and shorter hospital stay [14]. In addition, the authors of this chapter have presented data on four patients who underwent RPLND for para testicular rhabdomyosarcoma in a recent national meeting [15]. Procedures were performed using a Hidden incision endoscopic surgery technique with four robotic ports in line at the suprapubic region and an assisting port in the umbilicus. (Figs. 1 and 2) The average lymph node count was 23, and the average hospital stay was 3 days. Ultimately using MIS to perform RPLND surgery has shown great potential. Oncological outcomes may be similar, but there is a clear benefit in recovery.

## Future scenario

As the robotics techniques continue to evolve and with new platform coming to the market offering additional options including smaller trocars, the future for robotic surgery in children seems promising. Both pediatric urologist and pediatric surgeons could benefit from implementing these techniques for the treatment of a variety of cancers from

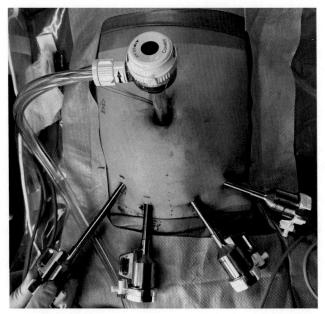

FIG. 1    Robotic RPLND port position.

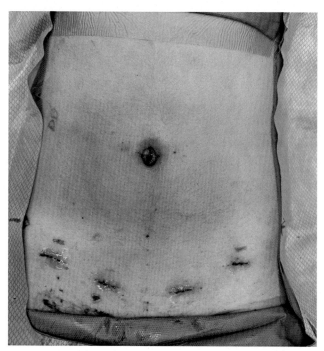

FIG. 2   RPLND immediate postoperative appearance.

intrabadominal to thoracic lesions. Robotics adds a significant upgrade to the laparoscopic technique that should not be overlooked even by experienced standard laparoscopic surgeons. Additionally, more plataforms will help in driving one of the main downsides of robotic surgery, the cost, down and help in making it more available worldwide.

## Conclusions

Despite widespread use in the adult oncological world, robotic surgery for pediatric tumors is still in its infancy. While mirroring the adult procedures, most techniques for tumor resection are feasible and safe for children. Furthermore, data do seem to suggest also equivalent oncological outcomes. However, objective data about the benefits enjoyed by the adult population are still lacking. Still, one could expect similar results based on other robotics and MIS procedures done in the pediatric population. Ultimately,the lack of a large volume of cases will continue to make it difficult to prove a real benefit for a while unequivocally. Nevertheless, robotics surgeons should continue to refine the technique and collect appropriate data to hopefully show that the benefits associated with the robotics approach also apply to oncological surgery in children.

## Key points

- Robotic surgery offers and upgrade to standard laparoscopic surgery treatment of cancer in children.
- A variety of solid intraabdominal as well as thoracic tumors can be successfully treated using the robotic platform.
- Even complex procedure, such as retroperitoneal lymph node dissection, can be achieved using the robotics technique, leading to significant improvement in recovery due to the well-known benefits of Minimally invasive surgery.

## References

[1] Varda BK, et al. Has the robot caught up? National trends in utilization, perioperative outcomes, and cost for open, laparoscopic, and robotic pediatric pyeloplasty in the United States from 2003 to 2015. J Pediatr Urol 2018;14(4):336 e1–8.
[2] Vartolomei MD, et al. Robot-assisted partial nephrectomy mid-term oncologic outcomes: a systematic review. J Clin Med 2022;11(20).

[3] Blanc T, et al. Robotic surgery in pediatric oncology: lessons learned from the first 100 tumors-a Nationwide experience. Ann Surg Oncol 2022; 29(2):1315–26.

[4] Simmons KL, et al. Open versus minimally-invasive surgical techniques in pediatric renal tumors: a population-level analysis of in-hospital outcomes. J Pediatr Urol 2021;17(4):534 e1–7.

[5] Warmann SW, et al. Minimally invasive nephrectomy for Wilms tumors in children - data from SIOP 2001. J Pediatr Surg 2014;49(11):1544–8.

[6] Bouty A, et al. Minimally invasive surgery for unilateral Wilms tumors: multicenter retrospective analysis of 50 transperitoneal laparoscopic total nephrectomies. Pediatr Blood Cancer 2020;67(5):e28212.

[7] Gavens E, Arul GS, Pachl M. A single Centre matched pair series comparing minimally invasive and open surgery for the resection of pediatric renal tumours. Surg Oncol 2020;35:498–503.

[8] Bouty A, et al. What is the risk of local recurrence after laparoscopic transperitoneal radical nephrectomy in children with Wilms tumours? Analysis of a local series and review of the literature. J Pediatr Urol 2018;14(4):327 e1–7.

[9] McKay KG, et al. Oncologic Fidelity of minimally invasive surgery to resect neoadjuvant-treated Wilms tumors. Am Surg 2022;88(5):943–52.

[10] Meehan JJ. Robotic surgery in small children: is there room for this? J Laparoendosc Adv Surg Tech A 2009;19(5):707–12.

[11] Mitra AP, et al. Robotic adrenalectomy in the pediatric population: initial experience case series from a tertiary center. BMC Urol 2020;20(1):155.

[12] Agarwal DK, et al. Pediatric robotic prostatectomy and pelvic lymphadenectomy for embryonal rhabdomyosarcoma. Urology 2018;119:143–5.

[13] Brown CT, et al. Utilization of robotics for retroperitoneal lymph-node dissection in pediatric and non-pediatric hospitals. J Robot Surg 2020; 14(6):865–70.

[14] Cost NG, et al. Robot-assisted laparoscopic retroperitoneal lymph node dissection in an adolescent population. J Endourol 2012;26(6):635–40.

[15] Ernst MEK, Ching C, Dajusta D. Robotic retroperitoneal lymph node dissection for paratesticular rhabdomyosarcoma in children. In: Society of Pediatric Urology Fall Meeting; 2022. Las Vegas, NV.

III. The current and future clinical applications of robotic surgery among medical specialties

# 65

# Robot-assisted kidney transplantation

Joao Manzi[a], Phillipe Abreu[b], and Rodrigo Vianna[a]

[a]Miami Transplant Institute, Jackson Memorial Hospital, University of Miami, Coral Gables, FL, United States [b]Division of Transplant Surgery, Department of Surgery, University of Colorado Anschutz Medical Center, Aurora, CO, United States

## Kidney transplantation

Chronic kidney disease (CKD) is a progressive, incurable disease with high morbidity and mortality that generally affects the adult population [1]. The condition is defined by a reduction in the functional capacity of the kidney, with an estimated Glomerular filtration rate of less than 60 mL/min per $1·73 m^2$, marker of kidney damage or abnormalities detected through laboratory or imaging testing, which is present for at least 3 months [2]. CKD is a significant public health problem worldwide, with an increasing and currently estimated prevalence of approximately 13.4% (11.7%–15.1%) [3]. In less than 20 years, in 2040, it is estimated that CKD will become the fifth leading cause of death globally, one of the most significant projected growths for causes of death [4].

When the filtration rate reaches less than 15 mL/min per $1·73 m^2$, it is then classified as End-stage renal disease (ESRD), in which the kidney function can no longer sustain life in the long term [2]. It is estimated that worldwide, the prevalence of ESRD is 4.9–7 million people [3]. The main signs and symptoms of kidney failure are progressive uremia, anemia, acidemia, volume overload, and electrolyte abnormalities [5]. Patients with ESRD depend on Renal replacement therapy or kidney transplantation to stay alive [6].

Most people with ESRD are treated with hemodialysis or peritoneal Dialysis [2]. This is mainly due to the impossibility of performing kidney transplantation, a therapeutic possibility that provides survival advantages over long-term Dialysis [7]. Approximately 56% of ESRD patients on Dialysis are actively waiting for a kidney transplant, but due to low availability, only 25% receive it, and 6% end up dying on the transplant waiting list every year [2]. Currently, more than 90.000 people are waiting for a kidney transplant in the United States. Last year, 4152 people died on the waiting list [8].

Long-term survival rates for kidney transplantation vary significantly by country and demographics [7]. In the United States, the 5-year graft survival for primary kidney transplants from deceased donors is 72%, while for living donors is 85%; these same values in Europe are 79% and 87%, respectively [9]. When comparing ethnicities in the United States, the 5-year graft survival is 71% for the White population, 43% for the Hispanic population, and 62% for African Americans [10].

These numbers represent a significant improvement over those seen in recent decades, mainly due to the evolution of immunosuppressive therapies, since the surgical technique has remained relatively stable, since the first long-term successful kidney transplant in 1954 [11–13]. Even with the introduction of the use of laparoscopy, technical difficulties in performing vascular anastomoses have hampered its utilization for transplantation, except for a few reported cases and small series. Laparoscopy has become the standard for live donor nephrectomy [14,15]. This scenario has changed significantly since 2002, when Hoznek et al. published the first case of Robot-assisted kidney transplantation (RAKT) [16].

## History of RAKT

The da Vinci robotic surgical system (Intuitive Surgical, Inc.) was introduced in the market in 2000, bringing significant improvements to the existing limitations in laparoscopy, such as three-dimensional view with stable camera guidance, a higher movement precision, and magnification options, which enabled its use in more complex surgeries [11]. In the same year, the world's first transabdominal hand-assisted robotic donor nephrectomy was performed at the University of Illinois at Chicago [17]. Two years later, Hoznek et al. published the first successful transplant using the da Vinci robotic surgical system in a 26-year-old male patient, using stereoscopic magnification and ultra-precise suturing techniques for the vascular anastomoses of the implant [16].

In the first years after its debut, the widespread adoption of RAKTs was hindered by robot-specific issues, such as the lack of fine haptic feedback, prolonged warm ischemia time, and high costs [18]. In 2011, Boggi et al. described the first entirely RAKT case in Europe, suggesting improvements compared to the first presented technique [19]. In the following years, Robot-assisted kidney transplantation gained strength, becoming an important alternative to open surgery, especially, for obese patients [20].

## Indications

In 2011, obese patients (BMI $\geq 30 \, kg/m^2$) comprised 20%–50% of patients on Dialysis [21]. These patients typically have longer waiting times, with a median of 39 months for patients with a BMI $<25 \, kg/m^2$ and 59 months for patients with a BMI $>40 \, kg/m^2$ [22]. The prolonged waiting times are mainly caused by the exclusion or hesitancy to include them on the transplant list due to high rates of perioperative complications, such as prolonged hospitalization, lower graft survival, delayed graft function, and higher rates of surgical site infections [23–28].

In 2009, the group from the University of Illinois at Chicago performed the first intra-abdominal robotic kidney transplant in an obese patient, introducing this technique as an option for these patients instead of a Conventional open kidney transplant (COKT) [29]. Since then, its use for obese patients has been adopted in various centers worldwide, showing lower complication rates when compared to COKT despite more prolonged warm ischemic times [30].

In a metaanalysis performed by Slagter et al. the Dutch group compared 482 RAKT procedures with 1316 COKT procedures, finding a lower risk of surgical infection with RAKT (risk ratio [RR] = 0.15, $P < 0.001$), less postoperative pain (mean difference [MD] = −1.38 points in the Visual Analogue Scale pain score, $P < 0.001$), and a shorter length of hospital stay (MD = −1.69 days, = 0.03) [31]. There were no differences in renal function, patient, and graft survival [31].

More recently, motivated by the good results found for obese patients, the European Robotic Urological Section group evaluated the expansion of RAKT use for nonobese patients [32]. Comparing obese patients ($\geq 30 \, kg/m^2$ BMI), overweight ($<30/\geq 25 \, kg/m^2$ BMI), and nonoverweight recipients ($<25 \, kg/m^2$ BMI), similar rates of minor and major postoperative complications were obtained with the use of robotics, suggesting a possible expansion of the results found in obese patients to these other groups [32].

## Technique: Robotic kidney transplantation

### Donor operation

Robot-assisted donor nephrectomy is a complex and advanced surgical procedure. In general, donors who undergo robot-assisted surgery are usually discharged in less than 24 h, which not only reduces admission costs and improves pain management in the postoperative period but also decreases the burden of the donation process to the donors.

The positioning of the patient is critical in robotic surgeries, given the minimal room for adjustment once the case is started. Adequate positioning ensures that the surgical team has sufficient access to the surgical field and can perform the procedure safely and effectively, taking full advantage of the potential of using the robot. The described techniques can be done using either the da Vinci Xi or the Si model, following the same step-by-step and port-positioning.

The left kidney is the most commonly side chosen to proceed with the donation, with longer vessels and less complicated anatomic relations to the surrounding organs. In these cases, the patient is positioned in right lateral decubitus, with full flex of the surgical table lowering the upper body and the lower limbs, exposing the left kidney (also known as the "Jackknife position"). The right arm is extended over an arm board for better fixation and balance. The left arm is attached to the side of the body to ensure good robot access. The main goal of this positioning is to increase the space between the left rib cage and the left anterior superior iliac spine. The patient is fixed to the operating table using safety straps in order to prevent any movements during the surgery (Fig. 1). After positioning, prepping, and draping the area to maintain sterile conditions, the robotic ports are inserted under direct view, using the lateral aspect of the rectus abdominis as the reference line (8 mm subcostal left, 8 mm para-rectal para-medial left, 12 mm para-rectal para-umbilical left, and 8 mm para-rectal supra-pubic—inside a Pfannenstiel incision) with an assistant 12 mm

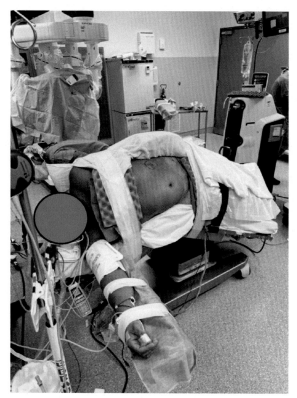

FIG. 1    Patient position for robotic-assisted left donor nephrectomy.

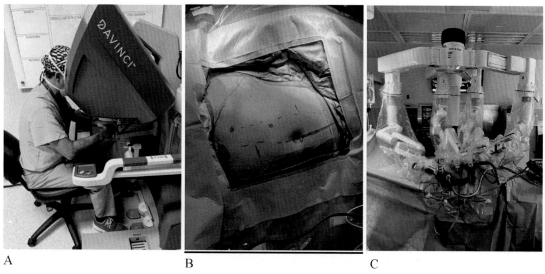

A                                    B                                    C

FIG. 2    (A) Surgeon positioning; (B) trocars positioning; (C) robotic arms positioning.

laparoscopic disposable port triangulating with the 2 upper robotic ports towards the midline. The robotic system is then docked behind the patient's back (Fig. 2). The Pfannenstiel incision is performed and dissected until the peritoneum layer (kept intact) to expedite the organ removal once the operation is completed. This incision is covered with a clean, dry lap and an adhesive drape to secure the para-median port and avoid air leaks. When the right kidney is used for donation, the positioning follows the same sequence mirroring the left side, with an extra 5 mm laparoscopic disposable port inserted in the subxiphoid region in order to retract the liver cephalad with a laparoscopic auto-static grasper.

III. The current and future clinical applications of robotic surgery among medical specialties

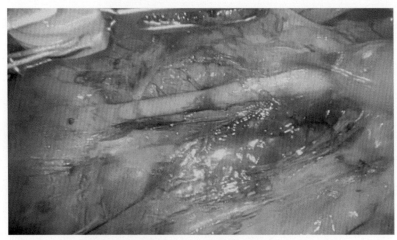

FIG. 3    The gonadal vein is then dissected until its origin in the left renal vein.

Didactically, we present the robot-assisted donor nephrectomy technique in 10 successive steps:

(1) The first step is the dissection of the left para-colic gutter promoting the medial rotation of the left colon (Mattox maneuver) exposing the left retroperitoneum (aorta and left kidney).

(2) After the first step is completed, the left ureter is identified and carefully dissected from the left gonadal vein all the way to the bifurcation of the iliac vessels.

(3) The gonadal vein is then dissected cephalad to its origin in the left renal vein (Fig. 3).

(4) At this moment, attention is turned to the upper pole of the left kidney, which is dissected from the left adrenal gland.

(5) As a fifth step, the renal vein is dissected medially towards the IVC to enable mobilization and exposure of the origins of the adrenal and gonadal veins.

(6) After dissecting the renal vein, the renal artery is then dissected medially from the surrounding tissues until its take-off from the aorta.

(7) The kidney is then mobilized from the retroperitoneum, and all collateral branches are ligated.

(8) Vessels are now well dissected, and the kidney is mobilized. A second time-out is performed to confirm ABO compatibility, and the vessels are stapled using the 45 mm robotic white load stapler, ensuring the division, and sealing of the vessels (Fig. 4).

(9) With total mobilization of the kidney, it is removed using a specimen retrieval Endo bag through the previously performed Pfannenstiel incision.

FIG. 4    The vessels are stapled using the robotic stapler.

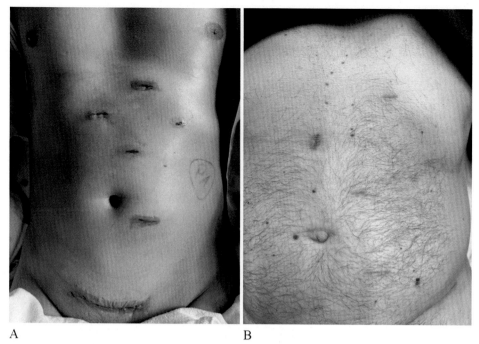

FIG. 5    Incision aspect post robot-assisted donor nephrectomy. (A) Immediately after surgery. (B) A month after surgery.

**(10)** Lastly, a careful review of hemostasis is performed, ensuring that any areas that need additional attention are addressed. The robotic system is then undocked, the ports are removed, and the abdominal wall is closed as the routine.

The final aspect is cosmetically very satisfying for the patients (Fig. 5).

### Recipient operation—Graft implantation

In the recipient operation, the patient is placed in a lithotomy position with the legs open so the robotic arms can be positioned in between them. As in the donor nephrectomy, this technique can be done using either the da Vinci Xi or the Si model. Four 8 mm robotic ports are inserted in line, approximately 12 cm from the right inguinal ligament. The first one inserted is the camera 8 mm robotic port in the midline, and then the far most left flank port is inserted under direct view. The remaining left port is placed between the previous ports, accommodating at least 8 cm of distance between the ports. Then, a 12 mm laparoscopic disposable port is inserted in the far most right flank for the assistant manipulation. Finally, an 8 mm robotic port is inserted in between the camera and the assistant port, also maintaining them 8 cm apart from each other. A 2–3 in. Pfannenstiel incision is performed, and a gel-port is placed to allow the allograft insertion in the abdominal cavity. The patient is then turned to a 20°–25° Trendelenburg to allow appropriate cephalad mobilization of the intestines and adequate exposure of the surgical target anatomy (iliac vessels).

The recipient operation can be didactically summarized in 10 steps:

**(1)** A very meticulous kidney preparation on the back table is performed as the first step. All sutures are trimmed and prepared to be ready for intra-corporeal use (Fig. 6). Routinely, #6–0 GoreTex sutures, TTc-9 needles are used for vascular anastomosis. #5–0 PDS suture and RB-1 needles are used for the ureteral anastomosis. A 7–8 cm double-arm suture is created by tying the suture ends and used to perform the venous anastomosis. A 6–7 cm double-arm suture is also created for the arterial anastomosis. The ureteral anastomosis is performed with 7–8 cm single-arm sutures. After the preparation of all sutures to be used during the implantation of the kidney, the graft is wrapped in a Raytec with an opening in the center to allow exposure of the hilum, with strategically placed stitches to help in identifying the orientation of the vessels. (Fig. 7). When shorter veins (e.g., right donor kidney used) or multiple arteries are found, careful reconstructions should be made during as part of the back-table procedure, allowing a single ostium and additional venous length (even extending it with donor vessels from cadaveric vessels when available). The ureter should also be trimmed and spatulated in the back-table (the length will influence the anastomotic site in the bladder).

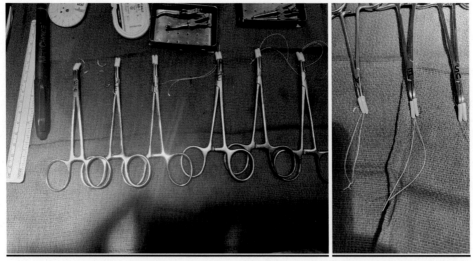

FIG. 6    All sutures are trimmed and prepared to be ready for intra-corporeal use. Routinely, #6-0 GoreTex sutures, TTc-9 needles are used for vascular anastomosis.

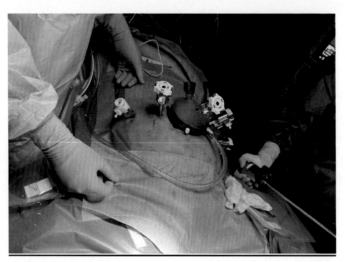

FIG. 7    The graft is introduced in the cavity through a Gelport:

(2) In the second step, a peritoneal flap is dissected in the right lower quadrant in order to expose the bladder and iliac vessels (similar to the dissection performed in a right inguinal hernia repair).

(3) The iliac vessels and the bladder are then dissected to prepare them for the implantation of the renal allograft. Dissection should only be performed with the minimal possible length for the clamps to be placed to avoid possible complications (for ex, lymphocele).

(4) The robotic arms are placed anteriorly, holding the abdominal wall, while the Gelport is opened for the graft insertion in the abdominal cavity. The positioning of the graft is a critical point, with the cortex facing the pelvis and the hilum facing the iliac vessels (the graft lays on top of the sigmoid colon). Ice slush is also directly applied to the peritoneum to decrease the warming time of the graft. The third arm can be used to assist in this step, pushing the kidney to the pelvis.

(5) Straight robotic bulldog vascular clamps are applied to the external iliac vein proximally and distally to the anastomotic site. An end-to-side running venous anastomosis is performed from the graft renal vein to the recipient external iliac vein (Fig. 8). At this point, another robotic bulldog vascular clamp is applied to the graft renal vein, and the iliac vein clamps are released, reestablishing the venous return of the patient without perfusion of the kidney graft.

(6) Straight robotic bulldog vascular clamps are applied to the external iliac artery proximally and distally to the anastomotic site. An end-to-side running arterial anastomosis is performed from the graft renal artery to the

FIG. 8 Arterial anastomosis end-to-side.

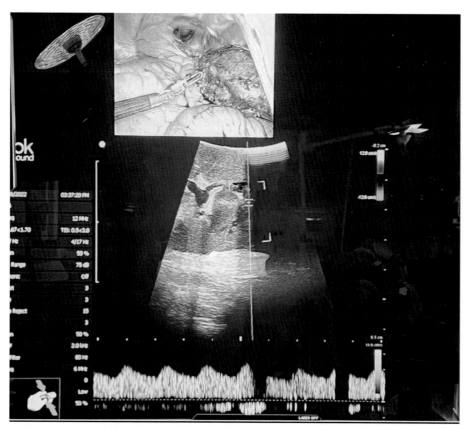

FIG. 9 An ultrasound may be used to evaluate the kidney.

recipient external iliac artery. Subsequently, another robotic bulldog vascular clamp is applied to the graft renal artery, and the iliac artery clamps are released, reestablishing perfusion of the donor leg.

(7) The robotic bulldog vascular clamps are removed from the renal vein and renal artery, reperfusing the organ. After adequate hemostasis, 2 cc of indocyanine green dye is injected to confirm kidney perfusion with the firefly technique, followed by an intraoperative ultrasound to check vascular patency (Fig. 9).

III. The current and future clinical applications of robotic surgery among medical specialties

**(8)** The graft is then rotated to the other side in order to be positioned inside the previously created peritoneal flap for better fixation and retroperitonealization of the kidney graft.

**(9)** An anastomosis with the graft ureter and the recipient's bladder is performed with minimal bladder dissection, placing a $4.5 \times 8$ cm urological stent.

**(10)** Finally, hemostasis is reviewed, instrument and gauze count are verified. The abdominal wall is closed.

## Future scenario

Transplantation has been a fundamental component in the care of patients with the ESRD since its development in 1954. Considering the increase in the number of patients with Chronic kidney disease, it will likely play an even more significant role in global health care. In this context, Robot-assisted kidney transplantation represents a new development leap, a new emerging and possibly safer procedure.

Robot-assisted surgery has evolved significantly in recent years, with the expansion of its applicability to different complex specialties, in which transplantation surgery will certainly play a relevant role. The initial experience has been showing promising and improved outcomes to donors and recipients, to the extent that it is fair to expect that open kidney transplants will perhaps no longer exist in the next few decades once robotic systems become widely available and more surgeons master the technique.

At the Miami Transplant Institute, robot-assisted donor nephrectomy has replaced the laparoscopic technique entirely. Patients have experienced a much faster recovery, with the majority of them being discharged within 24 h, largely due to diminished postoperative pain. As artificial intelligence progresses, robot-assisted surgery will become the standard for many procedures: not only will it offer more precise dissection of delicate structures, but it will also eliminate possible human errors.

## Key points

- Robot-assisted kidney transplantation (RAKT) has emerged as a favorable option compared to open surgery, with multiple studies demonstrating its benefits, such as reduced hospitalization, lower complication rates, and improved pain management.
- The indications for RAKT have expanded significantly in recent years, with its adoption extending beyond obese patients to a broader patient population, reflecting its increasing prevalence in the surgical field.
- Technical aspects of RAKT, including patient positioning and a strategic step-by-step, are essential for leveraging the full potential of robotic systems, highlighting the necessity for surgeons to master the intricacies of the technique.
- The prominence of robotic systems in surgery is anticipated to grow, indicating a future where proficiency in robotic procedures will be critical for surgical success and innovation in patient care.

## References

[1] Kalantar-Zadeh K, Jafar TH, Nitsch D, et al. Chronic kidney disease. The Lancet 2021;398:786–802. Elsevier B.V.

[2] Webster AC, Nagler EV, Morton RL, Masson P. Chronic kidney disease. Lancet 2017;389.

[3] Lv JC, Zhang LX. Prevalence and disease burden of chronic kidney disease. Adv Exp Med Biol 2019;1165:3–15.

[4] Foreman KJ, Marquez N, Dolgert A, et al. Forecasting life expectancy, years of life lost, and all-cause and cause-specific mortality for 250 causes of death: reference and alternative scenarios for 2016–40 for 195 countries and territories. Lancet 2018;392(10159).

[5] Zarantonello D, Rhee CM, Kalantar-Zadeh K, Brunori G. Novel conservative management of chronic kidney disease via dialysis-free interventions. Curr Opin Nephrol Hypertens 2021;30.

[6] Territo A, Mottrie A, Abaza R, et al. Robotic kidney transplantation: current status and future perspectives. Vol. 69. Minerva Urologica e Nefrologica; 2017.

[7] Hariharan S, Israni AK, Danovitch G. Long-term survival after kidney transplantation. N Engl J Med 2021;385(8).

[8] Bethesda M. United States renal data system. Incidence, prevalence, patient characteristics, and treatment modalities. National Institute of Diabetes and Digestive and Kidney Diseases; 2023. National Data—OPTN [Internet]. 2023 [cited 2023 Mar 14]; Available from: https://optn.transplant.hrsa.gov/data/view-data-reports/national-data/#.

[9] Wang JH, Skeans MA, Israni AK. Current status of kidney transplant outcomes: dying to survive. Adv Chronic Kidney Dis 2016;23.

[10] Gondos A, Dohler B, Brenner H, Opelz G. Kidney graft survival in europe and the United States: strikingly different long-term outcomes. Transplantation 2013;95(2).

[11] Pein U, Girndt M, Markau S, et al. Minimally invasive robotic versus conventional open living donor kidney transplantation. World J Urol 2020;38(3).

[12] Pai SA. Surgery of the soul: reflections on a curious career. BMJ 2002;324(7340).

[13] Murray JE, Merrill JP, Harrison JH. Kidney transplantation between seven pairs of identical twins. Ann Surg 1958;148(3).

[14] Greco F, Hamza A, Wagner S, et al. Hand-assisted laparoscopic living-donor nephrectomy versus open surgery: evaluation of surgical trauma and late graft function in 82 patients. Transplant Proc 2009;41(10).

[15] Hoda MR, Hamza A, Greco F, et al. Early and late graft function after laparoscopic hand-assisted donor nephrectomy for living kidney transplantation: comparison with open donor nephrectomy. Urol Int 2010;84(1).

[16] Hoznek A, Zaki SK, Samadi DB, et al. Robotic assisted kidney transplantation: an initial experience. J Urol 2002;167(4 I).

[17] Horgan S, Vanuno D, Sileri P, et al. Robotic-assisted laparoscopic donor nephrectomy for kidney transplantation. Transplantation 2002;73(9).

[18] Hameed AM, Yao J, Allen RDM, et al. The evolution of kidney transplantation surgery into the robotic era and its prospects for obese recipients. Transplantation 2018;102.

[19] Boggi U, Vistoli F, Signori S, et al. Robotic renal transplantation: first European case. Transpl Int 2011;24(2).

[20] Øyen O, Scholz T, Hartmann A, Pfeffer P. Minimally invasive kidney transplantation: the first experience. Transplant Proc 2006;38(9).

[21] Collins AJ, Foley RN, Chavers B, et al. United States Renal Data System 2011 Annual Data Report: Atlas of chronic kidney disease & end-stage renal disease in the United States. Am J Kidney Dis 2012;59.

[22] Segev DL, Simpkins CE, Thompson RE, et al. Obesity impacts access to kidney transplantation. J Am Soc Nephrol 2008;19(2).

[23] Nanni G, Tondolo V, Citterio F, et al. Comparison of oblique versus hockey-stick surgical incision for kidney transplantation. Transplant Proc 2005;37(6).

[24] Lynch RJ, Ranney DN, Shijie C, et al. Obesity, surgical site infection, and outcome following renal transplantation. Ann Surg 2009;250(6).

[25] Meier-Kriesche HU, Kaplan B. Waiting time on dialysis as the strongest modifiable risk factor for renal transplant outcomes: a paired donor kidney analysis. Transplantation 2002;74(10).

[26] Harris AD, Fleming B, Bromberg JS, et al. Surgical site infection after renal transplantation. Infect Control Hosp Epidemiol 2015;36(4).

[27] Modlin CS, Flechner SM, Goormastic M, et al. Should obese patients lose weight before receiving a kidney transplant? Transplantation 1997;64(4).

[28] Meier-Kriesche HU, Arndorfer JA, Kaplan B. The impact of body mass index on renal transplant outcomes: a significant independent risk factor for graft failure and patient death. Transplantation 2002;73(1).

[29] Giulianotti P, Gorodner V, Sbrana F, et al. Robotic transabdominal kidney transplantation in a morbidly obese patient. Am J Transplant 2010;10(6).

[30] Oberholzer J, Giulianotti P, Danielson KK, et al. Minimally invasive robotic kidney transplantation for obese patients previously denied access to transplantation. Am J Transplant 2013;13(3).

[31] Slagter JS, Outmani L, Tran KTCK, et al. Robot-assisted kidney transplantation as a minimally invasive approach for kidney transplant recipients: a systematic review and meta-analyses. Int J Surg 2022;99.

[32] Prudhomme T, Beauval JB, Lesourd M, et al. Robotic-assisted kidney transplantation in obese recipients compared to non-obese recipients: the European experience. World J Urol 2021;39(4).

# 66

# Robotic abdominal organ transplantation

*Celeste Del Basso$^a$, Fabio Antonellis$^b$, and Giovanni Battista Levi Sandri$^c$*

$^a$SCDU General Surgery, Surgical Oncology, Minimally Invasive, Robotic and HBP Surgery, AOUAL SS. Antonio e Biagio e Cesare Arrigo, Alessandria, Italy $^b$Department of General Surgery, Colleferro, Italy $^c$Digestive Surgery Unit, Fondazione Policlinico Universitario Agostino Gemelli IRCCS, Rome, Italy

## Introduction

Robotic surgery and abdominal transplantation are two cutting-edge medical advancements that have revolutionized the field of surgery. The minimally invasive approach, first with the laparoscopic approach and in the last 10 years with the robot-assisted procedures, has significantly improved patient outcomes, allowing for more precise and minimally invasive procedures.

Robotic surgery, also known as robot-assisted surgery, is a rapidly evolving field that combines the expertise of skilled surgeons with the precision and dexterity of robotic systems. It involves the use of robotic arms, equipped with surgical instruments and cameras, to perform various surgical procedures with enhanced precision and control. The surgeon operates the robotic system using a console, manipulating the instruments with exceptional accuracy. This technology has transformed the way surgeons approach complex procedures, allowing for smaller incisions, reduced scarring, and quicker recovery times for patients. Nowadays a few robotic systems are available with different abilities and costs for hospitals.

The convergence of robotic surgery and abdominal transplantation has brought numerous benefits to the field of transplantation surgery. Robotic assistance has allowed surgeons to perform abdominal transplant procedures with greater precision, enabling them to navigate intricate anatomical structures and perform meticulous suturing. The enhanced visualization provided by robotic systems aids surgeons in identifying blood vessels and nerves with greater accuracy, minimizing the risk of complications. Additionally, the minimally invasive nature of robotic surgery reduces postoperative pain, decreases the risk of infection, and accelerates the recovery process for transplant donor and recipients [1]. These aspects are essential for the living donor transplantation to develop in more Asian countries.

Most applications of robotic surgery in transplantation have involved kidney transplantation. In this chapter, we will explore the concepts of robotic surgery and abdominal transplantation, highlighting their key features, benefits, and impact on medical practice focusing on the liver, pancreas, and intestinal transplantation.

## Liver transplantation

Liver transplantation (LT) is a life-threatening procedure. The most frequent tumoral indication for LT is Hepatocellular carcinoma (HCC). The Minimally invasive liver surgery (MILS) has been described as a safe procedure for HCC resection in cirrhotic patients.

### Prior to liver transplant recipient surgery

Due to the organ shortages more frequently an HCC resection as a bridge to LT is required to avoid dropout patients from the waiting list. Panaro et al. first described a robotic liver resection as a bridge to LT I patient with an HCC in segment III. After this report, many centers described their own experience [2]. The role of MILS HCC resection prior to

LTwas investigated in a multicentric Italian study [3]. The study concludes that MILS seems to be protective over open surgery for the risk of delisting, posttransplant patient death, and tumor recurrence. A small series of seven patients transplanted after robotic liver resection was reported by Magistri et al. [4], the authors suggested an advantage of the robotic technique that seemed to provide easier access to the abdominal cavity, potentially reducing operative time and intraoperative complications during LT. The "Italian Group of Minimally Invasive Liver Surgery" analyzed the role of laparoscopy and robotic approach for repeat liver resection even in patients in waiting list [5]. Patients at LT surgery had minor adhesions rather than patients transplanted after open surgery. This translates into less blood loss, transfusion, and morbidity [3].

## Living donor

Living donation is the most intuitive tool to overpass the organ shortage. Moreover, this solution allows transplanting a high-quality liver to the recipient with a shorter cold ischemia time. The living liver donation is more developed in eastern countries due to local beliefs. Right liver lobe is generally the graft of choice for adult recipient, while the left lobe is preferred in pediatric transplant. In living donation the first goal is the safety of the donor, however, from 16% to 34% of donors had postoperative complications [6]. Specifically, in open donation 30%–50% of the complications are associated with trauma to the abdominal wall [7]. This is why any effort to decrease the donor morbidity is mandatory.

The first case of robotic living donor right hepatectomy was reported by Giulianotti in 2012 [8].

The main difference between the two operative techniques is the operative time, longer for the robotic approach. But this difference may decrease with the experience of the different Centers. In fact despite the previously mentioned difference, some studies showed even better results than laparoscopic approach in terms of blood loss, conversion rate, and morbidity rate [9]. On the other hand, no differences were observed for the blood loss and the graft function. If we compare the robotic to the pure laparoscopic approach, the reported difference between the two groups seems to be not advantageous [10]. A first robotic left liver donation was successfully described in 2014 [9]. To date, several centers have started a robotic liver living donor program (Saudi Arabia, Taiwan, and Korea) with encouraging results. Few years later, a comparative study between 25 robotic left lateral sectionectomy and laparoscopic approaches observed a lower blood loss, postoperative pain, and length of stay in the robotic group, without difference in morbidity and mortality [11]. A very early experience from Chen et al. comparing robotic and open right donor reported a higher operative time in the robotic group (596 vs 383 min) associated with minor donor morbidity. Broering et al. reported 35 cases of robotic right donor hepatectomy with higher operative time and warm ischemia time, but with lower donor morbidity [12]. Rho and colleagues reported a large series of robotic right liver donation (52 cases) compared to laparoscopic and open cases [13]. The authors observed a longer operative time, less blood loss, and postoperative pain. A subanalysis of the robotic cases, showed a decreased operative time in different surgical steps according to the surgeon's higher experience. In a metaanalysis, Yeow and colleagues reported interesting data [14]. Robotic approach was associated with longer operative time compared to all other approaches, the estimated blood loss was lesser in the laparoscopic and robotic groups. In terms of intraoperative blood transfusion, major complications, biliary complication, and the length of stay times were similar across all techniques [14]. The metaanalysis concludes that the robotic approach for living donor hepatectomy is a safe and feasible procedure when performed in high-volume centers. In 2021, the King Faisal Specialist Hospital from Riyadh (Saudi Arabia) published its experience of 318 robotic donor hepatectomies [15]. Of them, there were 132 right lobes, 113 left lateral lobes, and 73 left lobes. This is out of doubt the major world experience in robotic liver donation. The authors reported no mortality and an overall morbidity of 5.9%. The morbidity was higher in the right lobe group (9.1%) compared to the left lobe group (5.5%) and the left lateral group (2.7%). By note, all except one complication were classified as Clavien-Dindo I/II. Interestingly, Amma et al. described a donor's peak transaminase and serum bilirubin levels significantly lower in the robotic group despite a longer operative time [16]. The authors suggest a minor surgical trauma and metabolic stress sustained during the robotic procedure. However, the previously mentioned results are in contrast with previous results from different centers [13].

The intent of robotic liver donation is to reduce the donor morbidity, improving the postoperative outcome by avoiding bleeding, or bile duct injuries [17]. The preoperative donor imaging is mandatory as for the living donor a crucial step of the hepatectomy is the bile duct division. In open surgery an intraoperative cholangiogram allows in confirming the road map for bile duct division. In robotic surgery, actually, it is not possible to perform a cholangiogram [16].

A case of segment I liver harvesting by robotic procedure was reported for a pediatric living transplant [18]. A segment II monosegment graft was used for a 14-month-old. Donor and recipient were uneventful.

As for the laparoscopic approach, a learning curve is necessary to achieve excellent results. For the robotic approach this learning curve is shorter. Kim reporting the experience of the Yonsei University in Korea, described a transition from negative to a positive value in the CUSUM curve after the 16th case of robotic donor. This result was observed for the graft-out time and the total operative time [19].

A first hybrid laparoscopic/robotic graft implantation in living donor LT was reported in 2022 [20]. The authors from the Seoul National University College of Medicine used the robot after reperfusion. The robotic approach allowed elaborate suturing that is required for hepatic artery and bile duct anastomosis as these structures have a small diameter. One year later the same center reported the world's first totally robotic liver transplant [21]. The authors performed a total robot-assisted explant hepatectomy in a 57-year-old man with B+ blood type requiring living donor LT because of alcoholic liver cirrhosis with Hepatocellular carcinoma. The Robot-assisted explant hepatectomy was completed in 327 min and the follow-up robotic engraftment took 295 min.

## Postliver transplant recipient surgery

In order to gain full access to the liver during the transplant procedure, an extensive incision is usually required. The incisions commonly used are Mercedes, J shaped, and midline. The Mercedes incision is a transverse incision with median extension, both on the rectus abdominis muscle. As we are in a minimally invasive trend, the incision is more likely a J shaped or a midline one. Out of doubt, the transplant procedure leaves in the abdominal cavity signs of the procedure with visceral adhesions. Nevertheless, several surgical procedures have been successfully performed using the minimally invasive technique in transplant patients.

Rofaiel et al. described a robotic terminal jejunogastrostomy [22]. The indication of the surgery was a progressive biliary dilation about a year after the transplant with several episodes of cholangitis. Total docked time was 146 min with an estimated blood loss of <5 mL, no complications were reported. A short series of three robotic biliary revisions to hepaticojejunostomy was published in 2021 by Hawksworth et al. [23]. The authors compared the three robotic cases with four open cases. Median case time for biliary revision was longer in the robotic group, 373 min (286–373) compared to open group, 280 min (163–321). The median length of stay was shorter in the robotic group, 4 days [1–4], compared to open group, 7 days [4–10]. There were no mortalities in either group, or stricture recurrences seen during the follow-up period.

Two patients with de novo gastric cancer after LT were successfully treated with total gastrectomy plus D2 lymphadenectomy and robotic digestive diversion using a da Vinci robotic surgery system at Southwest Hospital (China) [24]. The duration of the surgery was 315 and 275 min, with the estimated blood losses of 145 and 125 mL. The patients were discharged on days seven and nine after surgery, and no complications occurred. Postoperative pathological stages were pT4aN3aM0, stage III B and pT4aN2M0, stage III A, respectively.

The group of Giulianotti published a case of sleeve gastrectomy after LT [25]. The operation was completed within 158 min with no postoperative complications.

## Pancreas transplantation

The pancreas is a single organ with two different functions, the endocrine function having above all the role of glycemic control by secreting insulin. The aim of pancreas transplantation is to restore physiological glycemic control by preventing and managing the dramatic complications of diabetes. Though pancreas transplantation is often performed with simultaneous kidney transplantation, in the present chapter we will discuss only the pancreas alone.

## Living donor

Being a unique organ, the only solution for a living donation is the partitioning of the pancreas [26]. This results in a donation of the distal pancreas. The consequence is a worldwide limited experience in living donation. The procedure consists of a distal pancreatectomy with spleen preservation. Nowadays, robotic distal pancreatectomy is commonly performed in many centers with a worldwide hospital mortality of 0.28% [27]. The preservation of the spleen is a procedure performed in 44.8% of cases. These last two data are related to oncological surgeries. As the pancreas transplant is a nonlife-saving transplant, for the living donation its widespread use is more limited. In 2007, Horgan et al. [28] reported the first robotic hand-assisted simultaneous nephrectomy and distal pancreatectomy. The robotic system was used in the pancreatic step to divide the splenic vessels and to complete the pancreatectomy. Oberholzer used the robotic system to achieve a simple dissection of the anterior surface of the splenic vein from the posterior surface of the neck of the pancreas [29]. Yet the robotic living donation for pancreas transplantation is more like a technical demonstration than an actual procedure.

## Recipient surgery

Boggi described the first world experience of a robot-assisted pancreas transplant [30,31]. the first case with simultaneous kidney transplant and two other cases of solitary pancreas. They had no complications. The aim of this report was to test the feasibility of the procedure using the robotic system. Yeh et al. reported the first case in a morbid obesity patient with type 1 diabetes [32]. The same center published the largest series of 10 patients transplanted using the robotic procedure [33]. The authors compared open vs robotic pancreas transplantations. The robotic group had 10 cases. Patients undergoing robotic surgery had a significantly higher BMI than those in the open approach (33.7 vs 27.1). Operative time was significantly longer in the robotic approach (7.6 vs 5.3h), while the estimated blood loss was lower in the robotic (150 vs 200). Complications between the two groups were similar. These important results confirmed the noninferiority of the robotic procedure for organ transplantation.

## Other organ transplantation

## Intestinal transplantation

Living donor-related small bowel transplantation is a surgical innovation to expand the pool of intestinal graft donors, and a standardized technique for laparoscopic living donor has been established. A segment of ileum, 180–200cm in length, is resected 15cm proximal to the ileocecal valve, which is always preserved [34]. A unique experience of five cases has been reported by the Fourth Military Medical University in China [35]. Total operative time ranged from 130 to 195min with a surgeon console time of 30–35min. None of the donors required blood transfusions during or after surgery. All five donors had mild diarrhea (three–five bowel movements per day) during the first month, which decreased to two–three times per day by the second month after the procedure. With regards to the intestinal transplant recipients, very often such patients have history of multiple previous surgeries as a consequence of their original disease and are expected to have hostile abdominal cavities due to adhesions, thus contraindicating a minimally invasive approach in bowel transplantation.

## Uterus transplantation

Noteworthy in this chapter, a group from China published the first case, a robotic uterus procurement [36]. The uterus transplantation procedure consisted of robot-assisted uterine procurement, orthotopic replacement and fixation of the retrieved uterus, revascularization, and end-to-side anastomoses of bilateral hypogastric arteries and ovarian-uterine vein to the bilateral external iliac arteries and veins. The surgery is similar to radical extensive hysterectomy. The robot-assisted donor uterus procurement surgery lasted 6h, and the recipient surgery lasted 8h, 50min. Blood loss of the donor was 100mL, and blood loss of the recipient was 490mL.

## Summary

The robotic system is more often used in abdominal transplantation. The most widely used is for LT. All surgeons and centers wishing to start a robotic transplantation program must have adequate experience in both living donation and minimally invasive surgery. The initial selection of easier cases is recommended to reduce the morbidity during the learning curve.

## Future scenario

The robotic surgery is presently for many benign and oncological surgical procedures. As for the laparoscopic technique, after a short period of skepticism the robot has gained its place in the operative room. The incredible dexterity of the robotic instrument allows the surgeon to overpass his personal precision limits. The minimally invasive surgery is out of doubt the new gold standard for many procedures and robot is the king in most of them. The minimizing of scars and postoperative pain with comparable or even less morbidity is the key to unrolling more living donors. The next step is full robotic procedures, procurement, and transplant.

# Key points

- Robotic living liver donation is safe and feasible.
- Robotic surgery allows pre and post liver transplant procedures.
- The role of robotic surgery for pancreas transplantation needs more investigation.
- The robotic system allows surgeons to perform abdominal transplants to overcome their limitations.

# References

[1] Levi Sandri GB, de Werra E, Mascianà G, Guerra F, Spoletini G, Lai Q. The use of robotic surgery in abdominal organ transplantation: a literature review. Clin Transplant 2017;31(1).
[2] Panaro F, Piardi T, Cag M, Cinqualbre J, Wolf P, Audet M. Robotic liver resection as a bridge to liver transplantation. JSLS 2011;15(1):86–9.
[3] Levi Sandri GB, Lai Q, Ravaioli M, Di Sandro S, Balzano E, Pagano D, et al. The role of salvage transplantation in patients initially treated with open versus minimally invasive liver surgery: an intention-to-treat analysis. Liver Transplant 2020;26(7):878–87.
[4] Magistri P, Olivieri T, Assirati G, Guerrini GP, Ballarin R, Tarantino G, et al. Robotic liver resection expands the opportunities of bridging before liver transplantation. Liver Transplant 2019;25(7):1110–2.
[5] Levi Sandri GB, Colasanti M, Aldrighetti L, Guglielmi A, Cillo U, Mazzaferro V, et al. Is minimally invasive liver surgery a reasonable option in recurrent HCC? A snapshot from the I go MILS registry. Updates Surg 2022;74(1):87–96.
[6] Kim SH, Kim YK. Improving outcomes of living-donor right hepatectomy. Br J Surg 2013;100(4):528–34.
[7] Abecassis MM, Fisher RA, Olthoff KM, Freise CE, Rodrigo DR, Samstein B, et al. Complications of living donor hepatic lobectomy--a comprehensive report. Am J Transplant 2012;12(5):1208–17.
[8] Giulianotti PC, Tzvetanov I, Jeon H, Bianco F, Spaggiari M, Oberholzer J, et al. Robot-assisted right lobe donor hepatectomy. Transplant Int 2012;25(1):e5–9.
[9] Wu YM, Hu RH, Lai HS, Lee PH. Robotic-assisted minimally invasive liver resection. Asian J Surg 2014;37(2):53–7.
[10] Daskalaki D, Gonzalez-Heredia R, Brown M, Bianco FM, Tzvetanov I, Davis M, et al. Financial impact of the robotic approach in liver surgery: a comparative study of clinical outcomes and costs between the robotic and open technique in a single institution. J Laparoendosc Adv Surg Tech A 2017;27(4):375–82.
[11] Troisi RI, Elsheikh Y, Alnemary Y, Zidan A, Sturdevant M, Alabbad S, et al. Safety and feasibility report of robotic-assisted left lateral Sectionectomy for pediatric living donor liver transplantation: a comparative analysis of learning curves and mastery achieved with the laparoscopic approach. Transplantation 2021;105(5):1044–51.
[12] Broering DC, Elsheikh Y, Alnemary Y, Zidan A, Elsarawy A, Saleh Y, et al. Robotic versus open right lobe donor hepatectomy for adult living donor liver transplantation: a propensity score-matched analysis. Liver Transplant 2020;26(11):1455–64.
[13] Rho SY, Lee JG, Joo DJ, Kim MS, Kim SI, Han DH, et al. Outcomes of robotic living donor right hepatectomy from 52 consecutive cases: comparison with open and laparoscopy-assisted donor hepatectomy. Ann Surg 2022;275(2):e433–42.
[14] Yeow M, Soh S, Starkey G, Perini MV, Koh YX, Tan EK, et al. A systematic review and network meta-analysis of outcomes after open, mini-laparotomy, hybrid, totally laparoscopic, and robotic living donor right hepatectomy. Surgery 2022;172(2):741–50.
[15] Broering D, Sturdevant ML, Zidan A. Robotic donor hepatectomy: a major breakthrough in living donor liver transplantation. Am J Transplant 2022;22(1):14–23.
[16] Amma BSPT, Mathew JS, Varghese CT, Nair K, Mallick S, Chandran B, et al. Open to robotic right donor hepatectomy: a tectonic shift in surgical technique. Clin Transplant 2022;36(9):e14775.
[17] Finotti M, D'Amico F, Mulligan D, Testa G. A narrative review of the current and future role of robotic surgery in liver surgery and transplantation. Hepatobil Surg Nutr 2023;12(1):568.
[18] Rela M, Rajalingam R, Shetty G, Cherukuru R, Rammohan A. Robotic monosegment donor hepatectomy for pediatric liver transplantation: first report. Pediatr Transplant 2022;26(1):e14110.
[19] Kim NR, Han DH, Choi GH, Lee JG, Joo DJ, Kim MS, et al. Comparison of surgical outcomes and learning curve for robotic versus laparoscopic living donor hepatectomy: a retrospective cohort study. Int J Surg 2022;108:107000.
[20] Suh KS, Hong SK, Lee S, Hong SY, Suh S, Han ES, et al. Purely laparoscopic explant hepatectomy and hybrid laparoscopic/robotic graft implantation in living donor liver transplantation. Br J Surg 2022;109(2):162–4.
[21] Lee KW, Choi Y, Lee S, Hong SY, Suh S, Han ES, et al. Total robot-assisted recipient's surgery in living donor liver transplantation: first step towards the future. J Hepatobiliary Pancreat Sci 2023.
[22] Rofaiel G, Martinez E, Pan G, Sossenheimer M, O'Hara R, Gallegos J, et al. Creation of a robotically assisted terminal Jejunogastrostomy is safe and effective in regaining Antegrade enteral bile duct access after live donor liver transplant with roux limb. Transplant Direct 2019;5(8):e476.
[23] Hawksworth J, Radkani P, Nguyen B, Aguirre O, Winslow E, Kroemer A, et al. Robotic Hepaticojejunostomy for late anastomotic biliary stricture after liver transplantation: technical description and case series. Ann Surg 2022;275(6):e801–3.
[24] Chen Q, Fan J, Liu J, Li C, Liu J, Qian F. Robotic surgical system completed radical gastrectomy for gastric cancer after liver transplantation: case report and systematic review. Transl Cancer Res 2020;9(6):4028–35.
[25] Elli EF, Masrur MA, Giulianotti PC. Robotic sleeve gastrectomy after liver transplantation. Surg Obes Relat Dis 2013;9(1):e20–2.

[26] Kirchner VA, Finger EB, Bellin MD, Dunn TB, Gruessner RWG, Hering BJ, et al. Long-term outcomes for living pancreas donors in the modern era. Transplantation 2016;100(6):1322–8.

[27] Levi Sandri GB, Abu Hilal M, Dokmak S, Edwin B, Hackert T, Keck T, et al. Figures do matter: a literature review of 4587 robotic pancreatic resections and their implications on training. J Hepatobiliary Pancreat Sci 2023;30(1):21–35.

[28] Horgan S, Galvani C, Gorodner V, Bareato U, Panaro F, Oberholzer J, et al. Robotic distal pancreatectomy and nephrectomy for living donor pancreas-kidney transplantation. Transplantation 2007;84(7):934–6.

[29] Oberholzer J, Tzvetanov I, Mele A, Benedetti E. Laparoscopic and robotic donor pancreatectomy for living donor pancreas and pancreas-kidney transplantation. J Hepatobiliary Pancreat Sci 2010;17(2):97–100.

[30] Boggi U, Signori S, Vistoli F, Amorese G, Consani G, De Lio N, et al. Current perspectives on laparoscopic robot-assisted pancreas and pancreas-kidney transplantation. Rev Diabet Stud 2011;8(1):28–34.

[31] Boggi U, Signori S, Vistoli F, D'Imporzano S, Amorese G, Consani G, et al. Laparoscopic robot-assisted pancreas transplantation: first world experience. Transplantation 2012;93(2):201–6.

[32] Yeh CC, Spaggiari M, Tzvetanov I, Oberholzer J. Robotic pancreas transplantation in a type 1 diabetic patient with morbid obesity: a case report. Medicine (Baltimore) 2017;96(6):e5847.

[33] Spaggiari M, Tulla KA, Okoye O, Di Bella C, Di Cocco P, Almario J, et al. The utility of robotic assisted pancreas transplants - a single center retrospective study. Transplant Int 2019;32(11):1173–81.

[34] Gruessner RW, Sharp HL. Living-related intestinal transplantation: first report of a standardized surgical technique. Transplantation 1997;64(11):1605–7.

[35] Wu G, Li Q, Zhao Q, Wang W, Shi H, Wang M, et al. Robotic-assisted live donor Ileal Segmentectomy for intestinal transplantation. Transplant Direct 2017;3(10):e215.

[36] Wei L, Xue T, Tao KS, Zhang G, Zhao GY, Yu SQ, et al. Modified human uterus transplantation using ovarian veins for venous drainage: the first report of surgically successful robotic-assisted uterus procurement and follow-up for 12 months. Fertil Steril 2017;108(2):346–356.e1.

# 67

# Robotic technology in orthopedic joint and hip surgery

*Pedro Debieux*[a,b]*, Carlos Leonardo Malta Braga*[c,d]*,*
*and Camila Cohen Kaleka*[a,b]

[a]Hospital Israelita Albert Einstein, São Paulo, SP, Brazil [b]Hospital Beneficência Portuguesa de São Paulo,
São Paulo, Brazil [c]Santa Casa de Belo Horizonte, Belo Horizonte, Brazil [d]Hospital Unimed, Belo Horizonte, Brazil

## Introduction

Robotic technology in orthopedic surgery has been used in the clinical field in recent decades [1]. The first robot used was the ROBODOC system (initially by Curexo Technology, Fremont, CA) in 1992, followed by the CASPER robot (Universal Robotic Systems Ortho, Germany) used in hip arthroplasties [1–4], which became one of the first medical specialties where this technology was used. The utilization of robotic technology has shown potential, increasing accuracy and reproducibility and, consequently, improving results. Orthopedic surgery benefits from the static nature of the skeletal anatomy [2,5]; it simplifies the generation of preoperative images and enhances the accuracy of registration and intraoperative computer navigation compared with soft tissue surgery, which requires more complex algorithms [5].

The robot technology used in orthopedics can be initially categorized as passive, semiactive, and active [1,4,6]. Passive robot technology is a system that relies on the direct and continuous control of the surgeon, whereas active robot technology performs a task independent of the surgeon's involvement, through preprogrammed algorithms and previously defined parameters. In semiactive systems, robots employ feedback to limit their manipulation, thereby increasing surgeon control and, in theory, operative safety. These semiactive systems are also referred to as "haptic systems" [1,6].

Unlike computer-assisted surgery, which only provides guidance and passive feedback, robots must, by definition, play some active role in surgery to be classified as such. They can also be differentiated based on their direct or indirect actions. In direct approaches, robots cut the bone into the desired final shape. In contrast, indirect methods involve the utilization of resources to enable the placement or maintenance of cutting templates [1,7]. Finally, the robotic cutting method can be categorized into three distinct groups. Autonomous robots are those that cut the bone without human control. Haptic robots involve human interaction to move the robot for cutting, milling, or drilling purposes, although robot's movement is constrained by tactile, sound, or image feedback. Finally, robots with boundary control require human interaction for movement; however, if the robot surpasses a set boundary, clipping is disabled or prevented by some means, even though the robot remains unrestricted in its movement [1,6].

According to these principles, robotic systems in orthopedics can be classified as follows [6]:

1. Direct and autonomous: Robots cut the bone as per a plan, without direct human guidance.
2. Direct and haptic: Robots cut the bone under the guidance of a human hand within a haptic boundary.
3. Direct and threshold control: Robots cut the bone under the guidance of a human hand and shut down when thresholds are surpassed.
4. Indirect: Robots do not touch the bone but hold the cutting jigs.
5. Indirect and haptic: Robots process features in the bone to receive cutting templates within a haptic boundary.
6. Indirect and limited control: Robots process bone features to receive cutting templates.

Robots used in orthopedic surgery rely on preoperative programming, which distinguishes them from other medical specialties. This enables the surgeon to have a surgical plan and understand the outcome before the procedure [1]. The programs can be based on the use of preoperative images, such as radiographs or computerized tomography scans, or they can be imageless. In imaging systems, the anatomy is confirmed by recording the mapping of relevant bone anatomical points, which are easily obtained by the surgeon during the surgical procedure. The combination of these procedures, including preoperative programming, preoperative imaging, and perioperative recording, enables the robot to assist the surgeon in achieving the desired outcome. Potential drawbacks of this system may include radiation exposure, increased cost, and the need to undergo preoperative examinations at a different location.

In systems that do not utilize preoperative images, the virtual model is obtained solely through preoperative recording, and the subsequent programming is based on the obtained information, along with the preoperative plan created on the system platform. The disadvantage of this system is the absence of true preoperative planning.

Another point of differentiation among orthopedic robotic systems is the distinction between open and closed platforms. Closed platforms refer to those in which the robotic system limits the use of one or more implants from a specific manufacturer, whereas open platforms are those that allow the use of different companies and implants. Open platforms offer surgeons greater flexibility in selecting implants, even allowing customization based on the specific needs of the patient. By contrast, closed systems provide greater specificity in terms of implant design and biomechanics, enhancing the predictability of the surgical outcome [1].

Although these various systems have differences, all those employed in orthopedic surgery follow a flow comprising four stages: setup, registration, planning, and execution [8]. The order in which these stages are implemented depends on the specific system and whether preoperative imaging exams are utilized. When imaging is used, the first step is planning, where the surgeon defines the surgical objective, ideal implant positioning, desired alignment, and other references according to the procedure. The next step is the setup, which refers to the process of positioning the robot in the operating room around the patient and fixing the trackers to the bone structures (Fig. 1). The third step is registration, which requires recording the relevant intraoperative bone anatomical references. The final stage is the execution of the procedure, wherein the level of the surgeon's involvement aligns with the type of robot and its autonomy, as described earlier. In systems that do not employ imaging, planning progresses from the first step to the third step, after registration [8].

Knee and hip arthroplasties are procedures that traditionally yield excellent outcomes in terms of alleviating pain, improving function, and enhancing patients' quality of life [7]. Nevertheless, a significant proportion of patients remains dissatisfied with the achieved results, even in the absence of technical justifications. Hence, robotic surgery emerges as a technical alternative, seeking to enhance the accuracy and precision of bone cutting and the alignment and positioning of implants. This approach has the potential to improve functional outcomes, increase long-term survival rates, and reduce the need for revision surgeries [2,9]. The number of procedures utilizing this technology has increased significantly in recent years, although the technology's use is still concentrated in services with a high volume of procedures per year [7].

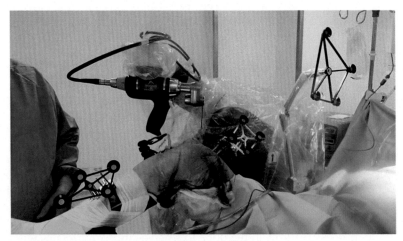

FIG. 1    Robotic arm (Mako) positioned next to a limb with its navtrackers installed.

# Robotic technology in total hip arthroplasty (THA)

The success of THA is influenced by multiple factors, and the entire manual procedure is highly dependent on the position of the pelvis on the surgical table and its stability throughout the procedure, as well as the surgeon's expertise. Reports indicate that the rate of unsatisfactory positioning of the acetabular component can range from 38% to 45% [2]. Robotic surgery aims to improve implant positioning, restore hip offset, and optimize implant sizing, which are important factors for reducing complications, such as dislocations and loosening. The planning begins by obtaining preoperative 3D tomographic images, enabling the study of the anatomy and its variations while reducing or eliminating inherent biases associated with 2D image templates traditionally used in this planning. This is particularly notable in terms of accurately assessing the depth of the acetabulum and the shape of the medullary canal [1,2]. The planning process is completed after the registration of the patient's anatomical points obtained during the preoperative period. The robotic arm assists the surgeon in accurately milling the acetabulum, ensuring appropriate acetabular orientation and center of rotation and adequate coverage [1–3]. In terms of femoral milling, the robot optimizes positioning by considering factors such as the femoral version, induced center of rotation by the planned nail, and comparative limb length [2,3] and reduces the risk of perioperative fractures [1]. Throughout the process, the goal is to preserve bone stock at each step.

Robotic hip surgery, however, has disadvantages, such as the potential increase in surgery time, prolonged learning curve, and increased cost and number of preoperative examinations. Although studies have shown improvements in radiographic alignment and positioning, long-term studies that conclusively establish enhancements in functional outcomes or implant survival are scarce [3].

# Robotic technology in total knee arthroplasty (TKA)

Robotic assistance has been implemented in TKA and unicompartmental knee arthroplasty (UKA) in addition to the aforementioned advancements in hip joint procedures. The desire to reduce complications, enhance patient satisfaction, and improve the reproducibility of results [1] has driven the investigation on the impact of robotics on knee joint procedures, tracing back to the early stages of this technique in orthopedics. Despite knee arthroplasty being an effective treatment modality, it presents a dissatisfaction rate of 8%–20% [10,11], and up to 13% of patients may require revision surgery for various reasons [10]. Notably, UKA tends to exhibit higher satisfaction rates than TKA, albeit with higher revision rates [10].

In an attempt to better understand the reason for this large proportion of dissatisfied patients, researchers have demonstrated a correlation between the long-term success of conventional knee arthroplasties and intraoperative factors, such as component alignment, limb alignment, implant size and positioning, maintenance of the joint line, and soft tissue balance [10,11].

Robotic TKA enhances the surgeon's preoperative planning capabilities and the real-time intraoperative dynamic reference (Figs. 2 and 3) to facilitate continuous assessments of the range of motion and ligament tension [12]. The

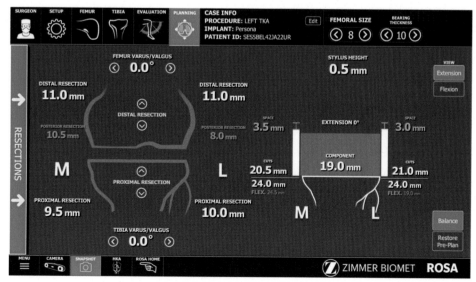

FIG. 2 Planning screen of the ROSA robotic system for performing a total knee arthroplasty.

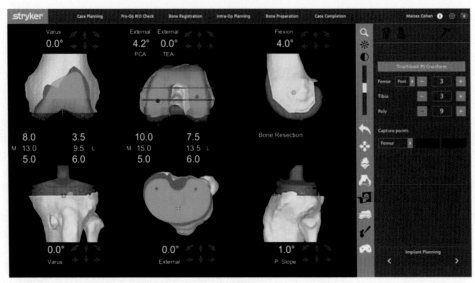

FIG. 3    Mako robotic system planning screen for total knee arthroplasty.

advantages of robotic TKA include improved accuracy in component positioning, high precision in implant and limb alignment, precise and economical bone cuts, and soft tissue balance [5,10,11], which are precisely the prerequisites for achieving favorable functional outcomes and long-term survival in TKA (Figs. 2 and 3).

Despite the literature consensus on the improvement in radiological standards, disagreement still exists regarding the improvement in clinical outcomes with the implementation of this technology. Preliminary studies are being conducted, showing short-term improvements in functional scores [6,9], although without significant findings in the medium or long term [10,11]. Chin et al. conducted a review of studies on robotic-assisted UKA and found improved clinical outcomes in the short-term follow-up [10]. These results must be interpreted in light of the intrinsic limitations and heterogeneity of the studies, as listed in a systematic review by Kort et al.: heterogeneity in follow-up periods, quality of research designs, commercial biases, and technology maturity comparing different generations of systems robotics, among others [11].

Despite the aforementioned factors, the utilization of robots in arthroplasties may entail potential adverse events, disadvantages, and complications. Among these, we can highlight the increase in surgical time, elevated costs, a steep learning curve, potential iatrogenic injuries, infection risks related to tracker pin placement, and the possibility of fractures in the tracker pin holes [10,12]. Longer surgical time is related to longer planning time, intraoperative recording, and the learning curve. As longer surgical time has been associated with an increased infection rate in conventional TKA [13], this extended duration could similarly affect robotic surgery. However, to date, no study has confirmed a higher infection rate in robotic surgery. The increased cost may be justified if a lower rate of complications and revisions is demonstrated in the future. Conversely, iatrogenic injuries have been more commonly associated with active and autonomous robots. Thus far, the analyzed results concerning surgical complications have not revealed significant differences compared with conventional TKA, indicating that robotics is at least as safe as conventional methods [10]. Robotic technology has been implemented in other fields of knee surgery, such as patellofemoral arthroplasty [4], osteotomies, and ligament reconstructions.

## Robotic technology in orthopedics

Robotic assistance in revision hip and knee arthroplasties has not yet been approved for clinical use by the Food and Drug Administration or the European Commission's Medical Device Coordination Group. Wu et al. conducted a metaanalysis to review studies on the use of robot-assisted technology in revision arthroplasty; their results indicate that the technology is not inferior to the conventional method, particularly in terms of accurate preoperative planning and guidance in operations [9]. In THA revisions, benefits include accurate acetabular widening and cup placement, balanced lower extremity length, offset restoration, and properly matched anteversion. In TKA revisions, benefits mainly include accurate bone cutting and the ability to locate component alignment and mechanical alignment, although further studies are needed.

Despite having potential promising results, robotic surgery in the field of foot and ankle surgery still has limited application when compared with knee and hip surgery [14]. In this context, the robotic technology application is described for various procedures, including intraarticular injections [5], fracture reduction, internal fixation in the calcaneus and tibial pilon, correction of pseudarthrosis, arthrodesis of the calcaneo-navicular coalition, and midfoot arthrodesis [6,14]. Owing to the high cost associated with the complexity and long operation durations, the more justifiable application of robotic technology is in high-precision procedures. In total ankle arthroplasty (TAA), robotic technology shows promising applicability if it can replicate the success achieved in TKA and THA [6]. The improved accuracy in bone resection, better balance of soft tissues, and enhanced positioning and alignment of implants have the potential to mitigate the relatively high failure rates observed in TAA when compared with those in THA and TKA [14].

In orthopedic trauma, robotic surgery can assist surgeons by improving limb alignment, reducing fracture rates [6], and optimizing the positioning of the synthesis material and the entry point of the intramedullary nails, as well as lowering the exposure to fluoroscopy [5,6,15]. Robotic assistance can also be beneficial in the repair of peripheral nerve injuries, as it combines virtual reality to enhance movements and robotic technology to filter out typical tremors of human inaccuracy [16], thus improving microsurgical precision.

The use of robots in upper limb surgery is still limited [14]. Cadaveric studies have been conducted in the field of arthroscopic shoulder surgery, where robot technology has demonstrated its potential as an alternative. With the development of specialized instruments, robots can enhance visualization and perform precise tasks in small spaces [17]. Additionally, the evaluation of robotic assistance in arteriography is being explored [18]. In shoulder arthroplasty, the positioning of the glenoid has been evaluated by applying computerized navigation, showing improved accuracy [19].

Robotics in orthopedic surgery has been available for more than 20 years, and its implementation has increased significantly [7]. Improving the precision of bone cutting, component alignment, and limb alignment already leads to better short-term clinical outcomes. However, there is little solid evidence currently available on long-term benefits [8]. Furthermore, the implementation of these technologies adds costs to the healthcare system [7–9]. New studies must be carried out to determine the efficiency and cost-benefit ratio of robotic surgery in orthopedics, especially in the long term [1,5,6,8,10,11].

## Future scenario

Robotic surgery in orthopedics has become an inevitable reality [1], following the path already established in other medical specialties. The number of procedures performed with robotic assistance is expected to increase in the coming years, both due to the greater number of robots and their decentralization, as well as the number of procedures possible to be performed with this technology. The future scenario for robotic surgery foresees, from a technical point of view, improvements in the data capture process, both in terms of quality and speed, which should become increasingly less invasive. The customization of implants and the possibility of adapting the robot to multiple implants would also be a useful addition. A better understanding of preoperative data would allow its use in revisions and complex deformities, a context in which robotic assistance may soon become indispensable. No technological innovation, however, would overcome the possibility of expanding the use of robotic surgery in more centers around the world.

## Key points

- Robotic surgery in orthopedics has become an inevitable reality.
- Robots in orthopedic surgery have a flow with four stages: setup, registration, planning, and execution.
- This approach has the potential to improve functional outcomes, increase long-term survival rates, and reduce the need for revision surgeries.
- Improving the precision of bone cutting, component alignment, and limb alignment already leads to better short-term clinical outcomes.
- Little solid evidence is currently available on long-term benefits.

## References

[1] Jacofsky DJ, Allen M. Robotics in arthroplasty: a comprehensive review. J Arthroplasty 2016;31(10):2353–63. https://doi.org/10.1016/j.arth.2016.05.026.

[2] Kouyoumdjian P, Mansour J, Assi C, Caton J, Lustig S, Coulomb R. Current concepts in robotic total hip arthroplasty. SICOT-J 2020;6.

[3] Fontalis A, Kayani B, Thompson JW, Plastow R, Haddad FS. Robotic total hip arthroplasty: past, present and future. Orthop Trauma 2022;36(1):6–13. https://doi.org/10.1016/j.mporth.2021.11.002.

[4] Wolf A, Jaramaz B, Lisien B, DiGioia A. MBARS: mini bone-attached robotic system for joint arthroplasty. Int J Med Robot Comput Assist Surg 2005;01(02):101. https://doi.org/10.1002/rcs.20.

[5] Karthik K, Colegate-Stone T, Dasgupta P, Tavakkolizadeh A, Sinha J. Robotic surgery in trauma and orthopaedics: a systematic review. Bone Joint J 2015;97(3):292–9. https://doi.org/10.1302/0301-620X.97B3.35107.

[6] Chen AF, Kazarian GS, Jessop GW, Makhdom A. Current concepts review: robotic technology in orthopaedic surgery. J Bone Jt Surg – Am 2018;100(22):1984–92. https://doi.org/10.2106/JBJS.17.01397.

[7] Boylan M, Suchman K, Vigdorchik J, Slover J, Bosco J. Technology-assisted hip and knee arthroplasties: an analysis of utilization trends. J Arthroplasty 2018;33(4):1019–23. https://doi.org/10.1016/j.arth.2017.11.033.

[8] Picard F, Deakin AH, Riches PE, Deep K, Baines J. Computer assisted orthopaedic surgery: past, present and future. Med Eng Phys [Internet] 2019;72:55–65. https://doi.org/10.1016/j.medengphy.2019.08.005.

[9] Wu X-D, Zhou Y, Shao H, Yang D, Guo S-J, Huang W. Robotic-assisted revision total joint arthroplasty: a state-of-the-art scoping review. EFORT Open Rev 2023;8(1):18–25. https://doi.org/10.1530/EOR-22-0105.

[10] Chin BZ, Tan SSH, Chua KCX, Budiono GR, Syn NLX, O'Neill GK. Robot-assisted versus conventional total and unicompartmental knee arthroplasty: a meta-analysis of radiological and functional outcomes. J Knee Surg 2021;34(10):1064–75. https://doi.org/10.1055/s-0040-1701440.

[11] Kort N, Stirling P, Pilot P, Müller JH. Robot-assisted knee arthroplasty improves component positioning and alignment, but results are inconclusive on whether it improves clinical scores or reduces complications and revisions: a systematic overview of meta-analyses. Knee Surgery Sport Traumatol Arthrosc [Internet] 2022;30(8):2639–53. https://doi.org/10.1007/s00167-021-06472-4.

[12] Nogalo C, Meena A, Abermann E, Fink C. Complications and downsides of the robotic total knee arthroplasty: a systematic review. Knee Surgery Sport Traumatol Arthrosc [Internet] 2023;31(3):736–50. https://doi.org/10.1007/s00167-022-07031-1.

[13] Pugely AJ, Martin CT, Gao Y, Schweizer ML, Callaghan JJ. The incidence of and risk factors for 30-day surgical site infections following primary and revision total joint arthroplasty. J Arthroplasty [Internet] 2015;30(9):47–50. https://doi.org/10.1016/j.arth.2015.01.063.

[14] Stauffer TP, Kim BI, Grant C, Adams SB, Anastasio AT. Robotic technology in foot and ankle surgery: a comprehensive review. Sensors 2023;23(2). https://doi.org/10.3390/s23020686.

[15] Garcia P, Rosen J, Kapoor C, Noakes M, Elbert G, Treat M, et al. Trauma pod: a semi-automated telerobotic surgical system. Int J Med Robot Comput Assist Surg 2009;5(2):136–46. https://doi.org/10.1002/rcs.238.

[16] Garcia JC, Lebailly F, Mantovani G, Mendonca LA, Garcia J, Liverneaux P. Telerobotic manipulation of the brachial plexus. J Reconstr Microsurg 2012;28(7):491–4. https://doi.org/10.1055/s-0032-1313761.

[17] Bozkurt M, Apaydin N, Işik Ç, Bilgetekin YG, Acar HI, Elhan A. Robotic arthroscopic surgery: a new challenge in arthroscopic surgery Part-I: robotic shoulder arthroscopy; a cadaveric feasibility study. Int J Med Robot Comput Assist Surg 2011;7(4):496–500. https://doi.org/10.1002/rcs.436.

[18] Monfaredi R, Iordachita I, Wilson E, Sze R, Sharma K, Krieger A, Fricke S, Cleary K. Development of a shoulder-mounted robot for MRI-guided needle placement: phantom study. Int J Comput Assist Radiol Surg [Internet] 2018;13(11):1829–41. https://doi.org/10.1007/s11548-018-1839-y.

[19] Nguyen D, Ferreira LM, Brownhill JR, King GJW, Drosdowech DS, Faber KJ, Johnson JA. Improved accuracy of computer assisted glenoid implantation in total shoulder arthroplasty: an in-vitro randomized controlled trial. J Shoulder Elbow Surg 2009;18(6):907–14. https://doi.org/10.1016/j.jse.2009.02.022.

# 68

# Robotic assistance in spine surgery

*Adam Vacek[a] and Ricardo B.V. Fontes[b]*

[a]University of Edinburgh Medical School, Edinburgh, United Kingdom [b]Rush University Medical Centre, Chicago, IL, United States

## Introduction

Around 4.83 million spinal surgeries occur annually worldwide, with approximately 1.34 million in the United States alone [1]. Spinal surgeons often must work with complex anatomy and master highly precise surgical techniques. With the recent advancements in medical technology, robotic systems have emerged with hopes of improving spinal surgery's safety, efficiency, clinical outcomes, and cost-effectiveness. Robotics was introduced to spinal surgery in the late 20th and early 21st centuries, with the first FDA (Food and Drug Administration)-approved spine robot introduced in 2004 [2]. The theoretical principles of stereotaxis and localization in the three-dimensional space, however, had been developed much earlier and were a direct neurosurgical contribution to spine surgery.

Current robotic systems are often divided into supervisor-controlled, telesurgical, and shared-control robots, with most currently available systems falling under the shared-control category [1]. These systems are increasingly being incorporated into spinal surgery to aid with navigation, trajectory determination, or screw implantation [1]. They are used in cases like percutaneous transforaminal interbody fusions, endoscopic surgery cases, and spine tumor surgery for assisting with implants, biopsies, and other percutaneous interventions and their application has been consistently expanded [4].

However, there are also potential pitfalls with robotics that represent new challenges in spine surgery. The basic principle of robotic surgery remains, as in surgical navigation or surgical augmentation, operating off a virtual spine in the robot's computer. At all times it must be ensured that this virtual spine does not deviate from the surgical reality, which can lead to frame shift errors. These errors represent surgical challenges not previously faced by surgeons: in pedicle screw positioning, these can lead to misplaced implants bilaterally and at multiple levels. There are multiple instances that can lead to these mismatches such as reference array vulnerabilities, soiling of tracking tools, and intraoperative changes in the spinal morphology, among others, and these must be carefully mitigated [5]. Additionally, the literature data on accuracy, radiation exposure, clinical outcomes, and operative times are still quite unclear in some instances and require further exploration.

This chapter provides an overview of robotic systems in spinal surgery and presents our sample protocol for robotic assistance utilizing a case example. We hope to share our experience and assist the reader in making an informed decision whether this technique is applicable to his or her practice.

## Historical overview

The use of robots in spine surgery depended on the sequential development of technologies and surgical techniques over the years. The basic concept is stereotaxis, a methodology utilizing external landmarks to localize human anatomy within the three-dimensional space. Stereotaxis itself is a concept more than 100 years old and has been broadly used with frame-based systems and simple radiographs in cranial neurosurgery. The initial advance that enabled the application of stereotaxis in spine surgery was the development and increased availability of computed tomography (CT) and magnetic resonance imaging (MRI) in the 1980s [2].

*Handbook of Robotic Surgery*
https://doi.org/10.1016/B978-0-443-13271-1.00042-X

The next two technical steps were the development of a robotic arm and efficient image guidance, which happened in parallel. The robotic arm was developed first and the first "robotic" spinal surgery was performed in 1985 with the PUMA 200 robotic arm used to place and stabilize a biopsy cannula [2]. Its cousin, the PUMA 560, was used for a neurosurgical brain biopsy [1]. In the mid-1990s, NASA (National Air and Space Administration) and the Ames Research Center created the AESOP robot that eliminated the need to hold the endoscopic camera manually, allowed much more stabilized operating field image, and contributed to fatigue reduction in surgeons [1,2]. These primitive robotic arms were little more than holders and were unable to provide guidance or to dynamically alter their trajectories.

Image-guidance in spine surgery required the development of an effective line-of-sight guidance system that could localize an instrument in space and enough computing power to match it to a CT image, MRI, or even a radiograph stored in the computer. The combination of these abilities in spine and cranial surgery was popularly called neuronavigation and the first application of this technique in spine surgery for pedicle screw placement was described by Girardi et al. in 1999 [2].

Once the concepts of stereotaxis, image-guidance, and robotic assistance could be combined, commercial systems were launched in the early 2000s. The Mazor SpineAssist (Mazor Robotics, Caesarea, Israel) was the first spine robot to gain U.S. FDA approval in 2004, improving on the previous issues with incorrect image synchronization and long calculation times [3,4]. The second generation of this system was the Mazor Renaissance, launched in 2011, which expanded its applicability to additional procedures in the spine [2,7].

At this point, surgical issues with image shift and loss of reference were becoming evident, which limited the applicability of robotic systems. The next generation of robots came into existence in the mid- to late-2010s and incorporated image-guidance and the ability to more effectively re-register surgical anatomy during the surgical case [5]. The ROSA OneSpine (MedTech, Montpellier, France), Mazor X (Medtronic Inc., Dublin, Ireland), and the ExcelsiusGPS (Globus Medical Inc., Audubon, PA, USA) are the more popular examples of this third generation of spinal robotic systems [3–5]. Other commercially available robotic systems are constantly being developed, which can be grouped into this third generation and can now be found worldwide.

## Surgical rationale and operating principles

Surgical robots can broadly be divided into supervisor-controlled, telesurgical, and shared-control systems and robot-assisted navigation systems. The supervisor-controlled systems enable the surgeon to plan the operation ahead of time and perform the operation themselves under the close supervision of the surgeon. Meanwhile, telesurgical systems let the surgeon remotely control the robot, its instruments, and their motions. The shared-control systems allow the surgeon and robot to simultaneously control the instruments and motions. Robot-assisted navigation systems utilize real-time imaging to aid the surgeon by guiding their instruments, but the robot has no control over the instruments [2,6,7]. We technically consider robot-assisted navigation system, as a part of neuronavigation and as described previously, is a concept incorporated in most new third generation robotic systems.

At the time this chapter was written, there were 18 commercially-available robotic systems identified worldwide, with three systems currently FDA-approved for use in the United States—the Mazor X Stealth Edition, the ROSA Spine robot, and the ExcelsiusGPS. All these systems utilize the shared-control approach [2]. These systems are currently FDA-cleared for open and percutaneous pedicle screw insertion, interbody device placement, and image-guidance in surgery. Their usage, however, is constantly being expanded and these systems may be used for additional procedures in different parts of the world with other regulatory constraints, such as for endoscopic spine surgery.

The two most often quoted benefits of robotic assistance in spine surgery remain increased accuracy of implant placement and decreased radiation exposure to the surgical team—not necessarily to the patient. For many years there has been a debate whether robotic assistance effectively more accurately delivered spinal implants. "Horror stories," stemming from loss of image reference and dissociation of the virtual spine from the real spine resulting in catastrophic neural injuries not normally seen in open freehand or real-time, fluoroscopy-guided pedicle screw placement, continuously hurt the reputation of both navigation and robotic systems. After over 20 years of navigation use and 10 years of robotic assistance in spine surgery, workflow protocols have been introduced that minimize these errors, namely, intraoperative CT- or cone beam-based 2D and 3D image acquisition in the prone position and improvement of re-registration procedures either with fluoroscopy or optical recognition (7D Surgical, North York, ON, Canada). Additionally, there is now at least one robotic system (Globus Excelsius) that incorporates two references in the patient, so that loss of accuracy (change in the relative position of the references to each other) can be detected intraoperatively.

The quality of the available literature is limited by the inherent retrospective nature of published studies, rapid technological advances in the field, and commercial and personal biases. Additionally, the technical difficulty of pedicle screw placement—the traditional means by which robotic accuracy is measured in spine surgery—varies tremendously according to indication: a series of pediatric congenital deformity cases cannot be directly compared to Minimally-invasive (MIS) adult degenerative cases. These issues notwithstanding, the largest literature metaanalysis was published by Naik et al. in 2022, which by then already included a significant number of cases performed by second and third generation spine robots. Naik et al. were able to identify >6200 patients and >31,000 pedicle screws across 78 studies. Robot-assisted placement was consistently more accurate than freehand (which included fluoroscopy guidance), 2D navigation, and preoperative CT navigation (based on a preoperative CT image obtained in the supine position), which was equivalent to intraoperative CT-based 3D navigation [8]. Reported accuracy for robotic- and 3D navigation is reported around 98%–99% and for freehand placement around 90%. Radiation exposure could not be analyzed but it is intuitive that if staff steps out of the operating room for the intraoperative CT, staff radiation exposure is limited to the registration of fluoroscopic images. Other secondary outcomes such as overall complications, intraoperative blood loss, and overall length of stay were also found to be generally in favor of robotic assistance, but the number of studies greatly decreased and as stated previously, surgical indication may greatly influence these secondary outcomes more so than the pedicle screw placement method [8].

Frequent, and pertinent criticisms of robotic assistance in spine surgery were also reflected in the Naik et al.'s metaanalysis, namely operative duration was significantly longer in the robotic and both navigation modalities. In shorter cases this may be an inconvenience and a cost factor, but in surgery for adult spinal deformity with long thoracolumbar fusions, this may make the difference between a one- or two-stage operation, or even in the amount of osteotomies to be performed. Another frequent criticism, especially, since the advent of parapedicular or "in-out-in" techniques for freehand screw placement in small thoracic pedicles, is that the vast majority of pedicle breaches during freehand placement are lateral and of no clinical consequence. On the other hand, navigation or robotic misplacements happen due to a mismatch between the real-world anatomy and the virtual spine stored in the computer interface. This can happen for obvious reasons, such as a loss of attachment of the reference array, or for subtle reasons that nonetheless result in inaccuracies in any given direction including medial with a much higher potential for neurological injury and to affect all pedicle screws placed after the change occurred [9]. This is a real concern and an optimized workflow with a surgical team used to image-guidance techniques and troubleshooting the robot is always necessary.

## Case example and sample workflow

A 58-year-old man presented with refractory low back and radiated bilateral leg pain in an L5 distribution. He was diagnosed with an L5-S1 spondylolisthesis secondary to L5 spondylolysis resulting in severe bilateral L5-S1 foraminal stenosis. Following failed nonoperative management, he was offered Minimally-invasive L5-S1 transforaminal interbody fusion (Fig. 1).

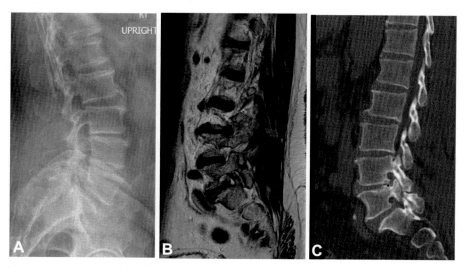

FIG. 1    Case example. L5-S1 spondylolisthesis with L5 spondylolysis and severe L5-S1 foraminal stenosis is visualized by XR (A), MRI (B), and CT (C).

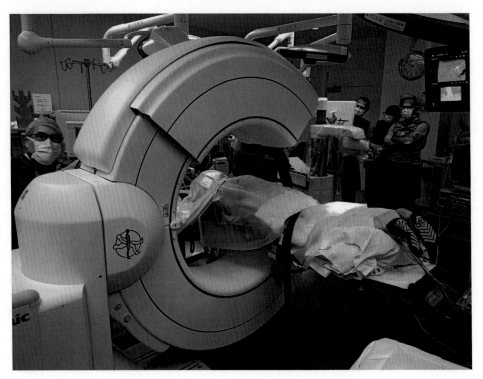

FIG. 2    Intraoperative image acquisition with the Medtronic O-arm (Medtronic Inc., Dublin, Ireland).

The steps in robotic assistance in spine surgery involve image acquisition, planning, and registration and the effective execution of the said plan. A robotic navigation system comprises a registration array, tracking cameras, patient reference arrays, and tool arrays [5]. Image acquisition can be performed with a preoperative CT image or after patient positioning (Fig. 2). Preoperative acquisition saves time in the operating room whereas imaging obtained in a prone position avoids potential inaccuracies from changing the lumbar lordosis from supine to prone positioning. Image acquisition can also be performed with a reference frame already positioned in the patient and thus obviates the step of registration—i.e., matching the stored CT image to the patient's anatomy. Insertion of a reference frame followed by acquisition may be considered the "standard" neuronavigation protocol, while most robotic protocols will use vertebra-by-vertebra registration since most robotic reference frames are bulkier and difficult to include in standard 9-in. fluoroscopy. At any point after the CT image is acquired, the surgeon can use the touch-screen interface to plan the screw placement (Fig. 3).

In this illustrative case, we are using the Globus ExcelsiusGPS robot. After the case is started and the patient is prepped and draped, reference frames for robotic assistance are inserted usually at the spinous processes for thoracic and upper lumbar cases, and in the posterior superior iliac spines for lumbar cases (Fig. 4). A frame is attached to the C-arm and is thus visible to the navigation camera (Fig. 5). The image-guidance computer can then see the patient (via a reference frame) and the C-arm in space. Anteroposterior (AP) and lateral images of each vertebra are obtained and based on the distances and calculations made on C-arm images, each vertebra is registered in space to the preoperative CT image. One additional advantage of this method in robotic acquisition is that registration is performed on a vertebra-per-vertebra basis, while in navigation acquisition with the frame already positioned it is the whole vertebral segment, usually around 5–7 vertebrae, which is registered and thus makes it more susceptible to changes in the anatomy in the interior of this segment.

The robot is then approximated, and the robotic arm will provide the planned trajectory to the surgeon for each sequential step of screw placement (Fig. 6). Intraoperative adjustments in trajectory can be performed and if there is suspicion of loss of accuracy, the spine can be registered again with a new set of AP and lateral radiographs.

Our group has preferred placing the pedicle screws as the initial step of the MIS-TLIF (minimally-invasive transforaminal interbody fusion) to increase accuracy. We have found that posted screw systems that allow connecting the tulip later thus remain out of the way of the interbody work. After decompression and interbody fusion, tulip heads are attached to the screws, rods passed and locked. Postoperative radiographs showed an excellent result and the patient's preoperative complaints were completely resolved by the sixth postoperative week (Fig. 7).

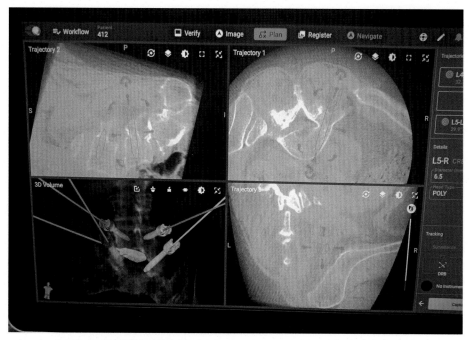

FIG. 3  Planning interface for the Globus ExcelsiusGPS robot.

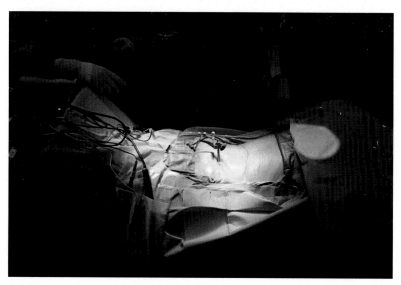

FIG. 4  Patient in prone position, head to the right, with the main reference array (patient's *right side*) and the surveillance markers (*left side*) inserted on both posterior superior iliac spines.

## Troubleshooting and tips

Troubleshooting and appropriate error management in robot-assisted spine surgery is of paramount importance, particularly because of its applicability to percutaneous, Minimally-invasive techniques in which the anatomy may not be visible and available for inspection. Familiarity with a rescue technique such as freehand, fluoroscopy-assisted or navigation-assisted implant placement is necessary. In our initial experience in establishing a robotic spine surgery protocol, we had a conversion rate of 10%–20%; after the initial 80 cases, conversion had become exceedingly rare. Part of avoiding conversion is recognizing those cases in which robotic assistance may not be the best technique: extreme spinal deformity, prior laminectomies, and extensive spinal instrumentation all make registration by the robot extremely difficult as the computer may be unable to visualize the spine among the instrumentation or an accurate AP

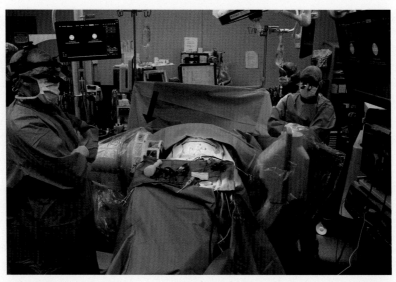

FIG. 5    Registration procedure with C-arm in the lateral position. *Black arrow* marks the C-arm reference, which includes IR-visible beads that are recognized by the robot's camera.

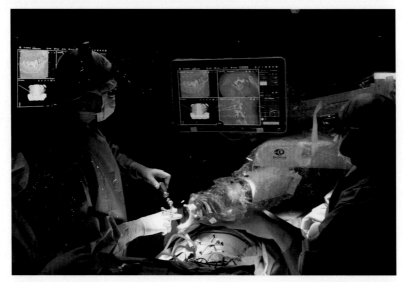

FIG. 6    Pedicle screw placement with the Excelsius robot and surgeon interface visible with real-time instrument guidance.

(antero-posterior) image may not be possible because of lack of the spinous process. In these cases, we have reverted to neuronavigation with segmental, en bloc registration thus obviating the need for XR-based registration.

All navigation and robotic errors can be broadly categorized into evident or silent. Line-of-sight issues, loss of visualization due to sphere damage or debris, inability to register or calculate trajectories, and gross inaccuracies of the reference ("bumps") are all examples of evident errors in which the robotic assistance is interrupted and the surgeon is immediately aware of a problem. The more familiar a team is with neuronavigation, the more readily these errors can be addressed: frequently cleaning the IR (infrared) spheres, redirecting operating room lighting away from the reference spheres, ensuring that the reference arrays are always safe and away from instruments, repositioning of the navigation camera, and the presence of a surveillance marker are all strategies that can minimize and correct these problems.

The silent errors are much more dangerous and need to be recognized as early as possible. Skiving of sharp tools along bone surfaces and missing the planned trajectory, for example, may be silent or obvious—use of sharp burs for the initial trajectory and appropriate tapping may minimize this phenomenon. Minor shifts in the reference frame, intervertebral motion, and addition of interbody implants may all incorporate small errors in accuracy that can be

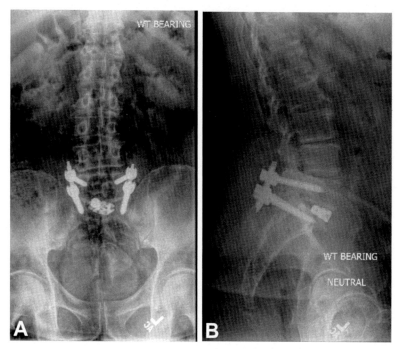

FIG. 7    Postoperative XR showing good radiological result—AP (A) and lateral (B) projections.

incremental. Our group has learned to minimize these small but potentially catastrophic errors by making the implant placement the first step of the planned procedure so as to introduce the least amount of bias. Repeating the registration procedure or reacquiring an initial CT image can solve a large portion of these errors and the team needs to be prepared for these alternatives. Above all, the surgical team needs to be prepared to use a bail-out technique if necessary and abandon the robot in select cases, which should decrease with experience.

## Future scenario

The additional literature is being produced, which further demonstrates the value of RSS across different countries, indications, and pathologies. While it may be argued that robotic assistance is equivalent clinically to freehand or fluoroscopy-assisted pedicle screw placement, there is not a single high-level publication that demonstrates a noninferior result. Development of real-time re-registration and a feedback loop so that the robot can detect loss of resistance that characterizes the transition from bone to soft tissue are the technical aspects that currently limit the applicability of robots in spine surgery. It is expected that once these issues are resolved, the predominant mode of use will switch from shared-control to supervisor-controlled mode in the medium to long term.

## Conclusions

It is an exciting time to become proficient in robot-assisted spine surgery. Third-generation robots that incorporate image-guidance have introduced a degree of familiarity with neuronavigation in addition to user-friendly interfaces that make it largely accessible to any surgeon who is used to a cell phone or tablet. While the learning curve is relatively steep, once optimized the time utilized is similar but marginally increased than conventional techniques, with a statistically significant increase in accuracy. Robotic assistance or neuronavigation is not yet able to be considered as a standard of care, but discharging the patient from the hospital with a misplaced, clinically-significant implant may be considered a breach, and a single return to the operating room in our institution costs around $25,000 US dollars in 2014 values, without accounting for implant costs [10]. Additional technical advances such as real-time, constant registration and re-registration, and computer communication with the surgical implants for localization in space and expanded applicability of robotic assistance to other spinal procedures will ensure that robots will become a part of an every contemporary spine operating room in the relatively short term.

## Key points

- With around 1.34 million spine surgeries performed in the United States every year, robotic systems are being increasingly implemented in more areas of spine surgery.
- Robotics is being developed to increase accuracy, reduce radiation exposure, and improve clinical outcomes in spine surgery.
- Robotic systems are used in pedicle screw placement to combat some of the downsides of freehand or fluoroscopic approaches but also they introduce several new problems, such as issues with reference and tracking arrays, displacement of the robotic arm, or tool skiving, which all need to be carefully mitigated.
- Most of the currently approved and commercially available systems fall under the shared-control category, allowing the surgeon and the robot to control the instruments and motions simultaneously.
- Studies exploring clinical results are often imperfect and present conflicting evidences. More high-quality research with a more robust methodology and longer follow-up times is required to fully evaluate the accuracy, safety, clinical outcomes, and cost-effectiveness of robotics in spine surgery.

## References

[1] Baker JD, Sayari AJ, Basques BA, DeWald CJ. Robotic-assisted spine surgery: application of preoperative and intraoperative imaging. Semin Spine Surg 2020;32, 100789. https://doi.org/10.1016/j.semss.2020.100789.

[2] Girardi FP, Cammisa FP, Sandhu HS, Alvarez L. The placement of lumbar pedicle screws using computerised stereotactic guidance. J Bone Joint Surg Br 1999;81:825–9. https://doi.org/10.1302/0301-620x.81b5.9244.

[3] Mao JZ, Agyei JO, Khan A, Hess RM, Jowdy PK, Mullin JP, et al. Technologic evolution of navigation and robotics in spine surgery: a historical perspective. World Neurosurg 2021;145:159–67. https://doi.org/10.1016/j.wneu.2020.08.224.

[4] Alluri RK, Avrumova F, Sivaganesan A, Vaishnav AS, Lebl DR, Qureshi SA. Overview of robotic technology in spine surgery. HSS J: Musculoskelet J Hosp Spec Surg 2021;17:308–16. https://doi.org/10.1177/15563316211026647.

[5] Kaushal M, Kurpad S, Choi H. Robotic-assisted systems for spinal surgery. In: Scerrati A, De Bonis P, editors. Neurosurg. Proced.—Innov. Approaches. IntechOpen; 2020. https://doi.org/10.5772/intechopen.88730.

[6] D'Souza M, Gendreau J, Feng A, Kim LH, Ho AL, Veeravagu A. Robotic-assisted spine surgery: history, efficacy, cost, and future trends. Robot Surg Res Rev 2019;6:9–23. https://doi.org/10.2147/RSRR.S190720.

[7] Vadalà G, Salvatore SD, Ambrosio L, Russo F, Papalia R, Denaro V. Robotic spine surgery and augmented reality systems: a state of the art. Neurospine 2020;17:88–100. https://doi.org/10.14245/ns.2040060.030.

[8] Naik A, Smith AD, Shaffer A, Krist DT, Moawad CM, MacInnis BR, et al. Evaluating robotic pedicle screw placement against conventional modalities: a systematic review and network meta-analysis. Neurosurg Focus 2022;52:E10. https://doi.org/10.3171/2021.10.FOCUS21509.

[9] Jeswani S, Drazin D, Hsieh JC, Shweikeh F, Friedman E, Pashman R, et al. Instrumenting the small thoracic pedicle: the role of intraoperative computed tomography image–guided surgery. Neurosurg Focus 2014;36:E6. https://doi.org/10.3171/2014.1.FOCUS13527.

[10] Fontes RBV, Wewel JT, O'Toole JE. Perioperative cost analysis of minimally invasive vs open resection of Intradural extramedullary spinal cord tumors. Neurosurgery 2016;78:531–9. https://doi.org/10.1227/NEU.0000000000001079.

# 69

# Cranial robotic surgery

*Ryan Kelly and Sepehr Sani*

Department of Neurosurgery, Rush University Medical Center, Chicago, IL, United States

## History

Science fiction has long portrayed robots as autonomous machines, either assisting or completing tasks too complex for their human counterparts. Although robotic brain surgery may seem futuristic, these machines have already begun to appear in the operating room. While modern robots did not appear until the end of the 20th century, their ancestors can be traced back to as early as 1908 with the development of the first stereotactic system by the neurosurgeon Victor Horsley and mathematician Robert Clarke [1] (Fig. 1). This Cartesian-based device allowed for strategic and accurate placement of cortical and subcortical electrodes. The Horsley-Clarke frame was further developed by the Canadian neuroanatomist Aubrey Mussen in 1918 as well as Ernst Spiegel and Henry Wycis at Temple University in Philadelphia who, in 1947, demonstrated its first clinical use in human subjects [1,2] (Fig. 2). This laid the foundation for stereotactic neurosurgery, and innovation rapidly progressed in this field over the next several decades. Lars Leksell from the Karolinska Institute in Sweden visited Speigel and Wycis in 1947 and began development on his own arc-based stereotactic frame. The Leksell frame saw its first clinical application with the stereotactic implantation of radioactive phosphorus for the treatment of a craniopharyngioma patient in 1948 (Fig. 3). The Leksell frame has since become the work horse for stereotactic neurosurgeons around the world and continues to remain in use to this day.

Generally credited with being the first robotic device used in surgery, Arthrobot (developed at the University of British Columbia in Vancouver) performed over 60 arthroscopic procedures within its first year of operation between 1983 and 1984 [3]. The robot was used to assist the surgeon by rigidly stabilizing the patient's leg during total hip arthroplasties while also acting as a surgical assistant. The development of a cranial specific device came one-year later in 1985. The Programmable Universal Machine for Assembly (PUMA) 200 was an industrial robotic arm developed by Victor Scheinman at the robot company Unimation [4,5] (Fig. 4). The PUMA 200 was originally designed for General Motors to assist in automated assembly and consisted of a single robotic arm with six degrees of freedom. However, its application in medical practice became apparent in 1985 when Kwoh et al. at Memorial Medical Center in Long Beach, CA, used it to hold a cannula during a neurosurgical biopsy of an intracranial lesion [6,7]. The robotic arm was able to navigate to the correct location by fixating it to a stereotactic frame attached to the patient's head. The PUMA 200 was able to interface with a CT image guidance interface, allowing the treating surgeon to identify the target lesion while a simple command to the robot allowed it to localize its end-effector probe to the target location. This allowed for a smoother, faster, and more accurate procedure compared to those using manual stereotactic frames [6].

In the mid-late 1990s, robotic applications in medicine began to accelerate. The da Vinci robot (Intuitive) began development in 1995 but would not become commercially available until 2000 for use in gynecologic, urologic, and laparoscopic procedures [4,8]. Neurosurgeons also began to develop their own robotic systems to tailor to the specific demands of cranial and spine surgery. One such robotic system was Minerva, which was developed at the University of Lausanne in Switzerland and performed the first stereotactic surgery in 1993 involving the aspiration of an intracranial cystic lesion [9] (Fig. 5). The Minerva system was designed to improve upon conventional stereotactic procedures by providing feedback to the surgeon throughout the operation. Minerva performed many of the repetitive tasks of the surgery including skin incision, calvarial drilling, dural opening, and probe manipulation while the surgeon performed the steps of the surgery (i.e., aspiration/biopsy). The system also worked within the CT scanner allowing for real-time imaging of the patient to determine the accuracy of the robot and the ability to adjust for

FIG. 1    Victor Horsley and Robert Clarke brass stereotactic frame (1900–1910).

FIG. 2    Aubrey Mussen's stereotactic head frame (1918).

FIG. 3    Lars Leksell arc-based stereotactic head frame first developed in 1949.

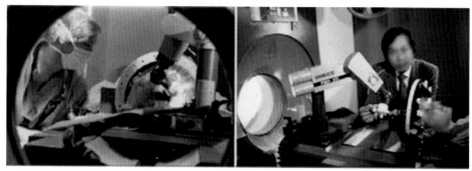

FIG. 4    PUMA 200 robotic arm with intraoperative CT scan.

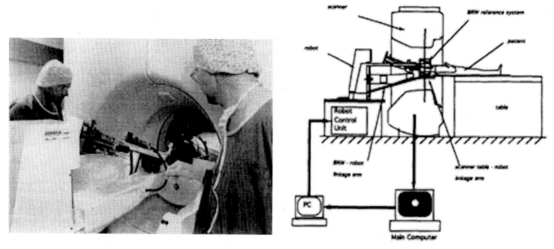

FIG. 5    Minerva robot with intraoperative CT (left) and schematic (right) of Minerva setup [9].

intraoperative brain shifts. Unfortunately, this led to large radiation doses to the patient on subsequent scans, as well as the cumbersome design of the system, which eventually led to the project's termination [7]. While some of these early robots had limited applications, the development of new robotic systems began to accelerate in the late 1990s with numerous devices being released throughout the turn of the 21st century.

## Modern robots

### NeuroMate

Many of the first neurosurgical robots were limited to rudimentary procedures such as intracranial biopsies and aspirations. However, toward the end of the 1990s, robots with more features began to emerge, allowing for their use in more complex surgeries. The NeuroMate platform started development in 1987 at Integrated Surgical Systems in Sacramento, CA but did not receive FDA approval until 1997. NeuroMate is a six-jointed robotic arm that allowed for either frame-based or frameless registration using either a conventional stereotactic localizer or an ultrasonic registration system. In the updated ultrasonic system, a detachable plate with fiducial markers is implanted into the patient's skull, and either a MRI or CT is completed prior to surgery for registration [10] (Fig. 6). One advantage of the NeuroMate system compared to prior robotic devices was that it was possible to incorporate preoperative imaging into the system. This allowed for new applications, including its use for deep brain stimulation (DBS). Several studies by Varma et al. demonstrated NeuroMate's success in the treatment of 106 patients with movement disorder pathologies, including Parkinson's disease, dystonia, and essential tremor [10]. The NeuroMate system also has found success in more conventional stereotactic procedures such as biopsies and endoscopic surgery [11]. NeuroMate remains commercially available with recent updates to its software having improved its safety features such as user-defined safety volumes, areas of reduced movement speed, and automatic termination of movement errors that exceed a pre-defined threshold [12].

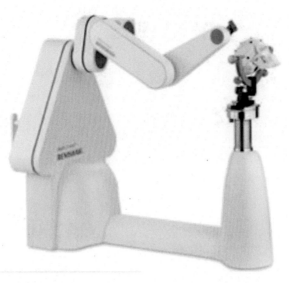

FIG. 6 NeuroMate robotic arm with cranial head-holder and fiducial marks.

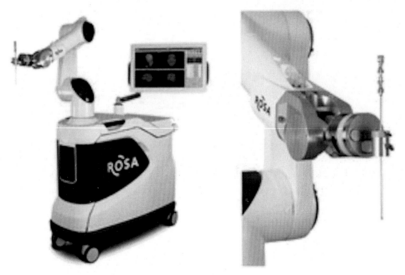

FIG. 7 ROSA One Brain robotic arm and CAN with cannula end effector.

## ROSA

The popularity of robotic-assisted devices rose in the early 2000s, and numerous medical companies began to develop their own systems. The Robotic Stereotactic Assistance (ROSA) was developed by Zimmer Biomet Robotics and received FDA approval for use in 2012 (Fig. 7). The system has been developed to provide either frame-based or frameless options depending on surgeon's preference and need. In the frame-based workflow, the robotic stand is attached to the head clamp intraoperatively, and the robotic arm registers points along the frame while intraoperative imaging is merged with a preoperative scan. With the frameless option, preoperative placement of fiducial markers or a laser-based facial matching software is used to register with preoperative MRI or CT scans to the session. The ROSA robot has found success in several neurosurgical procedures, including laser ablation thermal therapy, stereoelectroencephalography (sEEG) lead placement, DBS, endoscopic third ventriculostomy (ETV), and biopsies [13–17]. It has also become popular with pediatric neurosurgeons as the developing brain in this population poses additional challenges including smaller anatomy and increased vulnerability to injury necessitating accurate and minimally invasive approaches [17]. The most current iteration of this system is the ROSA ONE navigation system that offers tools to treat cranial, spine, and knee pathologies, making it the only robotic system on the market approved for use in these surgeries [18].

# Mazor

Several robotic systems have made the transition to cranial neurosurgical procedures after being developed for other applications. One such system is the Mazor Renaissance Guidance System developed by Mazor Robotics which received FDA approval in 2011 (Fig. 8). The Mazor Renaissance has been extensively employed in spine surgery with a systemic review conducted by Joesph et al. identifying 24 studies that have utilized the Mazor robotic system for this

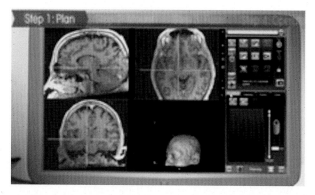

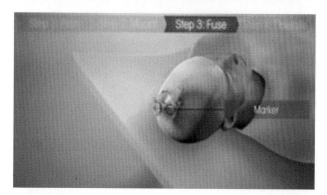

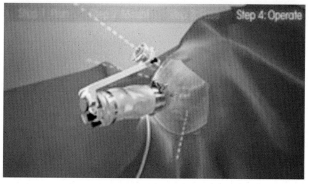

FIG. 8    Mazor Renaissance Guidance system with CAN planning, mounting of cranial fiducial with marker, and attachment of robotic device.

III. The current and future clinical applications of robotic surgery among medical specialties

application [19]. The Mazor Renaissance cranial system is a skull-mounted robotic-assisted device that allows for six degrees of freedom. It was the first system to allow for completely frameless cranial surgery as the skull-mounted device only required a single fiducial marker that was applied prior to intraoperative CT scan. Since its adaption for cranial use, several studies have demonstrated its effectiveness in reducing mean operative time and number of microelectrode recording passes as well as improved accuracy compared to traditional frame-based methods [20–22]. Recently, Medtronic bought Mazor Robotics in 2018 and began development on their own cranial-mounted device. The Stealth Autoguide is a cranial robotic guidance system that is currently being developed for use in sEEG lead placement and stereotactic biopsies (Fig. 9). While there is currently a paucity of literature on the use of the Mazor Renaissance system and Stealth Autoguide in cranial procedures, its application demonstrates promise that it can be used in other stereotactic neurosurgical operations [23–25].

## Globus ExcelsiusGPS

Similar to the Mazor Renaissance robot, the Globus ExcelsiusGPS was initially developed as a spine-focused robotic arm approved for use in 2017. The ExcelsiusGPS utilizes a floor-mounted base with the robotic arm and controller area network (CAN) interface attached. The robotic arm allows for six degrees of freedom, while the system platform provides real-time tracking to the surgeon while performing screw placement. The utilization of the ExcelsiusGPS in spine surgery has been well documented with several studies establishing its efficacy in thoracic and lumbar spine surgeries [26,27]. Globus has only recently developed a cranial workstation using the same robot that it uses with its spine system (Fig. 10). This new cranial workstation offers both frame-based and frameless registration options.

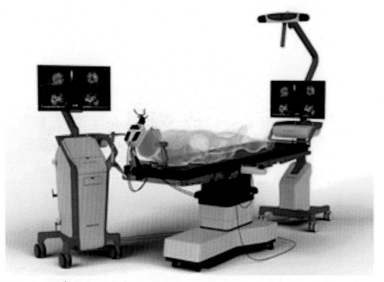

FIG. 9   Medtronic Stealth Autoguide workstation.

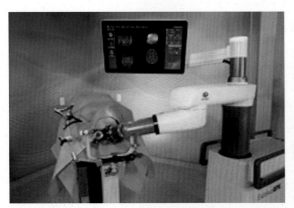

FIG. 10   Globus ExcelsiusGPS cranial workstation with frame-based and frame-less head attachments.

However, its use in cranial surgery has only recently been reported with the first DBS surgery using the ExcelsiusGPS robot being performed successfully in 2021 [28]. As of the time of this publication, this has thus far been the only documented use of the Globus system in a cranial neurosurgical procedure; however, the ExcelsiusGPS cranial workstation has been developed to perform biopsies, shunts, DBS, and sEEG placement [29].

## NeuroArm

While most of these robotic systems have been developed to have the surgeon be present in the operating room, there are some teleoperated systems that allow the operator to remotely control the machine during surgery. The DaVinci robotic system is probably the most documented of these teleoperated systems, but its uses are limited to general, gynecological, and urological procedures. In 2001, researchers at the University of Calgary and a Canadian aerospace engineering company (MacDonald, Dettwiler and Associates Ltd.) began development on a teleoperated robotic system with the goal of performing complex micro-neurosurgical operations while incorporating real-time MR imaging without disrupting the surgery. In 2012, the neuroArm became fully integrated in the clinical workflow at the University of Calgary and has since performed over 1000 neurosurgical procedures [7]. The neuroArm consists of two remote-controlled robotic arms each with six degrees of freedom and the ability to use both specifically designed and existing neurosurgical instruments [30] (Fig. 11). The surgeon sits at the control center that is located in a separate room and offers real-time three-dimensional (3-D) high-definition images via the robot along with the ability to integrate intraoperative MRIs. The robotic arms are manipulated using modified hand controllers that also provide advanced haptic feedback to the surgeon using two force sensors mounted on each arm [30]. A surgical assistant is present in the operating suite and acts in conjunction with the neuroArm. All members of the surgical team wear headsets allowing for rapid communication between surgeon assistant and support staff. One of the primary advantages of the neuroArm system is that it allows for constant up-to-date images to be obtained throughout the surgery allowing the surgeon to correct for brain shift and to determine the extent of current resection aiding in the complete removal of the pathology [30]. At present, this system is the only teleoperated neurosurgical robot currently in use and can only be found at two locations in Canada, the University of Calgary, and the Foothills Medical Centre also located in Calgary. The research into the neuroArm is still ongoing with new developments to improve the operating software, incorporate custom control algorithms and machine learning, and to enhance the integration between the control station and the robot to allow for more precise maneuverability to improve surgical outcomes [31].

## CorPath

At present, the use of robots in cranial neurosurgery has primarily been limited to functional and stereotactic procedures (i.e., DBS, sEEG, biopsies, ETV, etc.). However, advances in these machines have now expanded the use of

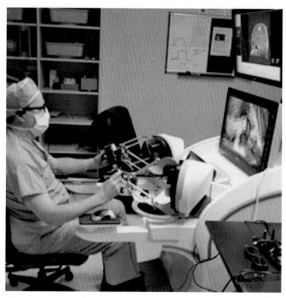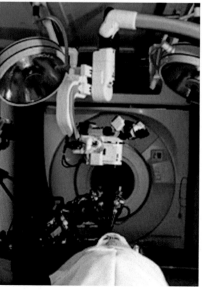

FIG. 11    NeuroArm remote workstation (left) and robotic arms with intraoperative MRI (right).

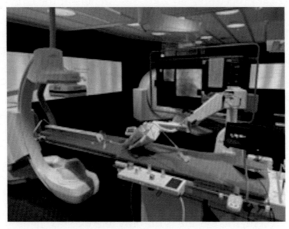

FIG. 12    CorPath GRX robotic arm (top) and intraoperative setup (bottom).

robots into the endovascular suite. The CorPath 200 robotic system is the first endovascular robot-assisted device and received FDA approval in 2012 (Fig. 12). The system consists of a robotic arm that houses a single-use cassette allowing the maneuverability of guide catheters mounted to the procedure table. The interventionalist is then able to control the cassette at a remote workstation outside the angiography suite, guiding that catheter to the desired location. One major advantage to this method is the significant reduction in radiation exposure to the treating physician while also allowing for accurate catherization and treatment of the pathology [32,33]. At present, the CorPath 200 and its successor the CorPath GRX are only FDA approved for use in cardiac catheterization procedures with several studies demonstrating the safety and efficacy of this device in human trials [34,35]. The reduction in radiation exposure is one major argument driving the increased interest from neuro-interventionalists in adopting this technology for intracranial use. Several studies have evaluated the use of the CorPath system for neuroendovascular procedures as an off-label device demonstrating its proof of concept and adaptability [36,37]. However, significant limitations of the device need to be remedied prior to its adoption in modern neuroendovascular treatment. Some of these include the need for physician-directed vascular access, manual operation of the endovascular table, the need for neuro-specific guidewires that are currently unable to be housed in the cassette, and lack of haptic feedback necessary for microcatheter insertion [37].

## Advantages

The use of robots in surgery has seen significant interest over the past decade, with one study demonstrating an increase in use of 13.3% from 2012 to 2018 in general surgery practices across the United States [38]. This number is only expected to increase as more surgical programs adopt robots for operative use. While neurosurgery has often led other fields in development, the adoption of robots for cranial surgery has shown slower progress compared to other surgical subspecialties. Within neurosurgery, robotic spine surgery has seen the most robust advancement and development with global trends anticipating an increase to $2.77 billion from $26 million over the course of 2016–2022 [39]. This significant increase in market-share will ultimately drive new innovation and research to improve these systems for more widespread adoption. Robotic devices offer several advantages compared to traditional surgical techniques, one of the most touted being its improved accuracy and reliability. A systemic review and meta-analysis conducted by in 2021 assessed mean target error (MTE) of DBS of sEEG lead placement across 37 articles from

2002 to 2018. This study demonstrated that over the 16-year period that the study was conducted, robotic-assistance achieved a 0.79 mm reduction in MTE in DBS cases and 0.74 mm reduction in MTE in sEEG cases compared to those that did not use robots. This study also demonstrated that there was no difference in accuracy with the use of either frame-based or frameless registration during robotic use. This has been replicated in other studies and phantom models demonstrating the increased accuracy of robots compared to standard surgical techniques [40].

When implemented correctly in a surgical workflow, robots have the potential to cut costs, including operative time, radiation exposure, and revision rates [41–43]. While operative times can vary depending on the institution, surgeon, and procedure, any reduction can potentially lead to better patient care and financial savings. While the initial operative time may be increased as surgeons are learning the device and how to best implement the robot into their surgery, over time, robotic use can approach traditional stereotactic methods and possibly lead to overall reductions in operative time. In a study by Vakharia et al. assessing the median time for sEEG electrode placement between two groups, they demonstrated a reduction in time in favor of the robotic cohort (6.36 min) compared to the manual cohort (9.06 min) [44]. This reduction in time per electrode placement translated to overall lower operative times again favoring the robotic cohort (176.4 (153.7–202.6) min) compared to the manual group (201.5 (175.5–231.3) min); however, this was not statistically significant. While these reductions may seem minimal, they can translate to significant savings while also reducing the anesthetic exposure to patients to improve postoperative recovery.

Increased radiation exposure from repeat fluoroscopic use can lead to a number of significant health complications including cancer, cataracts, and dysfunction in fertility [45,46]. The stochastic effects of ionizing radiation are measured in sieverts (Sv), with 1 Sv being associated with a 5.5% increase in developing cancer and recommendations of no more than 50 Sv in a single year and a maximum average of 20 Sv per year over 5 years [46]. A recent prospective randomized study by Kim et al. demonstrated a 30% decrease in total fluoroscopic time after their first eight cases while using robotic-assistance in posterior lumbar interbody fusion (PLIF) cases compared to conventional open PLIF techniques [47]. Another important consideration for cost reductions include revision surgery for either misplaced instrumentation or postoperative surgical site infection. Several studies have established that the use of robots in spine surgery can lead to overall reductions in revision surgery and operating costs at their hospitals [43,48,49]. Kantelhardt et al. found a 46% reduction in revision surgeries while using the SpineAsssit robot in their study, while Menger et al. found that the use of robotic assistance in surgery avoided 9.47 revision surgeries, with savings of $314,661 and secondary savings of $608,546 over a one-year period.

## Disadvantages

While neurosurgeons are some of the most aggressive adaptors of new technologies, the highly technical nature of some procedures can limit the transition of new instruments, devices, and methods into clinical practice. Surgical robots have been shown to improve accuracy and help with repetitive tasks; however, their implementation pose a significant demand on the surgeon and hospital using them. Several studies have shown that these machines take time to master, and consistent usage is important to improving clinical outcomes and decreasing operative time. Hu et al. demonstrated that over a two-year period in 150 robot-assisted cases, the accuracy of pedicle screw placement increased from 82% to 92%, while the rate of overall screw malposition was 0.7% and correlated to surgeon experience with the robot [50]. Abhinav et al. also showed that use of NeuroMate robot for sEEG placement led to an initial increase in operative times (5.6 h; 5.0–6.3 h) compared to conventional techniques (3.1 h; 1.7–4.5 h); however, larger case series have shown that as surgeons become more familiar with the system, the disparity in operative times reduces and approach similar results [51,52].

The implementation of surgical robots in neurosurgical practices remains limited with the top 30 neurosurgical programs accounting for 47.5% of all robotic spine and 60% of robotic cranial usage in the United States [53]. While these systems have many advantages, cost is a significant obstacle that limits its widespread adoption. In the long term, robots have been shown to reduce hospital and operative costs; however, their initial set-up are considered a significant investment for the hospital. The listing price for the Mazor X spine system, including the hardware and set-up costs, in 2018 was $1.1 million USD, not including disposable equipment, implants, and yearly maintenance fees [48]. While some large hospital organizations can afford this upfront cost, many smaller hospitals, especially those in less developed countries and regions, cannot devote the capital to secure and maintain this type of technology. As newer systems are developed and employed in hospitals that can afford them, previous iterations become more cost-effective and can be used in resource-limited healthcare systems. Additionally, some robotic companies have started to offer flat-rate packages that include maintenance, instruments, and assistants to reduce costs and make these systems more accessible [54]. However, at present, the high investment and operating cost of surgical robots only further the health disparities between those patients and hospitals who have the financial means to support and utilize these devices. Despite the financial barrier to many hospitals and regions, surgical robots will continue to be in high demand by both healthcare systems and patients given their perceived benefits and marketability.

## Future scenario

Surgical robots have only just started to find a foothold in neurosurgical practice with their adoption in spine, functional, and stereotactic surgeries. However, advances in this technology are rapidly progressing, and the integration of high-speed connectivity, artificial intelligence (AI), and augmented/virtual reality (AR/VR) is being incorporated into robotic system to improve and expand their uses [42]. The development of 5G technology has led to the emergence of the "Internet of Skills," or the ability to transmit technical skills and large amounts of information across digital interfaces effectively eliminating the perception of distance [55]. The expansion of this technology will allow users to become fully immersed in their environment, allowing for the immediate receiving and transmitting of information. The goal of this technology is to eventually perform complex procedures remotely, with delays of <10 ms at distances up to 1500 km away, allowing surgeons to reach patients who would otherwise not have access to surgical care [55]. The integration of AI into the robotic workflow also has the potential to further improve these machines and allow for increased responsiveness by predicting surgeon movements through pattern recognition while also processing large amounts of medical information assisting with diagnostics [54].

AR is the concept of displaying a virtual image on a real environment to enhance the users' experience. In neurosurgery, this can come in the form of overlaying a virtual display on a patient to improve visuospatial recognition of complex anatomy and reduce operative learning curves to improve education [56,57]. The use of AR technology has increased significantly over the past 10 years with one systemic review identifying only 13 articles related to AR in 2013 to over 70 articles published in 2021, with the majority of publications related to spine surgery and neuro-navigation [57]. The implementation of AR in the operating room can improve surgeon ergonomics by projecting images directly onto the patients instead of having the surgeon adjust their view to display monitors [58,59]. This technology can also be used to improve surgical education by allowing residents and surgeons to virtually practice skills such as pedicle screw placement, ventriculostomies, tumor resection, and aneurysm clippings before encountering these scenarios on real patients [59].

The integration of robots, high-speed connectivity, AI, and AR/VR will someday allow neurosurgeons to perform complex surgeries away from the sterile field. Future robots will be equipped with all of the tools necessary to perform skin incisions, craniotomies, biopsies, resections, and closure while the surgeon views the robotic arms at work through an AR/VR display that projects the surgical field to the surgeon along with intended trajectories and critical structures. The surgeon will then be able to control the robot's actions in real time through high-speed connectivity outside of the operating room thus reducing radiation exposure and limiting breaches in sterility [59]. While this degree of integration is decades away, the components are already being employed in neurosurgical practices around the country. Someday, these machines will expand the reach of surgeons to critical populations while improving the quality of life of the healthcare personal and patients who utilize them.

## Key points

- Neurosurgical robots can trace their roots to 1908 with the development of the stereotactic headframe developed by Victor Horsley and Robert Clarke.
- Most modern cranial neurosurgical robots are typically composed of a robotic arm with several degrees of freedom and allow for the integration of preoperative imaging utilizing either/both frameless or frame-based registration.
- There is a growing demand for the adoption of neurosurgical robots with several advantages, including improved accuracy/reliability and reductions in operative time, radiation exposure, and revision rates.
- While these machines have several benefits, their high start-up/maintenance costs and steep learning curve limit their widespread use to hospital systems with the financial means and resources to support these devices.

## References

[1] Serletis D, Pait TG. Early craniometric tools as a predecessor to neurosurgical stereotaxis. J Neurosurg 2016;124:1867–74.
[2] Spiegel EA, Wycis HT, Marks M, Lee AJ. Stereotaxic apparatus for operations on the human brain. Science 1947;106:349–50.

[3] Takács Á, Nagy D, Rudas I, Haidegger T. Origins of surgical robotics: from space to the operating room. Acta Polytech Hung 2016;13:13–30.

[4] Shah J, Vyas A, Vyas D. The history of robotics in surgical specialties. Am J Robot Surg 2014;1:12–20.

[5] Marcus HJ, Vakharia VN, Ourselin S, Duncan J, Tisdall M, Aquilina K. Robot-assisted stereotactic brain biopsy: systematic review and bibliometric analysis. Childs Nerv Syst 2018;34:1299–309.

[6] Kwoh YS, Hou J, Jonckheere EA, Hayati S. A robot with improved absolute positioning accuracy for CT guided stereotactic brain surgery. IEEE Trans Biomed Eng 1988;35:153–60.

[7] Mattei TA, Rodriguez AH, Sambhara D, Mendel E. Current state-of-the-art and future perspectives of robotic technology in neurosurgery. Neurosurg Rev 2014;37:357–66. discussion 366.

[8] George EI, Brand TC, LaPorta A, Marescaux J, Satava RM. Origins of robotic surgery: from skepticism to standard of care. JSLS 2018;22. e2018.00039.

[9] Glauser D, Fankhauser H, Epitaux M, Hefti JL, Jaccottet A. Neurosurgical robot minerva: first results and current developments. J Image Guid Surg 1995;1:266–72.

[10] Varma TRK, Eldridge P. Use of the NeuroMate stereotactic robot in a frameless mode for functional neurosurgery. Int J Med Robot 2006;2:107–13.

[11] Haegelen C, Touzet G, Reyns N, Maurage C, Ayachi M, Blond S. Stereotactic robot-guided biopsies of brain stem lesions: experience with 15 cases. Neurochirurgie 2010;56:363–7.

[12] Kajita Y, Nakatsubo D, Kataoka H, Nagai T, Nakura T, Wakabayashi T. Installation of a neuromate robot for stereotactic surgery: efforts to conform to Japanese specifications and an approach for clinical use—technical notes. Neurol Med Chir (Tokyo) 2015;55:907–14.

[13] Brandmeir N, Acharya V, Sather M. Robot assisted stereotactic laser ablation for a radiosurgery resistant hypothalamic hamartoma. Cureus 2016;8, e581.

[14] Chan AY, Tran DKT, Gill AS, Hsu FPK, Vadera S. Stereotactic robot-assisted MRI-guided laser thermal ablation of radiation necrosis in the posterior cranial fossa: technical note. Neurosurg Focus 2016;41:E5.

[15] González-Martínez J, Bulacio J, Thompson S, Gale J, Smithason S, Najm I, et al. Technique, results, and complications related to robot-assisted stereoelectroencephalography. Neurosurgery 2016;78:169–80.

[16] Liu L, Mariani SG, De Schlichting E, Grand S, Lefranc M, Seigneuret E, et al. Frameless ROSA® robot-assisted lead implantation for deep brain stimulation: technique and accuracy. Oper Neurosurg (Hagerstown) 2020;19:57–64.

[17] Nelson JH, Brackett SL, Oluigbo CO, Reddy SK. Robotic stereotactic assistance (ROSA) for pediatric epilepsy: a single-center experience of 23 consecutive cases. Children (Basel) 2020;7.

[18] Zimmer Biomet Holdings, Inc. Zimmer Biomet receives FDA clearance of ROSA® ONE spine system for robotically-assisted surgeries, https://www.prnewswire.com/news-releases/zimmer-biomet-receives-fda-clearance-of-rosa-one-spine-system-for-robotically-assisted-surgeries-300817471.html. [Accessed 24 January 2023].

[19] Joseph JR, Smith BW, Liu X, Park P. Current applications of robotics in spine surgery: a systematic review of the literature. Neurosurg Focus 2017;42:E2.

[20] VanSickle D, Volk V, Freeman P, Henry J, Baldwin M, Fitzpatrick CK. Electrode placement accuracy in robot-assisted asleep deep brain stimulation. Ann Biomed Eng 2019;47:1212–22.

[21] Ho AL, Pendharkar AV, Brewster R, Martinez DL, Jaffe RA, Xu LW, et al. Frameless robot-assisted deep brain stimulation surgery: an initial experience. Oper Neurosurg (Hagerstown) 2019;17:424–31.

[22] Liang AS, Ginalis EE, Jani R, Hargreaves EL, Danish SF. Frameless robotic-assisted deep brain stimulation with the mazor renaissance system. Oper Neurosurg (Hagerstown) 2022;22:158–64.

[23] Philipp L, Miller C, Wu C. Letter: placement of stereotactic electroencephalography depth electrodes using the stealth autoguide robotic system: technical methods and initial results. Oper Neurosurg (Hagerstown) 2022;23:e216–7.

[24] Mazur-Hart DJ, Yaghi NK, Shahin MN, Raslan AM. Stealth autoguide for robotic-assisted laser ablation for lesional epilepsy: illustrative case. J Neurosurg Case Lessons 2022;3.

[25] Tay ASS, Menaker SA, Chan JL, Mamelak AN. Placement of stereotactic electroencephalography depth electrodes using the stealth autoguide robotic system: technical methods and initial results. Oper Neurosurg (Hagerstown) 2022;22:e150–7.

[26] Huang M, Tetreault TA, Vaishnav A, York PJ, Staub BN. The current state of navigation in robotic spine surgery. Ann Transl Med 2021;9:86.

[27] Jiang B, Karim Ahmed A, Zygourakis CC, Kalb S, Zhu AM, Godzik J, et al. Pedicle screw accuracy assessment in ExcelsiusGPS® robotic spine surgery: evaluation of deviation from pre-planned trajectory. Chin Neurosurg J 2018;4:23.

[28] Behm C. Globus medical marks 1st cranial surgery with ExcelsiusGPS, https://www.beckersspine.com/robotics/52751-globus-medical-marks-1st-cranial-surgery-with-excelsiusgps.html. [Accessed 23 January 2023].

[29] ExcelsiusGPS® cranial solutions, https://www.globusmedical.com/musculoskeletal-solutions/excelsiustechnology/excelsiusgps/excelsiusgps-cranial-solutions/. [Accessed 23 January 2023].

[30] Sutherland GR, Wolfsberger S, Lama S, Zarei-nia K. The evolution of neuroArm. Neurosurgery 2013;72(Suppl 1):27–32.

[31] neuroArmPLUS-CellARM, https://neuroarm.org/neuroarmpluscellarm. [Accessed 23 January 2023].

[32] Goldstein JA, Balter S, Cowley M, Hodgson J, Klein LW. Occupational hazards of interventional cardiologists: prevalence of orthopedic health problems in contemporary practice. Catheter Cardiovasc Interv 2004;63:407–11.

[33] Jacob S, Boveda S, Bar O, Brézin A, Maccia C, Laurier D, et al. Interventional cardiologists and risk of radiation-induced cataract: results of a French multicenter observational study. Int J Cardiol 2013;167:1843–7.

[34] Weisz G, Metzger DC, Caputo RP, Delgado JA, Marshall JJ, Vetrovec GW, et al. Safety and feasibility of robotic percutaneous coronary intervention: PRECISE (percutaneous robotically-enhanced coronary intervention) study. J Am Coll Cardiol 2013;61:1596–600.

[35] Mahmud E, Dominguez A, Bahadorani J. First-in-human robotic percutaneous coronary intervention for unprotected left main stenosis. Catheter Cardiovasc Interv 2016;88:565–70.

[36] Britz GW, Panesar SS, Falb P, Tomas J, Desai V, Lumsden A. Neuroendovascular-specific engineering modifications to the CorPath GRX robotic system. J Neurosurg 2019;1–7.

[37] Sajja KC, Sweid A, Al Saiegh F, Chalouhi N, Avery MB, Schmidt RF, et al. Endovascular robotic: feasibility and proof of principle for diagnostic cerebral angiography and carotid artery stenting. J Neurointerv Surg 2020;12:345–9.

[38] Sheetz KH, Claflin J, Dimick JB. Trends in the adoption of robotic surgery for common surgical procedures. JAMA Netw Open 2020;3, e1918911.

III. The current and future clinical applications of robotic surgery among medical specialties

[39] Global surgical robots for the spine industry trend, growth. OpenPR; 2017, February 20. Available from: https://www.openpr.com/news/442943/global-surgical-robots-for-the-spine-industry-trend-growth-shares-strategy-and-forecasts-2016-to-2022.html. [Accessed 15 January 2023].

[40] Spyrantis A, Woebbecke T, Rueß D, Constantinescu A, Gierich A, Luyken K, et al. Accuracy of robotic and frame-based stereotactic neurosurgery in a phantom model. Front Neurorobot 2022;16:762317.

[41] Fomenko A, Serletis D. Robotic stereotaxy in cranial neurosurgery: a qualitative systematic review. Neurosurgery 2018;83:642–50.

[42] D'Souza M, Gendreau J, Feng A, Kim LH, Ho AL, Veeravagu A. Robotic-assisted spine surgery: history, efficacy, cost, and future trends. Robot Surg 2019;6:9–23.

[43] Menger RP, Savardekar AR, Farokhi F, Sin A. A cost-effectiveness analysis of the integration of robotic spine technology in spine surgery. Neurospine 2018;15:216–24.

[44] Vakharia VN, Rodionov R, Miserocchi A, McEvoy AW, O'Keeffe A, Granados A, et al. Comparison of robotic and manual implantation of intracerebral electrodes: a single-Centre, single-blinded, randomised controlled trial. Sci Rep 2021;11:1–10.

[45] Srinivasan D, Than KD, Wang AC, La Marca F, Wang PI, Schermerhorn TC, et al. Radiation safety and spine surgery: systematic review of exposure limits and methods to minimize radiation exposure. World Neurosurg 2014;82:1337–43.

[46] Narain AS, Hijji FY, Yom KH, Kudaravalli KT, Haws BE, Singh K. Radiation exposure and reduction in the operating room: perspectives and future directions in spine surgery. World J Orthop 2017;8:524–30.

[47] Kim H, Jung W, Chang B, Lee C, Kang K, Yeom JS. A prospective, randomized, controlled trial of robot-assisted vs freehand pedicle screw fixation in spine surgery. Int J Med Robot 2017;13.

[48] Fiani B, Quadri SA, Farooqui M, Cathel A, Berman B, Noel J, et al. Impact of robot-assisted spine surgery on health care quality and neurosurgical economics: a systemic review. Neurosurg Rev 2020;43:17–25.

[49] Kantelhardt SR, Martinez R, Baerwinkel S, Burger R, Giese A, Rohde V. Perioperative course and accuracy of screw positioning in conventional, open robotic-guided and percutaneous robotic-guided, pedicle screw placement. Eur Spine J 2011;20:860–8.

[50] Hu X, Lieberman IH. What is the learning curve for robotic-assisted pedicle screw placement in spine surgery? Clin Orthop Relat Res 2014;472:1839–44.

[51] Cardinale F, Cossu M, Castana L, Casaceli G, Schiariti MP, Miserocchi A, et al. Stereoelectroencephalography: surgical methodology, safety, and stereotactic application accuracy in 500 procedures. Neurosurgery 2013;72:353–66. discussion 366.

[52] Abhinav K, Prakash S, Sandeman DR. Use of robot-guided stereotactic placement of intracerebral electrodes for investigation of focal epilepsy: initial experience in the UK. Br J Neurosurg 2013;27:704–5.

[53] Singh R, Wang K, Qureshi MB, Rangel IC, Brown NJ, Shahrestani S, et al. Robotics in neurosurgery: current prevalence and future directions. Surg Neurol Int 2022;13.

[54] Aruni G, Amit G, Dasgupta P. New surgical robots on the horizon and the potential role of artificial intelligence. Investig Clin Urol 2018;59:221–2.

[55] Kim SSY, Dohler M, Dasgupta P. The internet of skills: use of fifth-generation telecommunications, haptics and artificial intelligence in robotic surgery. BJU Int 2018;122:356–8.

[56] Bernardo A. Virtual reality and simulation in neurosurgical training. World Neurosurg 2017;106:1015–29.

[57] Cannizzaro D, Zaed I, Safa A, Jelmoni AJM, Composto A, Bisoglio A, et al. Augmented reality in neurosurgery, state of art and future projections. A systematic review. Front Surg 2022;9.

[58] Yoon JW, Chen RE, Kim EJ, Akinduro OO, Kerezoudis P, Han PK, et al. Augmented reality for the surgeon: systematic review. Int J Med Robot 2018;14, e1914.

[59] Madhavan K, Kolcun JPG, Chieng LO, Wang MY. Augmented-reality integrated robotics in neurosurgery: are we there yet? Neurosurg Focus 2017;42:E3.

# Robotic reconstructive microsurgery

*Jefferson Braga Silva[a], Catarina Vellinho Busnello[b],*
*and Leandro Totti Cavazzola[c]*

[a]Hospital Moinhos de Vento, Porto Alegre, Brazil [b]Hospital de Clínicas Porto Alegre, Porto Alegre, Brazil
[c]Universidade Federal do Rio Grande do Sul, Porto Alegre, Brazil

## Introduction

The idea of robotic systems was introduced in 1980, first with industrial robots, used for stereotaxic biopsy[1]. Visceral robotic surgery was evidenced, then, during World War II and also by the National Aeronautics and Space Administration (NASA). Since then, especially, after the 1980s it is gaining huge space at surgical centers. The first models of robots were used for neurosurgery and orthopedics and now one of the most benefited by this technology has been microsurgery[1–3]. In 2007 the first microsurgical vascular anastomosis was performed using daVinci robotic assistance[4].

Over the last years, different models of robots have been created. The first one, used in 1985 in neurosurgery for biopsies, was the PUMA robot and then the ROBODOC system was used for orthopedics, in 1982. After, different models were created and the most known was the daVinci model, approved by FDA in 2000. In 2001 Zeus model was also approved, but the MUSA-2 robot (MicroSure, Eindhoven, The Netherlands) (Fig. 1), introduced in between 2020 and 2021, seemed to stand out in microsurgery [1,4–6]. Innovations keep coming. The Symani robot (Medical Microinstruments, MMI, Calci, Italy) (Fig. 2), a system with flexible arms, different from the others, seems to be the promise for performing microsurgery, already used for lymphatic and post traumatic upper limb reconstructions [4,5,7].

MUSA-2 robotic system was created in The Netherlands and it was considered the first robot dedicated for open microsurgery. This device has two fixed arms to which can be attached microinstruments and is capable of replicate movements commanded by the surgeon. MUSA-2 robot can be used in combination with a microscope or exoscope, however, the surgeon must stay in a fixed position. Symani robot, created in Italy, and MUSA-3 (Fig. 3) were designed to overcome this problem. MicroSure has acknowledged the user feedbacks of MUSA-2. Therefore, they created MUSA-3, a cart-based system with a surgeon console that can be positioned anywhere in the operating room. With flexible, wristed robotic instruments, Symani system enables the surgeon to access deeper structures, sitting in a comfortable position. Both of them have already proved to be feasible for micro and even supermicrosurgical anastomosis [7,8]. Besides that, there are the microsurgical handheld devices, like Micron device, a compact system also able to filter tremor [9].

Microsurgery is an important and still innovative technique, being increasingly used in several areas. Free flap reconstruction, nerve and lymphatic surgeries, and eye, hepatobiliary, transoral, urological, gynecological surgeries, and neuromicrosurgeries are some of the examples [6–10]. Since it requires extremely precise movements and techniques, robotic surgery has become the standard for reconstructive microsurgery, enabling the performance of challenging procedures [2,3,6]. Moreover, with the improvement of technology and microsurgery, nowadays supermicrosurgery has become a possibility, especially, for lymphatic surgery [11]. In this chapter we will present the current applications for reconstructive microsurgery using robotic assistance and its speculations for the future.

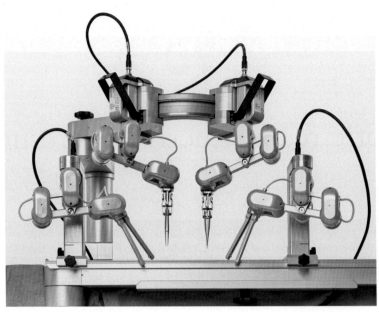

FIG. 1    MUSA-2 surgical system.

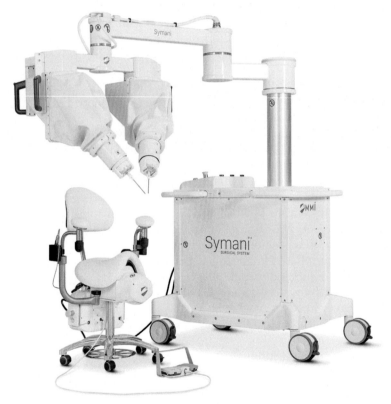

FIG. 2    Symani surgical system.

## Background

Microsurgery demands a high level of surgical skills, which can only be obtained with a lot of practice during years. Robot models are able to perform more precise movements and guarantee better techniques in narrow spaces [2–5]. Studies have shown that it may take longer to learn how to manipulate the robots and get used to them, but when it is

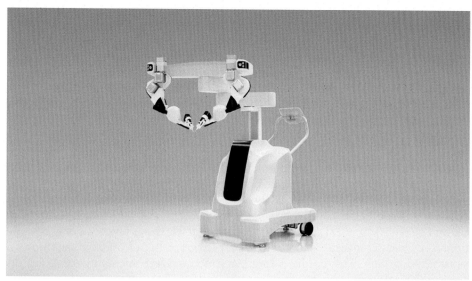

FIG. 3 MUSA-3 surgical system.

done, robot microsurgery is able to help with procedures that could not be done using conventional surgery [5]. Furthermore, after several surgeries there seems to be a technical improvement and time reduction [7].

Robotic microsurgery guarantees better vision and dexterity that are essential for reconstructive surgery. Therefore, there is a great room for robotics in microsurgery [2,7,12]. However, the first robotic systems introduced were not ideal for microsurgery, since they lacked needed micro instruments, quality of vision, and force control [6,13]. In 2014 a new robot, designed for microsurgery and supermicrosurgery, was projected by the Dutch technicians, the MUSA Micro-Sure Robot, and studies have demonstrated promising results [6]. Besides this, other microsurgery-assisted robots are also in tests, so that they can overcome robotic systems' limitations [13].

Nerve injuries due to tumor or trauma are very recurrent lesions at emergency centers and clinics, and surgical treatment is most time-requiring. However, relapse or motor and sensory sequel caused by conventional methods are not rare. Since microvessel anastomosis and nerve manipulation demand high precision, microsurgery with robotic assistance has been studied to overcome these treatment failures, with promising results. Minimally invasive neurosis, and nerve tumor resection using robots appear to be a possible key for these issues [3,14,15]. Besides these, the use of minimally invasive surgery using robots was used for intraneural perineurioma biopsy of brachial plexus with successful results [16].

The current literature already has described robot-assisted microsurgery for nerve reconstruction [17–19]. The high visual quality and precise movements provided by robotic systems enable procedures such as end-to-end nerve coaptation from difficultly accessed areas [17,18]. The reconstruction of a sympathetic trunk with a minimally invasive technique is a feasible procedure using robotic assistance, used for postoperative compensatory sweating treatment [17,18]. Moreover, neurolysis for treatment of meralgia also is described, with successful results, especially, due to the three-dimensional view made possible by robotic systems [19].

Brachial plexus reconstruction is another promising field for robotic microsurgery. This area comprehends a complex anatomy, with important vessels, nerves, and muscles, and conventional approaches require extensive dissection [12,20,21]. Robots could be used to preserve these structures and also avoid scarring and adherence [12]. Clinical studies have proved the feasibility of peripheral nerve repair using robotic assistance for brachial plexus neurorrhaphy, but it still needs some technical improvement to access difficult areas [15,20,21]. Exploration of the ulnar nerve for elbow paralysis treatment and also nerve transfer to deltoid muscle using the nerve to the long head of the triceps for shoulder abduction proved to be possible by telemicrosurgery, with better precision and less tremor than conventional surgery [22,23]. Furthermore, microsurgery and even supermicrosurgery have been described as innovative techniques for ischemic diabetic foot, preventing amputation in several cases [24,25].

Supermicrosurgey is an important field of reconstructive surgery, defined as anastomosis of vessels with less than 0.8 mm diameter. It can be applied for lymphedema treatment, digital reconstruction, fingertip replantation, nerve coaptation, and other tissue reconstructions. A high level of surgical skills are demanded for this technique, since it deals with such tiny and delicate structures. If supermicrosurgery has already gained this space, with robotic assistance and the capacity of improving surgical precision it must be the future of reconstructive surgery [11,26,27].

Supermicrosurgery is a well-established supermicrosurgery purpose in lymphedema treatment, mostly through lympho-venous anastomoses and vascularized lymph node transfer [27,28]. Since most collecting lymphatic vessels have a diameter of 0.5 mm or less, this technique has become the standard for lymphatic surgery. This represents a major milestone for micro procedures, but updates continue with robotic assistance. The first lymphatic robot microsurgery in humans was possible using the MUSA-2 system, proving to be a secure and effective robot for lympho-venous anastomoses, already used to treat breast cancer-associated lymphedema [28,29]. The second system available, enhanced with flexible wristed robotic instruments, is the Symani Surgical System and studies have shown promising outcomes, affirming that this system is safe for microsurgery of vessels smaller than 1 mm [5,28]. These systems provide not only better dexterity and vision, but also favorable ergonomic positions, preventing musculoskeletal pains, fatigue, and even tremor [28,30]. Moreover, a recent study described a microvascular reconstruction of a soft tissue with a free flap, for which MUSA-2 robot was used for end-to-end anastomosis and an excope (ORBEYE 4K 3D system) used for visual magnification, an important enhancement for microsurgery [8].

Supermicrosurgery is also very useful for soft tissue and tumor reconstruction, where micro vessels and nerve reconstruction are needed. Its competence in dissecting the most superficial perforators for flaps prevent donor-site morbidity, scarring, and sequels [11,26,27]. The use of robotics for all these procedures is not yet well established in the literature, but cases of nerve epineural coaptation, flaps, perforator-to-perforator anastomoses, and distal replantations have been described, which must be the future. As robot-assisted microsurgery scales down motions and tremor and allows access to locations not possible for hand surgery, these systems are ideal for supermicro anastomosis [7,27].

Although plastic surgery has been a resistant area for robotic microsurgery, it is gaining space [4,31–33]. Different studies have already proved the possible and safe application of robotics in plastic and reconstructive surgery, most of them are with animal models, but clinical human studies have also been published [32]. To date, the most related benefits of robotic microsurgery in plastic surgery are tremor cancellation and motion scale, which could provide better accuracy. However, studies demonstrated that there still have to be improvements in microsurgical instruments and techniques for reconstructive plastic surgical procedures [32,33]. The robotic instruments that are now on the market are not as delicate as needed for reconstructive surgery and there is still no robotic system that can handle different tissues at the same time. These limitations result in a more time-consuming surgery and a technique that is not indicated for some procedures, until then [32].

Regarding plastic and reconstructive surgery, breast and head and neck reconstructions are the ones that stand out the most. Robot-assisted microsurgery has shown to be effective in breast surgeries, reducing scars, enabling better hemostasis, and also improving flap pedicles. In these procedures, nipple-sparing mastectomies also have presented good outcomes regarding esthetic and vasculature preservation and robotic microsurgery could be effective for vessel anastomosis [28,34,35]. An important clinical study also demonstrated robot feasibility in harvesting the internal mammary vessels for breast reconstruction [36].

As reported, head and neck surgery is another area that has been positively affected by robotic reconstructive microsurgery. This new technology guarantees better outcomes not only in restoration but also functional preservation. Robotic systems can be used with modern microscopes and achieve even better visualization to perform microvascular anastomosis. RoboticScope (BHS Technologies, Innsbruck, Austria) is a good example, allowing surgeons to make different movements and stay in nonanatomical positions, with better dexterity and visual magnification [37].

Concerning transoral microsurgery, oral tumor resection and even subglottic stenosis have already been described, revealing a safe and feasible procedure [37–39]. Transoral reconstruction, due to oral tumor, had benefited from robotic assistance, mainly by microvessel anastomosis [39,40]. With high visual magnification and possibility to access narrow places, robotic systems are ideal for facial artery identification and manipulation [12]. Moreover, with the ability of reducing post op edema and injury during the procedure, robot microsurgery became one of the standard therapeutic modalities in this area [3].

Robotic microsurgery also played a good roll in neurosurgery. Tremor and fatigue control, essential for microneurosurgery, was achieved using robotic assistance, opening up opportunities for more complex procedures. Tumor resection and vascular disease treatment using robots are already a reality [41].

Men's infertility treatment is another area that has been studied and presented great outcomes with robotic microsurgery. Varicocele is a recurrent cause for men's infertility and studies have demonstrated that minimally invasive procedures are some of the best options, since they ensure testicular vessel preservation and a low rate of complication. Regarding hospitalization time, recurrence and complication, robotic assistance plays an important role [42]. Robot-assisted microsurgery also was showed to be feasible for vasectomy reversal. The abatement in surgeon's fatigue and procedure duration justifies the promise of this technology in urology [42].

Colorectal surgeons are also experiencing the benefits of robotic microsurgery, not only in colectomy but mainly in rectal lesions. Conservatively, these lesions were treated by open surgery or, when feasible, an endoscopic or laparoscopic technique. However, studies demonstrated that robot microsurgery could supply these procedures, with minimally invasive surgery, better visualization, and the possibility to work in smaller areas with more adequate ergonomic positions. The current literature revealed good outcomes in benign lesions and even malignant lesions in early stages [43,44].

Robotic surgery is consolidated in ophthalmology and now, micro and reconstructive robotic surgeries are showing promising results. The accurate movements needed in vitreoretinal treatments make it exciting to study robotic microsurgery for ophthalmology applications. Precise and large movement seems to be achieved with robotic assistance, allowing membrane peeling, core vitrectomy, and also microtube sutures [45]. Surgically treated retinal conditions are growing with advancing technology, however, studies emphasize the need for improvement in sense motion in robotic systems, so that they could supply new and overcome conventional techniques [46,47].

The following table shows some procedures already done by robotic microsurgery in humans with successful outcomes (Table 1).

TABLE 1 Procedures already done by robotic microsurgery in humans.

| Lindenblatt et al [5]. | Symani surgical system | Lympho-venous anastomosis |
|---|---|---|
| Barbon et al. [7] | Symani surgical system | Lympho-venous anastomosis, arterial anastomoses, lympho-lymphatic anastomosis, and epineural coaptations |
| Van Mulken et al. [8] | MUSA system and ORBEYE 4K 3D system | End-to-end anastomosis |
| Tigan et al. [14] | da Vinci surgical system | Neurolysis and nerve tumor resection |
| Lequint et al. [16] | da Vinci surgical system | Intraneural perineurioma biopsy of brachial plexus |
| Chang et al. [17] | da Vinci surgical system | Endoscopic thoracic sympathectomy reversal (end-to-end coaptation) |
| Chang et al. [18] | da Vinci surgical system | Suturing of sural nerve graft for sympathetic nerve reconstruction |
| Bruyere et al. [19] | Da Vinci surgical system | Neurolysis of lateral cutaneous nerve of thigh |
| Facca et al. [20] | da Vinci surgical system | Peripheral microneural repair |
| Garcia et al. [21] | da Vinci surgical system | Brachial plexus neurorrhaphy |
| Naito et al. [22]. | da Vinci surgical system | Restoration of elbow flexion using Oberlin technique |
| Miyamoto et al. [23] | da Vinci surgical system | Nerve transfer to deltoid muscle using the nerve to the long head of the triceps for shoulder abduction treatment |
| Van Mulken et al. [29] | MUSA system | Lympho-venous anastomosis |
| Kim et al. [34]. | da Vinci Xi surgical system | Nipple-sparing mastectomies |
| Boyd et al [36] | Aesop voice-activated robotic arm | Harvesting of internal mammary vessels in breast reconstruction |
| Boehm et al [37] | RoboticScope system | Transoral microvascular anastomosis |
| Fiorelli et al. [38] | AcuBlade robotic microsurgery system | Subglottic stenosis correction |
| Selber [39] | da Vinci surgical system | Transoral microvascular anastomosis |
| Lai et al. [40] | da Vinci surgical system | Microvascular anastomosis in free flap reconstruction |
| Goto et al. [41] | iArmS tool | Microscopic neurosurgery |
| Shu et al. [48] | da Vinci surgical system | Varicocelectomy |
| Parekattil and Gudeloglu [42] | da Vinci surgical system | Vasovasostomy and vasoepididymostomy |
| Lo et al. [44] | da Vinci surgical system | Rectal lesions |
| Edwards et al. [47] | Preceyes BV robotic system | Vitreoretinal procedures |

## Future scenario

Robot-assisted microsurgery has been witnessing a constant improvement over the past few years. As we present till now, different techniques, robotic systems, and procedures and opportunities are still in progress. Studies have shown excellent results and huge expectations for future outcomes. Among the upcoming scenarios, some limitations of the present robotic systems must be highlighted as promises to be enhanced. The next robot-assisted microsurgeries must count with systems with even better visual quality, the presence of haptic feedback, microinstuments appropriate for micro and supermicrosurgery, and the ability to perform more and more surgeries [28].

## Conclusions

This chapter brings up what is in evidence about microsurgery and also its evolution and upcoming scenarios. After an extensive review of the literature, we can summarize the benefits (and drawbacks) of robotic assistance for microsurgery as shown in Table 2.

Robotic microsurgery has been gaining enormous space until today. As stated throughout this chapter, it is already well established for several areas of surgery. One of them is for treating nerves, such as end-to-end coaptation, reconstruction of the brachial plexus and sympathetic trunk, nerve biopsy, and neurolysis [3,12,14–18,20,21]. Neurosurgery, urology, ophthalmology, transoral microsurgery, and proctology also benefit [37–48]. In plastic surgery, breast and head and neck reconstructions are in evidence and one of the most studied areas is the treatment of lymphedema, with lymphovenous anastomosis and vascularized lymph node transfer [11,26–30,34–40].

Despite so many benefits achieved by using robot microsurgery, it still has some disadvantages that need to be adjusted. Robotic systems lack tactile sensations that could cause thrombosis and intimal injury when vessels are manipulated. The need for a bigger surgical room, modern systems, and cost effectiveness are also some of the critics. Besides that, procedure time is a much discussed topic, it may take longer time to first perform a microanastomosis, but many studies have revealed that it must be due to the lack of familiarity with the new technology and practice could improve this. Furthermore, the learning curve for robotic surgery is shorter than for conventional surgery and it has fewer surgical complications, which could represent lower cost in the end [2,3,7,28,37,49]. In corroboration, The Structured Assessment of Robotic Microsurgical Skills, an instrument already validated, was used in one study that proved that dexterity, visuospatial ability, and operative flow were refined and time of procedure diminished with training and practice [49].

## Key points

- Robotic microsurgery will change the outcome of microsurgery outcomes.
- Robotic microsurgery will make super microsurgery affordable and change the paradigm of microsurgical anastomoses.
- Peripheral nerve surgery will have a technical improvement with robotic surgery.

TABLE 2   Advantages and disadvantages of robotic assistance for microsurgery.

| Advantages | Disadvantages |
| --- | --- |
| Three dimensional vision | Loss of sensation |
| No tremor | Cost |
| No psychological influence | Bigger surgical room |
| No tiredness | Lack of haptic feedback |
| Dexterity | Time |
| Precision | Microinstruments |
| Minimally invasive | |
| Ergonomic platform | |

# References

[1] Pugin F, Bucher P, Morel P. History of robotic surgery: from AESOP® and ZEUS® to da Vinci®. J Visc Surg 2011;148(5 Suppl):e3–8.

[2] Braga Silva JB, Busnello CV, Cesarino MR, Xavier LF, Cavazzola LT. Is there room for microsurgery in robotic surgery? Rev Bras Ortop (Sao Paulo) 2022;57(5):709–17.

[3] Saleh DB, Syed M, Kulendren D, Ramakrishnan V, Liverneaux PA. Plastic and reconstructive robotic microsurgery—a review of current practices. Ann Chir Plast Esthet 2015;60(4):305–12.

[4] Aitzetmüller MM, Klietz ML, Dermietzel AF, Hirsch T, Kückelhaus M. Robotic-assisted microsurgery and its future in plastic surgery. J Clin Med 2022;11(12):3378.

[5] Lindenblatt N, Grünherz L, Wang A, et al. Early experience using a new robotic microsurgical system for lymphatic surgery. Plast Reconstr Surg Glob Open 2022;10(1):e4013.

[6] van Mulken TJM, Boymans CAEM, Schols RM, et al. Preclinical experience using a new robotic system created for microsurgery. Plast Reconstr Surg 2018;142(5):1367–76.

[7] Barbon C, Grünherz L, Uyulmaz S, Giovanoli P, Lindenblatt N. Exploring the learning curve of a new robotic microsurgical system for microsurgery. JPRAS Open 2022;34:126–33.

[8] van Mulken TJM, Qiu SS, Jonis Y, Profar JJA, Blokhuis TJ, Geurts J, Schols RM, van der Hulst RRWJ. First-in-human integrated use of a dedicated microsurgical robot with a 4K 3D exoscope: the future of microsurgery. Life 2023;13(3):692.

[9] Zhang D, Si W, Fan W, et al. From teleoperation to autonomous robot-assisted microsurgery: a survey. Mach Intell Res 2022;19:288–306.

[10] van Mulken TJM, Schols RM, Qiu SS, et al. Robotic (super) microsurgery: Feasibility of a new master-slave platform in an in vivo animal model and future directions. J Surg Oncol 2018;118:826–31.

[11] Yamamoto T, Yamamoto N, Kageyama T, et al. Supermicrosurgery for oncologic reconstructions. Glob Health Med 2020;2(1):18–23.

[12] Ibrahim AE, Sarhane KA, Selber JC. New Frontiers in robotic-assisted microsurgical reconstruction. Clin Plast Surg 2017;44(2):415–23.

[13] Hangai S, Nozaki T, Soma T, et al. Development of a microsurgery-assisted robot for high-precision thread traction and tension control, and confirmation of its applicability. Int J Med Robot 2021;17(2):e2205.

[14] Tigan L, Miyamoto H, Hendriks S, Facca S, Liverneaux P. Interest of telemicrosurgery in peripheral nerve tumors: about a series of seven cases. Chir Main 2014;33(1):13–6.

[15] Chen LW, Goh M, Goh R, Chao YK, Huang JJ, Kuo WL, Sung CW, Chuieng-Yi LJ, Chuang DC, Chang TN. Robotic-assisted peripheral nerve surgery: a systematic review. J Reconstr Microsurg 2021;37(6):503–13.

[16] Lequint T, Naito K, Chaigne D, Facca S, Liverneaux P. Mini-invasive robot-assisted surgery of the brachial plexus: a case of intraneural perineurioma. J Reconstr Microsurg 2012;28(7):473–6.

[17] Chang TN, Daniel BW, Hsu AT, et al. Reversal of thoracic sympathectomy through robot-assisted microsurgical sympathetic trunk reconstruction with sural nerve graft and additional end-to-side coaptation of the intercostal nerves: a case report. Microsurgery 2021;41(8):772–6.

[18] Chang TN, Chen LW, Lee CP, Chang KH, Chuang DC, Chao YK. Microsurgical robotic suturing of sural nerve graft for sympathetic nerve reconstruction: a technical feasibility study. J Thorac Dis 2020;12(2):97–104.

[19] Bruyere A, Hidalgo Diaz JJ, Vernet P, Salazar Botero S, Facca S, Liverneaux PA. Technical feasibility of robot-assisted minimally-invasive neurolysis of the lateral cutaneous nerve of thigh: about a case. Ann Chir Plast Esthet 2016;61(6):872–6.

[20] Facca S, Hendriks S, Mantovani G, Selber JC, Liverneaux P. Robot-assisted surgery of the shoulder girdle and brachial plexus. Semin Plast Surg 2014;28(1):39–44.

[21] Garcia Jr JC, Lebailly F, Mantovani G, Mendonca LA, Garcia J, Liverneaux P. Telerobotic manipulation of the brachial plexus. J Reconstr Microsurg 2012;28(7):491–4.

[22] Naito K, Facca S, Lequint T, Liverneaux PA. The Oberlin procedure for restoration of elbow flexion with the da Vinci robot: four cases. Plast Reconstr Surg 2012;129(3):707–11.

[23] Miyamoto H, Leechavengvongs S, Atik T, Facca S, Liverneaux P. Nerve transfer to the deltoid muscle using the nerve to the long head of the triceps with the da Vinci robot: six cases. J Reconstr Microsurg 2014;30(6):375–80.

[24] Suh HS, Oh TS, Hong JP. Innovations in diabetic foot reconstruction using supermicrosurgery. Diabetes Metab Res Rev 2016;32(Suppl 1):275–80.

[25] Suh HS, Oh TS, Lee HS, et al. A new approach for reconstruction of diabetic foot wounds using the Angiosome and Supermicrosurgery concept. Plast Reconstr Surg 2016;138(4):702e–9e.

[26] Hong JPJ, Song S, Suh HSP. Supermicrosurgery: principles and applications. J Surg Oncol 2018;118(5):832–9.

[27] Badash I, Gould DJ, Patel KM. Supermicrosurgery: history, applications, training and the future. Front Surg 2018;5:23.

[28] Gousopoulos E, Grünherz L, Giovanoli P, Lindenblatt N. Robotic-assisted microsurgery for lymphedema treatment. *Plast Aesthetic* Res 2023;10(1):7.

[29] van Mulken TJM, Schols RM, Scharmga AMJ, et al. First-in-human robotic supermicrosurgery using a dedicated microsurgical robot for treating breast cancer-related lymphedema: a randomized pilot trial. Nat Commun 2020;11(1):757.

[30] Will PA, Hirche C, Berner JE, Kneser U, Gazyakan E. Lymphovenous anastomoses with three-dimensional digital hybrid visualization: improving ergonomics for supermicrosurgery in lymphedema. Arch Plast Surg 2021;48(4):427–32.

[31] Jimenez C, Stanton E, Sung C, Wong AK. Does plastic surgery need a rewiring? A survey and systematic review on robotic-assisted surgery. JPRAS Open 2022;26(33):76–91.

[32] Tan YPA, Liverneaux P, Wong JKF. Current limitations of surgical robotics in reconstructive plastic microsurgery. Front Surg 2018;5:22.

[33] Dobbs TD, Cundy O, Samarendra H, Khan K, Whitaker IS. A systematic review of the role of robotics in plastic and reconstructive surgery-from inception to the future. Front Surg 2017;4:66.

[34] Kim BS, Kuo WL, Cheong DC, Lindenblatt N, Huang JJ. Transcutaneous medial fixation sutures for free flap inset after robot-assisted nipple-sparing mastectomy. Arch Plast Surg 2022;49(1):29–33.

[35] Bishop SN, Selber JC. Minimally invasive robotic breast reconstruction surgery. Gland Surg 2021;10(1):469–78.

[36] Boyd B, Umansky J, Samson M, Boyd D, Stahl K. Robotic harvest of internal mammary vessels in breast reconstruction. J Reconstr Microsurg 2006;22(4):261–6.

III. The current and future clinical applications of robotic surgery among medical specialties

[37] Boehm F, Schuler PJ, Riepl R, Schild L, Hoffmann TK, Greve J. Performance of microvascular anastomosis with a new robotic visualization system: proof of concept. J Robot Surg 2022;16(3):705–13.

[38] Fiorelli A, Mazzone S, Costa G, Santini M. Endoscopic treatment of idiopathic subglottic stenosis with digital AcuBlade robotic microsurgery system. Clin Respir J 2018;12(2):802–5.

[39] Selber JC. Transoral robotic reconstruction of oropharyngeal defects: a case series. Plast Reconstr Surg 2010;126(6):1978–87.

[40] Lai CS, Lu CT, Liu SA, Tsai YC, Chen YW, Chen IC. Robot-assisted microvascular anastomosis in head and neck free flap reconstruction: preliminary experiences and results. Microsurgery 2019;39(8):715–20.

[41] Goto T, Hongo K, Ogiwara T, et al. Intelligent Surgeon's arm supporting system iArmS in microscopic neurosurgery utilizing robotic technology. World Neurosurg 2018;119:e661–5.

[42] Parekattil SJ, Gudeloglu A. Robotic assisted andrological surgery. Asian J Androl 2013;15(1):67–74.

[43] Watanaskul S, Schwab ME, Chern H, Varma M, Sarin A. Robotic transanal excision of rectal lesions: expert perspective and literature review [published online ahead of print, 2022 Oct 16]. J Robot Surg 2022. https://doi.org/10.1007/s11701-022-01469-8.

[44] Lo KW, Blitzer DN, Shoucair S, Lisle DM. Robotic transanal minimally invasive surgery: a case series. Surg Endosc 2022;36(1):793–9.

[45] Yang UJ, Kim D, Hwang M, et al. A novel microsurgery robot mechanism with mechanical motion scalability for intraocular and reconstructive surgery. Int J Med Robot 2021;17(3):e2240.

[46] Urias MG, Patel N, Ebrahimi A, Iordachita I, Gehlbach PL. Robotic retinal surgery impacts on scleral forces: in vivo study. Transl Vis Sci Technol 2020;9(10):2.

[47] Edwards TL, Xue K, Meenink HCM, et al. First-in-human study of the safety and viability of intraocular robotic surgery. Nat Biomed Eng 2018;2:649–56.

[48] Shu T, Taghechian S, Wang R. Initial experience with robot-assisted varicocelectomy. Asian J Androl 2008;10(1):146–8.

[49] Alrasheed T, Liu J, Hanasono MM, Butler CE, Selber JC. Robotic microsurgery: validating an assessment tool and plotting the learning curve. Plast Reconstr Surg 2014;134(4):794–803.

# 71

# Robotic plastic surgery

*Marco Faria-Correa[a] and Savitha Ramachandran[b]*

[a]Plastic Surgery, Mount Elizabeth Novena Specialist Centre, Singapore, Singapore [b]Department of Plastic and Reconstructive Surgery, Singapore General Hospital, Singapore, Singapore

## Introduction

The first reported use of the daVinci Surgical Robotic System (Intuitive Surgical, Sunnyvale, CA, United States) was in a robotic-assisted laparoscopic cholecystectomy [1]. Since then, Intuitive Surgical has become the leading force in surgical robotics. The daVinci robot has been widely implemented in many surgical specialties, from cardiac surgery to gynecology. Currently, 80% of radical prostatectomies are now being performed robotically in the United States. Addition to the daVinci robot, including a fourth instrument arm, facilitated broadening of robotic applications to new specialties such as colorectal surgery. However, the dominance of the daVinci system is beginning to be challenging with new competitors entering the market [2].

Plastic and reconstructive surgery is an innovative specialty, often at the forefront of the technical innovation within surgery. It is a unique specialty that collaborates with multiple surgical disciplines. Therefore, as robotic surgery becomes increasingly adopted by the various specialties, it is imperative for plastic surgeons to embrace this new surgical platform and explore potential uses for it within the field of plastic and reconstructive surgery. Robotic surgery should be eventually incorporated into the surgical armamentarium, of all plastic surgeons.

## History of robotic applications to plastic surgery

### Microsurgery

Although the daVinci robot was developed for general surgery, early adopters of this technology to plastic surgery identified microsurgery as an area of great potential application for robotic surgery. Katz et al. performed the first daVinci system-assisted anastomosis in a porcine model in 2005, closely followed by work in canine tarsal and femoral vessels [2,3]. In these studies, they concluded significant advantages such as the elimination of tremor at a microsurgical level. However, the lack of fine purpose-built microsurgical instruments turned out to be a major limitation, and hence, to date, it is not routinely adopted in clinical practice. Further animal and human cadaveric work has cemented the idea that robotically assisted microvascular surgery is both feasible and, in some instances, potentially beneficial, such as when working at depth and for surgeon comfort. The benefits of tremor elimination, motion scaling, and ergonomics in cadaver studies were mirrored in early clinical studies by Boyd et al. and Van der Hulst et al. [4,5]. However, both the authors alluded to increased time using the robot, compared to traditional techniques which reduced with increased experience and later in the learning curve. The learning curve for robotic microsurgery has also been established along with objective tools for the assessment of robotic microsurgery. To date, there are a total of 18 studies identified that discussed the use of robotics for a microsurgery application (Table 1). Eleven of these were preclinical studies in synthetic, animal, and cadaveric models while seven were clinical studies.

To address these concerns and lack of fine instrumentation in the daVinci robotic platform, in recent years, there have been newer robotic platforms developed specifically for the application of robotic microsurgery with delicate fine instrumentation designed for both microsurgery and even supermicrosurgery [19]. One of the most exciting additions to robotic microsurgery has been the debut of the Symani microsurgery robotic platform in 2018. The Symani Surgical

TABLE 1 Preclinical and clinical studies relating to the use of robotics in microvascular procedures.

| Reference year | Operations performed | Outcomes reported |
|---|---|---|
| Katz et al. (2005) [3] | Animal model. Arterial and venous anastomoses and free-flap transplantation (N = 1 pig) | All anastomoses grossly patent, confirmed by audible Doppler signals, visibly adequate perfusion of tissues, and arterial bleeding seen after incision distal to the anastomoses 4 h after the procedure |
| Knight et al. (2005) [6] | Animal model. Arterial end-to-end anastomoses (N = 31 vs N = 30 controls) | A remarkable degree of tremor filtration, but significantly slower operative time. All anastomoses were patent and nonleaking case controlled |
| Karamanoukian et al. (2006) [7] | Animal tissue samples. Slit arteriotomy and end-to-end arterial anastomoses in porcine hearts. The Zeus robotic system is a viable tool for microsurgical vascular reconstruction | It allows for precise movement, lack of hand tremor, enhanced microvascularization and improved ergonomics, compared to conventional human assistance. The major advantage is the ability of the robot to scale down the surgeon's movements to a microscopic level |
| Katz et al. (2006) [2] | Animal cadavers. Microvascular anastomoses of tarsal and superficial femoral vessels (N = 2 dog cadavers) | All anastomoses were successful and patent postoperatively |
| Taleb et al. (2009) [8] | Animal cadaver (N = 1, pig limb transplant) | Telesurgery could improve limb replantation and transplantation management, especially regarding operating gesture precision |
| Ramdhian et al. (2011) [9] | Animal tissue samples. Earthworm segment anastomoses (N = 15) | The high-quality 3D vision allowed by the robotic system was excellent and compensated for loss of tactile feedback. The robotic system eliminated physiological tremor. Motion scaling by the robot improved precision of the surgical gesture |
| Lee et al. (2012) [10] | Live animal models. Femoral artery end-to-end anastomoses (N = 20) | Generation of learning curves for robotic-assisted microvascular anastomosis. Important aspects of learning identified included starting level, learning plateau, and learning rate |
| Robert et al. (2013) [11] | Human cadaver. Radial/ulnar artery dissection and microvascular anastomoses (N = 2 cadavers, 4 anastomoses) | Successful anastomoses. The assembling and disassembling of the vascular clamp were time-consuming. In both cases (radial and ulnar arteries), the 10/0 needle was bent and a second suture had to be used |
| Alrasheed et al. (2014) [12] | Synthetic vessel models. Microvascular anastomoses (N = 50) | Successful validation of microsurgical assessment tool and characterization of learning curve. Proficiency gained by operators over five learning sessions |
| Selber and Alrasheed (2014) [13] | Synthetic models. Microvascular anastomoses (N = 5 per surgeon) | Definition of a learning curve in microsurgery and the development of a structured assessment of robotic microsurgical skills |
| Willems et al. (2016) [14] | Synthetic microvessel models. Microvascular anastomoses (N = 80, vs 80 control) | Manual surgery was superior to robotically assisted microsurgery in technically easy exposures. In difficult exposures (greater depth and lower sidewall angles), however, robotically assisted microsurgery had a shorter surgery time and a higher comfort rating. The Objective Structured Assessment of Technical Skills scores were similar to those assessing traditional microsurgery |
| van Mulken et al. (2018) [15] | Anastomoses on silicone vessels (N = 10) | It is feasible to complete anastomotic microsurgery on silicone vessels using the MicroSure robotic system |
| van Mulken et al. (2018) [16] | End-to-end anastomoses. Abdominal aorta and femoral vessels in the rat model (N = 7) | Feasibility of performing a microvascular anastomosis in a rat model using a first prototype of a new robotic platform, which is specifically designed for microsurgery |
| Clinical studies | | |
| Boyd et al. (2006) [4] | Pedicle was harvested with robotic-assisted technique. Microvascular anastomosis via standard technique | An average pedicle length of 6.7 cm is long enough to allow anastomosis without vein graft |

III. The current and future clinical applications of robotic surgery among medical specialties

TABLE 1  Preclinical and clinical studies relating to the use of robotics in microvascular procedures—cont'd

| Reference year | Operations performed | Outcomes reported |
|---|---|---|
| Van der Hulst et al. 2007 [5] | Case report. Breast reconstruction with muscle sparing-free TRAM-flap, using robotic arterial anastomosis ($N=1$) | The time to perform this anastomosis was about 40 min and significantly longer than the standard technique (around 15 min) |
| van Mulken et al. (2020) [17] | Robotic-assisted LVA ($N=8$) | At 3 months, the patient outcome improves. Furthermore, a steep decline in duration of time required to complete the anastomosis is observed in the robotic-assisted group (33–16 min). Here, we report the feasibility of robotic-assisted supermicrosurgical anastomosis in LVA, indicating promising results for the future of reconstructive supermicrosurgery |
| Giovanoli et al. (2022) [18] | Symani surgical system to perform lymphovenous and arterial anastomosis for lymphatic reconstruction ($N=10$) | Feasibility and safety of the robotic system to perform lymphatic surgery |

*Modified from Dobbs TD, Cundy O, Samarendra H, Khan K, Whitaker IS. A systematic review of the role of robotics in plastic and reconstructive surgery-from inception to the future. Front Surg 2017;4:66.*

System is a platform consisting of two robotic arms and a system that has 7–20X motion scaling with tremor filtration to address the demands and complexity of microsurgery and supermicrosurgery [18]. Since the establishment of their prototype in 2017, the developers of this robotic platform have dedicated themselves to optimizing this robotic platform for precision and control to suit requirements of delicate micro- and supermicrosurgical work. The Symani Surgical System saw its first clinical use in Careggi University Hospital in Florence, Italy, where both the first free flaps and LVA procedures were done [20]. In 2022, two more Symani Surgical Systems have been installed in Europe and are in use in the fields of plastic surgery and maxillofacial surgery at the University of Zurich and Salzburg, respectively.

## Muscle flap harvest

Muscle free-flaps are traditionally raised through a large incision overlying the muscle belly. With the advent of laparoscopic surgery, the clear advantages of shortened recovery time, and improved outcomes demonstrated by minimally invasive surgery, several surgeons developed laparoscopic approaches for muscle flaps for specific clinical indications [21]. As the field of minimally invasive surgery advanced towards robotic surgery, developing robotic approaches to muscle harvest for reconstruction was a natural step in the same direction. Laparoscopic harvesting is clearly documented in the literature, but with poor uptake due to difficulties with the visualization of the operative field and the inherent limitations of laparoscopic instruments. This is where robotic platforms can overcome limitations of laparoscopic surgery while simultaneously providing the obvious advantages of reduced scarring and faster recovery time, which ultimately leads to improved clinical outcomes [21].

To date, five human cadaveric studies and 17 clinical studies have been identified, describing the use of the robot for muscle flap harvest ranging from latissimus dorsi muscle flap harvest to the DIEP (deep inferior epigastric perforator) flap harvest (Table 2). In these clinical studies, it was demonstrated that the robot improves visualization, reduces the scar burden, and results in reduced postoperative pain and hospital stay [1,42,43]. However, it must be mentioned that the overall operative time was increased, especially at the early stages of the learning curve [1].

Majority of the publications about robotic-assisted muscle harvest are related to robotic latissimus dorsi muscle flap. Most publications report no significant postoperative complications, except a few cases of postoperative seromas, which were conservatively managed [43]. The robotic-assisted harvesting technique of the latissimus dorsi flap for breast reconstruction is safe and comparable to the conventional methods. Reduced hospital stays and superior aesthetic outcome have been widely reported, but increased cost and the steep learning curve have been cited as the main disadvantage of this modern and minimally invasive surgical approach [43].

As the robotic approach for latissimus dorsi flap for breast reconstruction evolved, the concept was applied to other forms of autologous breast reconstruction. It had already reported that the robotic rectus flap harvest was feasible [44], and the desire for improved muscle preservation and reduced donor site morbidity in DIEP flap harvest led to the

TABLE 2	Preclinical and clinical studies on robotic muscle harvest.

| Reference year | Operations performed | Outcomes reported |
|---|---|---|
| Selber (2011) [21] | Human cadaver. Latissimus dorsi muscle harvest ($N=10$ in 8 cadavers) | Successful harvest of all muscles |
| Patel and Pedersen (2012) [22] | Human cadaver. Rectus abdominis muscle dissection and harvest ($N=2$) | Successful harvest of all muscles |
| Selber et al. (2012) [23] | Human cadaver. Latissimus dorsi muscle harvest and transfer ($N=8$) | Successful harvest and transfer of all flaps that left no visible incisions, with no major complications |
| Honart et al. (2018) [24] | Human cadaver. DIEP flap harvest ($N=2$) | DIEP flap harvest with robotically assisted dissection of the deep inferior epigastric vessels was feasible in a cadaver. This technique might reduce donor site morbidity in comparison to the open approach, due to smaller fascial incision |
| Terzic et al. (2020) [25] | Cadaveric training model for DIEP flap harvest using two approaches: transabdominal preperitoneal (TAPP) and totally extraperitoneal (TEP) ($N=8$) | Both TAPP and TEP are feasible, with TEP less invasive, preserving the posterior rectus sheath, and decreasing complication risks. However, there is a steeper and longer learning curve for TEP |
| **Clinical studies** | | |
| Patel et al. (2012) [26] | Case report. Pedicled myocutaneous latissimus dorsi flap for shoulder reconstruction after sarcoma resection ($N=1$) | No objective outcomes reported flap successfully raised robotically. One of the limitations is the time/learning curve |
| Selber et al. (2012) [23] | Robotic latissimus dorsi muscle harvest ($N=7$) | Seven muscle flaps were harvested without converting to an open technique. Both free flaps were successfully transferred. All pedicled flaps resulted in successful breast reconstructions. Flap harvest complications included a single, temporary radial nerve palsy in the contralateral extremity, likely from positioning |
| Lazzaro et al. (2013) [27] | Case report. Intercostal muscle flap after lobectomy (done in conjunction with VATS) ($N=1$) | Success of surgery—no conversion to open procedures and both patients returned home 5 days postop |
| Ibrahim et al. (2014) [28] | Case series. Rectus abdominus muscle flap harvest ($N$ not reported) | Less tissue violation, compared to open technique, resulting in reduced postoperative pain, shorter duration of hospital stay, and more rapid functional recovery |
| Selber et al. (2014) [29] | Robotic latissimus dorsi flap for breast reconstruction ($N=17$) | Robotic-assisted harvest of the latissimus dorsi muscle is associated with a low complication rate and reliable results for delayed reconstruction of the irradiated breast while eliminating the need for a donor-site incision |
| Chung et al. (2015) [30] | Case series. Transaxillary gasless robotic-assisted latissimus dorsi muscle harvest (3 delayed reconstructions, 4 immediate after nipple sparing mastectomy, 5 corrections of deformity in Poland syndrome) ($N=12$) | Operating time, general satisfaction, cosmetic satisfaction, scar, and symmetry satisfaction were all outcomes measured via survey given to all patients with follow-up longer than months Robotic time decreases with experience |
| Singh et al. (2015) [31] | Case series and retrospective review. Extralevator abdominoperineal excision with robotic rectus abdominis flap harvest, for reconstruction after resection of distal rectal adenocarcinoma ($N=3$) | An incisionless robotic flap harvest with the preservation of the anterior rectus sheath obviates the risk of ventral hernia while providing robust tissue closure of the radiated abdominoperineal excision wound |

TABLE 2   Preclinical and clinical studies on robotic muscle harvest—cont'd

| Reference year | Operations performed | Outcomes reported |
|---|---|---|
| Delaney et al. (2018) [32] | Robotic DIEP flap harvest case report | Uneventful postop course. This technique enabled improved precision of flap harvest while also decreasing the donor-site morbidity by minimizing the incision length of the anterior rectus sheath |
| Cuylits et al. (2020) [33] | Robotic latissimus dorsi muscle harvest ($N = 6$) | Robotic-assisted latissimus dorsi muscle flap harvest is a safe, reproducible, and effective tool that offers precise dissection control and that leaves a minimal thoracic scar |
| Moon et al. (2020) [34] | Autologous chest reconstruction using a robotic-assisted LD muscle flap harvest ($N = 21$) | No serious complications such as flap loss were recorded for any patient. The time for robotic surgery markedly decreased as experience accumulated |
| Selber et al. (2021) [35] | Robotic harvest of the rectus abdominis muscle for reconstruction of pelvic defects ($N = 7$) | This study demonstrates the safety, efficacy, and reproducibility of robotic harvest of the rectus abdominis muscle in complex, multidisciplinary, minimally invasive pelvic surgery. The technique avoids violation of the anterior rectus sheath and wound complications related to open flap harvest, and early |
| Lee et al. (2021) [36] | Robotic DIEP flap extraperitoneal approach | |
| Selber et al. (2022) [37] | Robotic-assisted latissimus dorsi muscle harvest ($N = 15$) | Zero adverse events attributable to harvest and 100% muscle viability after harvest and zero conversions to open technique. There appears to be little to no long-term functional deficit or pain from muscle harvest. Given these results and their own prestudy guidelines, the robotic latissimus dorsi qualifies for 510(k) submission by Intuitive Surgical and approval by the U.S. FDA |
| Kim et al. (2022) [38] | Robotic DIEP flap for breast reconstruction ($N = 21$) | This study suggests that a robotic DIEP flap offers enhanced postoperative recovery, accompanied by a reduction in postoperative pain and hospital stay |
| Levine et al. (2022) [39] | Robotic DIEP flap for breast reconstruction ($N = 4$) | There were no flap failures, and no patient experienced abdominal wall donor site morbidity on physical exam |
| Selber et al. (2022) [40] | Robotic DIEP flap for breast reconstruction ($N = 21$) | The robotic deep inferior epigastric perforator flap is a safe and reliable technique that decreases the length of fascial incision and short-term complications associated with the open approach |
| Dusseldorp et al. (2023) [41] | $N = 1$ | Successful outcome |

*Modified from Dobbs TD, Cundy O, Samarendra H, Khan K, Whitaker IS. A systematic review of the role of robotics in plastic and reconstructive surgery-from inception to the future. Front Surg 2017;4:66.*

evaluation of robotic DIEP harvest [45]. Robotic DIEP harvest enables the longest possible pedicle harvest through the smallest possible fascial incision and hence reduces donor site morbidity [45].

To date, a review of Seven publications, detailing a total of 56 robotic-assisted DIEP flap harvest procedures analyzing technical feasibility of flap harvest in cadavers, viable flap harvest in live patients, harvest time and pedicle dissection time, pedicle length, fascial incision length, donor site pain, need for postoperative narcotic, donor site morbidity, and hernia formation, demonstrated successful DIEP flap harvesting without the need for conversion to the conventional open procedure [42]. Postoperative complications were minimal. Robotic DIEP flap harvest was shown to be safe, and there were no reports of donor-site morbidity. The main advantages of the robotic approach include decreased postoperative pain and length of hospital stay, along with improved aesthetic outcomes. The main disadvantages are increased operative time and cost [42]. Although at its current iteration, the robotic-assisted DIEP flap is feasible, and it may not be practical in all settings.

III. The current and future clinical applications of robotic surgery among medical specialties

## Robotic abdominoplasty

Robotic repair of rectus diastasis is a state-of-the-art concept that has evolved over 32 years from the new concept of miniabdominoplasty. In 1989, the senior author Dr Marco Faria-Correa performed a deep analysis of the miniabdominoplasty technique, originally described in the 1980s, and identified that for a select group of patients with good abdominal skin elasticity, without redundant folds of skin, it was unnecessary to have a long skin incision to perform the rectus plication.

He also noted that the rectus plication through the miniabdominoplasty technique was limited to the lower part of the abdomen and hence did not provide a full functional and cosmetic repair of rectus diastasis, because the rectus diastasis is simply not just limited to the lower abdomen and extends to the xiphoid process. Hence, these patients who had plication limited to lower abdomen had upper abdominal bulge due to ineffective abdominal muscle repair.

As such, he innovated a novel technique of abdominoplasty, using minimal incisions and a minimally invasive approach.

In 1991, he initiated a research project to adopt endoscopy to a gasless subcutaneous technique. He designed a set of instruments to facilitate this procedure (minimally invasive gasless abdominoplasty).

From 1989 to 2015, with a series of more than 600 cases, using this method, and patients of more than 20-years follow-up with successful lasting results, Dr. Marco was inspired to bring this technique to the next level. He was aware of the advantages of robotic surgery over the endoscopic technique.

In 29 April 2015, Dr. Marco did his inaugural case pioneering the use of robotic technology to gasless rectus plication. He also designed instruments to support this procedure and technique.

This minimally invasive lipoabdominoplasty (MILA) is becoming popular using both endoscopy and robotics and can be performed in both ways, gasless and with the use of $CO_2$. Using three keyhole incisions, at the bikini line, the central one for the camera, about 8 cm apart in each side, is able to dissect a V-shaped tunnel from the pubis to xiphoid, up to the sternal border of the rectus abdominal muscle. He then identified and repaired the umbilical, ventral hernia, and rectus diastasis using two layers of suture. The first layer is closed with ethilon 2.0 inverted buried stitch, 1 cm apart from each other. Followed by a second layer of suturing, using V lock barbed sutures 2.0. readjusting and tightening muscle aponeurosis of the core muscle in the midline. Liposuction can be performed in patients with lipodystrophy. Skin tightening technology can be used as an adjunct to treat skin laxity.

The use of suction drains is mandatory, to prevent the seroma. Compressive garment, taping, and lymphatic drainage massage are also helpful for faster recovery.

To strengthen the core muscles, physiotherapy should start 8–12 weeks after surgery.

A recent addition to robotic applications is the plastic surgery [46]. Riding on the similar advantages described for robotic applications to hernia repair, robotic rectus muscle harvest, robotic repair of rectus diastasis, and reconstruction of the abdominal wall, with small incisions and excellent visualization while avoiding the traditional tummy-tuck (abdominoplasty) horizontal incisions altogether. Thus, the result is reduced morbidity and improved aesthetic outcomes for the select group of patients with rectus diastasis without access to the abdominal skin, and this technique has been pioneered by the senior author and has shown good long-term outcomes.

| | | |
|---|---|---|
| Faria-Correa et al. (1992) [47] | Videoendoscopic abdominoplasty subcutaneouscopy | Endoscopic abdominoplasty is feasible with smaller scars |
| Faria-Correa et al. (2008) [48] | Videoendoscopic subcutaneous abdominoplasty | Endoscopic abdominoplasty is feasible with good long-term results |
| Faria-Correa et al. (2017) [49] | Minimally invasive robotic abdominoplasty | Robotic abdominoplasty is feasible with minimal scars and good results |

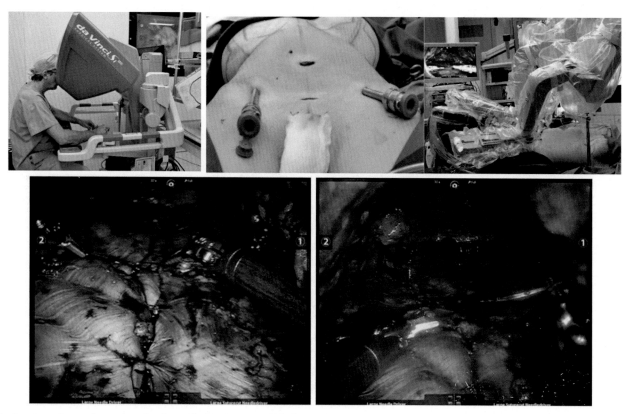

Robotic setup of ports for rectus plication and plication using V-lock sutures

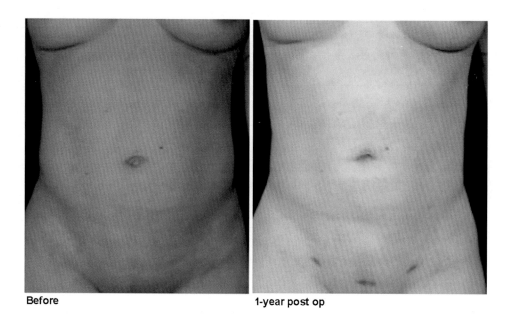

Before                              1-year post op

III. The current and future clinical applications of robotic surgery among medical specialties

## Nerve surgery

Given the clear advantages of superhuman precision, improved visualization, and small incisions, the application of robotic surgery is very difficult to reach areas such as the axilla for brachial plexus surgery which seems like a natural fit. A total of 19 studies including preclinical experimental researches and clinical reports were identified with the majority exploring the role of robotics in brachial plexus work into brachial plexus reconstruction [50]. The rest of the publications described robotic applications to peripheral nerve tumors management, peripheral nerve decompression or repair, peripheral nerve harvesting, and sympathetic trunk reconstruction. There were three animal studies, four cadaveric studies, eight clinical series, and four studies, demonstrating clinical, animal, or cadaveric studies simultaneously [50]. In out of total 53 clinical cases, only 20 (37.7%) cases were successfully approached with minimally invasive and intervened robotically; 17 (32.1%) cases underwent conventional approach and the nerves were intervened robotically; 12 (22.6%) cases were converted to open approach but still intervened the nerve by robot; and 4 (7.5%) cases failed to approach robotically and converted to open surgery entirely. Chang et al. concluded that although the robotic system improves surgery with virtual reality visualization, tremor filtration, minimally invasive approaches, and ergonomics, the clinical advantage in peripheral nerve surgery is still limited to specific clinical scenarios [50]. From the experiences of early adopters, some choose robotics instead of microscopic technique after open dissection because local anatomy can be clearly demonstrated, and iatrogenic injury of the adjacent tissue can be prevented.

## Transoral robotic surgery (TORS)

Transoral robotic surgery has revolutionized head and neck surgery with improved access afforded by the robotic instruments, visualization, and ergonomics for the surgeon for the treatment of malignant conditions of the oral cavity and oropharynx [51]. It facilitates access to the deep recess of the oral cavity while avoiding morbid traditional mandibulotomies and lip split approaches. If there is no communication between the oral cavity or oropharynx and neck dissection, then the defect could be left to heal by secondary intention. However, in more complex or advanced stages of disease, reconstruction using local flaps or free tissue transfer is required. Traditionally, mandibulotomies have to be performed, to gain access for satisfactory flap inset and reconstruction of the intraoral floor of mouth defects. The concept of using robotic approach for the intraoral flap inset was first introduced in 2009 by Mukhija et al., and since then, transoral robotic surgery has become the biggest area for robotic-assisted plastic surgery procedures, with two preclinical and 21 clinical studies identified [1,52]. Local reconstructive options include the use of the facial artery musculomucosal flap, commonly used in the reconstruction of the floor of the mouth, and soft palate has been described by Bonawitz and Duvvuri using the robot for raising and in-setting the flap with good results [53]. Others demonstrated that the use of the robot to perform a musculomucosal advancement flap pharyngoplasty gives good results, both in terms of orocutaneous fistula risk and functional outcomes. The commonest reported free-flap used following TORS resection is the radial forearm flap; however, the anterolateral thigh flap is also described. In the majority of cases, the robot was used for flap inset, with authors reporting good access and visualization that allowed a water-tight inset to be achieved and no flap complications despite the lack of a traditional jaw spilled. The robot has also been used in a number of studies to perform the vascular anastomosis.

## Transoral robotic cleft surgery (TORCS)

Transoral robotic cleft surgery is an evolving field with great potential and builds on the foundation and profile of transoral robotic surgery (TORS) [54]. To date, three preclinical and two clinical studies have been published in the literature [55–57]. Aside for robotic microsurgery, this is the only field of robotic plastic surgery in which the objective assessment of skills has been investigated using the hand motion analysis [58]. Future work is necessary to develop more suitable instrumentation and to compare outcomes of a full robotic cleft palate repair to the traditional approach in real patients before widescale implementation can be adopted. Importantly, training and validation of simulators for the purposes of robotic cleft surgery are needed to provide an assessment of surgeons who want to use surgical robots before it becomes more mainstream in cleft surgery [54].

## Future scenario

In the field of PRAS, robotic innovations will be focused on improved precision, development of instruments suitable for delicate tissue handling, and fine surgical movements. The advent of bid data collection platforms and artificial intelligence will lead to the future developments of the robotic platform, all aimed at reducing the learning curve, preventing errors, and eventually improving patient outcomes.

## Key points

- Plastic surgery is a newcomer to the Robotic Surgery's arena due to the pioneering efforts of innovative reconstructive surgeons, surgery being a growing field.
- Robotic applications to plastic/reconstructive surgery have gained traction in several subareas of the specialty as minimally invasive flap harvest for reconstruction, transoral robotic surgery for head and neck reconstruction/cleft palate repair, abdominal wall reconstruction, brachial plexus exploration/nerve repair, and microvascular surgery.
- Robotic applications can reduce scarring, in procedures with precision.
- The minimal access option of the robotic platform confers significant advantage for the patients who require reconstruction of soft tissue without skin.
- Robotic microsurgery has been the archetypal example of how these recent advancements and development of instruments suited for delicate tissue handling and fine surgical movements.

## References

[1] Dobbs TD, Cundy O, Samarendra H, Khan K, Whitaker IS. A systematic review of the role of robotics in plastic and reconstructive surgery-from inception to the future. Front Surg 2017;4:66.
[2] Katz RD, Taylor JA, Rosson GD, Brown PR, Singh NK. Robotics in plastic and reconstructive surgery: use of a telemanipulator slave robot to perform microvascular anastomoses. J Reconstr Microsurg 2006;22(1):53–7.
[3] Katz RD, Rosson GD, Taylor JA, Singh NK. Robotics in microsurgery: use of a surgical robot to perform a free flap in a pig. Microsurgery 2005; 25(7):566–9.
[4] Boyd B, Umansky J, Samson M, Boyd D, Stahl K. Robotic harvest of internal mammary vessels in breast reconstruction. J Reconstr Microsurg 2006;22(4):261–6.
[5] van der Hulst R, Sawor J, Bouvy N. Microvascular anastomosis: is there a role for robotic surgery? J Plast Reconstr Aesthet Surg 2007; 60(1):101–2.
[6] Knight CG, Lorincz A, Cao A, Gidell K, Klein MD, Langenburg SE. Computer-assisted, robot-enhanced open microsurgery in an animal model. J Laparoendosc Adv Surg Tech A 2005;15(2):182–5.
[7] Karamanoukian RL, Bui T, McConnell MP, Evans GR, Karamanoukian HL. Transfer of training in robotic-assisted microvascular surgery. Ann Plast Surg 2006;57(6):662–5.
[8] Taleb C, Nectoux E, Liverneaux P. Limb replantation with two robots: a feasibility study in a pig model. Microsurgery 2009;29(3):232–5.
[9] Ramdhian RM, Bednar M, Mantovani GR, Facca SA, Liverneaux PA. Microsurgery and telemicrosurgery training: a comparative study. J Reconstr Microsurg 2011;27(9):537–42.
[10] Lee JY, Mattar T, Parisi TJ, Carlsen BT, Bishop AT, Shin AY. Learning curve of robotic-assisted microvascular anastomosis in the rat. J Reconstr Microsurg 2012;28(7):451–6.
[11] Robert E, Facca S, Atik T, Bodin F, Bruant-Rodier C, Liverneaux P. Vascular microanastomosis through an endoscopic approach: feasibility study on two cadaver forearms. Chir Main 2013;32(3):136–40.
[12] Alrasheed T, Liu J, Hanasono MM, Butler CE, Selber JC. Robotic microsurgery: validating an assessment tool and plotting the learning curve. Plast Reconstr Surg 2014;134(4):794–803.
[13] Selber JC, Alrasheed T. Robotic microsurgical training and evaluation. Semin Plast Surg 2014;28(1):5–10.
[14] Willems JIP, Shin AM, Shin DM, Bishop AT, Shin AY. A comparison of robotically assisted microsurgery versus manual microsurgery in challenging situations. Plast Reconstr Surg 2016;137(4):1317–24.
[15] van Mulken TJM, Boymans C, Schols RM, Cau R, Schoenmakers FBF, Hoekstra LT, et al. Preclinical experience using a new robotic system created for microsurgery. Plast Reconstr Surg 2018;142:1367–76.
[16] van Mulken TJM, Schols RM, Qiu SS, Brouwers K, Hoekstra LT, Booi DI, et al. Robotic (super) microsurgery: feasibility of a new master-slave platform in an in vivo animal model and future directions. J Surg Oncol 2018;118:826–31.
[17] van Mulken TJM, Schols RM, Scharmga AMJ, Winkens B, Cau R, Schoenmakers FBF, et al. First-in-human robotic supermicrosurgery using a dedicated microsurgical robot for treating breast cancer-related lymphedema: a randomized pilot trial. Nat Commun 2020;11(1):757.
[18] Lindenblatt N, Grunherz L, Wang A, Gousopoulos E, Barbon C, Uyulmaz S, et al. Early experience using a new robotic microsurgical system for lymphatic surgery. Plast Reconstr Surg Glob Open 2022;10(1), e4013.
[19] Aitzetmuller MM, Klietz ML, Dermietzel AF, Hirsch T, Kuckelhaus M. Robotic-assisted microsurgery and its future in plastic surgery. J Clin Med 2022;11(12).
[20] Innocenti M. Back to the future: robotic microsurgery. Arch Plast Surg 2022;49(3):287–8.
[21] Selber JC. Robotic latissimus dorsi muscle harvest. Plast Reconstr Surg 2011;128(2):88e–90e.

[22] Patel NV, Pedersen JC. Robotic harvest of the rectus abdominis muscle: a preclinical investigation and case report. J Reconstr Microsurg 2012;
28(7):477–80.

[23] Selber JC, Baumann DP, Holsinger FC. Robotic latissimus dorsi muscle harvest: a case series. Plast Reconstr Surg 2012;129(6):1305–12.

[24] Struk S, Sarfati B, Leymarie N, Missistrano A, Alkhashnam H, Rimareix F, et al. Robotic-assisted DIEP flap harvest: a feasibility study on cadaveric model. J Plast Reconstr Aesthet Surg 2018;71(2):259–61.

[25] Manrique OJ, Bustos SS, Mohan AT, Nguyen MD, Martinez-Jorge J, Forte AJ, et al. Robotic-assisted DIEP flap harvest for autologous breast reconstruction: a comparative feasibility study on a cadaveric model. J Reconstr Microsurg 2020;36(5):362–8.

[26] Patel NP, Van Meeteren J, Pedersen J. A new dimension: robotic reconstruction in plastic surgery. J Robot Surg 2012;6(1):77–80.

[27] Lazzaro RS, Guerges M, Kadosh B, Gulkarov I. Robotic harvest of intercostal muscle flap. J Thorac Cardiovasc Surg 2013;146(2):486–7.

[28] Ibrahim AE, Sarhane KA, Pederson JC, Selber JC. Robotic harvest of the rectus abdominis muscle: principles and clinical applications. Semin Plast Surg 2014;28(1):26–31.

[29] Clemens MW, Kronowitz S, Selber JC. Robotic-assisted latissimus dorsi harvest in delayed-immediate breast reconstruction. Semin Plast Surg 2014;28(1):20–5.

[30] Chung JH, You HJ, Kim HS, Lee BI, Park SH, Yoon ES. A novel technique for robot assisted latissimus dorsi flap harvest. J Plast Reconstr Aesthet Surg 2015;68(7):966–72.

[31] Singh P, Teng E, Cannon LM, Bello BL, Song DH, Umanskiy K. Dynamic article: tandem robotic technique of extralevator abdominoperineal excision and rectus abdominis muscle harvest for immediate closure of the pelvic floor defect. Dis Colon Rectum 2015;58(9):885–91.

[32] Gundlapalli VS, Ogunleye AA, Scott K, Wenzinger E, Ulm JP, Tavana L, et al. Robotic-assisted deep inferior epigastric artery perforator flap abdominal harvest for breast reconstruction: a case report. Microsurgery 2018;38(6):702–5.

[33] Fouarge A, Cuylits N. From open to robotic-assisted latissimus dorsi muscle flap harvest. Plast Reconstr Surg Glob Open 2020;8(1), e2569.

[34] Moon KC, Yeo HD, Yoon ES, Lee BI, Park SH, Chung JH, et al. Robotic-assisted latissimus dorsi muscle flap for autologous chest reconstruction in poland syndrome. J Plast Reconstr Aesthet Surg 2020;73(8):1506–13.

[35] Asaad M, Pisters LL, Klein GT, Adelman DM, Oates SD, Butler CE, et al. Robotic rectus abdominis muscle flap following robotic extirpative surgery. Plast Reconstr Surg 2021;148(6):1377–81.

[36] Choi JH, Song SY, Park HS, Kim CH, Kim JY, Lew DH, et al. Robotic DIEP flap harvest through a totally extraperitoneal approach using a single-port surgical robotic system. Plast Reconstr Surg 2021;148(2):304–7.

[37] Shuck J, Asaad M, Liu J, Clemens MW, Selber JC. Prospective pilot study of robotic-assisted harvest of the latissimus dorsi muscle: a 510 (k) approval study with U.S. Food and Drug administration investigational device exemption. Plast Reconstr Surg 2022;149(6):1287–95.

[38] Lee MJ, Won J, Song SY, Park HS, Kim JY, Shin HJ, et al. Clinical outcomes following robotic versus conventional DIEP flap in breast reconstruction: a retrospective matched study. Front Oncol 2022;12, 989231.

[39] Daar DA, Anzai LM, Vranis NM, Schulster ML, Frey JD, Jun M, et al. Robotic deep inferior epigastric perforator flap harvest in breast reconstruction. Microsurgery 2022;42(4):319–25.

[40] Bishop SN, Asaad M, Liu J, Chu CK, Clemens MW, Kapur SS, et al. Robotic harvest of the deep inferior epigastric perforator flap for breast reconstruction: a case series. Plast Reconstr Surg 2022;149(5):1073–7.

[41] Dayaratna N, Ahmadi N, Mak C, Dusseldorp JR. Robotic-assisted deep inferior epigastric perforator (DIEP) flap harvest for breast reconstruction. ANZ J Surg 2023;93(4):1072–4.

[42] Khan MTA, Won BW, Baumgardner K, Lue M, Montorfano L, Hosein RC, et al. Literature review: robotic-assisted harvest of deep inferior epigastric flap for breast reconstruction. Ann Plast Surg 2022;89(6):703–8.

[43] Vourtsis SA, Paspala A, Lykoudis PM, Spartalis E, Tsourouflis G, Dimitroulis D, et al. Robotic-assisted harvest of latissimus dorsi muscle flap for breast reconstruction: review of the literature. J Robot Surg 2022;16(1):15–9.

[44] Pedersen J, Song DH, Selber JC. Robotic, intraperitoneal harvest of the rectus abdominis muscle. Plast Reconstr Surg 2014;134(5):1057–63.

[45] Selber JC. The robotic DIEP flap. Plast Reconstr Surg 2020;145(2):340–3.

[46] Bilezikian J, Durbin B, Miller J, Oyola AM, Hope WW. Robotic plication of rectus diastasis with associated hernias: a case series. Surg Technol Int 2022;41.

[47] Faria-Correa MA. Videoendoscopic abdominoplasty (subcutaneouscopy). Rev Soc Bras Cir Plast Est Reconstr 1992;7:32–4.

[48] Faria-Correa MA. Videoendoscopic subcutaneous abdominoplasty. In: Endoscopic plastic surgery 2nd ed. IV. Missouri: Quality Medical Publishing, Inc; 2008. p. 559–86.

[49] Faria-Correa MA. Minimally invasive robotic abdominoplasty. In: Advances in plastic & reconstructive surgery. Applis Publishers; 2017.

[50] Chen LW, Goh M, Goh R, Chao YK, Huang JJ, Kuo WL, et al. Robotic-assisted peripheral nerve surgery: a systematic review. J Reconstr Microsurg 2021;37(6):503–13.

[51] Selber JC. Transoral robotic reconstruction of oropharyngeal defects: a case series. Plast Reconstr Surg 2010;126(6):1978–87.

[52] Mukhija VK, Sung CK, Desai SC, Wanna G, Genden EM. Transoral robotic assisted free flap reconstruction. Otolaryngol Head Neck Surg 2009;140(1):124–5.

[53] Bonawitz SC, Duvvuri U. Robot-assisted oropharyngeal reconstruction with free tissue transfer. J Reconstr Microsurg 2012;28(7):485–90.

[54] Nadjmi N. Transoral robotic cleft palate surgery. Cleft Palate Craniofac J 2016;53(3):326–31.

[55] Selber JC. Discussion: infant robotic cleft palate surgery: a feasibility assessment using a realistic cleft palate simulator. Plast Reconstr Surg 2017;139(2). 466e-7e.

[56] Podolsky DJ, Fisher DM, Wong Riff KWY, Looi T, Drake JM, Forrest CR. Infant robotic cleft palate surgery: a feasibility assessment using a realistic cleft palate simulator. Plast Reconstr Surg 2017;139(2). 455e-65e.

[57] Khan K, Dobbs T, Swan MC, Weinstein GS, Goodacre TE. Trans-oral robotic cleft surgery (TORCS) for palate and posterior pharyngeal wall reconstruction: a feasibility study. J Plast Reconstr Aesthet Surg 2016;69(1):97–100.

[58] Ghanem A, Podolsky DJ, Fisher DM, Wong Riff KW, Myers S, Drake JM, et al. Economy of hand motion during cleft palate surgery using a high-fidelity cleft palate simulator. Cleft Palate Craniofac J 2019;56(4):432–7.

# 72

# Robotic ophthalmologic surgery

*Vagner Loduca Lima[a], Rafael Cunha de Almeida[a,b], Taurino dos Santos Rodrigues Neto[d], and Alexandre Antonio Marques Rosa[a,c]*

[a]Department of Ophthalmology, ABC Foundation School of Medicine, Santo André, Brazil [b]Department of Ophthalmology, University of Santo Amaro, School of Medicine, São Paulo, Brazil [c]Department of Ophthalmology, Federal University School of Pará, Pará, Brazil [d]Associate Physician in RetinaPro Ophthalmological Clinic, São Paulo, Brazil

## Introduction

The term robotics derives from the Czech word "robota" which means "servant" or "worker." [1] The use of robots in surgery can help improve the accuracy of tasks, reduce tremors, improve visualization, and better control from a distance. Robotics has been used in various medical areas for more than 20 years and has assisted doctors in operating rooms. As an example, the success of the da Vinci Surgical System (Intuitive Surgical Inc. Sunnyvaley CA.US) has become widely recognized and perhaps the most prevalent robotic surgical system in the world [2].

Robotic systems offer a promising solution to these human limitations due to the possibility of precise, tremor-free movements and offer greater depth resolution through a variety of integrated imaging modalities such as digital microscopy and optical coherence tomography (OCT). These advantages naturally increase the safety and effectiveness of surgical procedures [3].

However, despite the success of the da Vinci Surgical System in other specialties, the incorporation of surgical robotics in ophthalmic surgery has not had the same degree of success or growth.

Several obstacles remain before robotic surgery becomes popular in ophthalmology. A high cost, steep learning curve, and patient confidence present some of the challenges in robotics in ophthalmology. Currently, considerable studies and tests have been carried out for various robotic systems, only a few of them have reached the commercial stage and ophthalmology itself has a long way to go in robotic technology.

## Classification of surgical systems

One method for categorizing the range of intraocular robotic surgical systems is according to the degree of human versus robotic control. In traditional surgery, a human surgeon controls surgical tools and uses an optical microscope as a viewing tool. In turn, in robot-assisted surgery, the surgical tool itself is modified to be a miniature robotic system. The surgeon controls this tool to perform a hands-on surgical procedure, while the robotic tool offers tremor cancellation, depth lock, and other features. An example of this type is the portable robotic manipulator from the Robotics Institute of Carnegie Mellon University and Johns Hopkins University—"THE MICRON" [4].

In teleoperated robotic surgery, the surgeon controls a robotic system using joysticks and uses an optical microscope or a digital heads-up display to obtain a view of the surgical field. Joystick movement is mapped directly to robotic movement, and therefore advantages such as jitter filtering and motion scaling can be implemented. Examples of this type include the system from the department of robotics and embedded systems at the Technical University of Munich "RAM!S" and the system from the University of California "The intraocular Robotic Intervention Surgery System (IRISS)" [5].

In a cooperative robotic system, the surgeon holds and controls the surgical tool simultaneously with the robotic system and uses a microscope and/or OCT to obtain visualization of the surgical field. The surgeon maintains direct manual control over the movement of the instruments, while the robotic system provides assisted compensation for hand tremor and allows for prolonged immobilization of the surgical instruments. An example of this type is the system of the "Catholic University of Leuven" [3].

Finally, in a partially or fully automated system, the robotic system is fully integrated with the microscope and/or OCT to provide visualization of the surgical field and guidance to the movement commands of the robotic system, which directly holds and operates the surgical tools. Specific procedures or steps of a procedure are automatically performed by the robotic system while the surgeon supervises through the visual feedback provided. Replacement commands are commonly offered. An example of this type is IRISS [6].

## Ophthalmological surgical systems

Currently, there are only a few robotic surgical systems developed specifically for ophthalmological procedures. Some of these systems include:

MICRON was developed in 2010 through a collaboration between the Robotics Institute at Carnegie Mellon University and Johns Hopkins University. It consists of an active handheld micromanipulator that, through an image guidance approach, reduces hand tremor to provide smooth movement during surgical procedures. To date, all evaluation studies have been performed only on artificial or animal eye models [7].

The Robotic Retinal Dissection Device (R2D2): This system was developed by the Johns Hopkins University School of Medicine to assist surgeons in performing delicate and precise dissections of the retina. The R2D2 system uses a robotic arm to manipulate a microsurgical instrument, allowing the surgeon to make controlled incisions in the retina [8].

The da Vinci Surgical System: Although not designed specifically for ophthalmic surgery, the da Vinci Surgical System has been used in a limited capacity for some ophthalmic procedures, such as removing certain types of eye tumors. The system uses robotic arms and instruments to perform minimally invasive surgeries [2].

The Preceyes Surgical System: This is a robotic-assisted surgical system developed by the Dutch company Preceyes BV, and among all, it is the one with the largest number of works and manuscripts carried out and the most promising, especially in vitreoretinal surgeries and therefore it is worth a more detailed description about the system and its current use. The Preceyes Surgical System (PSS) is a robot designed for ophthalmological procedures, with a very high degree of precision and delicacy, especially for retinal surgeries and also intra-vitreal injections. The system consists of a console with control and a robotic arm, which is mounted on a tower above the surgical table. The control console is used by the surgeon to control the movement of the robotic arm inside the eye. One of the advantages, especially for retinal surgeries, is the high degree of precision and stability, thus preventing touches to the retina, or other damage that could be caused by inadvertent movements of the surgeon during the procedure. Another advantage would be the possibility of programming repeated movements, avoiding greater wear and tear on the surgeon [9,10].

The use of PSS was compared to surgery performed exclusively by surgeons experienced in vitreoretinal surgeries, demonstrating that the final surgical result was satisfactory, with no statistical difference between surgical complications or adverse effects [11]. Another study demonstrated safety in performing the steps of the surgery. Epiretinal membrane peeling using PSS and demonstrated that although surgical time was longer in the robot-assisted group (which probably results from learning and adaptation to the use of technology), during the study, surgical time in robot-assisted surgeries decreased from 72 to 46 min. On the other hand, the distance covered by the clamp was shorter in the group assisted by the PSS, which may suggest greater efficiency in the future with the help of this device. Central retinal thickness and best corrected visual acuity were equal in both groups in this study [4]. In another recent study, Cereda et al. evaluated the effectiveness of an instrument-integrated OCT-based distance sensor in retinal surgeries using the PSS and demonstrated that its use is reliable and could be a promising step in this type of surgery [6].

## Limiting factors

There are factors that limit the progress of robotic technology in ophthalmology. One of the limitations is the cost, as the technology is still quite expensive and may not be accessible to all patients. Furthermore, there is a significant learning and training process required for surgeons to be able to effectively utilize these robotic surgical systems, the learning curve is long in many cases and in high-precision, high-performance surgeries such as ophthalmology, where skilled surgeons achieve excellent results such as in cataract surgery, often with short surgical times, it is difficult for surgeons to accept and use these new technologies.

Another limitation is the lack of sufficient studies to evaluate the efficacy and safety of these technologies compared to conventional surgical techniques.

## Conclusion

Ophthalmic applications of robot-assisted surgery are still proving their identity and what they can accomplish. First evaluated for cataract surgery, these technologies have moved into vitreoretinal surgeries, with the first studies in humans. This new world of surgery presents numerous challenges for those developing robotic platforms. Surgeons deal with inadequate spatial resolution and depth perception of microstructures, natural hand tremor, and difficulty feeling force. Several barriers still need to be overcome before robotic technology reaches its full surgical potential in ophthalmology.

## Future scenario

The trend is that as research and studies advance, the use of robotics in ophthalmological surgeries tends to grow, especially in vitreoretinal surgeries, which are more delicate and require increasingly greater precision, especially due to the potential of the robotic arm to allow a high degree of precision and stability, reducing tremors and thus preventing touches to the retina, or other damage that could be caused by inadvertent movements of the surgeon during the procedure. In this sense, in the current scenario, the PSS is the one with the largest number of works carried out and the most promising.

Another trend is the integration with artificial intelligence, which can play a role in helping surgeons make real-time decisions based on advanced data and analysis, further improving outcomes.

## Key points

- Robot-assisted surgeries offer substantial improvements in terms of motion control, tremor cancellation, enhanced visualization, and better sense of distance and have enormous potential in ophthalmic surgeries due to the high degree of precision, delicacy, and motion control inside the eye.
- The high cost, limited availability, need for specialized microscopes and visualization systems challenges the application of robotics in ophthalmic surgery.
- The use of robotics in ophthalmological surgeries is more advanced in vitreoretinal surgeries, with the first studies on humans already carried out. In this sense, in the current scenario, the Preceyes Surgical System is the one with the largest number of works carried out and the most promising.

## References

[1] Hockstein NG, Gourin CG, Faust RA, Terris DJ. A history of robots: from science fiction to surgical robotics. J Robotic Surg 2007;1:113–8.
[2] Pandey SK, Sharma V. Robotics and ophthalmology: are we there yet? Indian J Ophthalmol 2019;67(7):988–94.
[3] Gerber MJ, Pettenkofer M, Hubschman JP. Advanced robotic surgical systems in ophthalmology. Eye 2020;34:1554–62.
[4] MacLachlan RA, Becker BC, Tabarés JC, Podnar GW, Lobes Jr LA, Riviere CN. Micron: an actively stabilized handheld tool for microsurgery. IEEE Trans Robot 2012;28(1):195–212. https://doi.org/10.1109/TRO.2011.2169634. PMID: 23028266. Epub 2011 Nov 18. PMCID: PMC3459696.
[5] Wilson JT, Gerber MJ, Prince SW, Chen CW, Schwartz SD, Hubschman JP, Tsao TC. Intraocular robotic interventional surgical system (IRISS): Mechanical design, evaluation, and master-slave manipulation. Int J Med Robot 2018;14(1). https://doi.org/10.1002/rcs.1842. Epub 2017 Jul 31 PMID: 28762253.
[6] Cereda MG, Parrulli S, Douven YGM, Faridpooya K, van Romunde S, Hüttmann G, Eixmann T, Schulz-Hildebrandt H, Kronreif G, Beelen M, de Smet MD. Clinical evaluation of an instrument-integrated OCT-based distance sensor for robotic vitreoretinal surgery. Ophthalmol Sci 2021;1(4), 100085. https://doi.org/10.1016/j.xops.2021.100085. PMID: 36246942. PMCID: PMC9560530.
[7] Faridpooya K, van Romunde SHM, Manning SS, van Meurs JC, Naus GJL, Beelen MJ, Meenink TCM, Smit J, de Smet MD. Randomised controlled trial on robot-assisted versus manual surgery for pucker peeling. Clin Exp Ophthalmol 2022;50(9):1057–64. https://doi.org/10.1111/ceo.14174. Epub 2022 Oct 17 PMID: 36177965.

[8] Vander Poorten CNE, RiviereJJ ACB, et al. Robotic retinal surgery. In: Handbook of robotic and image-guided surgery. Elsevier; 2020. p. 627–72.

[9] He B, Smet MD, Sodhi M, et al. A review of robotic surgical training: establishing a curriculum and credentialing process in ophthalmology. Eye 2021;35:3192–201.

[10] https://www.preceyes.nl/.

[11] Edwards TL, Xue K, Meenink HC, Beelen MJ, Naus GJ, Simunovic MP, et al. First-in-human study of the safety and viability of intraocular robotic surgery. Nat Biomed Eng 2018;2:649.

# Certification and credentialing for robotic surgery, a developed country scenario: United States

*Chandru P. Sundaram and Courtney Yong*

Indiana University School of Medicine, Indianapolis, IN, United States

## Introduction

Since its entry into the market in 2000, the use of robotic surgery continues to increase in many surgical specialties including general surgery, urology, and obstetrics and gynecology [1–3]. More recently, it has begun gaining traction in other specialties such as otolaryngology [4]. Expectedly, as the number of robotic surgeries performed rises, surgical training has increasingly included training in robotic surgery [5–7]. Despite this, the training and credentialing process for robotic surgery has lagged behind its widespread integration into practice and residency education [8]. Robotic surgery is included in society-specific curricula, such as the American Board of Surgery's (ABS) SCORE curriculum and the American Urological Associations (AUAs) Core Curriculum, and the American College of Obstetricians and Gynecologists (ACOG) has recommendations for training in robotic surgery and credentialing in new technologies [9–11]. However, there are no specific criteria for credentialing, and multiple needs assessments have shown no standardized robotic surgery curriculum in surgical residency programs across specialties [7,12]. This chapter outlines the existing methods of credentialing and granting privileges for robotics to surgeons and the future of robotic surgery training.

## The learning curve

The learning curve in surgery is a period during which a surgeon is still acquiring skills and experience in performing a procedure [13]. It has been documented that surgeons have worse outcomes early in their learning curve, and complication rates and surgical outcomes improve as the surgeon gains more experience [13,14]. Minimally invasive surgery has previously been under scrutiny for evaluation of the learning curve, and robotics is no exception [15]. One proposed benefit of robotics as compared to laparoscopy is a shortened learning curve partially due to ease of use [16]. The learning curve for several robotic procedures has been studied across specialties [15,17,18]. The goal of training in robotic surgery is to minimize the effects of the learning curve while in independent practice. Credentialing in robotic surgery is one way to maintain this baseline standard of training, with the goal to minimize serious adverse events.

Credentialing in robotic surgery became a focus of public attention in 2013, when the US Food and Drug Administration (FDA) noted an increased number of reports filed on medical devices including the surgical robot [19]. It was unclear whether this was related to a true increase in complication rates or an increase in device use and improved accuracy of reporting. At the same time, a quality improvement study found that 3% of robotic surgery complications were improperly reported to the FDA [20]. This expectedly led to the media and public to question the safety of robotic surgery as well as the reliability of training and validity of credentialing. In response, some state medical boards developed recommendations for robotic surgery training and credentialing [19].

Updated reports have shown that robotic surgery has comparable outcomes to other methods of surgery, and robotic surgery may even have advantages such as decreased blood loss and length of hospital stay [21–24]. With these findings, the use robotic surgery continues to grow in frequency and across specialties. However, the question regarding appropriate training and credentialing for robotic surgeons remains, and whether current credentialing requirements are enough to maintain a standard to minimize adverse patient events.

## Current board certification and credentialing requirements

As the use of robotics in surgery encompasses a wide variety of specialties, there are several specialty boards that certify surgeons that perform robotic surgery. Examples include national boards such as the American Board of Surgery (ABS), American Board of Urology (ABU), and the American Board of Obstetrics and Gynecology (ABOG). However, these boards simply certify surgeons to practice within their specialty and do not have specific recommendations for certification in robotic surgery. Within each specialty, there are societies dedicated to minimally invasive surgery and robotics such as the Society of American Gastrointestinal and Endoscopic Surgeons and Minimally Invasive Robotic Association (SAGES-MIRA), Society of Robotic Urologic Surgeons (SURS), and American Association of Gynecologic Laparoscopists (AAGL), and each offers and facilitates additional specialty training fellowships for interested surgeons. While surgeons can apply for membership to these societies, there are no true board certifications for robotic surgery.

Although there are no board certifications for robotic surgery, hospitals are required to assess practicing physicians in the credentialing process. The Joint Commission requires individual institutions to have specific credentialing requirements [25]. However, the specific components required for credentialing remain at the discretion of each individual institution. There have been attempts by both surgical societies and other groups of experts to standardize the credentialing requirements, but credentialing remains heterogeneous among institutions [25,26].

Of the various boards and societies, SAGES-MIRA, SURS, and the AUA have issued formal statements on robotic-specific surgery training and credentialing [27–29]. Both the SAGES-MIRA and SURS statements were published during the early years of robotic surgery and have not been updated since. Indeed, since the SAGES-MIRA and SURS statements were issued, four models of robots have been released by Intuitive Surgical (Sunnyvale, CA, USA), the Si, Xi, X, and SP. The AUA recommendations have been updated as recently as 2018 by the board of directors. A summary of each of these consensus statements is found in Table 1.

The SAGES-MIRA consensus puts forth minimum requirements for hospitals to grant privileges to surgeons to perform robotic surgery. These requirements are: (1) Satisfactory completion of an accredited surgical residency program, (2) Formal training either during residency or at a structured program after residency, and (3) Documentation of

TABLE 1　Recommended criteria for obtaining robotic surgery privileges, by society recommendations.

| Criteria | Included in recommendation | | |
| --- | --- | --- | --- |
| | **SAGES-MIRA** | **SURS** | **AUA** |
| Recommendations apply to all robotics procedures | X | | X |
| Completion of residency | X | X | X |
| Robotics experience documented in residency | X | X | X |
| Completion of online courses/modules for robotic surgery | | X | X |
| Case log of a certain number of robotics cases | X | X | X |
| Completion of Intuitive Surgical's robotics curriculum | | | X |
| Privileges for open surgery | X | | X |
| Hands-on dry lab experience | X | | X |
| Proctoring of initial cases | X | X | X |
| Qualifications of proctor defined in the recommendation | | | X |
| Discussion of legality of proctoring | | X | X |
| Informed consent from the patient | | X | X |
| Recommendations regarding how to maintain privileges | X | X | X |

practical experience including both submission of case logs and in-person proctoring [27]. The statement also recommends continued monitoring of performance through hospital-specific quality assurance mechanisms and continuing medical education programs [27]. It does not make specific recommendations regarding the fulfillment of these requirements.

The SURS statement is specific to credentialing for robotic-assisted radical prostatectomy. It recommends: (1) Completion of formal training either in residency or a "mini-residency program," (2) Proctoring during the "first few cases" of a robotic surgeon's career with an institution-defined role for the proctor, and (3) A ongoing maintenance of privileges by a robotic committee review of the performance of the surgeon [28]. However, similarly to the SAGES-MIRA statement, it does not make more detailed recommendations to institutions for surgeon credentialing and continued evaluation of surgeon performance.

The AUA statement, which was developed in 2009 and most recently updated in 2018, put forth similar requirements [29]. Specifically, they outline two pathways by which they recommend a surgeon obtain robotic surgery privileges. The first and most common method in the current training landscape is that a surgeon must have completed an accredited urology residency during which they have completed at least 20 cases and at least 10 of which they were on the console during "key portions of the procedure." With this method, the AUA recommends a letter from the program director to confirm adequate training for independent robotics practice. The second method outlined is for those surgeons who did not undergo formal training in robotic surgery. The AUA recommends that these surgeons complete the AUA's Fundamentals of Urologic Robotic Surgery module, one of the Basic Procedure Modules of the AUA's Robotic Surgery Online Course, and completion of Intuitive Surgical's Online System Training module, which is discussed in further detail later in this chapter. Additionally, the AUA recommends that these surgeons have privileges for an open surgical approach for any procedure they are planning on performing robotically, and that the surgeon is instructed in an in-person, hands-on experience, proctored by an experienced robotic surgeon, and finally undergo a review of surgical outcomes after their initial robotic experience with a panel of peers at their institution [29]. Regardless of the method of obtaining privileges, the AUA recommends maintaining privileges by monitoring outcomes and surgical volume as well as participating in continuing medical education [29].

Intuitive Surgical, the company responsible for the only FDA-approved robotic surgical system in the United States, also provides some recommended criteria for credentialing. While the Intuitive website has a specific disclaimer regarding medicolegal ramifications of training [30], there is also a published list of six criteria for obtaining a Certificate of da Vinci System Training for Residents and Fellows [31]. The requirements are a combination of online modules and in-person training with a representative from Intuitive (Table 2) [31].

There are ethical and legal questions that have been raised regarding the training of surgeons via the medical device manufacturer [8,32,33]. Despite this, many institutions include Intuitive Surgical's aforementioned Certificate to fulfill some or all of their credentialing requirements [8].

However, with several guidelines available but no governing body, credentialing requirements for robotic surgeons remains highly variable among institutions, as each institution sets its own credentialing requirements. The credentialing requirements at our institution can be found in Fig. 1 as an example, and generally, these requirements match the recommendations by SAGES-MIRA, SURS, and AUA. Our institution specifies the credentials of a proctor; a proctor

TABLE 2 Criteria for obtaining a "Certificate of da Vinci® System Training," according to Intuitive Surgical.

| Criteria | Method of completion |
|---|---|
| Criteria 1: Completion of da Vinci System Online Training, for Surgeons, Residents, and Fellows | Video-based training online covering system component overviews, instructions in the use of system components, instruments, accessories, and advanced technologies. |
| Criteria 2: Completion of da Vinci System Overview In-Service Training | Hands-on training with an Intuitive representative reviewing the topics covered in Criteria 1. |
| Criteria 3: Completion of da Vinci System Online Assessment | An online assessment to check the skills learned in Criteria 1 and 2. |
| Criteria 4: Performance of da Vinci Procedures in Primary Role of Console Surgeon (20 recommended) and Primary Role of Bedside Assistant (10 recommended) | Live surgery case log, intended to be performed under supervision while in residency. |
| Criteria 5: Solicit Letter of Verification of da Vinci System Training and Procedures Completed from Chief of Surgery or Program Director | Letter from the program director or other authority to endorse appropriate surgical training at the completion of residency. |
| Criteria 6: Submit Copy of Online Training Certificate, Case Log, and Letter of Verification to an Intuitive Representative | Obtain the certificate by completing the prior criteria. |

**ROBOTIC PRIVILEGE SECTION-** General Surgery, Bariatric, OB/GYN, & Urology
Previous Revisions May 2022 & July 2022
Latest Revision: September 2022

| | |
|---|---|
| **Education/Training** | **ALL APPLICANTS:** Completion of an ACGME or AOA accredited residency or fellowship training program in a surgical discipline.<br><br>**AND**<br><br>**Pathway #1:** For new surgeons with ACGME residency or fellowship training in robotics, a letter from the residency or fellowship program director of the trained specialty indicating competency in performing robotic procedures.<br><br>**OR**<br><br>**Pathway #2:** For transferring robotic surgeon currently performing robotic procedures, a letter from the service line leader of the trained specialty indicating competency in performing robotic procedures.<br><br>**OR**<br><br>**Pathway #3:** Surgeon must provide documentation of successful completion of the training course for (manufacturer's designed course) for the specific device. |
| **Clinical Experience (Initial)** | **ALL APPLICANTS:** Provisional privileges will be granted until 3 (or more) proctored cases are performed in the primary console role with the following expectations. 1) Meet all standard performance criteria for a minimum of the last two consecutive procedures performed. (See Appendix A for standard performance criteria.) 2) Optimal outcomes are achieved per review by the specialty service line leader at the primary facility the robotic procedures are performed.<br><br>**AND**<br><br>**Pathway #1:** For new surgeons with residency &/or fellowship training in robotics with 20 procedures performed in the primary console role within the training period on a similar platform (example Intuitive XI, Si, SP, etc.) being used at IU Health.<br><br>**OR**<br><br>**Pathway #2:** Transferring robotic surgeon currently performing robotic procedures with experience of a minimum of 20 procedures performed in the primary console role.<br><br>**OR**<br><br>**Pathway #3:** Currently credentialed surgeon interested in performing robotic procedures. |
| **Additional Qualifications** | Surgeon must have clinical privileges for the corresponding open and laparoscopic procedures that will be performed robotically.<br><br>A proctor for robotics procedures is required to perform a minimum of 40 procedures during a two-year appointment period. |

FIG. 1   Robotic credentialing requirements at Indiana University. The credentialing requirements for robotic surgeons at our institution include most of the recommended requirements from SAGES-MIRA, SURS, and AUA.

must have performed at least 40 procedures in a 2 year period. Additionally, our institution provides a specific evaluation form as a guide for the proctor to use to determine surgeon competency (Fig. 2).

The heterogeneity of institutional credentialing policies was highlighted in a study by Huffman et al., which reviewed the credentialing requirements for robotic general surgeons at 42 institutions from 24 different states: Only 38% of the institutions required applicants to have completed an accredited residency program and only 60% required board eligibility or certification [8]. Other discrepancies included whether institutions required proctoring and what constituted proctoring, and the requirements for maintaining privileges including average number of robotic cases and maximum allowable time gap between robotic cases [8]. Indeed, this heterogeneity among credentialing requirements at different institutions is seen in other specialties [34,35]. Therefore, the question remains: Is board certification in a given surgical specialty enough to provide baseline quality for robotic surgeons? This, of course, depends on the quality of surgical residency training.

## ROBOTICS PROCTOR REPORT

*NOTE: Once you have completed this form, please return it confidentially to __{to be determined by local privileging committee}__ .*

Proctoring Physician: _____

Primary Physician/Operator: _____

Procedure Performed: _____

Hospital: _____

Robotic Platform (e.g., da Vinci Si, Xi): _____

Please rate the console surgeon's performance in the following areas:

| | Meets Standards | More Practice Recommended | N/A |
|---|---|---|---|
| **Preparation and setup** | | | |
| • Positions patient safely in a manner appropriate for procedure | | | |
| • Configures ports to allow for appropriate spacing and inserts ports to appropriate depth avoiding inadvertent injury | | | |
| • Able to direct successful docking | | | |
| • Selects instruments appropriate for procedure | | | |
| **Skills** | | | |
| • Directs scope/camera to maintain safe view of the operative field | | | |
| • Safely handles tissue | | | |
| • Uses surgical energy safely | | | |
| • Sutures and/or ties knots safely and effectively | | | |
| • Independently able to troubleshoot common problems (ie instrument collisions, etc.) and direct solutions with effective team communication | | | |
| • Uses console adjustments and clutching to maintain appropriate ergonomic body positioning (surgeon hands /arms in comfortable position and not colliding with console) | | | |
| **Quality** | | | |
| • Performs all steps of the procedure safely and effectively | | | |
| • Performs all steps of the procedure with reasonable efficiency | | | |

## ROBOTICS PROCTOR REPORT cont.

**Is this surgeon ready to operate on their own?**
Yes, no need for additional proctoring          No, needs additional proctoring

**Feedback on areas where more practice is needed to meet standards (mandatory if No chosen for ready to operate on their own):**

I understand this information is confidential and will be used for the purpose of evaluating surgical skills and competency. This report will be maintained in the ____{to be determined by local privileging committee}____ .

**Proctoring Physician's Signature** _____          **Date** _____

FIG. 2  Proctor Evaluation Form at Indiana University. Proctors are given specific guidelines for determining surgical competency in robotic surgery.

## Residency curriculum and training in robotic surgery

For both general surgery and obstetrics and gynecology trainees, there is a set curriculum for laparoscopic skills as outlined in the Fundamentals of Laparoscopic Surgery (FLS) course [36]. As of 2008, the ABS has required residency graduates to pass the FLS exam, a test of basic laparoscopic skills, in order to be board-eligible [37]. By contrast, although many surgical specialties now use the robot, the amount of exposure to robotic surgery for surgical trainees is variable between specialties and among programs, and there is no universally adopted curriculum or assessment [38].

III. The current and future clinical applications of robotic surgery among medical specialties

Needs assessments performed in every surgical specialty have identified a significant gap between the adoption and now widespread use of robotic surgery in practice and the exposure and comfort of trainees with the platform [7,12,39–41]. While most surgical training programs across specialties include exposure to robotics, some residents graduate feeling underprepared to perform robotics procedures in independent practice [7,12,39,41]. Despite many programs reporting a robotic curriculum, there is both heterogeneity in the curriculum among programs as well as inconsistency of resident experience with the curriculum in a single program [7,12,39–41].

In 2011, the Fundamentals of Robotic Surgery (FRS) course, similar to the FLS curriculum, was developed by a panel of experts with the goal to fill the need for a standardized robotic surgery training program [42]. A study published in 2020 then validated its efficacy in a multispecialty, multi-institutional, randomized controlled trial [43]. However, there are no reports as to how many residency programs use the FRS course in practice, and other institutions and groups have both proposed and use independent robotics curricula [5]. Despite the efforts for its development and validation, the FRS course is not yet the robotics-equivalent of the FLS course for laparoscopy.

Besides FRS, several other independent courses have been developed including Intuitive's own robotic training curriculum, the Robotics Masters Series developed by SAGES, the Robotics Training Network curriculum, and the Fundamental Skills of Robotic-Assisted Surgery Training Program [38]. Despite the overall heterogeneity of the curricula used for robotic surgery training, there are three similar themes in all curricula: (1) Learning how to operate the robotic functions, (2) Simulation to develop robotic surgical skills, and (3) Supervised participation in live surgery. Learning how to operate the robot is taught in dry lab and is relatively homogeneous, often being facilitated by Intuitive Surgical [12]. However, both simulation and supervised operative time are both areas of continued study as to the optimal balance of education and patient safety.

There are four main robotic surgery simulators on the market. These are: (1) da Vinci Skills Simulator (dVSS, Intuitive Surgical, Sunnyvale, CA, USA), (2) Robotic surgery simulator (RoSS, Simulated Surgical Systems, San Jose, CA, USA), (3) dV trainer (Mimic, Seattle, WA, USA), and (4) RobotiX Mentor (3D Systems, Simbionix Products, Cleveland, OH, USA) [19,38]. However, these systems can be cost-prohibitive, with the dVSS costing $80,000, the RoSS costing $125,000, the dV trainer costing $110,000, and the RobotiX Mentor costing $137,000 [44,45]. These prices do not include cost of annual maintenance, estimated at $12,500, additional software and learning programs, or other costs of simulation and training labs such as equipment or animal laboratories [19,45]. Therefore, programs can also achieve simulation by using the live robot and objects such as pegboards, chicken skin, and hydrogel models [19].

Each of these simulation technologies and techniques has been studied in trainees across surgical specialties [46–51]. Although they have been compared to each other and each individually validated, there is no clear best system [46–50]. While it has been repeatedly confirmed that simulation is extremely important in developing robotic surgical skill, as previously mentioned, no standardized curriculum has emerged to guide the use of simulation in education. Therefore, the exact methods by which to use simulation in surgical education remain an area of interest.

The use of simulation in evaluation of surgical skill has also emerged as a topic of interest in surgical education literature [52]. Indeed, evaluation and assessment of surgical trainees remains an area of potential study. However, like credentialing, robotic surgery curricula, and simulation in education, the evaluation and assessment of robotic surgical skill is heterogeneous between training programs, and there is no standardized method of skills assessment [53]. Several proposed methods include manual evaluation of trainees by attendings using assessment scales, procedure-specific assessment to evaluate a trainee's competence in a specific portion of a procedure, and automated assessment including simulation performance [53]. However, none has emerged as a standardized method of evaluation and assessment of surgeon performance, and an objective measure of progression along the surgical learning curve during training and in early career remains elusive.

## The future scenario

## Future of robotic surgery: Training and credentialing new technology

Both education and credentialing practices have lagged behind the exponential growth of robotic surgery, but the field of robotic surgery continues to grow with the development of new techniques, applications, and technologies. New applications of existing robotic surgical technology have been developed and performed on live patients [54–57]. Besides the development of new surgical techniques using the robot, new robotic systems have been released onto the market, including the single port robot by Intuitive Surgical and other surgical robots such as the Hugo Robotic Assisted Surgery (Hugo RAS) system (Medtronic, Minneapolis, MN, USA) and Senhance robotic system (TransEnterix Surgical Inc., Morrisville, NC, USA). The single port robot has been used for lives surgeries in the United

States, while the other two robotic surgery systems have to date only been used abroad due to lack of FDA approval. Internationally, other robotics systems have also developed, including the Micro Hand S, a Chinese-produced surgical robot developed by Central South University (Changsha, Hunan, China) and Tianjin University (Tianjin, China), the Revo-I developed by Meerecompany (Yongin, Republic of Korea), the SSI Mantra (SS Innovations, Gurugram, Haryana, India), and the Versius Robot (CMR Surgical, Cambridge, United Kingdom) [58–64].

However, the development of new surgical techniques and technologies has brought forth ethical concerns regarding the use of these systems on patients [65,66]. This also brings into question the reliability of the already-inconsistent credentialing requirements for robotic surgeons and whether they truly protect against adverse events. Without well-defined credentialing requirements and little oversight by medical boards, surgeons are free to use medical devices in innovative ways. This can significantly advance the field of surgery and lead to technological and medical breakthroughs. It can also increase the risk to patients undergoing surgery, as the learning curve for these new surgeries is even less well defined than that of routine procedures [66,67].

Both SAGES and ACOG had the foresight to write about the consideration of privileges in the setting of future development of novel procedures and technologies. In their respective guidelines (SAGES) and guiding principles (ACOG) documents, written in 2015 and 2014 respectively, the societies outline how surgeons should engage with new technologies and how hospital systems might consider privileges for performing procedures or using technology that has not previously existed [68,69]. While the guidelines each go into different levels of depth, a few key themes emerge from both: (1) Patient selection and implementation of the new technique to benefit patient outcomes, (2) Informed consent, and (3) Maximizing surgeon education and training in the new technology or technique prior to use [68,69].

As new surgical technologies and techniques continue to emerge, the questions of privileges, credentialing, and training will further deepen. Currently, credentialing and surgical privileges are governed by individual institutions with only recommendations from surgical societies, but research continues to explore methods for better and more consistent training. Regardless, protecting patient safety and improving outcomes is the ultimate goal for the training, board certification, and credentialing processes.

## Key points

- While robotic surgical use and technology has experienced rapid growth and widespread use, the training and credentialing process has lagged.
- Professional societies in urology, obstetrics and gynecology, and general surgery have proposed general guidelines for the credentialing of robotic surgeons, but surgeon credentialing has been heterogeneous among institutions.
- With new robotic surgical technology in development, professional societies, medical boards, and individual institutions must continue to critically evaluate the credentialing and privileging process to protect patient safety and improve outcomes.

## References

[1] Sheetz KH, Claflin J, Dimick JB. Trends in the adoption of robotic surgery for common surgical procedures. JAMA Netw Open 2020;3(1), e1918911. https://doi.org/10.1001/jamanetworkopen.2019.18911 [published Online First: 03.01.2020].
[2] Shah AA, Bandari J, Pelzman D, et al. Diffusion and adoption of the surgical robot in urology. Transl Androl Urol 2021;10(5):2151–7. https://doi.org/10.21037/tau.2019.11.33.
[3] Sinha R, Sanjay M, Rupa B, et al. Robotic surgery in gynecology. J Minim Access Surg 2015;11(1):50–9. https://doi.org/10.4103/0972-9941.147690.
[4] Sharma A, Bhardwaj R. Robotic surgery in otolaryngology during the Covid-19 pandemic: a safer approach? Indian J Otolaryngol Head Neck Surg 2021;73(1):120–3. https://doi.org/10.1007/s12070-020-02032-3 [published Online First: 05.08.2020].
[5] George LC, O'Neill R, Merchant AM. Residency training in robotic general surgery: a survey of program directors. Minim Invasive Surg 2018;2018:8464298. https://doi.org/10.1155/2018/8464298 [published Online First: 08.05.2018].
[6] Merrill SB, Sohl BS, Thompson RH, et al. The balance between open and robotic training among graduating urology residents-does surgical technique need monitoring? J Urol 2020;203(5):996–1002. https://doi.org/10.1097/JU.0000000000000689 [published Online First: 11.12.2019].
[7] Vetter MH, Palettas M, Hade E, et al. Time to consider integration of a formal robotic-assisted surgical training program into obstetrics/gynecology residency curricula. J Robot Surg 2018;12(3):517–21. https://doi.org/10.1007/s11701-017-0775-0 [published Online First: 28.12.2017].
[8] Huffman EM, Rosen SA, Levy JS, et al. Are current credentialing requirements for robotic surgery adequate to ensure surgeon proficiency? Surg Endosc 2021;35(5):2104–9. https://doi.org/10.1007/s00464-020-07608-2 [published Online First: 06.05.2020].
[9] Arca MJ, Adams RB, Angelos P, et al. American Board of Surgery Statement on assessment and robotic surgery. Am J Surg 2021;221(2):424–6. https://doi.org/10.1016/j.amjsurg.2020.09.039 [published Online First: 17.10.2020].

[10] Committee opinion no. 628: robotic surgery in gynecology. Obstet Gynecol 2015;125(3):760–7. https://doi.org/10.1097/01. AOG.0000461761.47981.07.

[11] Gahan JC. AUA core curriculum: Laparoscopy and robotics, https://university.auanet.org/core/lap-robotics/laparoscopy-and-robotics/ index.cfm2022. [accessed 23 December 2022.

[12] Wang RS, Ambani SN. Robotic surgery training: current trends and future directions. Urol Clin North Am 2021;48(1):137–46. https://doi.org/ 10.1016/j.ucl.2020.09.014 [published Online First: 05.11.2020].

[13] Khan N, Abboudi H, Khan MS, et al. Measuring the surgical 'learning curve': methods, variables and competency. BJU Int 2014;113(3):504–8. https://doi.org/10.1111/bju.12197 [published Online First: 02.07.2013].

[14] Hatlie MJ. Climbing 'the learning curve'. New technologies, emerging obligations. JAMA 1993;270(11):1364–5.

[15] Pernar LIM, Robertson FC, Tavakkoli A, et al. An appraisal of the learning curve in robotic general surgery. Surg Endosc 2017;31(11):4583–96. https://doi.org/10.1007/s00464-017-5520-2 [published Online First: 14.04.2017].

[16] Flynn J, Larach JT, Kong JCH, et al. The learning curve in robotic colorectal surgery compared with laparoscopic colorectal surgery: a systematic review. Colorectal Dis 2021;23(11):2806–20. https://doi.org/10.1111/codi.15843 [published Online First: 15.08.2021].

[17] Mazzon G, Sridhar A, Busuttil G, et al. Learning curves for robotic surgery: a review of the recent literature. Curr Urol Rep 2017;18(11):89. https://doi.org/10.1007/s11934-017-0738-z [published Online First: 23.09.2017].

[18] Turner TB, Kim KH. Mapping the robotic hysterectomy learning curve and re-establishing surgical training metrics. J Gynecol Oncol 2021;32(4), e58. https://doi.org/10.3802/jgo.2021.32.e58 [published Online First: 12.04.2021].

[19] Bahler CD, Sundaram CP. Training in robotic surgery: simulators, surgery, and credentialing. Urol Clin North Am 2014;41(4):581–9. https:// doi.org/10.1016/j.ucl.2014.07.012 [published Online First: 22.08.2014].

[20] Cooper MA, Ibrahim A, Lyu H, et al. Underreporting of robotic surgery complications. J Healthc Qual 2015;37(2):133–8. https://doi.org/ 10.1111/jhq.12036.

[21] Jara RD, Guerron AD, Portenier D. Complications of robotic surgery. Surg Clin North Am 2020;100(2):461–8. https://doi.org/10.1016/j. suc.2019.12.008 [published Online First: 13.02.2020].

[22] Porpiglia F, Fiori C, Bertolo R, et al. Five-year outcomes for a prospective randomised controlled trial comparing laparoscopic and robot-assisted radical prostatectomy. Eur Urol Focus 2018;4(1):80–6. https://doi.org/10.1016/j.euf.2016.11.007 [published Online First: 23.11.2016].

[23] Prete FP, Pezzolla A, Prete F, et al. Robotic versus laparoscopic minimally invasive surgery for rectal Cancer: a systematic review and Meta-analysis of randomized controlled trials. Ann Surg 2018;267(6):1034–46. https://doi.org/10.1097/SLA.0000000000002523.

[24] Tsung A, Geller DA, Sukato DC, et al. Robotic versus laparoscopic hepatectomy: a matched comparison. Ann Surg 2014;259(3):549–55. https:// doi.org/10.1097/SLA.0000000000000250.

[25] Stefanidis D, Huffman EM, Collins JW, et al. Expert consensus recommendations for robotic surgery credentialing. Ann Surg 2022;276(1):88–93. https://doi.org/10.1097/SLA.0000000000004531 [published Online First: 17.11.2020].

[26] Bhora FY, Al-Ayoubi AM, Rehmani SS, et al. Robotically assisted thoracic surgery: proposed guidelines for privileging and credentialing. Innovations (Phila) 2016;11(6):386–9. https://doi.org/10.1097/IMI.0000000000000320.

[27] Herron DM, Marohn M, Group S-MRSC. A consensus document on robotic surgery. Surg Endosc 2008;22(2):313–25. discussion 11–2 https:// doi.org/10.1007/s00464-007-9727-5. [published Online First: 28.12.2007].

[28] Zorn KC, Gautam G, Shalhav AL, et al. Training, credentialing, proctoring and medicolegal risks of robotic urological surgery: recommendations of the society of urologic robotic surgeons. J Urol 2009;182(3):1126–32. https://doi.org/10.1016/j.juro.2009.05.042 [published Online First: 21.07.2009].

[29] American Urological Association. Robotic Surgery (Urologic) Standard Operating Procedure (SOP), https://www.auanet.org/guidelines-and-quality/guidelines/other-clinical-guidance/robotic-surgery-(urologic)-sop2018. [accessed 23 December 2022.

[30] Intuitive Surgical Inc. Product training disclaimer, https://www.intuitive.com/en-us/about-us/company/legal/product-training-disclaimer. [accessed 23 Decmeber 2022.

[31] Intuitive Surgical Inc. Certificate of da Vinci® system training for residents and Fellows, https://www.intuitive.com/en-us/-/Media/ISI/ Intuitive/pdf/equivalency-certificate-requirement-210352-.pdf2019. [accessed 23 December 2022].

[32] Fairhurst C. Taylor v Intuitive Surgical, https://www.courts.wa.gov/opinions/pdf/922101.pdf2017. [accessed 23 December 2022.

[33] McLean T. The complexity of litigation associated with robotic surgery and cybersurgery. Int J Med Robot 2007;3:23–9. https://doi.org/ 10.1002/rcs.121.

[34] Erickson BK, Gleason JL, Huh WK, et al. Survey of robotic surgery credentialing requirements for physicians completing OB/GYN residency. J Minim Invasive Gynecol 2012;19(5):589–92. https://doi.org/10.1016/j.jmig.2012.05.003 [published Online First: 06.07.2012].

[35] Administration FaD. Medical product safety network (MedSun) small sample survey – final report, topic: da Vinci surgical system, https:// www.fda.gov/media/87485/download2013. [accessed 23 December 2022.

[36] https://www.flsprogram.org [accessed 23 December 2022.

[37] Surgery ABo. ABS to require ACLS, ATLS and FLS for general surgery certification, https://www.absurgery.org/default.jsp?news_ newreqs2008. [accessed 23 December 2022.

[38] Chen R, Rodrigues Armijo P, Krause C, et al. A comprehensive review of robotic surgery curriculum and training for residents, fellows, and postgraduate surgical education. Surg Endosc 2020;34(1):361–7. https://doi.org/10.1007/s00464-019-06775-1 [published Online First: 05.04.2017].

[39] Okhunov Z, Safiullah S, Patel R, et al. Evaluation of urology residency training and perceived resident abilities in the United States. J Surg Educ 2019;76(4):936–48. https://doi.org/10.1016/j.jsurg.2019.02.002 [published Online First: 23.02.2019].

[40] Khalafallah YM, Bernaiche T, Ranson S, et al. Residents' views on the impact of robotic surgery on general surgery education. J Surg Educ 2021;78(3):1007–12. https://doi.org/10.1016/j.jsurg.2020.10.003 [published Online First: 20.10.2020].

[41] Shaw RD, Eid MA, Bleicher J, et al. Current barriers in robotic surgery training for general surgery residents. J Surg Educ 2022;79(3):606–13. https://doi.org/10.1016/j.jsurg.2021.11.005 [published Online First: 26.11.2021].

[42] Smith R, Patel V, Satava R. Fundamentals of robotic surgery: a course of basic robotic surgery skills based upon a 14-society consensus template of outcomes measures and curriculum development. Int J Med Robot 2014;10(3):379–84. https://doi.org/10.1002/rcs.1559 [published Online First: 26.11.2013].

[43] Satava RM, Stefanidis D, Levy JS, et al. Proving the effectiveness of the fundamentals of robotic surgery (FRS) skills curriculum: a single-blinded, multispecialty, Multi-institutional Randomized Control Trial. Ann Surg 2020;272(2):384–92. https://doi.org/10.1097/SLA.0000000000003220.

[44] Hertz AM, George EI, Vaccaro CM, et al. Head-to-head comparison of three virtual-reality robotic surgery simulators. JSLS 2018;22(1). https://doi.org/10.4293/JSLS.2017.00081.

[45] Rehman S, Raza SJ, Stegemann AP, et al. Simulation-based robot-assisted surgical training: a health economic evaluation. Int J Surg 2013;11(9):841–6. https://doi.org/10.1016/j.ijsu.2013.08.006 [published Online First: 27.08.2013].

[46] Walliczek-Dworschak U, Mandapathil M, Fortsch A, et al. Structured training on the da Vinci skills simulator leads to improvement in technical performance of robotic novices. Clin Otolaryngol 2017;42(1):71–80. https://doi.org/10.1111/coa.12666 [published Online First: 15.05.2016].

[47] Brinkman WM, Luursema JM, Kengen B, et al. da Vinci skills simulator for assessing learning curve and criterion-based training of robotic basic skills. Urology 2013;81(3):562–6. https://doi.org/10.1016/j.urology.2012.10.020 [published Online First: 04.01.2013].

[48] Kesavadas T, Stegemann A, Sathyaseelan G, et al. Validation of robotic surgery simulator (RoSS). Stud Health Technol Inform 2011;163:274–6.

[49] Yang K, Zhen H, Hubert N, et al. From dV-trainer to real robotic console: the limitations of robotic skill training. J Surg Educ 2017;74(6):1074–80. https://doi.org/10.1016/j.jsurg.2017.03.006 [published Online First: 24.04.2017].

[50] Leijte E, Claassen L, Arts E, et al. Training benchmarks based on validated composite scores for the RobotiX robot-assisted surgery simulator on basic tasks. J Robot Surg 2021;15(1):69–79. https://doi.org/10.1007/s11701-020-01080-9 [published Online First: 20.04.2020].

[51] Raison N, Gavazzi A, Abe T, et al. Virtually competent: a comparative analysis of virtual reality and dry-lab robotic simulation training. J Endourol 2020;34(3):379–84. https://doi.org/10.1089/end.2019.0541.

[52] Mills JT, Hougen HY, Bitner D, et al. Does robotic surgical simulator performance correlate with surgical skill? J Surg Educ 2017;74(6):1052–6. https://doi.org/10.1016/j.jsurg.2017.05.011 [published Online First: 13.06.2017].

[53] Chen J, Cheng N, Cacciamani G, et al. Objective assessment of robotic surgical technical skill: a systematic review. J Urol 2019;201(3):461–9. https://doi.org/10.1016/j.juro.2018.06.078.

[54] Batailler C, Hannouche D, Benazzo F, et al. Concepts and techniques of a new robotically assisted technique for total knee arthroplasty: the ROSA knee system. Arch Orthop Trauma Surg 2021;141(12):2049–58. https://doi.org/10.1007/s00402-021-04048-y [published Online First: 13.07.2021].

[55] Perrier ND, Randolph GW, Inabnet WB, et al. Robotic thyroidectomy: a framework for new technology assessment and safe implementation. Thyroid 2010;20(12):1327–32. https://doi.org/10.1089/thy.2010.1666.

[56] Usuda J, Inoue T, Sonokawa T, et al. New technique for introducing a surgical stapler during robot-assisted lobectomy for lung Cancer. J Nippon Med Sch 2022;89(2):169–75. https://doi.org/10.1272/jnms.JNMS.2022_89-211 [published Online First: 14.09.2021].

[57] Galfano A, Ascione A, Grimaldi S, et al. A new anatomic approach for robot-assisted laparoscopic prostatectomy: a feasibility study for completely intrafascial surgery. Eur Urol 2010;58(3):457–61. https://doi.org/10.1016/j.eururo.2010.06.008 [published Online First: 16.06.2010].

[58] Yi B, Wang G, Li J, et al. Domestically produced Chinese minimally invasive surgical robot system "Micro hand S" is applied to clinical surgery preliminarily in China. Surg Endosc 2017;31(1):487–93. https://doi.org/10.1007/s00464-016-4945-3 [published Online First: 18.05.2016].

[59] Lim JH, Lee WJ, Choi SH, et al. Cholecystectomy using the Revo-i robotic surgical system from Korea: the first clinical study. Updates Surg 2021;73(3):1029–35. https://doi.org/10.1007/s13304-020-00877-5 [published Online First: 16.09.2020].

[60] Banerjee I, Banerjee I, Banerjee S. Is robotics the real game changer for urological cancer care during COVID-19 crisis? Nepal J Epidemiol 2021;11(2):988–93. https://doi.org/10.3126/nje.v11i2.38133 [published Online First: 202106 30.06.2021].

[61] Bertolo R, Garisto J, Gettman M, et al. Novel system for robotic single-port surgery: feasibility and state of the art in urology. Eur Urol Focus 2018;4(5):669–73. https://doi.org/10.1016/j.euf.2018.06.004 [published Online First: 18.06.2018].

[62] Ragavan N, Bharathkumar S, Chirravur P, et al. Evaluation of Hugo RAS system in major urologic surgery: our initial experience. J Endourol 2022;36(8):1029–35. https://doi.org/10.1089/end.2022.0015 [published Online First: 08.03.2022].

[63] Samalavicius NE, Janusonis V, Siaulys R, et al. Robotic surgery using Senhance((R)) robotic platform: single center experience with first 100 cases. J Robot Surg 2020;14(2):371–6. https://doi.org/10.1007/s11701-019-01000-6 [published Online First: 12.07.2019].

[64] Rocco B, Turri F, Sangalli M, et al. Robot-assisted radical prostatectomy with the Versius robotic surgical system: first description of a clinical case. Eur Urol Open Sci 2023;48:82–3. https://doi.org/10.1016/j.euros.2022.11.019 [published Online First: 02.01.2023].

[65] Strong VE, Forde KA, MacFadyen BV, et al. Ethical considerations regarding the implementation of new technologies and techniques in surgery. Surg Endosc 2014;28(8):2272–6. https://doi.org/10.1007/s00464-014-3644-1 [published Online First: 25.06.2014].

[66] Geiger JD, Hirschl RB. Innovation in surgical technology and techniques: challenges and ethical issues. Semin Pediatr Surg 2015;24(3):115–21. https://doi.org/10.1053/j.sempedsurg.2015.02.008 [published Online First: 02.03.2015].

[67] Angelos P. Ethics and surgical innovation: challenges to the professionalism of surgeons. Int J Surg 2013;11(Suppl 1):S2–5. https://doi.org/10.1016/S1743-9191(13)60003-5.

[68] Stefanidis D, Fanelli RD, Price R, et al. SAGES guidelines for the introduction of new technology and techniques. Surg Endosc 2014;28(8):2257–71. https://doi.org/10.1007/s00464-014-3587-6 [published Online First: 18.06.2014].

[69] Committee Opinion No. 674: guiding principles for privileging of innovative procedures in gynecologic surgery. Obstet Gynecol 2016;128(3):e85–8. https://doi.org/10.1097/AOG.0000000000001646.

III. The current and future clinical applications of robotic surgery among medical specialties

# 74

# Fluorescence-guided robotic surgery

*Bruno Zilberstein[a], Raphael L.C. Araujo[b,c], Rubens A. Sallum[b], Samuel Aguiar, Jr.[d], Miguel Nacul[e], and Flavio Daniel Saavedra Tomasich[f]*

[a]Service of Digestive Surgery, Sao Leopoldo Mandic School of Medicine, Campinas, SP, Brazil [b]Department of Digestive Surgery, Universidade Federal de São Paulo, São Paulo, SP, Brazil [c]Department of Oncology, Hospital Israelita Albert Einstein, São Paulo, SP, Brazil [d]Department of Colorectal Surgery, Hospital AC Camargo, São Paulo, SP, Brazil [e]Service of Surgery, Hospital Moinhos de Vento, Porto Alegre, RS, Brazil [f]Department of Surgery, Federal University of Paraná, Curitiba, Paraná, Brazil

## Fundamentals of fluorescence imaging

Fluorescence is a photoluminescence that involves light emitted by an atom or molecule after the absorption of electromagnetic energy. The absorbed light promotes an electronic transition between the ground state and an excited state. After a short period, the electrons relax to their ground state. The wavelengths associated with absorption/fluorescence are in the range of 200–1000 nm which can be subdivided into the ultraviolet (200–400 nm), visible (400–600 nm), and near-infrared region (600–1000 nm) [1].

Fluorophores are molecules that yield fluorescence emission. Typically, a fluorophore contains merged or conjugated aromatic groups in its chemical structure. Minerals, organic molecules, transition metal complexes, and nanoparticles are types of luminescent materials currently available. The most widely available molecules are small fluorescent molecules. They offer high versatility regarding optical properties and chemical reactivity. Fluorescence intensity, emission wavelength, fluorescence lifetime, and fluorescence anisotropy or polarization are fundamental parameters of a fluorescent molecule exploited to get practical information about the microenvironment [1].

The emission of fluorescence is sensitive to the influence of many parameters of the microenvironment such as polarity, pH, viscosity, pressure, temperature, quenchers, ions, etc. Therefore, any variation of this signal would provide valuable temporal and spatial information about the surroundings of the fluorophore. When such a molecule allows exploring the structure and dynamics of a system, then it is named as "fluorescent probe." This fact defines the success of fluorescence techniques as a tool for studying the structure and dynamics of a given system with high sensitivity [1].

Fluorescence provides information in a broad range of areas, from materials technology (polymers, surfaces, etc.) to biological research such as the study of biological membranes, the interaction between biomolecules in different cellular events, observation of macroscopic tissue, and immunoassays [1].

The first described use of fluorescent imaging in surgery dates to 1948 when surgeons noticed fluorescein concentration increased in malignant tissue and used this property to identify and localize intracranial neoplasms during neurosurgery. Since then, additional fluorescent agents have been used for a variety of clinical and surgical applications [2]. In recent years, fluorescence has reached the medical arena with new technologies that allow the performance of fluorescence image-guided surgery. This technique based on molecular navigation using fluorescent probes increases the efficacy of surgical tissue resection allowing surgeons to have real-time visualization of the target structure by emitting part of this energy at a different wavelength [1]. Intraoperative fluorescence imaging offers the benefits of high contrast and sensitivity, low cost, ease of use, safety, and visualization of cells and tissues both in vitro and in vivo [3]. This technique combines a fluorescent contrast agent, for example, indocyanine green (ICG), and a near-infrared (NIR) imaging system [4].

White light (visible light spectrum) has classically been used as an aid during open surgery. In the case of laparoscopic surgery, this light is emitted by systems coupled to the camera. The use of this light allows the surgeon to visualize structures and planes to perform surgery [5]. When anatomical or planar identification is not possible, the use of contrast/stain markers has been used to aid this identification. Fluorescence near-infrared (FCLI) dyes, such as ICG, have the advantage that they do not affect the dissection of the surgical field during the use of white light [5].

## Indocyanine green

ICG was developed during the World War II as a photographic drug and tested in 1957 at the Mayo Clinic for use in human medicine. After it was approved, ICG was initially used mainly in the diagnosis of liver function and, later, in cardiology [6]. ICG is a tricarbocyanine that has strong absorption and maximum emission at ≈780 and ≈820 nm, respectively. ICG solution is prepared by diluting 25 mg of ICG in 10 mL of distilled water, to a final concentration of 2.5 mg/mL. Diluted and injected intravenously, it is invisible to the eye; therefore, it does not interfere with the surgical field [7]. It binds strongly to serum proteins such as human serum albumin which allows its angiographic use and its hepatic/biliary elimination (T1/2–3 min) [8,9]. The fluorophore is stable at room temperature and soluble in sterile water (>2.5 mg/mL). The standard dose for clinical use (0.1–0.5 mg/mL/kg) is well below the toxicity level. At high concentrations and in saline, ICG forms insoluble, nonfluorescent aggregates [8]. The preparation contains 5% iodinated sodium [5]. The ICG stability and degradation properties depend on several factors like ICG concentrations, solvent (water, methanol, or plasma), and incubation temperature. The ICG maintains stability for 3 days when diluted in water and stored at 4°C in the dark, with a loss of 20% of fluorescence intensity within this period, suggesting that ICG should be used within 1 or 2 days after dilution, under these conditions [10]. If injected outside the blood vessels, ICG binds to proteins and can be found in lymph nodes, reaching the nearest one in 15 min. After 1–2 h, it can be seen in regional lymph nodes and deposited within macrophages. Through its bright fluorescence, even lymphatic vessels and lymph nodes in dense fat can be easily visualized [10]. Tracer deposition can last for prolonged periods in the lymphatic vessels and lymph nodes (1–3 days), allowing both preoperative and intraoperative injection for lymphatic mapping [10].

In recent years, ICG angiography has been introduced into clinical practice with an emphasis on surgical procedures to assess organ perfusion in various conditions, especially for identification of liver and biliary tree anatomy, vascular anatomy, and visualization of perfusion of gastrointestinal anastomoses [11]. ICG is also routinely used endoluminal (intraureteric) and interstitial (for lymph node identification). However, this use is outside what is recommended by the company [5].

## Fluorescence-assisted robotic surgery

Technologically dependent surgeries such as video surgery represented a turning point in the history of surgery, establishing an era of dominance of minimally invasive procedures in a large part of surgical specialties. The introduction of robotic devices has brought a greater appeal for the performance of minimally invasive procedures. Robotic surgery maintains the benefits of minimally invasive surgery that include reduction of blood loss, postoperative pain, inflammatory response, complications of the surgical site (infections and incisional hernia), the possibility of an immediate return to daily activities, among others, increasing technology-based advantages that expand the technical possibilities of surgery [12].

The introduction of the da Vinci robotic system (Intuitive, Sunnyvale, CA, USA) increased the diversity and complexity of cases addressed using minimally invasive surgery. Within this evolution, reconstructive surgery represents a particularly challenging scenario. Along with the development of its platforms, this system has allowed the use of intraoperative images in real time to guide during surgery. One of the advantages of the robotic system is the ease with which you can switch between using regular white light and near-infrared fluorescence (NIRF) [5].

## IGC in robotic-assisted surgery

The major current uses of fluorescence-assisted robotic surgeries are in gastrointestinal, urologic, and gynecologic surgeries, mainly for surgical oncology. In this regard, an ideal probe provides the contrast required for the resection of an entire tumor or affected lymph node while sparing healthy tissue [13]. There are three evolving uses of fluorescent probes in surgery: biochemically activatable "smart" probes, fluorescent imaging of the lymphatic system, and fluorescence-assisted robotic surgery [14]. An additional technical challenge remains in the design and integration of new probes with surgical technology both existing and in development.

Fluorescent imaging has been used in a wide variety of applications to help guide robotic-assisted surgeries: visualize vascular and lymphatic anatomy, evaluate tissue perfusion, map biliary anatomy, identify lesions, and image metabolic activity [15]. Combining the minimally invasive approach of robotic surgery with the accuracy and precision of fluorescence has the potential to improve safety and outcomes in a wide range of treatments, with several studies already highlighting the benefits [15].

## Esophagus

Esophagectomy, especially for esophageal cancer: assessment of gastric conduit perfusion, including feasibility, creation of the gastroesophageal anastomosis, and qualification of perfusion, along with lymphatic mapping and identification of critical anatomy. These tools are uniquely leveraged using the robotic platform to standardize and quantify key technical aspects of the operation [16].

## Stomach

Gastric cancer (GC)/gastrectomy with lymphadenectomy: providing real-time anatomy assessment and intraoperative visualization of blood flow, lymph nodes, and lymphatic vessels (real-time lymphatic mapping) [10].

In GC, after inspection of the abdominal cavity and confirmation of the absence of metastasis or unresectability criteria, the ICG administration is performed. ICG and NIRF imaging may be used for sentinel lymph node biopsy and analysis, lymphadenectomy guidance and quality control, and localization of the tumor [17]. First, intraoperative endoscopic injection of 0.2 mL of ICG is injected into the submucosal layer at four points around the lesion. The sentinel lymph node is identified and removed for detailed analysis. Next, a robotic D2 gastrectomy is performed. For the determination of the lymphadenectomy extent, the ICG solution can be injected on the day before the surgery. It allows for intraoperative identification of the lesion location and margin check, lymphadenectomy guidance, and verification of its adequacy. After the end of the surgery, the fluorescence system is activated, for final control of the lymphatic drainage route assessment and/or identification of residual lymph nodes in the dissected area (Figs. 1–3). It often identified small vessels that would not have otherwise been visualized [15]. For the assessment of the anastomosis perfusion, a bolus is injected intravenously and an area that reveals good perfusion in the stomach is chosen to be sectioned [15] (Fig. 4). If there is any doubt about vascularization, the ICG solution can be injected intravenously

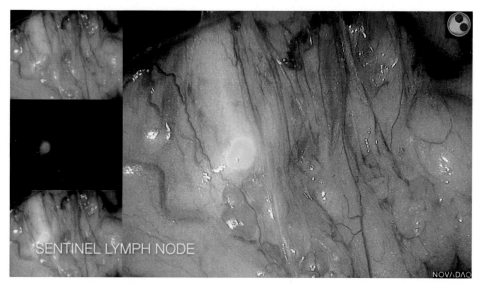

FIG. 1    After the end of the surgery, the fluorescence system is activated, for final control of the lymphatic drainage route assessment and/or identification of residual lymph nodes in the dissected area.

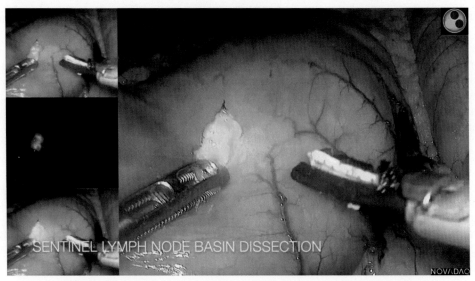

FIG. 2    Lymph node identified in Fig. 1, removed using fluorescence.

FIG. 3    Lymph node identified in Fig. 1, removed visualized with the firefly system.

FIG. 4    Assessment of the anastomosis perfusion: after a bolus was injected intravenously. An area revealing good perfusion in the stomach is chosen to be sectioned.

in a dose of 3.75–7.5 mg bolus to assess the blood supply. The perfusion of the stomach will be visualized and evaluated through the fluorescence system.

Chen et al. conducted a prospective randomized clinical trial that showed the safety and efficacy of ICG NIRF tracer-guided imaging during laparoscopic D2 lymphadenectomy in GC patients. It did not increase perioperative complications, suggesting that ICG NIRF tracer-guided imaging can be performed for routine lymphatic mapping during laparoscopic gastrectomy [18]. Fluorescent lymphography using ICG and NIRF imaging for lymph node navigation was also studied in robotic gastrectomy. Similar results, with more lymph nodes retrieved and reduced lymph node noncompliance, were found with fluorescent imaging [19].

## Colon

In colorectal surgery, especially in cancer, fluorescence has been used intraoperatively mainly to evaluate tissue perfusion [11]. ICG is provided as a freeze-dried powder sterile and water-soluble, which is injected intravenously, at a dose of 0.2 mg/kg, after sectioning the mesentery, but before proximal transection of the colon. Fluorescence is excited by exposure to an infrared light source, activated by the surgeon in the robot console. Vascularity of the colon or rectum is analyzed after 1 min of ICG dye injection and waited until a good perfusion signal is evident, to confirm the chosen level of colonic resection by standard visual inspection. Perfusion of the previously identified resection site is judged as "good" (representing even distribution of fluorescence at the chosen level of proximal colon resection), "poor" (uneven distribution of fluorescence at the chosen level of colon proximal to resection), or "absent" (if no fluorescence is observed in the 10 cm proximal to the chosen level of colonic resection). As it has a half-life of about 4 min, the ICG injection is repeated if the surgeon deems it necessary [20]. Diogenes et al. studied 75 patients in which the IGC was used to analyze the vascularization at the level of colorectal anastomosis, compared with 132 others in which the dye was not used, verify that the rate of fistulas in the first group was 4% and 19.7% when IGC was not used. It shows the importance of using this methodology to increase safety in cases of colorectal anastomosis [20].

Liu et al. published a systematic review and metaanalysis in 2021 that aimed to evaluate whether ICG-FA could prevent anastomotic leakage in colorectal surgery [21]. Thirteen studies of 4037 patients were included in the metaanalysis. The study included 1806 patients in the ICG group and 2231 patients in the control group. The pooled incidence of anastomotic leakage in the ICG group was 3.8% compared with 7.8% in the control group. There was a significant difference in the anastomotic leakage rate with or without the use of ICG-FA (OR 0.44; 95% CI 0.33–0.59; $P < 0.00001$). Reoperation rates were 2.6% and 6.9% in ICG and control groups, respectively. The application of intraoperative ICG-FA was associated with a lower risk of reoperation (OR 0.39; 95% CI 0.16–0.94; $P = 0.04$). The overall complication rate was 15.6% in the ICG group compared with 21.2% in the control group. Overall complications were significantly reduced when using ICG-FA (OR 0.62; 95% CI 0.47–0.82; $P = 0.0008$). The mortality rate was not statistically different with or without the use of ICG-FA (OR 1.22; 95% CI 0.20–7.30; $P = 0.83$). The authors conclude that ICG-FA reduced the risks of anastomotic leakage, reoperation, and overall complications for colorectal cancer patients undergoing colorectal surgery. Intraoperative fluorescence angiography with ICG is now widely applied to assess colonic vascular perfusion and may contribute to the prevention of secondary anastomotic fistula to perfusion deficit [22,23]. Also, due to its widespread use in surgery and its well-established safety profile, ICG may be a viable alternative to nanquim ink for tattooing tumors. ICG is cleared from circulation in less than 5 min when injected by intravenously due to the binding of proteins and phospholipids. When it is administered as an intratumoral injection, it persists for several days which can serve as a safe colon tumor marker [24].

## Liver and biliary tree

Since ICG is metabolized mainly by hepatic parenchymal cells and secreted into the bile, it has been used to visualize the biliary tree as a cholangiography to prevent accidental intraoperative bile duct injury during robotic cholecystectomy for gallstone disease. This application is safe, effective, and particularly helpful with obese patients or in cases of acute cholecystitis, two cases that make the surgery more challenging [15,25].

In liver surgery, a major challenge is performing radical resection with maximal preservation of the liver parenchyma and obtaining a low rate of complications. Despite the developments, visual inspection, palpation, and intraoperative ultrasound remain the most utilized tools during surgery today. In laparoscopic and robotic liver surgery, fluorescence-guided surgery (FGS) for benign and malignant hepatobiliary (HPB) neoplasms has significantly increased and improved imaging methods. It enables the identification of subcapsular liver tumors through the accumulation of ICG, after preoperative intravenous injection, in cancerous tissues of hepatocellular carcinoma and noncancerous hepatic parenchyma, around intrahepatic cholangiocarcinoma and liver metastases, and it can also be used for visualizing extrahepatic bile duct anatomy and hepatic segmental borders, increasing the accuracy and the easiness of open and minimally invasive hepatectomy [26].

A recent metaanalysis evaluating fluorescent imaging in the detection of hepatic tumors with 6 studies incorporating 587 patients demonstrated that complication rates were lower in the fluorescence-guided vs standard white light hepatectomy group. No serious reactions to ICG were reported [27]. Also, ICG fluorescence imaging has been studied in colorectal liver metastases resection. van der Vorst et al. concluded that ICG imaging significantly increases the number of intrahepatic colorectal liver metastases identified and reduces postoperative hospital stay and 1-year recurrence rate without increasing hepatectomy-related complications and mortality rates [28].

## Pancreas

In pancreatic surgery, Pessaux et al. employed a novel combination of fluorescence visualization and augmented-reality three-dimensional imaging to guide robotic duodenopancreatectomy in a patient with an intraductal papillary mucinous neoplasm [29].

## Urology

Robotic nephrectomy: identify renal vasculature to alert surgeons to abnormal vascular anatomy, avoiding unintended injuries. Differentiate tumors, also benign and malignant, from normal parenchyma in robotic partial nephrectomy for renal cortical tumors. Visual aid to reduce the overall time of ischemia after renal artery clamping [30]. It helps surgeons spare larger portions of the kidney in partial nephrectomy and reduce postoperative complications, blood loss, and metastasis [31].

Robotic partial adrenalectomy: differentiate between adrenal mass and normal parenchyma, helping to spare as much of the adrenal gland as possible while completely excising the tumor [32].

Robotic-assisted ureteral construction: intraureteric injection of ICG used to identify the ureter and locate ureteral strictures, enabling to precisely identify the extent of a ureteric stricture, bladder diverticula, and tumor locations [33].

Robotic radical prostatectomy: ICG can be injected into the prostate to mark prostatic tissue and map potential sentinel lymph nodes. Because ICG is not prostate-specific, new fluorescent tracers that achieve targeting via prostate-specific membrane antigens are being developed to enhance tissue contrast [34,35].

Robotic cystectomy: bladder cancer patients underwent an experimental technique in which ICG was injected both directly into the tumor and intravenously. This permitted tumor marking, sentinel lymph node detection, and mesenteric vasculature identification to be accomplished simultaneously [34].

## Gynecology

Robotic hysterectomy: detection of sentinel nodes for endometrial cancer [36]. Fluorescent lymph node mapping with robotic hysterectomy in patients with endometrial cancer can help accurately and safely avoid full lymphadenectomies in women with high-grade metastatic tumors, thus preventing secondary complications such as lymphedema [37].

## Thorax

Robotic thymectomy: identify the pericardiophrenic neurovascular bundle that could reduce surgery time and the risk of accidental nerve injury [38].

Robotic anatomic segmentectomy: demarcate boundaries of lung segments for surgical resection [39].

## Pediatric surgery

The clinical application of FGS in pediatrics is just in the initial phase. Surgical resection of hepatoblastoma and its metastasis, real-time imaging of the biliary tree, and urogenital system. Other current uses concern the assessment of blood perfusion (intestine, myocutaneous flap, transplanted liver, and lymphatic flow imaging), all with a very small number of cases [40].

## Consensus, limitations, and perspectives

In 2022, 19 international experts published a consensus about FGS based on a meeting that occurred in Frankfurt, Germany, in September 2019 [41]. The methodology used was a Delphi survey. The identified areas of consensus and nonconsensus evaluated were fundamentals, patient selection/preparation, technical aspects, and effectiveness and safety. The overall agreement was unanimous regarding the safety of the procedure, decreasing the overall cost of a patient's peri- and postoperative care. The experts also had 100% agreement that fluorescence imaging technology has the potential to dramatically facilitate many surgical procedures and to significantly enhance patient outcomes as an important tool for the evaluation of tissue perfusion. They see, over the next decade, the role of FGS in clinical practice and research increasing. They finally defined fluorescence imaging, with and without ICG, as useful for training surgical residents and for surgical quality control and should be part of the routine surgical practice [41].

Most of the surgeon's experts agreed that cost was not a significant barrier to using this technology, but whether the potential benefits of fluorescent imaging warrant the cost of adding it to a surgical field must be determined. More evidence confirming the reduction of postoperative complications could help validate the initial time and expense required [41].

Several other obstacles and limitations must be overcome for fluorescent imaging to gain even wider use than its current applications. The ICG preparation contains 5% iodinated sodium for which reactions can be seen in allergic patients [5]. Approximately 1% of the population is hypersensitive to iodine and cannot be given ICG. Before administering ICG, patients should be asked if they are allergic to iodine, but adverse reaction has been demonstrated to be very uncommon [41]. Another issue about iodine contrast is that it is rapidly excreted by the body and has a low penetrative ability. So, to achieve sufficient contrast, it must be administered very close to the time of surgery, or the dose must be increased [15]. The presence of inflammation, fibrosis, or excess fat also makes visualizing ICG more difficult. Elevated liver function tests are another contraindication. Alternative imaging molecules must be sought for these patients [15]. Newer detection methods could improve imaging by quantitatively assessing and mapping fluorescence intensity instead of relying on qualitative evaluation.

## Future scenario

The evolution of fluorescent imaging stands on the development of targeted fluorophores with higher specificity for tumors, including small molecules, peptides, and antibodies [15]. The increased sensitivity and specificity for cancer tissue will increase the ability of the surgeon to perform a more precise surgery. Future clinical trials with specific evidence-based indications and standardized techniques are also needed to implement this technology in the cancer surgical routine [42]. Recently developed fluorescently labeled peptides can specifically label degenerated nerve branches, which in the future could aid patients undergoing surgical nerve repair [43].

Also, newer detection methods could improve imaging by quantitatively assessing and mapping fluorescence intensity instead of relying on qualitative evaluation [15].

A great deal of current research and clinical trials are examining the potential use of quantum dots as an alternative fluorescent marker [44,45]. These molecules would overcome the obstacle of iodine allergy and can be easily modified to alter their biodistribution and fluorescence emission. They also allow quantitative detection, have high fluorescence intensity, and have a long emission lifetime.

The incorporation of artificial intelligence, machine learning, and big data associated with the evolution of the robotic platforms as fluorescence imaging becomes more sophisticated and targeted, will result in additional surgical applications benefiting both patients and physicians [12,15].

## Conclusions

ICG fluorescence imaging is a promising tool, allowing real-time image-guided surgery with wide application, especially in gastrointestinal, urologic, and gynecological oncologic surgery. Fluorescence imaging with ICG is useful for the visualization of critical anatomical structures such as arteries and veins, the assessment of tissue perfusion, the detection of cancerous lesions, the localization of sentinel lymph nodes, the visualization of segmented organs such as the lungs and liver, and the detection of small organs such as the parathyroid glands. The technology involved in robotic-assisted surgery is the most natural pioneering area of ICG fluorescence imaging. Although additional studies are needed to fully validate the utility of fluorescent imaging in robotic-assisted surgeries, a growing body of evidence indicates that coupling these technologies has the potential to further increase the efficacy and safety of these procedures. As fluorescence imaging becomes more sophisticated and targeted, additional surgical applications will follow, benefiting both patients and physicians and the interest in using ICG imaging in medical robotics is increasing quickly. In the future, this technology probably will be a necessity and a strong recommendation, especially in oncologic procedures.

## Key points

- Fluorescence near-infrared dyes, such as ICG, allows real-time image-guided surgery with wide application, especially in gastrointestinal, urologic, and gynecological oncologic surgery.
- Fluorescent imaging has been used in a wide variety of applications to help guide robotic-assisted surgeries: visualize vascular and lymphatic anatomy, evaluate tissue perfusion, map biliary anatomy, identify lesions, and image metabolic activity.
- In gastrointestinal surgery, the major current uses of fluorescence-assisted robotic surgeries are in gastric and colon cancer, liver, and biliary tract surgery.
- Combining the minimally invasive approach of robotic surgery with the accuracy and precision of fluorescence has the potential to improve safety and outcomes in a wide range of treatments as an important tool for the evaluation of tissue.

## Appendix: Supplementary material

Supplementary material related to this chapter can be found on the accompanying CD or online at https://doi.org/10.1016/B978-0-443-13271-1.00063-7.

## References

[1] Menéndez GO, Leskow FC, Spagnuolo CC. Basic concepts of fluorescence and fluorescent probes. In: Dip FD, Ishizawa T, Kokudo N, Rosenthal RJ, editors. Fluorescence imaging for surgeons. Cham: Springer International Publishing; 2015. p. 3–18. [cited 2023 Apr 17]. Available from: https://link.springer.com/10.1007/978-3-319-15678-1_1.

[2] Moore GE, Peyton WT, French LA, Walker WW. The clinical use of fluorescein in neurosurgery: the localization of brain tumors. J Neurosurg 1948;5(4):392–8.

[3] Alander JT, Kaartinen I, Laakso A, Pätilä T, Spillmann T, Tuchin VV, et al. A review of indocyanine green fluorescent imaging in surgery. Int J Biomed Imaging 2012;2012:1–26.

[4] Meijer RPJ, Faber RA, Bijlstra OD, Braak JPBM, Meershoek-Klein Kranenbarg E, Putter H, et al. AVOID; a phase III, randomised controlled trial using indocyanine green for the prevention of anastomotic leakage in colorectal surgery. BMJ Open 2022;12(4), e051144.

[5] Cadillo-Chávez R. Usefulness of indocyanine green (ICG) in robotic reconstructive surgery. Arch Esp Urol 2019;72(8):759–64.

[6] Fox IJ, Brooker LG, Heseltine DW, Essex HE, Wood EH. A tricarbocyanine dye for continuous recording of dilution curves in whole blood independent of variations in blood oxygen saturation. Proc Staff Meet Mayo Clin 1957;32(18):478–84.

[7] Escobedo JO, Rusin O, Lim S, Strongin RM. NIR dyes for bioimaging applications. Curr Opin Chem Biol 2010;14(1):64–70.

[8] van den Berg NS, van Leeuwen FWB, van der Poel HG. Fluorescence guidance in urologic surgery. Curr Opin Urol 2012;22(2):109–20.

[9] Luo S, Zhang E, Su Y, Cheng T, Shi C. A review of NIR dyes in cancer targeting and imaging. Biomaterials 2011;32(29):7127–38.

[10] Sakamoto E, Kodama Pertille Ramos MF, Dias AR, Safatle-Ribeiro AV, Zilberstein B, Nahas SC, et al. Indocyanine green imaging to guide lymphadenectomy in laparoscopic distal gastrectomy—with video. Ann Med Surg (Lond) 2021;69, 102657. https://doi.org/10.1016/j.amsu.2021.102657.

[11] Wada T, Kawada K, Hoshino N, Inamoto S, Yoshitomi M, Hida K, et al. The effects of intraoperative ICG fluorescence angiography in laparoscopic low anterior resection: a propensity score-matched study. Int J Clin Oncol 2019;24(4):394–402.

[12] Nacul MP. Laparoscopy & robotics: a historical parallel. Rev Col Bras Cir 2020;47, e20202811.

[13] Nguyen QT, Tsien RY. Fluorescence-guided surgery with live molecular navigation—a new cutting edge. Nat Rev Cancer 2013;13(9):653–62.

[14] Kobayashi H, Ogawa M, Alford R, Choyke PL, Urano Y. New strategies for fluorescent probe design in medical diagnostic imaging. Chem Rev 2010;110(5):2620–40.

[15] Landau MJ, Gould DJ, Patel KM. Advances in fluorescent-image guided surgery. Ann Transl Med 2016;4(20):392.

[16] Papageorge MV, Sachdeva UM, Schumacher LY. Intraoperative fluorescence imaging in esophagectomy and its application to the robotic platform: a narrative review. J Thorac Dis 2022;14(9):3598–605.

[17] Sakamoto E, Dias AR, Ramos MFKP, Safatle-Ribeiro AV, Zilberstein B, Ribeiro JU. Indocyanine green and near-infrared fluorescence imaging in gastric cancer precision surgical approach. Arq Gastroenterol 2021;58(4):569–70.

[18] Chen QY, Xie JW, Zhong Q, Wang JB, Lin JX, Lu J, et al. Safety and efficacy of indocyanine green tracer-guided lymph node dissection during laparoscopic radical gastrectomy in patients with gastric cancer: a randomized clinical trial. JAMA Surg 2020;155(4):300.

[19] Kwon IG, Son T, Kim HI, Hyung WJ. Fluorescent lymphography–guided lymphadenectomy during robotic radical gastrectomy for gastric cancer. JAMA Surg 2019;154(2):150.

[20] Diogenes I, Stevanato P, Bezerra T, Nakagawa W, Takahashi R, Kupper B, et al. Evaluation of anastomotic perfusion with indocyanine green angiography during left colon and rectal cancer surgeries. In: ePoster; 2023. p. 136. Boston, MA [cited Apr 17, 2023] https://doi.org/10.1245/s10434-023-13332-7.

[21] Liu D, Liang L, Liu L, Zhu Z. Does intraoperative indocyanine green fluorescence angiography decrease the incidence of anastomotic leakage in colorectal surgery? A systematic review and meta-analysis. Int J Colorectal Dis 2021;36(1):57–66.

[22] Hayami S, Matsuda K, Iwamoto H, Ueno M, Kawai M, Hirono S, et al. Visualization and quantification of anastomotic perfusion in colorectal surgery using near-infrared fluorescence. Tech Coloproctol 2019;23(10):973–80.

[23] Joosten JJ, Reijntjes MA, Slooter MD, Duijvestein M, Buskens CJ, Bemelman WA, et al. Fluorescence angiography after vascular ligation to make the ileo-anal pouch reach. Tech Coloproctol 2021;25(7):875–8.

[24] Garcia Badaracco A, Ward E, Barback C, Yang J, Wang J, Huang CH, et al. Indocyanine green modified silica shells for colon tumor marking. Appl Surf Sci 2020;499, 143885.

[25] Spinoglio G, Priora F, Bianchi PP, Lucido FS, Licciardello A, Maglione V, et al. Real-time near-infrared (NIR) fluorescent cholangiography in single-site robotic cholecystectomy (SSRC): a single-institutional prospective study. Surg Endosc 2013;27(6):2156–62.

[26] Rossi G, Tarasconi A, Baiocchi G, De' Angelis GL, Gaiani F, Di Mario F, et al. Fluorescence guided surgery in liver tumors: applications and advantages. Acta Biomed 2018;89(9-S):135–40.

[27] Qi C, Zhang H, Chen Y, Su S, Wang X, Huang X, et al. Effectiveness and safety of indocyanine green fluorescence imaging-guided hepatectomy for liver tumors: a systematic review and first meta-analysis. Photodiagnosis Photodyn Ther 2019;28:346–53.

[28] van der Vorst JR, Schaafsma BE, Hutteman M, Verbeek FPR, Liefers GJ, Hartgrink HH, et al. Near-infrared fluorescence-guided resection of colorectal liver metastases: fluorescence imaging of liver cancer. Cancer 2013;119(18):3411–8.

[29] Pessaux P, Diana M, Soler L, Piardi T, Mutter D, Marescaux J. Robotic duodenopancreatectomy assisted with augmented reality and real-time fluorescence guidance. Surg Endosc 2014;28(8):2493–8. https://doi.org/10.1007/s00464-014-3465-2.

[30] Borofsky MS, Gill IS, Hemal AK, Marien TP, Jayaratna I, Krane LS, et al. Near-infrared fluorescence imaging to facilitate super-selective arterial clamping during zero-ischaemia robotic partial nephrectomy: near-infrared fluorescence imaging in zero ischaemia RPN. BJU Int 2013;111(4):604–10.

[31] Harke N, Schoen G, Schiefelbein F, Heinrich E. Selective clamping under the usage of near-infrared fluorescence imaging with indocyanine green in robot-assisted partial nephrectomy: a single-surgeon matched-pair study. World J Urol 2014;32(5):1259–65.

[32] Manny TB, Pompeo AS, Hemal AK. Robotic partial adrenalectomy using indocyanine green dye with near-infrared imaging: the initial clinical experience. Urology 2013;82(3):738–42.

[33] Lee Z, Moore B, Giusto L, Eun DD. Use of indocyanine green during robot-assisted ureteral reconstructions. Eur Urol 2015;67(2):291–8.

[34] Manny TB, Patel M, Hemal AK. Fluorescence-enhanced robotic radical prostatectomy using real-time lymphangiography and tissue marking with percutaneous injection of unconjugated indocyanine green: the initial clinical experience in 50 patients. Eur Urol 2014;65(6):1162–8.

[35] Laydner H, Huang SS, Heston WD, Autorino R, Wang X, Harsch KM, et al. Robotic real-time near infrared targeted fluorescence imaging in a murine model of prostate cancer: a feasibility study. Urology 2013;81(2):451–7.

[36] Rossi EC, Jackson A, Ivanova A, Boggess JF. Detection of sentinel nodes for endometrial cancer with robotic assisted fluorescence imaging: cervical versus hysteroscopic injection. Int J Gynecol Cancer 2013;23(9):1704–11.

[37] Paley PJ, Veljovich DS, Press JZ, Isacson C, Pizer E, Shah C. A prospective investigation of fluorescence imaging to detect sentinel lymph nodes at robotic-assisted endometrial cancer staging. Am J Obstet Gynecol 2016;215(1). 117.e1–7.

[38] Wagner OJ, Louie BE, Vallières E, Aye RW, Farivar AS. Near-infrared fluorescence imaging can help identify the contralateral phrenic nerve during robotic thymectomy. Ann Thorac Surg 2012;94(2):622–5.

[39] Pardolesi A, Veronesi G, Solli P, Spaggiari L. Use of indocyanine green to facilitate intersegmental plane identification during robotic anatomic segmentectomy. J Thorac Cardiovasc Surg 2014;148(2):737–8.

[40] Paraboschi I, De Coppi P, Stoyanov D, Anderson J, Giuliani S. Fluorescence imaging in pediatric surgery: state-of-the-art and future perspectives. J Pediatr Surg 2021;56(4):655–62.

[41] Dip F, Boni L, Bouvet M, Carus T, Diana M, Falco J, et al. Consensus conference statement on the general use of near-infrared fluorescence imaging and indocyanine green guided surgery: results of a modified Delphi study. Ann Surg 2022;275(4):685–91.

[42] Marano A, Priora F, Lenti LM, Ravazzoni F, Quarati R, Spinoglio G. Application of fluorescence in robotic general surgery: review of the literature and state of the art. World J Surg 2013;37(12):2800–11.

[43] Hussain T, Mastrodimos MB, Raju SC, et al. Fluorescently labeled peptide increases identification of degenerated facial nerve branches during surgery and improves functional outcome. PLoS One 2015;10, e0119600.

[44] Fang M, Peng CW, Pang DW, et al. Quantum dots for cancer research: current status, remaining issues, and future perspectives. Cancer Biol Med 2012;9:151–63.

[45] Kamila S, McEwan C, Costley D, et al. Diagnostic and therapeutic applications of quantum dots in nanomedicine. Top Curr Chem 2016;370:203–24.

III. The current and future clinical applications of robotic surgery among medical specialties

# 75

# Robotic endoscopy

*Marcio Roberto Facanali Junior*[a] *and Everson Luiz de Almeida Artifon*[b,c]

[a]Department of Gastroenterology, Hospital das Clínicas of University of São Paulo Medical School, São Paulo, SP, Brazil [b]Surgery Department, Hospital das Clínicas of University of São Paulo Medical School, São Paulo, SP, Brazil [c]Department of Surgery Post-Graduate Program, São Paulo Medical School, São Paulo, SP, Brazil

## Introduction

The semi-flexible tube endoscope, first developed by Georg Kelling in 1898, who performed peritoneoscopy by placing a camera at the tip of the endoscope [1], improved by Rudolf Schindler in collaboration with Georg Wolf, who in 1932 invented the first semi-flexible gastroscope [2]. In 1957, Basil Hirschowitz produced a glass fiber gastroscope and transformed it into a fiber bundle called fiber optic endoscopy [3]. Afterward, television played a fundamental role in the technological evolution of gastroscopes [4], as they tried to introduce the technology for endoscopy from 1964, transferring the image captured by the fiberscope to a television monitor. After the invention of the fiberscope, another major advance was the invention of charge-coupled devices (CCD) in 1969 [5].

Initially, endoscopy began as a diagnostic method of the gastrointestinal tract; however, it has become an important method of treating gastrointestinal pathologies nowadays. In the last 20 years, the objectives of robotic technology for use in surgical procedures have changed its objective [6], initially developed to access geographically remote regions via telesurgery [7], and today, it is already used to optimize access in areas difficult to reach anatomically or to facilitate manual manipulations for inexperienced surgeons.

Robotic assistance can help shorten the learning curve, allowing less experienced surgeons to perform complex procedures such as endoscopic submucosal dissection (ESD) [8].

To perform transluminal endoscopic surgery through natural orifices (NOTES), there was also a need to develop platforms that would bring stability, in addition to allowing manipulation of tweezers that conventional equipment does not allow. In addition to the search for autonomy in performing examinations, generating the demand for equipment performs procedures that today only the endoscopist can perform.

Thus, the current robotic technologies incorporated in endoscopy can be divided between those developed for better maneuverability and stability of tweezers for resections and those with active movement in search of endoscopy autonomy.

## Flexible robotic endoscopy

Flexible endoscopy is widely used for diagnostic and therapeutic purposes of the gastrointestinal tract because it is minimally invasive and fast and can be performed by only one endoscopist and, in most cases, without general anesthesia [9]. Nonetheless, with the advancement of its therapeutic use, mainly in resections of lesions due to ESD, the time and complexity of the procedure has drastically increased, highlighting the operational limitations imposed by the flexible endoscope. Conventional flexible endoscopes have major limitations regarding the stability and freedom of movement of the tweezers, with little possible angulation.

To solve this problem, robotics is being incorporated to improve the accuracy of tissue triangulation, traction for dissection and in transluminal endoscopic surgery through natural orifices (NOTES) [9,10].

In addition, it is necessary to continue the search for lower rates of adverse events and greater patient acceptance by promoting less pain and discomfort during the examination.

Major research and investment in the introduction of robotic mechanisms in endoscopy is directed toward:

— Creation of platforms capable of handling tweezers with a wide degree of freedom for performing ESD and NOTES [11–13].
— Mechanisms for the active introduction of devices to reduce the influence of the endoscopist's manual skill, reduce the need for an endoscopist to be present in therapeutic procedures where more than one endoscopist is needed, and reduce the discomfort and pain reported by the patient [14,15].
— Evolution of endoscopic capsules with the aim of using them in the screening of stomach and colon diseases and also as a therapeutic method [16,17].

## General design

The design of a robotic system for gastrointestinal NOTES usually includes a master console and two mechanical arms. The end of the mechanical arm can be a gripper or an electrosurgical unit (ESU), inserted into an overtube.

The smaller the diameter of the overtube, the easier it becomes to pass through natural narrow and curved holes during the procedure [18].

The performance of the robotic system depends on the flexibility of the distal part of the mechanical arm, grip strength [18], and overtube diameter. The flexibility of the distal part of the mechanical arm refers to degrees of freedom (DOF). Grip strength is the maximum effort exerted by the end effector (claw), which is usually expressed as a unit of force (Newton (N)).

## Control-performance structure

The robotic system for gastrointestinal NOTES usually adopts a control-actuation structure.
The structure includes three parts:

### Master console

The master console collects the surgeon's control actions, converts them into angular data, and sends them to the motion controller. Manipulators are used as master console devices like joysticks.

### Montion controller

The motion controller interprets the angular data and sends the results to the mechanical arms [19].

### Mechanical drive arms

Then, the mechanical arms connected in series trigger the corresponding movements [20]. In the process of actuation, the antidither algorithm can effectively filter hand dithering and make the actuation more accurate and stable [21]. Therefore, actuation controlled by robots with mechanical arms can provide high precision [22].

## Robotic endoscopic platforms

Among the existing robotic endoscopic platforms, most consist of a flexible robotic endoscope that can be telemanipulated. Within the robotic endoscope, there are usually at least two articulated effectors capable of achieving triangulation, adequate tissue retraction, and optimal exposure of the operative field. Table 1 shows the robotic systems according to the year of publication [23].

## ViaCath system

Developed by Abbott et al. [24], it consists of a main console, two mechanical drive arms and a long, flexible instrument. Both the endoscope and the mechanical drive arms are introduced into the gastrointestinal tract through a 19-mm-diameter overtube in parallel. The end effectors of the two mechanical actuation arms are located at the front of the endoscope and can operate under camera monitoring. Each mechanical actuation arm has 8 DOF, and a gripper and the gripping force of the clamp is 3 Newton.

TABLE 1  Control-actuation robotic system for gastrointestinal NOTES [23].

| System | Diameter of overtube (mm) | DOF | Gripping force [N] | Research progress |
|---|---|---|---|---|
| ViaCath | 19 | (8+1)×2 | 3 | Animal in vivo |
| EndoSAMURAI | 18 | 5×2 | – | Animal in vivo |
| EndoMaster | 19.5 | 9 | 2.87 | Patient trial |
| Testbed (Zhao) | 12 | (5+1)×2 | – | Animal in vitro |
| Robot (Lau) | 18 | 10+2 | 0.47 | Animal in vitro |
| STRAS | 16 | 10+1 | 0.9 | Animal in vitro |
| CYCLOPS | 30 | (5+1)×2 | 19 | Vivo preclinical |
| EndoMaster EASE | – | 5×2 | 5.8 | Patient trial |
| K-FLEX | 17 | 14 | 2.94 | Animal ex vivo |

## EndoSAMURAI system

Developed by Olympus Corporation, the system includes a conventional endoscope, a main console and two mechanical drive arms on an 18-mm-diameter overtube. The end effectors are biopsy forceps with serrated jaws and an ESU. Each mechanical actuation arm has 5 DOF, and the gripping force value is not specified. Both mechanical drive arms are parallel and can be controlled using manipulators [12]. This system has been validated in ex vivo and in vivo experiments [25,26].

## EndoMaster system

Designed by Phee et al. and Ho et al. [11], in Singapore, at Nanyang Technological University, the system adopts a control-actuation structure, which comprises a main console, a motion controller, and two mechanical actuation arms. The diameter of the overtube is 19.5 mm, and there are two channels with diameters of 2.8 and 3.7 mm to pass through a gripper and an ESU. The master console responds to surgeon input; the motion controller controls the mechanical actuation arms, the gripper, and the ESU [11,27]. The system was validated with in vivo animal experiments and in five patients (ESD) [28,29].

## Testbed system

Zhao et al. [30] developed a mechanically and manually operated endoscopic test bench. Two mechanical actuation arms and a master console are integrated in an overtube with a diameter of 12 mm, and the mechanical actuation arm is equipped with a final actuator at the opposite end, having 5 DOF, without specification of gripping force.

## Two-arm robot system

Lau et al. [31] designed a two-arm robot for ESD, which adopts a control-actuation structure, with an overtube diameter of 18 mm. One of the two mechanical actuation arms has a surgical forceps that can lift the mucosa at the other end and the other end effector is a unipolar ESU. The two arms pass through two 6-mm-diameter instrument channels in the overtube, each of which has 5 DOF, and the gripping force of the forceps is 0.47 Newton. The robot was validated by in vitro animal experiments [31].

## STRAS/Anubiscope system

This system is the robotic version of Anubiscope, which was designed by Zorn et al. [13], and consists of a common endoscope, a master console and two mechanical drive arms. The overtube has a diameter of 16 mm, which allows it to accommodate two working channels, one 4.3 mm and the other 3.2 mm in diameter for mechanical drive arms. The

III. The current and future clinical applications of robotic surgery among medical specialties

overtube is equipped with a camera. The system has 10 DOF, and the gripper can apply a force of 0.9 Newton. STRAS has been validated for ESD by in vivo animal experiments [13].

## CYCLOPS system

Developed by Vrielink et al. [32], CYCLOPS adopts a control-actuation framework. The master console uses haptic devices with force feedback (Geomagic Touch) to indicate the boundaries of the workspace. There is also a detachable scaffold at the end of the overtube. When deployment is complete, the mechanical arms can move in all directions within DOF range. The scaffold needs to enter the digestive tract of the cannula, and its maximum width, in the folded state, is 30 mm. When the overtube is bent, the end-effector executive force in the X-, Y-, and Z-axis directions changes from 3.47 to 19.08 Newton. Since 2018, this system has been prepared for clinical validation [32].

## EndoMaster EASE system

EndoMaster EASE, the second-generation EndoMaster system, consists of master manipulators, actuation instruments, and a commercially customized endoscope. Two Omega7 haptic interfaces are used as the main manipulators. The robotic endoscope is customized to have three channels of instruments. One of the channels has a diameter of 2.5 mm for the passage of endoscopic tools, while the other two channels are where the 4.4-mm-diameter needle driver and the 4.2-mm-diameter forceps can be inserted. The actuation instruments are interchangeable, with 5 DOF each instrument. The gripping forces of the needle driver and tweezers can be 4.3 and 5.8 Newton [33].

## K-FLEX system

This platform was proposed by Huang et al. [34], for intracavitary surgery. The K-FLEX system consists of a master console and two mechanical drive arms. The robot has a total of 14 DOF and an overtube of 17 mm in diameter, and the clamping force of the mechanical arm is 2.94 Newton.

## Robotic flexible colonoscopes

Robotic-assisted colonoscopy aims to improve the patient's tolerance for the examination, promote cecal intubation regardless of the endoscopist's skill, and reduce possible risks such as intestinal perforation. Therefore, it is necessary for the colonoscope to have active movement and mold itself to the colon, assisted by a robot, with the doctor being able to manipulate the colonoscope remotely [35]. The existence disposals have different insertion tactics. The majority uses inchworm-like movements, or techniques derivate from balloon enteroscopy [9].

### Aer-O-scope system

A control station and disposable components make up this system. The disposable components are a rectal introducer, a power cable, and the optical capsule surrounded by a carrier balloon. The rectal introducer is a silicone tube with a balloon attached to prevent air loss [36]. After its introduction through the anal canal, the rest of the device is inserted. The two balloons are inflated, and the carbon dioxide ($CO_2$) is inflated between them. Pneumatic force applied to the bowel pushes the balloon forward, while the introducer balloon remains in the rectum. The pressures in the balloons and in the intestine (before and after the balloon) are constantly measured and transmitted to the workstation. A computer algorithm adjusts the three pressures to advance the vehicle's balloon and avoid punctures. Once the system is in the cecum, pressures are changed to keep the colon distended for evaluation and regression of the balloon to the rectum. The camera provides a 360° circumferential view [37] Gluck et al. [38] reported a cecal intubation rate of 98.2%.

### Endotics system

This system is based on the inchworm-like movement. A disposable probe has a movable tip manipulated and a flexible body controlled by a clinician at the workstation. The devices, proximal and distal, can attach to the mucosa, and an extension and retraction mechanism between them promotes the insertion of the instrument inchworm-like

[39]. More recent studies have reported a rate of 81.6%, even lower than that of control group of 94.3%, requiring minimal sedation and few reports of pain [14,39].

## NeoGuide endoscopy system

The system consists of an articulated colonoscope controlled by a computer console designed to maintain the natural loops of the colon during insertion. There are sensors on the tip and on the outside, which detect the position of the instrument. The tube segments are independent and mobile, electronically controlled. As the doctor inserts the tube, it is shaped by the computer console into the natural loops. Eickhoff et al. reported cecal intubation in 10 of 11 patients [40].

## Invendoscope system

It is a motorized, portable, single-use colonoscope. Eight wheels outside the patient drive the tube, which is controlled by a joystick and the doctor. The tube diameter is 10 mm, and the length can be from 170 to 210 cm depending on the version. An inverted double-sleeve protects the tube and unfolds during its insertion, serving as a propulsion. Groth et al. [15] reported a cecal intubation rate of 98.4% and a mean time of 15 min. Only three out of 61 patients required sedation.

## ColonoSight system

It is an automatic advancement system composed of a reusable colonoscope, with LEDs and a camera at the tip, covered by a disposable multilumen sheath wrapped with a working channel, to avoid contact of the endoscope with potentially infectious agents, eliminating the need for disinfection [41]. The device is powered by an electro-pneumatic unit that inflates the outer sheath to generate, progressively unfolding it, a forward force on the distal tip that allows pulling the endoscope through the colon, reducing the overall "push" force required to insert the device. A multicenter study showed a cecal intubation rate of 90% in an average time of 11 min. No complications were reported [41].

## Endoscopic capsule

Endoscopic capsules represent an attractive alternative to traditional endoscopic techniques for gastrointestinal screening due to the absence of discomfort and the need for sedation; however, current models are passive devices, and it is not possible to control the direction of the camera. They are currently used in the investigation of the small intestine in occult bleeding and have no therapeutic properties [42]. When applied to the study of the colon, robotic capsule endoscopy can overcome the disadvantages of pain and discomfort of conventional colonoscopy, but still lack reliability and diagnostic accuracy and fail to perform therapeutic functions simultaneously [42,43].

A robotic capsule endoscopy platform should consist of six modules: locomotion, localization, vision, telemetry, energy, and diagnostic and therapeutic tools. However, most capsules developed to date have only a few of these functions [42]. Active locomotion of the capsules can be carried out by the capsule itself (through the beating of tails, "legs," "oars," or propellers) or externally by magnetic energy [35,44]. One of the greatest difficulties in obtaining self-propelled capsules is the durability of their energy module, since the batteries need to be very small to fit inside the capsule [16,42]. Externally powered locomotion is more feasible and uses magnets to create force fields that interact with magnetic components inside the capsule, not requiring the presence of locomotion components in the capsule or batteries.

## NaviCam system

Platform applied only to the gastric district so far is assisted by robotics and capable of magnetically navigating a wireless endoscopic probe. The external static magnetic field generated by the NaviCam platform precisely controls, with 5 DOFs, an endoscopic capsule of 28 × 12 mm, which incorporates a CMOS camera with a field of view of 140° and a depth of field of 0 to 60 mm, LEDs, and a permanent magnet [45]. The platform received CFDA mark approval with a Class III medical device registration certificate entitled "Magnetically Controlled Robotic Capsule Endoscope."

## Future scenario

Flexible robotic systems are starting to appear on the market and their future looks promising, especially for growing applications in the anatomy of the gastrointestinal tract. More technological advances will be needed to make the entire GI tract accessible for therapeutic endoscopy.

Robotics, such as therapeutic endoscopy, is an exciting and ever-evolving field.

## Key points

- Robotic therapeutic endoscopy allows endoscopic-surgical procedures to be carried out safely, with a low risk of complications and favoring an early return to activity for the patient involved.
- The development of robotic endoscopy will be spread to the upper digestive tract endoscopy, bronchoscopy, colonoscopy, endoscopic capsules, and transluminal endoscopic surgery through natural orifices (NOTES).
- Robotic technology and balanced and precise triangulation raise the level of benefit in patient outcomes.
- More technological advances will be needed to make the entire GI tract accessible for therapeutic endoscopy.
- In the future, robotic endoscopy has the potential to enhance the capabilities of health professionals in the areas of telemanipulation, patient-specific image-based navigation, automated image analysis, diagnosis, and much more.
- The cost–benefit ratio is still unfavorable for the widespread use of this model in public health policies

## References

[1] Spaner SJ, Warnock GL. A brief history of endoscopy, laparoscopy, and laparoscopic surgery. J Laparoendosc Adv Surg Tech A 1997;7 (6):369–73. Available from: http://www.liebertpub.com/doi/10.1089/lap.1997.7.369.

[2] De Groen PC. History of the endoscope [scanning our past]. Proc IEEE 2017;105(10):1987–95. Available from: http://ieeexplore.ieee.org/document/8047436/.

[3] Davis CJ. A history of endoscopic surgery. Surg Laparosc Endosc 1992;2(1):16–23. Available from: http://www.ncbi.nlm.nih.gov/pubmed/1341495.

[4] Marsh BR. Historic development of bronchoesophagology. Otolaryngol Head Neck Surg 1996;114(6):689–716. Available from: http://www.ncbi.nlm.nih.gov/pubmed/8643291.

[5] Suzuki N. Teaching college students principle of endoscopes through an educational method of image processing. Int J Biomed Sci Eng 2015;3 (1):5. Available from: http://www.sciencepublishinggroup.com/journal/paperinfo.aspx?journalid=259&doi=10.11648/j.ijbse.20150301.12.

[6] Atallah S. Assessment of a flexible robotic system for endoluminal applications and transanal total mesorectal excision (taTME): could this be the solution we have been searching for? Tech Coloproctol 2017;21(10):809–14. Available from: http://link.springer.com/10.1007/s10151-017-1697-6.

[7] Marescaux J, Leroy J, Rubino F, Smith M, Vix M, Simone M, et al. Transcontinental robot-assisted remote telesurgery: feasibility and potential applications. Ann Surg 2002;235(4):487–92. Available from: http://journals.lww.com/00000658-200204000-00005.

[8] Mascagni P, Lim SG, Fiorillo C, Zanne P, Nageotte F, Zorn L, et al. Democratizing endoscopic submucosal dissection: single-operator fully robotic colorectal endoscopic submucosal dissection in a pig model. Gastroenterol Int 2019;156(6):1569–1571.e2. Available from: https://linkinghub.elsevier.com/retrieve/pii/S0016508519303671.

[9] Kume K. Flexible robotic endoscopy: current and original devices. Comput Assist Surg 2016;21(1):150–9.

[10] Kuriki P, Mota V, Moricz AD, Sassatani AS, Campos TD. NOTES / CETON – Cirurgia Endoscópica Transluminal por Orifício Natural : revisão de literatura. Arq Med Hosp Fac Cienc Med Santa Casa São Paulo 2008;53(3):118–24.

[11] Phee SJ, Low SC, Huynh VA, Kencana AP, Sun ZL, Yang K. Master and slave transluminal endoscopic robot (MASTER) for natural Orifice Transluminal Endoscopic Surgery (NOTES). In: 2009 Annual International Conference of the IEEE Engineering in Medicine and Biology Society; 2009. p. 1192–5.

[12] Spaun GO, Zheng B, Swanström LL. A multitasking platform for natural orifice translumenal endoscopic surgery (NOTES ): a benchtop comparison of a new device for flexible endoscopic surgery and a standard dual-channel endoscope. Surg Endosc 2009;2720–7.

[13] Zorn L, Nageotte F, Zanne P, Legner A, Dallemagne B, Marescaux J, et al. A novel Telemanipulated robotic assistant for surgical endoscopy: preclinical application to ESD. IEEE Trans Biomed Eng 2018;65(4):797–808.

[14] Tumino E, Sacco R, Bertini M, Bertoni M, Parisi G, Capria A. Endotics system vs colonoscopy for the detection of polyps. World J Gastroenterol 2010;16(43):5452–6.

[15] Groth S, Rex DK, Thomas R, Hoepffner N. High cecal intubation rates with a new computer- assisted colonoscope : a feasibility study. Am J Gastroenterol 2011;1075–80.

[16] Quirini M, Member S, Menciassi A, Scapellato S, Member S, Stefanini C, et al. Design and fabrication of a motor legged capsule for the active exploration of the gastrointestinal tract. IEEE/ASME Trans Mechatron 2008;13(2):169–79.

[17] Rey JF, Ogata H, Hosoe N, Ohtsuka K, Ogata N, Ikeda K, et al. Feasibility of stomach exploration with a guided capsule endoscope. Endoscopy 2010;42(7):541–5.

[18] Zhao J, Feng B, Zheng M-H, Xu K. Surgical robots for SPL and NOTES: a review. Minim Invasive Ther Allied Technol 2015;24(1):8–17. Available from: http://www.tandfonline.com/doi/full/10.3109/13645706.2014.999687.

[19] Ullah MI, Ajwad SA, Irfan M, Iqbal J. Non-linear control law for articulated serial manipulators: simulation augmented with hardware implementation. Elektron ir Elektrotechnika 2016;22(1). Available from: http://www.eejournal.ktu.lt/index.php/elt/article/view/14094.

[20] Alam W, Mehmood A, Ali K, Alharbi S, Iqbal J. Nonlinear control of a flexible joint robotic manipulator with experimental validation. Strojniški Vestn - J Mech Eng 2018;64(1). Available from: http://www.sv-jme.eu/article/nonlinear-control-of-flexible-joint-robotic-manipulator-with-experimental-validation/.

[21] Podobnik J, Munih M. Haptic interaction stability with respect to grasp force. IEEE Trans Syst Man Cybern Part C (Applications Rev) 2007;37 (6):1214–22. Available from: http://ieeexplore.ieee.org/document/4343998/.

[22] Zuo S, Wang S. Current and emerging robotic assisted intervention for Notes. Expert Rev Med Devices 2016;13(12):1095–105. Available from: https://www.tandfonline.com/doi/full/10.1080/17434440.2016.1254037.

[23] Du H, Liu X, Sun H, Zhu Q, Sun L. Progress in control-actuation robotic system for gastrointestinal NOTES development. Jamshed I, editor, Biomed Res Int 2022;30:1–9. Available from: https://www.hindawi.com/journals/bmri/2022/7047481/.

[24] Abbott DJ, Becke C, Rothstein RI, Peine WJ. Design of an endoluminal NOTES robotic system. In: 2007 IEEE/RSJ international conference on intelligent robots and systems. IEEE; 2007. p. 410–6. Available from: http://ieeexplore.ieee.org/document/4399536/.

[25] Ikeda K, Sumiyama K, Tajiri H, Yasuda K, Kitano S. Evaluation of a new multitasking platform for endoscopic full-thickness resection. Gastrointest Endosc 2011;73(1):117–22. Available from: https://linkinghub.elsevier.com/retrieve/pii/S0016510710020997.

[26] Fuchs K-H, Breithaupt W. Transgastric small bowel resection with the new multitasking platform EndoSAMURAI™ for natural orifice transluminal endoscopic surgery. Surg Endosc 2012;26(8):2281–7. Available from: http://link.springer.com/10.1007/s00464-012-2173-z.

[27] Ho K-Y, Phee SJ, Shabbir A, Low SC, Huynh VA, Kencana AP, et al. Endoscopic submucosal dissection of gastric lesions by using a master and slave transluminal endoscopic robot (MASTER). Gastrointest Endosc 2010;72(3):593–9. Available from: https://linkinghub.elsevier.com/retrieve/pii/S0016510710015233.

[28] Phee SJ, Reddy N, Chiu PWY, Rebala P, Rao GV, Wang Z, et al. Robot-assisted endoscopic submucosal dissection is effective in treating patients with early-stage gastric neoplasia. Clin Gastroenterol Hepatol 2012;10(10):1117–21. Available from: https://linkinghub.elsevier.com/retrieve/pii/S154235651200626X.

[29] Phee SJ, Ho KY, Lomanto D, Low SC, Huynh VA, Kencana AP, et al. Natural orifice transgastric endoscopic wedge hepatic resection in an experimental model using an intuitively controlled master and slave transluminal endoscopic robot (MASTER). Surg Endosc 2010;24 (9):2293–8. Available from: http://link.springer.com/10.1007/s00464-010-0955-8.

[30] Zhao J, Zheng X, Zheng M, Shih AJ, Kai X. An endoscopic continuum testbed for finalizing system characteristics of a surgical robot for NOTES procedures. In: 2013 IEEE/ASME international conference on advanced intelligent mechatronics. IEEE; 2013. p. 63–70. Available from: http://ieeexplore.ieee.org/document/6584069/.

[31] Lau KC, Leung EYY, Chiu PWY, Yam Y, Lau JYW, Poon CCY. A flexible surgical robotic system for removal of early-stage gastrointestinal cancers by endoscopic submucosal dissection. IEEE Trans Ind Inform 2016;12(6):2365–74. Available from: http://ieeexplore.ieee.org/document/7484685/.

[32] Vrielink TJCO, Zhao M, Darzi A, Mylonas GP. ESD CYCLOPS: A new robotic surgical system for GI surgery. In: 2018 IEEE international conference on robotics and automation (ICRA). IEEE; 2018. p. 150–7. Available from: https://ieeexplore.ieee.org/document/8462698/.

[33] Cao L, Li X, Phan PT, Tiong AMH, Kaan HL, Liu J, et al. Sewing up the wounds: a robotic suturing system for flexible endoscopy. IEEE Robot Autom Mag 2020;27(3):45–54. Available from: https://ieeexplore.ieee.org/document/8994188/.

[34] Hwang M, Kwon D. K-FLEX: A flexible robotic platform for scar-free endoscopic surgery. Int J Med Robot Comput Assist Surg 2020;16(2). Available from: https://onlinelibrary.wiley.com/doi/10.1002/rcs.2078.

[35] Li Z, Chiu PWY. Robotic endoscopy. Visc Med 2018;34(1):45–51.

[36] Colonoscope S, Vucelic B, Rex D, Pulanic R, Pfefer J, Hrstic I, et al. The Aer-O-scope : proof of concept of a pneumatic. Skill 2006;672–7.

[37] Gan R. Proof – of – concept study of the Aer – O – Scope omnidirectional colonoscopic viewing system in ex vivo and in vivo porcine models. Endoscopy 2007;412–7.

[38] Gluck N, Melhem A, Halpern Z, Mergener K, Goldfarb S, Santo E. Su1709 Aer-O-scope colonoscope system demonstrates efficacy and safety for colorectal cancer screening in humans. Gastrointest Endosc 2015;81(5). AB386.

[39] Trends N. Functional evaluation of the Endotics System, a new disposable self-propelled robotic colonoscope : in vitro tests and clinical trial. Int J Artif Organs 2009;32(8):517–27.

[40] Eickhoff A, Van DJ, Ph D, Jakobs R, Kudis V, Hartmann D, et al. Computer-assisted colonoscopy ( the NeoGuide endoscopy system ): results of the first human clinical trial ("PACE Study"). Am J Gastroenterol 2007;102(2):261–6.

[41] Shike M, Fireman Z, Eliakim R, Segol O, Sloyer A, Cohen LB, et al. Sightline ColonoSight system for a disposable, power-assisted, non-fiberoptic colonoscopy (with video). Gastrointest Endosc 2008;68(4):701–10. Available from: https://linkinghub.elsevier.com/retrieve/pii/S001651070800014X.

[42] Ciuti G, Caliò R, Camboni D, Neri L, Bianchi F, Arezzo A, et al. Frontiers of robotic endoscopic capsules: a review. J Micro-Bio Robot 2016;11 (1–4):1–18.

[43] Ciuti G, Menciassi A, Dario P. Capsule endoscopy : from current achievements to open challenges. IEEE Rev Biomed Eng 2011;4:59–72.

[44] Ciuti G, Donlin R, Valdastri P, Arezzo A, Menciassi A, Morino M, et al. Robotic versus manual control in magnetic steering of an endoscopic capsule. Endoscopy 2009;148–52.

[45] Liao Z, Hou X, Lin-Hu E-Q, Sheng J-Q, Ge Z-Z, Jiang B, et al. Accuracy of magnetically controlled capsule endoscopy, compared with conventional gastroscopy, in detection of gastric diseases. Clin Gastroenterol Hepatol 2016;14(9):1266–1273.e1. Available from: https://linkinghub.elsevier.com/retrieve/pii/S1542356516302002.

III. The current and future clinical applications of robotic surgery among medical specialties

# 76

# Energy sources and vessel sealers in robotic surgery

*Luiz Carlos Von Bahten[a], Leonardo Emilio da Silva[b,c], and Flavio Daniel Saavedra Tomasich[d,e,f,g]*

[a]Clinica Cirurgica, UFPR, Curitiba, Parana, Brazil [b]Department of Surgery, Faculty of Medicine, Federal University of Goias, Goiânia, Brazil [c]Robotic Surgery, Hospital Israelita Albert Einstein, Goiania, Goias, Brazil [d]Department of Surgery, Federal University of Paraná, Curitiba, Paraná, Brazil [e]National Directory, Brazilian College of Surgeons (CBC), Rio de Janeiro, Brazil [f]Committee on Minimally Invasive Surgery and Robotic Surgery, Brazilian College of Surgeons (CBC), Rio de Janeiro, Brazil [g]Department of Surgery, Erasto Gaertner Cancer Center, Curitiba, Paraná, Brazil

## Introduction

Robotic surgery has exponentially increased over the last decade. It has gained popularity due to its ability to provide surgeons with enhanced agility and visualization, improving surgical precision. The introduction of robotic technology in surgery has allowed for innovation, while creating several new challenges that teams must overcome to provide safe and efficient surgical care [1].

As minimally invasive surgical techniques progress, the demand for efficient, reliable methods for vascular ligation and tissue closure becomes pronounced. Consequently, energy sealing has advanced rapidly in the last decade, with many new devices becoming available. With the advent of new technology, it is crucial to understand how instruments work to utilize them and prevent injury entirely. Energy sources and vessel sealers are essential tools in achieving hemostasis and tissue dissection during robotic procedures and represent a significant advancement in robotic surgery, providing surgeons with an intuitive and reliable tool. Likewise, energy devices help in shortening the console time in robotic surgery [2]. This chapter will explore the different types of energy sources and vessel sealers commonly used in robotic surgery.

## Energy sources

### Electrosurgery

Tissue fusion is achieved by the introduction of heat to tissues via one or more energy delivery modalities, including monopolar or bipolar radio-frequency current, direct conduction via heated platens, and ultrasonically oscillated fusion "blades," or by the incidence of optical energy in the superficial or deep tissue layers, such as laser welding [3]. The selection of energy delivery modality may vary with the practitioner's preference of devices or with the intended application and desired tissue.

The term "electrocautery" pertains to an electrically heated wire contacting tissue without transferring electricity to the patient. "Electrosurgery" stems from Morton's observation in 1881 that alternating current of 100 kHz could safely and painlessly pass through the body without inducing muscle spasms or electrical burns; this paved the way for advances at the turn of the century [4].

In electrosurgery, energy is converted from electromagnetic to kinetic and finally to thermal to create several effects on cells and tissues. These effects are proportionate to the temperature of the tissues relative to the duration of the application and the compression applied. Below 40°C, reversible cell damage occurs. Above 49°C denaturation occurs with irreversible cell damage. Coagulation occurs at temperatures over 70°C, with collagen conversion to gelatin. Above 100°C, desiccation, or drying of cells, and gelatin stickiness occur, and above 200°C, carbonization occurs, with deep tissue necrosis observed as black eschar formation [4].

Electrosurgery refers to surgical techniques that use electrical currents to cut, coagulate, desiccate, or fulgurate tissues. These techniques are broadly classified into two types: monopolar and bipolar electrosurgery. They leverage electrical energy to generate heat, affecting tissues in various ways [5].

Electrosurgery systems exist in monopolar or bipolar forms. The tissues of the patient form part of the circuit in both. Monopolar systems generate a current that is passed through a metal-tipped probe. The current radiates out of this electrode as an electrical field, heating tissues in contact with it. An adhesive cutaneous grounding pad, or "indifferent plate," applied to an extremity connects the patient to the completed circuit. Bipolar methods use two electrodes within a single probe, between which a current is passed to cauterize tissues. The electrodes may be the two tines of a pair of forceps or the jaws of a laparoscopic grasper. Current administration to any vessel between bipolar electrodes will cause to shrink with luminal thrombus formation but not complete obliteration of the lumen [4,6–9].

## Monopolar electrosurgery

Monopolar energy remains the most widely used dissection tool in robotic surgery. This approach passes an electrical current from a generator to a monopolar instrument through the patient's body and back to the generator via a grounding pad. The high-frequency electrical current is concentrated at the instrument's tip, allowing precise cutting or coagulation of tissue. Monopolar electrosurgery is versatile and practical for a wide range of surgical procedures. A more appropriate rationale to support the designation "monopolar" is that the active electrode in monopolar electrosurgery contains only one of the poles in the circuit. In this construct, the patient is the other electrode [10].

Monopolar electrosurgery requires considerable knowledge, understanding, and vigilance of the operator to avoid the hazards of unintentional thermal injury using accidental visceral contact with active or heated electrodes; direct or capacitive coupling; insulation defects in instruments or connecting wires; damaged, faulty, or improper placement of the return electrode; and combustion of volatile substances [11].

Lack of basic electrosurgical knowledge or ignorance of principles of electrosurgery and equipment among surgeons has been reported. As a result, thermal injuries during laparoscopic electrosurgery occur, which frequently lead to significant morbidity and mortality rates and medicolegal actions. A survey of 506 attendants at the American College of Surgeons indicated that 85.6% of respondents used monopolar electrosurgery during laparoscopy, and 18% reported internal laparoscopic burns [12]. In another survey, 13% of attendants of the 1995 meeting of the Society of Laparoscopic Surgeons reported one or more ongoing litigations from laparoscopic burns [13,14].

## Bipolar electrosurgery

Bipolar electrosurgery, in contrast, involves the passage of current between two tips of a forceps-like instrument. Thus, bipolar electrosurgery consolidates an active electrode and returns the electric current into an instrument, separated by a small distance. Dissimilar to monopolar electrosurgery, the intervening tissue is part of the electrical circuit rather than the patient.

This technique confines the electrical current to the tissue between the instrument's tips, offering greater precision and reduced risk of electrical injury to surrounding tissues. Bipolar electrosurgery is particularly beneficial in delicate surgeries or in areas with a high density of critical structures.

Bipolar electrodes cannot cut tissues. Although a continuous ("cut") waveform is applied using bipolar instruments, cutting is inefficient because the amount of tissue involved is minimal, and vaporization is inefficient and cumbersome. Because the corona discharge travels in opposite directions along the two cables, it cancels itself out, and capacitive coupling does not occur. Furthermore, a return/dispersive electrode is not required, and the risk of dispersive electrode burns is also eliminated [5].

Despite the many technical advantages, tissue ends using bipolar electrosurgery are not always predictable; hemostasis often requires several contiguous applications, application duration per coaptation is entirely subjective, and thermal spread can significantly grow beyond the tissue pedicle. Finally, despite the surgeon's judgment that coagulation has been reached, unintentional transection of a patent vascular core cannot be eliminated. Opposed to the security of a suture ligature, coaptive hemostasis using conventional bipolar electrosurgery is accomplished by blood vessel shrinkage along with a proximal thrombus [15].

## Advantages and limitations

Advantages:

- Precision and Control: Both techniques offer high precision in cutting and coagulating tissues, which is essential for complex surgeries.
- The bidirectional current flow eliminates laparoscopic monopolar electrosurgery's direct and capacitive coupling risk.
- Reduced Blood Loss: Electrosurgery often results in less bleeding than traditional surgical methods.
- Versatility: A wide range of procedures can be performed, from minor outpatient surgeries to significant operations.

Limitations:

- Thermal Injury Risk: The heat generated can cause burns or damages to adjacent tissues if not used carefully.
- Contraindications: Certain patient conditions, such as the presence of a pacemaker, may limit the use of electrosurgery.
- Learning Curve: Surgeons require specific training and experience to master these techniques and understand their nuances.

# Ultrasonic energy

## Mechanism of action

As early as the 1970s, ultrasonic cauterization methods have been used for hemostasis by applying high vibrational frequency rather than electrical energy. Ultrasonic energy in robotic surgery involves the use of high-frequency sound waves. An ultrasonic transducer that converts electrical energy into mechanical vibrations generates this energy. These vibrations are transmitted to the surgical site, where they cause rapid movement of the surgical instrument's tip. An ultrasound transducer drives a vibrating blade and oscillates longitudinally against a nonvibrating pad, thus disrupting hydrogen bonds in proteins and forming a coagulum that subsequently seals the vessels. The friction created by this movement generates heat that can cut through or coagulate tissues, depending on the intensity and exposure time [16].

There are two types of ultrasonic dissectors. Low power dissectors that cleave water-containing tissues by cavitation and leave organized structures with low water content intact (e.g., vessels)—Cusa Cavitron Ultrasonic Surgical Aspirator, Valleylab, Boulder, CO, USA; SelectorH, Surgical Technical Group, Hampshire, GB. The other, is a high-power dissector that cleaves all the tissues—Harmonic devices [16].

Harmonic devices (Ethicon Inc., Cincinnati, OH, USA); Sonicision (Medtronic Inc., Minneapolis, MN, USA) curved jaw cordless ultrasonic dissection system have led the evolution of ultrasonic technology. They were developed and are capable of simultaneous cutting and coagulation using high-frequency vibrations in the range of 55,000 Hz. Numerous studies have shown that Harmonic devices, compared to conventional electrosurgery, are associated with superior coagulation with less thermal damage. It reduced surgical smoke production and improved surgical outcomes [17].

The consistent observations for reductions in pain and length of stay with Harmonic devices across many procedure types may be partially attributed to fewer thermal tissue damages associated with ultrasonic methods.

Monopolar electrosurgical devices cut and coagulate using a current to produce high temperatures (150–400°C), resulting in cell explosion and subsequent hemostasis [18].

Some data show significant cost advantages for Harmonic devices compared to conventional techniques [19,20].

## Advantages and limitations

Advantages:

- Precision: Allows for precise cutting and coagulation, minimizing damage to surrounding tissues.
- Reduced Blood Loss: Effective coagulation leads to reduced bleeding.
- Minimal Thermal Spread: Lower risk of collateral thermal damage compared to other energy sources.

Limitations:

- Learning Curve: Requires specific training to use effectively.
- Equipment Cost: High initial investment in specialized equipment.
- Tissue Specificity: This may not be as effective on all tissue types, requiring alternative methods and procedures.

## Laser energy

### Mechanism of action

Lasers in laparoscopic and robotic surgery are extremely limited. Available data consist primarily of small cohorts providing low evidence [21]. Laser energy in robotic surgery utilizes concentrated light beams. These lasers can be adjusted regarding wavelength, pulse duration, and energy output, allowing for targeted application. The laser energy is absorbed by tissues, causing localized heating. This can vaporize tissues for cutting or create a coagulative effect for sealing blood vessels.

Even though initial studies with currently available laser modalities demonstrated promising results, several drawbacks in each technique must be addressed before being widely accepted as standard care. Despite investigation, laser usage during laparoscopic and robotic urological procedures has not gained widespread acceptance and remains experimental [22].

### Advantages and limitations

Advantages:

- High Precision: Enables exact tissue targeting.
- Versatility: Effective on various tissues and for different surgical applications.
- Reduced Trauma: Minimizes physical contact, reducing tissue trauma.

Limitations:

- Cost: Involves significant investment in equipment and maintenance.
- Safety Concerns: Risk of eye damage and burns if not properly handled.
- Training Requirements: Demands high skill and training to use effectively.

## Vessel sealers

## Bipolar vessel sealers

True vessel wall fusion can be achieved using a combination of pressure and pulsed energy to denature collagen and elastin in tissue bundles, vessel walls, and lymphatics to reform into a permanent plastic-like seal that resists deformation with tensile strength of up to 3 times normal systolic pressure. Sealing tissue is grasped in the instrument's jaws, and a calibrated force is applied to the tissue, while temperature generation is minimized. The tissue type held in the forceps is diagnosed using proprietary generator technology, and the appropriate amount of energy is delivered to seal the tissue [23].

The clinical evaluation of new bipolar devices should reach well beyond their capacity to desiccate/coagulate and cut tissues. Other pivotal considerations include the relative responsiveness to variable tissue content, tissue tension, and efficacy in fat, as well as the propensity for smoke production and tissue sticking. Each instrument should also be evaluated as an operative instrument, including the relative ability to grasp, elevate, dissect, and coapt tissues of variable densities and perimeters [15].

The Vessel Sealer Extend (Intuitive Surgical, Inc., Sunnyvale, CA, USA) is a device designed specifically for robotic surgery, allowing for efficient and effective vessel sealing during surgical interventions. [24].

The Vessel Sealer Extend is one of the most potent hemostatic energy devices and is based on the principle of coagulation by bipolar electrodes. It performs coagulation and hemostasis with a pulsed output and transects the hemostatic tissue with an internal knife or incision electrode. Bipolar vessel sealers use electrical energy passed between two electrodes to coagulate and seal blood vessels. The energy denatures the collagen and elastin in vessel walls, causing them to fuse. This process creates a permanent seal that prevents bleeding.

The Vessel Sealer Extend is a versatile tool that seamlessly integrates with robotic surgical systems. It offers several key features that contribute to its effectiveness in vessel sealing.

The Vessel Sealer Extend utilizes advanced bipolar energy technology to seal and divide vessels up to 7 mm in diameter. This technology ensures precise and controlled energy delivery, minimizing thermal spread, and reducing the risk of collateral tissue damage.

The device has an articulating tip that enhances maneuverability and access to difficult-to-reach anatomical structures. This feature allows surgeons to navigate complex surgical sites easily, ensuring optimal sealing of vessels in challenging anatomical locations.

Real-time Feedback: The Vessel Sealer Extend provides real-time feedback to the surgeon, including audible and visual cues, ensuring accurate and reliable vessel sealing. This feedback system enhances surgical precision and promotes safer procedures.

## Benefits and clinical applications

The Vessel Sealer Extend offers numerous benefits in robotic surgery, making it an indispensable tool for surgeons. Some of the key advantages include:

- Enhanced Efficiency: The device enables faster vessel sealing, reducing surgical time. This efficiency translates into shorter patient anesthesia exposure, decreased blood loss, and improved surgical outcomes.
- Reduced Complications: The Vessel Sealer Extends advanced energy delivery system minimizes the risk of vessel leaks and bleeding, reducing the likelihood of postoperative complications such as hematoma formation and seroma.
- Versatility: The device is suitable for a wide range of surgical procedures, including but not limited to colorectal surgery, gynecological procedures, urological interventions, and general surgery. Its versatility allows surgeons to address various clinical scenarios with confidence.
- Cost-effectiveness: By reducing surgical time and minimizing postoperative complications, the Vessel Sealer Extend can contribute to cost-effectiveness in robotic surgery. It optimizes resource utilization and enhances patient satisfaction.

## Maryland bipolar forceps vessel sealing system

### Mechanism of action

In recent years, the concept of refined and tubeless surgery gradually emerged due to the requirement of rapid postoperative recovery [25]. Therefore, coarse, blunt, and thermally damaged electrocoagulation forceps cannot meet the requirements of robotic surgery. Consequently, medical centers are gradually replacing these instruments with bipolar Maryland Bipolar Forceps. The Maryland Bipolar Forceps (Medtronic Inc., Minneapolis, MN, USA) system combines bipolar electrical energy and pressure to seal vessels. It applies precise energy and pressure to denature collagen and elastin in the vessel walls, causing them to meld together and form a permanent seal. It is used for accurate dissection, similarly to monopolar scissors, and by closing the forceps to compress tissue and applying electric current, the more vital hemostatic ability can be achieved. These forceps are used at low- or high-voltage generator settings [26].

A notable difference between these forceps and Cadiere Forceps is that if the hand control in the surgeon's console is 30 degrees open, Cadiere Forceps would be 30 degrees available; Maryland Bipolar Forceps would be 45 degrees open, and long Maryland Bipolar Grasper would be 70 degrees open. In other words, the opening angle of hand control in the surgeon's console and the opening angle of most instruments in the patient cart are not the same, particularly in the case of the long Maryland Bipolar Grasper that would be open at an angle that is more than twice that of the hand control [24]. Maryland Bipolar Forceps are both safe and effective and have significant advantages in reducing operative time, intraoperative bleeding, and surgical trauma [27].

### Advantages and limitations

Advantages

- Versatility: Effective on a wide range of vessel sizes and types.
- Reduced Bleeding: Significantly minimizes intraoperative bleeding.
- Speed: Faster than traditional suturing methods.

Limitations

- Device Dependency: Relies on the availability and functionality of specific equipment.
- Cost: Higher initial and operational costs.
- Training: Requires specialized training for optimal use.

# SynchroSeal (Intuitive Surgical, Inc., Sunnyvale, CA)

### Features and functionality

SynchroSeal is a novel robotic-assisted bipolar vessel-sealing instrument with radio-frequency transection capability that can replace a mechanical knife, enabling simultaneous seal and transection of vessels and tissue bundles [28]. The multifunctional capabilities of SynchroSeal include grasping, blunt dissection, and tissue access.

In preclinical tests outcomes of a study comparing with Harmonic Ace + 7, SynchroSeal presented superior grasping capabilities, and was more efficient in sealing activation time and cooldown time, and has a safer profile in terms of jaw temperature [29]. It can successfully seal vessels at pressures significantly higher than the usual systolic pressure, and vessel can maintain their integrity in vivo at the time of the seal and during a postprocedure period of 3 weeks [28]. The SynchroSeal is a cutting-edge device that seamlessly integrates with robotic surgical systems, offering several key features that contribute to its effectiveness in tissue closure:

### Synchronized stapling and cutting

The SynchroSeal combines stapling and cutting functionalities in a single device, allowing for simultaneous tissue closure and division. This synchronized action streamlines the surgical workflow, reducing procedural time and enhancing efficiency.

### Adaptive firing technology

The device utilizes adaptive firing technology, automatically adjusting staple height based on tissue thickness. This feature ensures consistent staple formation and optimal tissue compression, minimizing the risk of postoperative complications such as bleeding and leaks.

### Intelligent feedback system

This incorporates an intelligent feedback system that provides real-time information to the surgeon. This includes audible and visual cues, ensuring accurate staple formation, and preventing potential errors during tissue closure.
Benefits and Clinical Applications [30]:
The Intuitive SynchroSeal offers benefits in robotic surgery. Some of the key advantages include:

- Enhanced Precision: The device's synchronized stapling and cutting mechanism allows for precise tissue closure, reducing the risk of tissue damage and ensuring optimal wound healing. This precision contributes to improved surgical outcomes and patient satisfaction.
- Reduced Complications: The adaptive firing technology of the SynchroSeal ensures consistent staple formation, minimizing the risk of postoperative complications such as bleeding, leaks, and strictures. This feature promotes safer procedures and reduces the need for additional interventions. The histology study evidenced a mean thermal spread of less than 2 mm for both Sync and Seal modes, which suggests minimal thermal spread adding to the safety profile of the tool [28].
- Versatility: The device suits various surgical procedures, including gastrointestinal surgeries, thoracic procedures, and general surgery. Its versatility allows surgeons to confidently address multiple clinical scenarios, promoting its widespread adoption in robotic surgery.
- Streamlined Workflow: The synchronized stapling and cutting functionality of the SynchroSeal streamlines the surgical workflow, reducing procedural time and optimizing resource utilization. This efficiency can minimize anesthesia exposure

## Safe selection and utilization of energy sources and vessel sealers

## Considerations for device selection

We reinforce that it is necessary to analyze several critical factors in selecting energy sources and vessel sealers for robotic surgery, including tissue type, surgical procedure specifics, and device capabilities. We emphasize the importance of matching the device properties with surgical needs.

## Considerations of safety precautions

We must insist on the daily practice to take account of comprehensively the safety measures for operating energy devices and vessel sealers. This includes the prevention of accidental burns, avoiding collateral damage to adjacent structures, and ensuring patient safety.

## Appropriate technique and training

It is necessary to explore deeply the requirements for training and for the refined techniques aiming at an effective and safe use of energy devices in robotic surgery. This includes mentoring, proctoring, hands-on training, simulation-based learning, and continuous education.

## Prevention of thermal injury

It is recommendable to keep in mind strategies and best practices to minimize risks of thermal injury during robotic procedures. This involves understanding device-specific heat profiles and avoiding prolonged tissue thermal exposure.

# Future scenario

## Advanced energy sources

The development of new energy sources for robotic surgery, including ultrasonic and laser technologies will progress more and more. These innovations probably will impact on surgical outcomes.

## Integration of imaging technologies

The integrating of advanced imaging technologies with robotic platforms, which enhances the surgeon's ability to visualize and target tissue with energy devices accurately, will certainly advance.

## Robotic feedback systems

Efforts in course for the haptic feedback systems in modern robotic surgery platforms, probably will provide surgeons with tactile sensations, and enhance their control and precision when using energy devices.

# Conclusion

During robotic surgery, energy sources and vessel sealers are vital for achieving hemostasis and tissue dissection. Understanding these devices' mechanisms of action, advantages, and limitations is crucial for surgeons to make informed decisions during procedures. Appropriate selection, utilization, and adherence to safety precautions are essential for optimal surgical outcomes and patient safety. As robotic surgery continues to evolve, advancements in energy sources and vessel sealers will further enhance the field, improving patient care and outcomes.

# Key points

- Electrosurgery in Robotic Surgery: we discuss electrosurgery, including monopolar and bipolar techniques, highlighting their precision, control, and versatility, along with the associated risks like thermal injury and the need for specific surgeon training.
- Ultrasonic Energy: it covers the use of ultrasonic energy in robotic surgery, emphasizing its precision, reduced blood loss, and minimal thermal spread, while also noting the challenges like the learning curve and equipment cost.

- Laser Energy: the use of laser energy is detailed, focusing on its high precision, versatility, and reduced trauma, as well as the associated costs and safety concerns requiring careful handling and skilled operation. It has not been used in daily practice.
- Vessel Sealers: the chapter explores various vessel sealers like bipolar, ultrasonic, and the vessel sealer systems, discussing their mechanisms, advantages like effective hemostasis and speed, and limitations including vessel size constraints and heat generation risks.
- Clinical Applications and Future Directions: the applications of these technologies tend to expand in almost all surgical specialties. Among several future advancements, in this field, we can mention the future energy sources, the integration of new imaging technologies, tactile feedbacks for surgeons, and artificial intelligence.

# References

[1] Clanahan JM, Awad MM. E615-623 Medicine and society: peer-reviewed article how does robotic-assisted surgery change OR safety culture? AMA J Ethics 2023;25.
[2] Kuroda K, Kubo N, Sakurai K, Tamamori Y, Hasegawa T, Yonemitsu K, et al. Comparison of short-term surgical outcomes of two types of robotic gastrectomy for gastric cancer: ultrasonic shears method versus the maryland bipolar forceps method. J Gastrointest Surg 2023;27(2):222–32. Available from: https://pubmed.ncbi.nlm.nih.gov/36376726/.
[3] Kramer EA, Rentschler ME. Energy-based tissue fusion for Sutureless closure: applications, mechanisms, and potential for functional recovery. Annu Rev Biomed Eng 2018;20:1–20. Available from: https://pubmed.ncbi.nlm.nih.gov/29865874/.
[4] Shabbir A, Dargan D. Advancement and benefit of energy sealing in minimally invasive surgery. Asian J Endosc Surg 2014;7(2):95–101. Available from: https://pubmed.ncbi.nlm.nih.gov/24754878/.
[5] Vilos GA, Rajakumar C. Electrosurgical generators and monopolar and bipolar electrosurgery. J Minim Invasive Gynecol 2013;20(3):279–87. Available from: https://pubmed.ncbi.nlm.nih.gov/23659748/.
[6] Reidenbach HD. Fundamentals of bipolar high-frequency surgery. Endosc Surg Allied Technol 1993;1(2):85–90.
[7] Schurr MO, Farin G, Reidenbach HD. Bipolar high-frequency. Endosc Surg Allied Technol 1993;1(2):115–6.
[8] Tucker RD, Hollenhorst MJ. Bipolar electrosurgical devices. Endosc Surg Allied Technol 1993;1(2):110–3.
[9] Mueller W. The advantages of laparoscopic assisted bipolar high-frequency surgery. Endosc Surg Allied Technol 1993;1(2):91–6.
[10] Diana M, Marescaux J. Robotic surgery. Br J Surg 2015;102(2):e15–28. Available from: www.bjs.co.uk.
[11] Soderstrom RM. Electrosurgical injuries during laparoscopy: prevention and management. Curr Opin Obstet Gynecol 1994;6(3):248–50.
[12] Tucker RD. Laparoscopic electrosurgical injuries: survey results and their implications. Surg Laparosc Endosc 1995;5(4):311–7.
[13] Sacks ES. Focus on laparoscopy. Health Devices 1995;24:4–5.
[14] Abu-Rafea B, Vilos GA, Al-Obeed O, AlSheikh A, Vilos AG, Al-Mandeel H. Monopolar electrosurgery through single-port laparoscopy: a potential hidden hazard for bowel burns. J Minim Invasive Gynecol 2011;18(6).734–40. Available from: https://pubmed.ncbi.nlm.nih.gov/21925969/.
[15] Brill AI. Bipolar electrosurgery: convention and innovation. Clin Obstet Gynecol 2008;51(1):153–8. Available from: https://pubmed.ncbi.nlm.nih.gov/18303509/.
[16] Entezari K, Hoffmann P, Goris M, Peltier A, van Velthoven R. A review of currently available vessel sealing systems. Minim Invasive Ther Allied Technol 2024;16(1):52–7. Available from: https://pubmed.ncbi.nlm.nih.gov/17365677/.
[17] Kloosterman R, Wright GWJ, Salvo-Halloran EM, Ferko NC, Mennone JZ, Clymer JW, et al. An umbrella review of the surgical performance of harmonic ultrasonic devices and impact on patient outcomes. BMC Surg 2023;23(1). Available from: https://pubmed.ncbi.nlm.nih.gov/37386399/.
[18] Wiatrak BJ, Willging JP. Harmonic scalpel for tonsillectomy. Laryngoscope 2002;112(8 Pt 2 Suppl 100):14–6. Available from: https://pubmed.ncbi.nlm.nih.gov/12172231/.
[19] Cheng H, Clymer JW, Qadeer RA, Ferko N, Sadeghirad B, Cameron CG, et al. Procedure costs associated with the use of Harmonic devices compared to conventional techniques in various surgeries: a systematic review and meta-analysis. Clinicoecon Outcomes Res 2024;10:399–412. Available from: https://pubmed.ncbi.nlm.nih.gov/30087572/.
[20] Ferko N, Wright GWJ, Syed I, Naoumtchik E, Tommaselli GA, Gangoli G. A device category economic model of electrosurgery technologies across procedure types: a U.S. hospital budget impact analysis. J Med Econ 2021;24(1):524–35. Available from: https://pubmed.ncbi.nlm.nih.gov/33851557/.
[21] Werkhaven J. Laser applications in pediatric laryngeal surgery. Otolaryngol Clin North Am 1996;29(6):1005–10.
[22] Pushkar DY, Kolontarev KB. Lasers in laparoscopic and robotic surgery: is there a need for them. Curr Opin Urol 2022;32(2):199–203. Available from: https://pubmed.ncbi.nlm.nih.gov/34954704/.
[23] Ortenzi M, Ghiselli R, Baldarelli M, Cardinali L, Guerrieri M. Is the bipolar vessel sealer device an effective tool in robotic surgery? A retrospective analysis of our experience and a meta-analysis of the literature about different robotic procedures by investigating operative data and post-operative course. Minim Invasive Ther Allied Technol 2018;27(2):113–8. Available from: https://pubmed.ncbi.nlm.nih.gov/28604140/.
[24] Hirahara N, Matsubara T, Hayashi H, Tajima Y. Features and applications of energy devices for prone robot-assisted minimally invasive esophagectomy: a narrative review. J Thorac Dis 2022;14(9):3606–12. Available from: https://pubmed.ncbi.nlm.nih.gov/36245588/.
[25] Ross T, Tolley NS, Awad Z. Novel energy devices in head and neck robotic surgery—A narrative review. Robot Surg 2020;7:25–39. Available from: https://pubmed.ncbi.nlm.nih.gov/32426397/.
[26] Milsom JW, Trencheva K, Momose K, Peev MP, Christos P, Shukla PJ, et al. A pilot randomized controlled trial comparing THUNDERBEAT to the Maryland LigaSure energy device in laparoscopic left colon surgery. Surg Endosc 2022;36(6):4265–74. Available from: https://pubmed.ncbi.nlm.nih.gov/34724584/.

[27] Hong Z, Bai X, Sheng Y, Cui B, Lu Y, Cheng T, et al. Efficacy of using Maryland forceps versus electrocoagulation hooks in da Vinci robot-assisted thoracoscopic mediastinal tumor resection. World J Surg Oncol 2023;21(1). Available from: https://pubmed.ncbi.nlm.nih.gov/37337217/.

[28] Ibanez Jimenez C, Lath A, Ringold F. Novel multifunctional robotically assisted bipolar instrument for simultaneous radiofrequency sealing and transection: preclinical and single-center experience. BMC Surg 2022;22(1). Available from: https://pubmed.ncbi.nlm.nih.gov/35109833/.

[29] Pellegrino F, Tin AL, Sjoberg DD, Benfante NE, Weber RC, Porwal SP, et al. The effect of the da Vinci® vessel sealer on robot-assisted laparoscopic prostatectomy complications. J Robot Surg 2023;17(4):1763–8. Available from: https://pubmed.ncbi.nlm.nih.gov/37043122/.

[30] Abaza R, Henderson SJ, Martinez O. Robotic vessel sealer device for lymphocele prevention after pelvic lymphadenectomy: results of a randomized trial. J Laparoendosc Adv Surg Tech A 2022;32(7):721–6. Available from: https://pubmed.ncbi.nlm.nih.gov/34677080/.

# 77

# Robotic-assisted endovascular surgery

*Andressa Cristina Sposato Louzada[a], Guilherme Yazbek[b], and Nelson Wolosker[a]*

[a]Albert Einstein Israeli College of Health Sciences (FICSAE), São Paulo, Brazil [b]Vascular and Endovascular Surgery Department at AC Camargo Cancer Center, São Paulo, Brazil

## Introduction

As technology advances, it must be incorporated into medicine to benefit patients first and foremost, but also, ideally, healthcare professionals. Over the last few decades, we surgeons have seen a shift toward minimally invasive techniques, providing a drop in surgical complications, such as wound complications and postoperative pain, subsequently reducing recovery time and hospital stays. In vascular surgery specifically, we have even observed that the endovascular approach may improve survival, as is the case with endovascular aortic repair for ruptured abdominal aortic aneurysms, considered the gold standard repair [1].

Even though endovascular surgery has already been assimilated by most vascular surgeons and both the technique and the materials are constantly improving, there are still some challenges to overcome, such as:

— Ionizing radiation exposure for patients and especially for the endovascular team
— Two-dimensional angiographic projection images
— Anatomical limitations
— Stability of catheters and guidewires
— Risk of distal embolization
— Risk of vessel or organ injury

Each of these limitations is addressed and may be overcome by endovascular robotics. Robotic surgery is the new vanguard of surgical innovation and, although it has been already consolidated in other specialties, such as urology and gastrointestinal surgery, most of the robots available represent an improvement on laparoscopy/video-assisted thoracoscopy. Still, this kind of approach is not really the most suitable for vascular surgery; therefore, robotic devices and platforms compatible with our minimally invasive surgery, endovascular surgery, are needed. This chapter will discuss the main robotic devices developed to assist endovascular surgeries.

## Steerable robotic catheters

Standard catheters have fixed distal tips single-axis rotation, often lack stability for target vessel catheterization, and their movements are highly dependent on the skills of the endovascular surgeon [2]. These characteristics increase the risks of unsuccessful cannulation or the need for multiple catheters, operator tremor interference, and vessel trauma, including dissection, perforation, and distal embolization. Robotic catheters are being designed to overcome these challenges. Examples of these robotic catheters are Magellan (Hansen Medical, Inc.), Niobe system (Stereotaxis, Inc.), Amigo (Catheter Precision, Inc.), and prototype CathROB [3]; these last two more are intended for cardiac electrophysiologists.

*Handbook of Robotic Surgery*
https://doi.org/10.1016/B978-0-443-13271-1.00065-0

## Magellan and Sensei X2 (Hansen Medical, Inc., Mountain View, CA, USA)

The Magellan system is the most widely studied and used catheter in endovascular treatments. It comprises a robotic arm at the patient table, including a single steerable catheter with two bending sections that can be controlled on a remote physician console, which can be placed outside the radiation source.

The steerable catheter can be remotely advanced, retracted, angulated, and rotated up to 360 degrees. It is compatible with 0.014″, 0.018″, and 0.035″ guide-wires and comes in 6 and 9 French diameters. Its distal tip can be precisely bent up to 90 degrees (9F version) and up to 180 degrees (6Fr) version. Its movements give it 7 degrees of freedom (Fig. 1).

Insertion can be either directly percutaneously or via a second short 9F or 6F sheath.

Acquisition of Sensei robotic system by Hansen Medical integrated cutting-edge catheter control with advanced 3D visualization (Fig. 2).

There are several reports of its feasibility and safety to treat vascular diseases, such as:

- Perera et al. [4] studied cerebral embolization during arch manipulation in 11 patients undergoing thoracic endovascular aortic repair, and compared procedures using manual and robotic catheters. Analyzing the number of high-intensity transient signals (HITS) detected by intraoperative transcranial Doppler, they observed a significantly lower number of HITS when using the robotic catheter.
- Riga et al. [5] successfully performed a robot-assisted three-vessel fenestrated endovascular aneurysm repair using Magellan to cannulate the left renal artery. Cochennec et al. [6] studied 16 fenestrated and/or branched stent-grafting procedures and 1 endovascular repair requiring the chimney technique for complex abdominal aortic aneurysm repair. Robotic navigation was attempted for access to 37 target vessels for a maximum of 15 min; manual catheters were used if cannulation was unsuccessful. Successful robotic cannulation was achieved for 30 (81%) target vessels. No intraoperative complications related to robotic navigation were observed.
- Jones et al. [7] performed 13 carotid artery stenting procedures using Magellan to navigate the arch and obtain a stable position in the common carotid artery. They described a technical success rate of 100% and no postoperative neurological complications.

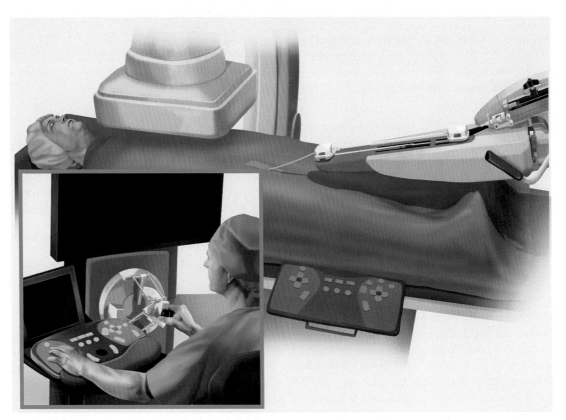

FIG. 1    Robotic arm at the patient table and the remote-control console.

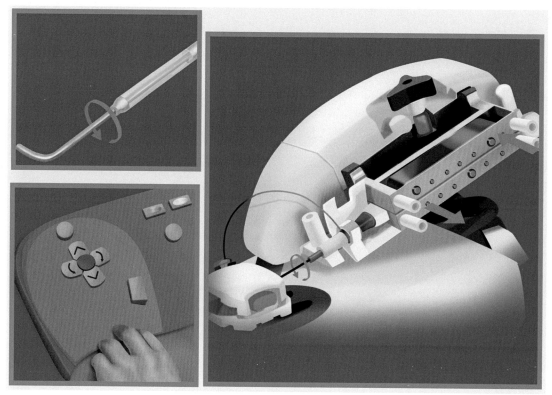

FIG. 2    At the *top left*, we observe the shaft and soft steerable catheter; at the right, the machine mirroring the catheter's movement; at the *bottom left*, the remote control.

- Rolls et al. [8] performed five bilateral uterine artery embolizations using the Magellan catheter, with a technical success rate of 100% and improvement in symptoms.
- Chinnadurai et al. [9] performed 2 carotid artery stenting, 11 visceral artery stenting, 1 hypogastric artery stenting, and 3 hypogastric artery embolization using a robotic catheter along with image fusion guidance. Five patients had failed manual catheterization attempts before. All target vessels were successfully cannulated with Magellan.
- Magellan was also used for challenging intravascular foreign body retrieval: Owji et al. [10] successfully used Magellan to retrieve an inferior vena cava filter that was placed 6 months earlier and presented with caval perforation from filter struts. Wolujewicz [11] successfully retrieved a transected port catheter embolized into the pulmonary artery with Magellan after a manual attempt failed.

In addition to the findings and impressions during its use in humans, there are also several studies on its use in animals and vascular models. The main benefits highlighted were:

- Less vessel trauma due to the ability to vary the distal shape, to bend up to 180 degrees, and adapt to the anatomical surroundings [2,12].
- Ease of use and high maneuverability [5,10].
- High stability [7,12]
- Less radiation exposure and overall fatigue, allowing the surgeon to be seated comfortably away from radiation source [2,5].
     Nevertheless, there are still some noteworthy limitations:
- Target vessels can be successfully achieved using Magellan, but the therapeutic devices must be delivered manually.
- Sheaths are large and single-use.
- They lack manual haptic feedback.

Despite its promising performance in endovascular treatments, in 2021, Hansen Medical was acquired by Auris Health, Inc. and Magellan ceased to be marketed.

III. The current and future clinical applications of robotic surgery among medical specialties

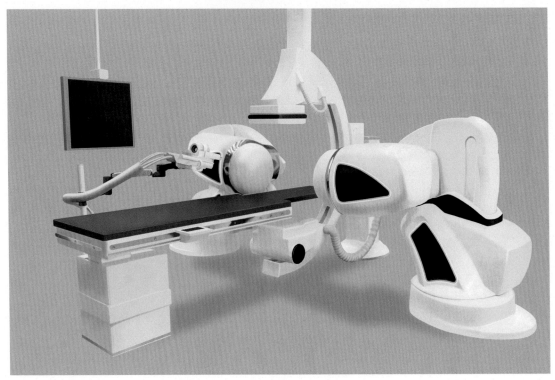

FIG. 3    Stereotaxis products.

## Niobe

The Stereotaxis Niobe Robotic Magnetic Navigation System (Stereotaxis, Inc., St. Louis, USA) consists of two robotically-controlled magnets next to the patient's table, creating a magnetic field used to remotely navigate the dedicated catheter.

The magnetic catheter tip can be controlled with 3 degrees of freedom and, compared with electromechanically controlled systems, is softer, potentially reducing the risk of vessel wall injury.

Recently, Stereotaxis, Inc. has incorporated two innovations to the Niobe system: the Genesis RMN System and Vdrive. Genesis RMN System uses smaller magnets that rotate along their center of mass, which may lead to higher responsiveness to physician control and faster procedures. The Vdrive Robotic Navigation System was designed aiming to improve navigation and stability for cardiac diagnosis and ablation (Fig. 3).

As the Amigo catheter, the Niobe system is more widely used and studied for cardiac interventions [13]. Still, it was also applied experimentally to endovascular cerebral intervention [14] and may have a potential role for endovascular peripheral interventions.

## Robotic-assisted platform: Corpath

In its latest version, the CorPath GRX robotic-assisted platform (Corindus, a Siemens Healthineers Company, Waltham, Massachusetts, USA) consists of a remote physician control console and a bedside robotic arm.

The bedside articulated robotic arm is attached to the patient's table and must be covered with a sterile cover, positioned toward the access vessel, and then receive a sterile single-use cassette, in which the endovascular material will be loaded. The platform is currently only compatible with 0.014″ guide-wires and rapid exchange therapeutic devices (balloon dilation catheters or stents) (Fig. 4).

The vessel access should be performed manually with a standard introducer sheath. A standard catheter should then be manually placed and must be connected to a hemostatic valve inside the cassette. At the tip of the cassette, an outer sheath is pulled over the standard catheter and must be attached to the standard introducer sheath, allowing the robot to move the catheter. After this, the guide-wire and therapeutic devices can be loaded into the cassette and progressed through the hemostatic valve and standard catheter (Fig. 5).

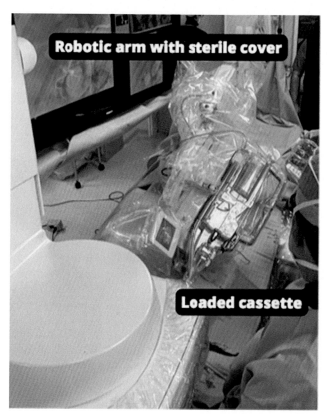

FIG. 4  The patient's bedside robotic arm already protected with a sterile cover and the loaded sterile cassette attached.

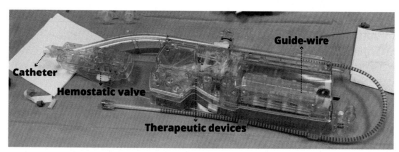

FIG. 5  Sterile cassette indicating where each endovascular device is loaded.

From the remote-control station, the endovascular surgeon can control the catheter and the guide-wire with three degrees of freedom, and the therapeutic device. This station should be placed far from the radiation source, protecting the physician. Part of the surgical team will still have to be at the patient's bedside, with sterile and radiation protection equipment, to perform the surgical steps that must be done manually, contrast injection, and load and unload devices into the cassette. Nonetheless, the surgical staff at the patient's bedside can move away from the main radiation source during the robotic-assisted surgical steps; thus, the whole team receives less ionizing radiation.

The standard endovascular catheter can be rotated, advanced, and retracted from the remote console's dedicated joystick. For advancing and retracting, it is possible to use touchscreen buttons that move 1 mm at a time instead of the joystick, and there is an average total movement limit of 20 cm, which is the length of the cassette's outer sheath that can be controlled. There is no haptic feedback per se, but the robot will signal whether the catheter is advancing through a critical stenotic area using an alarming sound and an image of a car going out of a highway. It can even halt its advance when it deems it incompatible, such as a total vessel occlusion.

The 0.014″ guide-wire can be rotated, advanced, and retracted from the remote console's dedicated joystick. As with the catheter, it is possible to use touchscreen buttons that move 1 mm at a time instead of the joystick for advancing and retracting. A software update included the technIQ technology, adding smart procedural automation features to the guide-wire control:

— Rotate on Retract: rotates the guidewire upon joystick retraction, which is especially useful when the guidewire enters a nontarget vessel, helping the operator to navigate in another direction toward a targeted lesion.
— Wiggle and spin: Both commands allow the physician to move the guide-wire while simultaneously rotating it 360 degrees, each at a different speed.
— Constant speed: Combining this command with the touchscreen buttons moving 1 mm at a time, it is possible to measure a lesion with high precision.

The therapeutic device, balloon dilation catheter or stent, can be advanced or retracted and controlled with its own dedicated joystick. Balloon inflation or stent deployment should be performed manually.

It is of note that the console allows the operator to deactivate or activate each of the joysticks, which improves stability and precision, especially during catheter or therapeutic device retraction or change, as the guide-wire command can be blocked so it stays still. Another feature is that the console has an accelerator button, indicated to speed up retractions (Fig. 6).

The latest software and hardware modifications included active device fixation (ADF) control software, permitting automated manipulation of the guidewire relative to the microcatheter, and a modified drive cassette suitable for neuroendovascular instruments [15].

The CorPath robotic-assisted platform has been successfully applied to a variety of endovascular procedures:

• Percutaneous coronary interventions (PCI): PRECISE study [16] demonstrated the safety and feasibility of robotic PCI using the older generation, CorPath 200, enrolling 164 patients. Technical success was achieved in 98.8% of the cases; clinical procedural success was achieved in 97.6%. There were no device-related complications, and they observed a reduction in radiation exposure. Brunner et al. [17] reported 86 coronary lesions with robotic-assisted repair, including complex and chronic total occlusion lesions. Partial manual assistance was used in 25.6%, and conversion to manual PCI was required in 5.8%.
• Peripheral vascular intervention (PVI): In the RAPID trials I [18] and II [19], authors demonstrated the safety and feasibility of robotic PVI using the older generation platform, CorPath 200, by treating patients with symptomatic peripheral artery disease and femoropopliteal lesions. Technical and clinical success rates were 100% in both studies. Behnamfar et al. [20] reported successful below-the-knee PVI as well.
• Neurointerventions: Weinberg et al. [21] conducted a retrospective comparison between 6 cases of robotic-assisted transradial carotid artery stenting (TR CAS) and 7 manual TR CAS, involving 13 consecutive patients. There were no technical or access-site complications and no catheter exchanges. Despite being significantly longer in duration,

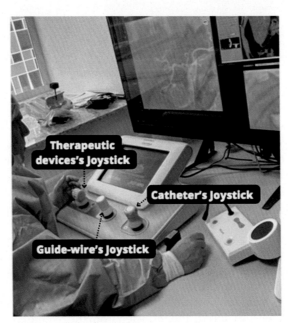

FIG. 6    Remote-control station, with each dedicated device's joysticks.

robotic-assisted TR CAS was no different from manual TR CAS in terms of postoperative complications or mortality. The authors' results suggest that RA TR CAS is feasible, safe, and effective. Nogueira et al. [22] reported four successful cases of CAS with microguidewire, emboli-protection, and angioplasty balloon navigation performed robotically except for navigation/deployment of the stent due to current incompatibility with the platform. Pereira et al. [23] and Cancelliere et al. [24] reported successful intracranial aneurysm repair with all intracranial steps, including stent placement and coil deployment, performed with assistance from CorPath.

The main benefits highlighted were:

- High stability.
- Less radiation exposure and overall fatigue allow the surgeon to be seated comfortably away from the radiation source.
- Better technical success, lower procedure times, and reduced fluoroscopy and contrast use for PVI.
- Compatibility with standard endovascular material.

However, there are some limitations:

- Therapeutic devices can be placed in the targets using the robot but must be delivered manually.
- Single-use cassette.
- Incompatibility with 0.035″ and 0.018″ guidewires.
- Poor haptic feedback.
- It is difficult to move the robotic arm during the procedure; sometimes, it is even needed to reset.

## Future scenario

In this chapter, we discussed the main robotic devices and platforms that have been more extensively studied in human subjects. Still, several prototypes are currently being developed with very promising features.

In addition to all the above-mentioned potential benefits, remotely performed robotic interventions can not only serve to reduce radiation exposure but can also be used to intervene in patients in isolation due to an infectious pathogen [25], as it would be very valuable if another pandemic breaks through. Furthermore, teleprocedures also have the potential to offer cutting-edge treatments in underserved areas [26].

Probably the most urgent improvement in robotic devices is incorporating of a reliable haptic feedback mechanism to reduce the risk of vessel lesions, such as dissection or perforation. Such tools are currently being tested [27,28]. Integrating 3D evaluation should also be an aim [9], to increase precision and not only to minimize vessel wall injury but also the use of contrast and radiation emission. Finally, compatibility with existing endovascular materials is crucial to cost reduction and should also considered by robotic device developers and marketers.

## Key points

- Robotic technology has several potential advantages:
  - The main endovascular surgeon, can be wholly distanced from the radiation source and comfortably seated while performing the robotic steps, improving well-being and reducing occupational hazards, which may reflect better performance.
  - The intervention team, who can move away from the radiation source during robotic navigation.
  - The patient, as robotic devices may offer several potential advantages over standard endovascular techniques, such as more precise navigation, enhanced safety, more accurate lesion treatment, reduction of procedure duration, contrast use, and radiation exposure.
- Endovascular surgeons must take a more active role in the development and study of robotic techniques and devices since most of the available literature comes from cardiac and neurointerventionalists.

## References

[1] Chaikof EL, Dalman RL, Eskandari MK, Jackson BM, Lee WA, Mansour MA, et al. The Society for Vascular Surgery practice guidelines on the care of patients with an abdominal aortic aneurysm. J Vasc Surg 2018;67(1):2–77.e2.
[2] Cruddas L, Martin G, Riga C. Robotic endovascular surgery: current and future practice. Semin Vasc Surg 2021;34(4):233–40.

[3] Cercenelli L, Bortolani B, Marcelli E. CathROB: a highly compact and versatile remote catheter navigation system. Appl Bionics Biomech 2017;2017:1–13.

[4] Perera AH, Riga CV, Monzon L, Gibbs RG, Bicknell CD, Hamady M. Robotic arch catheter placement reduces cerebral embolization during thoracic endovascular aortic repair (TEVAR). Eur J Vasc Endovasc Surg Off J Eur Soc Vasc Surg 2017;53(3):362–9.

[5] Riga CV, Bicknell CD, Rolls A, Cheshire NJ, Hamady MS. Robot-assisted fenestrated endovascular aneurysm repair (FEVAR) using the Magellan system. J Vasc Interv Radiol JVIR 2013;24(2):191–6.

[6] Cochennec F, Kobeiter H, Gohel M, Marzelle J, Desgranges P, Allaire E, et al. Feasibility and safety of renal and visceral target vessel cannulation using robotically steerable catheters during complex endovascular aortic procedures. J Endovasc Ther Off J Int Soc Endovasc Spec 2015;22 (2):187–93.

[7] Jones B, Riga C, Bicknell C, Hamady M. Robot-assisted carotid artery stenting: a safety and feasibility study. Cardiovasc Intervent Radiol 2021;44(5):795–800.

[8] Rolls AE, Riga CV, Bicknell CD, Regan L, Cheshire NJ, Hamady MS. Robot-assisted uterine artery embolization: a first-in-woman safety evaluation of the Magellan system. J Vasc Interv Radiol 2014;25(12):1841–8.

[9] Chinnadurai P, Duran C, Al-Jabbari O, Abu Saleh WK, Lumsden A, Bismuth J. Value of C-arm cone beam computed tomography image fusion in maximizing the versatility of endovascular robotics. Ann Vasc Surg 2016;30:138–48.

[10] Owji S, Lu T, Loh TM, Schwein A, Lumsden AB, Bismuth J. Robotic-assisted inferior vena cava filter retrieval. Methodist Debakey Cardiovasc J 2017;13(1):34–6.

[11] Wolujewicz M. Robotic-assisted endovascular pulmonary artery foreign body retrieval: a case report. Vasc Endovascular Surg 2016; 50(3):168–70.

[12] Duran C, Lumsden AB, Bismuth J. A randomized, controlled animal trial demonstrating the feasibility and safety of the Magellan™ endovascular robotic system. Ann Vasc Surg 2014;28(2):470–8.

[13] Kiemeneij F, Patterson MS, Amoroso G, Laarman G, Slagboom T. Use of the Stereotaxis Niobe® magnetic navigation system for percutaneous coronary intervention: results from 350 consecutive patients. Catheter Cardiovasc Interv 2008;71(4):510–6.

[14] Kara T, Leinveber P, Vlasin M, Jurak P, Novak M, Novak Z, et al. Endovascular brain intervention and mapping in a dog experimental model using magnetically-guided micro-catheter technology. Biomed Pap 2014;158(2):221–6.

[15] Britz GW, Panesar SS, Falb P, Tomas J, Desai V, Lumsden A. Neuroendovascular-specific engineering modifications to the CorPath GRX robotic system. J Neurosurg 2019;133(6):1830–6.

[16] Weisz G, Metzger DC, Caputo RP, Delgado JA, Marshall JJ, Vetrovec GW, et al. Safety and feasibility of robotic percutaneous coronary intervention: PRECISE (percutaneous robotically-enhanced coronary intervention) study. J Am Coll Cardiol 2013;61(15):1596–600.

[17] Brunner FJ, Waldeyer C, Zengin-Sahm E, Kondziella C, Schrage B, Clemmensen P, et al. Establishing a robotic-assisted PCI program: experiences at a large tertiary referral center. Heart Vessels 2022;37(10):1669–78.

[18] Mahmud E, Schmid F, Kalmar P, Deutschmann H, Hafner F, Rief P, et al. Feasibility and safety of robotic peripheral vascular interventions: results of the RAPID trial. JACC Cardiovasc Interv 2016;9(19):2058–64.

[19] Mahmud E, Schmid F, Kalmar P, Deutschmann H, Hafner F, Rief P, et al. Robotic peripheral vascular intervention with drug-coated balloons is feasible and reduces operator radiation exposure: results of the robotic-assisted peripheral intervention for peripheral artery disease (RAPID) study II. J Invasive Cardiol 2020;32(10):380–4.

[20] Behnamfar O, Pourdjabbar A, Yalvac E, Reeves R, Mahmud E. First case of robotic percutaneous vascular intervention for below-the-knee peripheral arterial disease. J Invasive Cardiol 2016;28(11):E128–31.

[21] Weinberg JH, Sweid A, Sajja K, Gooch MR, Herial N, Tjoumakaris S, et al. Comparison of robotic-assisted carotid stenting and manual carotid stenting through the transradial approach. J Neurosurg 2020;135(1):21–8.

[22] Nogueira RG, Sachdeva R, Al-Bayati AR, Mohammaden MH, Frankel MR, Haussen DC. Robotic assisted carotid artery stenting for the treatment of symptomatic carotid disease: technical feasibility and preliminary results. J NeuroInterv Surg 2020;12(4):341–4.

[23] Mendes Pereira V, Cancelliere NM, Nicholson P, Radovanovic I, Drake KE, Sungur JM, et al. First-in-human, robotic-assisted neuroendovascular intervention. J NeuroInterv Surg 2020;12(4):338–40.

[24] Cancelliere NM, Lynch J, Nicholson P, Dobrocky T, Swaminathan SK, Hendriks EJ, et al. Robotic-assisted intracranial aneurysm treatment: 1 year follow-up imaging and clinical outcomes. J Neurointerv Surg 2022;14(12):1229–33.

[25] Legeza P, Britz GW, Loh T, Lumsden A. Current utilization and future directions of robotic-assisted endovascular surgery. Expert Rev Med Devices 2020;17(9):919–27.

[26] Madder RD, VanOosterhout S, Parker J, Sconzert K, Li Y, Kottenstette N, et al. Robotic telestenting performance in transcontinental and regional pre-clinical models. Catheter Cardiovasc Interv Off J Soc Card Angiogr Interv 2021;97(3):E327–32.

[27] Li X, Guo S, Shi P, Jin X, Kawanishi M. An endovascular catheterization robotic system using collaborative operation with magnetically controlled haptic force feedback. Micromachines 2022;13(4):505.

[28] Zhang L, Gu S, Guo S, Tamiya T. A magnetorheological fluids-based robot-assisted catheter/guidewire surgery system for endovascular catheterization. Micromachines 2021;12(6):640.

# Index

Note: Page numbers followed by *f* indicate figures and *t* indicate tables.

## A

Printed in the United States
by Baker & Taylor Publisher Services